Aspirin and Related Drugs

Aspirin and Related Drugs

K.D. Rainsford

PhD, FRCP Edin, FRCPath, FRSC, FIBiol, FIBMS, Dr(hc)
Biomedical Research Centre
Sheffield Hallam University
Sheffield, UK

Taylor & Francis
Taylor & Francis Group

LONDON AND NEW YORK

First published 2004 by Taylor & Francis
11 New Fetter Lane, London EC4P 4EE

Simultaneously published in the USA and Canada
by Taylor & Francis Inc,
29 West 35th Street, New York, NY 10001

Taylor & Francis is an imprint of the Taylor & Francis Group

Typeset in Baskerville by Wearset Ltd, Boldon, Tyne and Wear
Printed and bound in the United States by Edwards Brothers Inc., Ann Arbor, Michigan

Every effort has been made to ensure that the advice and information in this book is true and accurate at the time of going to press. However, neither the publisher nor the authors can accept any legal responsibility or liability for any errors or omissions that may be made. In the case of drug administration, any medical procedure or the use of technical equipment mentioned within this book, you are strongly advised to consult the manufacturer's guidelines.

British Library Cataloguing in Publication Data
A catalogue record for this book is available from the British Library

Library of Congress Cataloging in Publication Data
A catalog record for this book has been requested

ISBN 0–7484–0885–1

'To those who have gone before'

In recognition of those who have contributed so much to the explosion of science and clinical practice on the actions and uses of aspirin and the salicylates as well as other related analgesic/anti-inflammatory/antipyretic drugs

Contents

4 PHARMACOKINETICS AND METABOLISM OF THE SALICYLATES 97

G.G. Graham, M.S. Roberts, R.O. Day and K.D. Rainsford

5 METABOLISM AND PHARMACOKINETICS OF IBUPROFEN 157

G.G. Graham and K.M. Williams

6 PHARMACOKINETICS AND METABOLISM OF PARACETAMOL (ACETAMINOPHEN) 181

G.G. Graham and M. Hicks

7 PHARMACOLOGY AND BIOCHEMISTRY OF SALICYLATES AND RELATED DRUGS   215

K.D. Rainsford

8 SIDE EFFECTS AND TOXICOLOGY OF THE SALICYLATES   367

K.D. Rainsford

9 REYE'S SYNDROME AND ASPIRIN   555

J.F.T. Glasgow and S.M. Hall

13 ASPIRIN AND NSAIDs IN THE PREVENTION OF CANCER, ALZHEIMER'S DISEASE AND OTHER NOVEL THERAPEUTIC ACTIONS 707

K.D. Rainsford

Contributors

W. Watson Buchanan MD, FRCP (Glas. & Edin.), FRCPC
Faculty of Health Sciences and Medical Centre, McMaster University, Hamilton, Ontario, Canada

Walter F. Kean MD, FRCP (Edn., Glas.), FRCPC
Faculty of Health Sciences and Medical Centre, McMaster University, Hamilton, Ontario, Canada

John G. Kelton MD, FRCPC
Professor and Dean of Health Sciences, McMaster University, Hamilton, Ontario, Canada

John F.T. Glasgow
Royal Belfast Hospital for Sick Children and Queen's University of Belfast, Institute of Clinical Science, Belfast, Northern Ireland, UK

Garry G. Graham MSc, PhD
School of Physiology and Pharmacology, University of NSW, and Department of Clinical Pharmacology and Toxicology, St Vincent's Hospital, Sydney, Australia

Susan M. Hall
Sheffield Children's Hospital and University of Sheffield, Sheffield, UK

Mark Hicks MSc, PhD
Department of Clinical Pharmacology and Toxicology, St Vincent's Hospital, Sydney, Australia

Jan R. McTavish PhD
Department of Social Sciences, Alcorn State University, Lorman, Mississippi, USA

Richard O. Day BS, MD, FRACP
Department of Clinical Pharmacology and Toxicology, St Vincent's Hospital, Sydney, Australia

Kim D. Rainsford PhD, FRCP Edin, FRCPath, FRSC, FIBiol, FIBMS, Dr(hc)
Biomedical Research Centre, Sheffield Hallam University, Sheffield, UK

Michael S. Roberts PhD, DSc
Department of Medicine, University of Queensland, Princess Alexandra Hospital, Brisbane, Australia

Kathryn E. Webert MD, FRCPC
Hematologist, McMaster University, Hamilton, Ontario, Canada

Kenneth M. Williams BSc, PhD
School of Physiology and Pharmacology, University of NSW, Sydney, and Department of Pharmacology, St Vincent's Hospital, Sydney, Australia

Preface

In my book *Aspirin and the Salicylates* (Butterworths, London, 1984) I wrote of my fascination with the salicylates. Now, two decades later, I can say that the fascination has grown, especially as we learn more about the mechanisms of action of these drugs and, through this, a wider range of uses for them. Advances in the past decade or so regarding the molecular biology of inflammatory diseases, the neurobiology of pain and fever and the physiopathology of thromboembolic diseases have given great insights into the molecular and cellular mechanisms of aspirin and salicylates, as well as the many non-steroidal anti-inflammatory and analgesic drugs – the latter often being compared with the salicylates for their actions, safety and efficacy. In fact, the salicylates are often used as a basis of comparison in laboratory as well as clinical studies. This reflects the extent of knowledge of these drugs and their past and present utility as antipyretic, analgesic, anti-inflammatory and, with some like aspirin, anti-thrombotic effects.

Since its introduction a century ago, aspirin has enjoyed waves of popularity in its use as well as in its application in new therapeutic areas. The remarkable thing about aspirin is that despite challenges from many other non-steroidal anti-inflammatory drugs and non-narcotic as well as weak narcotic analgesics, it is still extensively used to control pain, fever and inflammation. Its popularity in highly competitive markets where it is sold extensively on a non-prescription (or over-the-counter) basis, such as in the USA, is testimony to its recognition by the public as a very effective drug. Issues about its safety, especially regarding the gastrointestinal tract and kidneys, have long been recognised, but with careful use and the development of novel formulations these adverse reactions can be regarded as relatively rare – especially for occasional use by adults. The possible risk of Reye's syndrome in children is still a vexed question, as can be seen in Chapter 9. It still seems odd that the syndrome appears not to have been recognised or to have caused appreciable fatalities in children until several decades ago. No doubt the decline in fatalities associated with Reye's syndrome has reflected the reduction in the intake of aspirin, but there may be many other factors involved in this decline.

The popular use of aspirin as anti-thrombotic agent, which arose from the recognition that the antiplatelet aggregating effects of this drug could be responsible for gastrointestinal bleeding, was quickly turned to good therapeutic use. With careful studies to identify the requirement for low dosage to obtain selective effects on the platelets' thromboxane production without markedly influencing prostaglandin production by blood vessels, it was possible to obtain a degree of selectivity in therapy or prophylaxis of thromboembolic diseases. Thus, aspirin is undoubtedly the drug of choice in prophylaxis of cardiovascular conditions in subjects at potential risk of developing myocardial infarction. However, gastrointestinal bleeding can still present a problem at low doses of this drug, although for many this adverse reaction is of low grade and possibly an acceptable risk. Despite its popularity and wide acceptance, there is still research ongoing to identify safer anti-thrombotic agents. The major reason for the success of aspirin as a prophylactic or therapeutic agent for cardiovascular conditions is that it is cheap as well as effective. Cheapness and reliability also account for its widespread use by the lay public, although ibuprofen and paracetamol (acetaminophen) are potent competitors and both have lower risks for developing gastrointestinal and possibly renal adverse reactions. It is because of the competitive aspects of use of these analgesics and other non-steroidal and analgesic agents that aspects of comparison of their actions and use are considered in this book alongside those of aspirin and salicylates.

The major focus in this book is on the key aspects concerning the historical uses and developments of the salicylates, their chemistry and occurrence, pharmacological and toxicological effects, adverse drug reactions and clinical uses. The '...and Related Drugs' part of the title is intended to refer to comparisons with other analgesics, non-steroidal anti-inflammatory drugs (NSAIDs) or alternative classes of therapeutic agents that represent competitors, or where there are other important reasons for comparing their actions or adverse reactions. This book has to some extent been modelled on the format of my earlier book, *Aspirin and the Salicylates*, and a few sections in some of the chapters I have written have been taken, because of historical content, and updated from this book.

In attempting to present a comprehensive account of the salicylates I have enlisted the help of and contributions from leading researchers and physicians in the field. The content emphasis of their contributions has been largely their own, while I have attempted to give an overview. I accept there are some areas of overlap or even differing views; I believe that it is important to have the former because of the need for a comprehensive account within an area of review and discussion for the sake of readability. It is also important to present differing views because this reflects our state of knowledge and the dynamic state of the subject area. To have a homogeneous presentation might help some readers, but for those requiring a critical analysis it is imperative to bring out the contrasts and controversies.

Although the term 'related drugs' logically covers some of the commonly used analgesics, especially ibuprofen and paracetamol, there has been no attempt here to present a comprehensive account that includes these drugs, aside from chapters on their pharmacokinetics and comparisons with the salicylates and some historical aspects. The reader is referred to comparison monographs from the publisher – Professor Prescott's excellent book *Paracetamol. A Critical Bibliographic Review*, and *Ibuprofen. A Critical Bibliographic Review*, which was edited by myself and intended to be to some extent a comparison monograph. Perhaps the present volume will be regarded as completing the triad of the most popular or most extensively used analgesic, anti-inflammatory and antipyretic agents. It is recognised that there are some other significant drugs in this class that probably deserve a volume in themselves, but we have attempted broad comparisons with these and have included the newer non-steroidal anti-inflammatories as well.

To achieve a comprehensive account of many areas concerning the mode of action and therapeutics of the salicylates, it is important to cover much important earlier literature that was published in the first half or so of the twentieth century, during which there were important formative data and observations published on aspirin, the salicylates and related drugs. It would be easy simply to cover the most recent literature, but this ignores highly significant and important information that has led to modern development of these drugs. Indeed, there have been numerous examples where revisiting an earlier area of investigation has enabled a new view to be developed, based on this earlier information. For example, the importance of the mitochondrial actions of the salicylates in the development of induced cellular death (apoptosis) brings together the observations of the 1950s and 1960s regarding the effects of salicylates on the uncoupling of oxidative phosphorylation and their effects on intermediary metabolism, with more recent data on the activation of caspases and cytochrome c release from mitochondria. This may be important, along with the newer information on the actions of these drugs on oxyradical and cytokine-mediated signal transduction events, in understanding the protective effects of aspirin and related drugs in colon and other cancers as well as the mode of action of these drugs in the development of gastrointestinal ulceration and bleeding. As a further example, the long-debated and important therapeutic question regarding the efficacy of high-dose aspirin compared with that of salicylic acid, its dimer, salsalate, or the sodium salt in the long-term treatment of pain and inflammation in rheumatic diseases has been revisited again with evidence that the major anti-inflammatory and analgesic actions of aspirin reside in the salicylate that is produced therefrom. Recent studies on the molecular pharmacology of aspirin versus salicylate give further support to this view. The question is an important one therapeutically because the serious gastrointestinal adverse reactions from salicylate (and possibly certain formulations thereof) have long been recognised as less than those from aspirin. Yet why is it that aspirin is still, despite competition from paracetamol, ibuprofen and other NSAIDs (including the new COX-2 selective inhibitors), used so exclusively in self-medication of chronic, if episodic, arthritic disease as well as in acute states? The simple answer is that many consider that the drug works, and it is cheap. Maybe too many think only of its limitations concerning adverse effects in the major organ systems. We should not forget the old German adage 'Bitter im Mund, gesund im Korper', or 'Bitter in the mouth, healthy in the body' (from *Familiar Medical Quotations*, edited by M.B. Strauss, published by Little Brown, 1968) for this, as Professor Watson Buchanan has pointed out, is a reflection that you need to experience some adverse reactions to know that a drug works!

The previous question has relevance to the actions of competitors of aspirin, both new and old. Current interest in the COX-2 selective NSAIDs highlights important competitive issues for the salicylates – amongst the oldest of the analgesics. Aspirin and the salicylates have faced such competition in the past, including from ibuprofen and paracetamol. However, one single outstanding therapeutic action in the prevention of coronary vascular disease and stroke, which resulted from the discovery of the antiplatelet effects of the drug over three decades ago, has given aspirin a new lease of life. Intense interest in the mode of action of aspirin (which stems from recent understanding of the molecular biology of the cyclo-oxygenases, apoptosis and other components of the regulation of cell cycle and growth) coupled with clinico-epidemiological evidence that it might prevent colon and maybe other cancers now gives further scope for the therapeutic use of aspirin and others of its class.

With these new(er) discoveries has come recognition that the salicylates are employed as 'bench', 'gold' or 'clinical' standards for comparison with newer agents.

A number of outstanding books and reviews have been written about aspirin and the salicylates, and some of these are listed at the end of the preface. Many of the books have long been out of print, and their availability from libraries is limited or declining. Where possible, many of the points from earlier key literature have been included in this book so that it will provide an important source of information on the salicylates.

I would especially like to thank Mrs Veronica Rainsford-Köchli for her help with translating texts from German and French and for secretarial assistance in preparing this book, and my secretary, Mrs Marguerite Lyons, and Mrs Ann Shepherd for their secretarial support. My thanks too to Mr Richard Seabrook and Miss Kate Maybury, for assistance with obtaining references, and to the libraries of the Royal Society of Medicine and Sheffield Hallam University for the immense help given in obtaining articles and books – some of them from very remote locations.

This book is dedicated to the many colleagues and friends who have willingly shared their views and opinions with me over the years and who have given much valued advice in preparing this book. I especially appreciate the impartial advice, criticism and valuable comments from colleagues and friends, among them Professors Watson Buchanan, Richard Hunt and Walter Kean, (McMaster University, Canada), Dr Michael Whitehouse and Professor Michael Roberts (University of Queensland, Australia), Professor Garry Graham (University of New South Wales, Australia), Dr Michael Powanda (M/P Biomedical Consultants LLC, California, USA) and Dr Brian Callingham (University of Cambridge, UK). I am also most grateful to Dr Callingham for his valuable help in editing the text and for discussing some issues and controversies. This book is also dedicated to the memory of the late Professor Derek Willoughby, who did much pioneering work in the field of inflammation science, and who sadly passed away on 13 March 2004 just as this book was going to press.

Sheffield
April, 2004

USEFUL BOOKS ON ASPIRIN AND SALICYLATES

Barnett, H.J.M., Hirsh, J. and Mustard, J.F. 1982, *Acetylsalicylic Acid. New Uses for an Old Drug*. New York: Raven Press.

Bekemeier, H. (ed.) 1977, *100 Years of the Salicylic Acid as an Antirheumatic Drug*. Halle: Martin-Luther University.

Dale, T.L.C. (ed.) 1975, *Proceedings of the Aspirin Symposium*. London: Royal College of Surgeons.

Dixon, A.St.J., Martin, B.K., Smith, M.J.H. and Wood, P.H.N. (eds). 1963, *Salicylates. An International Symposium; Postgraduate Medical School London 13–15 Sept 1962*. London: J. & A. Churchill Ltd.

Düllmann, H. 1934, *Über die Wirkungsverstärkung des Pyramidons und Aspirins durch Dionin*. Münster: Werne-Lippe.

Feinman, S.E. (ed.) 1994, *Beneficial and Toxic Effects of Aspirin*. Boca Raton: CRC Press.

Forrestal, D.J. (ed.) 1977, *Faith, Hope and $5,000. The Story of Monsanto*. New York: Simon and Schuster.

Fryers, G. (ed.) 1990, *Aspirin – Towards 2000. Proceedings of an International Meeting of the European Aspirin Foundation. Brussels, 2–3 May 1989*. London: Royal Society of Medicine Services.

Gross, M. and Greenberg, L.A. (eds). 1948, *The Salicylates. A Critical Bibliographic Review*. New Haven: Hillhouse Press.

Hallam, J., Goldman, L. and Fryers, G.R. (eds). 1981, *Aspirin Symposium 1980. Proceedings of an International Symposium held by the Aspirin Foundation, Royal College of Surgeons, 5 June 1980*. London: Royal Society of Medicine.

Hanzlik, P.J. (ed.) 1927, *Actions and Uses of the Salicylates and Cincophen*. Baltimore: Williams and Wilkins Co.

Mann, C.C. and Plummer, M.L. (eds). 1991, *The Aspirin Wars. Money, Medicine, and 100 Years of Rampant Competition*. New York: Alfred A. Knopf.

Mielhke, K. 1978, *Diflunisal in Clinical Practice. Proceedings of a Special Symposium held at the XIV Congress of Rheumatology, San Francisco, USA, June 29 1977*. New York: Futura Publishing Co. Inc.

Morgan, B.S. 1959, *Apothecary's Venture: The Scientific Quest of the International Nicholas Organisation*. Slough: Aspro-Nicholas.

Rainsford, K.D. 1984, *Aspirin and the Salicylates*. London: Butterworths.

Schlenk, O. (ed.) 1947, *Die Salicylsäure*. Berlin: Verlag Dr Werner Saenger.

Smith, M.J.H. and Smith, P.K. (eds). 1966, *The Salicylates. A Critical Bibliographic Review*. New York: Interscience Publishers.

Smith, P.K., Kelley, V.C., Bunim, J. and Paul, W.D. 1956, *Aspirin: Recent Advances in its Pharmacology and Clinical Use*. Medical Research Symposium. St Louis, MO: Monsanto Chemical Company & St Louis Medical Society.

Smith, R.G. and Barrie, A. (eds). 1976, *Aspro – How a Family Business Grew Up*. Worcester: The Trinity Press.

Vane, J.R. and Botting, R.M. (eds). 1992, *Aspirin and Other Salicylates*. London: Chapman & Hall Medical.

IMPORTANT REVIEW ARTICLES

Adams, S.S. and Cobb, R. 1967, Non-steroidal anti-inflammatory drugs. *In:* G.P. Ellis and G.B. West (eds), *Progress in Medicinal Chemistry*, pp. 59–133. London: Butterworths.

Atkinson, D.C. and Collier, H.O.J. 1980, Salicylates: molecular mechanism of therapeutic action. *Advances in Pharmacology and Chemotherapy*, **17:** 233–288.

Babhair, S.A. 1984, Salicylamide. *Analytical Profiles of Drug Substances*, **13:** 521–551.

Collier, H.O.J. 1969, A pharmacological analysis of aspirin. *Advances in Pharmacology and Chemotherapy*, **7:** 333–405.

Collier, H.O.J. 1984, The story of aspirin. *In:* M.J. Parnham and J. Bruinvels (eds), *Discoveries in Pharmacology*, Vol. 2, p. 555. Amsterdam: Elsevier Science Publishers.

Domenjoz, R. 1955, *Pharmakotherapeutische Weiterentwicklung der Antipyretica-Analgetica. Naunyn-Schmiedeberg's Archive for Experimental Pathology and Pharmacology*, **225:** 14–44.

Fryers, G. 1990, Aspirin – towards 2000. *Royal Society of Medicine International Congress and Symposium Series*, **168**.

Hallam, J., Goldman, L. and Fryers, G.R. 1981, *Aspirin symposium 1980. Royal Society of Medicine International Congress and Symposium Series*, **39**.

Horsch, W. 1979, Die Salicylate. *Pharmazie*, **34:** 585–604.

Kim, D.H. 1978, Aspirin (1). Discovery, current and potential new therapeutic uses, and mechanism of action. *Archives of Pharmacological Research*, **1:** 41–54.

Rainsford, K.D. 1985, Salicylates. *In:* K.D. Rainsford (ed.), *Anti-Inflammatory and Anti-Rheumatic Drugs*, Vol. 1, pp. 109–147. Boca Raton: CRC Press.

Whitehouse, M.W. 1965, Some biochemical and pharmacological properties of anti-inflammatory drugs. *Progress in Drug Research*, **8:** 321–429.

Winter, C.A. 1966, Nonsteroidal anti-inflammatory agents. *Progress in Drug Research*, **10:** 139–203.

Website:
Aspirin Foundation:
www.aspirin-foundation.com

Abbreviations and Nomenclature

The term 'aspirin' is used, in accordance with its widespread generic usage throughout the world, as the name for the chemical, acetylsalicylic acid. In some European countries and Canada this name is still protected by trademark (to Bayer AG). Its use in this book recognises its convenience and widespread use in the scientific and medical community, and is in no way intended to denote use or endorsement of the trademark.

'Salicylates' is used to denote all drugs having the 2-hydroxybenzoic acid structure. When used in the general sense, it implies that, based on the current state of knowledge, it seems reasonable to employ this name to cover the actions or properties of all these compounds. Caution should, however, be expressed in such an extrapolation, and the reader must be mindful of this.

Standard chemical, biochemical and pharmacological abbreviations are employed and, where necessary, are defined at their first use in the text. They have, where possible, been derived from those detailed in *Units, Symbols and Abbreviations* (edited by D.N. Baron, 1979; Royal Society of Medicine, 1 Wimpole Street, London W1M 8AE).

The enzyme nomenclature employed is that described in *Enzyme Nomenclature: Recommendation of the Nomenclature Committee of the International Union of Biochemistry* (1978, Academic Press, New York), with some minor exceptions noted in the text in keeping with modern usage.

The following list of abbreviations is provided for convenient usage:

A23187 = calcium ionophore (Lilly) (= calimycin)
Acetyl-SCoA = acetyl-(S)Coenzyme A
ADP = adenosine diphosphate
ALT = alanine amino transferase (*syn* SGPT)
AMP = adenosine monophosphate
Ang = angiotensin
AOM = azoxymethane
APC = adenomatous polyposis coli
APP = amyloid precursor protein
ASA = aspirin (2-acetoxybenzoic acid) = acetylsalicylic acid
AST = aspartate amino transferase (*syn* SGOT)
ATP = adenosine triphosphate
AUC = area under the plasma concentration curve
BBN = N-butyl-N-(4-hydroxybutyl) nitrosamine
B-cell = bone marrow-derived lymphocytes
BW755c = 3-amino-1-[*m*-(trifluoromethyl)phenyl]-2-pyrazoline
COX = cyclo-oxygenase (see also PGHs below)
CuDIPS = copper 3,5′-diisopropylsalicylate
Cyclic AMP = adenosine cyclic 3′:5′-monophosphate
Cyclic GMP = guanosine cyclic 3′:5′-monophosphate
DEAE = diethylaminoethyl
DH = dimethylhydrazine
Diplosal = salicylsalicylic acid (i.e. salicyl ester of salicylic acid)
ED_{10} = effective dose required to produce 10 lesions to the gastric mucosa
ED_{50} = effective dose required to produce a response in 50 per cent of animals
EDTA = ethylethenediamine tetraacetic acid

Ent. cell = enterochromaffin cell

ER = endoplasmic reticulum

ESR = erythrocyte sedimentation rate

ETYA = 5,8,11,14-eicosatetraynoic acid

FANFT = N-[4-(5-nitro-2-furyl)2-thiazolyl]-formamide

GAGs = glycosaminoglycans

G-cell = gastrin cell

gentisic acid = 2.5-dihydroxybenzoic acid

GPs = glycoproteins

GSH = glutathione

GTP = guanosine triphosphate

H_1, H_2 = histamine type 1 and 2 receptors, respectively

Hb = haemoglobin

HETE(s) = hydroxyeicosatetraenoic acids (variously substituted)

HHT = 12-l-hydroxyheptadecatrienoic acid

HPETE(s) = hydroperoxyeicosatetraenoic acids (variously substituted)

HPLC = high-performance (or pressure) liquid chromatography

IC_{50} = inhibitory concentration required to produce 50 per cent reduction in response

IgG = immunoglobulin G

IL = interleukin

LCHAD = long-chain 3-hydroxyacyl-CoA-dehydrogenase

Log P = logarithm of the partition coefficient between n-octanol and an aqueous mixture

LTs (C_4, D_4, E_4) = leukotriene(s) (C_4, D_4, E_4, respectively)

MC = methylcholanthrene

MK-447 = 2-aminomethyl-4-*tert*-butyl-6-iodophenol

MNNG = N-methyl-N'-nitro-N-nitrosoguanidine

MOPS = multisubstrate oxidising peroxidases

NDGA = nordihydroguaiaretic acid

NFκB = nuclear factor kappa B

NFT = neurofibrillary tangles

NSAID = non-steroidal anti-inflammatory drug

OA = osteoarthritis

$O_2^{\cdot-}$ = superoxide anion

$[O]^{\cdot}$ = hypothetical oxygen radical species deriving from peroxidation of PGG_2 or HPETEs

OH• = hydroxyl radical

PAF = platelet aggregating factor = 1-*O*-alkyl-2-acetyl-*sn*-glyceryl-3-phosphorylcholine

PCO_2 = partial pressure of carbon dioxide

PGHS = prostaglandin G/H synthase; this defines the prostaglandin G to H conversion which includes both cyclo-oxygenase and peroxidase activities. The term 'COX' is often applied to these combined activities as a shorthand reference. It is recognised that the term COX is synonymous with that of a gene in mitochondria that codes for cytochrome oxidase, and that is not employed in this text. The use of COX recognises the widespread and convenient verbal and written use of this term to denote PGHS activity.

PER = peroxidative activity

PGDH = prostaglandin 16-hydroxydehydrogenase

PO_2 = partial pressure of oxygen

PrGns = proteoglycans

Pyrocatechoic acid = 2,3-dihydroxybenzoic acid

Pyroresorcylic acid = 2,6-dihydroxybenzoic acid

RA = rheumatoid arthritis

RNA and DNA = ribonucleic and deoxyribonucleic acids

SA = salicylic acid (2-hydroxybenzoic acid)

SAL = salicylate (anion)

Salol = phenol ester of salicylic acid

Salophen = phenetsal = paracetamol ester of salicylic acid

SASP = salicylazasulphapyridine = salazapyrin; sulphasalazine; 2-hydroxyl-5-[4-[2-pyridinylamino(sulphonyl)phenyl]azo]benzoic acid

Serotonin = 5-hydroxytryptamine

SGOT = serum glutamate oxaloacetate transaminases (*syn* AST)

SGPT = serum glutamate pyruvate transaminases (*syn* ALT)

SLE = systemic lupus erythematosus

SRS-A = slow reacting substance(s) in anaphylaxis (= $LTC_4 + LTD_4$)

T-cell = thymus-derived lymphocytes

Tx (A_2, B_2) = thromboxanes (A_2, B_2)

UDP = uridine diphosphate

V_D = volume of distribution

The following plates illustrate some scenes and artefacts related to the development and marketing of aspirin since its introduction over a century ago.

Plate 1   The Pharmacological Laboratory at Farbenfabriken vorm Friedr. Bayer & Co, Elberfeld (Germany), at the end of the 1890s, wherein it is presumed that the experiments leading to the discovery of aspirin took place. Seated second from the right is Prof. Dr Henrich Dreser, the Head of the Pharmacology Department. (Photograph courtesy of Bayer AG, Leverkusen.)

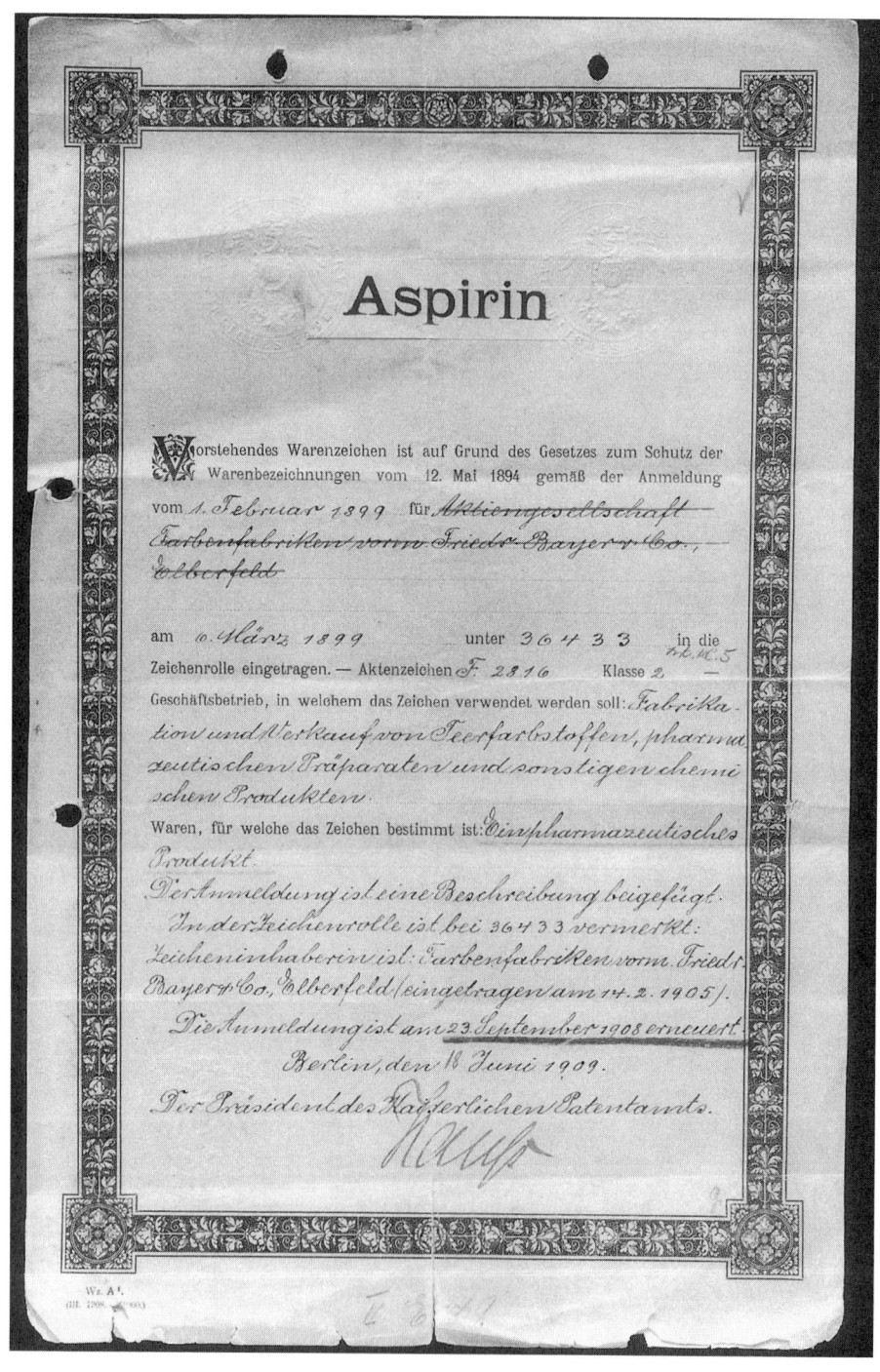

Plate 2 Aspirin trade mark registration document issued on March 6 1899 by the Imperial Patent Office, Berlin. (Photograph courtesy of Bayer AG, Leverkusen.)

Plate 3 First aspirin powder packed in a glass bottle. Before aspirin tablets became available, the drug was sold in powdered form. (Photograph courtesy of Bayer AG, Leverkusen.)

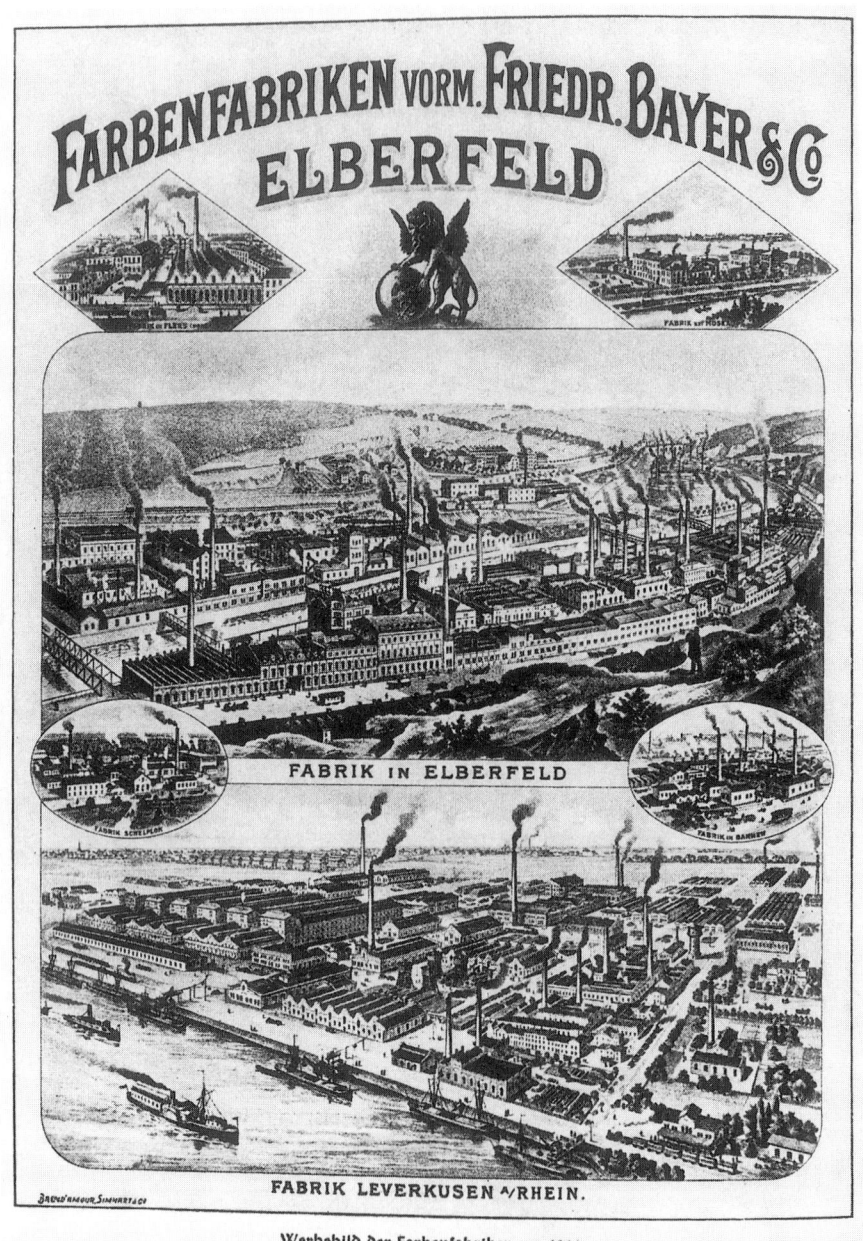

Werbebild der Farbenfabriken um 1900

Plate 4 Illustrated publicity depicting Farbenfabriken vorm Friedr. Bayer & Co, Elberfeld (Germany) at the turn of the nineteenth century. (Photograph courtesy of Bayer AG, Leverkusen.)

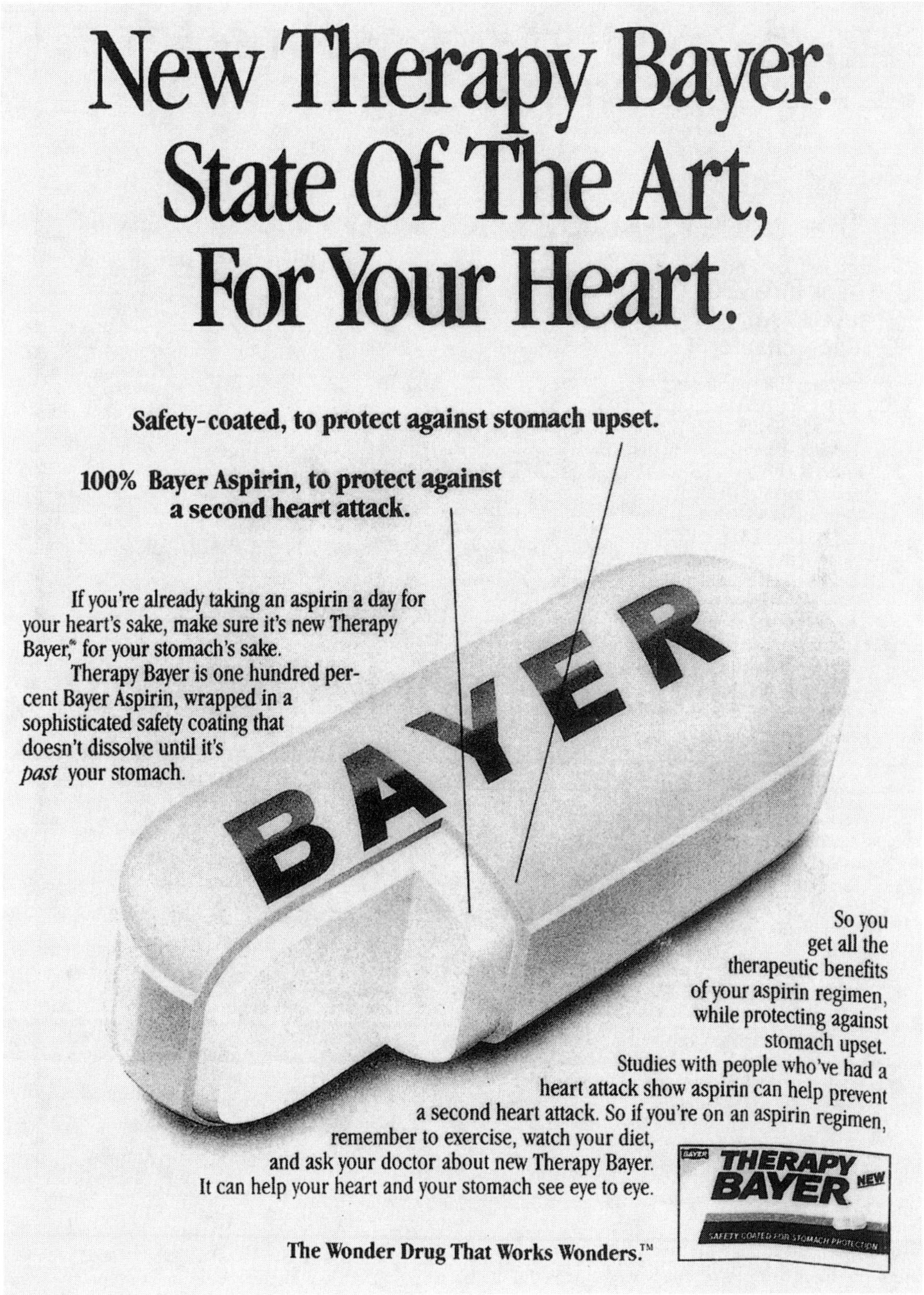

Plate 5 A Bayer advertisement of the 1970s highlighting a 'safety-coated' formulation to prevent against stomach upset while using aspirin to protect against a heart attack.

Today, most advances in medicines concern mode of delivery

One of the latest advances is total solubility of aspirin.

Aspirin: a 2300 year success story begins a new chapter.

The 2300 year history of salicylates, of which aspirin (Acetylsalicylic acid) is a member, begins with Hippocrates. It also includes the eminent names of Pliny, Celcus, Dioscorides, Galen and Rater, Sydenham, Piria, Von Gerhardt, Hofmann.

The efficacy of aspirin has always been held in high regard.

It is recognised as one of the safest and most effective analgesics and anti-inflammatory drugs. It is unique in possessing analgesic, anti-inflammatory and anti-pyretic activities.

What began as an infusion of willow bark, passed through powder forms to solid tablets.

Then came the first solubles – some still marketed – and finally, a most sophisticated formulation, a totally soluble tablet.

Aspro Clear.

To its principal advantages of speed of action and absence of residue, Aspro Clear adds another! It must be taken in water, in conformity with accepted medical practice.

Plate 6 An advertisement from the 1970s featuring 'Aspro-Clear'™ (from the Aspro-Nicholas Company – another producer of aspirin), which was a successful soluble form of aspirin with wide consumer acceptance because of its palatability in solution.

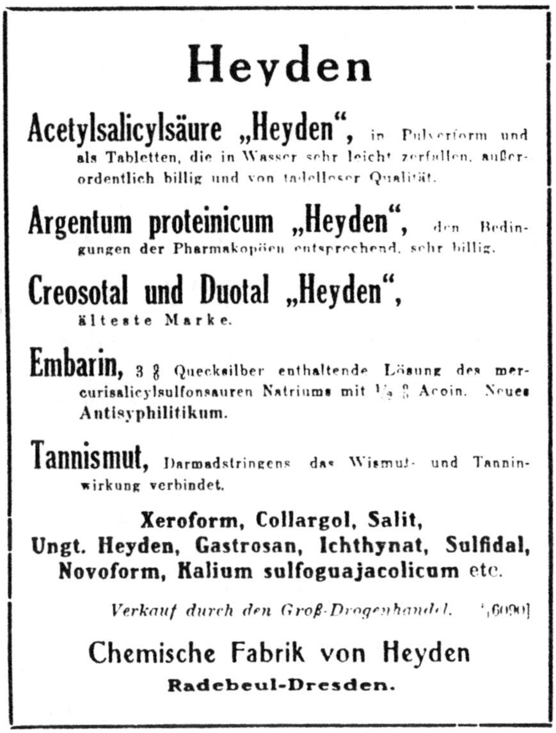

Plate 7 An advertisement in a German pharmaceutical journal in 1912 for Acetylsalicylsäure 'Heyden', produced by the Chemische Fabrik von Heyden (Dresden, Germany) – a competitor to Bayer. Clearly the chemical name used by Heyden was not as convenient as the trade mark name 'Aspirin', owned by Bayer. (Courtesy of Dr Jan McTavish.)

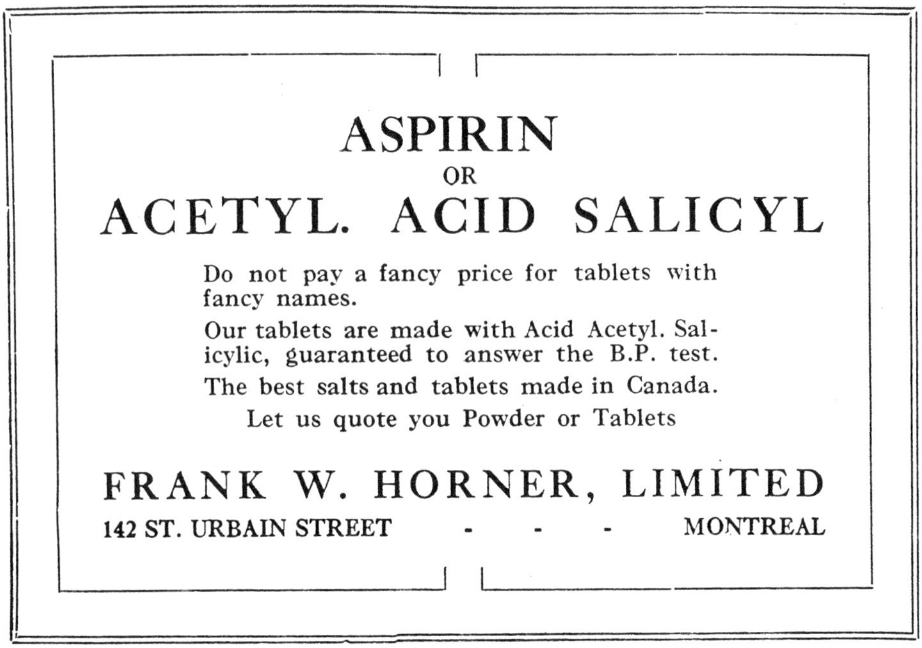

Plate 8 An advertisement in the *Canadian Pharmaceutical Journal* of 1917, which uses the trade mark name for Aspirin (now held by the Bayer company in that country) and implies that a less expensive product of the same name could be purchased from the company in Montreal. (Courtesy of Dr Jan McTavish.)

History and Development of the Salicylates

K.D. Rainsford

INTRODUCTION

The historical development of the salicylates is one of the classical stories in the field of pharmacognosy – i.e. of natural product-derived remedies giving rise to modern pharmacological agents. These drugs have proven an immensely successful class and have been a major impetus for the development of other anti-inflammatory, analgesic and antipyretic agents (Rainsford, 1984; Otterness, 1995).

This chapter reviews the historical and medico-scientific developments of the salicylates – some aspects of which have been reviewed previously (Rainsford, 1984). Chapter 2 reviews the development of the classical analgesics in the wider context and highlights some of the important industrial history of these drugs, especially of aspirin.

EARLY USE OF SALICYLATE-CONTAINING PLANTS

Hippocratic era

The history of aspirin and the present day salicylates has its origins in the use of various salicylate-containing plant extracts. About 2400 years ago, Hippocrates recommended juices of the poplar tree and willow bark for the treatment of eye diseases and pain in childbirth, respectively (Gross and Greenberg, 1948). In the monumental *Papyrus Ebers* (*circa* 1550 BC), it is stated that a remedy to expel rheumatic pains (phlegma) in the womb is the application of dried leaves of myrtle, which contain appreciable salicylates (Gross and Greenberg, 1948). This is prepared with an 'excellent' beer applied to the sacral and hypogastric regions (Ebbell, 1937). Thus, the analgesic and anti-inflammatory properties of plant extracts containing salicylates have been recognised from these early times.

Roman and Greek medicine

In AD 30, Aulus Cornelius Celsus, in one of his encyclopaedic works *De re medica*, recognised the four cardinal signs of inflammation: rubor, calor, dolor and tumor (i.e. redness, heat, pain and swelling) (Margotta, 1968). He stated that a boiled vinegar extract of willow leaves could be employed for the relief of pain owing to prolapse of the uterus and other conditions. Caius Plinus Secundus (AD 23–79), also known as Pliny the Elder, wrote in his massive treatise *Natural History* of the use of poplar bark infusions for pain in sciatica and poultices made from vinegar-soaked poplar bark for the treatment of gout. Pliny also recognised the keratolytic actions inherent in the salicylate-containing preparations, and recommended a paste made from the ash of willow bark for removing corns and callosities (Gross and Greenberg, 1948). In about AD 60 the famous army physician Dioscorides, of Silicia (now in Southern Turkey), wrote a compendium of the pharmaceutical properties of plants, the *Materica Medica*, and stated that a boiled aqueous extract of willow leaves or ash of willow bark could be used for treating corns, skin diseases, gout and earache (Gross and Greenberg, 1948).

In the second century AD, Galen employed the antiseptic properties of willow leaves (which are now known to contain salicylates) for the treatment of various skin conditions such as wounds, ulcers and erysipelas (Gross and Greenberg, 1948). Thus by Roman times there were numerous different therapeutic applications for salicylate-containing plants, many of which are appropriate by today's standards.

Asia and America

Salicylate-containing plants were used from the early days of civilisation, not only in Europe and the Middle East, but also in Asia (Perry and Metzger, 1980). It is, however, more difficult to date their introduction in this region. In China, preparations of the bark of the poplar tree (*Populus alba* L.) and decoctions of young shoots from *Salix babylonica* L. have been used for centuries for treatment of rheumatic fever, colds, haemorrhages and goitre, and as a general antiseptic for wounds and abscesses (Stuart, 1911; Hu, 1945). The bark of another willow species, *Salix purpura*, has been used in Burma for the treatment of rheumatism (Mosig and Schramm, 1955). Undated references also exist regarding the therapeutic uses of wintergreen as a treasured herb of the North American Indian (Levy, 1970).

European Middle Ages

There are frequent references in writings and pharmacopoeias of the Middle Ages and Renaissance periods to the therapeutic value of remedies including bark leaves and fruits that contain various salicylates (Gross and Greenberg, 1948; Lévesque and Lafont, 2000; Figures 1.1 and 1.2). In the Swiss herbal compendium *Herbarius zu Deutsch* (*circa* 1486; De Cuba and von Kaub, 1486–1490), the 'Master Serapio' is quoted as recommending plasters of burnt willow bark and leaves mixed with vinegar for the treatment of wounds and ulcers. A mixture of powdered or crushed willow leaves mixed with peppercorns and 'burnt' water (probably Schnapps or some such distilled spirit) is recommended for diarrhoea, and willow juice mixed with water is also suggested as being useful for menstrual bleeding and dysentery. Galenius is quoted in this herbal compendium as recognising that *Salix* flowers made into a plaster with rose oil made tissues that were initially hot and wet, cold and dry – tacit recognition of the anti-inflammatory properties present in extracts of these species. Similarly, alcohol mixtures or extracts of wintergreen and other salicylate-containing species were also recommended extensively for external and internal use by W.H. Ryff in his *Reformed German Apothecary* (Ryff, 1573; Figure 1.1).

In his *Dictionary of Drugs*, Nicolas Lemery (1759), from Rouen, apothecary of the Grand-Prévost (later the Medical Sciences Academy), stated that leaves of the *Salix* genera could be used to stop fevers, haemorrhages and vapours, and even for treating insomnia (Lévesque and Lafont, 2000). A 1760 manuscript in the museum Flaubert de l'Histoire de la Médecine et Pharmacie de Rouen gives numerous recipes based on acidic aqueous extracts of the leaves of reine-des-prés for the relief of fevers (Lévesque and Lafont, 2000). It appears that the antipyretic properties of leaves or other parts of *Salix* species were thus recognised by several French and German writers of the time (Gross and Greenberg, 1948; Lévesque and Lafont, 2000).

Another of the earlier references to the antipyretic actions of salicylate-containing preparations is attributed to the Reverend Edward Stone, of Chipping Norton in Oxfordshire, England (Figure 1.3). Stone is

Figure 1.1 Early documentation on the uses and actions of the salicylates from plant sources. Description and medical uses of the wintergreen plant in the *Reformed German Apotecken* of W.H. Ryff, published in 1573. The wintergreen (*Pyrola* sp.) is described by this author as being the most useful of the herbs for treating wounds. When used externally or internally a syrup, juice or 'Schnapps' (distillate) is prepared. Ryff points out that this has a rough taste on the tongue. He describes this herb as having extensive healing powers, and states that it can be employed to heal ruptures. The name 'wintergreen' is derived from the fact that it resists the frosts and cold of winter yet still remains green. (From Rainsford, 1984. Reproduced with permission of the publishers, Butterworths/Heinemann.)

accredited with serious study of this property. In a report to the President of the Royal Society (London) in 1763, he described what appears to have been the first clinical trial of a salicylate-containing preparation in some 50 subjects (Stone, 1763). In this trial, which extended over some 5 years, he employed the bark of the willow (*Salix alba*) as a replacement for Peruvian bark or Quinquina (a source of quinine as we know it today) for the treatment of paroxysms and fever from agues or malaria, which was then still endemic in Britain. Peruvian bark had been used in the treatment of this condition and also of rheumatism since its introduction to Europe in 1676 by Sir Thomas Sydenham, but its association with the Jesuits, its high cost, the short supply and the peddling of bogus preparations (Dewhurst, 1966) no doubt contributed to interest in a locally available replacement. The idea of employing this remedy came to Stone after he observed that willow bark had a similar bitter taste to that of Peruvian bark. Furthermore he stated (Stone, 1763):

> As this tree delights in a moist or wet soil, where agues chiefly abound, the general maxim that many natural maladies carry their cures along with them, or that their remedies lie not far from their causes, was so very apposite to this particular case, that I could not help applying it; and that this might be the intention of Providence here ...

THESAVRVS PHYTOLOGICVS
Das ist:
Neu-eröffneter und reichlich-versehener
Kräuter-Schatz,
Worinnen
Alle in der Artzney-Kunst gebräuchliche Gewächse/
welche in allen 4. Theilen der Welt/ sonderlich aber in Europa,
herfürkommen, nebst ihrer ausführlichen Beschreibung, Nahmen und
Beynahmen, in Teutsch-Lateinisch und Teutscher Sprache, ingleichen dererselben
sonderbaren Eigenschafften, Tugenden und fürtrefflichen Würckungen, samt beygefüg-
ter Art und Weise, wie solche herrliche und bewährte Artzney-Mittel in allerley
Zufällen und Kranckheiten an Menschen und Viehe ohn einige Gefahr
zu gebrauchen;
In zweyen Theilen befindlich/
Deren der erste die Kräuter und Blumen/ der andere die Bäume
und Stauden-Gewächse vorstellig machet.
Allen Aertzten/Wund-Aertzten/Apotheckern/Gärtnern/
Hauß-Vättern und Hauß-Müttern/ absonderlich auch denen
auf dem Lande wohnenden kranck darnieder liegenden Personen so wohl zur
Hülffe und Ergötzung/ als hauptsächlich zu Nutz und Erhaltung
guter Gesundheit mitgetheilet/
Zugleich mit dreyen vollständigen Registern versehen
Von
PETRO HOTTON,
Med. Doct. et Hort. Acad. Lugdun. Praefect.
Mit Röm. Kayserl. Majest. allergnädigstem PRIVILEGIO.
Nürnberg/
Verlegts Johann Leonhard Buggel und Johann Andreas Seitz,
Buchhändlere. 1738.

Natur/ Krafft und Würckung.

Diese zeitige Beerlein werden in den Apothecken fürnemlich gebraucht, haben eine Krafft den zähen Schleim und die wässerige Feuchtigkeiten aus dem Leib zu purgiren, dahero in der Wassersucht, Cachexia und Glieder-Kranckheit gegeben werden. Gebräuchlicher aber ist
der daraus bereitete Syrup, den man Syrupum de Spina cervina, sive
domesticum & familiarem nennet; so durch den Stulgang die Gall,
Schleim, und insonderheit die wässerige Feuchtigkeiten aus dem Leib
ohne sonderlichen Zwang, führet, dahero in der Wassersucht, Gelbsucht,
Fiebern, Frantzosen und andern Kranckheiten nützlich gebraucht wird.
Man kan auch ein Rob de Spina cervina, welches mit dem Syrup glei-
che Krafft und Würckung hat. Das Decoctum mit Alaun-Wasser
vermischt, widersteht der Mundfäule.

Weiden / Salix.

Weiden/Weiden-Baum/ weisse Weiden/ (als an welchen die
Blätter und Gerten weißlich,) Bruch-Weiden/ Seidelwei-
den/Band-Weiden/ Felbinger/ Lein-Weiden/ Graecè iris,
Latinè Salix, alba, arborescens, Casp. Bauhin. arborea, angustifolia,
alba vulgaris, maxima fragilis, alba hirsuta, Joh. Bauhin. Es gibt der
Weiden unterschiedene Arten, welche man aber in den Apothecken ohne
Unterschied braucht, und zwar die Blätter, Kätzlein, Rinde, Mistel,
und Schwämme.

Natur/ Krafft und Würckung.

Die Blätter haben einen bittern Geschmack, ziehen etwas zusam-
men, kühlen, und trocknen: in Wasser gesotten und getruncken, sind gut
vor das Brechen, Blut-speyen und Blutgang, verstopffen auch die
überflüssige Monat-Zeit, stillen das Nasenbluten, temperiren die Hitze
des Gebluts.

Die Rinde zu Aschen gebrannt, und mit Schaf-Koth und Essig
vermischt, und übergelegt, vertreibet die Wartzen und Hüneraugen.

Das Wasser/ so von den jungen Schößlein der weissen Weiden/
welche erst im May ausschlagen, destilliret wird, ist sehr gut für Grieß
und Stein, treibet solchen durch den Harn fort: Die Augen damit ge-
waschen, vertreibet und benimmt die Röthe derselben, und macht sie
hell und klar.

Aaa aaaa 2   Etliche

Figure 1.2 Description by P. Hotton (1738) of the nature and actions of extracts of *Salix* sp. (From Rainsford, 1984. Reproduced with permission of the publishers, Butterworths/Heinemann.)

Figure 1.3 The Parish Church of St Mary the Virgin, Chipping Norton, Oxfordshire, where the Reverend Mr Edward Stone was vicar during the mid-eighteenth century, when he studied the antipyretic activity of willow bark extracts. (Photograph kindly provided by Dr Brian Callingham, University of Cambridge, Cambridge, UK.)

Seemingly, the teleological thinking inherent here was common philosophy at the time in the form of the ancient Doctrine of Signatures.

The substitution of willow bark for Peruvian bark was so successful that in 1798 a Bath apothecary, Mr William White, was able to report that this had enabled a saving of at least £20 a year to the charity, The Bath City Infirmary and Dispensary (White, 1798). The therapeutic efficacy of willow bark was also endorsed by Wilkinson (1803) of Sunderland, UK. It is of interest that the weeping willow, *Salix babylonica*, was probably introduced into Britain from Eastern countries during the late seventeenth or early eighteenth centuries (Scaling, 1872).

In 1849 a wintergreen-like preparation of salicylates was isolated from the spruce, *Gaultheria procumbens*, by an American, William Procter, and was given the name 'Tea of Canada'. It was claimed to have antiseptic and antirheumatic properties.

An issue of particular interest is when development first took place of recognised or standardised procedures for preparing extracts of *Salix* species and oils of wintergreen (*Wintergrünöl*) (see also Figure 1.2). In an attempt to identify when these procedures were developed to an acceptable standard for inclusion in pharmacopoeias, several such texts have been consulted. Among the earliest of these was James' *Pharmacopoeia Universalis*, or *A New Universal English Dispensatory* (1764) which describes the use of the Common Willow that 'outwardly they are of service in Haemorrhages from wounds, or from the nosticles, that like disorders; and are of service in Baths for the feet, in order to procure sleep, and to cool the Heat of Fevers. The Ashes of the Bark of this Tree are effectual for extirpating Warts and Corns' (James, 1974). The *Pharmacopoeia Helvetica Basilea* (1771), which, although having a very detailed Index Morborum et Curationum recommending a wide range of natural medications or concoctions for the treatment of arthritic, inflammatory or painful conditions, makes no mention of the use of *Salix* preparations. Likewise, there is no mention of these preparations in the *Pharmacopoeia Austrico-Provincialis* of 1780 (Dr Gy Rádóczy, personal communication), the *Strasbourg Pharmacopoeia Generalis* (Spielman, 1783), or the first *Pharmacopoeia Helvetica* (1865) or the Altera edition (1872). The third edition of the *Pharmacopoeia Helvetica* of 1893 gives a detailed description of the solubility and general properties of *Acidium Salicylicum*, methods for (re)crystallisation and the preparation of various salts.

Of the German pharmacopoeias, the *Deutsche Pharmacopoeia* of 1872 (Hayer, 1872) has no mention of the *Salix* preparations. However, in the *Kommentar zur Pharmacopoeia Germanica* in 1874, Hayer (1874) gives a detailed description of the preparation of *Wintergrünöl* (*methyl salizylsäure*), and includes the following formula:

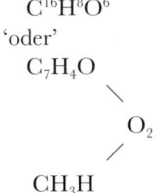

$C^{16}H^8O^6$
'oder'
C_7H_4O

O_2

CH_3H

He also gives the specific gravity as 1.18 and the distillation point as 225°C.

It would therefore appear that Hayer (1874) was the first to describe the preparation of methyl salicylate, while the preparation of salicylic acid was only described in 1893 despite it having been synthesised by Kolbe and Lautemann in 1860. The apparent lack of references to *Salix* preparations in pharmacopoeias in the late eighteenth and early nineteenth centuries does not mean that they were not recognised or produced until then. Recipes for preparing therapeutically active extracts or preparations were often employed regardless of their lack of citation in pharmacopoeias in this period. The first edition of the *Pharmacopoeia Hungarica* of 1883 gave details of salicylic acid and its sodium salt (Dr Gy Rádóczy, personal communication).

Nineteenth century

During the nineteenth century period there was an enormous upsurge in interest and therapeutic development of the salicylates. The main advances were:

1. The preparation of salicylic acid, first from natural sources of salicylate and later from chemical synthesis (Gross and Greenberg, 1948; Bekemeier, 1977).
2. Recognition of the therapeutic properties of salicylic acid, salicin (salicyl alcohol glycoside) and methyl salicylate; the latter two served as the initial sources of the salicyl moiety for preparation of the acid (Gross and Greenberg, 1948; Bekemeier, 1977).
3. Synthesis and manufacture of acetylsalicylic acid (aspirin) for clinical use (Alstaedter, 1985; Busse, 1989).

CHEMICAL DEVELOPMENT OF SALICYLATES

Initially salicylaldehyde was extracted by the Swiss pharmacist, Pagenstechser (1835), who obtained it from distillation of the flowers of *Spiraea ulmaria*. He subsequently transmitted this information to the German chemist K.J. Löwig, who then produced salicylic acid (which he called *Spirsäure*, or Spiria acid, an acid of the *Spiraea* species) by oxidation of salicylaldehyde (Flückinger, 1888). Between 1826 and 1829, an Italian pharmacist in Verona worked on the isolation of the glucoside, salicin, from willow (Lévesque and Lafont, 2000). Buchner, a German, isolated salicin from willow in 1828 (Lévesque and Lafont, 2000). Leroux, a pharmacist at Vitry-le-Francois, isolated and purified salicin in 1829 and later investigated its antipyretic properties (Leroux, 1830; Galmiche, 1957; Hedner and Everts, 1998; Lévesque and Lafont, 2000). He showed that salicin was the glucoside of salicyl alcohol, and developed a procedure for obtaining 30 g of salicin from 1.5 kg of willow bark (Lévesque and Lafont, 2000). In 1833, E. Merck of Damstadt (the forerunner to the chemical company of the same name) developed a procedure for extraction that was half as successful, and 2 years later Löwig crystallised *Spirsäure* (the acid of the *Spiraea* sp.) (Lévesque and Lafont, 2000). Thus, several attempts were being made at that time to isolate and purify the principle salicylate from *Salix* species.

Pure salicin was used extensively in the mid-eighteenth century for treatment of rheumatism (Gross and Greenberg, 1948; Galmiche, 1957; Bekemeier, 1977; Hedner and Everts, 1998; Lévesque and Lafont, 2000). In 1835 Piria, who was head of the Chemical Institute of Pisa, prepared salicylic acid from salicin (Galmiche, 1957; Hedner and Everts, 1998; Lévesque and Lafont, 2000). It was also prepared by the action of phosphorous perchloride on oil of wintergreen (*Gaultheria*) or methyl salicylate by the French chemist Cahours (1845), and by the Scottish chemist Couper (1858). Incidentally, Couper also elucidated the carbon bonding of aromatic compounds (including that of salicylic acid) before Kekulé, to whom this is often attributed.

In 1877, Benoit described the preparation of salicylic acid by treating anthranilic acid with nitric acid. However, the chemical synthesis of salicylic acid as employed today was pioneered in 1860 by Herman Kolbe (1818–1884) (Kolbe, 1860; Kolbe and Lautemann, 1860). Later refinements by Kolbe led to the development in 1874 of the famous procedure bearing his name, whereby sodium phenoxide is carboxylated with carbonic acid (see details in Chapter 3). Kolbe developed the first full-scale commercial synthesis in 1874 in the kitchen of one of his students, von Heyden, in Dresden (Kolbe, 1874; Bekemeier, 1977). Frederick von Heyden (1838–1926) was convinced by Kolbe to take up its manufacture, and he did this at his 'Salicylsäurefabrik Dr von Heyden', which was later renamed the 'Chemische Fabrik F. von Heyden AG' of Dresden-Radebeul (Bekemeier, 1977). The Kolbe procedure is essentially that used for the commercial synthesis of salicylic acid today.

The ready supply of salicylic acid soon enabled its use to be extended as an antiseptic – as suggested by Kolbe – to replace the more noxious compound phenol, which was originally pioneered by Lister (1827–1912) and used in surgery up until that time (Otterness, 1995).

CLINICAL OBSERVATIONS ON THE USE OF SALICYLATES IN THE NINETEENTH CENTURY

The use of salicin, salicylic acid and its sodium salt in the early eighteenth century appears to have been largely for the relief of fever (especially typhoid and acute rheumatic fever) as well as for various painful conditions (Enser, 1876; Maclagan, 1876; Gross and Greenberg, 1948; Goodwin and Goodwin, 1981; Figure 1.4). It is difficult to establish the exact timing of the acceptance or use of these drugs, or indeed the identity of the individuals who first established their efficacy. It is difficult to determine the impact of the Reverend Stone's findings in the mid-eighteenth century as well as the availability of salicin extracts or pure salicylic acid (or the sodium salts) in the mid-nineteenth century, which enabled trials of these agents for treatment of fever, pain and rheumatic conditions.

It appears that the Scottish Physician Thomas John Maclagan (1838–1903; Figure 1.5), Resident Superintendent of the Dundee Royal Infirmary, may have been the first person to recognise the effectiveness of salicin for the fever and articular pain in patients with 'acute rheumatism' (rheumatic fever)

Index Medicus Advertiser.

Figure 1.4 Advertisements from the *Index Medicus Advertiser* of 1880 attesting to the virtues of the salicylic acid and salicylin-containing preparations for treatment of a variety of conditions including rheumatism and gout. (From Rainsford, 1984. Reproduced with permission of the publishers, Butterworths/Heinemann.)

Figure 1.5 Dr Thomas John Maclagan (1838–1903), whose studies in Dundee on salicin and salicylic acid formed a basis for understanding their antipyretic and antirheumatic affects. (Reproduced from the paper by Stewart and Fleming (1987) in the *Scottish Medical Journal*, with permission of its publishers, Hermiston Publications Ltd, Whitekirk, East Lothian Scotland.)

in trials initiated in November 1874 but not published until 1876 (Maclagan, 1876a; 1876b; Stewart and Fleming, 1987). In an uncontrolled trial of eight patients he noted rapid reduction in fever and loss of articular pain in those that received salicin (Maclagan, 1876a). Maclagan was much concerned with febrile illnesses, as is evident from his considerable number of publications on this subject (Stewart and Fleming, 1987), and drew the analogy between fever from infectious diseases (e.g. malaria) and that in acute rheumatic fever. He recognised the need for a therapeutic agent for rheumatism that would have actions analogous to those of quinine in cinchona bark in relieving fever (Stewart and Fleming, 1987). He also recognised the association of rheumatism occurring in damp localities, and the growth of various willow species in these conditions so employing the Doctrine of Signatures in the same way as the Reverend Stone had a century earlier.

Antipyretic activity

The full therapeutic potential of pure salicylic acid as an antipyretic was first fully explored in 1875 by the young Swiss medical assistant Carl Emil Buss (1849–1878), at the St Gallen Cantonal Hospital (Buss, 1875; 1876; 1878; reviewed by Büss and Balmer, 1962). Following the submission of his 49-page thesis concerning the antipyretic actions of salicylates for his State Medical Examination at Basel (Buss, 1875), Buss moved to the Medical Clinic at the Bürgerspital (now Kantonspital) in Basel where he performed what must have been the first detailed experiments in fevered animals and patients (with typhoid fever and other afflictions) that demonstrated the antipyretic effects of salicylic acid (Buss, 1875; 1876; 1878; Büss and Balmer, 1962). These experiments were performed with remarkable scientific ability and insight. Buss may also have been the first to document detailed observations and experiments (many on himself) demonstrating the appearance of tinnitus and other side effects following oral ingestion of large quantities (up to 4 g) of salicylic acid (Buss, 1875; Büss and Balmer, 1962). He observed gastric irritation induced by salicylic acid in rabbits, and found that co-administration of sodium bicarbonate reduced this

(Büss and Balmer, 1962) – a procedure still employed with aspirin today. Buss also recognised the value of salicylic acid for the treatment of acute joint rheumatism (Büss and Balmer, 1962).

In December 1875, Riess (from the Berlin Municipal Hospital) published results showing the effects of salicylic acid in the treatment of fever in 400 patients with typhoid – a not inconsiderable number of patients (Riess, 1875; 1876). He claimed that the internal effects of salicylic acid 'energetically lowered' the fever within hours. He ascribed the effect to the antiseptic actions of salicylic acid, based on what Kolbe and co-workers had previously considered the anti-fermentation action on bacteria. Indeed, Kolbe had proposed that since carboxylation of phenol was used to produce salicylic acid, the latter would serve as a pro-drug and form phenol and so act as an internal antiseptic (Otterness, 1995). This antiseptic concept no doubt underlay the application of benzoic acid (or its sodium salt) in the studies of Senator (1879) regarding its antipyretic action in treating rheumatic polyarthritis. This author's results suggested that benzoate had equivalent antipyretic effects to that of salicylate.

Riess (1875) noted that dosages of 5 g of salicylic acid given 23 times to seven volunteers were well tolerated. He did, however, notice that even in these healthy individuals the temperature was lowered by an average of 0.9°C in 4–6 hours. He observed that when given as an aqueous suspension or an alcoholic solution (5 g salicylic acid, 20 parts spirit vin. Rect. and 30 parts glycerin) it was 'not pleasurable', and after repeated dosage led to irritation of the throat and oesophagus. Having had the idea that salicylic acid was absorbed as the sodium salt, he experimented with sodium phosphate and bicarbonate salts. He observed that most of the patients took these sodium salt formulations without difficulty although occasionally there was vomiting after the first dose, but this was not so bad that the salicylic acid couldn't take effect.

Riess (1876) observed that 2 g salicylic acid reduced fever in children between the ages of 6 and 12 years. This was probably the first report of the use of this drug as an antipyretic in children. Patients with cystitis were also found by Riess to have their fever lowered by administration of 5 g salicylic acid. He noted that the dose of 5 g salicylic acid had no influence on heart rate in febrile patients. This dose often produced a reduction in temperature of 2° to 6°C within 2 hours. The degree of reduction in fever was found to depend on 'the reason for the fever' as well as the time of dosage. There was some variability in the duration of effect – in a large number of cases it lasted for up to 24 hours. The condition in which the efficacy of salicylic acid was best demonstrated was ileo-typhoid.

A considerable number of studies were reported in the late 1870s and early 1880s attesting to the antipyretic effects of salicylates (Fürbringer, 1875; Blanchier and Bochefontaine, 1878; Beyer, 1880; Hallopeau, 1880; Kersch, 1880; Laborde, 1880; 1881; Livon, 1880; Löwitt, 1881).

Treatment of acute rheumatism (rheumatic fever)

Stricker (1876), of the Charité Berlin, reported the successful treatment of acute rheumatism with salicylic acid. He noted that the polyarthritis was relieved 48 hours after treatment with 5 g salicylic acid, and pericarditis had likewise disappeared. There was an interesting issue raised by Stricker concerning the purity of salicylic acid preparations then available. He noted that the yellow tint of some preparations suggested the presence of impurities such as carbolic acid, which when suspended in water made a cloudy mixture. He pointed out that when the formulation was recrystallised it had shiny white needles, was odourless, and completely dissolved in alcohol. Stricker's attention to these details is important and, the inference that impure preparations of salicylic acid were then available highlights the potential for the variations in results and side effects that may have been observed by other authors.

Most of the studies by Stricker were case reports and opinions of the author. He did express doubts about the use of salicylic acid in the treatment of rheumatic polyarthritis.

There was much interest and heated debate regarding the use of various salicylate preparations for treating rheumatic conditions in the London, German and French medical journals of the late 1870s and early 1880s (for example, see Anon., 1876; 1880a; 1880b; 1881; Kunze, 1876; Maclagan, 1876a; 1876b; 1879; 1880; Myers, 1876; Bälz, 1877; Hughes, 1877; Bartels, 1878; Boulonmié, 1879; Diesterweg, 1879; Downes, 1879; Finlay and Lucas, 1879; Senator, 1879; Bryden Hill, 1880; Ord, 1880;

Robertson, 1880; Taylor, 1880; Chateris, 1881; Fagge, 1881; Fowler, 1881; Hall, 1881; Hood, 1881; Investigator, 1881; Kemp, 1881; Owen, 1881; Warner, 1881). Among these were reports of clinical trials, a considerable number of which were performed in London hospitals. Donnelly (1880) in the USA reported about the efficacy of potassium salicylate in treating acute rheumatism.

As can be seen from these papers, the use of salicylates was not without its critics at this time. Myers (1876) expressed concern about the all too enthusiastic acceptance of Maclagan's work, especially as he pointed out that salicylic acid could cause haemorrhage of the mucosa in the stomach. Maclagan (1879) was obviously aware of the potential hazards of these preparations, but still considered that large doses of salicylic acid or salicin were equally efficacious in treating rheumatic fever. Finlay and Lucas (1879) also observed that a much higher relapse rate in rheumatic fever was observed with low doses of sodium salicylate than in patients given potassium bicarbonate or quinine. Maclagan was obviously aware of the potential cardiotoxicity of high doses of the salicylates, and warned that they could cause fatal depression of heart function in susceptible patients with rheumatic fever (Maclagan, 1880). Others (Huber, 1879; Akland, 1881; Apolant, 1881; Watts Parkinson, 1881) highlighted the cerebral complications (including oedema and delirium) that could occur with high doses of salicylates. Fowler (1881) raised serious concerns about the purity and price of salicylic acid preparations. He had Kolbe's salicylic acid preparations tested, and the results showed the presence of 'a considerable quantity of a substance not precipitated by nitrate of silver, and therefore not salicylic acid'. While Kolbe's preparation may have been cheaper, the implication was that it was not pure.

While salicylate was superior for treatment of 'acute rheumatism' or rheumatic fever was the subject of much debate (James, 1881). Maclagan (1880), an advocate of salicin because salicylic acid was regarded as being more irritant to the alimentary canal (Myers, 1876), considered also that 'salicylic acid, no matter whether given alone or in combination with soda, exercises depressing action on the heart'. He was especially concerned about this depressant action in subjects with rheumatic myocarditis: 'the depression is likely to be alarming, and may be fatal' (Maclagan, 1880). He was not the only person to observe cardiac effects of salicylates in this condition (Mahomed, 1880). Interestingly, he considered that cardiac inflammation could predispose the patient to dangerous side effects. Other authors (Fowler, 1881) found salicylic acid had no appreciable toxicity, although it did produce gastrointestinal (GI) and central nervous system (CNS) side effects.

Taylor (1880) considered that sodium salicylate improved the pericarditis that was evident upon admission. However, patients did appear to become 'unusually anaemic' and took a long time to recover. There was a relatively high frequency of relapses with the salicylates (Finlay and Lucas, 1879; Fagge, 1881). Tinnitus and deafness were also common, probably reflecting use of high doses of these drugs. James (1881) explored the potential of various salts (ammonium, potassium, lithium, lime) of salicylates and also, interestingly, the quinia and cinchonidia – the latter reflecting the potential of two drugs as salts of one another being employed as antipyretics.

A summary of the debate in the Medical Society of London regarding the use of salicin and salicylates in treating rheumatic fever, which included presentations and discussion by Drs Hilton Fagge, Isambard Owen, Thomas John Maclagan and Greenhow (quoted above), led the writer (Anon., 1881) to conclude:

A sufficient length of time has now elapsed since the introduction of these drugs into medicine to enable us to arrive at some definite conclusion concerning their action; and such a conclusion can only be arrived at impartially, freed from bias or prejudice, by carefully compiled statistics. For, in spite of the unfavourable conclusions which Dr Greenhow felt compelled to draw from his elaborate and critical analysis of his series of cases communicated to the Clinical Society two years ago, there is certainly no evidence that these remedies are losing ground in medical favour.

Few can question the potency of the remedy in cutting short the pyrexia and joint affection, although it does not appear to influence the cardiac manifestations. The open questions are the best mode of administration of the drug, the amount necessary to be given, the relation of relapse or 'recrudescence' to the treatment, if any relation there be, and the toxic effects generally ascribed to it, especially those involving the heart and the nervous system.

That 'statistics' of the outcomes should be considered was also emphasised. Some of the reports included a substantial number of cases (Fagge, 1881; Hood, 1881; Warner, 1881), but the term 'statistics' here is a misnomer since statistical analyses *per se* were not performed on these data at that time. None the less, authors published data regarding the results from salicylate treatments at Guy's and the London Hospital, and the East London Hospital for Children. The cases (totalling 415 patients) that received salicylates and were reported by Fagge (1881) mostly responded to salicylic acid or salicin, and these seemed to have improved more than patients treated with lemon juice, bicarbonate of potassium or free blistering; these latter treatments might be considered to approach the efficacy of placebos by present day standards. Of 355 patients treated at Guy's Hospital with salicylic acid or salicin, 128 appeared to have had one or more relapses (Fagge, 1881). Most appeared to have become free from pyrexia within 11 days of initiation of therapy. In another group, about one-third of patients appeared to have had relapses (a similar proportion to that in the aforementioned group), but most of these had an improved febrile state by 4 days after starting of therapy.

Dr Charles H. May, from the Mt Sinai Hospital New York, reported what was probably the largest set of data on the effects of salicylic acid compared with other popular remedies of the time (May, 1884). This study comprised 400 cases treated at the Roosevelt Hospital, New York. The patients had rheumatism (acute, subacute and chronic), but the study excluded those with muscular, syphilitic or gonorrhoeal rheumatism, and those with rheumatoid arthritis. These data are impressive, since they include full patient details, the time of year that patients were admitted, the duration of their treatments, their maximum temperatures, frequency of attacks, and relapses, and the frequency of cardiac valvular lesions, other complications and major outcomes. The patients received various remedies, among them: (a) salicylic acid as a solution initially, then as capsules, comprising 10 grains every 2 h at first, then at 3- to 4-hourly intervals; (b) Rochelle salt (potassium sodium tartrate), 1 drachm three times a day; (c) iodine of potassium 10 to 20 grains three times a day; and (d) 'wine of colchicum' and iodide of potassium 10 to 20 grains of each three times daily. Alkalis were commonly prescribed at this time, and no doubt the use of potassium iodide was based on the view that this would treat the infection that underlay the disease. Some patients also received atropine ($\frac{1}{120}$th grain t.i.d by injection). This was probably the most substantial compilation of data on the therapy of rheumatic fever at this time. The data summarised from an analysis of a subset of 271 cases from this study (Table 1.1) (May, 1884) were used as evidence for the efficacy of salicylic acid, which was probably superior to the other remedies.

Professor Germain Seé is claimed to have introduced salicylic acid to France and reported its use in acute and chronic rheumatism in 1877 (Editorial, 1877; Galmiche, 1957; Keitel, 1977; Lévesque and Lafont, 2000). He described the effectiveness of up to 10 g per day of sodium salicylate in the treatment of gout and noted the urinary excretion of uric acid with this therapy (Keitel, 1977). The editorial in the *British Medical Journal* noted that Seé was making much money from this therapy, for a gouty old gentleman once had to pay 2000 Francs (or £80) for a consultation (Editorial, 1877). Nonetheless, the use of salicylic acid, its sodium salt and salicin for the treatment of various rheumatic conditions, including gout, became quite extensive after this rather chequered history.

In a thesis for Doctorate of Medicine at the School of Medicine in Paris, Hogg (1877) described the results of treating 12 patients with 'rheumatism articulaire' having 'aigu or subaigu' using salicylic acid, sodium salicylate or salicin. He mentions in the thesis a trip to England, where he had heard about using salicylates for the treatment of typhoid and 'rheumatism articulaire febrile'. Thus the question of whether Professor Seé was responsible for the introduction of salicylates as therapy in France is debatable. The extensive number of citations in Gross and Greenberg (1948) attests to the extensive use of salicylates in treating rheumatic fever thereafter, although their efficacy and influence on the duration of hospitalisation and pericardial outcomes was often the subject of debate. The overall consensus has been that the use of high-dose salicylates in treating rheumatic fever during the pre-antibiotic period undoubtedly reduced pericardial complications, as well as being effective in reducing fever and time of hospitalisation (Gross and Greenberg, 1948).

CHAPTER 1

TABLE 1.1

Outcomes from treatment with salicylic acid and other remedies as reported by May (1884).

Treatment	Average duration of joint symptoms (days)* before hospital	No. of cases	Average duration of joint symptoms (days)* after hospital	No. of cases	Average duration of pyrexia (days)*	No. of cases	Average duration in hospital (days)*	No. of cases
Salicylic acid	17.2	110	12.5	113	4.9	78	24.8	113
Rochelle salt	37.1	21	26.8	21	8.7	10	34.3	21
Iodide of potassium	85.3	14	26.1	14	n.d.	n.d.	40.8	14
Iodide of potassium + colchicum	38.2	31	25.8	21	n.d.	n.d.	33.7	31
Salicylic acid, then iodide of potassium + colchicum	19.1	18	23.3	18	3.3	11	26.3	18
Atropia	10.6	8	21.4	8	7.3	6	40	8
Miscellaneous	23.4	65	27.4	66	12.4	33	34.5	66

*Data on duration rounded up to one decimal place from original data.

EARLY CLINICAL STUDIES WITH SALICYLATES IN THE TREATMENT OF RHEUMATOID ARTHRITIS

Goodwin and Goodwin (1981), in their comprehensive review on the failure to recognise the use of salicylates in the treatment of rheumatoid arthritis, considered this might have been related to the relatively recent origins of this disease – which perhaps only became evident in the past 300 years or so. Recent analysis of historical writings, palaeopathology records, paintings and the fine arts has, while reinforcing this view, given hints that there may have been some patients with this condition even back in Roman times (Buchanan and Kean, 2001). The difficulty has been accurate diagnosis of the condition in these subjects based on the historical information (Buchanan and Kean, 2001). Differentiation of rheumatoid arthritis from gout, rheumatic fever and degenerative arthropathies was not achieved until the earlier part of the nineteenth century, when a number of English and French physicians described and illustrated classical features of rheumatoid arthritis (Buchanan and Kean, 2001). However, it was Sir Alfred Baring Garrod who first introduced the term 'rheumatoid arthritis' and provided a clear clinical description of this disease (Garrod, 1859), while Sir George Frederick Still first described 'juvenile rheumatoid arthritis' (Still, 1896–1897) and thus the condition bears his name (Goodwin and Goodwin, 1981; Buchanan and Kean, 2001). A major problem highlighted by Ord (1880) was the wide range of synonyms employed to describe what was later accepted as rheumatoid arthritis. In this early period a wide range of remedies was employed for this rheumatic condition, among them cathartics, bleeding, sweating, quicksilver (mercury), and induced blisters (Goodwin and Goodwin, 1981). The famous English physician, William Heberden the Elder (Heberden, 1802) also recommended Peruvian Bark (the source of quinine and opium).

Gross and Greenberg (1948) mention that one of the earliest references to the use of salicylate-containing plants for the treatment of 'inflammatory rheumatism' (presumably rheumatoid arthritis) was in *Dr Chase's Recipes, or Information for Everybody*, which was printed in Ann Arbor, Michigan, in 1865. The extensive introduction of salicylates for the treatment of rheumatoid arthritis during the latter part of the nineteenth century came with the availability of salicylic acid. There were several reports (Compagnon, 1880; Stewart, 1901) of the marked effects that high doses of salicylates had in alleviating the pain, stiffness, swelling and malaise of rheumatoid arthritis. However, as Goodwin and Goodwin pointed out in their detailed analysis of the use of salicylates (and later aspirin) in treating this condition, there was neither ready acceptance nor recognition of the efficacy of high doses in treatment until the 1950s (Goodwin and Goodwin, 1981). Indeed, the famous Canadian physician Sir William Osler, who recommended high-dose salicylates for treating rheumatic fever, considered that they were of little use in rheumatoid arthritis, emphasising in 1900 that 'the salicylates are useless' (for this disease) (Goodwin and Goodwin, 1981). There was also a belief during the early part of the twentieth century that they could be addictive and create dependence (Goodwin and Goodwin, 1981). Although at this time salicylates could be useful for pain relief in rheumatoid arthritis, often the dosage was lower than that required for good control of inflammatory pain. In their careful reconstruction of the period when salicylates were ignored for their value in rheumatoid arthritis, Goodwin and Goodwin (1981) noted that there was a trend to increase the dose from around six tablets to nine tablets a day in the 1920s to 1930s, progressively 'pushing the dose to the limits of toxicity' (i.e tinnitus).

As seen in Figure 1.4, which is reproduced from the *Index Medicus Advertiser* of 1880, salicylate therapy was being widely exploited in the USA at this time. Supplies of these salicylates came from Europe, the main supplier of salicylic acid being Heyden's plant in Dresden.

It has only recently been shown that willow bark extracts have *proven* therapeutic effects in the treatment of musculoskeletal pain under properly controlled clinical trial conditions. In a randomised, double-blind, placebo-controlled trial, Chrubasik and co-workers (2000) found that 120 to 240 mg bark extract appeared to have pain-relieving effects in a dose-related manner in 191 patients with low back pain. Patients given 240 mg/d were pain-free for 39 per cent of the time, those given 120 mg/d for 21 per cent of the time, and those in the placebo group for 6 per cent of the time.

CHEMICAL DEVELOPMENT AND EARLY CLINICAL STUDIES WITH ASPIRIN

Synthesis and commercial development of aspirin

A French chemist, Charles Frederich von Gerhardt, had studied acetylating reactions and first produced aspirin (acetylsalicylic acid) in 1853 by the treatment of sodium salicylate with acetyl chloride. The product (which was later found by Kraut to be impure) was referred to as *wasserfreie salicylsäure-essigsäure* (acetosalicylic anhydride) (Galmiche, 1957). He also studied the alkaline hydrolysis of aspirin to acetic and salicylic acids and the reaction of the latter with silver oxide (Galmiche, 1957; von Gerhardt, 1843; 1853a; 1853b). At Innsbruch (Austria), von Gilm (1859) obtained pure aspirin by treating salicylic acid in the same way. Kraut (1869) obtained pure aspirin from Gerhardt's product by extraction with ether (Galmiche, 1957).

Full commercial exploitation of aspirin was not to come until over 40 years later, in Germany. While Heyden's plant was doing capacity business in salicylic acid production, the order went out from the director of pharmacological research at Friedrich Bayer and Company (Elberfeld, Germany), Professor Dr Heinrich Dreser (Figure 1.6a), for his chemists to synthesise a compound that was competitively superior to Heyden's (von Gilm, 1859). The high doses of sodium salicylate and salicylic acid employed at the time had been found to cause nausea, vomiting and other untoward gastrointestinal symptoms (Myers, 1876; Shaw, 1887; Gross and Greenberg, 1948), so there was good reason to look for a safer salicylate (still a problem today!). It is said that the father of one of the chemists at Farben-

(a)   (b)

Figure 1.6   (a) Professor Dr Heinrich Dreser, Head of the Pharmacology Department at Bayer, and (b) Dr Felix Hoffman, to whom the discovery of aspirin was originally attributed. (Photographs donated by Bayer AG, Leverkusen, Germany.)

fabriken Friedrich Bayer and Company suffered from severe rheumatoid arthritis and pleaded with his son, Felix Hoffmann (Figure 1.6b), to search for a less irritating drug than sodium salicylate. Hoffman searched through the stores of salicylate preparations that had been synthesised for other purposes years before, and had them tested (Hochwalt, 1957; Busse, 1989). As a result of these tests he decided that aspirin was the most satisfactory, so he prepared samples of pure aspirin and gave them to his father to try (Hochwalt, 1957). The idea of acetylating phenols or aminophenols to control their toxicity was not new, since this had been applied in the synthesis of phenacetin and other compounds before that time (Leake, 1957).

However, the story that Hoffman was the 'discoverer' of aspirin has recently been challenged by Dr Walter Sneader of Strathclyde University (Glasgow, UK) (*Royal Society of Chemistry Press Release*, 6–10 September 1999). Dr Sneader became suspicious of the story while researching for a lecture on aspirin and asked the Bayer Company to allow him to examine Dr Hoffman's notebooks, only to discover that Dr Hoffman's supervisor, Dr Arthur Eichengrün (Figure 1.7), may have been the originator. Eichengrün had previously developed methods to make a compound more tolerable. While Hoffman undoubtedly did synthesise aspirin in 1897, it was clear from Dr Sneader's investigations that Hoffman had methodically adapted Eichengrün's scheme to make salicylic acid more tolerable. It was not until 1934 that Hoffman actually claimed the work as his own initiative, but in 1949 Eichengrün published his version of the events in an article in *Die Pharmazie* 1 month before his death (Eichengrün, 1949). Dr Sneader's investigations suggest that Eichengrün had become a very successful owner of a chemical company and that, owing to the anti-semitic sentiments of the 1920s and 1930s, he was not in a position to challenge the events claimed by Hoffman. Although interned in a concentration camp for 14 months he survived to write his own version of events, including the fact that he never benefited financially from sales of aspirin. It has therefore emerged from Dr Sneader's investigations that Eichengrün is the person to whom credit should be given for the discovery of aspirin (DiscoveryHealth.com, 2001).

It is also said that Dreser realised that he had an outstanding drug in 1899, and subsequently the Bayer Company introduced aspirin in that year (Martin, 1963; Busse, 1989). He also recognised that its chemical name, acetylsalicylic acid, was far too difficult to pronounce and sounded too much like the salicylic acid it was designed to replace (Hochwalt, 1957). In devising the simpler name, he is said to have recalled that natural salicylic acid had been prepared from plants of the *Spiraea* family: Löwig had named the salicylic acid he produced *Spirsäure*. Thus, Dreser added the 'A' for acetyl to 'Spirin' from 'Spiraea' to make 'Aspirin'; this trademark name, combined with the patent protection, enabled the company to enjoy complete monopoly on the drug for 17 years (Hochwalt, 1957). There is also another, perhaps less convincing, story that the name was derived from the early Neapolitan Bishop, St Aspirinus, the patron saint of headaches! (Jourdier, 1999).

Dreser had reasoned that the pharmacological actions, including its effects on 'nerve working' and actions of the heart, were essentially due to salicylate. He considered that salicylate was formed after 'splitting' of the acetyl group of aspirin following its absorption from the stomach (Dreser, 1899; 1907). This is essentially in agreement with present day knowledge (except that certain actions of aspirin are due to acetylation of proteins, so specific actions can also be attributed to the acetyl group – e.g. inhibition of platelet aggregation, and effects on the prostaglandin cyclo-oxygenases). Nonetheless, Dreser recognised a pro-drug long before this concept had been exploited. The evidence for this pro-drug concept came from studies on: the hydrolysis of acetyl and other esters of salicylic acid (Dreser, 1907); and on the effects of sodium aspirin compared with sodium salicylate on the respiration of yeast (measured by CO_2 production) and fermentation (in stomach churd) of bacteria (now known to be *Escherichia coli*), these being indices of the disinfecting properties that were thought to underlie the pharmacological effects of the salicylates at that time (Dreser, 1899).

Aspirin was found to be more rapidly hydrolysed than the higher carbon acyl esters. Salicylate was also found to be more effective than its acetyl derivative in inhibiting yeast fermentation, so providing further evidence for the pro-drug concept (Dreser, 1899). However, Dreser appears to have passed over the equivalent effects of both these salicylates in inhibiting bacterial fermentation (due, no doubt, to the enzyme aspirin esterase in the stomach churd hydrolysing aspirin).

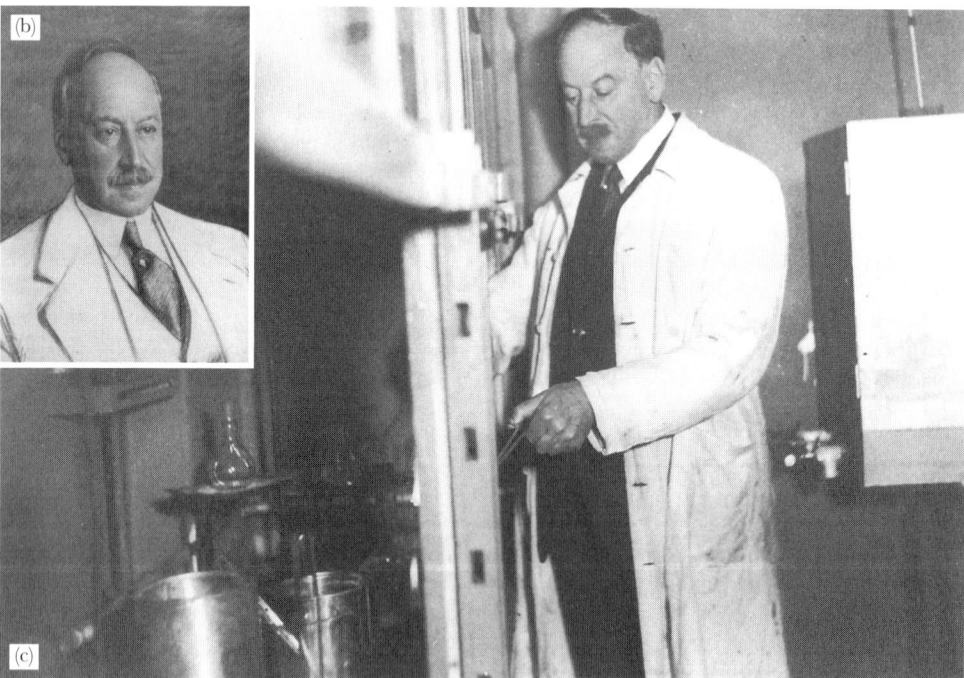

Figure 1.7 (a) Dr Arthur Eichengrün at his desk; (b) a portrait in later life; and (c) in the laboratory. He supervised Dr Hoffman at Bayer in the discovery of aspirin. (Photographs generously provided courtesy of Dr Eichengrün's grandson, Mr Ernst Eichengrün, Königswinter, Germany.)

Dreser reported in 1899, and later in 1907, a series of extraordinary experiments in fish (including goldfish) demonstrating that the inherent 'aggressive' actions of salicylic acid, which he considered were responsible for its irritant actions in the stomach, were not evident with the 'novaspirin' – the new aspirin. He dipped the tails of artificially ventilated fish into the drug solutions and found that aspirin did not produce the opacification or whitening of the skin seen with salicylic acid. Since no such effect was produced with dilute hydrochloric acid solutions, Dreser reasoned that the salicylic acid-induced opacification was not due to dissociation of hydrogen ions from the latter, even though he did not observe whitening with its sodium salt. It is curious that Dreser never performed experiments to compare the effects of the two salicylates on the stomachs of laboratory animals, even though Buss had performed such experiments back in 1875. Had he done so, he would have found what is known today – that aspirin is much more irritant to the stomach than either salicylic acid or sodium salicylate. Also, salicylic acid does not have the same nauseating effect as its sodium salt and is probably just as effective as aspirin in controlling pain in rheumatic conditions. Thus the whole origin of aspirin appears to have been built on the early successful promotion of a completely false premise about the drug's actions in the stomach, based on the wrong experiments!

Clinical trials

The first clinical trials of aspirin were performed by Witthauer in 1898 (Witthauer, 1899), and Wohl-gemuth (1899) at the University of Leyden Medical Clinic in Berlin. Wohlgemuth studied a total of 10 patients given aspirin 1 to 3 g supplied by the Elberfelders (Bayers) for the treatment of acute joint rheumatism, juvenile rheumatoid arthritis and a variety of other conditions. Some patients were given aspirin as an alcoholic solution (because of its insolubility in water). Remarkably, in the light of today's knowledge that alcohol markedly increases the irritancy of aspirin, no pain or other symptoms of gastric distress were reported in these patients (Wohlgemuth, 1899).

The early trials with aspirin were in patients with rheumatoid arthritis, yet as noted previously (Goodwin and Goodwin, 1981; see also Chapter 12) it took half a decade before identification of the necessity of high doses of this drug to achieve good antirheumatic effects.

ASPECTS OF THE COMMERCIAL DEVELOPMENT OF ASPIRIN

The Bayer Company enjoyed immense profitability from aspirin by careful protection of its patents until the beginning of the First World War (Hochwalt, 1957), and this and the important commercial history are analysed in Chapter 2. Some other aspects concerning the industrial history of the production and use of aspirin are, however, worth mentioning. Germany also had a virtual monopoly on the synthesis of many other synthetic drugs. In the USA there were many early attempts to wrest the fine chemical industry away from the European monopoly at the time. In 1901 John Queeny founded the Monsanto Chemical Works in the USA, initially to produce saccharin, but an interest soon developed in manufacturing aspirin (Hochwalt, 1957) – no doubt encouraged by the impending war. Late in 1912, Dr Gaston Dubois of Monsanto visited an unnamed chemist at Brugg in Switzerland and purchased from him, for 2000 Swiss Francs, a process for producing aspirin in one operation. Monsanto had also begun manufacture of salicylic acid in 1916, which placed it in a very favourable position for manufacturing the drug cheaply; however, this did not take place until 1917 (Richard L. Wasson, personal communication). A total of 11 other firms also began manufacturing aspirin at the time, although Monsanto appears initially to have taken the lead.

Following initial testing, Monsanto built a process plant in 1916 and by 1917 had sold 2368 pounds of aspirin. About this time, a battle ensued between Bayer and Monsanto: Bayer was obviously anxious to defend their patent rights vigorously. In November 1918, the US Patent Office cancelled Bayer's registered rights to the name of aspirin because they were thought to have been improperly and unlawfully registered (Hochwalt, 1957; Chapter 2). Furthermore, in a famous case for infringement of tradename rights the US Supreme Court ruled that Bayer's aspirin had been over-advertised to such an extent that it had become a common name, and thus Bayer's monopoly of aspirin had effectively been broken in the USA.

In 1920, Monsanto purchased a half-interest in the firm of R. Graesser Salicylates Ltd, then of Ruabon (Wales), which was manufacturing aspirin for the UK market (P.F. Carter, personal communication). This heralded the large-scale production of aspirin by this company in Europe.

An interesting situation developed about the same time in an emerging nation, Australia. The beginning of the First World War saw the interruption of supplies of aspirin from Germany, and the Australian Federal Government suspended Bayer's patent rights (Grenville-Smith and Barrie, 1976). Although the company had not taken up their option for these rights in Australia, the Government suspended their option as part of the wartime regulations and also to encourage local manufacture of this much-needed drug. A young pharmacist, George Nicholas, in collaboration with a chemist, Shmith, attempted to produce aspirin under extraordinarily primitive conditions in a shed outside Nicholas' Pharmacy in Flinder's Lane in Melbourne. By 1915 they had produced aspirin of sufficient purity to satisfy the government analyst that it complied with the standard of being 'dinkum' (local slang for genuine), and even drew the support of the then Attorney General and later Prime Minister, Mr Billy Hughes. After a dispute, Shmith withdrew from the partnership and George Nicholas joined with his brother and founded the firm Nicholas Proprietary Ltd (later Nicholas-Kiwi Ltd.). The Nicholas brothers were often subject to parliamentary attack regarding their propriety, some of this criticism having origins in British interests anxious to exploit the Australian market (Grenville-Smith and Barrie, 1976). Nicholas' registered their well-known trademark 'Aspro', 'As' being an abbreviation for aspirin and 'pro' being for propriety. Despite fluctuating fortunes and conflicts, their business grew immensely in the great influenza epidemics of 1919 and the early 1920s. Aspirin manufacturers in other countries likewise benefited considerably from this severe epidemic. The Nicholas company quickly developed commercial interests in Britain and later in other European countries (Grenville-Smith and Barrie, 1976). The profitable sales of aspirin by Beechams in Britain during this period were a great boost to the famous conductor, Sir Thomas Beecham. Although he was a part owner with his father of this company, he was not really interested in its running but did use its profits to fund his exploits into British Opera and the purchase of the Covent Garden Opera House in London; that at least we owe to aspirin! (Lazell, 1976).

The development of various formulations of aspirin (soluble, injectable, sustained-release, enteric-coated) has proceeded apace, especially since the Second World War. Among earlier developments was the highly successful Disprin™ (1948–1949), followed in quick succession by the incorporation of that formulation into the *British Pharmacopoeia* in 1952 (Smith, 1962; Dr N.C. Vary, personal communication). The development of these preparations was initially under very primitive conditions in the early 1940s, when the possibility of incorporating calcium carbonate with aspirin to overcome the gastrointestinal irritancy of aspirin was explored (G. Colman Green, unpublished paper *The Archeology and Social History of Aspirin*; Dr N.C. Vary, personal communication). What were eventually very successful formulations based on rapidly dissolving preparations were largely successful because of their palatability. These were also among the first preparations of aspirin to be marketed in aluminium foil.

Today, aspirin rates second to alcohol as the most consumed drug in the world. Annual production of aspirin in the USA is in excess of 100 million kilograms (Anon., 1980a). Actual demand is considerably short of this, probably being about 70 million kilograms. Demand for aspirin seems to have grown considerably over the years; in 1965 production in the USA was estimated at 13 million kilograms (Gottesman and Chin, 1968; Anon., 1980a). It appears also to have grown in other countries (Rainsford, 1975). It has been estimated that each year 50 billion tablets are taken worldwide (Jourdier, 1999). In Canada alone, where there has been extensive development of generic pharmaceuticals, over 130 aspirin-containing preparations have been identified (Brigden and Smith, 1997)! Despite intense competition, aspirin is still a well-known, effective and cheap analgesic available for prescription and over-the-counter sale today, and its popularity as a preventative against coronary vascular diseases, colon cancer and cataracts is now becoming legendary (Alstaedter, 1985).

Loss of the market share to paracetamol (acetaminophen), which occurred in the early 1970s, and the impact of recommendations in the 1980s that the drug should not be taken by children because of the risks of Reye's syndrome coupled with the marketing of newly developed non-steroid anti-inflammatory drugs (NSAIDs) has only had limited influence on the use of aspirin, the sale of which has been described as 'recession-proof' (Anon., 1980a). It is still described as 'A Miracle Drug' (Jourdier, 1999).

OTHER SALICYLATES

Aspirin was not the only salicylate ester developed in the last century, for acetyl esters of salicylaldehyde and salicylamide, salicyluric acid, salol (salicylphenyl ester) and salophen (salicyl paracetamol ester) were all known and some of these were used clinically long before aspirin was manufactured by Bayer (Liebig *et al.*, 1859; Flückinger, 1888; Geissler and Möller, 1889), Clever marketing obviously played a large part in the success of aspirin in those early and very crucial days of development (Chapter 2). As will be seen from subsequent chapters, an extraordinary variety of other salicylates has been developed in recent years, including a potent salicylate, diflunisal, which serves to illustrate the immense versatility of the class as remarkably effective therapeutic agents.

REFERENCES

Acland, T.D. 1881, Delirium following the treatment of acute rheumatism by salicylic acid. *British Medical Journal*, **i:** 337.

Alstaedter, R. (ed.), 1985, *Aspirin®. The Medicine of the Century*. Bayer AG, Germany.

Anon. 1876, Salicylic acid and salicin in acute rheumatism. *Guy's Hospital Gazette*, **1:** 84–85.

Anon. 1980a, *Chemical Marketing Reporter*, 29 Sep and 8 Dec.

Anon. 1980b, Nonsteroidal anti-inflammatory drugs for rheumatoid arthritis. *Medical Letter*, **22:** 29–31.

Anon. 1881, The debate on salicin and the salicylates in rheumatism. *Lancet*, **ii:** 1058.

Apolant, E. 1881, Symptoms of cerebral hyperaemia after large doses of salicylic acid. *Berliner Klinische Wochenschrift*, **18:** 82.

Bälz, E. 1877, Salicylsäure, salicylsaures Natron und Thymol in ihrem Einfluss auf Krankheiten. *Archiv für Heilkunde*, **18:** 61–81.

Barber, E. 1880, *A synopsis of the British Pharmacopoeia for the use of dispensers and students*. London: J & A Churchill.

Bartels, 1878, Über die therapeutische Verwerthung der Salicylsäure und ihres Natronsalzes in der Innere Medicin. *Deutsche Medizinsche Wochenschrift*, **4:** 399–403; 411–413; 423–425; 435–437.

Bekemeier, H. 1977, On the history of salicylic acid. *In:* H. Bekemeier (ed.), *100 Years of the Salicylic Acid as an Antirheumatic Drug*, pp. 6–13. Halle-Wittenberg: Martin-Luther Universität.

Benoit, P.E. 1877, *Description of Preparation of Sal Acid from Anthrombic Acid*. Thesis, Paris.

Beyer, H.G. 1880, A contribution to the knowledge of the physiological properties of salicylic acid. *Archives of Medicine, New York*, **iii:** 216–223.

Blanchier and Bochefontaine, 1878, Recherches expérimentales sur l'action physiologiques du salicylate de soude. *Comptes Rendes Societé de Biologie (Paris)*, **6s(v):** 287–291.

Boulonmié, P. 1879, La vérité ou le pou et le contre proprietés medicales au salicylate de soude. *Revue de literature Médecine de Paris*, **iv:** 488, 536, 564–566.

Brigden, M. and Smith, R.E. 1997, Acetylsalicylic acid-containing drugs and nonsteroidal anti-inflammatory drugs available in Canada. *Canadian Medical Association*, **156:** 1025–1028.

Bryden Hill, E. 1880, Acute rheumatism. *Lancet*, **i:** 227–228.

Buchanan, W.W. and Kean, W.F. 2001, Rheumatoid arthritis as seen through long-distance spectacles. *Inflammopharmacology*, **9:** 3–22.

Buss, C.E. 1875, *Über die Anwendung der Salicylsäure als Antipyreticum*. Leipzig: J.B. Hirschfeld.

Buss, C.E. 1876, *Zur Antipyretischen Bedeutung der Salicylsäure und des Neutralen Salicylsäuren Natron*. Stuttgart: F. Enke.

Buss, C.E. 1878, *Über Wesen und Behandlung der Fiebers. Klinisch-Experimentelle Untersuchungen*. Stuttgart: F. Enke.

Büss, H. and Balmer, H. 1962, Carl Emil Buss (1849–1878) and the beginning of salicylic acid therapy. *Gesnerus*, **19:** 130–154.

Busse, W.-D. 1989, History and philosophy of Bayer Pharmaceutical Research. *Artzneimittel Forschung*, **39:** 935–937.

Cahours, A. 1845, Recherches sur les acides volatiles à six atomes d'oxygène. *Annales de Chimie et de Physique*, **13:** 87–115.

Charteris, M. 1881, Salicin and salicylate in acute rheumatism. *British Medical Journal*, **i:** 229.

Chrubasik, S., Eisenberg, E., Balan, E., Weinberger, T., Luzzati, R. and Conradt, C. 2000, Treatment of low back pain exacerbations with willow bark extract: a randomised double-blind study. *American Journal of Medicine*, **109:** 9–14.

Compagnon, J. 1880, *De l'utilite du Salicylate de Soude dans le Traitement du Rheumatismé*. Paris.

Couper, A.S. 1858, *On a New Chemical Theory and Researches on Salicylic Acid*. Edinburgh: E.S. Livingstone, Alembic Club Reprints No. 21. Also in: *Edinburgh New Philosophy Journal*, **9:** 213–217.

De Cuba, J. and Von Kaub, J.W. 1486–1490, Salix rel Salamentum. *In:* J. Wonneke (ed.), *Herbarius zu Deutsch: 'Gart der Gesundheit'*. Basel: Michael Fueter.

Dewhurst, K. 1966, *Dr Thomas Sydenham (1624–1689). His Life and Original Writings*. London: The Wellcome Historical Medical Library.

Diesterweg, A. 1879, Zur Salicylbehandlung des acuten Gelenkrheumatismus. *Deutsche Medizinische Wochenschrift*, **v:** 551–555.

Discoveryhealth.com, 2001, Aspirin: A bitter pill. www.healthdiscovery.com/premiers/aspirin/aspirin.html. September 18, 2001.

Donnelly, M. 1880, The salicylate of potash in acute rheumatism and dyspepsia. *Medical Record NY*, **xvii:** 258.

Downes, C.H. 1879, Acute rheumatism and salicylic acid. *Guy's Hospital Gazette*, **4:** 78–81, 87–89.

Dreser, H. 1899, Pharmakologisches über Aspirin (Acetylsalicylsäure). *Archiv für die Gesammte Physiologie des Menschen und der Thiere (Pflügers)*, **76:** 306–318.

Dreser, H. 1907, Über modifizierte Salicylsäuren. *Medizinische Klinik*, **3:** 390–393.

Dumas, M. 1838, Note sur l'huile essentielle des fleurs de Reine des Prés (Spirea ulmaria). *Comptes Rendes Société de la Academie de Sciences (Paris)*, **7:** 940.

Ebbell, B. 1937, *The Papyrus Ebers*. Copenhagen: Levin & Munksgaard.

Editorial. 1877, Special Correspondence, Paris, M. Seé on Salicylic acid. *British Medical Journal*, **2:** 865.

Eichengrün, A. 1949, 50 Jahre Aspirin. *Die Pharmazie*, **4:** 582–584.

Ensor, F. 1876, The willow as a remedy for acute rheumatism. *Lancet*, **i:** 910.

Fagge, C.H. 1881, Remarks on the use of the salicylates in acute rheumatism. *Lancet*, **ii:** 1030–1033.

Finlay, D.W. and Lucas, R.H. 1879, Salicylate and alkaline treatment of acute rheumatism. *Lancet*, **2:** 420–421.

Flückinger, F.A. 1888, *Pharmazeutische Chemie*. Berlin: R. Gaertner's Verlag.

Fowler, J.K. 1881, On the treatment of acute rheumatism with salicylic acid. *Lancet*, **ii:** 1120–1121.

Fürbringer, P. 1875, Untersuchungen über die antifebrile Wirkung der Salicylsäure in Sonderheit über ihre temperaturherabsetzende Kraft bei septischem Fieber. *Centralblat. Medizin Wissenschaft*, **13:** 273–276.

Galmiche, P. 1957, Mais qui a inventé l'aspirine? *Le Presse Médicale*, **65:** 303.

Garrod, A.B. 1859, *The Nature and Treatment of Gout and Rheumatic Gout*. London: Walton and Maberly.

Geissler, E. and Möller, J. 1889, *Real Encyclopädie der gesamten Pharmacie*. Leipzig: Urban und Schwarzenberg.

Goodwin, J.S. and Goodwin, J.M. 1981, Failure to recognize efficacious treatments: a history of salicylate therapy in rheumatoid arthritis. *Perspectives in Biology and Medicine*, **25**: 78–92.

Gottesman, R.T. and Chin, D. 1968, Salicylic acid. *Kirk-Othmer Encyclopedia of Chemical Technology*. New York: Wiley.

Grenville-Smith, R. and Barrie, A. 1976, *Aspro. How a Family Business Grew Up*. Worcester: Ebenezer Bayliss. The Trinity Press.

Gross, M. and Greenberg, L.A. 1948, *The Salicylates. A Critical Bibliographic Review*. New Haven: Hillhouse Press.

Hall, De Havilland, 1881, The salicylate treatment of acute rheumatism. *Lancet*, **ii**: 1081–1082.

Hallopeau 1880, Du salicylate de soude dans le traitement de la fievre typhoide er de l'erysipèle. *Gazette Médicine de Paris*, **xxx**: 266.

Hayer, H. 1872, *Deutsche Pharmacopoeia*. Berlin: Verlag von Koeniglichen Geheimen ober Notbuch Druckerei.

Hayer, H. 1874, *Deutsche Pharmacopoeia*, 2nd edn. Berlin: Verlag von Julius Springer.

Hebenden, W. 1802, *Commentaries on the History and Cure of Diseases*. London: T. Payne, Newsgate.

Hedner, T. and Everts, B. 1998, The early clinical history of salicylates in rheumatology and pain. *Clinical Rheumatology*, **17**: 17–25.

Hochwalt, C.A. 1957, The story of aspirin. *Australian Journal of Pharmacy*, **38**: 771–772.

Hogg, W.D. 1877, *De l'Usage Therapeutique de L'Acide Salicylique ses Composés et accessoirement de la Salicine*. Thèse Pour le Doctorat de la Faculté de Médicine. Paris; V.A. Delahaye et Cie., Libraires Editeurs.

Hood, D.W.C. 1881, Statistics in connexion with the treatment of acute rheumatism by the salicylates. An analysis of 1200 cases at Guy's Hospital. *Lancet*, **ii**: 1119–1120.

Hu, S.Y. 1945, Medicinal plants of Chengtu herb shops. *Journal of the Western China Border Research Society*, **15B**: 97–177.

Huber, G. 1879, *Des Accidents Cérébraux concécutifs à l'administration du Salicylate de Soude*. Paris, 42pp.

Hughes, 1877, Acide salicylique et salicylates. *Nice-Médical*, **2**: 40–47.

Investigator, 1881, On the use of the salicylates in acute rheumatism. *Lancet*, **ii**: 1150.

James, P. 1881, The salicylates. *British Medical Journal*, **1**: 428.

James, R. 1764, *Pharmacopoeia Universalis: or A New English Dispensatory*. 3rd Edition. London: Printed for T. Osborne *et al.*

Jourdier, S. 1999, A miracle drug. www.chemsoc.org/chembytes/ezine/1999/jourdier.htm.

Keitel, W. 1977, 100 Jahre Anwendung der Salicylsäure und ihrer Derivate als Antirheumaticum. *In:* H. Bekemeier (ed.), *100 Years of Salicylic Acid as an Antirheumatic Drug*, pp. 39–50. Halle-Wittenberg: Martin-Luther Universitaet.

Kemp, G. 1881, On the use and abuse of salicylic acid. *British Medical Journal*, **i**: 510.

Kersch, 1880, Über Wirkung und Anwendung der Salicylsäure nach angestellten Versuchen an Thieren und am Krankenbette. *Memorabilien Heilbrund*, **xxv**: 433–438.

Kolbe, H. 1860, Über Synthese der Salicylsäure. *Liebig's Annalen der Chemie*, **113**: 125–127.

Kolbe, H. 1874, Über eine neue Darstellungsmethode und einige bemerkenswerte Eigenschaften der Salicylsäure. *Journal für Practische Chemie*, **10**: 89–112.

Kolbe, H. and Lautemann, E. 1860, Über die Constitution und Basicität der Salicylsäure. *Liebig's Annalen der Chemie*, **115**: 157–206.

Kraut, K. 1869, Über Salicylverbindungen. *Annalen Chemische Pharmazie*, **150**: 1–20.

Kunze, C.F. 1876, Über einige bisher noch ungekannte Wirkungen des salicylsauren Natrons. *Deutsche Zeitschrift für Praktische Medicin*, **28**: 323–324.

Laborde, J.V. 1880, L'action physiologique de l'acide salicylique et du salicylate de soude, du mechanism de cette action. *Tribune Médicine de Paris*, **xiii**: 53, 68.

Laborde, J.V. 1881, L'acide salicylique considéré comme antifermenmenteur et antiseptique. *Tribune Médicine de Paris*, **xiv**: 254–257.

Lazell, H.G. 1976, *From Pills to Penicillin. The Beecham Story*. London: Heinemann.

Leake, C.D. 1957, *An Historical Account of Pharmacology*. Springfield, Illinois: C.C. Thomas.

Leroux, M. 1830, Mémoir relatif à l'analyse de l'écorce de saule et à la découverte d'un principe immediate propre à remplacer le sulfate de quinine. *Journal Chimie Médicale*, **6**: 340–342.

Lévesque, H. and Lafont, O. 2000, L'aspirine à travers les siècles: rappel historique. *Rev. Med. Interne*, **21 (Suppl. 1):** 8–17.

Levy, J. de B. 1970, *The Illustrated Herbal Handbook*. London: Faber and Faber.

Liebig, J.V., Poggendorf, J.C., Woehler, F., Fehling, H.V. and Kolbe, H. 1859, *Handwoerterbuch der Reinen und Angewandten Chemie*. Braunschweig: Friedrich Viehweg.

Livon, Ch. 1880, De l'action physiologiques de l'acide salicylique et de salicylate de soude sur la respiration. Barlatier-Teissat père et fils. *Comp. Rend. Soc. Acad. De Sc. Paris*, **xc:** 321; also *Tribune Médicine de Paris*, **xiii:** 207–210.

Löwitt, M. 1881, Über den Einfluss von Salicylpräparaten auf die Temperaturcurve einiger Typhus- und Untermittelsfälle. *Medizinsch Chirurgie Centralblatt Wien*, **xvi:** 50, 62, 73, 85, 98, 110, 170, 182, 194.

Maclagan, T.J. 1876a, The treatment of rheumatism by salicin and salicylic acid. *British Medical Journal*, **1:** 627.

Maclagan, T.J. 1876b, Treatment of acute rheumatism by salicin. *Lancet*, **1:** 342–343, 383–385.

Maclagan, T.J. 1879, The treatment of acute rheumatism by salicin and salicylic acid. *Lancet*, **ii:** 875–877.

Maclagan, T.J. 1880, Note on the danger attending the use of salicylic acid in acute rheumatism. *Lancet*, **i:** 327.

Mahomed, F.A. 1880, The effect of salicylates on the heart. *Lancet*, **i:** 228.

Margotta, R. 1968, *An Illustrated History of Medicine* (P. Lewis, ed.). London: Paul Hamlyn.

Martin, B.K. 1963, Significant factors in the history of aspirin. In: A.St.J. Dixon, B.K., Martin, M.J.H. Smith and P.H.N. Wood (eds), *Salicylates. An International Symposium*, pp. 6–8. London: Churchill.

May, C.H. 1884, Statistics of four hundred cases of rheumatism, with especial reference to treatment. *Medical Record, New York*, **25:** 61–62, 87–92, 116–121, 173–178.

Mosig, A. and Schramm, G. 1955, Pharmazie. *Biehefte*, **4:** 1–71.

Myers, A.B.R. 1876, Salicin in acute rheumatism. *Lancet*, **ii:** 676–677.

Ord, W.M. 1880, Some of the conditions included under the general term 'rheumatoid arthritis'. *British Medical Journal*, **1:** 155–158.

Otterness, I.G. 1995, The discovery of drugs to treat arthritis: A historical view. In: V.J. Merluzzi and J. Adams (eds), Boston: Burkhäuser, 1–26.

Owen, I. 1881, The salicylate treatment of acute and subacute rheumatism. *Lancet*, **ii:** 1081.

Pagenstechser, F. 1835, Über das destillierte Wasser und Öel der Blüthen von Spirea Ulmaria. *Buchner's Repertorium f.d. Pharmazie*, **49:** 337–367.

Perry, L.M. and Metzger, J. 1980, *Medicinal Plants of East and South-East Asia: Attributed Properties and Uses*. Cambridge (USA): MIT Press.

Rainsford, K.D. 1975, Aspirin. Actions and uses. *Australian Journal of Pharmacy*, **56:** 373–382.

Rainsford, K.D. 1984, *Aspirin and the Salicylates*. London: Butterworths.

Riess, L. 1875, Über die innerliche Anwendung der Salicylsäure. *Berliner Klinische Wochenschrift*, **12:** 673–676.

Riess, L. 1876, Nachtrag zur innerlichen Anwendung der Salicylsäure, insbesondere bei dem akuten Gelenkrheumatismus. *Berliner Klinische Wochenschrift*, **13:** 86–89.

Robertson, Wm.H. 1880, Salicin and the salicylates in rheumatic fever. *Lancet*, **ii:** 192.

Royal Society of Chemistry, press release 1999, Jewish scientist's claim to discover aspirin denied by Nazis. www.rsc.org/pdf/pressoffice/1999/annconf99press3.pdf.

Ryff, W.H. 1573, *Reformierte Deutsche Apotect*. Strasbourg: Josiam Ribel.

Scaling, W. 1872, *The Salix or Willow: In a Series of Papers by William Scaling, Ten Years Basket Maker to Her Majesty and the Royal Family*. London: Simpkin, Marshall & Co.

Senator, H. 1879, Über die Wirkung der Benzoësäure bei der rheumatischen Polyarthritis. *Zeitschrift für Klinische Medizin (Berlin)*, **1:** 243–264.

Shaw, L.E. 1887, Cases of haemorrhage occurring during treatment by salicylate of soda. *Guy's Hospital Report*, **44:** 125–135.

Smith, S.E. 1962, The history and development of the Reckitt analgesic preparations. *Royal College of Health Magazine*, **1:** 19–25.

Spielmann, D.J.R. (ed.) 1783, *Pharmacopoeia Generalis*. Strasbourg: Johannis Georgh Treuttel.

Stewart, J. 1901, Chronic articular rheumatism and rheumatoid arthritis, and gout. *In:* H.M. Hare (ed.), *A System of Practical Therapeutics*, Philadelphia: Lea.

Stewart, W.J. and Fleming, L.W. 1987, Perthshire pioneer of anti-inflammatory agents. *Scottish Medical Journal*, **32:** 141–146.

Still, G.F. 1896–1897, On a form of chronic joint disease in children. *Medical Chirurgie Transactions*, **80:** 47–59.

Stone, E. 1763, An account of the success of the bark of the willow in the cure of agues. [In a letter to the Right Honourable George Earl of Macclesfield, President of the R.S.] *Philosophical Transactions of the Royal Society of London*, **53:** 195–200.

Stricker, 1876, Über die Resultate der Behandlung der Polyarthritic rheumatica mit Salicylsäure. *Berliner Klinische Wochenschrift*, **13:** 1–2, 15–16, 99–103.

Stuart, G.A. 1911, *Chinese Materia Medica*. Shanghai: Vegetable Kingdom.

Taylor, D.F. 1880, Ten cases of rheumatic fever treated with salicin. *The Medical Press and Circular, Transactions of Societies*, May **19:** 41–413.

von Gerhardt, C. 1843, Über die Zusammensetzung des Salicins, sowie über die Beziehungen zwischen der salicyl-, phenyl-, or Indigoreihe. *Liebigs Annalen der Chemie und Pharmacie*, **45:** 19–29.

von Gerhardt, C. 1853a, Recherches sur les acides organiques anhydrides. *Annales Chimie Serie*, **37:** 285–342.

von Gerhardt, C. 1853b, Untersuchungen über die Wasserfreien organischen Säuren. *Liebigs Annalen der Chemie und Pharmacie*, **87:** 149–179.

von Gilm, H. 1859, Acetylderivate der Phloretin- und Salicylsäure. *Liebigs Annalen der Chemie und Pharmacie*, **112:** 180–182.

Warner, F. 1881, Analysis of statistics illustrating the action of salicin compounds in the treatment of acute and subacute rheumatism. *Lancet*, **ii:** 1080.

Watts Parkinson, C.H. 1881, Delirium in acute rheumatism after salicylate of soda. *British Medical Journal*, **1:** 730.

White, W. 1798, *Observations and Experiments on the Broadleafed Willowbark*. Bath: Hazard.

Wilkinson, G. 1803, *Experiments and Observations on the Cortex Salix Latifoliae or Broadleafed Willow Bark*. Newcastle: Walker.

Witthauer, K. 1899, Aspirin, ein neues Salicylpräparat. *Therapeutische Monatschaft*, **13:** 330.

Wohlgemuth, J. 1899, Über Aspirin (Acetylsalicyl-säure). *Therapeutische Monatshefte*, **13:** 276–278.

CHAPTER 1

The Industrial History of Analgesics: the Evolution of Analgesics and Antipyretics

J.R. McTavish

RECOGNITION OF PAIN

For many centuries in the Western world, pain was thought to be unavoidable – either an affliction every mortal must endure or a punishment for sin. In more religious eras, all pain had meaning and purpose and was therefore not to be eradicated merely for the sake of greater personal comfort or convenience. Physicians, too, gave meaning to pain by identifying it as an important symptom that provided clues to the identity of a particular disease or warned of some significant danger to the body. Pain should not be masked, lest in ignoring it the victim fall prey to an even worse fate. Besides, doctors disdained the relief of symptoms as mere empiricism, scarcely better than quackery, and certainly unworthy of the intellectual legacy of their profession. Instead, physicians preferred to address the underlying *causes* of disease, usually basing their treatments on the concept of the four classical humours that were believed to form each person's constitution. It was the doctors' goal to maintain or restore a healthy humoral equilibrium by adjusting the balance between individuals and their environment. Pain, it was thought, would abate when the imbalance that caused it was corrected. It is therefore difficult to identify what substances were historically used as analgesics. Even the opiates were more often used to 'make a powerful impression on the nervous system, with a view of breaking up morbid action' than simply to relieve pain (Beck, 1851).

Thus, the history of analgesics does not really begin until the middle of the nineteenth century, when – as a result of the discovery of surgical anaesthesia – the issue of pain and its control was more fully addressed in medical practice. It was not until the mid-1840s that the English language even had a vernacular term for analgesic, the word first appearing in the registered trade name of a proprietary medication: 'Perry Davis' Painkiller' (Pernick, 1985). It is significant, however, that both the word and the concoction it represented were not introduced by orthodox or regular medical practitioners. In fact, the early history of painkilling and analgesic drugs demonstrates two important and interconnected features. First, the stimulus for analgesic medications came from the general public, not from the medical profession. And second, by the end of the First World War, the reaction to the public's demands resulted in the establishment of the very lucrative 'over-the-counter' (OTC) drug industry. The close relationship between pain and profits thus informs the development of analgesic drugs, and could even be said to be the model for many other aspects of the modern pharmaceutical industry.

THERAPEUTIC NIHILISM AND FEVERS

In the nineteenth century, orthodox medicine began to discard ancient humoral concepts and develop new, scientifically based medical disciplines such as neurology and biochemistry. New therapeutic principles, however, remained elusive. Practitioners remained wary of empirical or symptomatic approaches, and continued to rely on traditional drugs and treatments. Unfortunately, attempts to restore the patient's constitutional equilibrium seemed largely ineffective when dealing with the victims of diseases such as cholera and typhoid, which struck nineteenth century populations in massive epidemics. Even cocaine, morphine, strychnine, quinine and the other drugs newly isolated from vegetable sources had little impact in these situations. Besides, many people recovered from their illnesses whether treated or not. Therapeutics was in some sense futile. While no practitioner gave up treatment entirely, many, like Sir William Osler, adopted a sceptical attitude known as 'nihilism' and advocated more reliance on the healing power of nature (Paton, 1979). As before, this meant that symptomatic treatment was not endorsed: pain was simply the price of life.

This was not a view shared by the general public. Unable to find much help in orthodox medicine, many people turned to nostrums, which ignored the intricacies of medical theory and went straight to the heart of the matter, unequivocally promising relief from suffering. Because most of these concoctions contained fairly high amounts of narcotics and/or alcohol, they were in large measure able to live up to their guarantees. In specifically targeting pain, nostrum sellers had found a lucrative use for the morphine and cocaine that regular medicine still employed mostly for other reasons.

Sales of painkillers skyrocketed worldwide in the second half of the nineteenth century. Proprietaries could be obtained for a modest sum; they were nationally and internationally advertised with glowing testimonials from ordinary citizens, and were readily purchased without inconvenience or embarrassment (Young, 1961; Dukes, 1963). In this way, self-medication – always an option in the history of medicine – became an even more serious challenge to a medical profession struggling not only to assert its authority over disease, over homeopathy and other irregular practices, and over its own members, but also to overcome a public prejudice against doctors (Huerkamp, 1990; Bynum, 1994).

Despite these difficulties, science and medicine were nevertheless developing a better understanding of disease processes that potentially laid the foundation for more effective therapeutics. Before germ theory became respectable (in the 1880s), investigators concentrated on physical or chemical explanations of disease, focusing in particular on the common syndrome of high temperature, rigors, muscle wasting, rapid pulse, alterations of metabolism, and delirium generally known as fever (Rageth, 1964; McTavish, 1987a). Experience in treating malarial fevers since the seventeenth century with cinchona (Peruvian bark) led to the isolation of quinine from the bark in 1820. Although it was not effective in all febrile conditions, quinine was nevertheless thought to be the best available antipyretic, and chemists at mid-century eagerly sought either a synthetic version or something very closely related. The search had an unexpected outcome: in 1856 the English teenager William Henry Perkin, hoping to derive quinine from a coal-tar derivative, instead discovered aniline mauve and launched not a new line of pharmaceutical substances but the synthetic dyestuffs industry (Perkin, 1896). It would not be long, however, before drug making and dyes were again intimately linked.

THE ORGANIC CHEMICAL INDUSTRY

By the 1880s, the synthetic chemical production had shifted from Britain to the newly unified Germany, becoming the country's showpiece industry (Haber, 1958). From the start the German industry was both consciously scientific and highly commercial, noted for its impressive number of new organic compounds, its intense and often underhanded competitiveness, and its huge profits (Beer, 1959; Meyer-Thurow, 1982; Lenoir, 1988). Although most of the industry's chemists sought new dyes, the search for artificial quinine had not been abandoned – in fact it had greater hope of success now that the chemical industry was producing so many substances thought to be related to this compound.

Pyrazolones

One potential quinine substitute was hydrochloride of hydroxy-N-ethyltetrahydroquinoline, discovered in 1882 by a chemist at the University of Erlangen. The pharmacologist at this university identified its antipyretic properties. Given the trade name *Kairin* (from the Greek for 'timely'), it was manufactured by Hoechst (a major producer of dyestuffs) and used briefly as an antipyretic, but was generally not popular even in small doses because of its toxicity (Filehne, 1882; Fruitknight, 1886). In 1883 another German university chemist discovered and tested a compound that he identified as hydroxy-methylquinizine, which seemed to be a better antipyretic than Kairin, with fewer side effects. Hoechst marketed it under the trade name *Antipyrine* (Filehne, 1884). Cheaper than quinine and apparently just as effective, Antipyrine became popular with doctors and profitable for the manufacturer. (The compound was in fact a pyrazolone (phenyldimethylpyrazolone); it became the basis for further pharmaceutical research at Hoechst, resulting in dimethylamino-antipyrine or aminopyrine (Pyramidon) in 1893, and dipyrone (Novalgin) in 1920; Brune, 1986) Antipyrine, as the name suggests, was promoted in medical circles primarily as an antipyretic, and although it proved to be analgesic as well, it was prescribed for pain mostly in febrile conditions: doctors did not advocate it as a general painkiller. However, the public did, especially in the USA (Greenberg, 1950; McTavish, 1999a). Even the *New York Times* (29 January 1893) took note of the growing number of 'people whose habit it is to pay a visit to the nearest convenient drug store on the occurrence of a headache or temporary disability from some passing pain, more or less acute, and to partake freely of antipyrine or its associates, independent of a physician's prescription'. So popular was the drug with the laity that drugstores could not keep it in stock, despite the fact that Antipyrine as a patented chemical was quite expensive.

Acetanilid

In 1886 the medicinal properties of an unpatented chemical were accidentally discovered, which would have an enormous impact on the development of analgesics as a category distinct from antipyretics. According to a story that unfortunately cannot be substantiated, two Strasbourg physicians, A. Cahn and P. Hepp, while treating a patient for intestinal parasites, had ordered the vermifuge naphthalene from their pharmacist (Gross, 1946). An assistant mistakenly sent the aniline derivative acetanilid (n-phenylacetamide), which did not expel the worms but did reduce the fever from which the patient was also suffering. The doctors, convinced that this was an important discovery, then introduced the substance to the medical world under the trade name *Antifebrin*. It was manufactured by Kalle & Co., a small chemical firm later absorbed by Hoechst (Cahn and Hepp, 1886). Much less expensive than Antipyrine, Antifebrin soon outsold its rival two to one as a prescription antipyretic. However, it achieved distinction as an analgesic only when the even cheaper generic versions produced by other German companies were adopted by the nostrum trade for use in 'headache powders', a new category of patent medicine that seems to have been a direct response to the public's demand for general pain remedies that did not contain narcotics. Physicians continued to disapprove of this symptomatic application, but they were unable to prevent the increasing popularity of acetanilid as a proprietary ingredient (Hiss, 1899; Anon., 1911), despite the dangers associated with the drug. Cardiac depression and methaemoglobinaemia were noted almost immediately; agranulocytosis was recognised in 1922. Indeed, although it is not likely that the chemical was the culprit in all cases (Gross, 1946), many fatalities were attributed to acetanilid poisoning (Adams, 1912), self-prescribed headache powders taking most of the blame (Austin and Larrabee, 1906). Physicians, however, continued to emphasise that the proper course of action for such things as headache was not to mask the pain, but to eradicate its cause by means of cathartics, leeches, blisters, change in diet, abstinence from sex, or other unpleasant therapy, and so men and women who did not find comfort in these recommendations felt they had no choice but to turn to the proprietaries for relief.

Phenacetin

The popularity of acetanilid as a prescription antipyretic had prompted a young research chemist at the Farbenfabriken Bayer in Germany, a large dyestuffs maker and principal rival of Hoechst, to wonder if some profitable use might not be made of 30 000 kilos of paranitrophenol being stored as waste in the factory's cellars. Carl Duisberg (later an important director of Bayer) decided to convert this into paraphenetidin, then acetylated it to give acetophenetidin. In clinical trials its antipyretic properties exceeded even his expectations (Duisberg, 1923), so when Bayer introduced it as *Phenacetin* in 1887, the company had high hopes for its first pharmaceutical product. Bayer could only obtain a patent on the chemical in the USA, however, so it was very quickly produced by other German chemical manufacturers. Furthermore, Bayer did not take sufficient care to ensure that the trade name 'Phenacetin' remained its exclusive property. By the mid-1890s, the German courts had declared the word generic (Anon., 1896), although outside Germany Bayer energetically prosecuted trademark infringers and usually won its cases.

As with acetanilid and Antipyrine, the analgesic properties of phenacetin were initially of secondary interest to the medical world, yet were immediately appreciated by the public, who demanded the chemical from their druggists for the treatment of headache and other pain (Smith, 1958). Such a large market was developing for all the synthetic antipyretics, and so great were the profits from them, that by the middle of the 1890s Hoechst and Bayer had established dedicated pharmaceutical research and development facilities and were actively seeking novel medicaments. In the years before the First World War, novocaine, veronal, sulfonal, salvarsan and, of course, aspirin were the results of these endeavours.

THE SALICYLATES

The origins of aspirin (acetylsalicylic acid, ASA) are quite unrelated to those of the other antipyretic analgesics. It has often been claimed that aspirin is simply a modern version of ancient remedies, salicylate-bearing plants such as willow (*Salix alba*), meadowsweet (*Spirea ulmaria*), and wintergreen (*Gaultheria procumbens*) having been used as medicines since ancient times. When organic chemists of the early nineteenth century took an interest in these plants they were able to derive numerous products from them (Partington, 1964), but only one – salicin (salicyl alcohol glycoside, the discovery attributed to Leroux in 1830) – appeared with any frequency in the materia medica, alongside powdered willow bark (which had actually been in use only since the Reverend Edward Stone's employment of it to treat 'ague' in the previous century; Stone, 1763). Both items were described as having properties analogous to cinchona and quinine, and there was hope that salicin would prove as effective as quinine since it was considerably cheaper. However, most physicians found the salicylates fell short (Brande, 1833). The bark and the extract were deemed to be 'inferior in power' to cinchona and quinine, and the need to use larger doses to achieve the same effects rendered them more expensive in the long run (Beasley, 1865). Salicylates thus had limited clinical popularity, although in the 1830s chemists and physiologists continued to discover and study new versions of them, leading to the identification of salicylic acid in willow, meadowsweet and spirea by Piria, Löwig and Gerhardt among others (Gross and Greenberg, 1948). An Italian military doctor, for example, curious about the fate of salicylic acid in the body, examined his test subjects' urine for metabolic by-products (Bertagnini, 1855), and in 1859 the German chemist Hermann Kolbe synthesised salicylic acid from phenol as an interesting problem in organic chemistry (Kolbe, 1860). In neither case, however, were clinical applications important. For the next 15 years salicylic acid received little attention from physicians, remaining a quinine substitute of last resort.

In 1873, Kolbe, by now chemistry professor at Leipzig University, needed a large amount of salicylic acid in order to resolve a scientific dispute. Because the commercially available natural-source acid was expensive and of poor quality, he revived his synthesis of 1859, preparing salicylic acid from phenol. In fact, he refined his method to such an extent that he was able to produce quantities of pure acid efficiently and cheaply, and even helped set up a factory – the Salicylsäurefabrik Dr von Heyden, near Dresden – for large scale production. Nonetheless, he did not communicate his new method until well after he had obtained patents on it and the factory was already in place (Kolbe, 1874; Schlenk, 1934; Rocke, 1993). It is evident that Kolbe foresaw a profitable use for salicylic acid, but it was not the traditional uses of salicylates that aroused his interest. Rather, Kolbe was intrigued by his derivation of the acid from phenol (carbolic acid), which at that time the British surgeon Joseph Lister was using extensively to prevent postoperative infections. Lister used carbolic acid in the belief that it killed the micro-organisms responsible for surgical sepsis and infectious diseases in general. Although germ theory remained highly controversial for many years, Lister's antiseptic procedures were too successful to ignore, and in one form or another they became well established. Moreover, sepsis and the related phenomena of zymosis (infectious disease) and fermentation became attractive ideas in clinical medicine because they could account for so many different pathological processes, especially those associated with fevers. Heat, metabolic by-products and so forth could all be explained as the results of fermentation processes. Hence surgical or external antisepsis gave rise to the concept of *internal* antisepsis, the idea that infectious diseases could be inhibited at the sites where they did the most damage – in the blood and tissues of patients. It was a theory that might at last rescue therapeutics from its long-standing nihilism (Crellin, 1981). Carbolic acid having proved to be intensely irritating, its internal use did not seem feasible, so Kolbe, believing that salicylic acid might be a better-tolerated substitute, tested its ability to inhibit fermentation in milk, wine, beer and other mixtures *in vitro*. He was immensely pleased with the results. Convinced that it was indeed strongly antiseptic, he persuaded the Professor of Surgery at Leipzig University to employ salicylic acid in surgery, and reported excellent outcomes from these trials. Kolbe then ingested the acid himself without ill effect.

Having established salicylic acid's credentials as a non-poisonous external antiseptic, Kolbe finally published these findings and went on to advocate strenuously that it be tested internally, to see if it could inhibit scarlet fever, diphtheria, measles, smallpox, syphilis, dysentery, typhus, cholera, pyemia and perhaps even rabies. There were, in fact, very few ailments that Kolbe did not think would respond to the acid, especially if they fell in the 'fermentative' or 'zymotic' category (Kolbe, 1875). He also thought it would make a good toothpowder, mouthwash, foot antiperspirant and preservative for meat and barrels of water on ships. Kolbe was so impressed with the possibilities of salicylic acid as a panacea, he took a daily dose of it as a kind of prophylactic for the rest of his life (Lockemann, 1930).

Kolbe's colleagues at Leipzig similarly found much to praise, employing salicylic acid in diphtheria and 'intestinal fermentations' with favourable results, but it was only after Carl Emil Buss of Basel published his own clinical findings in mid-1875 that interest in the drug really escalated. Buss had begun his trials around January 1875, simply substituting salicylic acid for quinine. He found it effective as an antipyretic in all kinds of fevers, free from unpleasant side effects, costing far less than quinine (although he had to use larger doses) and especially effective in rheumatic conditions. He was inclined to agree with Kolbe that its actions were due to its antiseptic properties (Buss, 1875).

For the next several years, salicylates were one of the most discussed items in the medical world. Salicylic acid as an internal antiseptic had caught the medical imagination, and the initial reports of its successful use in myriad conditions, unmarred by appreciable side effects, had elevated it to almost miracle drug status (Geissler, 1876). In May 1880 *Scientific American* stated simply that salicylic acid was 'the most important antiseptic, antizymotic and antipyretic ever discovered'.

The enthusiastic use of salicylic acid in so many different diseases, however, quickly led to the discovery that it did indeed possess some unpleasant side effects – notably gastric irritation ranging from moderate discomfort to bleeding (Wolffberg, 1875). Kolbe took issue with these reports and maintained that only salicylic acid that was not produced by his method could cause problems (Kolbe, 1876). In

CHAPTER 2

attempting to buffer the acid, physicians began using sodium salicylate instead (Moeli, 1875; Riess, 1875). They found that although it was just as effective an antipyretic, analgesic, and antirheumatic as the pure acid, it did not have antiseptic properties. Its effectiveness could not be accounted for by the theory of internal antisepsis. This was a great disappointment, and the indiscriminate use of salicylates declined markedly.

Nevertheless, the dramatic ability of both salicylic acid and sodium salicylate to relieve pain, reduce swollen joints and lower febrile temperatures in rheumatic fever assured these drugs an important place in the materia medica as a specific for this disease (Riess, 1876; Stricker, 1876), although the salicylates were not recognised until much later as being particularly useful in rheumatoid arthritis (Goodwin and Goodwin, 1981), and do not appear to have been employed very frequently after 1880 in inflammatory conditions not associated with rheumatism. A few physicians claimed to have been treating rheumatic fever with salicin or other natural-source salicylates for years, but since they did not publish until after the success of the Kolbe acid their observations had had no widespread impact. For example, Scottish physician Thomas Maclagan maintained he had been using salicin in rheumatism since November 1874, basing his use on the view that providence had decreed the presence of salicin-bearing willow trees in damp areas where this disease was common, but he did not make this known for 2 years (Maclagan, 1876).

It is apparent that most of the medical world regarded salicylates as completely new medicines, and not as improved versions of older ones. They also came from a new and rather unlikely source – coal-tar. Unlike earlier pharmaceuticals such as quinine or salicin, which were prepared from plant sources in establishments that had evolved from traditional apothecary shops, salicylic acid and sodium salicylate were the synthetic creations of the organic chemical industry, an industry that at that time was without medical traditions or connections, and had a reputation for aggressive profit-seeking.

Moreover, Kolbe's product was protected by patents in Germany and elsewhere, which ethicists found troubling. 'Patent medicines' had a long association with quackery and charlatanism in the name of greed and profit; 'patented medicines' could likewise be seen as mercenary, monopolistic and contrary to the humanitarian ideals the medical profession was supposed to endorse. Kolbe, of course, argued that patents prevented counterfeit chemicals from contaminating the market, and pointed to the side effects of salicylic acid as proof that unpatented chemicals were dangerous. Later, when more and more synthetic drugs were patented, the manufacturers justified the temporary monopoly and the high prices they could charge because of it) as a fair way to compensate them for the costs of research.

There being so few industrially produced synthetic drugs in the late nineteenth century, however, their implications remained largely unaddressed at that time. Most of the materia medica was still in the public domain, known by official Latin names. The drugs were produced from natural sources by small ethical manufacturers who advertised and sold only to medical professionals, not to the public (Sonnedecker, 1976). ('Ethical' as used in the pharmaceutical industry in the late nineteenth century simply meant a company that did not advertise its products to the public.) This state of affairs by and large prevailed until the Second World War, but the impact of the industrial coal-tar producers nonetheless began to be felt before 1900 as the products of the industry became more numerous and useful. It was the successor to salicylic acid and sodium salicylate, aspirin, that is an especially prominent example of how commercial interests could come to dominate therapeutic or professional concerns.

Aspirin

In 1894 the Farbenfabriken Bayer hired a young chemist, Felix Hoffmann, whose father, it is said, suffered from both chronic rheumatism and the deleterious effects of treatment with sodium salicylate. The story has it that the younger Hoffmann was therefore personally motivated to find a better-tolerated version of the drug. By October 1897 his laboratory notebooks reveal he had found a way to modify salicylic acid through the process of acetylation: he substituted an acetyl group ($COCH_3$) for the hydrogen of the hydroxyl group (OH) in the salicylic acid molecule. This was accomplished by the action of acetic anhydride on salicylic acid, and the product was acetylsalicylic acid (ASA). In December 1897 he also prepared ASA from salicylic acid using acetyl chloride (Bayer, 1983).

Acetylsalicylic acid was not entirely new, however, having been anticipated (at least in theory) as early as 1853 when Strasbourg chemist, Charles Gerhardt, described the synthesis of *salicylate acétique* from sodium salicylate and acetylchloride (Gerhardt, 1853). In 1869 Karl Kraut described a process that Bayer's rivals constantly referred to as anticipating Hoffmann's (Kraut, 1869). Nevertheless Bayer claimed that any synthesis prior to Hoffmann's had not produced true ASA and therefore the company applied for patents: in Germany in February 1898 (never granted – Bayer's claim of novelty did not stand up); in Great Britain, granted December 1898; and in the USA, awarded 27 February 1900.

It is Hoffmann's colleague at Bayer, Professor Heinrich Dreser, head of the pharmacology department, who usually receives credit for introducing this chemical into practice. Bayer's own version of events describes Dreser as recognising very quickly that the new chemical was at least as efficacious as its predecessor, and that it possessed fewer side effects. Although the first clinical trials were not held until early 1899, they supposedly confirmed Dreser's laboratory findings. (Comprehensive testing of new medicines was not legally required by any country at this time; most drugs were introduced after minimal evaluation in the laboratory, where the absence of overt toxicity in test animals was generally all that was necessary before the substance was available for clinical trials. Efficacy and safety were determined 'in the field'.) However, according to Dr Arthur Eichengrün, head of Bayer's chemical research laboratories at the time, what actually happened to Hoffmann's ASA was not nearly so straightforward (Eichengrün, 1918; 1949). In 1949 Eichengrün published a memoir claiming that he, not Hoffmann, had come up with the idea of acetylating salicylic acid. (Acetylation was a process that Bayer used extensively to produce not only aspirin but also diacetyl morphine – heroin – in 1898, and which Eichengrün used to invent safety film after he left Bayer.) Eichengrün claimed that Dreser had opposed human use of acetylsalicylic acid because he had found that in the isolated frog-heart test (where the heart was perfused with a solution of the drug) the substance was 'a direct heart-poison'. Eichengrün, having more faith in the chemical, said he surreptitiously sent some acid of his own manufacture to a Bayer agent in Berlin, who persuaded physicians there to experiment with it on their patients. A dentist, so this story goes, had a patient with both toothache and a fever. He gave the patient a dose of the new drug to bring down the fever, but the patient then happily remarked that the toothache too had vanished, thereby revealing acetylsalicylic acid's analgesic properties. The Berlin doctors kept asking for more of the drug, so Eichengrün (apparently not on good terms with his colleague) went over Dreser's head and presented the Bayer directors with his findings. Dreser was still not convinced that ASA had any value at all until Duisberg forced him to test it further. Unfortunately, Eichengrün's story cannot be trusted entirely; he had spent the last 14 months of the war in Theresienstadt concentration camp, where he in fact wrote his account. He appears to have been more annoyed by the Nazis' failure to acknowledge his contribution to German glory (i.e. aspirin) than by their treatment of him during the war. It is also curious that Carl Duisberg does not mention aspirin, Bayer's most famous product, in his memoirs.

At any rate, in January 1899 Bayer needed to find a new name for the substance. Having learned from its experience with phenacetin, the company now chose its drug trade names with great care because if handled correctly these were legally private property and could be owned in perpetuity. On 23 January 1899 Bayer created the name that, according to the company, was derived from 'a' for acetyl and 'spir' for spiraeaic acid, chemically identical to salicylic acid but found in *Spiraea ulmaria* (Bayer, 1983). 'Aspirin' was then immediately registered as a trade name in every country where Bayer did business.

At about the same time two young German physicians conducted separate trials in rheumatic patients, using aspirin as a substitute for sodium salicylate. Both publications, appearing in the spring of 1899, stated plainly that the new drug was better than the older salicylates because it lacked the painful and deleterious side effects while retaining all the benefits. Both doctors claimed that aspirin quickly and dramatically reduced the swelling in the inflamed joints, lowered the temperature, and eased the pain of acute rheumatic attacks. One concluded: 'Aspirin is an improved substitute for sodium salicylate because it lacks the unpleasant side effects such as stomach upsets and loss of appetite'

CHAPTER 2

(Wohlgemuth, 1899). The other agreed: 'I believe that I can recommend Aspirin to my colleagues with all possible certainty' (Witthauer, 1899).

In June, Dreser published his pharmacological studies (Dreser, 1899). Whatever his original opinion of aspirin really was, this article gave the drug a glowing report. He included the physicians' clinical findings and went on to describe the properties of aspirin as observed in experiments *in vitro*, on frogs and rabbits, and upon himself. Testing his own urine for the presence of aspirin after ingesting a sample of the drug, Dreser found none although there was evidence of free salicylate. He concluded that aspirin passes through the stomach largely unchanged, which in his view explained the absence of gastric irritation. Aspirin was then decomposed in the intestines, where the free salicylic acid produced the pharmacologically significant effects.

By 1902, according to one calculation, some 160 articles on aspirin had already appeared, 'a literature . . . so voluminous that it is scarcely possible to review it' (Wohr, 1902). The majority of these articles pronounced the drug useful, and particularly recommended it as a substitute for sodium salicylate in the treatment of acute rheumatism. Side effects of therapeutic doses were minor and manageable, and included tinnitus, skin rashes, excessive sweating and some gastric irritation. All disappeared after a few hours; the stomach upsets could be avoided by having the patient drink sufficient water or by accompanying the drug with an acidic drink such as lemon juice, which was to prevent the aspirin from decomposing until it had reached the alkaline environment of the intestines. Dreser had shown, in fact, that alkalis decomposed aspirin into salicylic acid quite quickly. If this occurred in the stomach, the patient would suffer all the side effects for which salicylic acid was notorious.

Severe reactions to aspirin were rare (Winckelmann, 1903; Barnett, 1905; Dockray, 1905; Gilbert, 1911). Even the physicians who reported cases of toxicity had generally treated hundreds of patients without incident (Nusch, 1901; Lindsay and Bruce-Leckie, 1913). Although aspirin seemed to be virtually risk free, Bayer did not entertain the idea of advertising it directly to the public. Such an act would risk the company's status as an ethical manufacturer.

In addition to being used in rheumatic conditions, aspirin, like its predecessors, was found to be an antipyretic, reliably lowering the temperature in such conditions as tuberculosis, postpartum infections and typhoid. It was even thought to be of benefit in mild diabetes mellitus by reducing urinary sugar excretion; moreover, children took it willingly (Renon, 1900; Cybulski, 1902; Görges, 1903; Fink, 1911; Chambers, 1912). As an analgesic, aspirin was used for toothache, menstrual pain and even the pain of certain cancers (Witthauer, 1900; Burnet, 1905; Merkel, 1905; Chidichimo, 1906), although its application in these situations was always tempered by the need to relate treatment to the underlying causes of the pain. Its popularity as a general painkiller and headache remedy, therefore, was greater among patients than physicians – who tended to reserve aspirin for headaches stemming from 'gouty diathesis' or other 'appropriate' conditions (Tirard, 1905). Patients who were prescribed the drug for a gouty headache might, of course, discover that it also worked on a hangover headache. Because no prescription was legally required, anyone could purchase aspirin by simply asking a pharmacist. To Bayer's delight, the public seems to have asked rather often. In 1906, when aspirin was still primarily touted as an ethical 'antirheumatic', the company noted:

> Aspirin has in the decade [*sic*] since its introduction become so popular that it is unsurpassed by any other drug. Surely it is not an exaggeration to say that it is today the most used and beloved medicine we manufacture. For this we can thank Aspirin's relatively easy digestibility, its generally prompt and prolonged effects, and above all its inherent, notable analgesic effect on all sorts of conditions. By its numerous applications even in minor complaints, Aspirin has won the public's trust and has become a household remedy in the true sense of the word.

Yet modern opinion is that aspirin is more irritant to gastric mucosa than either salicylic acid or sodium salicylate, and some even suggest that aspirin's unprecedented success was due less to its actual properties than to the way it was marketed (Rainsford, 1984). Marketing, in fact, was crucially important in the aspirin story. Bayer's campaign on behalf of its product was a sore point amongst the world's physicians and pharmacists. It could quite possibly have led them to abandon the drug in disgust at the company's behaviour. Nevertheless they prescribed it enthusiastically, and did not often mention any reservations about its efficacy or its sequelae – highly suggestive of the value practitioners placed on it. That they did not seem to notice many side effects may say more about wishful thinking in the medical world than historians have heretofore explored. It is a curious irony, in fact, that paracetamol (acetaminophen), which at present is considered the analgesic and antipyretic equivalent of aspirin and its superior in terms of side effects, was rejected as ineffective and harmful when first investigated in 1893. It took another 50 years before it was rediscovered (Spooner and Harvey, 1976).

THE VALUE OF A GOOD NAME

Although as a medicine aspirin is not without its controversies, the drug is unique in that it has been used and abused by more people for more years with fewer harmful effects than any other substance. Nevertheless it is a powerful medication of which it has often been said that had it been discovered today it would have remained a prescription drug.

Yet aspirin has in fact become the quintessential over-the-counter remedy: cheap, ubiquitous, relatively harmless, and effective if used as directed – although its use today, at least as an analgesic, is rarely guided by physicians. Doctors in the early years of this century were opposed to the idea that aspirin or any other drug should become available to the public on demand. The story of how an ethical medication that seemed to provide physicians with a powerful therapeutic agent should have so quickly escaped professional control reveals how new forces, especially commercial concerns, were insinuating themselves into the traditional medical environment. Aspirin, and to a lesser extent the other analgesic antipyretics, provides one of the most illustrative examples of how these forces were beginning to affect medical practice.

Medical practitioners had long been accused of being venal frauds, but by the late nineteenth century progress in medical science had overcome much of this criticism. Physicians were riding on a new wave of respect and admiration. Professional medical and pharmaceutical organisations in both Europe and the USA worked hard to preserve the honour of the professions, and included in their ethical codes a disavowal of all commercial features, especially those related to the materia medica (Fishbein, 1947).

Drugs had long been held to be nature's gift to suffering humanity and to reside in the public domain. Their names, whether Latin or vernacular, likewise belonged to the public. In theory, a trade name for a legitimate drug, as distinct from a nostrum or so-called patent remedy, was acceptable if its purpose was to distinguish one manufacturer's brand of a product from another's, but in reality no true professional could countenance even this use. If the product in question was, say, tincture of opium, calling it anything else was confusing and pointless.

The wave of synthetic drugs ushered in by Kairin, and especially by Antipyrine, however, introduced absolutely new chemical entities without natural equivalents and with scientific names of impossible length and complexity. This immediately led to the much more convenient and memorable trade name standing in for the chemical or pharmaceutical designation. Believing that the trade name was the generic name, doctors used it in prescriptions. Pharmacists, however (better trained in chemistry), knew there was a difference, but were forced to dispense one particular manufacturer's product only, since substitution was actionable at law. To add to the confusion, a substance might also appear in the market with several different trade names but without any chemical synonym, leaving doctors completely ignorant of the drug's identity (Anon., 1907; Fiedler, 1979).

In the USA, where the patent laws meant that almost all the new medicinal chemicals were protected from generic competition for 17 years, there was the added fear that a legal monopoly on a potentially life-saving medication (such as Emil Behring's diphtheria antitoxin) was simply an 'attempt to blackmail suffering humanity in the interest of a foreign manufacturer' (Kiernan, 1898). Physicians and pharmacists complained for years about this situation (Meserve, 1897; Wilbert, 1903); the discontent came to head with Bayer's Aspirin (McTavish, 1987b; 1999b).

In principle, 'Aspirin' simply identified Bayer's brand of a particular chemical whose scientific name was 'acetylsalicylic acid'. Yet virtually all the clinical and pharmacological publications used the full chemical name only once, using 'Aspirin' everywhere else. Authors rarely mentioned that Aspirin was a trade name. One American physician in 1900 even said flatly that the product had been termed 'aspirin ... for the sake of convenience', apparently oblivious to the name's status as private property (Hewitt, 1900).

When competitive ASA first appeared in Germany in early 1901, its manufacturers and retailers identified their products as 'acetylsalicylic acid'. Bayer, however, continued to call Aspirin simply a 'salicylate' or even a 'substitute for salicylates', which resulted in German doctors being confused as to whether Aspirin and acetylsalicylic acid were medically or chemically equivalent (Anon., 1901). It had not been ASA but Aspirin, after all, that had received the good press. (Heyden, the chief rival manufacturer, did not invent its own trade name for ASA until 1913: Acetylin.) Bayer, of course, asserted that any side effects associated with ASA were due to a non-Bayer product, and impugned the quality of products from other sources.

Bayer also enhanced the attractions of its product by selling it in a standardised, consumer-ready form – the tablet (at least in Germany, where Aspirin was one of first drugs available in this form). It came from the factory in a handsome container, labelled prominently with the name Aspirin, the warning 'To be sold only in original package' and Bayer's logo, which by 1903 was the now famous Bayer Cross. Pharmacists complained that this prevented them from exercising their professional skills, and were angry that Bayer implied they were incapable of dispensing the correct medication (McTavish, 1987c).

Bayer's tactics annoyed physicians too, since factory packages allowed patients to learn the brand name of the drug, which they could then use whenever they wanted to ask a pharmacist directly, thus avoiding a return visit to the doctor. Consumer-ready cartons were also *de facto* advertisements, enabling a pharmaceutical manufacturer to publicise its wares without appearing to violate the rules for an 'ethical' drug company (Stephan, 1911).

German pharmacists tried to restore some control by deliberately, if illegally, dispensing Heyden or Hoechst ASA when Aspirin was ordered. However, Bayer's legal advisers diligently and indefatigably tracked down and prosecuted the offenders, even threatening action against pharmacists who sold the generic drug to walk-in customers who asked for 'aspirin'. In one case the customer turned out to be a Bayer employee (Buchholtz, 1911). When Bayer could prove intentional fraud it won its cases, but when an Aspirin prescription was filled with the non-Bayer drug and identified as acetylsalicylic acid, or even when it was labelled 'substitute for Aspirin', German judges found for the defendants and the generics had a small victory (Anon., 1905).

The habit of brand-name prescribing also led to unseemly rancour between German physicians and pharmacists. Doctors defended the practice by saying it guaranteed that the right product would be dispensed; druggists, insulted, accused doctors of being lazy, using trade names only 'because of their simplicity, deterred from the scientific names because of their excessive length'. (Anon., 1913). Various solutions were suggested, including the transfer of all trade names to the public domain, as well as the establishment of a national committee to evaluate all new drugs and verify claims for efficacy. The pharmaceutical industry, however, was a much more powerful force than either German pharmacists or doctors (Heubner, 1912; 1913). 'Aspirin' to this day remains the property of Bayer AG in Germany, and is as carefully guarded as ever.

In England, where a challenge from the Heyden Company in 1905 had resulted in the voiding of Bayer's acetylsalicylic acid patent (Patents, 1905; Report of Patent, Design and Trade Mark Cases, 1920), dissatisfaction with Bayer's behaviour was more muted than in Germany, but a brief spat between Bayer and readers of the *Pharmaceutical Journal* in March 1911 reveals that some enmity did exist. Bayer had written to the *Pharmaceutical Journal* to remind pharmacists that substituting the generic drug for Aspirin was illegal in Britain. Bayer was an honest firm, said its representative, that not only had a legal right to the profits from the trade name, but also deserved them. An outraged reader, however, responded that because Bayer's prices were extortionate and its behaviour unethical, the firm not only did not deserve the profits, it also 'may be said to be dishonest'. The solution, he suggested, was government regulation to curb trademark rights (Meldrum, 1911).

Some observers suggested that a convenient generic name such as 'salacetin' should be adopted for ASA, but the idea did not catch on, probably because 'aspirin' was already too familiar. The First World War finally provided the opportunity to deal with the situation and to strike a blow for King and Country at the same time; in early 1915, the courts removed Aspirin (owned by the enemy) from the UK Register of Trade Marks (Anon., 1915a). Under British laws relating to alien property, profits could be held in escrow until the end of the war, but the cancellation of the trade name of course eliminated an important source of those profits. The *Pharmazeutische Zeitung*, apparently forgetting its own disputes with Bayer, noted angrily: 'that the whole war is being conducted by the English with piratical intentions cannot be better indicated than by this incident'. (Anon., 1914). Despite some initial misgivings about its being a German name, 'aspirin' became and remained a legal generic term in Great Britain and Bayer's brand must compete with many others (although as a result of some convoluted negotiations 'Bayer' was co-owned by the German company and Sterling Products, an American firm, until 1994, when German Bayer bought Sterling outright and once again became sole owner of the Bayer name).

It was in the USA, however, that Aspirin was consciously seized upon as a symbol of unwanted industrial influence on the medical world, and where the future of the product as a lucrative over-the-counter consumer item was first realised. As part of its efforts to professionalise American physicians, the American Medical Association (AMA) decried the use of nostrums, trade names, and patented drugs, and encouraged educated prescribing of generic items in Latin (Anon., 1897; 1899; 1900; 1905b). Nevertheless, in 1909 a prescription ingredient survey revealed that of the 10 most prescribed items (which included sodium bicarbonate, distilled water and glycerin), two were patented, trade-named analgesic antipyretics from Bayer: Phenacetin and Aspirin (Gathercoal, 1933). (The US patent on phenacetin had expired in 1906, but the status of the name was still disputed.) Physicians wishing to treat patients with the newest wonder drugs had no choice but to prescribe proprietary items.

Realising that such drugs were here to stay, the AMA made the best of the situation by allowing those that met its strict criteria to be included in its *New and Non-official Remedies* (NNR), begun in 1907 as an annual guidebook to the proprietaries considered permissible in modern ethical practice. One of the rules for including a drug was that it never be advertised to the public. The AMA supported the contemporary campaign to curb the excesses of the patent medicine sellers who lured gullible Americans into buying nostrums loaded with alcohol, cocaine, morphine, opium, as well as acetanilid and other synthetics. Ultimately, said the AMA, no drug was safe unless prescribed or at least recommended by a physician. Nevertheless, nostrum makers had more influence in Washington than the AMA, and the Food and Drug Act of 1906 required only that specified drugs be listed on the product's label and not that they be made inaccessible to the public. Nevertheless the new law had some impact, and after 1906 many nostrums were reformulated or simply removed from trade, although headache treatments continued to be popular and profitable. Bayer, however, had no intention of allowing its products in the nostrum market (that is, under the name Bayer; the company did sell chemicals in bulk to proprietary manufacturers). Every year its branded products met the AMA's criteria for inclusion in the *NNR*. The company's medical and scientific credentials were regarded as impeccable.

However, American practitioners were not happy. There was already some suspicion that Bayer had exerted undue influence in the eighth revision of the *United States Pharmacopoeia* (USP VIII) (1900, published in 1905), the official guide to drug standards, including nomenclature. Aspirin had not yet been an issue, but Bayer's other popular drugs were. *USP VIII* was the first to address the problem of how to handle important commercial items and so, conscious of setting a precedent, it refused to include either trade names or substances 'controlled by unlimited proprietary or patent rights'. It did not escape notice, however, that four Bayer drugs, advertised to the medical world for years by trade names alone, appeared in the *USP* under their scientific names, two of which had been *invented* for use in the book: the sedatives Trional and Sulfonal were given the generic names sulphonethylmethane and sulphonmethane. They were cross-referenced in the index only as diethylsulphonmethylethyl-methane and diethylsulphondimethylmethane! 'Trional' and 'Sulfonal' were mentioned nowhere, yet the latter names were the only ones that doctors had previously encountered. Because it was unlikely that physicians would now use the complicated scientific terms, Bayer would continue to reap the benefits of trade-name prescribing. As one commentator asked: 'Surely pharmacists must have begun to wonder how such strange luck – and always for them good luck – follows this firm. Did they have a friend upon the committee of revision?' (Anon., 1905c). Aspirin would certainly be considered for *USP IX*. Would Bayer's luck continue to hold?

Although Bayer fought off all legal challenges to the acetylsalicylic acid patent in the USA and maintained its absolute monopoly on the chemical (Anon., 1909), the company was clearly worried that when the patent expired on 27 February 1917 Bayer's ownership of the trade name would expire with it. There was good reason to be concerned. Several important legal precedents had already been set: linoleum, for example, had once been a valuable trade name but was now in the public domain. Bayer's own Phenacetin, despite all the precautions the company took, became a generic by default when the patent expired in 1906 because Bayer had failed to provide the 'real' name – acetphenetidin – for the public to use as a synonym. This situation in fact had very little effect on Bayer because in 1906 few American chemical companies were capable of manufacturing the chemical and many saw little profit in trying. Bayer had lowered the price dramatically and had always had a reputation for quality, so the firm continued to supply the drug without much competition. Indeed, Bayer's phenacetin became a principal ingredient of the headache powders reformulated after the Food and Drug Act, even though the company still told American druggists that it would prosecute unauthorised use of the trade name.

However, Aspirin was by far Bayer's most popular product, and when the war in Europe prevented German raw materials from reaching the production facilities in the USA, American companies such as Monsanto saw an excellent opportunity to break the German stranglehold on medicinal chemicals (Haynes, 1945; Hochwalt, 1957). Bayer had always known that such competition was inevitable once the patent protection disappeared, but now it feared that 'Aspirin', too, might go the way of phenacetin. In 1906, to avoid the synonym problem, Bayer had begun identifying Aspirin by its 'public' name, 'the monoacetic acid ester of salicylic acid' – a term no one else ever used. This obfuscation only served to annoy the AMA and US pharmacists: there would be no sympathy for the company's claim that 'Aspirin' was its private property.

In 1916, therefore, the Bayer Company, Inc. (established in 1913 as an independent American firm separate from the German company) took an unprecedented step. Agonising over the prospect of losing the goodwill of the American medical establishment, Bayer nonetheless decided to promote Aspirin directly to the public by advertising in major newspapers around the country (McTavish, 1987b). However:

> No selling talk of any kind will be used. The reader will not be urged to buy anything. The product will not be suggested as a remedy for any ailment. The uses to which it can be put will not be mentioned. The sole object of the publicity is trade-mark identification.
>
> (Anon., 1916)

Considering how most other medicines were advertised to laymen at this time, Bayer was handling the campaign with 'extreme delicacy'. As they appeared in the *New York Times*, for example, the advertisements were indeed circumspect. Of modest size, there were many versions, rarely repeated, but all contained the words 'Bayer-Tablets of Aspirin' in large type with the Bayer Cross prominently displayed. Sometimes there was a drawing of an Aspirin bottle or tablets. Usually there was some statement advising readers that only *Bayer* was genuine Aspirin, and that all other imitations and substitutes were implicitly or expressly dangerous.

Newspaper articles suggest that the public was very interested in all of this. The *Philadelphia Ledger*, for example, foresaw the impending legal battle over the name aspirin as promising 'lively times in pharmaceutical circles'. But, as *The American Journal of Pharmacy* (Anon., 1917) pointed out, such advertising would also lead to the pernicious practice of 'self-medication on the part of the public'. With precious few effective medications available, professional medicine saw no advantage in allowing any to slip from its jurisdiction and did its best to discourage the self-prescribing. Nonetheless, it was becoming apparent that physicians had never really had control of the new synthetics. When the AMA dropped Bayer Aspirin from the *NNR* in 1917, the drug remained as popular as ever. Professional endorsement was, and had been, irrelevant.

Public advertising was not without its own hazards, however. Bayer no doubt realised that it would result in laypersons using 'aspirin' as a generic. Nevertheless the company fully intended to prosecute pharmacists who infringed on the trade name, and in fact brought suit against the United Drug Company of Boston in mid-1917 for this violation, the case not being resolved until 1921. In the meantime, however, American Bayer had a number of other difficulties.

Although American Bayer was ostensibly an independent firm, it maintained very close ties with the German parent company. There is even good evidence that the chief chemist, Dr Hugo Schweitzer (an American citizen of German birth) was a spy in the Kaiser's secret service (but see Reimer, 1996). At the very least he acted as the banker for the ring of spies and saboteurs, and in 1915 he was involved in a 'plot' to keep Thomas Edison from selling phenol to the British for making munitions (Mann and Plummer, 1991). So after America declared war on Germany in April 1917, the Alien Property Custodian did not hesitate to seize American Bayer's assets and hold them until the end of hostilities, when they were sold at auction. Yet so valuable was the ownership of Aspirin that all through the war the trustees who ran the Bayer company on behalf of the Custodian continued to advertise Aspirin in the newspapers in defiance of the AMA, in copy that impugned the quality of competitors' products and hinted at the deceitfulness of druggists who substituted.

As a German–American enterprise, Bayer's public advertising had been discreet and dignified, befitting an ethical firm courting the custom of physicians despite the AMA's disapproval. However, the new post-war owners, Sterling Products, Inc. (which had purchased Bayer from the Alien Property Custodian in December 1918 for $5 310 000 plus back taxes) was an established nostrum purveyor of the old school, dealing in Dodson's Liver Tone, Danderine Hair Tonic, and Cascarets Candy Cathartic. Sterling established the Winthrop Chemical Company to handle the ethical products, but aspirin was important enough to get its own company, The Bayer Company, Inc., which sold nothing but this drug and its combinations. In 1919 The Bayer Company, Ltd. was established at Windsor, Ontario, to handle the Canadian market.

Sterling undertook a massive advertising campaign, spending $1 000 000 in 1919 alone, but the expense seems to have been worthwhile, resulting in a net profit of $2 159 143 that year compared to a pre-war average of about $850 000 (McTavish, 1987b). The ownership of the trade name was still disputed until April 1921, however, when Judge Learned Hand decreed that aspirin could be employed as a generic term at the retail level, although not for wholesale quantities (Anon., 1921). The ruling was immediately ignored, and aspirin has been used ever since as a generic at all levels of trade. However, Sterling still owned Bayer and the Bayer Cross, and its efforts on behalf of trademark recognition were unequivocally successful. The Bayer Cross is today one of the most recognised proprietary symbols in the USA, even if the word 'aspirin' has now become virtually synonymous with any small white tablet used for headaches.

Sterling had more luck with the trade name in Canada, where the onset of the war had not resulted in the confiscation of Bayer's property: in 1914 the Bayer trademarks were technically owned by an American company. Canadian pharmacists and the small Canadian chemical industry nevertheless looked forward to the day when the term would be made generic, and several drug companies boldly advertised 'aspirin' during the war, despite threats of court action (Anon., 1915b; 1920). At the first post-war meeting of the Canadian Pharmaceutical Association in Winnipeg in 1919, where aspirin was a major topic of discussion, Canadian druggists vented their spleen at the company, convinced that Sterling Products was a thinly disguised German Bayer. Sterling's representative was actually told to leave the convention (Anon., 1919), an incident the local newspaper described in dramatic headlines: 'Oust German Supplies', and 'Druggists to Fight German Aggression'.

In 1923 a group of druggists petitioned to have 'Aspirin' removed from the register of trademarks in Canada on the grounds that it was a *de facto* generic; the Exchequer Court granted the petition but Sterling immediately appealed (Exchequer Court, 1923). In 1924, three of five judges of the Supreme Court of Canada reversed the lower court's decision, on the grounds that since the word had been properly registered to German Bayer on 28 April 1899, reassigned to American Bayer on 12 June

1913, and was now properly owned by Sterling, there was no legal reason for expunging the word. It became private property once again, and has remained so ever since (Supreme Court, 1924). Sterling was naturally happy with this arrangement; German Bayer was not.

In order to buy Bayer from the United States Alien Property Custodian in 1918, Sterling had had to demonstrate that it was an all-American company with all-American owners, and promise that it would have nothing to do with the German firm. German Bayer, however, was anxious to retrieve at least some of its property, for which it had received no compensation, and approached Sterling with a number of proposals about sharing royalties for old and new products. While Sterling negotiated agreements for a number of drugs and chemicals, the Americans adamantly refused to include aspirin in any of them (Mann and Plummer, 1991). In 1923, however, they did agree to pay German Bayer 50 per cent of the profits from the sale of Canadian Aspirin for the next 50 years. Even the Second World War did not stop the payments, although after September 1939 the money went to the custodian of enemy property in Ottawa. Canadian Bayer tried to void the contract in 1944, but the judge – calling IG Farben, of which Bayer was now a part, a probable cause of the war and 'an octopus with its tentacles spread out not only to all European countries, but also into the United States and other parts of the western hemisphere' – could not find a legal reason to let the Canadian company out of its obligations. Besides, he said, the money could be used to pay reparations to Canada when hostilities ended; the post-war world would need giants like IG to help restore the peace-time economy. Canadian Aspirin sales could help (*Dominion Law Review*, 1944). Perhaps they did. IG Farben no longer exists, but Bayer AG is still one of the most profitable chemical companies in the world, and apparently one of the most patient. In late 1994, 77 years after losing its American property, Bayer AG bought Sterling Drug of New York and is once again owner of the Bayer name in the USA, Canada and Great Britain.

CONCLUSION

The early history of the synthetic analgesics shows that a vast industry has come to play an important role in both professional and domestic medicine. Dissatisfaction with the traditional materia medica at the end of the nineteenth century had created a ready market for new drugs, which the German organic chemical firms stumbled into almost by accident. To avoid any association with the disreputable nostrum trade, the synthetic pharmaceutical industry stressed its scientific and ethical characteristics. Yet it promoted its drugs with a commercial enthusiasm and astuteness unprecedented in the legitimate drug field.

Furthermore, by allowing their products to become nostrum ingredients, the synthetic drug manufacturers helped fuel popular demand for effective, non-narcotic pain remedies. Fear of addiction was turning opiate use into a criminal act (Trebach, 1982). Legislation in many countries attempted to restrict narcotics to medically supervised applications. Discovering that acetanilid, phenacetin and similar chemicals could get rid of a headache or toothache without the danger of habituation, and that they could be used without a doctor's involvement, the public showed its appreciation by buying them in quantity. It was the marketplace that prompted American Bayer to consider profits more compelling than 'ethics' and contribute to the creation of the modern over-the-counter trade, where genuinely useful, professionally recommended medications were made available for home use. Physicians had originally been reluctant to approve of this, partly to protect their own incomes and partly because they believed it was dangerous; however, because most of the ailments for which these drugs were used were considered to be trivial or transient, doctors eventually conceded that some kinds of self-medication were appropriate and practical (*Annals of the New York Academy of Sciences, 1965*).

In any case, synthetic antipyretic analgesics have remained the cornerstone of the OTC market, with aspirin in the lead until challenged by paracetamol in the 1950s, and later by ibuprofen. Of the original nineteenth century products, only aspirin, despite some uneasiness about its adverse effects (Reye's syndrome and gastric ulcers), has remained in general use (now enhanced by its role in preventing thrombosis). Phenacetin, Pyramidon and the others were withdrawn even from prescription use, although they are still sometimes to be found in Third-World pharmacies (Silverman *et al.*, 1992).

In addition to their therapeutic importance, the history of these drugs also demonstrates the growing role of the pharmaceutical industry as a new, and often troubling, participant in health care. Despite its great satisfaction with items of real clinical merit, the medical world viewed the industry with some trepidation, and concerns about its activities persist to this day. Questions concerning the use of advertising to the public and to professionals (Lexchin, 1984; American College of Physicians, 1990), the safety of generics versus brand names (Shapiro *et al.*, 1982; Cohen, 2001), the cost of pharmaceuticals (United States Department of Health, Education and Welfare, 1969), the advisability of self-medication (Andrews and Levin, 1979) and many other issues, first addressed more than a century ago with the synthetic analgesics, still have not been answered according to the industry's numerous critics (Arledge, 1989). Yet it is also true that the pharmaceutical industry has been responsible for many of the therapeutic advances on which modern medicine prides itself – if not actually discovering a product, then certainly refining and perfecting it (Wainwright, 1990). The early history of the analgesic drugs reflects the sometimes uneasy relationship between the ideals of medicine and the realities of modern business. Headache sufferers around the world, however, are simply grateful that, whether by intention or accident, this relationship resulted in such wonderful conveniences as aspirin.

REFERENCES

Adams, S.H. 1912, *The Great American Fraud. Articles on the Nostrum Evil and Quackery*. Chicago: AMA.

Anon. 1896, Zum Wortschutz 'Phenacetin'. *Pharmazeutische Zeitung*, **41:** 307.

Anon. 1897, The ideals of the Association. *Journal of the American Medical Association*, **28:** 1198.

Anon. 1899, Should the physician prescribe patented drugs? *Journal of the American Medical Association*, **33:** 169.

Anon. 1900, Relations of pharmacy to the medical profession. *Journal of the American Medical Association*, **34:** 986–988, 1049–1051, 1114–1116, 1178–1179, 1327–1329, 1405–1407; **35:** 27–29, 89–91.

Anon. 1901, Acidum acetylosalicylicum. *Pharmazeutische Zentralhalle*, **42:** 311.

Anon. 1905, Trade-names in the new pharmacopoeia. *Druggists Circular*, **49:** 301–305.

Anon. 1905a, Aspirinprozess. *Pharmazeutische Zeitung*, **50:** 618.

Anon. 1905b, The secret nostrum vs. the ethical proprietary preparation. *Journal of the American Medical Association*, **44:** 718–719.

Anon. 1907, The nomenclature of synthetic medicines. *British Medical Journal*, **ii:** 398–399.

Anon. 1909, Aspirin–acetylsalicylic acid patent upheld. *Druggists Circular*, **53:** 496.

Anon. 1911, The abuse of drugs and the sale of patent medicines. *Pharmaceutical Journal*, **86:** 477–478.

Anon. 1913, Kann durch ärztliche Rezepte ein Vergehen gegen das Warenzeichengesetz begangen werden? *Pharmazeutische Zeitung*, **58:** 558.

Anon. 1914, Editorial. *Pharmazeutische Zeitung*, **59:** 946.

Anon. 1915a, The Aspirin trade-mark. *Chemist and Druggist*, **86:** 16–17.

Anon. 1915b, Editorial. *Canadian Pharmaceutical Journal*, **49:** 138.

Anon. 1916, Big campaign for aspirin to forestall expiration of patent. *Printers Ink*, **95:** 189–191.

Anon. 1917, Anent Aspirin. *American Journal of Pharmacy*, **89:** 129–130.

Anon. 1919, The CPA Winnipeg Convention. *Canadian Pharmaceutical Journal*, **53:** 103–104.

Anon. 1920, Canadian-made 'Aspirin'. *Canadian Chemical Journal*, **4:** 119.

Anon. 1921, Aspirin or acetylsalicylic acid – an important court decision. *Journal of the American Medical Association*, **76:** 1356.

Anon. 1965, Home medication and the public welfare. *Annals of the New York Academy of Sciences*, **120:** 807–1024.

Andrews, L. and Levin, L. 1979, Self-care and the law. *Social Policy*, **9:** 44–49.

American College of Physicians. 1990, Physicians and the pharmaceutical industry. *Annals of Internal Medicine*, **112:** 624–625.

Arledge, E. (producer). 1989, Prescriptions for profit. *Frontline*, Public Broadcasting System, (transcript).

Austin, C.E. and Larrabee, R.C. 1906, Acetanilid poisoning from the use of proprietary headache powders. *JAMA*, **46:** 1680–1681.

Barnett, H.N. 1905, Aspirin in rheumatism: a warning. *British Medical Journal*, **21**.

Bayer [Farbenfabriken vorm. Fr. Bayer & Co]. 1906, Aspirin. *In: Pharmazeutische Produkte*, p. 76. Elberfeld: Bayer.

Bayer [Aktiengesellschaft]. 1983, *Aspirin, ein Jahrhundertpharmakon. Daten, Fakten, Perspektiven*, p. 25. Leverkusen: Bayer.

Beasley, H. 1865, *The Book of Prescriptions*, p. 449. Philadelphia: Lindsay & Blakiston.

Beck, J. 1851, *Lectures on Materia Medica and Therapeutics*, pp. 374–375. New York: Samuel, S. & William Wood.

Beer, J.J. 1959, *The Emergence of the German Dye Industry*. Urbana: University of Illinois Press.

Bertagnini, C. 1855, Über das Verhalten einiger Säuren im thierischen Organismus. *Jahresbericht der Chemie* (Abstr.), 490–491.

Brande, W. 1833, *A Manual of Pharmacy*, 3rd edn, p. 188. London: Renshaw & Rush.

Brune, K. 1986, Knorr and Filehne in Erlangen. *In:* Brune, K. (ed.), *100 Years of Pyrazolone Drugs*. Basel: Birkhäuser Verlag.

Buchholtz. 1911, Apotheker und Grossindustrie. *Apotheker Zeitung*, **26:** 902–903.

Burnet, J. 1905, The therapeutics of aspirin and mesotan. *Lancet*, **i:** 1193–1196.

Buss, C.E. 1875, Über die Anwendung der Salicylsäure als Antipyreticum. *Deutsches Archiv für Klinische Medizin*, **15:** 457–501.

Bynum, W. 1994, *Science and the Practice of Medicine in the Nineteenth Century*, pp. 176–218. Cambridge: Cambridge University Press.

Cahn, A. and Hepp, P. 1886, Das Antifebrin, ein neues Fiebermittel. *Centralblatt für Klinische Medizin*, **7:** 561–564.

Chambers, G. 1912, On acetyl-salicylic acid, with special reference to its value in typhoid fever. *British Medical Journal*, **i:** 121–122.

Chidichimo, F. 1906, Die physiologische und therapeutische Wirkung des Aspirins, mit besonderer Berucksichtigung der Wirkung auf den Uterus. *Therapeutische Monatschaft*, **20:** 389–396.

Cohen, J.S. 2001, *The Case Against the Drug Companies. Prescription Drugs, Side Effects, and Your Health*. New York: Jeremy P. Tarcher/Putnam.

Crellin, J.K. 1981, Internal antisepsis or the dawn of chemotherapy? *Journal of the History of Medicine and Allied Sciences*, **36:** 9–18.

Cybulski, H. 1902, Aspirin in der Therapie der Lungentuberkulose. *Therapie der Gegenwart*, **43:** 424–425.

Dockray, J.S. 1905, Some toxic effects of aspirin. *British Medical Journal*, **ii:** 1692–1693.

Dominion Law Review. 1944, Bayer Company, Ltd. v. Farbenfabriken vorm. Bayer & Co. *et al.* 2 D.L.R. 616.

Dreser, H. 1899, Pharmakologisches über Aspirin (Acetylsalicylsäure). *Pflügers Archiv*, **76:** 306–318.

Duisberg, C. 1923, Zur Geschichte der Entdeckung des Phenacetins. *In:* C. Dreser (ed.), *Abhandlungen, Vorträge, und Reden aus den Jahren 1882–1921*. Berlin: Verlag Chemie.

Dukes, M.N.G. 1963, *Patent Medicines and Autotherapy in Society*, pp. 149–153. Den Haag: Drukkerij Pasmans.

Eichengrün, A. 1918, Pharmazeutisch-wissenschaftliche Abteilung. *In:* A. Eichengrün (ed.), *Geschichte und Entwicklung der Farbenfabriken vorm. Friedr. Bayer & Co. Elberfeld in den ersten 50 Jahren (1863–1913)*. Leverkusen: unpublished.

Eichengrün, A. 1949, 50 Jahre Aspirin. *Pharmazie*, **4:** 582–584.

Exchequer Court, Canada. 1923, In re 'Aspirin', American Druggists Syndicate Ltd. (Petitioner) and Bayer Company, Limited (Objecting Party) 1923 Ex.C.R. 65.

Fiedler, W. 1979, Antikamnia: the story of a pseudo-ethical pharmaceutical. *Pharmacy in History*, **21:** 59–72.

Filehne, W. 1882, Über neue Mittel, welche die fieberhafte Temperatur zur Norm bringen. *Berliner Klinische Wochenschrift*, **19**: 681–683.

Filehne, W. 1884, Über das Antipyrin, ein neues Antipyreticum. *Zeitschrift für Klinische Medizin*, **7**: 641–642.

Fink, G.H. 1911, High temperature after labour and its treatment with aspirin. *British Medical Journal*, **i**: 251–252.

Fishbein, M. 1947, *A History of the American Medical Association 1847 to 1946*, pp. 35–40. Philadelphia: Saunders.

Fruitknight, H. 1886, Kairin and Antipyrine. *Medical Record*, **29**: 646–648.

Gathercoal, E.N. 1933, *The Prescription Ingredient Survey*, p. 85. Chicago: AMA.

Geissler, A. 1876, Über die antipyretische Wirkung der Salicylsäure und der salicylsauren Salze. *Jahrbuch der in-u-ausländischen gesamten Medizin*, **172**: 185–202.

Gerhardt, C. 1853, Recherches sur les acides organiques anhydres. *Annales de Chimie*, **37**: 285–342.

Gilbert, B. 1911, Unusual idiosyncrasy to Aspirin. *Journal of the American Medical Association*, **56**: 1262.

Goodwin, J.S. and Goodwin, J.M. 1981, Failure to recognize efficacious treatments: a history of salicylate therapy in rheumatoid arthritis. *Perspectives in Biology and Medicine*, **25**: 78–92.

Görges. 1903, Über neuere Arzneimittel: Aspirin und Digitalis Dialysat. *Berliner Klinische Wochenschrift*, **39**: 753–755.

Greenberg, L. 1950, *Antipyrine: A Critical Bibliographic Review*, p. 41. New Haven: Hillhouse Press.

Gross, M. 1946, *Acetanilid: A Critical Bibliographic Review*, pp. 58–103. New Haven: Hillhouse Press.

Gross, M. and Greenberg, L. 1948, *The Salicylates: A Critical Bibliographic Review*, p. 5. New Haven: Hillhouse Press.

Haber, F. 1958, *The Chemical Industry During the Nineteenth Century*, pp. 170–179. Oxford: Oxford University Press.

Haynes, W. 1945, *The American Chemical Industry: A History*, Vol III, pp. 311–326. New York: Van Nostrand.

Heubner, W. 1912, Allerlei Heilmittel-Unheil. *Therapeutische Monatschaft*, **26**: 185–192.

Heubner, W. 1913, Der Entwurf eines neuen Warenzeichengesetzes. *Therapeutische Monatschaft*, **27**: 334–336, 678–684.

Hewitt, G.M. 1900, Aspirin: an improved salicylic acid. *Medical Bulletin*, **22**: 409–413.

Hiss, A.E. 1899, *Thesaurus of Proprietary Preparations and Pharmaceutical Specialities*, p. 23. Chicago: G.P. Engelhard.

Hochwalt, C.A. 1957, The story of Aspirin. *Chemistry*, **6**: 10–14.

Huerkamp, C. 1990, The making of the modern medical profession, 1800–1914: Prussian doctors in the nineteenth century. *In:* G. Cocks and K. Jarausch (eds), *German Professions 1800–1950*. New York: Oxford University Press.

Kiernan, J.A. 1898, The patenting of medical discoveries. *Medicine*, **4**: 801–805.

Kolbe, H. 1860, Über Synthese der Salicylsäure. *Liebigs Annalen*, **113**: 125–127.

Kolbe, H. 1874, Über eine neue Darstellungsmethode und einige bermerkenswerthe Eigenschaften der Salicylsäure. *Journal für Praktische Chemie*, **10**: 89–112.

Kolbe, H. 1875, Weitere Mittheilungen über Wirkungen der Salicylsäure. *Journal für Praktische Chemie*, **11**: 9–23.

Kolbe, H. 1876, Prüfung der Salicylsäure auf Reinheit. *Journal für Praktische Chemie*, **14**: 143–144.

Kraut, K. 1869, Über Salicylverbindungen. *Liebigs Annalen*, **150**: 1–20.

Lenoir, T. 1988, A magic bullet: research for profit and the growth of knowledge in Germany around 1900. *Minerva*, **26**: 66–88.

Lexchin, J. 1984, *The Real Pushers: A Critical Analysis of the Canadian Drug Industry*. Vancouver: New Star.

Lindsay, J. and Bruce Leckie, A.J. 1913, [letter] *British Medical Journal*, **i**: 1108.

Lockemann, G. 1930, Kolbe. *In:* G. Bugge (ed.), *Das Buch der Grossen Chemiker*, pp. 124–135.

Maclagan, T.J. 1876, The treatment of acute rheumatism by salicin. *Lancet*, **i**: 342–343, 383–384, 585.

Mann, C. and Plummer, M. 1991, *The Aspirin Wars. Money, Medicine, and 100 Years of Rampant Competition*, pp. 40–41. New York: Knopf.

McTavish, J.R. 1987a, Antipyretic treatment and typhoid fever 1860–1900. *Journal of the History of Medicine and Allied Sciences*, **42:** 486–506.

McTavish, J.R. 1987b, What's in a name? Aspirin and the American Medical Association. *Bulletin of the History of Medicine*, **61:** 342–366.

McTavish, J.R. 1987c, Aspirin in Germany: the pharmaceutical industry and the pharmaceutical profession. *Pharmacy in History*, **29:** 103–115.

McTavish, J.R. 1999a, The headache in American Medical Practice in the nineteenth century: a historical overview. *Headache*, **39:** 287–298.

McTavish, J.R. 1999b, What was Bayer doing before it had Aspirin? A glimpse into pharmaceutical marketing practices in America around the turn of the century. *Pharmacy in History*, **41:** 3–15.

Meldrum, M. 1911, Trade-mark rights. *Pharmaceutical Journal*, **86:** 425.

Merkel, F. 1905, Aspirin als Analgeticum in der Gynäkologie und Geburtshilfe. *Deutsches Archiv für Klinische Medizin*, **84:** 261–264.

Meserve, A.K.P. 1897, The use of advertised drugs by physicians. *Journal of Medicine and Science*, **4:** 271–277.

Meyer-Thurow, G. 1982, The industrialization of invention: a case study from the German chemical industry. *Isis*, **73:** 363–381.

Moeli, C. 1875, Über den Ersatz der Salicylsäure als Antifebrile durch das salicylsauren Natron. *Berliner Klinische Wochenschrift*, **12:** 517–519.

Nusch, A. 1901, Weitere Mitteilungen über therapeutischen Werth von Heroin und Aspirin. *Münchener Medizinische Wochenschrift*, **48:** 457–460.

Partington, J.R. 1964, *A History of Chemistry*. London: Macmillan.

Paton, W.D.M. 1979, Evolution of Therapeutics. Osler's therapeutic nihilism and the changing pharmacopoeia. *Journal of the Royal College of Physicians and Surgeons*, **13:** 74–83.

Perkin, W.H. 1896, The origins of the coal-tar colour industry. *Journal of the Chemical Society*, **69:** 596–637.

Pernick, M.S. 1985, *A Calculus of Suffering. Pain, Professionalism, and Anesthesia in Nineteenth Century America*. New York: Columbia.

Pharmacopoeia of the United States of America. 1905, 8th Revision, p. xli. Philadelphia.

Rageth, S. 1964, Die antipyretische Welle in der zweiten Hälfte des 19. Jahrhunderts. *Zürcher Medizingeschichtliche Abhandlungen*, **24:** 24.

Rainsford, K.D. 1984, *Aspirin and the Salicylates*, p. 9. London: Butterworths.

Reimer, T.M. 1996, *Bayer and Company in the United States: German Dyes, Drugs and Cartels in the Progressive Era*. PhD Dissertation, pp. 196–236, Syracuse University (New York).

Renon, L. 1900, Sur la valeur therapeutique de laspirine. *Bulletin de la Societé Medical des Hôpitaux de Paris*, **17:** 993–997.

Report of Patent, Design and Trade Mark Cases. 1920, Farbenfabriken vormals Friedrich Bayer & Co v. Chemische Fabrik von Heyden, **22:** 501–518.

Riess, L. 1875, Über die innerliche Anwendung der Salicylsäure. *Berliner Klinische Wochenschrift*, **12:** 297–298.

Riess, L. 1876, Nachtrag zur innerlichen Anwendung der Salicylsäure, insbesondere bei dem acuten Gelenkrheumatismus. *Berliner Klinische Wochenschrift*, **13:** 86–88.

Rocke, A.J. 1993, *The Quiet Revolution. Hermann Kolbe and the Science of Organic Chemistry*, pp. 304–309. Berkeley: University of California.

Schlenk, O. 1934, *Chemische Fabrik von Heyden Aktiengesellschaft, Radebeul-Dresden*. Radebeul: privately published.

Shapiro, L., Angorn, R. and McCormick, W. 1982, Look-alike prescription drugs. A historical perspective of the legal issues. *American Pharmacy*, **22(8):** 26–30.

Silverman, M., Lydecker, M. and Lee, P.R. 1992, *Bad Medicine. The Prescription Drug Industry in the Third World*. Stanford: Stanford University.

Smith, P.K. 1958, *Acetophenetidin: A Critical Bibliographic Review*, pp. 121–137. New York: Interscience Publishers.

Sonnedecker, G. 1976, *Kremers and Urdangs History of Pharmacy*, 4th edn. Philadelphia: Lippincott.

Spooner, J.B. and Harvey, J.G. 1976, The history and usage of paracetamol. *Journal of International Medical Research*, **4 (Suppl.)**: 1–6.

Stephan. 1911, Apotheker und Grossindustrie. *Apotheker Zeitung*, **26**: 903.

Stone, E. 1763, An account of the success of the bark of the willow in the cure of agues. *Philosophical Transactions of the Royal Society of London*, **53**: 195–200.

Stricker. 1876, Über die Resultate der Behandlung der Polyarthritis rheumatica mit Salicylsäure. *Berliner Klinische Wochenschrift*, **13**: 1–2, 15–16, 99–103.

Supreme Court, Canada. 1924, Bayer Company, Ltd. v. American Druggists Syndicate Ltd. 1924 S.C.R. 558.

Tirard, N. 1905, Some clinical observations with new remedies. *Lancet*, **i:** 83–84.

Trebach, A. 1992, *The Heroin Solution*, p. 46. New Haven: Yale University Press.

United States Department of Health, Education, and Welfare. 1969, *Task Force on Prescription Drugs, Final Report*. Washington, DC.

Wainwright, M. 1990, *Miracle Cure. The Story of Penicillin and the Golden Age of Antibiotics*. Oxford: Blackwell.

Wilbert, M.I. 1903, On the problem of proprietary and trade names. *Proceedings of the American Pharmaceutical Association*, **51**: 529–530.

Winckelmann, H. 1903, Aspirin-Nebenwirkung. *Münchener Medizinische Wochenschrift*, **50**: 1817.

Witthauer, K. 1899, Aspirin, ein neues Salicylpräparat. *Therapeutische Monatschaft*, **13**: 330.

Witthauer, K. 1900, Weitere Erfolge mit Aspirin. *Therapeutische Monatschaft*, **14**: 535.

Wohlgemuth, J. 1899, Über Aspirin (Acetylsalicylsäure). *Therapeutische Monatschaft*, **13**: 276–279.

Wohr, F. 1902, Observations of three hundred and sixty-two cases treated with Aspirin. *Medical Bulletin*, **24**: 274–276.

Wolffberg, S. 1875, Über die antipyretische Bedeutung der Salicylsäure. *Deutsches Archiv für Klinische Medizin*, **16**: 162–185.

Young, J.H. 1961, *The Toadstool Millionaires: A Social History of Patent Medicines in America before Federal Regulation*. Princeton: Princeton University Press.

■ CHAPTER 2 ■

3

Occurrence, Properties and Synthetic Developments of the Salicylates

K.D. Rainsford

INTRODUCTION

The salicylates are present in a wide variety of plants and microbial species. They are used extensively for production of drugs, and inspection under the heading of 2,5-dihydroxy-benzoic acid derivatives in *Chemical Abstracts* shows that their derivatives have wide application as synthetic precursors for the production of other compounds used in the chemical and pharmaceutical industries.

In this chapter the occurrence of salicylates, their formation in biological systems and their principal chemical and biological properties are reviewed. The medicinal chemistry and general chemical properties of the salicylates that are used pharmaceutically will also be considered.

NATURALLY OCCURRING SALICYLATES

Various salicylates are found in an immense variety of plant and bacterial species (see Table 3.1). Salicyl alcohol (or saligenin) and its glycoside (salicin) occur in the willow and poplar trees and the black haw (Figure 3.1; Gross and Greenberg, 1948). Methyl salicylate (oil of wintergreen, or Gaultheria) occurs in various species, ranging from trees (e.g. birch, myrtle and beech) to grasses (e.g. wheat, rye, sugar cane), legumes (e.g. peas, beans, clover), many common fruits (e.g. oranges, apples, strawberries, cherries, plums, raspberries and grapes) and exotic plants (e.g. Indian liquorice, ipecacuanha, feijoa fruits, teaberry and coffee) (Table 3.1; Gross and Greenberg, 1948; Mu and Young, 1966; De Alencar *et al.*, 1972; Starodubtseva *et al.*, 1971; Collins and Halim, 1972; Delaude *et al.*, 1974; Janssen *et al.*, 1996; 1997; Tamaki *et al.*, 2000; Tamaki *et al.*, 2000). The content of methyl salicylate present in conventional (black) tea has been found to be related to its flavour characteristics, the optimal methyl salicylate content being 7 to 10 μg/kg (Abraham *et al.*, 1976). Salicylaldehyde is produced by members of the *Spiraea* genus, *Filipendula* sp. (e.g. meadowsweet, bridal wreath).

Salicylic acid and its 3-hydroxy-and 6-methyl derivatives are produced in relative abundance by some moulds, several marine organisms and bacteria (Table 3.1).

Many of these naturally occurring salicylates (including salicylic acid itself) are therapeutically effective as weak anti-inflammatory, analgesic and/or antipyretic agents, although in some instances their potency is less than that of aspirin (Adams and Cobb, 1967; Whitehouse *et al.*, 1977; Borchers *et al.*, 2000). Various 6-n-alkylsalicylic acids (see Durrani and Tyman, 1979) are present in the shell liquid

TABLE 3.1

Some naturally occurring salicylate (2-hydroxybenzoic acid) derivatives.

Compound	Occurs as	Found in
Salicyl alcohol (2-hydroxybenzyl alcohol or saligenin)	Phenolic glucose (salicin) and free alcohol	Poplar, willow bark and leaves[1,] and *Gardenia jasmanoides*[2]
Salicylaldehyde	Free phenol or its glucoside (helicin)	*Spiraea* spp., *Filipendula* spp., e.g. meadowsweet, bridal wreath[3]
Methyl salicylate	Aglycone and various glycosides	Oils of wintergreen (*Gaultheria*), birch, myrtle, linden, madder, blackthorn, coffee, wild pansy, bay tree, Indian liquorice, soap berry, beech, wheat, rye, sugar cane, tea, coffee, cloves, olives, cassia, lily of the valley, *Camellia, Feijoa, Filipendula, Paederia, Parkia, Phellinus, Polygala Primula, Theaceae* sp.[1,4,5]
Salicylic acid	Free acid, glycosides or as mycobactins	*Mycobacterium* spp., *Pseudomonas* spp., barley[6], tobacco[7], *Arabidopsis* spp.[8], pea-roots[9], duckweed[10]
3-Hydroxysalicylic (2,3-dihydroxy-benzoic) acid	3-*O*-Glucoside, N-(lysine, glycine) or other conjugates or enterochelin	Coliform and other bacterial spp., periwinkle leaves[11].
6-Methyl- and various 5-*n*-alkyl-salicylic acids	Methyl ester or free acids	Ponerine ants, cashew nuts, *Pentaspadon* spp., Chrysanthemum spp.[12]
2-Hydroxy-acetophenone	Free phenol	*Chione glabra* (West Indies)[1]

[1]Gross and Greenberg, 1948; [2]Vallélian-Bindschelder *et al.*, 1998; [3]Saifullina and Kozhina, 1975; [4]Mu and Young, 1966; [5]Fujita *et al.*, 1974; [6]Weichert *et al.*, 1999; [7]Lee and Raskin, 1999; [8]Silva *et al.*, 1999; [9]Blilou *et al.*, 1999; [10]Mizukami *et al.*, 1986; [11]Tamaki *et al.*, 2000, [12]see text.

of the cashew nut (*Anacardium occidentale*) as a mixture comprising various anacardic acid derivatives (6-[$C_{15}H_{31-x}$], where x = 0, 2, 4 or 6). Anacardic acids have antifungal activity against many plant pathogenic fungi (Prithiviraj *et al.*, 1997). Other 6-*n*-substituted salicylic acids are found in various plant sources, e.g. ginkgolic acid (6-[$C_{15}H_{39}$]-salicylic acid) in *Ginko biloba*, pelandjauic acid in *Pentaspadon* sp., and others in *Chrysanthemum* sp. (Durrani and Tyman, 1979). 6-*n*-Pentadecylsalicylic acid and 6-*n*-tridecylsalicylic acids have also been isolated from various plants and marine organisms (Table 3.1; Durrani and Tyman, 1979) and their chemical synthesis achieved starting with n-alkyl-lithiums to give fluoroanisoles, followed by carbonation to 6-alkyl-methoxybenzoic acids and subsequent demethylation (Durrani and Tyman, 1979). Anacardic acids and analogues have been found to have inhibitory effects on the metabolism of Gram-positive and some Gram-negative bacteria (Gellerman *et al.*, 1969). Several isomers of 6-alkenyl-salicylic acids were isolated from *Spondias mombin* and were found to have antibacterial effects against *Bacillus cereas*, *Streptococcus pyrogenes* and *Mycobacterium fortuitum* (MIC 3 to 25 µg/m) and molluscocidal effects against the *Schistosome* harbouring snail, *Biophalaria glabrata* ($LD_{90} \approx 1$–3 ppm) (Corthout *et al.*, 1994). The latter property may prove important in the prevention of schistosomiasis (Bilharziaisis) (Corthout *et al.*, 1994). The reason why salicylates are produced in such relatively high abundance by plants is probably because of their roles as growth regulators and in host defence (Table 3.2; Raskin, 1992).

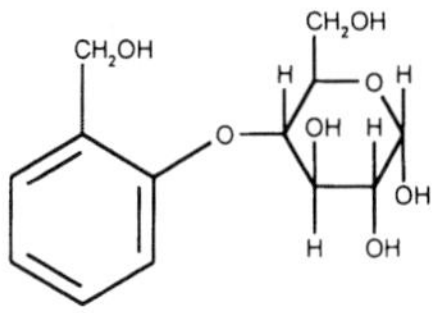

Salicin

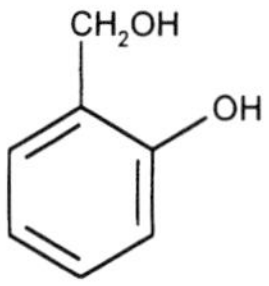

Salicyl Alcohol

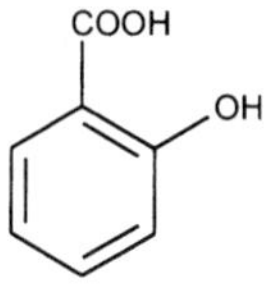

Salicylic acid

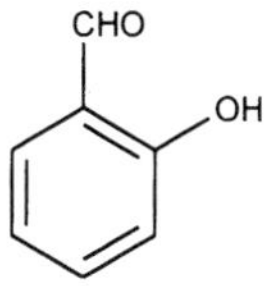

Salicylaldehyde

Figure 3.1 Chemical structures of some naturally occurring salicylates.

TABLE 3.2

Summary of the roles and mechanisms of salicylic acid (SA) in host defence in some plants.

Principal systems involved in mediating SA-related host defence systems:

- Signalling in plant defence against pathogens/injury by local and Systemic Acquired Resistance (SAR)
- SAR is mediated by SA-dependent and SA-independent pathways
- Induction of pathogenesis-related resistance systems (PRS)

Mechanisms:

- Gene expression of PRS range by SA from early, early to intermediate, or late responses
- Early responses involve transcription factor activation, i.e. signalling via MAP-type and other kinases→phosphorylation/dephosphorylation cycles and (NFκB/IκB)-type responses
- Increases in ion fluxes ($Ca^{++}H^{+}/K^{+}$); modulation by salicyl radical, nitric oxide (NO), H_2O_2 or lipid peroxides
- Localised cell death→containment of infection
- Late-response genes activated to produce PRs
- Activation of heat shock proteins (HSPs)
- Induction of 13-lipoxygenases
- Salicylate metabolised to methyl salicylate, salicyl alcohol and glucoside derivatives – serves to regulate amount of salicylate available for cell actions

Based on reviews by Raskin, 1992; Dempsey *et al.*, 1999; Lee and Raskin, 1999; Pieterse and vanLoon, 1999 and Cameron, 2000 and on original investigations by Mizukami *et al.*, 1986; Vallélian-Bindschedler *et al.*, 1998; Blatt *et al.*, 1999; Blilou *et al.*, 1999; Chong *et al.*, 1999; Conje and Bornman, 1999; Hugot *et al.*, 1999; Ikeda *et al.*, 1999; Jirage *et al.*, 1999; Kim *et al.*, 1999; Kock *et al.*, 1999; Lantin *et al.*, 1999; Manandhar *et al.*, 1999; Silva *et al.*, 1999; Thara *et al.*, 1999; Thomma *et al.*, 1999; Valkonen and Watanabe, 1999; Van Wees *et al.*, 1999; Venkatappa *et al.*, 1999; Weichert *et al.*, 1999; Berlanga and Vinas, 2000; Birkett *et al.*, 2000; Dellagi *et al.*, 2000; Greenberg, 2000; Greenberg *et al.*, 2000; Kachroo *et al.*, 2000; Kharat and Mahadeven, 2000; Kumar and Klessig, 2000; Mittler *et al.*, 2000; Nok *et al.*, 2000; Ohtake *et al.*, 2000; Somers *et al.*, 2000; Ton *et al.*, 2000; Weichert *et al.*, 2000.

Salicylic acid and its 3-hydroxy derivative are growth factors for certain bacteria, being required for iron transport (Young *et al.*, 1967; Ratledge and Hall, 1971; 1972; Ratledge *et al.*, 1974). Salicylic acid is produced in extracellular culture filtrates of several bacteria grown in iron-deficient media (Chipperfield and Ratledge, 2000). Salicylate is, however, not synthesised by *Mycobacterium* sp. when iron (Fe^{3+}) is present in abundance, there being feedback control by Fe^{3+} on salicylate synthesis (Young *et al.*, 1967; Ratledge and Hall, 1971; 1972; Ratledge *et al.*, 1974). Other aromatic acids (e.g. citrate, oxalate) cannot substitute for salicylate in its role on iron transport, but excess phosphate does cause suppression of salicylate-mediated iron transport (Ratledge and Hall, 1972). It has been suggested that salicylate may function as a siderophore and serve as an iron-solubilising agent in bacteria, but the insolubility of ferric ions in the presence of ubiquitous phosphate ions and theoretical calculations of the solubility of Fe(III) salicylate complexes make it unlikely that this function can be ascribed to salicylate itself (Chipperfield and Ratledge, 2000). However, salicylate does serve as a precursor for the biosynthesis of siderophores such as mycobactin S (Figure 3.2; Ratledge and Hall, 1970; Ratledge and Marshall, 1971; Marshall and Ratledge, 1972); yersinobactin of *Yersinia enterocolitica* (Chambers *et al.*, 1996); vulribactin of *Vibrio vulnificus*; and parabactin from *Paracoccus denitrificans* (Drechsel and Winkelmann, 1997). The 3-hydroxy analogue of salicylate serves as a precursor for the synthesis of enterochelin/enterobactin (Figure 3.2), which, like mycobactin S, is also required for iron transport in bacteria (Young *et al.*, 1967; Ratledge and Marshall, 1971; Dewick, 1997). It is possible that free salicylates may serve in the transport of ions (e.g. Fe^{3+}) in plants. The synthesis of the glycosides or derivatives of salicylates may also represent end (or inactivated) products.

In bacteria, the carbon atoms of salicylic acid and its 3-hydroxy derivative are synthesised from the glucose metabolites, phosphenolpyruvate and erythrose 4-phosphate via the shikimate–chorismate pathway (Figure 3.2; Young *et al.*, 1967; Marshall and Ratledge, 1972; Haslam, 1974; Dewick, 1997). In plants another branch of this pathway yields phenylalanine, which is metabolised either via cinnamate to salicylaldehyde or through benzoate to salicylate (Haslam, 1974; Raskin, 1992; Dewick, 1997; Figure 3.2). *O*-Glucosylation of salicylaldehyde forms its glycoside, helicin (Figure 3.2). Alternatively, salicylaldehyde may be oxidised to salicyl alcohol (saligenin), which subsequently combines with glucose to produce salicyl alcohol glucoside or salicin (Dewick, 1997; Figure 3.2).

Salicylic acid is also obtained from the microbial oxidation of naphthalene in a strain of *Pseudomonas* (Williams *et al.*, 1975; Filonov *et al.*, 2000). The gene coding for the enzyme salicylate hydroxylase, a flavoprotein enzyme that is responsible for the first step in degradation of salicylate to catechol, has been cloned from *Pseudomonas* (Kim and Tu, 1989), NMR properties studied (Vervoort *et al.*, 1991) and high expression systems prepared (Suzuki *et al.*, 2000). These developments and systems controlling dehydration of naphthalene to salicylate could be exploited for bioremediation. An entirely different series of reactions are involved in the biosynthesis of 6-methylsalicylic acid in *Penicillium* moulds (Birch *et al.*, 1955; Whitehouse *et al.*, 1977) and in some mycobacteria and fungi (Pettersson, 1966). In this case acetyl-S-CoA and malonyl-S-CoA serve as precursors, which enter the polypeptide pathway to form 6-methylsalicylic acid, which then serves as an intermediate in the synthesis of antibiotics (Figure 3.2).

There is interest in the biosynthesis of the salicylates in plants and micro-organisms, not only because of their biological importance but also as a practical means for obtaining these compounds as raw materials for the synthesis of more elaborate derivatives of this highly successful group of drugs (Birch *et al.*, 1955). It has been proposed that with increasing costs or possible shortages of petroleum and coal as starting materials for the synthesis of salicylic acid (from phenol), it may be more economic to biosynthesise salicylates by high-volume fermentation using selected mutant strains (Whitehouse *et al.*, 1977; Adilakshmi *et al.*, 2000) or plasmid-harbouring (Filonov *et al.*, 2000) strains of micro-organisms Alternatively, it may be more appropriate to exploit the energy derived directly from the sun and employ plant cell culture techniques and plasmid technology to grow plant cell strains with high salicylate-synthesising capacity.

Cell cultures of the leaves of the *Rubiaeceae* species *Gardenia jasmanoides*, have been shown to produce salicin and isosalicilin via glucosylation of salicyl alcohol (Mizukami *et al.*, 1986). These plant cell culture systems could afford a basis for producing salicylates naturally.

Salicylates are also found extensively in a variety of plant species, including the 'model' plant

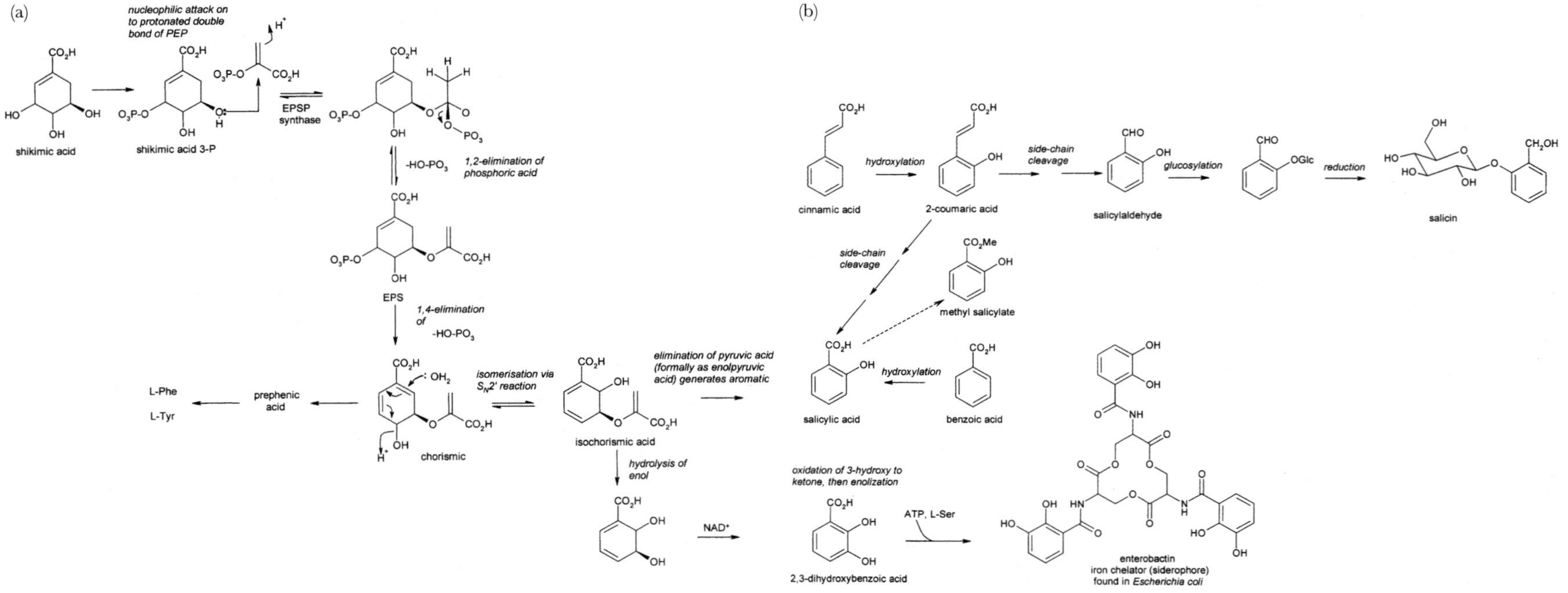

Figure 3.2 Biosynthetic routes to salicylate and derivatives in (a) bacteria (e.g. *Escherichia coli*) and (b) plant species (after Raskin, 1992; Dewick, 1997).

Arabidopsis thalania (Table 3.1), which is used for studies on growth regulation and genetic studies (its genetic structure having been recently reported by The Arabidopsis Genome Initiative, 2000). Salicylate produces disease resistance to pathogenic fungi, along with that of the gas methyl jasmonate, in *Arabidopsis* (Silva *et al.*, 1999; Thomma *et al.*, 2000). Various mutants have been developed in *Arabidopsis* that exhibit positive or negative regulation, and their actions linked to cell death and pathogen resistance (Greenberg, 2000). There are indications from studies of mutants of *Arabidopsis* that gene regulating accelerated cell death by salicylate (*acd5*) effects can be separated from those controlling Systemic Acquired Resistance (SAR), which are Nonexpressor of PR1/No Immunity (NPR1/NIM1) (Dong *et al.*, 1999; Greenberg *et al.*, 2000). Another gene system in *Arabidopsis* that controls the salicylate-mediated programmed cell death response, the Hypersensitive Response (HR), which is another component of disease resistance, has been identified as have mutants conferring variations in HR (Yu *et al.*, 2000). This system mediates the HR response by accumulation of reactive oxygen species (ROS) (Wolfe *et al.*, 2000). There is a variety of signalling pathways implicated in the development of HR and other components of the SAR in *Arabidopsis* (Norman-Setterblad *et al.*, 2000; Ohtake *et al.*, 2000), including those regulating ROS (Mackerness *et al.*, 1999; see also Table 3.2).

Figure 3.2 shows a summary of production of the salicylates in plant species and their properties. It has been suggested that the widespread ingestion of foods containing salicylates may have contributed to the reduction in cardiovascular mortality since the 1960s (Ingster and Feinleib, 1997; Janssen *et al.*, 1997). Similar statements have been made at international conferences and Internet sites, that salicylates in vegetables may account for the lower rate of colon cancer in vegetarians.

There has been much interest in the role of salicylic acid in inducing pathogen resistance in various plants (Raskin, 1992; Vallélian-Bindschedler *et al.*, 1998; Blilou *et al.*, 1999; Dempsey *et al.*, 1999; Dellagi *et al.*, 2000; Kachroo *et al.*, 2000; Klessig *et al.*, 2000; Niggeweg *et al.*, 2000; Quirino *et al.*, 2000; Schulzel-Lefert and Vogel, 2000), where its gene-regulated production modified by introduction of bacterial genes has been exploited as a means of enhancing this resistance (Dempsey *et al.*, 1999; Mur *et al.*, 2000; Verberne *et al.*, 2000). Complex mechanisms involving induction of intracellular signalling pathways mediate the pattern of resistance by salicylate (Table 3.2) as well as related phytohormones, including those relating to: (a) the production of H_2O_2 and H_2O_2-scavenging enzymes, catalase and ascorbate; (b) a high-affinity salicylic acid binding protein (SABP2); (c) a salicylic acid-inducible protein kinase (S1PK); and (d) a homologue to the IκBα (involved in mammalian signalling pathways (NPR1) and transcription factors (TGA/OBF family of the bZIP group; Dempsey *et al.*, 1999; Dat *et al.*, 2000; Klessig *et al.*, 2000). There may be other signalling pathways regulated by salicylic acid, e.g. in potatoes (Dellagi *et al.*, 2000) and in rice (Agrawal *et al.*, 2000; Table 3.2). There is also evidence of interplay between salicylic acid and nitric oxide-mediated pathways of host resistance (Klessig *et al.*, 2000). The development of host resistance by salicylate may be pathogen-specific (Vallélian-Bindschedler *et al.*, 1998).

A hormonal role for salicylic acid in plants is also evident in regulation of flowering (Tamot *et al.*, 1987; Raskin, 1992) and leaf senescence (Morris *et al.*, 2000), and in development of ripening in bananas (Srivastava and Dwivedi, 2000) and peas (McCue *et al.*, 2000). In barley there is an interesting metabolic pathway involving regulation by salicylate of lipid metabolism, where 13-lipoxygenase is induced by salicylate (together with the related phytohormone, jasmonate), leading to production of 13S; 9Z, 11E, 15Z)-13-hydroxy-9,11,15-octadecatrienoic acid (13-HOT), which in turn induces the expression of PR1b (Weichert *et al.*, 1999).

Oxidant status plays a key role in determining tolerance to stress, e.g. in tobacco (Dat *et al.*, 2000). Thus, heat shock can increase H_2O_2 content in the roots of tobacco treated with salicylic acid while the activity of enzymes involved in regulating redox activity declines (Dat *et al.*, 2000). There is a concentration dependence of these two components of the regulation of redox status combined with conversion at high concentrations of salicylic acid to its glucosylated metabolite (Dat *et al.*, 2000), thus showing a link between salicylate-related cellular control and metabolism of this regulator. Glucosylation of salicylic acid is pathogen-inducible (Lee and Raskin, 1999). One of the mitochondrial NAD(P)H-dependent oxidising pathways, a non-phosphorylating pathway, which is increased in beetroot with aging is increased by salicylate (Potter *et al.*, 2000). This site of action in mitochondria of salicylate may have significance in control of senescence and possible host resistance in plants.

Salicylates have also been identified in beaver castor (i.e. of scent glands) where it is secreted instead of via the usual urinary route. The salicylates are probably metabolic transformation products from vegetable sources in the diet of the beaver (Lederer, 1941). The methyl ester of 6-methylsalicylate is produced as a defence secretion in the ponerine ant (Duffield and Blum, 1978). Otherwise, salicylates do not appear to be abundant in animal tissues or secretions except from ingested foods (Ingster and Feinleib, 1997; Janssen et al., 1997).

CHEMICAL PROPERTIES

Summaries of the principal chemical and physical properties of salicylic acid, aspirin and other salicylates are shown in Tables 3.3 to 3.6 respectively.

Reactions of salicylic acid

Salicylic acid reacts in a manner reflecting a compound that has both an aromatic carboxylic group and a phenolic hydroxyl group. In aqueous solutions it gives a violet colour with ferric chloride. When heated quickly it rapidly sublimes, but when heated slowly it undergoes decarboxylation. However, when heated to 200°C it forms phenyl salicylate, probably by the combination of phenol (from decomposition) with salicylic acid. When the potassium salt of salicylic acid is heated to 230°C, p-hydroxybenzoic acid is formed. When reduced with sodium and isopentanol, salicylic acid causes opening of the aromatic ring to form the dicarboxylic acid, pimelic acid. Treatment of salicylic acid with bromine water results in replacement of the carboxyl group by bromine, and produces s-tribromo-phenol. Similar reactions occur when salicylic acid is treated with nitric acid, with resultant trinitro-phenol being formed (Finar, 1963). Salicylic acid undergoes electrophilic substitution reactions resulting in ring substitution (Gottesman and Chin, 1968).

Structure and reactions of aspirin

The crystal morphology of aspirin (Table 3.4) has been shown to be a dimer with the hydrogen bonds formed across a centre of symmetry (Wheatley, 1964; Umeyama et al., 1979). The length of each hydrogen bond is 2.645 Å (Wheatley, 1964). Quantum chemical calculations have shown that the dimer with two hydrogen bonds is more stable through two carboxyl groups (Umeyama et al., 1979). The interaction energy of the hydrogen bond dimer is due to charge transfer and electrostatic interaction terms (Umeyama et al., 1979). Various growth morphologies of aspirin have been investigated and the crystal forms modelled (Meenan, 1997), and the simulations thus obtained have been explained by hydrogen bonding, surface charge and steric considerations. The surface chemistry of aspirin crystals has been studied by dynamic force microscopy, from which it has been shown that the methyl groups have interactions at the (001) crystal planes while the carboxyl moieties have larger interactions with the (100) planes (Danesh et al., 2000).

The morphology of aspirin crystals has been shown to be markedly affected by the solvents used for recrystallisation (Summers et al., 1970; Meenan, 1997). Thus 'commercial' aspirin typically formed from aromatic, acyclic or chlorinated solvents has a larger depth than if recrystallised from acetone, methanol, dioxane, heptane or water (Meenan, 1997). These features of the crystal structure of aspirin are obviously of importance for formulation of the drug, markedly affecting dissolution (Kim et al., 1985) as well as absorption of oral forms of the drug (Martin, 1971).

Aspirin appears to have the characteristics of an acid anhydride (Davidson and Auerbach, 1953). It has been suggested by Davidson and Auerbach (1953) that the acetylating capacity of aspirin may be accounted for by assuming an equilibrium between aspirin and salicyloyl acetic anhydride.

CHAPTER 3

TABLE 3.3

Physical and chemical properties of salicylic acid.

Empirical formula	$C_7H_6O_3$
Formula weight	138.12
Crystal structure	
Space group	$P2_1/a$
Number of molecules per unit	4
Dimensions of unit cell[a]	a = 11.56 Å, b = 11.21 Å, c = 4.93 Å, β = 91°22′
Constants	
Melting point	157–159°C
	b_{20} 211°C
Sublimation point	76°C d_4^{20} 1.443
Heat of combustion	22.699 kJ/g at 15°C in air
	21.921 kJ/g absolute at 15°C in vacuo
Heat of solution	−26.57 kJ/mol at 15°C
Dissociation constants in:	
absolute ethanol	1.06×10^{-3} (25°C)
	1.13×10^{-3} (50°C)
ethanol (96%)	2.1×10^4 (25°C)
chloroform	1.55–1.56 (30.5°C)
carbon tetrachloride	0.35–0.36 (30.5°C)
benzene	1.001–1.021 (30.5°C)
water	1.06×10^{-3} (25°C)

Solution properties
- Gradually discolours in sunlight.
- When rapidly heated at atmospheric pressure it decomposes into phenol and CO_2.
- One gram dissolves in 460 ml of water, 15 ml boiling water, 2.7 ml alcohol, 3 ml acetone, 42 ml chloroform, 3 ml ether, 135 ml benzene, 52 ml oil turpentine, about 60 ml glycerol, about 80 ml fats or oils.
- Solubility in water is increased by sodium phosphate, borax, alkali acetates or citrates.
- pH of saturated aqueous solution 2.4.
- Salicylic acid is very easily soluble in liquid ammonia and is insoluble in liquid sulphur dioxide.
- Salicylic acid or its salts form reddish coloured solutions even by merest traces of ferric salts as a result of complexation providing useful method for detection. Incompatible with iron salts, spirit nitrous ether, lead acetate iodine.
- Slightly soluble in water (with heating)
- Very soluble in ethanol.

The solubility of salicylic in solvents [g/100 g of saturated solution (temp.)]
- methanol, 38.46 (21°C)
- ethyl alcohol (absolute) 34.87 (21°C)
- *n*-propyl alcohol, 27.36 (21°C)
- diethyl ether 24.4 (17°C)
- acetone 31.3 (23°C)

Other properties
Deliquescent in moist air. The aqueous solution is slightly acid to litmus.

Salts
Lithium salt $C_7H_5LiO_3$ or lithium salicylate. White or greyish white, odourless, sweetish powder.
Silver salt $C_7H_5AgO_3$ silver salicylate. White to reddish-white crystals. Slightly soluble in water and alcohol.

References: Gottesman and Chin, 1968; Merck Index, 1983; Kroschwitz and Howe-Grant, 1996; McBryde *et al.*, 1970.
[a]Crystal data from Cochran, 1953 quoted by Sundaralingham and Jensen, 1965 in their computerised refinement of the structure of salicylic acid.

TABLE 3.4

Physico-chemical properties of aspirin.

Empirical formula	$C_9H_8O_4$
Formula weight	180.15
	d 1.40
Melting point	135° (rapid heating) to 143°
	The melt solidifies at 118°
Crystal properties	
Appearance	Monoclinic $P2_1/c$ space group; four molecules per unit cell
Parameters	a = 11.433–11.446 Å,[a] b = 6.596 Å,[a]
	c = 11.388[a]–11.395 Å,[b] β = 95°33'[a]–95°68'[b]
Lattice energies (range)	25.87–33.89 Kcal/mol (calculated) [b]
General properties	
UV max. 0.1 $N H_2SO_4$	229 nm ($E^{1\%}$ 484) ($CHCl_3$)
	227 nm ($E^{1\%}$ 68)

Odourless, but in moist air it is gradually hydrolysed to acetic and salicylic acids

Stable in dry air   K at 25° = 3.27 × 10⁻⁴

Solubility

One gram dissolves in 300 ml water at 25°, in 100 ml water at 37°, in 5 ml alcohol or chloroform, in 10–15 ml diethyl ether. Less soluble in anhydrous diethyl ether

Pharmaceutical incompatibilities and stability
- Decomposed by boiling water or when dissolved in solution of alkali hydroxides and carbonates.
- Hydrolysis occurs in admixture with salts containing water of crystallisation.
- Aspirin forms a damp, pasty mass when titrated with acetanilide, phenacetin, antipyrine, aminopyrine, methenamine, phenol or phenyl salicylate.
- Powders containing aspirin with an alkali salt such as sodium bicarbonate become gummy on contact with atmospheric moisture.
- Solutions of the alkaline acetates and citrates, as well as alkalis themselves, dissolve aspirin, but the resulting solution hydrolyses rapidly to form salts of acetic and salicylic acids. Sugar and glycerol have been shown to hinder this decomposition.

References: Merck Index 10th edn, 1983; Crystal data from Wheatley, 1964,[a] Meenan, 1997.[b]

Thermal decomposition of aspirin leads to the formation not only of salicylic and acetic acids but also salicylsalicylic acid, acetyl-salicylsalicylic acid and cyclic polymers of salicylic acid. Thus, pyrolysis of aspirin with simultaneous distillation of products at 300 to 350°C (15 mm) produces these cyclic polymers termed salicylides (Reepmeyer, 1983). Reepmeyer (1983) produced highly linear oligomeric salicylate esters and their acetate derivatives by heating a solid mixture of aspirin and magnesium carbonate at 85°C for 2 hours. These reactions reflect the complex acetylation and acid anhydride characteristics of aspirin.

Miscellaneous physico-chemical properties of salicylic acid and its derivatives

The dissociation constants of a range of salicylic acid derivatives have been determined from potentiometric, spectrometric and kinetic analyses (Nishikawa *et al.*, 1983; Djurendic *et al.*, 1990; Aydin *et al.*, 1997). Humbert and co-workers (1998) have reported the infrared and Raman spectroscopic properties of salicylic acid and its derivatives. Mn (salicylate)$_2$(H$_2$O)$_2$ has been prepared and its EPR spectra studied (Alambar *et al.*, 1983)

TABLE 3.5

Physico-chemical properties of miscellaneous salicylate compounds.

I. Salicylsalicylic acid (Salsalate, Diplosal)	
Empirical formula	$C_{14}H_{10}O_5$
Formula weight	258.22
Crystal powder	
Crystalline properties[1]	Orthorhombic crystals
Space group	*Fdd2*
Unit cell dimensions	a = 12.9610 Å; b = 28.3230 Å; c = 12.9410 Å; α = 90°; β = 90°; γ = 90°
Volume	4750.6 Å³
Z	16
Solubility	Insoluble in water but gradually hydrolyses in it to salicylic acid. Soluble in ethanol and diethyl ether; sparingly soluble in benzene
II. Saligenen (salicyl alcohol)	
Empirical formula	$C_7H_8O_2$
Formula weight	124.13
Crystal powder or plates	
Density	1.16
Melting point	86–87°C (sublimes at 100°C)
Solubility	Soluble in 15 parts water, ethanol, chloroform, diethyl ether, benzene
Reactions	Gives red colour with H_2SO_4
III. Salicin (salicyl alcohol glucoside)	
Empirical formula	$C_{13}H_{18}O_7$
Formula weight	286.27
Crystalline properties	Orthorhombic crystals
Melting point	199–202°C
$[\alpha]_D^{25}$	−62° to −67° (c = 3)
$[\alpha]_D^{20}$	−45.6 (c = 0.6 in absolute ethanol)
Solubility	Soluble in 1 g/3 ml boiling water, 1 g/23 ml water, ethanol, alkalis, pyridine, glacial acetic acid. Almost insoluble in diethyl ether or chloroform
IV Salicylamide	
Empirical formula	$C_7H_7NO_2$
Formula weight	137.13
Melting point	140°C (metastable below melting point)
Crystalline properties	Monoclinic rod-like crystals of yellowish-white appearance
Cell dimensions	a = 12.93 Å, b = 5.02 Å, c = 24.80 Å
Solubility	Soluble in water (1 g/500 ml), propyleneglycol (1 g/20 ml), glycerol (with warming), hot water, ethanol (1 g/15 ml), chloroform (1 g/100 ml), diethyl ether (1 g/35 ml) Forms a water-soluble sodium salt at pH 9
V. Sulphasalazine/salicylazosulphapyridine (Salazopyrin)	
Empirical formula	$C_{18}H_{14}N_4O_5S$
Formula weight	398.39
Crystalline properties	Minute brownish-yellow crystals dec. 240–245°C
Solubility	Slightly soluble in ethanol, insoluble in water, benzene, chloroform, diethyl ether

From Merck Index, 10th edn., 1983, Babhair *et al.*, 1984. [1]Crystal structure described by Cox *et al.*, 2000.

TABLE 3.6

Physicochemical constants of some salicylates in various solvents.

Drug	pK$_a$	log P
Aspirin	3.5	−0.47
5-Chlorosalicylic acid	2.8	−0.13
Diflunisal	4.0	0.67
Salicylamide	8.1	1.26
Salicylic acid	2.9	−1.06

Values of pK$_a$ and log P from Hansch and Anderson, 1967; Weast and Astle, 1978; Babhair *et al.*, 1984; or determined experimentally (Rainsford *et al.*, 1979), the log P values being determined from partitioning of the drug between *n*-octanol/0.1 mol/l sodium phosphate buffer pH 7.4. The pK$_a$ values were obtained from titration of the drugs in various concentrations of either dioxane or ethanol in H$_2$O with corrections (by extrapolation to water values) for dielectric constant. Values for diflunisal and 5-chlorosalicylic acid determined experimentally.

Non-enzymic hydrolysis of aspirin

The spontaneous hydrolysis of aspirin varies markedly with pH (Edwards, 1950; 1952; Martin, 1971; Garrett, 1957; Figure 3.3) and the presence of counter ions (Moll and Stauff, 1985). Under the acidic conditions present in the stomach (i.e. pH 2 to 3), the rate of hydrolysis is much lower than at higher pH values (greater than 9 to 11), where the rate increases dramatically. The rate of hydrolysis of aspirin at pH 5 to 8 (such as it is in the upper intestinal tract) is about double that at pH 2, where it is at a minimum. The pH hydrolysis curve varies somewhat according to the buffer system employed (Jones *et al.*, 1978), but in general the pattern resembles that shown in Figure 3.3.

The rate of hydrolysis of aspirin increases markedly in the presence of 10 to 90 per cent aqueous ethanol mixtures (Garrett, 1957; Rainsford *et al.*, 1979). This is of relevance in view of the frequent consumption of alcohol with aspirin. Interestingly, no appreciable hydrolysis occurs in *absolute* methanol or propanol (Rainsford *et al.*, 1979), but does so in aqueous methanol mixtures where it occurs at a higher rate than that in water (Umeyama and Nakagawa, 1977). Hydrolysis of aspirin in aqueous media is reduced by the addition of sorbitol (Blang and Wesolowski, 1959). Ammonium ions and amines (e.g. histamine) and α-amino acids (at less than 20 mmol L per litre) also stimulate hydrolysis of aspirin, but porcine mucus (0.05 to 5.0 per cent) and pepsin (0.1 to 2.5 per cent) do not (Rainsford *et al.*, 1979). It appears, therefore, that amines and amino acids in the gastric juice could possibly

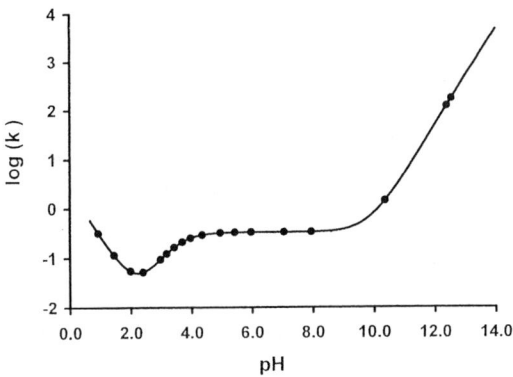

Figure 3.3 Rate of hydrolysis (log k) of aspirin in aqueous media (ionic strength 0.5 M) as a function of pH at 25°C (from studies by Some *et al.*, 2000 which is similar to data from Edwards, 1950; 1952 performed at 20°C). Reproduced from *International Journal of Pharmaceutics*, **198:** 39–49. (© 2000 with permission of Elsevier Science.)

enhance aspirin hydrolysis, whereas polyols having the properties of sorbitol may inhibit hydrolysis of the drug. The hydrolysis of aspirin and some derivatives has been found to be much lower in the pseudophase of micelles of cetyl-trimethyl ammonium bromide than in water (Broxton *et al.*, 1987). The significance of this is that the micelles like those employed in these studies by Broxton *et al.* (1987) may represent simple membrane models and so replicate or mimic situations found in membranes through which aspirin molecules will pass during absorptive and other phases or distribution. The lower rate of hydrolysis in membranes may be related to the orientation of the drug in the lipid phases of membranes.

The mechanism of hydrolysis of aspirin and its analogues in aqueous media has been investigated in considerable detail (Edwards, 1950; 1952; Garrett, 1957; Blang and Wesolowski, 1959; Fersht and Kirby, 1967a; 1967b; 1968a; 1968b; Bundgaard and Larsen, 1976; Umeyama and Nakagawa, 1977; Rainsford *et al.*, 1979; Alibrandi *et al.*, 1996). Using ^{18}O tracer studies, Fersht and Kirby, (1967a; 1967b; 1968a; 1968b) concluded that aspirin (and its analogues) are hydrolysed by a mechanism involving an intramolecular generalised basis catalysis involving formation of salicyl acetic anhydride (I) (Figure 3.4) as previously supposed. The mechanism as proposed by these authors is summarised in Figure 3.4. Interestingly, Fersht and Kirby commented that the mechanism approached the complexity of an enzymatic hydrolysis. Bundgaard and Larsen (1976) studied the hydrolysis of aspirin in non-hydroxylic (aprotic) solvents (benzene, ethyl acetate, chloroform, etc.) and established that an equivalent mixture of salicylic acid and acetylsalicylic acetic anhydride was formed. They postulated that aspirin is hydrolysed in aprotic solutions through the intramolecular addition of the carboxyl group to the ester carbonyl moiety to form the mixed salicyl acetic anhydride. This then reacts at the anhydride moiety with the carboxyl groups of a second aspirin molecular to produce either (a) salicylic acid and acetylsalicylic acetic anhydride, or (b) acetic acid and an aspirin–salicyl ester analogue (by another route). Clearly, this mechanism of hydrolysis in non-hydroxylic solvents is more complex than the base-catalysed hydrolysis in hydroxylic solvents. The hydrolysis in non-hydroxylic solvents is important because of the potential for the formation of impurities during manufacture, where such solvents are used extensively.

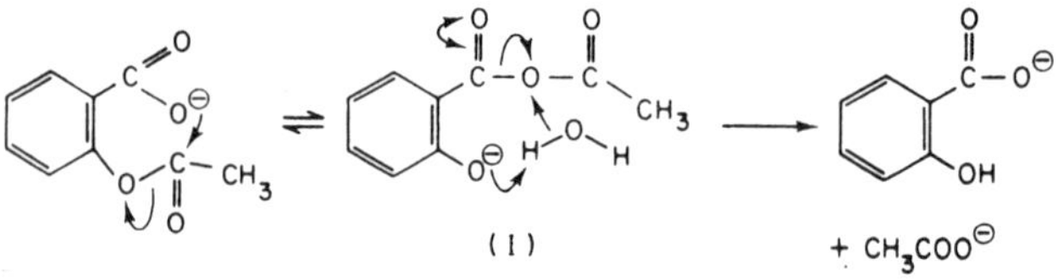

Figure 3.4   Mechanism of the hydrolysis of aspirin in aqueous media.

Solution chemistry of the salicylates

The salicylates are relatively lipophilic compounds, the Log $P_{n\text{-octanol/water}}$ value of aspirin (1.23) being lower than that of salicylic acid (2.26) (Hansch and Anderson, 1967; Table 3.6). Based on determinations of Log $P_{n\text{-octanol/water}}$ being lower than calculated using the additivity principle (i.e. the relative constant $\pi = \log P_x - \log P_H$, where H is the parent compound and X the derivative in a series), Hansch and Anderson (1967) proposed that a range of benzoyl and other derivatives, including salicylic acid, aspirin and salicylamide, exhibited intramolecular hydrogen bonding, which accounted for these differences in experimental compared with expected values of log P. Other studies have also confirmed the propensity of salicylic acid to form hydrogen bonds, in this case with heteroatomic systems (Berthelot *et al.*, 1996).

Based on fluorescence spectroscopy studies of salicylic acid, methyl salicylate and *o*-anisic acid, Kovi *et al.* (1972) concluded that intramolecular protolytic dissociations are formed during photo-tautomerism of these molecules. When salicylic acid is present in acidic or basic solutions, it is capable of undergoing two pathways of protolytic dissociations. The pK$_a$ values for the ground state of salicylic acid that were obtained spectroscopically by Kovl *et al.* (1972) were -8.0, 3.0 and 14.0, while those in the excited state by fluorometric titration, i.e. pK_a^*, were -7.0 and 16.0. The fluorescent behaviours of *o*-anisic acid (where there is no hydroxyl-proton) and methyl salicylate (where there is no carboxylic proton) supported the conclusion that there are two pathways of protolytic exchange in the intramolecular proton movements in salicylic acid.

Another interesting physical property of salicylic acid is its propensity to form transient free radicals (Borg, 1965). Thus, aqueous solutions of sodium salicylate or salicylaldehyde administered with the univalent oxidant, ceric sulphate, in $0.2\,M\,H_2SO_4$ instantly react to give olive-coloured products. Interactions with KM_nO_4 or $(NH_4)_2IrCl_6$ in neutral phosphate buffer give gradual changes in colour over a few minutes, with the former reagent in alkali producing a green product implying homolytic oxidation of salicylate with concomitant univalent reduction of the purple permanganate to green manganate (Borg, 1965). Potentiometric titrations of salicylate with the above oxidants quantitatively established the oxidation of salicylate. Electron paramagnetic resonance spectra using stop–flow techniques of salicylate oxidised with cerate or permanganate have shown the transient production of free radicals; similar observations have been made with salicylaldehyde and aspirin. The production of free radicals from salicylate may be an important pharmacological property, but the occurrence of this will depend on the redox state in cellular or physiological systems.

A series of studies by Hata and co-workers of the interactions of menadione with compounds having presumed electron-donor characteristics showed there are charge transfer interactions between salicylic acid and this drug (Hata and Tomioka, 1968a; 1968b; Hata *et al.*, 1968), Molecular orbital calculations confirmed these observations and established that, both theoretically and practically, salicylic acid can be regarded as an electron donor.

COMMERCIAL SYNTHESIS AND PROPERTIES OF THE SALICYLATES

Only those salicylates that have been developed or evaluated for clinical use as anti-inflammatory, analgesic and/or antipyretic agents will be considered here. The reader is referred to the useful review of Gottesman and Chin (1968) for details of other salicylates. A comparison of the principal therapeutic properties of the salicylates is shown in Table 3.7. Other more detailed descriptions are given in the subsequent chapters.

Salicylic and acetylsalicylic acids, their salts and esters

Essentially, the original method developed by Kolbe and Lautermann in 1874 is employed in the commercial production of salicylic acid today (Geissler and Möller, 1889; Gottesman and Chin, 1968). In the Kolbe reaction, alkali or alkaline earth phenoxides (prepared from middle petroleum oils or coaltar) are treated with carbon dioxide under high pressures and temperatures (120 to 170°C). The phenoxide ion is susceptible to electrophilic aromatic substitution by carbon dioxide (acting as the electrophile), as shown in Figure 3.5.

The original Kolbe reaction required 2 moles of sodium phenoxide being converted into 1 mole of phenol. It suggests that the phenolic hydroxyl group of the latter is more acidic than that in phenol (Fuson, 1962). The Schmitt modification avoided the low yields of the Kolbe reaction (which required relatively high temperature and pressure conditions) achieving this by lower temperatures and a longer time of reaction (Fuson, 1962). The modern method employs these milder conditions and is known as the Kolbe–Schmitt reaction.

The mechanisms of carbonation of phenol and related phenolic compounds have been investigated (Baine *et al.*, 1954), including the reaction conditions where the alkaline metal is changed or where

■ CHAPTER 3 ■

TABLE 3.7

Summary of the main pharmacological actions of some of the principal salicylates.

Drug	Actions			
	Anti-inflammatory[a]	Analgesic[b]	Antipyretic[c]	Anti-thrombotic[d]
Aspirin*	+++	+++	+++	+++
Benorylate	++	++	+	+
Choline salicylate	+++	++	+++	ND
5-Chlorosalicylic acid	++++	++	+++	+++[§]
2,3-Dihydroxybenzoic (=3-hydroxysalicylic) acid	+++	+++	+++	++
2,3-Diacetoxybenzoic acid (=3-acetoxy aspirin)	+++	+++	+++	++
Diflunisal	++++[†]	+++	++[†]	+
Salsalate (=salicylsalicylic acid; (Diplosal)	++	++	+++	0
Meseclazone	++++	+++	+++	+++
Phenyl salicylate (Salol)	+	++	+	0
Methyl salicylate[‡]	++	++	+++	+
Salicylaldehyde	++	++	++	+
Salicylamide	0	++	++	0
3-Methylsalicylic acid	++	++	+++	ND
Salicylic acid (or its sodium salt)	++	++	+++	0

Assessments graded on an arbitrary scale of 0 to 4 (i.e. increasing biological activity from assays of: (a) anti-inflammatory activity in the rat carrageenan paw oedema and adjuvant arthritis models; (b) analgesic activity in the mouse acetylcholine or phenyl quinone model; (c) antipyretic activity against yeast pyrogen in rats; and (d) inhibition of human platelet aggregation induced by ADP or collagen *in vitro*. (Further details in Chapter 7).

*Varies considerably according to formulation; [†]Prolonged action due to long half-life; [‡]Topically applied; [§]Variable according to aggregating agent; ND No data available. (From Rainsford (1984).)

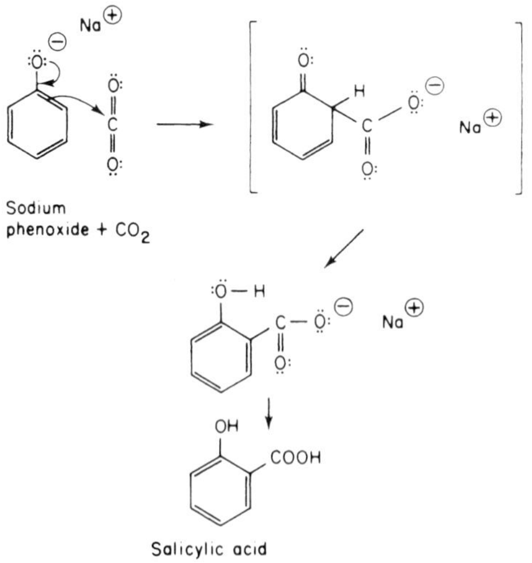

Sodium
phenoxide + CO_2

Salicylic acid

Figure 3.5 Synthesis of salicylic acid. (From Rainsford (1984). Reproduced with permission of the publishers, Butterworths/Heinemann.)

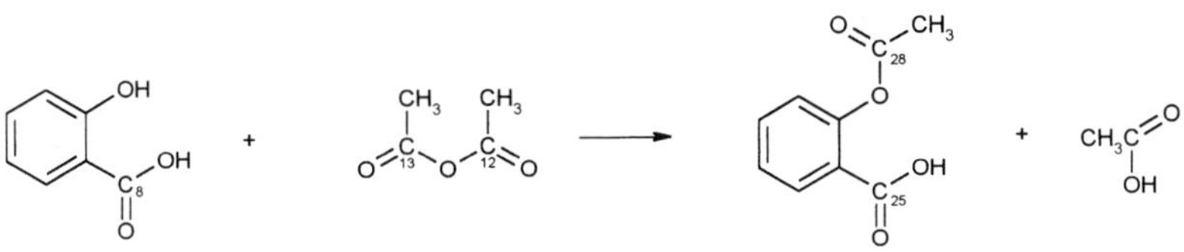

Figure 3.6 Synthesis of acetyl salicylic acid (aspirin). (From Rainsford (1984). Reproduced with permission of the publishers, Butterworths/Heinemann.)

carbonation is employed with the simpler Marassé procedure involving high temperature and pressure conditions. Using potassium phenoxide in place of the sodium salt, much lower yields of salicylic acid are obtained with proportionately greater yield of p-hydroxybenzoic acid (Baine et al., 1954; Fuson, 1962). The Marassé procedure does not achieve as high a yield as the Kolbe–Schmitt procedure (Baine et al., 1954).

Acetylsalicylic acid is subsequently produced by the acetylation of salicylic acid with acetic anhydride (Edmunds, 1966; Gottesmann and Chin, 1968; McKetta, 1977, Figure 3.6). The usual commercial production requires the recovery of the by-products acetic acid, excess acetic anhydride and unrecovered aspirin as salicylic acid (McKetta, 1977). A typical manufacturing process is shown in Table 3.8.

While the production of aspirin is a relatively old art, few major changes in the chemistry of the process have been made over the years. However, changes continue to be made in such areas as reaction cycles, temperatures, minimising economics and product quality (McKetta, 1977). In one commercial procedure, the reaction of salicylic acid is performed at elevated temperatures in the presence of fractional molar excess of acetic anhydride and acetic acid under controlled pressures and temperatures (Edmunds, 1966). Water and acetic acid are then added under reduced pressure. The yield claimed by using this procedure is greater than 99 per cent (Edmunds, 1966).

A convenient laboratory-based method for the production of aspirin on a milligram scale is shown in Table 3.9. This synthesis can also be applied as a convenient and simple general chemistry class experiment (Olmsted, 1998). The method can be adapted for large-scale laboratory production of aspirin or for the synthesis of either [^{14}C]-acetyl-labelled aspirin using [^{14}C]-acetic anhydride or [^{3}H], or [^{14}C]-salicylic acid from acetyl [^{3}H] or [^{14}C]-labelled salicylic acid. Production of [^{3}H]-labelled salicylic acid can be achieved by catalytic dehydrogenation with tritium gas (Michel and Truchot, 1963). The specific activity of the product obtained is sufficiently high to be useful in autoradiographic studies.

The alkaline salts (Na^+, K^+, Ca^{2+}) of salicylic acid are easily prepared by neutralising (and solubilising) the acid with the respective hydroxides or alkali salts (e.g. bicarbonate) (Geissler and Möller, 1989). However, care must be exercised in employing this procedure to form the salts of aspirin, since it is rapidly hydrolysed in aqueous solution to salicylic and acetic acids (see p. 55). The usual procedure in preparing these aspirin salts is to use equimolar solutions (with respect to the acid) of the bicarbonate salt so that the solutions are not made excessively alkaline, a procedure that would enhance the risk of hydrolysis.

The relative avidity of heavy metal ions for salicylates varies with the type of salicylate and the metal ion. Thus the chelate stability has been found to be salicylic greater than gentisic, which is

TABLE 3.8

Typical production of aspirin on a commercial scale. Basis: 100 lb bulk aspirin finished product, 90% yield on salicylic acid.

Raw material	lb/100 lb aspirin*
Salicylic acid	85
Acetic anhydride	65

*Recovered acetic acid amounts to 37 lb/100 lb aspirin.

TABLE 3.9

Preparation of aspirin on a milligram scale.

---

1. Dissolve 3.6 mg salicylic acid in 500 μl benzene to which 1 μl pyridine (as a catalyst) has been added.
2. Add 2.7 mg acetic anhydride in 500 μl benzene and allow the reaction to stand at room temperature for 18 h.
3. Purify the mixture by plating it out onto a preparative thin layer chromatography plate of Merck Kieselgel F_{254} fluorescent impregnated plates using 10 : 1 v/v petroleum ether (40–60°C bpt)/propionic acid. The zone corresponding to aspirin can be identified by UV light, scraped off and the aspirin extracted from the matrix with chloroform. Alternatively, the mixture may be separated by HPLC (e.g. using a system comprising a C-8 reverse-phase column, μ Bondipack, eluted with an isocratic gradient of 125 : 125; 250 : 0.5 v/v/v/v isopropyl alcohol/acetonitrile/water/0.1 μM ortho-phosphoric acid monitored at 215 nm (Altun *et al.*, 2001).
4. For the preparation of ^{14}C-acetyl labelled aspirin, ^{14}C-acetic anhydride may be added in place of acetic anhydride in no. 2 above.

Synthetic methods provided by D.R. Boreham, Nicholas Research Laboratories. Drug Metabolism and Kinetic Unit, Report No. 3, 3 April 1969.

greater than salicyluric acids (Pecci and Foye, 1960). These acids form stable complexes with Cu^{2+}, Fe^{3+} and Al^{3+} ions, but not with Co^{2+}, Ni^{2+}, Zn^{2+}, Mg^{2+}, Ca^{2+} or Ag^+ ions (Pecci and Foye, 1960; Suranyi *et al.*, 1995). These intrinsic affinities of the salicylates determine their ability to form pharmaceutically acceptable complexes (e.g. zinc aspirinate, technetium aspirin, vanadate aspirin; see Elshahawy *et al.*, 1993; Hartman and Vahrenkamp, 1994; Etcheverry *et al.*, 2000), as well as their relative biological significance in, for example, chelating reactions (Reid *et al.*, 1951) including those involving metal ions *in vivo* (Gaubert *et al.*, 2000).

In contrast to the poor solution stability of magnesium salicylate indicated above, a magnesium salicylate tetrahydrate tablet formulation has been prepared (Alam and Gregoriades, 1981). A manganese complex of salicylic acid, Mn (salicylate)$_2$(H$_2$O)$_2$, has been prepared and its EPR spectra studied (Alambar *et al.*, 1983).

Copper–salicylate complexes

There has been much interest in the potential pharmacological properties of copper complexes of salicylates (Lederle and Kollbrunner, 1980; Sorenson, 1982; Crouch *et al.*, 1985; Berthon, 1995), and these may possibly be more effective anti-inflammatory agents than the corresponding acids (see also Chapter 13). Copper salicylate in anhydrous and acidic, basic and mono-potassium salts of the anhydrous derivatives were originally prepared by Pickering (1929) using the basic salts, which were precipitated from salts of acids in various molar proportions. De Coninck (1915) described conditions for the preparation of copper salicylate by addition of salts of copper to the acid in ethanolic solution. The tetrahydrate obtained with excess ethanol yielded the basic monocuprous monohydrate and 'green needles with yellow reflex' were obtained upon action of hot water. The basic compound underwent decomposition and yielded copper oxide, phenol and CO_2 when heated.

Cupric and cuprous salts of salicylic and acetylsalicylic acids are readily prepared by treating the Na^+ or K^+ salts of the salicylates with cupric chloride solutions (at 4°C) in molar proportions of 1 : 2 of Cu^{2+} to salicylate or aspirin. The complexes formed are Cu(II)$_2$(salicylate)$_4$[Cu(II)$_2$(Sal)$_4$] or Cu(II)$_2$(acetylsalicylate)$_4$[Cu(II)$_2$(ASA)$_4$]. The 3,5-diisopropyl-salicylic acid (DIPS) complex were crystallised from dimethyl-formamide or diethyl ether and single crystal structures determined, the former having the chemistry of Cu(II)$_2$(3,5-DIPS)$_4$.DMF$_2$ (Morant *et al.*, 2000). Cu (DIPS) also has been reported to have anti-tumour and anticonvulsant activities (Sorenson, 1982; Crouch *et al.*, 1985). Cu (DIPS) and the Fe(III) and Mn(III) homologues have been shown to have radioprotective and radiorecovery effects (Sorenson, 1982; Irving *et al.*, 1996).

Recently, the Cu(II)–(salicylate)–(pyrazine) bridged polymer has been synthesised and its X-ray crystal structure analysed (Longguan *et al.*, 2000). Each salicylate ligand was found to connect with three copper centres, forming a novel rhombus-type two-dimensional coordination framework with angles of 60° and 120°. While no biological activity of this novel complex appears to have been reported, the polymeric feature may afford a unique means to deliver copper-salicylate into biological systems (e.g. membranes) and act as a superoxide mimetic or have other bioactivity (e.g. enzymic character).

Alkyl and aryl esters

The methyl and ethyl esters of salicylic or acetylsalicylic acid can be prepared by treating these acids with acidified (concentrated HCl or H_2SO_4) methanol or ethanol, respectively (Gottesmann and Chin, 1968). The methyl esters of these salicylic acids may also be prepared by treating them with ethereal diazomethane. Fan and co-workers (1998) have recently shown that a variety of salicylate esters may be prepared using microwave irradiation as an energy source at normal pressure. The products are pro-drugs, since they yield the pharmacologically active acids by hydrolysis following their absorption either from the gastrointestinal tract or through the skin (Cross *et al.*, 1998; 1999) (see p. 401).

The phenolic esters of salicylic or acetylsalicylic acid, e.g. salicylsalicylic acid (salsalate, diplosal) or phenyl salicylate (salol), are well-known pro-drugs that have low gastric irritation but may, in some cases, be therapeutically less effective compared with the parent drugs (see pp. 51, 401).

Phenyl salicylate is prepared by heating salicylic acid and phenol in the presence of phosphorous oxychloride (Fischer, 1893; Gottesmann and Chin, 1968). Salicylsalicylic (Figure 3.7) acid is prepared by treating salicylic acid (dissolved in benzene, pyridine or toluene) at low temperatures with phosphorous trichloride, phosphorous oxychloride, or thionyl chloride (Gottesmann and Chin, 1968). The corresponding *o*-acetyl derivatives can be prepared by treating these esters with acetic anhydride in pyridine (as described for the preparation of aspirin).

Salicylamide

This drug has been used historically as an antipyretic analgesic agent (Hart, 1946), but is not used appreciably today. It also has some modest sedative action. Salicylamide is prepared by reacting methyl salicylate with ammonia (Gottesmann and Chin, 1968) as shown in Figure 3.8.

Salicylamide has comparable analgesic and antipyretic effects to aspirin and salicylic acid, although it has less anti-inflammatory effects in animals (Ichniowski and Hueper, 1946; Gross and Greenberg, 1948; Krause, 1977; Whitehouse *et al.*, 1977; Table 3.7). It is regarded, however, as having antipyretic, analgesic and even anti-inflammatory activity in humans similar to that of salicylic acid (Insel, 1990). A

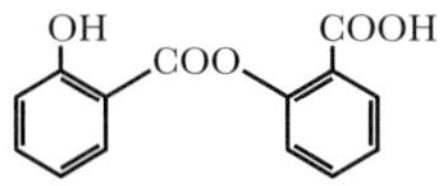

Figure 3.7   Salicylsalicylic acid [salsalate; diplosal].

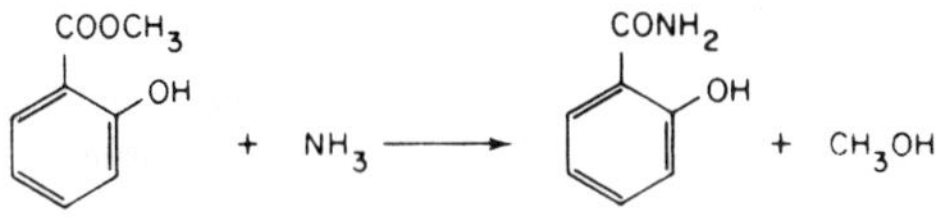

Figure 3.8   Synthesis of salicylamide. (From Rainsford (1984). Reproduced with permission of the publishers, Butterworths/Heinemann.)

double-blind study (one of the early trials performed with analgesic drugs) found that salicylamide was no better than placebo in relief of pain and in patient satisfaction in patients with osteoarthritis or other musculoskeletal conditions (Batterman and Grossman, 1955), whereas several other earlier studies quoted by these authors found that salicylamide had greater analgesic properties in rheumatic patients than aspirin.

The physicochemical and spectroscopic properties of salicylamide, as well as general analytical methods, have been reviewed and are described by Babhair *et al.* (1984) (See also Table 3.6.).

The morpholinomethyl-, *o*-acyloxymethyl, *o*-acyl and *N*-methyl derivatives have been developed as pro-drugs of salicylamide with the objective of reducing the presystemic metabolism of the latter (Bundgaard *et al.*, 1986; D'Souza *et al.*, 1986). No major clinical developments seem to have eventuated from studies with these pro-drugs.

Salicylate derivatives

A summary of the main pharmacological properties of some of the major salicylates and derivatives is shown in Table 3.7. There have been several comprehensive reviews of the structure–activity relationships of the therapeutic properties of the salicylates and their derivatives (Adams and Cobb, 1967; Scherrer, 1974; Hannah *et al.*, 1978; Jones *et al.*, 1978; Kim, 1979a; 1979b; Williamson, 1989). Based on this knowledge there have been considerable attempts made to produce more effective and safer drugs than aspirin or salicylic acid. Among the major factors affecting the development of anti-inflammatory/analgesic activity are ring substituents, especially where an electron-withdrawing ring substituent is added.

Chemically, substitutions are generally favoured at the 3- and 5- positions of salicylic acid (Gottesman and Chin, 1968). The electron-donating character of the phenolic group tends to increase the electron density at the 3- and 5- positions, while the electron-attracting nature of the carboxyl group decreases the electron density around the 4- and 6- positions (Gottesman and Chin, 1968). Thus, strong electrophiles (e.g. halogen, sulphate, nitrite and carbonate ions) tend to attack at the 3 or 5 positions. Steric hindrance around the 3- position favours the formation of 5- substituents. By these principles, the 5-chloro-, 5-bromo- and 3,5-dibromo-salicylic acids or their *o*-acetyl esters (see below) have been synthesised and are potent anti-inflammatory/analgesic and antipyretic agents; they even have some unique properties compared with salicylic acid or aspirin (Table 3.7).

Diflunisal and flufenisal

More complex and expensive procedures are required for the synthesis of these drugs than with the simpler salicylates, since there is addition required of aryl and heteroaryl groups to salicylic acids. Among the successful commercial efforts are diflunisal [5-(2,4-difluorophenyl)-salicylic acid] and flufenisal [*o*-acetyl-5-(4-fluorophenyl)-salicylic acid], which are potent salicylates (Table 3.7) developed in the Merck, Sharp and Dohm Laboratories (Rahway, New Jersey, USA) (Hannah *et al.*, 1978; Jones *et al.*, 1978). Commercial interest in flufenisal waned in favour of the more potent and less gastro-irritant and nephrotoxic analogue, diflunisal (Hannah *et al.*, 1978). This latter drug is synthesised (see Figure 3.9) by first forming the 2,4-difluorobiphenyl structure. This is achieved by reacting benzene with 2,4-difluoro-aniline or, alternatively, producing the 4-methoxy derivative by replacing benzene with 2,4-difluoro-anisole (Hannah, 1978). The former product (2,4-difluorobiphenyl (I)) is then used to form the 4-acetyl derivative (II) by Friedel–Crafts acylation and the corresponding acetoxy derivative (III) is then obtained by the Baeyer–Villiger reaction. The 4-hydroxy derivative (i.e. the phenol (IV)) is obtained by alkaline hydrolysis of the 4-acetyl derivative or by hydrolysing the 4-methoxy derivative (V) with acetic and hydro-iodic acids. The phenolic product is then carboxylated using a modification of the Kolbe method to yield diflunisal (VI) (Hannah *et al.*, 1978).

Diflunisal (Dolobid®, Merck, Sharpe and Dohme) is some seven to nine times more potent as an

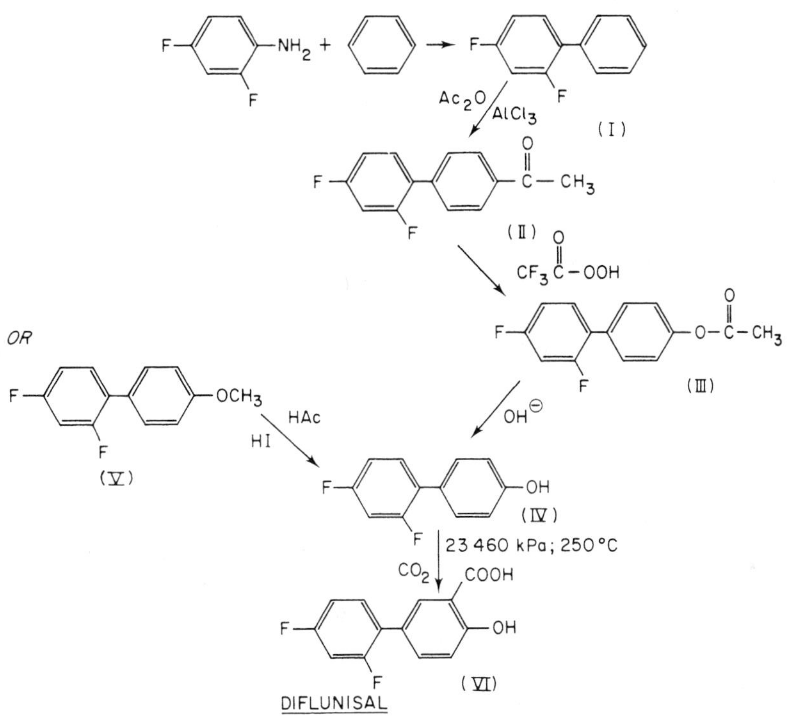

Figure 3.9 Synthesis of diflunisal. (From Rainsford (1984). Reproduced with permission of the publishers, Butterworths/Heinemann.)

anti-inflammatory agent than aspirin (compared on a weight basis), and is also quite a potent analgesic and antipyretic drug (Table 3.7; Hannah *et al.*, 1977; 1978; Jones, 1978; Dascombe, 1984; see p. 222). It is also appreciably less ulcerogenic than aspirin (Hannah *et al.*, 1977; 1978; Stone *et al.*, 1977; Jones *et al.*, 1978). Long-chain esters of diflunisal were prepared by Hung *et al.* (1997) from the respective acyl anhydride, using sulphuric acid as a catalyst. The aqueous hydrolysis, melting points and solubility of these esters decreased with increase in chain length, and the hydrolysis was reduced in the presence of albumin.

Benorylate

Of the immense number and variety of salicylate derivatives that have been tested (see Table 3.7, also Adams and Cobb, 1967; Scherrer, 1974; Whitehouse and Rainsford, 1982), only a few have achieved any appreciable clinical acceptance. Many of these drugs were developed with the specific objective of reducing the serious gastrointestinal side effects (e.g. ulceration and haemorrhage) inherent in aspirin and some other salicylic acids. Benorylate (WIN-11450, Benoral®, Sterling-Winthrop), is one such pro-drug, being the paracetamol ester of aspirin (i.e. 4-acetaminophenyl-2-acetoxybenzoate; Figure 3.10).

The synthesis of benorylate was accomplished by Robertson (1964) by treating paracetamol with NaOH and then adding acetyl-salicoyl chloride. The pharmacological properties were first described by Rosner and co-workers (1968). They found that the analgesic activity assayed using the Raddall–Selitto test of benorylate was slightly greater than that of aspirin or paracetamol on a dose-per-weight basis, and it also had a longer duration of action. Few or no acute gastric lesions were produced in fasted rats with benorylate, and the equimolar mixture of aspirin and paracetamol was more irritant, being only slightly less so than of aspirin.

Benorylate is hydrolysed to paracetamol, aspirin and salicylic acid *in vivo* following intestinal

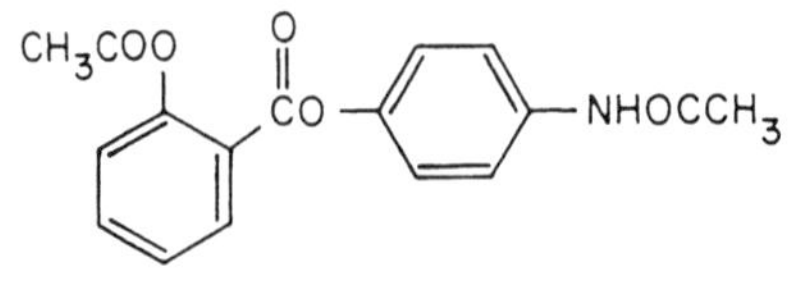

Figure 3.10   Benorylate.

absorption (Liss and Robertson, 1975). Thus, the pharmacological actions of this drug can be ascribed to the generation of the parent drug *in vivo*. The gastrointestinal side effects are certainly somewhat less than with aspirin (Whitehouse and Rainsford, 1982). It has been used in the treatment of rheumatic conditions (Berry *et al.*, 1981) and postoperative dental pain (Moore *et al.*, 1989) though its popularity has declined over the years.

Salicylazosulphapyridine (sulphasalazine) [SASP] and 5-aminosalicylic acids

Salicylazosulphapyridine (2-hydroxy-5-{[4-[(2-pyridinyl)amino]t-sulphonly]azo}benzoic acid or salazo-pyrin) was developed by Svartz at Pharmacia AB, Sweden, in the late 1930s (Svartz and Kallner, 1940; Svartz, 1941; 1942; 1948) from studies on a series of sulphonamide derivatives of the salicylates (Figure 3.11). It is now used as an anti-rheumatic agent (Bax, 1992). It was also developed for the treatment of ulcerative colitis and other inflammatory bowel diseases (Svartz, 1948).

It is interesting that this is a 5-substituted salicylate, since later work has shown that this is the most favourable position for adding aryl or heteroaryl substituents to enhance anti-inflammatory activity (Hannah *et al.*, 1978; Jones *et al.*, 1978). The sulphonamide group by itself does not appear to be associated with specific anti-inflammatory activity (Moore, 1974). The original notion was that the azo group of this compound would have a specific avidity for elastin-rich connective tissue. Later work has shown that the 5-aminosalicylic acid, formed from the splitting of the azo group of salicylazosulphapyridine *in vivo*, also has an affinity for this tissue (Svartz, 1941; Svartz and Kallner, 1941; Adams and Cobb, 1967).

There were originally rather variable reports about the effectiveness of salicylazosulphapyridine (SASP) in the treatment of rheumatoid arthritis, ankylosing spondylitis and, to a lesser extent, psoriatic arthritis (e.g. see Svartz, 1941; 1948; Sinclair and Duthie, 1949; Kuzell and Gardner, 1950; Rainsford and Buchanan, 1993). There is a suggestion of delayed action of the drug (Adams and Cobb, 1967; Svartz, 1941; 1948). Interest in this drug was revived during the 1980s and it has been shown to have 'disease-modifying' anti-rheumatic activity, as shown in both biochemical studies *in vitro* and in clinical analyses (McConkey *et al.*, 1978; 1980; Bird *et al.*, 1980; Bax, 1992; Gaginella and Walsh, 1992; Rainsford and Buchanan, 1993). While SASP might be somewhat less effective than the classical antirheumatic agents, i.e. gold salts and D-penicillamine, it is more effective than hydroxychloroquine and alclofenac (Bird *et al.*, 1980; 1982; Maetzel *et al.*, 2000). Thus, this drug may represent a unique salicylate with truly disease-modifying effects distinct from the symptomatic effects of salicylates and other NSAIDs.

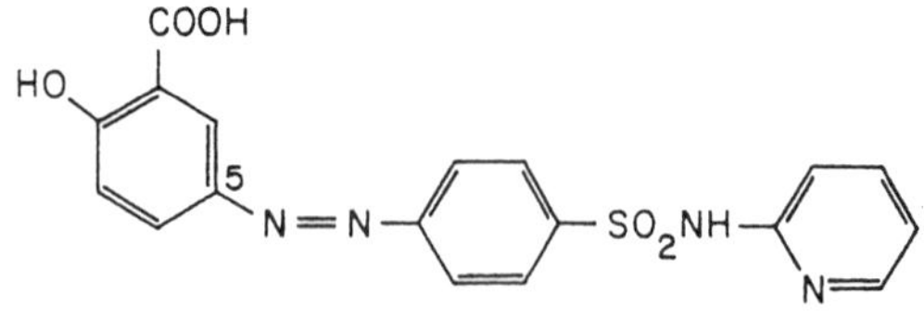

Figure 3.11   Salicylazosulphapyridine (sulphasalazine).

SASP is now recognised to be a very useful drug for the treatment of inflammatory bowel diseases such as ulcerative colitis and Crohn's disease (Klotz *et al.*, 1980; Klotz, 2000). The therapeutic efficacy of the 5-aminosalicylic acid (known as mesalazine), which is formed from salicylazosulphapyridine *in vivo*, means it is now extensively used for colitis (Klotz, 2000). It can, like the parent drug, affect prostaglandin metabolism by inhibiting both cyclo-oxygenase and lipoxygenase pathways, but also inhibits breakdown by dehydrogenases of prostaglandins (Hoult and Moore, 1980; Bakhle, 1981) and has other multiple actions on leucocyte activation and the production of reactive oxygen species (Gaginella and Walsh, 1992), as well as suppression of fatty acid oxidation (Roediger *et al.*, 1986), all of which are beneficial for the control of chronic inflammation. The effects of 5-ASA on the suppression of fatty acid oxidation appear unique to this drug (Roediger *et al.*, 1986).

Dai and co-workers (1998) have described a method for the synthesis of mesalazine.

The positional isomer of mesalazine, 4-aminosalicylic acid, is a tuberculostatic agent (Gross and Greenberg, 1948), and has been reported as being inactive in ultraviolet-induced erythema in guinea pig and mouse permeability assays (Adams and Cobb, 1967). These assays of anti-inflammatory activity may not represent the full spectrum of responses incurred in acute or chronic inflammation (Winder *et al.*, 1958; Whitehouse, 1963; 1964; 1965; see also Chapter 7). Also, 4-aminosalicylic acid is unstable in aqueous solution (where it decarboxylates to form 3-aminophenol), and thus this compound has little prospect of being effective *in vivo*.

5-aminomethylsalicylic acid has been shown to have anti-inflammatory and analgesic activity approximately equipotent with that of salicylate (Tamura, 1977a; 1977b; 1977c). This drug also exhibits uricosuric activity in laboratory animals. Finally, it is of interest that other sulphonamide derivatives of salicylates have been synthesised, although their biological activity has not been reported (El-Naggar *et al.*, 1975).

Olsalazine, the dimeric form of 5-ASA (sodium azosalicylate), was developed with the objective of achieving specific and localised delivery of 5-ASA in the colon, like that of SASP but without sulphapyridine, which has a number of the serious side effects associated with SASP (Truelove, 1988). A recent report by Yang *et al.* (1998) describes the synthesis of olsalazine.

EXPERIMENTAL DRUGS

Historically there has been an immense number and variety of salicylates developed for therapeutic use as substitutes for acetylsalicylic or salicylic acids; some have even been introduced clinically. While it can be said that few have surpassed the therapeutic efficacy of those salicylates discussed previously, it is important to understand the rationale behind some of these developments. Also, it may prove useful in the future to re-examine the properties of some of these using the newer biochemical or biological assays. Some may be regarded as useful tools for dissecting the biochemical/cellular actions of the anti-inflammatory/analgesic drugs, e.g. as slow or fast acetylators of proteins, including prostaglandin cyclo-oxygenase, for varying inhibition of platelet aggregation (Edmunds, 1966; Smith and Willis, 1971; Roth *et al.*, 1975; Siegel *et al.*, 1979; 1980). Others may prove to have specific biological properties worthy of development as unique therapeutic agents, e.g. 3,5-dibromo-aspirin and other diaspirins as anti-sickling agents (Walder *et al.*, 1977).

Nitric oxide-releasing aspirins (NO-aspirins)

Wallace, Del Soldato and co-workers have described the synthesis and action(s) of nitrate esters of aspirin nitrate; 2-acetoxy-benzoate-2-(2-nitroxy)-butyl ester (NCX-4215) and 2-acetoxy-benzoate-2-(nitroxy-methyl)-phenyl-ester (NCX-4016, or NOx aspirin) (see Figure 3.12) which has anti-thrombotic activity without the gastro-ulcerogenic activity of aspirin (Minuz *et al.*, 1995; Arena and Del Soldato, 1997; Wallace *et al.*, 1998; Del Soldata *et al.*, 1998; 1999; Tashima *et al.*, 2000); Chiroli *et al.*, 2003. The

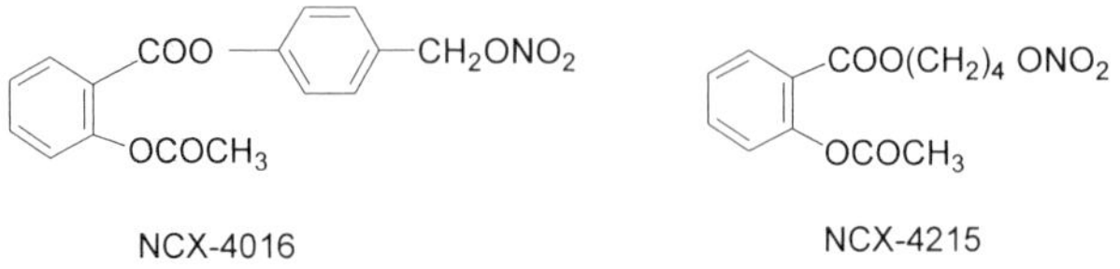

Figure 3.12 NO-donating aspirins.

potential advantage of these NO-aspirins would appear to be that the anti-thrombotic effects of aspirin (from inhibiting platelet aggregation and thromboxane production) are enhanced by the vasodilatory actions of NO released from the hydrolysed nitro-butoxyl moiety (Tagliaro *et al.*, 1997). While NO-aspirins inhibit platelet aggregation and thromboxane production, these effects are somewhat less potent than those of aspirin (Minuz *et al.*, 1995). NO-aspirins do, however, produce relatively potent relaxation of arteries, which was not evident with aspirin

Currently, NOx-aspirin is being developed by NicOx and is in phase II clinical trials for the prevention of cardiovascular disease (Scrip, 2001). Recently, NOx-aspirin 400 or 800 mg daily for 7 days in human volunteers was found to have less endoscopically observed gastro-duodenal injury than aspirin 200 and 420 mg daily, and no significant injury compared with placebo (Fiorucci *et al.*, 2003). There were no differences in platelet aggregation or TxB_s production following arachidonic acid stimulation of platelets, showing that it is still possible to have the antiplatelet effects of aspirin while not having mucosal damage by the NO-aspirins. Of the other nitrobutoxyl-derivatives, NO-naproxen (HCT-3012) was in phase II trials for the treatment of pain and inflammation but recently development of this was terminated. The development of other of these NO-releasing forms of salicylates and NSAID derivatives is awaited with much interest.

There has recently been much interest in the development of other novel NO-donating aspirins. Among these attempts have been furoxan and furazan derivatives (Cena *et al.*, 2003) and the isosorbide mononitrate ester of aspirin (Gilmer *et al.*, 2001).

There has been some concern about the chemical conditions under which NO is released from the nitrobutoxyl-moiety of aspirin and other NSAIDs, since the pH conditions in the stomach may vary, so affecting the redox potential that controls formulation of nitrites and NO from nitrates. An ingenious development by Endres *et al.* (1999) made an attempt to overcome some of these limitations, and was based on the fact that organic nitrates release NO when incubated with thiols that have a carbonyl group located two carbons from the third group in a co-planar orientation (as in cysteine, N-acetyl cysteine and thiosalicylate (Endres *et al.*, 1999). By combining the organic nitrate with a NO-liberating third of one molecule, a pro-drug may be obtained that exhibits more facilitated NO-release and reduces the possibility of nitrate tolerance. Using this approach, Endres *et al.* (1999) synthesised a range of s-nitro-oxyacylated esters and anilides of this salicylic acid and showed that these were chemically stable in phosphate buffered solutions, and exhibited relaxation of the thoracic aorta and activation of

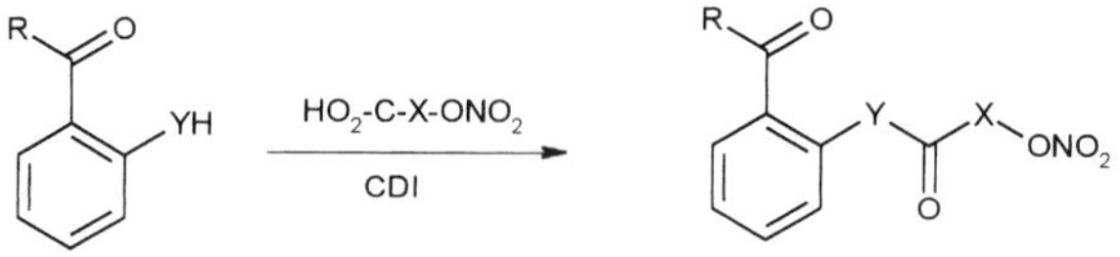

R = O-alkyl, N-(CH₃)₂

Y = S, O

X = C - spacers

CDI = carbodiimide

Figure 3.13 General synthetic route for the preparation of nitrate-thiol-hybrid salicylate pro-drugs (Endres *et al.*, 1999). (From Rainsford (1984). Reproduced with permission of the publishers, Butterworths/Heinemann.)

guanylate cyclase. The most active compound was SE-175 (methyl S-(4-nitro-oxymethyl-benzoyl)-thiosalicylate). Interestingly, these compounds could not be prepared using S-bromoacyl-thiosalicylates and silver nitric analogues to that used for NOx-NSAIDS (Arena and Del Soldato, 1997), but by acetylation of the (thio)-salicylates with nitroacids catalysed by carboxydiimidazole (CDT) (Figure 3.13).

Meanwhile, further understanding of the chemistry of nitrosothiols is being developed (Hogg, 2000; Wang *et al.*, 2000). Several nitrosothiol-derivatives of other NSAIDs are being developed (Bandarage *et al.*, 2000; Tam *et al.*, 2000) which may prove to have optimal biological activity.

Meseclazone

Meseclazone (W-2395, Wallace Laboratories, Cranbury, New Jersey, USA; Figure 3.14) was an ingenious pro-drug development (Sofia *et al.*, 1974; 1975; Edelson *et al.*, 1975; Sofia, 1978) in which the potent anti-inflammatory/analgesic drug 5-chlorosalicylic acid is cyclised to form an isoxazolabenzoxazone (7-chloro-3,3a-dihydro-2-methyl-2H,9H-isoxazolo-(3,2-b)(1,3)-benzoxazin-9-one).

Studies in laboratory animals and humans have shown that the pharmacologically active moiety 5-chlorosalicylic acid is produced by hydrolysis following intestinal absorption, the isoxazole moiety yielding 3-hydroxybutyrate and its tricarboxylic acid cycle metabolites following metabolism through this pathway (Edelson *et al.*, 1975; Dromgoole *et al.*, 1978). Unfortunately development of this compound did not proceed further because of toxicity, especially in the liver (D. Sofia, personal communication).

In rats the predominant products excreted in the urine are free and conjugated 5-chlorosalicylic acid (46.1 and 47.7 per cent respectively) with only traces of the glycine conjugate. In dogs the proportion of free 5-chlorosalicylic acid is greater (73.6 per cent) and that of the free and conjugated 5-chlorosalicylurate (about 4–5 per cent each) is greater than in rats (Edelson *et al.*, 1975).

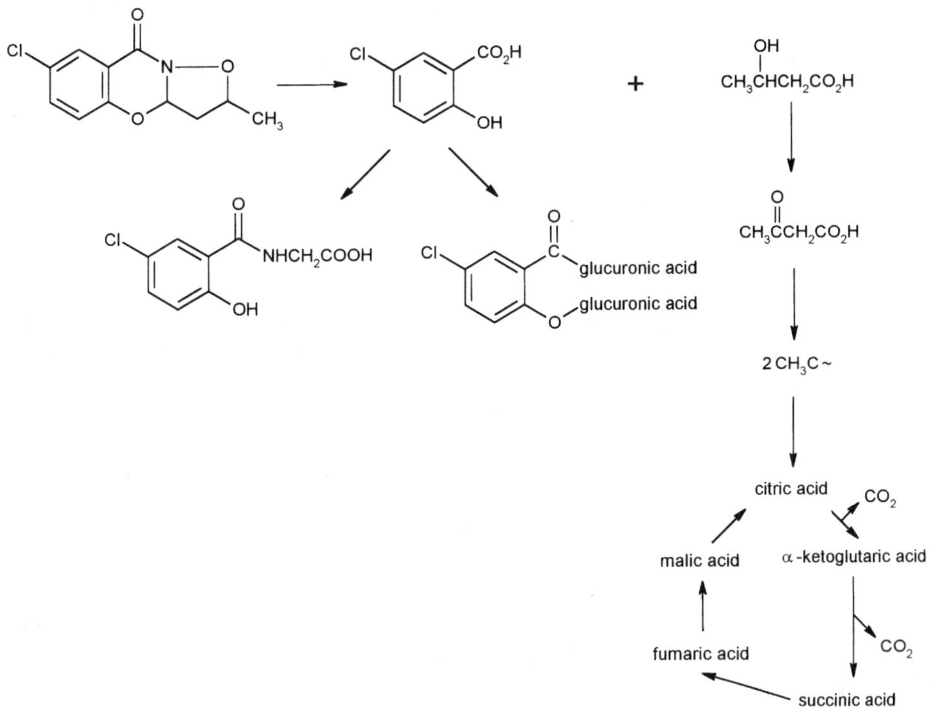

Figure 3.14 Meseclazone and its metabolism to 5-chlorosalicylic acid, acid or phenolic glucuronides and glycine conjugates of the latter and metabolites of the tricarboxylic acid pathway (Edelson *et al.*, 1975). (From Rainsford (1984). Reproduced with permission of the publishers, Butterworths/Heinemann.)

Seclazone

Seclazone (W-2354, Wallace Laboratories, Cranbury, USA) is actually the progenitor of meseclazone in which the methyl group is absent from the 2- position (of meseclazone). This drug showed activity comparable with that of meseclazone and, similarly, generates the pharmacologically active metabolite 5-chlorosalicylate *in vivo* upon intestinal absorption (Edelson *et al.*, 1973). Interest in this drug appears, however, to have declined since the initial studies in the early 1970s.

3-Methylsalicylic acid

3-Methylsalicylic acid (Lipterol®, AGN-356; Nicholas Pty. Ltd, Slough, UK) was selected after an exhaustive study of the alkyl (ring)-substituted salicylates by the former Nicholas company (James, 1975, personal communication) in the 1950s and 1960s. 3-Methylsalicylic acid is prepared from *o*-cresol by the Kolbe process. The *o*-cresol is dissolved at an equivalent amount of 70 per cent in an autoclave, the water evaporated off and CO_2 introduced under pressure, and the temperature increased (Aspro-Nicholas Ltd, 1966; Aspro-Nicholas Report, from James, personal communication, 1975). This drug was found to be relatively safe and as effective as aspirin. It is less gastrotoxic than either its parent acid or aspirin, and was shown in a limited trial to be as therapeutically effective as aspirin for the treatment of rheumatoid arthritis. Furthermore, it was found to have hypocholesteraemic activity (Lightbody and Reid, 1960); and it has slightly more analgesic activity than salicylic acid when given orally to rodents (Gorini *et al.*, 1963; Sievertsson *et al.*, 1970). The rate of elimination in man of 3-methylsalicylic acid (and its *o*-acetyl derivative) is much slower than that of either salicylic acid or aspirin (Cummings and Martin, 1965). This may be related to the greater percentage of the 3-methyl derivative bound to circulating albumin compared with salicylic acid itself (Stafford, 1962). 3-Methyl-aspirin (Amalin®, see Dyson and May, 1959) has a much slower rate of hydrolysis than aspirin ($K_1 = 1.73 \times 10^{-3}$/h and 6.80×10^{-3}/h, respectively at pH 7.4; Cummings and Martin, 1965).

Tachycardia and effects on cardiac conduction were noted in 1963 by Macdougall and Alexander (1963) in patients treated with 3-methylsalicylic acid. Stockman had also noted in 1912 that 3-methylsalicylic acid slows and depresses the heart rate when given to patients with 'rheumatic conditions' (Stockman, 1912). It is curious that commercial development proceeded with little consideration for establishing the reliability of these possible cardiac effects in an animal test system.

A report suggested that high doses of 3-methylsalicylic acid induced testicular atrophy in rats (James, 1975, personal communication). Thus interest waned, but the toxicological aspects may deserve more careful reinvestigation. In view of the current development of anti-inflammatory drugs with long half-lives, this drug or its acetyl derivative might prove useful either as an experimental tool or possibly where once or twice daily dosage is required in place of the more frequent administration of aspirin.

Aspirin anhydride

This drug was first considered in 1908 (Martin, 1971) as a possible precursor of aspirin in view of it being a dimeric product that would hydrolyse to aspirin (Figure 3.15) and so release aspirin *in vivo*. Some physicochemical evidence suggested that this drug might have significant advantages over aspirin (Garrett, 1957), but this later proved largely unfounded (Martin, 1971). An earlier report also claimed that this drug had fewer gastrointestinal side effects in man compared with aspirin (Kyriakopulos *et al.*, 1960), but later studies failed to substantiate these claims (Stubbe *et al.*, 1962; Wood *et al.*, 1962). This anhydride has low bioavailability in man (Martin, 1971) and is therapeutically less effective than aspirin in the treatment of rheumatoid arthritis (Wood *et al.*, 1962); it also shows considerable gastric irritancy in animals (Rainsford and Whitehouse, 1980a; 1980b).

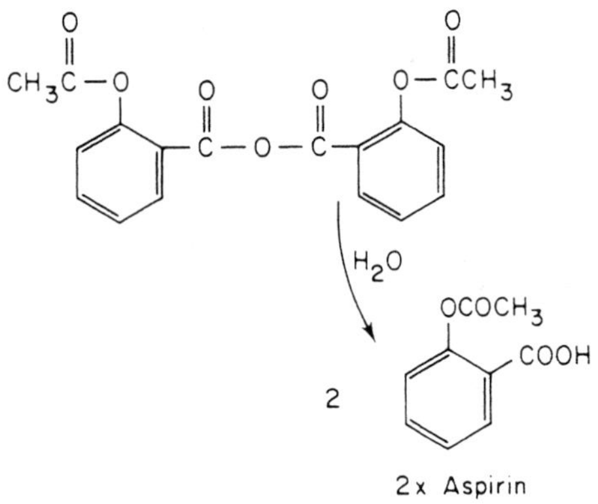

Figure 3.15 Aspirin anhydride. (From Rainsford (1984). Reproduced with permission of the publishers, Butterworths/Heinemann.)

Dihydroxy- and diacetoxybenzoic acids

These ring-substituted salicylates are of historical as well as pharmacological interest. *In vivo* formation of gentisic acid, i.e. 5-hydroxy-salicylic acid (2,5-dihydroxybenzoic acid) occurs and the contribution of metabolic transformation (in this case of ring hydroxylation (Chapter 4) to the activity of the parent (acetyl)-salicylic acids is considered.

Gentisic and pyrocatechoic (2,3-dihydroxybenzoic) acid were considered about 50 years ago as possibly more effective drugs than salicylate or aspirin (Meyer and Ragan, 1948; Clarke *et al.*, 1958). It was found, however, that neither drug proved effective in the treatment of rheumatoid arthritis, nor did they show appreciable anti-inflammatory properties in various animal models (see Bywaters, 1963; Collier, 1963; Adams and Cobb, 1967; for discussion). At this time, the notion that metal chelate formation was a desirable property for anti-inflammatory activity also led to the testing of the 6-hydroxy derivative of salicylic acid (i.e. γ-resorcylic or 2,6-dihydroxybenzoic acid), which was thought to be a more potent chelating agent than salicylic acid (Reid *et al.*, 1951; Clarke *et al.*, 1958). Later investigations proved this hypothesis was unfounded (for discussion see Bywaters, 1963; Collier, 1963; Adams and Cobb, 1967). Likewise, the other di- and trihydroxybenzoates proved ineffective in various animal test models (Adams and Cobb, 1967). The lack of effectiveness of these hydroxysalicylic acids is probably because they are unstable in aqueous solutions, being oxidised to the corresponding quinols or polymers thereof (Whitehouse *et al.*, 1977). This could account for the poor gastrointestinal absorption of these drugs *in vivo* (Austen, 1963). Clarke and Mosher (1953) observed its oxidation in the urine of human subjects following ingestion of the drug (they also noted that components in urine could apparently catalyse the oxidation and reduction of gentisate). One way of reducing or eliminating the possibility of oxidation (and thus poor absorption) of these hydroxysalicylates is to use their diacetyl derivatives (Whitehouse *et al.*, 1977). Thus, 2,3-diacetoxybenzoate (Moviren®) (Adams and Cobb, 1967) was reported to have favourable effects in the treatment of various rheumatic conditions (Ory, 1953; Venanzi, 1954). It shows anti-inflammatory activity almost equipotent with that of aspirin in the carrageenan-induced and adjuvant-arthritis models and, like aspirin, inhibits platelet aggregation (Whitehouse *et al.*, 1977). Furthermore, in contrast to aspirin, it has negligible gastric ulcerogenicity and also much lower general toxicity in rats (Whitehouse *et al.*, 1977). On this basis it has been proposed as an effective and safer alternative to aspirin. It could be synthesised by acetylating 2,3-dihydroxybenzoic acid derived from fermentation of bacterial sources (Whitehouse *et al.*, 1977; see p. 49) and so be an alternative to those salicylates derived from fossil fuels. The properties of various other methyl-substituted hydroxysalicylates (dihydroxybenzoates), and dihydroxybenzenes and their halogen derivatives have been critically reviewed by Adams and Cobb (1967). These authors concluded that although anti-inflammatory activity is evident with some members of this series, this property could not be divorced from the marked toxic effects exhibited by these drugs.

Ester pro-drugs

The term 'pro-drug' was first introduced in 1958 by Albert to indicate drug transformation prior to producing substances or metabolites that would interact with receptors. The term 'drug latentiation' was later developed by Harper in 1959 to describe the alteration of drugs to derivatives (Digenis and Swintowsky, 1975).

Interest has been shown in the development of ester pro-drugs of the salicylates almost since salicylic acid was first synthesised in the nineteenth and early part of the twentieth centuries. The concept of a pro-drug salicylate derivative that would be split in the body to deliver pharmacologically active products (Digenis and Swintowsky, 1975) was well known even before aspirin was synthesised. Phenyl salicylate (salol, salolum) was first made in 1886 by Nencki (1895), and Fischer (1893) described the use of this drug for treating joint rheumatics. The same author stated that salophen (phenetsal, i.e. the paracetamol ester of salicylate) was brought into production by the Bayer Company as a substitute for salol for use in the treatment of joint rheumatism. It was thought that this compound would not deliver the phenol (i.e. the toxic carbolic acid) on 'splitting', and indeed it was found to be much less toxic than salol (Fischer, 1893).

In the early part of this century, salicylsalicylic acid (originally known as Diplosal and later as Salsalate) was extensively used in the treatment of arthritic conditions, and it was known to be hydrolysed following absorption to salicylate (Tocco, 1912; Hanzlik and Presho, 1925). That these and other salicylate (carboxylic acid) esters might be less irritant to the mucous cells of the gastrointestinal tract was known (and the significance appreciated) in these early days. The misconception that aspirin, a phenolic ester, was therefore also less irritant is, of course, classical (Chapter 1). However, the search for other compounds proceeded because it was apparent that salol, diplosal and salophen were not very potent analgesics. In fact, some early authors stated that these compounds exhibited practically no activity (e.g. see Dyson and May, 1959), which may be a reflection of the slow release of active salicylate following hydrolysis *in vivo* (Hanzlik and Presho, 1925).

Eterylate

Eterylate (2-acetoxy-benzoic acid 2-(4-acetylamino)-phenoxyethyl ester; Laboratories Alter SA, Madrid, Spain) is an analogous derivative of benorylate in which an ethanolic group is linked between the paracetamol and aspirin (Sunkel *et al.*, 1978; Hopkins, 1979; Figure 3.16). It is prepared by reacting p-(2-hydroxyethoxy)acetanilide with o-acetoxybenzoic acid chloride in triethylamine (Sunkel *et al.*, 1978). Eterylate is appreciably more lipophilic than aspirin. Presumably it is hydrolysed after oral absorption to aspirin (salicylate) and paracetamol, and thus can be regarded as a pro-drug. Yang and co-authors (1998) have recently described a modified method for the synthesis of eterylate.

Acute studies in rats show that eterylate has activity comparable with aspirin in anti-inflammatory (carrageenan-foot oedema) and analgesic (the Randall–Selitto) assays (Sunkel *et al.*, 1978). It produces

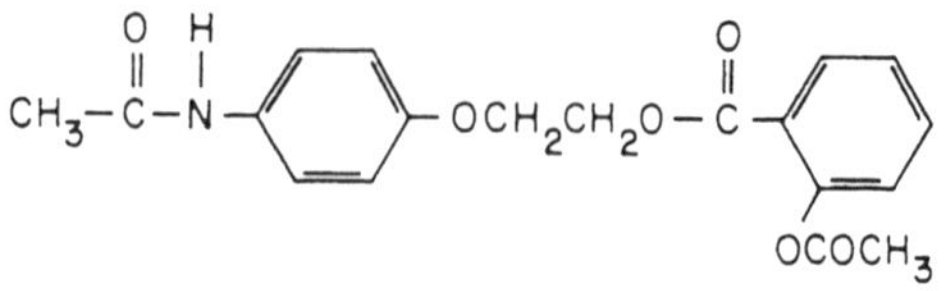

Figure 3.16   Eterylate.

less gastric damage and gastrointestinal blood loss following repeated dosage for 30 days to rats (Sunkel *et al.*, 1978). Clinical trials and pharmacokinetic studies in man have been undertaken (Hopkins, 1979).

Fosfosal

A novel *o*-phosphoryl ester of salicylic acid, fosfosal (2-phosphonobenzoic acid), was developed by J. Uriach & Cia, S.A. Spain (Garcia-Rafanell, 1980; Figure 3.17). In contrast to both aspirin and salicylic acid, this compound is highly soluble in water. This physicochemical property may account for the low gastric mucosal damage observed in laboratory animals and the fewer gastrointestinal side effects observed in a limited clinical study in man (Garcia-Rafanell, 1980). It appears to have the same therapeutic effectiveness as aspirin in laboratory animals and in a small clinical trial (Garcia-Rafanell, 1980), but more extensive studies are required before a full assessment of this drug can be made.

Triflusal

The 4-trifluoromethyl derivative of aspirin, (2-(acetyloxy)-4-(trifluoromethyl) benzoic acid or Triflusal™ is a potent inhibitor of platelet aggregation and which is categorised as an antithrombotic (Merck Index, 1983; Garcia Rafenell *et al.*, 1986) though it does not appear to have found wide acceptance. A new method of synthesising this drug from 3-fluoro-salicylic acid has been recently described (Micklatcher and Cuchman, 1999).

Alkyl esters

The simplest alkyl ester of aspirin, methyl-*o*-acetyl salicylate was first developed and patented by Thorpe in 1918. The process involved mixing methyl salicylate with a slight molar excess of acetic anhydride in the presence of an alkali metal acetate as a catalyst and, after heating to 90–100°C for 10 to 24 hours, precipitating the product with dilute alcohol, adding hot water until turbid, then cooling and filtering of the product.

The alkyl and aryl (carboxylic) esters of aspirin have markedly less gastrotoxicity than their acids (Rainsford and Whitehouse, 1976; 1980a; 1980b; Whitehouse *et al.*, 1977). Some of these derivatives still possess the full anti-inflammatory and other properties of the parent acids. Of these esters, the aspirin methyl ester has been found the most efficacious and shows the least toxicity of these compounds. Studies with radioactively-labelled aspirin methyl ester show that both aspirin and salicylate are generated *in vivo* (Rainsford *et al.*, 1980). Also, the pattern of biodisposition following oral absorption and subsequent uptake into inflammatory foci (e.g. carrageenan-inflamed paws of rats) is essentially the same as for aspirin. This derivative is probably one of the simplest, safest and least expensive of all the low gastrotoxic aspirin esters developed so far. The effectiveness of this pro-drug has still to be proven clinically.

Several long-chain alkyl esters of salicylic acid and diflunisal were prepared by the respective acyl anhydride (in the presence of H_2SO_4) or acyl chloride (with pyridine) and their solubility *in vitro* in rats of hydrolysis in buffer in the presence or absence of albumin and protein binding determined (Hung *et*

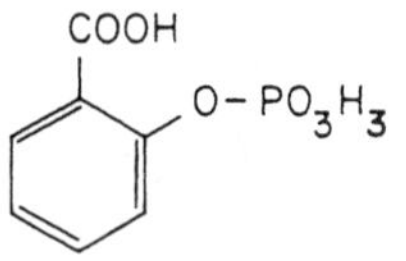

Figure 3.17   Fosfosal.

■ CHAPTER 3 ■

al., 1997). It is of interest that the diflunisal esters have antiplatelet and hydroxyl scavenging activities, and have appreciably lower gastrointestinal irritancy in rats than the parent acids (Hung *et al.*, 1998; Yung-Yu and Roberts, 1998a, 1998b).

The methods developed for the synthesis of alkyl and related esters in recent years have seen refinements in what have been well-established techniques. Thus, Chen and Zhang (1998) described the preparation of isopropyl salicylate using butyl titanate as a catalyst. Zhu *et al.* (1994) employed ultrasound for the synthesis of 3,5 di-isopropyl-salicylic acid. Jin *et al.* (1996) developed a catalytic system employing heteropoly phosphotungistic acid and silica for the preparation of isoamylsalicylate. Zhang *et al.* (1995) employed ferric chloride as a catalyst in the esterification reaction for preparing butyl p-hydroxybenzoate.

De Nil (1998) described a patent for the preparation of alkyl salicylate esters having antimicrobial activity, using alcohol esterification and azeotropic distillation. Chen and Zhao (1999) showed that isopropyl-salicylate could be prepared by esterification of salicylic acid using mixed acid catalysis. Using chloromethylsalicylates, Qaisi and Roth (1995) developed procedures for the synthesis of 5-alkoxymethylsaliclic acid. Sheng *et al.* (1999) described the synthesis of iso-amyl salicylate from salicylic acid using sulphonate polyaryletherketone resin as a catalyst. Another method describes the preparation of isoamyl-salicylate using heteropolyacid catalysis (Zhang *et al.*, 1999). Hu *et al.* (1997) claimed a method for preparing alkyl salicylates using esterification catalysis.

Other esters

Two novel triglycerides of aspirin have been developed by Kumar and Billimora (1978) and Paris *et al.* (1979). The idea behind their development is that aspirin would be released by the actions of (lipoprotein) lipases following absorption of the drug in the intestine. The bioavailability of the orally administered (1,3-didecanoyloxy)propyl derivative of aspirin (A-45474, Abbott Laboratories Ltd., North Chicago, USA – Figure 3.18) appears by comparison with the molar equivalent of aspirin to be greater than the 1,3-dipalmitoyl-glycerol derivative (Kumar and Billimora, 1978; Carter *et al.*, 1980; Paris *et al.*, 1979).

Also, the latter derivative is incompletely absorbed when given orally. Both drugs produce less gastric irritation following acute oral administration to rats (Kumar and Billimora, 1978; Paris *et al.*, 1979; Carter *et al.*, 1980). While data on the activity of the dipalmitoyl derivatives in animal tests are lacking, it appears from the bioavailability studies that this drug has much lower therapeutic potential than aspirin itself. The derivative also shows delayed onset of antipyretic activity (in the yeast-induced fever in rats), which is commensurate with its delayed absorption (Carter *et al.*, 1980). The drug may prove more effective on repeated dosage to animals with chronic inflammatory conditions (e.g. adjuvant arthritis), where sustained blood levels of salicylate(s) are desirable.

Based on the concept that the elevated hydrolytic activity of inflamed tissues might be a site for specifically releasing aspirin from pro-drugs, three aspirin esters, the phenylalinine ethyl ester and its amide, and phenyl-lactate ethyl esters (Figure 3.19) have been developed and tested for activity as substrates for the enzymes carboxypeptidase A and chymotrypsin (Glenn *et al.*, 1979; Banerjee and

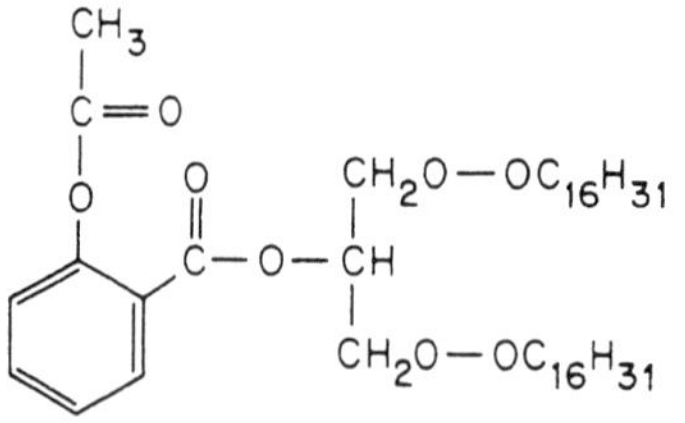

Figure 3.18 1,3-Dipalmitoyl-glycerol aspirin. (From Rainsford (1984). Reproduced with permission of the publishers, Butterworths/Heinemann.)

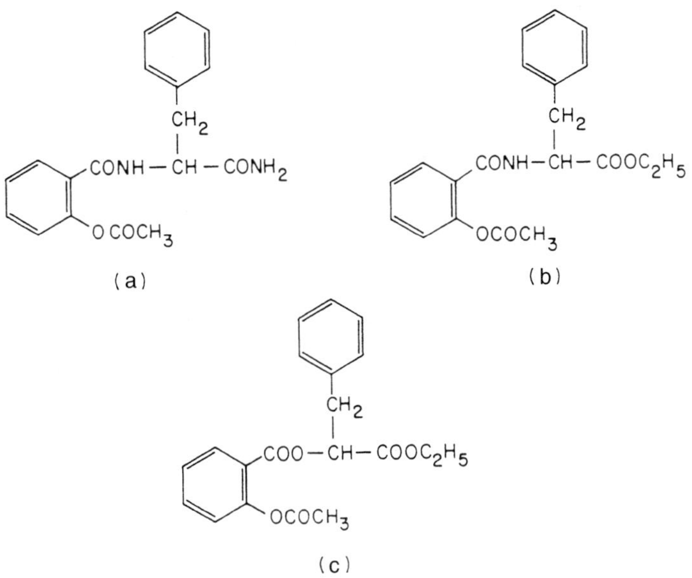

Figure 3.19 (a) Aspirin phenylalanine amide; (b) aspirin phenylalalinine ethyl ester; (c) aspirin phenyl-lactate ethyl ester. (From Rainsford (1984). Reproduced with permission of the publishers, Butterworths/Heinemann.)

Amidon, 1981a; 1981b; 1981c). The phenyl-lactate ethyl ester was the best substrate for carboxypeptidase A, but showed some product inhibition (Banerjee and Amidon, 1981c). This latter property may be of some therapeutic potential in controlling hydrolytic breakdown by protease in inflamed tissues, but this and other therapeutic and toxicological aspects have yet to be reported.

The methylthiomethyl-, methylsulphinylmethyl- and methylsulphonylmethyl esters (Figure 3.20) of aspirin have been synthesised and shown to generate aspirin rapidly following incubation with human plasma *in vitro* and salicylate following oral administration to beagle dogs (Loftsson *et al.*, 1981). These compounds appear to be pro-drugs capable of producing aspirin and salicylate *in vivo*, but their full pharmacological and toxicological properties have yet to be reported.

A considerable number of potentially interesting salicylate esters have been synthesised, some of which have been reported to have biological activity. Among these are the basic amino acid esters (e.g. aspirin-L-arginine-ethyl ester and aspirin-*p*-guanidino-L-phenylalanine ethyl ester); these were shown to generate salicylate or esters following hydrolysis with trypsin or chymotrypsin, depending on the amino acid substituent (Tsunematsu *et al.*, 1991).

A series of 2-, 3- or 4-formyl esters of aspirin that are highly lipophilic have been found to undergo

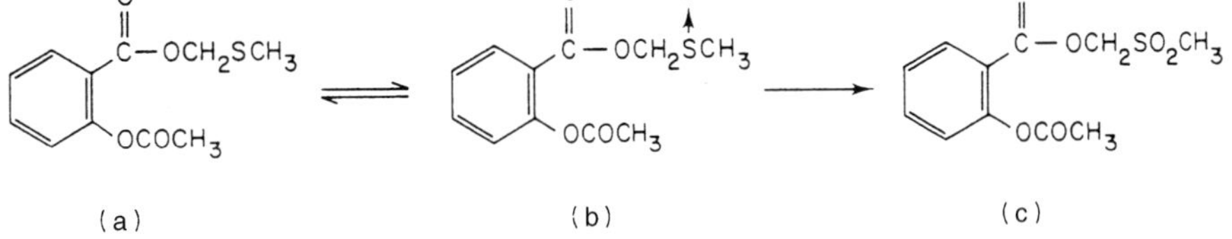

Figure 3.20 (a) Methylthiomethyl aspirin; (b) methylylsulphinylmethyl aspirin; (c) methylsulphonylmethyl aspirin. Possible metabolic interconversions are shown by the arrows (Loftsson *et al.*, 1981). (From Rainsford (1984). Reproduced with permission of the publishers, Butterworths/Heinemann.)

alkaline hydrolysis to aspirin (Bowden et al., 1997). The 2-, 3- or 4-formylphenyl derivative of aspirin had about double the anti-oedemic potency but the same course of effect in the rat carrageenan paw oedema assay. However, 2-formylphenyl derivatives with halogen, methyl-methoxy- or phenoxy-substituents at the 4 phenyl position failed to exhibit anti-oedemic activity in this assay even though they were all hydrolysed to aspirin (Bowden et al., 1997). The 2-formyl-phenyl-aspirin had appreciable inhibitory effects in vitro on prostaglandin production in rat whole blood stimulated with calcium ionophore. It also exhibited gastroduodenal ulcers and haemorrhages at an oral dose of 30 mg/kg after 24-h dosing to rats. Thus, although these esters have chemical novelty and are more potent as anti-inflammatory agents than aspirin, it is apparent they confer no benefit compared with the parent drug or alkyl esters.

Several phenolic esters of aspirin were synthesised by Cha and Lee (2000) using some natural derivatives with antioxidant activity, e.g. sesamol, eugenol, cinnamyl alcohol or 7-hydroxy-4-methyl-coumarin in the presence of 1,1-carbonyldiimidazole. The derivative of sesamol had strong antioxidant and anticoagulant effects, and was more potent as an antiplatelet aggregating agent than aspirin. The cinnamyl ester was a slightly less potent inhibitor of bleeding time in rats, while the other two derivatives were inactive in this assay. Thus, the strategy of combining antiplatelet activity (from the aspiryl moeity) and antioxidant actions (from the phenolic ester derivative) would appear to have been achieved by the sesamol derivative.

Morpholinomethyl-, o-acyloxymethyl-o-acyl and N-acyl-derivatives of salicylamide have been prepared as possible pro-drugs of salicylamide (Bundgaard et al., 1986; D'Souza et al., 1986). These exhibit wide variations in rates of hydrolysis. Some of these compounds may have utility as oral or rectal drug delivery systems, with the advantage of having protection against the metabolic conjugation of the phenolic group (Bundgaard et al., 1986).

Ethyl-carbamoyloxybenzoate (ECB) was developed as a pro-drug derivative of carsalam (or carbonylsalicylamide – a cyclised carbamyl derivative of salicylamide), which was shown by Kamal (1990) to form salicylamide, salicylic acid and carsalam when incubated with post-mitochondrial supernatants derived from the livers of rats, rabbits and dogs. Based on this the author concluded that ECB is a pro-drug of salicylates, although no in vivo data were provided to support this premise.

A hydroperoxy-cyclic derivative of aspirin, 3-hydroperoxy-3-methylphthalide (3-HYP), was synthesised and its pharmacological properties studied by Killackey et al. (1984), the development of which was based on the premise that there is a 'close association' of ASA (aspirin) with the production of pharmacologically active hydroperoxide metabolites of A1 (arachidonic acid). The synthesis of 3-HMP was undertaken by exposing the precursor 3-methyl phthalide (3-MP) to sunlight in a dessicator for 60 days (Killackey, 1982; Killackey et al., 1984). 3-MP was prepared from 2-acetyl benzoic acid by the action of sodium borohydride in methanol with 0.4 g/100 ml sodium hydroxide, and the mixture was extracted with chloroform and evaporated under vacuum to yield a clear oil with a characteristic 'pepper-like' amine (Killackey, 1982). The antiplatelet effects and production of thromboxane B_2 in human platelet lysates incubated with radiolabelled arachidonic acid of 3-HMP were studied in comparison with aspirin. It was found that 3-HMP had five times the inhibitory effects (IC_{50} 10 μM) compared with aspirin, and this compound was more intent as an inhibitor of PGI_2-generation by rabbit aorta rings (Killackey et al., 1984). Intravenous administration of 10 mg/kg 3-HMP inhibited PMN leucocyte accumulation into the plural cavity of carrageenan-injected rats whereas the 10-times higher dose of aspirin was without effects.

Two positional 3- or 5- isomeric phenyl esters of salicylate were prepared by Razzak (1979) and subsequently reduced by platinum or palladium to aminophenyl derivatives, which were then acetylated with acetic anhydride to form phenyl 3- or 5-acetamidosalicylic acids. Only limited biological activity was noted in these compounds. In the mouse phenyl p-quinone (i.p.) test, inhibition of writhing was obtained with the phenyl-5- (but not the phenyl-3) acetamido derivative with an IC_{50} (presumed orally administered) of 56 mg/kg. The toxic dose of the phenyl-5-derivative was 3 g/kg i.p. over 24 h. The analgesic activity of this derivative is unremarkable compared with aspirin (see Chapter 7), and the margin of toxicity in relation to therapeutic effects limited.

Swintowsky and co-workers (1984) prepared the hexylcarbonate ester of salicylate (SKF 26070;

HCSA) and showed that it had therapeutic effects of aspirin in laboratory animal models without the gastric ulcerogenicity. In bioavailability and kinetic studies in humans the plasma salicylate levels from a single dose of 480 mg HCSA taken orally in capsule or as a liquid dispersion were comparable in AUC values to those from 300 mg aspirin, but the peak values were achieved much later with HCSA (2.73, 3.55 h) than with aspirin (1.471 h), which is in accord with that expected from a pro-drug. No further development of HCSA appears to have been undertaken.

Inoue and co-workers (1979) prepared a range of C_3 to C_6 alkyl esters of salicylic acid and aspirin from which they compared their relative rates of hydrolysis by isolated preparations of human intestinal and liver esterases as a basis for a selection to study the pharmacokinetics of the esters in humans. Thus, n-propionyl and n-heptanoyl salicylic acid and n-pentyl-o-carboxyphenyl carbonate were selected and their plasma kinetics compared with aspirin. All the novel esters showed good absorption, but peak values were delayed by 1 to 2 hours compared with aspirin. The gastric ulcerogenic effects of the esters in acute ulcer assays in rats were however disappointing, since n-heptanoylsalicylic (UD_{50} 105 mg/kg) and phenyl-o-carboxyphenyl carbonate (UD_{50} 95 mg/kg) were more ulcerogenic than aspirin (UD_{50} 42 mg/kg). Even the ethyl and hexyl esters of aspirin were most ulcerogenic than aspirin, thus showing that not all alkyl esters have low ulcerogenicity (see Chapter 8).

Aonuma et al. (1980; 1981; 1982; 1983) prepared the isopropyl antipyrine ester of aspirin (AIA) and showed that it had the conventional anti-inflammatory and analgesic activities of the two analgesic components of this ester, but with very low gastro-ulcerogenic activity. AIA was found potently to inhibit platelet aggregation in vitro (Aonuma et al., 1981), it inhibited thrombus formation in an extra-corporeal shunt model in rats, and appeared to have little or no effect on production of PGI_2 by isolated rat aorta – in contrast to aspirin, which was markedly inhibitory (Aonuma et al., 1983). The radiolabelled drug underwent metabolism to salicyl-isopropylantipyrine and the glucuronide and sulphate esters, but the carboxylamide bond was not cleared in vivo (Aonuma et al., 1982). This indicates that the pharmacological activity of AIA rests either in this compound or in its de-acetylated derivative, which is unusual among the esters of aspirin, especially in comparison with drugs like benorylate. The lack of appreciable inhibitory effects on PGI_2 production in the vascular system is an advantage over aspirin, since the latter has the disadvantage of preventing production of the natural platelet anti-aggregatory PGI_2 (Chapter 7).

The hexylcarbonate (phenolic) ester of salicylic acid (SK&F-26070) was shown by Misher and co-workers (1968) to have antipyretic, analgesic and anti-inflammatory activity slightly less than that of aspirin (on a mg/kg and mmol/kg basis), but it was noticeably less ulcerogenic in rats and dogs. No further development of this compound appears to have occurred since these initial studies.

On the premise that modification of the carboxylic acid moiety of salicylic acid could reduce the gastric irritancy of salicylates, Wilder-Smith and co-workers (1963) developed several oxadiazo-2-ol compounds, including o-hydroxyphenyl-1,3,4-oxadiazo-2-ol (WS-132; Figure 3.21). While not strictly esters these compounds represent modifications of the carboxyl group, which is a major feature of carboxyl esters and may account for their low ulcerogenicity.

In addition to having tuberculostatic activity, this compound showed analgesic activity comparable with that of aspirin in the guinea pig using the tooth pulp stimulation assay. This compound was less active in antipyretic or anti-inflammatory (UV erythema and cotton pellet granuloma) assays (Wilder-Smith et al., 1963). Regrettably, no evidence of reduced ulcerogenicity (cf. aspirin) was reported.

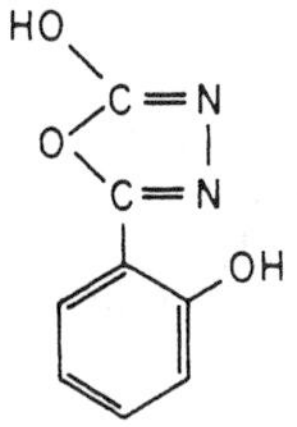

Figure 3.21 o-Hydroxyphenyl-1,3,4-oxadiazo-2-ol (WS-132). (From Rainsford (1984). Reproduced with permission of the publishers, Butterworths/Heinemann.)

One of the major issues concerning the developmentt of ester pro-drugs of aspirin as low gastro-toxic anti-platelet drugs is that many have a high rate of hydrolysis of the acetyl-moiety (Gilmer *et al.*, 2003). In attempts to reduce this aqueous hydrolysis of the acetyl group Gilmer *et al.* (2003) were led to synthesize isosorbide aspirin diester (ISDA) which they found has a U-shaped pH-dependent hydrolysis curve with low rates around pH 2–7.5. They also found that ISDA had a high degree of stability in human plasma and that it was hydrolysed by plasma butyryl-cholinesterase to aspirin. ISDA may, therefore, represent a model advance in the development of stable aspirin pro-drugs with low gastro-toxicity

Biphenyl aspirin

Biphenyl-aspirin (2′-acetoxy-biphenyl-2-carboxylic acid; Figure 3.22) and its amide derivative were reported by Gringauz to be as effective as aspirin (on a molar basis) in acute (carrageenan-oedema) and chronic (adjuvant-arthritis) inflammatory conditions in rats. These compounds were also found to have analgesic activity in mice, using the acetylcholine writing assay, comparable with that of aspirin (Gringauz, 1976). Biphenyl-aspirin does not induce any detectable gastric damage in the stress-sensitised rat, which is a good prediction of low ulcerogenic potential; see Chapter 8 (Rainsford, 1977). This certainly represents an interesting lead compound on which to increase potency and hopefully retain the low ulcerogenicity inherent in the parent compound.

Increasing the number of carbon atoms between the carboxylic acid moiety and benzene ring appears to be associated with reduction in potency of the aspirin. Furthermore, the vinyl analogues, a variety of cinnamates and propoxybenzoic acids are virtually inactive in animal tests (Gringauz, 1970). The aspirin homologue β-(*o*-acetoxy-phenyl)-propionic acid has shown analgesic activity in animal tests and in rheumatic patients, but also pronounced gastric irritancy (Bauer and Lasala, 1960).

Other compounds

Several marine organisms (coral, sponge and a brown alga) have been found to contain 6-*n*-tridecylsalicylic acid. This compound was found to have anti-inflammatory activity (against carrageenan oedema) in rats comparable with that of salicylic acid and aspirin, but with less gastric irritancy (Buckle *et al.*, 1980).

Further discussion of other salicylate derivatives with low anti-inflammatory/analgesic activity in animal tests can be found in the literature (Gross and Greenberg, 1948; Adams and Cobb, 1967; Walford *et al.*, 1971; Scherrer, 1974; Whitehouse *et al.*, 1977; Hannah *et al.*, 1978; Jones *et al.*, 1978; Glenn *et al.*, 1979; Whitehouse and Rainsford, 1982).

Acetyl-3,5-dibromosalicylic acid and dibromosalicyl-bis-fumarate ('dispirit')

These compounds have been developed to enable the synthesis of cross-linked haemoglobin as an oxygen carrier to replace whole blood (or red blood cells) in transfusion medicine (Klotz and Tam,

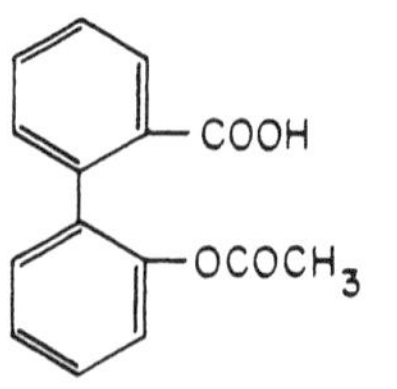

Figure 3.22 Biphenyl aspirin. (From Rainsford (1984). Reproduced with permission of the publishers, Butterworths/Heinemann.)

1973; Walder *et al.*, 1977; Chatterjee *et al.*, 1986; Winslow, 1995; Williams, 1996). The potential applications for these agents include restoration of oxygen delivery, prevention or reversal of hypovolaemia and organ failure, haemodilution in patients undergoing elective surgery, extra corporeal oxygenation during cardiopulmonary bypass, and in cardiogenic, septic or post-surgical shock. These potential uses of modified salicylates appear to have arisen from the observations that acetylation by aspirin of Lys-59 and Lys-144 of haemoglobin-S from patients with sickle-cell anaemia (HbS) might reduce the sickling of erythrocytes (ES) from patients with this disease, and improve oxygenation. Light-scattering studies under deoxygenated conditions have shown that there is some reduction in shape change of ES cells, but the same also occurs with naproxen (Rabinowitz *et al.*, 1974), thus raising questions about the role of acetylation by aspirin in this state. Attempts to improve the acetylating capacity of aspirin for HbS led to the development of acetyl-3,5-dibromo-salicylic acid or dibromoaspirin (Walder *et al.*, 1977) and other site-directed agents comprising certain meta-substituted amine derivatives of benzoic acid and prolyl-salicylic acid derivatives (Abraham *et al.*, 1984). The objective of the latter study was to correct the polymerisation and subsequent gelation of HbS that is due to the mutation of the hydrophilic Glu-6 residue to the hydrophobic Val-6 in HbS. The disubstituted benzoic acids were designed to interact with polar groups near the Val-6 mutation (donor) site, and had a hydrophobic group designed to occupy a non-polar area on the surface of the protein. The prolyl–salicylate compounds were designed to react covalently at the mutation acceptor area. Unfortunately these proved inactive as antigelling agents, but these attempts have been useful to understand some of the structure–action requirements for achieving correction of the HbS defects. The most effective of the benzoic acid amides was 5-(α-carboxy-*m*-anisic acid (Abraham *et al.*, 1984).

Other studies showed that dibromo-aspirin and the succinate and fumarate analogues had acute anti-inflammatory activity in the carrageenan paw oedema model at oral doses on a weight basis (200 mg/kg) comparable with aspirin (Thompson and Klotz, 1985). The diaspirin succinate had more potent acute analgesic activity in the phenylacetate abdominal writhing assay in mice. Hypothermic response were noted with dibromo-aspirin and the fumarate derivative (Thompson and Klotz, 1985). The acylating capacity of a series of diacyl hydroxy salicylate esters was explored by Massil *et al.* (1983) for anti-sickling activity. Of the compounds prepared the 2,5-diacetoxybenzoic acid was the most effective.

In structure–activity studies it was found that carboxyl substituents added to the salicylate ring increased the acetylating capacity of various aspirins and diaspirins in acetylating haemoglobin (Massil *et al.*, 1984). These authors developed the concept that salicylic acid esters with increased negative charge would be more effective as modifiers of haemoglobin. Following these observations these authors found that monoesters of dicarboxylic acid esters, bis(-5-carboxymethylsalicylic)-fumarate and the succinic acid homologue, were found to be most effective in modifying haemoglobin oxygenation/deoxygenation.

Dibromosalicyl-*bis*-fumarate (DSF, Dispirit™) is also referred to, probably inappropriately, as 'diaspirin'; the inaccuracy of this terminology is because this halogenated salicylate is not an acetylated ester like that of aspirin, and nor does it have similar properties. Nonetheless, this unfortunate term has become commonplace. In the preparation of DSF-linked haemoglobin (DSF-Hb) is produced by cross-linking the α–α subunits of the haemoglobin from outdated red blood cells is achieved using the diaspirin compound 3,5-dibromosalicyl-*bis* fumarate (Chatterjee *et al.*, 1986). This covalent cross-linking occurs between the *N*-terminal amino acids of the α chains at Lys-α-1 and Lys-α-2. The α–α cross-link stabilises the native haemoglobin, so resulting in increased vascular retention time, decreasing oxygen affinity and preventing the breakdown of haemoglobin to dimers and renal elimination. Because of its structural similarity to haemoglobin, it is assumed that DCF-Hb is metabolised in a similar manner to endogenous haemoglobin (Przybelksi *et al.*, 1996; Palaparthy *et al.*, 2000). It has been shown that DSF-Hb possesses an excellent oxygen carrying capacity in rats (Snyder *et al.*, 1987). The haemodynamic and cardiovascular effects of DCL-Hb have also been explored (Gulati and Rebello, 1994; Gulati *et al.*, 1994). Overall, however, clinical trials with diaspirin cross-linked haemoglobin preparations in blood transfusion and haemorrhagic shock have been disappointing (Przybelski *et al.*, 1999; Sloan *et al.*, 1999; Lamy *et al.*, 2000). The negative or poor outcomes in these studies may relate to the effects of haemoglobin in scavenging nitric oxide, participating in free-radical reactions, activation of the immune system and its neurotoxicity (Hess, 1996).

CHAPTER 3

Miscellaneous compounds with anti-microbial or anti-viral activity

Halogen derivatives of salicylates are generally regarded as being more potent than parent drugs. Troung *et al.* (1998) described the synthesis of some chlorosalicylamides and showed these had potent antibacterial and antifungal activities. A patent by De Nil (1998) has described the synthesis of some alkyl derivatives of salicylates and related compounds with anti-microbial activity.

Combrink *et al.* (2000) prepared a number of salicylamide derivatives in which quinolines were added at the amino-group of salicylamide. These were investigated for anti-influenza activity *in vitro*, and 2-methyl-*cis*-decahydroquinoline was found to be the most potent inhibitor (IC$_{50}$ 90 ng/ml).

Bacteriocides comprising *p*-hydroxybenzoic acid derivatives have been prepared by Castelum *et al.* (1997).

VARIOUS CHEMICAL APPLICATIONS

Measurement of oxyradicals

Oxyradical production can be measured *in vitro* and *in vivo* in some biological fluids by taking advantage of the potential for salicylate to be attacked by hydroxyl anions (OH•) under neutral pH conditions (pH 7.4) to give 2,3- and 2,5-dihydroxybenzoic acids (2,3-DHB and 2,5-DHB), and catechol (Kaur and Halliwell, 1996), the former two being metabolites of salicylate or aspirin ingestion in humans (Chapter 4). Whether or not 2,3-DHB or 2,5-DHB represent the *sole* products of oxyradical attack following ingestion of salicylate or aspirin (i.e. a natural end product of oxyradical actions *in vivo*) is a matter for debate. Procedures for determining OH• based on production of these salicylate metabolites and catechol have been developed in which the products are separated by HPLC on a column of Spherisorb 5ODS (25 cm × 4.6 mm fitted with a guard column), using electrochemical detection and elution with an isocratic gradient of methanol/34 mM sodium citrate; 27.7 mM sodium acetate buffer pH 4.75; methanol (97.2:2.8 v/v) and resorcinol as the internal standard (Kaur and Halliwell, 1996). The specificity for determining 2,3-DHB appears to underlay the production of oxyradicals, since 2,5-DHB is formed by cytochrome P$_{450}$.

To enable measurement of the hydroxyl radical *in vivo* Sasaki *et al.* (1999) synthesized the ^{11}C-derivatives of salicylic acid, aspirin and 2-methoxybenzoic acid and measured the decarboxylated [11]CO$_2$ in tissue which was trapped in liquid argon.

Salicylate – selective electrodes

Several attempts have been made to develop electrodes that may be selective for the detection of salicylate present in biological fluids, foods and preservatives (de Carvalho *et al.*, 2000; Shahrokhian *et al.*, 2000). Among the most recent of these was the polyvinyl pyrrolidine (PVC) membrane-bound matalions Al (III) and Sn (IV) salophons directly coated onto graphite electrodes to serve as ionophores (Shahrokhian *et al.*, 2000). These salicylate-sensitive electrodes were found to be sensitive to micromolar concentrations of salicylate, with detection possible over a wide range of pH values, from 3 to 8; good recoveries were found with salicylate present in a sample and in a synthetic serum sample, but not in natural serum or other biological material. De Carvalho *et al.* (2000) developed a micro-biosensor based on the reaction of salicylate hydrolase (from *Pseudomonas* species) which is covalently bound via carbodiimide to a carbon fibre electrode. The reaction of salicylate with salicylate hydroxylase involves the oxidature cleavage of the carboxyl group of salicylate in the presence of NADH to form catechol and release bicarbonate ions. A 2-electron transfer process then occurs in the conversion of catechol to its *ortho*-quinone, which enables the amperometric detection of the electrons generated in this reaction (de Carvalho *et al.*, 2000). It is claimed that this system has high sensitivity, low detection limits and good stability. It has the potential for analysis of salicylic acid and aspirin in pharmaceutical preparations over a wide range of pH values, from 6 to 8, with only minor variations according to the ionic strength of buffers (de Carvalho *et al.*, 2000). The potential for application in biological systems has yet to be explored. Given control of the redox conditions and availability of reduced adenine nucleotides (NADH), it may be possible to employ this system under *in vitro* conditions (e.g Ussing chambers, cell/organ cultures, organ bath systems). It may require more extensive modification to be employed for direct analysis of salicylate(s) in serum, plasma and urine samples where there are variations in redox state and potential interactions from other biomolecules.

Further development as robust 'at bedside' or laboratory methods is awaited, with much potential being evident for the application of such electrodes – principally, systems where they could be employed for 'in line' HPLC analysis of salicylates.

CONCLUSIONS

The salicylates, i.e. salicylic acid, saligenin and salicin, can be considered to be essentially natural drugs. With ingestion of salicylate-containing plants, mankind has no doubt developed an adaptation to these compounds. Many simple derivatives of the basic salicylate structure have been and still continue to be developed because these natural drugs have been so successful historically, and still represent a safe and effective group for therapy in these modern times.

The successful enhancement of the anti-inflammatory and analgesic potency of salicylic acid through the development of its biphenyl derivative, diflunisal, probably represents the most significant advance in recent times.

Of the immense efforts made, especially during the past 30 years, to produce more potent salicylate derivatives than aspirin or salicylate, it appears that only the 5-phenyl derivatives (i.e. the representatives being diflunisal and flufenisal) and 3- or 5-halogenated derivatives have been successful. This illustrates the particular molecular specificity for enhancement of anti-inflammatory and analgesic activities.

Other advances are being made in the development of an immense variety of pro-drugs of aspirin, which will be most useful in alleviating the gastrotoxicity inherent in it. The success of these pro-drugs will depend on their ratio of therapeutic gastrotoxic activities, this being related to their rate(s) of metabolism to aspirin and salicylate *in vivo*.

REFERENCES

Abraham, K.O., Shankaranarayana, M.L., Rahgavan, B. and Natarahan, C.P. 1976, Determination of methyl salicylate in black tea. *Microchemica Acta*, **1:** 11–15.

Abraham, D.J., Gazze, D.M., Kennedy, P.E. and Mokotoff, M. 1984, Design, synthesis, and testing of potential antisickling agents. 5. Disubstituted benzoic acids designed for the donor site and proline salicylates designed for the acceptor site. *Journal of Medicinal Chemistry*, **27:** 1549–1559.

Adams S.S. and Cobb, R. 1967, Non-steroidal anti-inflammatory drugs. *In:* Ellis, G.P. and West, G.B. (eds), *Progress in Medicinal Chemistry*, pp. 59–138. London: Butterworths.

Adilakshmi, T., Ayling, P.D. and Ratledge, C. 2000, Mutational analysis of a role for salicylic acid in iron metabolism of *Myobacterium smegmatis*. *Journal of Bacteriology*, **182:** 264–271.

Agrawal, G.K., Jwa, N.S. and Rakwal, S. 2000, A novel rice (*Oryza sativa* L.) acidic PR1 gene highly responsive to cut, phytohormones, and protein phosphatase inhibitors. *Biochemical and Biophysical Research Communications*, **274:** 157–165.

Alam, A.S. and Gregoriades, D. 1981, Physicochemical properties of magnesium salicylate. *Journal of Pharmaceutical Sciences*, **70:** 961–962.

CHAPTER 3

Alambar, E.S., Carlisle, J.A. and Carlisle, G.O. 1983, Synthesis and magnetic properties of manganese(II) salicylate dihydrate. *Inorganica Chimica Acta*, **78:** L65–L66.

Alibrandi, G., Micali, N., Trusso, S. and Villari, A. 1996, Hydrolysis of aspirin studied by spectrophotometric and fluorometric variable-temperature kinetics. *Journal of Pharmaceutical Sciences*, **85:** 1105–1108.

Altun, M.L., Ceyhan, T., Cartell, M., Atay, T., Ozdemir, N. and Cevheroglu, S. 2001, L.C. method for analysis of acetylsalicylic acid, caffeine, and codeine phosphate in pharmaceutical preparations. *Journal of Pharmaceutical and Biomedical Analysis*, **25:** 93–101.

Aonuma, S., Kohama, Y., Fujimoto, S. and Makino, T. 1981, Studies on aspirin derivatives with very little side-effect. II. Potent platelet anti-aggregant activity and no mutagenicity of aspirin-isopropylantipyrine (AIA). *Journal of Pharmaceutical Dynamics*, **4:** 803–811.

Aonuma, S., Kohama, Y. and Fujimoto, S. 1982, Studies on aspirin derivatives with very little side-effect. III. Absorption, distribution, excretion and metabolism of titrium-labeled aspirin-isopropylantipyrine (AIA). *Journal of Pharmaceutical Dynamics*, **5:** 252–258.

Aonuma, S., Kohama, Y., Fujimoto, S., Nomura, M. and Yamahata, E. 1983, Studies on aspirin derivatives with very little side-effect. IV. Inhibitory effect of aspirin-isopropylantipyrine (AIA). *Journal of Pharmaceutical Dynamics*, **6:** 9–17.

Aonuma, S., Kohama, Y., Komiyama, Y. and Fujimoto, S. 1980, Gastric, ulcerogenic and biological activities of N-3'-propylphenazonyl-2-acetoxybenzamide. *Chemistry and Pharmacology Bulletin*, **28:** 1237–1244.

Arena, B. and Del Soldato, P. 1997, Nitric esters having a pharmacological activity and process for their preparation. US Patent 5 621 000, 15 April, 1997.

Aspro-Nicholas Ltd. 1966, Report on Lipterol or AGN 356.

Austen, K.F. 1963, Discussion. *In:* A.St.J. Dixon, B.K. Martin, M.J.H. Smith and P.H.N. Woods (eds), *Salicylates. An International Symposium (Discussion)*, p. 64. London: Churchill.

Aydin, R., Ozer, U. and Turkel, N. 1997, Potentiometric and spectroscopic determination of acid dissociation constants of some phenols and salicylic acids. *Turkish Journal of Chemistry*, **21:** 428–436.

Babhair, S.A., Al-Badr, A.A. and Aboul-Enein, H.Y. 1984, Salicylamide. *Analytical Profiles of Drug Substances*, **13:** 521–551.

Baine, O., Adamson, G.F., Barton, J.W., Fitch, J.L., Swayampati, D.R. and Jeskey, H. 1954, A study of the Kolbe–Schmitt reaction II. The carbonation of phenols. *Journal of Organic Chemistry*, **19:** 510–514.

Bakhle, Y.S. 1981, Inhibition by clinically used dyes of prostaglandin inactivation in rat and human lung. *British Journal of Pharmacology*, **72:** 715–721.

Bandarage, U.K., Chen, L., Fang, X., Garvey, D.S., Glavin, A., Janero, D.R., Letts, L.G., Mercer, G.J., Saha, J.K., Schroeder, D., Shumway, M.J. and Tam, S.W. 2000, Nitrosothiol esters of diclofenac: synthesis and pharmacological characterization as gastrointestinal sparing prodrugs. *Journal of American Chemical Society*, **43:** 4005–4016.

Banerjee, P.K. and Amidon, G. 1981a, Physiochemical property modification strategies based on enzyme substrate specificities I: rationale, synthesis and pharmaceutical properties of aspirin derivatives. *Journal of Pharmaceutical Sciences*, **70:** 1299–1303.

Banerjee, P.K. and Amidon, G. 1981b, Physiochemical property modification strategies based on enzyme substrate specificities II: alpha-chymotrypsin hydrolysis of aspirin derivatives. *Journal of Pharmaceutical Sciences*, **70:** 1304–1306.

Banerjee, P.K. and Amidon, G. 1981c, Physiochemical property modification strategies based on enzyme substrate specificities III: carboxypeptidase A hydrolysis of aspirin derivatives. *Journal of Pharmaceutical Sciences*, **70:** 1307–1309.

Batterman, R.C. and Grossman, A.J. 1955, Effectiveness of salicylamide as an analgesic and antirheumatic agent. Evaluation of the double-blinded technique for studying analgesic drugs. *Journal of the American Medical Association*, **159:** 1619–1622.

Bauer, C.W. and Lasala, E.F. 1960, The preparation of properties of an aspirin homolog β-(o-acetoxy phenyl) propionic acid. *Journal of the American Pharmaceutical Association*, **49:** 48–53.

Bax, D.E. 1992, Sulfonamide. *In:* J.S. Dixon and D.E. Furst (eds), *Second-Line Agents in the Treatment of Rheumatic Diseases*, pp. 267–286. New York: Marcel Dekker.

Berlanga, M. and Vinas, M. 2000, Salicylate induction of phenotypic resistance to quinolones in *Serratia marcescens*. *Journal of Antimicrobial Chemotherapy*, **46:** 279–282.

Berry, H., Liyanage, S.P., Durance, R.A., Goode, J.D. and Swannell, A.J. 1981, A double-blind study of benorylate and chlormezanone in musculoskeletal disease. *Rheumatology Rehabilitation*, **1:** 46–49.

Berthelot, M., Laurence, C., Foucher, D. and Taft, R.W. 1996, Partition coefficients and intramolecular hydrogen bonding. 1. The hydrogen-bond basicity of intramolecular hydrogen-bonded heteroatoms. *Journal of Physical Organic Chemistry*, **9:** 225–261.

Berthon, G. (ed.) 1995, Handbook of Metal–Ligand Interactions in Biological Fluids. New York: Marcel Dekker.

Birch, A.J., Massey-Westropp, R.A. and Moyle, C.J. 1955, Studies in relation to biosynthesis III. 2-Hydroxy-6-methylbenzoic acid in Penicillium griseofuluum Dierckx. *Australian Journal of Chemistry*, **8:** 539–544.

Bird, H.A., Dixon, J.S. Pickup, M.E., Lee, M.R. and Wright, V. 1980, A biochemical comparison of alclofenac and D-penicillamine in rheumatoid arthritis. *Annals of the Rheumatic Diseases*, **39:** 281–284.

Bird, H.A., Dixon, J.S., Pickup, M.E., Rhind, V.M., Lowe, L.R., Lee, M.R. and Wright, V. 1982, A biochemical assessment of salphasalzine in rheumatoid arthritis. *Journal of Rheumatology*, **9:** 36–45.

Birkett, M.A., Campbell, C.A.M., Chamberlain, K., Guerrieri, E., Hick, A.J., Martin, J.L., Matthes, M., Napier, J.A., Pettersson, J., Pickett, J.A., Poppy, G.M., Pow, E.M., Pye, B.J., Smart, L.E., Wadhams, G.H., Wadhams, L.J. and Woodcock, C.M. 2000, New roles for *cis*-jasmone as an insect semichemical and in plant defense. *Proceedings of the National Academy of Sciences of the USA*, **97:** 9329–9334.

Blang, S.M. and Wesolowski, J.W. 1959, The stability of acetylsalicylic acid in suspension. *Journal of the American Pharmaceutical Association*, **48:** 691–694.

Blatt, M.R., Grabov, A., Brearley, J., Hammond-Kosack, K. and Jones, J.D.K. 1999, K$^+$ channels of cf-9 transgenic tobacco guard cells as targets for *Cladoporium fulvum* Avr9 elicitor-dependent signal transduction. *Plant Journal*, **19:** 453–462.

Blilou, I., Ocampo, J.A. and Garcia Garrido, J.M. 1999. Resistance of pea roots to endomycorrhizal fungus of Rhizobium correlates with enhanced levels of endogenous salicylic acid. *Journal of Experimental Botany*, **50:** 1663–1668.

Borchers, A.T., Keen, C.L., Stern, J.S. and Gershwin, M.E. 2000, Inflammation and Native American medicine: the role of botanicals. *American Journal of Clinical Nutrition*, **72:** 339–347.

Borg, D.C. 1965, Transient free radicals from salicylate. *Biochemical Pharmacology*, **14:** 627–631.

Bowden, K., Huntington, A.P. and Powell, S.L. 1997, Prodrugs – Part 1. Formylphenyl esters of aspirin. *European Journal of Medicinal Chemistry*, **32:** 987–993.

Broxton, T.J., Christie, J.R. and Sango, X. 1987, Micellar catalysis of organic reactions. 20. Kinetic studies of the hydrolysis of aspirin derivatives in micelles. *Journal of Organic Chemistry*, **52:** 4814–4817.

Buckle, P.J., Baldo, B.A. and Taylor, K.M. 1980, The anti-inflammatory activity of more natural products – 6n-tridecylsalicylic acid, flexibilide and dendalone 3-hydroxybutyrate. *Agents and Actions*, **10:** 361–367.

Bundgaard, H. and Larsen, C. 1976, Intramolecular and intermolecular transformations of aspirin in nonhydroxylic solvents. *Journal of Pharmaceutical Sciences*, **65:** 776–778.

Bundgaard, H., Klixbüll, U. and Falch, E. 1986, Prodrugs as drug delivery systems. 44. *O*-acyloxymethyl, *o*-acyl and *n*-acyl salicylamide derivatives as possible prodrugs for salicylamide. *International Journal of Pharmaceutics*, **30:** 111–121.

Bywaters, E.G.L. 1963, Discussion *In:* A.St.J. Dixon, B.K. Martin, M.J.H. Smith and P.H.N. Woods (eds), *Salicylates: An International Symposium*, p. 64. London: Churchill.

Cameron, R.K. 2000, Salicylic acid and its role in plant defense responses: what do we really know? *Physiological and Molecular Plant Pathology*, **56:** 91–93.

Carter, G.W., Young, P.R., Swett, L.R. and Paris, G.Y. 1980, Pharmacological studies in the rat with [2(1,3-didecanoyloxy)-propyl]2-acetyloxybenzoate (A-45474): an aspirin prodrug with negligible gastric irritation. *Agents and Actions*, **10:** 240–245.

Cena, C., Lolli, M.L., Lazzarato, L., Guaita, E., Morini, G., Coruzzi, G., McElroy, S.P., Megson, I.L.,

Fruttero, R. and Gasco, A. 2003, Antiinflammatory, gastrosparing, and antiplatelet properties of new NO-donor esters of aspirin. *Journal of Medicinal Chemistry*, **46:** 297–304.

Cha, B.C. and Lee, S.B. 2000, Synthesis and biological activity of aspirin derivatives. *Archives of Pharmaceutical Research*, **23:** 116–120.

Chambers, C.E., McIntyre, D.D., Mouch, M. and Sokol, P.A. 1996, Physical and structural characterization of Yersinophore, a siderophore produced by clinical isolates of *Yersinia enterocolitica*. *Biometals*, **9:** 157–167.

Chatterjee, R., Welty, E.V., Walder, R.Y., Pruitt, S.L., Rogers, P.H., Arnone, A. and Walder, J.A. 1986, Isolation and characterization of a new hemoglobin derivative cross-linked between the alpha chains (lysine 99 alpha1 Lysine 99 alpha 2). *Journal of Biological Chemistry*, **261:** 9929–9937.

Chen, H, and Zhao, D. 1999, Mixed acid catalyst for synthesis of isopropyl salicylate. *Riyong Huaxue Gongye*, **3:** 12–14.

Chen, Y. and Zhang, Y. 1998, Synthesis of isopropyl salicylate with butyl titanate as catalyst. *Xiandai Huagong*, **18:** 30–31.

Chipperfield, J.R. and Ratledge, C. 2000, Salicylic acid is not a bacterial siderophore: a theoretical study. *Biometals*, **13:** 165–168.

Chiroli, V., Benedini, F., Ongini, E. and Del Soldato, P. 2003, Nitric oxide-donating non-steroidal anti-inflammatory drugs: the case of nitroderivatives of aspirin. *European Journal of Medicinal Chemistry*, **38:** 441–446.

Chong, J., Baltz, R., Fritig, B. and Saindrenan, P. 1999, An early salicylic acid, pathogen- and elicitor-inducible tobacco glucosyltransferase: role in compartmentalization of phenolics and H_2O_2 metabolism. *FEBS Letters*, **458:** 204–208.

Clarke, N., Clarke, C.N. and Mosher, R.E. 1958, Phenolic compounds in chemotherapy of rheumatic fever. *American Journal of the Medical Sciences*, **235:** 7–22.

Clarke, N.E. and Mosher, R.E. 1953, Phenolic compounds in the treatment of rheumatic fever. II. The metabolism of gentistic acid and ethanolamide of gentisic acid. *Circulation*, **7:** 337–344.

Collier, H.O.J. 1963, Antagonism by aspirin and like-acting drugs of kinins and SRS-A in guinea pig lung. *In:* A.St.J. Dixon, B.K. Martin, M.J.H. Smith and P.H.N. Woods (eds), *Salicylates. An International Symposium*, pp. 120–126. London: Churchill.

Collins, R.P. and Halim, A.F. 1972, An analysis of the odorous constituents produced by various species of Phellinus. *Canadian Journal of Microbiology*, **18:** 65–66.

Combrink, K.D., Gulgeze, H.B., Yu, K.L., Pearce, B.C., Trehan, A.K., Wei, J.M., Deshpande, M., Krystal, M., Torri, A., Luo, G.X., Cianci, C., Danetz, S., Tiley, L. and Meanwell, N.A. 2000, Salicylamide inhibitors of influenza virus fusion. *Bioorganic and Medicinal Chemistry Letters*, **10:** 1649–1652.

Conje, M.J. and Bornman, L. 1999, Salicylic acid influences Hsp70/HSC70 expression in *Lycopersicon esculentum*: Dose- and time-dependent induction or potentiation. *Biochemical and Biophysical Research Communications*, **265:** 422–427.

Corthout, J., Pieters, L., Claeys, M., Geerts, S., Vanden-Berghe, D. and Vlietink, A. 1994, Antibacterial and molluscocidal phenolic acids from Spondias mombin. *Planta Medica*, **60:** 460–463.

Cox, P.J., Gilmour, G.I. and McManus, S.M. 2000, Hydrogen bonding in salicylsalicylic acid (Salsalate) crystals. *International Journal of Pharmaceutics*, **204:** 133–136.

Cross, S.E., Anderson, C. and Roberts, M.S. 1998, Topical penetration of commercial salicylate esters and salts using human isolated skin and clinical microdialysis studies. *British Journal of Pharmacology*, **46:** 29–35.

Cross, S.E., Megwa, S.A., Benson, H.A. and Roberts, M.S. 1999, Self promotion of deep tissue penetration and distribution of methyl salicylate after topical application. *Pharmaceutical Research*, **16:** 427–433.

Crouch, R.K., Kensler, T.W., Oberley, L.W. and Sorenson, J.R.J. 1985, Possible medicinal uses of copper complexes. *In:* K.D. Karlin and J. Zubieta (eds), *Biological and Inorganic Copper Chemistry*, pp. 139–157. New York: Adenine Press.

Cummings, A.J. and Martin, J. 1965, Systematic study of the elimination of 3-methyl salicylic acid and its acetyl derivative in man. *British Journal of Pharmacology*, **25:** 470A–480A.

Dai, G.-H., Fei, W. and Xu, Z.-P. 1999, Synthesis of mesalazine. *Zhongguo Yiyao Gongye Zazhi*, **29:** 443–444.

Danesh, A., Davies, M.C., Hinder, S.J., Roberts, C.J., Tendler, S.J.B., Williams, P.M. and Wilkins, M.J. 2000, Surface characterisation of aspirin crystal planes by dynamic chemical force microscopy. *Analytical Chemistry*, **72:** 3419–3422.

Dascombe, M.J. 1984, Effects of diflunisal on fever in the rabbit and the rat. *Journal of Pharmacy and Pharmacology*, **36:** 437–440.

Dat, J.F., Lopez Delgado, H., Foyer, C.H. and Scott, I. 2000, Effects of salicylic acid on oxidative stress and thermotolerance in tobacco. *Journal of Plant Physiology*, **156:** 659–665.

Davidson, D. and Auerbach, L. 1953, The acid anhydride character of aspirin. *Journal of the American Chemical Society*, **75:** 5984–5986.

De Alencar, R., Alves de Lima, R., Correa, R.G.C., Gottleib, O.R., Leao Da Silva, M., Marx, M.C., Maia, J.G.S., Magalhaes, M.T. and Assummpcao, R.M.V. 1972, Essential oils of Brazilian plants. *Anais da Academia Brasileirade Cienses*, **44 (Suppl.):** 312–334.

de Carvalho, R.M., de Oliviera Neto, G. and Kubota, L.T. 2000, Microbiosensor for salicylate based on modified carbon fibre. *Analytical Letters*, **33:** 425–442.

De Coninck, W.O. 1915, Basic copper salicylate. *Bulletin Societie Chimie*, **17:** 234–243 (quoted in *Chemical Abstracts*, **10:** 3221).

Delaude, C. 1974, Saponins in polygalaceae. Examination of saponisides of *Polygala usafeunsis*. *Bulletin de la Société Royale des Sciences de Liège*, **43:** 253–256.

Dellagi, A., Birch, P.R.J., Heilbronn, J., Avrova, A.O., Montesano, M., Palva, E.T. and Lyon, G.D. 2000, A potato gene, erg-1 is rapidly induced by *Erwinia carotovora* ssp. *Atroseptica*, *Phytophthora infestans*, ethylene and salicylic acid. *Journal of Plant Physiology*, **157:** 201–205.

De Nil, P. 1998, Method for the synthesis of antimicrobial alkyl hydroxybenzoate esters using alcohol esterification and azeotropic distillation. Patent: PCT International; WO 9856748 A1, 1998. Application No. WO 98BE84 (19980609) *BE 97491 (19970609). *Chemical Abstracts*, **130 (5):** 52239g.

Del Soldato, P. 1998, Nitric esters having anti-inflammatory and/or analgesic activity and process for their preparation. US Patent 5 780 495, 14 July 1998.

Del Soldata, P., Sorrentino, R. and Pinto, A. 1999, NO-aspirins: a class of new anti-inflammatory and antithrombotic agents. *Trends in Pharmacological Sciences*, **20:** 319–323.

Dempsey, D.A., Shah, J. and Klessig, D.F. 1999, Salicylic acid and disease resistance in plants. *Critical Reviews in Plant Sciences*, **18:** 547–575.

Dewick, P.M. 1997, *Medicinal Natural Products. A Biosynthetic Approach*. Chichester: Wiley.

Digenis, G.A. and Swintosky, J.V. 1975, Drug latentiation. *In:* J.R. Gillette and J.R. Mitchell (eds), *Concepts in Biochemical Pharmacology*, Part 3, pp. 86–111. New York: Springer-Verlag.

Djurendic, E., Suranyi, T.M. and Miljkovic, D.A. 1990, Synthesis of some salicylic acid derivatives and determination of their acidity constants. *Collection of Czechoslovac Chemical Communications*, **55:** 1763–1768.

Dong, H.S., D.P., Bauer, D.W. and Beer, S.V. 1999, Harpin induces disease resistance in *Arabidopsis* through the systemic acquired resistance pathway mediated by salicylic acid and the NIM1 gene. *Plant Journal*, **20:** 207–215.

Drechsel, H. and Winkelmann, G. 1997, Iron chelation and siderophores. *In:* G. Winkelmann and C.J. Carrano (eds), *Transition Metalelaney, Ts in Microbial Metabolism*, p. 149. Amsterdam: Harwood Academic Publishers.

Dromgoole, S.H., Nyman, K.E., Furst, D.E. and Kolman, W.A. 1978, Metabolism of meseclazone in man. *Drug Metabolism and Disposition*, **6:** 102–104.

D'Souza, M., Venkataramanan, R., D'Mello, A. and Niphadkar, P. 1986, An alternative prodrug approach for reducing presystemic metabolism of drugs. *International Journal of Pharmaceutics*, **31:** 165–167.

Duffield, R.M. and Blum, M.S. 1978, Methyl 6-methyl salicylate: identification and function in a ponerine ant (*Gnamptogenys pleurodon*). *Experientia (Basel)*, **31:** 466.

Durrani, A.A. and Tyman, J.H.P. 1979, Long chain phenols. Part 15. Synthesis of 6-n-alkylsalicylic acids (and isomeric acids) from fluoroanisoles with alkyl-lithium. *Journal of the Chemical Society*, 1979: 2079–2087.

Dyson, G.M. and May, P. 1959, *May's Chemistry of Synthetic Drugs*, pp. 110–141. London: Longmans.

CHAPTER 3

Edelson, J., Douglas, J.F. and Ludwig, B.J. 1973, Absorption, distribution and metabolic rate of 7-chloro-3,3a-dihydro-2H,9H-isoxazole(3,2-6)(1,3) benzoxazin-9-one). *Journal of Pharmaceutical Sciences*, **62:** 229–232.

Edelson, J., Douglas, J.F., Ludwig, B.J., Schuster, E.B. and Shahinian, S. 1975, Absorption, distribution, and metabolic fate of 7-chloro-3,3a-dihydro-2-methyl 2*H*, 9*H*-isoxazolo (3,2-b) (1,3-benzoxazine-9-one) in rats, dogs and humans. *Journal of Pharmaceutical Sciences*, **64:** 1316–1321.

Edmunds, R.T. 1966, US Patent 3 235 583. *Chemical Abstracts*, **64:** 17495a.

Edwards, L.J. 1950, Hydrolysis of aspirin. Determination of the thermodynamic dissociation constant and a study of the reaction kinetics by ultraviolet spectrophotometry. *Transactions of the Faraday Society*, **46:** 723–735.

Edwards, L.J. 1952, Hydrolysis of aspirin. II. *Transactions of the Faraday Society*, **48:** 696–699.

El-Naggar, A.M., Islam, A.M., Hannout, I.B. and Gaafer, M.M. 1975, Synthesis of some N-Salicylsulphonylamino acid derivatives. *Indian Journal of Chemistry*, **13:** 248–250.

Elshahawy, A.S., Mahfouz, R.M., Aly, A.A.M. and Elzohry, M. 1993, Technetium aspirin molecule complexes. *Journal of Chemical Technology and Biotechnology*, **56:** 227–231.

Endres, S., Hacker, A., Noack, E., Kojda, G. and Lehmann, J. 1999, NO-donors, part 3: nitrooxyacylated thiosalicylates and salicylates – synthesis and biological activities. *European Journal of Medicinal Chemistry*, **34:** 895–901.

Etcheverry, S.B., Williams, P.A., Barrio, D.A., Salice, V.C., Ferrer, E.G. and Cortizo, A.M. 2000, Synthesis, characterization and bioactivity of a new VO2+/aspirin complex. *Journal of Inorganic Biochemistry*, **80:** 169–171.

Fan, P., Ge, C.-H., Liu, K., Chen, Q.-H. and Sun, J.-Y. 1998, Synthesis of salicylate esters under the microwave irradiation and normal pressure. *Hecheng Huaxue*, **6:** 342–344.

Fersht, A.R. and Kirby, A.J. 1967a, Intramolecular nucleophilic catalysis of ester hydrolysis by the carboxylate group. *Journal of the American Chemical Society*, **89:** 5960–5961.

Fersht, A.R. and Kirby, A.J. 1967b, Intramolecular general acid catalysis of ester hydrolysis by the carboxylic acid group. *Journal of the American Chemical Society*, **89:** 5961–5962.

Fersht, A.R. and Kirby, A.J. 1968a, Intramolecular nucleophilic catalysis of ester hydrolysis by the ionized carboxyl group. The hydrolysis of 3,5-dinitroaspirin anion. *Journal of the American Chemical Society*, **90:** 5818–5826.

Fersht, A.R. and Kirby, A.J. 1968b, Intramolecular nucleophilic catalysis in the hydrolysis of substituted aspirin acids. *Journal of the American Chemical Society*, **90:** 5826–5832.

Filonov, A.E., Karpov, A.V., Kosheleva, I.A., Puntus, I.F., Balashova, N.V. and Boronin, A.M. 2000, The efficiency of salicylate utilization by *Pseudomonas putida* strains catabolizing naphthalene via different biochemical pathways. *Process Biochemistry*, **35:** 983–987.

Finar, I.L. 1963, *Organic Chemistry, Vol. 1, The Fundamental Principles*. London: Longmans.

Fiorucci, S., Santucci, L., Gresele, P., Faccino, R.M., Del Soldato, P. and Morelli, A. 2003, Gastrointestinal safety of NO-aspirin (NCX-4016) in healthy human volunteers: A proof of concept endoscopic study. *Gastroenterology*, **124:** 600–607.

Fischer, B. 1893, *Die Neueren Arzneimittel für Ärzte, Apotheker und Drogisten*. Berlin: Julius Springer Verlag.

Fujita, Y., Fujita, S. and Yoshikawa, H. 1974, Comparative biochemical and chemical-taxonomical studies on the plants of theaceae. I. Essential oils of *Camelilia sasanqua, C. japonica* and *Thea sinensis. Osaka Kogyo Gijutsu Shikenso Kiho*, **25:** 198–204.

Fuson, R.C. 1962, *Reactions of Organic Compounds*. New York: Wiley.

Gaginella, T.S. and Walsh, R.E. 1992, Sulfasalazine. Multiplicity of action. *Digestive Diseases and Sciences*, **37:** 801–812.

Garcia-Rafanell, J. 1980, Fosfosal. *Drugs of the Future*, **5:** 290–292.

Garcia-Rafanell, J., Ramis, J., Gomez, L. and Forn, J. 1986, Effect of triflusal and other salicylic-acid derivatives on cyclic-AMP levels in rat platelets. *Archives Internationales de Pharmacodynamie et de Therapie*, **284:** 155–165.

Garrett, E.R. 1957, The physical chemical evidence for aspirin anhydride as a superior form for the oral administration of aspirin. *Journal of the American Pharmaceutical Association, Scientific Edition*, **48:** 678–683.

Gaubert, S., Bouchaut, M., Brumas, V. and Berthon, G. 2000, Copper–ligand interactions and the physiological free radical processes. Part 3. Influence of histidine, salicylic acid and anthranilic acid on copper-driven Fenton chemistry in vitro. *Free Radical Research*, **32:** 451–461.

Geissler, E. and Möller, J. 1889, *Real Encyclopädie der Gesammter Pharmacie.* Leipzig: Urban und Schwarzenberg.

Gellerman, J.L., Walsh, N.J., Werner, N.K. and Schlenk, H. 1969, Antimicrobial effects of anacardic acids. *Canadian Journal of Microbiology*, **15:** 1219–1223.

Gilmer, J.F., Moriarty, L.M., Lally, M.N. and Clancy, J.M. 2002, Isosorbide-based aspirin prodrugs II. Hydrolysis kinetics of isosorbide diaspirinate. *European Journal of Pharmaceutical Sciences*, **16:** 297–304.

Gilmer, J.F., Moriarty, L.M., McCafferty, D.F. and Clancy, J.M. 2001, Synthesis, hydrolysis kinetics and anti-platelet effects of isosorbide mononitrate derivatives of aspirin. *European Journal of Pharmaceutical Sciences*, **14:** 221–227.

Glenn, E.M., Bowman, B.J. and Rohloff, N.A. 1979, Anomalous biological effects of salicylates and prostaglandins. *Agents and Actions*, **9:** 257–264.

Gorini, S., Valcavi, U. and Zonta-Bolego, N. 1963, Nuova derivati dell'acido O-cresotinico ad analgesia ed antiflogistica. *Bolletina della Societa Italiana de Biologia Spermentale*, **39:** 873–875.

Gottesmann, R.T. and Chin, D. 1968, *Kirk-Otmer Encyclopedia of Chemical Technology*, 2nd edn, Vol. 17, pp. 720–743. New York: Wiley.

Greenberg, J.T. 2000, Positive and negative regulation of salicylic acid-dependent cell death and pathogen resistance in *Arabidopsis lsd6* and *ssi1* mutants. *Molecular Plant–Microbe Interactions*, **13:** 877–881.

Greenberg, J.T., Silverman, F.P. and Liang, H. 2000, Uncoupling salicylic acid-dependent cell death and defense-related responses from diseases resistance in the *Arabidopsis* mutant *acd5*. *Genetics*, **156:** 341–350.

Gringauz, A. 1970, Certain desubstituted *o*-aminoacetoxy- and propoxybenzoic and cinnamic acids and their test butyl esters. *Journal of Pharmaceutical Sciences*, **59:** 422–425.

Gringauz, A. 1976, Synthesis of 2'-acetoxybiphenyl-2-carboxylic acid and its derivatives as potential anti-inflammatory and analgesic agents. *Journal of Pharmaceutical Sciences*, **65:** 291–294.

Gross, M. and Greenberg, L.A. 1948, *The Salicylates. A Critical Bibliographic Review.* New Haven: Hillhouse Press.

Gulati, A. and Rebello, S. 1994, Role of adrenergic mechanisms in the pressor effect of diaspirin cross-linked hemoglobin. *Journal of Laboratory and Clinical Medicine*, **124:** 125–133.

Gulati, A., Sharma, A.C. and Burhop, K.E. 1994, Effect of serum-free hemoglobin and diaspirin cross-linked hemoglobin on the regional circulation and systemic hemodynamics. *Life Sciences*, **55:** 827–837.

Hannah, J., Ruyle, W.V., Jones, H., Matzuk, A.R., Kelly, K.W., Witzel, B.E., Holtz, W.J., Houser, R.W., Shen, T.Y. and Sarret, L.H. 1977, Discovery of diflunisal. *British Journal of Clinical Pharmacology*, **4:** 7S–13S.

Hannah, J. Ruyle, W.V., Jones, H., Matzuk, A.R., Kelly, K.W., Witzel, B.E., Holtz, W.J., Howser, R.W., Shen, T.Y., Sarett, L.H., Lotti, V.J., Ridley, E.A., Van Arman, C.G. and Winter, C.A. 1978, Novel analgesic–antiflammatory salicylates. *Journal of Medicinal Chemistry*, **21:** 1093–1100.

Hansch, C. and Anderson, S.M. 1967, The effect of intramolecular hydrogen bonding on partition coefficients. *Journal of the American Chemical Society*, **32:** 2583–2586.

Hanzlik, P.J. and Presho, N.E. 1925, The Salicylates XVIII. Liberation of salicyl from and excretion of methyl salicylate, with a note on the irregular toxicity of the ester in man. *Journal of Pharmacology and Experimental Therapeutics*, **26:** 61–70.

Hart, R.E. 1946, The analgesic potency and acute toxicity of salicylamide and certain of its derivatives as compared with established analgesic-antipyretic drugs. *Federation of the American Association of Societies of Experimental Biology*, **5:** 182.

Hartman, U. and Vahrenkamp, H. 1994, Coordination chemistry of zinc with pharmaceutical ligands – the zinc–aspirin complex. *Bulletin of the Polish Academy of Sciences – Chemistry*, **42:** 161–167.

Haslam, E. 1974, *The Shikimate Pathway.* London: Butterworths.

Hata, S.-I. and Tomioka, S. 1968a, Charge transfer absorption bands of the complexes between various electron donors and menailione in aqueous solution. *Chemical and Pharmaceutical Bulletin*, **16:** 1397–1398.

Hata, S.-I. and Tomioka, S. 1968b, Effect of solvents on the charge transfer absorption bands of the complexes between various electron donors and menadione. *Chemical and Pharmaceutical Bulletin*, **16:** 2078–2079.

Hata, S.-I., Mizuno, K. and Tomioka, S. 1968, Effects of electron donors on the photodecomposition of menadione in aqueous solution III. The relation between the structure of the complex in electronic aspect and the rate of stabilization. *Chemical and Pharmaceutical Bulletin*, **16:** 1–5.

Hess, J.R. 1996, Alternative oxygen carriers. *Current Opinion in Hematology*, **3:** 492–497.

Hogg, N. 2000, Biological chemistry and clinical potential of S-nitrosothiols. *Free Radical Biology and Medicine*, **28:** 1478–1486.

Hopkins, S.J. 1979, Eterylate. *Drugs of the Future*, **6:** 577–579.

Hoult, J.R.S. and Moore, B.K. 1978, Sulphasalazine is a potent inhibitor of prostaglandin 15-hydroxy-dehydrogenase: possible basis for therapeutic action in ulcerative colitis. *British Journal of Pharmacology*, **64:** 6–8.

Hoult, J.R.S. and Moore, B.K. 1980, Effects of sulphasalazine and its metabolites on prostaglandin synthesis, inactivation and actions on smooth muscle. *British Journal of Pharmacology*, **68:** 719–730.

Hu, L., Yu, G., Lui, Z. and Yan, G. 1997, Esterification for synthesis of alkyl salicylate. Patent CN 1149576, 14 May 1997. *Chemical Abstracts*, **131 (25):** 336817q.

Hugot, H., Aime, S., Conrod, S., Poupet, A. and Galiana, E. 1999, Developmental regulated mechanisms affect the ability of a fungal pathogen to infect and colonize tobacco leaves. *Plant Journal*, **20:** 163–170.

Humbert, B., Alnot, M. and Quiles, F. 1998, Infrared and raman spectroscopical studies of salicylic and salicylate derivatives in aqueous solution. *Spectrochimica Acta, Part A*, **54A:** 456–476.

Hung, D.Y., Mellick, G.D., Prankerd, R.J. and Roberts, M.S. 1997, Synthesis, identification, characterization, stability, solubility, and protein binding of ester derivatives of salicylic acid and diflunisal. *International Journal of Pharmaceutics*, **153:** 25–39.

Hung, D.Y., Mellick, G.D., Masci, P.P., Whitaker, A.N., Whitehouse, M.W. and Roberts, M.S. 1998, Focused antithrombotic therapy: novel anti-platelet salicylates with reduced ulcerogenic potential and higher first-pass detoxification than aspirin in rats. *Journal of Laboratory and Clinical Medicine*, **132:** 469–477.

Ichniowski, C.T. and Hueper, W.C. 1946, Pharmacological and toxicological studies on salicylamide. *Journal of the American Pharmaceutical Association (Scientific Edition)*, **35:** 225–230.

Ikeda, Y., Koizumi, N., Kusano, T. and Sano, H. 1999, Sucrose and cytokinin modulation of *wpk4*, a gene encodin *a snf-1* related protein kinase from wheat. *Plant Physiology*, **121:** 813–820.

Ingster, L.M. and Feinleib, M. 1997, Could salicylates in food have contributed to the decline in cardiovascular disease mortality? A new hypothesis. *American Journal of Public Health*, **87:** 1554–1557.

Inoue, M., Morikawa, M., Tsuboi, M. and Sugiura, M. 1979, Studies of human intestinal esterase IV. Application to the development of ester prodrugs of salicylic acid. *Journal of Pharmacodynamics*, **2:** 229–236.

Insel, P.A. 1990, Analgesic–antipyretics and anti-inflammatory agents: drugs employed in the treatment of rheumatoid arthritis and gout. *In: Goodman and Gilman's The Pharmacological Basis of Therapeutics*, 8th edn, pp. 638–681. New York: Pergamon Press.

Irving, K.J., Wear, M.A., Simmons, H., Tipton, L.G., Tipton, J.B., Maddox, K.M., Willingham, W.M. and Sorenson, J.R.J. 1996, An examination of the radioprotective and radiorecovery activities of Fe(III)(3,5-diisopropylsalicylate)$_3$ and Mn(III)$_2$(II)(μ_3-O)(3,5-diisopropylsalicylate)$_6$. *Inflammopharmacology*, **4:** 309–321.

Janssen, P.L.T.M.K., Holman, P.C.H., Venema, D.P., Van Staveren, W.A. and Katan, M.B. 1996, Salicylates in foods. *Nutrition Reviews*, **54:** 357–359.

Janssen, P.L.T.M.K., Katan, M.B., Van Staveren, W.A., Hollman, P.C.H. and Venema, D.P. 1997, Acetylsalicylate and salicylate in foods. *Cancer Letters*, **114:** 163–164.

Jin, Z., Zhang, Q., Liu, L., Sun, S. and Zhang, S. 1996, Catalytic synthesis of isoamyl salicylate by heterogeneous supported heteropoly acid catalyst PW12/SiO$_2$. *Riyang Huaxue Gongye*, **1:** 21–22.

Jirage, D., Tootle, T.L., Reuber, T.L., Frost, L.N., Feys, B.J., Parker, J.E., Ausubel, F.M. and Glaze-brook, J. 1999, *Arabidopsis thaliana* pad4 encodes a lipase-like gene that is important for salicylic acid signalling. *Proceedings of the National Academy of Sciences of the USA*, **96:** 13583–13588.

Jones, H., Fordice, M.W., Greenwalt, R.B., Hannah, J. Ruyle, R.V., Walford, G.L. and Shen, T.Y. 1978, Synthesis and analgesic–anti-inflammatory activity of some 4- and 5-substituted heteroarylsalicylic acids. *Journal of Medicinal Chemistry*, **21:** 1100–1104.

Kachroo, P., Yoshioka, K., Shah, J., Dooner, H.K. and Klessig, D.F. 2000, Resistance to turnip crinkle virus in *Arabidopsis* is regulated by two host genes and is salicylic acid dependent, but *NPR1*, ethylene, and jasmonate independent. *The Plant Cell*, **12:** 677–690.

Kamal, A. 1990, Metabolism of ethyl 2-carbamoyloxybenzoate (4003.2), a prodrug of salicylic acid, carsalam and salicylamide. *Biochemical Pharmacology*, **40:** 1669–1671.

Kaur, H. and Halliwell, B. 1996, Salicylic acid and phenylalanine as probes to detect hydroxyl radicals. *In:* N.A. Punchard and F.J. Kelly (eds), *Free Radicals. A Practical Approach*, pp. 101–116. Oxford: Oxford University Press.

Kharat, A.S. and Mahadevan, S. 2000, Analysis of the beta-glucoside utilization (bgl) genes of *Shigella sonnei*: evolutionary implications for their maintenance in a cryptic state. *Microbiology*, **146:** 2039–2049.

Killackey, J.J.F. 1982, Studies of the pharmaceutical effects of benzoic acid analogs in thrombosis and inflammation. PhD Thesis, University of Western Ontario.

Killackey, J.J.F., Killackey, B.A., Cerskus, I. and Philp, R.B. 1984, Anti-inflammatory properties of a hydroperoxide compound structurally related to acetylsalicylic acid. *Inflammation*, **8:** 157–169.

Kim, D.H. 1979a, Aspirin (I) Discovery, current and potential new therapeutics uses, and mechanism of action. *Archives of Pharmacal Research*, **1:** 41–54.

Kim, D.H. 1979b, Aspirin (II) Structure activity relationship of salicylates and improvements of their therapeutic value through structural modification. *Archives of Pharmacal Research*, **2:** 71–78.

Kim, Y., Matsumoto, M. and Machida, K. 1985, Specific surface energies and dissolution behaviour of aspirin crystal. *Chemical and Pharmaceutical Bulletin*, **33:** 4125–4131.

Kim, Y. and Tu, S.-C. 1989, Molecular cloning of salicylate hydroxylase genes from *Pseudomonas cepacia* and *Pseudomonas putida*. *Archives of Biochemistry and Biophysics*, **269:** 295–304.

Kim, W.Y., Kim, C.Y., Cheong, N.E., Choi, Y.O., Lee, K.O., Lee, S.H., Park, J.B., Nakano, A., Bahk, J.D., Cho, M.J. and Lee, S.Y. 1999, Characterization of two fungal elicitor-induced rece cDNA's encoding functional homologues of the rab-specific GDP dissociation inhibitor. *Planta*, **210:** 143–149,

Klessig, D.F., Durner, J., Noad, R., Navarre, D.A., Wendehenne, D., Kumar, D., Zhou, J.M., Shah, J., Zhang, S., Kachroo, P., Trifa, Y., Pontier, D., Lam, E. and Silva, H. 2000, Nitric oxide and salicylic acid signaling in plant defense. *Proceedings of the National Academy of Sciences of the USA*, **97:** 8849–8855.

Klotz, I.M. and Tam, J.W.O. 1973, Acetylation of sickle cell hemoglobin by aspirin. *Proceedings of the National Academy of Sciences of the USA*, **70:** 1313–1315.

Klotz, U. 2000, The role of aminosalicylates at the beginning of the new millennium in the treatment of chronic inflammatory bowel disease. *European Journal of Clinical Pharmacology*, **56:** 353–362.

Klotz, U., Maier, K., Fischer, C. and Heinkel, K. 1980, Therapeutic efficacy of sulfasalazine and its metabolites in patients with ulcerative colitis and Crohn's disease. *New England Journal of Medicine*, **303:** 1499–1502.

Kock, J.L.F., Van Wyk, P.W.J., Venter, P., Coetzee, D.J., Smith, D.P., Viljoen, B.C. and Nigam, S. 1999, An acetylsalicylic acid-sensitive aggregation phenomenon in *Dipodascopsis uninucleata*. *Antonie van Leeuwenhoek*, **75:** 261–266.

Kovl, P.J., Miller, C.L. and Schulman, S.G. 1982, Biprotonic *versus* intramolecular photoautomerism of salicylic acid and some of its methylated derivatives in lowest excited single state. *Analytica Chimica Acta*, **61:** 7–13.

Krause, E. 1977, Analgesic activity of the salicylic acid and its derivatives. *In:* H. Bekemeier (ed.), *100 Years of the Salicylic Acid as an Anti-rheumatic Drug*, pp. 95–113. Halle-Wittenberg: Martin-Luther Universität.

Kroschwitz, J.I. and Howe-Grant, M. (eds). 1997, *Kirk-Othmer Encyclopedia of Chemical Technology*, Fourth Edition, pp. 601–619. New York: Wiley.

Kumar, R. and Billimora, J.D. 1978, Gastric ulceration and the concentration of salicylate in plasma

■ CHAPTER 3 ■

in rats after administration of ^{14}C-labelled aspirin and its triglyceride 1,3-dipalmitoyl-2(2'-acetoxy-[^{14}C] carboxylbenzoyl) glycerol. *Journal of Pharmacy and Pharmacology*, **30:** 754–758.

Kumar, D. and Klessig, D.F. 2000, Differential of tobacco MAP kinases by the defense signals nitric oxide, salicylic acid, ethylene and jasmonic acid. *Molecular Plant–Microbe Interactions*, **13:** 347–351.

Kuzell, W.C. and Gardner, G.M. 1950, Salicylazosulfapyridine (Salazopyrin or Azo Pyrin) in rheumatoid arthritis and experimental polyarthritis. *California Medicine*, **73:** 476–480.

Kyriakopoulos, A.A., Clarke, M.L., Mock, D.C. and Hagans, J.A. 1960, A comparative study of gastrointestinal bleeding incident to the administration of aspirin, aspirin anhydrides and a placebo. (Abstract).

Lamy, M.L., Daily, E.K., Brichant, J.F., Larbuisson, R.P., Demeyere, R.H., Vandermeersch, E.A., Lehot, J.J., Parsloe, M.R., Berridge, J.C., Sinclair, C.J., Baron, J.F. and Przybelski, R.J. 2000, Randomised trial of diaspirin cross-linked haemoglobin solution as an alternative to blood transfusion after cardiac surgery: The DCLHb Cardiac Surgery Trial Collaborative Group. *Anesthesiology*, **92:** 646–656.

Lantin, S., O'Brien, M. and Malton, D.P. 1999, Pollination wounding and jasmonate treatments induce the expression of a developmentally regulated pistil dioxygenase at a distance in the ovary, in the wild potato *Solanum chacoense bitt*. *Plant Molecular Biology*, **41:** 371–386.

Lederer, E, 1941, Constituents of beaver castor. *Travaux Membres de la Societé de Chimie Biologique*, **23:** 1457–1462.

Lederle, E. and Kollbrunner, F. 1980, Arzneimittel auf der Grundlage von Kupfer(II) – acetylsalicylat und seine Verwendung in der Therapie. Deutsches Patent No. DE 3033354 A1, 4th September 1980.

Lee, H. and Raskin, I. 1999, Purification, cloning and expression of a pathogen-inducible UDP-glucose: salicylic acid glucosyltransferase from tobacco. *Journal of Biological Chemistry*, **274:** 36637–36642.

Lightbody, T.D. and Reid, J. 1960, Ortho-cresotinate and diabetes mellitus. *British Medical Journal*, **2:** 1704–1707.

Liss, E. and Robertson, A. 1975, The distribution and elimination of radioactivity in the rat after administration of ^{14}C-4-acetamidophenyl-2-acetoxybenzoate (Benorylate). *Arzneimittel Forschung*, **25:** 1792–1793.

Loftsson, T., Kaminski, J.J. and Bodor, N. 1981, Improved delivery through biological membranes VIII: design, synthesis and *in vitro* testing of true prodrugs of aspirin. *Journal of Pharmaceutical Sciences*, **70:** 743–749.

Longguan, Z., Kitagawa, S., Kondo, M. and Miyasaka, H. 2000, Synthesis and crystal structure of new salicylate-bridged coordination polymer, [$Cu_2(sal)_2(pyz)(MeOH)_2$]$_n$. *Chemistry Letters*, **2000:** 536–537.

MacDougall, A.I. and Alexander, W.D. 1963, Effects of salicylates on cholesterol metabolism *In:* A.St.J. Dixon, B.K. Martin, M.J.H. Smith and P.H.N. Woods (eds), *Salicylates. An International Symposium*, pp. 92–96. London: Churchill.

Mackerness, S.A.-H., Surplus, S.L., Blake, P., John, C.F., Buchanan-Wollaston, V., Jordan, B.R. and Thomas, B. 1999, Ultraviolet-B-induced stress and changes in gene expression in *Arabidopsis thaliana*: role of signalling pathways controlled by jasmonic acid, ethylene and reactive oxygen species. *Plant, Cell and Environment*, **22:** 1413–1423.

Maetzel, A., Wong, A., Strand, V., Tugwell, P., Wells, G. and Bombardier, C. 2000, Meta-analysis of treatment termination rates among rheumatoid arthritis patients receiving disease-modifying antirheumatic drugs. *Rheumatology*, **39:** 975–981.

Manandhar, H.K., Mathur, S.B., Petersen, V.S. and Christensen, H.T. 1999, Accumulation of transcripts to pathogenesis-related proteins and peroxidase in rice plants triggered by *Pyricularia oryzae*, *Bipolaris sorokiniana* and UV light. *Physiological and Molecular Plant Pathology*, **55:** 289–295.

Marshall, B.J. and Ratledge, C. 1972, Salicylic acid biosynthesis and its control in *Mycobacterium smegmatis*. *Biochimica et Biophysica Acta*, **264:** 101–116.

Martin, B.K. 1971, The formulation of aspirin. *Advances in Pharmaceutical Sciences*, **3:** 107–171.

Massil, S.E., Shi, G.-Y. and Klotz, I.M. 1983, Acylation of hemoglobin by aspirin-like diacyl esters. *Journal of Pharmaceutical Sciences*, **73:** 1013–1014.

Massil, S.E., Shi, G.-Y. and Klotz, I.M. 1984, Electrostatic effects in acylation of hemoglobin by aspirins. *Journal of Pharmaceutical Sciences*, **73:** 1851–1853.

McBryde, W.A.E., Rohr, J.L., Penciner, J.S. and Page, J.A. 1970, Stability constants of three iron(III) salicylates. *Canadian Journal of Chemistry*, **48:** 2574.

McConkey, B., Amos, R.S., Butler, E.P., Crockson, R.A., Crockson, A.P. and Walsh, L. 1978, Salazopyrin in rheumatoid arthritis. *Agents and Actions*, **8:** 438–441.

McConkey, B., Amos, R.S., Durkham, S., Forster, P.J.G., Hubball, S. and Walsh, L. 1980, Sulphasalazine in rheumatoid arthritis. *British Medical Journal*, **280:** 442–444.

McCue, P., Zheng, Z.X., Pinkham, J.L. and Shetty, K. 2000, A model for enhanced pea seeding vigour following low pH and salicylic acid treatments. *Processes of Biochemistry*, **35:** 603–613.

McKetta, J.J. 1977, *Encyclopedia of Chemical Processing and Design*, Vol. 4, pp. 24–30. New York: Marcel Dekker.

Meenan, P. 1997, Crystal morphology predictive techniques to characterize crystal habit: application to aspirin ($C_9H_8O_4$). *Separataion and Purification by Crystallization*, **667:** 2–17.

Merck Index. An Encyclopedia of Chemicals, Drugs and Biologicals 1983, Tenth edn. (ed. Martha Windholz). Rahway: Merck & Co., Inc.

Meyer, K. and Ragan, C. 1948, The antirheumatic effect of sodium gentisate. *Science (Washington DC)*, 108, 281.

Michel, R. and Truchot, R. 1963, Synthesis of various ^{3}H-labelled radioisomers of salicylic acid. *Proceedings of the Conference on Methods of Preparing, Storing Marked Molecules. Euratom, Brussels*, p. 1171.

Micklatcher, M.L. and Cushman, M. 1999, An improved method for the synthesis of 3-fluorosalicylic acid with application to the synthesis of 3-(trifluoromethyl)salicylic acid. *Synthesis-Stuttgart*, **11:** 1878–1880.

Minuz, P., Lechi, C., Tommasoli, R., Gaino, S., Degan, M. Zuliani, V., Bonapace, S., Benoni, G., Adami, A., Cuzzolin, L. and Lechi, A. 1995, Antiaggregating and vasodilatory effects of a new nitroderivative of acetylsalicylic acid. *Thrombosis Research*, **80:** 367–376.

Misher, A., Adams, H.J., Fishler, J.J. and Jones, R.G. 1968, Pharmacology of the hexylcarbonate of salicylic acid. *Journal of Pharmaceutical Sciences*, **57:** 1128–1131.

Mittler, R., Herr, E.H., Orvar, B.L., Van Camp, W., Willekens, H., Inze, D. and Ellis, B.E. 2000, Transgenic tobacco plants with reduced capability to detoxify reactive oxygen intermediates are hyperresponsive to pathogen infection. *Proceedings of the National Academy of Sciences USA*, **96:** 14165–14170.

Mizukami, H., Terao, T., Amano, A. and Ohashi, H. 1986, Glucosylation of salicyl alcohol by *Gardenia jasminoides* cell cultures. *Plant Cell Physiology*, **27:** 645–650.

Moll, F. and Stauff, D. 1985, Wechselwirkungen zwischen Acetylsalicylsäure und Lysin in lösung. *Archiv der Pharmazie (Weinheim)*, **318:** 120–127.

Moore, G.G.I. 1974, Sulfonamides with Antiinflammatory Activity. *In:* R.A. Scherrer and M.W. Whitehouse (eds), *Antiinflammatory Agents, Chemistry and Pharmacology*, Vol. 2, pp. 159–177. New York: Academic Press.

Moore, U., Seymour, R.A., Williams, F.M., Nicholson, E. and Rawlins, M.D. 1989, The efficacy of benorylate in postoperative dental pain. *European Journal of Clinical Pharmacology*, **36:** 35–38.

Morant, G., Dung, N.H., Daran, J.C., Viossat, B., Labouze, X., Roch-Arveiller, M., Greenaway, F.T., Cordes, W. and Sorenson, J.R. 2000, Low-temperature crystal structures of tetrakis-μ-3,5-diisopropylsalicylatobis-dimethylformamidodicopper(II) and tetrakis-μ-3,5-diisopropylsalicylato-bis-diethyletherato-dicopper(II) and their role in modulating polymorphonuclear leukocyte activity in overcoming seizures. *Journal of Inorganic Biochemistry*, **81:** 11–22.

Morris, K., Mackerness, S.A.H., Page, T., John, C.F., Murphy, A.M., Carr, J.P. and Buchanan-Wollaston, V. 2000, Salicylic acid has a role in regulating gene expression during leaf senescence. *Plant Journal*, **23:** 677–685.

Mu, Q.-C. and Young, I.-J. 1966, Studies on essential oils of *Genus gaultheria L.* in yunnan and its applications. *Acta Pharmaceutica Sinica*, **13:** 453–455.

Mur, L.A.J., Brown, I.R., Darby, R.M., Bestwick, C.S., Bi, Y.M., Mansfield, J.W. and Draper, J. 2000, A loss of resistance to avirulent bacterial pathogens in tobacco is associated with the attenuation of a salicylic acid-potentiated oxidative burst. *Plant Journal*, **23:** 609–621.

Nencki, M. 1895, Eine Bemerckung, die Ausscheidung des Organismus fremder Stoffe in dem Magen betreffend. *Archive Experimental Pathology and Pharmacology*, **36:** 400–402.

Niggeweg, R., Thurow, C., Kegler, C. and Gatz, C. 2000, Tobacco transcription factor TGA1.1 is the main component of as-1-binding factor ASF-1 and is involved in salicylic acid- and auxin-inducible expression of as-1-containing target promoters. *Journal of Biological Chemistry*, **275:** 19897–19905.

Nishikawa, Y., Hiraki, K. and Maruyama, M. 1983, Studies on room temperature phosphorimetric analysis of several salicylic acid derivatives. *Bunseki Kageku*, **32:** 69–75.

Nok, A.J., Shuaibu, M.N., Bonire, J.J., Dabo, A., Wushishi, Z. and Ado, S. 2000, Triphenyltin salicylate – antimicrobial effect and resistance – the pyrophosphatase connection. *Journal of Enzyme Inhibition*, **15:** 411–420.

Norman-Setterblad, C., Vidal, S. and Palva, E.T. 2000, Interacting signal pathways control defense gene expression in *Arabidopsis* in response to cell wall-degrading enzymes from *Erwinia carotovora*. *Molecular Plant–Microbe Interactions*, **13:** 430–438.

Ohtake, Y., Takahashi, T. and Komeda, Y. 2000, Salicylic acid induces the expression of a number of receptor-like kinase genes in *Arabidopsis thaliana*. *Plant and Cell Physiology*, **41:** 1038–1044.

Olmsted, J. 1998, Synthesis of aspirin – a general chemistry experiment. *Journal of Chemical Education*, **75:** 1261–1263.

Ory, M. 1953, Essai du Pyrocatéylpyrocaté chique dam la Polyarthrite chronique évolutive. *Bruxelles-Medical*, **33:** 1556–1559.

Palaparthy, R., Kastrissios, H. and Gulati, A. 2000, Pharmacokinetics of diaspirin cross-linked haemoglobin in a rat model of hepatic cirrhosis. *Journal of Pharmacy and Pharmacology*, **53:** 179–185.

Paris, G.Y., Garmaise, D.L., Cimon, D., Swett, L.R., Carter, G.W. and Young, P.R. 1979, Glycerides as prodrugs. 1. Synthesis and anti-inflammatory activity of 1,3-bis(alkanoyl)-2-(o-acetylsalicyloyl) glycerines (aspirin triglycerides). *Journal of Medicinal Chemistry*, **22:** 683–687.

Pecci, J. and Foyle, W.O. 1960, The avidity of salicylic, gentisic, and salicyluric acids for heavy metal cations. *Journal of the American Pharmaceutical Association*, **49:** 411–414.

Pedersen, A.K. and FitzGerald, G.A. 1985, Preparation and analysis of deuterium-labeled aspirin: application to pharmacokinetic studies. *Journal of Pharmaceutical Sciences* **74:** 188–192.

Pettersson, G. 1966, On the role of 6-methylsalicylic acid in the biosynthesis of fungal benzoquinones. *Acta Chemica Scandinavica*, **20:** 151–158.

Pickering, S.U. 1927, Copper salts and their behavior with alkalis. *Journal of the Chemical Society*, **101:** 174–192.

Pieterse, C.M.J. and Van Loon, L.C. 1999, Salicylic acid-independent plant defence pathways. *Elsevier Science*, **1360:** 52–58.

Potter, F.J., Bennett, E. and Wiskich, J.T. 2000, Effects of ageing and salicylate on beetroot mitochondria. *Australian Journal of Plant Physiology*, **27:** 445–450.

Prithiviraj, B., Manickam, M., Singh, U.P. and Ray, A.B. 1997, Antifungal activity of anacardic acid, a naturally occurring derivative of salicylic acid. *Canadian Journal of Botany*, **75:** 207–211.

Przybelski, R.J., Daily, E.K., Kisicki, J.C., Mattia-Goldberg, C., Bounds, M.J. and Colburn, W.A. 1996, Phase I study of the safety and pharmacologic effects of diaspirin cross-linked hemoglobin solution. *Critical Care Medicine*, **24:** 1993–2000.

Przybelski, R.J., Daily, E.K., Micheels, J., Sloan, E., Mols, P., Koenigsberg, M.D., Bickell, W.H., Thompson, D.R., Harviel, J.D. and Cohn, S.M. 1999, A safety assessment of diaspirin cross-linked hemoglobin (DCLHb) in the treatment of hemorrhagic, hypovolemic shock. *Prehospital and Disaster Medicine*, **14:** 251–264.

Quaisi, A.M. and Roth, H.J. 1995, Investigation into the synthesis of 5-alkoxymethylsalicylic acid derivatives. *Bulletin of the Faculty of Pharmacy (Cairo University)*, **33:** 13–16.

Quirino, B.F., Noh, Y.S., Himelblau, E. and Amasino, R.M. 2000, Molecular aspects of leaf senescence. *Trends in Plant Science*, **5:** 278–282.

Rabinowitz, I.N., Wolf, P.L. and Berman, S. 1974, Light-scattering studies of retardation of sickling by aspirin-like drugs. *Research Communications in Chemical Pathology and Pharmacology*, **8:** 417–420.

Rainsford, K.D. 1977, The comparative gastric ulcerogenic activities of non-steroidal anti-inflammatory drugs. *Agents and Actions*, **7:** 573–577.

Rainsford, K.D. 1978, Structure-activity relationships of non-steroidal anti-inflammatory drugs I. Gastric ulcerogenic activity. *Agents and Actions*, **8:** 587–605.

Rainsford, K.D. and Buchanan, W.W. 1993, Sulphasalazine and aminosalicylates in rheumatoid and related arthropathies. *Inflammopharmacology*, **2:** 323–330.

Rainsford, K.D., Ford, N.L.V. and Watson, H.M. 1979, Unpublished studies.

Rainsford, K.D. Schweitzer, A., Green, P., Whitehouse, M.W. and Brune, K. 1980, Bio-distribution in rats of some salicylates with low gastric ulcerogenicity. *Agents and Actions*, **10:** 457–464.

Rainsford, K.D. and Whitehouse, M.W. 1976, Gastric irritancy of aspirin and its congeners: anti-inflammatory activity without this side-effect. *Journal of Pharmacy and Pharmacology*, **28:** 599–601.

Rainsford, K.D. and Whitehouse, M.W. 1980a, Anti-inflammatory anti-pyretic salicylic acid esters with low gastric ulcerogenic activity. *Agents and Actions*, **10:** 451–456.

Rainsford, K.D. and Whitehouse, M.W. 1980b. Are all aspirins really alike? A comparison of gastric ulcerogenicity with bio-efficacy in rats. *Pharmacological Research Communications*, **12:** 85–95.

Raskin, I. 1992, Role of salicylic acid in plants. *Annual Review of Plant Physiology and Plant Molecular Biology*, **43:** 439–463.

Ratledge, C. and Hall, M.J. 1970, Uptake of salicylic acid into mycobactin S by growing cells of *Mycobacterium smegmatis*. *FEBS Letters*, **10:** 309–312.

Ratledge, C. and Hall, M.J. 1971, Influence of metal ions on the formation of *Mycobacterium smegmatis* grown in static culture. *Journal of Bacteriology*, **108:** 314–319.

Ratledge, C. and Hall, M.J. 1972, Isolation and properties of auxotrophic mutants of *Mycobacterium smegmatis* requiring either salicylic or mycobactin. *Journal of General Microbiology*, **72:** 143–150.

Ratledge, C., Macham, P., Brown, K.A. and Marshall, B.J. 1974, Ion transport in *Mycobacterium smegmatis*: a restricted role for salicylic acid in the extracellular environment. *Biochimica et Biophysica Acta*, **372:** 39–51.

Ratledge, C. and Marshall, B.J. 1971, Ion transplant in *Mycobacterium Smegmatis* and the role of mycobactin. *Biochimica et Biophysica Acta*, **279:** 58–74.

Razzak, K.S.A. 1979, Synthesis of phenyl-3- and phenyl-5-acetamidosalicylates are potential analgesics. *Journal of Pharmaceutical Sciences*, **68:** 893–896.

Reepmeyer, J.C. 1983, Thermal decomposition of aspirin: formation of linear oligomeric salicylate esters. *Journal of Pharmaceutical Sciences*, **72:** 322–323.

Reid, J., Watson, R.D., Cockran, J.B. and Sproull, D.H. 1951, Sodium γ-resorcylate in rheumatic fever. *British Medical Journal*, **2:** 321–325.

Robertson, A. 1964, Amino phenol derivatives. British Patent No. 1,101,747. Also US patent No. 3,431,293, March 4, 1969.

Roediger, W., Schapel, G., Lawson, M., Radcliffe, B. and Nance, S. 1986, Effect of 5-aminosalicylic acid (5-ASA) and other salicylates on short chain fat metabolism in the colonic mucosa. Pharmacological implications for ulcerative colitis. *Biochemical Pharmacology*, **35:** 221–225.

Rosner, I., Malhie, P. and Mottot, G. 1968, Pharmacologie du-4-(acétamido-) phényl 2-acétoxybenzoate (Win-11 450). *Thérapie*, **23:** 525–534.

Roth, G.J., Stanford, N. and Majerus, P.W. 1975, Acetylation of prostaglandin synthase by aspirin. *Proceedings of the National Academy of Sciences of the USA*, **72:** 3073–3076.

Saifullina, N.A. and Kozhina, I.S. 1975, Composition of the essential oils from flowers of *Filipendula ulmaria*, *F. denuta* and *F. stepposa*. *Rastitel'nye Resursy*, **11:** 542–544.

Sasaki, T., Ogawa, K., Ishii, S. and Senda, M. 1999, Synthesis of [11C]salicylic acid and related compounds and their biodistribution in mice. *Applied Radiation Isotopes*, **50:** 905–909.

Scherrer, R.A. 1974. Aryl- and heteroarylalkanoic acids and related compounds. *In:* R.A. Scherrer and M.W. Whitehouse (eds), *Anti-inflammatory Agents*, Vol. 1, pp. 45–89. New York: Academic Press.

Schulzel-Lefert, P. and Vogel, J. 2000, Closing the ranks to attack by powdery mildew. *Trends in Plant Science*, **5:** 343–348.

Scrip. 2001, Feb. 9, No. 2616, p. 23.

Shahrokhian, S., Amini, M.K., Kia, R. and Tangestaninejad, S. 2000, Salicylate-selective electrodes based on Al(III) and Sn(IV) salophens. *Analytical Chemistry*, **72:** 956–962.

Sheng, S., Cai, M. and Song, C. 1999, Synthesis of isoamyl salicylate using sulfonated polyarylethylketone ketone (S-TEKK) resin as a catalyst. *Huaxue Shiji*, **21:** 377–378.

CHAPTER 3

Sievertsson, H. Nilsson, J.L.G. and Paalzow, L. 1970, The analgesic properties of methylsalicylic acids. *Acta Pharmaceutica Suecica*, **7:** 289–292.

Silva, H., Yoshioka, K., Dooner, H.K. and Klessig, D.F. 1999, Characterization of a new *Arabidopsis* mutant exhibiting enhanced disease resistance. *Molecular Plant–Microbe Interactions*, **12:** 1053–1063.

Sinclair, R.T.G. and Duthie, J.J.R. 1949, Salazopyrin in the treatment of rheumatoid arthritis. *Annals of the Rheumatic Diseases*, **8:** 226–231.

Sloan, E.P., Koenigsberg, M., Gens, D., Cipolle, M., Runge, J., Mallory, M.N., Rodman, Jr, G. (for the DCLHb Traumatic Hemorrhage Shock Study Group). 1999, Diaspirin cross-linked hemoglobin (DCLHb) in the treatment of severe traumatic hemorrhagic shock. *Journal of the American Medical Association*, **282:** 1857–1864.

Snyder, S.R., Welty, E.V., Walder, R.Y., Williams, L.A. and Walder, J.A. 1987, HbXL99 alpha: a hemoglobin derivative that is cross-linked between alpha subunits is useful as a blood substitute. *Proceedings of the National Academy of Sciences USA*, **84:** 7280–7284.

Sofia, R.D. 1978, Comparative antiphlogistic activity of mesaclazone, 5-chlorosalicylic acid, acetylsalicylic acid, phenylbutazone, indomethacin and hydrocortisone in various raw paw edema models. *Pharmacology*, **16:** 148–152.

Sofia, R.D., Diamantis, W., Gordon, R., Kletzkin, M., Berger, F.M., Edelson, J. Singer, H. and Douglas, J.F. 1974, Pharmacology of a new nonsteroidal anti-inflammatory agent, 7-chloro-3,3a-dihydro-2-methyl-2*H*, 9*H*-isoxazolo-(*3,2-b*)(*1,3*)-benzoxazin-9-one (W-2395). *European Journal of Pharmacology*, **26:** 51–62.

Sofia, R.D., Diamantis, W. and Ludwig, B.J. 1975, Comparative anti-inflammatory, analgesic and antipyretic activities of 7-chloro-3,3a-dihydro-2-methyl-2*H*, 9*H*-isoxazolo-(*3,2-b*)(*1,3*)-benzoxazin-9-one and 5 chlorosalicylic acid in rats. *Journal of Pharmaceutical Sciences*, **64:** 1321–1324.

Some, I.T., Bogaerts, O., Hanus, R., Hanocq, M. and Dubois, J. 2000, Improved kinetic parameter estimation in pH-profile data treatment. *International Journal of Pharmacy* 30, **198:** 39–49.

Somers, E., Keijers, V., Ptacek, D., Halvorsen Ottoy, M., Vanderleyden, J. and Faur, D. 2000, The salCAB operon of *Azospirillium irakense*, required for growth on salicin, is repressed by SalR, a transcriptional regulator that belongs to the Lacl/GalR family. *Molecular and General Genetics (Berlin)*, **263:** 1038–1046.

Sorenson, J.R.J. 1982, Copper complexes as the active metabolites of antiinflammatory agents. *In:* J.R.J. Sorenson (ed.), *Inflammatory Diseases and Copper. The Metabolic and Therapeutic Roles of Copper and Other Essential Metalloelements in Humans*. New Jersey: Humana Press.

Srivastava, M.K. and Dwivedi, U.N. 2000, Delayed ripening of banana fruit by salicylic acid. *Plant Science*, **158:** 87–96.

Stafford, W.L. 1962, The binding by bovine plasma and plasina fractions of salicylic acid and some of its 3-alkyl analogues. *Biochemical Pharmacology*, **11:** 685–692.

Starodubtseva, V.P., Kharabeva L.G. and Barbakadze. T.P. 1971, Effect of ecological factors on the composition of the essential oils of feijoa fruits. *Sub-tropicheskie Kul'tury*, **1:** 105–109.

Stockman, R. 1912, The therapeutic action of the cresotinic acids. *Journal of Pharmacology and Experimental Therapeutics*, **4:** 97–108.

Stone, C.A., Van Arman, C.G., Lotti, V.J., Minsker, D.H., Risley, E.A., Bagdon, W.J., Bokelman, D.L., Jensen, R.D., Mendlowski, B., Tate, C.L., Peck, H.M., Zwickley, R.E. and McKinney, S.E. 1977, Pharmacology and toxicology of diflunisal. *British Journal of Clinical Pharmacology*, **4(Suppl. 1):** 19S–29S.

Stubbe, L.Th.F.L., Pietersen, J.H. and Van Heulen, C. 1962, Aspirin preparations and their noxious effect on the gastrointestinal tract. *British Medical Journal*, **1:** 675–680.

Summers, M.P., Carless, J.E. and Enever, R.P. 1970, Polymorphism of aspirin. *Journal of Pharmacy and Pharmacology*, **22:** 615–616.

Sunkel, C., Cillero, F., Armijo, M., Pino, M. and Alonso, S. 1978, Synthesis and pharmacological properties of eterylate, a new derivative of acetyl-salicylic acid. *Arzneimittel Forschung*, **28:** 1692–1694.

Suranyi, T.M., Djurendic, E.A. and Vastag, G.G. 1995, Spectrophotometric study of the formation of iron(III) complexes with some salicylic-acid derivatives. *Collection of Czechoslovak Chemical Communications*, **60:** 464–472.

Suzuki, K., Asao, E., Nakamura, Y., Nakamura, M., Ohnishi, K. and Fukuda, S. 2000, Overexpression of salicylate hydroxylase and the crucial role of Lys[163] as its NADH binding site. *Journal of Biochemistry*, **128:** 293–299.

Svartz, N. 1941, The treatment of rheumatic polyarthritis with acidic azo compounds. *Rheumatism*, **4:** 56–60.

Svartz, N. 1942, Salazopyrin, a new sulfanilamide preparation. *Acta Medica Scandanavica*, **60:** 577–598.

Svartz, N. 1948, Elt nyft sulfonamidpreparat. Förelö pande meddelande. *Nordisk Medicin (Stockholm)*, **9:** 554.

Svartz, N. and Kallner, S. 1940, Fortsatta undersökningar över biverkningar vid behandling med sulfonamidpreparat samt en Kort redogörelse för nya terapeutiska försök. *Nordish Medicin (Stockholm)*, **45:** 1935–1940.

Swintowsky, J.V., Foster, T.S., Truelove, J.E. and Mizuno, N. 1984, A preliminary investigation of the kinetics and tolerance of the hexylcarbonate prodrug of salicylic acid in man. *Current Therapeutic Research*, **36:** 792–796.

Tagliaro, F., Cuzzolin, L., Adami, A., Scarcella, D., Crivellente, F. and Benoni, G. 1997, Pharmacokinetics of a new nitroderivative of acetylsalicylic acid after a single dose in rats. *Life Sciences*, **60:** 101–106.

Tam, S.W., Saha, J.K., Garvey, D.S., Schroeder, J.D., Shelekhin, T.E., Janero, D.R., Chen, L., Glavin, A. and Letts, L.G. 2000, Nitrosothiol-based NO-donors inhibit the gastrointestinal mucosal damaging actions of NSAIDs. *Inflammopharmacology*, **8:** 81–88.

Tamaki, A., Ide, T. and Otsuka, H. 2000, Phenolic glycosides from the leaves of *Alangium platanifolium* var. *platanifolium*. *Journal of National Products*, **63:** 1417–1419.

Tamot, B.K., Khurana, J.P. and Maheshwari, S.C. 1987, Obligate requirement of salicylic acid for short-day induction of flowering in a new duckweed, *Wolffiella hyalina* 7378. *Plant Cell Physiology*, **28:** 349–353.

Tamura, S. 1977a, Studies on analgesics. III. Pharmacological action related to analgesic and anti-inflammatory activities of 5-aminomethyl-salicylic acid. *Yakugaku Zasshi*, **97:** 289–294.

Tamura, S. 1977b, Studies on analgesics. IV. The fate of 5-aminomethylsalicylic acid in rabbits and rats. *Yakugaku Zasshi*, **97:** 388–392.

Tamura, S. 1977c, Studies on analgesics. V. Mechanism of anti-inflammatory action of 5-aminomethylsalicylic acid. *Yakugaku Zasshi*, **97:** 295–303.

Tashima, K., Fujita, A., Umeda, M. and Takeuchi, K. 2000, Lack of gastric toxicity of nitric oxide-releasing aspirin, NCX-4016, in the stomach of diabetic rats. *Life Sciences*, **67:** 1639–1652.

Thara, V.K., Tang, X.Y., Gu, Y.Q., Martin, G.B. and Zhou, J.M. 1999, *Pseudomonas syringae* pv tomato induces the expression of tomato EREBP-like genes Pti4 and Pti5 independent of ethylene, salicylate and jasmonate. *Plant Journal*, **20:** 475–483.

The Arabidopsis Genome Initiative 2000, Analysis of the genome sequence of the flowering plant *Arabidopsis thaliana*. *Nature*, **408:** 796–815.

Thomma, B.P.H.J., Eggermont, K., Tierens, K.F.M.J. and Broekaert, W.F. 1999, Requirement of functional ethylene-intensive 2 gene for efficient resistance of *arabidopsis* in infection by *Botrytis cinerea*. *Plant Physiology*, **121:** 1093–1101.

Thomma, B.P.H.J., Eggermont, K., Broekaert, W.F. and Cammue, B.P.A. 2000, Disease development of several fungi on *Arabidopsis* can be reduced by treatment with methyl jasmonate. *Plant Physiology and Biochemistry*, **38:** 421–427.

Thompkins, L. and Lee, K.H. 1975, Comparison of analgesic effects of isosteric variations of salicylic acid and aspirin (acetylsalicylic acid). *Journal of Pharmaceutical Sciences*, **64:** 760–763.

Thompson, E.B. and Klotz, I.M. 1985, Pharmacological actions of dispirins, potential antisickling agents: analgesic and anti-inflammatory effects. *Research Communications in Chemical Pathology and Pharmacology*, **48:** 381–388.

Thorp, L. 1918, Process forming methyl ester of acetyl salicylic acid. US Patent No. 1,255,950. Official Gazette, p. 282, US Patent Office, Feb. 12, 1918.

Thurston, J.H., Marlier, E.M. and Whitmire, K.H. 2002, Towards a molecular model for bismuth(III) subsalicylate. Synthesis and [Bi(Hsal) (sal) (1,10-phenanthroline) (C7H8)2. *Chemical Communications (Cambridge)*, **23:** 2834–2835.

Tocco, E.L. 1912, Richerche farmacologiche sull'etere salicilsalicilico (Diphosate). *Archivio di Farmacologia Sperimentale e Science Affini*, **13:** 567–584.

Ton, J., Pieterse, C.M.J. and Van Loon, L.C. 2000, Identification of a locus in arabidopsis controlling both the expression of rhizobacteria-mediated induced systemic resistence (ISR) and basal resistance against *Pseudomonas syringae* pv tomato. *Molecular Plant–Microbe Interactions*, **12:** 911–918.

Troung, P., Mai, P.M., Tran, T.D., Nguyen, D.N., Nguyen, T.V.H. and Nguy, T.T.N. 1998, Synthesis and biological properties of chlorsalicylamide derivatives. *Tap Chi Duoc Hoc*, **5:** 8–12.

Truelove, S.C. 1988, Evolution of olsalazine. *Scandinavian Journal of Gastroenterology*, **148** (**Suppl.**): 3–6.

Tsunematsu, H., Ishida, E., Yoshida, S. and Yamamoto, M. 1991, Synthesis and enzymatic-hydrolysis of aspirin-basic amino-acid ethyl-esters. *International Journal of Pharmaceutics*, **68:** 77–86.

Umeyama, H. and Nakagawa, S. 1977, Ab initio molecular orbital studies on aspirin solvolysis and ester hydrolysis. *Chemical and Pharmacological Bulletin (Tokyo)*, **25:** 1671–1677.

Umeyama, H., Nakagawa, S. and Moriguchi, I. 1979, Molecular orbital study of the cleavage of aspirin crystals. *Journal of Physical Chemistry*, **83:** 2048–2052.

Valkonen, J.P.T. and Watanabe, K.N. 1999, Autonomous cell death, temperature sensitivity and the genetic control associated with resistance to cucumber mosaic virus (CMV) in diploid potatoes. *Theoretical and Applied Genetics*, **99:** 996–1005.

Vallelian-Bindschedler, L., Metraux, J.-P. and Schweizer, P. 1998, Salicylic acid accumulation in barley is pathogen specific but not required for defense-gene activation. *Molecular Plant–Microbe Interactions*, **11:** 702–705.

Van Wees, S.C.M., Luijendijk, M., Smoorenburg, I., Van Loon, L.C. and Pieterse, C.M.J. 1999, Rhizobacteria-mediated induced systemic resistance (ISR) in arabidopsis is not associated with a direct effect on expression of known defense-related genes but stimulates the expression of the jasmonate-inducible gene afvsp upon challenge. *Plant Molecular Biology*, **41:** 537–549.

Venanzi, S. 1954, L'acido diacetil-pirocatecol-3-carbossiliro nel trattamento delle malattie reumatiche. *Minerva Medica (Torino)*, **45:** 229–233.

Venkatappa, K., Tang, X., Gu, Y.Q., Martin, G. and Zhou, J.-M. 1999, *Pseudomonas syringae* pv tomato induces the expression of tomato EREBP-like genes Pt14 and Pt15 independent of ethylene, salicylate and jasmonate. *Plant Journal*, **20:** 467–483.

Verberne, M.C., Verpoorte, R., Bol, J.F., Mercado Blanco, J. and Linthorst, H.J.M. 2000, Overproduction of salicylic acid in plants by bacterial transgenes enhances pathogen resistance. *Nature Biotechnology*, **18:** 779–783.

Vervoort, J., Van Berkel, W.J.H., Müller, F. and Moonen, C.T.W. 1991, NMR studies on *p*-hydroxybenzoate hydroxylase from *Pseudomonas fluorescens* and salicylate hydroxylase from *Pseudomonas putida*. *European Journal of Biochemistry*, **200:** 731–738.

Walder, J.A., Zaugg, R.H., Iwaoka, R.S., Watkin, W.G. and Klotz, I.M. 1977, Alternative aspirins as antisickling agents: acetyl-3,5-dibromosalicylic acid. *Proceedings of the National Academy of Sciences of the USA*, **74:** 5499–5502.

Walford, G.L., Jones, H. and Shed, T.Y. 1971, Aza analogs of 5-(*p*-fluorophenyl) salicific acid. *Journal of Medicinal Chemistry*, **14:** 339–344.

Walker, J.R. and Smith, M.J.H. 1979, Adrenocortical stimulation and the anti-inflammatory actions of salicylates. *Journal of Pharmacy and Pharmacology*, **31:** 640–641.

Wallace, J.L., Muscara, M.N., McKnight, W., Dicay, M., Del Soldato, P. and Cirino, G. 1998, *In vivo* antithrombotic effects of a nitric oxide-releasing aspirin derivative, NCX-4016. *Thrombosis Research*, **93:** 43–50.

Wang, K., Zhang, W., Xian, M., Hou, Y.-C., Chen, X.-C., Cheng, J.-P. and Wang, P.G. 2000, New chemical and biological aspects of S-nitrosothiols. *Current Medicinal Chemistry*, **7:** 821–834.

Wangel, A. and Klockars, M. 1977, Lymphocyte subpopulations in rheumatoid synovial tissue. *Annals of the Rheumatic Diseases*, **36:** 176–180.

Weast, R.C. and Astle, M.J. (eds) 1978, *CRC Handbook of Chemistry and Physics*. 59th edn. West Palm Beach, Fl.: CRC Press Inc.

Weichert, H., Stenzel, I., Berndt, E., Wastermack, C. and Feussner, I. 1999, Metabolic profiling of oxylipins upon salicylate treatment in barley leaves – preferential inductin of the reductase pathway by salicylates. *FEBS Letters*, **464:** 133–137.

Weichert, H., Kolbe, A., Wasternack, C. and Feussner, I. 2000, Formation of 4-hydroxy-2-alkenals in barley leaves. *Biochemical Society Transactions*, **28:** 850–851.

Wheatley, P.J. 1964, The crystal and molecular structure of aspirin. *Journal of the Chemical Society*, **1964:** 6036–6048.

Whitehouse, M.W. 1965, Some biochemical and pharmacological properties of anti-inflammatory drugs. *Progress in Drug Research*, **8:** 321–429.

Whitehouse, M.W. and Rainsford, K.D. 1982. Comparison of the gastric ulcerogenic activities of different salicylates. *In:* C.J. Pfeiffer (ed.), *Drugs and Peptic Ulcer*, pp. 127–141. Boca Raton: CRC Press.

Whitehouse, M.W., Rainsford, K.D., Ardlie, N.G., Young, I.G. and Brune, K. 1977, Alternatives to aspirin, derived from biological sources. *In:* K.D. Rainsford, K. Brune and M.W. Whitehouse (eds), *Aspirin and Related Drugs: Their Actions and Uses*, **(Suppl. 1):** 43–57.

Wieth, J.O. 1970, Paradoxical temperature dependence of sodium and potassium fluxes in human red cells. *Journal of Physiology*, **207:** 563–580.

Wilder-Smith, A.E., Frommel, E. and Radouco-Thomas, S. 1963, Preliminary screening of some new oxadiazol-2-ols with special reference to their antipyretic analgesic, and anti-inflammatory properties. *Arzneimittel Forschung*, **13:** 338–341.

Williams, M.J. 1996, Blood substitutes for transfusion. *Techniques in Urology*, **2:** 179–186.

Williams, P.A., Catterall, F.A. and Murray, K. 1975, Metabolism of naphthalene, 2-methylnaphthalene, salicylate, and benzoate by *Pseudomonas* P_G: regulation of tangential pathways. *Journal of Bacteriology*, **124:** 679–685.

Williamson, W.R.N. (ed.) 1987, *Anti-inflammatory Compounds*. New York: Marcel Dekker.

Winslow, R.M. 1995, Blood substitutes – a moving target. *Nature Medicine*, **1:** 1212–1215.

Wolfe, J., Hutcheon, C.J., Higgins, V.J. and Cameron, R.K. 2000, A functional gene-for-gene interaction is required for the production of an oxidative burst in response to infection with avirulent *Pseudomonas syringae* pv. *Tomato* in *Arabidopsis thaliana*. *Physiological and Molecular Plant Pathology*, **56:** 253–261.

Wood, P.H.N., Harvey-Smith, E.A. and Dixon, A.St.J. 1962, Salicylates and gastrointestinal bleeding. Acetylsalicylic acid and aspirin derivatives. *British Medical Journal*, **1:** 669–675.

Yang, N., Sheng, Q., Yuan, X., Xiao, M., Huang, Y. and Li Zhiliang, 1998, Study on method for synthesis of eterylate. *Jishou Daxue Xuebao, Ziran Kexueban*, **19:** 86–87.

Young, I.G., Cox, G.B. and Gibson, F. 1967, 2,3-Dihydroxybenzoate as a bacterial growth factor and its route of biosynthesis. *Biochimica et Biophysica Acta*, **141:** 319–331.

Yu, I.-C., Fengler, K.A., Clough, S.J. and Bent, A.F. 2000, Identification of *Arabidopsis* mutants exhibiting an altered hypersensitive response in gene-for-gene disease resistance. *Molecular Plant–Microbe Interactions*, **13:** 277–286.

Yung-Yu, H.D. and Roberts, M.S. 1998a, Anti-platelet and hydroxy radical-radical scavenging diflunisal esters and related compounds. Patent PCT International WO9846234 14 April 1998. *Chemical Abstracts*, **129 (24):** 310909e.

Yung-Yu, H.D. and Roberts, M.S. 1998b, Esters of salicylates. Patent: PCT International; WO 9846234 A1, Date: 1998. Application No. WO 98AU260 (19980414) *AU 976123 (19970411). CA: **129 (24):** 310909e.

Zhang, J. and Sun, Z. 1995, Synthesis of butyl p-hydroxybenzoate by $FeCl_3$ catalysis. *Xiandai Huagong*, **15:** 33–34.

Zhang, J., Chen, T., Huag, T. and Lai, Y. 1999, Synthesis of isoamyl 2-hydroxybenzoate by heteropoly-acid catalysis. *Xiamen Daxue Xuebao, Ziran Kexueban*, **38:** 573–577.

Zhu, J.-L., Qin, W.-R. and Duyang, P.K. 1994, Ultrasound in synthesis of 3,5-diisopropyl-salicylic acid. *Nanjing Huagong daxue Xubao, Ziron Kexueban*, **23:** 71–73.

Pharmacokinetics and Metabolism of the Salicylates

G.G. Graham, M.S. Roberts, R.O. Day and K.D. Rainsford

INTRODUCTION

Many aspects of the handling of the salicylates by the body are rarely found with other drugs. These unusual features include the covalent binding of metabolites of aspirin and salicylate to tissues, the autoinduction of the metabolism of salicylate, and the saturable nature of both the protein binding and metabolism of salicylate. Despite these complexities of the disposition of the salicylates, they have been frequently used as model drugs because of the ease of assay of salicylate itself and the acceptability of large doses to volunteers. Although there has been a large amount of research on the salicylates, uncertainties still remain about the details and the clinical significance of several aspects of their absorption, distribution and elimination.

The disposition of the salicylates in animals is also of considerable interest. Not only are they used in veterinary practice, but both are also widely used in experimental animals in investigational studies to establish the principles of pharmacokinetics. Aspirin and salicylate are also widely used in pharmacological studies in animals, often as controls for other anti-inflammatory drugs. Knowledge of the absorption, distribution and elimination of these drugs in animals assists the design and interpretation of such studies.

The disposition of both aspirin and salicylate in the body is considered together in this chapter since the two are closely interrelated through the rapid conversion of aspirin to salicylate and the importance of salicylate in the pharmacological effects of aspirin. It should be noted, however, that salicylate should not always be considered as simply a metabolite of aspirin. Salicylate has been administered as a salt with sodium and a double salt with choline and magnesium, while several other pro-drugs of salicylate, such as salsalate (salicylsalicylic acid or disalcid) and benorylate, have been used. These and other pro-drugs, as well as some salicylate derivatives, are discussed separately in later sections of this chapter.

ABSORPTION OF ASPIRIN AND SALICYLATE

Aspirin is quite lipid-soluble in the un-ionised form with a log P value (logarithm of partition coefficient of the un-ionised form between octanol and water) of 1.19, while salicylic acid is much more lipid-soluble with a log P value of 2.26 (Drayton, 1990). Given that the pK_a values of aspirin and salicylic acid are 3.5 and 2.97, respectively, the un-ionised species are major forms only in the stomach and in the upper small intestine.

Once in solution, both aspirin and salicylate are totally absorbed from the gastrointestinal tract although several factors influence the rate of absorption of both aspirin and salicylate. The effective absorption of aspirin is, however, incomplete due to first-pass metabolism in the liver. Its effective absorption from solution is about 50 to 70 per cent in man, and is constant over a wide range of doses (Rowland *et al.*, 1972; Pedersen and Fitzgerald, 1984).

Various factors affect the rate of absorption of aspirin and salicylate. These include the physico-chemical properties of the compounds, the pH of the gastrointestinal lumen, the surface area of the tract, the rate of gastric emptying, and intestinal transit times. The rate and extent of absorption is also greatly affected by the pharmaceutical formulation, which, together with the pH of the immediate environment, controls the dissolution of the salicylates within the gastrointestinal tract,

The gastric absorption of aspirin is limited, despite the high proportion of the un-ionised aspirin present, its absorption being restricted by the surface area of the mucosa. Thus, only about 12 per cent of the mass of aspirin is absorbed from an unbuffered solution after 10 minutes in the stomach (Cooke and Hunt, 1970), and the extent of absorption decreases to about 1 per cent if gastric pH is increased to above 6, because most of the aspirin is then in the less permeable ionised form. However, the slower gastric absorption obtained after increasing the pH of gastric contents is not reflected in slower overall absorption *in vivo*, since buffered solutions of aspirin are quickly emptied into the small intestine, where absorption is rapid (Dotevall and Ekenved, 1976)

On an empty stomach, solutions of aspirin salts are absorbed quite quickly in man with a half-life of absorption ranging from 5 to 16 minutes (Rowland *et al.*, 1972). With a short half-life of elimination, the rate of absorption of aspirin profoundly affects the pattern of plasma concentrations after its oral administration. In particular, the peak levels of aspirin are markedly reduced with slowing the rate of absorption (Bochner *et al.*, 1988), which occurs particularly if aspirin is administered with meals, or as sustained release or enteric-coated preparations. For acute conditions, such as acute pain or as an acute antiplatelet agent after myocardial infarction, aspirin should be administered in solution or by chewing buffered tablets (Feldman and Cryer, 1999).

Salts of salicylic acid, such as sodium salicylate and choline magnesium salicylate, are now little used in clinical practice, but the oral absorption of salicylate has been widely studied, particularly as a model drug in early studies on the mechanism of the gastrointestinal absorption of drugs. Early studies showed that the rate of absorption of salicylate decreases with increasing pH in the lumen of small intestine, indicating that the un-ionised species diffuses more rapidly through the intestinal mucosa than the ionised form (Hogben *et al.*, 1959; Doluisio *et al.*, 1969). A similar dependence upon pH is seen in the stomach, although the rate of absorption is slower. While the rate of absorption of salicylate decreases with increasing pH, the fall is not as great as may be expected from the changing fraction un-ionised. Either the bulk pH does not reflect the pH at the membrane surface, or there is also diffusion of the salicylate anion (Hogben *et al.*, 1959; Doluisio *et al.*, 1969; O'Driscoll and Corrigan, 1983). Being somewhat less lipid-soluble than salicylic acid, aspirin is more slowly absorbed from the isolated intestine of the rat (Hogben *et al.*, 1959), although the absorption of aspirin is still very rapid in man.

Presystemic metabolism

Aspirin is stable in gastric and duodenal fluids, and is therefore absorbed by the gastrointestinal tract as unchanged aspirin. While esterases are present in the gut wall and liver (see p. 107), the major site of presystemic metabolism of aspirin in man is in the liver (Rowland *et al.*, 1972).

In the dog, there is presystemic metabolism of aspirin in the gastrointestinal tract but somewhat more in the liver. This is shown by comparison of the oral availability (45 per cent) after oral dosage compared to 64 per cent after infusion into the hepatic portal vein (Harris and Riegelman, 1969). The availability is about 75 per cent after infusion into the portal vein in the sheep (Cossum *et al.*, 1986). Only about one-quarter of an oral dose of aspirin is absorbed intact in the rat, but in this species the major site of first-pass metabolism is the gastrointestinal tract (Wientjes and Levy, 1988). Extraction by the liver is also important in the rat, with from 20 to 50 per cent being extracted by the liver (Wientjes and Levy, 1988; Mellick and Roberts, 1992). The extraction decreases with increasing perfusion rate, indicating at least one cause of variation in the extent of presystemic metabolism of aspirin. *In vivo* in the rat there is uptake of aspirin and salicylate into the stomach mucosa, with the acetyl moiety of aspirin binding covalently to proteins and other molecules in the stomach wall, indicating some presystemic metabolism in the stomach in this species (Morris *et al.*, 1973; Rainsford *et al.*, 1983). This gastric metabolism of aspirin may not be significant in its overall metabolism and absorption but, as discussed below, is consistent with its gastric toxicity.

During the absorption phase, the blood concentrations of aspirin should not be consistent in all blood vessels. The concentration of aspirin must be highest in the portal circulation due to absorption of drug from the small intestine. The concentration of aspirin should then be decreased, in order, by dilution by blood from the hepatic artery, by first-pass metabolism in the liver, by blood in the inferior vena cava, and then in the right atrium. The consequence is that the rate of reaction of aspirin with cyclo-oxygenase may be greatest with platelets in the portal circulation. This dilution effect may be the reason why antiplatelet effects of aspirin are seen before the drug is detectable in the systemic circulation (Pedersen and FitzGerald, 1984), and may be responsible, in part, for the selectivity of aspirin for platelets; aspirin having a predominant effect on the cyclo-oxygenase of platelets with much lesser influence on endothelial cyclo-oxygenase responsible for the synthesis of the antiplatelet mediator, prostacyclin. The acute dilution effect is decreased with a sustained release preparation of aspirin (Bochner *et al.*, 1989), but even sustained release preparations of aspirin show marked selectivity for platelets (Vial *et al.*, 1990). Selectivity for platelets in the portal circulation may even be greater for analogues of aspirin with an acyl group larger than the acetyl group (see section on O-acyl derivatives of salicylate).

The absorption of unchanged aspirin may also be a significant aspect of its analgesic activity because the analgesic activity of aspirin increases with increasing availability of intact aspirin (Seymour *et al.*, 1984). This interesting finding requires confirmation, however.

Enteric-coated formulations

Many enteric formulations of aspirin have been prepared in order to decrease its upper gastrointestinal toxicity. Protection of the gastrointestinal tract is, however, only partial. Enteric coatings are applied to whole tablets or to granules that are presented in capsules. In the past the gastrointestinal absorption of such formulations was inconsistent and often incomplete, but modern coatings appear to provide more reliable release of aspirin. The rate of absorption of enteric-coated tablets of aspirin is still variable, largely due to the retention of intact tablets in the stomach, because absorption only occurs after passage of the enteric-coated tablets into the small intestine. Considerable numbers of intact enteric-coated tablets have been recovered from patients with pyloric obstruction (Bogacz and Caldron, 1987). Such patients should not be given enteric-coated tablets of aspirin or any other drug. Capsules containing enteric-coated granules of drug are safer in this condition.

Localisation in the stomach

Gastrointestinal discomfort and damage are clinically significant side effects of aspirin and other non-steroidal anti-inflammatory drugs. As is well recognised, the therapeutic and side effects of drugs are dependent upon their uptake or binding at sites of action. With regard to the gastrointestinal toxicity of the salicylates and related drugs, it is of note that salicylate is concentrated in the parietal cells of the stomach and it has been suggested that the high concentrations of aspirin and salicylate in the parietal cell may initiate damage to the gastric mucosa (Martin, 1963; Rainsford and Brune, 1976). The localisation of aspirin and salicylate in the parietal cells is due to the high pH gradient at the parietal cell

CHAPTER 4

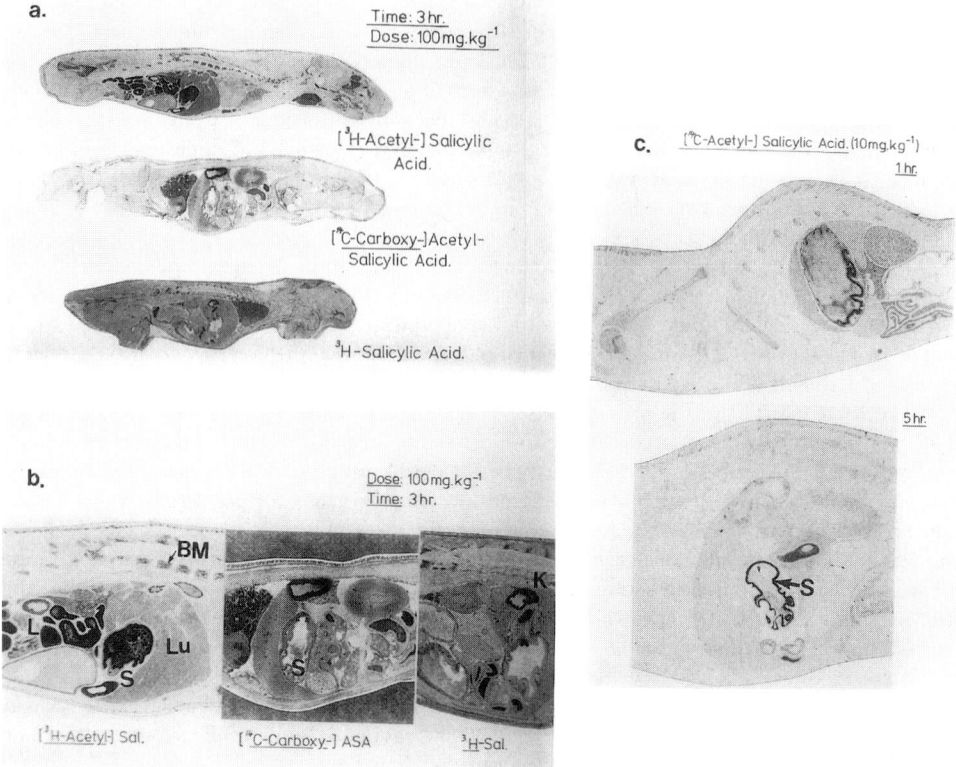

Figure 4.1 Autoradiographs of rats treated with ^{3}H-aspirin. Lighter areas indicate the more intense areas of radioactivity. The high uptake of label in the liver (Li) and kidney (K) is probably related to the metabolism and renal excretion of aspirin. Note the higher uptake of label in an inflamed paw than in a non-inflamed paw, and the high uptake in the stomach. (From Brune *et al.*, 1980.)

wall. In fact, the highest pH gradient in the body occurs adjacent to the parietal cell. At this cell, the exterior gastric pH is highly acidic and the interior about neutral. If the cell membrane is permeable predominantly to the un-ionised form of these acidic drugs, then the pH partition hypothesis predicts that high concentrations of the drugs accumulate in the parietal cells, as is the case (Figure 4.1; Rainsford *et al.*, 1981; 1983).

Percutaneous absorption

Salicylate can be applied to the skin in three forms: as un-ionised salicylic acid, as a salicylate salt (most commonly triethanolamine salicylate), as esters such as methyl salicylate (see p. 125) and, rarely, as aspirin.

Salicylic acid is often applied topically. Absorption appears to be slow and dependent on the concentration and formulation, but ultimately quite high proportions are absorbed. After single application to human skin, the percentage absorbed ranges from about 20 per cent when the skin is uncovered (Feldman and Maibach, 1970), to about 60 per cent after application in propylene glycol/ethanol and occlusion (Taylor and Halprin, 1975). In rats, salicylic acid penetrates only to a depth of 3 to 4 mm. The drug in deeper tissues is derived predominantly from the systemic absorption of the drug, particularly at longer times after application (Singh and Roberts, 1993a).

Low concentrations of salicylic acid (1 to 2 per cent) in ointments are often applied for a keratoplastic effect (slowing skin turnover), while higher concentrations (4 to 6 per cent) have keratolytic activity. Over a period of 3 days, applications of the higher concentrations increase the rate of uptake of salicylic acid but there is a subsequent decline (Roberts and Horlock, 1978).

Because of the quite high absorption of salicylate, toxicity must be anticipated after the topical application of large amounts of salicylic acid, particularly if the skin is damaged. Toxicity has been reported from salicylic acid after its topical application to large areas of skin of patients with psoriasis and icthyosis (Davies *et al.*, 1979). Even death has occurred due to high cutaneous exposure in psoriasis (von Weiss and Lever, 1964). The topical application of very large amounts of salicylic acid requires close clinical monitoring, including the measurement of the plasma concentrations of salicylate.

Aspirin is also absorbed slowly from the skin (Rougier *et al.*, 1987), and it appears to be usefully applied to the skin in the treatment of herpes zoster and post-herpetic neuralgia, although further evaluation is required (Bareggi *et al.*, 1998). The very low levels of aspirin in blood (about $2\,\mu M$) resulting from cutaneous application are still sufficient to reduce prostaglandin synthesis in the gastrointestinal tract with consequent gastric damage (Cryer *et al.*, 1999).

The percutaneous absorption of a salicylate salt (triethanolamine salicylate) is only about 1 per cent, and the concentrations in the dermis and subcutaneous tissue are much lower than after application of methyl salicylate (Morra *et al.*, 1996; Cross *et al.*, 1997; 1998).

The diffusion of salicylate salts from skin into synovial fluid has only been studied in the knee, where the concentrations are very low (Rabinowitz *et al.*, 1982) and appear far too low for significant anti-inflammatory effect. Studies on bilateral effusions conducted with topical preparations of other non-steroidal anti-inflammatory drugs, such as diclofenac (Radermacher *et al.*, 1991), indicate that topical NSAIDs largely gain access to synovial fluid of the knee from the systemic blood supply, rather than from diffusion from overlying skin. Overall, direct diffusion of salicylate and other non-steroidal anti-inflammatory drugs from skin to synovial fluid appears insignificant, although studies in joints other than the knee are still required. Preparations containing these drugs may, however, be useful for soft tissue rheumatism, which is frequently superficial.

Iontophoresis enhances the rate of permeation of salicylate through the stratum corneum, with maximal penetration occurring at higher pH values where salicylate is in the ionised form. The extent of penetration into deep tissues after iontophoresis is essentially the same as that obtained after application of salicylate solutions to the dermis (Singh and Roberts, 1993b).

DISTRIBUTION OF ASPIRIN AND SALICYLATE

Aspirin and salicylate distribute widely throughout the body, although their volumes of distribution during the elimination phase are only about 10 litres after low doses in adults (Rowland and Riegelman, 1968). Although aspirin and salicylate are bound to plasma proteins to a lesser extent than other non-steroidal anti-inflammatory drugs, their low volumes of distribution indicate that binding to tissue constituents is still lower than to plasma proteins. The binding of salicylate to plasma albumin has been examined in considerable detail, and many of the principles of drug distribution have been developed from studies on salicylate.

Binding to plasma proteins

Methodology

As is commonly found, the measured binding of both aspirin and salicylate may differ according to the method used. Higher unbound fractions are recorded after ultracentrifugation than after equilibrium dialysis (Verbeeck and Cardinal, 1985), although ultrafiltration and equilibrium dialysis give similar binding results (Spector *et al.*, 1972). *In vivo* ultrafiltration also gives similar results to those from *in vitro* ultrafiltration (Ghahramani *et al.*, 1998). The former is an indirect method based on the changed plasma concentrations of a drug when the plasma concentration of protein is increased after obstruction of a vein by a sphygmomanometer cuff.

As with other acidic drugs, albumin is the principal protein binding aspirin and salicylate in plasma. There have been many studies on the binding of salicylate to purified albumin, but a problem is the contrasting results with different batches of albumin (Aarons *et al.*, 1980). Much of this variation may be due to differing amounts of fatty acids, which displace salicylate from binding to albumin (Ashton *et al.*, 1989). Fatty acids may also decrease the binding of aspirin to albumin. The binding of salicylate to albumin decreases slightly with increasing pH (Moran and Walker, 1968) but quite markedly with increasing temperature (Zarolinski *et al.*, 1974), and, as is the case with other drugs, it is important to control the temperature in studies on the binding of salicylate if the results are to be correlated with conditions *in vivo*. An aspect of protein-binding methodology that is of surprisingly little concern is the volume of ultrafiltrate that is collected when this technique is used. Large volumes of plasma may be ultrafiltered without altering the concentration of salicylate in the filtrate (Whitlam and Brown, 1981).

Binding of aspirin to plasma proteins

Unchanged aspirin binds both irreversibly and reversibly to plasma proteins. Irreversible binding involves the transfer of the acetyl group to bind covalently to the plasma proteins (Pinckard *et al.*, 1970; see p. 108). The precise extent of reversible binding of aspirin to plasma albumin is difficult to determine, because of both the covalent binding and hydrolysis to salicylate, but the reversible binding of aspirin is about 60 per cent (Ghahramani *et al.*, 1998), with the bulk of the binding being to albumin.

Binding of salicylate to plasma proteins

Approximately 2 to 10 per cent of salicylate is free in plasma at low concentrations (less than 100 mg/l) in man (Bochner *et al.*, 1981; Shen *et al.*, 1991a). The unbound fraction increases approximately linearly throughout the therapeutic range, reaching about 25 per cent at 300 mg/l (Figure 4.2), which is often considered as the upper limit of the therapeutic range of plasma concentrations in the treatment of rheumatoid arthritis. This corresponds to a maximal unbound concentration of salicylate in plasma of 0.55 mM. Even higher concentrations may be unbound after overdoses (Alvan *et al.*, 1981).

Although most of the binding of salicylate is due to albumin (Reynolds and Cluff, 1960), the unbound proportion of salicylate in plasma is greater than in solutions of purified albumin (Costello *et al.*, 1982). Salicylate binds to multiple sites on serum albumin, but only one site need be considered at therapeutic doses. In a comprehensive study of the pharmacokinetics of salicylate, binding to only one site was sufficient to model the kinetics of distribution and elimination of the drug (Shen *et al.*, 1991a).

Apart from the concentrations of salicylate, several other factors influence its binding to plasma proteins. Most importantly, decreasing concentrations of albumin are associated with increased unbound fractions in plasma (Reynolds and Cluff, 1960; Yacobi and Levy, 1977). A variable level of albumin is the major cause of the intersubject differences in the protein binding in the plasma of healthy subjects (Yacobi and Levy, 1977) and in patients with chronic liver disease and chronic respiratory insufficiency (Perez-Mateo and Erill, 1977), while the lower binding of salicylate in pregnancy is associated with a lower concentration of plasma albumin (Yoshikawa *et al.*, 1984a). Due to the decreased serum albumin and increased fatty acids, the free fraction is increased four-fold in kwashiorkor (Ashton *et al.*, 1989). The fraction unbound is also generally increased by single mutations in various amino acids of albumin (Kragh-Hansen *et al.*, 1990). Some elderly patients may also have low protein binding of salicylate.

The binding of salicylate to plasma albumin is decreased in various acute infections (Reynolds and Cluff, 1960) and, along with other acidic drugs, by uraemia due to the accumulation of endogenous

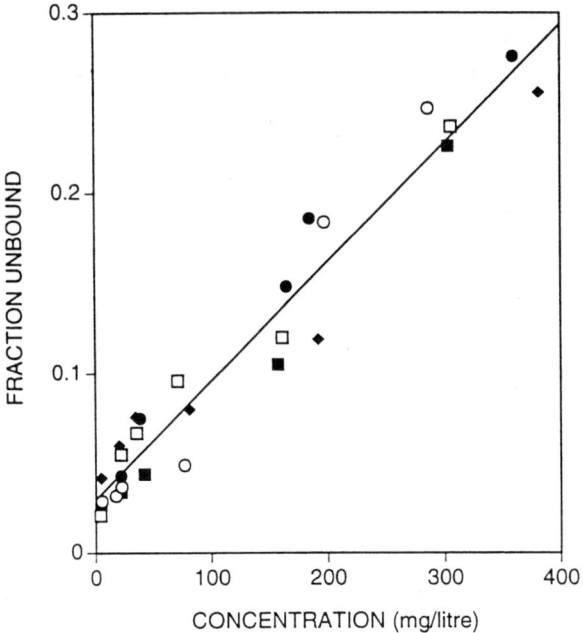

Figure 4.2 The linear relationship between the unbound fraction and the total plasma concentrations of salicylate in healthy subjects. The varying symbols refer to the different subjects in the study. The relationship between the unbound fraction and total plasma concentrations is non-linear. (Redrawn from Shen *et al.*, 1991a.)

■

CHAPTER 4

■

displacing agents, possibly furancarboxylic acids or indoxyl sulphate (Niwa *et al.*, 1988a; 1988b), both of which are strongly bound to plasma proteins.

Variable results have been presented on the plasma binding in patients with rheumatoid arthritis. Normal binding in rheumatoid patients without liver or renal disease has been reported (Gurwich *et al.*, 1984), while decreased binding has been associated with increasing severity of the disease (Netter *et al.*, 1984).

There is a marked species-dependence in the binding of salicylate to serum proteins, with high binding in man, rhesus monkey, rabbit and guinea pig, while several other species, including the rat, mouse and dog, have much lower binding (Sturman and Smith, 1967).

When salicylate binds to plasma albumin there is an associated displacement of some drugs, particularly other non-steroidal anti-inflammatory drugs. The displacement of another drug or metabolite does not usually lead to potentiation of the displaced compound because the unbound concentration remains essentially unchanged (Sellers, 1979; see p. 120). The interpretation of the total plasma concentration is, however, altered in order to take the higher unbound proportion into account. Endogenous substances such as tryptophan are also displaced by salicylate (McArthur *et al.*, 1971). The displacement of endogenous tryptophan has been linked to some of the pharmacological effects of salicylate (McArthur *et al.*, 1971; Badaway, 1982), but definitive evidence is lacking.

Tissue uptake of aspirin and salicylate

Volume of distribution

Salicylate diffuses into tissues, but the extent of tissue uptake is limited by its binding to plasma albumin. This is indicated by the increasing volume of distribution with decreasing albumin concentrations in man

(Ho *et al.*, 1985) and also in rats, where there is a greater volume of distribution and tissue uptake of salicylate in hypo-albuminaemic than in normal animals (Hirate *et al.*, 1989). Because of the saturable binding of salicylate to plasma proteins, the volume of distribution of salicylate increases with increasing body load of salicylate in children (Levy and Yaffe, 1974). In adults the volume of distribution is about 10 litres at about 50 mg/l (Rowland and Riegelman, 1968), but increases to about 16 litres (Rubin *et al.*, 1983) at a total plasma concentration of 300 mg/l, the approximate upper limit of the therapeutic range.

According to a model of drug distribution in the body there is slight overall binding of salicylate to tissues (Rubin *et al.*, 1983) and, consistent with this kinetic analysis, salicylate binds to constituents of skin (Walter and Kurz, 1988), liver and kidney, but not to brain, heart or skeletal muscle (McArthur *et al.*, 1970).

Tissue uptake and effects of the aspirin and salicylate

Uptake of any drug at its site of action is clearly a prerequisite for its therapeutic or toxic effects, and with aspirin and similar drugs their anti-inflammatory effect is consistent with their more marked uptake in inflamed than in non-inflamed joints of the rat (Brune *et al.*, 1980). Their toxic effects on the liver and kidney have also been related to their uptake in these tissues (Brune *et al.*, 1980; Rainsford *et al.*, 1983; Figure 4.1). Both aspirin and salicylate also permeate into the eye and persist in aqueous and vitreous humours for a longer time than in plasma (Valeri *et al.*, 1989). The presence of significant concentrations of aspirin may be responsible for its reported prevention of cataract formation through local acetylation of lens protein or other constituents of the eye (Rao *et al.*, 1985).

Kinetics of tissue uptake

While salicylates are widely distributed, uptake into tissues may be slow. For example, the concentrations of salicylate in perilymph (Jastreboff *et al.*, 1986), ascitic fluid and cells (Raghoebar *et al.*, 1987), muscle (Chen *et al.*, 1978), lymph (Sudo *et al.*, 1989), synovial fluid and inflammatory exudates (see p. 106) generally increase slowly after dosage, indicating slow transfer of salicylate into and out of the tissues (Graham, 1988). The slow transfer of salicylate into the brain (Brodie *et al.*, 1960; Chen *et al.*, 1978; Bannwarth *et al.*, 1986; Jastreboff *et al.*, 1986) is of particular note, with the half-life of permeation of salicylate from plasma to cerebrospinal fluid being nearly 2 hours in the dog (Brodie *et al.*, 1960). This slow uptake is probably due first to the high degree of ionisation and consequent low lipid solubility of salicylate at physiological pH, limiting to the diffusion of salicylate across the blood–brain barrier. The rate and extent of uptake of salicylate into the central nervous system is also limited by the binding to plasma albumin (Reed and Palmisano, 1975). Slow tissue uptake may account in part for the delayed toxicity of salicylate after overdoses. Slow onset of central analgesic and antipyretic effects are also predicted. By contrast to the slow uptake into several tissues, the uptake of salicylate into the liver is rapid (Chen *et al.*, 1978) and does not limit its rate of metabolism.

pH and uptake by cells

Salicylate is more than 99.99 per cent ionised at physiological pH values, but appears to diffuse through cell membranes mainly in the un-ionised form, salicylic acid. There may be some intestinal absorption of salicylate in the ionised form, but the increasing excretion with increasing urinary pH indicates resorption mainly in the un-ionised form. Alkalosis also leads to a lowered ratio of the intracellular to the extracellular concentrations of salicylate (Hill, 1971). Systemic acidosis has the reverse effects, and therefore should be avoided in salicylate overdose. Increased toxicity of salicylate is shown if a carbonic anhydrase inhibitor is administered because of the acidosis produced (Cowan *et al.*, 1984).

A carbonic anhydrase inhibitor does, however, increase urinary pH, and therefore increases the renal clearance of salicylate. For this reason acetazolamide has been used in treatment of salicylate overdose, but the consequent systemic acidosis must be controlled with sodium bicarbonate. Frequent determination of the pH and carbon dioxide content of blood is mandatory if carbonic anhydrase inhibitors are ever used in this situation (Schwartz *et al.*, 1959). Salicylate also inhibits the renal clearance of the carbonic anhydrase inhibitor, acetazolamide (Sweeney *et al.*, 1986). This renal interaction may increase the toxicity of acetazolamide, as well as increasing the effect of acetazolamide on the cellular uptake of salicylate.

The uptake of salicylate by isolated cells is also facilitated by acidosis. For example, the uptake by red and white blood cells is greater if extracellular pH is lowered (Garcia-Sancho and Sanchez, 1978; Joy and Cutler, 1987; Raghoebar *et al.*, 1988). Consistent with the effect of extracellular pH on the cellular uptake of the drug, salicylate decreases the killing of *Candida albicans* by neutrophils to a greater extent with decreasing pH of the medium (Brune and Graf, 1978).

Foetal uptake

Salicylate crosses the placenta and is widely distributed throughout the foetus and extra-foetal fluids, but, as is the case in other tissues, equilibrium is achieved slowly. In sheep, distributional equilibrium is achieved after about 40 minutes with a foetal:maternal ratio of 0.4 (Thiessen *et al.*, 1984). The lower levels in foetal plasma are not due to differences in the binding to plasma protein but rather appear to be due to the lower pH of foetal blood, and the distribution is thus generally consistent with the pH partition hypothesis (Varma, 1988). Because of its lipid solubility, aspirin should also cross the placenta and produce pharmacological effects in the foetus.

Breast milk

Salicylate is present in breast milk although at concentrations considerably lower than in plasma (Pütter *et al.*, 1974), consistent with the passive diffusion of salicylate into milk because the concentration of albumin and the pH of breast milk are both lower than the corresponding values of blood. The ingestion of salicylate in breast milk may lead to substantial levels in infants. For example, one mother had a plasma concentration of 230 mg/l, leading to a plasma concentration of 65 mg/l in the infant (Unsworth *et al.*, 1987). Because of concern about the possibility of salicylate causing Reye's syndrome in the infants, it is recommended that salicylates should not be taken by breast feeding mothers.

Synovial fluid

The site of the anti-inflammatory action of the salicylates and other non-steroidal anti-inflammatory drugs is probably synovial tissue. This tissue is not easily sampled, but synovial fluid (particularly from effusions of the knee) is relatively easy to obtain and is considered more closely to reflect the concentrations of salicylates and other non-steroidal anti-inflammatory drugs at their site of action than do plasma concentrations.

Although the kinetics of uptake of salicylate into synovial fluid have not been studied in detail, the rate of influx appears to be slow (Soren, 1973). The rate of loss of salicylate from the knee joints has been examined in detail. After its intra-articular administration the mean half-life of loss from the knee is 2.4 hours, considerably shorter than the half-life of loss of albumin from the knee (mean 13 hours), indicating that salicylate largely diffuses out of synovial fluid in the protein-free form (Owen *et al.*, 1994). When steady state is established the unbound concentrations in plasma and synovial fluid are equal (Rosenthal *et al.*, 1964), confirming that salicylate diffuses between plasma and synovial fluid in the unbound form. The total levels of salicylate in plasma exceed those in synovial fluid because of the lower binding to albumin in synovial fluid. The lower level of albumin in synovial fluid is a major, but not the only, reason for the lower binding to proteins in synovial fluid. This is because the protein binding of salicylate in plasma is greater than in synovial fluid, even when plasma is diluted to the same albumin concentration as synovial fluid (Trnavska and Trnavsky, 1980).

Aspirin also partitions between plasma and synovial fluid. After oral dosage the peak concentrations of the unchanged drug in synovial fluid occur after those in plasma, again indicating relatively slow diffusion into synovial fluid (Sholkoff et al., 1967).

Experimental inflammatory sites

The non-steroidal anti-inflammatory drugs, including aspirin, are localised to a greater extent in inflamed tissues in experimental animals than in normal tissues (Brune et al., 1980; Figure 4.1). There are several reasons for this finding, including the presence of the binding protein albumin and the larger volume of synovial fluid in inflamed joints. In addition, the greater cellular uptake of salicylate should be favoured by the often lower pH of synovial fluid in inflamed than in non-inflamed joints.

The kinetics of transfer of both salicylate and aspirin into experimental exudates have been studied but differing results have been obtained, no doubt due to variations in the diffusional barrier and differing protein concentrations developed at the site of inflammation. In rats, salicylate diffuses rapidly into and out of implanted sponges – more rapidly than albumin or naproxen (Doherty et al., 1977), a nonsteroidal anti-inflammatory drug that is more strongly bound to plasma proteins than salicylate. The different behaviour of salicylate and naproxen indicates that, in this system, salicylate diffuses into the sponges primarily in the unbound form (Graham, 1988). By contrast, Higgs et al. (1987) found that salicylate and aspirin diffuse slowly between plasma and exudates in polyester sponges soaked in carrageenan. Varying rates of diffusion into and out of inflammatory sites should be considered in experiments on the anti-inflammatory effects of the salicylates and related drugs.

Carrier-mediated transport into tissues

There are some data that indicate that carrier mechanisms may be involved in the cellular transport of salicylate. An anion transport inhibitor reduces, but does not abolish, the transport of salicylate into the red blood cell, an indication that it is transported by both an anion channel and by passive diffusion (Minami and Cutler, 1992). Salicylate is also transported out of cerebrospinal fluid at the choroid plexus by an active transport process (Lorenzo and Spector, 1973). Further, the uptake of salicylate by the brain is decreased by a variety of monocarboxylic acids, including self-inhibition by salicylate (Kang et al., 1990), while the concentrations of salicylate in brain, heart and skeletal muscle are increased by insulin (Wisniewski and Zarebski, 1968; Zarebski, 1973), observations that are consistent with carrier-mediated transport.

ELIMINATION OF ASPIRIN

The major mode of elimination of aspirin is by hydrolysis to salicylate (Figure 4.3). Because of its rapid hydrolysis only small amounts of aspirin are excreted unchanged in urine, and essentially all the aspirin is eliminated in urine as salicylate and its further metabolites.

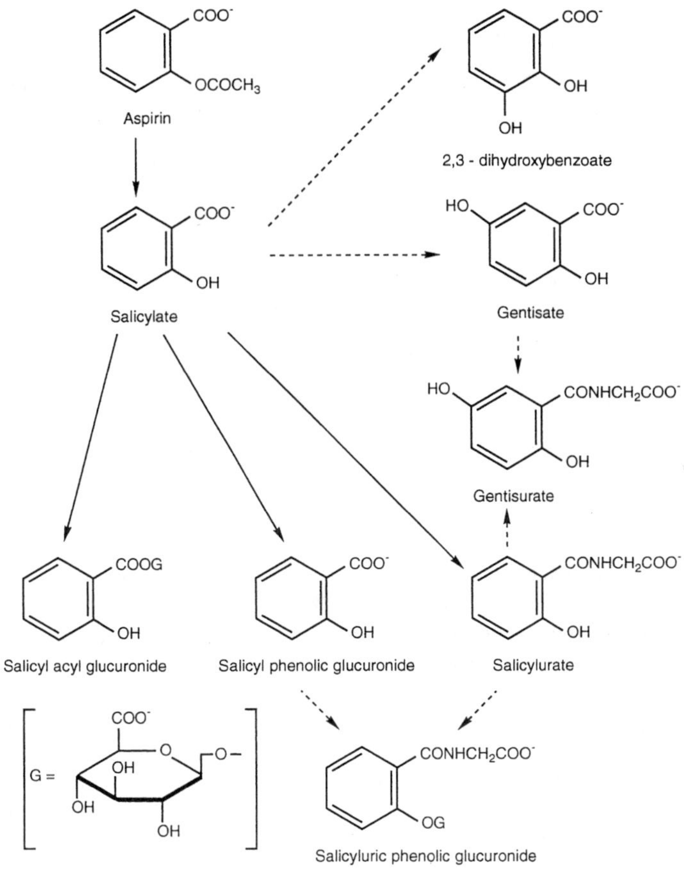

Figure 4.3 Pathways of metabolism of aspirin and salicylate. Major pathways are shown as solid arrows and minor pathways as broken. The major ionisation state at physiological pH is shown. The glucuronide group (G) is also largely ionised at physiological pH. The acetyl group of aspirin is in part transferred to proteins, while the acyl glucuronide is unstable under physiological conditions, undergoing hydrolysis, rearrangement and reaction with proteins. Apart from the metabolites shown, free radicals decarboxylate salicylate to catechol and possibly other metabolites also convert salicylate to reactive products. Salicylate is also partly excreted unchanged, dependent upon urinary pH (Figure 4.4).

Hydrolysis of aspirin

Aspirin undergoes both spontaneous and enzymatic hydrolysis to salicylate, but the spontaneous hydrolysis is insignificant in the body. This is shown by the half-life of 15.5 hours in physiological buffered saline (Rowland *et al.*, 1972). The half-life should be little different or even longer at any point within the pH range of 1.2 to 8.0 (Edwards, 1952). This half-life of spontaneous hydrolysis is far longer than the half-life of elimination (about 10 minutes) and the half-life in plasma *in vitro* (about 30 minutes) (Rowland and Riegelman, 1968).

Enzymatic hydrolysis of aspirin occurs in a variety of tissues, including the liver (Cossum *et al.*, 1986, Williams *et al.*, 1989a), gastrointestinal tract (Spenney and Nowell, 1979), kidney (Gaspari *et al.*, 1989), hind limbs (Cossum *et al.*, 1986) and blood. The major enzymes hydrolysing aspirin in human plasma are probably cholinesterases, since the hydrolysis of aspirin is inhibited by classical anticholinesterases

■ CHAPTER 4 ■

■ 107

such as physostigmine (Rainsford *et al.*, 1980). The esteratic activity is dependent upon calcium ions, and maintenance of physiological concentrations of calcium is important in *in vitro* studies on serum esterase. Significant aspirin esterase activity is also present in red blood cells, but it appears not to be a cholinesterase as it is resistant to physostigmine (Rylance and Wallace, 1981). In the intact red blood cell, the activity of aspirin esterase is controlled by the extracellular concentration of unbound aspirin and is consequently reduced by binding to the extracellular protein, albumin. Consequently, salicylate and fatty acids increase the activity of the esterase in red blood cells due to displacement of aspirin from extracellular albumin (Costello and Green, 1987). Aspirin also acetylates a variety of proteins in various tissues and components of blood, particularly albumin (see p. 102), and salicylate is released from this acetylation. This reaction resembles an esteratic process, but only a small portion of the aspirin hydrolysis in plasma can be attributed to albumin (Rainsford *et al.*, 1980).

The level of aspirin esterase activity in plasma shows considerable intersubject and also some inter-racial variation with lower esterase activity in Ghanaians than in British subjects (Williams *et al.*, 1986a). No other racial pattern in the elimination of aspirin has been reported. An interesting correlation occurs between the activity of aspirin esterase in plasma and the rate of elimination of salicylate, with higher esterase activity in subjects who eliminate salicylate most rapidly (Trnavsky and Zachar, 1975). This is a surprising correlation because the two drugs are eliminated by very different processes; aspirin by hydrolysis and salicylate mainly by conjugative processes as well as renal excretion (see p. 110).

Carboxylesterases are mainly responsible for the hydrolysis of aspirin in the liver (Inoue *et al.*, 1980). Aspirin esterase activity of liver is found both in the cytosol and microsomal fractions, with, in man, the greater activity in the microsomal fraction (Williams *et al.*, 1989a). Interestingly, there is a positive correlation between the total esterase activity in human liver and that in plasma.

There are considerable interspecies differences in the activity of plasma aspirin esterase, with cats and rabbits showing approximately the same esteratic activity as humans while rats have a higher and dogs a lower activity than man (Morgan and Truitt, 1965). The physiological role of the aspirin esterases is unclear. In the guinea pig, a carboxylesterase that hydrolyses aspirin resembles lysophospholipase. A contributing factor to the substrate activity of aspirin would appear to be that the drug, like the phospholipid substrates of lysophospholipase, is an anion at physiological pH values (White and Hope, 1984).

Acetylation of proteins by aspirin

A minor, although pharmacologically important, mode of metabolism of aspirin is by acetylation of a variety of proteins. A large number of proteins are acetylated *in vivo* (Rainsford *et al.*, 1983) and *in vitro*, including plasma albumin, other plasma proteins, various enzymes (Pinckard *et al.*, 1968), membranes of red blood cells (Green and Jung, 1981), haemoglobin (Bridges *et al.*, 1975) and renal proteins (Caterson *et al.*, 1978). The acetylation of albumin inhibits the reaction between glucose and albumin (glycosylation) and, to a lesser extent, the reaction between glucose and haemoglobin (Rendell *et al.*, 1986). The clinical significance of this latter reaction is not known at this stage, in contrast to the more detailed knowledge about the acetylation of prostaglandin endoperoxide synthase-1 synthase-2 (COX 1 and 2) by aspirin. Acetylation occurs at one serine of both these enzymes (Lecomte *et al.*, 1994), and is at least partly responsible for the pharmacological properties of aspirin. In particular, the irreversible acetylation of cyclo-oxygenase of platelets (Green and Jung, 1981) causes the prolonged inhibition of platelet function by aspirin.

Aspirin generally reacts with proteins through the ϵ-amino groups of lysine side chains (Hawkins *et al.*, 1969), whereas acetylation of a serine residue is responsible for the inactivation of COX 1 and 2 (Lecomte *et al.*, 1994). Acetylation with the lens gamma-crystallins may be responsible for the potential anti-cataract effect of aspirin, although there is argument as to whether the acetylation occurs with cysteine or lysine residues on gamma-crystallin (Cherian and Abraham, 1993; Qin *et al.*, 1993). RNA and DNA are also acetylated *in vitro*, but it is not known if this occurs *in vivo*.

The reaction between aspirin and a small number of small molecules has also been observed. In particular, aspirin interacts with carcinogenic N-hydroxy arylamines to form intermediates that readily acetylate DNA (Minchin *et al.*, 1992). Aspirin thus has the potential to influence the carcinogenic activity of N-hydroxy arylamines, but it is not known if sufficient concentrations of aspirin are present *in vivo* to influence the development of tumours by this mechanism.

Pharmacokinetics of aspirin

Unchanged aspirin can be detected in plasma for about 1 hour after its intravenous or oral administration. Following its intravenous administration in man, it has a distribution half-life of about 3 minutes, an elimination half-life of 10 minutes and a clearance of about 800 ml blood/min (Rowland and Riegelman, 1968; Figure 4.4). Aspirin is hydrolysed enzymatically in blood, but its clearance in blood accounts for only about 15 per cent of the total body clearance of the drug and the bulk of the clearance is considered to occur in the liver (Rowland *et al.*, 1972). By contrast, the clearance of aspirin in the rat is dose-dependent and at a low dose (40 mg/kg) is slightly greater than hepatic blood flow, indicating significant extrahepatic hydrolysis (Wientjes and Levy, 1988).

ELIMINATION OF SALICYLATE

A dose of aspirin or salicylate is recovered almost totally in urine as salicylate itself and its metabolites. The proportions excreted as salicylate and the various metabolites show considerable intersubject and

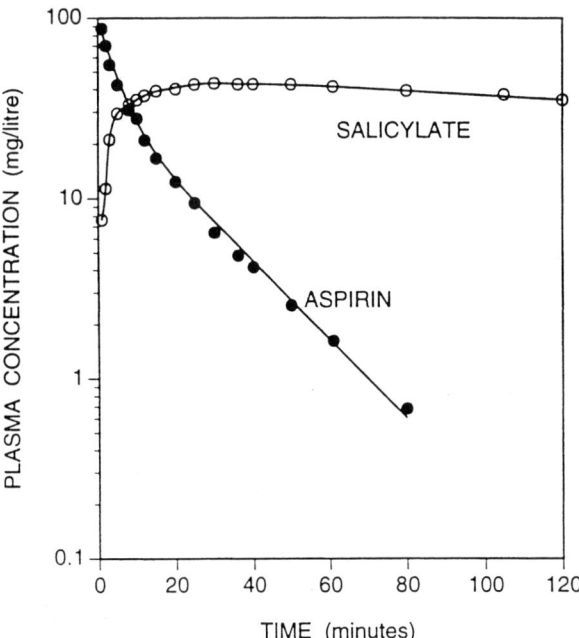

Figure 4.4 Time courses of plasma concentrations of aspirin and salicylate following a single intravenous dose of 650 mg aspirin. After an oral dose of aspirin, the plasma concentrations of salicylate are similar but the peak plasma concentrations of aspirin are lower because of incomplete bioavailability and hydrolysis to salicylate during the absorption phase. (Redrawn from Rowland and Riegelman, 1968.)

interspecies variation. As discussed below, small amounts of salicylate are also excreted in saliva and are subsequently swallowed and reabsorbed. While insignificant in the overall pattern of distribution and metabolism of salicylate, the salivary levels have been utilised in studies on the bioavailability of salicylates.

Urinary excretion of salicylate

The renal clearance of salicylate depends upon a number of processes; glomerular filtration, proximal active secretion, and passive resorption of the lipid-soluble un-ionised form (MacPherson *et al.*, 1955). As is the case with diflunisal (see p. 132), resorption can take place in the bladder (Au *et al.*, 1991), as well as from the renal tubules.

In the proximal tubule, resorption appears to occur by both a carrier-mediated mechanism and passive diffusion of the un-ionised species (Chatton and Roch-Ramel, 1992). Like other lipid-soluble carboxylic acids, the renal clearance of salicylate increases with urinary pH. The relationship between renal clearance and pH is non-linear, with the renal clearance of salicylate increasing disproportionately with pH (Smith *et al.*, 1946; Figure 4.5). This is of clinical significance during treatment with large doses of aspirin, since the steady state plasma concentrations of salicylate decrease substantially with even small increases in urinary pH. For example, small doses of sodium bicarbonate increase urinary pH from the range 5.6–6.1 to 6.2–6.9 and almost halve the average steady state concentrations of salicylate in plasma (Levy and Leonards, 1971). The influence of urinary pH on the excretion is of clinical importance, since antacids and buffers are often combined with aspirin in tablets or administered at the same time as aspirin in order to improve its gastrointestinal tolerance. Antacids such as aluminium and magnesium hydroxides, which are normally considered as non-systemic, may alkalinise urine to a point where the renal clearance of salicylate is increased and the plasma concentrations decreased (Gibaldi *et al.*, 1975). Care should be taken when patients on high-dose salicylates commence treatment with

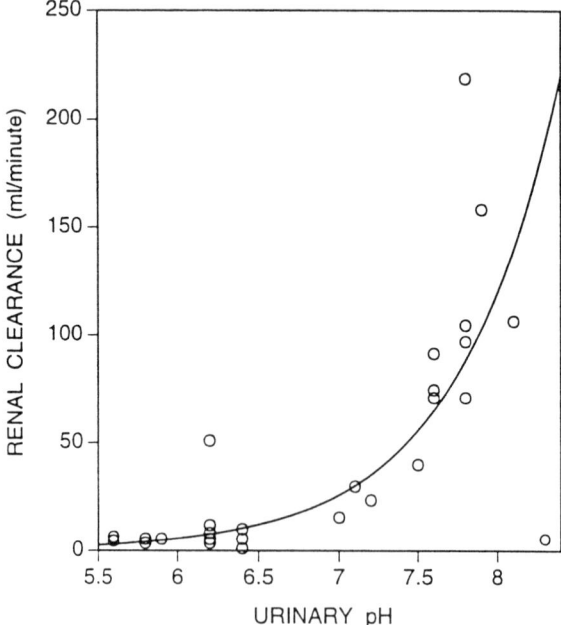

Figure 4.5 Relationship between the renal clearance of salicylate and urinary pH. (Redrawn from Smith *et al.*, 1946.)

antacids, since subtherapeutic plasma concentrations of salicylate may result. Conversely, sudden cessation of treatment with antacids may elevate plasma concentrations and lead to toxicity.

The increased renal clearance of salicylate under alkaline conditions is utilised in the treatment of overdosage (Gordon *et al.*, 1984). Alkalinisation of urine is more important than increased urine flow in overdose, and excessive urine flow is not required (Henry and Volans, 1984). Systemic alkalosis is also useful in the treatment of overdose because it decreases the cellular uptake of salicylate (see p. 104).

The renal clearance of salicylate is decreased by probenecid, presumably by inhibition of the proximal secretion of salicylate (Gutman *et al.*, 1955). This interaction, however, should rarely lead to significant accumulation of salicylate, since the urinary excretion of salicylate is minor unless the urine is alkaline. Salicylate also antagonises the uricosuric activity of probenecid. These interactions are of little clinical significance because probenecid is now little used for the treatment of gout, and salicylates are generally avoided in gouty patients because of the urate-retaining activity of low levels of salicylate.

Because of the saturable metabolism of salicylate, it has been suggested that the percentage urinary recovery of salicylate should increase with increasing doses. Overdoses of salicylates do lead to increased proportion of the dose being excreted as salicylate (Patel *et al.*, 1990a) but, within the range of therapeutic dosage, the already low renal clearance of salicylate decreases markedly from a mean of 1.9 ml/min at 1 g of aspirin daily to 0.3 ml/min at 4 g of aspirin daily, and the proportion excreted as salicylate actually falls (Bochner *et al.*, 1987). The contrasting results may be due to acidification of urine and hence greater resorption of salicylate at therapeutic doses. More recently, it has been suggested that the fall in renal clearance may be due to saturation of its secretion (Dubovska *et al.*, 1995).

Elimination of salicylate by repeated doses of charcoal

Salicylate can diffuse from the circulation into the small intestine, the rate being enhanced with increasing intestinal pH (Blair and Huang, 1992). This process, termed exsorption, has been utilised with several drugs in accelerating their removal from the body by binding the exsorbed drug to charcoal. Data on salicylate is, however, equivocal. There are case reports indicating that charcoal leads to faster salicylate elimination after overdose (Hillman and Prescott, 1985), although controlled studies with therapeutic doses indicate no significant effect on the elimination of salicylate (Ho *et al.*, 1989; Mayer *et al.*, 1992).

Salivary excretion of salicylate

As is the case with several acidic drugs, the salivary concentrations of salicylate are lower than in plasma in man, although there is a good correlation between the two (Graham and Rowland, 1972). The concentration is independent of the flow, and the salivary concentrations of salicylates have been used in studies on the bioavailability and drug interactions of salicylates, although the utility of salivary measurements of salicylate is limited by intersubject and intrasubject variation in the ratio of salivary to plasma concentrations (Khan and Aarons, 1989). Such variations may be very much dependent upon the reproducibility of the assay. Another practical problem is that mixed saliva may be contaminated by the orally administered drug retained in the mouth. It is probable that the salivary concentrations are more closely related to the unbound concentrations than to the total concentrations in plasma, but definitive proof is lacking. While there is rapid equilibration between plasma and salivary concentrations of salicylate in man, this is not the case in the rat, at least in saliva collected from the submaxillary gland (Putney and Borzelleca, 1972).

METABOLIC PATHWAYS OF SALICYLATE

Salicylate is converted to a variety of metabolites (Figure 4.3). In man, three conjugates are formed directly; salicylurate, a glycine conjugate which is the major metabolite, and two glucuronides – an acyl glucuronide linking the glucuronyl group to the carboxyl group of salicylate, and a phenolic glucuronide linked to the phenolic hydroxyl of salicylate. Two of these conjugative metabolic processes are saturable at therapeutic doses; namely the syntheses of salicylurate and salicyl phenolic glucuronide. In addition, salicylate undergoes non-saturable oxidation to two dihydroxybenzoates; gentisate (2,5-dihydroxybenzoate) and 2,3-dihydroxybenzoate. There is also some secondary metabolism, with the production of gentisurate, the glycine conjugate of gentisate, and the biosynthesis of the phenolic glucuronide of salicylurate. The main site of metabolism of all the major metabolic pathways is probably the liver, but other tissues, particularly the kidney, may also be involved.

Salicylurate

This metabolite is synthesised in a two-step process involving co-enzyme A (CoA), adenosine triphosphate (ATP) and glycine (Tishler and Goldman, 1970; Forman *et al.*, 1971).

Salicylate + ATP + CoA → Salicyl-CoA + pyrophosphate + AMP
Salicyl-CoA + glycine → salicylurate + CoA

Gentisurate is also formed by the mitochondrial fraction of liver, presumably by the same mechanism (Wilson *et al.*, 1978).

The site of metabolism of salicylate is contentious. No metabolism was detected in the perfused rat liver (Shetty *et al.*, 1994), but the technique (single-pass perfusion) may not have been adequate to detect the metabolism of a low-clearance drug like salicylate. The formation of salicylurate and the loss of salicylate should be more apparent if the perfusing medium is recirculated through the liver. The synthesis of salicylurate occurs in the mitochondrial fraction of bovine liver (Forman *et al.*, 1971), while the hepatic production of salicylurate in sheep is evident by slightly higher concentrations of salicylurate in the hepatic vein than in the portal vein (Cossum *et al.*, 1986). The production of salicylurate occurs in the isolated rat kidney (Bekersky *et al.*, 1980), but the kidney is probably not an important site of formation of salicylurate in man because the production of salicylurate appears normal in anephric patients (Lowenthal *et al.*, 1974). Similarly, impaired renal function in man leads to higher plasma concentrations of salicylurate (Bochner *et al.*, 1981) while renal failure in the rabbit causes an increased proportion excreted as metabolites (Laznicek *et al.*, 1989), both results indicating that the kidney is mainly involved in the elimination of salicylurate rather than its synthesis.

Salicylurate is bound to plasma proteins, although to a lesser extent than the parent, salicylate. During treatment with salicylate, about 20 per cent of the salicylurate is unbound (Ashton *et al.*, 1989). Salicylurate may be further metabolised to a phenolic glucuronide (Figure 4.3), but this pathway appears negligible except in renal impairment (Zimmerman *et al.*, 1981). Small amounts of salicylurate may also be converted to salicylamide through the intermediate formation of N-salicyl-α-hydroxyglycine (DeBlassio *et al.*, 2000). This two-step reaction involves the liberation of glyoxylate with both reactions catalysed by peptidylglycine α-amidating monooxygenase, an oxidative enzyme that is responsible for the formation of several endogenous amides. At this stage the synthesis of salicylamide from salicylurate has only been examined *in vitro* and its importance *in vivo* is unknown, but it can only be a minor metabolite because a dose of salicylate can be recovered in urine almost completely as the sum of free salicylate, salicylurate, the two glucuronides and gentisate.

Low concentrations of salicylurate, possibly with its glucuronide and other phenolic acids, may be present in the plasma and urine of patients who are not receiving salicylates therapeutically, and detectable levels in plasma have been associated with various conditions; in children with gastrointestinal disorders (Finnie *et al.*, 1976) and in uraemic patients (Lichtenwalner *et al.*, 1983). The source of salicylurate in the absence of therapeutic doses of salicylates is unclear, but may include the bacterial metabolism of L-tryptophan in the colon. This is, however, a contentious area, and there is one report of no assayable levels of salicylurate except when salicylate was present, indicating that salicylurate is only present after the ingestion of aspirin or other salicylates (Gulyassy *et al.*, 1987).

Kinetics of synthesis and elimination of salicylurate

The synthesis of salicylurate, in contrast to the formation of most drug metabolites, is saturable. As applied to the kinetics of drugs *in vivo*, two parameters describe the rate of metabolism of saturable processes. These are the maximal rate of metabolism and the Michaelis constant, K_m, the latter being determined as either the plasma concentration or body load of drug at which the rate of metabolism is 50 per cent of the maximum. After single doses of various salicylates, the maximal rate of conversion of salicylate to salicylurate is about 40 mg/h, with the various studies on the kinetics of this saturable pathway yielding relatively consistent estimates of the maximal rate (Levy *et al.*, 1972; Bochner *et al.*, 1981; Gunsberg *et al.*, 1984; Owen *et al.*, 1989; Shen *et al.*, 1991a; Table 4.1). The rate of conversion of salicylate to salicylurate is reasonably related to the unbound concentrations. From this analysis, the K_m is about 1.5 mg/l (Table 4.1), equivalent to a total plasma concentration of about 30 mg/l. Allowing for the volume of distribution of about 10 l, this corresponds to a total body load of about 300 mg of salicylate.

Because of the saturable metabolism of salicylate to salicylurate, the urinary output of salicylurate, as a fraction of the dose, falls with increasing dosage. This occurs after both therapeutic and overdoses of salicylates. For example, with increasing therapeutic daily doses of aspirin from 1 to 4 g daily, the output of salicylurate decreases from about 80 to 58 per cent of the daily dose (Bochner *et al.*, 1987). The relative output of salicylurate decreases further after overdosage (Patel *et al.*, 1990a).

Salicylurate itself is eliminated in a biphasic fashion, with an initial half-life of about 17 minutes in man (Levy *et al.*, 1969; Bochner *et al.*, 1981). Its elimination is dose-dependent in rats, although the half-life is still under 15 minutes (Morris, 1990). Salicylurate is cleared renally by proximal secretion (Knoefel *et al.*, 1962) with highly variable clearance values ranging from 103 to 893 ml/min during treatment with salicylates (Gunsberg *et al.*, 1984; Ho *et al.*, 1985; Owen *et al.*, 1989). Although its clearance is high, the renal secretion of salicylurate is less than that of other glycine conjugates, such as the well-known marker of renal secretion, para-aminohippurate. Not surprisingly, the renal clearance of salicylurate decreases with decreasing renal function and is inhibited by inhibitors or substrates of the secretory process, such as probenecid (Hekman and van Ginneken, 1983) and phenolsulphonphthalein (Russel *et al.*, 1987).

The clearance of salicylurate is very much higher than that of salicylate, and as a result its plasma concentrations are considerably below those of the parent salicylate and also are sustained because of

TABLE 4.1

Kinetic parameters of conversion of salicylate to its major metabolites after single* doses of aspirin or sodium salicylate (in relation to unbound concentrations in plasma; mean + SD). From data of Levy *et al.*, 1972; Bochner *et al.*, 1981; Gunsberg *et al.*, 1984; Owen *et al.*, 1989; Shen *et al.*, 1991a.

Metabolite	Kinetic parameters
Salicylurate	V_{max} 43 + 15 mg/h
	K_m 1.5 + 1.1 mg/l
Acyl glucuronide	CL 0.64 + 0.26 l/h
Phenolic glucuronide	V_{max} 25 + 6 mg/h
	K_m 3 + 2 mg/l
Gentisate	CL 0.38 + 0.08 l/h

*Maximal rates of conversion of salicylate to salicylurate and the phenolic glucuronide increase with repeated doses.

the saturable metabolism of salicylate. After single therapeutic doses of salicylates the plasma concentrations of salicylurate plateau at about 5 mg/l, but are higher in patients with poor renal function (Bochner *et al.*, 1981).

Small amounts of circulating salicylurate are hydrolysed back to salicylate in the dog (Nakamura *et al.*, 1989) and rat (Nakamura *et al.*, 1988), but not in man (Boxenbaum *et al.*, 1979). In the rat, some of the hydrolysis to regenerate salicylate occurs in the kidney (Bekersky *et al.*, 1980), while in both the rat and rabbit salicylate is reformed by micro-organisms in the large intestine (Shibasaki *et al.*, 1985; Nakamura *et al.*, 1989).

Salicylate glucuronides

The two glucuronides of salicylate are formed in lesser amounts than salicylurate. After a single large dose of aspirin or sodium salicylate, the urinary recoveries of the phenolic and acyl glucuronides account for about 20 and 10 per cent of the dose, respectively. The synthesis of the acyl glucuronide follows non-saturable kinetics, but, like the synthesis of salicylurate, the conversion of salicylate to the phenolic glucuronide follows saturable kinetics. Despite this saturable pattern, the urinary recovery of salicyl phenolic glucuronide actually increases with increasing dose rates. At doses of aspirin ranging from 1 to 4 g daily, the urinary output of this glucuronide increased from 5.2 to 10.5 per cent (Bochner *et al.*, 1987). This increase may be due, first, to induction of the phenolic glucuronide pathway on repeated dosage, since the percentage increase in the activity of the phenolic glucuronide pathway appears slightly greater than that of the salicylurate pathway (Day *et al.*, 1988a; see p. 119). Second, the K_m of the phenolic glucuronide pathway is higher than that of salicylurate (Table 4.1), making the synthesis of the phenolic glucuronide relatively less saturated than the salicylurate pathway at low plasma concentrations. In various species the two glucuronides of salicylate are formed in the intestine, liver and kidney (Schachter *et al.*, 1959), but data are lacking in man.

The phenolic glucuronide is linked to salicylate through an ether bond, which is anticipated to be stable under physiological conditions. However, the acyl glucuronide contains an ester grouping and, like other acyl glucuronides, hydrolyses under physiological conditions. Salicyl acyl glucuronide is broken down with a half-life of 1.4 hours at 37°C and pH 7.4 (Dickinson and Baker, 1991). In addition to its hydrolysis back to salicylate, rearrangement of the ester glucuronide also occurs (Bradow *et al.*, 1989). As is the case with several other acyl glucuronides, the glucuronides react with albumin *in vitro* to form a salicylate–protein adduct (Dickinson and Baker, 1991; Liu *et al.*, 1996). The occurrence of salicylate–protein adducts has not yet been examined in man, but low concentrations of adducts of salicylate with plasma proteins have been detected in rats with very considerable increases if renal failure is induced, due to retention of the acyl glucuronide (Liu *et al.*, 1996). Because of the instability of the salicyl acyl glucuronide, decomposition may occur in the bladder and during assay of urine samples. Consequently, its formation may be underestimated and the elimination of salicylate overestimated.

Hydroxylated salicylates: gentisate and 2,3-dihydroxybenzoate

Gentisate (2,5-dihydroxybenzoate; Figure 4.3) is a minor metabolite of salicylate, usually accounting for less than 5 per cent of the elimination of the drug, and is present in plasma at markedly lower levels than salicylate (Levy *et al.*, 1972; Bochner *et al.*, 1981; Shen *et al.*, 1991a). Gentisate is also present in synovial fluid at concentrations similar to those in plasma (Cleland *et al.*, 1985; Sitar *et al.*, 1985). Unlike two other pathways of metabolism of salicylate, gentisate is formed by non-saturable processes (Table 4.1).

Salicylate is converted to gentisate by enzymatic and possibly also by non-enzymatic free radical processes. Gentisate is produced by two cytochrome P450s, CYP4502E1 and CYP4503A4 (Dupont *et al.*, 1999), both of which are induced by alcohol.

Gentisate is also synthesised in a variety of free radical systems. Thus, it is synthesised by hydroxyl or similar highly reactive radicals in chemical systems (Grinstead, 1960), and is also produced during the reperfusion of several tissues after a period of ischaemia (Das *et al.*, 1991; Ophir *et al.*, 1993), possibly due to non-enzymatic reactions with the free radicals produced during reperfusion. Activated neutrophils also convert salicylate to gentisate (Davis *et al.*, 1989), the process requiring the enzyme myeloperoxidase, as well as the free radical, superoxide (Kettle and Winterbourn, 1994). The formation of gentisate from salicylate is low, only 55 ng from 10^6 neutrophils and 10 mM salicylate (Davis *et al.*, 1989). The formation of gentisate by free radicals may, however, be the cause of the reported increased excretion of this metabolite in patients with rheumatic fever (Kapp and Coburn, 1942), although this latter observation requires confirmation with modern specific techniques. Small quantities of gentisate, together with somewhat lower levels of salicylate, are also formed from mesalazine by free radicals produced by activated neutrophils (Dull *et al.*, 1987a; Liu *et al.*, 1995).

An isomer of gentisate, 2,3-dihydroxybenzoate (Figure 4.3), is also formed from salicylate *in vivo* but is present in lower concentrations in plasma and urine than gentisate (Grootveld and Halliwell, 1988). Unlike gentisate it is not produced by native cytochrome P450 systems, but may be solely produced by free radical mechanisms; by neutrophils (Davis *et al.*, 1989), by xanthine oxidase (Richmond *et al.*, 1981) and, like gentisate, during reperfusion of tissues after ischaemia (Das *et al.*, 1991) and in fatigued muscle (Hasegawa *et al.*, 1997). The plasma levels of 2,3-dihydroxybenzoate *in vivo* may be an indicator of the formation of hydroxyl radical or similar highly reactive radicals, such as peroxynitrite, in the body (Grootveld and Halliwell, 1986; Narayan *et al.*, 1997). Consistent with this hypothesis, the plasma concentration of 2,3-dihydroxybenzoate is increased by conditions that may be associated with increased free radicals; by hyperoxia in rats (O'Connell and Webster, 1990), in alcoholism (Thome *et al.*, 1997), by the administration of paraquat to mice (Kim and Wells, 1996), and in diabetes (Ghiselli *et al.*, 1992). A more recent finding is the increased formation of 2,3-dihydroxybenzoate after exposure to a low level of ozone, which is considered to increase the concentration of hydroxyl radical or a similar reactive radical *in vivo* (Liu *et al.*, 1997). Possible artefacts should, however, be noted. The formation of 2,3-dihydroxybenzoate from salicylate on metal surfaces (Montgomery *et al.*, 1995) and in the purely chemical system of the Fenton system (hydrogen peroxide and an iron salt) means that careful controls are required to ensure that the formation of 2,3-dihydroxybenzoate is a real biological phenomenon.

Free radicals, particularly the hydroxyl radical, are highly reactive, but salicylate is one of the few compounds that may possibly attain levels that are sufficiently high to scavenge reactive radicals significantly in biological systems and to yield measurable levels of the resulting products. Presumably because of its scavenging activity, salicylate reduces experimental reperfusion injury of the liver (Colantoni *et al.*, 1998) and myocardium (Das *et al.*, 1991), but the concentration (2 mM) used in the protein-free medium exceeds the unbound concentration in plasma *in vivo* and at this stage it is not known if therapeutic dosage yields sufficient salicylate to reduce this reperfusion injury *in vivo*.

The formation of gentisate and 2,3-dihydroxybenzoate may also be significant aspects of the toxicology and pharmacology of salicylate. While gentisate and, to a lesser extent, 2,3-dihydroxybenzoate may be nephrotoxic (McMahon *et al.*, 1991), both are potent scavengers of free radicals (Betts *et al.*, 1985) and useful tissue protective activity could result from the scavenging of hydroxyl or other highly reactive free radicals by these metabolites. For example, the scavenging activity of gentisate leads to inhibition of the depolymerisation of hyaluronic acid caused by free radicals (Carlin *et al.*, 1985). Furthermore, gentisate inhibits several functions of neutrophils (Lorico *et al.*, 1986), while both gentisate and its quinhydrone oxidation product interact with prostaglandin synthase in a fashion resembling paracetamol, being either activators or inhibitors, depending on the experimental conditions (Blackwell *et al.*, 1975; Holmes *et al.*, 1984; 1985). The other hydroxylated product, 2,3-dihydroxybenzoate, inhibits cyclo-oxygenase and has anti-inflammatory activity in experimental inflammation (Whitehouse *et al.*, 1976). Additionally, experimental lung injury (Baldwin *et al.*, 1985) and gentamicin-induced renal injury (Walker and Shah, 1988) are both decreased by 2,3-dihydroxybenzoate.

The question is whether dosage with aspirin or salicylate produces concentrations of gentisate or 2,3-dihydroxybenzoate that are sufficiently high to produce toxicological or pharmacological effects. The concentrations of gentisate in both plasma and synovial fluid observed during treatment with aspirin are about 5 to 10 µmol/l (Cleland *et al.*, 1985; Sitar *et al.*, 1985), while the plasma concentrations of

CHAPTER 4

2,3-dihydroxybenzoate appear to be lower. Present data indicate that there is little quenching of oxidising radicals at these concentrations, although quenching has been reported at gentisate concentrations that may be obtained clinically by dosage with this preformed metabolite (Betts *et al.*, 1985). An additional consideration is that gentisate has low lipid solubility, with consequent poor diffusion into cells. It has been suggested that the oxidised metabolites may have pharmacological activity within inflammatory cells (such as neutrophils or monocytes) that produce these metabolites (Haynes *et al.*, 1993), but this hypothesis has not been tested.

By itself, gentisate is an inhibitor of prostaglandin synthesis with an IC_{50} value of about $50\,\mu M$, a concentration about five times higher than produced during high-dose treatment with aspirin or salicylate (Hinz *et al.*, 2000). Gentisate has been used in the past for the treatment of rheumatic fever, and 2,3-dihydroxybenzoate was reported to be anti-inflammatory in an open trial in rheumatic fever (Clarke *et al.*, 1958). Clearly, the clinical activity of both compounds requires reassessment in modern clinical trials. The value of gentisate is, however, somewhat doubtful because of the arthritis that occurs in association with the accumulation of a close analogue, homogentisate, in the inherited metabolic disease alkaptonuria. The arthritis of alkaptonuria may be due to the production of free radicals during the auto-oxidation of homogentisate (Martin and Batkoff, 1987). Interestingly, homogentisate accumulates during treatment with aspirin (Montgomery and Mamer, 1978), possibly due to inhibition of homogentisate oxidase by gentisate.

Some aspects of the metabolism and excretion of gentisate are known, but the handling of 2,3-dihydroxybenzoate has not been studied. Gentisate is mainly excreted unchanged, with some conjugation with glycine to yield gentisurate (Wilson *et al.*, 1978), which may, like salicylurate, be further metabolised to gentisamide (DeBlassio *et al.*, 2000). The renal clearance of gentisate is about $60\,ml/min$, considerably lower than that of salicylurate (Gunsberg *et al.*, 1984) and lower than its predicted glomerular filtration rate, since it is largely unbound in plasma. Consistent with its passive resorption, the renal excretion may be pH-dependent (Batterman and Sommer, 1953), although further studies are required with modern methodology to delineate its pharmacokinetics when administered by itself.

Salicylate is also oxidised by products other than to gentisate and 2,3-dihydroxybenzoate. It is decarboxylated by a very reactive radical, possibly hydroxyl, produced by neutrophils and by xanthine oxidase (Sagone and Husney, 1987). Another decarboxylated and oxidised product has recently been identified. This is catechol, which is increased in an animal model of Parkinsonism (Sam *et al.*, 1998).

Reactive metabolites

Salicylate is metabolised to reactive metabolites in rat kidney mitochondria and, to a lesser extent, in liver mitochondria (Kyle and Kocsis, 1986a). With increasing age, there is increased covalent binding of salicylate to the mitochondria of renal cortex. Pretreatment of rats with piperonyl butoxide, an inhibitor of the cytochrome P450 system, decreases both the renal toxicity and covalent binding of salicylate equivalents to renal mitochondria in rats (Kyle and Kocsis, 1986b). This correlation indicates that reactive oxidative metabolites of salicylate are responsible for its nephrotoxicity, at least in this acute state seen in rats. Any relationship to the longer-term analgesic nephropathy associated particularly with combinations of analgesics is, however, not known.

Comparative metabolic pathways of salicylate

The pathways of elimination of salicylate are generally similar in all species examined, although the relative amounts of the metabolites vary. The glucuronide and glycine conjugates of salicylate are found in the rat (Nelson *et al.*, 1966; Patel *et al.*, 1990b), dog (Alpen *et al.*, 1951) and rabbit (Short *et al.*, 1991), but salicylurate is the only conjugate of salicylate detected in the urine of goats and cattle (Short *et al.*, 1990). In all these species, a higher proportion of the dose is excreted in urine as free salicylate than in man. This may possibly be due to the high doses that have often been administered, leading to saturation of the salicylurate pathway, lower maximal velocities of the salicylurate pathway, and/or an alkaline urinary pH, which increases the renal clearance of salicylate.

Cattle and goats show a variable pattern of excretion of salicylate. In both species, more salicylurate and lesser amounts of salicylate are found in urine after oral dosage than after intravenous dosage (Short *et al.*, 1990). This may be due to the saturable conversion of salicylate to salicylurate, because slow absorption after oral dosage should lead to lower initial plasma concentrations than after intravenous dosage, and hence lesser saturation of the salicylurate pathway.

FACTORS CONTROLLING THE ELIMINATION OF ASPIRIN AND SALICYLATE

One of the features of the clinical pharmacology of salicylate is the marked interpatient variation in the amounts of the various metabolites and salicylate excreted (Hutt *et al.*, 1986) and in the pharmacokinetics of the drug (Paulus *et al.*, 1971; Graham *et al.*, 1977). Consequently, high doses of any salicylate require careful monitoring to optimise dosage. At one stage the oral absorption of the salicylates was thought to be incomplete in some patients, but apart from incomplete absorption from some slow-release or enteric-coated formulations and possible incomplete absorption in Kawasaki disease, it is now realised that the variable plasma concentrations result from interpatient differences in the rate of elimination of salicylate. The influence of urinary pH on the elimination of salicylate is, as discussed above, relatively straightforward, but control over the metabolism of this drug is complex (Table 4.2). Furthermore, some areas of the literature on the pharmacokinetics of salicylate are inconsistent, making it difficult to summarise the control over the metabolism of the salicylates. Interpatient differences in the kinetics of elimination of aspirin have also been observed, although variation in kinetics of salicylate is the more important because of its slower elimination and its greater accumulation during long-term therapy.

Genetics

Studies in twins indicate that the conjugation of salicylate to salicylurate is under genetic control, with other conjugations probably controlled in a similar fashion (Furst *et al.*, 1977). No clinically important polymorphism in the elimination has been detected, although patients with the rare Crigler–Najjar syndrome have a deficiency of glucuronyl transferase and produce much less salicylate glucuronides than normals (Childs *et al.*, 1959). Genetic control over the metabolism of salicylate may be the cause of higher ratios of salicylurate to salicyl glucuronides in Caucasians than in Nigerians (Emudianughe *et al.*, 1986).

TABLE 4.2

Summary of factors altering the rate of elimination of salicylate.

Factor	Change in clearance of salicylate
Urinary pH	Increased clearance in alkaline urine
Corticosteroids	Possible induction of metabolism
Circadian pattern	Lowest clearance in early morning
Sex	Possible greater clearance in males
Age	Decreased clearance in elderly
Liver disease	Decreased clearance of unbound salicylate
Previous ingestion of salicylates	Auto-induction of clearance

Corticosteroids

From case studies, corticosteroids appear to induce the elimination of salicylate. Decreases in the plasma concentrations of salicylate consistent with induction of metabolism have been reported during dosage with corticosteroids, even after intra-articular injections of corticosteroids (Klinenberg and Miller, 1965; Graham et al., 1977; Edelman et al., 1986), but by contrast corticosteroids do not alter the rate of salicylate elimination after single doses of aspirin or salicylate, either in man (Day et al., 1988b) or in the dog (Day et al., 1976). Possibly, corticosteroids enhance the auto-induction of salicylate metabolism or induce the activity of minor metabolic pathways that are more significant at higher doses of salicylate because of its saturable metabolism.

Circadian variations

There are circadian rhythms in the pharmacokinetics of salicylate, with the highest plasma concentrations (Markiewicz and Semenowicz, 1979) and, correspondingly, the slowest urinary elimination after early morning dosage (Reinberg et al., 1967). The assays of salicylate in urine detected both salicylate and salicylurate, and it is not known whether the circadian rhythm is due to metabolism or excretion of salicylate. The pattern in the kinetics of salicylate does not parallel the endogenous hydocortisone rhythm, since this steroid shows the reverse pattern with the highest plasma concentrations in the early morning and lowest about midnight. However, delayed response to hydrocortisone is possible.

Sex

Another contentious area is the influence of sex on the pharmacokinetics of salicylate. Several groups have reported that the clearance of salicylate is higher in males than in females (Graham et al., 1977; Trnavska and Trnavsky, 1983; Ho et al., 1985; Miners et al., 1986), although there is considerable overlap. Other groups have, however, found little or no such difference (Greenblatt et al., 1986; Montgomery et al., 1986; Aarons et al., 1989). Higher metabolic clearances to salicylurate have been found in males (Ho et al., 1985; Miners et al., 1986), although in Nigerians there is lesser salicylurate excretion but greater excretion of salicyl glucuronides in males than in females (Emudianughe et al., 1986). An hormonal influence on the kinetics of salicylate is indicated by the increased clearance of salicylate in women who are taking oestrogen-containing oral contraceptives (Gupta et al., 1982; Miners et al., 1986).

The limited studies in experimental animals support an influence of sex on the pharmacokinetics of salicylate. Except in very young rats, the clearance of salicylate is consistently larger in male than in female rats, the difference correlating with a greater output of glucuronides in the male rats (Varma and Yue, 1984). Ovariectomy of female rats increases salicylate clearance but castration of male rats is without effect, while the sex of rats alters the pattern of urinary metabolites, particularly after the induction of diabetes (Emudianughe, 1990). Overall, the higher body weight of males may correlate with their higher clearance, but an hormonal influence on the metabolism of salicylate is probable.

Males have also been reported to have lower plasma concentrations of unchanged aspirin after oral doses of the drug (Ho et al., 1985), but there are conflicting data on the activity of aspirin esterase activity in males and females. Greater esterase activity has been found in the blood of males in two studies (Menguy et al., 1972; Miners et al., 1986), with other reports of no sex-related difference in serum or red blood cell aspirin esterase (Rainsford et al., 1980; Rylance and Wallace, 1981). While there may be sex-related differences in the elimination of aspirin, it should be noted that aspirin is still rapidly eliminated.

Age

The elimination of aspirin is not slowed in old age (Roberts *et al.*, 1983) but, as with other factors, there are conflicting findings about the influence of age on the kinetics of elimination of salicylate in man. In some studies in elderly subjects, either a decreased clearance of salicylate (Salem and Stevenson, 1977) or a decreased maximal rate of synthesis of salicylurate was observed (Ho *et al.*, 1985), but there were no such changes in other studies (Roberts *et al.*, 1983; Netter *et al.*, 1985; Montgomery *et al.*, 1986). Despite the conflicting results, it is reasonably concluded that a proportion of the elderly may be slow eliminators of salicylate and that elderly taking anti-inflammatory doses of salicylates should be carefully monitored for side effects due to excessive accumulation (Grigor *et al.*, 1987). Because of their impaired renal function, higher plasma concentrations of the metabolites salicylurate and gentisate are seen in the elderly than in younger adults (Montgomery and Sitar, 1981).

In neonates salicylate is mainly eliminated by conversion to conjugates, as is the case in adults, but the capacity of these pathways is markedly lower than in adults (Levy and Garrettson, 1974). Age-dependent metabolism is clearly seen in rats, where the clearance of salicylate reaches a maximum at about 3 weeks and then declines with increasing age.

A rare affliction in children is Kawasaki disease, which is an acute febrile illness with effects on many organs, including the gastrointestinal tract. High doses of aspirin are the mainstay of treatment of the acute phases of the disease, but the plasma concentrations of salicylate are low, probably due to low binding to plasma proteins and, possibly, incomplete oral absorption (Koren *et al.*, 1988).

Pregnancy

From limited data, it appears that the clearance of salicylate is increased during pregnancy (Amon *et al.*, 1977). Conflicting data have been published on the effect of pregnancy on the elimination of salicylate in the rat, with two reports of no change in the half-life of salicylate in the rat (Varma and Yue, 1984; Yoshikawa *et al.*, 1984b) and one report of a 50 per cent increase in the half-life (Dean *et al.*, 1989).

Liver disease

Although highly variable, the average concentrations of unbound salicylate in plasma are considerably increased in alcoholic liver disease although, because of the decreased albumin levels and low plasma binding, the total concentrations (free + bound) are not increased (Roberts *et al.*, 1983). The total concentrations of unchanged aspirin are also not increased by liver disease, but the unbound levels have not been recorded.

Auto-induction of metabolism of salicylate

A feature of the metabolism of salicylate is that it increases during continued dosage and the plasma concentrations correspondingly fall. For example, the plasma levels at day 21 of daily dosage with 60 mg/kg of aspirin were on average, only 48 per cent of those achieved on day 7 of treatment (Müller *et al.*, 1975). The decrease is due to increased conversion of salicylate to salicylurate and salicyl phenolic glucuronide (Furst *et al.*, 1977; Day *et al.*, 1983; 1988b; Owen *et al.*, 1989). The plasma concentrations of salicylate may of course also fall because of poor compliance, but auto-induction of its metabolism is now well established.

Other drugs

There are few drugs that influence the elimination of the salicylates. As discussed above (see p. 107), the enzymes hydrolysing aspirin interact with the classical inhibitors of these enzymes. These interactions are not of clinical interest, but may be utilised in investigational studies in experimental animals.

There are few recorded effects of other drugs on the elimination of salicylate. The H_2 blocker and inhibitor of the hepatic metabolism of many drugs, cimetidine, inhibits the elimination of salicylate (Trnavska et al., 1985), while the synthesis of gentisurate is increased by pretreatment with the well-known inducer of the cytochrome P450 system, phenobarbitone (Wilson et al., 1978). No effect of phenobarbitone on the synthesis of the major metabolite, salicylurate, has been reported.

Benzoate inhibits the synthesis of salicylurate by acting as a competitive substrate (Levy and Amsel, 1966), and the block in the synthesis of salicylurate has been used to investigate the pharmacokinetics of this metabolite (Levy et al., 1969). The formation of salicylurate is also inhibited by the solvent m-xylene through the same general mechanism, involving first the conversion of m-xylene to m-methylbenzoic acid, which is the active inhibitor (Campbell et al., 1988).

Glycine has contrasting effects on the synthesis of salicylurate and hippurate, the glycine conjugate of benzoate. Only the synthesis of hippurate is increased by glycine. The rate of synthesis of salicylurate is not (Amsel and Levy, 1969). Glycine levels in plasma are, however, depleted by salicylate, but only after overdoses (Patel et al., 1990c).

Several other factors may affect the pharmacokinetics of salicylate, including bed rest, which may lead to lower plasma concentrations of salicylate (Bayles, 1963).

Effects of salicylates on the disposition and pharmacological effects of other drugs

Salicylates affect the pharmacokinetics of a variety of other drugs, most commonly either decreasing the renal excretion or binding to plasma albumin (Table 4.3). The displacement of drugs from plasma

TABLE 4.3

Effects of aspirin and salicylate on the actions of other drugs.

Alcohol	High concentrations of alcohol increase gastrointestinal bleeding
Antihypertensives and diuretics	Decreased antihypertensive and natriuretic actions possible; check for partial loss of activity when commencing treatment with aspirin
Lithium	No interaction in patients with good renal function, but a decreased renal clearance and increased plasma lithium is considered possible in patients with impaired renal function; monitor plasma concentrations of lithium
Methotrexate	Decreased renal clearance increases risk of toxicity of cytotoxic doses of methotrexate
Other NSAIDs	Displacement from binding to plasma proteins and decreased total plasma concentrations of several other NSAIDs; not of clinical significance
Phenytoin	Displacement of phenytoin from binding to plasma proteins; does not increase the activity of phenytoin, but a doubling of % unbound must be considered during high dosage with aspirin when monitoring the plasma concentrations of phenytoin
Probenecid	Decreased hyperuricaemic effect of probenecid; avoid aspirin apart from occasional doses
Sulphonylureas	Possible enhanced hypoglycaemic response of oral antidiabetic agents; monitor blood glucose
Valproate	Decreased clearance and increased unbound concentrations of valproate; use an alternative NSAID or analgesic
Warfarin	Potential for increased bleeding due to antiplatelet effect of aspirin; risk of bleeding, particularly from gastrointestinal tract; avoid aspirin wherever possible

Interactions summarised from Rainsford, 1996; Griffin and D'Arcy, 1997 and Hansten and Horn, 2000.

proteins does not lead to potentiation of the actions of the other drugs unless the clearance of the other drugs is inhibited. A good example is the displacement of phenytoin from its binding to plasma albumin. The percentage unbound increases, but the unbound concentrations are unchanged (Paxton, 1980). Most drug interactions that have been associated with aspirin therapy are probably due to salicylate, and therefore will be produced by salicylate salts as well as aspirin. A major exception is the increased risk of bleeding during treatment with warfarin or other oral anticoagulants (Table 4.3). This interaction is related to inhibition of platelet aggregation by aspirin, and does not occur with salicylate.

KINETICS OF ELIMINATION OF SALICYLATE

The time course of plasma concentrations of salicylate after a single dose of a salicylate follows a complex pattern because of the saturable metabolism and binding of the drug. Somewhat surprisingly the unbound concentrations decrease in nearly a log-linear fashion, but the total concentrations (free + bound) decrease, on a semi-logarithmic plot, in a more markedly non-linear fashion (Shen *et al.*, 1991a; Figure 4.6) because of the saturable binding of salicylate to plasma proteins. The half-life of elimination of total salicylate in plasma is about 2 to 3 hours after low doses (100 to 300 mg), but the initial half-life is considerably longer after larger doses.

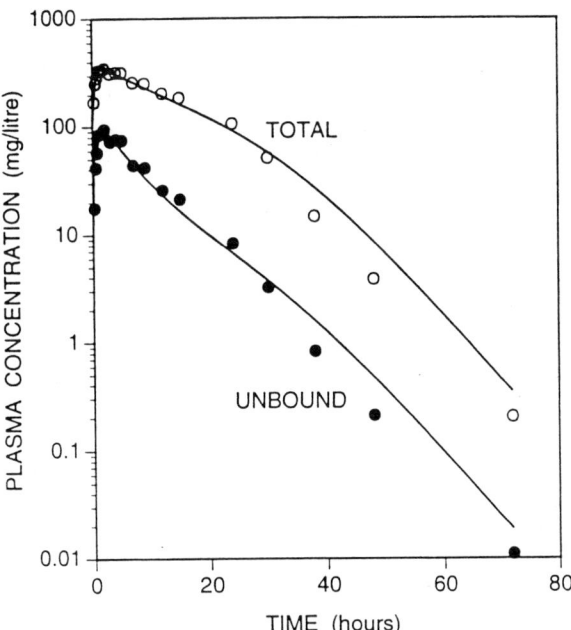

Figure 4.6   Time courses of plasma concentrations of unbound and total salicylate following an oral dose of sodium salicylate equivalent to 3 g salicylic acid. The curves are modelled on the saturable binding of salicylate to plasma proteins, its saturable metabolism to salicylurate and salicyl phenolic glucuronide, and non-saturable elimination by all other pathways. Because of the saturable binding to plasma proteins, the curvature in the semi-logarithmic plots of the plasma concentrations versus time is more marked than is the case with the unbound concentrations. (Redrawn from Shen *et al.*, 1991a.)

Accumulation of salicylates

Because of its short half-life, aspirin does not accumulate significantly in plasma during its long-term administration, although since it irreversibly acetylates cyclo-oxygenases its effects may outlast the transient appearance of unchanged aspirin. Salicylate, however, does accumulate since it is eliminated considerably more slowly, and it accumulates for at least 2 days during multiple dosage regimens. Two loading doses of 1300 mg aspirin 4 hours apart produce high initial plasma concentrations of salicylate, and may be considered for severe inflammation (Talbert *et al.*, 1979). Similar plasma concentrations of salicylate are produced by aspirin or simple salts of salicylate during long-term therapy and, because of the considerable interindividual differences in the pharmacokinetics of salicylate, the degree of accumulation of salicylate is highly variable (Paulus *et al.*, 1971; Graham *et al.*, 1977), whether or not the dosage is given at a rate proportional to body weight. For example, the plasma concentrations of salicylate varied from 44 to 330 mg/l when aspirin was administered at 50 mg/kg daily (Gupta *et al.*, 1975), even with the urine acidified to minimise the excretion of unchanged salicylate. In this case, the interpatient differences in the plasma concentrations are related to the rate of metabolism of salicylate to salicylurate, but, as outlined above, the renal clearance may also greatly influence the plasma concentrations of salicylate and the interpatient variations may be even more marked when urinary pH is not controlled. In clinical practice, non-compliance must always be considered when monitoring the plasma concentrations of salicylate in the decreasing number of patients who are treated with high doses of aspirin.

Non-linear accumulation of salicylate

Because of the saturable metabolism of salicylate, it may be expected that its plasma concentrations should increase disproportionately with any increase in dosage. In fact, however, the total plasma concentrations increase little more than in direct proportion to the dose of salicylate (Furst *et al.*, 1979; Figure 4.7). There are two reasons for this behaviour; first, the metabolism of salicylate to both salicyl phenolic glucuronide and salicylurate is induced with repeated dosage, and second, the binding of salicylate to plasma proteins decreases with increasing total concentrations in plasma. By contrast, the unbound concentrations increase more steeply with any increase in dosage (Figure 4.7).

The fluctuations in salicylate concentrations over a dosage interval depend both on the dosage interval and on the plasma concentrations achieved. At the high concentrations (above 150–200 mg/l) required for anti-inflammatory activity, the plasma concentrations generally fluctuate little (Cassell, 1979; Pachman *et al.*, 1979), as predicted by Levy and Giacomini (1978). There are, however, considerable fluctuations at lower concentrations because of the shorter half-life of salicylate at low plasma concentrations. The relationship between fluctuations and plasma salicylate concentration are of consideration when monitoring its plasma concentrations. Thus, a high level (over about 150 mg/l) indicates little change over a 6 to 12 hourly dosage interval, but lower concentrations indicate that the plasma concentrations may be very variable over the dosage interval (Levy, 1980).

Comparative pharmacokinetics of salicylate

The half-lives of elimination of salicylate from the rat and rabbit are approximately 6 and 5 hours (Nelson *et al.*, 1966; Short *et al.*, 1991). In these species, pharmacokinetics has not been studied over a wide range of doses and saturable elimination has therefore not been examined. In the dog the half-life of unbound salicylate appears constant at about 10 hours over a wide range of doses, but the total drug shows the saturable pattern seen in man (McCann and Palmisano, 1973).

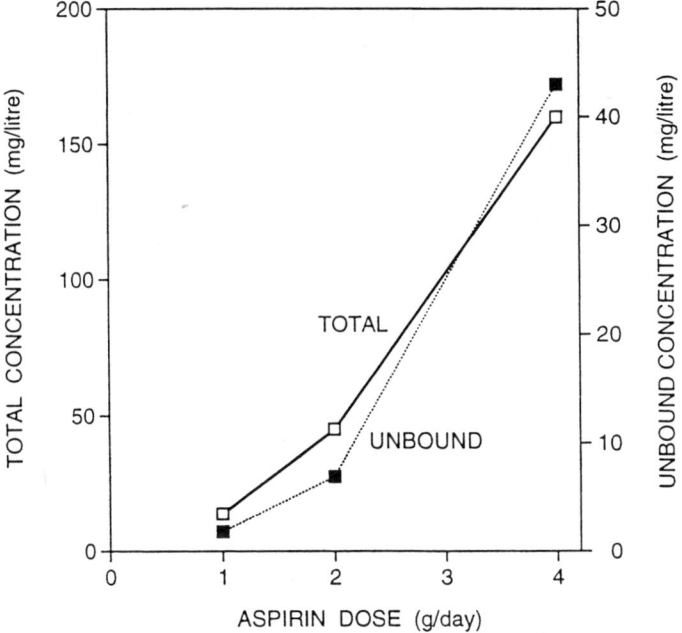

Figure 4.7   Relationship between unbound (broken line) and total plasma (unbroken line) concentrations of salicylate and dosage with aspirin for one week. Note that the relatively greater accumulation of unbound salicylate. The plasma concentrations during longer-term treatment may decrease due to auto-induction of the metabolism of salicylate. (Drawn from the data of Bochner *et al.*, 1985.)

PRO-DRUGS

Apart from aspirin many other salicylate pro-drugs have been synthesised, but only two, benorylate and salsalate, have been used orally in recent years. Their metabolism, together with that of two other pro-drugs, salicin and methyl salicylate, is described below. Salicin is the active component of willow bark, while methyl salicylate is also a natural product and is the major constituent of some essential oils. It is the major salicylate ester used topically.

Benorylate

Benorylate is the paracetamol ester of aspirin and paracetamol. It is thus a diester of salicylic acid (see diesters, below). Benorylate is well tolerated by the gastrointestinal tract, and is popular in some countries for paediatric and geriatric patients. Because of its low water solubility and slow dissolution within the gastrointestinal tract, benorylate is absorbed more slowly than either aspirin or paracetamol (Robertson *et al.*, 1972).

The major pathway of metabolism of benorylate is through the intermediate, phenetsal (the paracetamol ester of salicylate), which is hydrolysed rapidly to salicylate and paracetamol (Williams *et al.*, 1989b; Figure 4.8). Benorylate is completely hydrolysed during passage through the gastrointestinal mucosa of the rat (Humphreys and Smy, 1975), and is also hydrolysed more rapidly than aspirin by rat hepatocytes (Williams *et al.*, 1991). Overall, it appears that benorylate undergoes essentially total presystemic metabolism to salicylate and paracetamol, and that no other hydrolytic products are present in blood.

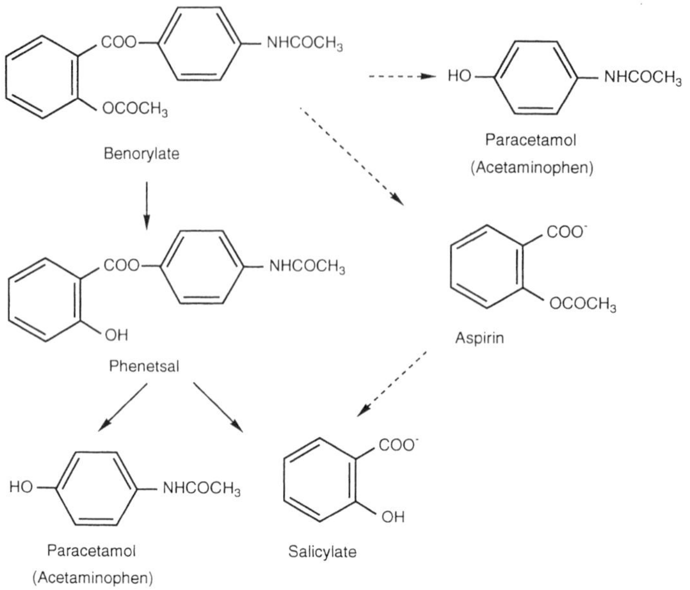

Figure 4.8 Metabolic pathways of benorylate. Phenetsal is the major intermediate in human liver and plasma *in vitro*, with the other intermediate, aspirin, being present at much lower levels. Paracetamol and salicylate are the only metabolites in peripheral blood. Paracetamol is metabolised to glucuronide, sulphate and mercapturate conjugates, while salicylate is metabolised and excreted as shown in Figure 4.3.

It has been found in clinical trials that dosage with benorylate yielded higher plasma concentrations of salicylate than the equivalent dosage with aspirin. Two reasons may be suggested for this surprising finding. First, benorylate is better tolerated orally than aspirin and its compliance may be better than aspirin, resulting in higher plasma concentrations, and second, in the trials where this phenomenon was found, benorylate was administered before aspirin (Aylward, 1973). Thus, auto-induction of the metabolism of salicylate may have occurred during the dosage period with benorylate resulting in the production of lower plasma concentrations during the subsequent dosage with aspirin.

Salsalate (salicylsalicylic acid)

This is the salicylic ester of salicylic acid (Figure 4.9). It is quite popular in North America, where it is often preferred to aspirin in the treatment of rheumatoid arthritis. Because of its low water solubility salsalate is absorbed more slowly than aspirin, although it is still well absorbed, as indicated by its low excretion in faeces and its almost complete recovery in urine (Dromgoole *et al.*, 1984).

Salsalate is detectable in plasma after its oral administration, and disappears with a half-life of 1.1 hours (Harrison *et al.*, 1981). Salsalate is largely but not completely hydrolysed to salicylate (Figure 4.9). Less than 1 per cent is excreted unchanged, but up to 13 per cent is excreted as glucuronides of the parent drug. Because of its incomplete conversion to salicylate, salsalate treatment yields plasma levels of salicylate that are about 15 to 25 per cent lower than those produced by either aspirin or a plain salicylate salt (Cohen, 1979; Dromgoole *et al.*, 1983).

There has been little study of the factors controlling the pharmacokinetics of salsalate. The absorption of salsalate is slowed by food, but not sufficiently to alter the plasma concentrations of the metabolite salicylate (Harrison *et al.*, 1992), while renal failure results in lower plasma concentrations of salsalate but higher concentrations of the metabolite salicylate (Williams *et al.*, 1986b). The elevated

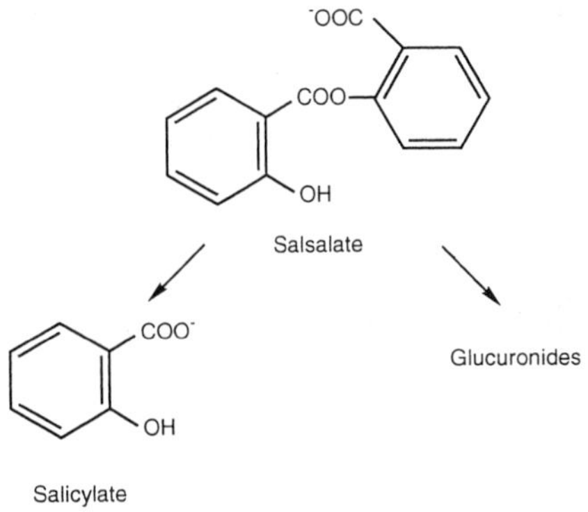

Figure 4.9 Metabolic pathways of salsalate (salicylsalicylate). The structures of the glucuronide conjugates of salsalate are not known; the formation of phenolic, acyl or both glucuronides is possible. Small amounts of salsalate are excreted unchanged. Some salicylate is excreted in urine while most is further metabolised as shown in Figure 4.3.

concentrations of salicylate are in contrast to the lack of effect of renal failure on the pharmacokinetics of salicylate administered alone (Lowenthal *et al.*, 1974), but the cause of these contrasting results is not known.

O-acyl esters of salicylate

O-acyl esters of salicylate are esterified through the phenolic hydroxyl group, and the acidic carboxylate group is unaltered. The best known example is, of course, aspirin, but recently, homologues have been synthesised with the acetyl group of aspirin replaced with acyl groups containing up to eight carbons. The salicylate esters are more lipophilic with increasing length of the carbon chain, and the higher homologues are extracted almost completely in one pass through rat liver (Hung *et al.*, 1998a). Like aspirin, the higher homologues have antiplatelet activity but have lesser gastrointestinal toxicity than aspirin in rats (Hung *et al.*, 1998b). Potentially, these esters could produce their antiplatelet activity within the portal circulation and thus be even more selective for platelets than aspirin. Another ester, octyl salicylate, has been applied to the skin as a sunscreen. Its systemic absorption is low (less than 2 per cent; Walters *et al.*, 1997).

Methyl salicylate

Methyl salicylate is the ester of salicylate in which the acidic carboxylate moiety is methylated but the phenolic hydroxyl is unchanged. It is therefore a neutral compound, which is used only for topical application in liniments and creams.

Methyl salicylate is readily hydrolysed to salicylate, although some methyl salicylate is found in blood. Humans hydrolyse methyl salicylate more slowly than rats and dogs, and only negligible amounts are found in the circulation in these latter species (Davison *et al.*, 1961). In agreement with the relatively slow hydrolysis in humans, some unchanged methyl salicylate is excreted in urine (Castagnou, 1952). Hydrolysis of methyl salicylate occurs in several tissues of the rat, with the greatest activity in the liver (Davison *et al.*, 1961). Hydrolysis also occurs in rabbit skin, with dermally administered methyl salicylate being absorbed and appearing in the venous drainage as both salicylate and unchanged ester (Behrendt and Kampffmeyer, 1989). In humans, unchanged methyl salicylate has not been detected in the dermis and subcutaneous tissue (Cross *et al.*, 1998).

The absorption of methyl salicylate from the skin depends on a number of factors, including the pharmaceutical formulation (i.e. the type and quantity of vehicle), the area covered, the time and site of application, skin blood flow, temperature and hydration, but in all cases studied the absorption of salicylate has been incomplete. For example, when creams and an ointment containing methyl salicylate were applied to 50 cm² areas of the forearm, only 12 to 20 per cent of the methyl salicylate was absorbed over a 10-hour period, even though the skin was covered (Roberts *et al.*, 1982). The plasma levels of salicylate resulting from the continuous application of these preparations was low, at less than 10 mg/l. In clinical use methyl salicylate may be applied to larger areas, proportionally increasing the amount absorbed, but the skin is usually uncovered leading to lesser absorption. Thus, when 5 g of pure methyl salicylate was applied to the skin of the chest and back, plasma concentrations peaked at less than 20 mg/l (Danon *et al.*, 1986), a concentration achieved by an oral dose of about 250 mg aspirin. Nevertheless, substantial amounts of methyl salicylate are absorbed when applied to large areas of skin. Further, methyl salicylate appears to promote its own absorption (Cross *et al.*, 1999).

Following the application of 20 per cent methyl salicylate, the estimated concentrations of salicylate in the dermis and subcutaneous tissue peak at about 4 mg/l (Cross *et al.*, 1998; 1999). This concentration appears low, but the concentrations of salicylate in the dermis and subcutaneous are 30-fold higher than in plasma, indicating that it was derived from the locally applied material and not from the circulation.

Diesters

These are derivatives of salicylate in which both the carboxylate and the phenolic group are esterified. The best known example is benorylate (see above), but the simplest is the methyl ester of aspirin, which shows lesser gastric toxicity than aspirin in rats (Rainsford and Whitehouse, 1980) but has not been evaluated in man. A more complex diester is the ethylcarbonate ester of methyl salicylate, which has been used as a model for studies on diffusion and metabolism within skin (Guzek *et al.*, 1989).

Salicin

Salicin is the glycoside of salicyl alcohol, which is hydrolysed and oxidised, finally to yield salicylate. It is thus a pro-drug of salicylate. After its oral administration to the rat, however, it yields less than a molecular equivalent of salicylate (Fotsch and Pfeiffer, 1990), due either to incomplete absorption or to incomplete conversion to salicylate.

ANALYTICAL METHODS FOR THE SALICYLATES

A large variety of analytical methods has been developed for determining the concentrations of salicylates in body tissues and fluids. Table 4.4 and the following review summarise the salient features of the different methods, in order to provide a guide for their application. The assayable limits for the general type of assays in Table 4.4 are approximate, because these limits vary with the quantities of plasma taken, the extraction procedure, and the performance of the equipment. A more detailed guide to the assays is provided by Stewart and Watson (1987).

As is generally the case, the high specificity and sensitivity of gas chromatographic (GC) or high performance liquid chromatographic (HPLC) assays makes them the preferred methods in pharmacokinetic and metabolic studies. In clinical work, chromatographic assays are of lesser use than colorimetric, fluorometric or immunoassay methods, because of the time taken to set up the chromatographic procedures. Possible interference by other drugs and their metabolites should also be considered, particularly when chromatographic methods are not being used. The chromatography of salicylate may be required to confirm the results of the other methods, particularly in forensic work. Further

TABLE 4.4

Research and clinical chemistry assays of aspirin and salicylate.

Technique	Drug	Use/fluid	Approximate minimal assayable limit
Colorimetric – Fe^{3+}	Salicylate	Clinical/plasma	50 mg/l
Fluorometric – terbium	Salicylate	Clinical/plasma	2 mg/l
Fluorescence	Salicylate	Clinical/plasma	1 mg/l
Fluorescence polarisation	Salicylate	Clinical	5 mg/l
High performance liquid chromatography	Salicylate	Research/plasma	50 μg/l
	Metabolites	Urine	2 mg/l
	Aspirin	Research/plasma	5 μg/l
Gas chromatography/mass spectrometry	Aspirin	Research/plasma	10 μg/l

specificity in chromatography can be achieved by the use of mass spectroscopy or, with HPLC, by ultraviolet spectroscopy of the peaks. With the availability of modern HPLC equipment, thin-layer chromatography is less used than previously in pharmacokinetic or metabolic studies on salicylates, although it may still be useful as part of a screen for overdoses.

Colorimetric and spectrophotometric procedures

The classical methods for determining salicylate have involved the purple complex formed with ferric ions in weakly acid solution. The most widely used has been the method of Trinder (1954), in which mercuric chloride in the reagent solution precipitates proteins, and the ferric chloride simultaneously complexes the salicylate. The result is an optically clear solution, the absorption being read at 540 nm.

The major problem with the colorimetric methods is the interference by other drugs and by endogenous compounds. Comparison with results obtained by chromatographic techniques show that there is no significant interference with the Trinder method by the concentrations of salicylurate or other metabolites in plasma (Jarvie *et al.*, 1987). Ketone bodies interfere with the colorimetric methods, but can be overcome by boiling the biological samples (Stewart and Watson, 1987).

A colorimetric method in which the purple ferric–salicylate complex is broken down by the addition of phosphoric acid allows an estimation of the blanks of individual patients, and appears at least as accurate as the Trinder method (Jarvie *et al.*, 1987). It also obviates the need for the toxic mercuric chloride used in Trinder's method. Overall, the colorimetric methods are sufficiently specific and sensitive for clinical use but not for pharmacokinetic or metabolic studies.

The high concentrations of salicylurate in urine prevent the assay of urinary salicylate, because this metabolite also gives a purple colour with ferric ions and the Trinder or related methods cannot be used with urine, except as a qualitative test for salicylate. The separate colorimetric assay of salicylate and salicylurate is, however, possible by the extraction of urine with ethylene dichloride and carbon tetrachloride followed by back extraction into an aqueous ferric nitrate solution, the principle being that carbon tetrachloride extracts salicylate with little salicylurate, whereas both salicylate and salicylurate are extracted by ethylene dichloride (Levy and Procknal, 1968). This method has been extensively used in studies on the pharmacokinetics of salicylate, and the results with this method have recently been confirmed with specific chromatographic assays (Shen *et al.*, 1991b). The total urinary output of salicylate and its conjugates, salicylurate and the two glucuronides, are assayed after hydrolysis with hydrochloric acid, and extraction of the total salicylate released (Levy and Procknal, 1968). A difficulty with the extraction procedures is that toxic chlorinated solvents such as carbon tetrachloride have often been used to extract the total salicylate. Dibutyl ether is an excellent alternative solvent (Page *et al.*, 1974).

Colour development with Folin–Ciocalteau reagent has also been used for the assay of salicylate, based on the reducing properties of the phenolic group (Smith, 1951). All the colorimetric assays detect not only both salicylate and salicylurate but also other salicylates, most importantly diflunisal. Unchanged aspirin does not give a colour with ferric ions, and aspirin can be measured by difference before and after hydrolysis; however, except at the earlier times after dosage the plasma concentrations of salicylate are considerably greater, and difference assays for unchanged aspirin are impractical.

Spectrophotometric assays have also been used based on the absorbance of salicylate in the ultraviolet range (Stevenson, 1960), although blanks are still appreciable and interference from a variety of other drugs is possible. More specific spectrophotometric or polarographic assays have been developed based on the oxidation of salicylate by salicylate hydroxylase (You, 1985; Morris *et al.*, 1990).

Fluorescence assays

The fluorescence of salicylate allows its assay at lower concentrations than is possible with the colorimetric assays, with amounts to low microgram quantities being easily assayed. The fluorescence of salicylate has been utilised after simple dilution of plasma, but blanks are considerable (Øie and Frislid, 1971). Lesser blanks are found after extraction of plasma by ether (Rowland and Riegelman, 1967) or by dialysis of plasma allowing the automated assay of salicylate (Hill and Smith, 1970). As in the colorimetric assay salicylurate also interferes with the fluorometric assay, but except in renal failure the plasma concentrations of salicylurate are low and its fluorescence does not interfere significantly with that of salicylate when the fluorometer settings are set at the maxima for salicylate. Furthermore, salicylurate can be differentiated from salicylate by the increasing fluorescence of salicylurate in the range pH 6 to 11, compared to the constant fluorescence of salicylate in the same pH range (Truitt *et al.*, 1955), or by the longer wavelength of maximal excitation of salicylurate (Pütter, 1975), particularly through the use of synchronous fluorescence in which the excitation and emission wavelengths are changed simultaneously (de la Pena *et al.*, 1988). Interference from salicylurate can be removed entirely by the formation of the extremely fluorescent ternary complex with the lanthanide, terbium, and EDTA, a method that allows the assay of salicylate in as little as 10 μl of plasma (Bailey *et al.*, 1987). However, several substituted salicylates, including mesalazine, also yield fluorescent products (Huang and Gao, 1989).

Aspirin is not fluorescent at the neutral to alkaline pH values at which the fluorescence of salicylate is usually measured, but its weak fluorescence in acetic acid–chloroform solutions (Miles and Schenk, 1970) allows its determination, simultaneously with salicylic acid, by synchronous fluorescence spectrometry without chromatography (Konstantianos *et al.*, 1991).

Neither the colorimetric nor the fluorometric assays allow direct determination of the salicylate glucuronides. These can be assayed colorimetrically or fluorometrically in urine after the conversion of the acyl glucuronide to the hydroxamate by the addition of hydroxylamine and the acid-catalysed hydrolysis of the phenolic glucuronide (Schachter and Manis, 1958). The same techniques have been used in a chromatographic assay (Mallikaarjun *et al.*, 1989).

Gas chromatography

This technique has been used to separate and assay aspirin in the presence of high concentrations of salicylate. Salicylate can be assayed at the same time. Most methods involve the conversion of aspirin and salicylate to their more volatile trimethylsilyl derivatives (Rowland and Riegelman, 1967; Thomas *et al.*, 1973), but the two functional groups on salicylate, the phenolic and carboxyl groups, lead to the possibility of a mixture of trimethylsilyl derivatives, and care must be taken to convert the salicylate to a single derivative. The trimethylsilyl derivatives, can be detected by flame detectors or by mass spectroscopy, but a silicon selective detector has also been used (Osman and Hill, 1982). The coupling of mass spectrometers to gas chromatographs greatly increases the sensitivity and specificity of drug assays, and aspirin has been assayed in such a coupled system (Pedersen and Fitzgerald, 1985).

High performance liquid chromatography (HPLC)

The specificity, sensitivity and relative ease of HPLC assays has made this the most frequently used method in pharmacokinetic and metabolic studies in recent years. Several systems have been developed to assay aspirin, salicylate, salicylurate and gentisate in both plasma and urine (Cham *et al.*, 1979; Reidl, 1983; Shen *et al.*, 1990), although a common problem has been an endogenous compound with a similar retention time to gentisate (Aghabeigi and Henderson, 1992). The use of electrochemical detection with HPLC allows the sensitive assay of the gentisate and 2,3-dihydroxybenzoate (Coudray *et al.*, 1995). This technique is useful in studies on the interaction between salicylate and free radicals.

In metabolic studies, all the urinary products of salicylate, including the glucuronides, can now be assayed by HPLC in the one system, making this the optimal procedure for the evaluation of the kinetics of the known metabolites (Shen *et al.*, 1991b). The salicylates are usually detected by absorption in the ultraviolet range, the wavelength used depending upon the material being assayed. Scanning the HPLC peaks by the photodiode array detector increases the specificity of the ultraviolet detection (Hill and Langner, 1987), while the fluorescence detector increases the sensitivity and specificity of the HPLC assays of salicylate and salicylurate. Aspirin is only weakly fluorescent under normal HPLC conditions, and post-column hydrolysis to salicylate increases the sensitivity of its HPLC assay (Siebert and Bochner, 1987). The use of HPLC requires initial extraction or protein precipitation, and several procedures are described.

Recovery

Various procedures have been used to prevent the hydrolysis of unchanged aspirin before extraction from plasma or blood. Rapid cooling and the addition of protein precipitants is probably the best technique to measure the whole blood concentrations of unchanged aspirin (Marzo *et al.*, 1992). When assaying the plasma concentrations of aspirin, inhibitors of the enzymatic hydrolysis of aspirin have been added to prevent its hydrolysis during centrifugation. Fluoride has been used as an inhibitor (Rowland and Riegelman, 1967), although there is some doubt about its efficacy (Marzo *et al.*, 1992), and physostigmine has also been used (Cham *et al.*, 1980). The treated plasma can be separated and stored at $-80°C$, although there is some hydrolysis, possibly during the thawing phase (Reidl, 1983).

Another problem that must be considered in the procedures involved in the preparation of extracts for chromatography is the possible loss of salicylic acid by sublimation when extracts are concentrated. One procedure to overcome this problem is the evaporation of the solvent at $0°C$ in long tubes (15 cm) under a gentle stream of nitrogen (Goehl *et al.*, 1981).

Immunoassay

This technique has become very popular in clinical laboratories in recent years because of the speed, specificity and the absence of an extraction procedure. A commercial fluorescent polarisation assay shows little cross-reactivity with aspirin, salsalate and salicylurate, but gentisate shows about the same response as salicylate (Koel and Nebinger, 1989). Because of the low plasma concentrations of gentisate, this should not produce significant interference. There is, however, considerable reactivity from diflunisal, sulphasalazine and its metabolite mesalazine, and the method is not applicable if the patient is taking these drugs.

Nuclear magnetic resonance spectroscopy (NMR)

NMR has been applied in the identification of various compounds, including salicylate and its metabolites, in urine (Vermeersch *et al.*, 1988), particularly after chromatographic clean up (Wilson *et al.*, 1988).

CHAPTER 4

The application of NMR is, however, limited by its low sensitivity and the specialised equipment and expertise required, but it may be useful in some cases as an initial screening procedure for salicylates and their metabolites.

SALICYLATE DERIVATIVES

Mesalazine

Mesalazine (5-aminosalicylate, mesalamine) is a salicylate derivative that is not an analgesic but is used in the treatment of ulcerative colitis and Crohn's disease. It is most commonly used either by itself in sustained release formulations or as pro-drugs, particularly sulphasalazine (Salazopyrine) or olsalazine (Azodisalicylate), which release mesalazine in the large intestine. At most, only very small amounts of mesalazine are converted further to salicylate. Sulphasalazine is also used as a slow-acting antirheumatic drug.

Oral absorption

Mesalazine is rapidly and at least 75 per cent absorbed from the upper gastrointestinal tract after the administration of suspensions or plain capsules (Nielsen and Bondesen, 1983; Myers et al., 1987), although its absorption is decreased by a meal through binding to constituents in food (Yu et al., 1990). The absorption of mesalazine is considerably lower when delivered directly to the large intestine or rectum (Bondesen et al., 1988; Grisham and Granger, 1989) than when released in the upper gastrointestinal tract. The capacity of the colon to absorb mesalazine is even further reduced by inflammation (Campieri et al., 1985).

Mesalazine is one of the few drugs where the concentrations at the sites of absorption have been estimated. In the cat colon perfused with 10 mM mesalazine, the interstitial concentration of the drug was estimated as 130 µM (20 mg/l; Grisham and Granger, 1989), sufficient for it to act, as described below, as a scavenger of free radicals.

Of the pro-drugs of mesalazine, the most widely used is sulphasalazine, which is poorly absorbed from the small intestine but metabolised in the large intestine to sulphapyridine and mesalazine. The released sulphapyridine is almost totally absorbed, in contrast to the poor absorption of mesalazine (Bondesen et al., 1986). The cleavage of sulphasalazine in the large intestine means that there is a delay of about 4 hours before any mesalazine or sulfapyridine is absorbed. This delay before the detection of sulphapyridine in blood has, in fact, been used as a test for the transit time to the large intestine (Kennedy et al., 1979). Of more clinical importance is that much of the haematological toxicity of sulphasalazine is due to the metabolite sulphapyridine (Pirmohamed et al., 1991). Olsalazine is, like sulphasalazine absorbed poorly from the small intestine (Sandberg-Gertzen et al., 1983), and is cleaved to yield two molecules of mesalazine in the large intestine. At least 30 per cent of an oral dose is absorbed as mesalazine and its metabolites (Willoughby et al., 1982), but only about 2 per cent of an oral dose of olsalazine is absorbed intact (Ryde and Ahnfelt, 1988). The conversion of the two pro-drugs to mesalazine and the subsequent absorption of the active form is very similar (Stretch et al., 1996).

Comparison between the coated tablets and the pro-drugs is of interest in the therapeutic properties of mesalazine. In subjects with normal transit times, the presently available coated tablets release more mesalazine in the small intestine and less into the large intestine than the two pro-drugs (Rijk et al., 1988; Stoa-Birketvedt and Florholmen, 1999). For both the sustained release preparations and the pro-drugs, a decreased transit time through the gastrointestinal tract reduces the delivery of mesalazine to the large intestine, although the reduction is greatest for the two pro-drugs (Christensen et al., 1987; Rijk et al., 1989).

Mesalazine has been administered as retention enemas. The formulation of these preparations has a marked effect on the spread and retention time in the colon, which are important factors in the efficacy of the enemas (Otten et al., 1997).

Elimination

Mesalazine is primarily eliminated by metabolism to N-acetyl-5-aminosalicylate (Figure 4.10). Mesalazine is not only acetylated systemically, presumably in the liver, but is also acetylated locally by intestinal flora (Dull *et al.*, 1987b) and colonic epithelial cells (Ireland *et al.*, 1990).

Three other metabolites have recently been identified (Figure 4.10); N-formyl-5-aminosalicylate (Tjornelund *et al.*, 1991), N-butyryl-5-aminosalicylate, and the conjugate with glucose, N-β-D-glucopyranosyl-5-aminosalicylate (Tjornelund *et al.*, 1989). All are present in human plasma, but only trace amounts of the glucose conjugate are found in urine because of its instability under the pH conditions and temperature of urine. Like other aromatic amines mesalazine reacts non-enzymatically with glucose, but the amounts of the glucose conjugate in plasma are too great to be formed in this manner from the available concentrations of the drug and glucose. All the pathways of elimination have not yet been identified, and there are unidentified compounds in faeces and urine (van Hogezand *et al.*, 1988).

As is the case with salicylate, mesalazine is chemically modified by free radicals and the reactions may be related to its therapeutic activity in ulcerative colitis. Thus, faeces from such patients treated with sulphasalazine contain some of the same products that are produced by reaction of mesalazine with free radicals in the Fenton system (Ahnfelt-Ronne *et al.*, 1990). These products are not present in faeces from rheumatoid patients treated with sulphasalazine, indicating that the local metabolism of mesalazine is specific to ulcerative colitis.

Mesalazine is metabolised by activated neutrophils and monocytes to yield salicylate and gentisate (Davis *et al.*, 1989), as well as to compounds in which the amino group is modified. Hypochlorous acid is formed by activated neutrophils, and several products have been identified or suggested from its reaction with mesalazine. A possible product is 5-nitrososalicylic acid (Williams and Hallett, 1989). Intermediates in the production of gentisate are the potentially toxic iminoquinone and quinone derivatives, which react with thiols in a similar fashion to the reactive intermediate formed by the oxidation of paracetamol (Liu *et al.*, 1995). The scavenging of hypochlorite may protect against the potentially

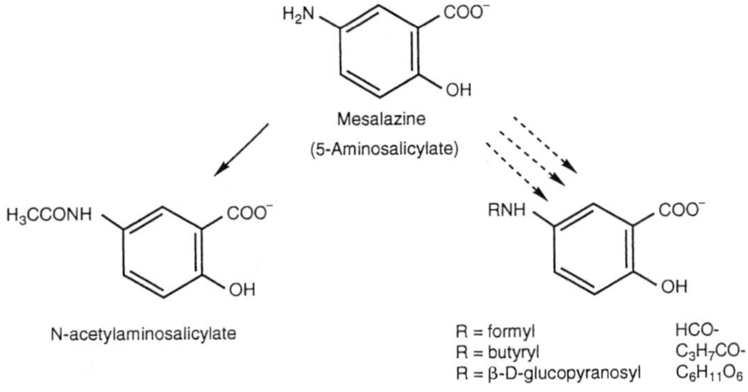

Figure 4.10 Pathways of metabolism of mesalazine (5-aminosalicylate) and its precursor, olsalazine. Sulphasalazine, a diazo conjugate of mesalazine with sulphapyridine, is cleaved to yield mesalazine and sulphapyridine. Sulphapyridine is then metabolised by acetylation and hydroxylation.

■ CHAPTER 4 ■

damaging effects of the activated cells in the inflammatory bowel disease, but could also decrease the antimicrobial activity of the white blood cells or lead to toxicity.

Mesalazine is eliminated rapidly, with a half-life of about 50 minutes. A distributional phase with a half-life of 17 minutes is also seen when the drug is administered intravenously (Myers *et al.*, 1987). Plasma concentrations of the acetyl metabolite are initially lower than those of mesalazine after intravenous dosage or after absorption from the upper gastrointestinal tract, but may exceed the lower plasma concentrations of unchanged mesalazine after release in the large intestine, consistent with the acetylation of the drug at this site. During long-term treatment with sulphasalazine, the plasma concentrations of the acetyl metabolite of mesalazine are markedly lower than after a single dose (Taggart *et al.*, 1992). The cause of this phenomenon has not been established. During treatment with sulphasalazine the plasma levels of mesalazine are increased by age but are not altered by the classical acetylator status, indicating that mesalazine can only be metabolised to a minor extent by the NAT2 enzyme responsible for the acetylation of isoniazid and the sulphonamides (Taggart *et al.*, 1992).

Although poorly absorbed, unchanged sulphasalazine and olsalazine are detectable in plasma after oral administration. During daily treatment with 2 g sulphasalazine, the average concentrations of the unchanged drug are about 5 mg/l (Taggart *et al.*, 1992). The half-life of sulphasalazine in young patients with rheumatoid arthritis is about 6 hours, increasing to 10 to 20 hours in the elderly, although the plasma concentrations during long-term therapy in the elderly are not significantly higher than in young patients. Olsalazine has a shorter half-life of elimination, at about 55 minutes, while its sulphate conjugate has the surprisingly long half-life of about 7 days (Ryde and Ahnfelt, 1988).

Assays

Mesalazine, its acetylated metabolite and pro-drugs have been assayed by a variety of HPLC techniques. The most sensitive involve fluorescence (Lee and Ang, 1987; Bystrowska *et al.*, 2000) or electrochemical detection (Nagy *et al.*, 1988). A method in which mesalazine and all the presently known metabolites are chromatographed appears to be very suitable for detailed pharmacokinetic studies (Tjornelund and Hansen, 1991).

DIFLUNISAL

There are some similarities between the elimination and pharmacokinetics of diflunisal and salicylate, but dissimilarities are also noted.

Absorption

Diflunisal is less water-soluble than either aspirin or salicylate, and its oral absorption is consequently slower (Nuernberg and Brune, 1989). Thus, peak concentrations are attained at 2 to 3 hours after dosage.

Distribution

Diflunisal is bound to plasma albumin much more strongly than is salicylate. However, like salicylate, the unbound fraction of diflunisal increases with the concentration of diflunisal, and over the therapeutic concentrations (50 to 250 mg/l) the unbound fraction in human plasma nearly doubles from about 0.1 to just under 0.2 per cent (Verbeeck *et al.*, 1990). Like several strongly protein-bound drugs, the uptake into synovial fluid is slow and lags behind the absorption of the drug (Nuernberg *et al.*, 1991).

Elimination

Diflunisal is metabolised to acyl and phenolic glucuronides (Figure 4.11), as is the case with salicylate, but the glycine conjugate (corresponding to salicylurate) has not been detected in man or other species. Another difference is that the sulphate conjugate of diflunisal accounts for about 10 per cent of the urinary metabolites in man (Loewen *et al.*, 1986), whereas the sulphate adduct of salicylate has not been identified in man. A minor metabolite is 3-hydroxydiflunisal, which is excreted as unidentified conjugate (Macdonald *et al.*, 1989).

As is the case with other acyl glucuronides, including that of salicylate, the acyl glucuronide of diflunisal is reactive. Three types of reactions occur; it is hydrolysed back to diflunisal (see below), rearranged (Dickinson and King, 1989), and covalent adducts of diflunisal are produced with proteins (Watt and Dickinson, 1990). The adducts with plasma proteins accumulate during treatment, and concentrations equivalent to about 3 mg/l (about 1 to 2 per cent of the levels of unchanged diflunisal) have been reported, although further accumulation is possible. The plasma protein adduct is eliminated in a biphasic fashion with a terminal half-life of about 10 days (McKinnon and Dickinson, 1989), possibly related to the turnover of protein. In the rat covalent binding slowly occurs in a variety of tissues (King and Dickinson, 1993), particularly in the bladder because of its exposure to high concentrations of the acyl glucuronide and a rearranged product of the glucuronide (Dickinson and King, 1993). The total resorption of diflunisal acyl glucuronide from the bladder is, however, insignificant, although the resorption of diflunisal occurs readily (Dickinson and King, 1996).

The covalent binding of drugs to proteins is considered as a potential cause of tissue toxicity and immunogenic reactions. The injection of adducts of diflunisal with albumin leads to the production of circulating antibodies in the rat (Worrall and Dickinson, 1995) although only mild hypersensitivity reactions are associated with diflunisal therapy (Cook *et al.*, 1988).

After single doses, the half-life of elimination of diflunisal is about 10 hours. The clearance of diflunisal, like that of salicylate, is dose-dependent, and the accumulation is greater than predicted from a single dose. During multiple dosage the total plasma concentrations are about double that predicted,

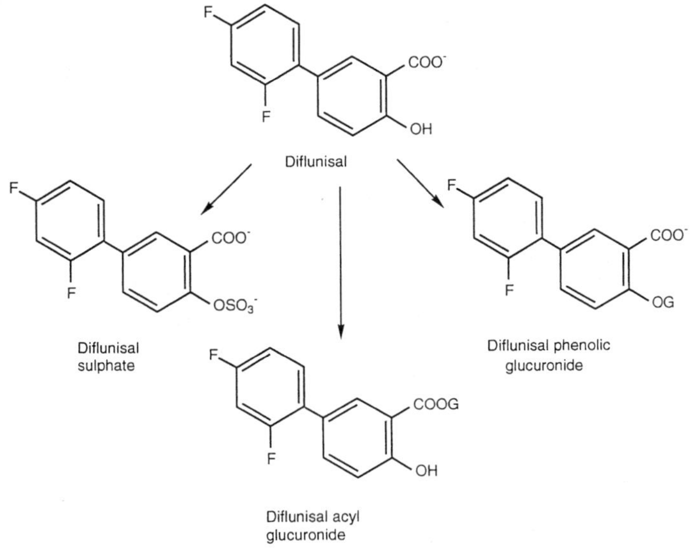

Figure 4.11 Metabolic pathways of diflunisal. G = glucuronide group. The acyl glucuronide is unstable, being hydrolysed back to diflunisal, forming rearrangement products and covalent adducts with plasma proteins. The hydroxylated metabolite is largely excreted as an as yet unidentified conjugate.

■ CHAPTER 4 ■

but because of the saturable protein binding the unbound plasma concentrations increase to a greater extent. Thus, during dosage with 500 mg diflunisal twice daily, the unbound plasma concentrations are about 350 per cent of the plasma concentrations predicted from a single dose (Verbeeck *et al.*, 1990). Because of the saturable metabolism and slow absorption the plasma concentrations are well sustained during long-term therapy, and the plasma concentrations during dosage with 1000 mg once a day are little different from the concentrations at 500 mg twice a day (Mojaverian *et al.*, 1985).

Sulphation is normally considered to be a high-affinity, low-capacity pathway, but in the case of diflunisal the sulphation pathway is not saturable at therapeutic doses in man. For diflunisal both glucuronidation pathways are saturable (Verbeeck *et al.*, 1990), as opposed to saturation of only the phenolic glucuronide pathway of salicylate.

The plasma clearance of diflunisal decreases with decreasing renal function (Verbeeck *et al.*, 1979; Meffin *et al.*, 1983). In rats preformed acyl glucuronide is rapidly hydrolysed, indicating that retention of the ester glucuronide and its subsequent hydrolysis back to diflunisal is the cause of the apparently reduced clearance (Dickinson *et al.*, 1989). By comparison, the phenolic glucuronide is stable and thus does not undergo this futile cycling (Brunelle and Verbeeck, 1997).

There have been few studies on the effects of other factors or drugs on the pharmacokinetics of diflunisal. Parallel with salicylate, there may be the influence of sex hormones on the kinetics of diflunisal. The clearance of diflunisal is about 60 per cent higher in men than in women, and is elevated to a similar extent in women by oral contraceptives (Macdonald *et al.*, 1990). Smoking increases the clearance of diflunisal by about 35 per cent, possibly by induction of the minor hydroxylation pathway. In cirrhosis, the unbound clearance of diflunisal is decreased by about one-third because of impairment of the glucuronidation pathways (Macdonald *et al.*, 1992).

Pro-drugs of diflunisal

As is the case with salicylate, acyl esters of diflunisal have been prepared with the aim of decreasing the gastrointestinal toxicity of diflunisal and producing selective inhibitors of platelet aggregation. As with the salicylate esters, all the diflunisal esters are hydrolysed readily with the lipophilic ester, the acetyl ester showing the greatest extraction by the rat liver (Hung *et al.*, 1998c).

Assays of diflunisal

Diflunisal can be assayed by a variety of HPLC techniques, involving absorbance or fluorometric detection (Meffin *et al.*, 1983; Ray and Day, 1983; Dickinson and King, 1989). The stability of the conjugates of diflunisal must be considered in the assay of these conjugates. The phenolic glucuronide is stable except at very low pH, but the acyl glucuronide is unstable at pH values above 4.5, with rearrangement and hydrolysis occurring (Hansen-Moller *et al.*, 1987; Dickinson and King, 1989; Watt and Dickinson, 1990). The sulphate conjugate is stable at physiological pH, but unstable below about 4.5. The optimal pH for the stability of these two conjugates during the preparation of samples for HPLC samples is 4 to 5, although there is some loss of the phenolic glucuronide due to trapping in the protein pellet from the plasma (Dickinson and King, 1989).

SALICYLAMIDE

Salicylamide is the amide of salicylic acid and has analgesic, antipyretic and sedative activities, although these activities are at best weak. Although salicylamide has little pharmacological activity, it has been used as model drug in pharmacokinetic and metabolic studies. Salicylamide is not hydrolysed to salicylate in the body and must be considered largely as a distinct drug, although small amounts of salicylamide may be formed from salicylate. The plasma concentrations of salicylamide are considerably lower than those achieved by equivalent doses of sodium salicylate because of the first-pass metabolism of salicylamide and its more extensive tissue distribution resulting from lesser binding to plasma proteins (Seeberg *et al.*, 1951).

Absorption

Although well absorbed from the gastrointestinal tract, salicylamide undergoes a variable degree of first-pass metabolism. The extent of first-pass metabolism is dose-dependent, with high first-pass metabolism and very low plasma concentrations after small doses and substantive plasma concentrations after doses of 1 to 2g (Barr *et al.*, 1973). Under such circumstances the formulation becomes a critical variable, because a slow-release form should provide lower concentrations at first-pass metabolic sites than those produced by a rapid-release preparation. The result should therefore be a greater first-pass metabolism after the slow-release form. Consistent with this prediction, the plasma concentrations of salicylamide after an oral dose of 1.3g are about five times higher following administration of the drug in solution than if given as a suspension or tablet, whereas the total metabolite concentrations are nearly equal for the three preparations (Barr *et al.*, 1973). The extent of first-pass metabolism also depends on the liver blood flow, the fractional absorption decreasing with increasing perfusion.

Elimination

The amide group of salicylamide is unaffected by metabolism and therefore salicylate is not a metabolite, but the aromatic ring and phenolic group are modified in a similar fashion to salicylate (Figure 4.12). Salicylamide is mainly eliminated in man by transformation to the phenolic glucuronide and sulphate (Becher *et al.*, 1952; Foye *et al.*, 1975). The phenolic glucuronide is analogous to the phenolic glucuronide of salicylate, but the sulphate conjugate of salicylate has never been definitely identified in

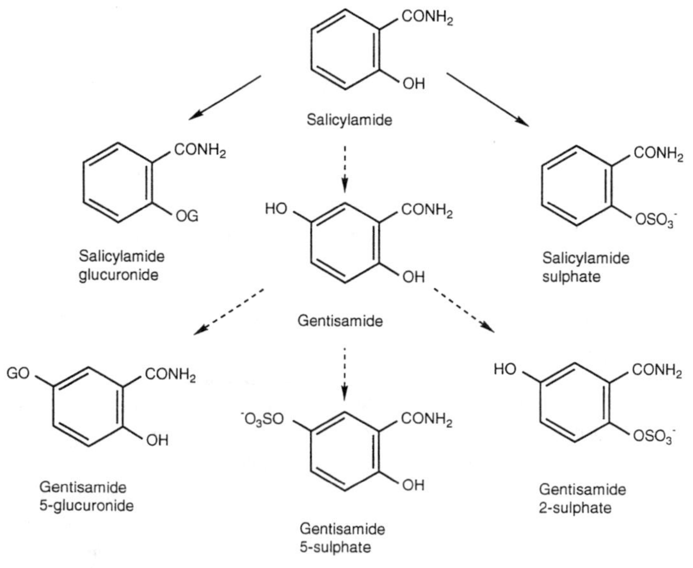

Figure 4.12 Pathways of metabolism of salicylamide. G = glucuronide group. A metabolite hydroxylated in the 3 position, 2,3-dihydroxybenzamide, has also been identified in the mouse.

man although the sulphate conjugate of diflunisal is formed in man. As is the case with salicylate, hydroxylation of salicylamide also occurs, but only small proportions of the drug are converted to gentisamide (the analogue of gentisate) and from this to its glucuronide (Song *et al.*, 1972). Hydroxylation to the 2,3-dihydroxybenzamide followed by glucuronidation has also been identified in mice (Howell *et al.*, 1988).

Like salicylate, the metabolism of salicylamide is saturable and the proportion of minor metabolites increases as the major pathways become saturated; thus the production of gentisamide is significant in man only when the formation of the sulphate is saturated (Levy and Matsuzawa, 1967). The availability of inorganic sulphate controls the relative proportions of the metabolites. Both in man *in vivo* and in isolated rat hepatocytes, sulphoconjugation is enhanced while glucuronide formation is suppressed if the availability of sulphate is increased (Levy and Matsuzawa, 1967; Koike *et al.*, 1981).

From studies in rat hepatocytes, the relative production of the sulphate conjugate is favoured at low concentrations because of the relevant kinetic parameters of this pathway. In these cells the apparent K_m for sulphoconjugation is 0.006 mM, while the V_{max} of sulphate formation is dependent on the supply of sulphate, varying between 0.5 and 2.1 nmoles/min per million cells at 0.5 and 1.2 nM inorganic sulphate. The K_m value for the glucuronidation of salicylamide is much higher at 0.19 mM, while the V_{max} is of the same order at 1.28 nmoles/minute per million cells in the absence of sulphate. Because of the lower K_m value for the sulphoconjugation, this process is favoured over glucuronidation at low salicylamide concentrations.

An interesting aspect of the metabolism of salicylamide has been demonstrated in the perfused rat liver. The sequential metabolism of salicylamide is not identical to the metabolic pattern of a preformed metabolite. As discussed above, salicylamide is hydroxylated to gentisamide and then glucuronidated. By comparison, preformed gentisamide is primarily sulphated at both the 2- and 5-positions to yield gentisamide, and is glucuronidated only when the higher-affinity sulphation pathway is saturated (Xu *et al.*, 1990). Possibly, the glucuronidation of the metabolically formed gentisamide is favoured because of the occurrence of both oxidation and glucuronidation processes on the endoplasmic reticulum in the liver.

Apart from its biotransformation in the liver, salicylamide is extensively metabolised in the intestine in the rabbit (Barr and Riegelman, 1970; Barr *et al.*, 1973), rat (Shibaski *et al.*, 1981) and dog (Gugler *et al.*, 1975). Data obtained in the rat indicate that significant first-pass metabolism of salicylamide occurs both in the small intestine and liver, with the greater extraction in the liver (Xu *et al.*, 1989). Metabolism also occurs in the kidneys, lung and forelimbs of dogs, particularly at low doses, and the lung may contribute to the high first-pass metabolism of low doses of salicylamide (Fielding *et al.*, 1986).

Like salicylamide, ascorbic acid is metabolised in part to a sulphate conjugate, and ascorbic acid has therefore been used to explore the determinants of the formation of salicylamide sulphate. Orally administered ascorbic acid markedly decreases the formation of salicylamide sulphate in man and rats, whereas intravenous administration of the vitamin has little or no effect in rats (Houston and Levy, 1976). Similarly, the oral administration of sulphate, but not its intravenous infusion, decreases the oral availability of salicylamide in the dog, indicating a local effect of sulphate on the intestinal first-pass metabolism of salicylamide (Waschek *et al.*, 1985).

Interactions between salicylamide and salicylates have been detected, although the mechanisms are unclear. The formation of both the sulphate and glucuronide conjugates of salicylamide is inhibited by salicylate and paracetamol (Levy and Yamada, 1971). Salicylamide significantly increases the plasma concentrations of aspirin (Abdel-Rahman *et al.*, 1991), although the mechanism is again unknown.

Apart from the availability of sulphate, several factors affecting the metabolism of salicylamide have been identified. For example, age affects the metabolic profile, the sulphate conjugate being the predominant metabolite in children whereas the glucuronide is the major metabolite in adults (Alam *et al.*, 1977). Pyrogen induced fever reduces the formation of the glucuronide metabolite in man (Song *et al.*, 1972), while in acute intermittent porphyria the half-life of salicylamide is increased by 43 per cent due to impairment of the formation of the sulphate and hydroxylated metabolites (Song *et al.*, 1974).

REFERENCES

Aarons, L., Clifton, P., Fleming, G. and Rowland, M. 1980, Aspirin binding and the effect of albumin on spontaneous and enzyme-catalysed hydrolysis. *Journal of Pharmacy and Pharmacology*, **32:** 537–543.

Aarons, L., Hopkins, K., Rowland, M., Brossel, S. and Thiercelin, J.-F. 1989, Route of administration and sex differences in the pharmacokinetics of aspirin administered as its lysine salt. *Pharmaceutical Research*, **6:** 660–666.

Abdel-Rahman, M.S., Reddi, A.S., Curro, F.A., Turkall, R.M., Kardy, A.M. and Hansrote, J.A. 1991, Bioavailability of aspirin and salicylamide following oral co-administration in human volunteers. *Canadian Journal of Physiology and Pharmacology*, **69:** 1436–1442.

Aghabeigi, B. and Henderson, B. 1992, Observations on the measurement of 2,3- and 2,5-dihydroxy-benzoic acid using high performance liquid chromatography. *Inflammopharmacology*, **1:** 363–368.

Ahnfelt-Ronne, I., Nielsen, O.H., Christensen, A., Langholz, E., Binder, V. and Riis, P. 1990, Clinical evidence supporting the radical scavenger mechanism of 5-aminosalicylate. *Gastrenterology*, **98:** 1162–1169.

Alam, S.N., Roberts, R.J. and Fischer, L.J. 1977, Age-related differences in salicylamide and acet-aminophen conjugation in man. *Journal of Pediatrics*, **90:** 130–135.

Alpen, E.L., Mandel, H.G., Rodwell, V.W. and Smith, P.K. 1951, The metabolism of C^{14} carboxyl salicylic acid in the dog and man. *Journal of Pharmacology and Experimental Therapeutics*, **102:** 150–155.

Alvan, G., Bergman, U. and Gusttafsson, L.L. 1981, High unbound fraction of salicylate in plasma during intoxication. *British Journal of Clinical Pharmacology*, **11:** 625–626.

Amon, I., Amon, K., Zschiesche, M. and Hueller, H. 1977, Kinetics and distribution of acetylsalicylic acid in pregnant and nonpregnant women. *Zentralblatt für Pharmazie, Pharmakotherapie und Laboratoriums-diagnostik*, **116:** 385–391.

Amsel, L.P. and Levy, G. 1969, Drug biotransformation interactions in man. II: A pharmacokinetic study of the simultaneous conjugation of benzoic and salicylic acids with glycine. *Journal of Pharmaceutical Sciences*, **58:** 321–326.

Ashton, J.M., Bolme, P. and Zerihun, G. 1989, Protein binding of salicylic acid and salicyluric acid in serum from malnourished children: the influence of albumin, competitive binding and non-esterified fatty acids. *Journal of Pharmacy and Pharmacology*, **41:** 474–480.

Au, J.L.-S., Dalton, J.T. and Wientjes, M.G. 1991, Evidence of significant absorption of sodium salicy-late from urinary bladder of rats. *Journal of Pharmacology and Experimental Therapeutics*, **258:** 357–364.

Aylward, M. 1973, Toxicity of benorylate. *British Medical Journal*, **3:** 347–348.

Badaway, A.A.-B. 1982, Mechanisms of elevation of rat brain tryptophan concentration by various doses of salicylate. *British Journal of Pharmacology*, **76:** 211–213.

Bailey, M.P., Rocks, B.F. and Riley, C. 1987, Rapid spectrofluorimetric determination of plasma sali-cylate with EDTA and terbium. *Analytica Chimica Acta*, **201:** 335–338.

Baldwin, S.R., Simon, R.H., Boxer, L.A., Till, G.O. and Kunkel, R.G. 1985, Attenuation by 2,3-dihy-droxybenzoate of acute lung injury induced by cobra venom factor in the rat. *American Review of Respiratory Disease*, **132:** 1288–1293.

Bannwarth, B., Netter, P., Gaucher, A. and Royer, R.-J. 1986, Delayed appearance of salicylate in cerebrospinal fluid. *Journal of Rheumatology*, **13:** 993–994.

Bareggi, S.R., Pirola, R. and De Benedittis, G. 1998, Skin and plasma levels of acetylsalicylic acid: a comparison between topical aspirin/diethyl ether and oral aspirin in acute herpes zoster and post-herpetic neuralgia. *European Journal of Pharmacology*, **54:** 231–235.

Barr, W.H., Aceto, T., Chung, M. and Shukur, M. 1973, Dose dependent drug metabolism during the absorptive phase. *Revue Canadienne de Biologie*, **32 (Suppl.):** 33–42.

Barr, W.H. and Riegelman, S. 1970, Intestinal drug absorption and metabolism. II Kinetic aspects of intestinal glucuronide conjugation. *Journal of Pharmaceutical Sciences*, **59:** 164–168.

Batterman, R.C. and Sommer, E.M. 1953, Fate of gentisic acid in man as influenced by alkalinization and acidification. *Proceedings of the Society for Experimental Biology and Medicine*, **83:** 376–379.

■ CHAPTER 4 ■

Bayles, T.B. 1963, Discussion on plasma salicylate levels in rheumatoid arthritis. *In:* A.St.J. Dixon, B.K. Martin, M.J.H. Smith and P.H.N. Wood (eds). Salicylates: An International Symposium. London: Churchill.

Becher, A., Miksch, J., Rambacher, P. and Schäfer, A. 1952, Behaviour of salicylamide in human metabolism. *Klinische Wochenschrift*, **30:** 913–918.

Behrendt, H. and Kampffmeyer, H.G. 1989, Absorption and cleavage of methyl salicylate by skin of single-pass perfused rabbit ears. *Xenobiotica*, **19:** 131–141.

Bekersky, I., Colburn, W.A., Fishman, L. and Kaplan, S.A. 1980, Metabolism of salicylic acid in the isolated perfused rat kidney. *Drug Metabolism and Disposition*, **8:** 319–324.

Betts, W.H., Whitehouse, M.W., Cleland, L.G. and Vernon-Roberts, B. 1985, *In vitro* antioxidant properties of potential biotransformation of salicylate, sulphasalazine and amidopyrine. *Journal of Free Radicals in Biology and Medicine*, **1:** 273–280.

Blackwell, G.J., Flower, R.J. and Vane, J.R. 1975, Some characteristics of the prostaglandin synthesizing system in rabbit kidney microsomes. *Biochimica et Biophysica Acta*, **398:** 178–190.

Blair, C.H. and Huang, J.D. 1992, Effect of theophylline on the intestinal clearance of drugs in rats. *Journal of Pharmacy and Pharmacology*, **44:** 483–486.

Bochner, F., Graham, G.G., Cham, B.E., Imhoff, D.M. and Haavisto, T.M. 1981, Salicylate metabolite kinetics after several salicylates. *Clinical Pharmacology and Therapeutics*, **30:** 266–275.

Bochner, F., Graham, G.G., Polverino, A., Imhoff, D.M., Tregenza, R.A., Rolan, P.E. and Cleland, L.G. 1987, Salicyl phenolic glucuronide pharmacokinetics in patients with rheumatoid arthritis. *European Journal of Clinical Pharmacology*, **32:** 153–158.

Bochner, F., Siebert, D.M., Rodgers, S.E., McIntosh, G.H., James, M.J. and Lloyd, J.V. 1989, Measurement of aspirin concentrations in portal and systemic blood in pigs: effect on platelet aggregation, thromboxane and prostacyclin production. *Thrombosis and Haemostasis*, **61:** 211–216.

Bochner, F., Williams, D.B., Morris, P.M.A., Siebert, D.M. and Lloyd, J.V. 1988, Pharmacokinetics of low-dose oral modified release, soluble and intravenous aspirin in man, and effects on platelet function. *European Journal of Clinical Pharmacology*, **35:** 287–294.

Bogacz, K. and Caldron, P. 1987, Enteric-coated aspirin bezoar: elevation of serum salicylate level by barium study. *American Journal of Medicine*, **83:** 783–786.

Bondesen, S., Nielsen, O.H., Schou, J.B., Jensen, P.H., Lassen, L.B., Binder, V., Krasilnikoff, P.A., Dano, P., Hansen, S.H., Jacobsen, O., Rasmussen, S.N. and Hvidberg, E.F. 1986, Steady-state kinetics of 5-aminosalicylic acid and sulphapyridine during sulfasalazine prophylaxis in ulcerative colitis. *Scandinavian Journal of Gastroenterology*, **21:** 693–700.

Bondesen, S., Schou, J.B., Pedersen, V., Rafiolsadat, Z., Hansen, S.H. and Hvidberg, E.F. 1988, Absorption of 5-aminosalicylic acid from colon and rectum. *British Journal of Clinical Pharmacology*, **25:** 269–272.

Boxenbaum, H., Bekersky, I., Jack, M.L. and Kaplan, S.A. 1979, Influence of gut microflora on bioavailability. *Drug Metabolism Reviews*, **18:** 259–279.

Bradow, G., Kan, L. and Fenselau, C. 1989, Studies of intramolecular rearrangements of acyl-linked glucuronides using salicylic acid, flufenamic acid, and (*S*)- and (*R*)-benzoxaprofen and confirmation of isomerization in acyl-linked $\Delta^9$11-carboxytetrahydrocannabinol. *Chemical Research in Toxicology*, **2:** 316–324.

Bridges, K.R., Schmidt, G.J., Jensen, M., Cerami, A. and Bunn, H.F. 1975, The acetylation of hemoglobin by aspirin. *Journal of Clinical Investigation*, **56:** 201–207.

Brodie, B.B., Kurz, H. and Schanker, L.S. 1960, The importance of dissociation constant and lipid-solubility in influencing the passage of drugs into the cerebrospinal fluid. *Journal of Pharmacology and Experimental Therapeutics*, **130:** 20–25.

Brune, K. and Graf, P. 1978, Non-steroidal anti-inflammatory drugs: influence of extra-cellular pH on biodistribution and pharmacological effects. *Biochemical Pharmacology*, **27:** 525–530.

Brune, K., Rainsford, K.D. and Schweitzer, A. 1980, Biodistribution of mild analgesics, *British Journal of Clinical Pharmacology*, **10:** 279S–284S.

Brunelle, F.M. and Verbeeck, R.K. 1997, Conjugation–deconjugation cycling of diflunisal via beta-glucuronidase catalyzed hydrolysis of its acyl glucuronide in the rat. *Life Sciences*, **60:** 2013–2021.

Bystrowska, B., Nowak, J. and Brandys, J. 2000, Validation of a LC method for the determination of 5-aminosalicylic acid and its metabolite in plasma and urine. *Journal of Pharmaceutical and Biomedical Analysis*, **22:** 341–347.

Campbell, L., Wilson, H.K., Samuell, A.M. and Gompertz, D. 1988, Interactions of m-xylene and aspirin metabolism in man. *Journal of Indian Medicine*, **45:** 127–132.

Campieri, M., Lanfranchi, G.A., Boschi, S., Brignola, C., Bazzocchi, G., Gionchetti, P., Minguzzi, M.R., Belluzi, A. and Labo, G. 1985, Topical administration of 5-aminosalicylic acid enemas in patients with ulcerative colitis. Studies on rectal absorption and excretion. *Gut*, **26:** 400–405.

Carlin, G., Djursater, R., Smedgard, G. and Gerdin, B. 1985, Effect of anti-inflammatory drugs on xanthine oxidase and xanthine induced depolymerization of hyaluronic acid. *Agents and Actions*, **16:** 377–384.

Cassell, S., Furst, D.E., Dromgoole, S. and Paulus, H. 1979, Steady-state serum salicylate levels in hospitalized patients with rheumatoid arthritis. *Arthritis and Rheumatism*, **22:** 384–388.

Castagnou, R., Larcebau, S. and Queyment, A. 1952, Toxicology of methyl salicylate. *Bulletin de la Societe de Pharmacie Marseille*, **3:** 30–34.

Caterson, R.J., Duggin, G.G., Horvath, J., Mohandas, J. and Tiller, D. 1978, Aspirin, protein transacetylation and inhibition of prostaglandin synthetase in the kidney. *British Journal of Pharmacology*, **64:** 353–358.

Cham, B.E., Johns, D., Bochner, F., Imhoff, D.M. and Rowland, M. 1979, Liquid-chromatographic procedure for the simultaneous quantitation of salicylic acid, salicylurate and gentisic acid in plasma. *Clinical Chemistry*, **25:** 1420–1425.

Cham, B.E., Ross-Lee, L. and Bochner, F. 1980, Measurement and pharmacokinetics of acetylsalicylic acid by a novel high performance liquid chromatographic assay. *Therapeutic Drug Monitoring*, **2:** 365–372.

Chatton, J.-Y. and Roch-Ramel, F. 1992, Transport of salicylic acid through monolayers of a kidney epithelial cell line (LLC-PK1). *Journal of Pharmacology and Experimental Therapeutics*, **261:** 518–524.

Chen, C.N., Coleman, D.L., Andrade, J.D. and Temple, A.R. 1978, Pharmacokinetic model for salicylate in cerebrospinal fluid, blood, organs and tissues. *Journal of Pharmaceutical Sciences*, **67:** 38–45.

Cherian, M. and Abraham, E.C. 1993, In vitro glycation and acetylation (by aspirin) of rat crystallins. *Life Sciences*, **52:** 69–70.

Childs, B., Sidbury, J.B. and Migeon, C.J. 1959, Glucuronic acid conjugation by patients with familial nonhemolytic jaundice and their relatives. *Pediatrics*, **23:** 903–913.

Christensen, L.A., Slot, O., Sanchez, G., Boserup, J., Rasmussen, S.N., Bondesen, S., Hansen, S.H. and Hvidberg, E.F. 1987, Release of 5-aminosalicylic acid from Pentasa during normal and accelerated intestinal time. *British Journal of Clinical Pharmacology*, **23:** 365–369.

Clarke, N.E., Clarke, C.H. and Mosher, R.E. 1958, Phenolic compounds in chemotherapy of rheumatic fever. *American Journal of Medical Sciences*, **235:** 7–22.

Cleland, L.G., Lowthian, P.J., Imhoff, D., Bochner, F., Betts, W.H. and O'Callaghan, J. 1985, Plasma and synovial fluid gentisate in patients receiving salicylate therapy. *Journal of Rheumatology*, **12:** 136–139.

Cohen, A. 1979, A comparative blood salicylate study of two salicylate tablet formulations utilising normal volunteers. *Current Therapeutic Research*, **23:** 772–778.

Colantoni, A., de Maria, N., Caraceni, P., Bernardi, M., Floyd, R.A. and Van Thiel, D.H. 1998, Prevention of reoxygenation injury by sodium salicylate in isolated-perfused rat liver. *Free Radical Biology and Medicine*, **25**, 87–94.

Cook, D.J., Achong, M.R. and Murphy, F.R. 1988, Three cases of diflunisal hypersensitivity. *Canadian Medical Association Journal*, **138:** 1029–1030.

Cooke, A.R. and Hunt, J.N. 1970, Absorption of acetylsalicylic acid from unbuffered and buffered gastric contents. *American Journal of Digestive Diseases*, **15:** 95–102.

Cossum, P.A., Roberts, M.S., Kilpatrick, D. and Yong, A.C. 1986, Extrahepatic metabolism and distribution of aspirin in vascular beds of sheep. *Journal of Pharmaceutical Sciences*, **75:** 731–737.

Costello, P.B. and Green, F.A. 1987, The extracellular control of intracellular aspirin hydrolysis. *Arthritis and Rheumatism*, **30:** 412–418.

Costello, P.B., Green, F.A. and Jung, C.Y. 1982, Free versus bound salicylate in adults and children with chronic inflammatory joint disease. *Arthritis and Rheumatism*, **25:** 32–37.

Coudray, C., Talla, M., Martin, S., Fatome, M. and Favier, A. 1995, High-performance liquid chromatography–electrochemical determination of salicylate hydroxylation products as an in vivo marker of oxidative stress. *Analytical Biochemistry*, **227:** 101–111.

Cowan, R.A., Hartnell, G.G., Lowdell, C.P., Baird, I.M. and Leak, A.M. 1984, Metabolic acidosis induced by carbonic anhydrase inhibitors and salicylates in patients with normal renal function. *British Medical Journal*, **289:** 347–348.

Cross, S.E., Anderson, C. and Roberts, M.S. 1998, Topical penetration of commercial salicylate esters and salts using human isolated skin and clinical microdialysis studies. *British Journal of Clinical Pharmacology*, **46:** 29–35.

Cross, S.E., Anderson, C., Thompson, M.J. and Roberts, M.S. 1997, Is there tissue penetration after application of topical salicylate formulations? *Lancet*, **350:** 636.

Cross, S.E., Megwa, S.A., Benson, H.A.E. and Roberts, M.S. 1999, Self promotion of deep tissue penetration and distribution of methylsalicylate after topical application. *Pharmaceutical Research*, **16:** 427–433.

Cryer, B., Kliewer, D., Sie, H., McAllister, L. and Feldman, M. 1999, Effects of cutaneous aspirin on the human stomach. *Proceedings of the Association of American Physicians*, **111:** 448–456.

Danon, A., Ben-Shimon, S. and Ben-Zvi, Z. 1986, Effect of exercise and heat exposure on percutaneous absorption of methyl salicylate. *European Journal of Clinical Pharmacology*, **31:** 49–52.

Das, D.K., Cordis, G.A., Rao, P.S., Liu, X. and Maity, S. 1991, High-performance liquid chromatographic detection of hydroxylated benzoic acids as an indirect measure of hydroxyl radical in heart: its possible link with the myocardial reperfusion injury. *Journal of Chromatography*, **536:** 273–282.

Davies, M.G., Briffa, D.V. and Greaves, M.W. 1979, Systemic toxicity from topically applied salicylic acid. *British Medical Journal*, **1:** 661.

Davis, W.B., Mohammed, B.S., Mays, D.C., She, Z.-W., Mohammed, J.R., Husney, R.M. and Sagone, A.L. 1989, Hydroxylation of salicylate by activated neutrophils. *Biochemical Pharmacology*, **38:** 4013–4019.

Davison, C., Zimmerman, E.F. and Smith, P.K. 1961, On the metabolism and toxicity of methyl salicylate. *Journal of Pharmacology and Experimental Therapeutics*, **132:** 207–211.

Day, R.K., Williamson, H.E. and Roberts, R.J. 1976, Failure of prednisolone to alter plasma salicylate concentrations in dogs. *Journal of Pharmaceutical Sciences*, **65:** 773–774.

Day, R.O., Furst, D.E., Dromgoole, S.H. and Paulus, H.E. 1988a, Changes in salicylate serum concentration and metabolism during chronic dosing in normal volunteers. *Biopharmaceutics and Drug Disposition*, **9:** 273–283.

Day, R.O., Harris, G., Brown, M., Graham, G.G. and Champion, G.D. 1988b, Interaction of salicylate and corticosteroids in man. *British Journal of Clinical Pharmacology*, **26:** 334–337.

Day, R.O., Shen, D.D. and Azarnoff, D.L. 1983, Induction of salicylurate formation in rheumatoid arthritis patients treated with salicylates. *Clinical Pharmacokinetics*, **8:** 263–271.

Dean, M., Penglis, S. and Stock, B. 1989, The pharmacokinetics of salicylate in the pregnant rat. *Drug Metabolism and Disposition*, **17:** 87–90.

DeBlassio, J.L., deLong, M.A., Glufke, U., Kulathila, R., Merkler, K.A., Vederas, J.C. and Merkler, D.J. 2000, Amidation of salicylauric acid and gentisuric acid: a possible role for peptidyl α-amidating monooxygenase in the metabolism of aspirin. *Archives of Biochemistry and Biophysics*, **383**, 46–55.

de la Pena, A.M., Salinas, F. and Meras, I.D. 1988, Simultaneous determination of salicylic and salicyluric acids in urine by first-derivative synchronous fluorescence spectroscopy. *Analytical Chemistry*, **60:** 2493–2496.

Dickinson, R.G. and Baker, P.V. 1991, Studies on the reactivity of salicyl acyl glucuronide. *Clinical and Experimental Pharmacology and Physiology*, (**Suppl. 18**): 13.

Dickinson, R.G. and King, A.R. 1989, Reactivity considerations in the analysis of glucuronide and sulphate conjugates of diflunisal. *Therapeutic Drug Monitoring*, **11:** 712–720.

Dickinson, R.G. and King, A.R. 1993, Studies on the reactivity of acyl glucuronides V. Glucuronide-derived covalent binding of diflunisal to bladder tissue of rats and its modulation by urinary pH and beta-glucuronidase. *Biochemical Pharmacology*, **46:** 1175–1182.

Dickinson, R.G. and King, A.R. 1996, Vesico–hepato–renal cycling of acidic drugs via their reactive acyl glucuronides metabolites? Studies with diflunisal in rats. *Clinical and Experimental Pharmacology and Physiology*, **23:** 665–668.

Dickinson, R.G., King, A.R. and Verbeeck, R.K. 1989, Elimination of diflunisal as its acyl glucuronide, phenolic glucuronide and sulphate conjugates in bile-exteriorized and intact rats. *Clinical and Experimental Pharmacology and Physiology*, **16:** 913–924.

Doherty, N.S., Anttila, M. and Dean, P.B. 1977, Penetration of naproxen and salicylate into inflammatory exudates in the rat. *Annals of the Rheumatic Diseases*, **36:** 244–248.

Doluisio, J.T., Billups, N.F., Dittert, L.W., Sugita, E.T. and Swintosky, J.V. 1969, Drug absorption I: an *in situ* rat gut technique yielding realistic absorption rates. *Journal of Pharmaceutical Sciences*, **58:** 1196–1200.

Dotevall, G. and Ekenved, G. 1976, The absorption of acetylsalicylic acid from the stomach in relation to intragastric pH. *Scandinavian Journal of Gastroenterology*, **11:** 801–806.

Drayton, C.J. 1990, *Comprehensive Medicinal Chemistry*, Vol. 6. Oxford: Pergamon Press.

Dromgoole, S.H., Cassell, S., Furst, D.E. and Paulus, H.E. 1983, Availability of salicylate from salsalate and aspirin. *Clinical Pharmacology and Therapeutics*, **34:** 539–545.

Dromgoole, S.H., Furst, D.E. and Paulus, H.E. 1984, Metabolism of salsalate in normal subjects. *Journal of Pharmaceutical Sciences*, **73:** 1657–1659.

Dubovska, D., Piotrovskij, V.K., Gajdos, M., Krivosikova, Z., Spustova, V. and Trnovec, T. 1995, Pharmacokinetics of acetylsalicylic acid and its metabolites at low doses; a compartmental modeling. *Methods and Findings in Experimental and Clinical Pharmacology*, **17:** 67–77.

Dull, B.J., Salata, K., Langenhove, A.V. and Goldman, P. 1987a, 5-aminosalicylate oxidation by activated leukocytes and protection of cultured cells from oxidative damage. *Biochemical Pharmacology*, **36:** 2467–2472.

Dull, B.J., Salata, K. and Goldman, P. 1987b, Role of the intestinal flora in the acetylation of sulfasalazine metabolites. *Biochemical Pharmacology*, **36:** 3772–3774.

Dupont, I., Berthou, F., Bodenez, P., Bardou, L., Guirriec, C., Stephan, N. *et al.* 1999, Involvement of cytochromes P-450 2E1 and 3A4 in the 5-hydroxylation of salicylate in humans, *Drug Metabolism and Disposition*, **27:** 322–326.

Edelman, J., Potter, J.M. and Hackett, L.P. 1986, The effect of intra-articular steroids on plasma salicylate concentration. *British Journal of Clinical Pharmacology*, **21:** 301–307.

Edwards, L.J. 1952, The hydrolysis of aspirin. *Transactions of the Faraday Society*, **48:** 696–699.

Emudianughe, T.S. 1990, Sex differences in salicylic acid metabolism in streptozotocin induced diabetes in rats. *Fundamental and Clinical Pharmacology*, **4:** 483–489.

Emudianughe, T.S., Oduleye, S.O., Ebadan, E.E. and Eneji, S.D. 1986, Sex difference in salicylic acid metabolism in Nigerian subjects. *Xenobiotica*, **16:** 177–179.

Feldman, M. and Cryer, B. 1999, Aspirin absorption rates and platelet inhibition times with 325-mg buffered aspirin tablets (chewed or swallowed intact) and with buffered aspirin solution. *American Journal of Cardiology*, **84:** 404–409.

Feldman, R.J. and Maibach, H.I. 1970, Absorption of some organic compounds through the skin in man. *Journal of Investigative Dermatology*, **54:** 399–404.

Fielding, R.M., Waschek, J.A., Effeney, D.J., Pogany, A.C., Pond, S.M. and Tozer, T.N. 1986, Extrahepatic extraction of salicylamide in dogs. *Journal of Pharmacology and Experimental Therapeutics*, **235:** 97–102.

Finnie, M.D.A., Ersser, R.S., Seakins, J.W.T. and Snedden, W. 1976, The occurrence and identification of O-hydroxyhippuric acid (salicyluric acid) in the urine of sick children. *Clinica Chimica Acta*, **70:** 171–178.

Forman, W.B., Davidson, E.D. and Webster, L.T. 1971, Enzymatic conversion of salicylate to salicylurate. *Molecular Pharmacology*, **7:** 247–259.

Fotsch, G. and Pfeiffer, S. 1990, Comparative study of serum levels of salicylic acid after oral administration of salicin and sodium salicylate in rats. *Pharmazie*, **45:** 535–536.

Foye, W.O., Duvall, R.N., Lange, W.E., Talbot, M.H. and Prien, E.L. 1975, Urinary metabolites of salicylamide in stone formers. *Journal of Pharmacology and Experimental Therapeutics*, **195:** 416–423.

Furst, D.E., Gupta, N. and Paulus, H.E. 1977, Salicylate metabolism in twins. Evidence suggesting a genetic influence and induction of salicylurate formation. *Journal of Clinical Investigation*, **60:** 32–42.

Furst, D.E., Tozer, T.N. and Melmon, K.L. 1979, Salicylate clearance, the resultant of protein binding and metabolism. *Clinical Pharmacology and Therapeutics*, **26:** 380–389.

Garcia-Sancho, J. and Sanchez, A. 1978, Use of salicylic acid to measure the apparent intracellular pH in the Ehrlich ascites-tumor cells and *Escherichia coli*. *Biochimica et Biophysica Acta*, **509:** 148–158.

Gaspari, F., Perico, N., Locatelli, M., Corna, D., Remuzzi, G. and Garattini, S. 1989, Renal handling of aspirin in the rat. *Journal of Pharmacology and Experimental Therapeutics*, **251:** 295–304.

Ghahramani, P., Rowland-Yeo, K., Yeo, W.W., Jackson, P.R. and Ramsay, L.E. 1998, Protein binding of aspirin and salicylate measured by in vivo ultrafiltration. *Clinical Pharmacology and Therapeutics*, **63:** 285–295.

Ghiselli, A., Laurenti, O., De Mattia, G., Maiani, G. and Ferro-Luzzi, F. 1992, Salicylate hydroxylation as an early marker of in vivo oxidative stress in diabetic patients. *Free Radical Biology and Medicine*, **13:** 621–626.

Gibaldi, M., Grundhofer, B. and Levy, G. 1975, Time course and dose dependence of antacid effect on urine pH. *Journal of Pharmaceutical Sciences*, **64:** 2003–2004.

Goehl, T.J., DeWoody, C.T. and Sundaresan, G.M. 1981, Sublimation losses of salicylic acid from plasma during analysis. *Clinical Chemistry*, **27:** 776.

Gordon, I.J., Bowler, C.S., Coakley, J. and Smith, P. 1984, Algorithm for modified alkaline diuresis in salicylate poisoning. *British Medical Journal*, **289:** 1039–1040.

Graham, G.G. 1988, Kinetics of non-steroidal anti-inflammatory drugs in synovial fluid. *Agents and Actions*, (**Suppl. 20**): 66–75.

Graham, G. and Rowland, M. 1972, Application of salivary salicylate data to biopharmaceutical studies of salicylates. *Journal of Pharmaceutical Sciences*, **61:** 1219–1222.

Graham, G.G., Champion, G.D., Day, R.O. and Paull, P.D. 1977, Patterns of plasma concentrations and urinary excretion of salicylate in rheumatoid arthritis. *Clinical Pharmacology and Therapeutics*, **22:** 410–420.

Green, F.A. and Jung, C.Y. 1981, Acetylation of erythrocytic membrane peptides by aspirin. *Transfusion*, **21:** 55–58.

Greenblatt, D.J., Abernethy, D.J., Boxenbaum, H.G., Matlis, R., Ochs, H.R., Harmatz, J.S. and Shader, R.I. 1986, Influence of age, gender, and obesity on salicylate kinetics following single doses of aspirin. *Arthritis and Rheumatism*, **29:** 971–980.

Griffin, J.P. and D'Arcy, P.F. 1997, *A Manual of Drug Interactions*. Amsterdam: Elsevier.

Grigor, R.R., Spitz, P.W. and Furst, D.E. 1987, Salicylate toxicity in elderly patients with rheumatoid arthritis. *Journal of Rheumatology*, **14:** 60–66.

Grinstead, R.R. 1960, Oxidation of salicylate by a model peroxidase catalyst iron-ethylenediaminotetracetato-iron(III) acid. *Journal of the American Chemical Society*, **82:** 3464–3471.

Grisham, M.B. and Granger, D.N. 1989, 5-Aminosalicylic acid concentration in mucosal interstitium of cat small and large intestine. *Digestive Diseases and Sciences*, **34:** 573–578.

Grootveld, M. and Halliwell, B. 1986, Aromatic hydroxylation as a potential measure of hydroxyl radical formation in vivo. *Biochemical Journal*, **237:** 499–504.

Grootveld, M. and Halliwell, B. 1988, 2,3-dihydroxybenzoic acid is a product of human aspirin metabolism. *Biochemical Pharmacology*, **37:** 271–280.

Gugler, R., Lain, P. and Azarnoff, D.L. 1975, Effect of portacaval shunt on the disposition of drugs with and without first-pass effect. *Journal of Pharmacology and Experimental Therapeutics*, **195:** 416–423.

Gulyassy, P.F., Jarrard, E. and Stanfel, L.A. 1987, Contributions of hippurate, indoxyl sulfate, and *o*-hydroxyhippurate to impaired ligand binding by plasma in azotemic humans. *Biochemical Pharmacology*, **36:** 4215–4220.

Gunsberg, M., Bochner, F., Graham, G., Imhoff, D., Parsons, G. and Cham, B. 1984, Disposition of and clinical response to salicylates in patients with rheumatoid disease. *Clinical Pharmacology and Therapeutics*, **35:** 585–593.

Gupta, K.C., Joshi, J.V., Hazari, K., Pohujani, S.M. and Satoskar, R.S. 1982, Effect of low oestrogen combination oral contraceptive on metabolism of aspirin and phenylbutazone. *International Journal of Clinical Pharmacology, Therapy and Toxicology*, **20:** 511–513.

Gupta, N., Sarkissian, E. and Paulus, H.E. 1975, Correlation of plateau serum salicylate level with rate of salicylate metabolism. *Clinical Pharmacology and Therapeutics*, **18:** 350–355.

Gurwich, E.L., Raees, S.M., Skosey, J. and Niazi, S. 1984, Unbound plasma salicylate concentration in rheumatoid arthritis patients. *British Journal of Rheumatology*, **23:** 66–73.

Gutman, A.B., Yu, T.F. and Sirota, J.H. 1955, A study, by simultaneous clearance techniques, of salicylate excretion in man. Effect of alkalinisation of the urine by bicarbonate administration; effect of probenecid. *Journal of Clinical Investigation*, **34:** 711–721.

Guzek, D.B., Kennedy, A.H., McNeill, S.C., Wakshull, E. and Potts, R.O. 1989, Transdermal drug transport and metabolism. I. Comparison of *in vitro* and *in vivo* results. *Pharmaceutical Research*, **6:** 33–39.

Hansen-Moller, J., Dalgaard, L. and Hansen, S.H. 1987, Reversed-phase high-performance liquid chromatography assay for the simultaneous determination of diflunisal and its glucuronides in serum and urine. Rearrangement of the 1-O-acylglucuronide. *Journal of Chromatography*, **420:** 99–109.

Hansten, P.D. and Horn, J.R. 2000, *Drug Interactions*. St Louis: Facts and Comparisons.

Harris, P.A. and Riegelman, S. 1969, Influence of the route of administration on the area under the plasma concentration-time curve. *Journal of Pharmaceutical Sciences*, **58:** 71–75.

Harrison, L.I., Funk, M.L., Re, O.N. and Ober, R.E. 1981, Absorption, biotransformation, and pharmacokinetics of salicylsalicylic acid in humans. *Journal of Clinical Pharmacology*, **21:** 401–404.

Harrison, L.I., Riedel, D.J., Armstrong, K.E., Goldlust, M.B. and Ekholm, B.P. 1992, Effect of food on salsalate absorption. *Therapeutic Drug Monitoring*, **14:** 87–91.

Hasegawa, A., Suzuki, S., Matsumoto, Y. and Okubo, T. 1997, In vivo fatiguing contraction of rat diaphragm produces hydroxyl radicals. *Free Radical Biology and Medicine*, **22:** 349–354.

Hawkins, D., Pinckard, R.N., Crawford, I.P. and Farr, R.S. 1969, Structural changes in human serum albumin induced by ingestion of acetylsalicylic acid. *Journal of Clinical Investigation*, **48:** 536–542.

Haynes, D.R., Wright, P.F.A., Gadd, S.J., Whitehouse, M.W. and Vernon-Roberts, B. 1993, Is aspirin a pro-drug for antioxidant and cytokine modulating oxymetabolites? *Agents and Actions*, **39:** 49–58.

Hekman, P. and van Ginneken, C.A.M. 1983, Simultaneous kinetic modelling of plasma levels and urinary excretion of salicyluric acid, and the influence of probenecid., *European Journal of Drug Metabolism and Pharmacokinetics*, **8:** 239–249.

Henry, J. and Volans, G. 1984, Analgesic poisoning. I – Salicylates. *British Medical Journal*, **289:** 820–822.

Higgs, G.A., Salmon, J.A., Henderson, B. and Vane, J.R. 1987, Pharmacokinetics of aspirin and salicylate in relation to inhibition of arachidonate cyclooxygenase and antiinflammatory activity. *Proceedings of the National Academy of Sciences of the USA*, **84:** 1417–1420.

Hill, D.W. and Langner, K.J. 1987, HPLC photodiode array UV detection for toxicological drug analysis. *Journal of Liquid Chromatography*, **10:** 377–409.

Hill, J.B. 1971, Experimental salicylate poisoning; observations on the effects of altering blood pH on tissue and plasma salicylate concentrations. *Pediatrics*, **47:** 658–665.

Hill, J.B. and Smith, R.M. 1970, An automated microfluorometric determination for salicylate in body fluids and tissue extracts. *Biochemical Medicine*, **4:** 24–35.

Hillman, R.J. and Prescott, L.F. 1985, Treatment of salicylate poisoning with repeated oral charcoal. *British Medical Journal*, **291:** 1492.

Hinz, B., Kraus, V., Pahl, A. and Brune, K. 2000, Salicylate metabolites inhibit cycloxygenase-2-dependent prostaglandin E$_2$ synthesis in murine macrophages. *Biochemical and Biophysical Research Communications*, **274:** 197–202.

Hirate, J., Kato, Y., Horikoshi, I., Nagase, S. and Ueda, C.T. 1989, Further observations on the disposition characteristics of salicylic acid in analbuminemic rats. *Biopharmaceutics and Drug Disposition*, **10:** 299–309.

Ho, J.L., Tierney, M.G. and Dickinson, G.E. 1989, An evaluation of the effect of repeated doses of oral activated charcoal on salicylate elimination. *Journal of Clinical Pharmacology*, **29:** 366–369.

Ho, P.C., Triggs, E.J., Bourne, D.W.A. and Heazlewood, V.J. 1985, The effects of age and sex on the disposition of acetylsalicylic acid. *British Journal of Clinical Pharmacology*, **19**: 675–684.

Hogben, C.A.M., Tocco, D.J., Brodie, B.B. and Schanker, L.S. 1959, On the mechanism of absorption of drugs. *Journal of Pharmacology and Experimental Therapeutics*, **125**: 275–282.

Holmes, T.J., John, V., Vennerstrom, J., Kulmacz, R.J. and Lands, W.E.M. 1984, Stimulation of the cyclooxygenase activity of prostaglandin H synthase by salicylate-derived quinhydrones. *Prostaglandins*, **28**: 711–716.

Holmes, T.J., Vennerstrom, J.L. and John, V. 1985, Inhibition of cyclooxygenase mediated by electrochemical oxidation of gentisic acid. *Journal of Biological Chemistry*, **260**: 14092–14095.

Houston, J.B. and Levy, G. 1976, Effect of route of administration on competitive drug biotransformation interaction: salicylamide-ascorbic interaction in rats. *Journal of Pharmacology and Experimental Therapeutics*, **198**: 284–294.

Howell, S.R., Kotkoskie, L.A., Dills, R.L. and Klaasen, C.D. 1988, 3-Hydroxylation of salicylamide in mice. *Journal of Pharmaceutical Sciences*, **77**: 309–313.

Huang, H. and Gao, J. 1989, Fluorometric determination of terbium and dysprosium with a salicylic acid derivative-EDTA system. *Bunseki Kagaku*, **38**, 361–367.

Humphreys, K.J. and Smy, J.R. 1975, The absorption of benorylate from everted sacs of rat intestine. *Journal of Pharmacy and Pharmacology*, **27**: 962–964.

Hung, D.Y., Mellick, G.D., Anissimov, Y.G., Weiss, M. and Roberts, M.S. 1998a, Hepatic structure-pharmacokinetic relationships: the hepatic disposition and metabolite kinetics of a homologous series of O-acyl derivatives of salicylic acid. *British Journal of Pharmacology*, **124**: 1475–1483.

Hung, D.Y., Mellick, G.D., Masci, P.P., Whitaker, A.N., Whitehouse, M.W. and Roberts, M.S. 1998b, Focused antithrombic therapy: novel anti-platelet salicylates with reduced ulcerogenic potential and higher first-pass detoxification than aspirin in rats. *Journal of Laboratory and Clinical Medicine*, **132**: 469–477.

Hung, D.Y., Mellick, G.D., Anissimov, Y.G., Weiss, M. and Roberts, M.S. 1998c, Hepatic disposition and metabolite kinetics of a homologous series of diflunisal esters. *Journal of Pharmaceutical Sciences*, **87**: 943–951.

Hutt, A.J., Caldwell, J. and Smith, R.L. 1986, The metabolism of aspirin in man: a population study. *Xenobiotica*, **16**: 239–249.

Inoue, M., Morikawa, M., Tsuboi, M., Ito, Y. and Sugiura, M. 1980, Comparative study of human intestinal and hepatic esterases as related to enzymatic properties and hydrolysing activity for ester type drugs. *Japanese Journal of Pharmacology*, **30**: 529–535.

Ireland, A., Priddle, J.D. and Jewell, D.P. 1990, Acetylation of 5-aminosalicylic acid by isolated colonic epithelial cells. *Clinical Science*, **78**: 1105–1111.

Jarvie, D.R., Heyworth, R. and Simpson, D. 1987, Plasma salicylate analysis: a comparison of colorimetric, HPLC and enzymatic techniques. *Annals of Clinical Biochemistry*, **24**: 364–373.

Jastreboff, P.J., Hansen, R., Sasaki, P.G. and Sasaki, C.T. 1986, Differential uptake of salicylate in serum, cerebrospinal fluid, and perilymph. *Archives of Otolaryngology – Head and Neck Surgery*, **112**: 1050–1053.

Joy, M.M. and Cutler, D.J. 1987, On the mechanism of transport of salicylate and *p*-hydroxybenzoic acid across human red cell membranes. *Journal of Pharmacy and Pharmacology*, **39**: 266–271.

Kang, Y.S., Terasaki, T. and Tsuji, A. 1990, Acidic drug transport *in vivo* through the blood–brain barrier. A role of the transport carrier for monocarboxylic acids., *J Pharmacobio-Dynamics*, **13**: 158–163.

Kapp, E.M. and Coburn, A.F. 1942, Urinary metabolites of sodium salicylate. *Journal of Biological Chemistry*, **145**: 549–565.

Kennedy, M., Chinwah, P. and Wade, D.N. 1979, A pharmacological method of measuring mouth caecal transit time in man. *British Journal of Clinical Pharmacology*, **8**: 372–373.

Kettle, A.J. and Winterbourn, C.C. 1994, Superoxide-dependent hydroxylation by myeloperoxidase. *Journal of Biological Chemistry*, **269**: 7146–7151.

Khan, A. and Aarons, L. 1989, A note on the use of salicylate saliva concentration in clinical pharmacokinetic studies. *Journal of Pharmacy and Pharmacology*, **41**: 710–711.

Kim, P.M. and Wells, P.G. 1996, Phenytoin-initiated hydroxyl radical formation: characterization by enhanced salicylate hydroxylation. *Molecular Pharmacology*, **49:** 172–181.

King, A.R. and Dickinson, R.G. 1993, Studies on the reactivity of acyl glucuronides IV Covalent binding of diflunisal to tissues of the rat. *Biochemical Pharmacology*, **45:** 1043–1047.

Klinenberg, J.R. and Miller, F. 1965, Effects of corticosteroids on blood salicylate concentration. *Journal of the American Medical Association*, **196:** 601–604.

Knoefel, P.K., Huang, K.C. and Jarboe, C.H. 1962, Renal disposal of salicyluric acid. *American Journal of Physiology*, **203:** 6–10.

Koel, M. and Nebinger, P. 1989, Specificity data of the salicylate assay by fluorescent polarisation immunoassay. *Journal of Analytical Toxicology*, **13:** 358–360.

Koike, M., Sugeno, K. and Hirata, M. 1981, Sulphoconjugation and glucuronidation of salicylamide in isolated rat hepatocytes. *Journal of Pharmaceutical Sciences*, **70:** 308–311.

Konstantianos, D.G., Ioannou, P.C. and Efstathiou, C.E. 1991, Simultaneous determination of acetylsalicylic and salicylic acids in human serum by second-derivative synchronous spectrometry. *Analyst*, **116:** 373–378.

Koren, G., Schaffer, F., Silverman, E., Walker, S., Duffy, C., Stein, L., Suria, D., Schue, S., Thiessen, J.J., Gelfand, E. and Laxer, R. 1988, Determinants of low serum concentrations of salicylates in patients with Kawasaki disease. *Journal of Pediatrics*, **112**, 663–667.

Kragh-Hansen, U., Brennan, S.O., Galliano, M. and Sugita, O. 1990, Binding of warfarin, salicylate, and diazepam to genetic variants of human serum albumin with known mutations. *Molecular Pharmacology*, **37:** 238–242.

Kyle, M.E. and Kocsis, J.J. 1986a, Metabolism of salicylate by isolated kidney and liver mitochondria. *Chemico-Biological Interactions*, **59:** 325–335.

Kyle, M.E. and Kocsis, J.J. 1986b, The effect of mixed function oxidase induction and inhibition on salicylate-induced nephrotoxicity in male rats. *Toxicology and Applied Pharmacology*, **84:** 241–249.

Laznicek, M., Melicharova, L., Laznikova, A. and Kvetina, J. 1989, Pharmacokinetics of salicylate in acute kidney failure. *Ceskoslovenska Farmacie*, **38:** 343–347.

Lecomte, M., Laneuville, O., Ji, C., DeWitt, D.L. and Smith, W.L. 1994, Acetylation of human prostaglandin endoperoxide synthase-2 (cyclooxygenase-2) by aspirin. *Journal of Biological Chemistry*, **267:** 13207–13215.

Lee, E.J.D. and Ang, S.B. 1987, Simple and sensitive high-performance liquid chromatographic assay for 5-aminosalicylic acid and acetyl-5-aminosalicylic acid in serum. *Journal of Chromatography*, **413:** 300–304.

Levy, G. 1980, Clinical pharmacokinetics of salicylates. A re-assessment. *British Journal of Clinical Pharmacology*, **10:** 285S–290S.

Levy, G. and Amsel, L.P. 1966, Kinetics of competitive inhibition of salicylic acid conjugation with glycine in man. *Biochemical Pharmacology*, **15:** 1033–1038.

Levy, G., Amsel, L.P. and Elliott, H.V. 1969, Kinetics of salicyluric elimination in man. *Journal of Pharmaceutical Sciences*, **58:** 827–829.

Levy, G. and Garrettson, L.K. 1974, Kinetics of salicylate elimination by newborn infants of mothers who ingested aspirin before delivery. *Pediatrics*, **53:** 201–210.

Levy, G. and Giacomini, K.M. 1978, Rational aspirin dosage regimens. *Clinical Pharmacology and Therapeutics*, **23:** 247–252.

Levy, G. and Leonards, J.R. 1971, Urine pH and salicylate therapy. *Journal of the American Medical Association* **217:** 81.

Levy, G. and Matsuzawa, T. 1967, Pharmacokinetics of salicylamide elimination in man. *Journal of Pharmacology and Experimental Therapeutics*, **156:** 285–293.

Levy, G. and Procknal, J.A. 1968, Drug biotransformation in man. I. Mutual inhibition in glucuronide formation of salicylic acid and salicylamide in man. *Journal of Pharmaceutical Sciences*, **57:** 1330–1335.

Levy, G., Tsuchiya, T. and Amsel, L.P. 1972, Limited capacity for salicyl phenolic glucuronide and its effect on the kinetics of salicylate elimination in man. *Clinical Pharmacology and Therapeutics*, **13:** 258–268.

Levy, G. and Yaffe, S.J. 1974, Relationship between dose and apparent volume of distribution of salicylate in children. *Pediatrics*, **54:** 713–717.

Levy, G. and Yamada, H. 1971, Drug biotransformation interactions in man. III Acetaminophen and salicylamide. *Journal of Pharmaceutical Sciences*, **60:** 215–221.

Lichtenwalner, D.M., Suh, B. and Lichtenwalner, M.R. 1983, Isolation and chemical characterization of 2-hydroxybenzoylglycine as a drug binding inhibitor in uremia. *Journal of Clinical Investigation*, **71:** 1289–1296.

Liu, J.H., Malone, R.S., Stallings, H. and Smith, P.C. 1996, Influence of renal failure in rats on the disposition of salicyl acyl glucuronide and covalent binding of salicylate to plasma proteins. *Journal of Pharmacology and Experimental Therapeutics*, **278:** 277–283.

Liu, L., Leech, J.A., Urch, R.B. and Silverman, F.S. 1997, In vivo salicylate hydroxylation: a potential biomarker for assessing acute ozone exposure and effects in humans. *American Journal of Respiratory and Critical Care Medicine*, **156:** 1405–1412.

Liu, Z.C., McClelland, Z.C. and Uetrecht, J.P. 1995, Oxidation of 5-aminosalicylic acid by hypochlorous acid to a reactive iminoquinone. Possible role in the treatment of inflammatory bowel diseases. *Drug Metabolism and Disposition*, **23:** 246–250.

Loewen, G.R., McKay, G. and Verbeeck, R.K. 1986, Isolation and identification of a new major metabolite of diflunisal in man: the sulfate conjugate. *Drug Metabolism and Disposition*, **14:** 127–131.

Lorenzo, A.V. and Spector, R. 1973, Transport of salicylic acid by the choroid plexus. *Journal of Pharmacology and Experimental Therapeutics*, **184:** 465–471.

Lorico, A., Masturzo, P., Villa, S., Salmona, M., Semeraro, N. and de Gaetano, G. 1986, Gentisic acid: an aspirin metabolite with multiple effects on human blood polymorphonuclear leukocytes. *Biochemical Pharmacology*, **35:** 2243–2245.

Lowenthal, D.T., Briggs, W.A. and Levy, G. 1974, Kinetics of salicylate elimination by anephric patients. *Journal of Clinical Investigation*, **54:** 1221–1226.

Macdonald, J.I., Herman, R.J. and Verbeeck, R.K. 1990, Sex-difference and the effects of smoking and oral contraceptive steroids on the kinetics of diflunisal. *European Journal of Clinical Pharmacology*, **38:** 175–179.

Macdonald, J.I., Reid, R.S., Edom, R.W. and Verbeeck, R.K. 1989, Identification of a hydroxylated metabolite of diflunisal in man. *FASEB Journal*, **3:** A424.

Macdonald, J.I., Wallace, S.M., Mahachi, V. and Verbeeck, R.K. 1992, Both phenolic and acyl glucuronidation pathways are impaired in liver cirrhosis. *European Journal of Clinical Pharmacology*, **42:** 471–474.

MacPherson, C.R., Milne, M.D. and Evans, B.M. 1955, The excretion of salicylate. *British Journal of Pharmacology*, **10:** 484–489.

Mallikaarjun, S., Wood, J.H. and Karnes, H.T. 1989, High-performance liquid chromatographic method for the determination of salicylic acid and its metabolites in urine by direct injection. *Journal of Chromatography*, **493:** 93–104.

Markiewicz, A. and Semenowicz, K. 1979, Time-dependent changes in the pharmacokinetics of aspirin. *International Journal of Clinical Pharmacology and Biopharmacy*, **17:** 409–411.

Martin, B.K. 1963, Accumulation of drug anions in gastric mucosal cells. *Nature*, **198:** 896–897.

Martin, J.P. and Batkoff, B. 1987, Homogentisic acid autoxidation and oxygen radical generation: implications for the etiology of alkaptonuric arthritis. *Free Radical Biology and Medicine*, **3:** 241–250.

Marzo, A., Mancinelli, A., Cardace, G., Monti, N. and Martelli, E.A. 1992, NaF and two other esterase inhibitors unaffect acetyl salicylic acid enzyme hydrolysis. *Journal of Pharmacy and Pharmacology*, **44:** 786.

Mayer, A.L., Sitar, D.S. and Tenenbein, M. 1992, Multiple dose charcoal and whole bowel irrigation do not increase clearance of absorbed salicylate. *Archives of Internal Medicine*, **152:** 393–396.

McArthur, J.N., Dawkins, P.D. and Smith, M.J.H. 1970, The relation between circulating and tissue concentrations of salicylate in the mouse *in vivo*. *Journal of Pharmacy and Pharmacology*, **22:** 801–805.

McArthur, J.N., Dawkins, P.D., Smith, M.J.H. and Hamilton, E.B.D. 1971, Mode of action of antirheumatic drugs. *British Medical Journal*, **2:** 677–679.

McCann, W.P. and Palmisano, P.A. 1973, Salicylate pharmacokinetics in the dog. *Research Communications in Chemical Pathology and Pharmacology*, **5:** 17–25.

McKinnon, G.E. and Dickinson, R.G. 1989, Covalent binding of diflunisal and probenecid to plasma protein in humans: persistence of the adducts in the circulation. *Research Communications in Chemical Pathology and Pharmacology*, **66:** 339–354.

McMahon, T.F., Stefanski, S.A., Wilson, R.E., Blair, P.C., Clark, A.M. and Birnbaum, L.S. 1991, Comparative acute nephrotoxicity of salicylic acid, 2,3-dihydroxybenzoic acid and 2,5-dihydroxybenzoic acid in young and middle aged Fischer 344 rats. *Toxicology*, **66:** 297–311.

Meffin, P.J., Brooks, P.M., Bertouch, J., Veenendaal, J.R. and Harrington, B.J. 1983, Diflunisal disposition and hypouricaemic response in osteoarthritis. *Clinical Pharmacology and Therapeutics*, **33:** 813–821.

Mellick, G.D. and Roberts, M.S. 1992, Disposition of aspirin and salicylate in the isolated perfused rat liver. *Clinical and Experimental Pharmacology and Physiology*, (**Suppl. 21**): 47.

Menguy, R., Desbaillets, L., Masters, Y.F. and Okabe, S. 1972, Evidence for a sex-linked difference in aspirin metabolism. *Nature*, **239:** 102–103.

Miles, C.I. and Schenk, G.H. 1970, Fluorescence of acetylsalicylic acid in solution and its measurement in presence of salicylic acid. *Analytical Chemistry*, **42:** 656–659.

Minami, T. and Cutler, D.J. 1992, A kinetic study of the role of band 3 anion transport protein in the transport of salicylic acid and other hydroxybenzoic acids across the human erythrocyte membrane. *Journal of Pharmacy and Pharmacology*, **81:** 424–427.

Minchin, R.F., Ilett, K.F., Teitel, C.H., Reeves, P.T. and Kadlubar, F.F. 1992, Direct O-acetylation of N-hydroxy arylamines by acetylsalicylic acid to form carcinogen-DNA adducts. *Carcinogenesis*, **13:** 663–667.

Miners, J.O., Grgurinovich, N., Whitehead, A.G., Robson, R.A. and Birkett, D.J. 1986, Effect of gender and oral contraceptive steroids on the metabolism of salicylic acid and acetylsalicylic acid. *British Journal of Clinical Pharmacology*, **22:** 135–142.

Mojaverian, P., Rocci, M.L., Swanson, B.N., Vlasses, P.H., Chremos, A.N., Lin, J.H., Yeh, J.C. and Ferguson, R.K. 1985, Steady state disposition of diflunisal: once- versus twice-daily administration. *Pharmacotherapy*, **5:** 336–339.

Montgomery, J., Ste-Marie, L., Boismenu, D. and Vachon, L. 1995, Hydroxylation of aromatic compounds as indices of hydroxyl radical production: a cautionary note revisited. *Free Radical Biology and Medicine*, **19:** 927–933.

Montgomery, J.A. and Mamer, O.A. 1978, Profiles in altered metabolism. II Accumulation of homogentisic acid in serum and urine following acetylsalicylic acid ingestion. *Biomedical Mass Spectrometry*, **5:** 331–333.

Montgomery, P.R., Berger, L.G., Mitenko, P.A. and Sitar, D.S. 1986, Salicylate metabolism: effects of age and sex in adults., *Clinical Pharmacology and Therapeutics*, **39:** 571–576.

Montgomery, P.R. and Sitar, D.S. 1981, Increased serum salicylate metabolites with age in patients receiving chronic acetylsalicylic acid therapy. *Gerontology*, **27:** 329–333.

Moran, C.J. and Walker, W.H.C. 1968, The binding of salicylate to human serum. *Biochemical Pharmacology*, **17:** 153–156.

Morgan, A.M. and Truitt, E.B. 1965, Evaluation of acetylsalicylic acid esterase in aspirin metabolism. *Journal of Pharmaceutical Sciences*, **54:** 1640–1646.

Morra, P., Bartle, W.R., Walker, S.E., Lee, S.N., Bowles, S.K. and Reeves, R.A. 1996, Serum concentrations of salicylic acid following topically applied salicylate derivatives. *Annals of Pharmacotherapy*, **30:** 935–940.

Morris, C.H., Christian, J.E., Landolt, R.R. and Hansen, W.G. 1973, Metabolism of aspirin in rumen and corpus tissues of rat stomach during first four minutes after administration. *Journal of Pharmaceutical Sciences*, **62:** 1017–1018.

Morris, H.C., Overton, P.D., Ramsay, J.R., Campbell, R.S., Hammond, P.M., Atkinson, T. and Price, C.P. 1990, Development and validation of an automated, enzyme-mediated colorimetric assay of salicylate in serum. *Clinical Chemistry*, **36:** 131–135.

Morris, M.E. 1990, Pharmacokinetics and protein binding of salicylate metabolites in rats. *Drug Metabolism and Disposition*, **18:** 809–811.

Müller, F.O., Hundt, H.K.L. and de Kock, A.C. 1975, Decreased steady-state salicylic acid plasma levels associated with chronic aspirin ingestion. *Current Medical Research and Opinion*, **3:** 417–422.

Myers, B., Evans, D.N.W., Rhodes, J., Evans, B.K., Hughes, B.R., Lee, M.G., Richens, A. and Richards, D. 1987, Metabolism and urinary excretion of 5-amino salicylic acid in healthy volunteers when given intravenously or released for absorption at different sites in the gastrointestinal tract. *Gut*, **28:** 196–200.

Nagy, E., Csipo, I., Degrell, I. and Szabo, G. 1988, High-performance liquid chromatographic assay of 5-aminosalicylic acid and its acetylated metabolite in biological fluids using electrochemical detection. *Journal of Chromatography*, **425:** 214–219.

Nakamura, J., Shiota, H., Sasaki, H. and Shibasaki, J. 1988, Hydrolysis of salicyluric acid in intestinal microorganisms and prolonged blood concentration of salicylic acid following rectal administration of salicyluric acid in rats. *Journal of Pharmacobio-Dynamics* **11:** 625–629.

Nakamura, J., Shiota, H., Sasaki, H. and Shibasaki, J. 1989, Sustained blood concentration of salicylic acid following rectal administration of salicyluric acid in dogs. *Chemical and Pharmaceutical Bulletin*, **37:** 2537–2538.

Narayan, M., Berliner, L.J., Merola, A.J., Diaz, P.T. and Clanton, T.L. 1997, Biological reactions of peroxynitrite: evidence for an alternate pathway of salicylate hydroxylation. *Free Radical Research*, **27:** 63–72.

Nelson, E., Hanano, M. and Levy, G. 1966, Comparative pharmacokinetics of salicylate elimination in man and rats. *Journal of Pharmacology and Experimental Therapeutics*, **153:** 159–166.

Netter, P., Faure, G., Regent, M.C., Procknal, J.A. and Levy, G. 1985, Salicylate kinetics in old age. *Clinical Pharmacology and Therapeutics*, **38:** 6–11.

Netter, P., Monot, C., Stalars, M.C., Mur, J.M., Royer, R.J., Faure, G., Pourel, J., Martin, J. and Gaucher, A. 1984, Decrease of in vitro serum protein binding of salicylate in rheumatoid arthritis. *European Journal of Drug Metabolism and Pharmacokinetics*, **9:** 109–116.

Nielsen, O.H. and Bondesen, S. 1983, Kinetics of 5-aminosalicylic acid after jejunal instillation in man. *British Journal of Clinical Pharmacology*, **16:** 738–740.

Niwa, T., Takeda, N., Maeda, K., Shibata, M. and Tatematsu, A. 1988a, Accumulation of furancarboxylic acids in uremic serum as inhibitors of drug binding, *Clinica Chimica Acta*, **173:** 127–138.

Niwa, T., Takeda, N., Tatematsu, A. and Maeda, K. 1988b, Accumulation of indoxyl sulfate, an inhibitor of drug-binding, in uremic serum as demonstrated by internal-surface reversed-phase liquid chromatography. *Clinical Chemistry*, **34:** 2264–2267.

Nuernberg, B. and Brune, K. 1989, Buffering the stomach content enhances the absorption of diflunisal in man. *Biopharmaceutics and Drug Disposition*, **10:** 377–387.

Nuernberg, B., Koehler, G. and Brune, K. 1991, Pharmacokinetics of diflunisal in patients. *Clinical Pharmacokinetics*, **20:** 81–89.

O'Connell, M.J. and Webster, N.R. 1990, Hyperoxia and salicylate metabolism in rats. *Journal of Pharmacy Pharmacology*, **42:** 205–206.

O'Driscoll, K.M. and Corrigan, O.I. 1983, The influence of perfusion rate on salicylate absorption in the rat., *Journal of Pharmacy and Pharmacology*, **35:** 814–815.

Øie, S. and Frislid, K. 1971, A fluorometric method for direct determination of total salicylate in plasma. *Pharmaceutica Acta Helvetiae*, **46:** 632–636.

Ophir, A., Berenshtein, E., Kitrossky, N., Berman, E.R., Photiou, S., Rothman, Z. and Chevion, M. 1993, Hydroxyl radical generation in the cat retina during reperfusion following ischaemia. *Experimental Eye Research*, **57:** 351–357.

Osman, M.A. and Hill, H.H. 1982, Silicon-selective detection after gas chromatography for the determination of silylated salicylic acid in urine. *Journal of Chromatography*, **232:** 430–434.

Otten, M.H., De Haas, G. and Van den Ende, R. 1997, Colonic spread of 5-ASA enemas in healthy individuals, with a comparison of their physical and chemical characteristics. *Alimentary Pharmacology and Therapeutics*, **11:** 693–697.

Owen, S.G., Francis, H.W. and Roberts, M.S. 1994, Disappearance kinetics of solutes from synovial fluid after intra-articular injection. *British Journal of Clinical Pharmacology*, **38:** 349–355.

Owen, S.G., Roberts, M.S., Friesen, W.T. and Francis, H.W. 1989, Salicylate pharmacokinetics in patients with rheumatoid arthritis, *British Journal of Clinical Pharmacology*, **28:** 449–461.

Pachman, L.M., Olufs, R., Procknal, J.A. and Levy, G. 1979, Pharmacokinetic monitoring of salicylate therapy in children with juvenile rheumatoid arthritis. *Arthritis and Rheumatism*, **22:** 826–831.

Page, M.A., Anderson, R.A., Brown, K.F. and Roberts, M.S. 1974, The availability of sodium salicylate from enteric coated tablets. *Australian Journal of Pharmaceutical Science*, **NS3:** 95–99.

Patel, D.K., Hesse, A., Ogunbona, A., Notarianni, L.J. and Bennett, P.N. 1990a, Metabolism of aspirin after therapeutic and toxic doses. *Human and Experimental Toxicology*, **9:** 131–136.

Patel, D.K., Notarianni, L.J. and Bennett, P.N. 1990b, Comparative metabolism of high doses of aspirin in man and rat. *Xenobiotica*, **20:** 847–854.

Patel, D.K., Ogunbona, A., Notarianni, L.J. and Bennett, P.N. 1990c, Depletion of plasma glycine and effect of glycine by mouth on salicylate metabolism during aspirin overdose. *Human and Experimental Toxicology*, **9:** 389–395.

Paulus, H.E., Siegel, M., Mongan, E., Okun, R. and Calabro, J.J. 1971, Variations of serum concentrations and half-life of salicylate in patients with rheumatoid arthritis. *Arthritis and Rheumatism*, **14:** 527–532.

Paxton, J.W. 1980, Effects of aspirin on salivary and serum phenytoin kinetics in healthy subjects. *Clinical Pharmacology and Therapeutics*, **27**, 170–177.

Pedersen, A.K. and FitzGerald, G.A. 1984, Dose-related kinetics of aspirin. Presystemic acetylation of platelet cyclooxygenase. *New England Journal of Medicine*, **311:** 1206–1211.

Pedersen, A.K. and FitzGerald, G.A. 1985, Preparation and analysis of deuterium labelled aspirin: application to pharmacokinetic studies. *Journal of Pharmaceutical Sciences*, **74:** 188–192.

Perez-Mateo, M. and Erill, S. 1977, Protein binding of salicylate and quinidine in plasma from patients with renal failure, chronic liver disease and chronic respiratory insufficiency. *European Journal of Clinical Pharmacology*, **11:** 225–231.

Pinckard, R.N., Hawkins, D. and Farr, R.S. 1968, In vitro acetylation of plasma proteins, enzymes and DNA by aspirin. *Nature*, **219:** 68–69.

Pinckard, R.N., Hawkins, D. and Farr, R.S. 1970, The inhibitory effect of salicylate on the acetylation of human albumin by acetylsalicylic acid. *Arthritis and Rheumatism*, **13:** 361–368.

Pirmohamed, M., Coleman, M.D., Hussain, F., Breckenridge, A.M. and Park, B.K. 1991, Direct and metabolism-dependent toxicity of sulphasalazine and its principal metabolites towards human erythrocytes and leucocytes. *British Journal of Clinical Pharmacology*, **32:** 303–310.

Putney, J.W. and Borzelleca, J.F. 1972, Mechanisms of ^{14}C-salicylic acid excretion by the rat submaxillary gland. *Journal of Pharmacology and Experimental Therapeutics*, **182:** 515–521.

Pütter, J. 1975, Quantitative determination of the main metabolites of acetylsalicylic acid. 1. A method for the quantitative determination of salicylic acid and its metabolites. Studies in healthy individuals. *Arzneimittel-Forschung*, **25:** 941–944.

Pütter, J., Satravaha, P. and Stockhausen, H. 1974, Quantitative analysis of the main metabolites of acetylsalicylic acid. 3. Comparative analysis in the blood and milk of lactating women. *Zeitschrift für Geburtshilfe und Perinatologie*, **178:** 135–138.

Qin, W., Smith, J.B. and Smith, D.L. 1993, Reaction of aspirin with cysteinyl residues of lens-crystallins: a mechanism for the proposed anti-cataract effect of aspirin. *Biochimica et Biophysica Acta*, **1181:** 103–110.

Rabinowitz, J.L., Feldman, E.S., Weinberger, A. and Schumacher, H.R. 1982, Comparative tissue absorption of oral ^{14}C-aspirin and topical triethanolamine ^{14}C-salicylate in human and canine knee joints. *Journal of Clinical Pharmacology*, **22:** 42–48.

Radermacher, J., Jentsch, D., Scholl, M.A., Lustinetz, T. and Frölich, J.C. 1991, Diclofenac concentrations in synovial fluid and plasma after cutaneous application in inflammatory and degenerative joint disease. *British Journal of Clinical Pharmacology*, **31:** 537–541.

Raghoebar, M., van den Berg, W.B., Huisman, J.A.M. and van Ginneken, C.A.M. 1987, The cellular association of sodium salicylate and indomethacin in peritoneal fluid of ascites bearing mice. *Agents and Actions*, **22:** 314–323.

Raghoebar, M., van den Berg, W.B. and van Ginneken, C.A.M. 1988, Mechanisms of cell association of some non-steroidal anti-inflammatory drugs with isolated leukocytes. *Biochemical Pharmacology*, **37:** 1245–1250.

Rainsford, K.D. 1996, *Anti-inflammatory and Antirheumatic Therapy*. Boca Raton: CRC Press.

Rainsford, K.D. and Brune, K. 1976, Role of the parietal cell in gastric damage induced by aspirin and related drugs; implications for safer therapy. *Medical Journal of Australia*, **1:** 881–883.

Rainsford, K.D., Ford, N.L.V., Brooks, P.M. and Watson, H.M. 1980, Plasma aspirin esterases in normal individuals, patients with alcoholic liver disease and rheumatoid arthritis: characterization and the importance of the enzymic components. *European Journal of Clinical Investigation*, **10:** 413–420.

Rainsford, K.D., Schweitzer, A. and Brune, K. 1981, Autoradiographic and biochemical observations on the distribution of non-steroidal anti-inflammatory drugs. *Archives Internationales de Pharmacodynamie et de Therapie*, **250:** 180–194.

Rainsford, K.D., Schweitzer, A. and Brune, K. 1983, Distribution of the acetyl compared with the salicyl moiety of acetylsalicylic acid. Acetylation of biomolecules in organs in which side effects are manifest. *Biochemical Pharmacology*, **32:** 1301–1308.

Rainsford, K.D. and Whitehouse, M.W. 1980, Anti-inflammatory/anti-pyretic salicylic esters with low gastric ulcerogenic activity. *Agents and Actions* **10:** 451–456.

Rao, G.N., Lardis, M.P. and Cotlier, E. 1985, Acetylation of lens crystallins. A possible mechanism by which aspirin could prevent cataract formation. *Biochemical and Biophysical Research Communications*, **128:** 1125–1132.

Ray, J.E. and Day, R.O. 1983, High-performance liquid chromatographic analysis of diflunisal in plasma and urine: application to pharmacokinetic studies in two normal volunteers. *Journal of Pharmaceutical Sciences*, **72:** 1403–1405.

Reed, J.R. and Palmisano, P.A. 1975, Central nervous system salicylate. *Clinical Toxicology*, **8:** 623–631.

Reidl, U. 1983, Determination of acetylsalicylic acid and metabolites in biological fluids by high-performance liquid chromatography. *Journal of Chromatography*, **272:** 325–331.

Reinberg, A., Zagulla-Mally, Z.W., Ghata, J., Halberg, F. 1967, Circadian rhythm in duration of salicylate excretion referred to phase of excretory rhythms and routine. *Proceedings of the Society for Experimental Biology and Medicine*, **124:** 826–832.

Rendell, M., Nierenberg, J., Brannan, C., Valentine, J.L., Stephen, P.M., Dodds, S., Mercer, P., Smith, P.K. and Walder, J. 1986, Inhibition of glycation of albumin and hemoglobin by acetylation in vitro and in vivo. *Journal of Laboratory and Clinical Medicine*, **108:** 286–293.

Reynolds, R.C. and Cluff, L.E. 1960, Interaction of serum and sodium salicylate: changes during acute infection and its influence on pharmacological activity. *Bulletin of the Johns Hopkins Hospital*, **107:** 278–290.

Richmond, R., Halliwell, B., Chauhan, J. and Darbre, A. 1981, Superoxide-dependent formation of hydroxyl radicals: detection of hydroxyl radicals by the hydroxylation of aromatic compounds. *Analytical Biochemistry*, **118:** 328–335.

Rijk, M.C.M., van Hogezand, R.A., van Schaik, A. and van Tongeren, J.H.M. 1989, Disposition of 5-aminosalicylic acid by 5-aminosalicylic acid-delivering drugs during accelerated intestinal transit in healthy volunteers. *Scandinavian Journal of Gastroenterology*, **24:** 1179–1185.

Rijk, M.C.M., van Schaik, A. and van Tongeren, J.H.M. 1988, Disposition of 5-aminosalicylic acid by 5-aminosalicylic acid-delivering compounds. *Scandinavian Journal of Gastroenterology*, **53 (Suppl. 148):** 54–59.

Roberts, M.S., Favretto, W.A., Meer, A., Reckman, M. and Wongseelashote, T. 1982, Topical bioavailability of methyl salicylate. *Australian and New Zealand Journal of Medicine*, **12:** 303–305.

Roberts, M.S. and Horlock, E. 1978, Effect of repeated skin applications of percutaneous absorption of salicylic acid. *Journal of Pharmaceutical Sciences*, **67:** 1685–1687.

Roberts, M.S., Rumble, R.H., Wanwimolruk, S., Thomas, D. and Brooks, P.M. 1983, Pharmacokinetics of aspirin and salicylate in elderly subjects and in subjects with alcoholic liver disease. *European Journal of Clinical Pharmacology*, **25:** 253–261.

Robertson, A., Glynn, J.P. and Watson, A.K. 1972, The absorption and metabolism in man of 4-acetamidophenyl-2-acetoxybenzoate (Benorylate). *Xenobiotica*, **2:** 339–347.

Rosenthal, R.K., Bayles, T.B. and Fremont-Smith, K. 1964, Simultaneous salicylate concentrations in synovial fluid and plasma in rheumatoid arthritis. *Arthritis and Rheumatism*, **7:** 103–109.

Rougier, A., Lotte, C. and Maibach, H.I. 1987, In vivo percutaneous penetration of some organic compounds related to anatomic site in humans: predictive assessment by the stripping method. *Journal of Pharmaceutical Sciences*, **76:** 451–454.

Rowland, M. and Riegelman, S. 1967, Determination of acetylsalicylic acid and salicylic acid in plasma. *Journal of Pharmaceutical Sciences*, **56:** 717–720.

Rowland, M. and Riegelman, S. 1968, Pharmacokinetics of acetylsalicylic acid and salicylic acid after intravenous administration in man. *Journal of Pharmaceutical Sciences*, **57:** 1313–1319.

Rowland, M., Riegelman, S., Harris, P.A. and Sholkoff, S.D. 1972, Absorption kinetics of aspirin in man following oral administration of an aqueous solution. *Journal of Pharmaceutical Sciences*, **61:** 379–385.

Rubin, G.M., Tozer, N. and Øie, S. 1983, Concentration-dependence of salicylate distribution. *Journal of Pharmacy and Pharmacology*, **35:** 115–117.

Russel, F.G.M., Wouterse, A.C. and van Ginneken, C.A.M. 1987, Physiologically based pharmacokinetic model for the clearance of salicyluric acid and the interaction with phenolsulphonphthalein. *Drug Metabolism and Disposition*, **15:** 695–701.

Ryde, E.M. and Ahnfelt, N.-O. 1988, The pharmacokinetics of olsalazine sodium in healthy volunteers after a single i.v. dose and after oral doses with and without food. *European Journal of Clinical Pharmacology*, **34:** 481–488.

Rylance, H.J. and Wallace, R.C. 1981, Erythrocyte and plasma esterase. *British Journal of Clinical Pharmacology*, **12:** 436–438.

Sagone, A.L. and Husney, R.M. 1987, Oxidation of salicylates by stimulated granulocytes: evidence that these drugs act as free radical scavengers in biological systems. *Journal of Immunology*, **138:** 2177–2183.

Salem, S.A.M. and Stevenson, I.H. 1977, Absorption kinetics of aspirin and quinine in elderly. *British Journal of Clinical Pharmacology*, **4:** 397P.

Sam, E., Sarre, S., Michotte, Y. and Verbeke, N. 1998, Cathechol is the major product of hydroxylation in 1-methyl-4-phenylpyridium ion treated rats. *European Journal of Drug Metabolism and Pharmacokinetics*, **23:** 137–142.

Sandberg-Gertzen, H., Ryde, M. and Jarnerot, G. 1983, Absorption and excretion of single 1-g dose of azodisal sodium in subjects with ileostomy. *Scandinavian Journal of Gastroenterology*, **18:** 107–111.

Schachter, D., Kass, D.J. and Lannon, J. 1959, The biosynthesis of salicyl glucuronides by tissue slices of various organs. *Journal of Biological Chemistry*, **234:** 201–205.

Schachter, D. and Manis, J.G. 1958, Salicylate and salicyl conjugates: fluorometric estimation, biosynthesis and renal excretion in man. *Journal of Clinical Investigation*, **37:** 800–806.

Schwartz, R., Fellers, F.X., Knapp, J., Yaffe, S. 1959, Renal response to administration of acetazolamide (Diamox) during salicylate intoxication. *Pediatrics*, **23:** 1103–1114.

Seeberg, V.P., Hansen, D. and Whitney, B. 1951, Absorption and distribution of salicylamide. *Journal of Pharmacology and Experimental Therapeutics*, **101:** 275–282.

Sellers, E.M. 1979, Plasma protein displacement interactions are rarely of clinical significance. *Pharmacology*, **18:** 225–227.

Seymour, R.A., Williams, F.M., Ward, A. and Rawlins, M.D. 1984, Aspirin metabolism and efficacy in postoperative dental pain. *British Journal of Clinical Pharmacology*, **17:** 697–702.

Shen, J., Wanwimolruk, S., Clark, C.R. and Roberts, M.S. 1990, A sensitive assay for aspirin and its metabolites using reversed-phase ion-pair high-performance liquid chromatography. *Journal of Liquid Chromatography*, **13:** 751–761.

Shen, J., Wanwimolruk, S., Purves, R.D., McQueen, E.G. and Roberts, M.S. 1991a, Model representation of salicylate pharmacokinetics using unbound plasma salicylate concentrations and metabolite urinary excretion rates following a single oral dose. *Journal of Pharmacokinetics and Biopharmaceutics*, **19:** 575–595.

Shen, J., Wanwimolruk, S. and Roberts, M.S. 1991b, Novel direct high-performance liquid chromatographic method for determination of salicylate glucuronide conjugates in human urine. *Journal of Chromatography*, **565:** 309–320.

Shetty, B.V., Badr, M. and Melethil, S. 1994, Evaluation of hepatic metabolism of salicylic acid in perfused rat liver. *Journal of Pharmaceutical Sciences*, **83:** 807–808.

Shibasaki, J., Inone, Y., Kadosaki, K. and Nakamura, J. 1985, Hydrolysis of salicyluric acid in rabbit intestinal microorganisms. *Journal of Pharmacobio-Dynamics*, **8:** 989–995.

Shibaski, J., Konishi, R., Koike, M., Imamura, A. and Sueyasu, M. 1981, Some quantitative evaluation of first pass metabolism of salicylamide in rabbit and rat. *Journal of Pharmacobio-Dynamics*, **4:** 91–100.

Sholkoff, S.D., Eyring, E.J., Rowland, M. and Riegelman, S. 1967, Plasma and synovial fluid concentrations of acetylsalicylic acid in patients with rheumatoid arthritis. *Arthritis and Rheumatism*, **10:** 348–351.

Short, C.R., Hsieh, L.C., Malbrough, M.S., Barker, S.A., Neff-Davis, C.A., Davis, L.E., Koritz, G. and Bevill, R.F. 1990, Elimination of salicylic acid in goats and cattle. *American Journal of Veterinary Research*, **51:** 1267–1270.

Short, C.R., Neff-Davis, C.A., Hsieh, L.C., Koritz, G.D., Malbrough, M.S., Barker, S.A. and Davis, L.E. 1991, Pharmacokinetics and elimination of salicylic acid in rabbits. *Journal of Veterinary Pharmacology and Therapeutics*, **14:** 70–77.

Siebert, D.M. and Bochner, F. 1987, Determination of plasma aspirin and salicylic acid concentrations after low aspirin doses by high-performance liquid chromatography with post-column hydrolysis and fluorescence detection. *Journal of Chromatography*, **420:** 425–431.

Singh, P. and Roberts, M.S. 1993a, Dermal and underlying tissue pharmacokinetics of salicylic acid after topical application. *Journal of Pharmacokinetics and Biopharmaceutics*, **21:** 337–373.

Singh, P. and Roberts, M.S. 1993b, Iontophoretic delivery of salicylic acid and lidocaine to local subcutaneous structures. *Journal of Pharmaceutical Sciences*, **82:** 127–131.

Sitar, D.S., Chalmers, I.M. and Hunter, T. 1985, Plasma and synovial fluid concentrations of salicylic acid and its metabolites in patients with joint effusions. *Journal of Rheumatology*, **12:** 134–135.

Smith, M.J.H. 1951, Plasma-salicylate concentrations after small doses of acetylsalicylic acid. *Journal of Pharmacy and Pharmacology*, **3:** 409–414.

Smith, P.K., Gleason, H.L., Stoll, C.G. and Orgorzalek, S. 1946, Studies on the pharmacology of salicylates. *Journal of Pharmacology and Experimental Therapeutics*, **87:** 237–246.

Song, C.S., Bonkowsky, H.L. and Tschudy, D.P. 1974, Salicylamide metabolism in acute intermittent porphyria. *Clinical Pharmacology and Therapeutics*, **15:** 431–435.

Song, C.S., Gelb, N.A. and Wolff, S.M. 1972, The influence of pyrogen-induced fever on salicylamide metabolism in man. *Journal of Clinical Investigation*, **51:** 2959–2966.

Soren, A. 1973, Transport of salicylates from blood to joint fluid. *Archives of Internal Medicine*, **132:** 668–672.

Spector, R., Korkin, D.T. and Lorenzo, A.V. 1972, A rapid method for the determination of salicylate binding by the use of ultrafilters. *Journal of Pharmacy and Pharmacology*, **24:** 786–789.

Spenney, J.G. and Nowell, R.M. 1979, Acetylsalicylate hydrolase of rabbit gastric mucosa. Isolation and purification. *Drug Metabolism and Disposition*, **7:** 215–219.

Stevenson, G.W. 1960, Rapid ultaviolet spectrophotometric determination of salicylate in blood. *Analytical Chemistry*, **32:** 1522–1525.

Stewart, M.J. and Watson, I.D. 1987, Analytical reviews in clinical chemistry: methods for the estimation of salicylate and paracetamol in serum, plasma and urine. *Annals of Clinical Biochemistry*, **24:** 552–565.

Stoa-Birketvedt, G. and Florholmen, J. 1999, The systemic load and efficient delivery of active 5-aminosalicylic acid in patients with ulcerative colitis on treatment with olsalazine or mesalazine. *Alimentary Pharmacology and Therapeutics*, **13:** 357–361.

Stretch, G.L., Campbell, B.J., Dwarakanath, A.D., Yaqoob, M., Stevenson, A., Morris, A.I. *et al.* 1996, 5-aminosalicylic acid absorption and metabolism in ulcerative colitis patients receiving maintenance sulphasalazine, olsalazine or mesalazine. *Alimentary Pharmacology and Therapeutics*, **10:** 941–947.

Sturman, J.A. and Smith, M.J.H. 1967, The binding of salicylate to plasma proteins in different species. *Journal of Pharmacy and Pharmacology*, **19:** 621–623.

Sudo, L.S., Almeida, M.G., Yasaka, W. and Garcia-Leme, J. 1989, Lymphatic transport of salicylates in dogs. *General Pharmacology*, **20:** 779–783.

Sweeney, K.R., Chapron, D.J., Brandt, J.L., Gomolin, I.H., Feig, I.H. and Kramer, P.A. 1986, Toxic interaction between acetazolamide and salicylate: case reports and a pharmacokinetic interaction. *Clinical Pharmacology and Therapeutics*, **40:** 518–524.

Taggart, A.J., McDermott, B.J. and Roberts, S.D. 1992, The effect of age and acetylator phenotype on the pharmacokinetics of sulfasalazine in patients with rheumatoid arthritis. *Clinical Pharmacokinetics*, **23:** 311–320.

Talbert, R.L., Ludden, T.M. and West, R.E. 1979, Rapid establishment of therapeutic serum concentrations of salicylates. *Journal of Clinical Pharmacology*, **19:** 108–112.

Taylor, J.R. and Halprin, K. 1975, Percutaneous absorption of salicylic acid. *Archives of Dermatology*, **111:** 740–743.

Thiessen, J.J., Salama, R.B., Coceani, F. and Olley, P.M. 1984, Placental transfer in near-term ewes: acetylsalicylic acid and salicylic acid. *Canadian Journal of Physiology and Pharmacology*, **62:** 441–445.

Thomas, B.H., Solomonraj, G. and Coldwell, B.B. 1973, The estimation of acetylsalicylic acid and salicylate in biological fluids by gas-liquid chromatography. *Journal of Pharmacy and Pharmacology*, **25:** 201–204.

Thome, J., Zhang, J., Davids, E., Foley, P., Weijers, H.G., Wiesbeck, G.A. *et al.* 1997, Evidence for increased oxidative stress in alcohol-dependent patients provided by quantification of in vivo salicylate hydroxylation products. *Alcoholism: Clinical and Experimental Research*, **21:** 82–85.

Tishler, S.L. and Goldman, P. 1970, Properties and reactions of salicyl-coenzyme A. *Biochemical Pharmacology*, **19:** 143–150.

Tjornelund, J. and Hansen, S.H. 1991, High-performance liquid chromatographic assay of 5-aminosalicylic acid (5-ASA) and its metabolites N-β-D-glucopyranosyl-5-ASA, N-acetyl-5-ASA, N-formyl-ASA and N-butyryl-5-ASA in biological fluids. *Journal of Chromatography*, **570:** 109–117.

Tjornelund, J., Hansen, S.H. and Cornett, C. 1989, New metabolites of the drug 5-aminosalicylic acid. I. N-β-D-glucopyranosyl-5-aminosalicylic acid. *Xenobiotica*, **19:** 891–899.

Tjornelund, J., Hansen, S.H. and Cornett, C. 1991, New metabolites of the drug 5-aminosalicylic acid. II. N-formyl-5-aminosalicylate. *Xenobiotica*, **21:** 605–612.

Trinder, P. 1954, Rapid determination of salicylate in biological fluids. *Biochemical Journal*, **57:** 301–303.

Trnavsky, K. and Zachar, M. 1975, Correlation of serum aspirin esterase activity and half-life of salicylic acid. *Agents and Actions*, **5:** 549–552.

Trnavska, Z. and Trnavsky, K. 1980, Characterization of salicylate binding to synovial fluid and plasma protein in patients with rheumatoid arthritis. *European Journal of Clinical Pharmacology*, **18:** 403–406.

Trnavska, Z. and Trnavsky, K. 1983, Sex differences in the pharmacokinetics of salicylates. *European Journal of Clinical Pharmacology*, **25:** 679–682.

Trnavska, Z., Trnavsky, K. and Smondrk, J. 1985, The effect of cimetidine on the pharmacokinetics of salicylic acid. *Drugs under Experimental and Clinical Research*, **11:** 703–707.

Truitt, E.B., Morgan, A.M. and Little, J.M. 1955, Determination of salicylic acid and two metabolites in plasma and urine using fluorimetry for directly measuring salicyluric acid. *Journal of the American Pharmaceutical Association Scientific Edition*, **44:** 142–148.

Unsworth, J., d'Assis-Fonseca, A., Beswick, D.T. and Blake, D.R. 1987, Serum salicylate levels in a breast fed infant. *Annals of the Rheumatic Diseases*, **46:** 638–639.

Valeri, P., Romanelli, L., Martinelli, B., Guglielmotti, A. and Catanese, B. 1989, Time-course of aspirin and salicylate in ocular tissues of rabbits. *Lens and Eye Toxicology Research*, **6,** 465–475.

van Hogezand, R.A., van Schaik, A., van Hees, P.A.M. and van Tongeren, J.H.M. 1988, Stability of disodium azodisalicylate (olsalazine) and metabolites in urine and faeces stored at different temperatures. *European Journal of Drug Metabolism and Pharmacokinetics*, **13:** 261–265.

Varma, D.R. 1988, Modification of transplacental distribution of salicylate by acidosis and alkalosis. *British Journal of Pharmacology*, **93:** 978–984.

Varma, D.R. and Yue, T.L. 1984, Influence of age, sex, pregnancy and protein-calorie malnutrition on the pharmacokinetics of salicylate in rats. *British Journal of Pharmacology*, **82:** 241–248.

Verbeeck, R.K. and Cardinal, J.-A. 1985, Plasma protein binding of salicylic acid, phenytoin, chlorpromazine and pethidine using equilibrium dialysis and ultracentrifugation. *Arzneimittel-Forschung*, **35:** 903–906.

Verbeeck, R.K., Loewen, G.R., Macdonald, J.I. and Herman, R.J. 1990, The effect of multiple dosage on the kinetics of glucuronidation and sulphation of diflunisal in man. *British Journal of Clinical Pharmacology*, **29:** 381–389.

Verbeeck, R.K., Tjandramaga, T.B., Mullie, A., Verbesselt, R., Verberckmoes, R. and De Schepper, P.J. 1979, Biotransformation of diflunisal and renal excretion of its glucuronides in renal insufficiency. *British Journal of Clinical Pharmacology*, **7:** 273–282.

Vermeersch, G., Marko, J., Cartigny, B., Leclerc, F., Roussel, P. and Lhermitte, M. 1988, Salicylate poisoning detected by ^{1}H NMR spectroscopy. *Clinical Chemistry*, **34:** 1003–1004.

Vial, J.H., Roberts, M.S., McLeod, L.J. and Seville, P.R. 1990, Selective inhibition of platelet cyclooxygenase with controlled release, low-dose aspirin. *Australian and New Zealand Journal of Medicine*, **20:** 652–656.

von Weiss, J.F. and Lever, W.F. 1964, Percutaneous salicylic acid intoxication in psoriasis. *Archives of Dermatology* **90:** 614–619.

Walker, P.D. and Shah, S.V. 1988, Evidence suggesting a role for hydroxyl radical in gentamicin-induced acute renal failure in rats. *Journal of Clinical Investigation*, **81:** 334–341.

Walter, K. and Kurz, H. 1988, Binding of drugs to human skin: influencing factors and the role of tissue lipids. *Journal of Pharmacy and Pharmacology*, **40:** 689–693.

Walters, K.A., Brain, K.R., Howes, D., James, V.J., Kraus, A.L., Teetsel, N.M. *et al.* 1997, Percutaneous penetration of octyl salicylate from representative sunscreen formulations through human skin in vitro. *Food and Chemical Toxicology*, **35:** 1219–1225.

Waschek, J.A., Fielding, R.M., Pond, S.M., Rubin, G.M., Effeney, D.J. and Tozer, T.N. 1985, Dose-dependent sulphoconjugation of salicylamide in dogs; Effect of sulphate depletion or co-administration. *Journal of Pharmacology and Experimental Therapeutics*, **234:** 431–434.

Watt, J.A. and Dickinson, R.G. 1990, Reactivity of diflunisal acyl glucuronide in human and rat plasma and albumin solutions. *Biochemical Pharmacology*, **39:** 1067–1075.

White, K.N. and Hope, D.B. 1984, Partial purification and characterization of a microsomal carboxylesterase specific for salicylate esters from guinea-pig liver. *Biochimica et Biophysica Acta*, **785:** 138–147.

Whitehouse, M.W., Rainsford, K.D., Ardlie, N.G., Young, I.G. and Brune, K. 1976, Alternatives to aspirin derived from biological sources. *Agents and Actions*, (**Suppl. 1**): 43–57.

Whitlam, J.B. and Brown, K.F. 1981, Ultrafiltration in serum protein binding determinations. *Journal of Pharmaceutical Sciences*, **70:** 146–150.

Wientjes, M.G. and Levy, G. 1988, Non-linear pharmacokinetics of aspirin in rats. *Journal of Pharmacology and Experimental Therapeutics*, **245:** 809–815.

Williams, F.M., Moore, U., Seymour, R.A., Mutch, E.M., Nicholson, E., Wright, P., Wynne, H., Blain, P.G. and Rawlins, M.D. 1989b, Benorylate hydrolysis by human plasma and human liver. *British Journal of Clinical Pharmacology*, **28:** 703–708.

Williams, F.M., Mutch, E. and Blain, P.G. 1991, Esterase activity in rat hepatocytes. *Biochemical Pharmacology*, **41:** 527–531.

Williams, F.M., Mutch, E.M., Nicholson, E., Wynne, H., Wright, P., Lambert, D. and Rawlins, M.D. 1989a, Human liver and plasma aspirin esterase. *Journal of Pharmacy and Pharmacology*, **41:** 407–409.

Williams, F.M., Nicholson, E.N., Woolhouse, N.W., Adjepon-Yamoah, K.K. and Rawlins, M.D. 1986a, Esterase activity in plasma from Ghanian and British subjects. *European Journal of Clinical Pharmacology*, **31:** 485–489.

Williams, J.G. and Hallett, M.B. 1989, The reaction of 5-amino-salicylic acid with hypochlorite. Implications for its mode of action in inflammatory bowel disease. *Biochemical Pharmacology*, **38:** 149–154.

Williams, M.E., Weinblatt, M., Rosa, R.M., Griffin, V.L., Goldlust, M.B., Shang, S.F., Harrison, L.I. and Brown, R.S. 1986b, Salsalate kinetics in patients with chronic renal failure undergoing hemodialysis. *Clinical Pharmacology and Therapeutics*, **39:** 420–424.

Willoughby, C.P., Aronson, J.K., Agback, H., Bodin, N.O. and Truelove, S.C. 1982, Distribution and metabolism in healthy volunteers of disodium azodisalicylate, a potential therapeutic agent for ulcerative colitis. *Gut*, **23:** 1081–1087.

Wilson, I.D. and Nicholson, J.K. 1988, Solid phase extraction chromatography and NMR-spectroscopy (SPEC-NMR) for the rapid identification of drug metabolites in urine. *Journal of Pharmaceutical and Biomedical Analysis*, **6:** 151–165.

Wilson, J.T., Howell, R.L., Holladay, M.W., Brilis, G.M., Chrastil, J., Watson, J.T. and Taber, D.F. 1978, Gentisuric acid: metabolic formation in animals and identification as a metabolite of aspirin in man. *Clinical Pharmacology and Therapeutics*, **23:** 635–643.

Wisniewski, K. and Zarebski, M. 1968, Effect of insulin on the transport and the analgesic action of sodium salicylate. *Metabolism*, **17:** 212–217.

Worrall, S. and Dickinson, R.G. 1995, Rat serum albumin modified by diflunisal acyl glucuronide is immunogenic in rats. *Life Sciences*, **56:** 1921–1930.

Xu, X., Hirayama, H. and Pang, K.S. 1989, Studies in the once through vascularly perfused rat intestine–liver preparation. *Drug Metabolism and Disposition*, **17:** 556–563.

Xu, X., Tang, B.K. and Pang, K.S. 1990, Sequential metabolism of salicylamide exclusively to gentisamide-5-glucuronide and not gentisamide sulphate conjugates in single-pass in situ perfused rat liver. *Journal of Pharmacology and Experimental Therapeutics*, **253:** 965–973.

Yacobi, A. and Levy, G. 1977, Intraindividual relationships between serum protein binding of drugs in normal human subjects, and patients with impaired renal function and rats. *Journal of Pharmaceutical Sciences*, **66:** 1285–1288.

Yoshikawa, T., Sugiyama, Y., Sawada, Y., Iga, T., Hanano, M., Kawasaki, S. and Yanagida, M. 1984a, Effect of late pregnancy on salicylate, diazepam, warfarin, and propranolol binding: use of fluorescent probes. *Clinical Pharmacology and Therapeutics*, **36:** 201–208.

Yoshikawa, T., Sugiyama, Y., Sawada, Y., Iga, T. and Hanano, M. 1984b, Effect of pregnancy on tissue distribution of salicylate in rats. *Drug Metabolism and Disposition*, **12:** 50–505.

You, K. 1985, Polarographic quantification of salicylate in serum by salicylate hydroxylase. *Clinica Chimica Acta*, **149:** 281–284.

Yu, D.K., Elvin, A.T., Morrill, B., Eichmeier, L.S., Lanman, R.C., Lanman, M.B. and Giesing, D.H. 1990, Effect of food coadministration on 5-aminosalicylic acid oral suspension bioavailability. *Clinical Pharmacology and Therapeutics*, **48:** 26–33.

Zarebski, M. 1973, The effect of insulin on acute toxicity and transport of sodium salicylate into brain tissue. *Polish Journal of Pharmacology and Pharmacy*, **25:** 127–134.

Zarolinski, J.F., Keresztes-Nagy, S., Mais, R.F. and Oester, Y.T. 1974, Effect of temperature on the binding of salicylate by human serum albumin. *Biochemical Pharmacology*, **23:** 1767–1776.

Zimmerman, L., Jorhvall, H., Bergstrom, J., Furst, P. and Sjovall, J. 1981, Characterization of a double conjugate in uremic body fluids. *FEBS Letters*, **129:** 237–240.

CHAPTER 4

Metabolism and Pharmacokinetics of Ibuprofen

G.G. Graham and K.M. Williams

INTRODUCTION

The metabolism and pharmacokinetics of ibuprofen are of considerable interest because of the widespread use of the drug and its stereospecific metabolism. Being an asymmetric molecule it exists as an enantiomeric pair, like the other propionates that are used as non-steroidal anti-inflammatory drugs (NSAIDs). Ibuprofen is generally used as the racemic mixture of the two enantiomers, although S-ibuprofen is now available in some countries. There are several differences in the metabolism and pharmacokinetics of the two enantiomers, but the most significant aspect is the conversion of the R enantiomer to the pharmacologically active S enantiomer. Interactions of R-ibuprofen with lipid metabolism also occur, although significant clinical consequences of such interactions have not been demonstrated. Like aspirin and paracetamol, ibuprofen is administered in large doses, and it has been used as a prototype drug to investigate general principles of drug metabolism. Ibuprofen has also been used as a substrate to develop methods of drug metabolism, examples being the use of nuclear magnetic spectroscopy to study its metabolism (Shieh *et al.*, 1993; Spraul *et al.*, 1993).

ABSORPTION

Ibuprofen is essentially totally absorbed after its oral absorption with a bioavailability of 80 to 90 per cent (Martin *et al.*, 1990; Ceppi Monti *et al.*, 1992; Cheng *et al.*, 1994). The high oral absorption is due to the high lipid solubility of the un-ionised form, allowing its passive diffusion from the gastrointestinal tract, and to low first-pass metabolism in the liver. Its clearance is well below the hepatic flow rate of plasma, and its first-pass metabolism therefore should be low (see p. 166). The lipid solubilities and dissociation constants of the enantiomers are identical and, although membranes contain chiral molecules, there is no evidence that enantiomers have different rates of absorption once they are in solution. The rate of dissolution of the racemic mixture may, however, differ from that of the active S enantiomer if used by itself (Dwivedi *et al.*, 1992).

As is the case with other drugs, two aspects of the absorption of ibuprofen are of clinical importance. The extent of absorption is the more important in the long-term use of the drug, but the rate of absorption is important when rapid onset of action is required. Although the activity of ibuprofen lags behind changes in plasma concentrations (see p. 170), the onset of analgesic activity is still shorter with a soluble salt of ibuprofen than with conventional tablets containing the un-ionised acid form of the drug (Ceppi Monti *et al.*, 1992). Because of the short half-life of ibuprofen, sustained-release tablets are marketed for its long-term use, but at present definite clinical advantages over conventional tablets have not been established. Gastrointestinal tolerance may be increased (Valle-Jones *et al.*, 1984), but this requires more detailed study.

The rate of absorption of ibuprofen is slowed by food, although the total absorption is unaffected (Levine *et al.*, 1992). The absorption of ibuprofen is also slowed by oral surgery (Jamali and Kunz-Dober, 1999), presumably due to decreased parasympathetic and increased sympathetic responses. As with other analgesics, it is preferable to administer ibuprofen before surgery. Ibuprofen is very rapidly absorbed from solutions according to first order kinetics, but the dissolution of tablets yields a slower and more complex pattern of absorption (Gillespie *et al.*, 1982; Wagner *et al.*, 1984). Antacids generally have no significant effect on the total absorption of ibuprofen (Gontarz *et al.*, 1987) although magnesium hydroxide increases the rate of absorption of the drug from plain tablets (Neuvonen, 1991). As expected, charcoal decreases the absorption of ibuprofen, but the effect of charcoal is not increased by gastric lavage (Lapatto-Reiniluoto *et al.*, 1999).

Ibuprofen is largely present as an anion in the small intestine, and it is not surprising that cholestyramine, an ion-exchange resin, decreases the bioavailability of the drug. The effect is, however, relatively small, with only a 26 per cent reduction in absorption (al-Meshal *et al.*, 1994). This interaction should be further decreased by taking the ibuprofen and cholestyramine at widely different times.

Although the first-pass metabolism of ibuprofen is low, it has been suggested that there is inversion of R-ibuprofen to the S enantiomer by metabolism in the gastrointestinal tract, particularly if absorption is slow (Jamali *et al.*, 1988). The enzyme system capable of inverting R-ibuprofen to S-ibuprofen is present in the gastrointestinal tract (Jamali *et al.*, 1992), but kinetic analysis indicates that no significant inversion occurs if the absorption of ibuprofen is rapid (Hall *et al.*, 1993; Cheng *et al.*, 1994). Inversion also does not occur in the small intestine of the rat (Jeffrey *et al.*, 1990). Studies on animal skin also indicates that there is no inversion of R-ibuprofen during its absorption through skin (Millership and Collier, 1997).

Ibuprofen is absorbed through the skin, and peak plasma concentrations up to about one-fifth those after oral dosage have been reported after occlusive application in a preparation containing isopropyl alcohol and propylene glycol, which enhance topical absorption (Kleinbloesem *et al.*, 1995). Enhancement of skin permeation is also produced by the use of supersaturated solutions in which crystallisation is prevented by colloidal additives (Iervolino *et al.*, 2001). Topical application of ibuprofen yields substantial concentrations in subcutaneous tissue and muscle, even though less than 1 per cent is absorbed through uncovered skin from gel preparations (Dominkus *et al.*, 1996; Tegeder *et al.*, 1999). This absorption indicates that topical preparations should be useful for soft tissue rheumatism or when pain arises from damage to dermal or subcutaneous tissue or peripheral muscle. The dermal absorption is, however, slow, and the concentrations in muscle are very low in about 50 per cent of subjects, indicating that the relief of muscle pain may be small in some patients (Tegeder *et al.*, 1999). The importance of local concentrations in the epidermis at least is shown in an increasing local anti-inflammatory effect of ibuprofen with increasing uptake (Treffel and Gabard, 1993). Although local absorption into subcutaneous tissue and muscle occurs, work with other NSAIDs indicates that little drug absorbed through the skin permeates directly into the synovial fluid of the knee (Dawson *et al.*, 1988; Radermacher *et al.*, 1991). The extent of direct permeation into smaller joints is not known.

Like other NSAIDs, ibuprofen is taken up by the cornea with greater absorption at pH 6.4 than at slightly alkaline pH values (Gupta and Majumdar, 1997). This result is consistent with passive diffusion of the un-ionised form through the epithelium of the cornea.

DISTRIBUTION

Volume of distribution

The volume of distribution of both enantiomers of ibuprofen is small, being only about 0.1 to 0.2 l/kg (Table 5.1; Geisslinger *et al.*, 1990; Smith *et al.*, 1994). The small volume of distribution is the result of the very high binding to plasma albumin and low tissue binding. Because of the saturable binding to plasma albumin, the volume of distribution of racemic ibuprofen increases by nearly 50 per cent after doses increasing from 400 mg to 1200 mg in the elderly, although the increase is less marked in younger healthy volunteers (Albert *et al.*, 1984).

Binding to plasma proteins

Ibuprofen binds both reversibly and covalently to plasma proteins, most notably to albumin. The reversible binding involves the unchanged enantiomers, and the covalent binding is mediated through the reactivity of the glucuronides (see p. 162).

The reversible binding of ibuprofen is stereoselective, with only about 0.3 and 0.5 per cent of the R and S enantiomers unbound at low concentrations (Paliwal *et al.*, 1993). The unbound concentrations, however, increase at higher concentrations, and there is the further complication of mutual competition between the enantiomers causing displacement from binding sites (Evans *et al.*, 1989a; 1989b; Paliwal *et al.*, 1993). The result is that the unbound proportions of R- and S-ibuprofen are approximately 0.4 and 0.7 per cent after 600 to 800 mg doses of racemic ibuprofen, respectively. The enantiomer–enantiomer interaction and saturable binding accounts for the increased clearance of the total drug from plasma when administered at increasing doses of the racemic drug (Lockwood *et al.*, 1983) and for the increased clearance of the R enantiomer when administered as the racemate when compared to its ingestion as single enantiomers (Lee *et al.*, 1985). The clearances of the unbound enantiomers, however, are constant at doses in the therapeutic range, indicating that there is saturation only in the binding to the plasma proteins, and not in metabolic activity of the liver or other organs where metabolism occurs (Evans *et al.*, 1990; Paliwal *et al.*, 1993; Smith *et al.*, 1994).

Tissue distribution

Ibuprofen shows some reversible binding to tissue proteins, particularly to actin (Menzel and Kolarz, 1994), but the binding is still weaker than to plasma albumin. Ibuprofen forms covalent adducts with 60- and 110-kDa proteins in the mouse liver, but the production of the 110-kDa adduct is much lower

TABLE 5.1

Pharmacokinetic parameters of R- and S-ibuprofen. The parameters were determined after the administration of 300 mg of the individual enantiomers (from Evans *et al.*, 1990; Geisslinger *et al.*, 1990; Smith *et al.*, 1994).

Parameter	R-ibuprofen	S-ibuprofen
Unbound drug		
Clearance (l/min)	21	11
Half-life (h)	1.7	1.7
Total (bound + unbound)		
Clearance (ml/min)	76	77
Volume of distribution (l)	13	13
Half-life (h)	2.0	1.9

CHAPTER 5

than that of two potentially hepatotoxic NSAIDs, diclofenac and sulindac (Wade *et al.*, 1997). Ibuprofen is not considered to be hepatotoxic, and this lack of toxicity may be due to the low levels of the 110-kDa adduct. Ibuprofen is, however, incorporated covalently into lipids, unlike other NSAIDs (see p. 162).

After prolonged treatment of rats with ^{14}C-labelled ibuprofen substantial levels of radioactivity are found in several tissues, including the adrenals, ovaries, thyroid, skin and adipose tissues (Adams *et al.*, 1969a). The slow uptake and loss of labelled ibuprofen from these tissues indicates that the ibuprofen is incorporated in the tissues, possibly as hybrid triglycerides. The structure and function of these tissues does not appear to be affected.

Brain

Following oral dosage, the concentrations of both enantiomers of ibuprofen in cerebrospinal fluid are higher and more sustained than the unbound concentrations in plasma (Bannwarth *et al.*, 1995). Because of the high binding of ibuprofen to plasma proteins, the total concentrations in plasma are, however, much higher than in cerebrospinal fluid because of the absence of the binding proteins from normal cerebrospinal fluid. The slow diffusion of ibuprofen into and out of cerebrospinal fluid is consistent with the delayed antipyretic effects of ibuprofen (see p. 170).

Synovial fluid and other peripheral sites

Ibuprofen is typical of many highly protein-bound drugs that appear to diffuse into and out of synovial fluid largely in the unbound form, although the protein-bound form may contribute significantly to egress of ibuprofen from the joint. The features of the concentrations of ibuprofen in synovial fluid are (Figure 5.1; Day *et al.*, 1988; Walker *et al.*, 1993):

1. Over a dosage interval the mean concentrations of both enantiomers are lower than in plasma because the concentration of the binding protein, albumin, is lower in synovial fluid.
2. The concentrations of S-ibuprofen are on average about twice those of the R enantiomer. S-ibuprofen diffuses more rapidly than the R enantiomer, both into and out of synovial fluid. This difference correlates with the higher proportion of the S enantiomer that is unbound in plasma and synovial fluid.
3. The concentrations of both enantiomers peak later than in plasma and fluctuate to a much lesser degree in synovial fluid during multiple dosage. Kinetic analysis indicates that the reason for this pattern is that the half-lives of diffusion into and out of synovial fluid are of the same order or longer than the half-life of elimination (Graham, 1988).
4. During multiple dosage the trough concentrations in synovial fluid are higher than in plasma, particularly in patients in whom the rate of diffusion into and out of synovial fluid is very slow (Day *et al.*, 1988).
5. There are marked interpatient differences in the kinetics of diffusion into and out of synovial fluid (Day *et al.*, 1988).

It has been proposed that the sustained concentrations of S-ibuprofen in synovial fluid may correlate with its long duration of action, but several other possibilities are discussed below (see p. 170).

Ibuprofen also diffuses slowly into and out of other peripheral sites. For example, both enantiomers diffuse slowly into and out of blister fluid, and on average the time course in blister fluid resembles the time course in synovial fluid (Walker *et al.*, 1993). However, the kinetics of transfer in blister fluid correlate poorly with the kinetics in synovial fluid in individual patients (Seideman *et al.*, 1994). A similar delayed and sustained pattern of concentrations of ibuprofen in muscle and subcutaneous tissue is seen when the tissue concentrations are measured by the microdialysis technique (Tegeder *et al.*, 1999). The average concentrations are, however, of a similar order to the unbound concentrations in plasma. The

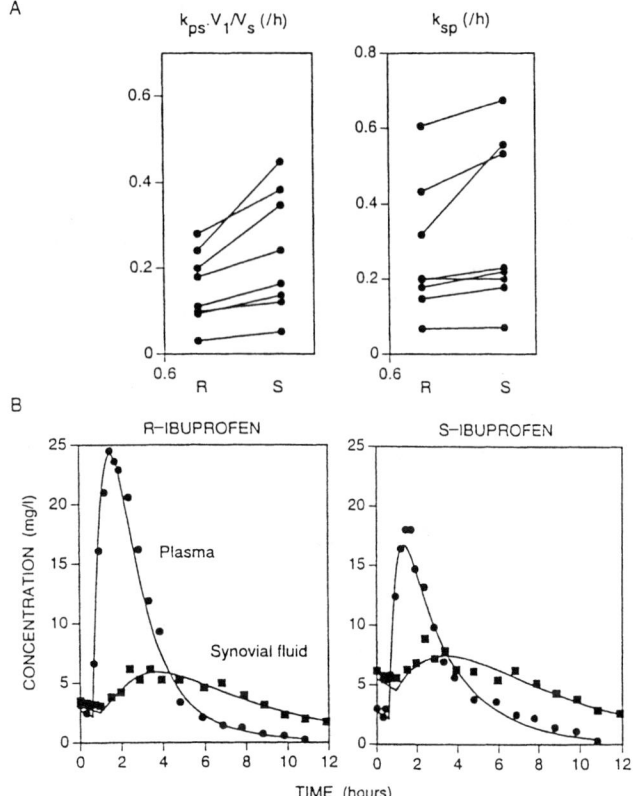

Figure 5.1 (a) Rate constants of transfer of R- and S-ibuprofen into and out of synovial fluid of patients with knee effusions. k_{ps} and k_{sp} are the first-order constants of transfer into and out of synovial fluid; V_1 and V_s are the volumes of the central compartment and synovial fluid, respectively. The quotient of the constants $k_{ps} \cdot V_1 / k_{ps} \cdot V_1$ gives the ratio of mean concentrations in synovial fluid to the mean concentrations in plasma. This ratio is approximately 0.6 and 0.7 for R- and S-ibuprofen, respectively. The faster transfer of S-ibuprofen into and out of synovial fluid correlates with the higher unbound proportions in plasma and synovial fluid. (b) Time courses of concentrations of R- and S-ibuprofen in synovial fluid in a patient who was receiving 800 mg of racemic ibuprofen every 8 hours. (Data of Day et al., 1988.)

uptake of ibuprofen into the heart is also slow, but follows a bi-exponential pattern, possibly reflecting uptake into extracellular and intracellular compartments (Askholt and Nielsen-Kudsk, 1985).

S-ibuprofen has been used as a model drug to determine the value of the site-specific delivery. A particular site utilised is the air pouch produced under rat skin. The air pouch resembles the synovial cavity, and is an easily accessible inflammatory compartment. The administration of S-ibuprofen into air pouches reduces the dose required for local drug exposure to well below that required by systemic administration (Martin et al., 1995).

Lipids

A feature of ibuprofen and some chemically related drugs is their stereospecific covalent incorporation into triglycerides (Fears, 1985; Williams et al., 1986; Sallustio et al., 1988). These products, termed hybrid lipids, are present in both plasma and adipose tissue. The formation of the hybrid lipids is dependent on the formation of the co-enzyme A thioester of ibuprofen. This intermediate is only formed from R-ibuprofen, and thus only R-ibuprofen leads to the formation of hybrid lipids. The co-enzyme A thioester of R-ibuprofen is, however, inverted to S-ibuprofen derivative (see p. 164), and therefore the hybrid lipids contain both R- and S-ibuprofen.

The hybrid lipids are formed by hepatocytes (Moorhouse *et al.*, 1991), and possibly in other tissues. In plasma, the hybrid lipids are present in concentrations equivalent to 3.5 to 9 per cent of the total plasma concentrations of ibuprofen (i.e. unbound and reversibly bound to albumin). The half-lives of disappearance of the triglycerides containing R- and S-ibuprofen from plasma are about 5 and 8 hours, respectively (Johnson *et al.*, 1995). These long half-lives may contribute to the slow terminal phase of elimination of ibuprofen by hydrolysis of the hybrid lipids back to the parent drug.

In adipose tissue in man the hybrid triglycerides accumulate to levels of about $10\,\mu g/g$ (Williams, 1991), and thus account for only a small percentage of the total triglycerides, but concentrations in adipose tissue still greatly exceed the unbound concentrations in plasma. The turnover of the hybrid triglycerides is slow. The rate of release of ibuprofen from adipose tissue is not known in humans, but the half-life of labelled ibuprofen is about 7 days in rats (see p. 160; Adams *et al.*, 1969a).

Ibufenac, an ibuprofen analogue that was withdrawn because of occasional hepatotoxicity, provides an interesting contrast with ibuprofen. Initial studies showed that the uptake of ibufenac into adipose tissue was about twenty-fold greater than that observed with ibuprofen (Adams *et al.*, 1970). At the time that these data were published it was not clear why a single methyl group, the only difference between the two drugs, should have led to such a change in the tissue distribution of ibufenac. The difference is inexplicable on the basis of the relative lipid solubilities of the two drugs. A logical explanation is that more ibufenac than ibuprofen is taken up into hybrid lipids, although this has not been tested. The acyl glucuronide is also more reactive than that of ibuprofen (see below).

Breast milk

The concentrations of ibuprofen in milk are very low (Walter and Dilger, 1997), and the dose to babies appears insignificant.

ELIMINATION

Ibuprofen is eliminated primarily by metabolism, with the metabolites being excreted subsequently in urine. A feature of the metabolism of ibuprofen is the conversion of the R enantiomer to the S enantiomer, a process that is responsible for the activation of the R enantiomer.

Glucuronidation

About 10 to 15 per cent of doses of ibuprofen are converted directly to the acyl glucuronides. There is very marked stereoselectivity with the glucuronidation of S-ibuprofen being far greater than that of the R enantiomer (Lee *et al.*, 1985; Geisslinger *et al.*, 1993; Rudy *et al.*, 1995). The glucuronides of ibuprofen are detectable in plasma, the concentrations being about 4 per cent of the concentrations of ibuprofen, although because of the high protein binding of the parent drug the concentrations of the glucuronides exceed the unbound concentrations of unchanged ibuprofen (Castillo *et al.*, 1995). Ibuprofen glucuronides are more stable than the glucuronides of other NSAIDs, and while they react with plasma proteins to produce covalent ibuprofen–protein adducts the levels of these adducts are lower than for several other NSAIDs, particularly zomepirac and tolmetin (Castillo *et al.*, 1995). Although the formation of covalent adducts of proteins is widely considered to have toxic potential, the formation of protein adducts has not been associated with any recognised toxic reaction to ibuprofen.

Ibufenac, the hepatotoxic analogue of ibuprofen, also forms an acyl glucuronide which reacts with protein faster than the acyl glucuronide of ibuprofen. Again this is suggestive of the toxicity of the acyl glucuronides, but definitive proof is lacking.

Metabolic oxidation

The isobutyl side chains of both enantiomers of ibuprofen are oxidised through the activity of cytochromes P4502C8 and particularly P4502C9 (Smith and Jones, 1992; Leemann *et al.*, 1993; Hamman *et al.*, 1997). Overall, the total rate of oxidation is slightly greater for the S enantiomer (Rudy *et al.*, 1991). Hydroxylation occurs at all three possible positions on the isobutyl side chain, although little of the 1-hydroxy metabolite is formed. The 3-hydroxy metabolite is readily oxidised to the 3-carboxy derivative, but the 2-hydroxy metabolite, being a tertiary alcohol, is not readily oxidised. Thus, the major oxidised metabolites are the 2-hydroxy derivative and 3-carboxyibuprofen, which are usually termed simply hydroxyibuprofen and carboxyibuprofen (Figure 5.2; Adams *et al.*, 1967;

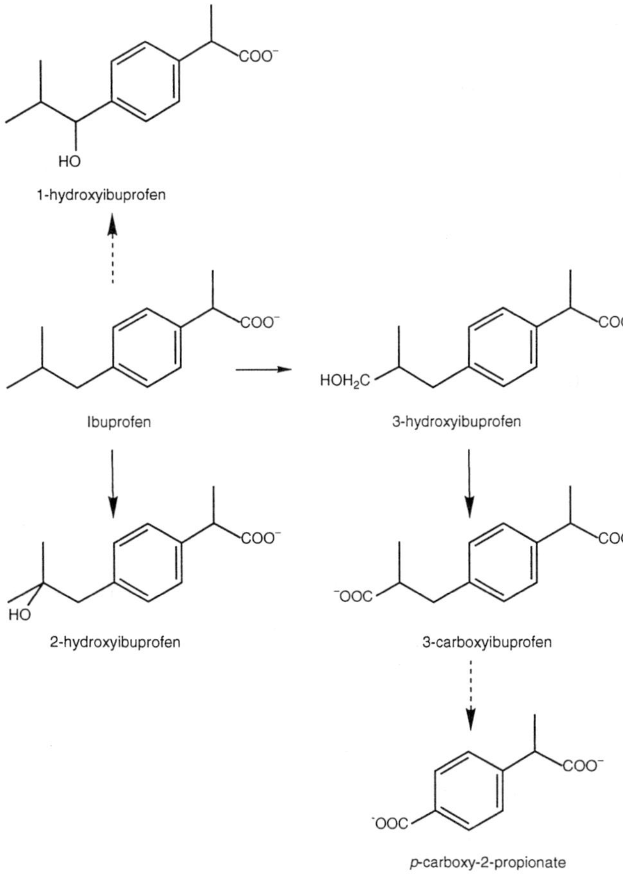

Figure 5.2 Oxidative pathways of the metabolism of ibuprofen. Minor pathways are shown as broken arrows. Three hydroxy metabolites have been identified. A major metabolite, 2-hydroxyibuprofen, is commonly termed hydroxyibuprofen, and 3-carboxyibuprofen is termed carboxyibuprofen. The oxidised metabolites and some unchanged ibuprofen are excreted largely as their acyl glucuronide conjugates linked through the propionyl carboxylate groups. Small amounts of unchanged ibuprofen are also conjugated with taurine.

Pettersen *et al.*, 1978; Kepp *et al.*, 1997). Cytochrome P4502C8 enzyme favours the formation of hydroxy-ibuprofen from the R enantiomer and the 2C9 enzyme catalyses the hydroxylation of S-ibuprofen at both the 2 and 3 positions (Hamman *et al.*, 1997). In humans, several variants of the 2C9 are well recognised. The metabolic clearance of S-ibuprofen is not altered by a form of CYP4502C9 in which arginine 144 is replaced by cysteine (*2) but is decreased by approximately 50 per cent in patients who are homozygous for the abnormal enzyme where isoleucine is replaced by leucine at codon 359 (termed *3/*3; Kirchheiner *et al.*, 2002). The clearance is reduced by about 25 per cent in heterozygotes (*1/*3). The clearance of ibuprofen is not altered by another form of CYP4502C9 in which arginine 144 is replaced by cysteine (*2). The *3 allele occurs in about 5 per cent of Caucasians and in lower proportions in other ethnic groups and, therefore, the *3/*3 homozygotes are very rare. Ibuprofen is a well tolerated drug and, although its antiplatelet effect is prolonged in the *3/*3 homozygotes, it is unclear if the slower metabolism is of clinical significance in the homozygotes. The decreased clearance in the heterozygotes is probably clinically insignificant. The clearance of S-ibuprofen potentially could be decreased by inhibitors of cytochrome P4502C9, such as amiodarone and fluconazole, but there are no reports on these potential interactions.

None of the oxidised metabolites have anti-inflammatory activity (Adams *et al.*, 1969b), and they are largely excreted as their conjugates with glucuronic acid (Mills *et al.*, 1973; Lockwood and Wagner, 1982; Kepp *et al.*, 1997). Together with the acyl glucuronides of ibuprofen, slightly less than 90 per cent of an oral dose of ibuprofen can be accounted for in urine as the total of the oxidised metabolites and their glucuronides (Evans *et al.*, 1989a).

The formation of hydroxyibuprofen does not introduce a further chiral centre, but the formation of the 3-hydroxyibuprofen and the subsequent production of the carboxy metabolite do introduce a second chiral centre. Thus two hydroxyibuprofens are formed, one each from R- and S-ibuprofen. Four carboxyibuprofens are formed, two each from R- and S-ibuprofen (Kaiser *et al.*, 1976; Tan *et al.*, 1997).

Inversion of R- to S-ibuprofen and related metabolism

The outstanding feature of the metabolism of ibuprofen is its intriguing chiral inversion (Kaiser *et al.*, 1976; Lee *et al.*, 1985; Fornasini *et al.*, 1997). In humans approximately 60 per cent of the R-ibuprofen is inverted to the S enantiomer, and there is no detectable inversion in the opposite direction (i.e. from S-ibuprofen to R-ibuprofen).

There is little interpatient variation in the amounts of R-ibuprofen converted to the S form (Rudy *et al.*, 1992; Geisslinger *et al.*, 1993). Thus R-ibuprofen is a reliable source of S-ibuprofen, although there is still considerable variation in the ratio of the plasma concentrations of S- to R-ibuprofen during long-term dosage (Oliary *et al.*, 1992), presumably as a result of interpatient differences in the oxidative metabolism of the S enantiomer.

The mechanism of the inversion has been studied in considerable detail. Work on the inversion started with the observation that racemic ibuprofen was excreted in urine as dextrorotatory metabolites (Adams *et al.*, 1967; Kaiser *et al.*, 1976). The first step is possibly the conversion of R-ibuprofen to its adenylate conjugate (Menzel *et al.*, 1994) and thence to the co-enzyme A (CoA) thioester (Knihinicki *et al.*, 1989; Figure 5.3). The overall production of the CoA thioester is known as the ligase reaction and is mediated by long chain acyl CoA synthetase (Bruggera *et al.*, 1996). This step is followed by the reversible inversion (epimerisation) of the R enantiomer CoA thioester to the CoA thioester of S-ibuprofen. The CoA thioesters of both enantiomers are hydrolysed rapidly back to the enantiomers of ibuprofen itself (Tracy and Hall, 1991; 1992). The hydrolysis of the CoA thioester of S-ibuprofen yields the S enantiomer, thus completing the inversion of the R enantiomer. The S enantiomer is not a substrate for the ligase enzyme system in humans, and consequently, inversion of the S to the R enantiomer does not occur (Figure 5.3).

The proposed mechanism of the inversion of the CoA thioester of R-ibuprofen to the S CoA thioester is shown in Figure 5.3. It involves the production of a planar intermediate that allows equilibration between the R and S CoA thioesters (Nakamura *et al.*, 1981; Chen *et al.*, 1991; Sanins *et al.*, 1991). The equilibrium constant is not unity, and the equilibrium lies slightly towards the CoA

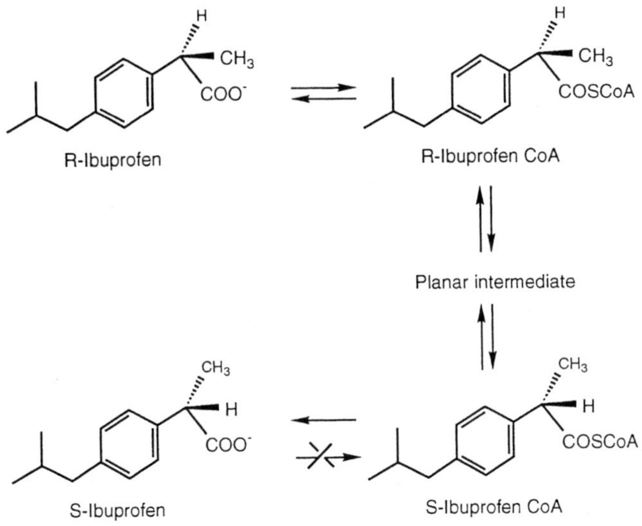

Figure 5.3   The metabolic inversion of R- to S-ibuprofen. The intermediate is considered to be the planar, although its exact nature is unclear. The inversion is unidirectional because only the R-enantiomer is converted to the CoA derivative.

thioester of S-ibuprofen (Chen *et al.*, 1991). The epimerase enzyme has considerable homology with carnitine dehydratase, an enzyme involved in lipid metabolism (Reichel *et al.*, 1997). The epimerase may thus have a normal biological function, although further work is still required to identify the enzyme involved in the epimerase step.

The enzyme system responsible for the inversion of R-ibuprofen is present in the liver (Cox *et al.*, 1985). It is retained in hepatoma cells (Menzel-Soglowek *et al.*, 1992) and is also present in several other organs, particularly the kidney. Enzymes in the lung are capable of metabolising the drug, including its inversion. However, albumin very much limits the extraction, and thus the lung is not a significant site of metabolism of ibuprofen *in vivo* (Hall *et al.*, 1992). The inversion system is also present in the gastrointestinal tract (Mehvar and Jamali, 1988), where it may lead to limited first-pass inversion of ibuprofen.

Interestingly, several fungi selectively invert the R enantiomer or its further metabolites, and these fungi may be useful in the preparation of the pure enantiomers of ibuprofen or related drugs (Hutt *et al.*, 1993).

The conversion of R-ibuprofen to CoA thioesters has several consequences, apart from the formation of hybrid triglycerides. If R-ibuprofen is converted to CoA thioesters, it would be expected that this enantiomer may inhibit the conversion of fatty acids to their CoA thioesters by acting as an alternative substrate. Inhibition of the formation of palmitoyl CoA does occur, but at concentrations well above the unbound concentrations in plasma, and effects on lipid metabolism are unlikely *in vivo* (Knights and Jones, 1992). Similarly, inhibition of the inversion of R-ibuprofen by the S enantiomer does not occur at the concentrations achieved *in vivo* (Smith *et al.*, 1994). R-ibuprofen produces a variety of effects on the liver – increasing the ratio of NADH:NAD, uncoupling oxidative phosphorylation and decreasing the concentrations of free CoA – but again the concentrations are supratherapeutic (Knights and Drew, 1992). The mucosa of the gastrointestinal tract is, however, exposed to high concentrations, which could lead to interactions with local metabolism (Evans, 1996).

Conjugates of acidic compounds with taurine require the prior formation of the CoA thioester, and it is therefore not surprising that ibuprofen is metabolised to its taurine conjugate. This is still a minor metabolite, accounting for only about 1.5 per cent of a dose of ibuprofen (Shirley *et al.*, 1994). The isomeric composition of the taurine conjugate indicates that, unexpectedly, S-ibuprofen is converted to the CoA thioester and some S-ibuprofen is converted to the R enantiomer (Shirley et al., 1994). The formation of glycine conjugates of drugs also requires the intermediate formation of the CoA thioester intermediates, but no glycine conjugate of ibuprofen has been found.

The sum of the biological interactions of the CoA thioesters of ibuprofen is suggestive of toxicity (Williams et al., 1993), but ibuprofen is still a very well-tolerated drug (Wechter, 1994). Possibly the only present conclusion is that several of the biochemical events that occur after the administration of R-ibuprofen could be avoided by the use of other NSAIDs or by dosage with pure S-ibuprofen, which is eliminated solely by conjugation and oxidative metabolism. While S-ibuprofen undergoes no significant conversion to the CoA thioester the S enantiomer is more readily converted to the glucuronide, which more readily forms an adduct with plasma albumin (see p. 162).

Urinary excretion

The excretion of unchanged ibuprofen is insignificant (Lockwood et al., 1983). Initial reports of excretion of the unchanged drug (Adams et al., 1969b) probably reflect the instability of the glucuronide in urine. The high protein-binding of both enantiomers of ibuprofen limits filtration at the glomerulus, and the high lipid solubility of the un-ionised drug ensures resorption of most of the filtered or secreted drug (Ahn et al., 1991a; Cox et al., 1991). As with many NSAIDs, however, ibuprofen blocks the secretion of other anions by the active anionic transport pathway. In particular, the secretion of methotrexate may be reduced by ibuprofen, an interaction that has potentially fatal consequences if large doses of methotrexate are administered. In this respect the use of S-ibuprofen alone may be somewhat safer than the racemic ibuprofen because of the lower total dose and plasma concentrations of the S enantiomer.

Although the urinary excretion of ibuprofen is extremely low, the metabolites are excreted readily in urine, both free and as the glucuronide conjugates (Geisslinger et al., 1993). All the free metabolites are acids, and they interfere with the profile of urinary organic acids – an important screen in paediatrics (Bennett et al., 1992). Those conducting this test should be aware of the interference.

Biliary excretion

There has been some interest in the biliary excretion of the NSAIDs because of the potential for enterohepatic cycling to contribute to the gastrointestinal toxicity of this group of drugs. Biliary excretion is a major mode of elimination of ibuprofen and its metabolites in rats (Dietzel et al., 1990) but excretion by this route in humans is much more limited, and less than 1 per cent of a dose of ibuprofen is excreted in bile in humans (Schneider et al., 1990). Therefore, this neither represents a major route of excretion of ibuprofen nor, if inhibited, is it likely to contribute to the toxicity of ibuprofen.

HALF-LIFE AND CLEARANCE

The majority of studies of ibuprofen characterising its pharmacokinetics are based on non-stereoselective techniques, although recent work has been concentrated on the kinetics of the enantiomers. Ibuprofen is regarded as a short half-life NSAID, having a half-life of about 2 hours in humans (Table 5.1), but a late slow phase is seen, indicating slow diffusion of drug from some tissues back to blood or slow release from covalent species such as hybrid lipids or protein adducts. The late phase does not lead to significant accumulation of the drug with repeated dosage (Aarons et al., 1983a; Rudy et al., 1995). After its intravenous injection a short-lived distribution phase is seen, but this phase is not seen after oral administration of the drug.

The clearance of both enantiomers is low at less than 100 ml/min (Table 5.1). This is much less than the plasma flow through the liver, indicating that first-pass clearance should be low, and is consistent with the high oral bioavailability of ibuprofen.

FACTORS AFFECTING THE ELIMINATION OF IBUPROFEN

There are substantial interpatient differences in the pharmacokinetics of ibuprofen, and there has been considerable effort to find factors that are responsible for these interpatient differences. Many studies have, however, been conducted with non-stereoselective assays of ibuprofen, and stereoselective interactions may have been missed. For example, if the clearance of S-ibuprofen is unchanged, a 50 per cent decrease in the clearance of R-ibuprofen is required to reduce the clearance of total ibuprofen by 20 per cent (Cox *et al.*, 1985). Knowledge of differences in the unbound clearance of the active enantiomer are also important because any change in this parameter alters the mean unbound plasma concentration of the active drug.

Gender

There are no significant gender differences in the elimination of racemic ibuprofen (Greenblatt *et al.*, 1984; Knights *et al.*, 1995; Rudy *et al.*, 1995).

Age

There is no clear effect of old age on elimination of total ibuprofen (R- plus S-ibuprofen) by humans (Albert *et al.*, 1984; Greenblatt *et al.*, 1984), the only change reported being a slower clearance of both enantiomers in elderly patients with renal impairment.

Intravenous ibuprofen has been under clinical trial in premature neonates for the treatment of intraventricular haemorrhage and closure of the ductus arteriosus. The clearance is very low, indicating gross immaturity of the cytochrome P450 systems (Aranda *et al.*, 1997). The half-life is correspondingly very long, the mean half-life being 30 hours. The consequent sustained plasma concentrations may be useful in the treatment of these cardiac conditions in the neonates.

On a body-weight basis, the clearance and volume of distribution in children aged 3 to 10 years with fever are the same as in adults (Nahata *et al.*, 1991). The plasma concentrations of active S-ibuprofen are slightly lower than those of the R enantiomer in infants aged 6 to 18 months, but it is not known if this is due to more rapid metabolism or lower binding of the S enantiomer to plasma proteins (Rey *et al.*, 1994). At this stage, the dosage of ibuprofen in children, on a body-weight basis, is recommended to be the same as in adults.

Treatment with ibuprofen appears beneficial for children with cystic fibrosis (Konstan *et al.*, 1995), and in such children the mean clearance and volume of racemic ibuprofen are both about 80 per cent greater than in normal children, the result being that the half-life is the same as in normals (Konstan *et al.*, 1991). Decreased plasma albumin occurs in nearly 50 per cent of patients with cystic fibrosis (Benabdeslam *et al.*, 1998), and the consequent lesser binding to plasma proteins at least partly explains the parallel increase in volume of distribution and clearance. The clearance is, however, highly variable, with nearly a six-fold interpatient difference (Murry *et al.*, 1999). A peak plasma concentration of at least 50 mg/l has been suggested for cystic fibrosis, and the dosage adjusted to obtain peak concentrations in the range of 50 to 100 mg/l (Konstan *et al.*, 1991). The ratio of R- to S-ibuprofen is normal in patients with cystic fibrosis, indicating that the relatively simple achiral assays of ibuprofen are sufficient for monitoring plasma concentrations of ibuprofen in these patients (Dong *et al.*, 2000). The more complex chiral assays are not required.

CHAPTER 5

Liver disease

The elimination of total ibuprofen is about normal or slightly impaired in patients with liver failure (Adamska-Dyniewska et al., 1982; Juhl et al., 1983), although the conversion of R- to S-ibuprofen is impaired in cirrhosis (Li et al., 1993). In isolated fatty livers of rats the change in the clearance is in the same direction, with the clearance of R-ibuprofen being decreased by about 30 per cent while the clearance of S-ibuprofen is unaffected (Cox et al., 1985).

Renal failure

The half-life of racemic ibuprofen is unaltered by nephrectomy in experimental animals (Au et al., 1984) or patients with severe renal failure maintained on dialysis (Antal et al., 1986). The clearance and volume of distribution in the renal failure patients are very variable, but are increased at least in some of the dialysis patients (Ochs et al., 1985). The free fraction of racemic ibuprofen in the plasma of patients with renal failure is also very variable, but is on average about three times that in healthy subjects. The result is that the clearance and volume of distribution of ibuprofen are decreased in some patients. Correspondingly, the mean unbound concentrations in many patients are therefore higher than in healthy subjects

Similar results are seen in the pharmacokinetics of the active S-ibuprofen in elderly patients with milder renal impairment. Again the pharmacokinetic parameters are very variable, but the unbound clearance is decreased and the unbound plasma concentrations are increased substantially in some (but not all) patients (Bradley et al., 1992; Chen and Chen, 1994; Rudy et al., 1995). NSAIDs have been associated with the development of reversible renal failure, and it is suggested that patients may be at greater risk of further impairment of renal function if the metabolic clearance of ibuprofen is decreased.

The hydroxy and carboxy metabolites and their glucuronides are excreted in renal failure and consequently do accumulate in severe cases, but are eliminated by haemodialysis (Antal et al., 1986).

Cardiovascular disease

Higher plasma concentrations of the active S enantiomer, but not the R enantiomer, have been reported in patients with a variety of cardiovascular diseases (Chen and Chen, 1995). Elevated concentrations of S-ibuprofen may contribute to the greater risk of renal toxicity in patients with coronary artery disease (Murray et al., 1990).

Obesity

Both the clearance of ibuprofen and volume of distribution of ibuprofen increase with increasing body weight in obese subjects. The result is that the half-life is unchanged (Abernethy and Greenblatt, 1985).

Rheumatoid arthritis

The pharmacokinetic behaviour of the active S-ibuprofen is not altered by rheumatoid arthritis although there is considerable interpatient variation, as is common for other NSAIDs (Geisslinger et al., 1993). The binding of racemic ibuprofen to plasma proteins is slightly decreased in rheumatoid arthritis (Aarons et al., 1983b), possibly due to a fall in plasma albumin (Skeith and Jamali, 1991).

Species

As is the case in man, other animals convert R-ibuprofen to the S enantiomer. The most widely studied is the rat which converts about 50 per cent of R-ibuprofen to the S enantiomer (Knihinicki *et al.*, 1990), a slightly lower proportion than in man. Rat liver *in vitro* has been widely used in the study of the inversion of the R enantiomer. In the microsomal fraction of rat liver, selectivity of the ligase enzyme for the R ibuprofen is shown at low concentrations, but there is some conversion of S ibuprofen to its CoA thioester conjugate at very high concentrations, which may not be achievable *in vivo* (Knights *et al.*, 1992).

The inversion of R-ibuprofen in the dog is similar to that in man, with about 70 per cent inversion of the R to the S enantiomer (Ahn *et al.*, 1991b; Beck *et al.*, 1991). Only about 30 per cent of the R enantiomer is inverted in the rabbit, lower than in other species, as a result of the very rapid clearance by oxidative pathways (Williams *et al.*, 1991). The conversion of the S enantiomer to the R form is insignificant in man, but there may be a very small amount of inversion of the S to the R enantiomer in both rats and rabbits (Chen *et al.*, 1991). The guinea pig is quite different to all other species examined, with rapid inversion that proceeds in both directions (Chen *et al.*, 1991).

Other drugs

The total plasma concentrations of ibuprofen are reduced when it is taken with aspirin, due to displacement of bound ibuprofen from binding to albumin (Grennan *et al.*, 1979; Aarons *et al.*, 1983b). The half-life is, however, unchanged. Simultaneously, there is a weak additive effect between aspirin and ibuprofen, as is expected from the anti-inflammatory actions of the two drugs. Cimetidine, a well-known inhibitor of the metabolism of many drugs, does not inhibit the elimination of R- or S-ibuprofen (Evans *et al.*, 1989a), in line with the oxidative metabolism of ibuprofen by cytochrome 2C9, which is resistant to cimetidine. There is an increased bleeding tendency in haemophiliacs receiving zidovudine (AZT) and ibuprofen, although zidovudine does not alter the clearance of ibuprofen (Ragni *et al.*, 1992).

A recent finding is that clofibrate greatly increases the clearance of R-ibuprofen although there is no significant effect on the total extent of conversion of the R to the S enantiomer (Scheuerer *et al.*, 1998a). From studies in rats, the mechanism is an increased expression of long chain acyl CoA synthetase (Scheuerer *et al.*, 1998b) or increased availability of CoA (Vos *et al.*, 1996). Induction of the synthesis of the CoA thioester of ibuprofen may not only increase the rate of conversion of R- to S-ibuprofen but also increase the amount of ibuprofen incorporated into long-lived hybrid lipid stores.

Effects of ibuprofen on the clearance of other drugs

Ibuprofen has little influence on the elimination of other drugs, and only supratherapeutic doses provide weak induction of the activity of cytochrome P4A1 in rats (Rekka *et al.*, 1994). Ibuprofen may decrease the clearance of digoxin in some patients, although indomethacin has a greater effect (Quattrochi *et al.*, 1983; Jørgensen *et al.*, 1991). Ibuprofen may also decrease the renal clearance of methotrexate, with increased toxicity when large doses of the cytotoxic agent are used or when smaller anti-rheumatic doses are used in patients with pre-existing renal impairment. Like other NSAIDs, ibuprofen potentially decreases the renal clearance of the lithium cation, and consequently the plasma concentrations of lithium should be monitored in lithium-treated patients who commence or stop taking ibuprofen.

PHARMACOKINETICS, PLASMA CONCENTRATIONS AND CLINICAL RESPONSE

Both enantiomers of ibuprofen have short half-lives, but dosage need not be as frequent as the short half-life may indicate. While the half-life is about 2 hours, ibuprofen is usually administered about every 6 to 8 hours. Even dosage every 12 hours yields activity similar to that when ibuprofen is administered every 6 hours (Brugueras *et al.*, 1978). Although clinical trials are often conducted on small numbers of patients and quite large differences may not be detected, the duration of action consistently appears to be longer than the short half-life. Several reasons for the prolonged action have been suggested.

First, the concentrations at sites of action may lag behind those in blood. The concentrations at potential sites of action, synovial fluid (Figure 5.1) and brain (see p. 160) are more sustained than in plasma. Consistent with slow tissue uptake into tissues, both the peak analgesic (Figure 5.4; Laska *et al.*, 1986; Suri *et al.*, 1997) and antipyretic effects (Brown *et al.*, 1992; Kelley *et al.*, 1992; Garg and Jusko, 1994; Troconiz *et al.*, 2000) occur after peak plasma concentrations have been achieved – in fact when the plasma concentrations are actually declining. Thus, the peak analgesic and antipyretic effects occur at about 2 hours after dosage, whereas peak plasma concentrations occur at about 1 hour or less after dosage with conventional or soluble tablets. From pharmacokinetic–pharmacodynamic analysis of the kinetics in blood and the antipyretic effect, the half-life of loss from the antipyretic site of action is about 80 minutes (Kelley *et al.*, 1992). Modelling of the antipyretic effect of ibuprofen is, however, difficult, because the body temperature cannot fall instantaneously and the slow decline in body temperature must be due (at least in part) to the time taken for heat loss, as well as the time taken for access of ibuprofen to its site of action. The delayed diffusion to its sites of action, together with slow absorption when the patient is in pain (Jamali and Kunz-Dober, 1999) means that wherever possible ibuprofen and other analgesics should be administered before pain develops.

Second, in some pain states high dosage may produce near maximal effects, and falling plasma concentrations may therefore produce little decrease in effect. A good example involves the antiplatelet effect of ibuprofen. As an inhibitor of the synthesis of thromboxane A2 by platelets, the plasma concentrations of ibuprofen are sufficient to produce a near maximal effect for several hours after standard

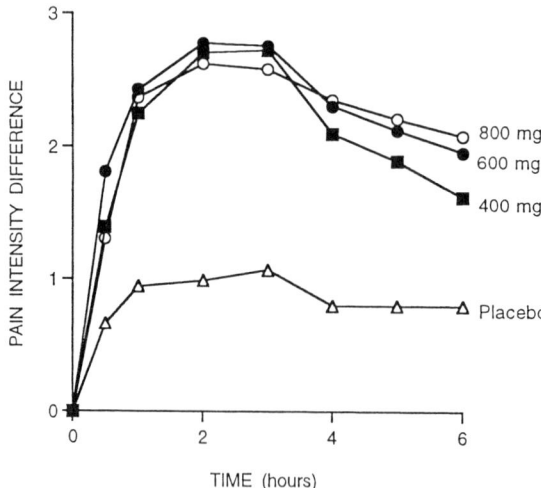

Figure 5.4 Time courses of analgesic effect of 400, 600 and 800 mg oral doses of racemic ibuprofen. Pain was measured on a 0 to 3 ordinal scale and subtracted from the baseline score. Note that the analgesic effect of the different doses is very similar, and that the effect decays very little over the period of 2 to 6 hours when the plasma concentrations are declining with a half-life of approximately 2 hours. (Redrawn from Laska *et al.*, 1986.)

oral doses, despite the large decrease in plasma concentrations over this time period (Longenecker *et al.*, 1985; Evans *et al.*, 1991; Villanueva *et al.*, 1993). The same phenomenon may be responsible for the near invariant time course of analgesic response after different doses of ibuprofen (Figure 5.4) or the lack of any appreciable change in analgesia between 1 and 5 hours after dosage despite a substantial fall in plasma concentrations over this time (Jones *et al.*, 1997).

Third, ibuprofen may affect just one aspect of a cascade. The observed pharmacological effect of the drug may therefore be delayed and may not cease immediately after inhibition of the synthesis of prostaglandins – the major mechanism of action of the NSAIDs such as S-ibuprofen.

An aspect of the clinical pharmacology of ibuprofen related to the duration of action is the correlation between effect and plasma concentrations. Such correlations may assist in the development of optimal dosage schedules. In the treatment of osteoarthritis with ibuprofen some measures of improvement correlate with the area under the time course of plasma concentrations, and it is concluded that some of the interpatient variation in response is due to pharmacokinetic differences (Bradley *et al.*, 1992); however, no correlations have been detected between plasma concentrations and anti-inflammatory effect in the treatment of rheumatoid arthritis (Grennan *et al.*, 1983). The pharmacokinetics of ibuprofen have several features that make correlations between effect and plasma concentrations difficult. These include the short half-life, the low and variable protein binding, and the variable average ratio of the concentrations of the S and R enantiomers (Oliary *et al.*, 1992; Rudy *et al.*, 1992). A correlation may only be found with the plasma concentrations of S-ibuprofen. A further problem may be the high peak concentrations, which exceed those for maximal efficacy.

In some cases, *in vitro* effects of ibuprofen are found only at concentrations well above those found during therapy. For example, both enantiomers inhibit the production of superoxide by neutrophils in a protein-free medium, the effects becoming statistically significant at 30 to 100 μmol/l (Villanueva *et al.*, 1993). Furthermore, both the aggregation and release of lysozyme of neutrophils is inhibited with an IC_{50} value of 40 μmol/l (Kaplan *et al.*, 1984). The concentrations are far in excess of the unbound concentrations in plasma, and would appear to indicate that the effect of ibuprofen on neutrophils is irrelevant *in vivo*. Neutrophils isolated from subjects treated with ibuprofen have, however, impaired aggregation and release of lysozyme (Kaplan *et al.*, 1984), although even the peak concentrations of unbound ibuprofen in plasma are only about 1 μmol/l. Furthermore, neutrophils are washed during their isolation and therefore much of the ibuprofen present in such cells should be removed during their separation from other cells.

ANALYTICAL METHODS

Non-stereoselective

Total ibuprofen has been analysed by both gas chromatography and high performance liquid chromatography. These methods were the basis for the early pharmacokinetic data derived on the total drug (R- plus S-ibuprofen). The total drug can be assayed by gas chromatography after conversion to a variety of esters. Mostly the methyl ester has been used (Kaiser and VanGiessen, 1974), with the electron capturing pentafluorobenzyl ester giving higher sensitivity (Kaiser and Martin, 1978).

Stereoselective

Recent analytical methods have focused on the chiral nature of the drug. Both standard approaches to the analysis of enantiomers have been applied to ibuprofen; namely achiral chromatography after chemical modification to produce a pair of diastereomerically related products, and also chromatography on chiral columns.

Using the reactivity of the carboxylate function of ibuprofen, a variety of derivatives have been used to produce diastereoisomeric derivatives. These include esters with enantiomeric alcohols (Lee *et al.*, 1984; Zhao *et al.*, 1994) and amides with enantiomeric amines (Mehvar *et al.*, 1988; Evans *et al.*, 1989a; Rudy *et al.*, 1990). This approach requires an enantiomerically pure reagent and a chemical reaction to be carried out. The chemical reaction must be conducted under mild conditions in order to prevent racemisation of the enantiomers. However, despite these limitations most methods for the assay of ibuprofen enantiomers are based on chemical modification to form diastereoisomers.

Chromatography in an asymetric environment can also be used to separate enantiomers without their prior derivatisation to disatereoisomers. One method involves the use of mobile phases containing chiral compounds, such as quinine or quinidine. This system has been used to separate compounds related to phenylpropionates (Schill, 1989), and may also be applicable to the enantiomers of ibuprofen. A more widely used technique is chromatography on a chiral column, which has been used to separate R- and S-ibuprofen (Li *et al.*, 1989; Menzel-Soglowek *et al.*, 1990; Noctor *et al.*, 1991; Naidong and Lee, 1994). With some systems, prior derivatisation to simple amides (Pirkle and Murray, 1990) or ureides (Ahn *et al.*, 1994) is required. The CoA thioesters of the enantiomers have also been separated on chiral materials (Tracy and Hall, 1991). The use of chiral columns appears simpler than chromatography of diastereoisomers, but this convenience is offset by the expense of the chiral columns, their shorter life expectancies and, to some degree, the intercolumn variability in resolving power, although there has been improvement in recent years. Separation of both the enantiomers of ibuprofen and some of its metabolites can now be obtained by electrophoresis (Bjornsdottir *et al.*, 1998).

Two of the stereoselective techniques, diastereoisomer derivatives and chiral chromatography, have also been used to assay the two enantiomeric forms of the major hydroxy metabolite and the four carboxy metabolites (Figure 5.2; Kaiser *et al.*, 1976; Rudy *et al.*, 1990; Tan *et al.*, 1997).

REFERENCES

Aarons, L., Grennan, D.M., Rajapakse, C., Brinkley, J., Siddiqui, M., Taylor, L. and Higham, C. 1983a, Anti-inflammatory (ibuprofen) drug therapy in rheumatoid arthritis – rate of response and lack of time dependency of plasma pharmacokinetics. *British Journal of Clinical Pharmacology*, **15:** 387–388.

Aarons, L., Grennan, D.M. and Siddiqui, M. 1983b, The binding of ibuprofen to plasma proteins. *European Journal of Clinical Pharmacology*, **25:** 815–818.

Abernethy, D.R. and Greenblatt, D.J. 1985, Ibuprofen disposition in obese individuals. *Arthritis and Rheumatism*, **28:** 1117–1121.

Adams, S.S., Cliffe, E.E., Lessel, B. and Nicholson, J.S. 1967, Some biological properties of 2-(4-isobutylphenyl)-propionic acid. *Journal of Pharmaceutical Sciences*, **56:** 1686.

Adams, S.S., Bough, R.G., Cliffe, E.E., Lessel, B. and Mills, R.F.N. 1969a, Absorption, distribution and toxicity of ibuprofen. *Toxicology and Applied Pharmacology*, **15:** 310–330.

Adams, S.S., McCullough, K.F. and Nicholson, J.S. 1969b, The pharmacological properties of ibuprofen, an antiinflammatory analgesic and antipyretic agent. *Archives Internationales de Pharmacodynamie et de Thérapie*, **178:** 115–129.

Adams, S.S., Bough, R.G., Cliffe, E.E., Dickinson, W., Lessel, B., McCullough, K.F. *et al.* 1970, Some aspects of the pharmacology, metabolism, and toxicology of ibuprofen. *Rheumatology and Physical Medicine*, **10 (Suppl.):** 9–22.

Adamska-Dyniewska, H., Tkaczewski, W. and Gajewska, B. 1982, Ibuprofen pharmacokinetics in patients with liver cirrhosis. *Wiadomosci Lekarskie*, **35:** 609–613.

Ahn, H.Y., Amidon, G.L. and Smith, D.E. 1991b, Stereoselective systemic disposition of ibuprofen enantiomers in the dog. *Pharmaceutical Research*, **8:** 1186–1190.

Ahn, H.Y., Jamali, F., Cox, S.R., Kittayanond, D. and Smith, D.E. 1991a, Stereoselective disposition of ibuprofen enantiomers in the isolated perfused rat kidney. *Pharmaceutical Research*, **8:** 1520–1524.

Ahn, H.Y., Shiu, G.K., Trafton, W.F. and Doyle, T.D. 1994, Resolution of the enantiomers of ibuprofen; comparison study of diasterisomeric method and chiral stationary phase method. *Journal of Chromatography B*, **653:** 163–169.

Albert, K.S., Gillespie, W.R., Wagner, J.G., Pau, A. and Lockwood, G.F. 1984, Effects of age on the clinical pharmacokinetics of ibuprofen. *American Journal of Medicine*, **77:** 47–50.

al-Meshal, M.A., el-Sayed, Y.M., al-Balla, S.R. and Gouda, M.W. 1994, The effect of colestipol and cholestyramine on ibuprofen bioavailability in man. *Biopharmaceutics and Drug Disposition*, **15:** 463–471.

Antal, E.J., Wright, C.E., Brown, B.L., Albert, K.S., Aman, L.C. and Levin, N.W. 1986, The influence of hemodialysis on the pharmacokinetics of ibuprofen and its major metabolites. *Journal of Clinical Pharmacology*, **26:** 184–190.

Aranda, J.V., Varvarigou, A., Beharry, K., Bansal, R., Bardin, C., Modanlou, H., Papageorgiou, A. and Chemtob, S. 1997, Pharmacokinetics and protein binding of intravenous ibuprofen in the premature newborn infant. *Acta Paediatrica*, **86:** 289–293.

Askholt, J. and Nielsen-Kudsk, F. 1985, Ibuprofen, pharmacokinetics and pharmacodynamics in the isolated rabbit heart. *Acta Pharmacologica et Toxicologica*, **56:** 99–107.

Au, D.S., Kuo, T.H., Mederski-Samoraj, B. and Lee, C.-S. 1984, Disposition of ibuprofen in nephrectomized dogs. *Journal of Pharmaceutical Sciences*, **73:** 705–708.

Bannwarth, R., Lapicque, F., Pehourcq, F., Gillet, P., Schaeverbeke, T., Laborde, C. *et al.* 1995, Stereoselective disposition of ibuprofen enantiomers in human cerebrospinal fluid. *British Journal of Clinical Pharmacology*, **40:** 266–269.

Beck, W.S., Geisslinger, G., Engler, H. and Brune, K. 1991, Pharmacokinetics of ibuprofen enantiomers in dogs. *Chirality*, **3:** 165–169.

Benabdeslam, H., Garcia, I., Bellon, G., Gilly, R. and Revol, R. 1998, Biochemical assessment of the nutritional status of cystic fibrosis patients treated with pancreatic enzyme extracts. *American Journal of Clinical Nutrition*, **67:** 912–918.

Bennett, M.J., Sherwood, W.G., Bhala, A. and Hale, D.E. 1992, Identification of urinary metabolites of $(+/-)$-2-(p-isobutylphenyl)propionic acid (Ibuprofen) by routine organic acid screening. *Clinica Chimica Acta*, **210:** 55–62.

Bjornsdottir, I., Kepp, D.R., Tjornelund, J. and Hansen, S.H. 1998, Separation of the enantiomers of ibuprofen and its phase I major metabolites in urine using capillary electrophoresis. *Electrophoresis*, **19:** 455–460.

Bradley, J.D., Rudy, A.C., Katz, B.P., Ryan, S.I., Kalasinski, L.A., Brater, D.C. *et al.* 1992, Correlation of serum concentrations of ibuprofen stereoisomers with clinical response in the treatment of hip and knee osteoarthritis. *Journal of Rheumatology*, **19:** 130–134.

Brown, R.D., Wilson, J.T., Kearns, G.L., Eichler, V.F., Johnson, V.A. and Bertrand, K.M. 1992, Single-dose pharmacokinetics of ibuprofen and acetaminophen in febrile children. *Journal of Clinical Pharmacology*, **32:** 231–241.

Bruggera, R., Reichel, C., Alia, B.G., Brune, K., Yamamoto, T., Tegeder, I. and Geisslinger, G. 2001, Expression of rat liver long-chain acyl-CoA synthetase and characterization of its role in the metabolism of R-ibuprofen and other fatty acid-like xenobiotics. *Biochemical Pharmacology*, **61:** 651–656.

Brugueras, N.E., LeZotte, L.A. and Moxley, T.E. 1978, Ibuprofen: a double-blind comparison twice-a-day therapy with four-times-a-day therapy. *Clinical Therapeutics*, **2:** 13–21.

Castillo, M., Lam, F., Dooley, M.A., Stahl, E. and Smith, P.C. 1995, Disposition and covalent binding of ibuprofen and its acyl glucuronide in the elderly. *Clinical Pharmacology and Therapeutics*, **57:** 636–644.

Ceppi Monti, N., Gazzaniga, A., Gianesello, V., Stroppolo, F. and Lodola, E. 1992, Activity and pharmacokinetics of a new oral dosage form of soluble ibuprofen. *Arzneimittel-Forschung*, **42:** 556–559.

Chen, C.-S., Shieh, W.-R., Lu, P.-H., Harriman, S. and Chen, C.-Y. 1991, Metabolic stereoisomeric inversion of ibuprofen in mammals. *Biochimica et Biophysica Acta*, **1078:** 411–417.

Chen, C.-Y. and Chen, C.-S. 1994, Stereoselective disposition of ibuprofen in patients with renal dysfunction. *Journal of Pharmacology and Experimental Therapeutics*, **268:** 590–594.

Chen, C.-Y. and Chen, C.-S. 1995, Stereoselective disposition of ibuprofen in patients with compromised renal haemodynamics. *British Journal of Clinical Pharmacology*, **40:** 67–72.

CHAPTER 5

Cheng, H., Rogers, J.D., Demetriades, J.L., Holland, S.D., Seibold, J.R. and Depuy, E. 1994, Pharmacokinetics and bioinversion of ibuprofen enantiomers in humans. *Pharmaceutical Research*, **11**: 824–830.

Cox, J.W., Cox, S.R., VanGiessen, G. and Ruwart, M.J. 1985, Ibuprofen stereoisomer hepatic clearance and distribution in normal and fatty *in situ* perfused rat liver. *Journal of Pharmacology and Experimental Therapeutics*, **232**: 636–643.

Cox, P.G.F., Moons, W.M., Russel, F.G.M. and van Ginneken, C.A.M. 1991, Renal handling and effects of S(+)-ibuprofen and R(−)-ibuprofen in the rat isolated perfused kidney. *British Journal of Pharmacology*, **103**: 1542–1546.

Dawson, M., McGee, C.M., Vine, J.H., Watson, T.R. and Brooks, P.M. 1988, The disposition of biphenylacetic acid following topical application. *European Journal of Clinical Pharmacology*, **33**: 639–642.

Day, R.O., Williams, K.M., Graham, G.G., Lee, E.J., Knihinicki, R.D. and Champion, G.D. 1988, Stereoselective disposition of ibuprofen enantiomers in synovial fluid. *Clinical Pharmacology and Therapeutics*, **43**: 480–487.

Dietzel, K., Beck, W.S., Schneider, H.-T., Geisslinger, G. and Brune, K. 1990, The biliary and enterohepatic circulation of ibuprofen in rats. *Pharmaceutical Research*, **7**: 87–90.

Dominkus, M., Nicolakis, M., Kotz, R., Wilkinson, F.E., Kaiser, R.R. and Chlud, K. 1996, Comparison of tissue and plasma levels of ibuprofen after oral and topical administration. *Arzneimittel-Forschung*, **46**: 1138–1143.

Dong, J.Q., Ni, L., Scott, C.S., Retsch-Bogart, G.Z. and Smith, P.C. 2000, Pharmacokinetics of ibuprofen enantiomers in children with cystic fibrosis. *Journal of Clinical Pharmacology*, **40**: 861–868.

Dwivedi, S.K., Sattari, S., Jamali, F. and Mitchell, A.G. 1992, Ibuprofen racemate and enantiomers: phase diagram, solubility and thermodynamic studies. *International Journal of Pharmaceutics*, **87**: 95–104.

Evans, A.M. 1996, Pharmacodynamics and pharmacokinetics of the profens: enantioselectivity, clinical implications, and special reference to S(+)-ibuprofen. *Journal of Clinical Pharmacology*, **36**: 7S–15S.

Evans, A.M., Nation, R.L. and Sansom, L.N. 1989a, Lack of effect of cimetidine on the pharmacokinetics of R(−)- and S(+)-ibuprofen. *British Journal of Clinical Pharmacology*, **28**: 143–149.

Evans, A.M., Nation, R.L., Sansom, L.N., Bochner, F. and Somogyi, A.A. 1989b, Stereoselective plasma protein binding of ibuprofen enantiomers. *European Journal of Clinical Pharmacology*, **36**: 283–290.

Evans, A.M., Nation, R.L., Sansom, L.N., Bochner, F. and Somogyi, A.A. 1990, The relationship between the pharmacokinetics of ibuprofen and the dose of racemic ibuprofen in humans. *Biopharmaceutics and Drug Disposition*, **11**: 507–518.

Evans, A.M., Nation, R.L., Sansom, L.N., Bochner, F. and Somogyi, A.A. 1991, Effect of racemic ibuprofen dose on the magnitude and duration of platelet cyclo-oxygenase inhibition: relationship between inhibition of thromboxane production and the plasma unbound concentration of S(+)-ibuprofen. *British Journal of Clinical Pharmacology*, **31**: 131–138.

Fears, R. 1985, Lipophilic xenobiotic conjugates: the pharmacological and toxicological consequences of the participation of drugs and other foreign compounds as substrates in lipid biosynthesis. *Progress in Lipid Research*, **24**: 177–195.

Fornasini, G., Monti, N., Brogin, G., Gallina, M., Eandi, M., Persiani, S. *et al.* 1997, Preliminary pharmacokinetic study of ibuprofen enantiomers after administration of a new oral formulation (ibuprofen arginine) to healthy male volunteers. *Chirality*, **9**: 297–302.

Garg, V. and Jusko, W.J. 1994, Pharmacodynamic modeling of non-steroidal anti-inflammatory drugs. *Clinical Pharmacology and Therapeutics*, **55**: 87–88.

Geisslinger, G., Schuster, O., Stock, K.-P., Loew, D., Bach, G.L. and Brune, K. 1990, Pharmacokinetics of S(+)- and R(−)-ibuprofen in volunteers and first clinical experience of S(+)-ibuprofen in rheumatoid arthritis. *European Journal of Clinical Pharmacology*, **38**: 493–497.

Geisslinger, G., Stock, K.-P., Loew, D., Bach, G.L. and Brune, K. 1993, Variability in the stereoselective disposition of ibuprofen in patients with rheumatoid arthritis. *British Journal of Clinical Pharmacology*, **35**: 603–607.

Gillespie, W.R., DiSanto, A.R., Monovich, R.E. and Albert, K.S. 1982, Relative bioavailability of commercially available ibuprofen oral dosage forms in humans. *Journal of Pharmaceutical Sciences*, **71**: 1034–1038.

Gontarz, N., Small, R.E., Comstock, T.J., Stalker, D.J. and Willis, H.E. 1987, Effect of antacid suspension on pharmacokinetics of ibuprofen, *Clinical Pharmacy*, **6:** 413–416.

Graham, G.G. 1988, Kinetics of non-steroidal anti-inflammatory drugs in synovial fluid. *Agents and Actions Supplements*, **24:** 66–75.

Greenblatt, D.J., Abernethy, D.R., Matlis, R., Harmatz, J.S. and Shader, R.I. 1984, Absorption and disposition of ibuprofen in the elderly. *Arthritis and Rheumatism*, **27:** 1066–1069.

Grennan, D.M., Aarons, L., Siddiqui, M., Richards, M., Thompson, R. and Higham, C. 1983, Dose–response study with ibuprofen in rheumatoid arthritis: clinical and pharmacokinetic findings. *British Journal of Clinical Pharmacology*, **15:** 311–316.

Grennan, D.M., Ferry, D.G., Ashworth, M.E., Kenny, R.E. and MacKinnon, M. 1979, The aspirin–ibuprofen interaction in rheumatoid arthritis. *British Journal of Clinical Pharmacology*, **8:** 497–503.

Gupta, M. and Majumdar, D.K. 1997, Effect of concentration, pH, and preservative on *in vitro* transcorneal permeation of ibuprofen and flurbiprofen from non-buffered aqueous drops. *Indian Journal of Experimental Biology*, **35:** 844–849.

Hall, S.D., Hassanzadeh-Khayyat, M., Knadler, M.P. and Mayer, P.R. 1992, Pulmonary inversion of 2-arylpropionic acids: influence of protein binding. *Chirality*, **4:** 349–352.

Hall, S.D., Rudy, A.C., Knight, P.M. and Brater, D.C. 1993, Lack of presystemic inversion of (R)- to (S)-ibuprofen in humans. *Clinical Pharmacology and Therapeutics*, **53:** 393–400.

Hamman, M.A., Thompson, G.A. and Hall, S.D. 1997, Regioselective and stereoselective metabolism of ibuprofen by human cytochrome P450 2C. *Biochemical Pharmacology*, **54:** 33–41.

Hutt, A.J., Kooloobandi, A. and Hanlon, G.W. 1993, Microbial metabolism of 2-arylpropionic acids: chiral inversion of ibuprofen and 2-phenylpropionic acid. *Chirality*, **5:** 596–601.

Iervolino, M., Cappello, B. Raghavan, S.L. and Hadgraft, J. 2001, Penetration enhancement of ibuprofen from supersaturated solutions through human skin. *International Journal of Pharmaceutics*, **212:** 131–141.

Jamali, F. and Kunz-Dober, C.M. 1999, Pain-mediated altered absorption and metabolism of ibuprofen: an explanation for decreased serum enantiomer concentration after dental surgery. *British Journal of Clinical Pharmacology*, **47:** 391–396.

Jamali, F., Mehvar, R., Russell, A.S., Sattari, S., Yakimets, W.W. and Koo, J. 1992, Human pharmacokinetics of ibuprofen enantiomers following different doses and formulations: intestinal chiral inversion. *Journal of Pharmaceutical Sciences*, **81:** 221–225.

Jamali, F., Singh, N.N., Pasutto, F.M., Russell, A.S. and Coutts, R.T. 1988, Pharmacokinetics of ibuprofen enantiomers in humans following oral administration of tablets with different absorption rates. *Pharmaceutical Research*, **5:** 40–43.

Jeffrey, P., Tucker, G.T., Bye, A., Crewe, H.K. and Wright, P.A. 1990, The site of inversion of R(−)-ibuprofen: studies using rat in-situ isolated perfused intestine/liver preparations. *Journal of Pharmacy and Pharmacology*, **43:** 715–720.

Johnson, J.L., Brater, D.C. and Hall, S.D. 1995, Formation of 'hybrid' glycerolipids in normal volunteers. *Clinical Pharmacology and Therapeutics*, **57:** 213.

Jones, K., Seymour, R.A. and Hawkesford, J.E. 1997, Are the pharmacokinetics of ibuprofen important determinants for the drug's efficacy in postoperative pain after third molar surgery?. *British Journal of Oral and Maxillofacial Surgery*, **35:** 173–176.

Jørgensen, H.S., Christensen, H.R. and Kampmann, J.P. 1991, Interactions between digoxin and indomethacin or ibuprofen. *British Journal of Clinical Pharmacology*, **31:** 108–110.

Juhl, R.P., Van Thiel, D.H., Dittert, L.W., Albert, K.S. and Smith, R.B. 1983, Ibuprofen and sulindac kinetics in alcoholic liver disease. *Clinical Pharmacology and Therapeutics*, **34:** 104–109.

Kaiser, D.G. and Martin, R.S. 1978, Electron-capture GLC determination of ibuprofen in serum. *Journal of Pharmaceutical Sciences*, **67:** 627–630.

Kaiser, D.G. and VanGiessen, G.J. 1974, GLC determination of ibuprofen [(±)-2-(p-isobutyl)propionic acid] in plasma. *Journal of Pharmaceutical Sciences*, **63:** 219–221.

Kaiser, D.G., VanGiessen, G.J., Reischer, R.J. and Wechter, W.J. 1976, Isomeric inversion of ibuprofen (R)-enantiomer in humans. *Journal of Pharmaceutical Sciences*, **65:** 269–273.

Kaplan, H.B., Edelson, H.S., Korchak, H.M., Given, W.P., Abramson, S. and Weissmann, G. 1984,

Effects of non-steroidal anti-inflammatory agents on human neutrophil functions *in vitro* and *in vivo*. *Biochemical Pharmacology*, **33:** 371–378.

Kelley, M.T., Walson, P.D., Edge, J.H., Cox, S. and Mortensen, M.E. 1992, Pharmacokinetics and pharmacodynamics of ibuprofen isomers and acetaminophen in febrile children. *Clinical Pharmacology and Therapeutics*, **52:** 181–189.

Kepp, D.R., Sidelmann, U.G. and Hansen, S.H. 1997, Isolation and characterization of major phase I and II metabolites of ibuprofen. *Pharmaceutical Research*, **14:** 676–680.

Kirchheiner, J., Meineke, I., Freytag, G., Meisel, C., Roots, I. and Bröckmoller, J. 2002, Enantio-specific effects of cytochrome P450 2C9 amino acid variants on ibuprofen pharmacokinetics and on the inhibition of cyclooxygenases 1 and 2. *Clinical Pharmacology and Therapeutics*, **72:** 62–75.

Kleinbloesem, C.H., Ouwerkerk, M., Spitznagel, W., Wilkinson, F.E. and Kaiser, R.R. 1995, Pharmacokinetics and bioavailability of percutaneous ibuprofen. *Arzneimittel-Forschung*, **45:** 1117–1121.

Knights, K.M. and Drew, R. 1992, The effects of ibuprofen enantiomers on hepatocyte intermediary metabolism and mitochondrial respiration. *Biochemical Pharmacology*, **44:** 1291–1296.

Knights, K.M. and Jones, M.E. 1992, Inhibition kinetics of hepatic microsomal long chain fatty acid-CoA ligase by 2-arylpropionic acid non-steroidal anti-inflammatory drugs. *Biochemical Pharmacology*, **43:** 1465–1471.

Knights, K.M., McLean, C.F., Tonkin, A.L. and Miners, J.O. 1995, Lack of effect of gender and oral contraceptive steroids on the pharmacokinetics of (R)-ibuprofen in humans. *British Journal of Clinical Pharmacology*, **40:** 153–156.

Knights, K.M., Talbot, U.M. and Baillie, T.A. 1992, Evidence of multiple forms of rat liver microsomal coenzyme A ligase catalysing the formation of 2-arylpropionyl-coenzyme A thioesters. *Biochemical Pharmacology*, **44:** 2415–2417.

Knihinicki, R.D., Day, R.O., Graham, G.G. and Williams, K.M. 1990, Stereoselective disposition of ibuprofen and flurbiprofen in rats. *Chirality*, **2:** 134–140.

Knihinicki, R.D., Williams, K.M. and Day, R.O. 1989, Chiral inversion of 2-arylpropionic acid non-steroidal anti-inflammatory drugs – 1 *In vitro* studies of ibuprofen and flurbiprofen. *Biochemical Pharmacology*, **38:** 4389–4395.

Konstan, M.W., Byard, P.J. and Hoppel, C.L. 1995, Effect of high-dose ibuprofen in patients with cystic fibrosis. *New England Journal of Medicine*, **332:** 848–854.

Konstan, M.W., Hoppel, C.L., Chai, B.L. and Davis, P.B. 1991, Ibuprofen in children with cystic fibrosis: pharmacokinetics and adverse effects. *Journal of Pediatrics*, **118:** 956–964.

Lapatto-Reiniluoto, O., Kivisto, K.T. and Neuvonen, P.J. 1999, Effect of activated charcoal alone or given after gastric lavage in reducing the absorption of diazepam, ibuprofen and citalopram. *British Journal of Clinical Pharmacology*, **48:** 148–153.

Laska, E.M., Sunshine, A., Marrero, I., Olson, N., Siegel, C. and McCormick, N. 1986, The correlation between blood levels of ibuprofen and clinical analgesic response. *Clinical Pharmacology and Therapeutics*, **40:** 1–7.

Lee, E.J.D., Williams, K.M., Graham, G.G. and Champion, G.D. 1984, Liquid chromatographic determination and plasma concentration profile of the optical isomers of ibuprofen in man. *Journal of Pharmaceutical Sciences*, **73:** 1542–1545.

Lee, E.J.D., Williams, K.M., Graham, G.G., Day, R.O. and Champion, G.D. 1985, Stereoselective disposition of ibuprofen enantiomers in man. *British Journal of Clinical Pharmacology*, **19:** 669–674.

Leemann, T.D., Transon, C., Bonnabry, P. and Dayer, P. 1993, A major role for cytochrome P450TB (CYP2C subfamily) in the actions of non-steroidal antiinflammatory drugs. *Drugs under Experimental and Clinical Research*, **19:** 189–195.

Levine, M.A., Walker, S.E. and Paton, T.W. 1992, The effect of food or sucralfate on the bioavailability of S(+) and R(−) enantiomers of ibuprofen. *Journal of Clinical Pharmacology*, **32:** 1110–1114.

Li, G., Treiber, G. and Klotz, U. 1989, The ibuprofen-cimetidine interaction. Stereochemical investigations. *Drug Investigation*, **1:** 11–17.

Li, G., Treiber, G., Maier, K., Walker, S. and Klotz, U. 1993, Disposition of ibuprofen in patients with liver cirrhosis. Stereochemical considerations. *Clinical Pharmacokinetics*, **25:** 154–163.

Lockwood, G.F., Albert, K.S., Gillespie, W.R., Bole, G.G., Harkom, T.M., Szpunar, G.J. and Wagner, J.G. 1983, Pharmacokinetics of ibuprofen in man. I. Free and total area/dose relationships. *Clinical Pharmacology and Therapeutics*, **34:** 97–103.

Lockwood, G.F. and Wagner, J.G. 1982, High performance liquid chromatographic determination of ibuprofen and its major metabolites in biological fluids. *Journal of Chromatography*, **232:** 335–343.

Longenecker, G.L., Swift, I.A., Bowen, R.J., Beyers, B.J. and Shah, A.K. 1985, Kinetics of ibuprofen effect on platelet and endothelial prostanoid release. *Clinical Pharmacology and Therapeutics*, **37:** 343–348.

Martin, S.W., Stevens, A.J., Brennan, B.S., Rowland, M. and Houston, J.B. 1995, Pharmacodynamic comparison of regional drug delivery for non-steroidal anti-inflammatory drugs, using the rat air-pouch model of inflammation. *Journal of Pharmacy and Pharmacology*, **47:** 458–461.

Martin, W., Koselowske, G., Töberich, H., Kerkmann, T., Mangold, B. and Augustin, J. 1990, Pharmacokinetics and absolute bioavailability of ibuprofen after oral administration of ibuprofen lysine in man. *Biopharmaceutics and Drug Disposition*, **11:** 265–278.

Mehvar, R. and Jamali, F. 1988, Pharmacokinetic analysis of the enantiomeric inversion of chiral non-steroidal antiinflammatory drugs. *Pharmaceutical Research*, **5:** 76–79.

Mehvar, R., Jamali, F. and Pasutto, F.M. 1988, Liquid-chromatographic assay of ibuprofen enantiomers in plasma. *Clinical Chemistry*, **34:** 493–496.

Menzel, E.J. and Kolarz, G. 1994, Binding capacity of ibuprofen to muscle proteins. *Arzneimittel-Forschung*, **44:** 341–343.

Menzel, S., Waibel, R., Brune, K. and Geisslinger, G. 1994, Is the formation of R-ibuprofenyl-adenylate the first stereoselective step of chiral inversion? *Biochemical Pharmacology*, **48:** 1056–1058.

Menzel-Soglowek, S., Geisslinger, G. and Brune, K. 1990, Stereoselective high-performance liquid chromatographic determination of ketoprofen, ibuprofen and fenoprofen in plasma using a chiral α-acid glycoprotein column. *Journal of Chromatography*, **532:** 295–303.

Menzel-Soglowek, S., Geisslinger, G., Mollenhauer, J. and Brune, K. 1992, Metabolic chiral inversion of 2-arylpropionates in rat H411E and human Hep G2 hepatoma cells. Relationship to *in vivo* metabolism. *Biochemical Pharmacology*, **43:** 1487–1492.

Millership, J.S. and Collier, P.S. 1997, Topical administration of racemic ibuprofen. *Chirality*, **9:** 313–316.

Mills, R.F.N., Adams, S.S., Cliffe, E.E., Dickinson, W. and Nicholson, J.S. 1973, The metabolism of ibuprofen. *Xenobiotica*, **3:** 589–598.

Moorhouse, K.G., Dodds, P.F. and Hutson, D.H. 1991, Xenobiotic triacylglycerol formation in isolated hepatocytes. *Biochemical Pharmacology*, **41:** 1179–1185.

Murray, M.D., Brater, D.C., Tierney, W.M., Hui, S.L. and McDonald, C.J. 1990, Ibuprofen-associated renal impairment in a large general internal medicine program. *American Journal of Medical Science*, **299:** 222–229.

Murry, D.J., Oermann, C.M., Ou, C.N., Rognerud, C., Seilheimer, D.K. and Sockrider, M.M. 1999, Pharmacokinetics of ibuprofen in patients with cystic fibrosis. *Pharmacotherapy*, **19:** 340–345.

Nahata, M.C., Durrell, D.E., Powell, D.A. and Gupta, N. 1991, Pharmacokinetics of ibuprofen in febrile children. *European Journal of Clinical Pharmacology*, **40:** 427–428.

Naidong, W. and Lee, J.W. 1994, Development and validation of a liquid chromatographic method for the quantitation of ibuprofen enantiomers in human plasma. *Journal of Pharmaceutical and Biomedical Analysis*, **12:** 551–556.

Nakamura, Y., Yamaguchi, T., Takahashi, S., Hashimoto, S., Iwatami, K. and Nakagawa, Y. 1981, Optical isomerization mechanisms of R(−)-hydratropic acid derivatives. *Journal of Pharmacobio-Dynamics*, **4:** S1.

Neuvonen, P.J. 1991, The effect of magnesium hydroxide on the oral absorption of ibuprofen, ketoprofen and diclofenac. *British Journal of Clinical Pharmacology*, **31:** 263–266.

Noctor, T.A.G., Felix, G. and Wainer, I.W. 1991, Stereochemical resolution of enantiomeric 2-aryl propionic acid non-steroidal anti-inflammatory drugs on a human serum albumin based high-performance liquid chromatographic chiral stationary phase. *Chromatographia*, **31:** 55–59.

Ochs, H.R., Greenblatt, D.J. and Verburg-Ochs, B. 1985, Ibuprofen kinetics in patients with renal insufficiency who are receiving maintenance hemodialysis. *Arthritis and Rheumatism*, **28:** 1430–1434.

Oliary, J., Tod, M., Nicolas, P., Petitjean, O. and Caillé, G. 1992, Pharmacokinetics of ibuprofen enantiomers after single and repeated doses in man. *Biopharmaceutics and Drug Disposition*, **13:** 337–344.

Paliwal, J.K., Smith, D.E., Cox, S.R., Berardi, R.R., Dunn-Kucharski, V.A. and Elta, G.H. 1993, Stereoselective, competitive, and non-linear plasma protein binding of ibuprofen enantiomers as determined *in vivo* in healthy subjects. *Journal of Pharmacokinetics and Biopharmaceutics*, **21:** 145–161.

Pettersen, J.E., Ulsaker, G.A. and Jellum, E. 1978, Studies on the metabolism of 2,4'-isobutylphenyl-propionic acid (ibuprofen) by gas chromatography and mass spectrometry. Dialysis fluid, a convenient medium for studies on drug metabolism. *Journal of Chromatography*, **145:** 413–420.

Pirkle, W.H. and Murray, P.G. 1990, The separation of the enantiomers of a variety of non-steroidal anti-inflammatory drugs (NSAIDs) as their anilide derivatives using a chiral stationary phase. *Journal of Liquid Chromatography*, **13:** 2123–2134.

Quattrochi, F.P., Robinson, J.D., Curry, R.W., Grieco, M.L. and Schulman, S.G. 1983, The effect of ibuprofen on serum digoxin concentrations. *Drug Intelligence and Clinical Pharmacy*, **17:** 286–288.

Radermacher, J., Jentsch, D., Scholl, M.A., Lustinetz, T. and Frölich, J.C. 1991, Diclofenac concentrations after cutaneous application in inflammatory and degenerative joint disease. *British Journal of Clinical Pharmacology*, **31:** 537–541.

Ragni, M.V., Miller, B.J., Whalen, R. and Ptachcinski, R. 1992, Bleeding tendency, platelet function, and pharmacokinetics of ibuprofen and zidovudine in HIV(+) hemophilic men. *American Journal of Hematology*, **40:** 176–182.

Reichel, C., Brugger, R., Bang, H., Geisslinger, G. and Brune, K. 1997, Molecular cloning and expression of a 2-arylpropionyl-coenzyme A epimerase: a key enzyme in the inversion metabolism of ibuprofen. *Molecular Pharmacology*, **51:** 576–582.

Rekka, E., Ayalogu, E.O., Lewis, D.F., Gibson, G.G. and Ioannides, C. 1994, Induction of hepatic microsomal CYP4A activity and of peroxisomal beta-oxidation by two non-steroidal anti-inflammatory drugs. *Archives of Toxicology*, **68:** 73–78.

Rey, E., Pariente-Khayat, A., Gouyet, L., Vauzelle-Kervroedan, F., Pons, G., D'Athis, P. *et al.* 1994, Stereoselective disposition of ibuprofen enantiomers in infants. *British Journal of Clinical Pharmacology*, **38:** 373–375.

Rudy, A.C., Anliker, K.S. and Hall, S.D. 1990, High-performance liquid chromatographic determination of the stereoisomeric metabolites of ibuprofen. *Journal of Chromatography*, **528:** 395–405.

Rudy, A.C., Bradley, J.D., Ryan, S.I., Kalasinski, L.A., Xiaotao, Q. and Hall, S.D. 1992, Variability in the disposition of ibuprofen enantiomers in osteoarthritis patients. *Therapeutic Drug Monitoring*, **14:** 464–470.

Rudy, A.C., Knight, P.M., Brater, D.C. and Hall, S.D. 1991, Stereoselective metabolism of ibuprofen in humans: administration of R-, S- and racemic ibuprofen. *Journal of Pharmacology and Experimental Therapeutics*, **259:** 1133–1139.

Rudy, A.C., Knight, P.M., Brater, D.C. and Hall, S.D. 1995, Enantioselective disposition of ibuprofen in elderly persons with and without renal impairment. *Journal of Pharmacology and Experimental Therapeutics*, **273:** 88–93.

Sallustio, B.C., Meffin, P.J. and Knights, K.M. 1988, The stereospecific incorporation of fenoprofen into rat hepatocyte and adipocyte triacylglycerols. *Biochemical Pharmacology*, **37:** 1919–1923.

Sanins, S.M., Adams, W.J., Kaiser, D.G., Halstead, G.W., Hosley, J., Barnes, H. and Baillie, T.A. 1991, Mechanistic studies on the metabolic chiral inversion of R-ibuprofen in the rat. *Drug Metabolism and Disposition*, **19:** 405–410.

Scheuerer, S., Hall, S.D., Williams, K.M. and Geisslinger, G. 1998a, Effect of clofibrate on the chiral inversion of ibuprofen in healthy volunteers. *Clinical Pharmacology and Therapeutics*, **64:** 168–176.

Scheuerer, S., Williams, K.M., Brugger, R., McLachlan, A.J., Brune, K., Day, R.O. and Geisslinger, G. 1998b, Effect of clofibrate on the chiral disposition of ibuprofen in rats. *Journal of Pharmacology and Experimental Therapeutics*, **284:** 1132–1138.

Schill, G. 1989, High performance ion-pair chromatography. *Journal of Biochemical and Biophysical Methods*, **18:** 249–270.

Schneider, H.T., Nürnberg, B., Dietzel, K. and Brune, K. 1990, Biliary elimination of non-steroidal anti-inflammatory drugs in patients. *British Journal of Clinical Pharmacology*, **29:** 127–131.

Seideman, P., Lohrer, F., Graham, G.G., Duncan, M.W., Williams, K.M. and Day, R.O. 1994, The stereoselective disposition of the enantiomers of ibuprofen in blood, blister and synovial fluid. *British Journal of Clinical Pharmacology*, **38:** 221–227.

Shieh, W.R., Gou, D.M., Liu, Y.C., Chen, C.-S. and Chen, C.-Y. 1993, A 13C NMR study on ibuprofen metabolism in isolated rat liver mitchondria. *Analytical Biochemistry*, **212:** 143–149.

Shirley, M.A., Guan, X., Kaiser, D.G., Halstead, G.W. and Baillie, T.A. 1994, Taurine conjugation of ibuprofen in humans and in rat liver *in vitro*. Relationship to metabolic chiral inversion. *Journal of Pharmacology and Experimental Therapeutics*, **269:** 1166–1175.

Skeith, K.J. and Jamali, F. 1991, Clinical pharmacokinetics of drugs used in juvenile arthritis. *Clinical Pharmacokinetics*, **21:** 129–149.

Smith, D.A. and Jones, B.C. 1992, Speculations on the substrate structure–activity relationship (SSAR) of cytochrome P450 enzymes. *Biochemical Pharmacology*, **44:** 2089–2098.

Smith, D.E., Paliwal, J.K., Cox, S.R., Berardi, R.R., Dunn-Kucharski, V.A. and Elta, G.H. 1994, The effect of competitive and non-linear plasma protein binding on the stereoselective disposition and metabolic inversion of ibuprofen in healthy subjects. *Biopharmaceutics and Drug Disposition*, **15:** 545–561.

Spraul, M., Hofmann, M., Dvortsak, P., Nicholson, J.K. and Wilson, I.D. 1993, High-performance liquid chromatography coupled to high-field proton nuclear magnetic resonance spectroscopy: application to the urinary metabolites of ibuprofen. *Analytical Chemistry*, **65:** 327–330.

Suri, A., Grundy, B.L. and Derendorf, H. 1997, Pharmacokinetics and pharmacodynamics of enantiomers of ibuprofen and flurbiprofen after oral administration. *International Journal of Clinical Pharmacology and Therapeutics*, **35:** 1–8.

Tan, S.C., Jackson, S.H., Swift, G.G. and Hutt, A.J. 1997, Stereospecific analysis of the major metabolites of ibuprofen in urine by sequential achiral-chiral high-performance liquid chromatography. *Journal of Chromatography B*, 53–63.

Tegeder, I., Muth-Selbach, U., Lötsch, J., Rüsing, G., Oelkers, R., Brune, K. *et al.* 1999, Application of microdialysis for the determination of muscle and subcutaneous tissue concentrations after oral and topical ibuprofen administration. *Clinical Pharmacology and Therapeutics*, **65:** 357–368.

Tracy, T.S. and Hall, S.D. 1991, Determination of the epimeric composition of ibuprofenyl-CoA. *Analytical Biochemistry*, **195:** 24–29.

Tracy, T.S. and Hall, S.D. 1992, Metabolic inversion of (R)-ibuprofen. Epimerization and hydrolysis of ibuprofenyl-coenzyme A. *Drug Metabolism and Disposition*, **20:** 322–327.

Treffel, P. and Gabard, B. 1993, Feasibility of measuring the bioavailability of topical ibuprofen in commercial formulations using drug content in epidermis and a methyl nicotinate skin inflammation assay. *Skin Pharmacology*, **6:** 268–275.

Troconíz, I.F., Armenteros, S., Planelles, M.V., Benítez, J., Calvo, R. and Domínguez, R. 2000, Pharmacokinetic–pharmacodynamic modelling of the antipyretic effect of two formulations of ibuprofen. *Clinical Pharmacokinetics*, **38:** 505–518.

Valle-Jones, J.C., Smith, J. and Rowley-Jones, D. 1984, A comparison in general practice of once and twice daily sustained-release ibuprofen and standard ibuprofen in the treatment of non-articular rheumatism. *British Journal of Clinical Practice*, **38:** 353–358.

Villanueva, M., Heckenberger, R., Strobach, H., Palmér, M. and Schrör, K. 1993, Equipotent inhibition by R(−)-, S(+)- and racemic ibuprofen of human polymorphonuclear cell function *in vitro*. *British Journal of Clinical Pharmacology*, **35:** 235–242.

Vos, R.M., Mayer, J.M., Etter, J.C. and Testa, B. 1996, Clofibric acid increases the unidirectional chiral inversion of ibuprofen in rat liver preparations. *Xenobiotica*, **26:** 571–582.

Wade, L.T., Kenna, J.G. and Caldwell, J. 1997, Immunochemical identification of mouse hepatic protein adducts derived from the non-steroidal anti-inflammatory drugs diclofenac, sulindac and ibuprofen. *Chemical Research in Toxicology*, **10:** 546–555.

Wagner, J.G., Albert, K.S., Szpunar, G.J. and Lockwood, G.F. 1984, Pharmacokinetics of ibuprofen in man. IV: Absorption and disposition. *Journal of Pharmacokinetics and Biopharmaceutics*, **12:** 381–399.

Walker, J.S., Knihinicki, R.D., Seideman, P. and Day, R.O. 1993, Pharmacokinetics of ibuprofen enantiomers in plasma and suction blister fluid in healthy volunteers. *Journal of Pharmaceutical Sciences*, **82:** 787–790.

Walter, K. and Dilger, C. 1997, Ibuprofen in human milk. *British Journal of Clinical Pharmacology*, **44:** 211–212.

Wechter, W.J. 1994, Drug chirality: on the mechanism of R-aryl propionic acid class NSAIDs. Epimerization in humans and the clinical implications for the use of racemates. *Journal of Clinical Pharmacology*, **34:** 1036–1042.

Williams, K.M. 1991, Molecular asymmetry and its pharmacological consequences. *Advances in Pharmacology*, **22:** 57–135.

Williams, K.M., Day, R.O. and Breit, S.N. 1993, Biochemical actions and clinical pharmacology of anti-inflammatory drugs. *Advances in Drug Research*, **24:** 121–198.

Williams, K.M., Day, R.O., Knihinicki, R.D. and Duffield, A. 1986, The stereoselective uptake of ibuprofen enantiomers into adipose tissue. *Biochemical Pharmacology*, **35:** 3403–3405.

Williams, K.M., Knihinicki, R.D. and Day, R.O. 1991, Pharmacokinetics of the enantiomers of ibuprofen in the rabbit. *Agents and Actions*, **34:** 381–386.

Zhao, M.J., Peter, C., Holtz, M.C., Hugenell, N., Koffel, J.C. and Jung, L. 1994, Gas chromatographic-mass spectrometric determination of ibuprofen enantiomers using R(−)-2,2,2-trifluoro-1-(9-anthryl)ethanol as derivatizing agent. *Journal of Chromatography B*, **656:** 441–446.

Pharmacokinetics and Metabolism of Paracetamol (Acetaminophen)

6

G.G. Graham and M. Hicks

INTRODUCTION

The metabolism and pharmacokinetics of paracetamol are very important aspects of the clinical pharmacology of this drug. Even the development of paracetamol was based upon studies on drug metabolism and pharmacokinetics. It was found that the old analgesics, acetanilide and phenacetin, are both converted largely to paracetamol (Lester and Greenberg, 1947; Brodie and Axelrod, 1949). The subsequent confirmation of earlier work showing that paracetamol is an analgesic (Flinn and Brodie, 1948) and questions about the renal toxicity of phenacetin led to the cessation of the use of phenacetin and its replacement by paracetamol.

Paracetamol achieves high concentrations in plasma and urine. It is assayed easily, and is eliminated by several metabolic pathways. Therefore, like salicylate, paracetamol is used widely as a model drug in studies on the principles of drug metabolism, pharmacokinetics and drug interactions. Examples are the examination of the metabolism of paracetamol by microdialysis probes in the liver, radiochemical procedures, mass spectrometry and nuclear magnetic resonance (Nicholson *et al.*, 1985; Dawson *et al.*, 1991; 1992; Stenken *et al.*, 1998). An older example is the study of the pharmacokinetics of paracetamol through the rate of excretion of paracetamol and its conjugated metabolites (Cummings *et al.*, 1967).

ABSORPTION

Paracetamol lacks the gastrointestinal toxicity of aspirin, and consequently it is usually administered as conventional, rapidly disintegrating tablets. It is only moderately lipid-soluble, with an octanol/water partition coefficient of 3.2 (Craig, 1990). Paracetamol is a very weak acid (pK$_a$ 9.7), and is therefore almost totally un-ionised at all physiological pH values. It is absorbed passively from the small intestine (Swaan *et al.*, 1994). The rates of absorption of paracetamol in different parts of the small intestine are similar, and are approximately proportional to the transmucosal water fluxes (Gramatte and Richter, 1994). Absorption from all parts of the small intestine is much faster than from the stomach, and consequently the rate of absorption of paracetamol depends upon the rate of gastric emptying and can be used as a measure of this physiological process (Heading *et al.*, 1973). The absorption is faster when the subjects are walking than when lying down (Rumble *et al.*, 1991) and, because of the more rapid

TABLE 6.1

Pharmacokinetic parameters of paracetamol (Benet *et al.*, 1996).

Fractional absorption	$88 \pm 15\%$
Urinary excretion	$3 \pm 1\%$
Binding to plasma proteins	0
Total clearance	5.0 ± 1.4 ml/min per kilogram
Volume of distribution	0.95 ± 0.12 l/kg
Half-life	2.0 ± 0.4 hours

passage of gastric contents through the pylorus when individuals are standing or lying on the right side, the rate of absorption of paracetamol is correspondingly faster than when lying on the left side (Renwick *et al.*, 1992; Vance *et al.*, 1992). Lying on the left side also decreases aspiration and is recommended during transport and initial examination if toxic doses of paracetamol and other drugs are suspected (Vance *et al.*, 1992). Recently, the impaired absorption of paracetamol has been shown to be an early marker of rejection of transplanted small bowel in the rat (Miyauchi *et al.*, 1997).

The total absorption of paracetamol is high at about 88 per cent (Table 6.1; Benet *et al.*, 1996). The mean total absorption is slightly higher than that predicted from the metabolic clearance of paracetamol and the hepatic blood flow. Considering that the metabolic clearance of paracetamol is about 350 ml/min and that hepatic blood flow is approximately 1400 ml/min, the predicted first-pass extraction of paracetamol should be approximately 20 per cent and the total absorption should therefore be 80 per cent, slightly lower than the actual figure. The greater effective absorption of paracetamol may be due either to saturable first-pass metabolism or to extrahepatic metabolism. Both hypotheses are supported; by studies in the rat where saturable first-pass metabolism in either the gastrointestinal tract and liver occurs (Bhargava and Hirate, 1989), and also by the observation that paracetamol may be metabolised in the kidney, which cannot play any part in first-pass metabolism (Bock *et al.*, 1993).

The total absorption of paracetamol is not affected by food, although the rate of absorption is slowed (Robertson *et al.*, 1991). The rate of absorption of larger doses is slowed by the rate of dissolution, the rate of absorption of a 2 g dose of paracetamol being slower than that of 0.5 or 1 g doses (Rawlins *et al.*, 1977). Because of its short half-life of elimination (Table 6.1), slow release capsules and tablets of paracetamol have been prepared (Ström *et al.*, 1990) and may usefully prolong the dosage interval from 6 to about 8 hours. Prolonged release over longer periods of time is not considered useful because the dose, and therefore the tablets, would then be very large and difficult to swallow.

The absorption of paracetamol after its rectal administration has also been examined, particularly in neonates and after surgery. In general, its rectal absorption has been slow and incomplete (Gaudreault *et al.*, 1988; Anderson *et al.*, 1995; 1999; van Lingen *et al.*, 1999) and a loading dosage schedule has been proposed for children (Anderson and Holford, 1997).

Measures to limit the absorption of overdoses

Charcoal binds paracetamol and limits the gastrointestinal absorption of overdoses of paracetamol if the charcoal is taken within 1 hour of taking the overdose (Chamberlain *et al.*, 1993). Not surprisingly, charcoal with a high surface area is more effective than charcoal with a lower surface area (Roberts *et al.*, 1997). The binding of paracetamol to multiple doses of charcoal in the gastrointestinal tract also leads to an increased clearance of paracetamol from the circulation of pigs, and should usefully lower the drug concentrations after overdosage in man (Chyka *et al.*, 1995). Multiple doses of charcoal should, however, be considered as secondary to the primary antidote, N-acetylcysteine.

The absorption of overdoses of paracetamol is also limited by other measures. Vomiting, such as that induced by ipecacuanha (Saincher *et al.*, 1997), decreases the absorption of paracetamol, but any significant effect on its absorption is lost within half an hour and induction of vomiting is not generally practicable in the treatment of an overdose. Absorption of a large dose is also decreased by about 40 per cent by purging with polyethylene glycol/electrolyte solution (Hassig *et al.*, 1993).

DISTRIBUTION

Paracetamol has a moderate volume of distribution (Table 6.1). Its volume of distribution (about 0.95l/kg) is just less than body weight, although it is still about 30 per cent larger than total body water. Studies with tritiated paracetamol show that the drug is distributed throughout the body, with the highest concentrations of the label in the liver, gastrointestinal tract and kidney, probably related to the metabolism and excretion of the drug (Brune *et al.*, 1980).

The distribution of paracetamol into anatomically large tissues, such as muscle, adipose tissue and liver, appears quite rapid, but for some tissues, particularly the brain, the transfer is much slower (see below). The contrasting rates of uptake into tissues are largely the result of the limited lipid-solubility of paracetamol. It is suggested that the drug readily diffuses from the fenestrated blood vessels in muscle but diffuses slowly though the blood–brain barrier, which is the intact endothelial lining of blood vessels in the central nervous system. Paracetamol is also taken up rapidly by red blood cells, such that the ratio of the concentrations in red blood cell water is slightly higher than plasma water (Pang *et al.*, 1995).

Central nervous system

Several experimental findings indicate slow transfer of paracetamol into and out of the central nervous system where paracetamol is considered to produce its major therapeutic actions:

1. The peak antipyretic effect in children occurs at about 2.5 hours after peak plasma concentrations are attained (Brown *et al.*, 1992; Kelley *et al.*, 1992). This delay may indicate, in part, slow equilibration at its central site of action, but the delay must also be due to the time taken for body temperature to fall. This cannot be instantaneous.
2. The peak analgesic effect of paracetamol is delayed, occurring at about 1.5 to 2 hours after the intravenous administration of paracetamol (Seymour and Rawlins, 1981; Piletta *et al.*, 1991) and at about 2.5 hours after an oral dose (Flinn and Brodie, 1948).
3. From pharmacokinetic–pharmacodynamic analysis, the equilibration half time between its effect compartment and blood is 1.6 hours (Anderson *et al.*, 1999). This mean estimate, however, has a very large experimental error.
4. The concentrations in cerebrospinal fluid lag behind those in plasma (Figure 6.1). According to the compartmental model used, the equilibration half time from cerebrospinal fluid in adults is one to two hours (Bannwarth *et al.*, 1992; Moreau *et al.*, 1993) but generally 1 hour or less in children (Anderson *et al.*, 1998).
5. Analgesic activity is lost slowly (Quiding *et al.*, 1984; Cooper *et al.*, 1989), particularly after a second dose (Levy, 1987).

Because of its delayed effect and the time taken for oral absorption, it is reasonable that paracetamol should be administered before pain is expected. Thus, it should be administered before short surgical procedures or before full recovery from longer procedures. In the latter case slow oral absorption may prevent good analgesia, and in the countries where it is available propacetamol, an injectable precursor of paracetamol, should be useful (Figure 6.1).

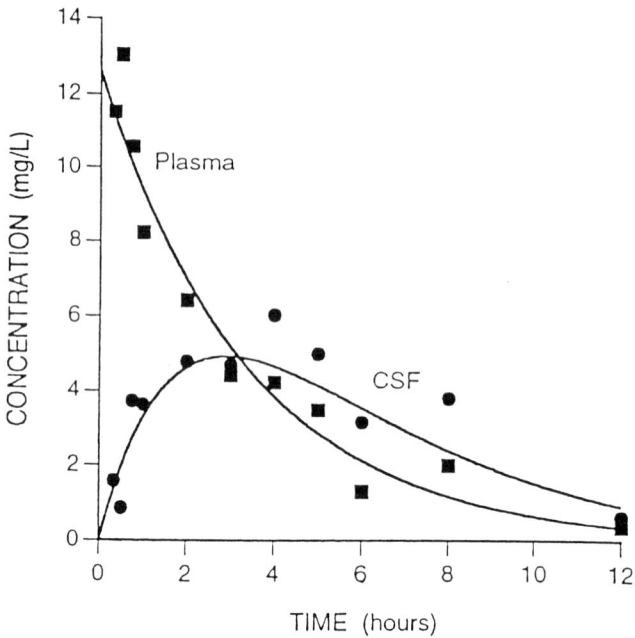

Figure 6.1 Time courses of concentrations of paracetamol in plasma and cerebrospinal fluid after an intravenous dose of 2 g of propacetamol. Propacetamol is the dimethylglycine ester of paracetamol, and is completely hydrolysed to paracetamol within 7 minutes. The half-life of loss of paracetamol from cerebrospinal fluid is 1.8 hours, compared to the elimination half-life of 2.8 hours. Note the significant delay before the maximum concentrations are produced in plasma. (From the data of Bannwarth *et al.*, 1992.)

Synovial fluid

Although the pharmacological activities of paracetamol are widely considered to derive from actions within the central nervous system, some of its actions may be in peripheral tissues, and the kinetics of the drug in joints is of some note. The half-life of loss of paracetamol from synovial fluid of the knee joints of rheumatoid patients is about 1 hour. By comparison, the rate of loss of salicylate from the knee is considerably slower ($t_{1/2}$ of loss = 2.5 hours; Owen *et al.*, 1994). The faster loss of paracetamol from the knee joint is probably due to its lack of binding to plasma proteins compared to the high binding of salicylate. The relatively short half-life of loss of paracetamol from the knee joint means that the concentrations in the knee should lag behind the concentrations in plasma, but to a lesser extent than salicylate and other non-steroidal anti-inflammatory drugs.

The classical NSAIDs, such as aspirin, salicylate and ibuprofen, are concentrated to a greater extent in the inflamed paws of rats, but by contrast no such concentration is seen with paracetamol (Figure 6.2). The difference between paracetamol and the NSAIDs probably results from paracetamol being essentially a neutral compound at physiological pH values, and its negligible binding to plasma albumin (see Chapter 4).

Muscle and adipose tissue

As determined by microdialysis probes, the concentrations of paracetamol in the interstitial fluid of muscle are very similar to those in plasma, while the same technique shows that the concentrations in

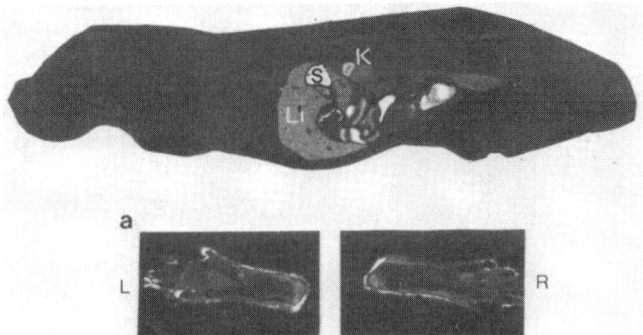

Figure 6.2 Autoradiographs demonstrating the distribution of ^{3}H-paracetamol in rats 3 hours after its oral administration. The lighter regions correspond to the more intense areas of radioactivity. Inflammation in the left paw only was produced by the local injection of carrageenan. Inflammation does not affect the uptake of ^{3}H-paracetamol into the paw, in contrast to the greater concentration of acidic NSAIDs in inflamed paws (Brune *et al.*, 1980).

adipose tissue are slightly lower than in plasma (Müller *et al.*, 1995). The concentrations in both these tissues follow the concentrations in plasma quite closely, and do not show the considerable lag seen with cerebrospinal and synovial fluids.

Eye

The transfer of paracetamol into and out of the lens is of interest because of the possible anti-cataract activity of paracetamol. The transfer appears slow because of the avascular nature of the lens (Shyadehi and Harding, 1991) and prolonged levels should be produced by dosage at wide dosage intervals. Transfer into and out of aqueous humour is, however, very rapid, and the concentrations in this compartment of the eye approximately equal those in plasma (Romanelli *et al.*, 1991).

Liver

Paracetamol enters hepatocytes by both a saturable mechanism and passive diffusion (McPhail *et al.*, 1993). Paracetamol distributes to a greater extent in the lysosomal fraction than in other fractions of the liver (Studenberg and Brouwer, 1993a), but there is no significant accumulation of unchanged paracetamol in hepatocytes as all the paracetamol entering the cells is metabolised (McPhail *et al.*, 1993). The kinetics of the transport processes in isolated hepatocytes can, however, be separated from the kinetics of the metabolic processes (Studenberg and Brouwer, 1993b).

Placenta

Paracetamol readily crosses the placenta. When administered at the second stage of labour, the concentrations of paracetamol in the plasma of the baby at birth equal the concentrations in the mother (Naga Rani *et al.*, 1989). From studies in the sheep, the placental transfer of paracetamol from the foetus back to the mother is the major mode of removal from the foetus (Wang *et al.*, 1990). There is limited oxidative metabolism and sulphation of paracetamol in the human foetal liver, although the production of the hepatotoxic metabolite indicates a risk of toxicity there (Rollins *et al.*, 1979). The metabolism of paracetamol in the neonate is slow and is described below.

Breast milk

Paracetamol concentrations in milk are about the same or slightly lower than in plasma, and decline in parallel with the concentrations in plasma (Bitzen *et al.*, 1981; Notarianni *et al.*, 1987). Treatment of mothers with paracetamol is, however, acceptable because only low doses will be given to the baby by breast feeding.

ELIMINATION

Oral doses of paracetamol are almost quantitatively recovered in urine as the sum of the metabolites, with only a small amount being excreted as unchanged paracetamol. The metabolism of paracetamol can be divided into two general routes, non-oxidative and oxidative (Figure 6.3).

Non-oxidative metabolism

Most paracetamol is metabolised by non-oxidative pathways, with approximately 50 and 30 per cent being eliminated in urine as the glucuronide and sulphate conjugates, respectively (Epstein *et al.*, 1991), and only a small amount of the glucuronide being excreted in bile (Jayasinghe *et al.*, 1986). The two pathways of conjugation appear saturable. In terms of the body content of paracetamol, the Michaelis constants of elimination are approximately 13 and 7 g for the formation of the glucuronide and sulphate, respectively (Slattery and Levy, 1979). These levels are well above single or multiple therapeutic doses, but are relevant to the kinetics after overdose. There are several phenol UDP-glucuronosyltransferases, but one transferase, UGT1A6, is very largely responsible for the formation of paracetamol glucuronide (de Wildt *et al.*, 1999), with glucuronidation activity in both the human liver and kidney (Bock *et al.*, 1993).

Although the sulphation appears saturable, a contributing factor is the limited availability of sulphate (Slattery *et al.*, 1987). It may therefore be expected that paracetamol should decrease the plasma concentrations of sulphate but conflicting data has been obtained, with both decreases in sulphate or sulphate donor (Slattery *et al.*, 1987; Hoffman *et al.*, 1990; Kim *et al.*, 1992) and increases (Blackledge *et al.*, 1991) being reported. The different results may be partly due to the consumption of sulphate. In a comparative study in two Canadian towns with greatly different levels of sulphate in the drinking water, low intake of sulphate was associated with slightly lower serum concentration of sulphate and a fall in sulphate concentrations and in the excretion of the sulphate conjugate of paracetamol during multiple dosage (Hindmarsh *et al.*, 1991). The differences in the kinetics of elimination of paracetamol were, however, small.

There is substantial variation in the extent of glucuronidation and sulphation of paracetamol between individuals. Glucuronidation appears bimodal, and the sulphation correlates negatively with glucuronidation (Patel *et al.*, 1993).

Oxidative metabolism by cytochrome P450s

The amount of paracetamol metabolised by oxidation is small but, because of the hepatotoxicity of one metabolite, the oxidative metabolism is of great clinical importance (Mitchell *et al.*, 1973a). Two primary oxidative metabolites are formed; N-acetyl-p-benzoquinone imine (NAPQI, the hepatotoxic metabolite) and 3-hydroxyparacetamol (Figure 6.3). NAPQI can be formed by several isoforms of cytochrome P450, with the principal cytochrome in man being CYP2E1 (Manyike *et al.*, 2000).

The human isoform CYP2A6 (Chen *et al.*, 1998), and in the rat CYP2B1, are the major cytochromes involved in the production of 3-hydroxyparacetamol. With CYP2B1, 3-hydroxyparacetamol is formed as a result of a different approach of paracetamol to the haem iron than in other

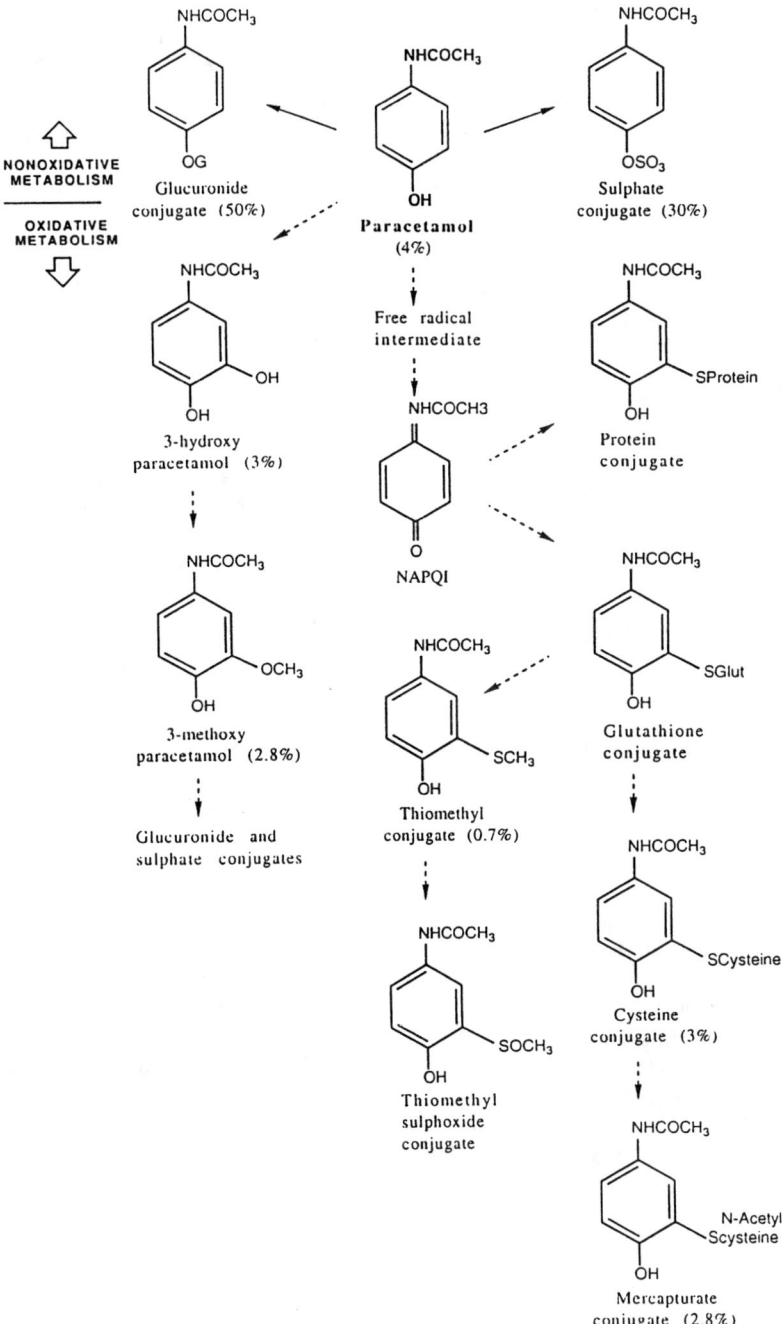

Figure 6.3 Metabolic pathways of paracetamol. Major pathways are shown as solid arrows and minor pathways as broken arrows. The thiomethyl and thiomethyl conjugates are converted to their sulphate and glucuronide conjugates. The pathways of metabolism by neutrophils and prostaglandin H synthase are shown in Figure 6.4.

isoforms responsible for the formation of NAPQI (Myers *et al.*, 1994). The two major metabolites are formed by the two-electron oxidative reactions, but it is proposed that the process consists of two sequential one-electron oxidative steps, the first step being the formation of a radical, possibly a phenoxyl radical (Koymans *et al.*, 1993) or a caged oxygen centred radical (Hoffmann *et al.*, 1990). The second electron may come from NADH whereas the first electron comes from NADPH (Sato and Marumo, 1991).

Oxidation to NAPQI occurs both in the liver and in the kidney (Hart *et al.*, 1994; Hoivik *et al.*, 1995). NAPQI is a strong electrophile, and reacts chemically with glutathione and other thiol compounds. The high intracellular concentrations of glutathione favour the reaction with glutathione, which appears to be purely chemical and not enzymatic (Henderson *et al.*, 2000). The glutathione conjugate is hydrolysed to the cysteine adduct and acetylated to yield the mercapturic acid derivative (conjugate with N-acetyl-cysteine; Figure 6.3). At most only 20 per cent of a dose of paracetamol is excreted in urine as the total of these metabolites in man, but the sum is widely used as an indicator of the conversion of paracetamol to the hepatotoxic metabolite. The ratio of the mercapturate to the total glutathione conjugates appears bimodal, indicating heterogeneity in the acetylation of the cysteine conjugate to the mercapturate. The urinary excretion of glutathione-derived conjugates of paracetamol is highly variable, with a 60-fold range being reported (Critchley *et al.*, 1986). The availability of glutathione is decreased with increasing doses (Jollow *et al.*, 1974; Gemborys *et al.*, 1981), commencing within the range of therapeutic doses in man (Slattery *et al.*, 1987). Most of the glutathione-derived metabolite fraction is excreted in urine, with about 20 per cent excretion in bile (Jayasinghe *et al.*, 1986).

Apart from metabolism of the glutathione conjugate to the mercapturate derivative, glutamate is also conjugated with the initial glutathione conjugate to form mono and bis glutamates, which are minor metabolites in the rat (Mutlib *et al.*, 2000).

NAPQI is widely considered to mediate the hepatic and renal toxicity of paracetamol. When the hepatic glutathione is depleted by about 75 per cent, NAPQI is available for reaction with the proteins mainly through their thiols (cysteine residues), although reactions with other amino acid side chains are possible (Streeter *et al.*, 1984). The result is cell damage, with centrilobular necrosis being the characteristic hepatic lesion (Mitchell *et al.*, 1973b). Not surprisingly, the hepatic distribution of CYP2E1 is also centrilobular (Hinson *et al.*, 1981; Tsutsumi *et al.*, 1989; Anundi *et al.*, 1993). The oxidative metabolism is more significant in hamsters and mice than in rats, rabbits and guinea pigs, possibly explaining the greater susceptibility of hamsters and mice to paracetamol-induced hepatotoxicity (Gregus *et al.*, 1988; Liu *et al.*, 1991). Although the reaction of NAPQI with glutathione appears to be purely chemical, glutathione S-transferase, which can catalyse the reaction between NAPQI and glutathione, actually decreases the hepatotoxicity of paracetamol in mice (Henderson *et al.*, 2000). The mechanism of this surprising finding is unknown.

After high doses, NAPQI binds to more than 20 hepatic proteins (Landin *et al.*, 1996), and an adduct with an hepatic protein is also found in plasma (Hinson *et al.*, 1990; Roberts *et al.*, 1991). Plasma levels of the hepatic enzyme alanine aminotransferase are proportional to the measured levels of the paracetamol–protein adducts and, taken together, are indicative of the risk of hepatotoxicity (Hinson *et al.*, 1990). Small quantities of NAPQI appear to diffuse out of hepatocytes or other cells forming NAPQI, since some paracetamol is covalently bound to haemoglobin. As expected this occurs, again through the cysteinyl residues (Axworthy *et al.*, 1988). The extent of this binding to haemoglobin is, however, about one-tenth of that to liver proteins.

It has been proposed that paracetamol would be unduly toxic in patients with low levels of glutathione. Thus, malnourished patients, or patients with hepatitis C, cirrhosis or AIDS were considered to be at great risk of hepatotoxicity, even from therapeutic doses of paracetamol. However, this does not appear to be the case (Lauterburg, 2002).

Overdoses of paracetamol may damage tissues other than the liver and kidney. In rodents, paracetamol is also cytotoxic to the olfactory mucosa due to its oxidative metabolism by cytochromes P4502A5 and P4502G1 in this tissue (Genter *et al.*, 1998). High doses of paracetamol also lead to depletion of glutathione in the mouse lung, indicating the local formation of the reactive metabolite of paracetamol (Chen *et al.*, 1990). Consistent with this finding, a high incidence of severe lung injury occurs in patients with fulminant hepatic failure due to paracetamol overdose (Baudouin *et al.*, 1995).

The cysteinyl groups of proteins and glutathione are not the only sites of covalent binding. Covalent binding to DNA has been detected in mice (Rogers *et al.*, 1997), particularly after depletion of GSH by the thiol reagent diethyl maleate (Hongslo *et al.*, 1994).

Oxidative metabolism by peroxidases (including prostaglandin H synthase)

Paracetamol is oxidatively metabolised by a variety of peroxidases, as well as by isoforms of cytochrome P450. The well known peroxidase, horse radish peroxidase, has been used widely in investigational studies, with the two major products being NAPQI, the hepatotoxic metabolite, and polymers of paracetamol (Potter and Hinson, 1986) through the intermediate formation of free radical species (West *et al.*, 1984; Fischer *et al.*, 1985; Mason and Fischer, 1986). More importantly, paracetamol interacts in a complex fashion with myeloperoxidase, a major enzyme of neutrophils involved with the oxidative burst of these cells (Figure 6.4). Paracetamol is a substrate for myeloperoxidase, competing with chloride for the active site of the enzyme and also binding irreversibly to myeloperoxidase (van Zyl *et al.*, 1989). Hypochlorite, the major product of myeloperoxidase, also reacts with

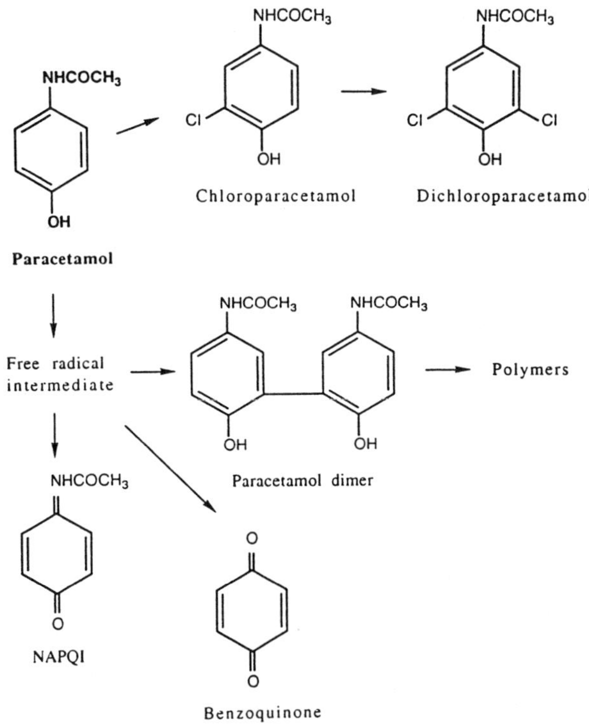

Figure 6.4 Pathways of metabolism of paracetamol by neutrophils due to the activity of myeloperoxidase. The reaction between hypochlorite yields chloroparacetamol and dichloroparacetamol in addition to the other products, but *in vivo* hypochlorite should be largely scavenged and the formation of the chlorinated products is unlikely. The peroxidase activity of prostaglandin H synthase leads to the production of NAPQI and polymers (mainly paracetamol dimer). NAPQI is also formed by cytochrome P450 2E1 in the liver.

paracetamol to produce the reactive intermediates, NAPQI and benzoquinone, together with the non-reactive chlorinated derivatives, 3-chloroparacetamol and 3,5-dichloroparacetamol (O'Brien *et al.*, 1990). Although these reactions between paracetamol and hypochlorite can be demonstrated *in vitro*, it is probable that albumin and other endogenous scavengers react preferentially with hypochlorite. Essentially, endogenous compounds should outscavenge paracetamol (Graham *et al.*, 1997).

Although paracetamol is probably not a significant biological scavenger of hypochlorite, it clearly does inhibit myeloperoxidase and reduce the production of hypochlorite by neutrophils (Shalabi, 1992; Graham *et al.*, 1997). There are several potential consequences. The decreased hypochlorite may reduce the tissue damage of inflammation and also interfere with antibacterial activity of neutrophils (van Zyl *et al.*, 1989). However, neutrophils possess other mechanisms that lead to antibacterial activity, and a significant decrease of antibacterial activity is unlikely. Considerable amounts of paracetamol derivatives bind to both DNA and RNA of neutrophils (Corbett *et al.*, 1989), indicating the possibility of toxic reactions other than hepatotoxicity. The reaction with nucleic acids may well be responsible for the genotoxic and teratogenic effects of paracetamol in experimental systems (Dybing *et al.*, 1984; Kociscova *et al.*, 1988; Fort *et al.*, 1992). Even agranulocytosis may be associated with paracetamol dosage in rare individuals (Mason and Fischer, 1992).

Prostaglandin H synthase (cyclo-oxygenase) has peroxidase activity, and it is not surprising that there is metabolism of paracetamol to polymers and to NAPQI by this enzyme (Potter and Hinson, 1987; Figure 6.4). Paracetamol increases the activity of prostaglandin H synthase in broken cell preparations but is an inhibitor in intact cells stimulated by a variety of factors, such as cytokines. It is, however, not known if there is any causal relationship between the inhibition of prostaglandin synthesis and the metabolism of paracetamol in intact cells (Graham *et al.*, 1999). Metabolism of paracetamol by prostaglandin H synthase may be a cause of its renal toxicity in overdose since high levels of this enzyme and low levels of cytochrome P450 occur in the renal inner medulla, where paracetamol-induced necrosis occurs (Smith and Wilkin, 1977).

Overall, it is now clear that free radical species and NAPQI are formed in a variety of cells in the body either by the cytochrome P450 systems or by peroxidases. Although paracetamol is a very safe drug at therapeutic doses, the oxidative metabolism of paracetamol is potentially important in view of the widespread use of paracetamol as an analgesic and antipyretic, especially in children and during pregnancy.

An interesting application of the metabolism of paracetamol by a peroxidase is in the assay of glycerol by a microdialysis enzyme electrode (Murphy and Galley, 1994). Paracetamol, along with ascorbate and urate, interferes with the assay but can be removed by a peroxidase.

Transacetylation

A recently discovered metabolic pathway of paracetamol is transacetylation. In the rat, about 10 per cent of the sulphate conjugate of paracetamol contains a different acetyl group from that in the administered drug, indicating removal of the acetyl group followed by reacetylation (Nicholls *et al.*, 1995). There is concern about this pathway of metabolism because the intermediate should be the renal toxin, 4-aminophenol (para-aminophenol), and there is some evidence that both oxidative metabolism to NAPQI and deacetylation to 4-aminophenol contribute to paracetamol-induced renal toxicity in the rat (Mugford and Tarloff, 1997). The level of this futile deacetylation appears to be about 1 to 2 per cent in man (Nicholls *et al.*, 1997). Thus the extent of this pathway appears low in man, but paracetamol is administered in large doses and a minor pathway represents several milligrams per dose. This futile cycle should be considered when measuring the urinary metabolites resulting from the oxidation of paracetamol. Peroxidase activity converts 4-aminophenol to a species that forms a glutathione conjugate (Eyanagi *et al.*, 1991). This conjugate and further metabolites, such as the mercapturic acid (Figure 6.3), should be distinguished from the corresponding conjugates of paracetamol. Not surprisingly, small amounts of 4-aminophenol are metabolised to paracetamol and its conjugates by rat hepatocytes (Yan *et al.*, 2000).

Urinary clearance

Only about 3 per cent of a dose of paracetamol is excreted unchanged (Table 6.1). Because of its low molecular weight and lack of protein binding, paracetamol is freely filtered at the glomerulus but resorption is limited owing to its only moderate lipid-solubility. The result is that the renal clearance of unchanged paracetamol is quite high, at about 20 ml/min, and is reduced in proportion to the decreased creatinine clearance in renal failure (Kietzman et al., 1990). Because paracetamol is almost completely un-ionised at physiological pH values its renal clearance is not influenced by urinary pH but is increased by higher urine flows, maintaining the urine: plasma concentration ratio at approximately 9.8 in subjects with normal renal function (Prescott, 1980; Kietzman et al., 1990).

Paracetamol is used as a test compound for the non-invasive determination of several metabolic pathways. One particular measure is the metabolic ratio, which is the ratio of the amount of metabolite (glucuronide, sulphate and cysteine conjugates) excreted in urine to the amount of unchanged drug. The value of the metabolic ratio of metabolites is, however, questionable if the renal clearance of the parent drug varies with urine output (Kietzman et al., 1990; Miners et al., 1992), as is the case with paracetamol. The glucuronide and sulphate conjugates are secreted and have much higher urinary clearances than paracetamol (Duggin and Mudge, 1975), but the total body clearance of paracetamol is greater than the major metabolites (see p. 192). Paracetamol sulphate shares an hepatic transporter with several organic sulphates and sulphonates (Sakuma-Sawada et al., 1997), but it is not known whether the same transporter is involved in the renal secretion of this paracetamol metabolite.

Salivary secretion

The clearance of paracetamol in saliva is small compared to its metabolic clearance and, furthermore, the swallowed drug should be resorbed, making salivary secretion an insignificant aspect of the elimination of paracetamol. The concentrations in saliva are approximately equal to those in plasma because paracetamol is not bound to plasma proteins (Glynn and Bastain, 1973), although some poor correlations between salivary and plasma concentrations have been reported (Smith et al., 1991), particularly in patients on dialysis (Lee et al., 1996). The salivary concentrations may be dependent upon salivary flow (Kamali et al., 1992). Despite these problems, the time course of salivary concentrations of paracetamol has been used in studies on the kinetics of the drug (Mucklow et al., 1980; Miners et al., 1983; Sahajwalla and Ayres, 1991; Ray et al., 1993; Al-Obaidy et al., 1995). Although pharmacokinetic studies using saliva are convenient because the collection of blood is avoided, the method must still be validated in the type of subjects who are being studied.

HALF-LIFE AND CLEARANCE

The half-life of elimination of paracetamol is about 2 hours after therapeutic doses (Table 6.1). After about 8 hours there is a slow terminal phase of elimination, with a half-life of about 4 to 5 hours in healthy subjects (Baraka et al., 1990; Figure 6.5). This terminal half-life is much longer in patients with renal failure (Prescott et al., 1989). The late slow phase of elimination is consistent with the slow uptake and loss from several other tissues, including the brain, but the reasons for the long terminal half-life in patients with renal failure is unclear. One possibility is that the parent drug is regenerated from a metabolite, as is now well recognised for the glucuronides of several other drugs.

The total clearance of paracetamol is high, with a mean of 5 ml/min per kilogram (Table 6.1). By comparison, the renal clearances of the glucuronide and sulphate conjugates are 2 ml/min per kilogram (Slattery et al., 1987). Thus, the renal clearances of the major metabolites are lower than the total clearance of the parent drug. Consequently, the glucuronide and sulphate conjugates attain plasma concentrations that are similar to those of unchanged paracetamol (Prescott, 1980; Slattery et al., 1987;

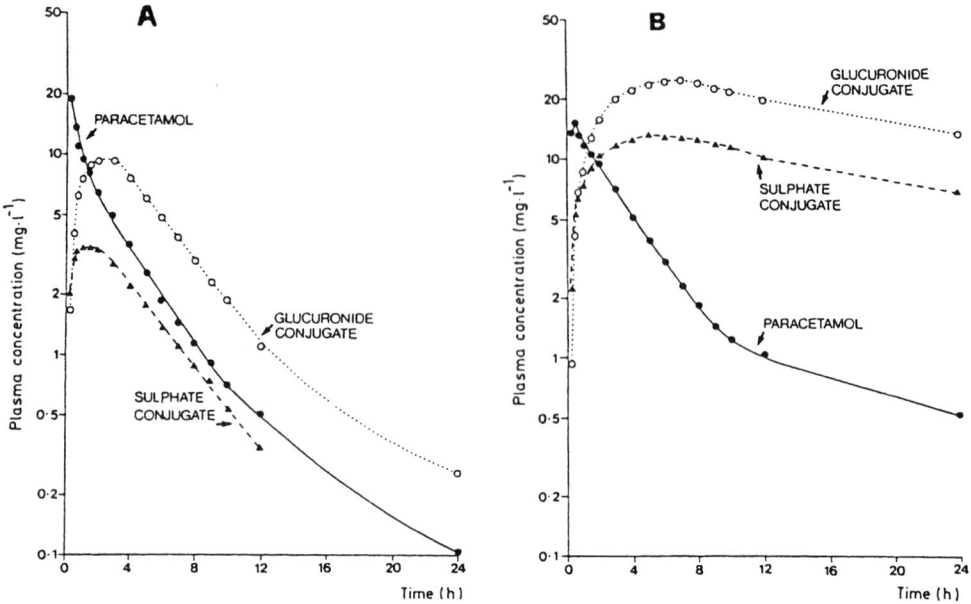

Figure 6.5   Time course of mean plasma concentrations of paracetamol and its conjugates after oral dose of 1 g of paracetamol to healthy subjects (A) and patients with moderate renal failure (B). The half-life of elimination of paracetamol is about 2 hours between 2 and 8 hours after dosage, but is longer after 8 hours after dosage, markedly so in the patients with renal failure. (Prescott *et al.*, 1992.)

Prescott *et al.*, 1989; Figure 6.5). The half-lives of elimination of the metabolites derived from glu-tathione follow a similar pattern, although the concentrations are lower. The methylthio conjugate peaks at a much later time (about 14 hours) than the parent drug and its major metabolites, possibly due to its formation after excretion into the intestine (Slattery *et al.*, 1987).

Dosage with paracetamol is high, but the total clearance is constant up to about 20 mg/kg. The clearance decreases and half-life of elimination increases, however, after higher or toxic doses (Slattery and Levy, 1979; Prescott, 1980; Clements *et al.*, 1984; Slattery *et al.*, 1987; Sahajwalla and Ayres, 1991), consistent with saturation of the most important metabolic processes or limited availability of necessary co-substrates such as sulphate and glutathione.

Because of the widespread use and toxicity of paracetamol, its pharmacokinetics have been studied by a variety of methods. Following intravenous dosage, the half-life can be determined with reasonable accuracy by measurement at only two times (2 and 6 hours, or 3 and 6 hours; Scavone *et al.*, 1990a). Of greater application is that Bayesian forecasting, following the measurement of an initial plasma concen-tration, allows the accurate prediction of plasma concentrations within the first 8 hours of an overdose, and therefore a reasonable prediction of the degree of toxicity (Gentry *et al.*, 1994).

FACTORS AFFECTING THE ELIMINATION OF PARACETAMOL

Other drugs

Many drugs and other chemicals are now known to affect the elimination of paracetamol (Tables 6.2–6.4). Interactions with experimental drugs and volatile anaesthetics are shown in Tables 6.3 and

TABLE 6.2

Effects of other drugs on the elimination of paracetamol (the effects of the volatile anaesthetics are listed in Table 6.4).

Interacting drug	Effect on paracetamol	Species	Reference
Antiepileptic drugs	Reduced bioavailability, increased glucuronidation and oxidative metabolism (all small effects)	Man	Perucca and Richens, 1979 Miners *et al.*, 1984 Bock *et al.*, 1987
Caffeine	Increased or decreased oxidative metabolism dependent upon pretreatment (see text)	Rat	Lee *et al.*, 1996
Cimetidine	Decreased hepatotoxicity	Rat, mouse	Mitchell *et al.*, 1981 Rudd *et al.*, 1981
	No effect on oxidative metabolism or the development of hepatotoxicity	Man	Critchley *et al.*, 1983 Vendemiale *et al.*, 1987 Slattery *et al.*, 1989 Burkhart *et al.*, 1995
Heparin	Half-life of paracetamol and plasma concentrations of sulphate and glucuronide conjugates increased	Rat	Scott *et al.*, 1991
Isoniazid	Induced metabolism and increased hepatotoxicity	Rat	Ryan *et al.*, 1986
	Induced oxidative metabolism which is inhibited by about 60% while dosage with isoniazid continues	Man	Epstein *et al.*, 1991 Park *et al.*, 1993 Zand *et al.*, 1993 Chien *et al.*, 1997
Methoxsalen	Decreased oxidative metabolism	Man	Amouyal *et al.*, 1987
Phenobarbitone	Increased metabolism and hepatotoxicity	Mouse	Mitchell *et al.*, 1973a
Probenecid	Markedly decreased clearance and increased half-life	Man	Abernethy *et al.*, 1985 Kamali, 1993 von Moltke *et al.*, 1993
	Non-competitive inhibition of glucuronidation *in vitro*	Rat	Turner and Brouwer, 1997
	Decreased renal secretion of sulphate conjugate but not glucuronide	Dog	Duggin and Mudge, 1975
Propranolol	Small decrease in clearance and increase in half-life	Man	Baraka *et al.*, 1990
Rifampicin	Slightly shorter half-life and increased glucuronidation	Man	Prescott *et al.*, 1981
	Oxidative metabolism not increased	Man	Manyike *et al.*, 2000
Sulfinpyrazone	Increased clearance	Man	Miners *et al.*, 1984
Zidovudine	No effect on clearance	Man	Burger *et al.*, 1994

6.4. Paracetamol is a well-tolerated drug at therapeutic doses, and in most cases the interactions with other drugs are not clinically significant – particularly if there are changes only in the non-oxidative processes. Furthermore, the effects of other drugs generally only account for a small proportion of the total variance in the clearance and half-life of paracetamol (Mucklow *et al.*, 1980). Effects on the oxidative metabolism of paracetamol are, however, of considerable interest because of potential changes in the degree of hepatotoxicity. The importance of any effect on the oxidative metabolism of paracetamol was shown in early hepatotoxicity studies in which an inducer of oxidative drug metabolism, such as phenobarbitone, increases the hepatic necrosis through increased conversion to NAPQI (Table 6.2).

CHAPTER 6

Conversely, inhibitors of oxidative drug metabolism, such as piperonyl butoxide and cimetidine, decrease the conjugation of paracetamol with glutathione and the hepatotoxicity of paracetamol in experimental animals (Table 6.2). There is, however, no significant effect in man and cimetidine is not useful in the treatment of paracetamol overdosage (Table 6.2).

Inducers of cytochrome P450 systems, the anticonvulsants phenytoin and carbamazepine, have been of particular concern because of the possibility of greater hepatotoxicity of paracetamol. The total clearance and the oxidative clearance of paracetamol are increased to approximately the same extent in patients taking anticonvulsants (Miners *et al.*, 1984; Table 6.2). Essentially, this means that the rate of the oxidative pathway is normal despite the decreased plasma concentrations. Thus, the fraction of a paracetamol dose that is oxidised is normal (approximately 10 per cent). In Chinese patients treated with phenytoin the percentage converted to oxidised metabolites actually decreases, and is normal in patients taking carbamazepine (Tomlinson *et al.*, 1996). On this basis, increased hepatotoxicity is not predicted in patients taking anticonvulsants. Nevertheless, anticonvulsants are considered to worsen the outcome of patients overdosed with paracetamol (Bray *et al.*, 1992), but it has been recommended that patients who are taking well-recognised inducers of drug metabolism and who have taken an overdose of paracetamol should be treated with N-acetyl cysteine if the plasma concentrations are 70 per cent or more of the levels that are considered predictive of severe toxicity (Proudfoot, 1982). This recommendation, made in 1982, still appears prudent.

Rifampin is an inducer of several cytochrome P450s, particularly cytochrome P3A4, but this cytochrome is not significantly involved in the oxidative metabolism of paracetamol. Rifampin therefore does not induce the oxidative metabolism of paracetamol (Manyike *et al.*, 2000).

The antituberculous drug isoniazid interacts with the oxidative metabolism of paracetamol in a complex fashion, and in a very similar fashion to alcohol (see p. 196; Table 6.4). Isoniazid apparently induces CYP2E1, although the effect is more likely to be due to stabilisation of the enzyme than to increased synthesis of the enzyme (Park *et al.*, 1993). Isoniazid, however, inhibits the oxidation of paracetamol while the administration of isoniazid is continued, and the induction of the metabolism of paracetamol is masked (Epstein *et al.*, 1991). After the cessation of treatment with isoniazid the inhibition is relieved by the elimination of isoniazid after about 1 day (Zand *et al.*, 1993), but at under 12 hours in fast acetylators of isoniazid (Chien *et al.*, 1997; Table 6.2). At this time, the oxidative metabolism is approximately doubled in the fast acetylators. There are several case reports of patients indicating that either isoniazid alone (Murphy *et al.*, 1990; Moulding *et al.*, 1991; Crippin, 1993) or combination therapy of isoniazid, rifampicin and pyrizinamide (Nolan *et al.*, 1994) potentiates the hepatotoxicity of paracetamol. Possibly, paracetamol may produce hepatotoxicity during the short period immediately after cessation of treatment with isoniazid when the oxidative metabolism of paracetamol is induced. Another possibility is that hepatotoxicity could be due to summation of the hepatotoxicity of isoniazid and paracetamol.

The effects of caffeine on the metabolism of paracetamol are complex and controversial. Caffeine has been reported both to decrease (Iqbal *et al.*, 1995) and to increase (Rainska-Giesek, 1995) the total body clearance of paracetamol. In the rat, caffeine has little or no effect on the oxidative metabolism in control animals, increases oxidative metabolism in phenobarbitone-treated animals, but inhibits this pathway in rats induced with the inducer methylcholanthrene (Lee *et al.*, 1991; Table 6.2). The action of caffeine appears to result from a direct activation of the cytochrome P450 systems, and is not due to induction through an increase in the amount of the enzymes. The interactions of caffeine with paracetamol require further evaluation in order to determine their clinical significance and also to investigate the mechanism of direct activation of the cytochrome P450 systems.

Probenecid profoundly blocks the metabolism of paracetamol in man, doubling its half-life of elimination (Table 6.2), and consequently the dosage of paracetamol should be reduced or the dosage interval increased in patients taking probenecid. Surprisingly probenecid does not inhibit the renal secretion of the glucuronide, although it does block the secretion of the sulphate conjugate.

Another marked inhibition of the clearance of paracetamol has been shown with heparin. In the rat, heparin approximately doubles the half-life of paracetamol (Table 6.2). This potential interaction has not been studied in man, but clearly requires further study.

TABLE 6.3

Effects of chemicals and experimental drugs on the elimination of paracetamol.

Interacting compound	Effect on paracetamol	Species	Reference
Dimethyl sulphoxide	Decreased oxidative metabolism and hepatotoxicity	Mouse	Jeffery *et al.*, 1988
Propylene glycol	Decreased oxidative metabolism and hepatotoxicity	Mouse	Thomsen *et al.*, 1995
Molybdate	Decreased sulphation due to depletion of sulphate and inhibition of enzymes involved in sulphate metabolism	Mouse	Gregus *et al.*, 1994
Oleanolic acid	Inhibited oxidative metabolism and hepatotoxicity	Mouse	Liu *et al.*, 1993

TABLE 6.4

Effects of environmental factors, smoking, volatile anaesthetics, alcohol and related compounds on the elimination of paracetamol.

Interacting compound	Effect on paracetamol	Species	Reference
Acute alcohol	Decreased oxidative metabolism *in vitro*	Rat, mouse, rabbit	Thummel *et al.*, 1988; Sato *et al.*, 1991
	Decreased oxidative metabolism *in vivo*	Man	Banda and Quart, 1982
Chronic alcohol	Increased total clearance	Man	Girre *et al.*, 1994
	Increased oxidative metabolism after elimination of alcohol	Man	Slattery *et al.*, 1996 Thummel *et al.*, 2000
Acetaldehyde	Decreased clearance *in vitro*	Rat, mouse, rabbit	Sato *et al.*, 1991
Disulfiram	Conflicting results possibly due to different dosage	Man	
	No effect on oxidative metabolism		Poulsen *et al.*, 1991
	Decreased oxidative metabolism		Manyike *et al.*, 2000
Flavonoids	Decreased oxidative metabolism *in vitro*	Rat	Li *et al.*, 1994
Polyvinyl chloride (environmental)	Increased clearance	Man	Smilgin *et al.*, 1993
Smoking	Increased clearance in some but not in other studies	Man	Mucklow *et al.*, 1980
		Man	Miners *et al.*, 1984 Scavone *et al.*, 1990b
	Increased glucuronidation in heavy smokers	Man	Bock *et al.*, 1987
Trichloroethylene	Increased clearance	Man	Ray *et al.*, 1993
Nitrous oxide/halothane	Unchanged clearance	Man	Lewis *et al.*, 1991
Watercress (phenethyl isothiocyanate)	Oxidative metabolism decreased by 25%	Man	Chen *et al.*, 1996

Chemicals and experimental drugs

The common solvents dimethylsulfoxide and propylene glycol inhibit the oxidative metabolism and hepatotoxicity of paracetamol (Table 6.3). These interactions should be considered in experimental studies where non-aqueous solvents are used as solvents for paracetamol which has very limited solubility in water. An interesting interaction has been found with oleanolic acid, a triterpenoid used in Chinese medicine to treat hepatitis. In mice, oleanolic acid inhibits the hepatic production of the reactive metabolite of paracetamol, NAPQI, and as a result decreases the hepatotoxicity of paracetamol (Table 6.3).

Environmental factors, smoking, alcohol and related compounds

Exposure to polyvinyl chloride and to other chlorinated compounds may increase the total clearance of paracetamol, although the pharmacokinetics of paracetamol are unchanged after recovery from anaesthesia with nitrous oxide and halothane (Table 6.4). Possibly, some chlorinated compounds in the environment may increase the conjugation of paracetamol.

The interaction between alcohol and paracetamol is very complex, and resembles the interaction with isoniazid. It has been stated that chronic alcoholics are more sensitive to overdoses of paracetamol and may even have liver damage after therapeutic doses. These claims are supported by induction of cytochrome P4502E1 by the ingestion of alcohol, but increased oxidative metabolism of paracetamol is only seen after the alcohol has been eliminated and slight induction of hepatic oxidative metabolism remains (Slattery et al., 1996; Thummel et al., 2000). While substantial levels of alcohol are present, the oxidative metabolism of paracetamol is inhibited (Table 6.4). The result is decreased hepatotoxicity, which has been observed in mice and rats (Banda and Quart, 1984; Thummel et al., 1988). At this stage, hepatotoxicity is considered possible in alcoholics at therapeutic doses of paracetamol, but 'there is insufficient evidence to support the alleged major toxic interaction' (Prescott, 2000).

There are conflicting data on the effect of smoking on the total body clearance of paracetamol, although any effect may depend upon the number of cigarettes smoked per day (Table 6.4). It has been concluded, however, that smoking and moderate chronic intake of alcohol, together with a variety of other environmental factors, including oral contraceptives, account for only about 30 per cent of the variance of paracetamol clearance. These factors 'are unlikely to be major determinants of paracetamol elimination in man' (Mucklow et al., 1980). Although smoking does not alter the pharmacokinetics of paracetamol to a great extent, it does have a much greater effect on the oxidation of other drugs through effects on cytochrome P450s other than CYP2E1.

Several flavonoids variously inhibit or induce the oxidative metabolism of paracetamol in vitro, but it is not known if sufficient are present in the diet to alter the metabolism of paracetamol. A foodstuff that does decrease the oxidative metabolism of paracetamol is watercress (Table 6.4). The active constituent is probably phenethyl isothiocyanate.

There are no reports of significant interactions of paracetamol on the clearance of other drugs. Inhibition of the metabolism of other drugs by paracetamol is, however, suggested by its plasma concentrations being close to the K_m of its oxidation by CYP3A4, and paracetamol may be an as yet unrecognised inhibitor of the oxidative metabolism of other drugs (Thummel et al., 1993).

Steroids and cytokines

Sex hormones have a clear effect on the elimination of paracetamol. Paracetamol clearance is about 20 per cent higher in males than in females, even after correction for the higher body weight of males (Wojcicki et al., 1979; Mucklow et al., 1980; Miners et al., 1983; Bock et al., 1994), but this small difference is unlikely to be clinically significant. Of greater potential significance is the approximately 50 per cent greater total clearance in females who are using oral contraceptive steroids than in females who do not take oral contraceptives (Mucklow *et al.*, 1980; Abernethy *et al.*, 1982; Miners *et al.*, 1983; Mitchell *et al.*, 1983). The difference has been ascribed to a marked increase in glucuronidation and a lesser increase in oxidative metabolism. These effects of oestrogens may have clinical consequences because of effects both on the duration of action of paracetamol and on its hepatotoxicity. There is also an interaction in the reverse direction, with the availability of ethinyloestradiol being increased by about 20 per cent due to paracetamol-induced inhibition of its sulphation (Rogers *et al.*, 1987).

Prednisolone stimulates the synthesis of glutathione after toxic doses of paracetamol in mice, and thus increases the amounts of glutathione conjugates and decreases the hepatotoxicity of paracetamol (Speck *et al.*, 1993). This observation has potential clinical significance, but has not been investigated.

Interferons have a complex effect on the metabolism of paracetamol, similar to that of alcohol. In mice the oxidative metabolism is inhibited within days of the administration of induction of interferon synthesis, but the oxidative metabolism is subsequently induced (Kalabis and Wells, 1990).

Other drug interactions

Apart from metabolic drug interactions, paracetamol has very few interactions with other drugs. It has been reported to increase slightly the prothrombin time of patients treated with warfarin (Antlitz *et al.*, 1968) but there has been considerable recent controversy about this interaction. Paracetamol has little effect on platelets at therapeutic doses, and increased bleeding with warfarin is unlikely.

Age

The clearance of paracetamol in old age is similar to that of young adults (Divoll *et al.*, 1982; Miners *et al.*, 1988; Herd *et al.*, 1991), but its clearance *in vivo* is lower in frail elderly patients (Wynne *et al.*, 1990). The latter patients should receive low doses of paracetamol but, judged from kinetic data, decreased dosage in fit elderly patients is not necessary. A case report indicates that advancing age with multiple organ insufficiency is associated with low metabolism and consequent toxicity from paracetamol (Bonkovsky *et al.*, 1994).

The metabolic clearance of paracetamol also appears low in the foetus. Glucuronidation is very low, although sulphation and oxidative activities are present. The presence of oxidative activity indicates the risk of hepatotoxicity to the foetus as well as to the mother and foetus if a pregnant woman takes an overdose of paracetamol (Rollins *et al.*, 1979).

Glucuronidation continues to be low in the neonate and does not mature for several years, while sulphation and oxidative activities are well developed (Levy *et al.*, 1975; Miller *et al.*, 1976). Total clearance is low in neonates and is particularly low in preterm neonates (28–32 weeks). Thus the mean half-life is 12 hours and is very variable in premature neonates, and consequently the dosage interval should be at least 8 hours (van Lingen *et al.*, 1999). In near- or full-term neonates less than 10 days old, the half-life is about 3 to 6 hours (Autret *et al.*, 1993; van Lingen *et al.*, 1999). It is recommended that neonates under 10 days receive a lower dosage (30 mg paracetamol/kg per day) than older infants in the treatment of fever. Older children should receive paracetamol only on weight- and age-based recommendations in order to prevent excessive plasma concentrations and possible hepatotoxicity.

Pregnancy

The clearance of paracetamol is increased during pregnancy and increases from the first to the third trimester of pregnancy, possibly because of the increasing oestrogen levels (Miners *et al.*, 1986; Rayburn *et al.*, 1986; Beaulac-Baillargeon and Rocheleau, 1994). The reported increase in clearance is up to 60 per cent but the oxidative metabolism is increased to an even greater extent (90 per cent) and the yield of oxidised metabolites is therefore increased by about 50 per cent (Miners *et al.*, 1986).

Disease states

Patients with rheumatoid arthritis may have lesser capacity to form the sulphate and glucuronide conjugates, and consequently it is suggested that they may be more at risk of hepatotoxicity than other patients (Bradley *et al.*, 1991). Confirmatory work is, however, required. One unusual rheumatoid patient was claimed to have ingested 15 to 20 g of paracetamol daily for 5 years without liver damage (Tredger *et al.*, 1995). The clearance to the sulphate conjugate was about twice normal in this patient, and the clearance by glucuronidation was lower than normal. The oxidative metabolism was slightly low, about 30 per cent lower than in normals. With the very high plasma concentrations the rate of production of oxidative metabolites must have been very high, and the lack of hepatotoxicity in this patient is not understood. In another inflammatory disease, ulcerative colitis, the sulphation of paracetamol is greatly reduced in the colonic mucosa (Ramakrishna *et al.*, 1991) but the systemic sulphation of paracetamol is not altered (Haderslev *et al.*, 1998).

An unexpected finding is that the metabolic clearance of paracetamol is increased by approximately 50 per cent by cystic fibrosis (Hutabarat *et al.*, 1991). The total clearance of paracetamol is about twice as high in patients with stomach cancer as in healthy subjects (Gawronska-Szklarz *et al.*, 1988), and the oxidative clearance is similarly higher in patients with hepatocellular (Leung and Critchley, 1991) or bladder carcinomas (Dolara *et al.*, 1988). There is no indication that paracetamol is carcinogenic in such patients, although it has been suggested 'that the use of paracetamol may contribute to an increase in the total burden of genotoxic damage in man' (Rannug *et al.*, 1995). This is, however, a contentious topic, and no genotoxic or carcinogenic risk was detected in another study (Kirkland *et al.*, 1992), although paracetamol, at concentrations slightly above therapeutic, inhibits DNA synthesis in a cell line (Richard *et al.*, 1991) – possibly through oxidation of paracetamol and subsequent reaction of NAPQI. This reaction with DNA has been detected in the livers and kidneys of mice (Hongslo *et al.*, 1994; Rogers *et al.*, 1997; see above).

The metabolic clearance of paracetamol is lowered in a variety of disease states; in hypothyroidism (Sonne *et al.*, 1990), in obstructive jaundice, and in severe but not in lower grades of liver disease in adults (Andreasen and Hutters, 1979; Forrest *et al.*, 1979; Brodie *et al.*, 1981; Froomes *et al.*, 1999) or children (Al-Obaidy *et al.*, 1995). At least occasional doses of paracetamol appear safe in patients with mild to moderate liver disease. The clearance of paracetamol is decreased by about 50 per cent in patients with cirrhosis, and is not increased by oxygen supplementation (Froomes *et al.*, 1999). The oxidative and sulphation pathways of paracetamol metabolism are higher than normal in advanced HIV infections, possibly a compensatory mechanism due to decreased glucuronidation (Esteban *et al.*, 1997), although they are not altered by zidovudine (AZT) (Table 6.2). The clearance of paracetamol appears normal in Down's syndrome (Griener *et al.*, 1990).

Because paracetamol is largely eliminated by hepatic metabolism its initial half-life and plasma concentrations are little changed in renal failure, although the late phase of elimination is very much slowed by renal failure (Figure 6.5), possibly because of regeneration of paracetamol from the glucuronide and sulphate conjugates that are retained in renal failure and consequently achieve much higher plasma concentrations than in normal patients (Prescott *et al.*, 1989; Martin *et al.*, 1993). These conjugates are, however, readily removed during dialysis (Øie *et al.*, 1975).

It is anticipated that patients with Gilbert's syndrome, a deficiency in the glucuronidation of bilirubin mainly due to an abnormality of the glucuronyl transferase UGT1A1, should have normal glucuronidation and therefore normal oxidative metabolism. Some patients, however, produce less glucuronide and, correspondingly, more oxidised metabolites than normal. The results in different studies are conflicting, but some patients with Gilbert's syndrome show decreased glucuronidation of paracetamol (de Wildt *et al.*, 1999; Esteban and Perez-Mateo, 1999). As discussed below, the oxidative pathway is associated with the hepatotoxicity of paracetamol. It is considered that Gilbert's syndrome should not increase the risk of hepatotoxicity after therapeutic dosage but may increase the severity of hepatic damage after overdoses (de Morais *et al.*, 1992a). The increase in oxidative pathways of metabolism is clear in cats and Gunn rats, who have a deficiency of glucuronidation (Savides *et al.*, 1984; de Morais *et al.*, 1992b; Court and Greenblatt, 1997).

Race

There are conflicting data on the oxidative metabolism of paracetamol in various racial groups. In a large study, the mean conversion of paracetamol to its oxidised metabolites in Ghanaian and Kenyan Africans was found to be 50 per cent lower than in Caucasians, but there were very large intersubject variations in both groups (Critchley *et al.*, 1986). There is a slightly lower (20 per cent) mean clearance in Asians than in Caucasians and, corresponding, a slightly longer mean half-life in Asians, but again the intersubject variations are large (Mucklow *et al.*, 1980). The total clearance is the same in Caucasians and Chinese, but the conversion to the oxidised metabolites is variously reported to be either the same (Osborne *et al.*, 1991) or lower (Lee *et al.*, 1992) in Chinese than in Caucasians. It is possible that the conflicting results are due to environmental factors, since the studies were conducted in different countries. A subset of both Caucasians (20 per cent) and Orientals (33 per cent) displayed extensive glucuronidation (Patel *et al.*, 1993).

Species

As is the case with several drugs, cats and Gunn rats convert little paracetamol to its glucuronide (Savides *et al.*, 1984; de Morais *et al.*, 1992b; Court and Greenblatt, 1997). Consequently, they produce large amounts of the oxidised metabolites and are therefore very susceptible to paracetamol-induced hepatotoxicity (see discussion above on Gilbert's syndrome).

Individualisation of dosage

A repeating theme in any examination of factors affecting the clearance of paracetamol is that many factors are known to alter its metabolism and pharmacokinetics, but there is still much unexplained variation. It may be possible to improve the dosage of paracetamol by individualising its dosage, but this could only be useful for patients who are taking high doses of paracetamol for the treatment of chronic pain, such as the pain of osteoarthritis. Individualising long-term dosage with the aid of measurements of plasma or salivary concentrations appears reasonable, but its value has not been tested. In the relief of the pain and discomfort of tonsillectomy in children, the EC_{50} (the plasma concentration associated with 50 per cent of maximal effect) is $3.4\,\mu g/ml$ (Anderson *et al.*, 1999), and this concentration has been chosen as the minimal target plasma concentration in the development of a sustained-release tablet of paracetamol.

METABOLISM OF PRO-DRUGS

Paracetamol has replaced the older drug phenacetin, which is rapidly oxidised (including first-pass in the liver) to paracetamol (Kuntzman *et al.*, 1977; Inaba *et al.*, 1979). Therefore, little phenacetin appears in the circulation. Phenacetin is still used experimentally as a marker of cytochrome P450IA2, and its conversion to paracetamol is induced by cigarette smoking (Kuntzman *et al.*, 1977; Kahn *et al.*, 1985). Many but not all poor metabolisers of the marker drug debrisoquine are also poor metabolisers of phenacetin (Sesardic *et al.*, 1988), showing that metabolism by cytochrome P4502D6 may be important in some subjects. The old analgesic acetanilide is hydroxylated to paracetamol (Lester and Greenberg, 1947) but, like phenacetin, is no longer used.

Several esters of paracetamol with aspirin have been prepared and tested. All are hydrolysed rapidly. Benorylate, an open ester of paracetamol and aspirin, is used clinically, while a cyclic ester has been also been synthesised and tested in rats. Both esters are rapidly hydrolysed to paracetamol and salicylate (Williams *et al.*, 1989; Marzo *et al.*, 1990). Esters have also been prepared with the non-steroidal anti-inflammatory drugs tolmetin and indomethacin. Both are hydrolysed rapidly to their constituent drugs in experimental animals (Cociglio *et al.*, 1991; Sabater *et al.*, 1993), but their value is dubious, particularly because of the differing dosage of the constituents.

Paracetamol is only slightly soluble in water, and its soluble ester with diethyl glycine (propacetamol) has been used by intravenous injection. According to the manufacturer (Laboratories Upsa), it is hydrolysed by plasma esterases within 7 minutes (see Figure 6.1).

ANALYTICAL METHODS

Paracetamol concentrations in plasma and urine have been assayed by a number of techniques (Stewart and Watson, 1987). Several methods are based on the absorbance of paracetamol in the ultra-violet region, or reactions yielding coloured products. A colorimetric qualitative method for the detection of paracetamol in urine is useful as an initial screen for overdoses (Simpson and Stewart, 1973). It is very rapid, simple, and determines whether the patient has taken any paracetamol. If it is negative, no plasma assay is required. The most widely used quantitative colorimetric method is one in which paracetamol is nitrated, and this method has been modified for use in the centrifugal analyser where small volumes of plasma or serum are used (Shihabi and David, 1984). The problem with spectro-photometric and colorimetric assays is, of course, interference by other drugs, but three more specific methods are now available in clinical laboratories; enzyme immunoassay (Hepler *et al.*, 1984), fluorescence polarisation immunoassay, and enzyme assay in which paracetamol is hydrolysed enzymatically and the resulting p-aminophenol detected colorimetrically (Hammond *et al.*, 1981).

In research on the metabolism and pharmacokinetics of paracetamol, liquid chromatographic methods for the determination of the drug and its metabolites are now widely used (Wilson *et al.*, 1982; Lau and Critchley, 1994) with one method allowing the separation of 13 urinary metabolites (Aguilar *et al.*, 1988). The chromatographic assay of the metabolites is difficult because of the lack of suitable standards but the glucuronide and sulphate conjugates are easily determined after hydrolysis by beta-glucuronidase and sulphatase (Nakamura *et al.*, 1987). In recent years, the development of mass spectrometry interfaces with HPLC has greatly assisted the detailed examination of the metabolites of paracetamol (Teffera and Abramson, 1994).

REFERENCES

Abernethy, D.R., Divoll, M., Ochs, H.R., Ameer, B. and Greenblatt, D.J. 1982, Increased metabolic clearance of acetaminophen with oral contraceptives use. *Obstetrics and Gynecology*, **60:** 338–341.

Abernethy, D.R., Greenblatt, D.J., Ameer, B. and Shader, R.I. 1985, Probenecid impairment of acetaminophen and lorazepam clearance: direct inhibition of ether glucuronide formation. *Journal of Pharmacology and Experimental Therapeutics*, **234:** 345–349.

Aguilar, M.I., Hart, S.J. and Calder, I.C. 1988, Complete separation of urinary metabolites of paracetamol and substituted paracetamols by reversed-phase ion-pair high-performance liquid chromatography. *Journal of Chromatography*, **426:** 315–333.

Al-Obaidy, S.S., McKiernan, P.J., Po, A.L.W., Glasgow, J.F.T. and Collier, P.S. 1996, Metabolism of paracetamol in children with chronic liver disease. *European Journal of Clinical Pharmacology*, **50:** 69–76.

Al-Obaidy, S.S., Po, A.L.W., McKiernan, P.J., Glasgow, J.F. and Millership, J. 1995, Assay of paracetamol and its metabolites in urine, plasma and saliva of children with chronic liver disease. *Journal of Pharmaceutical and Biomedical Analysis*, **13:** 1033–1039.

Amouyal, G., Larrey, D., Letteron, P., Geneve, J., Labbe, G., Belghiti, J. and Pessayre, D. 1987, Effects of methoxsalen on the metabolism of acetaminophen in humans. *Biochemical Pharmacology*, **36:** 2349–2352.

Anderson, B.J. and Holford, N.H. 1997, Rectal paracetamol dosing regimens: determination by computer simulation. *Paediatric Anaesthesia*, **7:** 451–455.

Anderson, B.J., Holford, N.H., Woollard, G.A. and Chan, P.L.S. 1998, Paracetamol plasma and cerebrospinal fluid pharmacokinetics in children. *British Journal of Clinical Pharmacology*, **46:** 237–243.

Anderson, B.J., Holford, N.H., Woollard, G.A., Kanagasundaram, S. and Mahadevan, M. 1999, Perioperative pharmacodynamics of acetaminophen analgesia in children. *Anesthesiology*, **90:** 411–421.

Anderson, B.J., Woollard, G.A. and Holford, N.H. 1995, Pharmacokinetics of rectal paracetamol after major surgery in children. *Paediatric Anaesthesia*, **5:** 237–242.

Andreasen, P.B. and Hutters, L. 1979, Paracetamol (acetaminophen) clearance in patients with cirrhosis of the liver. *Acta Medica Scandinavica Supplement*, **624:** 99–105.

Antlitz, A.M., Mead, J.A. and Tolentino, M.A. 1968, Potentiation of oral anticoagulant therapy by acetaminophen. *Current Therapeutic Research*, **10:** 501–507.

Anundi, N., Lahteenmaki, T., Rungren, M., Moldeus, P. and Lindros, K.O. 1993, Zonation of acetaminophen metabolism and cytochrome P450 2E1-mediated toxicity studied in isolated periportal and perivenous hepatocytes. *Biochemical Pharmacology*, **45:** 1251–1259.

Autret, E., Dutertre, J.P., Breteau, M., Jonville, A.P., Furet, Y. and Laugier, J. 1993, Pharmacokinetics of paracetamol in the neonate and infant after administration of propacetamol chlorohydrate. *Developmental Pharmacology and Therapeutics*, **20:** 129–134.

Axworthy, D.B., Hoffmann, K.-J., Streeter, A.J., Calleman, C.J. Pascoe, G.A. and Baillie, T.A. 1988, Covalent binding of acetaminophen to mouse hemoglobin. Identification of major and minor adducts formed in vivo and implications for the nature of the arylating metabolites. *Chemico-Biological Interactions*, **68:** 99–116.

Banda, P.W. and Quart, B.D. 1982, The effect of mild alcohol consumption on the metabolism of acetaminophen in man. *Research Communications in Chemical Pathology and Pharmacology*, **38:** 57–70.

Banda, P.W. and Quart, B.D. 1984, The effect of alcohol on the toxicity of acetaminophen in mice. *Research Communications in Chemical Pathology and Pharmacology*, **43:** 127–138.

Bannwarth, B., Netter, P., Lapicque, F., Gillet, P., Pere, P., Boccard, E., Royer, R.J. and Gaucher, A. 1992, Plasma and cerebrospinal fluid concentrations of paracetamol after a single intravenous dose of propacetamol. *British Journal of Clinical Pharmacology*, **34:** 79–81.

Baraka, O.Z., Truman, C.A., Ford, J.M. and Roberts, C.J.C. 1990, The effect of propranolol on paracetamol metabolism in man. *British Journal of Clinical Pharmacology*, **29:** 261–264.

Baudouin, S.V., Howdle, P., O'Grady, J.G. and Webster, N.R. 1995, Acute lung injury in fulminant hepatic failure following paracetamol poisoning. *Thorax*, **50:** 399–402.

Beaulac-Baillargeon, L. and Rocheleau, S. 1994, Paracetamol pharmacokinetics during the first trimester of human pregnancy. *European Journal of Clinical Pharmacology*, **46:** 451–454.

Benet, L.Z., Øie, S. and Schwartz, J.B. 1996, Design and optimization of dosage regimens; pharmacokinetic data. *In:* J.G. Hardman, L.E. Limbird, P.B. Molinoff and R.W. Ruddon (eds), *Goodman and Gilman's The Pharmacological Basis of Therapeutics*, 9th edn, p. 1712. New York: Pergamon Press.

Bhargava, V.O. and Hirate, J. 1989, Gastrointestinal, liver, and lung extraction ratio of acetaminophen in the rat after high dose administration. *Biopharmaceutics and Drug Disposition*, **10:** 389–396.

Bitzen, P.O., Gustafsson, B., Jostell, K.G., Melander, A. and Wahlin-Boll, E. 1981, Excretion of paracetamol in human breast milk. *European Journal of Clinical Pharmacology*, **20:** 123–125.

Blackledge, H.M., O'Farrell, J., Minton, N.A. and McLean, A.E. 1991, The effect of therapeutic doses of paracetamol on sulphur metabolism in man. *Human and Experimental Toxicology*, **10:** 159–165.

Bock, K.W., Forster, A., Gschaidmeier, H., Brück, M., Münzel, P., Schareck, W., Fournel-Gigleux, S. and Burchell, B. 1993, Paracetamol glucuronidation by recombinant rat and human phenol UDP-glucuronosyltransferases. *Biochemical Pharmacology*, **45:** 1809–1814.

Bock, K.W., Schrenk, D., Forster, A., Griese, E.U., Morike, K., Brockmeier, D. and Eichelbaum, M. 1994, The influence of environmental and genetic factors on CYP2D6, CYP1A2 and UDP-glucuronyltransferases in man using sparteine, caffeine, and paracetamol as probes. *Pharmacogenetics*, **4:** 209–218.

Bock, K.W., Wiltfang, J., Blume, R., Ullrich, D. and Bircher, J. 1987, Paracetamol as a test drug to determine glucuronide formation in man. Effects of inducers and of smoking. *European Journal of Clinical Pharmacology*, **31:** 677–683.

Bonkovsky, H.L., Kane, R.E., Jones, D.P., Galinsky, R.E. and Banner, B. 1994, Acute hepatic and renal toxicity from low doses of acetaminophen in the absence of alcohol abuse or malnutrition: evidence for increased susceptibility to drug toxicity due to cardiopulmonary and renal insufficiency. *Hepatology*, **19:** 1141–1148.

Bradley, H., Waring, R.H., Emery, P. and Arthur, V. 1991, Metabolism of low-dose paracetamol in patients with rheumatoid arthritis. *Xenobiotica*, **21:** 689–693.

Bray, G.P., Harrison, P.M., O'Grady, J.G., Tredger, J.M. and Williams, R. 1992, Long-term anticonvulsant therapy worsens outcome in paracetamol-induced fulminant hepatic failure. *Human and Experimental Toxicology*, **11:** 265–270.

Brodie, B.B. and Axelrod, J. 1949, The fate of acetophenetidin (phenacetin) in man and methods for the estimation of acetophenetidin and its metabolites in biological material. *Journal of Pharmacology and Experimental Therapeutics*, **97:** 58–67.

Brodie, M.J., Boobis, A.R., Hampden, C., McPherson, G.A., Benjamin, I.S. and Blumgart, L.H. 1981, Antipyrine and paracetamol metabolism in obstructive jaundice. *British Journal of Clinical Pharmacology*, **13:** 277P–278P.

Brown, R.D., Wilson, J.T., Kearns, G.L., Eichler, V.F., Johnson, V.A. and Bertrand, K.M. 1992, Single-dose pharmacokinetics of ibuprofen and acetaminophen in febrile children. *Journal of Clinical Pharmacology*, **32:** 231–241.

Brune, K., Rainsford, K.D. and Schweitzer, A. 1980, Biodistribution of mild analgesics. *British Journal of Clinical Pharmacology*, **10:** 279S–284S.

Burger, D.M., Meenhorst, P.L., Underberg, W.J., van der Heijde, J.F., Koks, C.H. and Beijnen, J.H. 1994, Short-term, combined use of paracetamol and zidovudine does not alter the pharmacokinetics of either drug. *Netherlands Journal of Medicine*, **44:** 161–165.

Burkhart, K.K., Janco, N., Kulig, K.W., Rumack, B.H. 1995, Cimetidine as adjunctive treatment for acetaminophen overdose. *Human and Experimental Toxicology*, **14:** 299–304.

Chamberlain, J.M., Gorman, R.L., Oderda, G.M., Klein-Schwartz, W. and Klein, B.L. 1993, Use of activated charcoal in a simulated poisoning with acetaminophen: a new loading dose for N-acetylcysteine? *Annals of Emergency Medicine*, **22:** 1398–1402.

Chen, L., Mohr, S.N. and Yang, C.S. 1996, Decrease of plasma and urinary oxidative metabolites of acetaminophen after consumption of watercress by human volunteers. *Clinical Pharmacology and Therapeutics*, **60:** 651–660.

Chen, T.S., Ritchie, J.P. and Lang, C.A. 1990, Life span profiles of glutathione and acetaminophen detoxification. *Drug Metabolism and Disposition*, **18:** 882–887.

Chen, W., Koenigs, L.L., Thompson, S.J., Peter, R.M., Rettie, A.E., Trager, W.F. and Nelson, S.D. 1998, Oxidation of acetaminophen to its toxic quinone imine and nontoxic catechol metabolites by baculovirus-expressed and purified human cytochromes P450 2E1 and 2A6. *Chemical Research in Toxicology*, **11:** 295–301.

Chien, J.Y., Peter, R.M., Nolan, C.M., Wartell, C., Slattery, J.T., Nelson, S.D., Carithers, R.L. and Thummel, K.E. 1997, Influence of polymorphic N-acetyltransferase phenotype on the inhibition and induction of acetaminophen bioactivation with long-term isoniazid. *Clinical Pharmacology and Therapeutics*, **61:** 24–34.

Chyka, P.A., Holley, J.E., Mandrell, T.D. and Sugathan, P. 1995, Correlation of drug pharmacokinetics and effectiveness of multiple-dose activated charcoal therapy. *Annals of Emergency Medicine*, **25:** 356–362.

Clements, J.A., Critchley, J.A.J.H. and Prescott, L.F. 1984, The role of sulphate conjugation in the metabolism and disposition of oral and intravenous paracetamol in man. *British Journal of Clinical Pharmacology*, **18:** 481–485.

Cociglio, M., Bres, J., Sauvaire, D., Alric, R. and Richard, M. 1991, Pharmacokinetics of an indomethacin pro-drug: apyramide after intravenous administration in dog. *European Journal of Drug Metabolism and Pharmacokinetics*, **16:** 275–280.

Cooper, S.A., Schachtel, B.P., Goldman, E., Gelb, S. and Cohn, P. 1989, Ibuprofen and acetaminophen in the relief of acute pain: a randomized, double-blind, placebo-controlled study. *Journal of Clinical Pharmacology*, **29:** 1026–1030.

Corbett, M.D., Corbett, B.R., Hannothiaux, M.-H. and Quintana, S.J. 1989, Metabolic activation and nucleic acid binding of acetaminophen and related arylamine substrates by the respiratory burst of human granulocytes. *Chemical Research in Toxicology*, **2:** 260–266.

Court, M.H. and Greenblatt, D.J. 1997, Molecular basis for deficient acetaminophen glucuronidation in cats. An interspecies comparison of enzyme kinetics in liver microsomes. *Biochemical Pharmacology*, **53:** 1041–1047.

Craig, P.N. 1990, Drug compendium. *In:* C. Hansch, P.G. Sammes and J.B. Taylor (eds), *Comprehensive Medicinal Chemistry*, p. 245. Oxford: Pergamon.

Crippin, J.S. 1993, Acetaminophen hepatotoxicity: potentiation by isoniazid. *American Journal of Gastroenterology*, **88:** 590–592.

Critchley, J.A.J.H., Dyson, E.H., Scott, A.W., Jarvie, D.R. and Prescott, L.F. 1983, Is there a place for cimetidine or ethanol in the treatment of paracetamol poisoning? *Lancet*, **1:** 1375–1376.

Critchley, J.A.J.H., Nimmo, G.R., Gregson, C.A., Woolhouse, N.M. and Prescott, L.F. 1986, Intersubject and ethnic differences in paracetamol metabolism. *British Journal of Clinical Pharmacology*, **22:** 649–657.

Cummings, A.J., King, M.L. and Martin, B.K. 1967, A kinetic study of drug elimination: the excretion of paracetamol and its metabolites in man. *British Journal of Pharmacology and Chemotherapy*, **29:** 150–157.

Dawson, J., Knowles, R.G. and Pogson, C.I. 1991, Quantitative studies of sulphate conjugation by isolated rat liver cells using [35S]sulphate. *Biochemical Pharmacology*, **42:** 45–49.

Dawson, J., Knowles, R.G. and Pogson, C.I. 1992, Measurement of glucuronidation by isolated rat liver cells using [14C] fructose. *Biochemical Pharmacology*, **43:** 971–978.

de Morais, S.M.F., Uetrecht, J.P. and Wells, P.G. 1992a, Decreased glucuronidation and increased bioactivation of acetaminophen in Gibert's syndrome. *Gastroenterology*, **102:** 577–586.

de Morais, S.M.F., Uetrecht, J.P. and Wells, P.G. 1992b, Biotransformation and toxicity of acetaminophen in congenic RHA rats with or without a hereditary deficiency in bilirubin UDP-glucuronsyltransferase. *Toxicology and Applied Pharmacology*, **117:** 81–87.

de Wildt, S.N., Kearns, G.L., Leeder, J.S. and van den Anker, J.N. 1999, Glucuronidation in humans. Pharmacogenetic and development aspects. *Clinical Pharmacokinetics*, **36:** 439–452.

Divoll, M., Abernethy, D.R., Ameer, B. and Greenblatt, D.J. 1982, Acetaminophen kinetics in the elderly. *Clinical Pharmacology and Therapeutics*, **31:** 151–156.

Dolara, P., Lodovici, M., Salvadori, M., Saltutti, C., Delle Rose, A., Selli, C. and Kriebel, D. 1988, Variations of cortisol hydroxylation and paracetamol metabolism in patients with bladder carcinoma. *British Journal of Urology*, **62:** 419–426.

Duggin, G.G. and Mudge, G.H. 1975, Renal tubular transport of paracetamol and its conjugates in the dog. *British Journal of Pharmacology*, **54:** 359–366.

Dybing, E., Holme, J.A., Gordon, W.P., Soderlund, E.J., Dahlin, D.C. and Nelson, S.D. 1984, Genotoxicity studies with paracetamol. *Mutation Research*, **138:** 21–32.

Epstein, M.M., Nelson, S.D., Slattery, J.T., Kalhorn, T.F., Wall, R.A. and Wright, J.M. 1991, Inhibition of the metabolism of paracetamol by isoniazid. *British Journal of Clinical Pharmacology*, **31:** 139–142.

Esteban, A. and Perez-Mateo, M. 1999, Heterogeneity of paracetamol metabolism in Gilbert's syndrome. *European Journal of Drug Metabolism and Pharmacokinetics*, **24:** 9–13.

Esteban, A., Perez-Mateo, M., Boix, V., Gonzalez, M., Portilla, J. and Mora, A. 1997, Abnormalities in the metabolism of acetaminophen in patients infected with the human immunodeficiency virus (HIV). *Methods and Findings in Experimental and Clinical Pharmacology*, **19:** 129–132.

Eyanagi, R., Hisanari, Y. and Shigematsu, H. 1991, Studies of paracetamol/phenacetin toxicity: isolation and characterization of p-aminophenol-glutathione conjugate. *Xenobiotica*, **21:** 793–803.

Fischer, V., West, P.R., Harman, L.S. and Mason, R.P. 1985, Free radical metabolites of acetaminophen and a dimethylated derivative. *Environmental Health Perspectives*, **64:** 127–137.

Flinn, F.B. and Brodie, B.B. 1948, The effect on the pain threshold of N-acetyl-p-aminophenol, a product derived in the body from acetanilide. *Journal of Pharmacology and Experimental Therapeutics*, **94:** 76–77.

Forrest, J.A., Adriaenssens, P., Finlayson, N.D. and Prescott, L.F. 1979, Paracetamol metabolism in chronic liver disease. *European Journal of Clinical Pharmacology*, **15:** 427–431.

Fort, D.J., Rayburn, J.R. and Bantle, J.A. 1992, Evaluation of acetaminophen-induced developmental toxicity using FETAX. *Drug and Chemical Toxicology*, **15:** 329–350.

Froomes, P.R., Morgan, D.J., Smallwood, R.A. and Angus, P.W. 1999, Comparative effects of oxygen supplementation on theophylline and acetaminophen clearance in human cirrhosis. *Gastroenterology*, **117:** 1257–1259.

Gaudreault, P., Guay, J., Nicol, O. and Dupuis, C. 1988, Pharmacokinetics and clinical efficacy of intrarectal solution of acetaminophen. *Canadian Journal of Anaesthesia*, **35:** 149–152.

Gawronska-Szklarz, B., Gabriel, J. and Wojcicki, J. 1988, Pharmacokinetics of paracetamol in patients with stomach cancer. *Polish Journal of Pharmacology and Pharmacy*, **40:** 41–45.

Gemborys, M.W. and Mudge, G.H. 1981, Formation and disposition of the minor metabolites of acetaminophen in the hamster. *Drug Metabolism and Disposition*, **9:** 340–351.

Genter, M.B., Liang, H.C., Gu, J., Ding, X., Negishi, M., McKinnon, R.A. and Nebert, D.W. 1998, Role of CYP2A5 and 2G1 in acetaminophen metabolism and toxicity in the olfactory mucosa of the CYP1a2(−/−) mouse. *Biochemical Pharmacology*, **55:** 1819–1826.

Gentry, C.A., Paloucek, F.P. and Rodvold, K.A. 1994, Prediction of acetaminophen concentrations in overdose patients using a Bayesian pharmacokinetic model. *Journal of Toxicology – Clinical Toxicology*, **32:** 17–30.

Girre, C., Hispard, E., Palombo, S., N'Guyen, C. and Dally, S. 1994, Increased metabolism of acetaminophen in chronically alcoholic patients. *Alcoholism, Clinical and Experimental Research*, **17:** 170–173.

Glynn, J.P. and Bastain, W. 1973, Salivary excretion of paracetamol in man. *Journal of Pharmacy and Pharmacology*, **25:** 420–421.

Graham, G.G., Milligan, M.K., Day, R.O., Williams, K.M. and Ziegler, J.B. 1997, Therapeutic considerations from pharmacokinetics and metabolism: ibuprofen and paracetamol. *In:* K.D. Rainsford and M.C. Powanda (eds), S*afety and Efficacy of Non-prescription (Over-the-Counter) Analgesics and NSAIDs.* pp. 72–92. Dordrecht: Kluwer Press.

Graham, G.G., Day, R.O., Milligan, M.K., Ziegler, J.B. and Kettle, A.J. 1999, Current concepts of the actions of paracetamol (acetaminophen) and NSAIDs. *Inflammopharmacology*, **7:** 255–263.

Gramatte, T. and Richter, K. 1994, Paracetamol absorption from different sites in the human small intestine. *British Journal of Clinical Pharmacology*, **37:** 608–611.

Gregus, Z., Madhu, C. and Klaassen, C.D. 1988, Species variation in toxication and detoxication of acetaminophen in vivo: a comparative study of biliary and urinary excretion of acetaminophen metabolites. *Journal of Pharmacology and Experimental Therapeutics*, **244:** 91–99.

Gregus, Z., Oguro, T. and Klaassen, C.D. 1994, Nutritionally and chemically induced impairment of sulfate activation and sulfation of xenobiotics in vivo. *Chemico-Biological Interactions*, **92:** 169–177.

Griener, J.C., Msall, M.E., Cooke, R.E. and Corcoran, G.B. 1990, Noninvasive determination of acetaminophen disposition in Down's syndrome. *Clinical Pharmacology and Therapeutics*, **48:** 520–528.

Haderslev, K.V., Sonne, J., Poulsen, H.E. and Loft, S. 1998, Paracetamol metabolism in patients with ulcerative colitis. *British Journal of Clinical Pharmacology*, **46:** 513–516.

Hammond, P.M., Scawen, M.D. and Price, C.P. 1981, Enzyme-based paracetamol estimation. *Lancet*, **1:** 391–392.

Hart, S.G., Beierschmitt, W.P., Wyand, D., Khairallah, E.A. and Cohen, S.D. 1994, Acetaminophen nephrotoxicity in CD-1 mice. I. Evidence of a role for in situ activation in selective covalent binding and toxicity. *Toxicology and Applied Pharmacology*, **126:** 267–275.

Hassig, S.R., Linscheer, W.G., Murthy, U.K., Miller, C., Banerjee, A., Levine, L., Wagner, K. and Oates, R.P. 1993, Effects of PEG-electrolyte (Colyte) lavage on serum acetaminophen concentrations. A model for treatment of acetaminophen overdose. *Digestive Diseases and Sciences*, **38:** 1395–1401.

Heading, R.C., Nimmo, J., Prescott, L.F. and Tothill, P. 1973, The dependence of paracetamol absorption on the rate of gastric emptying. *British Journal of Pharmacology*, **47:** 415–421.

Henderson, C.J., Wolf, C.R., Kitteringham, N., Powell, H. and Park, B.K. 2000, Increased resistance to acetaminophen hepatotoxicity in mice lacking glutathione S-transferase. *Proceedings of the National Academy of Sciences USA*, **97:** 12741–12745.

Hepler, B., Weber, J., Sutheimer, C. and Sunshine, I. 1984, Homogenous enzyme immunoassay of acetaminophen in serum. *American Journal of Clinical Pathology*, **81:** 602–610.

Herd, B., Wynne, H., Wright, P., James, O. and Woodhouse, K. 1991, The effect of age on glucuronidation and sulphation of paracetamol by human liver fractions. *British Journal of Clinical Pharmacology*, **32:** 768–770.

Hindmarsh, K.W., Mayers, D.J., Wallace, S.M., Danilkewich, A. and Ernst, A. 1991, Increased serum sulfate concentrations in man due to environmental factors: effects on acetaminophen metabolism. *Veterinary and Human Toxicology*, **33:** 441–445.

Hinson, J.A., Pohl, L.R., Monks, T.J. and Gillette, J.R. 1981, Acetaminophen-induced hepatotoxicity. *Life Sciences*, **29:** 107–116.

Hinson, J.A., Roberts, D.W., Benson, R.W., Dalhoff, K., Loft, S. and Poulsen, H.E. 1990, Mechanism of paracetamol toxicity. *Lancet*, **1:** 732.

Hoffman, D.A., Wallace, S.M. and Verbeeck, R.K. 1990, Circadian rhythm of serum sulfate levels in man and acetaminophen pharmacokinetics. *European Journal of Clinical Pharmacology*, **39:** 143–148.

Hoffmann, K.J., Axworthy, D.B. and Baillie, T.A. 1990, Mechanistic studies on the metabolic activation of acetaminophen in vivo. *Chemical Research in Toxicology*, **3:** 204–211.

Hoivik, D.J., Manautou, J.E., Tveit, A., Hart, S.G., Khairallah, E.A. and Cohen, S.D. 1995, Gender-related differences in susceptibility to acetaminophen-induced protein arylation and nephrotoxicity in the CD-1 mouse. *Toxicology and Applied Pharmacology*, **130:** 257–271.

Hongslo, J.K., Smith, C.V., Brunborg, G., Soderlund, E.J. and Holme, J.A. 1994, Genotoxicity of paracetamol in mice and rats. *Mutagenesis*, **9:** 93–100.

Hutabarat, R.M., Unadkat, J.D., Kushmerick, P., Aitken, M.L., Slattery, J.T. and Smith, A.L. 1991, Disposition of drugs in cystic fibrosis. III. Acetaminophen. *Clinical Pharmacology and Therapeutics*, **50:** 695–701.

Inaba, T., Mahon, W.A. and Stone, R.M. 1979, Phenacetin concentrations in portal and hepatic venous blood in man. *International Journal of Clinical Pharmacology and Biopharmacy*, **17:** 371–374.

Iqbal, N., Ahmad, B., Janbaz, K.H., Gilani, A.U. and Niazi, S.K. 1995, The effect of caffeine on the pharmacokinetics of acetaminophen in man. *Biopharmaceutics and Drug Disposition*, **16:** 481–487.

Jayasinghe, K.S.A., Roberts, C.J.C. and Read, A.E. 1986, Is biliary excretion of paracetamol significant in man?' *British Journal of Clinical Pharmacology*, **22:** 363–366.

Jeffery, E.H., Arndt, K. and Haschek, W.M. 1988, Mechanism of inhibition of hepatic bioactivation of paracetamol by dimethyl sulfoxide. *Drug Metabolism and Drug Interactions*, **6:** 413–424.

Jollow, D.J., Thorgeirsson, S.S., Potter, W.Z., Hashimoto, M. and Mitchell, J.R. 1974, Acetaminophen-induced hepatic necrosis. VI. Metabolic disposition of toxic and nontoxic doses of acetaminophen. *Pharmacology*, **12:** 251–271.

Kahn, G.C., Boobis, A.R. and Brodie, M.J. 1985, Phenacetin O-de-ethylase: an activity of a cytochrome P450 showing genetic linkage with that catalysing the 4-hydroxylation. *British Journal of Clinical Pharmacology*, **20:** 67–76.

Kalabis, G.M. and Wells, P.G. 1990, Biphasic modulation of acetaminophen bioactivation and hepatotoxicity by pretreatment with the interferon inducer polyinosinic-polycytidylic acid. *Journal of Pharmacology and Experimental Therapeutics*, **255:** 1408–1419.

Kamali, F. 1993, The effect of probenecid on paracetamol metabolism and pharmacokinetics. *European Journal of Clinical Pharmacology*, **45:** 551–553.

Kamali, F., Edwards, C. and Rawlins, M.D. 1992, The effect of pirenzipine on gastric emptying and salivary flow rate: constraints on the use of saliva paracetamol concentrations for the determination of paracetamol pharmacokinetics. *British Journal of Clinical Pharmacology*, **33**: 309–312.

Kelley, M.T., Walson, P.D., Edge, J.H., Cox, S. and Mortensen, M.E. 1992, Pharmacokinetics and pharmacodynamics of ibuprofen isomers and acetaminophen in febrile children. *Clinical Pharmacology and Therapeutics*, **52**: 181–189.

Kietzmann, D., Bock, K.W., Krahmer, B., Kettler, D. and Bircher, J. 1990, Paracetamol test: modification by renal function, urine flow and pH. *European Journal of Clinical Pharmacology*, **39**: 245–251.

Kim, H.J., Rozman, P., Madhu, C. and Klaassen, C.D. 1992, Homeostasis of sulphate and 3′-phosphoadenosine 5′-phosphosulfate in rats after acetaminophen administration. *Journal of Pharmacology and Experimental Therapeutics*, **261**: 1015–1021.

Kirkland, D.J., Dresp, J.H., Marshall, R.R., Baumeister, M., Gerloff, C. and Gocke, E. 1992, Normal chromosomal aberration frequencies in peripheral lymphocytes of healthy human volunteers exposed to a maximum daily dose of paracetamol in a double blind trial. *Mutation Research*, **279**: 181–194.

Kociscova, J., Rossner, P., Binkova, B., Bavorova, H. and Sram, R.J. 1988, Mutagenicity studies on paracetamol in human volunteers: I. Cytogenetic analysis of peripheral lymphocytes and lipid peroxidation in plasma. *Mutation Research*, **209**: 161–165.

Koymans, L., Donné-op den Kelder, G.M., te Koppele, J.M. and Vermeulen, N.P.E. 1993, Generalized cytochrome P450-mediated oxidation and oxygenation reactions in aromatic substrates with activated N-H, O-H, or S-H substituents. *Xenobiotica*, **23**: 633–648.

Kuntzman, R., Pantuck, E.J., Kaplan, S.A. and Conney, A.H. 1977, Phenacetin metabolism: effect of hydrocarbons and cigarette smoking. *Clinical Pharmacology and Therapeutics*, **22**: 757–764.

Landin, J.S., Cohen, S.D. and Khairallah, E.A. 1996, Identification of a 54-kDa mitochondrial acetaminophen-binding protein as aldehyde dehydrogenase. *Toxicology and Applied Pharmacology*, **141**: 299–307.

Lau, G.S. and Critchley, J.A.J.H. 1994, The estimation of paracetamol and its major metabolites in both plasma and urine by a single high-performance liquid chromatographic assay. *Journal of Pharmaceutical and Biomedical Analysis*, **12**: 1563–1572.

Lauterburg, B.H. 2002, Analgesics and glutathione. *American Journal of Therapeutics*, **9**: 225–233.

Lee, C.A., Thummel, K.E., Kalhorn, T.F., Nelson, S.D. and Slattery, J.T. 1991, Activation of acetaminophen-reactive formation by methylxanthines and known cytochrome P450 activators. *Drug Metabolism and Disposition*, **19**: 966–971.

Lee, H.S., Ti, T.Y., Koh, Y.K. and Prescott, L.F. 1992, Paracetamol elimination in Chinese and Indians in Singapore. *European Journal of Clinical Pharmacology*, **43**: 81–84.

Lee, H.S., Ti, T.Y., Lye, W.C., Khoo, Y.M. and Tan, C.C. 1996, Paracetamol and its metabolites in saliva and plasma in chronic dialysis patients. *British Journal of Clinical Pharmacology*, **41**: 41–47.

Lester, D. and Greenberg, L.A. 1947, The metabolic fate of acetanilid and other aniline derivatives: II. Major metabolites of acetanilid appearing in the blood. *Journal of Pharmacology and Experimental Therapeutics*, **90**: 68–75.

Leung, N.W.Y. and Critchley, J.A.J.H. 1991, Increased oxidative metabolism of paracetamol in patients with hepatocellular carcinoma. *Cancer Letters*, **57**: 45–48.

Levy, G. 1987, Pharmacokinetic analysis of the analgesic effect of a second dose of acetaminophen in humans. *Journal of Pharmaceutical Sciences*, **76**: 88–89.

Levy, G., Khanna, N.N., Soda, D.M., Tzuzuki, O. and Stern, L. 1975, Pharmacokinetics of acetaminophen in the human neonate: formation of acetaminophen glucuronide and sulfate in relation to plasma bilirubin concentration and D-glucaric acid excretion. *Pediatrics*, **55**: 818–825.

Lewis, R.P., Dunphy, J.A. and Reilly, C.S. 1991, Paracetamol metabolism after general anaesthesia. *European Journal of Anaesthesiology*, **8**: 445–450.

Li, Y., Wang, E., Patten, C.J., Chen, L. and Yang, C.S. 1994, Effects of flavonoids on cytochrome P450-dependent acetaminophen metabolism in rats and human liver microsomes. *Drug Metabolism and Disposition*, **22**: 566–571.

Liu, J., Liu, Y., Madhu, C. and Klaasen, C.D. 1993, Protective effects of oleanolic acid on acetaminophen-induced hepatotoxicity in mice. *Journal of Pharmacology and Experimental Therapeutics*, **266**: 1607–1613.

Liu, J., Sato, C. and Marumo, F. 1991, Characterization of the acetaminophen-glutathione conjugation reaction by liver microsomes: species difference in the effects of acetone. *Toxicology Letters*, **56**: 269–274.

Manyike, P.T., Kharasch, E.D., Kalhorn, T.F. and Slattery, J.T. 2000, Contribution of CYP2E1 and CYP3A to acetaminophen reactive metabolite formation. *Clinical Pharmacology and Therapeutics*, **67**: 275–282.

Martin, U., Temple, R.M., Winney, R.J. and Prescott, L.F. 1993, The disposition of paracetamol and its conjugates during multiple dosing in patients with end-stage renal failure maintained on haemodialysis. *European Journal of Clinical Pharmacology*, **45**: 141–145.

Marzo, A., Quadro, G., Treffner, E., Ripamonti, M., Meroni, G. and Lucarelli, C. 1990, High-pressure liquid chromatographic evaluation of cyclic paracetamol-acetylsalicylate and its active metabolites with results of a comparative pharmacokinetic investigation in the rat. *Arzneimittel-Forschung*, **40**: 813–817.

Mason, R.P. and Fischer, V. 1986, Free radicals of acetaminophen: their subsequent reactions and toxicological significance. *Federation Proceedings*, **45**: 2493–2499.

Mason, R.P. and Fisher, V. 1992, Possible role of free radical formation in drug-induced agranulocytosis. *Drug Safety*, **7 (Suppl. 1)**: 45–50.

McPhail, M.E., Knowles, R.G., Salter, M., Dawson, J., Burchell, B. and Pogson, C.I. 1993, Uptake of acetaminophen (paracetamol) by isolated rat liver cells. *Biochemical Pharmacology*, **45**: 1599–1604.

Miller, R.P., Roberts, R.J. and Fischer, L.J. 1976, Acetaminophen elimination kinetics in neonates, children and adults. *Clinical Pharmacology and Therapeutics*, **19**: 284–294.

Miners, J.O., Attwood, J. and Birkett, D.J. 1983, Influence of sex and oral contraceptive steroids on paracetamol metabolism. *British Journal of Clinical Pharmacology*, **16**: 503–509.

Miners, J.O., Attwood, J. and Birkett, D.J. 1984, Determinants of acetaminophen metabolism: effect of inducers and inhibitors of drug metabolism on acetaminophen's metabolic pathways. *Clinical Pharmacology and Therapeutics*, **35**: 480–486.

Miners, J.O., Osborne, N.J., Tonkin, A.L. and Birkett, D.J. 1992, Perturbation of paracetamol urinary metabolic ratios by urine flow rate. *British Journal of Clinical Pharmacology*, **34**: 359–362.

Miners, J.O., Penhall, R., Robson, R.A. and Birkett, D.J. 1988, Comparison of paracetamol metabolism in young adult and elderly males. *European Journal of Clinical Pharmacology*, **35**: 157–160.

Miners, J.O., Robson, R.A. and Birkett, D.J. 1986, Paracetamol metabolism in pregnancy. *British Journal of Clinical Pharmacology*, **22**: 359–362.

Mitchell, J.R., Jollow, D.J., Potter, W.Z., Davis, D.C., Gillette, J.R. and Brodie, B.B. 1973a, Acetaminophen-induced hepatic necrosis. I. Role of drug metabolism. *Journal of Pharmacology and Experimental Therapeutics*, **187**: 185–194.

Mitchell, J.R., Jollow, D.J., Potter, W.Z., Gillette, J.R. and Brodie, B.B. 1973b, Acetaminophen-induced hepatic necrosis. IV. Protective role of glutathione. *Journal of Pharmacology and Experimental Therapeutics*, **187**: 211–217.

Mitchell, M., Hanew, T., Meredith, C.G. and Schenker, S. 1983, Effects of oral contraceptive steroids on acetaminophen metabolism and elimination. *Clinical Pharmacology and Therapeutics*, **34**: 48–53.

Mitchell, M.C., Schenker, A., Avant, G.R. and Speeg, K.V. 1981, Cimetidine protects against acetaminophen hepatotoxicity in rats. *Gastroenterology*, **81**: 1052–1060.

Miyauchi, T., Ishikawa, M., Tashiro, S., Hisaeda, H., Nagasawa, H. and Himeno, K. 1997, Acetaminophen absorption test as a marker of small bowel transplant rejection. *Transplantation*, **63**: 1179–1182.

Moreau, X., Le Quay, L., Granry, J.C., Boishardy, N. and Delhumeau, A. 1993, Pharmacokinetics of paracetamol in the cerebrospinal fluid in the elderly. *Therapie*, **48**: 393–396.

Moulding, T.S., Redeker, A.G. and Kanel, G.C. 1991, Acetaminophen, isoniazid and hepatic toxicity. *Annals of Internal Medicine*, **114**: 431.

CHAPTER 6

Mucklow, J.C., Fraser, H.S., Bulpitt, C.J., Kahn, C., Mould, G. and Dollery, C.T. 1980, Environmental factors affecting paracetamol metabolism in London factory and office workers. *British Journal of Clinical Pharmacology*, **10:** 67–74.

Mugford, C.A. and Tarloff, J.B. 1997, The contribution of oxidation and deacetylation to acetaminophen nephrotoxicity in female Sprague–Dawley rats. *Toxicology Letters*, **93:** 15–22.

Müller, M., Schmid, R., Georgopoulos, A., Buxbaum, A., Wasicek, C. and Eichler, H.G. 1995, Application of microdialysis to clinical pharmacokinetics in humans. *Clinical Pharmacology and Therapeutics*, **57:** 371–380.

Murphy, L.J. and Galley, P.T. 1994, Measurement in vitro of human plasma glycerol with a hydrogen peroxide detecting microdialysis enzyme electrode. *Analytical Chemistry*, **66:** 4345–4353.

Murphy, R., Swartz, R. and Watkins, P.B. 1990, Severe acetaminophen toxicity in a patient receiving isoniazid. *Annals of Internal Medicine*, **113:** 799–800.

Mutlib, A.E., Shockcor, J., Espina, R., Graciani, N., Du, A. and Gan, L.S. 2000, Disposition of glutathione conjugates in rats by a novel glutamic acid pathway: characterization of unique peptide conjugates by liquid chromatography/mass spectrometry and liquid chromatography/NMR. *Journal of Pharmacology and Experimental Therapeutics*, **294:** 735–745.

Myers, T.B., Thummel, K.E., Kalhorn, T.F. and Nelson, S.D. 1994, Preferred orientations in the binding of 4′-hydroxyacetanilide (acetaminophen) to cytochrome P450 1A1 and 2B1 isoforms as determined by 13C- and 15N-NMR relaxation studies. *Journal of Medicinal Chemistry*, **37:** 860–867.

Naga Rani, M.A., Joseph, T., Narayanan, R. 1989, Placental transfer of paracetamol. *Journal of the Indian Medical Association*, **87:** 182–183.

Nakamura, J., Baba, S., Nakamura, T., Sasaki, H. and Shibasaki, J. 1987, A method for the preparation of calibration curves for acetaminophen glucuronide and acetaminophen sulfate in rabbit urine without the use of authentic compounds in high-performance liquid chromatography. *Journal of Pharmacobio-Dynamics*, **10:** 673–677.

Nicholls, A.W., Caddick, S., Wilson, I.D., Farrant, R.D., Lindon, J.C. and Nicholson, J.K. 1995, High resolution NMR spectroscopic studies on the metabolism and futile deacetylation of 4-hydroxy acetanilide (paracetamol) in the rat. *Biochemical Pharmacology*, **49:** 1155–1164.

Nicholls, A.W., Farrant, R.D., Shockcor, J.P., Unger, S.E., Wilson, I.D., Lindon, J.C. and Nicholson, J.K. 1997, NMR and HPLC-NMR spectroscopic studies of futile deacetylation in paracetamol metabolites in rat and man. *Journal of Pharmaceutical and Biomedical Analysis*, **15:** 901–910.

Nicholson, J.K., Timbrell, J.A., Bales, J.R. and Sadler, P.J. 1985, A high resolution proton nuclear magnetic resonance approach to the study of hepatocyte and drug metabolism. Application to acetaminophen. *Molecular Pharmacology*, **27:** 634–643.

Nolan, C.M., Sandblom, R.E., Thummel, K.E., Slattery, J.T. and Nelson, S.D. 1994, Hepatotoxicity associated with acetaminophen usage in patients receiving multiple drug therapy for tuberculosis. *Chest*, **105:** 408–411.

Notarianni, L.J., Oldham, H.G. and Bennett, P.N. 1987, Passage of paracetamol into breast milk and its subsequent metabolism by the neonate. *British Journal of Clinical Pharmacology*, **24:** 63–67.

O'Brien, P.J., Khan, S. and Jatoe, S.D. 1990, Formation of biological reactive intermediates by peroxidases: halide mediated acetaminophen oxidation and cytotoxicity. *Advances in Experimental Medicine and Biology*, **283:** 51–64.

Øie, S., Lowenthal, D.T., Briggs, W.A. and Levy, G. 1975, Effect of hemodialysis on kinetics of acetaminophen by anephric patients. *Clinical Pharmacology and Therapeutics*, **18:** 680–686.

Osborne, N.J., Tonkin, A.L. and Miners, J.O. 1991, Interethnic differences in drug glucuronidation: a comparison of paracetamol metabolism in Caucasians and Chinese. *British Journal of Clinical Pharmacology*, **32:** 765–767.

Owen, S.G., Francis, H.W. and Roberts, M.S. 1994, Disappearance kinetics of solutes from synovial fluid after intra-articular injection. *British Journal of Clinical Pharmacology*, **38:** 349–355.

Pang, K.S., Barker, F., Simard, A., Schwab, A.J. and Goresky, C.A. 1995, Sulfation of acetaminophen by the perfused rat liver: effect of red blood cell carriage. *Hepatology*, **22:** 267–282.

Park, K.S., Sohn, D.H., Veech, R.L. and Song, B.J. 1993, Translational activation of ethanol-inducible cytochrome P450 (CYP2E1) by isoniazid. *European Journal of Pharmacology*, **248:** 7–14.

Patel, M., Tang, B. and Kalow, W. 1993, Variability of acetaminophen metabolism in Caucasians and Orientals. *Pharmacogenetics*, **2:** 38–45.

Perucca, E. and Richens, A. 1979, Paracetamol disposition in normal subjects and in patients treated with antiepileptic drugs. *British Journal of Clinical Pharmacology*, **7:** 201–206.

Piletta, P., Porchet, H.C. and Dayer, P. 1991, Central analgesic effect of acetaminophen but not of aspirin. *Clinical Pharmacology and Therapeutics*, **49:** 350–354.

Potter, D.W. and Hinson, J.A. 1986, Mechanisms of acetaminophen oxidation to N-acetyl-P-benzo-quinone by horseradish peroxidase and cytochrome P-450. *Journal of Biological Chemistry*, **262:** 966–973.

Potter, D.W. and Hinson, J.A. 1987, The 1- and 2-electron oxidation of acetaminophen catalyzed by prostaglandin H synthase. *Journal of Biological Chemistry*, **262:** 974–980.

Poulsen, H.E., Ranek, L. and Jorgensen, L. 1991, The influence of disulfiram on acetaminophen metabolism in man. *Xenobiotica*, **21:** 243–249.

Prescott, L.F. 1980, Kinetics and metabolism of paracetamol and phenacetin. *British Journal of Clinical Pharmacology*, **10:** 291S–298S.

Prescott, L.F. 2000, Paracetamol, alcohol and the liver. *British Journal of Clinical Pharmacology*, **49:** 291–301.

Prescott, L.F., Critchley, J.A.J.H., Balali-Mood, M. and Pentland, B. 1981, Effects of microsomal enzyme induction on paracetamol metabolism in man. *British Journal of Clinical Pharmacology*, **12:** 149–153.

Prescott, L.F., Speirs, G.C., Critchley, J.A.J.H., Temple, R.M. and Winney, R.J. 1989, Paracetamol disposition and metabolite kinetics in patients with chronic renal failure. *European Journal of Clinical Pharmacology*, **36:** 291–297.

Proudfoot, A.T. 1982, *Diagnosis and Treatment of Acute Poisoning*. Oxford: Blackwell.

Quiding, H., Oikarinen, V., Sane, J. and Sjöblad, A.-M. 1984, Analgesic efficacy after single and repeated doses of codeine and acetaminophen. *Journal of Clinical Pharmacology*, **24:** 27–34.

Rainska-Giezek, T. 1995, Influence of caffeine on the toxicity and pharmacokinetics of paracetamol. *Annales Academiae Medicae Stetinensis*, **41:** 69–85.

Ramakrishna, B.S., Roberts-Thomson, I.C., Pannall, P.R. and Roediger, W.E. 1991, Impaired sulphation of phenol by the colonic mucosa in quiescent and active ulcerative colitis. *Gut*, **32:** 46–49.

Rannug, U., Holme, J.A., Hongslo, J.K. and Sram, R. 1995, An evaluation of the genetic toxicity of paracetamol. *Mutation Research*, **327:** 179–200.

Rawlins, M.D., Henderson, D.B. and Hijab, A.R. 1977, Pharmacokinetics of paracetamol (acetaminophen) after intravenous and oral administration. *European Journal of Clinical Pharmacology*, **11:** 283–286.

Ray, K., Sahana, C.C., Chaudhuri, S.B., De, G.C. and Chatterjee, K. 1993, Effects of trichloroethylene anaesthesia on salivary paracetamol elimination. *Indian Journal of Physiology and Pharmacology*, **37:** 79–81.

Rayburn, W., Shukla, U., Stetson, P. and Piehl, E. 1986, Acetaminophen pharmacokinetics: comparison between pregnant and nonpregnant women. *American Journal of Obstetrics and Gynecology*, **155:** 1353–1356.

Renwick, A.G., Ahsan, C.H., Challenor, V.F., Daniels, R., Macklin, B.S., Waller, D.G. and George, C.F. 1992, The influence of posture on the pharmacokinetics of orally administered nifedipine. *British Journal of Clinical Pharmacology*, **34:** 332–336.

Richard, A., Hongslo, J.K., Boone, P.F. and Holme, J.A. 1991, Structure–activity of paracetamol analogues: inhibition of replicative DNA synthesis in V79 Chinese hamster cells. *Chemical Research in Toxicology*, **4:** 151–156.

Roberts, D.W., Bucci, T.J., Benson, R.W., Warbritton, A.R., McRae, T.A., Pumford, N.R. and Hinson, J.A. 1991, Immunohistochemical localization and quantification of the 3-(cystein-S-yl)-

acetaminophen protein adduct in acetaminophen hepatotoxicity. *American Journal of Pathology*, **38:** 359–371.

Roberts, J.R., Gracely, E.J. and Schoffstall, J.M. 1997, Advantage of high-surface-area charcoal for gastrointestinal decontamination in a human acetaminophen ingestion model. *Academic Emergency Medicine*, **4:** 167–174.

Robertson, D.R., Higginson, I., Macklin, B.S., Renwick, A.G., Waller, D.G. and George, C.F. 1991, The influence of protein containing meals on the pharmacokinetics of levodopa in healthy volunteers. *British Journal of Clinical Pharmacology*, **31:** 413–417.

Rogers, L.K., Moorthy, B. and Smith, C.V. 1997, Acetaminophen binds to mouse hepatic and renal DNA at human therapeutic doses. *Chemical Research in Toxicology*, **10:** 470–476.

Rogers, S.M., Back, D.J., Stevenson, P.J., Grimmer, S.F.M. and Orme, M.L'E. 1987, Paracetamol interaction with oral contraceptive steroids: increased plasma concentrations of ethinyloestradiol. *British Journal of Clinical Pharmacology*, **23:** 721–725.

Rollins, D.E., von Bahr, C., Glaumann, H., Moldeus, P. and Rane, A. 1979, Acetaminophen: potentially toxic metabolite formed by human fetal and adult liver microsomes and isolated fetal liver cells. *Science*, **205:** 1414–1416.

Romanelli, L., Valeri, P., Morrone, L.A. and Pimpinella, G. 1991, Ocular disposition of acetaminophen and its metabolites following intravenous administration in rabbits. *Journal of Ocular Pharmacology*, **7:** 339–350.

Rudd, G.D., Donn, K.H. and Grisham, J.W. 1981, Prevention of acetaminophen-induced hepatic necrosis by cimetidine in mice. *Research Communications in Chemical Pathology and Pharmacology*, **32:** 369–372.

Rumble, R.H., Roberts, M.S. and Denton, M.J. 1991, Effects of posture and sleep on the pharmacokinetics of paracetamol (acetaminophen) and its metabolites. *Clinical Pharmacokinetics*, **20:** 167–173.

Ryan, D.E., Koop, D.R., Thomas, P.E., Coon, M.J. and Levin, W. 1986, Evidence that isoniazid and ethanol induce the same microsomal P-450 in rat liver, an isozyme homologous to rabbit liver cytochrome P-450 isoenzyme 3a. *Archives of Biochemistry and Biophysics*, **246:** 633–644.

Sabater, J., Domenech, J. and Obach, R. 1993, Pharmacokinetic study of 4'-acetamidophenyl-2-(5'-p-toluyl-1'-methylpyrrole)acetate in the rat. *Arzneimittel-Forschung*, **43:** 154–159.

Sahajwalla, C.G. and Ayres, J.W. 1991, Multiple-dose acetaminophen pharmacokinetics. *Journal of Pharmaceutical Sciences*, **80:** 855–860.

Saincher, A., Sitar, D.S. and Tenenbein, M. 1997, Efficacy of ipecac during the first hour after drug ingestion in human volunteers. *Journal of Toxicology*, **35:** 609–615.

Sakuma-Sawada, N., Iida, S., Mizuma, T., Hayashi, M. and Awazu, S. 1997, Inhibition of the hepatic uptake of paracetamol sulphate by anionic compounds. *Journal of Pharmacy and Pharmacology*, **49:** 743–746.

Sato, C., Liu, J., Miyakawa, H., Nouchi, T., Tanaka, Y., Uchihara M. and Marumo, F. 1991, Inhibition of acetaminophen activation by ethanol and acetaldehyde in liver microsomes. *Life Sciences*, **49:** 1787–1791.

Sato, C. and Marumo, F. 1991, Synergistic effect of NADH on NADPH-dependent acetaminophen activation in liver microsomes and its inhibition by cyanide. *Life Sciences*, **48:** 2423–2427.

Savides, M.C., Oehme, F.W., Nash, S.L. and Leipold, H.W. 1984, The toxicity and biotransformation of single doses of acetaminophen in dogs and cats. *Toxicology and Applied Pharmacology*, **74:** 26–34.

Scavone, J.M., Greenblatt, D.J., Blyden, G.T., Luna, B.G. and Harmatz, J.S. 1990a, Validity of a two-point acetaminophen pharmacokinetic study. *Therapeutic Drug Monitoring*, **12:** 35–39.

Scavone, J.M., Greenblatt, D.J., LeDuc, B.W., Blyden, G.T., Luna, B.G. and Harmatz, J.S. 1990b, Differential effect of cigarette smoking on antipyrine oxidation versus acetaminophen conjugation. *Pharmacology*, **40:** 77–84.

Scott, D.O., Sorenson, L.R., Steele, K.L., Puckett, D.L. and Lunte, C.E. 1991, In vivo microdialysis sampling for pharmacokinetic investigations. *Pharmaceutical Research*, **8:** 389–392.

Sesardic, D., Boobis, A.R., Edwards, R.J. and Davies, D.S. 1988, A form of cytochrome P450 in man, orthologous to form d in the rat, catalyses the O-dethylation of phenacetin and is inducible by cigarette smoking. *British Journal of Clinical Pharmacology*, **26:** 363–372.

Seymour, R.A. and Rawlins, M.D. 1981, Pharmacokinetics of parenteral paracetamol and its analgesic effects in post-operative dental pain. *European Journal of Clinical Pharmacology*, **20:** 215–218.

Shalabi, E.A. 1992, Acetaminophen inhibits the human polymorphonuclear function in vitro. *Immunopharmacology*, **24:** 37–46.

Shihabi, Z.K. and David, R.M. 1984, Colorimetric assay for acetaminophen in serum. *Therapeutic Drug Monitoring*, **6:** 449–453.

Shyadehi, A.Z. and Harding, J.J. 1991, Investigations of ibuprofen and paracetamol binding to lens proteins to explore their role against cataract. *Biochemical Pharmacology*, **42:** 2077–2084.

Simpson, E. and Stewart, M.J. 1973, Screening for paracetamol poisoning. *Annals of Clinical Biochemistry*, **10:** 171.

Slattery, J.T. and Levy, G. 1979, Acetaminophen kinetics in acutely poisoned patients. *Clinical Pharmacology and Therapeutics*, **25:** 184–195.

Slattery, J.T., McRorie, T.I., Reynolds, R., Kalhorn, T.F., Kharasch, E.D. and Eddy, A.C. 1989, Lack of effect of cimetidine on acetaminophen disposition in humans. *Clinical Pharmacology and Therapeutics*, **46:** 591–597.

Slattery, J.T., Nelson, S.D. and Thummel, K.E. 1996, The complex interaction between ethanol and acetaminophen. *Clinical Pharmacology and Therapeutics*, **60:** 241–246.

Slattery, J.T., Wilson, J.M., Kalhorn, T.F. and Nelson, S.D. 1987, Dose-dependent pharmacokinetics of acetaminophen: evidence of glutathione depletion in humans. *Clinical Pharmacology and Therapeutics*, **41:** 413–418.

Smilgin, Z., Drozdzik, M., Gawronska-Szklarz, B., Wojcicki, J., Tustanowski, S. and Gornik, W. 1993, Pharmacokinetics of acetaminophen in individuals occupationally exposed to polyvinyl chloride modified with plasticizers. *Medycyna Pracy*, **44:** 423–429.

Smith, M., Whitehead, E., O'Sullivan, G. and Reynolds, F. 1991, A comparison of serum and saliva paracetamol concentrations. *British Journal of Clinical Pharmacology*, **31:** 553–555.

Smith, W.L. and Wilkin, G.P. 1977, Immunochemistry of prostaglandin endoperoxide-forming cyclooxygenase: the detection of the cyclooxygenase in rat, rabbit and guinea pig kidney by immunofluorescence. *Prostaglandins*, **13:** 873–892.

Sonne, J., Boesgaard, S., Poulsen, H.E., Loft, S., Hansen, J.M., Døssing, M. and Andreasen, F. 1990, Pharmacokinetics and pharmacodynamics of oxazepam and metabolism of paracetamol in severe hypothyroidism. *British Journal of Clinical Pharmacology*, **30:** 737–742.

Speck, R.F., Schranz, C. and Lauterburg, B.H. 1993, Prednisolone stimulates hepatic glutathione synthesis in mice. Protection by prednisolone against acetaminophen hepatotoxicity in vivo. *Journal of Hepatology*, **18:** 62–67.

Stenken, J.A., Stahle, L., Lunte, C.E. and Southard, M.Z. 1998, Monitoring in situ liver metabolism in rats using microdialysis. Comparison of microdialysis mass-transport model predictions to experimental metabolite generation data. *Journal of Pharmaceutical Sciences*, **87:** 311–320.

Stewart, M.J. and Watson, I.D. 1987, Analytical reviews in clinical chemistry: methods for the estimation of salicylate and paracetamol in serum, plasma and urine. *Annals of Clinical Biochemistry*, **24:** 552–565.

Streeter, A.J., Dahlin, D.C., Nelson, S.D. and Baillie, T.A. 1984, The binding of acetaminophen to protein. Evidence for cysteine residues as major sites of arylation in vitro. *Chemico-Biological Interactions*, **48:** 349–366.

Ström, C., Forsberg, O., Quiding, H., Engevall, S. and Larsson, O. 1990, Analgesic efficacy of acetaminophen sustained release. *Journal of Clinical Pharmacology*, **30:** 654–659.

Studenberg, S.D. and Brouwer, K.L. 1993a, Hepatic disposition of acetaminophen and metabolites. Pharmacokinetic modeling, protein binding and subcellular distribution. *Biochemical Pharmacology*, **46:** 739–746.

Studenberg, S.D. and Brouwer, K.L. 1993b, Effect of phenobarbital and p-hydroxyphenobarbital glucuronide on acetaminophen metabolites in isolated rat hepatocytes: use of a kinetic model to examine the rates of formation and egress. *Journal of Pharmacokinetics and Biopharmaceutics*, **21:** 175–194.

Swaan, P.W., Marks, G.J., Ryan, F.M. and Smith, P.L. 1994, Determination of transport rates for arginine and acetaminophen in rabbit intestinal tissues in vitro. *Pharmaceutical Research*, **11:** 283–287.

Teffera, Y. and Abramson, F. 1994, Application of high-performance liquid chromatography/chemical reaction interface mass spectrometry for the analysis of conjugated metabolites: a demonstration using deuterated acetaminophen. *Biological Mass Spectrometry*, **23:** 776–783.

Thomsen, M.S., Loft, S., Roberts, D.W. and Poulsen, H.E. 1995, Cytochrome P4502E1 inhibition by propylene glycol prevents acetaminophen (paracetamol) hepatotoxicity in mice without cytochrome P4501A2 inhibition. *Pharmacology and Toxicology*, **76:** 395–399.

Thummel, K.E., Lee, C.A., Kunze, K.L., Nelson, S.D. and Slattery, J.T. 1993, Oxidation of acetaminophen to N-acetyl-p-aminobenzoquinoneimine by human CYP3A4. *Biochemical Pharmacology*, **45:** 1563–1569.

Thummel, K.E., Slattery, J.T. and Nelson, S.D. 1988, Mechanism by which ethanol diminishes the hepatotoxicity of acetaminophen. *Journal of Pharmacology and Experimental Therapeutics*, **245:** 129–136.

Thummel, K.E., Slattery, J.T., Ro, H., Chien, J.Y., Nelson, S.D., Lown, K.E. and Watkins, P.B. 2000, Ethanol and production of the hepatotoxic metabolite of acetaminophen in healthy adults. *Clinical Pharmacology and Therapeutics*, **67**, 591–599.

Tomlinson, B., Young, R.P., Ng, M.C.Y., Anderson, P.J., Kay, R. and Critchley, J.A.J.H. 1996, Selective liver enzyme induction by carbamazepine and phenytoin in Chinese epileptics. *European Journal of Clinical Pharmacology*, **50:** 411–415.

Tredger, J.M., Thuluvath, P., Williams, R. and Murray-Lyon, I.M. 1995, Metabolic basis for high paracetamol dosage without hepatic damage: a case study. *Human and Experimental Toxicology*, **14:** 8–12.

Tsutsumi, M., Lasker, J.M., Shimizu, M., Rosman, A.S. and Lieber, C.S. 1989, The intralobular distribution of ethanol-inducible P450IIE1 in rat and human liver. *Hepatology*, **10:** 437–446.

Turner, K.C. and Brouwer, K.L. 1997, In vitro mechanisms of probenecid-associated alterations in acetaminophen glucuronide hepatic disposition. *Drug Metabolism and Disposition*, **25:** 1017–1021.

van Lingen, R.A., Deinum, J.T., Quak, J.M., Kuizenga, A.J., van Dam, J.G., Anand, K.J., Tibboel, D. and Okken, A. 1999, Pharmacokinetics and metabolism of rectally administered paracetamol in preterm neonates. *Archives of Disease in Children Fetal and Neonatal Edition*, **80:** F59–F63.

van Zyl, J.M., Basson, K. and van der Walt, B. 1989, The inhibitory effect of acetaminophen on the myeloperoxidase-induced antimicrobial system of the polymorphonuclear leukocyte. *Biochemical Pharmacology*, **38:** 161–165.

Vance, M.V., Selden, B.S. and Clark, R.F. 1992, Optimal patient position for transport and initial management of toxic ingestions. *Annals of Emergency Medicine*, **21:** 243–246.

Vendemiale, G., Altomare, E., Trizio, T., Leandro, G., Manghisi, O.G. and Albano, O. 1987, Effect of acute and chronic cimetidine administration on acetaminophen metabolism in humans. *American Journal of Gastroenterology*, **82:** 1031–1034.

von Moltke, L., Manis, M., Harmatz, J.S., Poorman, R. and Greenblatt, D.J. 1993, Inhibition of acetaminophen and lorazepam glucuronidation in vitro by probenecid. *Biopharmaceutics and Drug Disposition*, **14:** 119–130.

Wang, L.H., Rudolph, A.M. and Benet, L.Z. 1990, Comparative study of acetaminophen disposition in sheep at three developmental stages; the fetal, neonatal and adult periods. *Developmental Pharmacology and Therapeutics*, **14:** 161–179.

West, P.R., Harman, L.S., Josephy, P.D. and Mason, R.P. 1984, Acetaminophen: enzymatic formation of a transient phenoxyl free radical. *Biochemical Pharmacology*, **33:** 2933–2936.

Williams, F.M., Moore, U., Seymour, R.A., Mutch, E.M., Nicholson, E., Wright, P., Wynne, H., Blain, P.G. and Rawlins, M.D. 1989, Benorylate hydrolysis by human plasma and human liver. *British Journal of Clinical Pharmacology*, **28:** 703–708.

Wilson, J.M., Slattery, J.T., Forte, A.J. and Nelson, S.D. 1982, Analysis of acetaminophen metabolites in urine by high performance liquid with UV and amperometric detection. *Journal of Chromatography*, **227:** 453–462.

Wojcicki, J., Gawronska-Szklarz, B., Kazimierczyk, J., Baskiewicz, Z. and Raczynski, A. 1979, Comparative pharmacokinetics of paracetamol in men and women considering follicular and luteal phases. *Arzneimittel-Forschung*, **29:** 350–352.

Wynne, H.A., Cope, L.H., Herd, B., Rawlins, M.D., James, O.F. and Woodhouse, K.W. 1990, The association of age and frailty with paracetamol conjugation in man. *Age and Ageing*, **19:** 419–424.

Yan, Z., Nikelly, J.G., Killmer, L. and Tarloff, J.B. 2000, Metabolism of para-aminophenol by rat hepatocytes. *Drug Metabolism and Disposition*, **28:** 880–886.

Zand, R., Nelson, S.D., Slattery, J.T., Thummel, K.E., Kalhorn, T.F., Adams, S.P. and Wright, J.M. 1993, Inhibition and induction of cytochrome P4502E1-catalyzed oxidation by isoniazid in humans. *Clinical Pharmacology and Therapeutics*, **54:** 142–149.

■
CHAPTER 6
■

Pharmacology and Biochemistry of Salicylates and Related Drugs

7

K.D. Rainsford

INTRODUCTION

Aspirin and salicylate have been studied extensively to understand their pharmacological and toxicological effects. In many respects the salicylates have been used not only as standards of comparison but also as prototypes for the development of NSAIDs and other analgesic anti-inflammatory agents. It is remarkable that hardly a month goes by without reports appearing showing that aspirin and other salicylates have effects on newly discovered biochemical or cellular systems that are involved in some process relating to the classical activities as anti-inflammatory, analgesic or antipyretic activities. Some of the novel activities that are being explored for potential therapeutic activities – e.g. prevention of cancer growth and development, regulation of apoptosis or against Alzheimer's disease. These observations then logically lead to studies of the effects of other anti-inflammatory agents, or the development of new agents that target some of these processes. Recent examples include the effects of aspirin and salicylates on signal transduction pathways that control the production of the inducible cyclo-oxygenase 2 (COX-2), metalloproteinases, cytokines and other mediators of inflammation or cellular activation.

Philosophically speaking, aspirin has been described as an 'anti-defensive' drug by the late Harry Collier. This concept, which Collier (1969) first proposed, was probably one of the first attempts at formulating the fundamental actions of aspirin in a single concept. It is easy to appreciate how this concept has evolved, since inflammation is fundamentally regarded as a defensive response by the body to insult or infection. The anti-defensive nature of aspirin clearly reflected the combined anti-inflammatory, analgesic and antipyretic activities inherent in this drug. The subsequent discovery of newer pharmacological properties (e.g. the prevention of colon cancer and lymphocyte activation) may necessarily lead to a challenge to this concept. This is because it is apparent that studies of the prevention of colon cancer, for instance, imply that aspirin, like other NSAIDs, is probably aiding the offensive processes of the body to overcome tumour growth and spread throughout the body. Part of these actions may reside in influences on lymphocyte functions and thus the immune surveillance, as well as direct effects on tumour growth and metastasis. Therefore we may have to rethink our concepts of the actions of aspirin according to the disease state for which it is being applied.

Fundamental to the understanding of the mode of action of aspirin, as with any drug, is appreciation of the patterns of its absorption, distribution metabolism and elimination (ADME; see Chapter 4). One of the main features that distinguishes aspirin from all other NSAIDs and analgesics is the ability of this drug to acetylate a variety of biomolecules – an effect that precedes the production of salicylates and thus is integrally part of the metabolism of the drug. The sites of acetylation on the body vary considerably (Brune *et al.*, 1977; Rainsford *et al.*, 1980; 1981), the highest degree occurring in the upper gastrointestinal tract at sites where the drug is present in highest concentration following absorption, i.e. the gastrointestinal mucosa, as well as in the circulation that passes through this region of the GI tract. Recognition that acetylation of the cyclo-oxygenases is part of the irreversible action of aspirin in blocking the activity of these enzymes, coupled with many observations that salicylate that is produced from aspirin has unique pharmacological properties of its own, has led to the concept that there may be two drugs in one in terms of pharmacological activity. In essence, therefore, aspirin may be regarded as 'a bifunctional drug'. Thus, the fate of the acetyl group being in part to form covalent acetylated biomolecules, including acetylation of the region near the active site of cyclo-oxygenases, and the salicylate half of the molecule having its own unique pharmacological actions emphasises the dual role of aspirin. Some may regard aspirin as a 'pro-drug' of salicylate, especially when viewing the use of aspirin in the treatment of chronic inflammatory conditions such as rheumatoid arthritis, where high doses of aspirin are given. In these states salicylate or its dimer, salsalate or diplosal, are roughly equipotent in controlling joint manifestations, including soft tissue swelling, and in pain relief. Thus it could be argued under these circumstances that the contribution of inhibition of cyclo-oxygenase to the therapeutic effects of the drug in controlling pain and joint swelling may be minor compared with the inherent actions of salicylate (including that produced from the hydrolysis of the salicylate dimer, salsalate).

In contrast to this, it is now well established (see Chapter 11) that only low doses of aspirin are required for prevention or treatment of thromboembolic diseases. In this case the acetylation of COX-1 in platelets is one of the principal pharmacological actions of aspirin, there being apparently few or no effects of salicylate in the platelet-mediated thromboembolic responses.

It can therefore be seen from these two examples of the treatment of chronic disease that the dose of aspirin can determine its pharmacologically important therapeutic effects.

The concept that dosage of the drug may be the major feature which combined with its unique pharmacokinetics, sets aspirin aside from many NSAIDs and analgesics.

Figure 7.1 shows the dynamic interplay between the pharmacokinetics of aspirin and its underlying pharmacological activities, emphasising the role of dosage in control of thromboembolic conditions compared with that of chronic inflammatory diseases.

There are several overall principles concerning the mode of pharmacological actions of the salicylates. While these generalisations are largely directed towards the effects of these drugs as anti-inflammatory, analgesic, antipyretic and, with aspirin, antithrombotic agents (the principal actions of this group), they are probably applicable to a large extent to the other actions that can be attributed to individual members of the salicylate family, as well as to other NSAIDs and analgesics. The principles underlying the therapeutic actions of these drugs can be stated as follows.

1. The salicylates inhibit a variety of enzymes and other biochemical processes, the net effects of which are to (a) block the production of inflammatory mediators; (b) depress cellular functions and intracellular control mechanisms; and (c) affect enzymes responsible for tissue destruction processes (Whitehouse, 1965; Smith and Smith, 1966; McArthur *et al.*, 1971a; Smith and Dawkins, 1971; Mizushima *et al.*, 1975; Amann and Peskar, 2002; Warner and Mitchell, 2003).

2. Individual salicylates can exert specific activities on cellular and biochemical processes involved in inflammation according to their physicochemical and structural properties (Whitehouse, 1965; Smith and Smith, 1966; Smith *et al.*, 1975; 1979; Stone *et al.*, 1977; Whitehouse *et al.*, 1977; Sofia, 1978; Glenn *et al.*, 1979; Walker *et al.*, 1976; Brune *et al.*, 1981). This specificity can be attributed to the following:

   • The propensity of these acidic drugs to accumulate in diseased/inflamed tissues or compartments (Brune *et al.*, 1977; Rainsford *et al.*, 1980; 1981). Extravasated plasma proteins carrying protein (albumin)-bound drugs accumulate in inflamed sites wherein pain is manifest as a result of vascular damage (Brune *et al.*, 1977). Also, the low intracellular/extracellular pH (approximately 5 to 6.8) that is evident in inflamed tissues may favour accumulation of these organic (i.e. lipophilic) acids into inflamed compartments according to the pH dependence of the drug-partitioning hypothesis (Brune *et al.*, 1977; Rainsford *et al.*, 1980; 1981) as well as from the leakage of plasma protein-bound drug through damaged or dilated post-capillary venules and other blood vessels.

- Enhanced activity of disease-perturbed biochemical and degradative processes in inflamed tissues means that many of the inhibitory effects of salicylates on these processes may be more readily expressed at a given concentration of the drug, compared with the same processes in non-inflamed tissues. This arises simply because the activity of the processes in inflamed tissues is often greater than those in non-inflamed tissues. The inhibitory effects of a drug on a biochemical reaction are more apparent quantitatively in inflamed tissues, where the reaction proceeds at a much faster rate than observed in non-inflamed 'normal' tissue.

- In addition to a quantitative difference in the biochemical process, there are also qualitative differences in various tissues of the response to the drug action. This can be illustrated by the case of differing specificity of action of anti-inflammatory/analgesic drugs on enzyme reactions, e.g. induced cyclo-oxygenase-2 (or COX-2) and nitric oxide synthase (iNOS), as well as lysosomal and metalloproteinase enzymic activities in different tissues (Flower and Vane, 1972; Bhatterjee and Eakins, 1974; Baenziger *et al.*, 1977; Atkinson and Collier, 1980; Amin *et al.*, 1995; 1999; Vane, 1998; Moncada *et al.*, 1999).

- Analgesic activity of the salicylates is predominantly peripheral in location, and is in part related to anti-inflammatory activity of these drugs (Guzman *et al.*, 1964; Lim *et al.*, 1964; 1969; Smith and Smith, 1966). There is also an important component of drug action on the central nervous system (Dubas and Parker, 1971; Paalzow, 1973; Tagliamonte *et al.*, 1973; Guerinot *et al.*, 1974).

- Antipyretic effects of the salicylates are primarily due to hypothalamic effects of these drugs on the actions of endogenous pyrogen on temperature receptor/regulatory areas in this region of the central nervous system (Cranston *et al.*, 1971a; Beckman and Rozkowska-Ruttiman, 1974; Perlow *et al.*, 1975; Schoener and Wang, 1975; 1976; Avery and Penn, 1976; Cranston *et al.*, 1976; Splawinski *et al.*, 1977; Dinarello and Wolff, 1978).

It has now become recognised that a major part of the anti-inflammatory, analgesic and antipyretic effects of the salicylates, like that of many of the NSAIDs, is due to effects on the inhibition of COX-2 and iNOS induced during inflammation, the induction of pain stimuli, and the response at the level of the dorsal horn and centres in the central nervous system and in the hypothalamus during the development of fever (Crofford, 1997; Dubois *et al.*, 1998; Vane *et al.*, 1998; Kam and See, 2000, Kam *et al.*, 2000). Although the inhibitory effects of salicylates on COX-2 activity undoubtedly plays a major part in the regulation of prostaglandin-mediated inflammatory and other responses, there is increasing evidence that COX-1 inhibition may also contribute to the anti-inflammatory and possibly the analgesic actions of these drugs (see later in this chapter). Furthermore there is some evidence that, while COX-2 may 'drive' the inflammatory processes, in the stages of repair COX-1 may be responsible for modulating the chronic inflammation, the selective inhibition of which may control granulomatous inflammation (Parnham, 1998).

The role of nitric oxide (NO), another important inflammatory mediator, in the activity of cyclo-oxygenases also represents a major target for the effects of salicylates (Amin *et al.*, 1995), although the effects of aspirin and salicylate on this pathway might be less potent than on the induction and activity of COX-2 and activity of COX-1. A recent study (Clancy *et al.*, 2000) suggests that nitric oxide stimulates the activity of the constitutive COX-1 isoenzyme but inhibits the activity of COX-2 during inflammation. As the production of nitric oxide is largely controlled by the activity of the inducible nitric oxide synthase during inflammation, any effects of the NSAIDs on the production of nitric oxide synthase or of nitric oxide itself mediated by this enzyme will have indirect effects on the regulation of the prostaglandin production by the two different isoforms of COX. Thus the high concentration of aspirin-induced inhibition of iNOS production (Amin *et al.*, 1995; 1999; Kwan *et al.*, 1997) but not that by salicylate may lead to reduction in the effects of stimulation by nitric oxide on COX-1 produced prostaglandins, and dysinhibition of the effects of nitric oxide on COX-2 derived prostaglandins (Clancy *et al.*, 2000). As it is now well established that aspirin and salicylate inhibit the production of COX-2 enzyme and iNOS proteins (Sanduja *et al.*, 1990; Amin *et al.*, 1995; Rainsford *et al.*, 1995; Xu *et al.*, 1999), coupled with the recent recognition that the NFκB-mediated transcription of the synthesis of

CHAPTER 7

DIFFERENTIAL EFFECTS OF ASPIRIN AND SALICYLATE ON INFLAMMATORY REACTIONS, PAIN, FEVER AND THROMBOSIS

A. Inflammation
Aspirin (High Dose)

Circulating Leucocytes/Endothelial Cells

Inflamed Joints and Tissues

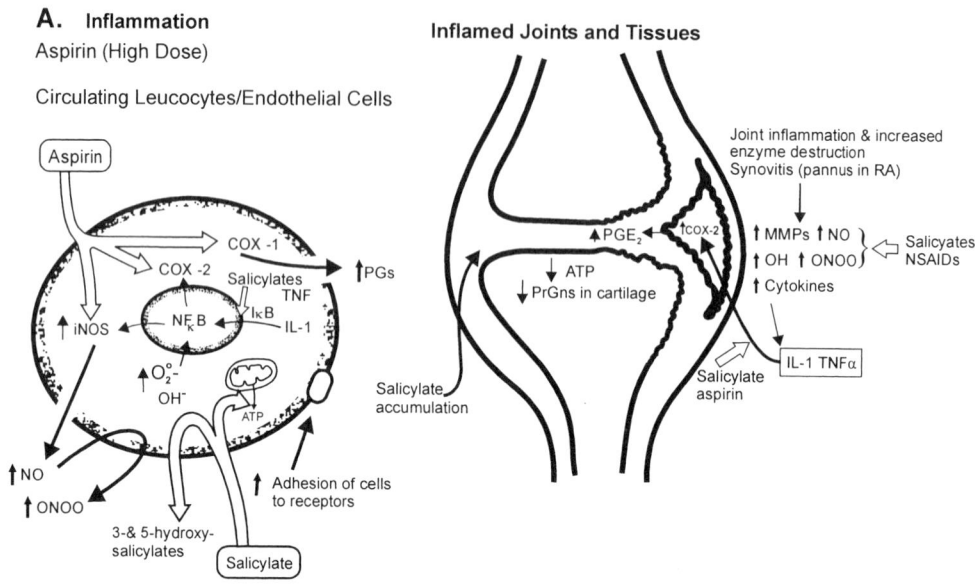

B. Pain (Moderate - High dose)

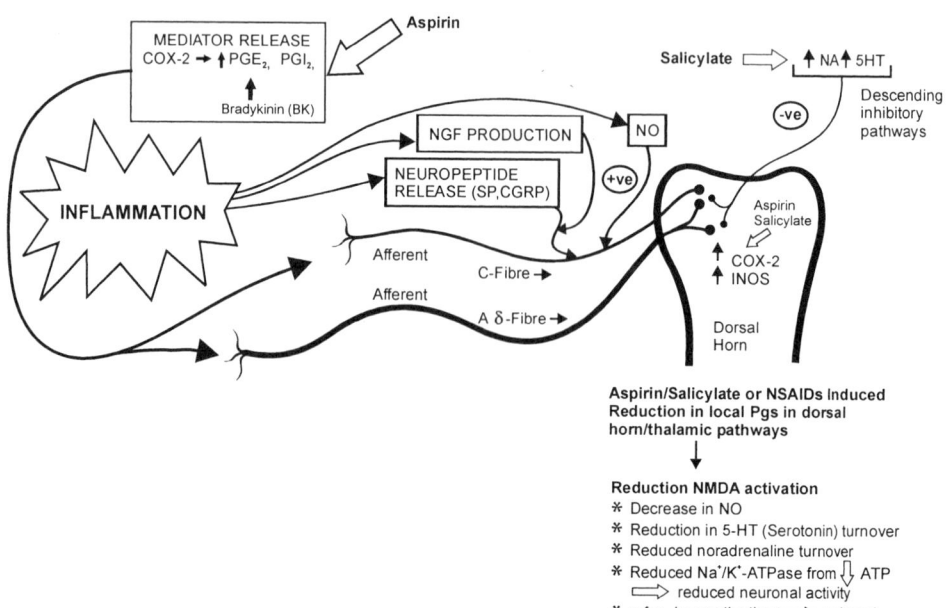

Aspirin/Salicylate or NSAIDs Induced
Reduction in local Pgs in dorsal
horn/thalamic pathways

Reduction NMDA activation
* Decrease in NO
* Reduction in 5-HT (Serotonin) turnover
* Reduced noradrenaline turnover
* Reduced Na⁺/K⁺-ATPase from ⇩ ATP
  ⟹ reduced neuronal activity
* c-fos, krox activation ⟹ reduced
  neuronal activation

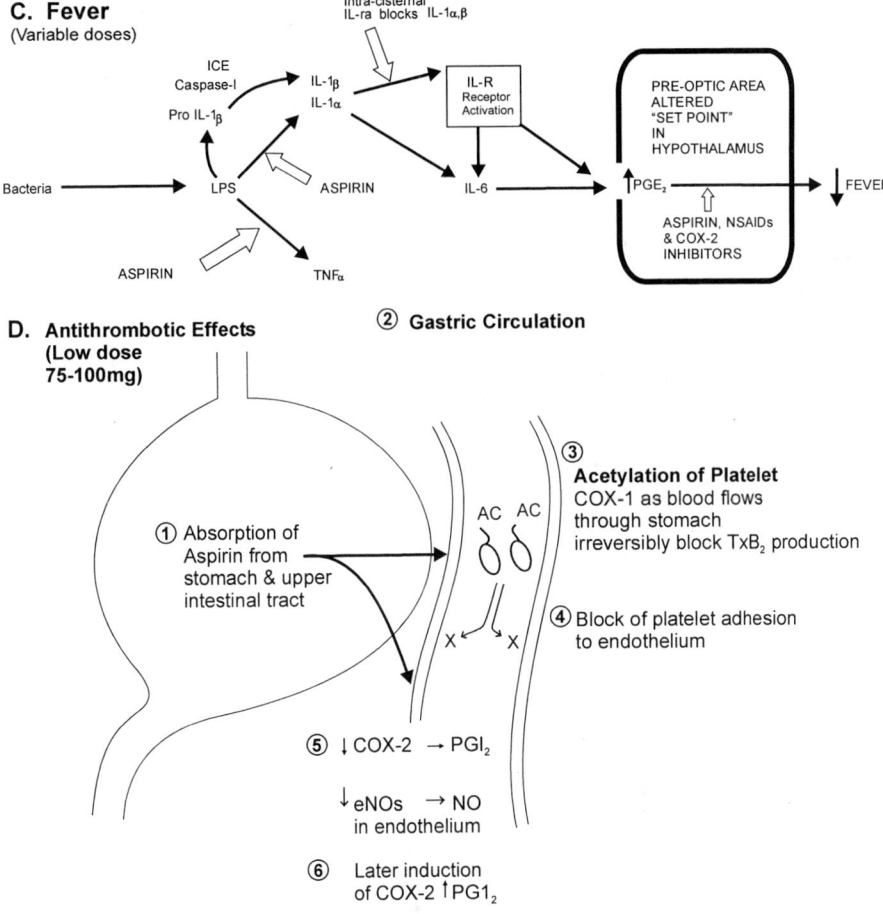

C. Fever
(Variable doses)

Intra-cisternal
IL-ra blocks IL-1α,β

ICE
Caspase-I → IL-1β
Pro IL-1β   IL-1α

IL-R
Receptor
Activation

PRE-OPTIC AREA
ALTERED
"SET POINT"
IN
HYPOTHALAMUS

Bacteria → LPS → ASPIRIN → IL-6 → ↑PGE₂ → ↓ FEVER

ASPIRIN, NSAIDs
& COX-2
INHIBITORS

ASPIRIN   TNFα

D. Antithrombotic Effects
(Low dose
75-100mg)

② **Gastric Circulation**

AC AC

③ **Acetylation of Platelet
COX-1 as blood flows
through stomach
irreversibly block TxB₂ production**

① Absorption of
Aspirin from
stomach & upper
intestinal tract

④ Block of platelet adhesion
to endothelium

X ← → X

⑤ ↓ COX-2 → PGI₂

↓ eNOs → NO
in endothelium

⑥ Later induction
of COX-2 ↑PG1₂

Figure 7.1 Integrated pharmacokinetic and pharmacodynamic concept for the actions of aspirin in inflammation (A), pain (B), fever (C) and thrombosis (D). There is marked dose-, drug-metabolic and site-dependency of effects. Clinically in inflammation aspirin and salicylate/salicylic acid are only effective at high doses e.g. 3–4 g per day), while anti-thrombotic effects are evident only with aspirin at low daily doses (e.g. 75–100 mg per day). Mild to moderate pain is controlled at doses of aspirin ranging from 650–1950 mg. Antipyretic effects are dependent on the extent of the febrile state. (A) *In inflammation* salicylates exert effects mainly at high doses by accumulating in inflamed tissues (e.g. as shown here in diarthrodal joints) where they exert *multiple inhibitory actions* on leucocyte and synovial production of inflammatory mediators including cyclo-oxygenase (COX)-derived prostaglandins, nitric oxide (NO), oxyradicals (OH'), the induction of COX-2, and possibly the production of pro-inflammatory cytokines (IL-1, TNFα) with consequent reduced expression of the production of leucocyte adhesion molecules. Control of NFκB activation (via IKK-kinases) by salicylates leads to reduction of early response genes among them COX-2, metalloproteinases (MMPs), IL-1 and TNFα. (B) *In pain* salicylates at moderate to high doses act: (a) peripherally by control of inflammation as well as the production during pain of neural peptides that exert inflammatory activities substance P (Sub P), calcitonin gene-related peptide (CGRP), bradykinin (BK), nerve growth factor (NGF), as well as prostaglandins (PGs); (b) in the dorsal horn by reducing the expression COX-2 and production of NO; and (c) in activating downward descending pathways from thalamic regions to increase production of noradrenaline (NA) and 5-hydroxytryptamine (5-HT, serotonin) that mediate inhibition of pain fibres in the dorsal horn. (C) *In control of fever* with moderate to high doses of aspirin and salicylate exert antipyretic effects by controlling the production of pro-inflammatory cytokines produced by bacterial activation of leucocytes (IL-1, ?TNFα), then IL-6 and the subsequent production of PGE₂ in the hypothalamic region. Altered set point occurs as the central response in fever. (D) *Anti-thrombotic effects* of aspirin (but not salicylate) are evident and are evident at low doses of the drug. During absorption and uptake into circulation through the stomach the acetyl (Ac) group of aspirin covalently acetylates the Ser[530] in the active site of platelet COX-1 irreversibly blocking its capacity to undertake catalytic oxidation of arachidonic acid to form thrombogenic, thromboxane A₂ in response to platelet activation – this block is exerted for the entire lifespan of the platelet (half-life 5 days approximately) since this cell is incapable of renewing production of COX-1. While low doses of aspirin may inhibit production of COX-2-derived prostacyclin (PGI₂) especially in the endothelia of blood vessels of the gastro-intestinal circulation this may be replenished by renewed synthesis of COX-2. Nitric oxide production is also inhibited by aspirin reducing the vascular dilating effects of this mediator.

cyclo-oxygenase-2 is inhibited by both aspirin and salicylate (Shi *et al.*, 1999; Yin *et al.*, 1998; 1999), it would appear that there are important intracellular transductional systems that regulate the production of these enzymes as well as others that may regulate the inflammatory process at other sites. Thus in viewing the mechanisms of action of the salicylates in relation to their therapeutic actions, the following key biochemical actions should be considered:

1. The acetylation by aspirin of COX-2 near the active site leads to a switch in enzyme activity to cause production of 15-lipoxygenase products and formation of lipoxins, with consequent inhibition of production of prostaglandins. This effect does not appear to be evident with salicylate.
2. Both aspirin and salicylate are potent inhibitors of the induction of COX-2 enzyme protein, and this may represent their major mode of action *in vivo*. The mechanism of this action may be related to the inhibition of the signal transduction system, possibly that mediated by NFκB, although other signalling systems (e.g. AP-1; Huang *et al.*, 1997) may also be involved.
3. Effects on nitric oxide production may indirectly influence the actions of prostaglandins produced by COX-1 and COX-2.
4. Inhibition of the accumulation and activation of leucocytes at inflamed sites, and especially of the production of oxygen radicals during phagocytosis or activation of leucocytes, represents a major target for the effects of salicylate produced from aspirin.

Earlier studies in the 1950s and 1960s on the mode of action of salicylates concentrated on defining their actions on enzyme systems that were then becoming recognised as being important in connective tissue metabolism. The studies by Professors Michael Whitehouse (Whitehouse, 1965; 1968) and Mervyn Smith (Smith, 1966; Smith and Dawkins, 1971), as well as several other groups were formative in focusing attention on the biochemical mechanisms of effects of the salicylates in comparison with other anti-inflammatory drugs. Advances in the understanding of biochemical, molecular and cell biology of cellular reactions to inflammatory or noxious stimuli have often been quickly followed by investigations of the mechanisms of drug effects on these systems.

In the same way, the pioneering work in 1971 of the Novel Laureate, Professor Sir John Vane, FRS showed that the actions of aspirin and other anti-inflammatory/antipyretic drugs could be explained by their ability to block production of inflammatory prostaglandins (Vane, 1971; 1978). This work had important foundations in the fundamental studies by Professor von Euler regarding the discovery and identification chemically of prostaglandins in seminal fluids.

The elegant chemical structural analysis of prostaglandins by Professors Bengt Samuelsson and Sune Bergstrom at the Karolinska Institute, Stockholm, set the basis for Vane and his colleagues to study the effects of aspirin and related drugs on prostaglandin production for which Vane, Samuelsson and Bergstrom received their Nobel Prize in 1982.

ANTI-INFLAMMATORY EFFECTS

Studies in animal models

Willoughby and Flower (1992) extensively reviewed the anti-inflammatory effects of salicylates in various animal models. These authors emphasised the differing roles of specific inflammatory mediators in the expression of anti-inflammatory effects of the salicylates *in vivo*.

The *in vivo* responses of the salicylates to inflammatory stimuli vary enormously according to (a) the type of inflammagen and the site(s) of its application, (b) the species and strain of animal, (c) variability in laboratory conditions (e.g. housing, density, handling temperature, humidity), and (d) experimental techniques (Domenjoz, 1955; Whitehouse, 1965; Winter, 1966; Green *et al.*, 1971; Swingle, 1974a; Otterness and Bliven, 1985).

Over 40 different inflammagens have been employed in laboratory animals for determining the anti-inflammatory activities of the salicylates (Domenjoz, 1955; Winder *et al.*, 1958; Whitehouse, 1965; Pearson, 1966; Winter, 1966; Adams and Cobb, 1967; Sancilio, 1969; Green *et al.*, 1971; Swingle, 1974a; Sofia, 1978; Otterness *et al.*, 1979; Carlson *et al.*, 1985; Otterness and Bliven, 1985; Cordeiro *et al.*, 1986). The dose range of response varies considerably according to the type of inflammagen employed (Tables 7.1 to 7.3). The effective dose for aspirin is in the order of 50 to 300 mg/kg in some acute models, the higher doses being in the toxicological range at which anti-inflammatory activity could be influenced by stimulation of the adrenocortical axis (Whitehouse, 1965; Green *et al.*, 1971). Most studies of the acute anti-inflammatory effects of salicylates and related drugs have been performed in either carrageenan, kaolin or urate-crystal induced foot-paw oedema in rats. The polyarthritis induced in rats by subcutaneous or subplantar injections of heat-killed *Mycobacterium tuberculosis* or *M. butyricum* and the carrageenan granuloma models have been amongst the most popular models of chronic inflammation (Pearson, 1966; Swingle, 1974b).

Acute inflammatory models

ULTRAVIOLET ERYTHEMA

Until the mid-1960s, ultraviolet-induced skin erythema in guinea pigs was a popular method for screening the acute anti-inflammatory activity of newly developed salicylate compounds (Pearson, 1966; Adams and Cobb, 1967; Swingle, 1974a), but this was found to have excessive variability even in tests in the same laboratory (Winder *et al.*, 1958). The impression from these early studies is that compounds were selected in the UV erythema model with a spectrum of activity akin to that of aspirin, ibuprofen or phenylbutazone (Winder *et al.*, 1958; Adams and Cobb, 1967), which are prostaglandin synthesis inhibitors.

The exposure of skin to UV light leads to localised reddening (erythema) and increased skin temperature (local hyperthermia), which are two of the cardinal signs of inflammation. Erythematous reaction can be discriminated following the application of UV to the shaven skin of guinea pigs (Otterness and Bliven, 1985).

The procedure for this model involves exposing the depilated skin of guinea pigs or rats (see Table 7.1) briefly (80 to 90 s) to UV-B irradiation (wavelength 280 to 320 nm), a major cause of sunburn in humans (Winder *et al.*, 1958; Otterness and Bliven, 1985). Adams and co-workers (1969; 1970) considerably modified the procedures employed in this assay as established by Wilhelmi (1949), and this has important consequences for establishment of the utility of the model. In the early stages of the development of the acute inflammation there is increased vascular permeability and mast cell degranulation with accompanying release of histamine and 5-hydroxytryptamine (5-HT, serotonin) within 10–15 minutes. During the period of 2 to 6 hours there is considerable production of prostaglandins (PGs) in the UV-irradiated skin. Thus, it is not surprising that inhibitors of PG production are detected in this model (Otterness and Bliven, 1985).

In early studies up until the mid-1960s, the ultraviolet-induced skin erythema in guinea pigs was a popular method for screening the acute anti-inflammatory activity of newly developed salicylate compounds (Pearson, 1966; Adams and Cobb, 1967; Swingle, 1974a).

PAW OEDEMA ASSAYS

Subplantar injections in rats of carrageenan (a mixture of sulphated polysaccharides from Irish moss, *Chondrus crispus*) elicits a biphasic swelling of the hind paws (Winter *et al.*, 1962; Vinegar *et al.*, 1969; 1976; Swingle, 1974b; Roch-Arveiller and Giroud, 1979). The initial phase of paw swelling begins immediately on injection and tapers off after 20 to 60 minutes. Peak paw swelling during the second phase occurs at 3 h, the swelling being some three to four times greater than that in the first phase. A rapid rise in paw temperature occurs initially and this then drops before increasing again during the second phase (Vinegar *et al.*, 1976).

TABLE 7.1

Anti-inflammatory activities of salicylates, compared with NSAIDs and non-narcotic analgesics in various animal models of inflammation.

	Drug ED_{50} mg/kg or % inhibition at dose (mg/kg); Confidence Interval (CI)				
Model/inflammagen	Aspirin	Benorylate	5-Chloro-salicylate	Diclofenac sodium	Diflunisal
Acute inflammation					
UV erythema: Guinea pigs[1]	75–115	–	–	2.1–9.0	–
Rats, oral[2]	148	–	–	–	–
Rats, topical[2]	9.2				
Paw oedema (rats), induced by	70–89[3]	94	37	2.02	6.2–9.8
Carrageenan[3,4]	69.9 (33.4–146)[4]				
Formalin[5]	17.8% @ 490	–	–	–	–
Dextran[5]	38.8% @ 490	–	–	–	–
Polymyxin B[5]	300	–	–	–	–
Mycobacterium tuberculosis (18 h)[6]	217	–	–	–	49
Kaolin[7]	35	–	87	–	–
Sodium urate crystals[7]	93	–	IA @ 200	–	–
Histamine[7]	115	–	IA @ 200	–	–
Serotonin[7]	IA @ 400	–	IA @ 200	–	–
Bradykinin[7]	349	–	IA @ 200	–	–
Chronic inflammation					
Adjuvant arthritis (rats)[8]	72–78	–	–	8.0	10
Mycobacterium (established disease)[b]					
Cotton-pellet granuloma (rats) ED_{50}	69	–	–	2.0	–
Croton-oil granuloma (rats)	>480	–	–	17.7	–

[1]Winder *et al.* (1958); Adams and Cobb (1967); Otterness *et al.* (1979)
[2]Law and Lewis (1977)
[3]Hannah *et al.* (1978); Sofia (1978); Sunkel *et al.* (1978); Sofia *et al.* (1979); Stone *et al.* (1977)
[4]Shimizu *et al.* (1975)
[5]Bertelli and Soldani (1979)
[6]Sofia *et al.* (1979)
[7]Sofia (1978)
[8]Sofia *et al.* (1974); Hannah *et al.* (1978)
[9]Nakamura and Shimizu (1979)
Abbreviation
IA = inactive; @ ≥ dose.
Notes
[a]Meseclazone is an experimental pro-drug of 5-chlorosalicylic acid.
[b]Adjuvant arthritis induced by subplantar injection in rats foot of heat-killed *Mycobacterium* species (Freund's adjuvant) in mineral oil.
Test drugs given for 14 d from 14-days post injection except meseclazone which was given from day 21 post-induction.

Flufenamic acid	Flurbiprofen	Ibuprofen	Indomethacin	Ketoprofen	Meseclazone[a]	Salicylate sodium
–	0.2	2.5–5.0	2.0–6.0	8.9–13.00	–	100–200
36.5–38.5	–	–	1.3	–	–	17.2
10.9	0.28	–		–	–	5.6
9.1	–	24.3	2.7	–	53	–
(4.0–20.7)		(10.6–39.2)	3.3 (2.1–7.3)			
–			12.9% @ 10			
–			1.7 @ 10			
–		150	3–9	–		
						–
–			2.9			72
–	–	–	1.2	–	40	–
–	–	–	1.6	–	100	–
–	–	–	IA @ 32	–	84	–
–	–	–	IA @ 32	–	IA @ 400	–
–	–	–	IA @ 32	–	IA @ 400	–
29.3	–	>45.0	0.3–1.7		145[b]	–
12	–	–	2.5	–	–	–
58.0	–	–	–	–	–	–

CHAPTER 7

TABLE 7.2

Anti-oedemic effects in response to various inflammagens of salicylates compared with other analgesics and anti-inflammatory agents (from Domenjoz, 1955).

Drug	Dose (s.c. mg/kg)	Per cent reduction			Dextran oedema	Hyaluronic acid oedema
		Formalin oedema				
		Normal	Hypo-physectomised	Adrenalectomised		
Aspirin	500	48	70	6	38	3
p-Aminosalicylic acid	500	28	41	−14	17	46
Sodium salicylate	500	10	5	−17	53	24
Aminopyrine	200	49	74	13	49	59
Antipyrine/Dipyrone	200	3	15	1	34	42
Metamizole	200	30	24	0	23	32
Phenylbutazone	200	64	62	42	38	61
ACTH	4 × 2.5	44	19	3	18	48
Cortisone	2 × 10	32	−8	20	38	51
Morphine HCl	10	37	34	8	46	21

TABLE 7.3

Comparison of potencies of salicylates and indomethacin in various animal models with their therapeutic doses employed in humans.

Drug	Carrageenan oedema (rat paw)		Adjuvant poly-arthritis (rats)				Therapeutic dose range in man[2]	
	mg/kg	mmol/kg	mg/kg	mmol/kg	mg/kg per day	(Potency)*	mmol/kg per day	(Potency)*
Aspirin	1.0	1.0	1.0	1.0	43–57	(1.0)	0.24–32	(1.0)
Benorylate	0.75	0.48	–	–	57+	(0.9)	0.18+	(3.5)
Diflunisal	11	16	7.5	10	7–11	(5.6)	0.028–0.044	(7.8)
Indomethacin	44	87	289	570	1.0	(47)	0.003	(100)
Salicylate	0.9	0.8	–	–	57+	(0.8)	0.36+	(0.8)

*Potencies calculated with respect to aspirin = 1.0; for clinical dosage, mid-range values chosen. The potencies from animal studies were calculated from the approximate mid-range values shown in Table 7.1.

[1]Data from Hannah et al. (1978); [2]Data from Miehlke (1978) and Otterness et al. (1979).

The first phase is accompanied by the release of histamine and 5-hydroxytryptamine (serotonin) from the mast cells of guinea pigs or rats respectively, and kinins generated in the bloodstream, all of which cause disruption of blood vessels in the inflamed tissue, so allowing extravasation of plasma proteins and penetration of inflammatory cells into the damaged site during the second phase (Vinegar et al., 1969; 1976; Di Rosa et al., 1971a; Garcia Leme et al., 1973; Bolam et al., 1974; Roch-Arveiller and Giroud, 1979). The amines and kinins also continue being released in the second phase, and contribute to the expression of inflammatory events in this phase (Vinegar, 1969; Bolam et al., 1974; Ferreira et al., 1974).

Release of lysosomal enzymes is evident within the first 10 to 30 minutes following subplantar injection of carrageenan, and progressively increases through to a maximum at 3 hours (i.e. during the second phase) (Anderson, 1970; Anderson et al., 1971; Weischer and Anda, 1975; Davies, 1977). These enzymes initiate tissue destruction and, accompanying this, production of tissue-destructive free oxygen radicals, which cause formation of highly reactive lipid peroxides (Oyanagui, 1976; McCord et al., 1979). These lipid peroxides appear to stimulate activity of phospholipases on phospholipids, so causing release of arachidonic acid and the generation of prostaglandins (PGs). These PGs are evident in large quantities during the second phase (i.e. 1 to 3 hours after carrageenan injection) (Willis, 1969; 1970; Kuehl et al., 1977), and reflect local induction in the inflamed paw of the COX-2 isoenzyme (Seibert et al., 1994) as well as that from accumulated leucocytes in which induction of this enzyme also occurs. Reactive oxygen species (ROS) or free oxygen radicals, possibly including singlet oxygen, produced as a consequence of the prostaglandin endoperoxide synthase (that catalyses the reaction of PGG_2 to PGH_2) and lipoperoxide metabolism also appear further to potentiate the cycle of the free radical-induced tissue damage, lipid peroxidation and prostaglandin production (Kuehl et al., 1977; 1979; De Vries and Verboom, 1980; Bragt and Bonta, 1981; Milanino and Velo, 1981). Hydroxyl ions produced by the iron-catalysed Fenton or Haber-Weiss reaction on H_2O_2 combine with NO to produce probably the most powerful oxidant peroxynitrite ($ONOO^-$).

A rapid influx of polymorphonuclear (neutrophil) leucocytes is most evident during the second phase (Di Rosa et al., 1971b; Vinegar et al., 1976) and these cells, together with aggregated platelets present around damaged endothelial tissues in the blood vessels, probably account for an appreciable amount of the eicosanoids (i.e. prostaglandins, thromboxanes and leukotrienes) that are produced in inflamed paws (Mustard and Packham, 1975; Vargaftig, 1977; Smith, 1979; Bray et al., 1981a). COX-2-derived prostaglandins are produced from neutrophils, monocytes/macrophages and endothelial cells, while platelet COX-1 also contributes to local prostanoid production (Cryer and DuBois, 1998; Kam and See, 2000).

That neutrophils are important in the second phase is shown by the fact that rats rendered deficient in these cells (i.e. granulocytopenic) show much reduced paw oedema during the second (but not the first) phase of the reaction (Vinegar et al., 1969; 1976; Di Rosa et al., 1971a). Furthermore, the degree of swelling in this phase is proportional to the number of neutrophils in the circulation (Vinegar et al., 1976). Migration of monocytes is only evident much later in the second phase, after about 3 hours, and progresses over the 5 to 24 hours after injection of carrageenan (Vinegar et al., 1969; Di Rosa et al., 1971a). The migration of both neutrophils and macrophages is controlled by the chemotactic factors (e.g. leukotrienes, chemokines and cytokines, complement components, endotoxins) that they produce, and these cells also produce enzymes that activate the complement pathway (Vinegar et al., 1976; Smith, 1979; Bray et al., 1980b). The migration of these cells is crucial to the development of the inflammatory response, since they produce a whole range of active oxygen species (superoxide anion, hydroxyl anions and hydrogen peroxide; Weissmann et al., 1978) and autolytic enzymes (Baggiolini et al., 1978), which cause release of more eicosanoids (Smith, 1979; Atkinson and Collier, 1980; Doig and Ford-Hutchinson, 1980), which are also evident in inflamed paws (Seibert et al., 1994). Thus there is enormous amplification of the initial inflammatory events by subsequent induction and activity of key enzymes involved in the inflammatory process. It should be noted that, by comparison with models of chronic inflammatory disease (e.g. rat adjuvant arthritis), there is little if any involvement of lymphocytic and humoral components of the immunological reactions during the acute paw oedema responses elicited following carrageenan injection.

Systemic responses to subplantar carrageenan are probably important in relation to the development of the inflammatory responses (Delhon *et al.*, 1977). In particular, the reductions in the blood leucocyte count, fibrinogen content and plasma proteins evident by 4 hours are of considerable significance (Strubelt and Zetler, 1980). Increases in the erythrocyte sedimentation rate and platelet aggregation are, however, only manifest at 24 hours after injection and thus do not contribute to the stages of the inflammatory reaction that cause the maximal oedema response (Strubelt and Zetler, 1980). While the paw oedema induced by subplantar carrageenan is essentially similar in adrenalectomised compared with normal (intact) rats (Atkinson and Leach, 1976; Strubelt and Zetler, 1980), there are indications that acute inflammatory reactions often elicit adrenocortical stimulation (Winter *et al.*, 1968) and this is associated with impairment of drug metabolism (Whitehouse and Beck, 1973; Whitehouse, 1977a). The carrageenan-induced paw oedema induced a small but significant increase in lipid peroxidation in the rat liver at 3 hours but not 24 hours after injection of the inflammagen (Robak, 1978). This may have relevance in the production of prostaglandins by way of a generalised systemic response. Elevated plasma levels of copper and the antioxidant protein ceruloplasmin have been observed 22 hours after injection of carrageenan, but not at earlier times (Conforti *et al.*, 1982; 1983a; 1983b). This may be relevant to the involvement of copper in free-radical generated inflammatory reactions (Oyanagui, 1976; McCord *et al.*, 1979; Milanino and Velo, 1981).

Otterness *et al.* (1979) and Otterness and Bliven (1985) compared the dose responses of a variety of anti-inflammatory agents tested in the carrageenan and UV erythema models with a clinical dose required for the effective treatment of rheumatoid arthritis. They concluded that the carrageenan oedema in rats is by far the best model of the two for predicting anti-inflammatory activity of these drugs in humans.

Inflammagens other than carrageenan sometimes elicit a different pattern of inflammatory responses to that observed with subplantar injection of carrageenan. For instance, with kaolin the paw oedema increases progressively up to 24 hours after injection (Gemmell *et al.*, 1979), whereas by this time the carrageenan-oedema has declined to almost half the peak response seen at 3 to 4 hours (Vinegar *et al.*, 1969; 1976). The kaolin oedema primarily involves the participation of kinins and prostaglandins, while histamine, serotonin and complement components only contribute in a very minor way to the response (Noordhoek *et al.*, 1977; Gemmell *et al.*, 1979). In contrast, amine liberation together with potent activation of complement pathways occurs in the immunological-type paw-oedema models (e.g. following injection of zymosan, antigen–antibody-type reactions; Gemmell *et al.*, 1979). The paw swelling induced in these acute immunological models clearly differs from both the carrageenan- and kaolin-induced oedema in that a broad monophasic response is obtained at about 3 hours after injection of the inflammogen (Gemmell *et al.*, 1979). Subplantar injection of some of the mediators of the acute inflammatory response (e.g. histamine, 5-HT, bradykinin or prostaglandins E_2 and I_2) also elicits paw swelling (Ferreira *et al.*, 1974; Ford-Hutchinson *et al.*, 1978; Higgs *et al.*, 1978). Generally, the quantity of these mediators required to elicit an oedematous reaction is above their concentration present in inflamed tissue; there is also evidence that some of the mediators tend to potentiate one another. These mediators elicit a generalised irritant reaction in the paw, setting in train the cascade of events that are quantitatively seen in the response to irritants such as carrageenan. Hence it is not possible to single out any one mediator as being the most important in acute inflammation, as was thought to be the case some years ago (Vane, 1971; Flower *et al.*, 1972; Vane, 1973; Flower, 1974). Furthermore, as discussed later, the evidence from studies with various drugs on isolated enzyme and cellular systems also mitigates against this concept.

PLEURISY, AIR-BLEB AND SPONGE-IMPLANTATION MODELS

These techniques have been developed to allow for easy and precise quantitative measurement of the chemical mediators and cellular components involved in the acute inflammatory response to various irritants, and the influence of drugs on these mediators and cells (Hurley *et al.*, 1968; Ishikawa *et al.*, 1968; Di Rosa *et al.*, 1971b; Willis *et al.*, 1972; Ammendola *et al.*, 1975; Williams and Johnson, 1976; Gilchrest and Watnick, 1977; Ford-Hutchinson *et al.*, 1978). In the pleurisy and air-bleb models inflammatory reactions are elicited by injection of irritants (such as carrageenan, turpentine or carboxymethylcellulose) into the pleural cavity or into a subcutaneous air bleb. This leads to a whole variety of reactions which, individually, can be difficult to distinguish in a temporal sense with the possible exception of the prolonged effects of cell migration and prostaglandin production (Ford-Hutchinson *et al.*, 1978; Horokova *et al.*, 1980). The sponge-implantation model, in which sponges impregnated with carrageenan are implanted (up to eight per rat) along the mid-line, allows for the temporal relationship between chemical mediators and cellular events to be determined, so that the effects of drugs can be established quantitatively (Ford-Hutchinson *et al.*, 1978). Quantitative production of prostanoids derived from COX-2 into the air pouch is possible when this is injected with carrageenan (Masferrer *et al.*, 1994). Vascular changes accompanying the development of pleural oedema can be measured using the Evan's blue dye effusion model (Sancilio, 1969).

It is, however, as well to remember that all these models suffer from the defect that the site of inflammation is well away from sites where there is involvement of cartilage and connective tissues of the peripheral joints, as in rheumatoid disease. There are indications that the responses elicited in the pleurisy models are quite different to those observed in the paw-oedema reactions, e.g. neutrophil migration and histamine involvement (Gilchrest and Watnick, 1977; Horokova *et al.*, 1980). Furthermore, the effects of some anti-inflammatory drugs on leucocyte emigration into the rat pleural cavity vary enormously according to seasonal conditions, suggesting that this may not be a very desirable model (Warne and West, 1978). The sponge-implantation model probably provides a reasonable basis for comparing some of the events involved in the early responses to carrageenan with those in the paw. It is clear that the time-course of responses in the sponge-implantation model are much slower than in the paw, so the sponge model does have the advantage in enabling clear time-dependent discrimination of the roles of chemical mediators and their effects on inflammatory cells.

Actions of salicylates in acute inflammatory models: comparative effects of different drugs

Comparison of the effects of salicylates in different models of acute inflammation is shown (with indomethacin as a standard) in Tables 7.1 to 7.3. From this, the following are evident:

1. The potencies of individual compounds vary considerably according to the type of irritant employed.
2. The drugs generally do not exert a profound effect when mediators (e.g. histamine, serotonin or bradykinin) are employed individually to elicit the paw inflammation.
3. Salicylate has about the same anti-oedemic activity in the carrageenan oedema as that of aspirin (Tables 7.1 and 7.3; see also Higuchi *et al.*, 1985; but see Chiabrando *et al.*, 1989). However, in the formalin-induced oedema aspirin is more potent than salicylic acid, and in the dextran and hyaluronic acid-induced oedemas aspirin has the lower potency of these two drugs (Table 7.2).
4. Diflunisal is the only salicylate in clinical use that is more potent than aspirin in the carrageenan paw oedema (Tables 7.1 and 7.3). Salicylate and benorylate are of the same order of potency as aspirin when compared on a milligram dose-for-body weight basis.
5. With the exception of diflunisal, the potencies (relative to aspirin) of these salicylates correspond quite well with the daily therapeutic dose range at which these drugs are employed in the clinic (Table 7.3). Based on the clinical potencies and dose responses in animal models (Tables 7.1 and 7.2), it appears that diflunisal, like indomethacin (Swingle, 1974b), is overestimated in potency in the carrageenan-oedema assay.
6. Generally NSAIDs inhibit both phases of the carrageenan oedema, but all, including the salicylates, are much less potent in the first compared with the second phase (Vinegar *et al.*, 1969; Swingle, 1974a). It should be noted that the carrageenan-oedema data in Table 7.1 are peak values recorded in the second phase of the oedema.
7. Salicylate has been shown to antagonise the anti-oedemic effects of aspirin in the carrageenan foot paw model (Telias *et al.*, 1985). It was suggested by the authors of this study that the mechanism for

this effect may be the interference by salicylate with the cyclo-oxygenase inhibitory actions of aspirin.

Role of mediators of the inflammatory response

The weak activity of the salicylates listed in Table 7.1 against the paw oedema elicited by the mediators of the first phase of the carrageenan-induced oedemas (during which histamine, serotonin and bradykinin are produced) is evidence for the fact that these drugs do not antagonise any one of the responses elicited by the mediators alone. While the formation of these mediators is affected by some salicylates (see p. 267 *et seq.*), it is apparent that the drug effect on any one of these substances is alone insufficient to attenuate the expression of the inflammatory reaction.

The relative significance of the influence of salicylates on prostaglandin production during the second phase of paw oedema has been the subject of some debate (e.g. see Smith, 1978; Bonta *et al.*, 1977; Vane, 1978; Higgs *et al.*, 1987). There are essentially two lines of evidence suggesting that inhibition of prostaglandin production, while having an important role, is not the only factor involved in the anti-oedemic activity of these and related drugs seen during the second phase.

First, aspirin and salicylate are equally effective in inhibiting the paw oedema, although under *in vitro* conditions salicylate is virtually ineffective as an inhibitor of prostaglandin synthesis whereas aspirin is a potent inhibitor (Smith *et al.*, 1975a; Chiabrando *et al.*, 1989). These differential effects of aspirin and salicylates are also observed *in vivo* when the drugs are applied locally in carrageenan-soaked sponges (Smith *et al.*, 1979). Since aspirin is metabolised rapidly to salicylate in rats (Hatori *et al.*, 1984) very little of the parent drug actually reaches the inflamed tissues in rats (Rainsford *et al.*, 1981; Higgs *et al.*, 1987), and so the effects of aspirin at the inflamed site must be due to systemically generated salicylate, which is the main metabolite accumulating at this site (Rainsford *et al.*, 1981). One of the clues to the anti-inflammatory equipotency of aspirin and salicylate in this model comes from the observations that both drugs are equally effective in inhibiting the migration of inflammatory leucocytes into the 9-hour carrageenan sponge *in vitro* (Walker *et al.*, 1976). This effect may be one of the most significant actions of the salicylates since, as already mentioned, early depletion of leucocytes (notably polymorphs) causes a marked reduction in the paw oedema and other acute inflammatory reactions. Another clue is that salicylate and aspirin are equi-active in reducing PGE_2 concentrations by 50 to 70 per cent in inflammatory exudates after oral administration of 200 mg/kg of the drugs. Thus it appears that the effect of aspirin in reducing prostaglandin production is related to salicylate that appears at inflamed sites (Higgs *et al.*, 1987). It is possible that salicylate is acting to inhibit the induction of the COX-2 isoenzyme (Wu *et al.*, 1991; Seibert *et al.*, 1994; Xu *et al.*, 1999).

The *O*-acetyl moiety of aspirin may exert specific effects on the early stages of prostaglandin-mediated oedema (e.g. acetylation of COXs and inhibition of 12-hydroxytetraenoic acid generation – see p. 264). This suggestion comes from the observations that aspirin (but not salicylate) inhibits the early stage of the arachidonic acid-potentiated carrageenan-paw oedema, an experimental model where prostaglandin production is deliberately enhanced during the first phase by co-injection of arachidonate with the carrageenan (Smith *et al.*, 1979). These studies show that the drug effects on leucocyte emigration and function may be of major significance in the acute inflammatory reactions, such as the carrageenan-induced oedema, apart from influences on prostaglandin production. Locally generated prostaglandins at inflamed sites are, in combination with histamine, bradykinin and leukotrienes, responsible for the enhanced permeability of damaged tissues. The effects of drugs such as aspirin on the generation of these mediators are distinct from those drug effects (e.g. phagocytosis, lysosomal enzyme release etc.) in the nonvascular compartments of damaged tissues (Williams and Peck, 1977; Ferreira, 1979; Westwick, 1979). Enhanced prostaglandin production from stimulated leucocytes (mainly polymorphs) (Walker *et al.*, 1976) and platelets (Smith *et al.*, 1976; Vincent *et al.*, 1978) that accumulate in inflamed sites also contributes to the cycle of prostaglandin-induced vascular changes and modulation of cell functions of these tissues. Thus by inhibiting leucocyte uptake into inflamed tissues the salicylates effectively prevent the accumulation of the sources of generation not only of

CHAPTER 7

prostaglandins but also of the reactive oxygen species (ROS), nitric oxide, leukotrienes and other inflammatory mediators that have such potent effects on the cycle of inflammatory responses.

A second line of evidence suggesting that salicylate effects on prostaglandin production may only partly explain the actions of these drugs in acute inflammation comes from studies in rats that have been depleted of prostaglandin precursors by being placed (prior to the induction of the oedema) on an essential fatty acid (EFA)-free diet (Bonta et al., 1977). Animals on such a diet have a deficiency in the linoleic (C18:2, omega 6), arachidonic (C20:4, omega 6) and other fatty acid precursors required for the biosynthesis of prostaglandins of the PGE_2 series (Van Dorp, 1971; 1978; Rivers and Frankel, 1981). EFA rats have a reduced paw oedema from carrageenan or kaolin injection (to about 60 per cent) compared with those on a normal (or EFA replete) diet (Bonta et al., 1974; 1976; 1977). This gives some indication of the quantitative involvement of prostaglandins in these models, and shows that prostaglandin-independent responses are of particular significance in mediating the inflammatory responses to carrageenan.

Bonta and co-workers found that aspirin exhibits an equal suppression of the carrageenan-induced oedema in rats on an EFA-deficient diet compared with that observed in control animals. They concluded from these studies that a major part of the anti-inflammatory activity of this drug must involve a mechanism *independent* of the inhibition of prostaglandin biosynthesis (Bonta et al., 1977). They suggested that drug effects on leucocyte migration could contribute to this prostaglandin-independent mechanism of the action of aspirin. Both dexamethasone and indomethacin exhibited anti-oedemic effects in EFA rats, so the prostaglandin-independent reduction in inflammation in this model is apparently common to all anti-inflammatory drugs (Bonta et al., 1976; 1977). While it is not possible to ascribe NSAID effects on the production of COX-2-derived prostaglandins from their effects on leucocyte emigration and their activation to produce ROS, proteases and more eicosanoids, it is probably a reasonable assumption that most NSAIDs (including the more potent salicylates) predominantly act in the acute phase (2 to 6 h) of the carrageenan model by affecting both prostaglandin generation and leucocyte emigration/activation.

Another important component of the inflammatory responses is part of the immunological system, the complement activation pathway. The protein components of this pathway initiate (a) increased blood vessel permeability; (b) tissue swelling; (c) infiltration of polymorphs; (d) activation of the kinin-forming system; and (e) lysis of cells involved in antigen–antibody reactions (Ward et al., 1979). *In vivo* and *in vitro* activation of the alternate pathway (involving properdin) can be produced by zymosan, a polysaccharide from yeast cell walls (Allison and Davies, 1974). Zymosan is phagocytosed by polymorphs and macrophages, and elicits production of prostaglandins (Davies et al., 1979; Gemmell et al., 1979; Smith, 1979). On injection into rat paws, zymosan causes a rapid increase in paw swelling at 0.5 h, which persists for up to 5 h after injection (Ford-Hutchinson et al., 1977; Gemmell et al., 1979). Appreciable prostaglandin production and cell migration is evident in zymosan-soaked polyvinyl sponges implanted for 5 and 16 hours in rats (Ford-Hutchinson et al., 1977). Like other NSAIDs, aspirin 100–300 mg/kg (given orally 1 h before the inflammagen) inhibits the zymosan-induced paw oedema in rats (Gemmell et al., 1979). Studies have shown that aspirin 0.4 mmol/1 (but not salicylate) inhibits complement activation *in vitro* by blocking both the classical and the alternate (properdin) pathways (Di Perri and Auteri, 1973; Voigtländer et al., 1980). Hence, the inhibition by aspirin of the zymosan-induced paw oedema could also be due to anti-complement activity (van Oss et al., 1961; Voigtländer et al., 1980), as well as other effects upon prostaglandin synthesis and cell migration.

It has been suggested that some NSAIDs exert their anti-inflammatory actions by stimulating the pituitary–adrenal system, so causing production of anti-inflammatory corticosteroids by affecting prostaglandin production (Gryglewsky et al., 1975). In drug screening it is usually considered mandatory to establish evidence of this by performing tests for anti-inflammatory activity assays in steroid-replete adrenalectomised animals (e.g. see Di Pasquale et al., 1975). Comparison of published data by Domenjoz (1955, Table 7.2) and Glenn et al. (1973; 1979) suggests that aspirin is much less potent in the formalin or carrageenan-induced paw oedema in adrenalectomised rats compared with that in normal rats. This indicates that aspirin may stimulate adrenocortical function, especially when high doses (>200 mg/kg) are used to produce anti-inflammatory effects. However, the results on hypophysectomised rats injected with formalin (Domenjoz, 1955; Table 7.2) indicated that aspirin is a more potent inhibitor in these animals, suggesting that hormones from the pituitary may partly blunt the anti-inflammatory effects of aspirin. The weak response produced in the carrageenan-induced paw oedema by aspirin in adrenalectomised rats is interesting in relation to the objection raised by Smith and Smith (1966) that the adrenocortical stimulation by the salicylates is a toxicological manifestation only observed at high doses of this drug. Further evidence to support this view comes from observations that the dose required for aspirin to stimulate release of free plasma corticosterone or depress adrenal ascorbate (an index of adrenal corticosteroid release) is much higher than that required to elicit anti-oedemic activity in rats (Cronheim *et al.*, 1952; Engelhart, 1978). Also, the effects of oral aspirin or salicylate on prostaglandin content and total leucocyte counts in 9 hour carrageenan-sponge exudates are identical in adrenalectomised as compared with normal rats (Bruni *et al.*, 1980). Other more potent NSAIDs (e.g. indomethacin, diclofenac, piroxicam) do, however, stimulate free plasma corticosterone levels in rats at doses that fall within the potency range required to affect anti-oedemic activity (Engelhart, 1978). Thus aspirin may differ from the more potent drugs by having effects independent of those on the adrenal gland.

In most studies in the carrageenan-induced paw oedema model, aspirin and other NSAIDs are given 0.5–1 h before injection of the irritant. The reason for this is that adequate blood and tissue levels of the drug are required for full expression of anti-inflammatory activity. In one sense, this type of assay design gives a measure of prophylactic effects of some NSAIDs. Therapeutic effects of such drugs are determined by giving the drugs *after* the injection of the inflammogen (Walker and Smith, 1979). This prior dosing has one advantage from the mechanistic point of view in that delayed effects of the drug (e.g. due to the delay in formation of an active metabolite or enterohepatic recirculation of the drug) or influences on the monocyte component of the inflammatory response (evident between 5 and 24 h) can be readily discriminated. In such therapeutic assays, aspirin 50 to 200 mg/kg has been found to exert effects only when given 2 to 24 h after the irritant, with the oedema being measured at short times (3 to 5 h) but not at longer times (24 to 26 h) after drug administration (Walker and Smith, 1979). No dose–response effects have been observed at 2 h, but there are direct dose–response effects apparent when the drug is given 6 h after the irritant (Walker and Smith, 1979). In view of the fact that blood and tissue levels of salicylate and aspirin decline appreciably by 24 h, it would seem that the therapeutic effects are either due to residual acetylation of proteins or other macromolecules, or to drug effects on monocytes (Militzer and Hirche, 1981).

Structure–activity relationships

To identify those structural components of the salicylates that are important for the acute anti-inflammatory activity (in the carrageenan-induced paw oedema), it is useful to examine the variation in anti-inflammatory activity with addition of various chemical substituents to the basic salicylate structure. In Table 7.4 a list has been compiled from data in three different laboratories (Whitehouse *et al.*, 1977; Hannah *et al.*, 1978; Glenn *et al.*, 1979; Rainsford and Whitehouse, 1980b). These data would appear to be reasonably self-consistent, and indeed the most extensive available in the literature.

The most significant contributions of specific chemical moieties to the acute anti-inflammatory activity of the salicylates (on a molar basis) are as follows:

1. A carboxylic acid group at the 1-position of the benzene nucleus (either unsubstituted or esterified with a readily hydrolysable alcohol or phenol moiety), or an aldehyde or oxide, *both combined* with a phenol (unsubstituted or esterified with a readily hydrolysable acyl or aryl moiety) at the 2-position, are apparently basic prerequisites for the potential expression of anti-inflammatory activity of the salicylates (compare compounds 1–5, 21–28, 30–32 and others in Table 7.4). Benzoic acid itself is ineffective in the carrageenan-induced oedema assay, illustrating the requirement for the 2-hydroxy substituent (Rainsford, 1981, unpublished results). However, thiosalicylic acid has potent anti-inflammatory activity in this model, thus indicating that sulphur may exchange with oxygen at the

CHAPTER 7

TABLE 7.4

Anti-inflammatory activity in the carrageenan paw oedema assay of various salicylate derivatives.

General structure

No.	Drug/compound	Substituents			Potency	
		R_1	R_2	R_3	(mg/kg)	(mmol/kg)
1	Aspirin	—COOH	—OCOCH$_3$	—	1.0	1.0
2	— Methyl ester	—COOCH$_3$	—OCOCH$_3$	—	0.8	0.9
3	— Ethyl ester	—COOC$_2$H$_5$	—OCOCH$_3$	—	0.8	0.9
4	— Phenyl ester (salol acetate)	—COOAr	—OCOCH$_3$	—	0.6	0.9
5	— Salicyl ester (diplosal acetate)	—CO—O—Ar$_2$COOH	—OCOCH$_3$	—	0.9	1.5
6	3-Acetoxy-aspirin (2,3-diacetoxybenzoic acid)	—COOH	—OCOCH$_3$	—OCOCH$_3$	1.0	1.3
7	Aspirin anhydride	—CO—O—OCAr—2OAc	—OCOCH$_3$	—	0.8	1.6
8	5-Chloro-aspirin	—COOH	—OCOCH$_3$	5-Cl	1.2	1.4
9	— Methyl ester	—COOCH$_3$	—OCOCH$_3$	5-Cl	0.7	IA
10	5-Bromo-aspirin	—COOH	—OCOCH$_3$	5-Br	1.2	1.7
11	— Methyl ester	—COOCH$_3$	—OCOCH$_3$	5-Br	0.7	1.1
12	3-Fluoro-aspirin	—COOH	—OCOCH$_3$	3-F	0.3	0.4
13	4-Trifluoromethyl-aspirin	—COOH	—OCOCH$_3$	3-CF$_3$	0.6	0.8
14	3,5-Dichloro-aspirin	—COOH	—OCOCH$_3$	3,5-Cl$_2$	0.9	1.3
15	5-(4-Fluorophenyl)-aspirin (flufenisal)	—COOH	—OCOCH$_3$	5-Ar-4F	3.7	5.8
16	5-Phenyl-aspirin	—COOH	—OCOCH$_3$	5-Ar	0.5	0.8
17	5-(2,4-Difluorophenyl)-aspirin	—COOH	—OCOCH$_3$	5-Ar	0.5	0.8
18	5-(2,4-Difluorophenyl)-o-ethylsalicylic acid	—COOH	—OCOC$_2$H$_5$	5-Ar-2,4-F$_2$	3.0	5.2
19	5-(2,4-Difluorophenyl)-o-butylsalicylic acid	—COOH	—OCOC$_3$H$_7$	5-Ar-2,4-F$_2$	7.4	13.5
20	5-(4-Fluorophenyl)-o-butylsalicylic acid	—COOH	—OCOC$_3$H$_7$	5-Ar-4F	3.5	6.1
21	Salicylic acid	—COOH	—OH	—	0.9	0.7
22	— Methyl ester	—COOCH$_3$	—OH	—	0.8	0.7
23	— Ethyl ester	—COOC$_2$H$_5$	—OH	—	0.8	0.8
24	— Phenyl ester (salol)	—COOAr	—OH	—	0.6	0.8
25	— Salicyl ester (diplosal or salicylsalicylic acid)	—COOAr—2OH	—OH	—	0.9	1.3
26	5-Chlorosalicylic acid	—COOH	—OH	5-Cl	1.8	1.7
27	— Methyl ester	—COOCH$_3$	—OH	5-Cl	0.8	0.9

No.	Compound				Potency	Potency
28	5-Bromosalicylic acid	–OH	–COOH	5-Br	1.5	1.8
29	3,5-Dichlorosalicylic acid	–OH	–COOH	3,5-Cl$_2$	0.9	1.1
30	Salicylaldehyde	–OH	–CHO	—	0.8	0.7
31	Salicylamide	–OH	–COHN$_2$	—	IA	IA
32	Salicylaldoxime	–OH	–CH=NOH	—	0.9	0.8
33	2,2′-Dithiosalicylic acid	–OH	–CSSH	—	IA	IA
34	3-Fluorosalicylic acid	–OH	–COOH	3-F	1.0	0.9
35	4-Fluorosalicylic acid	–OH	–COOH	4-F	0.2	0.2
36	4-Trifluoromethylsalicylic acid	–OH	–COOH	4-CF$_3$	IA	IA
37	5-Fluorosalicylic acid	–OH	–COOH	5-F	0.7	0.6
38	6-Fluorosalicylic acid	–OH	–COOH	6-F	IA	IA
39	5-Hexylsalicylic acid	–OH	–COOH	5-C$_6$H$_{11}$	IA	IA
40	3-Phenylsalicylic acid	–OH	–COOH	3-Ar	0.7	0.8
41	4-Phenylsalicylic acid	–OH	–COOH	4-Ar	0.6	0.7
42	5-Phenylsalicylic acid	–OH	–COOH	5-Ar	1.7	2.0
43	6-Phenylsalicylic acid	–OH	–COOH	6-Ar	IA	IA
44	3-Methyl-5-phenylsalicylic acid	–OH	–COOH	3CH$_3$,5-Ar	1.5	1.9
45	5-(4-Chlorophenyl)-salicylic acid	–OH	–COOH	5Ar-4Cl	3.7	5.3
46	5-(4-Fluorophenyl)-salicylic acid (desacetyl flufenisal)	–OH	–COOH	5Ar-4F	6.4	18.4
47	5-(2-Fluorophenyl)-salicylic acid	–OH	–COOH	5Ar-2F	4.5	1.7
48	5-(3-Fluorophenyl)-salicylic acid	–OH	–COOH	5Ar-3F	2.2	3.0
49	5-(4-Iodophenyl)-salicylic acid	–OH	–COOH	5Ar-I	0.7	0.2
50	3-Methyl-5-(4-fluorophenyl)-salicylic acid	–OH	–COOH	3Me,5Ar-4F	1.5	2.1
51	3-Propyl-5-(4-fluorophenyl)-salicylic acid	–OH	–COOH	3Pr,5Ar-4F	IA	IA
52	3-Fluoro-5-phenylsalicylic acid	–OH	–COOH	3F,5Ar	0.8	1.1
53	5-(2,4-Difluorophenyl)-salicylic acid (diflunisal)	–OH	–COOH	5-Ar-2,4-F$_2$	9.1	13.0
54	5-(2-Methyl-4-fluorophenyl)-salicylic acid	–OH	–COOH	5-Ar-2Me,4F	1.1	1.6
55	5-(3-Methyl-4-fluorophenyl)-salicylic acid	–OH	–COOH	5-Ar-3Me,4F	IA	IA
56	5-(3-Methoxy-4-fluorophenyl)-salicylic acid	–OH	–COOH	5-Ar-3MeO,4F	IA	IA
57	5-(2,3,4,5,6-Pentafluorophenyl)-salicylic acid	–OH	–COOH	5-Ar-2,3,4,5,6-F$_5$	4.5	13.0
58	3-(4-Chlorophenyl)-salicylic acid	–OH	–COOH	3Ar-4Cl	IA	IA
59	3-Methyl-5-phenylsalicylic acid	–OH	–COOH	3-Me,5Ar	1.5	1.9
60	3-Methyl-5-(4-fluorophenyl)-salicylic acid	–OH	–COOH	3Me,5Ar-4F	1.5	1.4
61	3-Propyl-5-(4-fluorophenyl)-salicylic acid	–OH	–COOH	3Pr,5Ar-4F	IA	IA
62	5-(N-Pyrryl)-salicylic acid	–OH	–COOH	5-(N-C$_4$H$_4$N)	1.8	2.0
63	4-(N-Pyrryl)-salicylic acid	–OH	–COOH	4-(N-C$_4$H$_4$N)	0.5	0.5
64	5-(N-Imidazolyl)-salicylic acid	–OH	–COOH	4-(N-C$_3$H$_3$N$_2$)	1.2	1.4
65	5-(4-Thiazolyl)-salicylic acid	–OH	–COOH	5-(4-C$_3$H$_2$NS)	1.0	1.2
66	4-(4-Thiazolyl)-salicylic acid	–OH	–COOH	4-(C$_3$H$_2$NS)	1.3	1.6
67	5-(2-Thienyl)-salicylic acid	–OH	–COOH	5-(2-C$_4$H$_3$S)	1.5	1.8

Potency values calculated from data published by Jones et al. (1978), Hannah et al. (1978), Glenn et al. (1979), Rainsford and Whitehouse (1980b) with potencies being with respect to aspirin = 1.0. All values <1.0 indicate potencies less than aspirin, and those >1.0 indicate potencies greater than aspirin. IA indicates inactive. (From Rainsford (1984).)

2-position but not at the carbon-1 position (compare compound 33 with salicylic acid, Table 7.4). Carboxyl and phenolic esters appear to exert their effects following hydrolysis *in vivo* (see Rainsford *et al.*, 1980, and Chapter 4).

2. Acetyl esters of non(ring)-substituted acids or their alkyl or aryl esters are slightly more active than their corresponding phenols (compare compounds 1–5 with 21–25 in Table 7.4). Addition of electron-withdrawing substituents (e.g. halogens) at positions 3 to 5 of the ring reduces the potency of these acetyl esters compared with the corresponding phenols (compare compounds 8, 10 and 12 with 26, 28 and 34, respectively, Table 7.4).

3. Addition of electron-withdrawing (and more lipophilic) substituents (e.g. halogens or phenyl substituents) has little effect on the anti-inflammatory activity of salicylic acid, with the exception that the 5-chloro- or 5-phenyl-analogues are much more effective than salicylic acid, the 5-bromo- or 5-hexyl-derivatives (compare compounds 21 with 26, 28, 29, 34–43, Table 7.4; Sofia *et al.*, 1974; 1975; 1979). As evident above, this relationship does not hold for the aspirin analogues substituted in the same way. The ineffectiveness of the 5-hexyl- compared with that of 5-phenylsalicylic acid) suggests that the mere provision of a bulky substituent at the 5-position of salicylic acid (as in the 5-hexyl compared with that of 5-phenylsalicylic acid) suggests that the mere provision of a bulky substituent at the 5-position of salicylic acid (as in the 5-hexyl-derivative does not confer any specific advantages. Clearly, the electronic charge of the second aromatic ring added at the 5-position (such as might be required for a charge transfer reactions with mediators of the inflammatory process) confers specific advantages over other substituents. Similar influences from the addition of aromatic substituents at the 5-position (of salicylic acid) can be seen with the 5-(*N*-pyrryl)-, 5-(*N*-imidazolyl)-, and 5-(2-thienyl)-derivatives (compounds 62–64 and 67, Table 7.4). The effect of adding the thiazolyl group at the 5-position creates a more potent compound than the 5-derivative (compare compound 64 with 65, Table 7.4).

4. Addition of a fluorine group at the 4-position of the second ring (i.e. the 5-(4-fluorophenyl)-derivative) markedly enhances the potency of the 5-phenyl-substituted salicylic acids (compare compound 42 with 46, Table 7.4). Furthermore, the type of halogen and its position of substitution on the second ring appears to be quite specific in its effects. Substitution of a chloro- or iodo- group and/or placing these at other positions in the second ring reduces their potency compared with the 5-(4-fluorophenyl)-derivative (compare compounds 45, 47–49, 52 and 58 with 46, Table 7.4). Addition of alkyl substituents to the 5-(4-fluorophenyl)-salicylic acid actually reduces the potency of this compound (compare compounds 50, 51, 60 and 61 with 46, Table 7.4). There is obviously no advantage in adding extra fluorine atoms on the ring of the 5-(4-fluorophenyl)-derivative in terms of molar potency (compare compounds 53 (diflunisal) and 57 with 46, Table 7.4). However it should be noted that, when considered on a body weight basis, compound 53 (diflunisal) is more potent than compound 46 (desacetyl-flufenisal).

5. The marked reduction in potency caused by addition of 5-phenyl- or 5-(2,4-difluorophenyl) groups to aspirin contrasts markedly with the enhancement of activity observed with addition of these substituents to salicylic acid (compare compounds 16 and 17 with 1, 42 and 53, Table 7.4). This reduction in potency of the 5-phenyl- or 5-(2,4-difluorophenyl) aspirin derivatives is not evident with addition of higher carbon alkyl esters (compare compounds 17–19, Table 7.4). Indeed, the butyl ester of diflunisal (compound 20) is slightly more potent (on both a molar or weight basis) than diflunisal itself (compare compound 19 with 53, Table 7.4). This suggests that there may be influences of aromatic substitution on alkyl hydrolysis (e.g. by esterases, see Chapter 4).

The impression given by the above comparison is that the key to enhancing the potency of salicylates lies in addition of certain 5-substituents (Hannah *et al.*, 1978). Dearden and George (1979) have however suggested, on the basis of a Hansch-type structure–activity analysis (Hansch, 1971; Hansch *et al.*, 1973; Hansch and Leo, 1979) of paw oedema data of Winter *et al.* (1962), that it is not the 5-substitution that enhances potency, but rather the 4-substitution that lowers potency relative to all other substitutions. Regrettably, these authors (Dearden and George, 1979) obtained estimates of the lipophilicity (i.e. log P) of the derivatives they studied by measuring the partition coefficients (P) of each of their compounds in octanol-'aqueous buffer pH 1.1'. Since the buffer system is at such an acidic pH, these acids would be completely non-ionised (protonated), creating an essentially non-physiological pH situation (normally pH 7.4 is used). The whole concept of using the partition coefficient as a measure of the movement of drugs through membranes and interaction with putative receptor sites, as ascribed by theory (Hansch, 1971), is completely ignored in the studies by these authors (Dearden and George, 1979). Thus, their conclusions have little validity or applicability to the *in vivo* situation. The general conclusions about the enhancing effect of 5-substitution stated above appear, therefore, to be justified (Hannah *et al.*, 1978).

Using quantitative structure–activity relationship (QSAR) methods based on Hansch parameters, as well as free-energy (Wilson) and Fijita–Ban analysis, Gombar *et al.* (1983) compared the physicochemical contributions of each of the substituents of salicylic acid with their anti-oedemic activity (using the data of Hannah *et al.*, 1978). Notable among these observations was that the 5-phenyl-substituent increased the biological activity, and this was enhanced when electron-withdrawing ring substituents were added on this ring. These molecular data give important insights to help the understanding of the chemical properties of the salicylates, underlying their biological activity, and give further support for the chemical rationale for the development of diflunisal.

There are two main outcomes from this comparison of chemical structure with anti-inflammatory activities of the salicylates:

1. It provides the rational basis for the development of diflunisal (Hannah *et al.*, 1977), which is clearly more potent than aspirin or salicylic acid.
2. With knowledge of the potency characteristics *in vivo* it is possible to:
   • Establish which chemical groups contribute most significantly to the anti-inflammatory activity, and from this design hypothetical receptors and/or binding sites to aid in discriminating biochemical and cellular actions of these drugs *in vitro* and *in vivo*.
   • Exploit particular compounds as tools for unravelling which moieties are involved in the expression of specific biochemical and cellular functions in the anti-inflammatory actions of these drugs.

RESPONSES TO ESTERS OF SALICYLATES

Topical application of liposoluble esters of salicylates is a well-established and effective means of controlling acute local inflammation in humans. Application of a 1 per cent methyl salicylate poultice (cataplasm) to rats' paws at the same time or up to 1 h after carrageenan injection at this site induces a sustained anti-inflammatory effect that lasts for up to 72 h after injection of the inflammagen (Hattori, 1978a). This 'therapeutic' anti-inflammatory effect of topical methyl salicylate is the same in adrenalectomised as in intact animals, indicating that there is no adrenocortical component in the anti-inflammatory activity of the drug (Hattori, 1978b). Some anti-inflammatory effects of methyl salicylate are evident at 24 to 72 h when the time between subcutaneous carrageenan injection and application of the methyl salicylate poultice is increased to 2 to 3 h (Hattori, 1978a), suggesting that there is an effect of the drug on the later components (e.g. polymorph and monocyte infiltration).

OTHER INFLAMMATORY MODELS

The responses of aspirin and other salicylates in acute paw oedema elicited by other inflammagens (Tables 7.1 and 7.2) varies considerably according to the type of agent employed. With kaolin and sodium urate the potency of aspirin falls within the range seen with carrageenan, suggesting the effects of the drug are predominantly on COX-2 and NO but not on immune pathways (e.g. complement activation). Interestingly, 5-chlorosalicylic acid a uricosuric agent is inactive in the urate model and shows about half the activity of aspirin in the kaolin model; indomethacin, in comparison, is equi-active in all three models. Moreover, aspirin (as well as 5-chlorosalicylic acid) is inactive against 5-HT (serotonin)- and bradykinin-induced paw oedemas whereas indomethacin is a potent inhibitor in both these models.

Aspirin is also a weak inhibitor of the polymyxin-B-, formalin- and dextran-induced paw oedemas,

CHAPTER 7

while indomethacin is a potent inhibitor of polymyxin-B but not the other two models. Aspirin and salicylate inhibit PAF-induced paw oedema whereas many other NSAIDS are inactive in this model (Cordieto *et al.*, 1986). These results reflect differing modes of action of these drugs on immune compared with inflammatory mediator – (i.e. prostaglandin/NO) based pathways of inflammation.

Domenjoz (1955) observed that high doses of aspirin (500 mg/kg) and other analgesics given s.c. to rats exhibited variable anti-oedemic activity in the formalin-, dextran- and hyaluronic acid-induced paw oedema models (Table 7.2). Thus aspirin, aminopyrine, dipyrone (metamizole) and phenylbutazone exhibited some anti-inflammatory effects in the formalin- and dextran-induced oedema models, but aspirin was without effect in the hyaluronic acid oedema model in which these drugs also had effects. Sodium salicylate 500 mg/kg had no effect in the formalin-induced oedema model, whereas this drug did exhibit anti-inflammatory effects in the dextran-induced oedema model. These variable effects of aspirin and salicylate might be explained in the relation to the particular mediators released during the oedemic response in these different models.

It is of interest that adrenalectomy largely abolished the anti-inflammatory effects of most of those drugs that showed anti-oedemic effects in the formalin-induced oedema model, with the exception of phenylbutazone and cortisone. With the latter corticosteroid the positive anti-inflammatory effect is to be expected as a result of replenishing stores of this hormone. The negative effects of drugs other than phenylbutzone suggests that there may be adrenocortical stimulation as part of the responses of the analgesics other than phenylbutazone in this model. Removal of the hypophysis did not, however, result in any appreciable difference in anti-inflammatory effects of salicylates or other analgesics in this model. In comparison with the responses in adrenolectomised animals this suggests that the effects of these drugs is not at the level of the pituitary–hypophyseal system but at the level of the adrenal glands. Domenjoz (1955) showed that most of the drugs depleted ascorbate stores in the adrenal glands of normal but not hypophysectomised animals, indicating that there may be a mechanism of anti-oedemic activity mediated via the adrenal gland and involving release of ascorbate possibly via the hypophyseal–pituitary system.

These early studies, while performed with high doses of parenterally-administered drugs, are interesting regarding the potential of salicylates to affect the pituitary–adrenal system via ascorbate. These aspects deserve more detailed investigation in relation to current understanding of the mechanisms of these drugs on hormonal and physiological pathways that are regulated by hormones.

Other features of interest found in these studies are that (a) *p*-aminosalicylate had anti-oedemic effects in the hyaluronate-induced oedema model but less so in the other models, suggesting this agent may have some protective effects in tissue destruction; (b) morphine HCl had anti-oedemic effects in all models, implying that opiate pathways mediate neuro-immuno-inflammatory mechanisms leading to expression of anti-inflammatory responses; and (c) aminopyrine had greater anti-oedemic actions in the formalin-induced oedema model than antipyrine, the latter being generally regarded as having only analgesic effects.

A topical inflammatory model that has been very popular in recent years involves the application of a topical irritant such as croton oil, arachidonic acid or other irritants in dimethylsulphoxide or another organic solvent (Tubaro *et al.*, 1985; Gábor, 2000). The degree of inflammation can be quantified by excision and weighing of the ears to which the irritants have been applied, by measuring the accumulation of radiolabelled leucocytes, or by histology. A large number of agents have been tested in these models (Carlson *et al.*, 1985; Gábor, 2000). When tetradecanoylphorbol acetate (TPA) or arachidonic acid are used as topical irritants it is possible to detect the activity of orally administered combined cyclooxygenase-lipoxygenase inhibitors or 5-lipoxygenase inhibitors (Carlson *et al.*, 1985). Some other non-anti-inflammatory drugs have the capacity to reduce the inflammation, so it is important to perform control experiments. It is, however, notable from the studies by Carlson *et al.* (1985) that the arachidonic-induced inflammation is less likely to detect non-anti-inflammatory compounds than that of TPA.

In the arachidonic acid- and TPA-induced oedemas high doses (200 mg/kg p.o.) of aspirin only exhibits weak anti-inflammatory activity in mice that is in many respects characteristic of that of other NSAIDs with the notable exception of zomepirac, the corticosteroids and the immunomodulator, dapsone.

Chronic polyarthritis in rats

The chronic polyarthritis induced in rats by the injection of heat-killed *Mycobacterium tuberculosis* (Freund's adjuvant) or *M. butyricum* has been extensively used to study the mode of action of NSAIDs, including the salicylates. The clinical and pathological features of the model (Glenn and Gray, 1965a, b; Fujihira *et al.*, 1970; Amkraut *et al.*, 1971; Beck and Whitehouse, 1974; Swingle, 1974b; Whitehouse *et al.*, 1974; Mohr and Wild, 1976; Van Arman, 1976; Harris, 1981) resemble rheumatoid arthritis, although it does have some features in common with Reiter's syndrome – a seronegative arthritic condition of the lower vertebrae, knee and ankle, prevalent mostly in males and accompanied by diarrhoea (initially), urethritis, conjunctivitis and prostatitis (Dick, 1972a; Dieppe *et al.*, 1985). Nonetheless, there are very useful features of this model that enable exploration of the mechanisms of drug action in respect of *individual* pathological changes relevant to rheumatoid arthritis. As long as it is clear what pathological features are prevalent in adjuvant disease and are relevant to rheumatoid arthritis, then it is possible to make useful extrapolations to the human disease from studies in this model. The specific limitations and advantages of this model will become evident in the following discussion.

The experimental results obtained from studies in adjuvant disease sometimes vary considerably in different laboratories according to (a) the preparation, source and particle size of *Mycobacterium* sp; (b) the suspending agent for preparation of the adjuvant; (c) the route of injection (i.e. s.c. in tail base or subplantar region) plus the age, strain, sex and housing conditions of the rats; and (d) the timing of drug administration in relation to induction of the disease (Fujihira *et al.*, 1970; Amkraut *et al.*, 1971; Beck and Whitehouse, 1974; Swingle, 1974b; Whitehouse and Beck, 1974; Whitehouse *et al.*, 1974; Liyanage *et al.*, 1975; Van Arman, 1976). These factors must be appreciated in analysing the data obtained by different authors.

Adjuvant disease is apparently specific to rats, although a weak reaction to the adjuvant may be elicited in mice and man (Graeme *et al.*, 1966; Mohr and Wild, 1976; Metzke, 1977). That the disease is specific to rats may be due to some immunological, physiological or other biological feature of this species (Pearson, 1966; Swingle, 1974b; Watnick, 1975; Mohr and Wild, 1976; Van Arman, 1976). The mycobacterial (i.e. Freund's) adjuvant is only one capable of inducing the disease, there being a whole host of agents that produce stimulation of different immunological responses in a variety of animals (see Whitehouse, 1977b). A variety of Gram-positive bacteria or extracts therefrom can induce adjuvant arthritis in rats (Waksman *et al.*, 1960; Newbold, 1964; Pearson and Wood, 1964; Glenn and Gray, 1965; Whitehouse *et al.*, 1969; Glenn *et al.*, 1973; Koga and Pearson, 1973; Koga *et al.*, 1976a; 1976b; Kohashi *et al.*, 1976; Vernon-Roberts *et al.*, 1976; Whitehouse, 1978). Specific accumulation of the bacilli particles, or of the arthritogenic peptidoglycan which is present in the bacilli in the lymph nodes is mandatory for the development of the disease (Glenn and Gray, 1965a; Newbold, 1974; Koga *et al.*, 1976c).

The development of arthritis is thought to be a T-cell mediated delayed-type hypersensitivity reaction (Whitehouse *et al.*, 1969; Kourounakis and Kapusta, 1976; Chang, 1977a; 1977b; Kayashima *et al.*, 1978; Gans *et al.*, 1980) analogous to that occurring in rheumatoid arthritis (Harris, 1981). This may result from (a) activation of a latent and as yet unidentified arthritogenic virus; (b) formation of an immunogenic component of host tissue (i.e. due to the effects of the adjuvant); or (c) combination of these two processes (Chang *et al.*, 1980; Harris, 1981). The latent virus hypothesis has been supported by observations that the antiviral agent interferon, or agents that stimulate endogenous interferon production, markedly suppress the development of adjuvant polyarthritis (Kapusta and Mendelson, 1967; Chang, 1977b; Chang and Hoffman, 1977). Interferon may also inhibit cell proliferation of lymphocytes (Evinger *et al.*, 1981) as well as virus replication, so that evidence of viral involvement in adjuvant diseases with this agent could be equivocal. However, adjuvant disease can be induced by injection (in mineral oil) of a compound (N,N-dioctadecyl-N',N'-bis(2-hydroxyethyl)propane diamide) that is devoid of any of the immunological properties of the peptidoglycan arthritogens, suggesting that the disease is triggered by relatively non-specific agents such as might be produced by virus activation (Chang *et al.*, 1980).

The histopathological and fine structural (i.e. electron microscopic) changes seen in the joints of the hindlimbs and tail of adjuvant-diseased animals varies with time depending on whether the adjuvant is injected into the base of the tail or into a rear foot pad (Pearson and Wood, 1964; Glenn and Gray,

■ CHAPTER 7 ■

1965a; Watnick, 1975), If the adjuvant is injected into the tail base, the pattern of inflammatory cell infiltration and synovial changes appears to be about the same as that seen in the non-(adjuvant) or vehicle-injected site (Pearson and Wood, 1963; Glenn and Gray, 1965b; Mori et al., 1970; Zurier and Sallas, 1973; Militzer, 1975; Watnick, 1975; Mohr and Wild, 1976; Nusbickel and Troyer, 1976). In the injected paw the oedematous swelling (within hours of injection) is accompanied by inflammatory cell infiltration, which later progresses in intensity as the disease proceeds into the chronic phase (Watnick, 1975; Mohr and Wild, 1976; Van Arman, 1976).

The sequence of events observed histologically in the non-injected hindlimb joints or in tail-injected rats begins after 5 to 10 days with the appearance of fibrin-like material in the articular cavity, which probably comes from the exudation of blood (Pearson and Wood, 1963; Mohr and Wild, 1976; Evinger et al., 1981). This appears to induce a rapid proliferation (synovitis) in close proximity to the fibrous tissue (Pearson and Wood, 1963; Glenn and Gray, 1965; Mori et al., 1970; Zurier and Sallas, 1973; Militzer, 1975; Watnick, 1975; Mohr and Wild, 1976; Nusbickel and Troyer, 1976). The first sign of a local inflammatory process in the paw is an increase in the number of mast cells (Mohr and Wild, 1976; Van Arman, 1976). This is followed (by day 10 post-inoculation) by infiltration of poly-morphs and monocytes into tibial bone marrow, synovium, tendon sheaths, bursae and periosteal sur-faces (Pearson and Wood, 1963; Mohr and Wild, 1976; 1977; Glenn et al., 1977). Lymphocytes also begin to accumulate in the synovial tissue (Pearson and Wood, 1963; 1964; Glenn and Gray, 1965; Watnick, 1975; Mohr and Wild, 1976).

The pattern of fibrin accumulation and leucocyte infiltration into the synovium and marrow in adjuvant disease is similar to that observed in rheumatoid arthritis (Pearson and Wood, 1963; Glenn and Gray, 1965a; Mori et al., 1970; Mohr and Wild, 1976). However, true lymphoid follicles (contain-ing T and B lymphocytes and plasma cells), which appear in the synovial tissue in rheumatoid disease and are considered pathognomic of the disease (Fassbender, 1976; Harris, 1981), are not frequently observed in the synovium of adjuvant-arthritic rats (Mohr and Wild, 1976). In adjuvant disease, however, there is a marked increase in the size and weight of the popliteal lymph nodes (maximal at 14 to 21 days), specifically those in the ipsilateral paw and/or in the lymphatics draining the site of tail injection (Watnick, 1975). Proliferation of T-cells and other lymphocyte populations is an important consequence of the effects of mycobacterial adjuvant (Kourounakis and Kapusta, 1974; 1976).

Histological changes in the lymph nodes of the hindlimb are evident 2 days after injection into the tail base of the adjuvant (Mori et al., 1970). A rapid influx of plasma cells is evident by this time, fol-lowed by the appearance of granulomatous cysts and foci of necrotic tissue in lymph nodes and spleen at 6 to 7 days (Mori et al., 1970). Later the tissue adjacent to tibio-tarsal joints proliferates and lipid vac-uoles appear (Glenn and Gray, 1965; Mori et al., 1970). Occasionally giant cells are seen in the lym-phoid tissue of adjuvant-arthritic rats (Mohr and Wild, 1976; Nusbickel and Troyer, 1976). This type of cell is seen in large numbers in the synovial tissue in rheumatoid arthritis, and is often considered a distinctive feature of this disease in humans (Mohr and Wild, 1976; Harris, 1981). Thus the histological changes seen in the hindlimb lymph nodes in adjuvant-arthritic rats are similar to the lymphoid nodes that appear in rheumatoid arthritis, so the distinctions between the disease in both species may not be so marked. It may be that anatomical restrictions prevent lymphoid follicles developing in the synovial membrane of adjuvant-treated rats.

In addition to the increase in proliferation of the synovial cells and fibroblasts that is evident in the early stages of the disease, there is also marked proliferation of the endothelial and other cells lining the vasculature (as in rheumatoid arthritis). The A (or macrophage-like) and B (actively protein synthesis-ing) cells that appear in the rheumatoid synovium are also evident in affected synovial tissue of adjuvant-arthritic rats (Muirden and Peace, 1969). Ferritin deposits (derived from extravasated blood) also appear in the lysosomes of macrophages in both diseases (Muirden and Peace, 1969), but the pathological consequences of this are unknown.

By about days 5 to 10 post-inoculation, a granulomatous outgrowth of tissue from the synovium (pannus) occurs into the joint space (Pearson and Wood, 1963; Mohr and Wild, 1976), which is also a striking feature of rheumatoid arthritis (Pearson and Wood, 1963; Fassbender, 1976; Harris, 1981). Accompanying the invasion of pannus is progressive destruction of cartilage mucopolysaccharides as well as collagen and other components of the underlying bony matrix (Nusbickel and Troyer, 1976). This damage appears to be due to the release of lysosomally derived hydrolytic enzymes from both the pannus tissue and those polymorphs that have migrated into the synovium and bone marrow (Mohr and Wild, 1976; Nusbickel and Troyer, 1976). Lymph node cells also participate in forming granulomatous pannus tissue (Parnham, 1980). In the final stages of the adjuvant disease, destruction of bony tissue near joints leads to their fusion (fibrotic ankylosis) and immobilisation (Pearson and Wood, 1963; Mohr and Wild, 1976), a similar event to that seen in rheumatoid arthritis (Pearson and Wood, 1963; Fassbender, 1976; Mohr and Wild, 1976). Outgrowths of bone from periosteum (exostoses) and extensive tendon damage also become evident (Pearson and Wood, 1963; Mohr and Wild, 1976). Similar pathological features are evident in the tail and lower spine to those seen in synovial joints of the hindlimb, even though articulations in the spine are of a fibrocartilagenous type (Pearson and Wood, 1963).

Extra-articular or systemic changes accompanying the development of adjuvant disease are very marked, and involve a variety of organs in the body. Some of these changes are similar to those observed in rheumatoid disease (Glenn and Gray, 1965; Van Arman, 1976). Changes in the circulating blood components include the following:

1. An initial decrease in the content of haemoglobin and iron, followed later by an anaemia resulting from depression in the number of red blood cells (Mori et al., 1970).
2. An increase in the platelet count with an accompanying increase in thrombus formation (Görög and Kovaks, 1976; Van Arman, 1976).
3. An increase in the red cell sedimentation rate (Watnick, 1975; Görög and Kovacs, 1970).
4. Increases in the plasma levels of acute phase proteins (haptoglobin, seromucoid), certain liver-derived enzymes (e.g. transaminases), collagenase, protease, lysozyme, orosomucoid, alkaline and acid hosphatases, α-, β- γ-globulins, gastrin and fibrinogen ('inflammation units') (Glenn and Gray, 1965; Piliero and Colombo, 1969; Mori et al., 1970; Collins and Lewis, 1971; Walz et al., 1971; Sternberg et al., 1975; Van Arman, 1976; Parrott and Lewis, 1977; Rooney et al., 1978; Kuberasampath and Rose, 1980).
5. Decreased serum albumin content (Van Arman, 1976).
6. Increases in the content of all leucocytes (Van Arman, 1976; Parnham, 1980)

While all these changes have some parallels in rheumatoid arthritis, there is one important difference between adjuvant and rheumatoid arthritis in that there are no demonstrable antigen–antibody complexes combined with complement present in blood or synovial fluid of rats with adjuvant disease (Van Arman, 1976). These immune complexes (including rheumatoid factor) are an important diagnostic and indeed immunopathological feature of rheumatoid arthritis (Van Arman, 1976). There are also extensive changes in the reticulo-endothelial system in adjuvant disease (Fujihira et al., 1970), which include alterations in the phagocytic activity (Parrott and Lewis, 1977). Also the mitogenic activity of lymphocyte populations is variously impaired or enhanced depending on the tissues from which these cells were derived and on the presence of serum (inhibitory) factors (Kourounakis and Kapusta, 1974; Kahan et al., 1975; Binderup et al., 1976).

Marked impairment of liver metabolism accompanies the development of adjuvant disease, and includes (a) reduced ability of this organ to detoxify foreign compounds (Morton and Chatfield, 1970; Whitehouse and Beck, 1973; Mathur et al., 1978; (b) increases in lysosomal hydrolases (Arrigoni-Martelli and Restelli, 1972); and (c) defective calcium-controlled mitochondrial functions (Barrit and Whitehouse, 1977).

Local biochemical changes in the paws or connective tissues in adjuvant disease include:

1. Increased lysosomal and collagenolytic activity (Anderson, 1970).
2. Increased levels of PGE_2 (but not the F-type prostaglandins) (Parnham et al., 1978) that are probably derived from COX-2, although some COX-1 derived PGs from monocytes and platelets would also be expected. It is noteworthy that both COX-1 and COX-2 are expressed by T cells (Pablos et al., 1999), being cell types that are pivotal in control of the cell-mediated processes underlying inflammation.

3. Decreased content of cyclic AMP (Parnham *et al.*, 1978).
4. Increases in the turnover of sulphated proteoglycans and content of mucopolysaccharides (Exer *et al.*, 1976).

These detailed considerations of the development of the chronic adjuvant polyarthritis and comparison of this with rheumatoid arthritis are important to show the specific biochemical and cellular effects of the salicylates and to compare these actions with those of other NSAIDs. Without a definite basis on which to identify and compare their individual actions, it is impossible to establish the relevance of the drug effects to their therapeutic effects on the pathology of the disease process.

Comparative effects of salicylates and other NSAIDs in adjuvant arthritis

In Table 7.3, it can be seen that the therapeutic potencies of aspirin and diflunisal in adjuvant disease compare well with the relative dose ranges for these drugs employed in the treatment of arthritis in man (i.e. on a milligram/weight or molar basis). From these data it can be seen that (a) the effective dose (ED_{50}) ranges reported for therapeutic effects of aspirin have been variously quoted from 78 to 279 mg/kg per day (Martel *et al.*, 1974; Hannah *et al.*, 1978), and (b) diflunisal is about seven to ten times more potent than aspirin and three to four times more potent than flufenamic acid, ibuprofen and naproxen. It should be emphasised that these drugs are given in the so-called therapeutic treatment, i.e. at a time approaching or after the disease has become manifest ('blown') (Van Arman, 1976).

Aspirin usually inhibits the so-called primary paw swelling (i.e. in the injected paw) within the dose range of 100 to 250 mg/kg (Graeme *et al.*, 1966; Wax *et al.*, 1975). There does not appear to be much difference in the effective dose for inhibiting the paw swelling following tail injection (Wax *et al.*, 1975; Stone *et al.*, 1977). In the primary paw swelling, aspirin is 0.087 times as effective as phenylbutazone and 53 times less active than indomethacin (Graeme *et al.*, 1966).

Van Arman observed an interesting drug antagonism when aspirin was given at a low dose of 3 mg/kg per day on day 4 prior to the induction of adjuvant disease, followed by administration of a prophylactic dose of aspirin 150 mg/kg per day (Van Arman, 1976). This pretreatment was sufficient completely to prevent the anti-inflammatory effects normally seen with aspirin 150 mg/kg per day given from 1 day either before or after the injection of the adjuvant (Watnick, 1975). No explanation for this effect has been advanced by this author (Van Arman, 1976), but it could be due to a wide variety of subtle sensitising actions of the drug on cellular and immunological responses controlling the responses to inflammatory insult.

Regrettably, very few reports exist on the quantitative responses (in the form of ED_{50} values) of the effects of the other salicylates in adjuvant arthritis. Sodium salicylate has been found to be as effective as aspirin (on a weight basis) when administered orally to rats in which the disease was induced by paw injection of the mycobacteria (Newbold, 1963). This is of particular interest, since clinical reports (see Chapter 12) suggest that there is little difference in analgesic effects in rheumatic patients. Clearly this aspect requires further study in adjuvant rats to (a) establish the therapeutic significance of the *o*-acetyl group of aspirin, and (b) determine whether it would be more advantageous to recommend the use of salts or esters of salicylic acid in place of aspirin for treatment of rheumatic conditions in view of the lower gastrotoxicity of the former group.

Meseclazone and its active metabolite (5-chlorosalicylic acid) have both been found to have a prophylactic effect when incorporated into the diet at doses of 0.1 and 0.2 per cent (approximately 75 and 150 mg kg per day), the swelling being reduced in both the injected and non-injected paws (Sofia *et al.*, 1975). The experimental drug 5-(*N*-pyrryl)-salicylic acid has a therapeutic ED_{50} only 1.5 times greater than aspirin (although the authors of this work claimed this drug was appreciably more potent than aspirin (Jones *et al.*, 1975). The methyl and phenyl esters of aspirin, as well as 2,3-diacetoxybenzoic acid (3-acetoxy-aspirin), are equipotent or slightly more potent than aspirin in the tail-injected model of adjuvant arthritis (Whitehouse *et al.*, 1977; Rainsford and Whitehouse, 1980a). Furthermore, these esters of aspirin are noticeably less toxic than aspirin in terms of systemic toxicity, as well as gastrotoxicity (Whitehouse *et al.*, 1977; Rainsford and Whitehouse, 1980a).

Generally, aspirin and salicylate are regarded as relatively weak anti-inflammatory drugs in the various forms of the adjuvant arthritis considered to date, especially by comparison with potent drugs such as indomethacin, the fenamates, phenylacetates, phenylpropionates and pyrazolones, which are traditionally regarded as potent anti-inflammatory agents (Swingle, 1974a; Watnick, 1975; Van Arman, 1976). Some drugs that are thought to modify (albeit slowly) the rheumatoid disease process (e.g. D-penicillamine, levamisole, gold salts and anti-malarials) are relatively weak or inactive, or may even exhibit pro-inflammatory effects in the adjuvant-arthritis model, depending on the timing and quantity of drug dosed (Newbold, 1963; Graeme *et al.*, 1966; Swingle, 1974a; Jones *et al.*, 1975; Sofia *et al.*, 1975; Watnick, 1975; Wax *et al.*, 1975; Van Arman, 1976; Arrigoni-Martelli, 1977). This has led many workers to question the value of the adjuvant-arthritis model as a procedure for detecting drug activity that could be regarded as 'disease modifying' or 'anti-arthritic'. The central problem is that hardly any attempt has been made to examine the individual components of the inflammatory responses in adjuvant disease. Few if any reports exist on the histopathological, cellular or biochemical changes elicited by the salicylates in the hindpaws of adjuvant-diseased rats. Only occasionally have scattered reports begun to appear on histopathological and X-ray morphometric data on the effects of anti-inflammatory drugs on the hindpaw region of adjuvant-arthritic rats, and none have included salicylates even by way of comparison (Ackerman *et al.*, 1979; Bensley and Nickander, 1982). Also, there is an important question of the time over which responses to drugs should be examined in the adjuvant-arthritic rat. In one study Graeme and co-workers (1966) found that aspirin 200 mg/kg per day p.o. was as effective in suppressing the swelling in the injected hindpaws of adjuvant-arthritic rats as was chloroquine 10 mg/kg per day p.o. when the period of observation was extended to 180 days. In this stage, chloroquine typically aggravated the swelling during the early stages (up to 40 days) whereas aspirin and other anti-inflammatory drugs suppressed the inflammation (Graeme *et al.*, 1966). The prolonged administration of aspirin under conditions where the disease has settled into a genuine chronic phase may thus reveal more significant anti-inflammatory actions than previously supposed (Graeme *et al.*, 1966). Antirheumatic compounds (e.g. chloroquine, gold salts, D-penicillamine) are generally slow in producing disease-modifying changes in man, and this is exemplified in the study quoted above in the case of chloroquine. It is possible that the responses to salicylates (or indeed other anti-inflammatory drugs) may likewise be slow in onset in order to achieve modification. albeit mild, of the disease in the adjuvant-arthritic rat as well as in man.

In a variant of the chronic adjuvant disease model in which acute responses are examined, the oral pretreatment (for 1 h) with aspirin, diflunisal and other NSAIDs inhibits the primary swelling of the adjuvant-injected paw at 18 to 40 hours following injection (Bhargava, 1971; Sofia *et al.*, 1979). The ED_{50} values of these drugs in this acute phase of the disease are somewhat higher than observed following therapeutic dosing in chronically inflamed animals (Sofia *et al.*, 1979; see Table 7.1). Diflunisal has been shown to have accelerated rate of uptake into the hindpaws and tails of adjuvant-injected (compared with non-injected) rats during the acute phase (16 to 20 h) of the disease (Winter *et al.*, 1979).

Biochemical and cellular responses to salicylates in arthritic disease in rats

In terms of overall responses to the effects of NSAIDs in adjuvant disease, perhaps one of the most significant systemic changes reflecting a positive improvement to this drug is seen in the changes in body weight. Salicylates, in common with other NSAIDs, restore the loss in weight that occurs during the disease, although the normal rate of increase body weight is still below that seen in healthy (non-diseased) rats (Newbold, 1963; Graeme *et al.*, 1966; Rosenthale *et al.*, 1974; Sofia *et al.*, 1974; Watnick, 1975; Wax *et al.*, 1975).

Chronic oral administration of aspirin 30 to 175 mg/kg per day or meseclazone 30–175 mg/kg per day decreases the erythrocyte sedimentation rate (ESR) in adjuvant disease, in common with that observed with other NSAIDs (Watnick, 1975; Sofia *et al.*, 1979). The reduction in ESR is regarded as possibly giving some indication of improvement in rheumatoid arthritis in man (Dick, 1972a; Dieppe *et al.*, 1985). Other authors have reported reduction in ESR in arthritic rats with high doses (400 mg/kg per day) of aspirin and also high doses of other NSAIDs, although the dose–response effects from indomethacin indicate that this drug is more potent in reducing ESR (Sloboda *et al.*, 1988). Of the other haematological parameters affected by salicylate administration, the total and individual counts of circulating white blood cells (WBC) are restored to normal by prophylactic treatment of arthritic rats with aspirin 150 to 300 mg/kg per day; other NSAIDs are likewise effective in restoring WBC counts (Rosenthale *et al.*, 1974). Serum ratios of albumin/globulins also return to normal values in arthritic rats treated prophylactically with aspirin 150 to 300 mg/kg per day or meseclazone 175 mg/kg per day (Sofia *et al.*, 1979). Plasma levels of β-glucuronidase and lysozyme are reduced by prophylactic treatment with meseclazone 175 mg/kg per day (Sofia *et al.*, 1979). These results suggest that aspirin and meseclazone do partly restore the blood and plasma components influenced by adjuvant disease. The drugs appear to be effective in preventing the abnormally high systemic generation of hydrolytic enzymes (including those of lysosomal origin), and thus may have some effects in controlling degradative processes manifest in adjuvant disease.

Little other information appears to be available on the therapeutic effects of the other salicylates on these blood parameters. It should also be noted that most of the data mentioned above are derived from prophylactic and not therapeutic treatment of adjuvant-diseased rats, so that it is hard to determine the exact significance of these results in therapeutic terms.

The weights of the main immunological organs (thymus, spleen and popliteal lymph nodes) are restored to normal by oral aspirin administration to polyarthritic rats (Rosenthale *et al.*, 1974; Watnick, 1975). The response is not as marked as that obtained with more potent anti-inflammatory or anti-arthritic drugs. Nonetheless, these data, combined with the restoration of abnormal white cell count in adjuvant disease, suggest that aspirin and possibly other salicylates may have some, albeit limited, effects on the immunological manifestations in adjuvant disease. Some *in vitro* effects of salicylates on the lymphocyte and macrophage data discussed later (p. 276 *et seq.*) also support this thesis. More potent salicylates (e.g. diflunisal) may even have more marked effects.

Adrenal weights are increased in adjuvant-arthritic rats (Rosenthale *et al.*, 1974), and this is often considered to reflect enhanced adrenocortical activity, which mediates stress responses (Selye, 1976). Chronic prophylactic administration of aspirin 75 to 400 mg/kg per day, like other NSAIDs, rapidly reduces the adrenal hypertrophy in adjuvant arthritis (Rosenthale *et al.*, 1974). Aspirin also reduces adrenal weight in normal rats, together with some other parameters indicative of pituitary–adrenal activity (Chitwood *et al.*, 1976). This suggests that the drug may have beneficial effects through a slight depression in adrenocortical function.

The accelerated catabolism of collagen observed in adjuvant disease (and indeed other chronic degradative–inflammatory diseases) has been found to be prevented by the therapeutic daily oral administration of sodium salicylate 300 mg/kg (Trnavska and Trnavsky, 1974). Furthermore, this treatment also normalises the conversion of soluble to insoluble collagen, a process that is depressed in adjuvant disease. Salicylate administration appears to slow the catabolism and enhance the formation of collagen. This has not been observed with chloroquine, and decreased catabolism occurs with phenylbutazone (Trnavska and Trnavsky, 1974). Hence, this potentially beneficial effect of salicylate may not be shared by other anti-inflammatory or anti-rheumatic drugs.

Actions of salicylates in other chronic models of inflammation

Aspirin and salicylate are relatively weak inhibitors of various types of granuloma formation established in laboratory animals, compared with the more potent NSAIDs (Whitehouse, 1965; Kulonen and Potilo, 1975; Nakamura and Shimizu, 1979). The therapeutic effects of NSAIDs on the formation of granulation tissue are of relevance to the pathogenesis of chronic inflammation, since granulomatous tissue is present in pannus, skin lesions and lymphoid follicles in both experimentally-induced and rheumatoid arthritis. The development of this tissue is obviously linked with the severity of joint degeneration and immobility and the extra-articular manifestations of the arthritis.

Wilhelmi (1963) showed that the granulation tissue and leucocyte infiltration that develops during the repair following excision of the dorsal skin of rats, was inhibited by repeated oral dosage of (a) salicylic acid 300 mg/kg per day for 12 days; (b) intraperitoneal injection of sodium salicylate 60 mg/kg per day for 12 days; or (c) topical application of salicyl vaseline 10 or 20 per cent twice daily for 5 or 12 days. Oral administration of aspirin 300 mg/kg per day for 5 days has been found to be equipotent with phenylbutazone 100 mg/kg per day p.o. for 5 days, but less effective than indomethacin or prednisolone in inhibiting a cellulose sponge granuloma injected for 7 to 21 days in rats (Kulonen and Potila, 1975). Aspirin, in common with the above-mentioned anti-inflammatory drugs, significantly inhibited the synthesis of collagen and decreased the content of RNA in the granulomas (Nakamura and Shimuzu, 1979). However, aspirin was found to be less effective than the other anti-inflammatory agents in inhibiting the biosynthesis of non-collagenous proteins, DNA and acid mucopolysaccharides, whose synthesis is normally enhanced during granuloma growth.

Electron-microscopic observations have shown that sodium salicylate 200 mg/kg per day s.c. was as effective as phenylbutazone 8 mg/kg per day s.c. or prednisolone 5 mg/kg per day s.c. in reducing the content of endoplasmic reticulum in developing fibroblasts in granulomas induced in rats (Jørgensen, 1964). In the same study, it was found that both salicylate and phenylbutazone induced extensive dilation of the mitochondria accompanied by extreme shortening (or even disappearance) of cristae in mitochondria of the developing fibroblasts in granulomas induced in rats. These studies suggest that salicylates are capable of inhibiting the biosynthesis of many cellular macromolecules and the functional capacity of mitochondria to produce cellular energy (ATP) in the rapidly proliferating cells of granulomas. This metabolic effect of the salicylates may well provide the basis for the observed reduction in granuloma weight observed following administration of these drugs.

Antigen-induced monoarticular arthritis in rabbits is another model of chronic inflammation, developed by Dumonde and Glynn (1962), which has been claimed by these and other workers strongly to resemble rheumatoid arthritis in man in regard to chronicity and histopathology (Dumonde and Glynn, 1962; Suguro, 1977). However, the developing polyarthritis of peripheral joints, which is characteristic of both rheumatoid and Freund's adjuvant-induced arthritis does not develop in the Dumonde and Glynn model in rabbits. The latter has a strictly monoarticular involvement, and hence its relevance, especially in terms of the systemic involvement seen in rheumatoid arthritis, is a limitation to the use of this model. Given this limitation, studies have shown that chronic oral dosage of aspirin reduces the content of synovial cells in the injected joint, but does not affect the joint swelling or cause any improvement in the joint histology as seen with more potent anti-inflammatory/anti-rheumatic agents (Blackham and Radziwonik, 1977). In a surgical model in rabbits salicylate was not found to affect the joint histology, activities of lysosomal enzymes in the joints or the indices of cartilage proteoglycan synthesis (Gold et al., 1976).

A variant of the adjuvant-induced arthritis model that has in recent years become very popular for the investigation of the actions of antirheumatic agents involves the use of injections of native type II collagen mixed in incomplete Freund's adjuvant and injected intradermally into the tail base of rats (Otterness and Bliven, 1985; Slaboda et al., 1988). A booster injection is given in the tail base on day 14, and by day 21 it is possible to measure hindpaw swelling and ankle joints in the afflicted rats (Sloboda et al., 1988). As in the adjuvant-induced arthritis, it is possible to quantify the joint damage by using a high resolution X-ray system (Sloboda et al., 1988). A number of authors have reported effects of various NSAIDs (including aspirin) in this model, although as in adjuvant arthritis the effects of aspirin are not particularly potent as compared with indomethacin. It is also possible to show that ESR is reduced by aspirin to an equivalent extent as that seen with other NSAIDs and corticosteroids or immunosuppressive agents (Sloboda et al., 1988). The advocates of the collagen II model in rats and mice claim that it has a greater relevance to the development of the autoimmune disease typified by rheumatoid arthritis in humans.

CHAPTER 7

BIOCHEMICAL EFFECTS OF SALICYLATES IN RELATION TO THEIR ANTI-INFLAMMATORY ACTIONS

As mentioned earlier in this chapter the salicylates, in common with other NSAIDs, have multiple actions on a whole range of biochemical actions, and no one single action is sufficient to account alone for the therapeutic effects of these drugs as anti-inflammatory agents. Several excellent reviews have been published over the years covering different aspects of the biochemical and cellular actions of the salicylates (Whitehouse, 1965; Smith and Smith, 1966; Collier, 1969; Arrigoni-Martelli, 1977; Hirschelman and Bekemeier, 1977; Ferriera, 1979; Atkinson and Collier, 1980; Packham, 1982; Vane and Botting, 1995; 1996; Amin *et al.*, 1999). Each of these is probably a reflection of current thinking at that time. As indicated at the beginning of this chapter, there has been much focus on the 'central role' of the effects of salicylates on the production of prostaglandins, nitric oxide, leukotrienes and, more recently, the lipoxins produced from metabolism of arachidonic acid. Other actions (e.g. on lympho-cyte functions, leucocyte migration and superoxide production) are also known to be important sites of action of the salicylates, especially in relation to their anti-inflammatory activity.

Effects on the synthesis and metabolism of prostaglandins, thromboxanes and leukotrienes

Metabolic pathways

An outline of the pathways of the biosynthesis of these eicosanoids from phospholipid-derived arachi-donic acid (C20:4) is shown in Figure 7.2. Arachidonate is the specific precursor for the synthesis of the eicosanoids, and is derived from dietary sources of the essential fatty acid, linoleate (C18:2) (Kinsella *et*

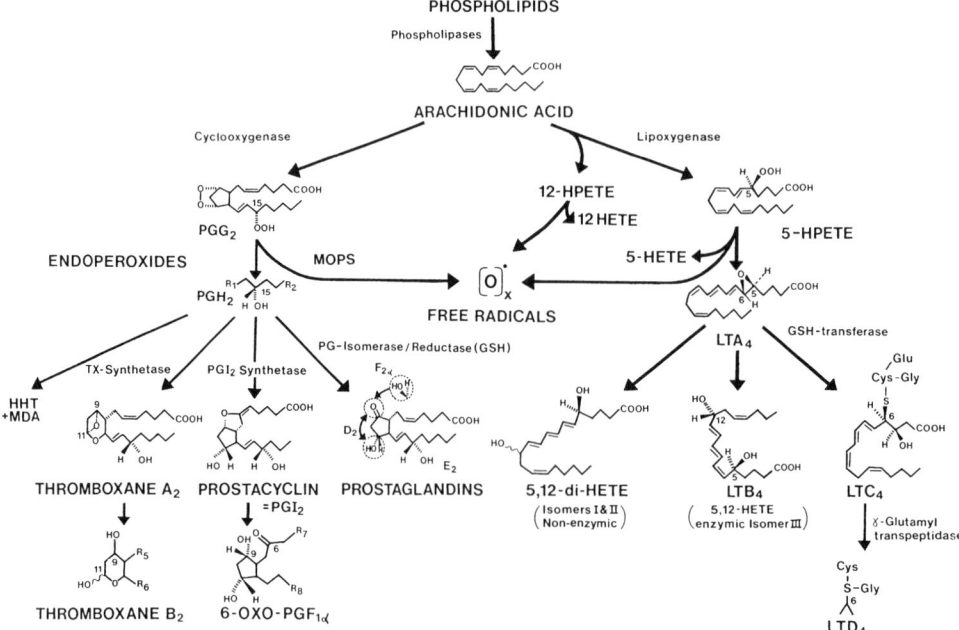

Chemical: HPETE = hydroperoxyeicosatetraenoic acid; LT = Leukotriene; PG = Prostaglandin

Figure 7.2 Pathways of the formation of eicosanoids from arachidonic acid. (From Rainsford, 1984. Reproduced with permission of the publishers, Butterworths/Heinemann.)

al., 1979; Lands, 1979; Deby, 1988). It is released from esterified precursors (phospholipids, triglycerides and cholesterol esters), by the action of phospholipase A_2 and C_1 and, possibly, cholesterol esterases (Lands, 1979). Arachidonic acid release is blocked by anti-inflammatory glucocorticoids (Lewis and Piper, 1975; Hong and Levine, 1976), cholesterol (Vigo *et al.*, 1980), indomethacin (Kaplan-Harris and Elsbach, 1980) and nicotinic acid (Bowery and Lewis, 1973).

Prostaglandin synthetase/cyclo-oxygenase pathways

The pathway for the synthesis of prostaglandins and thromboxanes proceeds through the combination of the cyclo-oxygenase/peroxidase reactions of the prostaglandin (PG) endoperoxide synthetase enzymes (PGHSs; EC 1.14.99.1) (Hamberg and Samuelsson, 1974; Miyamoto *et al.*, 1976; Yamamoto, 1988). These enzymes catalyse (a) the activation of molecular oxygen; (b) the addition in two steps of two molecules of dioxygen at the 11 and 15 positions, respectively, of arachidonic acid; followed by (c) abstraction of protons concomitant with rearrangement of the double bonds at the 11–12 position. Prostaglandin G_2 (PGG$_2$) is thus produced in which dioxygen is linked in an endoperoxide bridge across the 9–11 position of peroxidase activity at the 15 position (Miyamoto *et al.*, 1976; Lands, 1979; Yamamoto, 1988). Subsequent peroxidase activity at the 15-hydroperoxy group of PGG$_2$ liberates an oxygen radical and forms PGH$_2$, which is a central precursor for the subsequent formation of prostaglandins and thromboxanes (Figure 7.2).

A haem prosthetic group is present in the cyclo-oxygenase enzyme whose function appears to involve the activation of oxygen via the formation of perferryl ion Fe(IV)O^{2+} (Kuehl *et al.*, 1979a; Peterson and Gerrard, 1979; Rainsford and Swann, 1984; Yamamoto, 1988; Tang *et al.*, 1997) as well as orientation of the fatty acid (Yamamoto, 1988).

The PG endoperoxide synthetase also catalyses the peroxidase-like conversion of PGG$_2$ to PGH$_2$ (Miyamoto *et al.*, 1976; Yamamoto, 1988). As a consequence, potent oxidising radical species comprising a tyrosyl and an [O$^•$]$_x$ is released (Egan *et al.*, 1976; Kuehl *et al.*, 1979; 1980; Peterson and Gerrard, 1979; Tsai *et al.*, 2002). The [O$^•$]$_x$ species is extremely reactive and irreversibly inhibits the cyclo-oxygenase enzyme (Egan *et al.*, 1976), so its production is an important regulator of the activity of this enzyme *in vitro*.

PGH$_2$ serves as the central point for the subsequent formation of prostaglandins D_2, E_2, $F_{2\alpha}$ and I_2 (prostacyclin), and of thromboxane A_2 (Figure 7.2). Prostaglandins D_2, E_2 and $F_{2\alpha}$ are formed by the actions of specific isomerase/reductase enzymes on PGH$_2$, whereas prostacyclin (PGI$_2$) and thromboxane A_2 are formed by the actions of their respective synthetase enzymes on this intermediate. The biological activities of these inflammatory mediators have been reported extensively elsewhere (Ford-Hutchinson *et al.*, 1977; Vane and Ferreira *et al.*, 1979; Kuehl *et al.*, 1980; Curtis-Prior, 1988; Davies and MacIntyre, 1992; Lam and Austen, 1992) and a brief summary of their actions as they are important in the mediation of inflammatory reactions and pain is shown in Figure 7.3. Specific actions of these mediators on cells, especially as the salicylates influence them, will be discussed later. The important general points to note here are:

1. The potential for antagonism of the actions of one prostanoid on that of another, e.g. platelet aggregation promoted by thromboxane A_2 derived from platelets, which is antagonised by the PGI$_2$ produced by the vascular endothelium.

2. The ultimate expression of the activities of prostaglandins, thromboxanes and the leukotrienes depends on their individual effects upon the respective EP, FP, IP, DP-receptors for the prostaglandins of the E, F, I and D series, TP for thromboxane A_2, BLT for leukotriene B_4, and CysLT$_1$ and CysLT$_2$ for the cysteinyl-leukotrienes (*The IUPHAR Compendium of Receptor Characterisation and Classification*, 1998; Sugimoto *et al.*, 2000).

The receptor-mediated activities of prostaglandins, thromboxane (and possibly the leukotrienes as well) appear to depend on their transductional effects, which lead to increased effects upon the activities of the adenylate and guanylate cyclases and activation of the inositol–phosphate metabolism. There is evidence that the opposing actions of prostaglandins may be attributed to their respective actions on

CHAPTER 7

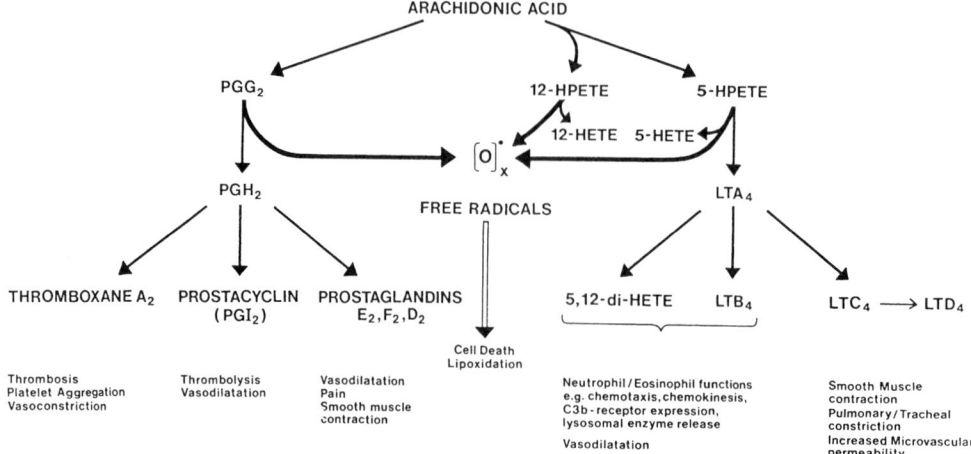

Chemical: HPETE = hydroperoxyeicosatetraenoic acid; LT = Leukotriene; PG = Prostaglandin

Figure 7.3 Principal effects of the eicosanoids with respect to inflammation. (From Rainsford, 1984. Reproduced with permission of the publishers, Butterworths/Heinemann.)

these membrane-bound enzymes that are involved in the production of putative second messengers of intracellular control, adenosine cyclic 3′,5′-monophosphate (cyclic AMP) and guanosine cyclic 3′,5′-monophosphate (cyclic GMP) (Willoughby and Dieppe, 1976; Bertelli and Schinelli, 1979). For instance, PGE_2 is considered a pro-inflammatory agent, whose function is to stimulate cyclic AMP production (Vane and Ferreira, 1979). Opposing this is $PGF_{2\alpha}$ (which has anti-inflammatory actions) stimulating guanylate cyclase to produce cyclic GMP (Willoughby and Dieppe, 1976). Willoughby considers that this may actually represent a futile system, since there may be antagonism of these cyclic nucleotides by one another (Willoughby and Dieppe, 1976). However, the possibility of understanding further modes of action of prostaglandins rests in part on studying their effects on cyclic nucleotide production. Also, as will be seen later, the salicylates (in common with other NSAIDs) can directly influence cyclic nucleotide production (Bertelli and Schinelli, 1979; Vane and Ferreira, 1979), e.g. the prostaglandin-E-mediated activity of adenylate cyclase. Thus these drugs appear to regulate prostaglandin-mediated cyclic nucleotide activity quite separately from their effects on the production *per se* of these inflammatory mediators.

The rate of catabolism of the prostaglandins and thromboxanes occurs rapidly *in vivo* and obviously influences their rate of production (i.e. by mass-action phenomena) and that of the tissue destructive free radicals (e.g. $[O']_x$), and the amount of these prostaglandins available for expression of biological activity. The principal features of the degradation of prostaglandins are as follows (Oates *et al.*, 1980):

1. Dehydrogenation of the 15-hydroxyl group (in lungs) to form a 15-keto derivative.
2. Reduction of the 13,14 double bond (in lungs) to yield the 13,14-dihydro derivatives (i.e. the 15-keto-13,14-dihydroprostaglandins). The levels of these products are usually measured to give an indication of the extent of catabolism *in vivo*.
3. There follows a two-step removal of two carbon units at each step from the carboxyl end to yield the 16-carbon derivative (by β-oxidation), hydroxylation of the C-20 methyl to an alcohol (by ω-hydroxylation), then oxidation of this (by ω-oxidation) to form a carboxylic acid. The major urinary product of PGE_2 in man is the 16-carbon dicarboxylic acid so formed.
4. Prostaglandins E_2 and $F_{2\alpha}$ may be interconverted by a $NADP^+$-linked 9-keto-reductase enzyme.
5. Prostacyclin (PGI_2) is non-enzymatically reduced to 6-keto-$PGF_{1\alpha}$ both *in vivo* and in aqueous media *in vitro*, which then undergoes a two-step β-oxidation (as for PGE_2 and $F_{2\alpha}$) and oxidative decarboxylation to yield pentanor-$PGF_{1\alpha}$. 6-Keto-PGF can also form 6-keto-$PGE_{1\alpha}$ (via the 9-keto-

reductase) and then proceeds to ω-hydroxylation (as in (3) above). 6-keto-$PGE_{1\alpha}$ is usually measured *in vivo* to give an indication of PGI_2 production.

Overall, the most significant enzyme system involved in the degradation of the prostaglandins E and F that has been studied most for drug effects on prostaglandin degradation is the 15-hydroxydehydrogenase (PDGH).

Lipoxygenase pathways

In addition to being susceptible to attack by oxygen at the 11 and 15 positions (as in the cyclo-oxygenase pathway), there are other positions where carbon atoms exist in positions that are activated for oxygenation. Principally these include the 5,12 positions, and the resultant monohydroperoxy-eicosatetranoic acids (HPETEs) are the first products formed in the lipoxygenase pathways (Borgeat and Samuelsson, 1979; Borgeat and Sirois, 1981; Narumiya *et al.*, 1981; Peters-Golden, 1998). The lipoxygenase (EC 1.13.11.12) enzymes catalyse the oxidation of unsaturated fatty acids (or esters thereof) having a 1,4-*cis,cis*-pentadiene system (Holman, 1951; Nugteren, 1975). 5-Lipoxygenase (5-LO) is involved in the first step of reactions leading from arachidonic acid to 5-(S)-HPETE and subsequent cyclicisation to LTA_4 (Provost *et al.*, 1999). Subsequently, the pathway divides to produce either LTB_4 by hydrolysis or the peptido-leukotrienes (LTC_4, D_4, E_4) (Rouzer, 1980; Rådmark, 2000). 5-LO is largely restricted to cells of the myeloid lineage.

12-Lipoxygenase has been isolated and characterised from platelets (where 12-H(P)ETE is the principal product (Nugteren, 1975).

These leukotrienes are extremely potent, vasoactive and leucotactic substances that are in some respects more inflammagenic or potent than the prostaglandins, although they have a different spectrum of activity compared with those products of the cyclo-oxygenase pathway (Figure 7.3; Haegsstrom, 1999).

Among these are the dihydroxy-products of the hydrolysis of LTA_4, in the 5-lipoxygenase pathway which results in the formation of LTB_4 (Rådmark, 1999). Production of this leukotriene is induced by pro-inflammatory cytokines (e.g. $TNF\alpha$, GM-CSF) as a consequence of gene activation of the 5-LO enzyme in neutrophils and macrophages-monocytes (Ford-Hutchinson *et al.*, 1980; Rådmark, 2000). The major effects of LTB_4 are in the recruitment of leucocytes to inflamed sites, degranulation and lysosomal release in neutrophils, promotion of neutrophil-dependent hyperalgesia and induction of leucocyte adhesion molecules on endothelia, and subsequent exudation into the extravascular space and plasma leakage (Hägsstrom, 1999). Inhibition of this pathway by drugs that control 5-LO, and particularly of the production of LTB_4, has been shown to be of importance in controlling inflammation (Higgs *et al.*, 1979; Dawson, 1980).

The other pathway from 5-LO involved in the formation of peptido- or cystenyl-leukotrienes (LTC_4, LTD_4 and LTE_4 – the components of what was formerly SRS-A) required addition of glutathione to the epoxy-groups of LTA_4 by LTC_4 synthase (Penrose, 1999), and is found in mast cells, eosinophils and monocytes-macrophages (Hedqvist *et al.*, 2000). LTA_4 may be released by PMNs and monocytes and taken up for transcellular synthesis by adjacent cells devoid of 5-LO but expressing either LTA_4 hydrolase and/or synthase, e.g. endothelial cells, lymphocytes, platelets (to produce 12-HETE products) or red blood cells (Hedqvist *et al.*, 2000). Cellular co-operativity in transcellular synthesis is also manifest in the production of lipoxins from 15-LO of mucosal cells and 5-LO in PMNs and monocytes (Serhan *et al.*, 1999).

The initial step in LT biosynthesis is brought about by the receptor-mediated influx of Ca^{2+}, translocation of cytosolic phospholipase A_2 and subsequent release of arachidonic acid – a system identical to that involved in prostaglandin production (Hedqvist *et al.*, 2000). Activation of 5-LO requires the presence of 5-LO-activating protein (FLAP) in a membrane-bound complex (Gillard *et al.*, 1992).

LTB_4 and LTC_4 are transported out of cells by a carrier-mediated system, the latter ATP-dependent. Extracellular LTC_4 is converted to LTD_4 and LTE_4 by γ-glutamyl-transpeptidase and a dipeptidase respectively (Penrose *et al.*, 1999).

■ CHAPTER 7 ■

The LTs act on specific receptors. LTB_4 acts on BLT receptors and those for the peptido-LTs on Cys LT_1 or Cys LT_2 (*The IUPHAR Compendium of Receptor Characterisation and Classification*, 1998; Nicosia, 1999; Alexander and Peters, 2000). These are G-protein coupled receptors (Alexander and Peters, 2000). Leukotrienes can also bind to peroxisomal proliferator-activated receptor (PPAR) transcription factors, which makes interpretation of their sites of action complicated (Alexander and Peters, 2000).

There is also evidence of co-operation or synergism between the products of these pathways, e.g. the $(LTB_4 + PGE_2$ or $PGI_2)$ induced vasodilatation and leucocyte accumulation/activation (e.g. Ford-Hutchinson *et al.*, 1981; Haeggstrom, 1999).

Cyclo-oxygenases and inflammation

The early studies (Willis, 1970; Smith and Willis, 1971; Vane, 1971; Ferriera *et al.*, 1971) showing the production of prostaglandins in inflammation and the effects of aspirin and related anti-inflammatories and analgesics on the production of prostaglandins set the stage for the view that these drugs exerted their effects by inhibiting the enzyme cyclo-oxygenase (COX) or prostaglandin G/H synthase (PGHS).

For the best part of two decades thereafter, the cyclo-oxygenase theory served as a foundation for understanding the actions of these drugs and in the development of new agents. Inhibition of prostaglandin production was considered a key target for discovering new drugs. With the conventional NSAIDs and non-narcotic analgesics, understanding their actions in relationship to the inhibition of prostaglandins in various organ systems while overall showing general relationships also revealed that there were some non-prostaglandin-related actions of these drugs that needed to be understood. Furthermore it became clear, especially during the 1980s, that increases in the levels of prostaglandins in inflamed tissues and in neural pathways involved in pain mediation were markedly increased in the presence of specific stimuli, e.g. cytokines, neurokinins and other inflammagens. Meanwhile, protein chemistry and molecular biological tools were being applied to understand the sequence structure and functions of what was then perceived to be *the* cyclo-oxygenase. Inconsistencies were revealed when the protein structure of cyclo-oxygenases from different sources and attribution to specific gene sites on chromosomes highlighted the existence of two isoforms, PGHS1 and PGHS2 or, in the shorthand version to describe the central actions of the NSAIDs in blocking the activity of the cyclo-oxygenase component of these proteins, COX-1 and COX-2 respectively (Vane and Botting, 1995; Crofford *et al.*, 1994; Crofford, 1997; Vane *et al.*, 1998). The concept has emerged from the studies on the structure and functions of COX-1 and COX-2 that COX-1 principally mediates physiological functions while COX-2 is principally involved in mediation of inflammation, pain, fever, and other general responses associated with the development of inflammation, a summary of these properties is shown in Figure 7.4. Being two different gene products, it has been possible to use molecular biological techniques to detect both the production of the messenger RNA specific for each of these types as well as the enzymes' proteins. Several lines of evidence have, however, given rise to the possibility that there may be a third type of cyclo-oxygenase activity tentatively described as COX-3 (Balsinde *et al.*, 2000; Willoughby *et al.*, 2000). The observation that a third cyclo-oxygenase may be involved in inflammation has been highlighted by the lack of effects of COX-2 inhibitors in modifying an experimentally-induced inflammatory response after 20 to 48 hours (Gilroy *et al.*, 1999). A second line of reasoning is that it has long been postulated that paracetamol may be acting on a specific cyclo-oxygenase system in the central nervous system in order to understand the actions of this drug in controlling pain since Flower and Vane (1972) showed that paracetamol had a more potent effect on cyclo-oxygenase derived from brain compared with that from spleen. In formalin-treated rats anti-nociceptive doses of paracetamol (100 to 300 mg/kg i.p.) have been shown markedly to reduce PGE_2 production in the spinal cord (Muth-Celbach *et al.*, 1999). This suggests that paracetamol does have effects on spinal production of prostaglandins in contrast to the less potent effects on other cyclo-oxygenase systems, except those that have been specifically stimulated, e.g. with phorbol ester. A macrophage-monocyte cell line J744.2 treated with diclofenac,

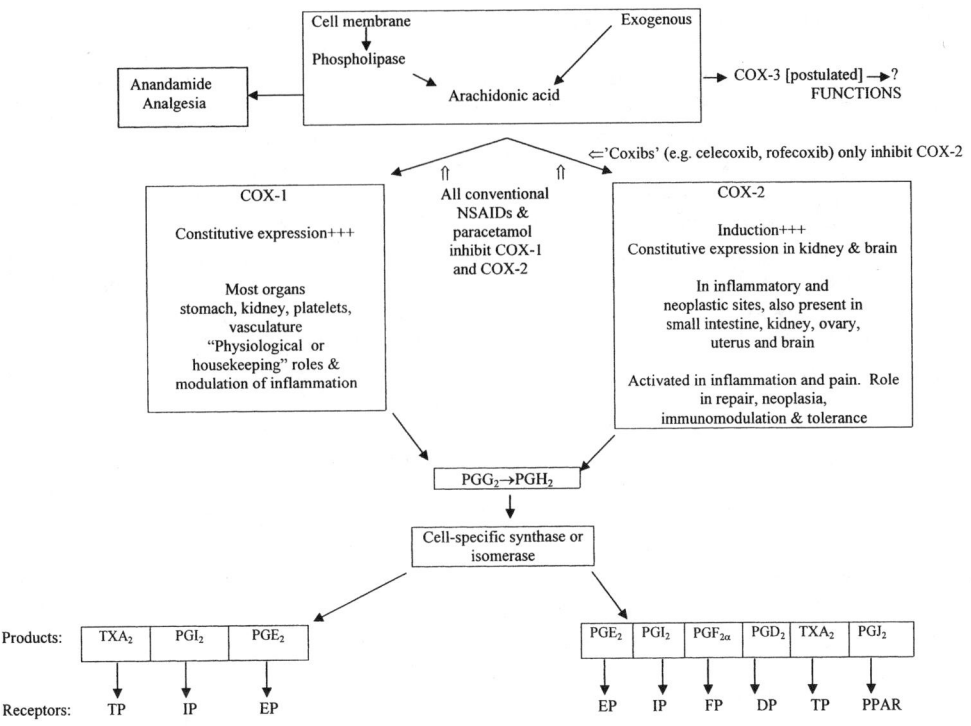

Figure 7.4 COX pathways. Each receptor-activated response leads to a specific physiolo-pathological response to the prostaglandin (PG) or thromboxane (TXA$_2$) receptor that is activated.

a potent pseudo-irreversible inhibitor of COX-2 activity, induced production of a cyclo-oxygenase that appeared to be more sensitive to inhibition with paracetamol than when treated with the inducing agent lipopolysaccharide (Simmons *et al.*, 1999).

PGHSs are membrane-bound glycoproteins with haem prosthetic groups. They have two homo-dimers each with three distinct domains – an N-terminal epidermal growth factor (EGF), a membrane-bound motif and the C-terminal catalytic domains (Picot *et al.*, 1994; Crofford, 1997). The active site of both COX-1 and COX-2 comprises a long, narrow, hydrophobic channel extending upwards from the membrane-binding domain to a catalytic domain. Comparison of the X-ray crystal structures of COX-2 and COX-1 reveals that many amino acids can be superimposed except for two small variations (referred to in COX-2 by their equivalent positions in COX-1) comprising those adjacent to the substrate-binding channel where valine substitutes for isoleucine at positions 434 and 523 respectively. As valine is a smaller amino acid than isoleucine this substitution effectively opens a wider side 'pocket' in the region adjacent to the main substrate channel in COX-2 compared with that in the same channel in COX-1. The effect of these substitutions is to increase the binding of larger COX-2 inhibitory NSAIDs, and this has been exploited for the design of COX-2 selective NSAIDs (Dannhardt and Laufer, 2000; Dannhardt *et al.*, 2000).

COX-2 is coded by a class of early–immediate response genes in which there is a binding site for the transcription factor NFκB in the promoter region. This transcription factor is activated by several cytokines (e.g. IL-1, TNFα), oxidants, endotoxin/LPS, PAF and phorbol esters, and controls the synthesis of pro-inflammatory proteins (including iNOS, IL-1, TNFα, ICAM-1, E-selectin and IL-8). Thus NFκB represents a central target for the control of several key inflammatory mediators, as well as being a target for inhibition by salicylates.

Nitric oxide and COX- and LOX-cross talk

Nitric oxide (NO), first recognised as endothelium-derived relaxation factor (EDRF), is a free radical with short half-life that is synthesised from L-arginine by either a calmodulin and Ca^{2+}-constitutive NO synthase (NOS) or an inducible iNOS, which is calmodulin/Ca^{2+}-independent enzyme (Moncada *et al.*, 1991; Nathan and Xie, 1994). There are two constitutive forms; type III or endothelial NOS (eNOS), and type I or neuronal NOS (nNOS). Of particular interest in inflammation is the iNOS that is found following stimulation by pro-inflammatory cytokines or LPS (endotoxin) in neutrophils, macrophages, microglia, endothelial cells, hepatocytes, and smooth and myocardial muscle (Moncado *et al.*, 1991; Dinnerman *et al.*, 1993; Amin *et al.*, 1995; Swierkosz *et al.*, 1995; Ikeda *et al.*, 1996; Kepka-Lenhart *et al.*, 1996; Kim *et al.*, 1998; Moncada, 1999), and contributes to the inflammatory events (Amin *et al.*, 1999). It has been suggested by Amin and co-workers (1999) that downregulation by aspirin of the expression and activity of iNOS contributes to the pleotropic actions of aspirin in the control of inflammation, pain and other activities of this drug.

Several lines of evidence suggest that there is regulation by NO of the activities of both COX-1 and LOX activities. While COX-2 and iNOS are co-induced by LPS and IL-1 (Swierkosz *et al.*, 1995; Sánchez de Miguel *et al.*, 1999; Wang and Brecher, 1999), it appears that in NO may regulate prostaglandin production (Busija and Thore, 1997). Thus, in macrophages NO inhibits both COX-2 activity and induction (Swierkosz *et al.*, 1995) but activates COX-1 activity (Clancy *et al.*, 2000), possibly by the actions of the product formed by the contribution of superoxide and nitric oxide (i.e. peroxynitrite), the latter of which has been shown to activate PGHS-1 or COX-1 (Upmacis *et al.*, 1999). NO appears to counteract the superinduction of COX-2 in cartilage from osteoarthritis patients (Amin *et al.*, 1997). Recent studies have illustrated the potential significance of iNOS in regulating the development of antigen-induced arthritis and its role in leucocyte–endothelial interactions in mice (Veihelmann *et al.*, 2001). In iNOS gene-knockout mice (iNOS−/−) leucocyte–endothelial cell interactions, expression of I-CAM and P-selectin and infiltration of leucocytes into joints were enhanced. The joint swelling was doubled compared with that in wild type (iNOS+/+) animals. These studies suggest that NO derived from iNOS has a negative regulatory role on the development of inflammation, especially at the level of leucocyte–endothelial and leucocyte-activation reactions.

A considerable body of data from studies in various isolated cell systems shows that both aspirin and salicylate inhibit the induction of iNOS by pro-inflammatory cytokines or LPS (Amin *et al.*, 1995; Farivar and Brecher, 1996; Farivar *et al.*, 1996; Kepka-Lenhart *et al.*, 1996; Sakitani *et al.*, 1997; Kim *et al.*, 1998; Kimura *et al.*, 1998; Katsuyama *et al.*, 1999; Sánchez de Miguel *et al.*, 1998; Chung *et al.*, 2000; Ryu *et al.*, 2000; Shimpo *et al.*, 2000). Recently, Moon *et al.* (2004) showed that the paralysis in experimental auto-immune encephalomyelitis (EAE) in rats was reduced by 200 mg/kg sodium salicylate given 13 days post-induction of the disease, coincident with reduction in the expression of both COX-2 and iNOS. This may reflect actions of salicylate or the induction of iNOS as well as COX-2, which are distinct from the drug effects on the activities of these enzymes. Clearly, further *in vivo* studies are necessary to define the effects of the salicylates on the nitric oxide system in relation to their anti-inflammatory and analgesic activities.

Effects of salicylates on prostaglandin production

The salicylates, in common with many acidic NSAIDs, inhibit the production of prostaglandins and thromboxanes *in vivo* and *in vitro* (Ferreira *et al.*, 1974; Walker *et al.*, 1976; Bonta *et al.*, 1977; Whitehouse *et al.*, 1977; Williams and Peck, 1977; Smith, 1978a; 1979; Vane, 1978; Ferreira, 1979; Westwick, 1979; Atkinson and Collier, 1980; Kuehl *et al.*, 1980; Brune *et al.*, 1981; Vane and Botting, 1995). While many of these drugs inhibit cyclo-oxygenase activity, the mechanism of their inhibitory actions is not the same for all. The salicylates, like the other NSAIDs, can also inhibit other enzymes involved in breakdown of prostaglandins, but at concentrations that are higher than that for the inhibition of the cyclo-oxygenase enzyme systems (Flower, 1974). Furthermore, the fact that the salicylates are phenolic compounds raises the possibility that they may be free radical scavengers (or acquire this scavenging ability during their metabolism *in vivo*) and hence influence the production of prostaglandins by 'drawing off' of free radicals. Recent studies suggest that salicylates like other NSAIDs may inhibit transport of prostanoids from cells (Warner and Mitchell, 2003) thus representing another site of action of these drugs on prostanoid production.

Differential effects of NSAIDs on cyclo-oxygenases

It is now well-established from a large number of studies that conventional NSAIDs, including aspirin, have differential effects on COX-1 and COX-2 activities (Vane and Botting, 1995; Vane et al., 1998; Blanco et al., 1999; Brooks et al., 1999; Warner et al., 1999). A considerable number of studies have shown that the IC_{50} values for the inhibition of COX-1 and COX-2 vary considerably according to whether the activities have been assayed with pure isolated enzymes, human recombinant enzymes, isoenzymes transfected into various cell lines, or in whole human blood ex vivo or in vitro (Table 7.5).

The relative potencies of aspirin, other salicylates and NSAIDs vary considerably in COX-2 compared with COX-1 inhibition studies over a wide range of concentration ranges (Laneuville et al., 1994; Gierse et al., 1995; Kargman et al., 1996; Young et al., 1996; Kirchner et al., 1997; Mancini et al., 1997; Warner et al., 1999; Kulkarni et al., 2000; see also Tables 7.5 and 7.6). This makes the determination of the relative selectivity for COX-1 and COX-2 particularly difficult to determine (Brooks et al., 1999). Each type of assay methodology has its own specific requirements. Thus studies in isolated enzymes, including those that are from human recombinant sources or in transfected cell lines, have advantages

TABLE 7.5

Comparison of the inhibitory activities of aspirin with other NSAIDs on COX-1 and COX-2 isoenzymes using human recombinant enzymes (transfected into CHO Chinese ovary, COS monkey kidney or S19 insect cells) and in whole human blood in vitro.

Drug	Human recombinant enzymes			Whole human blood		
	COX-1, IC_{50} (µM)	COX-2, IC_{50} (µM)	COX-2 selectivity COX-1/ COX-2	COX-1, IC_{50} (µm)	COX-2, IC_{50} (µM)	COX-2 selectivity COX-1/ COX-2
Aspirin	4.45	16	0.28	4.00	81	0.05
Celecoxib	0.32	0.002	160	6.3	0.99	6.36
Diclofenac sodium	0.0034	0.0023	1.47	0.14	0.06	2.33
Etodolac	50	0.041	1219	11.7	2.60	4.5
Fenoprofen calcium	1.96	26.3	0.07	2.73	4.03	0.68
Flurbiprofen	0.0015	0.0026	0.58	0.59	3.46	0.17
Ibuprofen	1.53	11.8	0.13	4.88	22.4	0.22
Ketoprofen	0.006	0.119	0.05	0.11	0.88	0.13
Ketorolac	0.083	0.012	6.9	0.32	0.88	0.36
Mefenamic acid	0.15	0.28	0.54	1.94	0.16	12.1
Meloxicam	1.95	0.06	33	1.87	0.54	3.46
Naproxen sodium	0.23	3.36	0.07	11.3	52.3	0.22
Nimesulide	2.1	0.31	6.7	4.1	0.31	13.2
Oxaprozin	14.9	>1000	<0.02	14.6	36.7	0.40
Piroxicam	2.52	0.39	6.5	2.62	5.35	0.49
Rofecoxib	26.3	0.34	77	19	0.51	37.2
Sulindac	NR	NR	NR	41.3	24.9	1.66
Tenoxicam	17.8	36.9	0.48	2.3	14.2	0.16
Tolmetin	0.58	1.44	0.40	1.23	7.09	0.17

From Kulkarni et al. (2000), with data derived from references therein and Prous Science MFLine® Database.
Note
NR=Not recorded

in order to determine the mechanistic actions and specificity of various drugs under controlled *in vitro* conditions. The argument for determining the clinical significance of the differential inhibition of the COX isoenzymes was brought to particular notice following The International Consensus Meeting of the Mode of Action of COX-2 Inhibition, which brought together experts in the different fields of arthritis, gastroenterology and pharmacology on the 5–6 December 1997 (Brooks *et al.*, 1999). Here the issues concerning the determination of assay specificity that is relevant to clinically observed levels of drugs were considered in detail. The conclusions from this meeting were that it was recommended that the human whole blood assay be used to determine COX specificity *in the clinical context*. Furthermore, the conclusion that a drug was COX-2 specific depended on whether the drug inhibited COX-2 only but not COX-1 across the entire therapeutic dosage range.

Critical issues concerning the evaluation of *in vitro* assays using isolated cell systems have shown COX-2 activity can vary considerably with the nature of the stimulus employed (Hulkower *et al.*, 1997). Furthermore, COX-1 activity as well as that of COX-2 can vary according to the presence of varying concentrations of exogenous or endogenous arachidonic acid (Chulada and Langenbach, 1997). Thus, in the presence of $10\,\mu\text{mol}/l$ exogenous arachidonic acid the COX-1 to COX-2 ratio in PGHS2 and PGHS1 murine transfected cell lines had a ratio of 10, with 10T1/2 mouse embryo fibroblasts and AS52 Chinese hamster ovary cells where the ratio was 5. In contrast in the same cell lines, when calcium ionophore was used to release endogenous arachidonic acid the ratios fell to 0.77 and 1.14 respectively (Chulada and Langenbach, 1997). In the same cell lines the COX-2 selective experimental drug NS-398 showed the predicted COX-2 selectivity, while indomethacin had a slightly greater selectivity for COX-2 than COX-1 in the calcium ionophore stimulated cells compared with that where exogenous arachidonic acid was added where the drug showed COX-1 activity preferentially. Some stimuli can influence the uptake of drug into cells. Thus, Raghoebar and co-workers (1988) showed that phorbolmyristate acetate inhibited the uptake of aspirin whereas it stimulated the uptake of sodium salicylate, thus highlighting the importance of determining drug concentrations within cells where different stimuli are employed to activate prostaglandin production.

An approach to achieving information regarding the effects of NSAIDs on the COX-isoenzymes that can be related to the concentrations of drugs *in vivo* and to allow full expression of the drug effects on the enzyme systems has been made by Warner *et al.* (1999), data for which are shown in Table 7.6. These data show the effects of a number of salicylates most of which, aside from aspirin, are weak inhibitors of COX-1 and have variably moderate potency as inhibitors of COX-2. Salicylate itself is a weak direct inhibitor of both enzymes and like aspirin and diflunisal is an unselective inhibitor of both isoenzymes. It has been well-known since the earlier work of Rome and Lands (1975) that there is a time-dependence with some NSAIDs for the full inhibitory effects of these drugs to be expressed either in isolated enzyme or isolated cell systems (Vane and Botting, 1996; Lora *et al.*, 1997; 1998; Gierse *et al.*, 1999). Thus, Lora *et al.* (1998) showed that aspirin exerted virtually no instantaneous inhibition of human recombinant COX-1 and COX-2 while several other NSAIDs that were studied as well as the COX-2 inhibitor NS398 were appreciably less potent inhibitors as compared with the same NSAIDs when they were pre-incubated with the enzyme and arachidonic acid for 20 minutes at 37°C. The ratios of the inhibitory concentrations required for instantaneous inhibition compared with that from pre-incubated enzyme was greater with aspirin and indomethacin and less so with mefenamic acid and NS398 for human recombinant COX-2, whereas the difference was greater for the latter drugs for COX-1. These results show that it is essential to pre-incubate enzymes or cells with aspirin and related drugs for the full expression of their inhibitory effects. Gierse and co-workers (1999) also came to similar conclusions, but showed that there were variations among a wider variety of NSAIDs and the COX-2 selective drug, celecoxib.

The acetylation of the serine residues in a region near to the entrance of the active site of the COX isoenzymes is a key feature that is important in understanding the inhibitory effects of aspirin, and it accounts for the irreversible inhibition of the enzyme (DeWitt *et al.*, 1990; Wennogle *et al.*, 1995; Mancini *et al.*, 1997; Kalgutkar *et al.*, 1998; Pan *et al.*, 1998; Hochgesang *et al.*, 2000). Analysis of the crystal structures of COX-1 with acidic NSAID inhibitors (like the salicylic acids) shows that the binding of these to the Arg[120] depends on the hydrophobic binding channel; this interaction occurs by hydrogen bonding. In COX-2, the binding by Arg[120] has only an accessory function. (Kalgutkar *et al.*, 1998).

In COX-1 derived from seminal vesicle preparations the Ser[530] residue and in human preparation Ser[529] are selectively and irreversibly acetylated in a location at the entrance of the arachidonic acid-binding channel (DeWitt *et al.*, 1990; Kalgutkar *et al.*, 1998). The acetylation of an analogous amino acid residue Ser[516] is achieved with aspirin in COX-2 during which Tyr[385] is involved as part of the mechanism (Hochgesang *et al.*, 2000). The consequences of the inhibition of COX-2 by aspirin are, however, profound in that there is a conversion of the enzymic activity of COX-2 to form 15(*R*)-hydroperoxy-eicosatetraenoic acid (15(*R*)-HPETE) (Figure 7.5). (Holtzman *et al.*, 1992; Lecomte *et al.*, 1992; Shimokawa and Smith, 1992; Mancini *et al.*, 1994; 1997; O'Neill *et al.*, 1994; Tang *et al.*, 1997; Schneider and Brash, 2000), a product that is important for the synthesis of aspirin-triggered epi-lipoxins (Serhan, 1997; see p. 260).

Kalgutkar *et al.* (1998) undertook a structure–activity analysis of some aspirin derivatives to establish if it was possible to convert these into COX-2 selective agents. They found that *O*-(acetophenyl)-hept-2-nyl sulphide (APHS) was a potent and selective inhibitor of this enzyme, and that it exerted its inhibitory effects by acetylation of the same Ser[516] as affected by aspirin. Thus by extending the alkyl side chain of the aspirin ester group it has been possible to create a selective COX-2 inhibitor.

An ester of salicylate, valeryl-salicylate, in contrast to APHS is a more selective inhibitor of COX-1 than aspirin (Bhattacharya *et al.*, 1995), although it is not as potent as the latter for inhibition of COX-1 (Table 7.6).

A number of salicylates that were found in early studies (prior to identification of COX isoenzymes) to inhibit the production of prostaglandins *in vivo* or in various cellular and subcellular systems. Apart from aspirin, diflunisal and 5-chlorosalicylic acid, salicylate and salicylsalicylic acid are relatively weak inhibitors of prostaglandin production, compared to other NSAIDs (Rainsford 1984). There has been some debate about the inhibitory effects *in vivo* and *in vitro* of salicylate itself. In platelets (which have COX-1), salicylate fails to inhibit the production of thromboxane A$_2$ *ex vivo* following administration to rats (Vargaftig, 1978a), whereas the production of prostaglandins in inflamed tissues is inhibited following oral administration of the drug to rats (Smith *et al.*, 1979; Eakins *et al.*, 1980). There are, however, clear differences

■ CHAPTER 7 ■

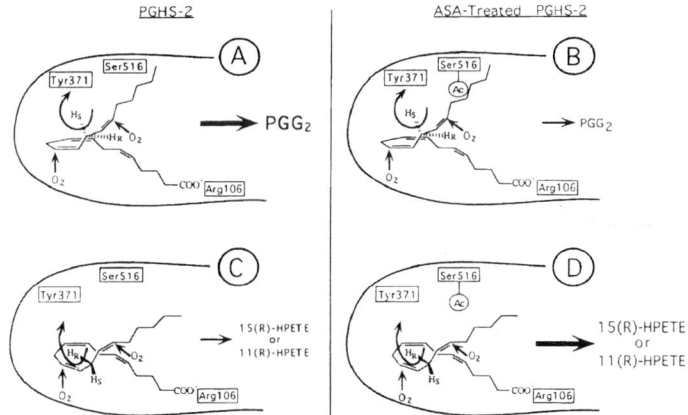

Figure 7.5 Conversion by aspirin of PGHS-2 (COX-2) to form 15- or 11-hydroperoxy eicosatetraenoic acids (15-HPETE, 11-HPETE) respectively.

This pathway is of particular interest, since in inflammatory states where there has been extensive production of PGHS-2/COX-2 following induction by pro-inflammatory cytokines the acetylation of COX-2 by aspirin leading to formation of 15-(R)-HETE (b) may contribute to the *anti-inflammatory* activity of aspirin, a route which is unique to this drug among the NSAIDs and analgesics. The mechanism of action of PGHS-2/COX-2 in the formation of PGG$_2$ and 15(R)-HPETE (as well as 11-(R)-HPETE) is shown in the two left-hand panels. The effect of acetylation by aspirin of the serine[516] to convert it to a form that results in formation of 15(R)-HPETE and 11-(R)-HPETE but not production of PGG$_2$ is shown in the two right-hand panels (Xiao *et al.*, 1997), the former being the substrate for 5-LO in leucocytes as shown in (a). (Reproduced from Xiao *et al.* (1997), with permission from the publishers, The American Chemical Society.)

TABLE 7.6

Potencies of compounds as inhibitors of prostanoid formation determined in the various COX-1 and COX-2 assay.

	WBA-COX-1 IC$_{50}$ (µM)	WBA-COX-2 IC$_{50}$ (µM)	WHMA-COX-2 IC$_{50}$ (µM)	IC$_{50}$ ratios WBA COX-1	IC$_{50}$ ratios WHMA COX-1	Ranking at IC$_{50}$ ratios WBA COX-1	Ranking at IC$_{50}$ ratios WHMA COX-1
Saliclates							
Aspirin	1.7	>100	7.5	>100	4.4	34	23
5-aminosalicylic acid	410	61	n.d.	0.15	n.d.	n.d.	n.d.
Diflunisal	113	8.2	134	0.1	1.2	9	14
Salicin	>100	>100	n.d.	–	n.d.	–	n.d.
Salicylaldehyde	>100	>100	n.d.	–	n.d.	–	n.d.
Sodium salicylate	4956	34440	482	6.9	0.10	16	15
Sulphasalazine	3242	2507	n.d.	0.8	n.d.	15	n.d.
Valeryl salicylate	42	2.3	n.d.	0.053	n.d.	–	n.d.
Other conventional NSAIDs							
Diclofenac	0.075	0.038	0.020	0.5	0.3	10	9
Flufenamate	3.0	9.3	n.d.	3.1	n.d.	13	n.d.
Flubiprofen	0.075	5.5	0.77	73	10	31	27
Ibuprofen	7.6	7.2	20	0.9	2.6	14	20
Indomethacin	0.013	1.0	0.13	80	10	29	24
Ketoprofen	0.047	2.9	0.24	61	5.1	31	25
Meclofenamate	0.22	0.7	0.2	3.2	0.91	22	11
Mefenamic acid	25	2.9	1.3	0.11	0.049	–	–
Naproxen	9.3	28	35	3.0	3.8	18	22
Piroxicam	2.4	7.9	0.17	3.3	0.1	17	13
Sulindac sulphide	1.9	55	1.21	29	0.64	20	10
Sulindac	>100	>100	58	–	–	–	–
COX-2 inhibitors							
Celecoxib	1.2	0.83	0.34	0.7	0.3	8	7
Etodolac	12	2.2	0.94	0.2	0.1	6	5
Meloxicam	5.7	2.1	0.23	0.37	0.040	11	6
Nimesulide	10	1.9	0.39	0.19	0.038	7	8
L745,337	>100	8.6	1.3	<0.01	<0.01	1=	1=
NS398	6.9	0.35	0.042	0.051	0.0061	5	4
Rofecoxib	63	0.84	0.31	0.013	0.0049	4	3
SC58125	>100	2.0	n.d.	>0.01	n.d.	1=	n.d.
Analgesics (non-narcotic)							
Aminopyrone	55	203	85	3.7	1.5	24	19
Paracetamol	>100	49	64	–	–	–	–

From Warner et al. (1999).

Notes

n.d. = not done; WBA = whole blood assay; WHMA = William Harvey Modified Assay in the potency of salicylate compared to aspirin, with aspirin always being much more potent as an inhibitor of prostaglandin production *in vitro* than salicylate (Flower, 1974; Smith *et al.*, 1979; Eakins *et al.*, 1980).

Fjalland (1974) found that the *in vitro* production of prostaglandin endoperoxides in guinea pig lungs was inhibited by salicylate and aspirin at concentrations (ID_{50} 660 µmol/l and 8.5 µmol/l, respectively) comparable with that of the total production of all prostaglandins (ID_{50} 0.66 µmol/l and 10.7 µmol/l, respectively). This suggests that the inhibitory effects of these drugs on prostaglandin production are evident at the same site, i.e. the prostaglandin cyclo-oxygenase. At this level aspirin would be expected to inhibit the enzyme activity irreversibly by acetylating the enzyme protein (Roth *et al.*, 1975), a feature that may be common to other *O*-acetylated salicylates, but not to non-acetylated salicylates or other NSAIDs. The inhibition of the cyclo-oxygenase by aspirin is prevented by salicylate (Roth *et al.*, 1975) and also by indomethacin and phenanthrolines (Roth *et al.*, 1975; Vargaftig, 1978b), suggesting a common site of interaction of NSAIDs with the position(s) of acetylation at the PG cyclo-oxygenase (COX-1) active site. It is also interesting that salicylate is able to interact with the same site on the cyclo-oxygenase as aspirin, especially since the *in vivo* generation of salicylate from aspirin (Chapter 4) would appear to determine the bioefficacy of aspirin in relation to its prostaglandin-mediated actions.

The weak activity of salicylate as an inhibitor of prostaglandin synthesis could be due to (a) its weak binding to the cyclo-oxygenase active site, and (b) the possibility that, as with all phenols, it could be acting as a scavenger of free radicals (Kuehl *et al.*, 1977; 1978; 1979a; Kuehl and Egan, 1978; Peterson and Gerrard, 1979), so enhancing the conversion of $PGG_2 \rightarrow PGH_2$, and thus partly overriding the tendency of this drug to inhibit the cyclo-oxygenase enzyme.

A similar mode of action has been attributed to diflunisal (Kuehl and Egan, 1978), which, although this drug is a more potent inhibitor of prostaglandin synthesis than salicylate (Kuehl and Egan, 1978; Table 7.5), also promotes conversion of hydroperoxy fatty acids to their corresponding hydroxy fatty acids (Kuehl and Egan, 1978), thus being a presumptive free radical scavenger (Kuehl and Egan, 1978).

Recent studies indicate that inhibition of COX's by salicylate is dependent on the oxidative state of the enzyme (Aronoff *et al.*, 2003). This may account for the inhibition of LPS-induced elevation of 6-keto-PGF in rats that appears to occur without affecting COX-2 expression (Giuliano *et al.*, 2000).

Studies by Kuehl and Egan (1978) have suggested that there are two sites of inhibition of prostaglandin cyclo-oxygenase (probably COX-1) by NSAIDs – a catalytic site and a 'supplementary' site. Drugs such as diflunisal and salicylate, which interact at the 'supplementary' site, are capable of reversing the inhibition of cyclo-oxygenase by indomethacin (Humes *et al.*, 1981). Rome and Lands observed that aspirin and other NSAIDs exhibit a time-dependent destruction of the cyclo-oxygenase (probably COX-1) enzyme *in vitro* (Rome and Lands, 1975). This was not evident with the methyl esters of these acids, suggesting that there is some requirement for the presence of a free carboxyl group for this drug-induced destruction of the enzyme activity. Many of the methyl esters of the NSAIDs have the same inhibitory effects as their parent acids on substrate binding (Rome and Lands, 1975). Contrary to what these authors stated, their data actually showed there was a large reduction in the inhibition constants (K_i) with the methyl esters of aspirin and indomethacin compared to their respective acids. These methyl esters would appear to be more tightly bound to the enzyme active site and so might exhibit more potent effects as inhibitors of the enzyme activity. Hence, the destruction of the enzyme by the NSAIDs is somehow associated with a reduction in the ability of these drugs to exhibit full inhibitory activity. The enzyme destruction by salicylates and indomethacin may be linked to the capacity of these drugs to actually *produce* free radicals *in vitro* (Borg, 1965). The latter possibility certainly clouds the concept of these free radical scavenging activity of salicylate, and obviously this aspect requires clarification.

Earlier studies (reviewed by Flower, 1974) suggested that the mechanism of inhibition (probably COX-1) by aspirin was a non-reversible type. The studies by Rome and Lands (1975) and the observed capacity of aspirin to acetylate the cyclo-oxygenase (Roth *et al.*, 1975) suggest that the basis for the non-reversibility of inhibition by aspirin resides with effects of both the carboxyl and the acetyl groups of this drug. However, the contrary conclusion that the inhibition by aspirin is competitive is rather more questionable, because (a) destruction of the enzyme occurs during the assay of its activity, and (b) there are marked effects of cofactors on the enzyme activity (Saeed and Warren, 1973; Flower, 1974; Saeed and Cuthbert, 1977), none of which have been reconciled with their *in vivo* availability.

Gierse *et al.* (1999) examined the mechanisms of the effects of the COX-2 selective drug celecoxib, in contrast to that of a range of other NSAIDs, on the kinetics of the isolated COX isoenzymes. Celecoxib was found to have typical competitive inhibition of COX-1 but, in contrast, competitive inhibition was initially followed by time-dependent inactivation of the COX-2 enzyme. The kinetic analysis of the effects of NSAIDs showed that these drugs exhibit one of four modes of action: (1) covalent irreversible (i.e. aspirin); (2) reversible (e.g. ibuprofen); (3) weak-binding and time-dependent (e.g. piroxicam, naproxen); and (4) tight-binding time-dependent (e.g. indomethacin). There was some variability in the drug effects on the two COX isoforms. The authors concluded that the use of IC_{50} values was inherently inaccurate because of the differing kinetic mechanisms involved in the inhibitory activity of the drugs.

The IC_{50} values of aspirin in the earlier studies that were reported for the inhibition of prostaglandin production by various microsomal or other *in vitro* preparations vary enormously. Reported values range from 0.7 to 23 200 µmol/l (Fjalland, 1974; Flower, 1974; Taylor and Salata, 1976). As mentioned previously, there are marked effects of cofactors on the IC_{50} values obtained with NSAIDs (Saeed and Warren, 1973; Flower, 1974; Saeed and Cuthbert, 1977) and the reported assay methodologies vary enormously (Flower *et al.*, 1972; Flower, 1974). Interpretation of these studies may be possible now in the light of knowledge of the presence of COX isoenzymes in these various systems.

Another approach towards resolving this problem of relating *in vitro* to *in vivo* effects is to examine the effects of NSAIDs on prostaglandin release by those cells mediating inflammatory reactions, e.g. cultured macrophages (Brune *et al.*, 1981). It is presumed that this type of assay of prostaglandin release accurately reflects the biosynthesis of prostaglandins from COX-2, and also avoids some of the difficulties regarding optimising substrate concentration and content of natural radical scavengers, PG cyclo-oxygenase enzyme 'protectants' etc., within the cell. Accepting these assumptions and given the potential difficulties of extrapolating reactions in one inflammatory cell system to all *in vivo* reactions, it has been found that the rank order for the inhibitory activities of NSAIDs on PG synthesis by macrophages, corresponds well to their respective potencies in anti-inflammatory assays in laboratory animals, and their respective therapeutic dose ranges employed in the treatment of arthritic conditions (Brune *et al.*, 1981). The IC_{50} values for aspirin and diflunisal for inhibiting macrophage prostaglandin production have been reported as 6.6 and 1.4 µmol/l, respectively (Brune *et al.*, 1981). These concentrations are well below the concentrations of these drugs in the therapeutic dose range (see Chapter 4) (compare these values with the IC_{50} for indomethacin of 0.0017 µmol/l in the same system. Other non-cyclo-oxygenase actions of aspirin could affect production of prostaglandins including effects on lipid metabolism (Bozza *et al.*, 1996).

The *in vivo* production of prostaglandins could also be affected through salicylate by impairment of the degradation of prostaglandins via 15-hydroxydehydrogenase (PGDH) (Flower, 1974; Hansen, 1974). Many NSAIDs inhibit this enzyme (Hansen, 1974; Taylor and Salata, 1976), but the effects of salicylate on this enzyme do not appear to have been examined *in vitro* or *in vivo*. This may be important because salicylate and its congeners are in general relatively potent inhibitors of many NAD^+-linked dehydrogenases (Smith, 1966). In view of the relative inactivity of salicylate as an inhibitor of cyclo-oxygenase activity, the inhibition of PGDH activity could assume greater importance than has been generally recognised.

Structure–activity studies – prostaglandin production

Using the same phorbol-ester stimulated production of PGE_2 in macrophages, Habicht and Brune (1983) compared the structure–activity relationships (SAR) of a number of substituted benzoates phenols and salicylates. While it is now evident that this system is complex in combining a number of drug effects that influence the actions in a cellular system, it has been useful in highlighting major chemical properties that determine the actions of these salicylates.

The assay probably largely reflects COX-2 activity, since the mouse peritoneal cells have been pre-incubated for 16 hours in a medium containing 10 per cent normal calf serum; they are then stimulated with the phorbol ester, TPA or PMA, which causes release of arachidonic acid. There are several important qualitative observations that can be made from the SAR of Habicht and Brune (1983) including the following:

1. The 5-cyclohexyl and phenyl-salicylic acids are the most potent salicylate inhibitors but the 4-phenyl-phenol is 10 times more potent, with the 2- and 3-phenyl, 4-methyl and 4-isopropyl phenols being about equipotent with 5-phenyl salicylic acid. All these compounds are the most potent in the series of salicylates, phenols or benzoic acid derivatives that were investigated. The results suggest that 4- or 5-substitutions with bulky substituents having electronic activity on salicylic acids or phenols confer increasing activity. Likewise, 5-chloro and 5-bromo and, to a lesser extent, 5-fluoro-derivatives increased the potency of salicylic acid preferentially above that of 3-, 4- or 6-halogens. 4-Methyl and 5-isopropyl salicylic acids were also more potent than 3- or 6- alkyl-substituted derivatives. These observations confirm the advantages of 5- or 4- bulky substituents, and especially those with large electron withdrawing groups such as 5-chloro-derivatives of salicylates, in enhancing their PG synthesis inhibitory effects.

2. The 3- and 5- but not the 4- or 6- hydroxy derivatives of salicylic acid are about 10 times more potent than salicylic acid itself. The enhanced activity of the 5- hydroxy- and inactivity of the 6-hydroxy-salicylic acids has also been confirmed recently by Hinz et al. (2000) using the LPS-stimulated RAW264.7 macrophage cell line, in which it was shown that 5-hydroxysalicylic acid, salicylic acid and even aspirin did not inhibit the induction of COX-2 mRNA, thus implying that the drug effects are on the activity of COX enzymes in this system.

3. Amino-substituted salicylic acids are not more potent than the parent acid, but amino-substituted phenols like that of the halogen and derivatives all uniformly resulted in approximately a 10-fold increase in potency.

4. Aside from 3- and 4- phenyl benzoic acids (which were about 10 times more potent than salicylic acid), alkyl-substituted benzoic acids were not more potent and some were less potent than salicylic acid.

5. It is clear that the phenolic group confers benefit for the potency of salicylic acid derivatives. The presence of groups on the 4- or 5-positions of the aromatic ring structure containing phenolic groups that distorts the electron cloud over the benzene ring opposite to that of phenolic and carboxylate moieties clearly influences the activity of the salicylic acid derivatives as COX-2 derived PG synthesis inhibitors.

Using physicochemical and electronic parameters, Habicht and Brune (1983) attempted QSAR of their data and concluded that molar refractivity contributed most to the physicochemical properties underlying the activity of salicylic acid and the other derivatives they studied, the contribution of lipophilicity being less pronounced. An analysis by Dean (1996) of the QSAR of a range of NSAIDs in the mouse macrophage assay suggested that inhibitory potency does not correlated with log P or pK_a (unless at extremes) but with hP, a parameter composed of hydrophobic contributions that includes atoms or molecules of carbon, hydrogen or halogens together with steric factors.

A further quantum chemical study based on this data by Mehler and Gerhards (1987) attempted to 'map' the drug actions to postulated receptor sites on cyclo-oxygenase akin to the early attempts (Gund and Shen, 1977). The authors observed that there was a good correlation between the IC_{50} of these compounds and the difference between the highest occupied and lowest unoccupied molecular calculated orbital values. This, combined with the electrostatic potential of the inhibitor, led to the proposal of a two-site model for inhibition of the enzyme. Given the complexities of this cellular system and the lack of consideration of the degree of ionisation of the various molecular species in the cell systems, this theoretical model may be useful for further understanding of the mechanisms of inhibitory effects of salicylates. More recent attempts by Zoete et al. (1999) to model structure–activity relationships based on the same set of data on prostaglandin E_2 synthesis developed by Habicht and Brune (1983) showed that there was a correlation between the reciprocal of the IC_{50} values and the energy of the highest occupied molecular orbitals (E-HOMO) of salicylic acid derivatives but not the amino derivatives, these having their own correlation.

Using calculated log P and pK_a values, correlations were observed with the E-HOMO and log P

7.2 values for both non-amino and acidic compounds. No indication of correlations with pK_a and Log P 7.2 were noted by these authors. Another limitation to this theoretical analysis was the use of calculated log P values, which are often suspect.

It is of interest in relation to the actions of salicylic acid metabolites that Hinz *et al.* (2000) found that the salicyl-CoA precursor to conjugation with glycine to form salicylic acid was a potent inhibitor of PGE_2 production in LPS-stimulated RAW 264.7 macrophages, but salicyluric acid itself was inactive. Presumably, the activation of salicylic acid by coupling with S-CoA conferred increased potency compared with salicylic acid.

Effects on PGHS-2 (COX-2) Induction

Aside from the direct effects of aspirin on the enzymatic activity of COX-2 and COX-1, it appears that this drug and its metabolite, salicylate, both have direct effects on the synthesis of the inducible PGHS-2 or COX-2 enzyme protein (O'Neill *et al.*, 1987; Sanduja *et al.*, 1990; Mitchell *et al.*, 1997; Xu *et al.*, 1999; Amann *et al.*, 2001). First indications that aspirin might have effects on the induction of COX-2 were provided by the studies of O'Neill *et al.* (1987). These authors showed that interleukin-1 (IL-1) treatment of human synovial cells *in vitro* stimulated PGE_2 production, and that this could be blocked by pretreatment with aspirin. Using ^{14}C-acetyl-labelled aspirin, they showed that there was acetylation of the IL-1-induced enzyme protein. While the exact identity of the two different forms of the COX enzymes was not known at this stage, these were the first indications of the existence of an inducible enzyme that was susceptible to the effects of aspirin during its induction.

Using monolayers of human umbilical vein endothelial cells (HUVEC), Sanduja and co-workers (1990) showed that the IL-1β-(10 U/ml) induced PGHS protein was inhibited equally by both aspirin and sodium salicylate with an IC_{50} of 60 nM. Moreover, Northern blot analysis of the 2.7 kb mRNA that codes for PGHS-2 showed this was inhibited by aspirin with the same IC_{50} (60 nM) as observed for the inhibition of the enzyme protein. Unfortunately, the authors did not report the effects of salicylate on the production of the mRNA for PGHS-2. The IC_{50} for the inhibition of both message and PGHS-2 protein for aspirin suggests that the primary site for the inhibition of PGHS-2 production is at the transcriptional level. These results are of particular interest, since the range of inhibition of the production of PGHS-2 protein is well below that for the inhibition of COX-2 activity (Sanduja *et al.*, 1990).

In mouse peritoneal macrophages, Tordjman *et al.* (1995) found that 50 μM salicylate and paracetamol reduced COX-2 protein, but not the mRNA for COX-2, while aspirin at the same concentration increased COX-2 protein; all these drugs inhibited PGE_2 production. The effects of the low concentrations of aspirin in blocking PGE_2 production in the wake of stimulation of COX-2 protein are interesting and suggestive of a positive feedback control which must be at the translational level since the mRNA for COX-2 was unaffected by aspirin. This might involve the potentiation of the stimulatory effects of soluble pancreatic secretory phospholipase A_2 receptor in inducing COX-2 (Yuan *et al.*, 2000).

The variable effects of aspirin and salicylate on induction of COX-2 in these various cellular systems may relate to the experimental conditions.

As noted previously, salicylate is a very weak inhibitor of activity of isolated or purified cyclooxygenases *in vitro* although it inhibits prostaglandin production in intact cells (Wu *et al.*, 1991). Thus the possibility has been raised that the reason for the apparent similarity in potency as antiinflammatory agents of aspirin and salicylate, especially in chronic conditions such as arthritic diseases, might be due to their potential as equipotent inhibitors of production of PGHS-2/COX-2 protein.

Sanduja *et al.* (1990) also observed that paracetamol (25 μg/ml) and naproxen (30 μg/ml) inhibited the synthesis of PGHS-2 while indomethacin 50 μM did not. The difference in effects of aspirin and indomethacin on production of immunoreactive PGHS-2 was also shown by Rainsford *et al.* (1995) in pig gastric mucosal tissues in organ culture. The observation that paracetamol inhibited production of PGHS-2 is of particular interest in view that paracetamol, like salicylate, is a weak inhibitor of COX activity. This suggests that (a) one of the sites of action of paracetamol is controlling the production of prostaglandin E_2 induced by endotoxin in the hypothalamus that is induced by endotoxin during fever,

(b) the production of PGs in the spinothalamic pathways, the COX-2 for which is induced during painful stimuli, might also represent the site for action of this analgesic. The higher concentration of this analgesic required for inhibition of PGHS-2 protein compared with that of salicylate and aspirin might explain the relatively weak effects of paracetamol on inflammation-mediated PG production.

In contrast to these observations in a primary cell line system, O'Sullivan *et al.* (1993) reported that *E. coli* lipopolysaccharide (LPS)-induced PGHS-2 in rabbit alveolar macrophages was unaffected by an unstated concentration of aspirin, whereas thromboxane B_2 production was reduced by aspirin at about 10^{-4}M. These results are difficult to reconcile with previous and more recent studies, especially since the lack of information on the concentrations of the drugs used in the Western blot analysis of PGHS-2 makes it difficult to interpret these results. In contrast, Amann *et al.* (2001) found that salicy-late 1.0–3.0 mmol/l increased expression of COX-2 protein while reducing PGE_2 production by human mononuclear cells. Aspirin and salicylate, but not 5-aminosalicylate have also been found to enhance IL-1 induced COX-2 expression by a mechanism that involves enhanced stabilization of mRNA for the enzyme (Mifflin *et al.*, 2004).

Salicylate and aspirin (1 nM-1 μM) were found by Xu *et al.* (1999) to inhibit production of PGHS-2 protein induced by the phorbol ester PMA (phorbol-12-myristate, 13-acetate) in HUVEC and human foreskin fibroblasts. As with IL-1β the mRNA for PGHS-2 induced by PMA was inhibited equally by aspirin and salicylate, though the concentration required for inhibition of mRNA PGHS-2 was some-what higher (0.1–10 μM) than that required for inhibition of the PGHS-2 protein. Salicylate did not affect the stability of mRNA but did affect the promotor, suggesting that the primary effect of the drug was on the synthesis of the message rather than control of its degradation.

Mitchell *et al.* (1997) investigated the mechanism of the effects of salicylate on the induction by IL-1β of PGHS-2 protein and PGE_2 production in human A549 cells. In particular, these authors exam-ined the relationship of these effects of salicylate to the previously reported effects of high concentrations of this drug on the activity of the nuclear factor, NFκB. While salicylate inhibited IL-1β-induced PGE_2 release with an IC_{50} of 31 μM, there was no effect on NFκB activation (see also next section). Thus, it appears that the IL-1β – stimulated PGE_2 production proceeded by a mechanism that was independent of COX-2 transcription or translation. Possible factors accounting for the differ-ences in responses to salicylate on PGHS-2 regulation observed by these authors compared with those of Xu *et al.* (1999) are that the former used established cell lines and incubated them for 24 hours, whereas Xu *et al.* (1999) used primary cells and incubated these for shorter periods (6 hours).

Thus, it would appear that there are differences in the actions of salicylates on the induction by IL-1β in different cell systems, and this could be due to variations in the controls of pathways involved in signal transduction in these cells.

Signal transduction and NFκB / IκB Interactions

NFκB is important in the induction of COX-2 by cytokines (Nakao *et al.*, 2000) and thus any effects of sal-icylates on this system will have consequences for the control of COX-2 mediated PGs (O'Neill, 1998).

Activation of the NFκB transcription factor has been considered by several authors as a site of action of salicylates since the original publication of Kopp and Ghosh (1994) reported that high con-centrations of aspirin and salicylate could inhibit its activity. However, these studies were criticised because the effects were only apparent at supratherapeutic (millimolar) concentrations of these drugs (Frantz and O'Neill, 1995). While this has been confirmed by several groups (Gautam *et al.*, 1995; Oeth and Mackman, 1995; Weber *et al.*, 1995; Grilli *et al.*, 1996; Pierce *et al.*, 1996; Takashiba *et al.*, 1996; Dodel *et al.*, 1999), there have also been some exceptions (Farivar and Brecher, 1996; Sakitani *et al.*, 1997). As indicated above, these differences may relate to the regulatory controls of NFκB in different cell systems as well as experimental conditions.

Yin *et al.* (1998) showed that the primary effect of salicylate and aspirin is to inhibit IκB kinase (IKK)-β activity, thus preventing the cytokine-induced phosphorylation of IκB, subsequent dissociation of this from NFκB. However, later studies by Alpert and Vilcek (2000) showed that this *in vitro* effect

CHAPTER 7

does not reflect the drug effects in intact cells. They found that salicylate inhibited TNFα- but not IL-1β-induced IKK activity. These results may have importance for the potential therapeutic effects of salicylate on the actions of TNFα *in vivo*. Using COX-2 or p105 (a precursor of NFκB) knock-out mice Weissman *et al.* (2002) showed that anti-inflammatory effects of aspirin or sodium salicylate are independent of COX-2 and NFκB. They showed that both drugs inhibited the Mak/Erk pathway and integrin-dependent aggregation of neutrophils. These studies provide further evidence for prostaglandin-independent actions of aspirin and salicylate.

Regulation by salicylates of signal transduction events downstream in the region of the immediate–early response gene transcription was considered by Stevenson *et al.* (1999) and Saunders *et al.* (2001). Stevenson *et al.* (1999) found that salicylate and other NSAIDs suppress the phosphorylation of the ribosomal S6 kinase 2 (RSK2) as well as the phosphorylation of RSK-2 substrates, cAMP-response element binding protein (CREB) and IκB. This is of particular interest in view of the effect of cAMP in upregulating PGHS-2 gene expression and PG production (Lo *et al.*, 2000; Kirtikara *et al.*, 2001). By inhibiting the CREB phosphorylation, salicylate may block the NFκB–IκB system that may involve the antioxidant effects of these drugs. Thus, Shi *et al.* (1999) found that aspirin inhibited NFκB activation induced by the $Fe(II) + H_2O_2$ production of hydroxyl radicals ($OH^{\cdot}$-) or hydrogen peroxide (H_2O_2). They also showed by ESR that aspirin was a superior antioxidant to ascorbate, glutathione and cysteine.

Other salicylates have been investigated for their actions on the NFκB–IκB signal transduction pathway (Hernandez *et al.*, 2001). These may also affect NFκB activation, with consequent effects on regulation of PGHS-2 or other proteins. Liptay *et al.* (1999) found that sulphasalazine but not its two major metabolites, 5-amino-salicylic acid or sulphapyridine (up to 5 mM), inhibited NFκB activation in the murine T-lymphocyte cell line, RBL5. The IC_{50} following 4-hour incubation with sulphasalazine was 0.625 mM, which is rather high by comparison with the previously mentioned studies with salicylate and aspirin. It is interesting from a structure–activity viewpoint that the 5-amino-salicylate (5-ASA) was without effect, especially as salicylate itself appears a potent inhibitor of NFκB activation in the micromolar range in most systems. Perhaps the added 5-amino moiety of 5-ASA reduces the antioxidant potential of salicylate, which may in turn limit the oxyradical mediated NFκB induction that is inhibited by salicylate.

Lipoxins and 15-epi-lipoxins

The discovery of lipoxins was pioneered by Professor Charles Serhan (now of Harvard Medical School) with Hamberg and Samuelsson of the Karolinska Institute (Serhan *et al.*, 1984; 1985). The term lipoxin refers to **lipox**ygenase **in**teraction products (Serhan *et al.*, 1985). These compounds are products formed from activated leucocytes; in the initial studies addition of 15(S)-HPETE led to formation of lipoxins (Serhan *et al.*, 1985). The involvement of both 5- and 15-lipoxygenases was foreseen and the products stimulated superoxide anion and lysosomal enzyme release. Later, 12-LO was also shown to be involved in their formation. They initially presented a dichotomous view to the prevailing and predominating views that the cyclo-oxygenase derived prostanoids and 5-, 12-, and 15-lipoxygenase-derived leukotrienes and respective HETES were the main arachidonate-derived mediators involved in inflammation and asthma. In this case the emphasis was on demonstrating their pro-inflammatory roles (although as discussed previously the contrasting roles of PGE_2 and $PGF_{2\alpha}$ were also recognised). The early work showing production of lipoxin A_4 (LXA_4) and LXB_4 by stimulated polymorphs (Serhan *et al.*, 1984; 1985) was puzzling in that here were lipid mediators 'looking for a role' in inflammation. However, the persistent work of Serhan and his colleagues as well as that inspired in others now enables a view of lipoxins as products of cellular co-operation (transcellular biosynthesis) that have important acute anti-inflammatory effects but also, paradoxically, stimulate monocytes and myeloid bone marrow-derived progenitor cells (Serhan *et al.*, 1999; Table 7.7A and 7.7B). They also affect natural killer cell activity (Ramstedt *et al.*, 1985) and so could be important in tumour surveillance.

The pathways for the biosynthesis of the lipoxins are shown in Figure 7.6. Normally lipoxin (LX) A_4 and LXB_4 are formed, but when exposed to aspirin two isomers of lipoxins are formed, the 15-epi-LXs,

TABLE 7.7

The occurrence in tissues of patients with various immuno-inflammatory diseases, vascular and leucocyte actions of the lipoxins. (From Serhan, 1997; Serhan *et al.*, 1999.)

A. Lipoxins in tissues and diseases
Asthma
Aspirin-sensitive asthma
Angioplasty-induced plaque rupture
Bone marrow generation
Lipoxin biosynthesis – defect in chronic myeloid leukemia
Glomerulonephritis
Sarcoidosis
Pneumonia
Nasal polyps
Rheumatoid arthritis

B. Vascular and smooth muscle actions of lipoxins
Vasodilatory
Relax aorta and pulmonary artery
Reverse $PGF_{2\alpha}$ and endothelin (ET-1) contraction
Stimulate nitric oxide generation
Stimulate endothelial prostacyclin formation
Stimulate $cPLA_2$-dependent arachidonate release and conversion

C. Selective actions of lipoxins on human leucocytes

Cell type	Lipoxin	Action
PMN	LXA_4 and LXB_4	Blocks emigration
		Transmigration and chemotaxis; downregulates CD11/18 and intracellular IP_3, Ca^{2+}; inhibit PMN-endothelial cell and epithelial cell interactions
Eosinophils	LXA_4	Inhibit chemotaxis to PAF and FMLP
Monocytes	LXA_4 and LXB_4	Stimulate chemotaxis and adherence; stimulate myeloid bone marrow-derived progenitors

15-epi-LXA_4 and 15-epi-LXB_4, these being referred to as 'aspirin-triggered 15-epi-lipoxins' (Figure 7.6; Serhan *et al.*, 1999), the 15-HPETE for this being derived from the activity of aspirin-treated PGHS-2.

The occurrence of lipoxins in tissues of patients with various immuno-inflammatory states, as well as the vascular and leucocyte actions of these mediators, is shown in Table 7.7. The signalling responses mediated by receptors can be summarised as follows (Serhan, 1997; Serhan *et al.*, 1999).

1. Lipoxins act at their own specific cell surface receptors (i.e. LXA_4-specific and separate LXB_4 receptor, which are distinct from one another)
2. LXA_4 interacts with a subclass of LTD_4 receptors that also bind LXA_4;
3. 15-Epi-LXA_4 acts as the same receptor as LXA_4
4. Lipoxins can act at intracellular targets after lipoxin transport and uptake or within their cells of origin
5. Lipoxins inhibit or attenuate receptor mediated actions of leukotrienes and block the chemotaxis, emigration and transmigration of granulocytes
6. Lipoxins down-regulate expression of adhesion molecules, intracellular calcium and phosphoinositide (IP-3) signals
7. Lipoxins stimulate monocyte cell activation
8. Aspirin-triggered 15-epi-LXs exhibit even greater potency in blocking the above-mentioned neutrophil functions and inhibit cell proliferation, thus constituting potent anti-inflammatory mediators, or 'chalones' (Serhan *et al.*, 1999; Fierro *et al.*, 2003).

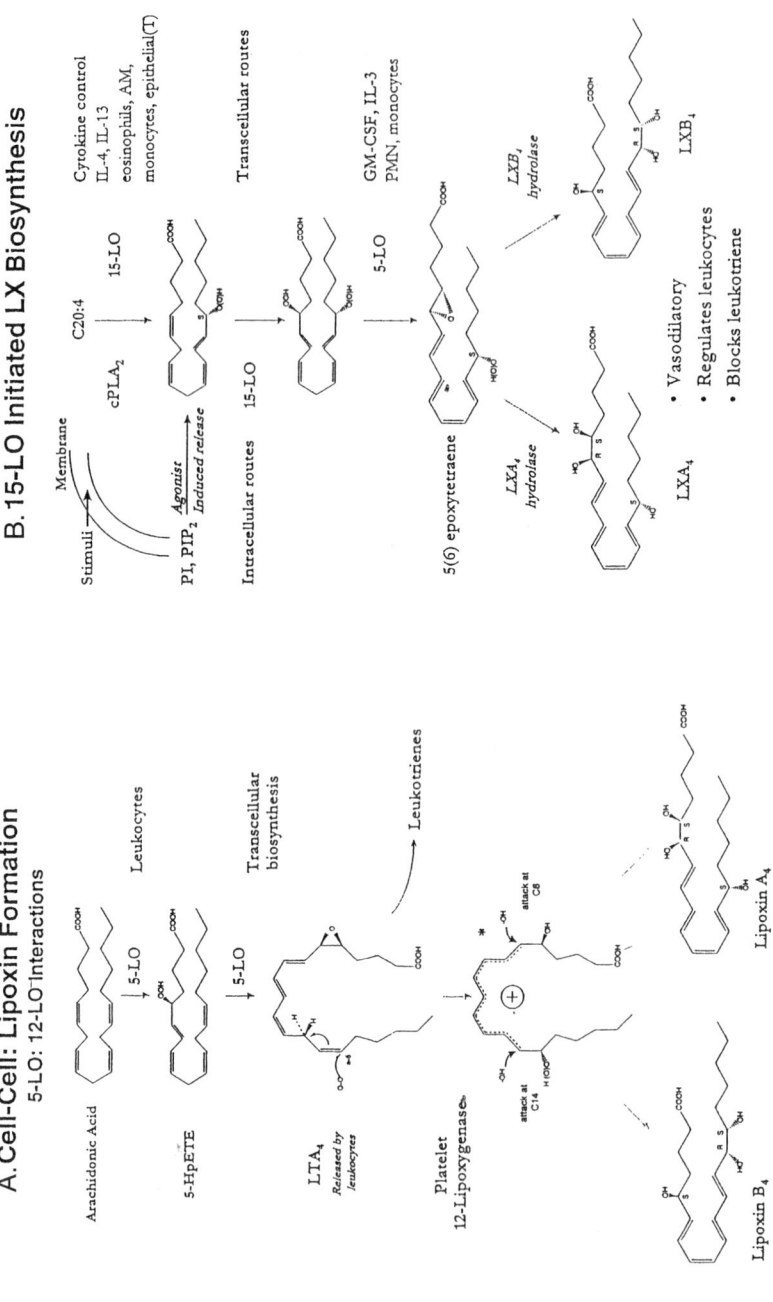

Figure 7.6 Pathways for the biosynthesis of the lipoxins (LX). There are two pathways involved in the production of LXs:

A. The 5-lipoxygenase pathway (5-LO) occurs in the cells of the vasculature and blood-borne cells (neutrophils, platelets), leucocytes with their 5-LO producing the leukotriene (LTA₄), which then is taken up by platelets to undergo formation of LXA₄ and LXB₄ by platelet 12-LO.

B. The 15-LO pathway in eosinophils, macrophages, monocytes and endothelial cells is under cytokine control, being negatively regulated primarily by IL-4 and IL-13 and amplified by IL-3 and GM-CSF of monocytes and polymorphs. 15-H(P)ETE produced by 15-LO in the former cells is metabolised by 5-LO in the latter group to form 5(6)-epoxytetraene and then LTA₄ hydrolase and LTB₄ hydrolase to form LXA₄ and LXB₄, respectively.

(Reproduced from Serhan (1997), by permission from the publishers, Elsevier Science.)

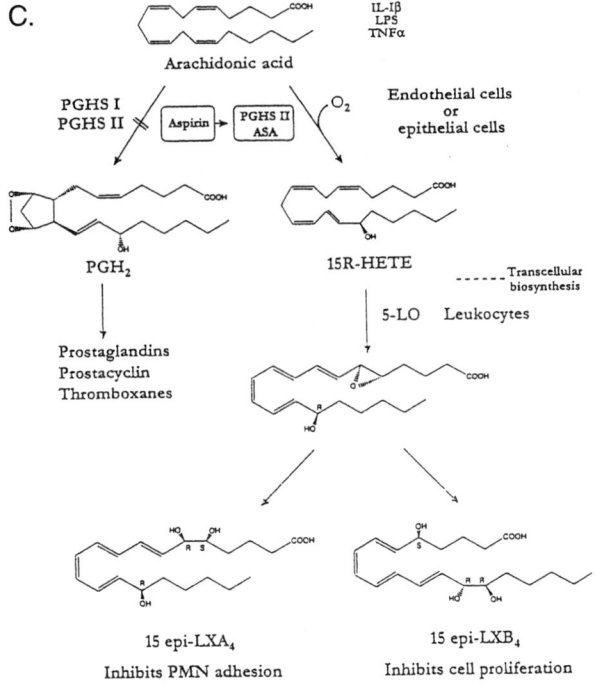

C.

Arachidonic acid

IL-Iβ
LPS
TNFα

PGHS I
PGHS II → | Aspirin | → | PGHS II ASA |

O₂

Endothelial cells
or
epithelial cells

PGH₂

15R-HETE
------- Transcellular biosynthesis

Prostaglandins
Prostacyclin
Thromboxanes

5-LO   Leukocytes

15 epi-LXA₄
Inhibits PMN adhesion

15 epi-LXB₄
Inhibits cell proliferation

Figure 7.6 *(contd)* (C) Formation of epi-lipoxins (epi-LXA₄ and epi-LXB₄) via the conversion of PGHS-2 or COX-2 by aspirin to produce 15-(R)-HETE (Figure 7.5) by cytokine-activated endothelial or epithelial cells. This then undergoes transcellular biosynthesis by 5-LO in leucocytes to produce epi-LXA₄ and epi-LXB₄ which inhibit polymorph adhesion and cell proliferation respectively. (Reproduced from Serhan (1997), with permission from the publishers, Elsevier Science).

Sulphasalazine (SASP, or salicylazosulphapyridine) and aminosalicylates

Of the other salicylates investigated, much interest has been shown in the actions of salicylazo-sulphapyridine (SASP) or sulphasalazine and its metabolites (especially 5-aminosalicylic acid) on eicosanoid metabolism, and how these effects relate to its mode of action (Gaginella and Walsh, 1992; Travis and Jewell, 1994). Waller (1973) suggested that because prostaglandins are produced in ulcerative colitis, SASP might act by inhibiting their production in controlling inflammation. Collier *et al.* (1976) recognised the important pharmacokinetic aspect that metabolism in the intestinal tract of SASP to 5-aminosalicylic acid (5-AS) and sulphapyridine (SP) raises the possibility that those metabolites may exert effects on PG metabolism. Using bovine and ram seminal vesicle preparations (probably COX-1 systems) they showed that SASP was about equipotent with aspirin (IC_{50} in range of 0.47–0.66 mM) in causing inhibition of bioassayable PGs (rat stomach strip), while 5-AS and SP, though inhibiting PG production, were about 10 times less potent – even less so than sodium salicylate. NSAIDs in microsomal systems often produce IC_{50} values that are higher than in cell-based systems, probably because of the varying co-factor and substrate requirements and redox balance. However, the data reviewed by Gaginella and Walsh (1992) showed that SASP and 5-AS inhibit the formation of COX products and metabolism of arachidonic acid to PGs in a variety of cell/tissue systems in the same mM concentration range as observed by Collier *et al.* (1976). Results by different authors regarding the effects of SP in these systems are more variable, with some reporting negative results.

Moore and Hoult (1982) showed that SASP was about 10 times more potent in inhibiting PG breakdown by the 15-hydroxy-prostaglandin dehydrogenase (PGDH) system in rabbit colon post-microsomal supernatants. This has been confirmed in human colonic systems (Peskar, 1985).

Inhibition of PGDH activity now appears important in the actions of SASP for the treatment of Crohn's disease and ulcerative colitis. Reduction in prostaglandin production by this drug in the lower bowel is associated with significant improvement in the manifestations of these diseases (Collier $et\ al.$, 1976; Hoult and Moore, 1978; 1980). Low concentrations of SASP ($IC_{50} = 50\,\mu$mol/l) totally block $PGF_{2\alpha}$ breakdown (Hoult and Moore, 1978; 1980). SASP (but not its metabolite sulphapyridine or 5-aminosalicylic acid) specifically inhibits PGDH instead of the other enzymes involved in prostaglandin catabolism (Hoult and Moore, 1980). While SASP only exhibits very weak activity as a cyclo-oxygenase inhibitor, it does seem to be amongst the most specific inhibitors of PGDH activity. Certainly the principal effects of this drug on overall prostaglandin metabolism might be ascribed to this specific drug action on PGDH (Hoult and Moore, 1980).

SASP, 5-AS and olsalazine also inhibit synthesis of peptido-leukotrienes and leukotriene B_4, in some assays the potency of the former being about 10 times that of 5-ASA (Horn $et\ al.$, 1991; Gaginella and Walsh, 1992; Oketano $et\ al.$, 2001). The inhibition of the production of peptido-leukotrienes by SASP is probably due to its inhibitory effects on LTC synthetase (IC_{50} 0.4 mM in RBL cells) as well as being a competitive inhibitor of glutathione S-transferase (K_i 0.21–0.46 μM in various enzyme fractions from rat liver; Bach $et\ al.$, 1985).

Using an $in\ situ$ equilibrium dialysis system to measure luminal concentrations of eicosanoids in patients with ulcerative colitis, Lauritsen $et\ al.$ (1986) observed that although topical 5-AS treatment (enemas) for 4 weeks or until remission did not cause reduction of PGE_2 or LTB_4 as a whole, there was a positive correlation between disease activity and concentrations of these eicosanoids. Schmidt $et\ al.$ (1995) found that sulphasalazine and 5-aminosalicylic acid treatment in biopsies of 107 patients with inflammatory bowel disease reduced production of both LTB_4 and PGE_2. The differences in results obtained by these authors and Lauritsen $et\ al.$ (1986) is probably due to the methods of sampling and tissue source.

Cronstein $et\ al.$ (1999) showed that 100 mg/kg SASP given to NFκB knock-out (p105−/−) mice exhibited the same inhibitory effects on leucocyte accumulation in air pouches as in wild-type mice. They claimed that this implied that SASP had anti-inflammatory effects independent of the inhibition of NFκB activation, but the leucocyte accumulation in this acute model is not totally representative of the effects on the entire acute inflammatory process (swelling etc.).

Actions of salicylates on 12-HETE and leukotriene production

Siegel and co-workers (1979; 1980) reported that aspirin, salicylate and indomethacin, but significantly not all NSAIDs, block the conversion of 12-HPETE to 12-HETE in platelets at plasma concentrations that are well within their therapeutic range. While the exact significance of these drug effects (especially in respect of their actions on platelet aggregation) has yet to be determined, it is possible that the effects of salicylates on this system could be of potential significance in relation to the $in\ vivo$ actions of this drug.

In contrast to the situation with 12-HETE, the role of leukotrienes as inflammatory mediators is clearer (Figure 7.3; Klickstein $et\ al.$, 1980; Borgeat and Sirios, 1981; Ford-Hutchinson $et\ al.$, 1980; 1981; Davidson $et\ al.$, 1982). 5-HETE and leukotriene B_4 (LTB_4) production in rat polymorphs where increased when stimulated by calcium ionophore (Bray $et\ al.$, 1980b). Diflunisal is active in this system at a concentration of 100 μmol/l, which is within its range for pharmacological activity (Bray $et\ al.$, 1980b). Salicylate is inactive, but nordihydroguaiaretic acid (NGDA), eicosatetraynoic acid (ETYA), BW-755 (3-amino-1-[m-trifluoromethyl-phenyl]-2-pyrazoline) and benzoxaprofen, which are all lipoxygenase inhibitors, are active at concentrations in the range 10 to 100 μmol/l (Baumann $et\ al.$, 1980; Bray $et\ al.$, 1980b).

In contrast to these observations on rat and guinea pig polymorphs, the same cells derived from the peritoneal fluids of rabbits exhibit 5-lipoxygenase inhibition by aspirin 0.05 to 1.0 mmol/l, but not by salicylate (Walker and Harvey, 1983). It appears that the different effects of aspirin on rabbit (compared with guinea pig polymorphs) may be due to the capacity of rabbit cells to metabolise arachidonate by the 5-lipoxygenase, compared with cyclo-oxygenase, pathways (Walker and Harvey, 1983). The rabbit cells are notably more active in producing lipoxygenase products, compared to cyclo-oxygenase products, than polymorphs from guinea pigs (Walker and Harvey, 1983). Rat peritoneal polymorphs may resemble those from guinea pigs in this regard. Aspirin-induced inhibition of 5-lipoxygenase may only be manifest in those cellular systems with high activity of this enzyme (e.g. monosodium urate-stimulated polymorphs (Rae et al., 1982). This inhibitory effect of aspirin may only be important pharmacologically in those conditions where there is considerable production of leukotrienes (e.g. 'aspirin-sensitive' asthma, gout) but not in other conditions where leukotriene production is low (e.g. rheumatoid arthritis).

Effects of salicylates on free radical production and actions

Free radicals are produced during tissue damage and phagocytosis (e.g. in inflammation), and cause the production of highly reactive peroxidation products (chiefly lipids) (Oyanagui, 1976; 1978; Kuehl et al., 1977; 1979a; Lunec and Dormandy, 1979; Bragt et al., 1980; Dormandy, 1980; Bragt and Bonta, 1981; Milanino and Velo, 1981; Walker and Harvey, 1983). These radicals contribute further to the cycle of tissue injury, stimulation of prostaglandin production and leucocyte infiltration (Kuehl et al., 1977; 1979a; Lunec and Dormandy, 1979; Bragt et al., 1980; Dormandy, 1980; Bragt and Bonta, 1981; Lunec et al., 1981; Milanino and Velo, 1981). The radicals produced during phagocytosis are most probably the superoxide ($O_2^{\cdot-}$) type whose destruction occurs naturally via superoxide dismutase to produce H_2O_2, which is then converted to H_2O by catalase (Fridovich, 1975; McCord et al., 1979; Dormandy, 1980; Braght and Bonta, 1981). However, the identity of those radical species generated during tissue injury is still debatable, both superoxide ($O_2^{\cdot-}$) and hydroxyl radicals ($OH^{\cdot}$) being likely candidates (Fridovich, 1975; Oyanagui, 1976; Kuehl et al., 1977; 1979; McCord et al., 1979; Bragt and Bonta, 1981; Milanino and Velo, 1981; Walker and Harvey, 1983). The more favourable procedure for measuring free radical production, aside from that of superoxide production (Oyanagui, 1978), has been to measure the more stable lipid peroxidation productions formed as a consequence of radical attack (Lunec and Dormandy, 1979; Lunec et al., 1981; Walker and Harvey, 1983).

Several lines of evidence indicate that salicylates act as oxyradical scavengers (Haggag et al., 1986; Haynes et al., 1993; Kataoka et al., 1997) or inhibit their actions in polymorphs that may also include the inhibition of surface ligand receptor interactions (Minta and Williams, 1985). However, the results depend on the type of system employed. Early in vivo studies showed that some NSAIDs, but notably not aspirin, were found to inhibit the production of $O_2^{\cdot-}$ in (a) paraffin oil elicited and non-stimulated guinea pig macrophages (Oyanagui, 1978); (b) IgG- or chemotactic factor-stimulated human monocytes (Lehmeyer and Johnstone, 1978; Oyanagui, 1978); or (c) by the hypoxanthine–xanthine oxidase system (De Alvare et al., 1976; Puig-Paradella and Planas, 1978). Salicylate may, however, be effective as a superoxide scavenger, possibly through its transformation to a hydroxylated metabolite (Betts et al., 1981; Carlin et al., 1985; Haynes et al., 1993). Chemically salicylate undergoes hydroxylation in the Fenton reaction at the 3 and 5 positions to produce 2,5-dihydroxybenzoic (gentisic) acid (DHB) (Grootveld and Halliwell, 1986; Maskos et al., 1990; Bailey et al., 1997). It has been suggested that this may be due to chelation of iron in the Fenton reaction (Cheng et al., 1996). Electron spin resonance (ESR) studies showed that aspirin was a more effective inhibitor of the Fenton-generated $OH^{\cdot}$ reaction than endogenous antioxidants such as ascorbate, cysteine and glutathione (Shi et al., 1999). The production of oxyradicals has been assayed with salicylate as a chemical probe (Bailey et al., 1997; Liu et al., 1997; see also Chapter 3).

Salicylate and aspirin induced scavenging of $OH^{\cdot}$ in human granulocytes was considered not to be related to effects on myeloperoxidase (Sagone and Husney, 1987) but possibly to the actions of HOCl (Wahn and Hammerschmidt, 1998) – although the taurine chlorination reaction was only weakly inhibited by aspirin 2 mM, other NSAIDs being more potent inhibitors of this reaction (Neve et al., 2001). The salicylate-induced increase in LDL oxidation appears to be antagonised by 2,5-DHB (Hermann et al., 1999a; 1999b) thus showing that the oxyradical scavenging effects of salicylates in vivo are complex but involve the DHB metabolites. Endothelial cell damage by high concentrations of

H_2O_2 0.65 mM are inhibited by pre-treatment with 3 to 30 μM aspirin; combination with vitamin E potentiating further the protective effects of aspirin (Podhaisky *et al.*, 1997).

In the xanthine–xanthine oxidase reaction, aspirin and salicylate are decarboxylated (Sagone and Husney, 1987). However, it appears that the principal actions of the salicylates involve their metabolism to dihdroxybenzoic acids.

Sodium salicylate 100 mg/kg i.p. given prior to induction of ischaemia of the rat intestine reduced the reperfusion injury; DHB metabolites were detected during injury and were thus presumed to have been generated during this injury (Udassin *et al.*, 1991). High doses of aspirin 500 mg/kg for 5 days increased liver glutathione peroxidase activity in rats (Kirkova *et al.*, 1995), and this may represent another mechanism for the antioxidant effects of this drug (Steer *et al.*, 1997).

Haynes *et al.* (1993) have proposed that aspirin is a pro-drug for the antioxidant and cytokine modulating effects of its oxymetabolites. In particular 2,5-DHB and 2,3-DHB are inhibitors of H_2O_2 production by human neutrophils, with 2,5-DHB inhibiting this reaction independent of the effects on PGE_2 production (Haynes *et al.*, 1993). Aspirin 100 mg/d for 8 days reduced the $O_2^{\cdot-}$ release from platelets stimulated by fMLP (Reinisch *et al.*, 2000). Sodium salicylate 2 mM prevented the reoxygenation injury induced in the isolated reperfused rat liver (Colantoni *et al.*, 1998). De Maria *et al.* (1997) showed that this inhibitory effect of salicylate on reperfusion injury was due to blockade of NFκB and iNOS. Shi *et al.* (1999) showed that aspirin 0.36–2.88 mM exerted a protective effect on the activation of NFκB by silica, LPS or the Fenton reaction in the mouse macrophage cell line, RAW 264.7.

The copper complexes of aspirin and several other salicylates have been found to be more effective superoxide scavengers *in vitro* than the parent drugs (De Alvare *et al.*, 1976; Betts *et al.*, 1981; Weser and Schubotz, 1981). There is a suggestion that the copper complexes of salicylates (and other acidic NSAIDs) may be formed *in vivo* and thus be a bioactive form of these drugs (Sorenson, 1982; see also p. 722, Binderup *et al.*, 1976), by acting as superoxide scavengers *in vivo*.

Other studies have suggested that neither superoxide, singlet oxygen, hydroxyl radical, nor hydrogen peroxide are involved in the microsomal (prostaglandin synthetase) oxidation of arachidonic acid (De Vries and Verboom, 1980). It would therefore appear that direct lipid peroxidation involves fatty acid systems other than the prostaglandin-synthesising systems (Porter *et al.*, 1980). The hydroxyl radical-mediated lipid peroxidation may be of particular significance in relation to the action of salicylates, since salicylate itself has been found to react with hydroxyl radicals (Amphlett *et al.*, 1968). Clearly in view of the potential importance of radical control in understanding the anti-inflammatory actions of the salicylates, especially with potent radical scavengers such as diflunisal, this aspect of the action of salicylates should receive intensive study.

Oxyradical-scavenging effects of salicylates may in part be responsible for the reported effects of these drugs on the binding or activity of heat shock proteins (Jurivich *et al.*, 1992; Koo *et al.*, 2000).

Peroxisomal metabolism

Peroxisomes (or formerly microbodies) that have functions in catabolic reactions (e.g. β-oxidation of fatty acids, H_2O_2-based respiration) and anabolic functions (e.g. plasminogen and bile acid biosynthesis) (Valle and Gärtner, 1993) are interesting and important sites of the actions of salicylates, like other NSAIDs that induce proliferation of this organelle. Recent studies have shown that NSAIDs, including aspirin, act at the nuclear receptors for perioxisosomal proliferators (PPAR), especially the subtype PPARγ, which is implicated in regulating the differentiation cells of the adipose lineage (Sarraf *et al.*, 1998). PPARγ has been shown also to be expressed in the colon during carcinogenesis and following ligand activation of PPARγ-receptors (PPAR-R) reduce tumour growth and gene expression associated with colon cancer (Sarraf *et al.*, 1998). In isolated microglial cells as well as cultured glial cells the prostaglandin analogue, 15-deoxy-$\Delta^{12,14}$-PGJ, (15d PGD_2), which is a PPAR, suppresses production of pro-inflammatory cytokines IL-1β and TNFα, iNOS and COX-2 (Koppal *et al.*, 2000). 15d PGD_2 also induces antioxidant reactions in glial cells, including increased glutathione levels and upregulation of haeme oxygenase-1 (Koppal *et al.*, 2000). These and other cellular controls by PPAR agonists make them important targets for regulating gene expression during differentiation and malignancy. This includes the actions of NSAIDs in inhibiting the APC regulated PPARδ gene (He *et al.*, 1999).

Early phase amines and kinins

Salicylates inhibit the production and/or actions of early phase acute inflammatory mediators (Hebborn and Shaw, 1963; Whitehouse, 1965; Skidmore and Whitehouse, 1966; 1967; Smith and Smith, 1966; Petillo *et al.*, 1969; Rosenior and Tonks, 1974; Rothschild *et al.*, 1975; Inoki *et al.*, 1977; 1978; Atkinson and Collier, 1980; Sharma *et al.*, 1980). Individually, these inhibitory effects of salicylates are of relatively low to moderate potency. As the release or actions of mediators is inter-related, both with one another and with the prostaglandin system (Lewis, 1977; 1979; Williams and Peck, 1977; Movat, 1979; Westwick, 1979; Williams, 1979), it is apparent that a concerted inhibitory attack by the salicylates on all these systems may well be responsible for the overall effects of these drugs on the mediators of acute inflammation. As will be apparent later, there are some more complex effects of the salicylates on the systemic production of histamine (Lewis and Whittle, 1977) which raises the possibility of contrasting effects in the various organs competing with or even antagonising those histamine-mediated events in inflamed tissues. Such systemic effects may also be apparent with the other mediators of the inflammatory responses, e.g. from the feedback modulation of the prostaglandin system. Thus a caveat must be invoked that the *in vivo* effects of salicylates on these components of the inflammatory response *per se*, revealed by examination in animal models, may not allow for full understanding *in toto*, especially in man. This situation has been well illustrated by Atkinson and Collier (1980), who pointed out that aspirin exerts a capricious effect in antagonising the actions of bradykinins in different species.

Histamine

Smith and Smith (1966) reviewed the early studies of the effects of salicylate on liberation and actions of histamine and concluded that salicylate exerted rather weak and unspecific effects on this system. Shortly afterwards, Skidmore and Whitehouse (1966) reported that salicylate, in common with other NSAI drugs, causes the inhibition *in vitro* of the enzyme, histidine decarboxylase, which is involved in histamine formation from histidine. This was soon confirmed by other authors (Radwan and West, 1968; Petillo *et al.*, 1969). The mechanism of this inhibitory effect probably involves salicylate competing with the cofactor, pyridoxal 5-phosphate, required by this enzyme (Skidmore and Whitehouse, 1966), a feature exhibited by this drug on other decarboxylases. Relatively high concentrations of salicylate (7.3–10.8 mmol/l) and aspirin (2.8–5.7 mmol/l), in common with other NSAI drugs, have been found to inhibit the release of histamine from rat peritoneal mast cells *in vitro* and *in vivo* (Yamasaki and Saeki, 1967; Champion *et al.*, 1977). It has been suggested that the mechanism of this effect may be related to the actions of these drugs in reducing ATP levels (Yamasaki and Saeki, 1967; Champion *et al.*, 1977) by uncoupling of oxidative phosphorylation (see p. 280).

If these effects of salicylates on histamine synthesis and release do in fact occur in inflamed tissues *in vivo*, then how is it that such weak effects are exhibited by these drugs on histamine release/actions in the same tissues? Several possibilities may account for these disparate effects:

1. Salicylates in high concentration stimulate histamine synthesis and release to the circulation from some tissues, such as the stomach (Petillo *et al.*, 1969), which may lead to enhanced concentrations in and around the inflamed site.
2. A similar increase in circulating histamine may arise from inhibition of histamine and histidine catabolism, as has been shown *in vitro* and *in vivo*; the inhibition, specific to salicylate and aspirin, is not shown by other NSAIDs or salicylamide and probably involves blockade of the conjugation of ribose with imidazole-acetic acid (Beaven *et al.*, 1974; Moss *et al.*, 1976).
3. These above-mentioned inhibitory effects of salicylates on histamine synthesis/release might only

■ CHAPTER 7 ■

occur in relatively intact cellular systems (i.e. where there is integrity of mast-cell populations); thus in inflamed tissues, these cells may be so damaged that the drugs are incapable of exerting appreciable effects therein.

5-Hydroxytryptamine (5HT, serotonin)

Early studies suggested that the serotonin (5-hydroxytryptamine) released from blood platelets accumulating at inflamed sites or from these tissues could be an important mediator of inflammation (Whitehouse, 1965; Skidmore and Whitehouse, 1966). While salicylates and other NSAIDs inhibit the enzyme 5-hydroxytryptophan decarboxylase, which is involved in serotonin production (Skidmore and Whitehouse, 1966), this effect is opposed by a stimulation of serotonin production by salicylates displacing its metabolic precursor, tryptophan, from its binding sites on albumin. This latter property is relevant to the analgesic actions of salicylate, and is discussed in further detail on p. 291.

Kinins

Of the three mediators discussed in this section, there is no doubt that the major effects of the salicylates are on the kinin system. Apart from inhibiting the bradykinin-induced prostaglandin production, both salicylate and aspirin directly inhibit the release and production of prostaglandin (Atkinson and Collier, 1980) the release and actions of bradykinin (Smith and Smith, 1966; Rosenior and Tonks, 1974; Rothschild et al., 1975; Inoki et al., 1977; 1978; Atkinson and Collier, 1980; Sharma et al., 1980). Early studies (see Smith and Smith, 1966) suggested that salicylates did not affect kinin formation by blockade of kallikrein activity; neither did it appear that enzymes involved in degradation of kinins (kininases) would be affected by these drugs. However, subcutaneous perfusion of aspirin into the paws of rats (previously exposed to noxious stimuli) inhibits both the release of a bradykinin-like substance and those enzymes involved in the synthesis of these mediators (Inoki et al., 1977; 1978). An apparent paradoxical effect of aspirin compared to salicylate was observed on plasma kininogen consumption in rats given adrenaline (Rothschild et al., 1975). While aspirin was found to inhibit kininogen consumption, the prior treatment with salicylate or indomethacin prevented this response (Rothschild et al., 1975). Thus, the reduced conversion of this plasma kinin precursor by aspirin appears to be due to blockade by this drug of kinin-forming enzymes. It also appears that these effects could be blocked by salicylate formed in vivo. This likely paradox still does not explain the earlier disparate results obtained by various authors (Hebborn and Shaw, 1963; Smith and Smith, 1966), though later studies (Rosenior and Tonks, 1974) have shown that salicylate blocks plasminogen activation in vitro. Even more puzzling is the observation of Castania and Rothschild (1974) that aspirin blocked the fall in plasma kininogen content induced by adrenaline. Sharma and co-workers (1980) have shown that ingestion of aspirin 3.9 g by patients with rheumatoid arthritis reduced their supranormal plasma levels of kininogens by 31 per cent. These results illustrate the clinical relevance of drug effects on the kinin-forming system, and also suggest that the drug has effects other than blocking kininogen consumption. Perhaps aspirin affects the synthesis of kininogen per se. The consequences of aspirin-induced reduction in kinin output are linked to the inhibitory effects of this drug on prostaglandin production. Both the hyperproduction of E_2- and I_2-type prostaglandins and bradykinin are necessary to elicit plasma exudation (as seen in inflammation), with bradykinin increasing vascular permeability (with only trace plasma exudation) and the prostaglandins E_2/I_2 potentiating this effect by their vasodilator effects (Williams and Peck, 1977; Westwick, 1979; Williams, 1979). So, by blocking kinin release, aspirin can prevent both of these actions directly, as well as the stimulation by kinins of prostaglandin (E_2/I_2) release. Aspirin might also be expected to inhibit bradykinin-induced increase in blood flow and its stimulation of glucose uptake (Dietze et al., 1978).

Lysosomal and other hydrolases

These hydrolytic enzymes are released in response to direct injury of cells resident in tissues, as well as from leucocytes that have emigrated into the inflamed areas. Thus the first release would be expected to occur during the early stages of an acute inflammatory reaction, whereas the leucocyte-derived lysosomal enzymes would be released at later stages of the acute inflammatory response or when polymorphs and monocytes have become established in chronically inflamed tissues. They may also release factors that influence vascular permeability (Anderson, 1968).

Studies of the effects of salicylates on enzyme release from isolated lysosomes *in vitro* (usually from rodent livers) is critically dependent upon the procedures used to prepare these subcellular fractions and the pH and other conditions employed during the incubations with the drugs (Harford and Smith, 1970). Salicylate and aspirin have been variously reported to prevent the release of lysosomal enzymes, presumably by stabilising the lysosomal membrane (Miller and Smith, 1966; Tanaka and Iizuka, 1968; Grisolia *et al.*, 1969), to have no effects (Ennis *et al.*, 1968; Weissman, 1968; Robinson and Wilcox, 1969; Harford and Smith, 1970), or to increase the liberation of these enzymes (Brown and Schwartz, 1969; Lee and Spencer, 1969) *in vitro*. Furthermore, aspirin inhibits lysosomal acid phosphatase and cathepsin activities, whereas the fenamates, indomethacin and phenylbutazone fail to do so but do inhibit hyaluronidase and β-glucuronidase enzymes from this organelle (Anderson, 1968). Thus when determining the influences of NSAIDs on the activities of individual enzymes released from within lysosomes, it is imperative to establish first the inhibitory effects of these drugs on the individual enzymes. Also, clearance between the homogeniser tube and pestle employed in tissue homogenisation influences the liberation of the lysosomal enzymes and their total activities, factors which are especially important when considering the drug effects *in vivo* (see below). Harford and Smith (1970) also found that salicylate 4 mmol/l at pH 7.0 exerted no discernible effects upon the release of lysosomal acid phosphatase, but as the pH was reduced to 5.0 the time of incubation required to elicit release of this enzyme was progressively less. The lability of the lysosomal membrane towards high concentrations of salicylates at low pH may be relevant to the mildly acidic environment in inflamed tissues.

Ignarro (1972) found that, in common with other anti-inflammatory drugs, oral or parenteral administration of aspirin 25 to 100 mg/kg to rats inhibited the release of β-glucuronidase from isolated liver lysosomes. The same effect was achieved in adrenalectomised animals, thus ruling out any effects of endogenous corticosteroids upon lysosomal stability (Lewis, 1970).

The relevance of studying drug effects on enzyme release from liver lysosomes is a rather vexed question. It is true that this aspect might be important in respect of those liver manifestations of arthritic disease (where there is a demonstrable increase in the plasma of lysosomal enzymes derived from the liver). In inflamed tissues it is possible that there may be a whole range of lysosome populations, all with varying membrane stability. Also, the enzyme types is in their susceptibility towards salicylates. Thus, Hayasaka and Sears (1978) have found that aspirin 10 to 100 μmol/l inhibits the activity of rabbit iridial but not bovine testicular hyaluronidases. Moreover, the inhibitory effects of aspirin on the activity of rat liver lysosomal acid phosphatase (Anderson, 1968) have not been confirmed with the same enzyme derived from the rabbit iris (Hayasaka and Sears, 1978). The differences in membrane-stabilising effects of aspirin (or salicylate) between different cells types *in vivo* or in isolated cells may not be important, since aspirin inhibits release of lysosomal enzymes from the rat retina *in vivo* (Dewar and Barron, 1977) and, most importantly, from polymorphs and macrophages, this being a property shared with other NSAI drugs (Ignarro and Colombo, 1972; Dowes *et al.*, 1975; Finlay *et al.*, 1975; Goto *et al.*, 1977; Northover, 1977; Smith, 1977; 1978b; Werb, 1978; Mikulikova and Trnavsky, 1980).

The role of calcium in the maintenance of the integrity of polymorph lysosomal membranes and its susceptibility to salicylate and other NSAIDs has been investigated in detail by Northover (1977). He found that the Ca^{2+}-dependent release of β-glucuronidase was inhibited by these drugs, but when Ca^{2+} was absent the otherwise low enzyme release was unaffected. The release of lysozyme was also inhibited by salicylate independent of the presence or absence of calcium ions (Northover, 1977).

The cellular (i.e. leucocytic) or *in vivo* effects of salicylates upon lysosomal membrane stability and enzyme release appear to be due to a variety of mechanisms. The *in vitro* studies cited above suggest direct effects of the drugs on both the membrane and enzymes of this organelle. In cellular systems there are, however, influences on lysosomal enzyme release from the drug effects on (a) cyclic nucleotide activity (Ignarro and Cech, 1975; Davies, 1977); (b) production of mediators from both the cyclo-oxygenase and lipoxygenase pathways of arachidonic acid metabolism (Smith, 1977; Hemler *et al.*, 1978; Goetzl, 1980; Mikulikova and Trnavsky, 1980; Walenga *et al.*, 1980; Rae and Smith, 1981); and (c) responses following ingestion of phagocytic ligands (i.e. immune complexes and complement C3b-opsonised particles) (Weissmann *et al.*, 1978; Davies *et al.*, 1980). Thus there is extensive interaction of the lysosomal system with other systems. It should also be noted that, upon phagocytosis or reactivation, the polymorph and macrophage will generate 'active' oxygen derivatives ($O_2^\cdot-$, $OH^\cdot$ and H_2O_2) (Weissman *et al.*, 1978; Rainsford and Swann, 1984) and coincidentally will release, through stimulated phospholipase activity, products of arachidonate metabolism (Goto *et al.*, 1977; Smith, 1977; Weissman *et al.*, 1978; Davies *et al.*, 1980; Mikulikova and Trnavsky, 1980). Consequently there is potential for cyclic activation of these systems, so that any drug influences on these components will affect the total responses by these leucocytes *in toto*. With the cyclic nucleotide regulation of lysosomal enzyme activity of phagocytic leucocytes the drugs or agents that increase cyclic AMP content will inhibit enzyme release, while the converse occurs with agents that decrease cyclic GMP levels (Ignarro and Cech, 1975; Davies, 1977; Weissmann *et al.*, 1978; Davies *et al.*, 1980). Since, in other cellular systems, aspirin and salicylate have been found to directly stimulate adenylate cyclase activity and production of cyclic AMP *in vitro* and *in vivo* (Mangla *et al.*, 1974; Mitznegg *et al.*, 1977), it seems reasonable to conclude that part of the stabilising action of these drugs on the lysosomal membrane *in vivo* (Ignarro, 1972) or intact cells (Ignarro and Columbo, 1972; Dowes *et al.*, 1975; Finlay *et al.*, 1975; Goto *et al.*, 1977; Northover, 1977; Smith, 1977; 1978; Werb, 1978; Mikulikova and Trnavsky, 1980) could be due to this elevation in cyclic AMP production.

The effects of salicylates on prostaglandin/leukotriene production could also directly influence lysosomal stability *in vivo*, either by affecting the prostaglandin/leukotriene-mediated cyclic nucleotide production or by actions of these eicosanoids acting as ionophores upon the lysosomal membrane *per se*. Thus if these drugs should act to decrease the proportion of $PGF_{2\alpha}$ relative to that of PGE_2, they could, by effects on cyclic nucleotide production, cause a reduction in lysosomal enzyme release. This is because, in polymorphs, PGE_2 (and PGE_1) enhance the levels of cyclic AMP that prevent enzyme release, whereas $PGF_{2\alpha}$ enhances cyclic GMP production, so stimulating the release of lysosomal enzymes (Zurier *et al.*, 1974; Rivkin *et al.*, 1975; Davies, 1977; Hatch *et al.*, 1977; Weissmann *et al.*, 1978). This cyclic nucleotide-mediated action of the prostaglandins could account for the anti-inflammatory effects of PGE_1 in the rat granuloma and arthritis models (Bonta *et al.*, 1979). An imbalance of endogenous PGE_2 relative to $PGF_{2\alpha}$ created by the exogenous PGE_1, or by the selective inhibitory effects of aspirin on the production of PGE_2 relative to $PGF_{2\alpha}$, could in part account for the cyclic nucleotide-mediated lysosomal enzyme release (i.e. in addition to direct effects of the drug on adenylate cyclase (Mangla *et al.*, 1974; Mitznegg *et al.*, 1977)).

The relevance of the inhibitory effects of aspirin on the release of chondrolytic lysosomal enzymes from leucocytes on the degradation by the latter of cartilage macromolecules has been investigated by several authors. Perper and Orowsky (1974) showed that aspirin 0.1 to 1.0 mmol/l inhibited the leucocyte enzyme-mediated degradation of sulphated macromolecules in cartilage, coincidentally with reduction in the release of lysosomal β-glucuronidase. This reduction in the degradation of cartilage sulphated macromolecules was evident when aspirin was incubated with intact human leucocytes as well as with granular lysates from these cells (Perper and Orowsky, 1974). Wojtecka-Lukasik and Dancewicz (1974) have furthermore established that aspirin 0.1 to 10.0 mmol/l, in common with other NSAIDs, inhibits the activity of partially purified collagenase preparation isolated from human leucocytes. No such effects were observed with hydrocortisone or cyclophosphamide, illustrating the specificity for these drug effects (Wojtecka-Lukasik and Dancewicz, 1974). The release of both collagenase and the neutral proteases from cultured chondrocytes activated by a macrophage factor has been found to be inhibited in a dose-related fashion by aspirin 0.01 to 100 mmol/l (Phadke *et al.*, 1979). These inhibitory effects were also observed with other anti-inflammatory drugs (Phadke *et al.*, 1979).

Overall, it appears that aspirin and possibly salicylate can inhibit the release of lysosomal enzymes in a wide variety of cellular systems *in vitro* and in isolated cells. These inhibitory effects may account for reduction in the diminished destruction of cartilage and surrounding connective tissues by lysosomal enzymes released from either leucocytes infiltrating into inflamed joint tissues by lysosomal enzymes or by resident (activated) chondrocytes.

Endothelial damage, leucocyte migration and platelet accumulation

Exudation of blood constituents into tissues is a major factor in the development of inflammation. The migration of leucocytes and thrombocytes into these sites is followed by their release of inflammatory mediators and, in arthritis, the proliferation of macrophages and lymphocytes. Enhanced vascular permeability (induced by prostaglandins, histamine and kinins released from these cells) can be the beginning of further cycles of cell emigration into inflamed sites, because they release soluble factors (complement components, e.g. C5a, peptides) as well as leukotrienes and prostaglandins – which, acting as chemotaxins, attract more cells into these sites and, in the case of the leukotrienes, prostaglandins E_2/I_2 and kinins, enhance vascular permeability (Grega *et al.*, 1981; Wedmore and Williams, 1981). Recent studies by Talbodec *et al.* (2000) show that high concentrations (3 to 30 mM) inhibit of aspirin the contractile effects of endothelin-1 in rat aorta and human mammary arteries by a mechanism involving the competitive inhibition of the binding of endothelin-1 to its receptors. These studies highlight the potential for another target for salicylates in control of vascular functions.

Clearly, any effects of salicylates on the production of these inflammatory mediators by resident and colonising cells, surrounding tissues and the endothelium could be major factors in attenuating leucocyte emigration and tissue oedema. There is evidence for co-operation between polymorphs in platelet–endothelial interactions that involve lipoxygenase pathways (Hernandez *et al.*, 1993) and the roles of platelet anti-aggregatory actions of NO released by polymorphs (Lopez-Farre *et al.*, 1995), the latter of which can be inhibited by aspirin or salicylate in endothelial cells but not when neutrophils are present (Sánchez de Miguel *et al.*, 1998). These factors may be important for measurement of the antiplatelet effects of aspirin and other NSAIDs *in vivo* using established models (e.g. Gryglewsky *et al.*, 1978).

Aside from these indirect drug effects on mediator production, it appears that aspirin could have a more direct 'stabilising' action upon the endothelium (Hladovec, 1979; Rosenblum *et al.*, 1980). Without doubt, the inhibitory effects of aspirin on platelet aggregation and thrombus formation and on fibrin deposition are major factors in preserving vascular integrity.

Platelet aggregation and thrombus formation

Platelet accumulation in inflamed sites (e.g. carrageenan-impregnated sponges in rats) can be prevented by salicylate 200 mg/kg p.o. However, prior intravenous administration of aspirin 30 to 100 mg/kg increased both platelet accumulation and fibrin accretion in the mechanically injured carotid arteries of rabbits (Buchanan *et al.*, 1981). These two apparently disparate observations highlight the difficulties in understanding the effects of aspirin and salicylate on platelet and vascular functions (Bertelé *et al.*, 1983), especially in quite different models of inflammation.

First, salicylate does not cause any significant effects on platelet aggregation, as does aspirin, when the aggregation is induced by ADP, collagen or arachidonic acid (O'Brien, 1968; Mills *et al.*, 1974a; Vargaftig, 1978b; Bertelé *et al.*, 1983). However, salicylate will inhibit aggregation induced by carrageenan in rabbit, but not in human, platelets (Vargaftig *et al.*, 1981). Thus the nature of the aggregating agent and species origin of the platelets can be crucial to the drug response, and must reflect the intrinsic differences in the platelet-aggregating mechanism affected by these drugs (Vargaftig *et al.*, 1981; Philp *et al.*, 1983).

The second factor is that aspirin inhibits the production of the potent anti-(platelet) aggregatory

CHAPTER 7

prostaglandin, PGI_2 (prostacyclin), which is largely generated by endothelial cells (Korbut and Moncada, 1978; MacIntyre *et al.*, 1978; Czervionke *et al.*, 1979; Fuccella, 1979; Preston *et al.*, 1981). The efficacy of aspirin as an inhibitor of platelet aggregation *in vivo* will therefore depend upon the relative inhibitory effects of this drug on the production of (a) the pro-aggregatory agent thromboxane A_2 by platelets; (b) the anti-aggregatory prostaglandin I_2 by endothelial cells; and (c) the production of anti-(platelet) aggregatory nitric oxide (Fuccella, 1979; Haslam and Wechsler, 1981; Boger *et al.*, 1996; Failli *et al.*, 1998; Shimpo *et al.*, 2000). The effects of NO may be of importance for the actions of NO-releasing aspirin analogues (Grosser and Schröder, 2000).

A third aspect is that salicylate can prevent the inhibitory effects of aspirin on platelet aggregation induced by arachidonic acid, the calcium ionophore (A23187), thrombin or collagen (Vane, 1979). Thus the amount of salicylate present in the circulation (relative to that of aspirin) following aspirin ingestion will determine the total aggregation response *in vivo*. The prevention by salicylate of aspirin-induced platelet aggregation appears to be due to competition by salicylate for the aspirin binding site on the cyclo-oxygenase enzyme (Roth *et al.*, 1975).

MECHANISMS OF INHIBITION OF PLATELET AGGREGATION BY SALICYLATES

Platelet aggregation appears to proceed by at least three routes:

1. Activation of phospholipase A_2 by thrombin, collagen, bradykinin, peptidoleukotrienes (slow-reacting substance of anaphylaxis, SRS-A) or intracellular calcium ions (i.e. artificially induced by Ca^{2+} ionophore A23187)
2. The ADP-dependent release reaction, which can be induced by thromboxane A_2 (TXA_2; Bertelé *et al.*, 1983) thrombin and collagen
3. Induction of platelet aggregating factor (PAF acether), which may in part be due to phospholipase C activity (Vargaftig *et al.*, 1981).

Since ADP is released by TXA_2 and PAF and, furthermore, ADP is involved in the so-called 'second wave' of platelet aggregation and recruitment of more thrombocytes, it is clear that regulation of platelet aggregation can be achieved by modulation of adenine nucleotide production. Moreover, elevation of cyclic AMP or PGI_2 levels will inhibit PAF production (Vargaftig *et al.*, 1981).

There is clear evidence that aspirin has actions independent of, as well as dependent on, the COX-1 enzyme system (Buchanan *et al.*, 1982; Violi *et al.*, 1989; Ferri *et al.*, 1994). The cyclo-oxygenase-independent actions of aspirin could be due to (a) receptor modification of thrombin binding sites (Buchanan *et al.*, 1982) or suppression of thrombin–antithrombin complexes (Hong *et al.*, 1999); (b) acetylation of thrombin by aspirin, leading to its inability to activate platelets (Han and Ardlie, 1974); (c) inhibition of the peroxidation of 12-HPETE (Siegel *et al.*, 1979; 1980); (d) the inhibition by aspirin of platelet glycolysis leading to reduced levels of ATP that is required for subsequent formation of cyclic AMP and ADP (Doery *et al.*, 1969); and/or (e) stimulation of cyclic AMP production (which could be counteracted by effects of the drug on ATP production). The interaction of platelets with von Willebrand factor is insensitive to aspirin (Ferri *et al.*, 1994).

There is no clear evidence that the thrombin receptor is modified by aspirin, despite the fact that the results of Buchanan and co-workers (1982) could be interpreted as such. It is conceivable that the inhibitory effects of aspirin on thrombin function could simply be due to acetylation by aspirin of thrombin itself (Han and Ardlie, 1974) without affecting the thrombin receptor.

Suggestions that aspirin could interfere with either the production or the functions of PAF seem unlikely in the light of certain studies (Nunn, 1981). However, it is of interest that aspirin 200 mg/kg i.p. and other NSAIDs can block the effects of PAF in inducing paw oedema in rats (Vargaftig and Ferreira, 1981), so this factor is potentially an important inflammatory mediator released from platelets present in inflamed tissues.

Inhibition of the peroxidative conversion of 12-HEPTE to 12-HETE could reduce the production of free oxygen radicals. It is notable that salicylate is a potent inhibitor of this reaction (Siegel *et al.*,

1979; 1980; Radomski *et al.*, 1986; Tremoli *et al.*, 1986), and this could explain in part the inhibitory action of salicylate upon platelet accumulation in the carrageen and sponge system (Smith *et al.*, 1976).

One of the products released from activated platelets, serotonin, in turn also influences platelet function. Malmgren and co-workers (1979) examined the possibility that aspirin blocked the uptake of serotonin by platelets, but failed to establish any inhibitory effects of the drug.

Aspirin does not appear to affect the release of alpha granules from platelets (Rinder *et al.*, 1993). The possibility that aspirin could influence ADP production by effects on 'ecto-ATPase' activity was considered by Yue and Davis (1978). These authors found that oral intake of aspirin 320 mg for 7 days did not significantly alter the activity of this enzyme in man, although the 'second-phase' of ADP-induced aggregation was inhibited. Also, Crutchley and co-workers (1980) found that aspirin 1 mmol/l does not influence the catabolism of ADP to AMP, adenosine, hypoxanthine and inosine. Thus the only direct effect of the drug on adenine nucleotide status appears to be on the production of ATP (Han and Ardlie, 1974). This may not be a dominant effect, since salicylate (which would also be expected to inhibit platelet glycolysis and ATP production) is virtually ineffective as an inhibitor of platelet aggregation.

In summary, the mechanisms of the actions of aspirin on platelets appear to involve inhibitory effects on (a) the production of the pro-aggregatory prostanoid, thromboxane A_2; (b) the membrane-activating functions of thrombin; and (c) the availability of ADP from ATP. The platelet adhesion to normal and damaged capillaries (or other blood vessels) is a major factor to be considered in the examination of potential beneficial effects of aspirin upon the microvasculature. It is widely assumed that by preventing adhesion of platelets to damaged capillaries, arterioles or venules in inflamed tissues, aspirin (and indeed other NSAIDs) would confer benefit by preventing ischaemia-mediated tissue damage and also remove a rich source of oxygenated eicosanoids produced from the resident thrombocytes.

Studies have been reported showing that the platelets in aspirin-treated animals adhere tenaciously to damaged blood vessels (Tschopp, 1979; Buchanan *et al.*, 1981), but other results have not always been in agreement (Sheppard, 1972; Cazenave *et al.*, 1977). One view is that platelet adhesion to endothelial cells and connective tissue components (e.g. collagen) can be promoted by high but not low doses (concentrations) of aspirin, because of the differential effects of this drug on platelet thromboxane A_2 production and endothelial cell synthesis of prostacyclin (MacIntyre *et al.*, 1978; Preston *et al.*, 1981; McAdam *et al.*, 1999). Related to this discussion is the possibility of there being differences in the activity or recovery of the apparent irreversible inhibition by aspirin of cyclo-oxygenase enzymes in endothelial cells, compared with those in thrombocytes. While it is possible under some conditions *in vivo* selectively to inhibit platelet thromboxane A_2 production without affecting the synthesis of prostacyclin with low doses of aspirin, some authors have failed to confirm such a selective inhibitory effect (Preston *et al.*, 1981). Furthermore, studies with human (umbilical) endothelial cell cultures have shown that the endothelial cell cyclo-oxygenase is as sensitive to aspirin as is the platelet cyclo-oxygenase (Jaffe and Weksler, 1979). Thus the widely held view now (Schror, 1997) is that the differences observed in platelet compared with endothelial cell prostanoid production *in vivo* are due to differences in effects of drug concentrations and intrinsic susceptibility of the cyclo-oxygenases from these two sources (Vane, 1979), which would appear to be related to differences in the two isoforms of the cyclo-oxygenases – COX-1 in platelets (Schror, 1997) and COX-2 in endothelial cells – that produce prostacyclin (Holtzman *et al.*, 1992; McAdam *et al.*, 1999). It is possible that the differences in effects of aspirin observed *in vivo* are a reflection of differences in the kinetics of aspirin and salicylate uptake into these two cellular systems. The ingestion of low doses of aspirin may achieve selective inhibition of platelet cyclo-oxygenase because of the greater likelihood of platelets coming into contact with high concentrations of aspirin after its passage through the gastrointestinal circulation. Relatively lower concentrations of aspirin would be expected to reach endothelial cells in extra-gastrointestinal organs and, in proportion to the levels of aspirin, much higher concentrations of salicylate would reach these cells, so allowing salicylate to prevent or even reverse the inhibitory effects of aspirin on endothelial cyclo-oxygenases. When larger doses of aspirin are ingested, proportionally greater quantities of this drug could reach the vascular tissues compared with salicylate, so that inhibition of endothelial cell prostacyclin would ensue. It should be noted that prostacyclin from vascular tissues probably does not

CHAPTER 7

persist in circulation under normal physiological conditions (Haslam and McClenaghan, 1981). Thus any inhibitory effects of aspirin on endothelial cell production of prostacyclin will have an immediate local effect, since there will be little if any prostacyclin persisting in the circulation.

While platelet COX-1 appears to be inhibited irreversibly by aspirin for the lifetime of these cells, it is possible that the inhibitory effects of the drug on endothelial cells (Jaffe and Weksler, 1979) can be reversed by the *de novo* protein synthesis of new COX-2 isoenzyme (McAdam *et al.*, 1999). Thus the overall *in vivo* effects of aspirin on synthesis of prostanoids and NO by these two cells will depend on the dose and pharmacokinetics of the drug and the capacity of endothelial cells to adapt reversibly to the inhibitory effects of the drug (Warhurst *et al.*, 1987; Wallis and Simkin, 1993; O'Kane *et al.*, 2003). There may also be a time factor in determining the effects of aspirin-mediated cyclo-oxygenase inhibition, especially in relation to platelet adherence to the damaged endothelium. Thus, Schaub and co-workers (1979) have found that repeated oral dosing of aspirin 325 mg to dogs for 33 days led to a reduction in the adhesion of platelets to damaged capillaries, whereas after a few days increased adhesion of thrombocytes was evident. Recent studies suggest that there are at least two polymorphisms of COX-1 that may be important in the sensitivity of this enzyme to inhibition by aspirin (Haluska *et al.*, 2003).

The importance of dosage of the drug in relation to platelet reactions at the damaged endothelium has also been highlighted by the work of Wu and co-workers (1981). They observed, by combined electron microscopy and uptake of ^{111}In-labelled platelets, that fewer platelets accumulated in the injured blood vessels of those rabbits dosed orally with aspirin 30 mg/kg compared with those given 150 mg/kg salicylate. These results could be interpreted in relation to the pharmacokinetics of aspirin and salicylate, as mentioned above.

COMPARISON OF THE ANTI-AGGREGATORY PROPERTIES OF DIFFERENT SALICYLATES: IMPORTANCE OF CHEMICAL STRUCTURE

The requirement for an acyl ester group at the 2-hydroxy position of salicylate, not necessarily an *O*-acetyl moiety, to inhibit platelet aggregation has been illustrated by the structure–activity studies reported by Mills *et al.* (1974a; 1974b), Whitehouse *et al.* (1977), Rotilio *et al.* (1984) and Violi *et al.* (1989). Some interesting observations can be made by considering the effects of adding an extra *O*-acetyl moiety to that of aspirin. Thus the 2,3-, 2,4- and 2,6-diacetoxybenzoic acids all had progressively lower activity (in that order) than that observed with aspirin (Mills *et al.*, 1974a; 1974b; Whitehouse *et al.*, 1977). If the notion that the presence of labile acetyl group(s) confers inhibitory activity on platelet aggregation by salicylates is correct, then it might be expected that the presence of an additional *O*-acetyl group would confer added benefits, whereas in fact this is not the case. It is conceivable that the presence of a second *O*-acetyl substituent on the aspirin somehow slows or prevents appreciable acetylation of platelet cyclo-oxygenase.

The requirement for a 2-acetoxy group on benzoic acid for optimal anti-aggregatory effects has been examined further by Cerskus and Philp (1981). Thus, 2-acetylbenzoic acid (which presumably exhibits poor lability of the acetyl moiety) is a very weak inhibitor of platelet aggregation and prostaglandin synthesis in this cell, the IC_{50} concentration for inhibition of platelet aggregation by collagen being 100 mmol/l.

The internal anhydride of 2-acetylbenzoic acid, 3-methylphthalide, was found to be more potent than 2-acetylbenzoic acid as an inhibitor of platelet aggregation (although it was somewhat less so than aspirin) and, interestingly, did not inhibit platelet prostaglandin synthesis *in vitro*. 2-Acetylbenzoic acid was moderately active *in vivo* after oral administration (300 mg/kg) to rats. These results suggest two important features: first, it is probably not necessary to have inhibition of prostaglandin biosynthesis in platelets to prevent their aggregation by these benzoic acid derivatives; and second, a labile acetyl moiety, while enhancing the potency of salicylates (i.e. in aspirin), is not an absolute requirement for platelet anti-aggregatory activity (Cerskus and Philp, 1981). Furthermore, the 2-propionyloxy derivative of benzoic acid has been found by Cerskus and Philp to have about the same anti-aggregatory activity as aspirin, thus illustrating that any labile acyl moiety will enhance the potency of benzoic or salicylic acid (i.e. not just an acetyl group). However, this acyl moiety must be at the 2-position for full activity since 3-propionyloxybenzoic acid is much less potent than its 2-substituted analogue (Cerskus and Philp, 1981).

The effects of adding lipophilic substituents to aspirin in the platelet aggregation *in vivo* in the rabbit have been studied by Dearden and George (1977). In a brief report these authors stated that increasing the lipophilicity caused an increase in the potency of inhibitors of platelet aggregation to a maximum, which fell off slightly as lipophilicity was further increased. The correlation between anti-aggregatory activity and lipophilicity (established by way of a Hansch-type analysis and expressed in the form of a quadratic equation) was not improved by adding steric or electronic terms (Dearden and George, 1977). This latter point is surprising, especially if lability of the acetyl moiety is a requirement for potency by aspirin, since addition of bulky lipophilic groups at the 3 position would be expected at least to attenuate this lability.

Specific effects on the anti-aggregatory activity of aspirin or salicylate may occur by adding lipophilic and/or electron-withdrawing substituents (Casadebaig *et al.*, 1991). Thus ring iodination markedly enhances the anti-aggregatory activity of aspirin (Mende, 1972), an effect that could be explained by the electron-withdrawing effects of the iodine moiety. In like manner, addition of trifluoromethyl (or 5-chloro) substituents markedly potentiates the anti-aggregatory activity of salicylates (Diamantis *et al.*, 1978; Garcia Rafenell *et al.*, 1979). There are indications that the mode of action of these halosalicylic acids could be quite different to that of aspirin. Thus, the trifluoromethylsalicylic acid, trifusal, is 60 per cent less potent than aspirin as an inhibitor of platelet cyclo-oxygenase, but is five times more potent than aspirin as an inhibitor of phosphodiesterase activity (Garcia Rafenell *et al.*, 1979). While inhibitory effects of aspirin on platelet phosphodiesterase may be of questionable significance because of the relatively high concentrations required to achieve such an effect, it is most likely that inhibition of this enzyme is important in the mode of action of triflusal (Garcia Rafenell *et al.*, 1979). In the presence of neutrophils, triflusal, but not its main metabolite 2-hydroxy-4-trifluoromethyl-benzoic acid, exerts a greater anti-aggregatory action than that from aspirin (Sánchez de Miguel *et al.*, 1998).

5-Chlorosalicyclic acid and its cyclic pro-drug, meseclazone, are both potent anti-aggregatory agents, but their effects, compared with those of aspirin, vary markedly according to the aggregating agent (Diamantis *et al.*, 1978). For example, meseclazone is almost equipotent with aspirin in anti-aggregatory activity *in vitro* when ADP is employed to induce aggregation, whereas 5-chlorosalicylate is about 27 times less potent (Diamantis *et al.*, 1978). In contrast, when thrombin is employed as the pro-aggregant, 5-chlorosalicylic acid is equipotent with aspirin as an anti-aggregatory drug, but mezecla-zone is almost inactive (Diamantis *et al.*, 1978). These results suggest that these drugs may have different modes of action in inhibiting platelet aggregation.

Chronic ingestion of the salicylic acid dimer, diplosal, failed to affect platelet aggregation in the presence of collagen, arachidonate, ADP or adrenaline (Estes and Kaplan, 1980). Likewise, the choline magnesium salt of salicylate is ineffective as an inhibitor of platelet aggregation (Start and Pisko, 1981).

Of the other salicylates employed therapeutically, diflunisal taken orally as either single or twice daily doses of 500 mg does not alter either ADP-induced platelet aggregation, platelet disaggregation or affect prothrombin and bleeding times (Tempero *et al.*, 1978). The lack of effects of this drug on platelet aggregation contrasts with the reported potency of this drug upon platelet cyclo-oxygenase activity (Majerus and Stanford, 1977). Thus it appears that inhibition of platelet cyclo-oxygenase activity may not, *per se*, be a prerequisite for preventing platelet aggregation.

Because the antiplatelet effects of aspirin are limited owing to its gastrotoxicity and inhibitory effects on vascular prostacyclin production, Hung *et al.* (1998) examined a series of *O*-acyl esters of salicylic acid and diflunisal for their antiplatelet, gastroirritant effects and hepatic extraction. The *O*-acyl derivatives all showed less gastro-ulcerogenicity in rats, while at the same time retaining inhibitory effects on arachidonic acid platelet aggregation and thrombin-induced serum thromboxane A$_2$ production by rat platelets comparable with that of aspirin. The effects of these esters on prostacyclin production were not, however, studied. Thus these *O*-acyl esters may offer potential as low ulcerogenic novel antiplatelet agents.

THROMBOLYTIC/FIBRINOLYTIC ACTIVITY

The regulation of fibrinolysis (Collen, 1980) is another potentially important action of the salicylates in thrombus formation, is their capacity to dissolve intravascular clots by the activation of endogenous

CHAPTER 7

fibrinolytic activity (Von Kaulla and Ens, 1967; Ballie and Sim, 1972; Andreenko and Karabasova, 1977; Moroz, 1977). The importance of these drug effects has been overshadowed by the attention devoted in recent years to the antiplatelet effects of the salicylates.

Early studies by Meyer and Howard (1943) suggested that hypoprothrombinaemia could be induced by high doses of salicylates, and that this could account for the hypocoagulability in aspirin-treated individuals. Later, Von Kaulla and co-workers (Von Kaulla and Ens, 1967; Hansch and Von Kaulla, 1970) studied the fibinolytic activity of over 800 ring-substituted salicylic acid derivatives. Salicylate was found to be a relatively weak fibrinolytic agent, but its potency could be markedly increased by the addition of bulky alkyl- or halo-substituents at the 3, 4 or 5 positions of salicylic acid (Von Kaulla and Ens, 1967; Hansch and Von Kaulla, 1970). Thus the introduction of methyl, iodine or bromine groups at these positions improves the activity of salicylate by some three to eight times (Von Kaulla and Ens, 1967). Double substitutions of halogens boost activity by over 100-fold (Von Kaulla and Ens, 1967). A structure–activity analysis by Hansch and Von Kaulla (1970), using Hansch's analytical procedure (Hansch, 1971; Hansch et al., 1973), showed the role that increasing lipophilicity, combined with electronic character of the substituent and its steric nature, play in enhancing the fibrinolytic activity of salicylate. The mode of action of salicylate-induced fibrinolysis appears to be due to reduction of plasminogen activator activity in plasma (Von Kaulla and Ens, 1967).

The mechanism of induction of the salicylate-induced hypoprothrombinaemia has been investigated by Park and Leck (1981). These authors found that methyl-salicylate induced hypoprothrombinaemia was induced in the rabbit in a manner resembling that of the coumarins, by blocking the activity of vitamin K_1-epoxide reductase, so preventing the regeneration of vitamin K_1 – an essential cofactor in the γ-carboxylation of glutamate residues in clotting factors II, VII and X (Park and Leck, 1981).

CHEMOTAXIS OF POLYMORPHS AND MONOCYTES

In addition to influencing the extravasation of leucocytes into inflamed tissues by actions at the vascular level, the salicylates, in common with other NSAIDs, have been considered for their potential to affect the migration of these cells into damaged tissues (Smith et al., 1975b; Rumore et al., 1987). These effects could be due to the drug-induced interference in the formation or actions of fibrin, kinins, complement factors, peptides, etc. The effects of these drugs on production of E-type prostaglandins and thromboxane A_2 may only have minor consequences in the attraction of leucocytes (Smith et al., 1975b; Smith, 1979). However, a powerful synergistic action of leukotriene B_4 (LTB_4) and blockade of the synthesis of this and other leukotrienes prevents the chemoattraction and aggregation of polymorphs (Bray et al., 1980b; Schiffmann, 1981). Thus, the inhibitory effects of aspirin and other NSAIDs upon leucocyte migration into inflamed tissues (Di Rosa, 1974; Smith et al., 1975; Borel and Feurer, 1978; Warne and West, 1978; Smith, 1979; Dawson, 1980) could be due to the actions on a combination of chemoattractant factors. Since the evidence for a primary chemoattractant action of the E-type prostaglandins and thromboxanes is weak (Smith et al., 1975) and aspirin is not an inhibitor of LTB_4 synthesis, it appears that interference with neither the lipoxygenase nor prostanoid synthesis can explain the in vivo effects of aspirin or salicylate on cell migration.

The effects of aspirin and salicylate on polymorph chemotaxis and migration in vitro or in vivo have however been variable, and many of the studies indicate these effects are only evident in some studies at high, supratherapeutic concentrations, or when concentrations of chemoattractants are relatively low (Spisani et al., 1979; Turner et al., 1983; Wildfeuer, 1983; Matzner et al., 1984; Pham Huy et al., 1985; Shimanuki et al., 1985; Zimmerman et al., 1985). Thus, aspirin 150 μg/ml did not alter the chemoattraction of human polymorphs towards denatured γ-globulin (Pecoud et al., 1980). Rat polymorphs stimulated with calcium ionophore A23187 to produce chemokinetic and aggregating responses were unaffected by 0.5 mM sodium salicylate or aspirin (Bray et al., 1980b). The activation of rat peritoneal leucocytes by complement with zymosan-activated serum was unaffected by salicylates (Dawson, 1980).

In contrast to these observations, Smith and Walker (1980) and Abramson (1987) have shown that aspirin 2 to 10 mM or salicylate 0.25 to 10 mM inhibits the aggregation of human polymorphs induced by fMLP. It may be that the differences in effects of salicylates may be related to the stimulus employed, the presence of protein components of plasma (Smith and Walker, 1980), and the biological effects that are produced. It is also possible that the mechanisms of drug uptake may vary among leucocyte populations and be affected by the stimulus (Raghoebar *et al.*, 1988).

Some other NSAIDs inhibit the chemoattraction of polymorphs (Dawson, 1980; Pecoud *et al.*, 1980), and these actions have been ascribed to their capacity to inhibit 5-lipoxygenase activity (Pecoud *et al.*, 1980). If this is an important event then diflunisal, which has some inhibitory effects on 5-lipoxygenase, could inhibit the chemotactic activity of leucocytes by this action (Baumann *et al.*, 1980). The significance of the role of the 5-lipoxygenase pathway in leucocyte adhesion has been shown from the observations of Kurose *et al.* (1996) showing that LTB_4 antagonists could additionally block the leakage of albumin in rat mesenteric venules, more so than NSAIDs.

Adhesion of human polymorphs and monocytes to human saphenous vein preparations, observed by phase-contrast microscopy *in vitro*, has been shown to be reduced by exposure of the endothelium to 10 mM aspirin, without affecting the production of nitric oxide (Fricchione *et al.*, 1998). Using a model of human polymorph transmigration through HUVEC monolayers, Pierce *et al.* (1996) found that relatively high concentrations of sodium salicylate inhibited their migration but had no effects on the adhesion of these cells. Salicylate inhibited the TNFα-induced increase in mRNA for adhesion molecules and the TNFα upregulation of vascular cell adhesion molecule-1 (VCAM-1), which functions in lymphocyte recruitment, e-selectin and endothelial cell leucocyte adhesion molecule-1 (Weber *et al.*, 1995). These effects of salicylate have been ascribed to the inhibitory effects on NFκB gene activation in response to inflammatory cytokines (Weber *et al.*, 1995; Pierce *et al.*, 1996, Lee and Burkart, 1998). Other studies indicate that inhibition by salicylate of integrin-mediated adhesion of neutrophils is dependent on Erk activity (Pillinger *et al.*, 1998).

Chemotaxis of neutrophils by agents that stimulate superoxide production does not appear to be affected by aspirin, but ibuprofen and indomethacin are inhibitory (Simchowitz *et al.*, 1979). Overall these studies suggest that the differences in responsiveness of polymorphs to salicylates may depend on the activating agent and the nature of the process (i.e. adhesion, migration and activation) affected.

The migration and activation of polymorphs from the synovial fluid but not those from the peripheral blood of patients with rheumatoid arthritis appears different to that from healthy controls (Kemp *et al.*, 1980). The chemotactic response of neutrophils from the peripheral blood of rheumatoid arthritis patients who took aspirin was increased, which is a paradoxical observation and may relate to their responsiveness to stimuli and drugs. The accumulation of polymorphs and expression of adhesion molecules in response to endotoxin in human volunteers has been shown to be refractory to aspirin (Jilma *et al.*, 1999). These results suggest that there may be important differences in the action of aspirin in different pathophysiological states.

Copper complexes of salicylates and other NSAIDs were found to be more potent than their parent acids in decreasing random migration and chemotaxis of polymorphs stimulated with fMLP and complement components *in vitro* and after serum-induced pleural inflammation in rats *in vivo* (Auclair *et al.*, 1980; Roch-Arveiller *et al.*, 1990). If the concept is that NSAIDs exert part of their pharmacological activities (see Chapter 13) via the copper-complexed forms, then these observations of the effects of copper NSAIDs could be relevant in showing that they affect polymorph functions only when they have been complexed with copper ions.

Chemotactic activity of monocytes appears to be affected differently by drugs compared with that of polymorphs, and may also vary according to the species from which the cells have been derived and the chemoattractant employed *in vitro*. Thus in contrast to the negative inhibitory effects of aspirin obtained in the chemotaxis of polymorphs, Borel and Feurer (1975) have observed inhibitory effects on migration of rabbit macrophages in the presence of very high concentrations (1000 μg/ml) of salicylate. Similar inhibitory effects were also obtained with relatively high concentrations of some (but not all) anti-inflammatory drugs (Borel and Feurer, 1975). Kemp and Smith (1982) observed that sodium salicylate 5.8 and 120 μM inhibited migration of human monocytes, but not polymorphs. The aspirin-induced inhibition of the migration of monocytes might be related to the effects of the drug on NFκB-regulated expression of VCAM-1 and e-selectin (Weber *et al.*, 1995). Salicylates may also regulate leucocyte migration by influencing lymphocyte functions (Cologniet *et al.*, 1977).

CHAPTER 7

In summary, aspirin and salicylate may inhibit specific phases involved in the cytokine-mediated migration of leucocytes into inflamed tissue (sites) by influencing vascular permeability and the actions and/or production of non-eicosanoid derived inflammatory mediators.

Displacement of albumin-bound peptides and tryptophan

McArthur and co-workers (McArthur and Dawkins, 1969; McArthur et al., 1971b; McArthur and Smith, 1972) suggested that one of the anti-inflammatory actions of the salicylates and related NSAIDs could be related to their effects in displacing tryptophan and/or some of its peptides from binding sites on circulating plasma proteins. It was suggested (McArthur and Dawkins, 1969) that one of the anti-inflammatory actions of the tryptophan so displaced might be to inhibit the infiltration of leucocytes into inflamed areas (Davis et al., 1968). The principal tryptophan-binding protein in plasma is albumin, and, of all the physiological amino acids in the circulation, only tryptophan is bound to plasma proteins (i.e. albumin; McArthur and Dawkins, 1969). Thus the effects of salicylates are clearly specific to this amino acid. The concept proposed by McArthur and co-workers (McArthur et al. (1971b) also suggests that the tryptophan displacement effect resembles that of certain peptides that protect susceptible tissues against chronic inflammatory insults. It is interesting that, based on this work, these authors patented a series of tryptophan and related dipeptides with anti-inflammatory activity (McArthur et al., 1975). In addition to anti-inflammatory actions, the salicylate-induced displacement of tryptophan could lead to increases in serotonin in the circulation and later in the brain, which could explain the analgesic activity of salicylates (see also p. 291).

Connective tissue metabolism

The inhibitory effects of the salicylates (notably aspirin and salicylate) upon the stimulated metabolism in inflamed connective tissues have been extensively investigated (see reviews of the early work by Whitehouse, 1965; 1968; Smith, 1966; Smith and Dawkins, 1971; Trnavsky, 1974; Brandt and Palmoski, 1984; Dingle, 1999). Some details of the biochemical effects on cellular metabolism are summarised in the Appendix to this chapter.

Fundamental to the understanding of the beneficial effects of salicylates in controlling connective tissue metabolism is the fact that it is markedly elevated in inflammatory states and, if it is sustained, can be undesirable, leading to fibrotic or scar tissue formation. A wealth of studies have been published showing that there is increased connective tissue metabolism in inflamed tissues (Trnavsky, 1974). Thus in synovial tissues of inflamed joints, e.g. in rheumatoid patients, there are increases in:

1. Glucose uptake, glycolytic activity and lactate production (Castor, 1971; Lindy et al., 1971; Gotze and Rossel, 1972; Henderson et al., 1979; Mossman et al., 1981)
2. Glucose 6-phosphate dehydrogenase activity (Butcher et al., 1973; Henderson and Glynn, 1981), i.e. the first step in glucose 6-phosphate catabolism through the hexose monophosphate shunt (pentose pathway)
3. The cytochrome oxidase activity of mitochondria (Henderson et al., 1978)
4. The synthesis of hyaluronate (Castor, 1971)
5. The ratio of sulphydryl/disulphide groups of the proteins of synoviocytes (Butcher et al., 1973)
6. The fibronectin content of synovial fluid and localisation of this glycoprotein, which is involved in cellular adhesion (Yamada and Olden, 1978), within the intimal cell layer (Revell et al., 1983; Scott et al., 1983)
7. Fibroblast proliferation with development of marked cellular independence (Clarris et al., 1977; Anastassiades et al., 1978).
8. Production and actions of pro-inflammatory cytokines (Dingle and Shield, 1990; Dingle, 1991; 1992; 1999; Rainsford, 1987; Rainsford et al., 1989; 1992)

There is also a low Po_2 inversely correlated with raised Pco_2, reflecting high oxidative metabolism (Lund-Olesen, 1970). The content of acidic glycosaminoglycans (GAGs) (Greiling and Kleesiek, 1982), especially polysulphates, may decline as a consequence of enhanced lysosomal enzyme activity (Chayen et al., 1971; Greiling and Kleesiek, 1982). Increased activity of some key enzymes involved in both the anaerobic and aerobic metabolism of glucose, cartilage mucopolysaccharide synthesis, as well as lysosomal enzyme activities, has also been observed in the synovial tissues of animals with experimentally induced arthritis (Mohr and Wild, 1976; Nusbickel and Troyer, 1976; Henderson and Glynn, 1981). Increased synthesis of collagen and sulphated GAGs occurs in the joints of osteoarthritic patients (Collins and McElligott, 1960; Lipiello et al., 1977), although the synthesis of sulphated GAGs is not correlated specifically with degree of osteoarthrosis (McKenzie et al., 1977).

Thus the consequence of employing salicylates for anti-inflammatory therapy is to reduce the metabolism of glucose to energy-yielding intermediates (ATP), the turnover of GAGs and collagenous proteins, and cell growth in inflamed tissues. Figure 7.7 shows a summary of the principal metabolic effects of the salicylates as they primarily affect cartilage and synovial tissue metabolism. It is apparent

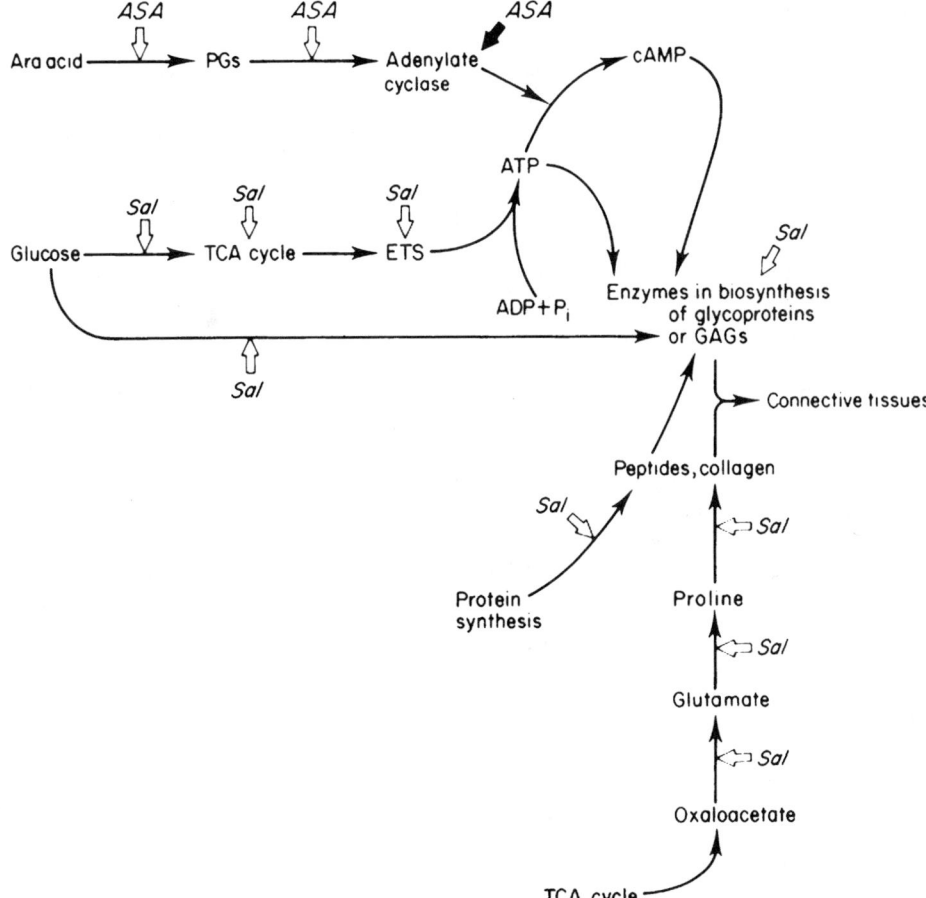

Figure 7.7 Principal effects of salicylates on connective tissue anabolism. Sites of inhibition (open arrows) or activation (solid arrows) by aspirin (ASA) or salicylate (SAL) on enzymes and regulators of synthesis of precursors or intermediates of glycosaminoglycans (GAGs) and subsequently proteoglycans. Abbreviations: cAMP, cyclic adenosine-3′,5′-monophosphate; ATP, adenosine triphosphate; Ara, arachidonic acid; TCA, tricarboxylic acid acid (cycle); PGs, prostaglandins; ETS, electron transport system in mitochondria. (From Rainford, 1984.)

CHAPTER 7

that these drugs inhibit practically every major step involved in the metabolism of glucose, GAGs, collagen and some essential proteins and energy-yielding molecules (e.g. ATP) involved in cell growth. These conclusions have largely been based on studies on the *in vitro* effects of drugs upon the activities of isolated enzymes or production of intermediates or products of various metabolic pathways by connective tissue preparations. This therefore represents specific actions on individual components that control connective tissue metabolism. With the interest in the role of prostaglandins in inflammation, there has been a tendency to ascribe the stimulatory effects on GAG metabolism of low concentrations of some prostaglandins to a prime involvement of these mediators in GAG synthesis (Kleine and Jungmann, 1977; Murota *et al.*, 1977). Some authors have claimed that the inhibitory effects of salicylates and other NSAI drugs on PG production is sufficient reason to account for the inhibition of the synthesis of GAGs (Kofoed *et al.*, 1973). Clearly this is only one possible action of the salicylates in isolated systems *in vitro*. Briefly, the mechanism of actions of the salicylates on the metabolism of glucose, GAGs and collagen can be summarised as below.

Metabolism of glucose

There are several mechanisms by which salicylates induce marked alterations in glucose metabolism, and these ultimately result in an acceleration of glucose utilisation and lactate production (Sturman and Smith, 1969; Davies *et al.*, 1970; Wolna and Inglot, 1973; Worathumrong and Grimes, 1975; Figure 7.8). The reformation of glucose (i.e. gluconeogenesis) is also inhibited by salicylate (Krebs and De Gasquet, 1964; Woods *et al.*, 1974). The other principal sources of glucose in liver and muscle glycogen are also affected by salicylate in normal (but not diabetic) individuals through mechanisms that involve decreased ATP levels and increased phosphorylase activity in those tissues (Smith and Smith, 1966; Vercesi and Focesi, 1977). Consequently, there is an apparent increase in blood glucose (Smith and Smith, 1966). The supply of this nutrient may not be a critical limitation for production of lactate. The accelerated conversion of glucose to lactate by salicylate occurs despite inhibitory effects of these drugs upon the dehydrogenases and 6-phosphofructokinase enzymes involved in the glycolytic (Embden–Meyerhoff) and pentose pathways of glucose metabolism (Whitehouse, 1965; Smith and Smith, 1966; Sturman and Smith, 1966; Smith and Sturman, 1967; McCoubrey *et al.*, 1970; Kaplan, 1973; Einarsson *et al.*, 1974; Cheshire and Park, 1977; Rainsford, 1982 (unpublished); see also Figure 7.6). As a consequence of salicylates depressing ATP levels (see below) there is less ATP available to cause the usual negative feedback control of 6-phosphofructokinase activity, i.e. one of the rate-limiting enzymes in glycolysis. The reduction in levels of glucose and also in the liver levels of certain gluconeogenic amino acids, as a consequence of inhibitory effects of salicylates on aminotransferase activities (Smith, 1966; Smith and Dawkins, 1971), could influence the availability of hexoses and amino sugars necessary for synthesis of connective tissue glycoproteins and GAGs.

Production of ATP

While accelerated glucose metabolism to pyruvate (and lactate) would be expected to yield slightly more than the normal ATP than is utilised, this is more than counteracted by the profound inhibitory effects of the salicylate anion on mitochondrial ATP production (Whitehouse, 1965; Smith and Smith, 1966; Smith and Dawkins, 1971; Galen *et al.*, 1974a; 1974b; Glenn *et al.*, 1979; Petrescu and Tarba, 1997). Appreciably more ATP is produced from the mitochondrial oxidation of the tricarboxylate cycle intermediates and 3-hydroxybutyrate than is generated from the anaerobic metabolism of glucose, so the effect of salicylates on uncoupling oxidative phosphorylation and, to some extent, the inhibition of the mitochondrial dehydrogenases will be more profound. The net effect of these mitochondrial effects of salicylates, as with some other NSAIDs, is to reduce the ATP content of cartilage (Roger and Kalbhen, 1962) and other cellular systems *in vitro* and *in vivo* (Smith and Jeffrey, 1956; Smith, 1966; Bullock *et al.*, 1970; Smith and Dawkins, 1971; Jørgensen *et al.*, 1976). Reduction in ATP can lead to

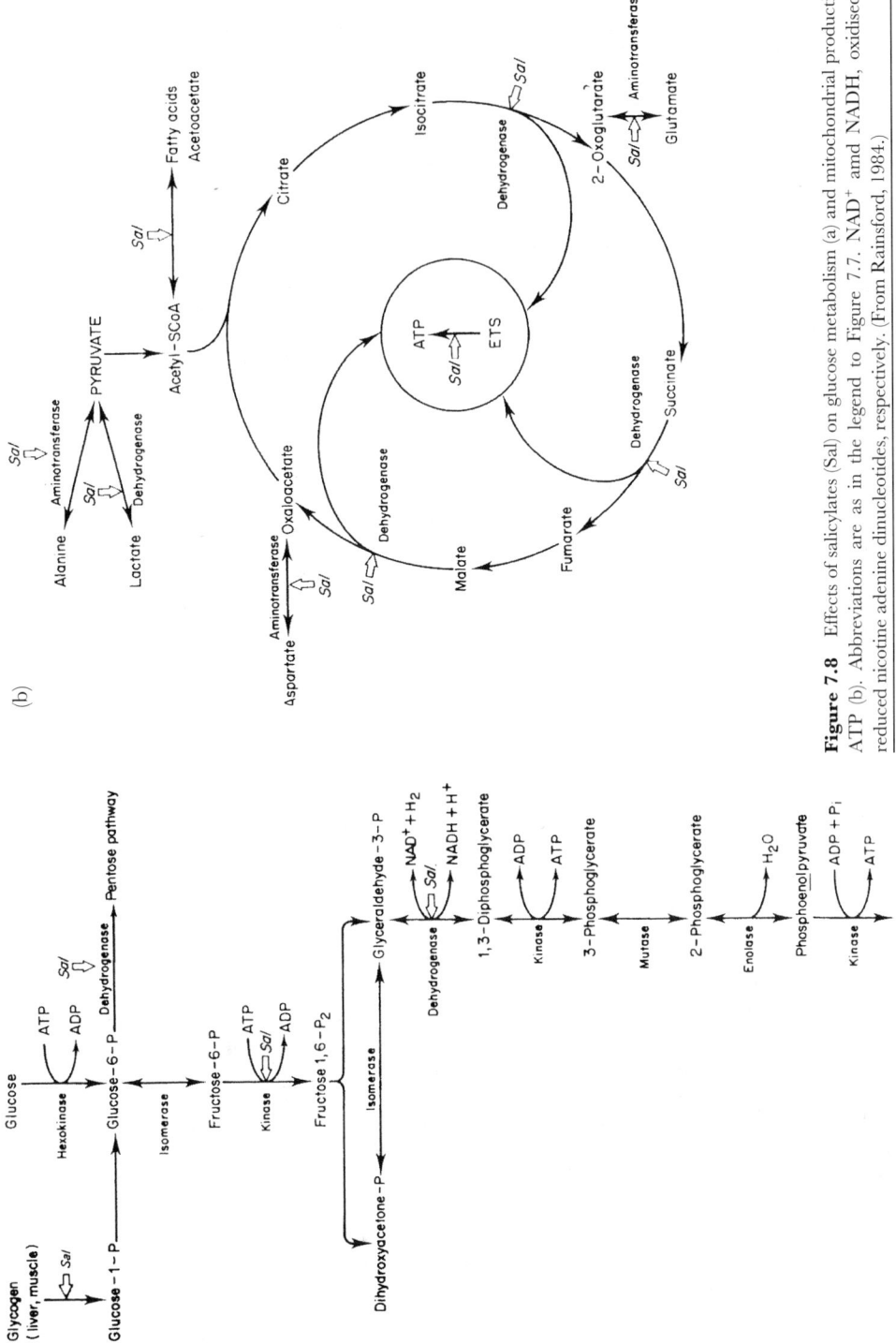

Figure 7.8 Effects of salicylates (Sal) on glucose metabolism (a) and mitochondrial production of ATP (b). Abbreviations are as in the legend to Figure 7.7. NAD⁺ and NADH, oxidised and reduced nicotine adenine dinucleotides, respectively. (From Rainsford, 1984.)

increase in adenosine and this may result in anti-inflammatory effects by activation of adenosine receptors (Pillinger *et al.*, 1998; Cronstein *et al.*, 1999; Sawynok and Liu, 2003). The uncoupling activity *in vitro* also occurs with a wide variety of salicylate derivatives (e.g. diflunisal; Baños and Reyes, 1989) and other NSAIDs (Whitehouse, 1965; Baños and Reyes, 1989; Petrescu and Tarba, 1997) and is evident at therapeutic doses of aspirin without structural changes in mitochondria or livers of rats given repeated daily doses of the drug (Tomoda *et al.*, 1994).

The potency of salicylate itself is enhanced by: (a) addition of halogens and aromatic subtituents on the benzene ring; (b) the substitution of a thiol group for the phenolic hydroxyl; or (c) the selected replacement of carboxylic acid with an amide (Whitehouse, 1964; Burke and Whitehouse, 1965; Whitehouse and Dean, 1965; Leader and Whitehouse, 1966; Karler *et al.*, 1968; Galen *et al.*, 1974a; 1974b). Under *in vitro* conditions aspirin is ineffective as an uncoupler of oxidative phosphorylation, i.e. it has to be hydrolysed to salicylate to achieve this effect (Thompkins and Lee, 1969). However, uncoupling is achieved in mitochondria from the livers of rats given repeated oral doses of aspirin (Mehlman *et al.*, 1972), i.e. after biotransformation to salicylate *in vivo*. Evidence for the uncoupling action of salicylate and aspirin occurring *in vivo* comes from (a) morphological evidence of alterations in liver and granulation tissue from animals given salicylates (Jørgensen, 1964; Gutowska-Grzegorczyk and Kalzak, 1968); (b) the reduced ATP levels in tissues of salicylate-treated animals (Smith and Jeffrey, 1956; Roger and Kalbhen, 1962; Smith and Smith, 1966; Bullock *et al.*, 1970; Smith and Dawkins, 1971; Jørgensen *et al.*, 1976); (c) effects on oxygen consumption and lactate/pyruvate in isolated perfused rat livers; and (d) measurement of ^{13}C-α-ketoisocaproic acid decarboxylation to $^{13}CO_2$ in the breath of humans following oral intake of 30 mg/kg aspirin (Lauterburg *et al.*, 1995). Oxidative phosphorylation *in vitro* is decreased by prior treatment with aspirin and salicylate *in vivo* when the isolated mitochondria are incubated with the usual substrates (2-oxoglutarate, succinate etc.); (Smith, 1966; Gutowska-Grzegorczyk and Kalzak, 1968; Thompkins and Lee, 1969; Dawkins and Smith, 1970; Smith and Dawkins, 1971; Mehlman *et al.*, 1972; Pocwiardowska, 1976a; 1976b). There are important effects of aspirin on Ca^{2+} flux and *vice versa*, with addition of Ca^{2+} overcoming the effect of aspirin *in vitro* (Tomoda *et al.*, 1994). Cyclosporin A and Mg^{2+} in the presence of low Ca^{2+} block the effect of aspirin, suggesting effects of aspirin on anion channels (Biban *et al.*, 1995; Venerado *et al.*, 1996). Aspirin treatment *in vivo* does not stimulate ATPase activity (Mehlman *et al.*, 1972), but salicylate has been reported to stimulate mitochondrial ATPase activity *in vitro* (Charnock and Opit, 1962; Chatterjee and Stefanovich, 1976, although the results have not always been in agreement (Pocwiardowska, 1976a; 1976b). The terminal oxidase of the mitochondrial respiratory chain is most sensitive to effects of salicylate (Penniall, 1958). This and other evidence (Smith and Dawkins, 1971) suggest direct actions of the drug on mitochondrial enzymes involving the coupling of electron transport along the respiratory chain (i.e. oxidation) to phosphorylation of ADP. As with other uncoupling agents, the salicylates may also exert their uncoupling effects by promoting electrogenic hydrogen ion transport across the inner membrane of the mitochondria (Cunarro and Weiner, 1975), thereby discharging the energised state essential for energy conservation.

Salicylates also inhibit some of the mitochondrial dehydrogenases located at the beginning of the respiratory chain, although at higher concentrations than required for uncoupling actions (Smith, 1966). These concentration-dependent effects on uncoupling/dehydrogenases is reflected in changes in the rate of respiration of both isolated tissues or cells and in whole animal systems. At low concentrations, salicylates stimulate oxygen consumption primarily as a result of uncoupling and raising ADP concentrations, whereas at higher concentrations respiration is depressed as a consequence of inhibiting key dehydrogenases of the tricarboxylate cycle (Smith and Smith, 1966; Smith and Dawkins, 1971).

Reduction in mitochondrial ATP production has far-reaching consequences for those synthetic and other reactions that utilise ATP either as a source of energy or for production of regulators (e.g. cyclic AMP) a variety of anabolic reactions involved in the biosynthesis of GAGs, glycoproteins and collagen in connective tissues. Reduction in mitochondrial ATP may contribute to apoptosis and inhibition of rheumatoid synovial cells observed with millimolar concentrations of aspirin and salicylate (Yamazaki *et al.*, 2002).

Synthesis of connective tissue proteoglycans and glycoproteins

Salicylates directly affect the biosynthesis of proteoglycans by:

1. Inhibiting the enzymes involved in the synthesis of the component sugars, sulphate moieties and peptides of proteoglycans
2. Depressing the availability of ATP (as noted above)
3. Direct effects on the production of cyclic nucleotides
4. Indirect effects on cyclic nucleotide production as a consequence of inhibiting prostaglandin synthesis.

There may also be indirect effects of salicyates on the control of proteoglycan synthesis mediated via effects on prostaglandin and nitric oxide production (Amin *et al.*, 1997; 1998; 1999).

Salicylates inhibit the biosynthesis of liver proteins following administration of high doses (500 to 600 mg/kg) of this drug to mice (Dawkins *et al.*, 1971), as well as after the incubation *in vitro* of liver microsomal preparations with pharmacological concentrations (1 to 3 mmol/l) of this and related salicylates (Dawkins *et al.*, 1966; Reunanen *et al.*, 1967; Burleigh and Smith, 1970; 1971). The sites of inhibition by salicylate of protein biosynthesis are at the transfer of certain (but not all) amino acids from their tRNA-linked precursors to the growing peptide chain (Burleigh and Smith, 1971). Some of the aminoacyl-tRNA-synthetases catalysing these reactions (e.g. those for glutamate, aspartate and histamine) are more sensitive to salicylate than others. Some of the aminoacyl-tRNA-synthetases catalysing these reactions (e.g. those for glutamate, aspartate and histamine) are more sensitive to salicylate than others. Furthermore, salicylate also inhibits the ATP-pyrophosphate exchange reaction (Burleigh and Smith, 1971). Thus, in addition to the general depressant effects on protein/polypeptide biosynthesis caused by lowering of ATP levels and the pyrophosphate exchange reaction, the salicylates also have specific effects on the incorporation of certain amino acids into polypeptides. They may also cause reduction in protein synthesis by inhibiting the synthesis of RNA (Janakidevi and Smith, 1970a), or by alterations in the distribution of amino acids (Dawkins and Smith, 1970).

Another inhibitory effect of salicylates *in vitro* on proteoglycan (PrGn) synthesis is the inhibition of the synthesis and incorporation of many of the neutral, acidic and amino sugars that are attached to or associated with the polypeptides of cartilage proteoglycans. The salicylates inhibit those enzymes involved in the incorporation of monosaccharides and sulphate moieties into polysaccharide components of proteoglycans (PrGns), GAGs or glycoproteins (GPs), as shown in Figure 7.9. It can be seen that nearly all the major enzyme steps appear to be inhibited by the salicylates. The evidence for these multiple sites of inhibition of GAGs and PrGns comes from studies on the effects of salicylates on (a) the uptake of radioactively labelled precursors *in vitro* or *in vivo* into isolated GAGs or PrGns, and/or (b) the activities of isolated enzymes *in vitro*. It should be emphasised that enzymes involved in the GAG biosynthesis have not been fully evaluated in cartilage or synovial tissues from either normal or arthritic animals or man. Thus the synopsis shown in Figure 7.9 indicates that salicylate inhibits the following enzymes *in vitro* (at the 5 mmol/l limit of pharmacological concentrations; Bannwarth and Dehais, 1991):

1. L-Glutamine-D-fructose-6-phosphate aminotransferase (EC 5.3.1.19) (Bollet, 1961; Jacobson and Boström, 1964; Schönhöfer, 1967; Schönhöfer and Perry, 1967; Perry, 1968; Anastassiades, 1973; Chan and Lee, 1975; Hugenberg *et al.*, 1993), which is involved in the formation of the amino sugar, glucosamine 6-phosphate. This is also a precursor for forming *N*-acetylamino sugars, *N*-acetylglucosamine and *N*-acetylgalactosamine, respectively. Salicylates also inhibit aminotransferases (Gould *et al.*, 1966) so that the supply of NH_3 for the synthesis of glutamine from its respective 2-oxoacid may be impaired. It is notable that although the majority of authors found effects of salicylate, Hugenberg *et al.* (1993) failed to observe any inhibitory effects of aspirin or salicylate up to 10 μmol/l on the bovine articular enzyme.

■ CHAPTER 7 ■

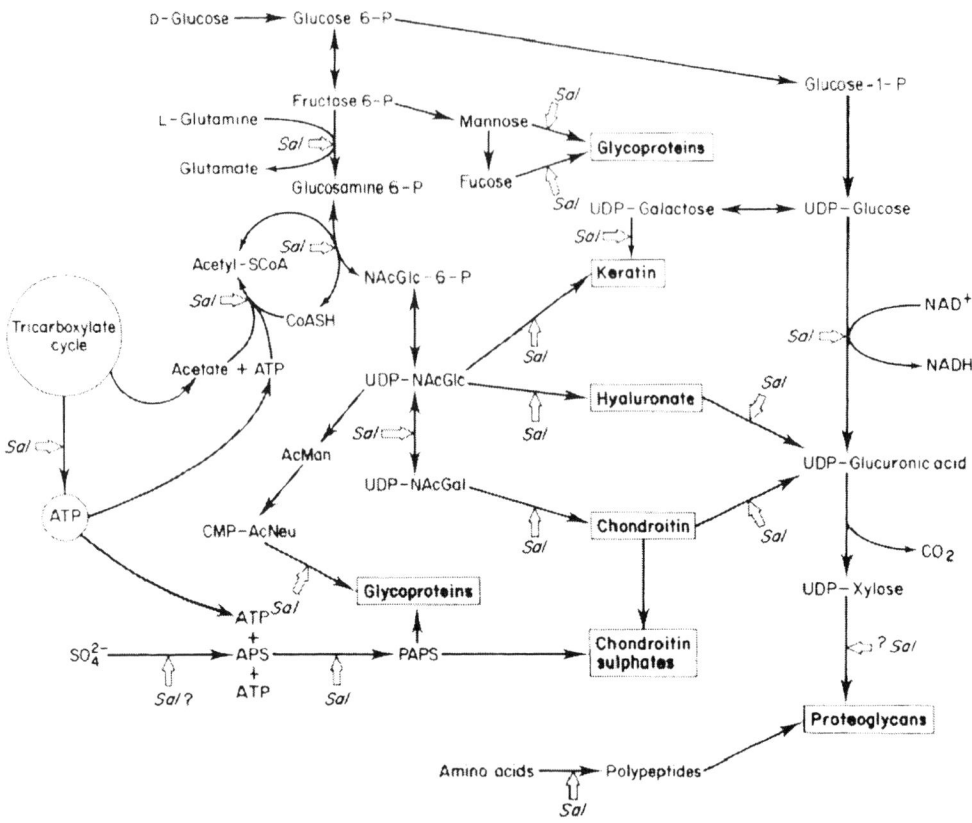

Figure 7.9 Sites of inhibition by salicylate (Sal) on the pathways of the formation of glycosaminoglycans (GAGs) and proteoglycans/glycoproteins. Abbreviations as in the legends to Figures 7.7 and 7.8. UDP, uridine diphosphate (derivatives); APS, adenosyl-5′-phosphosulphate; PAPS, 3′-phosphoadenosyl-5′-phosphosulphate; UDP-NAcGlc, UDP-*N*-acetyl-glucosamine; UDP-NAcGal, UDP-*N*-acetyl-galactosamine; CMP, cytosine monophosphate; AcMan, acetyl-mannosamine; NacGlc-6-P, *N*-acetyl-glucosamine-6-phosphate. (From Rainsford, 1984.)

2. Acetyl-SCoA synthetase (EC 6.2.1.1) involved in the formation of 'active' acetate prior to its incorporation into amino sugars to form *N*-acetylated derivatives (Kent and Allen, 1968).
3. UDP-*N*-acetylglucosamine 4-epimerase (EC 5.1.3.7; Nakagawa and Bentley, 1971).
4. UDP-glucose dehydrogenase (EC 1.1.1.22), which is involved in the formation of UDP-glucuronic acid from its glucose precursors (Lee and Spencer, 1969; Hugenberg *et al.*, 1993). High concentrations of aspirin and salicylate (10 μmol/l) were only found to be inhibitory by the latter authors.
5. UDP-glucuronosyltransferase and most of the enzymes involved in the transfer of *N*-acetylamino sugars and other hexosamines or hexoses (Kent and Allen, 1968; Musil *et al.*, 1968; Reed *et al.*, 1968; Rokosova-Cmuchalova and Bentley, 1968; Lee and Spencer, 1969; Kauke *et al.*, 1971; Nakagawa and Bentley, 1971; Lukie and Forstner, 1972; Palmoski and Brandt, 1979; David *et al.*, 1992; Hugenberg *et al.*, 1993).

It is evident from *in vitro* studies with isolated enzyme and cell preparations that there are marked differences in the concentrations of salicylate required to inhibit these enzymic reactions. Amongst the most sensitive is the acetyl-SCoA synthetase reaction, which has an approximate IC$_{50}$ of 250 μmol/l for salicylate (Kent and Allen, 1968). Many of the other reactions are sensitive to millimolar concentrations of salicylate, but the incorporation of [U-^{14}C]glucose into isolated GAGs was found by Nakagawa and Bentley (1971) to be inhibited by salicylate 20 to 60 μmol/l. Kent and Allen (1968) suggested that, since the most sensitive site for inhibition by salicylate is the formation of acetyl-SCoA, the formation of N-acetylglucosamine 6-phosphate may play a central role in the overall regulation of proteoglycan synthesis.

The effects of NSAIDs including salicylate on the activity of glycosyltransferases in centrifuged (1000 g) homogenates of human articular cartilage were studied by David et al. (1992). Salicylate at 300 μg/ml (2 μmol/l) but not 100 μmol/l (0.5 μmol/l) was found to inhibit xylosyl-, glucuronyl- and N-acetylgalactosaminyl transferases; there was no effect at either of these drug concentrations on glucuronyltransferases. High concentrations of indomethacin, ketoprofen, ibuprofen, paracetamol and tiaprofenic acid variously inhibited the incorporation of radiolabelled sugars involved in the above transferases, but salicylate was not studied (David et al., 1992).

Another important step in the synthesis of connective tissue GPs and GAGs that is affected by the salicylates is the transfer of sulphate moieties to the proteoglycans (e.g. chondroitin and keratin) to form sulphate esters. These sulphated PrGns and GAGs have important mechanical and constructive properties that are essential in maintaining joint function and the overall physiological properties of connective tissues.

In a pioneering series of papers Boström et al. (1963; 1964) and Whitehouse and co-workers (1961; 1962; 1963; 1964; 1965) showed that salicylate in common with other anti-inflammatory compounds (and even antirheumatic drugs, e.g. chloroquine), inhibited both the in vitro and in vivo incorporation of radioactively labelled sulphate into the GAGs isolated from rat cartilage or following in vitro incubation of bovine heart valves. These inhibitory effects have since been confirmed in cultures of aged human articular cartilage (McKenzie et al., 1976) and other cartilage systems in vitro (Palmoski and Brandt, 1979; Palmoski et al., 1980). Boström and co-workers (1964) claimed that the inhibitory effects of salicylate and other anti-inflammatory agents on cartilage metabolism could not be attributed to competition of the drug for 'active sulphate', i.e. 3'-phosphoadenosyl-5'-phosphosulphate (PAPS) (see Figure 7.9), but they did show the excretion of presumptive esters of these drugs in the urine of rats given oral doses of the drugs. With pure radioactively labelled PAPS available, it has been shown that an apparent competition does occur for PAPS at least in the synthesis of GPs of the gastric mucosa (Rainsford, 1978). Moreover, Foye and co-workers (1981) have shown that some salicylates inhibit the mucopolysaccharide sulphotransferase of bovine corneal tissue in vitro. Since the formation of PAPS is unaffected by salicylate or other anti-inflammatory drugs, it appears that, independent of actions of these drugs on ATP production (which would also induce a secondary depression of PAPS levels), they inhibit the sulphotransferase reaction. The possibility exists that sulphate esters of drugs may form as a consequence of this reaction. Salicylate-induced reduction in endogenous sulphate has been found to be related to the reduction in sulphated GAGs in murine patella cartilage (de Vries et al., 1985). Similar reduction in sulphate has been found in rats given paracetamol 200 mg/kg for up to 9 weeks, with consequent reduction in sulphated GAGs (van der Kraan et al., 1990). Alternatively, the metabolism of salicylates may occur by liver phenoltransferase activity, so causing the formation of the urinary sulphate esters that were observed by Boström and co-workers (1964).

The synthesis of GAGs from glucosamine precursors has been shown to be inhibited by salicylate and other NSAIDs in fibroblasts (Kalbhen et al., 1967).

Salicylates may influence the synthesis of GAGs by affecting the production or activity of the 'second messenger', cyclic AMP, by direct effects on adenylate cyclase activity, production of those prostaglandins that influence activity of this enzyme, or blocking the expression of cyclic AMP-mediated enzymic activities

Peters and co-workers (1975) found that salicylate 1 to 10 mmol/l inhibited the production of cyclic AMP by embryonic mouse fibroblasts incubated in culture in the presence or absence of PGE$_1$ and, in parallel, inhibited the production of GAGs by these cells. Dibutyryl-cyclic AMP or theophylline (a drug inhibiting breakdown of this cyclic nucleotide) both stimulate the synthesis of sulphated GAGs by fibroblasts (Coggins et al., 1972), thus illustrating the central role played by cyclic AMP on synthesis of GAGs. Since PGE$_1$ 0.1–100 μmol/1 has been found by Peters and co-workers (1975) to stimulate both cyclic

CHAPTER 7

AMP levels and GAG secretion by fibroblasts, it appears that the salicylate-induced reduction in the PGE_1-regulated cyclic AMP production may occur at the level of the PGE receptor on the adenylate cyclase. This concept is supported by the studies of Ortmann and Perkins (1977), who found that aspirin (no doubt hydrolysed to salicylate) and other NSAIDs inhibited the PGE_1-stimulated cyclic AMP accumulation in human astrocytoma cells in culture, though the effect of aspirin was, by comparison, rather weak. The inhibition of cyclic AMP production by salicylate in these culture systems should be contrasted with the aforementioned stimulatory effects of aspirin and salicylate on cyclic AMP production in gastric mucosal tissues (Rainsford, 1978). Salicylate 0.1 to 20 mmol/1, and to a lesser extent aspirin, have been found to inhibit the activity of cyclic AMP-dependent protein kinase by the drugs affecting the regulatory site on the enzyme (Dinnendahl et al., 1973). It is conceivable that the expression of cyclic-AMP-controlled GAG secretory processes could be affected by salicylate and aspirin at the level of the regulation of such kinase-type enzyme(s). Similar effects have been observed on cAMP-protein kinases in chronrocytes with other NSAIDs (Malemud et al., 1994).

A considerable number of studies have been performed examining the effects of salicylates and other NSAIDs on the in vitro production of GAGs and PrGns in isolated chondrocytes from chick embryonic sternum (Fujii et al., 1989), rabbits (Bjelle and Eronon, 1991), cattle (Anastassiades et al., 1998; Attur et al., 1998), dogs (Anderson et al., 1999) and humans (Bassleer et al., 1992; 1997; Henrotin et al., 1992). Bjelle and Eronen (1991) found that the total uptake of ^{35}S into GAGs in the matrix and cultures of rabbit chondrocytes was inhibited by 200 and 1000 μg/ml salicylic acid as well as by high concentrations of several other NSAIDs. Inhibition of the synthesis of GAGs or PrGns by aspirin and some other NSAIDs in chondrocytes from humans (Fujii et al., 1989; Bassleer et al., 1992; 1997; Henrotin et al., 1992), dogs (Anderson et al., 1999), cattle (Anastassiades et al., 1998) and chickens (Fujii et al., 1989).

Similar inhibitory effects of aspirin or other anti-inflammatory drugs on synthesis of PrGns from $^{35}SO_4^{2-}$-salicylate have been observed in human porcine purine or canine articular cartilage explants in organ cultures, although the concentration–effect relationships have been variable (Anastassides, 1973; McKenzie et al., 1976; Dekel et al., 1980; Palmoski et al., 1980; Brandt and Palmoski, 1984; de Vries et al., 1985; 1988; Carney, 1987; Rainsford et al., 1989; 1992; Dingle and Shield, 1990; Dingle, 1991; 1999). The inhibition of PrGn synthesis has been shown to be related to the increased release of IL-1 or catabolin-stimulated degraded GAGs (Rainsford et al., 1989; 1992; Dingle and Shield, 1990). The concentrations of salicylates required for these effects are relatively high, and in some studies might exceed the plasma-levels of these drugs encountered in therapy.

The synthesis of proteoglycans and degradation of PrGns has also been shown to be inhibited by some other NSAIDs that are potent inhibitors of COX-1/2 enzymes. However, some weak inhibitors of COX-1/2, e.g. azapropazone, do not affect PrGn synthesis or GAG release (Rainsford et al., 1989; 1992). Since salicylate is a weak inhibitor of COX-1/2, this suggests that the effects on this drug PrGn synthesis might not be related to influence on prostaglandin production.

The effects of salicylates and other NSAIDs can vary considerably according to the source of cartilage, i.e. from OA, RA or normal humans or animals. The metabolism of proteoglycans and other components of cartilage varies considerably in different arthritic states (Colins and McElligott, 1960; Oegema and Thompson, 1986; Lafeber et al., 1992), especially in respect of synthesis of proteoglycans. Thus the degree of inhibitory effects of NSAIDs may vary considerably depending on the pathological state of the tissues.

The recovery of cartilage following exposure to IL-1 (which has been sufficient to cause appreciable loss of PrGns) has been found to be retarded by aspirin (30 μg/ml) and ibuprofen (50 μg/ml) but not diclofenac (1 μg/ml) (Dingle, 1992).

Reversal of salicylate-1 mM induced inhibition of ^{35}S-GAG synthesis in canine femoral cartilage by addition of β-D-xyloside 1 μm was observed by Palmoski and Brandt (1982a). Similar effects were observed on the synthesis of GAGs from ^3H-glucosamine, suggesting that the effect of xyloside was at the level of oligosaccharide addition. These results could be important for development of potential cartilage protective agents.

Another component of the processes involved in cartilage degradation is the release of cartilage-degrading activity (CDA), catabolism, pro-inflammatory cytokines (that degrade cartilage, e.g. IL-1, TNF_α, IL-6) and metalloproteinases (Jiang et al., 2001).

The catabolin-like activity that is released from synovial tissues from either human subjects with osteo or rheumatoid arthritis or from pig articular synovium induces degradation of cartilage explants in culture (Herman *et al.*, 1987; 1989; Rainsford, 1987; Dingle and Shield, 1990). Some cyclo-oxygenase inhibitors are effective in preventing the catabolin-induced cartilage degradation, but their effects are essentially weak (Herman *et al.*, 1987; 1989; Rainsford, 1987). Piroxicam has been claimed to be an effective inhibitor of cartilage degrading activity, whereas indomethacin and sodium salicylate at high concentrations are not (Rainsford, 1987; Herman *et al.*, 1989). Aspirin is only a very weak inhibitor and does not cause concentrated related changes in the range of 10 to 100 μmol/l that are consistent with a pharmacologically acceptable effect of this drug.

In addition to influencing the synthesis of GAGs and PrGns, the salicylates, in common with other NSAIDs, could also affect the *in vivo* degradation of these macromolecules. Thus, Comper and co-workers (1981) found that salicylate 0.5 to 3.0 mmol/l inhibited the breakdown of ^{3}H-labelled proteoglycans of rabbit articular cartilage in organ culture. Presumably these effects are related to drug effects on the release and/or activities of lysosomal hydrolases (see p. 269). In other studies, aspirin caused an increase in proteoglycans after initially being treated with the cytokine degrading complex, catabolin, or interleukin-1 *in vitro* (Strathy and Gorski, 1987). Suppression by salicylates of cytokine-induced elevations of matrix metalloproteinases and other proteolytic enzymes (Oeth *et al.*, 1995; Nose *et al.*, 1997; Liptay *et al.*, 1999; Kang *et al.*, 2000; Jiang *et al.*, 2001), nitric oxide (see p. 250), oxyradicals (see p. 265) and signal transduction pathways (see p. 259) may be important in controlling the joint destruction in arthritic diseases.

The question of the significance of these various effects of salicylates on proteoglycan synthesis has been the subject of debate (Brandt and Palmoski, 1984; Herman and Hess, 1984; Tenenbaum, 1984; Cooke, 1985; Burkhardt and Ghosh, 1987; Doherty, 1989; Ghosh *et al.*, 1990). While there have been suggestions that aspirin in arthritic patients might protect organs and against cartilage degeneration (Chrisman *et al.*, 1972) other studies have suggested that aspirin may exacerbate joint damage in OA (Serup and Ovesen, 1981). A considerable number of studies have shown that the inhibition of proteoglycan synthesis *in vivo* relates to acceleration of cartilage degradation (Brandt and Palmoski, 1984; de Vries *et al.*, 1985), although the synthesis of sulphated proteoglycans might be due to the effect of salicylate in reducing endogenous sulphate (de Vries *et al.*, 1985).

In the cruciate ligament sectioned model of dog osteoarthritis, 150 mg/kg per day aspirin (a dose that produced serum salicylate levels averaging 20 to 25 mg/dl) was found to accelerate the degradation of articular cartilage in those joints in which osteoarthritis was evident, although it had no effect on cartilage from contralateral knees (Palmoski and Brandt, 1982b). These studies and others performed by Palmoski and Brandt were all performed in relatively small numbers of animals under conditions where the joints had already exhibited substantial injury. The results of this and other studies from Brandt's group (Palmoski and Brandt, 1984; 1985; Slowman-Kovacs *et al.*, 1989) consistently showed that there was reduction in the synthesis of proteoglycans and reduction in proteoglycan concentrations in the cartilage of animals given aspirin. In a model in which the limbs were immobilised or loaded it was found that proteoglycan depletion was evident with aspirin or salicylate. These variants of canine osteoarthritis are complex models, and the interpretation of whether or not aspirin and other salicylates have effects on accelerating cartilage destruction depends on the stage and intensity of the osteoarthritis in these models.

Since paracetamol has been regarded as an agent unlikely to accelerate cartilage or bone degradation in osteoarthritis (see Chapter 12), the observations of von der Kraan *et al.* (1990) that treatment of rats with paracetamol 200 mg/kg twice daily for 2 to 9 weeks led to diminution of circulating sulphate in the serum and reduction in cartilage GAGs implying that paracetamol may not be without effects on cartilage proteoglycan metabolism.

Other investigators (Chrisman and Snook, 1968; Marcelon *et al.*, 1976) have found little or no effects of salicylates (Pelletier, 1999).

The other major component of connective tissues influenced by the salicylates is the biosynthesis of collagen. While there is experimental evidence that salicylate and other NSAIDs do not alter the content of total collagen fractions, it appears that the content of neutral salt-soluble and alkali-soluble fractions (representing newly synthesised collagen) is increased. These changes are only manifest in animals with defective collagen metabolism (e.g. in experimental lathyrism – a condition induced by

CHAPTER 7

administration of β-aminopropionitrile, which causes the breakdown of inter- and intramolecular bonding of collagen).

Studies by Trnavsky and co-workers (1968a; 1968b; 1974) have shown the importance of the selective effects of salicylate on collagen metabolism in adjuvant-arthritic or lathyritic rats. Chronic oral feeding of salicylate 300 mg/kg per day to normal rats causes an increase compared with control animals in the specific activities of [^{14}C]hydroxyproline in both the skin and urine, reflecting an accelerated metabolic turnover of collagen. In contrast, this same salicylate treatment of lathyritic or arthritic rats restored synthesis of collagen and its accelerated breakdown to near normal levels (Trnavska et al., 1968a; 1968b; Trnavska and Trnavsky, 1974). These results suggest that salicylate normalises the collagen turnover in inflammatory conditions where there is abnormal collagen metabolism. Modest inhibition of bone resorption has been observed in a femoral head bone transplantation model in rats (Solheim et al., 1986a; 1986b; 1986c), and also in another model of bone resorption in rats (Gugiyasu, 1976; Yazdi et al., 1992).

While salicylate administration has similar beneficial effects on metabolic turnover of collagen of the granulation tissue induced in rats (Kulonen and Potila, 1975), electron-microscopic observations of fibroblast collagen in such tissues suggests that neither the number nor the morphology of collagen fibrils is altered by the drug, although the protein-synthetic and mitochondrial structures are markedly altered. It may be that the morphological appearance of collagen in granuloma tissue does not correlate with the biochemical changes. Further studies are needed to establish whether morphological changes suggestive of normalisation of collagen structure do occur as suggested from the biochemical evidence. Such changes may also be important for the abnormal calcifying connective tissue reaction, which has been reported to be prevented by salicylates (Cotty and Harris, 1968; Kulonen and Potila, 1975). Heterotropic bone formation has also been shown to be reduced by aspirin (Solheim et al., 1987) and this drug may reduce bone resorption (Herman et al., 1994). These results contrast with reports that aspirin may induce abnormalities in bone growth and connective tissue calcifying reactions in experimentally-induced bone destruction, including arthritis in rats (Carroll and Melfi, 1972; Gugiyasu, 1976; Wilhelmi, 1978).

Eicosanoids exert a variable effect on the resorption of bone in the mouse calvaria model according to the type of eicosanoid and concentration (Katz et al., 1981). Because of these regulatory effects of eicosanoids there has been interest in the effects of aspirin and other NSAIDs on bone resorptive activity in various model systems. Thus, Solheim (1986a; 1986b; 1986c) showed in a range of studies that aspirin administration to rats for various periods ranging from 9 to 18 days at doses of 115 mg/kg or lower decreased collagen synthesis and bone mineralisation and decreased the formation of bone. Similar effects were noted with naproxen 20 mg/kg. In a follow-up study, Solheim et al. (1987) showed that aspirin decreased osteogenesis in bone transplants in which the bone had been demineralised.

In a study of the effects of NSAIDs on demineralised bone induced bone formation using a similar system to that described by Solheim and co-workers in 1987, there was an enhancement of bone formation in rats given aspirin 75 mg/kg, ibuprofen 15 mg/kg and paracetamol 50 mg/kg. In contrast, indomethacin 5 mg/kg and piroxicam 4 mg/kg had inhibitory effects and flurbiprofen did not affect bone formation significantly. It appears therefore that in these models of bone resorption there may be variations in the effects of aspirin and other NSAIDs according to the type of drug and the model system employed. The increase in bone formation may be related to the development of ectopic bone formation, and may be considered to be an undesirable event.

In a canine model of osteoporosis induced by fibreglass cast immobilisation of the right hindlimb for 4 weeks, the *in vitro* release of prostaglandin PGE_2 from the calcaneus tibule cortical bone and tibule cancellous bone and ilium were found to be enhanced approximately two-fold (Waters et al., 1991). Aspirin 25 mg/kg every 8 h resulted in a 65 per cent reduction in bone PGE and a 13 per cent sparing of bone mass, implying that aspirin may have some effect in reducing bone loss in osteoporosis. The potential beneficial effects of aspirin and other NSAIDs in osteoporosis were examined by Lane et al. (1997), who found that daily intake of aspirin or other NSAIDs was associated with reduced urinary excretion of N-telopeptide cross-link, which, though small (12.5 per cent), was statistically significant. As aspirin use with COX-2 inhibitors may increase bone mineral density these drugs may protect against bone loss in osteoporosis and arthritic conditions (Carbone et al., 2003).

Investigations of the effects of aspirin and various fatty acid derivatives on rat calvary disks showed that aspirin inhibited the increased uptake induced by *cis*-vaccenic acid while prostacyclin increased the uptake of calcium (Walenga and Bergström, 1985). Ueno *et al.* (1980) and Kawashima *et al.* (1985) showed that aspirin 20 to 100 mg/kg p.o. inhibited the hypercalcaemia induced by 1α-hydroxyvitamin D_3, but did not affect the intestinal calcium absorption. Aspirin also abolished the hypercalcaemic action of parathyroid hormone. The authors suggested that aspirin may inhibit bone resorption.

In the mouse calvary model, thrombin-induced resorption was inhibited by 5 μmol/l aspirin with a concentration dependent on addition of calcium release. Acemetacin and indomethacin also inhibited the bone resorption in this system in concentration ranges that were lower than that observed with aspirin (Hoffman *et al.*, 1985).

The effects of sodium salicylate and indomethacin on the recruitment of osteoclast cells was investigated by Soekanto (1994) in bone marrow cells cultured for 8 days with 1α, 25-dihydroxyvitamin D_3, prostaglandin E_2 and recombinant interleukin 1. Sodium salicylate 1 to 100 μmol/l caused a progressive reduction in the number of tartrate-resistant acid phosphatase positive multinucleated cells that are probably osteoclasts. Similar effects were noted with 0.01 to 1 μmol/l indomethacin. The effects of both these drugs appear to be related to effects on prostaglandin production; the effect of sodium salicylate in depleting prostaglandin production appeared to be somewhat less than that of indomethacin. Sodium salicylate did not show any cytotoxic effects in this system (Soekanto *et al.*, 1994), and the results suggested that the inhibitory effect of sodium salicylate on the recruitment of osteoclasts was greater during the latter stage of bone marrow culture.

Overall, therefore, it would appear that aspirin and salicylate have inhibitory effects on bone resorption in some systems although there may be some considerable variability according to the experimental conditions. A study of the resorption of the lining membranous tissue obtained from endosteal curettage of the femoral prosthetic bed in patients with OA showed that sodium salicylate 100 μg/ml and indomethacin 1.5 μg/ml did not inhibit pseudomembrane-induced bone resorption and cytokine production, whereas piroxicam 10 μg/ml did inhibit this process (Herman *et al.*, 1994). Thus, pathological tissue may not show the apparent inhibitory effects on bone resorption as are observed in experimental models. It is clear that further investigations are required in order to clarify the effects of salicylates on bone resorption.

Among the significant potential mediators of inflammation in the control of cartilage and bone destruction is nitric oxide (Amin *et al.*, 1998). In osteoarthritis cartilage from patients undergoing knee replacement surgery, the production of nitric oxide (as nitrite) was inhibited at concentrations of 2 and 3 mmol/l of aspirin and sodium salicylate coincident with reduction in the production of prostaglandin E_2 (Attur *et al.*, 1998). Dexamethasone and indomethacin failed to inhibit the production of nitric oxide, although there was reduction in prostaglandin E_2 as expected. Interleukin-1-induced production of nitric oxide was also suppressed by aspirin (2 mmol/l). Similar effects on nitric oxide production were observed with 1 to 3 mmol/l aspirin or sodium salicylate in rat chondrosarcoma cells *ex vivo*, the Swarm rat chondrosarcoma having previously been implanted into rat so that it induced formation of a well-differentiated human chondrosarcoma. The exact significance of these effects on nitric oxide production has not been defined. It can only be speculated that as nitric oxide like prostaglandins may be important in the regulation of cartilage proteoglycan and other metabolic reactions, there may be important regulatory effects that occur as a consequence of the effects of salicylates on nitric oxide production.

Anti-proliferative activities

Oral administration of aspirin 300 mg/kg per day for 5 days induces a slight reduction in the content and rates of synthesis of DNA in sponge granulomas induced in rats (Samogyi *et al.*, 1969). This may be related to the reported inhibitory effects of salicylate on DNA and RNA synthesis *in vitro* (Janakidevi and Smith, 1970a; 1970b). Aspirin, salicylate and salicylamide have, however, been found to cause a slight increase at low concentrations, followed at high millimolar concentrations by a profound inhibition of cell growth, protein and nucleic acid synthesis of human fibroblast and rat hepatoma cell cultures (Hail *et al.*, 1977). Moderate to weak anti-proliferative and cytostatic effects of aspirin and salicylate have been observed on *in vitro* cultures of Ehrlich ascites tumour and hamster kidney cells (Warnecke and Seeber, 1969; Karzel *et al.*, 1973; Rubenstein *et al.*, 1976) and in other tumour cell models (See Chapter 13). It appears, therefore, that aspirin and salicylate have moderate growth inhibitory effects that may have benefit in the proliferation of cells in inflammatory foci.

Lymphocyte transformation and functions

It is well established that T and B lymphocyte functions are abnormal in arthritic diseases, especially in those lymphocyte subpopulations present in synovial tissues (see p. 237, and Wangel and Klockars, 1977; Meijer *et al.*, 1980).

Earlier studies (Austen, 1963) showed that salicylates suppress antigen–antibody reactions. Among the systems affected by aspirin are the systemic anaphylaxis and reverse passive Arthus reaction in rabbits and guinea pigs. Salicylates at high concentrations partially dissociated preformed complexes *in vitro*. Clinically aspirin can be a cause of hypersensitivity (Duncan *et al.*, 1977; see Chapter 8), so the effects of these drugs in controlling hypersensitivity reactions clearly have a different basis to the development of this adverse reaction.

Opelz and co-workers (1973) showed that aspirin in pharmacological ranges (0.56 to 2.2 mmol/l) inhibited the incorporation of [^{3}H]-thymidine in lymphocytes transformed by exposure to the mitogen, phyto-haemagglutinin (PHA). While these authors claimed that effects of aspirin on such lymphocyte transformations had not, to their knowledge, been published, there had in fact been at least two reports published previously by Griswold and Ujeki (1969) and Schneider and co-workers (1971) showing such inhibitory effects of aspirin and salicylate on lymphocyte transformability *in vitro*. There followed a series of papers confirming the inhibitory effects of aspirin or salicylate on transformation of peripheral lymphocytes from adjuvant-arthritic rats (Loveday and Eisen, 1973) and on the response to various mitogens in those from man (Panush and Anthony, 1976; Cologniet *et al.*, 1977; Dewse, 1977; Gabourel *et al.*, 1977a; Egorin *et al.*, 1978) and rodents (Gabourel *et al.*, 1977b; Mobarok Ali and Morley, 1980; Mullink and Von Blomberg, 1980). While there have been a few negative reports (Smith *et al.*, 1975a; Egorin *et al.*, 1978) these differences may be due to minor technical aspects (Smith *et al.*, 1975a); certainly the numbers of reports would appear to support the reality of the original observations.

The mechanism(s) of this suppressive action could involve effects of the drugs on DNA synthesis (Janakidevi and Smith, 1970a; Hail *et al.*, 1977; Schneider *et al.*, 1972), effects on sulphydryl reactivity, ion transport and/or membrane functions (Famaey and Whitehouse, 1973; 1975). It is also possible that the many complex effects of salicylates on the production of pro- and anti-inflammatory cytokines and NFκB/IκB or other signalling pathways (Takashiba *et al.*, 1996; Bitko *et al.*, 1997; Shackleford *et al.*, 1997; Schwenger *et al.*, 1998; Chen *et al.*, 1999; Lemay *et al.*, 1999; Daun *et al.*, 2000; Rossi *et al.*, 2000; Cianferoni *et al.*, 2001), hexokinase activity (Schneider *et al.*, 1972), and prostaglandin (Ceuppens and Goodwin, 1985) and cyclic AMP levels (Snider and Parker, 1975) could also contribute to the actions of salicylates on immune functions. Whatever the mechanism, it appears that this drug may have beneficial effects since the pro-inflammatory effects of lymphocytes in the acute and chronic inflammatory reactions in the rat are attenuated by aspirin, in common with other anti-inflammatory drugs (Winchurch *et al.*, 1974; Leme *et al.*, 1977), perhaps by repressing lymphocyte cytotoxicity (Winchurch *et al.*, 1974). Salicylate, in common with other NSAIDs, has important physical effects on lymphocyte membranes (Inglot and Wolna, 1968; Görög and Kovaks, 1970; Famaey and Whitehouse, 1973; 1975; Schwoch *et al.*, 1974; Mizushima *et al.*, 1975). Famaey and Whitehouse (1973; 1975) have furthermore provided some evidence suggesting that the NSAIDs stimulate sulphhydryl/disulphide groups in lymphocytes, although the exact meaning of this observation has yet to be established.

It should be noted that the presence of monocytes can influence the actions of salicylates and other NSAIDs on lymphocytes (Mullnik and Von Blomberg, 1980), perhaps through the effects of these drugs on prostaglandin production by the former cells. There may also be secondary effects of the salicylate on macrophage- or granulocyte-mediated interactions with lymphocytes as a result of inhibitory effects of these drugs on granulopoiesis (Gabourel et al., 1977b).

The established roles of prostaglandins, especially PGE_2, in regulating the functions of T- and B-cells, macrophage functions and antibody production have been shown to form the basis of immunological effects of NSAIDs (Ceuppens and Goodwin, 1985). Additionally, more recent studies have shown that aspirin and other salicylates have effects on the production of cytokines that regulate immune functions and the importance of the effects of salicylates on the actions of transcription factors involved in their regulation (Chen et al., 1999; Perez et al., 2002). Thus, human peripheral blood mononuclear cells (PBMC) stimulated with phytohaemagglutinin (PHA) show increased production of γ-interferon (IFN-γ) with aspirin 150 to 600 mol/l (Hsai and Tang, 1992). Other authors have reported stimulation of IL-2 as well as IFN-γ production by aspirin in various model systems (Zatz et al., 1985; Yousefi et al., 1987; Takaoki et al., 1988; Cesario et al., 1989; Hsai et al., 1989a; 1989b). Hsai and Tang (1992) investigated the ex vivo production of IL-2 and IFN-γ in volunteers, who took 325 mg aspirin. Peak levels of IL-2 (about $2 \times$ baseline) occurred at 12 h after ingestion, while those of IFN-γ peaked at 24 h and declined thereafter.

In a placebo-controlled trial of rhinovirus infection in normal volunteers, subjects took 325 mg aspirin every other day for three doses, or identical placebo (Hsia and Tang, 1992). Production ex vivo in PBMCs of IFN-γ and IL-2 was significantly increased in subjects who took aspirin compared with placebo controls. Unfortunately, in rhinovirus-inoculated subjects aspirin treatment did not affect the progress of the infection. In Balb/c mice given influenza virus vaccine, the immune response to the vaccine and subsequent virus challenge appeared to be enhanced by treatment on days 1 to 7 or 5 to 11 after injection with 2.5 or 4.7 mg/kg aspirin, compared with controls.

Jäpel et al. (1994) showed that lysine aspirin (75 or 100 μg/ml) did not affect the antibody-dependent and antibody-independent cytotoxicity of mouse peritoneal macrophages. In contrast, previous studies by Kleinerman et al. (1981) showed that both aspirin 50 to 200 μg/ml and salicylic acid 100 μg/ml enhanced human peripheral monocyte mediated cytotoxicity of ^{51}Cr-labelled chicken red cells. They showed that the effect of salicylic acid was independent of effects on PGE_2 production, thus implying that there is a PG-independent mechanism of cytotoxicity.

Other immunological actions of salicylates are discussed in Chapter 13.

ANALGESIC ACTIVITY

Induction, modulation and perception of pain

Current concepts of the mechanisms of nociceptive pain indicate that:

1. Pain initiated in the periphery generally involves a varying degree of local inflammation
2. Pain is mediated by sensory afferent pathways
3. Modulation and integration of afferent pathways occurs at the level of the spinal cord, brainstem, reticular formation, thalamus and ultimately in the sensory–motor cortex
4. Efferent responses from the CNS involve neural activation in the periaqueductal grey (PAG) and are mediated by downward neural activation in brainstem and spinal systems, ultimately integrating signals principally involving 5-HT, NA and endogenous opioid neurones (Wills, 1985; Field, 1987; Forrest, 1998; Scadding, 1998; Schaible, 1998; Julius and Basbaum, 2001; see Figure 7.10).

Neural impulses produced by stimulation of the nociceptors are transmitted through Aδ and C fibres to the spinal cord and brainstem. Brief, sharp, pricking pain is usually localised to the epidermis and the GI and other mucosa tissues. This type of pain is carried along thin myelinated Aδ fibres at a velocity of 2.5 to 3.0 m/s, giving fast transmission of the pain stimulus. Deep aching, burning or itching is more diffuse and longer lasting with C than with Aδ fibres. This pain is transmitted by unmyelinated C fibres

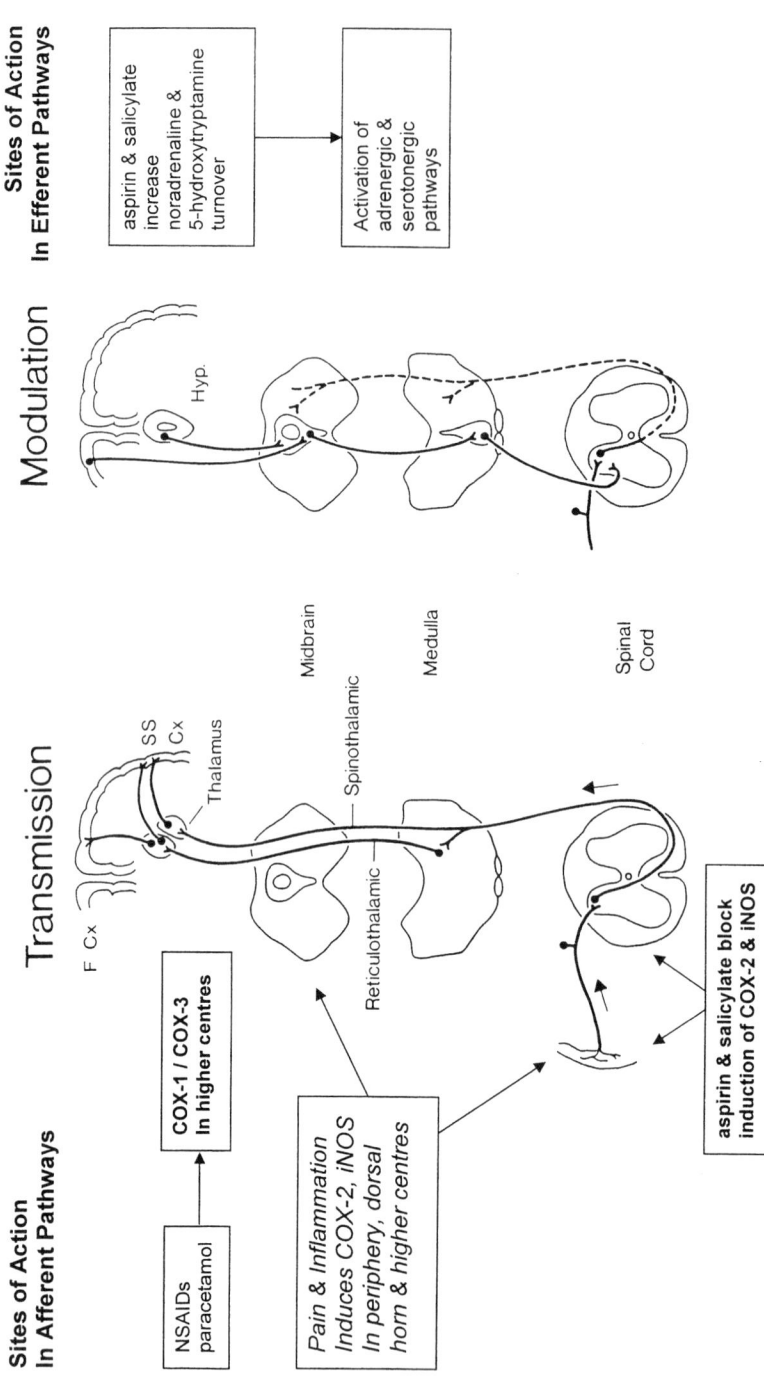

Figure 7.10 Diagram showing principal pain pathways mediating afferent (ascending pain) and efferent (modulating) pain pathways with sites of COX-2 and nitric oxide (NO) production from NO synthase in the dorsal shown and spinothalamic tracts. Aspirin and other NSAIDs affect the activities of COX-2 and NOS in the dorsal horn, spinothalamic tracts as well as in the periphery. Aspirin, salicylate and paracetamol cause increase in the turnover of 5-hydroxytryptamine (5-HT, serotonin) in serotonergic pathways and noradrenaline in downward projecting fibres so causing modulation of pain response, in a manner similar to, but not identical to, that of opioids. Based on Figure 1.4 in Fields (1987), and modified to show the COX-2, NOS, serotonergic and nor-adrenergic sites of effects of the salicylates and other drugs. (Reproduced from Fields, H.L. 1987, *Pain, Mechanics and Management* with permission of the McGraw-Hill companies).

at a velocity of less than 2.5 m/s, due to the size of the fibre and lack of myelin sheath. These afferent pain fibres terminate in the dorsal horn. Secondary neurones transmit the impulse from the laminae to the ventral and lateral horns, crossing to the opposite side of the cord. The ascending sensory tracts relay impulses towards the supraspinal targets through chains of successive neurones. In general, neural transmission is conveyed along six main pathways on each side of the spinal cord. The dorsal horn is involved at the first level in pain processing, involving the receipt of transmitted nociceptive impulses, sensory processing including local abstraction, integration selection, and subsequent output to high centres of sensory impulses to the spinoreticular and spinothalamic tracts that ascend the anterolateral quarter of the spinal cord. Spinothalamic tracts transmit pain information to the supraspinal targets, with the acute pain being transmitted by the neospinothalamic tract to the midbrain, postcentral gyrus and cortex, while the paleospinothalamic tract transmits dull, burning, itching pain to the reticular formation, pons, limbic system and midbrain (Fields, 1987; Forrest, 1998).

The transmission of pain signals by neurones in the thalamus and cortex evokes the sensory, emotional and cognitive components of the pain response. Activation of neurones in other supraspinal sites – posterior thalamus, intralaminar nuclei of the thalamus, hypothalamus and limbic system – are involved in arousal reactions, which constitute the affective aspects of pain and pain memory.

The efferent components in pain pathways are contained in the Autonomic Nervous System (ANS), which involves visceral motor fibres that control the activity of cardiac and smooth muscle and the glands – functions that are essentially involuntary. The ANS comprises sympathetic and parasympathetic nervous systems, which essentially have opposing effects on the activities of the same site in the nervous system. Efferent impulses from the brain are transmitted to the spinal cord through two classes of descending tracts, the pyramidal pathway and extrapyramidal pathways. The efferent pathway is responsible for the regulation of the afferent pain signals. In the brain the PeriAqueductal Grey (PAG) and the Nucleus Raphe Magnus (NRM) are interconnected and form descending pathways that transmit impulses down the spine to the dorsal horn. Transmission through these fibres leads to a component of analgesia mediated by opiate-releasing interneurones, which inhibit afferent transmission of pain signals. Neurotransmitters in these efferent pathways – endorphins and enkephalins – are released upon noxious stimuli and act as effective endogenous analgesics by inhibiting pain sensation through a variety of mechanisms. The endorphins and enkephalins act by activating opiate receptors located on the afferent neural cells. The action of endorphins/enkephalins on opioid receptors results in suppression of the release of excitatory neurotransmitters, preventing the transmission of painful impulses. These pathways mediate inhibition of descending neurones in the spinal cord, resulting in the suppression of nociceptive signals in spinal cord neurones.

Acute, chronic and neuropathic pain

Pain is protective by alerting the individual to a potentially harmful experience. There are three types of *acute pain*:

1. Somatic pain arising from the skin, joints and muscles, which may be superficial fleeting, sharp, or stabbing localised pain transmitted by $A\delta$ fibres, or prolonged pain that has deep, itching, aching perception transmitted by C fibres
2. Visceral pain from the internal organs, abdomen or skeleton, which is transmitted by visceral afferent C fibres to the dorsal horn
3. Referred pain is from the conveyance of the first two pathways sending impulses to the brain from somatic and visceral pathways, which consequently may cause confusion to the brain regarding localisation of the pain.

Chronic pain generally results from persistent stimulation, e.g. from long-term activation of T lymphocyte driven immuno-inflammatory pathways or from neuro-immunological activation. It is generally regarded as a persistent, prolonged painful experience, which may be debilitating and last over many

TABLE 7.8

Characteristics of acute, chronic and neuropathic pain.

Characteristic	Acute pain	Chronic pain	Neuropathic pain
Onset	Usually sudden onset	May be sudden or develop insidiously	May be rapid after injury, but often delayed
Duration	Transient; up to 6 months	Prolonged; months→years	
Pain identification	Painful and non-painful areas well identified	Painful and non-painful areas, changes in sensation less well identified	
Clinical signs	Typical response pattern with more visible signs; increased heart rate respiratory rate and blood pressure, pallor, flushing, dilated pupils, diaphoresis, elevated blood sugar, decreased gastric acid secretion and mobility, and increased blood flow to the organ and skin, nausea and vomiting	Response patterns vary; fewer overt signs and adaptation over a chronic course	Allodynia – the sensation of pain is caused by a stimulus that would not usually cause pain Hyperalgesia – inappropriate severity of pain experience from noxious stimuli Continuous background pain is interrupted by spontaneous, paroxysmal pain such as burning or stabbing skin sensations, or cramping, aching deeper musculoskeletal pain Hyperpathia – a delayed onset of the pain experience, which may only develop after repeated stimuli Sympathetic abnormalities – local sympathetic abnormalities are common, particularly with peripheral nerve lesions, but also occur with root, cord or thalamic lesions Changes in emotions and tiredness can affect the intensity of the pain over a short period of time, therefore psychological factors feature in neuropathic conditions.
Significance	Significant (informs person something is wrong)	Person looks for significance	
Pattern	Self-limiting or readily corrected	Relapsing or remitting, intensity may vary over time or remain constant	
Course	Suffering usually decreases over time	Suffering usually increases over time	
Actions	Leads to actions to relieve pain	Leads to action to modify pain experience	
Prognosis	Likelihood of eventual complete relief	Complete relief not usually possible	
Causes	Acute injury (minor cuts and bruises), toothache, headache, insect stings	Low back pain, neuralgias, hypersthesias, myofacial pain syndrome, haemoagneusias and may be associated with diseases such as cancer and AIDS	

months, the cause of which is often not always apparent. It may not respond to conventional analgesics, thus being more difficult to manage compared with acute pain.

Neuropathic pain is pain from damage to the nervous system. It is less common than nociceptive pain and is more difficult to treat because it mostly results from irreversible damage to the peripheral or central nervous system. The pain from injury to the nociceptive pathways may occur peripherally or centrally. It may involve damage to the receptors, peripheral nerves, posterior roots, the spinal cord or central regions of the brain (McNaughton, 1998). The underlying mechanisms of neuropathic pain are less clearly understood compared with those of nociceptive pain (Scadding, 1998).

Table 7.8 shows the characteristics of acute, chronic and neuropathic pain, and serves for understanding the pathophysiological basis of the actions of aspirin and related analgesics. It is worth highlighting the limitations of the procedures used in animal and human experimental models for determining the analgesic actions, since many of the subjective behavioural responses become apparent in more complex states in chronic and neuropathic pain.

Mediators of pain

Tissue damage caused by injury, disease or inflammation results in the release of endogenous algogenic or algesic chemicals, and these may reach the extracellular fluid that surrounds nociceptors, so mediating information about painful stimuli, leading to hyperalgesia. Among these mediators are mast cell amines, kinins, substance P and neurokinins, prostaglandins E_2 and I_2, nitric oxide, platelet-activating factor adenosine and cytokines, all of which are released during inflammatory responses accompanying tissue damage (Julius and Basbaum, 2001; Vanegas and Schaible, 2001; Sawynok and Liu, 2003). The target of the effects of NSAIDs has been controlling the neural cell production of COX-1 and COX-2 derived prostaglandins ($PGF_{2\alpha}$, PGE_2; Ballou *et al.*, 2000) acting on EP1, EP3, EP4 and IP-receptors in afferent pathways (Sugimoto *et al.*, 2000), and of nitric oxide (Figure 7.10), although indirect effects occur from these drugs on the production of mediators that underlie the inflammatory reactions in the periphery, resulting in nerve sensitisation and initiation of pain responses via afferent stimulation.

Analgesic activity of salicylates

A major part of the analgesic mechanism of the salicylates is due to the peripheral anti-inflammatory actions of these drugs (Dubas and Parker, 1971). Evidence for such a component has come from the elegant cross-perfusion experiments in dogs of Lim and co-workers (1964; 1969). Later studies by the same group established the role of bradykinin in the pain response elicited by electrical stimulation of the footpad of cats, and the effects of morphine (a centrally acting analgesic) compared with aspirin on the central nervous system (CNS) pathways associated with pain response (Guzman *et al.*, 1964). These latter studies implied that there could be a small CNS effect of aspirin, although the peripheral mode was obviously a dominant feature. Studies by Dubas and Parker (1971) on the effects of salicylate on the peripheral (footpad) and central (i.e. hypothalamic) stimulation (from implanted electrodes) in the rat showed that there could be an appreciable central component in the analgesic effects of the salicylates. This and later evidence (see p. 301), and especially the role played by prostaglandins E_2 and I_2 in eliciting the central (presumably vascular) component of analgesia, suggests that these agents could mediate certain CNS responses and that aspirin could attenuate some of these responses (Ferreira, 1980; Mense, 1983; Ferreira *et al.*, 1997). Further important evidence is accruing that aspirin and salicylate may have distinct effects in the CNS mediated by non-prostaglandin mechanisms (Brune *et al.*, 1991; Bannwarth *et al.*, 1995; Pini *et al.*, 1997; Greffrath *et al.*, 2002; Genc *et al.*, 2003), in part by causing alterations in the brain levels of 5-HT (serotonin) (see p. 305), a neurotransmitter that is known to have effects in mediating pain in central nervous pathways (Mayer and Price, 1976; Messing and Lytle, 1977; Shyu *et al.*, 1984). Since aspirin is known to inhibit platelet aggregation and platelets are a rich source of serotonin, it is possible that by preventing accumulation of thrombocytes (which occurs in the CNS during certain pain-eliciting states such as headaches), aspirin could directly influence specific types of pain response involving platelets.

■ CHAPTER 7 ■

The clinical aspects of the analgesic actions of the salicylates are discussed in Chapters 10 and 12. Here, aspects of the structure–activity and mechanistic responses in analgesia by the salicylates are considered. Before considering these aspects it is useful briefly to review the mechanism of pain responses.

Mechanisms of pain responses

The main component of what is described as inflammatory pain is termed 'hyperalgesia', and involves the sensitisation of specialised peripheral nerve endings, nociceptors, to mechanical and/or chemical stimulation (Dalessio, 1972; Dykes, 1975; Mayer and Price, 1976; Messing and Lytle, 1977; Belcher, 1979; Mense, 1983). Pathways of pain transmission involve the somatosensory system (Dykes, 1975; Ferreira, 1980; Ferreira *et al.*, 1997). This has two components, namely the lemniscal (which involves rapid neurotransmission to the cortex and medulla with consequent rapid adaptation), and the anteriolateral system (which informs the animal and possibly the human patient about the state of tissue injury by thermal, mechanical or chemical stimuli; Dykes, 1975; Ferreira, 1980). Some of the afferent pathways involved in pain signal transmission in the brainstem and spinal cord are serotoninergic (Messing and Lytle, 1977), and hence the interest in effects of aspirin and salicylate on serotonin production in the CNS.

Comparative effects of salicylates in pain models in animals

Animal models employed in the assay of NSAIDs and analgesics have been reviewed elsewhere (Bliven and Otterness, 1985; Chan, 1989). Table 7.9 shows data from standard analgesic assays in a variety of rodent models comparing aspirin and diflunisal with a range of other NSAIDs and some analgesics. The potencies of aspirin and diflunisal in these assays depends on the type of analgesic agent employed. Even potencies of salicylates relative to other NSAIDs vary considerably within the one assay (e.g. phenylquinone writhing), although there is rough agreement of 'orders of magnitude' of potency in relation to clinically accepted potency of these drugs in acute and chronic painful conditions.

Among factors accounting for variability in response is the timing of drug assay. Thus, Ohdo *et al.* (1995) showed that there was significant chronological variation in effects of aspirin on the latency of response in the hotplate assay (which paralleled hypothermic responses), not to plasma salicylate levels.

It has long been considered that aspirin has a greater acute analgesic activity than salicylate. The essential quantitative differences depend on the nature of the pain-eliciting stimulus, the route of administration of both the drug and pain-producing substance, and the species of animal employed. Nonetheless, the differences in analgesic activity of these drugs in different models are not entirely clear. In the cross-perfusion experiments of Lim and co-workers (1964; 1966; 1969), where bradykinin was used as the pain-eliciting agent aspirin had four times the analgesic activity of salicylate (assessed by the peripheral assay technique where the drug is given into the splenic artery (Lim *et al.*, 1964). Similar results have been found in the responses from the intra-arterial injection of bradykinin into rats (Thompkins and Lee, 1975). When given by the intraperitoneal route, the ED_{50} values for aspirin and salicylic acid are 64.9 and 138 mg/kg, respectively (Thompkins and Lee, 1975). Both these studies were performed when the drugs were gives parenterally. The differences become much less pronounced when the drugs are given orally (Thompkins and Lee, 1975), presumably because of the appreciable conversion of aspirin to salicylate.

It appears that the overall responses to analgesic drugs of intraperitoneal bradykinin in mice are better than those towards phenylquinone (Emele and Shanaman, 1963). Aspirin given orally has an ED_{50} of 21 mg/kg in the former system, whereas it has a value of 150 mg/kg in the latter (Emele and Shanaman, 1963). By comparison, salicylamide has an ED_{50} of 225 mg/kg in response to bradykinin (Emele and Shanaman, 1963), showing that this is a relatively weak analgesic. Also, orally administered paracetamol has been claimed to be 70 per cent less potent than oral aspirin in the phenylquinone-inducing writhing assays in mice (Siegmund *et al.*, 1957). However, the data of Römer (1980) has shown the reverse, since the ED_{50} in the same assay system for oral aspirin is 240 mg/kg and for oral

TABLE 7.9

Analgesic activity of aspirin compared with other NSAIDs and analgesics in rodent analgesia models.

Drug	Mice (PBQ) ED_{50} mg/kg* (95% C.I.)					Rat (Randal–Selitto) ED_{50} mg/kg (95% C.I.)		
Aspirin	181 (124–267)	182	120	224 (137–368)		32		276 (127–601)
Diclofenac sodium	17.3 (8.6–34.9)		156					
Diflunisal		55.6						
Fenoprofen calcium		3.7						
Flurbiprofen								
Ibuprofen		82.2	14.7		14.7	3.2	37	
Indomethacin	0.85 (0.53–1.35)	19.0	0.4	3.8 (1.3–11.0)	0.6	1.0	9	5.6 (2.6–12)
Ketoprofen	13.3 (6.25–28.3)							
Ketorolac								
Meclofenamate sodium		9.6						
Mefenamic acid	45.9 (26.8–78.6)	20.7						
Naproxen	13.3 (6.73–26.1)	24.1			9.4		14	
Oxaprozin				29 (14–58)				174 (91–330)
Phenylbutazone		129	1.9	244 (176–337)	32			112 (54–285)
Piroxicam		0.44				1.7		
Sulindac		7.2			18			
Zomepirac		0.7	0.6		216	3.2		
Paracetamol			>100		50			
Dipyrone			>400		21			
Codeine phosphate	8.2 (4.64–14.5)		5.6					
Author	Nakamura et al. (1982)	Pong et al. (1985)	Otterness and Bliven (1985)	Amanuma et al. (1984)	Griswold et al. (1991)	Otterness and Bliven (1985)	Griswold et al. (1991)	Amanuma et al. (1984)

TABLE 7.9 *continued*

Drug	Rat (AA flexion) ED$_{50}$ mg/kg (95% C.I.)	Mouse HAc ED$_{50}$ mg/kg p.o.	Mouse HAc ED$_{50}$ mg/kg p.o. (95% C.I.)	Rat HAc ED$_{50}$ mg/kg p.o. (95% C.I.)	Mice ACh ED$_{50}$ mg/kg p.o. (95% C.I.)
Aspirin	180	227 (121–425)	239 (172–408)	23.4 (11.6–47.6)	23 (11–45)
Diclofenac sodium			2.1 (1.2–3.8)	1.25 (0.72–2.18)	
Diflunisal	6				
Fenoprofen calcium					
Flurbiprofen				0.08 (0.04–0.18)	
Ibuprofen					7.2 (4.3–12.1)
Indomethacin	1.8	11.7 (5.3–26)	0.7 (0.4–1.0)	0.44 (0.22–0.87)	
Ketoprofen				0.86 (0.52–1.41)	
Ketorolac					
Meclofenamate sodium					
Mefenamic acid				30.9 (17.8–53.4)	
Naproxen			78 (62–114)	0.82 (0.48–1.42)	
Oxaprozin	89 (35–222)	45 (18–112)			
Phenylbutazone	12 (5.28)	300 (174–519)	36 (25–70)		
Piroxicam	7				
Sulindac	2.0				
Zomepirac					
Paracetamol	216				
Dipyrone					
Codeine phosphate	23			16.7 (10.6–26.4)	
Author	Otterness and Bliven (1985)	Amanuma *et al.* (1984)	Maeda *et al.* (1977)	Nakamura *et al.* (1982)	Amanuma *et al.* (1984)

TABLE 7.9 *continued*

Drug	Tail pressure mice ED$_{50}$ mg/kg p.o. (95% C.I.)	Rat AgNO$_2$ arth. ED$_{50}$ mg/kg p.o. (95% C.I.)	Rat formalin** ED$_{50}$ mg/kg p.o. (95% C.I.) after	
			Intrathecal admin.	Intraperitoneal admin.
Aspirin	>400	440 (244–792)	27 (18.41)	8 (514–12)
Diclofenac sodium				
Diflunisal				
Fenoprofen calcium				
Flurbiprofen				
Ibuprofen	319 (184–547)	115 (58–230)	18.9 (9.38)	
Indomethacin			2.1 (1–4.3)	3.1 (2.3–4)
Ketoprofen				
Ketorolac	>20	>20	1.9 (1.2–4)	2.6 (1.3–5)
Meclofenamate sodium			5.2 (3–8)	3.0 (2–4)
Mefenamic acid				
Naproxen				
Oxaprozin				
Phenylbutazone				
Piroxicam	306 (184–509)	306 (184–509)		
Sulindac	>300	>300		
Zomepirac				
Paracetamol			5.9 (4–9)	5.5 (2–14)
Dipyrone			257 (163–405)	6.0 (0.8–46)
Codeine phosphate				
Authors		Malmberg and Yaksh (1992a; 1992b)		

PBQ, Phenyl benzoquinone writhing assay; HAc, acetic acid writhing; AA flexation, Adjuvant arthritis flexion text; ACh, Acetyl choline writhing.

*Drugs given 30 min p.o. before 0.3% phenylquinone i.p.

**Drugs given 2 min before s.c. injection of 5% formalin

95% CI = 95% confidence interval

paracetamol is 44 mg/kg. Similar differences in oral potency of these two drugs have been observed in the Randall–Sellito test in the inflamed paws of rats (Rohdenwald et al., 1982) and in the lame-walking test in adjuvant arthritic rats (Higuchi et al., 1986). However, the two drugs are equipotent in the arthritic pain test, which involves measurement of the pain on flexion of the hindlimb of adjuvant-arthritic rats (Römer, 1980). The latter assay represents a pain response to a chronic inflammatory insult, and is considered a clinically-relevant model of human chronic pain (Besson and Guilbaud, 1988), whereas the other assays are clearly acute pain models. Studies with a similar (but of shorter time interval) arthritic pain test model in rats have shown that orally administered diflunisal has 26 times the potency of aspirin p.o., with paracetamol p.o. having the same potency as aspirin (Winter et al., 1979). The ED_{50} values at peak time of response (3 h) were 7.18, 200 and 180 mg/kg for diflunisal, aspirin and paracetamol, respectively (Winter et al., 1979).

Structure–activity studies of certain alkyl esters of O-diphenyl-acetylsalicylic acids have shown that the parent acid is the most effective, having analgesic activity in the mouse phenylquinone model roughly comparable with that of aspirin (Weaver et al., 1961). Ring alkyl substitution confers little advantage on the analgesic activity of either aspirin or salicylate, though clearly in the case of diflunisal there are added benefits from the 2,4-difluorophenyl substitution at the 5 position of salicylic acid (Stone et al., 1977). Likewise, the addition of the 4-fluorophenyl moiety at the 5 position of aspirin improves the analgesic activity of the latter (Stone et al., 1977).

Various alkyl and vinyl phenolic esters of salicylamide have been tested for analgesic activity (given i.p.) with very little enhancement in analgesia except for a slight enhancement noted with a few vinyl derivatives (Table 7.10; Adams and Cobb, 1967). Of experimental interest in relation to structure–activity analysis are the data from Gorino and co-workers (quoted by Adams and Cobb, 1965) on the analgesic compared with anti-inflammatory activities of some salicylamide and cresotino-mide derivatives (Table 7.10). While these do not have full dose–response data they are of interest in showing that (with a few exceptions) analgesic activity of the derivatives does have some relationship to anti-inflammatory activity, and that ortho- and meta-substituted derivatives show no marked increase in potency compared with salicylamide.

Several isosteres of aspirin were tested for analgesic activity against bradykinin in rats, but none, including the 2-thio- derivative of aspirin, were found to have activity even approaching that of aspirin or salicylic acid (Thompkins and Lee, 1975). The aspirin derivative, eterylate (see Chapter 3) has been found to have analgesic activity comparable with that of aspirin (Sunkel et al., 1978).

Overall, the only major improvement in analgesic activity of the salicylates has come with the development of diflunisal.

Dissociation of analgesic from anti-inflammatory activities of aspirin has been shown in the formalin hindpaw behaviour assay in mice (Hunskaar et al., 1986). Aspirin produced less licking of the injected paws during both the early (0 to 5 min) and late (20 to 30 min) phases of high licking activity. In contrast, indomethacin only affected the latter phase. Similar results with aspirin were observed in the orofacial pain response to formalin in rats (Clavelou et al., 1989).

The effects of aspirin have been studied in other less conventional models of analgesia in laboratory animals. Among these, Vinegar et al. (1990) claimed that the subplantar trypsin (250 µg)-induced hyperalgesia model, which gives a biphasic response with peaks at 10 and 150 min in response to pressure on the hind foot, gave ED_{50} values in response to aspirin as well as some NSAIDs and analgesics in the late phase that correlated with their doses required to produce analgesia in humans. The ED_{50} values for aspirin and paracetamol were 10 and 17 mg/kg, respectively.

A grid-shock model in mice that produces vocalisation was found to respond to a variety of opioid receptor agonists, α_2- and 5-HT_{1B}-agonists, muscarinic agonists, and a cholinesterase inhibitor, but not to aspirin (Swedberg, 1994). This would indicate that aspirin has no apparent opioid-like central effects, and this has been claimed for studies using the hotplate test in mice (Ballou et al., 2000). However, other studies have shown that aspirin has effects in this model in mice (Ohdo et al., 1995).

Intrathecal administration of aspirin and other NSAIDs has been found to exhibit analgesic activity (Antignoni, 1992; Jurna et al., 1992; Malmberg and Yaksh, 1992a; 1992b), but the response may be overcome by increasing the intensity of the noxious stimuli (Antignoni, 1992).

Intraperitoneal administration of ATP (160 μg) was employed by Gyires and Torma (1984) to test the responses to writhing of some NSAIDs, including aspirin, and morphine. The relative potencies of all these drugs were about the same, in contrast to results obtained with acetylcholine (see also Table 7.9). Both these agonists were affected by the PG receptor antagonist SC-19220, suggesting that they are PG-mediated responses. The reason for the lack of differences among the NSAIDs and morphine in these models was not apparent from these studies.

Topically applied salicylates, especially methyl salicylate, have long had widespread application for pain relief (Gross and Greenberg, 1948; Hattori, 1978b). Yet until recently, as pointed out by Day *et al.* (1999), relatively few concentration–time data in synovial fluids or other inflamed compartments related the actions of topically-applied NSAIDs to their relative efficacy. Using microdialysis techniques, Cross *et al.* (1998) showed that methyl salicylate penetrated the dermal layers of the skin while the triethanolamine salt of salicylate showed no significant penetration. The pain relief following topical application of a methyl salicylate also paralleled skin uptake of drug (Cross *et al.*, 1999; Figure 7.11). This study shows that skin penetration of methyl salicylate is directly related to pain relief. The selectivity of local penetration of the drug has relationships to the absorption of salts, counter ions and lipophilicity of drugs and complexes, as well as blood supply and recirculation (Roberts and Cross, 1999). Thus any improvements in the pain relief from topical salicylates or other NSAIDs will require assessment of the impact of these parameters and the measurement of skin penetrance and pain relief against the 'bench standard', methyl-salicylate.

Mechanisms of analgesia

Responses in central and peripheral nerves

The evidence shows that salicylates have both central as well as peripheral effects in mediating analgesic responses (Dugas and Parker, 1971; Fink and Irwin, 1982; Jurna *et al.*, 1992; Björkman, 1995; Pini *et al.*, 1997), despite the fact that earlier studies focused on the peripheral modes of action. Thus in these early studies Lim and co-workers, in 1969, showed that aspirin 100 mg/kg i.v. blocked the inhibitory effects of bradykinin on cortical and subcortical non-specific sensory potentials evoked by electrical stimulation of the footpad of chloralose-anaesthetised cats (Lim *et al.*, 1969). The increased electrical activity in the centrum medianum of the thalamus produced by intra-arterial injection of bradykinin 20 to 50 μg was inhibited by morphine 1 mg/kg i.v. (Lim *et al.*, 1969). However, aspirin 80 mg/kg i.v. actually enhanced firing in this region at various times following administration of this drug, which, it was suggested, could be due to effects of the drug on the peripheral bradykinin receptor (Lim *et al.*, 1969).

Sulc and Brozek (1972) and Sulc *et al.* (1973) showed there were small alterations in the EEG of normal subjects following intake of 0.5–1.95 g aspirin. Salicylate 200 mg/kg i.v. inhibited the depressant effects of morphine on EEG in rabbits (Paeile *et al.*, 1974), suggesting there may be effects of salicylates on CNS functions.

Rohdenwald and co-workers (1982) then showed that the cortical evoked potentials elicited following tooth stimulation in healthy human volunteers recorded by electroencephalography (EEG) (from Cz to F8) were markedly reduced following oral ingestion of aspirin 0.5 to 1.0 g in a dose-related fashion. The evoked potentials discriminated between placebo, aspirin and morphine treatments in response to painful stimuli (Buchsbaum *et al.*, 1981). The response to aspirin and other mild analgesics (paracetamol, metizimole) correlated very well with both the subjective measurements of pain (pain report and threshold of stimulation) and drug levels present in the saliva (Rohdenwald *et al.*, 1982). Differentiation of dose-effects of aspirin on EEG in normal men was shown by Fink and Irwin (1982); a dose of 1.95 g aspirin having greater quantitative effects on EEG and symptoms than the lower dose of 0.65 g. The response to a single dose of aspirin 0.6 g was slightly superior to that from paracetamol 1.0 g when measured by EEG dolomimetry and pain scores following dental or gynaecological surgery (Rohdenwald *et al.*, 1982).

In chronic adjuvant-arthritic rats, an intravenous injection of aspirin 50 mg/kg (as the lysine salt)

CHAPTER 7

■ 301

TABLE 7.10

Anti-inflammatory and analgesic activities of salicylamide and cresotinamide derivatives.

Compound number	R^1	R^2	R^3	R^4	LD_{50} i.p. mg/kg	Anti-inflammatory activity		Analgesic activity	
						Dose mg/kg i.p.	Percentage inhibition	Dose mg/kg i.p.	Percentage activity
1	CO.NH₂	OH			660	215	33.30	135	43
2	CO.NH₂	OEt			495	160	30.95	100	91
3	CO.NH₂	O.CH₂.CH=CH₂			320	110	44.28	65	50
4	CO.NH₂	O.CH₂.COOH			>4000	1000	54	1000	10
5	CO.NH₂	O.CH₂.CH=CHPh			>2000	500	28	300	9
6	CO.NH₂	O.CH₂.CH=CHMe			450	150	27	90	16–60
7	CO.NH₂	O.CH₂.CH=CH₂		MeCO.NH	440	184	28.55	104	38.33
8	H	O.CH₂.CH=CH₂		MeCO.NH	480	160	62.32	100	70
9	CO.NH₂	OH	Me		570	190	44.48	115	60
10	CO.NH₂	O.CH₂.CH=CH₂	Me		420	140	45.12	85	84
11	CO.NH₂	OEt	Me		510	170	27.83	100	23
12	CO.NH₂	O.CH₂.CH=CHPh	Me		645	215	34.74	130	24
13	CO.NH₂	O.CH₂.CMe=CH₂	Me		375	125	45.77	75	8
14	CO.NH₂	O.CH₂.CH=CHMe	Me		625	205	40.58	125	44
15	CO.NH₂	OH	Me	CH₂=CH.CH₂	460	155	39	95	57
16	CO.NH₂	OH	Me	MeCO.NH	525	175	51	105	5

From: Gorini, Valcavi and Zonta-Bolego as reviewed by Adams and Cobb (1967).

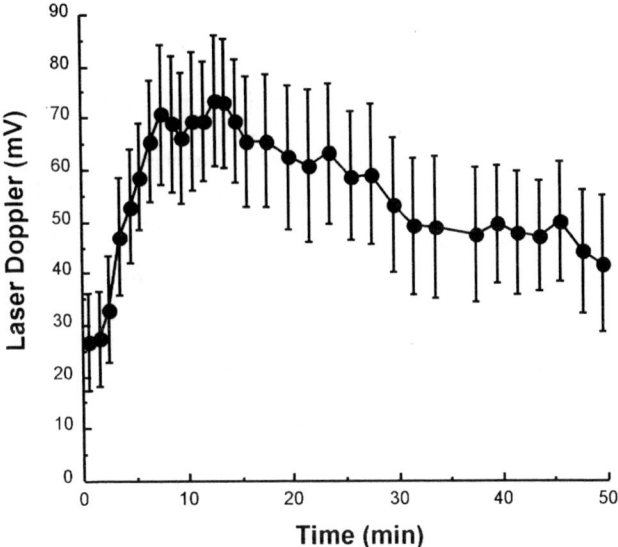

Figure 7.11 Blood flow measured by doppler flow techniques accompanying pain relief following application of a methyl salicylate formulation to human volunteers. (Figure kindly provided by Professor M.S. Roberts and Ms Sheree Cross, Department of Medicine, Princess Alexandra Hospital and University of Queensland, Brisbane, Queensland, Australia.)

CHAPTER 7

progressively reduced the number of discharges recorded (on 13 neurones) in the ventrobasal region of the thalamus, following pain elicited by mild lateral pressure or mobilisation of the affected hindlimb joints (Guilbaud *et al.*, 1982). The initial response was evident at 10 min, with the maximum (35 per cent of control) at 30 min after the administration of the drug, and this persisted for about 45 min thereafter (Guilbaud *et al.*, 1982). The lower dose of aspirin 25 mg/kg i.v. only produced a transient decrease in response at the ventrobasal nucleus, while no effects were apparent with 12.5 mg/kg of the drug (Guilbaud *et al.*, 1982).

Aspirin 100 mg/kg i.v. reduced the firing by nociceptive dorsal horn interneurones of anaesthetised cats, following application of heat to the peripheral receptive field. While this is a near-toxic parenteral dose of the drug in cats, it does illustrate peripheral responses of the drug in addition to the aforementioned central actions.

Some indications of a local anaesthetic-like action of aspirin on peripheral nerves were shown in the studies by Andrews and Orbach (1973). Both aspirin 1 mg/ml and paracetamol, like the local anaesthetics procaine and lignocaine, inhibited the action potentials elicited by acetylcholine in the perfused rabbit-liver preparation (Andrews and Orbach, 1973). Morphine, in contrast, failed to cause any effects, further illustrating the specificity of this by comparison with that of the weak non-narcotic analgesics.

A considerable amount of evidence suggests that aspirin and paracetamol have central effects that constitute an important part of their anti-nociceptive activity. A variety of behavioural electrophysiological and biochemical evidence exists that confirms this. Hunskaar and co-workers (1986) showed that aspirin and paracetamol exhibited a dose-related response in the hotplate test in mice, which was suggestive of central effects. Furthermore, a formalin-induced mouse model in which there is a wind up following spontaneous activity of C fibres in the second phase was found to be responsive to aspirin and paracetamol (Hunskaar *et al.*, 1985). Björkman and co-workers (Björkman *et al.*, 1994a; Björkman, 1995) showed that the writhing induced by visceral stimulation could be inhibited by subcutaneous administration of sodium salicylate and several other non-steroidals and paracetamol. In the case of sodium salicylate this could be blocked by naloxone, antagonist. This suggested that there might be some effects on opioid transmission systems. Spinal administration of aspirin antagonised the hyperalgesia caused by the activation of spinal glutamate or substance P receptors (Björkman *et al.*, 1994b).

In the formalin test in rats, the behavioural response elicited by flinching of the injected paw following injection of formalin was found to be inhibited by aspirin with an ID_{50} of 27 mg/kg intrathecal and by paracetamol with approximately 10-fold greater ID_{50} (Malmberg and Yaksh, 1992a). Malmberg and Yaksh (1992b) also showed that the thermal hyperalgesia induced by intrathecal administration in rats of NMDA, AMPA or substance P was blocked by aspirin with an ID_{50} of 100 nmol.

Aspirin 10 to 500 µg given intrathecally to rats was found to decrease in a dose-related fashion C fibre reflex elicited by electrical stimulation within the territory of the sural nerve recorded from the ipsilateral biceps femoralis muscle in anaesthetised animals (Bustamante *et al.*, 1997). Similar effects were noted with other NSAIDs. It was concluded from these studies that NSAIDs have significant antinociceptive effects at the spinal level are mediated by NMDA receptors (Björkman *et al.*, 1994a; Björkman, 1995).

In general there is convincing evidence that aspirin, like other NSAIDs, has effects at spinal or higher levels of the central nervous system (Fink and Irwin, 1982), and it is likely that supraspinal mechanisms play a significant role in analgesic activity that may or may not be related to effects on prostaglandins as well as on serotonergic and noradrenergic pathways (Bannwarth *et al.*, 1995; Björkman, 1995; Sandrini, 1999). There is evidence that the antinociceptive effects of NSAIDs including aspirin can be differentiated from their anti-inflammatory effects, providing further evidence for their central antinociceptive effects (McCormack and Brune, 1991). There is also considerable evidence that NMDA receptors may be implicated in the action of NSAIDs.

An interesting effect of calcium in antagonising the aspirin analgesia was observed in an ultrasonic algesic test performed in human volunteers (Misawa *et al.*, 1985). The authors showed that the threshold pain relief elicited by aspirin increased progressively for a period up to 90 minutes after administration of 1 g of aspirin and then progressively declined for the next 90 minutes. At all points where the pain threshold was measured after administration of calcium gluconate 1.5 g at 30 minutes and another 1.5 g at 90 minutes it was completely suppressed, suggesting that calcium totally blocked the analgesic effect of aspirin. The authors showed that there were no significant effects of calcium gluconate administration on the plasma levels of aspirin, and indeed the salicylate concentration even progressively increased for up to 180 min after aspirin administration in both aspirin only as well as aspirin plus calcium gluconate treated individuals. The mechanism of this antagonistic effect of calcium on aspirin analgesia is not known.

Prostaglandins

Actions of the salicylates in relation to release and effects of bradykinin and production of COX-2 derived prostaglandins in peripheral tissue have been discussed previously (p. 250). Figure 7.10 shows the possible sites of action of aspirin and other analgesics on various afferent and efferent pathways of pain mediation. The roles of COX-2 as well as COX-1 in mediating or modulating pain pathways are now well established (Ballou *et al.*, 2000).

The prostaglandin involvement in analgesic activity and related drugs received impetus during the earlier part of the 1970s, particularly following the studies of Collier and Schneider (1972) and Ferreira (1972). However, it became apparent that several complexities existed in the prostaglandin hypothesis, which could not be entirely accounted for by the inhibition of prostaglandin-mediated pain by salicylates. First, there was the problem that in some types of experimental pain (especially the electrical stimulation of teeth of anaesthetised cats; Haegerstam and Edwall, 1977) there is a clear prostaglandin-independent component. Also, it became apparent that prostaglandin $F_{2\alpha}$ could antagonise the pain enhancing actions of PGE_2 (Juan and Lembeck, 1977). Thus, it could be argued that drugs that inhibit the synthesis of both PGE_2 and $PGF_{2\alpha}$ might not have quite the benefits that might be expected, or at least not as much as selective inhibitors of PGE_2 production. Studies by Ferreira (1972) suggest that PG-mediated alterations in nucleotide production may more satisfactorily explain the effects of the prostaglandin component and actions of aspirin and related drugs thereon. This could be an important part of the central actions of aspirin and other salicylates in analgesia (Ferreira, 1972).

Inhibition of PGE_2 production in microglial cells by aspirin has been shown to be potentiated by caffeine and paracetamol (Fiebich *et al.*, 2000), highlighting the potential for drug interactions between these agents in mediating their analgesic activity.

Serotonin production in the CNS

The analgesic activity of morphine has been shown in part to be related to its effects in stimulating the production of serotonin (5-hydroxytryptamine) in the CNS (Messing and Lytle, 1977; Botting and Morrinan, 1982; Pini *et al.*, 1997). In addition, other studies (Collier and Schneider, 1972; Ferreira, 1972) suggest that the manipulation of serotonin levels in the brain (e.g. by dietary variations) induces parallel responses in pain susceptibility. The synthesis of serotonin depends on the tryptophan available in the brain, which in turn depends on the free levels of the amino acid in the circulation (i.e. from plasma) (Moir and Eccleston, 1968; De Montis, 1977). Since salicylate displaces tryptophan from its binding sites on albumin, it is possible that the increased free tryptophan in the circulation caused by salicylate-induced displacement of albumin-bound tryptophan and inhibition of the liver tryptophan-metabolising enzyme, tryptophan pyrrolase (Badawy and Smith, 1972; Badawy, 1982), could cause an increase in the level of this amino acid in the brain, so increasing serotonin production in the CNS. This could represent an efferent pathway of modulation of pain responses by the salicylates and other analgesics (Figure 7.10).

Evidence in support of this concept has come from studies by Tagliamonte and co-workers (1973) showing that salicylate 50 to 450 mg/kg i.p. administered to rats resulted in a dose-related increase in brain tryptophan with an increase in both serotonin and its metabolite 5-hydroxyindole acetic acid levels, with most marked changes occurring in the hippocampus, hypothalamus and brainstem. Later studies confirmed the salicylate-induced rise in brain levels of tryptophan, serotonin and 5-hydroxy-indole acetic acid, accompanied by an increase in serum tryptophan (Iwata *et al.*, 1975).

Enhanced urinary output of tryptophan metabolites has been shown both in arthritic patients and to be related to therapy with aspirin or cortisone (Bruckner *et al.*, 1972). Also, free tryptophan levels in the plasma of patients with fibrositis on a balanced diet of this amino acid were found to correlate with the severity of subjective pain (Maldofsky and Warsh, 1978).

Actions on nerve membranes and functions

Levitan and Barker (1972a; 1972b; 1972c) reported that a wide variety of salicylates exhibited effects on the permeability of the large molluscan buccal ganglionic neurones. These drugs caused a reversible, dose-dependent decrease in Na^+, Li^+ and Cl^- ions, and an increase in K^+ conductance (Levitan and Barker, 1972a; 1972b; 1972c). The effects on K^+ and Cl^- permeability were responsible for an increase in the membrane potential. Similar changes in permeability to K^+ and Cl^- ions by salicylate had been previously shown in erythrocytes (Wieth, 1969). This suggested that the selective effects of salicylate on permeability to these ions might be a generalised membrane disruption phenomena. Shortly after publication of the studies of Levitan and Barker, McLaughlin (1973) reported that salicylate, when added to phospholipid layers, decreased the conductance of negative permiant species and increased that of positive permanent species. This could be the basis for the selective permeability of Cl^- and K^+ ions observed in the molluscan neurones (McLaughlin, 1973). The effects on membrane permeability were found to be independent of the phospholipid composition of the bilayer. McLaughlin advanced a molecular mechanism for these drug effects suggesting that salicylate undergoes a hydrophobic adsorption on the insulating bilayer with concomitant production of a negative surface potential. The actions of salicylates on membrane conductance appear to be due to the ionised form of the drug, since similar effects were observed at pH 5.0 and 7.0 (Levitan and Barker, 1972a). These

CHAPTER 7

effects of the salicylates were found to be related to both lipophilicity and pK_a, and were essentially independent of steric effects (Levitan and Barker, 1972a). Direct membrane effects of salicylates on membranes (Li *et al.*, 1999) represents one effect of these drugs.

The relative potencies of different salicylates in causing changes in membrane potential were correlated with their analgesic activities (Levitan and Barker, 1972a). These authors concluded that the enhanced membrane potential and conductance due to increased K^+ and decreased Cl^- permeability caused by the salicylates could decrease the probability of generating action potentials in nerves, and hence reduce the effectiveness of synaptic output. More recent studies on the rabbit vagus (Anner *et al.*, 1970; Neto, 1980), frog node of Ranvier (Attwell *et al.*, 1979), frog sciatic (Ricciopo Neto, 1980) and the giant squid and crayfish (Neto and Tarashi, 1976) axons have provided further evidence for direct inhibitory actions of the salicylates on nerve impulse conduction, although the mechanisms of the drug effects vary with individual preparations as well as with the drug employed.

Anner and co-workers (1970) found that salicylate 2 to 10 mmol/1 block the conduction of B and C fibres in the rabbit desheathed vagus preparation *in vitro*, coincidental with a loss in intracellular potassium concentration. Changes in permeability to K^+ ions have, however, been suggested as being responsible for a hyperpolarisation seen at concentrations of 2 to 10 mmol/l; salicylate and its 5-bromo derivative induce a pronounced depolarising action (Neto, 1980). This effect has been ascribed to the drug-induced inhibition of the Na^+/K^+ electrogenic pump. A salicylate-induced inhibition of this pump has been shown from radiotracer studies of Na^+ and K^+ movements in frog nerves (Hurlbut, 1965). Neto (1980) suggests that this could be due to effects of the drug on uncoupling oxidative phosphorylation, so reducing ATP supply for operations of this pump – evidence is supported by changes observed by Anner and co-workers (1970) in the levels of inorganic phosphate induced by salicylate in rabbit vagus nerves at concentrations that coincided with reduction in action potential. The potent uncoupler, 2,4-dinitrophenol, has also been found to induce pronounced depolarisation, thus providing further support for the relationship between uncoupling activity and the effects of salicylate-like drugs on electrogenic pump activity (Neto, 1980). Inhibitory effects of 5-bromosalicylate 0.1 to 1.5 mmol/l and salicylate 10 mmol/l on both peak transient Na^+ conductance and steady-state K^+ conductance have been observed in voltage clamp studies in the crayfish and giant squid axons (Neto and Tarashi, 1976). The blockade in conductance by 5-bromosalicylate in these preparations was reversible, since internal washing of the drug-treated preparation restored nerve activity to normal (Neto and Tarashi, 1976). The results of washing studies in these experiments suggested that 5-bromosalicylate acted on a membrane site more accessible from the inside than from the outside of the nerve. However, Attwell and co-workers (1979) showed that salicylate had equal effects on sodium current when applied externally or internally to the frog Ranvier node preparation. These authors (Attwell *et al.*, 1979) were unable to observe any effects of salicylate 5 to 40 mmol/l on the amplitude of the action potential in this preparation as observed in other preparations by different authors (Anner *et al.*, 1970; Schorderet and Straub, 1971; Neto and Tarashi, 1976; Neto, 1980), although they did observe a slowing of the falling phase of the action potential. Aside from this, the evidence clearly indicated that salicylates in moderately high concentrations inhibit nerve transmission. Such effects are manifest at the membrane levels and resemble the actions of local anaesthetics, and could account for some of the analgesic actions of these drugs.

Reduction in ATP accompanied by increase in extracellular adenosine may inhibit pain signalling activation by adenosine of its receptors in pain pathways (Sawynok and Liu, 2003).

Further evidence for the possibility of effects of aspirin or salicylate on the CNS comes from studies of their actions on electroencephalography (EEG). Salicylate 200 mg/kg i.v. has been found to antagonise the depressant effects of morphine on the EEG in rabbits. Since aspirin 5.0 to 10.0 mg/kg s.c. has also been shown to antagonise a small part of the antinociceptive actions of morphine in mice, it appears that some of the actions of aspirin may be common with those of morphine at the level of the CNS and, in turn, this might be reflected in the drug-induced alterations in the EEG. Some (albeit small) alterations have been observed in the EEG of man following oral ingestion of aspirin 0.5 to 1.95 g, but, as with the aforementioned studies in animals, the exact meaning of these observations *vis-à-vis* the analgesic actions of the drugs is uncertain.

Other biochemical changes

Administration of salicylic acid 300 or 450 mg/kg p.o. to mice has been found to reduce (by about 33 to 36 per cent) the levels of noradrenaline in the brain, but only after inhibition by disulfiram of the rate-limiting enzyme, dopamine β-hydroxylase, involved in the synthesis of these catecholamines (Pfeiffer *et al.*, 1967). This reduction in noradrenaline levels was correlated with analgesic activity, suggesting that the drug may affect noradrenaline turnover in a manner analogous with that of morphine (Paalzow, 1973). This may also be an action of salicylates, like that of paracetamol, in analgesia via activation of efferent pathways (Figure 7.10).

Salicylates have been shown to inhibit glutamate decarboxylase from *Escherichia coli* and rat brain (Gould *et al.*, 1963; Smith *et al.*, 1963; Gould and Smith, 1965; McArthur and Smith, 1969). Since this enzyme catalyses the formation of γ-aminoisobutyric acid (GABA) and both this and glutamate are, respectively, inhibitory and stimulatory neurotransmitters, it is possible that salicylates could influence functions in the CNS by alterations in the production of these neurotransmitters. Salicylates could also affect the actions of glutamate (Huang *et al.*, 2004).

Salicylates may also influence the activities of the cholinesterases, especially the pseudo-cholinesterases, which have important functions in the CNS. Aspirin 2 mg/ml has been found to inhibit the histochemically determined activity of pseudocholinesterases in the rat brain *in vitro* (Delbarre *et al.*, 1971). Moreover, salicylate has been shown to inhibit human erythrocyte cholinesterase *in vitro* in concentrations ranging from 6.1 to 300 mmol/l (Shukuya, 1954). This inhibitory effect on cholinesterase activity may explain the apparent parasympathomimetic actions of aspirin (e.g. on mouse pupil, guinea pig bronchi), which could be of relevance to the analgesic actions of the drug.

There may also be components of the neuroimmunoinflammatory responses that are affected by salicylates that contribute to pain control. Thus, Schmelz *et al.* (2000) showed that topically-applied acetylsalicylate to the skin of rats attenuated neurogenic vascularisation.

ANTIPYRESIS

Antipyresis is a generalised response of inflammatory reactions. Current concepts of the regulation of fever responses focus on the view that the initiating events involve the release, following stimulation of leucocytes with microbial organisms or in response to injury, of what was originally described as endogenous pyrogen but is now known to comprise several cytokines (IL-1α, IL-1β, TNFα, IL-6, IL-18, IFN-γ). These act on receptors in the region close to the organum vasculosum laminae terminalis (OVTL), adjacent to the hypothalamus, and stimulates the synthesis of PGE$_2$ (Hellon *et al.*, 1991; Cao *et al.*, 1998; Kluger *et al.*, 1998) and release of glutamate (Huang *et al.*, 2001). The PGE$_2$ acts on EP$_3$ receptors, an effect that has been shown to be absent in mice lacking the gene for these receptors. Since COX-2-selective drugs (e.g. rofecoxib) appear effective as antipyretics (Schwartz *et al.*, 1999) it is possible that the PGE$_2$ stimulated by pyrogens probably is generated via COX-2.

Studies of COX-2 knock-out mice and the effects of the highly selective COX-2 inhibitor rofecoxib indicate that LPS and pyrogenic cytokines probably mediate their actions by upregulating COX-2 production of PGE$_2$.

Comparative efficacy of different salicylates

The moderately potent temperature-lowering effects of salicylates have been well known for centuries (Chapter 1; Hanzlik, 1927; Gross and Greenberg, 1948; Gorini *et al.*, 1963; Cranston, 1979; Milton, 1992). The first experimental evidence of the antipyretic effects of salicylate in man was obtained by Stone in 1763, and was later shown by Buss in 1876 in rabbits and patients with typhoid fever (see Chapter 1). Orally ingested aspirin has 1.5 times the antipyretic activity of salicylate (on a molar basis)

CHAPTER 7

in man (Seed, 1965). The differences are more marked in laboratory animals (Rainsford and Whitehouse, 1980a; Table 7.11), and it should be noted that the potency of aspirin depends on the type of formulation employed (Rainsford and Whitehouse, 1980b), no doubt reflecting differences in pharmacokinetics of these preparations. Diflunisal has about 1.5 times the antipyretic activity of aspirin in rats and rabbits injected with bacterial endotoxin, while clinical studies indicate about a similar potency in humans (Tempero *et al.*, 1978; Dascombe, 1984). Benorylate and the phenyl esters of aspirin have much lower antipyretic activity in rats compared with that of aspirin, but the phenyl ester of salicylic acid is actually more potent than its parent acid (Rainsford and Whitehouse, 1980b).

Studies in laboratory animals are usually acute experiments – i.e. the drug response to a pyrogen, usually yeast, injected parenterally is followed for several hours (Loux *et al.*, 1977; Rainsford and Whitehouse, 1980b: Otterness and Bliven, 1985). Some typical data on the effects of salicylates and other antipyretic drugs are shown in Table 7.11.

Mode of actions of salicylates

Sofia and co-workers (1973) have investigated the effects of chronic aspirin administration in lowering elevated hindpaw temperature of adjuvant-arthritic rats. This is a chronic model of pyresis that probably has appreciable local as well as central components, since the uninjected hindpaw has an appreciable thermal output as well as the adjuvant-injected paw. This study illustrates an aspect of temperature regulation: that the salicylates exert peripheral effects by way of effects on components of tissue respiration as well on the hypothalamus (Smith, 1966; Cranston *et al.*, 1971b; Beckman and Rozkowska-Ruttiman, 1974; Perlow *et al.*, 1975; Schoener and Wang, 1975; 1976; Avery and Penn, 1976; Cranston *et al.*, 1976; Spławinski *et al.*, 1977), both of which contribute to the total antipyretic actions of these drugs. Clearly the model chosen, the route of administration and the type of pyrogen injected influence the response to the salicylates.

Milton (1976; 1992), Cooper and co-workers (1982), Done (1983), Hellon *et al.* (1991), van Arman *et al.* (1991) and Aronoff and Neilson (2001) have reviewed the pathogenesis of fever and the mode of action of antipyretic agents. Until about a decade ago, the view was that prostaglandin E_2 is a mediator of pyrogen fever, and that antipyretic agents such as aspirin exert their effects in pyrogen fever primarily through their inhibitory actions on PGE_2 production. This view has been refined since, but the central role of the stimulation of PGE_2 production by pyrogens is in essence the currently held view.

Based on studies principally with salicylate (which though ineffective *in vitro* as an inhibitor of PG synthesis does have weak effects *in vivo*) and some selective inhibitors of PG synthesis, Cranston and co-workers (1971a; 1971b; 1975; 1979) concluded that E-type prostaglandins do not alone play a significant part in pyrexia or hypothalamic thermoregulation. Studies by these authors with 2,4-dinitrophenol and with salicylate analogues have shown that (a) uncoupling of oxidative phosphorylation is not a mode of action of these compounds (since intraventricular injections of 2,4-dinitrophenol fail to induce significant anti-pyresis), and (b) aspirin and salicylate are the only two agents of various benzoic acids studied that exhibit antipyretic activity (Cranston *et al.*, 1971a).

It is possible that interpretations of the actions of salicylates on hypothalamic PGE_2 synthesis by these authors may be model dependent and could imply that there are non-prostaglandin as well as prostaglandin-dependent systems mediating the antipyretic effects of these agents.

Aspirin has been shown to abolish or reverse the fever induced by injection of *E. coli* endotoxin or endogenous leucocytic pyrogen (EP) into the anterior region of either conscious rats (Spławinski *et al.*, 1977) or urethane-anaesthetised cats (Schoener and Wang, 1975; 1976). The studies indicate that both the pyrogen and aspirin exert their respective effects in the pre-optic/anterior hypothalamic area, but that antipyresis is not due to competition by aspirin with 'receptor' sites for leucocyte pyrogen (Schoener and Wang, 1975; 1976). The receptors on thermoregulatory neurones in this region of the brain are directly antagonised by aspirin when the drug is given before the pyrogen (Schoener and Wang, 1975). The evidence from these authors and from subsequent studies indicates that the site of action of aspirin and salicylate is in the pre-optic/anterior hypothalamic area (Beckman and Rozkowska-Ruttiman, 1974; Avery and Penn, 1976; Nakashima *et al.*, 1985). Thus Beckman and

TABLE 7.11

Effects of aspirin compared with other NSAIDs on yeast-induced hyperthermia in rats.

Drug	ED_{50} mg/kg, p.o. (85% C.I.)		
Aspirin	36.7 (19.0–95.7)	78.3 (30.6–200)	36.7 (19.0–95.7)
Aminopyrine			7.35 (4.24–12.8)
Flufenamic acid	0.68 (0.21–1.34)		
Ibuprofen	0.43 (0.25–0.70)	5.53 (0.32–94.6)	
Indomethacin	0.31 (0.06–1.80)	1.21 (0.41–3.53)	0.23 (0.1–0.48)
Phenylbutazone	9.13 (2.48–33.7)	12.4 (5.63–27.4)	
Tolmetin, Na	7.16 (2.52–13.9)		
Authors	Shimizu *et al.* (1975)	Atkinson and Leach (1976)	Nakamura *et al.* (1982)

Fever induced by 15 ml/kg 15% bakers' dry yeast s.c. 17 h before drug administration. Peak reduction of rectal temperature is at 2 h for ibuprofen and some other NSAIDs, but e.g. is 3 h for indomethacin and 2–4 h for tolmetin.

Rozkowska-Ruttiman (1974) have shown that iontophoretic injection of salicylate or acetylsalicylate into the hypothalamic region caused excitation of the warm-sensitive neurones, which are the cells that produce hypothermia in response to neurotransmitters. Avery and Penn (1976) have shown that injection of serotonin into the hypothalamus induces an increase in the colonic temperature of rats. Since salicylate induces an increase in levels of this transmitter in the hypothalamus (as a consequence of displacement of tryptophan from albumin – see p. 278 and Francesconi and Mager, 1975), it would seem to mitigate against salicylate-induced changes in serotonin levels being of importance in inducing hypothermia, and possibly antipyresis, as suggested by Avery and Penn (1976).

Certain endogenous peptides (arginine vasopressin, neurotensin) have been found to exhibit pyretic activity in the CNS (Cooper *et al.*, 1982; Mason *et al.*, 1982; Kluger *et al.*, 1998). The hyperthermia induced by intracisternal administration of neurotensin 10 μg to mice maintained at 6°C was not blocked by aspirin 100 mg/kg s.c., but was affected by indomethacin 5 mg/kg s.c. (Mason *et al.*, 1982). Mason *et al.* (1982) concluded that, because indomethacin inhibits synthesis of prostaglandins within the CNS as well as in peripheral organs whereas aspirin only acts in the periphery (Abdel-Halim *et al.*, 1978), it appears that the neurotension-induced hyperthermia in a cold environment is enhanced by reduction of prostaglandins in the CNS.

In conclusion, while salicylates affect the cytokine-mediated production in the hypothalamus of PGE_2 as a major part of their mechanism of antipyresis possibly by blocking actions of $NF_\kappa B$ (Grilli *et al.*, 1996) it is not possible to state precisely what other mechanisms are involved in the antipyretic actions of the salicylates. The reversal of PGE-mediated hyperthermia (e.g. from prostaglandins arriving in the blood at the hypothalamus from inflamed sites) may form a partial action of aspirin by means of influences of the drug in inflamed tissues. Salicylates may also bring about effects on the hypothalamus by influencing noradrenaline and 5-HT turnover (Francesconi and Mayer, 1975; Avery and Penn, 1976; Pini *et al.*, 1997) as well as suppression of glutamate in the OVTL (Huang *et al.*, 2004).

REFERENCES

Abdel-Halim, M.S., Sjoquist, B. and Anggord, E. 1978, Inhibition of prostaglandin synthesis in rat brain. *Acta Pharmacologica et Toxicologica (Kobenhavn)*, **43:** 266–272.

Abramson, S.B. 1987, Inhibition of neutrophil activation by NSAIDs. EULAR *Journal of Education and Information in Rheumatology*, **16:** 15–17.

Ackerman, N.R., Rooks, W.H. II, Schott, L., Genant, H., Maloney, P. and West, E. 1979, Effects of naproxen on connective tissue changes in the adjuvant arthritic rat. *Arthritis and Rheumatism*, **22:** 1365–1374.

Adams, S.S., Bough, R.G., Cliffe, E.E. *et al.* 1970, Some aspects of the pharmacology, metabolism, and toxicology of ibuprofen. *Rheumatology and Physical Medicine*, **11:** 9–22.

Adams, S.S. and Cobb, R. 1967, Non-steroidal anti-inflammatory drugs. *In:* G.P. Ellis and G.B. West (eds), *Progress in Medicinal Chemistry* pp. 59–138. London: Butterworths.

Adams, S.S., McCullough, K.F. and Nicholson, J.S. 1969, The pharmacological properties of ibuprofen, an anti-inflammatory, analgesic and antipyretic agent. *Archives Internationale Pharmacodynamie et de Thèrapie*, **178:** 115–129.

Alexander, S.P.H. and Peters, J.A. 2000, *TiPS Receptor and Ion Channel Nomenclature. Supplement 2000*. London: Elsevier/Current Trends.

Allison, A.C. and Davies, P. 1974, Lysosomes. *In:* G.P. Velo, D.A. Willoughby and J.P. Giroud (eds), *Future Trends in Inflammation*, Vol. 1, pp. 449–480. Padua: Piccin Books.

Alpert, D. and Vilcek, J. 2000, Inhibition of IκB kinase activity by sodium salicylate in vitro does not reflect its inhibitory mechanism in intact cells. *Journal of Biological Chemistry*, **275:** 10925–10929.

Amann, R. and Peskar, B.A. 2002, Anti-inflammatory effects of aspirin and sodium salicylate. *European Journal of Pharmacology*, **447:** 1–9.

Amann, R., Egger, T., Schuligoi, R., Heinemann, A. and Peskar, B.A. 2001, Sodium salicylate enhances the expression of cyclooxygenase-2 in endotoxin-stimulated human mononuclear cells. *European Journal of Pharmacology*, **433:** 129–134.

Amanuma, F., Okuyama, S. and Orikasa, S. 1984, The analgesic and antipyretic effects of a nonsteroidal anti-inflammatory drug, oxaprozin, in experimental animals. *Folia Pharmacologia Japonica*, **83:** 345–354.

Amin, A.R., Attur, M. and Abramson, S.B. 1998, Regulation of nitric oxide and inflammatory mediators in human osteoarthritis-affected cartilage: implication for pharmacological intervention. *In:* G.W. Rubanyl (ed.), *The Pathophysiology and Clinical Application of Nitric Oxide*, pp. 397–412. Harvard Academic Publishers.

Amin, A.R., Attur, M., Patel, R.N., Thakker, G.D., Marshall, P.J., Rediske, J., Stuchin, S.A., Patel, I.R. and Abramson, S.B. 1997, Superinduction of cyclooxygenase-2 activity in human osteoarthritis-affected cartilage. *Journal of Clinical Investigation*, **99:** 1231–1237.

Amin, A.R., Attur, M.G., Pillinger, M. and Abramson, S.B. 1999, The pleiotropic functions of aspirin: mechanisms of action. *Cellular and Molecular Life Science*, **56:** 305–312.

Amin, A.R., Vyas, P., Attur, M., Leszczynska-Piziak, J., Patel, I.R., Weissmann, G. and Abramson, S.B. 1995, The mode of action of aspirin-like drugs: effect on inducible nitric oxide synthase. *Proceedings of the National Academy of Sciences USA*, **92:** 7926–7930.

Amkraut, A.A., Solomon, G.F. and Kramer, H. 1971, Stress, early experience and adjuvant-induced arthritis in the rat. *Psychosomatic Medicine*, **33:** 203–214.

Ammendola, G., Di Rosa, M. and Sorrentino, L. 1975, Leukocyte migration and lysosomal enzymes release in rat carrageenan pleurisy. *Agents and Actions*, **5:** 250–255.

Amphlett, C.B., Adams, G.E. and Michael, B.D. 1968, *In:* R.F. Gould (ed.), *Advances in Chemistry*, Series No. 81, pp. 231–250. Washington: American Chemical Society.

Anastassiades, T.P. 1973, Effect of a synthetic hexosamine derivative on mucopolysaccharide synthesis by human capsule and synovium. *Biochemcial Pharmacology*, **22:** 3013–3023.

Anastassiades, T.P., Chopra, R., Law, C. and Wong, E. 1998, *In vitro* suppression of transforming growth factor-beta induced stimulation of glycosaminoglycan synthesis by acetylsalicylic acid and its reversal by misoprostol. *Journal of Rheumatology*, **25:** 1962–1967.

Anastassiades, T.P., Ley, J., Wood, A. and Irwin, D. 1978, The growth kinetics of synovial fibroblastic cells from inflammatory and noninflammatory arthropathies. *Arthritis and Rheumatism*, **21:** 461–466.

Anderson, A.J. 1968, The effect of anti-inflammatory drugs on the enzymic activity of a rat liver granular fraction which increases vascular permeability. *Biochemical Pharmacology*, **17:** 2253–2264.

Anderson, A.J. 1970, Lysosomal enzyme activity in rats with adjuvant-induced arthritis. *Annals of the Rheumatic Diseases*, **29:** 307–313.

Anderson, A.J., Brocklehurst, W.E. and Willis, A.J. 1971, Evidence for the role of lysosomes in the formation of prostaglandins during carrageenan-induced inflammation in the rat. *Pharmacological Research Communications*, **3:** 13–19.

Anderson, C.C., Cook, J.L., Kreeger, J.M., Tomlinson, J.L. and Wagner-Mann, C.C. 1999, *In vitro* effects of glucosamine and acetylsalicylate on canine chondrocytes in three-dimensional culture. *American Journal of Veterinary Research*, **60:** 1546–1551.

Andreenko, G.V. and Karabasova, M.A. 1977, Effect of acetylsalicylic acid on fibrinolysis under experimental conditions. *Farmakologiya i Toksikologiya*, **40:** 310–313.

Andrews, W.H. and Orbach, J. 1973, A study of compounds which initiate and block nerve impulses in the perfused rat liver. *British Journal of Pharmacology*, **49:** 192–204.

Anner, B., Ferrero, J.D. and Schorderet, M. 1970, Effects of salicylic and acetylsalicylic acid in inorganic phosphates and impulse conduction in mammalian nerve fibres. *Agents and Actions*, **1:** 249–251.

Antognini, J.F. 1993, Intrathecal acetylsalicylic acid and indomethacin are not analgesic for a supramaximal stimulus. *Anesthesia and Analgesia*, **76:** 1079–1082.

Aronoff, D.M. and Neilson, E.G. 2001, Antipyretics: mechanisms of action and clinical use in fever suppression. *American Journal of Medicine*, **111:** 304–315.

Aronoff, D.M., Boutaud, O., Marnett, L.J. and Oates, J.A. 2003, Inhibition of prostaglandin H2 synthases by salicylate is dependent on the oxidative state of the enzymes. *Advances in Experimental Biology and Medical Sciences*, **525:** 125–128.

Arrigoni-Martelli, E. 1977, *Inflammation and the Anti-inflammatories*. New York: Spectrum.

Arrigoni-Martelli, E. and Restelli, A. 1972, Release of lysosomal enzymes in experimental inflammations: effects of anti-inflammatory drugs. *European Journal of Pharmacology*, **19:** 191–198.

Atkinson, D.C. and Collier, H.O.J. 1980, Salicylates: molecular mechanism of therapeutic action. *Advances in Pharmacology and Chemotherapy*, **17:** 233–288.

Atkinson, D.C. and Leach, E.C. 1976, Anti-inflammatory and related properties of 2-(2,4-dichlorophenoxy)phenylacetic acid (fenclofenac). *Agents and Actions*, **6:** 657–666.

Attur, M.G., Patel, R., DiCesare, P.E., Steiner, G.C., Abramson, S.B. and Amin, A.R. 1998, Regulation of nitric oxide production by salicylates and tenidap in human OA-affected cartilage, rat chondrosarcomas and bovine chondrocytes. *Osteoarthritis and Cartilage*, **6:** 269–277.

Attwell, D., Bergman, C. and Ojeda, C. 1979, The action of salicylate ions on the frog node of Ranvier. *Journal of Physiology*, **295:** 69–81.

Auclair, C., Gautero, H. and Boivin, P. 1980, Effects of salicylate–copper complex on the metabolic activation in phagocytozing granulocytes. *Biochemical Pharmacology*, **29:** 3105–3109.

Austen, K.F. 1963, Immunological aspects of salicylate action – a review. *In:* A.St.J. Dixon, B.K. Martin, M.J.H. Smith and P.N.H. Wood (eds), *Salicylates. An International Symposium*, pp. 161–169. London: Churchill.

Avery, D.D. and Penn, P.E. 1976, Interaction of salicylate and body temperature changes caused by injections of neurohumours into the anterior hypothalamus: possible mechanisms in salicylate antipyresis. *Neuropharmacology*, **15:** 433–438.

Bach, M.K., Brashler, J.R. and Johnson, M.A. 1985, Inhibition by sulfasalazine of LTC synthetase and of rat liver glutathione S-transferases. *Biochemical Pharmacology*, **34:** 2695–2704.

Badawy, A.A.-B. 1982, Mechanisms of elevation of brain tryptophan concentration by various doses of salicylate. *British Journal of Pharmacology*, **76:** 211–213.

Badawy, A.A.-B. and Smith, M.J.H. 1972, Changes in liver tryptophan and tryptophan pyrrolase activity after administration of salicylate and tryptophan to the rat. *Biochemical Pharmacology*, **21:** 97–101.

Baenziger, N.L., Dillender, M.J. and Majerus, P.W. 1977, Cultured human skin fibroblasts and arterial cells produce a labile platelet-inhibitory prostaglandin. *Biochemical and Biophysical Research Communications*, **78:** 294–301.

Baggiolini, M., Bretz, U., Dewald, B. and Beigenson, M.E. 1978, The polymorphonuclear leukocyte. *Agents and Actions*, **8:** 3–10.

Bailey, S.M., Fauconnet, A.L. and Reinke, L.A. 1997, Comparison of salicylate and D-phenylalanine for detection of hydroxyl radicals in chemical and biological reactions. *Redox Report*, **3:** 17–22.

Ballie, A.J. and Sim, A.K. 1972, The effects of some synthetic compounds on in vitro fibrinolytic activity measured by different methods and the relevance to activity in vivo. *Thrombosis et Diathesis Haemorrhagica*, **28:** 351–358.

Ballou, L.R., Botting, R.M., Goorha, S., Zhang, J. and Vane, J.R. 2000, Nociception in cyclooxygenase isozyme-deficient mice. *Proceedings of the National Academy of Science USA*, **97:** 10272–10276.

Balsinde, J., Balboa, M.A. and Dennis, E.A. 2000, Identification of a third pathway for arachidonic acid mobilization and prostaglandin production in activated P388D(1) macrophage-like cells. *Journal of Biological Chemistry*, **275:** 22544–22549.

Bannwarth, B. and Dehais, J. 1991, Concentration des anti-inflammatoires non stéroidiens dans le cartilage articulaire chez l'homme. *Revue du Rhumatism*, **58:** 879–882.

Bannwarth, B., Demotes-Mainard, F., Schaeverbeke, T., Labat, L. and Dehais, J. 1995, Central analgesic effects of aspirin-like drugs. *Fundamental and Clinical Pharmacology*, **9:** 1–7.

Baños, G. and Reyes, P.A. 1989, A comparative study of the effect of ten non-steroidal anti-inflammatory drugs (NSAIDs) upon some mitochondrial and platelet functions. *International Journal of Biochemistry*, **21:** 1387–1394.

Barritt, G.J. and Whitehouse, M.W. 1977, Pathobiodynamics: effect of extrahepatic inflammation on calcium transport and drug metabolism by rat liver mitochondria in vitro. *Biochemical Medicine*, **17:** 99–115.

Bassleer, C., Magotteaux, J., Geenen, V. and Malaise, M. 1997, Effects of meloxicam compared to acetylsalicylic acid in human articular chondrocytes. *Pharmacology*, **54:** 49–56.

Bassleer, C.T., Henrotin, Y.E., Reginster, J.L. and Franchimont, P.P. 1992, Effects of tiaprofenic acid and acetylsalicylic acid on human articular chondrocytes in 3-dimensional culture. *Journal of Rheumatology*, **19:** 1433–1438.

Baumann, J., von Bruchhausen, F. and Wurm, G. 1980, Flavonoids and related compounds as inhibition of arachidonic acid peroxidation. *Prostaglandins*, **20:** 627–639.

Beaven, M.A., Horakova, Z. and Keiser, H. 1974, Inhibition by aspirin of ribose conjugation in the metabolism of histamine. *European Journal of Pharmacology*, **29:** 138–146.

Beck, F.W.J. and Whitehouse, M.W. 1974, I. Drug sensitivity of rat adjuvant arthritis, induced with 'adjuvants' containing no mineral oil components. *Proceedings of the Society for Experimental Biology and Medicine*, **146:** 665–669.

Beckman, A.L. and Rozkowska-Ruttiman, E. 1974, Hypothalamic and septal neuronal responses to iontophoretic application of salicylate in rats. *Neuropharmacology*, **13:** 393–398.

Belcher, G. 1979, The effects of intra-arterial bradykinin, histamine, acetylcholine and prostaglandin E1 on nociceptive and non-nociceptive dorsal horn neurones of the cat. *European Journal of Pharmacology*, **56:** 385–395.

Bensley, D.N. and Nickander, R. 1982, Comparative effects of benoxaprofen and other anti-inflammatory drugs on bone damage in the adjuvant arthritic rat. *Agents and Actions*, **12:** 313–319.

Bertelé, V., Falanga, A., Tomasiak, M., Dejana, E., Cerletti, C. and De Gaetano, G. 1983, Platelet thromboxane synthetase inhibitors with low doses of aspirin: possible resolution of the aspirin dilemma. *Science (Washington DC)*, **220:** 517–519.

Bertelli, A. and Schinelli, M.L. 1979, Activity of tolmetin on levels of cyclic nucleotides in experimental pleurisy. *Arzneimittel Forschung*, **29:** 779–781.

Bertelli, A. and Soldani, G. 1979, Polymyxin B-induced oedema in the hind paw of the rat as an assay for anti-inflammatory drugs. *Arzneimittel Forschung*, **29:** 777–778.

Besson, J.-M. and Guilbaud, G. (eds) 1988, The arthritic rat as a model of clinical pain? *Proceedings of the International Symposium on The Arthritic Rat as a Model of Clinical Pain, Saint-Paul de Vence, France, 6–8 June 1988.* Amsterdam-New York-Oxford: Excerpta Medica.

Betts, H., Cleland, L. and Whitehouse, M.W. 1981, *In vitro* studies of anti-inflammatory copper complexes: some difficulties in their chemical interpretation. *In:* J.R.J. Sorenson (ed.), *Inflammatory Diseases and Copper*, pp. 553–562. New Jersey: Humana Press.

Bhargava, A.S. 1971, Effect of anti-inflammatory agents on adjuvant-induced edema modified for primary test. *Pharmacological Research Communications*, **3:** 83–91.

Bhattacharya, D.K., Lecomte, M., Dunn, J., Morgans, D.J. and Smith, W.L. 1995, Selective inhibition of prostaglandin endoperoxide synthase-1 (cyclooxygenase-1) by valerylsalicylic acid. *Archives of Biochemistry and Biophysics*, **317:** 19–24.

Bhatterjee, P. and Eakins, K.E. 1974, Inhibition of the prostaglandin synthetase systems in ocular tissues by indomethacin. *British Journal of Pharmacology*, **50:** 227–230.

Biban, C., Tassani, V., Toninello, A., Siliprandi, D. and Siliprandi, N. 1995, The alterations in the energy-linked properties induced in rat-liver mitochondria by acetylsalicylate are prevented by cyclosporine-A or Mg^{2+}. *Biochemical Pharmacology*, **50:** 497–500.

Binderup, L., Bramm, E. and Arrigoni-Martelli, E. 1976, Actinomycin D peritonitis in rats: a tool for the study of early events in inflammation. *Experientia (Basel)*, **33:** 390–391.

Bitko, V., Velazquez, A., Yang, L., Yang, Y.C. and Barik, S. 1997, Transcriptional induction of multiple cytokines by human respiratory syncytial virus requires activation of NF-kappa B and is inhibited by sodium salicylate and aspirin. *Virology*, **232:** 369–378.

Bjelle, A. and Eronen, I. 1991, The *in vitro* effect of six NSAIDs on the glycosaminoglycan metabolism of rabbit chondrocytes. *Clinical and Experimental Rheumatology*, **9:** 369–374.

Björkman, R. 1995, Central antinociceptive effects of non-steroidal anti-inflammatory drugs and paracetamol. Experimental studies in the rat. *Acta Anaesthesiologica Scandinavica*, **39:** (**Suppl. 103**).

Björkman, R., Hallman, K., Hedner, J., Hedner, T. and Henning, M. 1994a, Hyperalgesia induced by intrathecal NMDA and substance P and AMPA is modulated by non-steroidal anti-inflammatory drugs. *Pain*, **57:** 259–264.

Björkman, R., Hallman, K., Hedner, J., Hedner, T. and Henning, M. 1994b, Acetaminophen blocks spinal hyperalgesia induced by NMDA and substance P. *Pain*, **58:** 265–272.

Blackham, A. and Radziwonik, H. 1977, The effect of drugs in established rabbit monoarticular arthritis. *Agents and Actions*, **7:** 473–480.

Blanco, F.J., Guitian, R., Moreno, J., De Toro, F.J. and Galdo, F. 1999, Effect of anti-inflammatory drugs on COX-1 and COX-2 activity in human articular chondrocytes. *Journal of Rheumatology*, **26:** 1366–1373.

Boger, R.H., Bode-Boger, S.M., Kramme, P., Tsikas, D., Gutzki, F.M. and Frolich, J.C. 1996, Effect of captopril on prostacyclin and nitric oxide formation in healthy human subjects: interaction with low dose acetylsalicylic acid. *British Journal of Clinical Pharmacology*, **42:** 721–727.

Bolam, J.P., Elliot, P.N.C., Ford-Hutchinson, A.W. and Smith, M.J.H. 1974, Histamine, 5-hydroxytryptamine, kinins and the anti-inflammatory activity of human plasma fraction in carrageenan-induced paw oedema in rats. *Journal of Pharmacy and Pharmacology*, **26:** 434–440.

Bollet, A.J. 1961, Inhibition of glucosamine 6-PO_4 synthesis by salicylate and other anti-inflammatory agents in vitro. *Arthritis and Rheumatism*, **4:** 624–631.

Bonta, I.L., Adolfs, M.J. and Parnham, M.J. 1979, Distribution and further studies on the activity of prostaglandin E in chronic granulomatous inflammation. *In:* K.D. Rainsford and A.W. Ford-Hutchinson (eds), *Prostaglandins and Inflammation*, pp. 121–132. Basel: Birkhäuser.

Bonta, I.L., Bult, H., Van Den Ven, L.L.M. and Noordhoek, J. 1976, Essential fatty acid deficiency: a condition to discriminate prostaglandin and non-prostaglandin mediated components of inflammation. *Agents and Actions*, **6:** 154–158.

Bonta, I.L., Bult, H., Vincent, J.E. and Zijlstra, F.J. 1977, Acute anti-inflammatory effects of aspirin and dexamethasone in rats deprived of endogenous prostaglandin precursors. *Journal of Pharmacy and Pharmacology*, **29:** 1–7.

Bonta, I.L., Chrispijin, H., Nordhock, J. and Vincent, J.E. 1974, Reduction in prostaglandin-phase in hind-paw inflammation and partial failure of indomethacin to exert anti-inflammatory effects in rats on essential fatty acid deficient diet. *Prostaglandins*, **5:** 495–503.

Borel, J.F. and Feurer, C. 1975, Chemotaxis of rabbit macrophages in vitro: inhibition by drugs. *Experientia (Basel)*, **31:** 1437–1439.

Borel, J.F. and Feurer, C. 1978, In vivo effects of anti-inflammatory and other drugs on granulocyte emigration in the rabbit skin collection chamber. *Journal of Pathology*, **124:** 85–93.

Borg, D.C. 1965, Transient free radicals from salicylate. *Biochemical Pharmacology*, **14:** 627–631.

Borgeat, P. and Samuelsson, B. 1979, Metabolism of arachidonic acid in polymorphonuclear leukocytes. Structural analysis of novel hydroxylated compounds. *Journal of Biological Chemistry*, **254:** 7865–7869.

Borgeat, P. and Sirios, P. 1981, Leukotrienes: a major step in understanding of immediate hypersensitivity reactions. *Journal of Medicinal Chemistry*, **24:** 121–126.

Boström, H., Bernsten, K. and Whitehouse, M.W. 1964, Biochemical properties of anti-inflammatory drugs. II. Some effects on sulphate-^{35}S metabolism in vivo. *Biochemical Pharmacology*, **13:** 413–440.

Boström, H., Moretti, A. and Whitehouse, M.W. 1963, Studies on the biochemistry of heart valves. I. On the biosynthesis of mucopolysaccharides in bovine heart valves. *Biochimica Biophysica Acta*, **74:** 213–221.

Botting, R. and Morinan, A. 1982, Involvement of 5-hydroxytryptamine in the analgesic action of pethidine and morphine in the mouse. *British Journal of Pharmacology*, **75:** 579–585.

CHAPTER 7

Bowery, B. and Lewis, G.P. 1973, Inhibition of functional vasodilatation and prostaglandin formation in rabbit adipose tissue by indomethacin and aspirin. *British Journal of Pharmacology*, **47:** 305–314.

Bozza, P.T., Payne, J.L., Morham, S.G., Langenbach, R., Smithies, O. and Weller, P.F. 1996, Leukocyte lipid body formation and eicosanoid generation: cyclooxygenaase-independent inhibition by aspirin. *Proceedings of the National Academy of Sciences USA*, **93:** 11091–11096.

Braht, P.C. and Bonta, I.L. 1981, Role of trace elements in hepatic changes during inflammatory conditions. *In:* K.D. Rainsford, K. Brune and M.W. Whitehouse (eds), *Trace Elements in the Pathogenesis and Treatment of Inflammation*, pp. 231–239. Basel: Birkhäuser.

Braht, P.C., Brausberg, J.I. and Bonta, I.L. 1980, Antiinflammatory effects of free radical scavengers and antioxidants: further support for proinflammatory roles of endogenous hydrogen peroxide and lipid peroxides. *Inflammation*, **4:** 289–299.

Brandt, K.D. and Palmoski, M.J. 1984, Effects of salicylates and other nonsteroidal anti-inflammatory drugs on articular cartilage. *American Journal of Medicine*, **77:** 65–69.

Bray, M.A., Ford-Hutchinson, A.W. and Smith, M.J.H. 1980a, *SRS-A and Leukotrienes* (ed. P.J. Piper), pp. 253–270. Chichester: Research Studies Press, John Wiley.

Bray, M.A., Ford-Hutchinson, A.W., Shipley, M.E. and Smith, M.J.H. 1980b, Calcium ionophore A23187 induces release of chemokinetic and aggregating factors from polymorphonuclear leucocytes. *British Journal of Pharmacology*, **71:** 507–512.

Brooks, P., Emery, P., Evans, J.F., Fenner, H., Hawkey, C.J., Patrono, C., Smolen, J., Breedveld, F., Day, R., Dougados, M., Ehrich, E.W., Gijon-Banos, J. and Kvien, T.K. 1999, Interpreting the clinical significance of the differential inhibition of cyclooxygenase-1 and cyclooxygenase-2. *Rheumatology*, **38:** 779–788.

Brown, J.H. and Schwartz, N.L. 1969, Interaction of lysosomes and anti-inflammatory drugs. *Proceedings of the Society for Experimental Biology and Medicine*, **131:** 614–620.

Bruckner, F.E., Smith, H.G., Lakatos, C. and Chamberlain, M.A. 1972, Tryptophan metabolism in rheumatoid neuropathies. *Annals of the Rheumatic Diseases*, **31:** 311–315.

Brune, K., Beck, W.S., Geisslinger, G., Menzel-Soglowek, S., Peskar, B.M. and Peskar, B.A. 1991, Aspirin-like drugs may block pain independently of prostaglandin synthesis inhibition. *Experientia*, **47:** 257–261.

Brune, K., Graf, P. and Rainsford, K.D. 1977, A pharmacokinetic approach to the understanding of therapeutic effects and side effects of salicylates. *In:* K.D. Rainsford, K. Brune and M.W. Whitehouse (eds), *Aspirin and Related Drugs. Their Actions and Uses*, pp. 9–26. Basel: Birkhäuser.

Brune, K., Rainsford, K.D., Wagner, K. and Peskar, B.A. 1981, Inhibition of anti-inflammatory drugs of prostaglandin production in cultured macrophages. *Naunyn-Schmiedebergs Archives of Pharmacology*, **315:** 269–276.

Bruni, G., Dal Pra, P., Dotti, M.T. and Segre, G. 1980, Effect of various antiinflammatory drugs on plasma (ACTH) and cortisol levels in rats. *Pharmacological Research Communications*, **12**: 349–357.

Buchanan, M.R., Dejana, E., Gent, M., Mustard, J.F. and Hirsch, J. 1981, Enhanced platelet accumulation onto injured carotid arteries in rabbits after aspirin treatment. *Journal of Clinical Investigation*, **67:** 503–508.

Buchanan, M.R., Rischke, J.A. and Hirsch, J. 1982, Aspirin inhibits platelet function independent of the acetylation of cyclo-oxygenase. *Thrombosis Research*, **25:** 363–373.

Buchsbaum, M.S., Davies, G.C., Coppola, R. and Naber, D. 1981, Opiate pharmacology and individual differences. I. Psychophysical pain measurements. *Pain*, **10:** 357–366.

Bullock, G.R., Delaney, V.B., Sawyer, B.C. and Slater, T.F. 1970, Biochemical and structural changes in rat liver resulting from the parenteral administration of a large dose of sodium salicylate. *Biochemical Pharmacology*, **19:** 245–253.

Burke, J.F. and Whitehouse, M.W. 1965, Some biochemical properties of thio analogues of salicylic acid. *Biochemical Pharmacology*, **14:** 1039–1048.

Burkhardt, D. and Ghosh, P. 1987, Laboratory evaluation of antiarthritic drugs as potential chondroprotective agents. *Seminars in Arthritis and Rheumatism*, **17 (Suppl. 1):** 3–34.

Burleigh, M. and Smith, M.J.H. 1970, The site of the inhibitory action of salicylate on protein biosynthesis *in vitro*. *Biochemical Journal*, **117:** 68P.

Burleigh, M. and Smith, M.J.H. 1971, The site of the inhibitory action of salicylate on protein biosynthesis in vitro. *Journal of Pharmacy and Pharmacology*, **23:** 519–527.

Busija, D.W. and Thore, C. 1997, Modulation of prostaglandin production by nitric oxide in astroglia. *Prostaglandins, Leukotrienes and Essential Fatty Acids*, **56:** 355–359.

Buss, C.E. 1876, *Zur Antipyretischen Bedeutung der Salicylsäure und des Neutralen Salicylsauren Natrons*. Stuttgart: Enke.

Bustamante, D., Paeile, C., Willer, J.-C. and Le Bars, D. 1997, Effects of intrathecal or intracerebroventricular administration of nonsteroidal anti-inflammatory drugs on a C-fiber reflex in rats. *Journal of Pharmacology and Experimental Therapeutics*, **281:** 1381–1391.

Butcher, R.G., Bitensky, L., Cashman, B. and Chayen, J. 1973, Differences in the redox balance in human rheumatoid and non-rheumatoid synovial lining cells. *Beitraege zur Pathologie*, **148:** 265–274.

Cao, C., Matsumura, K., Yamagata, K. and Watanabe, Y. 1998, Cyclooxygenase-2 is induced in brain blood vessels during fever evoked by peripheral or central administration of tumor necrosis factor. *Mol Brain Res*, **56:** 45–56.

Carbone, L.D., Tylavsky, F.A., Cauley, J.A., Harris, T.B., Lang, T.F., Bauer, D.C., Barrow, K.D. and Kiritchevsky, S.B. 2003, Association between bone mineral density and the use of nonsteroidal anti-inflammatory drugs and aspirin: impact of cyclooxygenase selectivity. *Journal of Bone and Mineral Research*, **18:** 1795–1802.

Carlin, G., Djursäter, G., Smedegård, G. and Gerdin, B. 1985, Effect of anti-inflammatory drugs on xanthine oxidase and xanthine oxidase induced depolymerization of hyaluronic acid. *Agents and Actions*, **16:** 377–384.

Carlson, R.P., O'Neill-Davis, L., Chang, J. and Lewis, A.J. 1985, Modulation of mouse ear edema by cyclooxygenase and lipoxygenase inhibitors and other pharmacologic agents. *Agents and Actions*, **17:** 197–204.

Carney, S.I. 1987, A study of the effects of NSAIDs on proteoglycan metabolism in cartilage explant cultures. *In:* E.C. Huskisson and Y. Shiokawa (eds), *New Trends in Rheumatology*, pp. 24–34. Amsterdam: Excerpta Medica.

Carroll, P.B. and Melfi, R.C. 1972, The histologic effect of topically applied acetylsalicylic acid on bone healing in rats. *Oral Surgery*, **33:** 728–735.

Casadebaig, F., Fupin, J.P., Gravier, D., Hou, G., Daret, D., Bernard, H., Larrue, J. and Boisseau, M. 1991, Action of some salicylate derivatives on *in vitro* platelet aggregation inhibitory and inhibition antagonistic effects. *Thrombosis Research*, **64:** 631–636

Castania, A. and Rothschild, A.M. 1974, Lowering of kininogen in rat blood by adrenaline and its inhibition by sympatholytic agents, heparin and aspirin. *British Journal of Pharmacology*, **50:** 375–381.

Castor, C.W. 1971, Connective tissue activation. II. Abnormalities of cultured rheumatoid synovial cells. *Arthritis and Rheumatology*, **14:** 55–66.

Cazenave, J.P., Guccione, M.A., Packham, M.A. and Mustard, J.F. 1977, Inhibition of platelet adherence to damaged vessel wall by drugs. *Advances in Experimental Medicine and Biology*, **82:** 228–230.

Cerskus, I. and Philp, R.B. 1981, Relationship of inhibition of prostaglandin synthesis in platelets to anti-aggregatory and anti-inflammatory activity of some benzoic acid derivatives. *Agents and Actions*, **11:** 281–286.

Cesario, T.C., Yousefi, S. and Carandang, G. 1989, The regulation of interferon production by aspirin, other inhibitors of the cyclooxygenase pathway and agents influencing calcium channel flux. *Bulletin of the New York Academy of Medicine*, **5:** 26–35.

Ceuppens, J.L. and Goodwin, J.S. 1985, Immunological responses in treatment with non-steroidal anti-inflammatory drugs, with particular reference to the role of prostaglandins. *In:* K.D. Rainsford (ed.), *Anti-Inflammatory and Anti-Rheumatic Drugs*, Volume 1. *Inflammation Mechanisms and Actions of Traditional Drugs*, pp. 89–105. Boca Raton: CRC Press.

Champion, G.D., Day, R.O., Ray, J.E. and Wade, D.N. 1977, The effect of non-steroidal anti-inflammatory drugs on adenosine triphosphate content and histamine release from rat peritoneal cell suspensions rich in mast cells. *British Journal of Pharmacology*, **59:** 29–33.

Chan, J. and Lee, K.H. 1975, Effect of uridine diphospho-N-acetylglucosamine and sodium salicylate on L-glutamine-D-fructose aminotransferase activity from rat gastric mucosa. *Journal of Pharmaceutical Sciences*, **64:** 1182–1185.

Chan, T.T. 1989, Analgesic testing in animal models. *In:* J.Y. Chang and A.J. Lewis (eds), *Pharmacological Methods in the Control of Inflammation*, pp. 195–212. New York: Alan R. Liss.

Chang, Y.H. 1977a, Adjuvant polyarthritis I: incorporation of quantitative measurements of humoral and cellular immune response. *Journal of Pharmacology and Experimental Therapeutics*, **201:** 1–7.

Chang, Y.H. 1977b, Adjuvant polyarthritis. II. Suppression by tilorone. *Journal of Pharmacology and Experimental Therapeutics*, **203:** 156–161.

Chang, Y.H. and Hoffman, W. 1977, Adjuvant polyarthritis III. Evidence in support of viral etiology. *Arthritis and Rheumatism*, **20:** 1507–1513.

Chang, Y.H., Pearson, C.M. and Abe, C. 1980, Adjuvant polyarthritis. IV. Induction by a synthetic adjuvant: immunologic, histopathologic, and other studies. *Arthritis and Rheumatism*, **23:** 62–71.

Charnock, J.S. and Opit, L.J. 1962, The effect of salicylates on adenosine-triphosphatase activity of rat liver mitochondria. *Biochemical Journal*, **83:** 596–602.

Chatterjee, S.S. and Stefanovich, V. 1976, Effect of anti-inflammatory agent on L-glutamine-D-fructose-6-phosphate transaminase. *Arzneimittel Forschung*, **26:** 499–502.

Chayen, J., Bitnesky, L., Butcher, R.G. and Cashman, B. 1971, Evidence for altered lysosomal membranes in synovial lining cells from human rheumatoid joints. *Beiträge zur Pathologie*, **142:** 137–149.

Chen, L.-C., Kepka-Lenhart, D., Wright, T.M. and Morris, S.M. 1999, Salicylate-enhanced activation of transcription factors induced by interferon-γ. *Biochemical Journal*, **342:** 503–507.

Cheng, I.F., Zhao, C.C.P., Amolins, A., Galazka, M. and Doneski, L. 1996, A hypothesis for the in vivo antioxidant action of salicylic acid. *Biometals*, **9:** 285–290.

Cheshire, R.M. and Park, M.V. 1977, The inhibition of lactate dehydrogenase by salicylate. *International Journal of Biochemistry*, **8:** 637–643.

Chiabrando, C., Castelli, M.G., Cozzi, E., Fanelli, R., Campoleoni, A., Balotta, C., Latini, R. and Garattini, S. 1989, Antiinflammatory action of salicylates: aspirin is not a prodrug for salicylate against rat carrageenin pleurisy. *European Journal of Pharmacology*, **159:** 257–264.

Chitwood, S.J., Goldman, M. and Blum, S.L. 1976, Effect of aspirin on adrenal weight in female rats. *Toxicology and Applied Pharmacology*, **35:** 397–400.

Chrisman, O.D. and Snook, G.A. 1968, Studies on the protective effect of aspirin against degeneration of human articular cartilage. A preliminary report. *Clinical Orthopedics*, **56:** 77–82.

Chrisman, O.D., Snook, G.A. and Wilson, T.C. 1972, The protective effect of aspirin against degeneration of human articular cartilage. *Clinical Orthopedics*, **84:** 193–196.

Chulada, P.C. and Langenbach, R. 1997, Differential inhibition of murine prostaglandin synthase-1 and -2 by nonsteroidal anti-inflammatory drugs using exogenous and endogenous sources of arachidonic acid. *Journal of Pharmacology and Experimental Therapeutics*, **280:** 606–613.

Chung, C.K., Koo, H.N., Chung, K.Y., Shin, T., Kim, H.R., Chae, H.J., An, N.H., Kim, C.H. and Kim, H.M. 2000, Inhibitory effect of sodium salicylate on nitric oxide production from TM4 sertoli cells. *International Journal of Immunopharmacology*, **22:** 685–692.

Cianferoni, A., Schroeder, J.T., Kim, J., Schmidt, J.W., Lichtenstein, L.M., Georas, S.N. and Casolaro, V. 2001, Selective inhibition of interleukin-4 gene expression in human T cells by aspirin. *Blood* **97:** 1742–1749.

Clancy, R., Varenika, B., Huang, W.Q., Ballou, L., Attur, M., Amin, A.R. and Abramson, S.B. 2000, Nitric oxide synthase/COX cross-talk: nitric oxide activates COX-1 but inhibits COX-2-derived prostaglandin production. *Journal of Immunology*, **165:** 1582–1587.

Clarris, B.J., Fraser, J.R., Moran, C.J. and Muirden, K.D. 1977, Rheumatoid synovial cells from intact joints. Morphology, growth and polykaryocytosis. *Annals of the Rheumatic Diseases*, **36:** 293–301.

Clavelou, P., Pajot, J., Dallel, R. and Raboisson, P. 1989, Application of the formalin test to the study of orofacial pain in the rat. *Neuroscience Letters*, **103:** 349–353.

Coggins, J.F., Johnson, G.S. and Pastan, I. 1972, The effect of dibutyryl cyclic adenosine monophosphate on synthesis of sulfated acid mucopolysaccharides by transformed fibroblasts. *Journal of Biological Chemistry*, **247:** 5759–5764.

Colantoni, A., De Maria, N., Caraceni, P., Bernardi, M., Floyd, R.A. and Van Thiel, D.H. 1998, Prevention of reoxygenation injury by sodium salicylate in isolated-perfused rat liver. *Free Radical Biology and Medicine*, **25:** 87–94.

Collen, D. 1980, On the regulation and control of fibrinolysis. Edward Kowalski Memorial Lecture. *Thrombosis and Haemostasis*, **43:** 77–89.

Collier, H.O.J. 1969, A pharmacological analysis of aspirin. *Advances in Pharmacology and Chemotherapy*, **7:** 333–405.

Collier, H.O.J., Francis, A.A., McDonald-Gibson, W.J. and Saeed, S.A. 1976, Inhibition of prostaglandin biosynthesis by sulphasalazine and its metabolites. *Prostaglandins*, **11:** 219–225.

Collier, H.O.J. and Schneider, C. 1972, Nociceptive response to prostaglandins and analgesic actions of aspirin and morphine. *Nature New Biology*, **236:** 141–143.

Collins, A.J. and Lewis, D.A. 1971, Lysosomal enzyme levels in the blood of arthritic rats. *Biochemical Pharmacology*, **20:** 251–253.

Collins, D.H. and McElligott, T.F. 1960, Sulphate ($^{35}SO_4$) uptake by chondrocytes in relation to histological changes in osteoarthritic human articular cartilage. *Annals of the Rheumatic Diseases*, **19:** 318–322.

Cologniet, E., Bendtzen, K., Soeberg, B. and Bendixen, G. 1977, Leucocyte migration inhibitory activity of concanavalin-A-stimulated lymphocytes. In vivo and in vitro modifications with dipyridamole and acetylsalicylic acid. *Acta Medica Scandinavica*, **201:** 197–201.

Comper, W.D., De Witt, M. and Lowther, D.A. 1981, Effects of anti-inflammatory drugs on proteoglycan degradation as studied in rabbit articular cartilage in organ culture. *Biochemical Pharmacology*, **30:** 459–468.

Conforti, A., Franco, L., Menegale, G., Milanino, R., Piemonte, G. and Velo, G.P. 1983a, Serum copper and ceruloplasmin levels in rheumatoid arthritis and degenerative joint disease and their pharmacological implications. *Pharmacological Research Communications*, **15:** 859–867.

Conforti, A., Franco, L., Milanino, R., Totorizzo, A. and Velo, G.P. 1983b, Copper metabolism during acute inflammation: studies on liver and serum copper concentrations in normal and inflamed rats. *British Journal of Pharmacology*, **79:** 45–52.

Conforti, A., Franco, L., Milanino, R. and Velo, G.P. 1982, Copper and ceruloplasmin (Cp) concentrations during the acute inflammatory process in the rat. *Agents and Actions*, **12:** 303–307.

Cooke, T.D.V. 1985, Mechanisms of cartilage degradation: relation to choice of therapeutic agent. *Seminars in Arthritis and Rheumatism*, **15:** 16–23.

Cooper, K.E., Neale, W.L. and Kasting, N.W. 1982, Temperature regulation, fever and antipyretics. *In:* H.J.M. Barnett, J. Hirsch and J.F. Mustard (eds), *Acetylsalicylic Acid: New Uses for an Old Drug*, pp. 153–160. New York: Raven Press.

Cordeiro, R.S., Silva, P.M., Martins, M.A. and Vargaftig, B.B. 1986, Salicylates inhibit PAF-acether-induced rat paw oedema when cyclooxygenase inhibitors are ineffective. *Prostaglandins*, **32:** 719–727.

Cotty, V.F. and Harris, A.F. 1968, The effect of acetylsalicylic acid (ASA) on calvification. *Archives Internationales de Pharmacodynamie et de Thérapie*, **176:** 28–31.

Cranston, W.I. 1979, Prostaglandins as mediators of pyrexia. *Agents and Actions* **(Suppl. 6):** 79–82.

Cranston, W.I., Duff, G.W., Hellon, R.F., Mitchell, D. and Townsend, Y. 1976, Evidence that brain prostaglandin synthesis is not essential in fever. *Journal of Physiology*, **259:** 239–249.

Cranston, W.I., Hellon, R.F. and Mitchell, D. 1975, A dissociation between fever and prostaglandin concentration in cerebrospinal fluid. *Journal of Physiology*, **253:** 583–592.

Cranston, W.I., Luff, R.H. and Rawlins, M.D. 1971a, Antipyretic properties of some metabolic and structural analogues of sodium salicylate. *Journal of Physiology*, **216:** 81P–82P.

Cranston, W.I., Luff, R.H., Rawlins, M.D. and Wright, V.A. 1971b, Influence of the duration of experimental fever on salicylate antipyresis in the rabbit. *British Journal of Pharmacology*, **41:** 344–351.

Crofford, L.J. 1997, COX-1 and COX-2 tissue expression: implications and predictions. *Journal of Rheumatology*, **24:** 15–19.

Crofford, L.J., Wilder, R.L., Ristimäki, A.P., Sano, H., Remmers, E.F., Epps, H.R. and Hla, T. 1994, Cyclooxygenase-1 and -2 expression in rheumatoid synovial tissues. *Journal of Clinical Investigation*, **93:** 1095–1101.

Cronheim, G., King, J.S., Jr. and Hyder, N. 1952, Effect of salicylic acid and similar compounds on the adrenal–pituitary system. *Proceedings of the Society for Experimental Biology and Medicine*, **80:** 51–55.

Cronstein, B.N., Montesinos, M.C. and Weissmann, G. 1999, Salicylates and sulfasalazine, but not glucocorticoids, inhibit leukocyte accumulation by an adenosine-dependent mechanism that is independent of inhibition of prostaglandin synthesis and p105 of NFκB. *Proceedings of the National Academy of Sciences USA*, **96:** 6377–6381.

Cross, S.E., Anderson, C. and Roberts, M.S. 1998, Topical penetration of commercial salicylate esters and salts using human isolated skin and clinical microdialysis studies. *British Journal of Pharmacology*, **46:** 29–35.

Cross, S.E., Megwa, S.A., Benson, H.A. and Roberts, M.S. 1999. Self promotion of deep tissue penetration and distribution of methyl salicylate after topical application. *Pharmaceutical Research*, **16:** 427–433.

Crutchley, D.J., Ryan, U.S. and Ryan, J.W. 1980, Effects of aspirin and dipyridamole on the degradation of adenosine diphosphate by cultured cells derived from bovine pulmonary artery. *Journal of Clinical Investigation*, **66:** 29–35.

Cunarro, J. and Weiner, W.M. 1975, Mechanisms of action of agents which uncouple oxidative phosphorylation; direct correlation between proton-carrying and respiratory-releasing properties using rat liver mitochondria. *Biochimica Biophysica Acta*, **387:** 234–240.

Curtis-Prior, P.B. (ed.) 1988, *Prostaglandins. Biology and Chemistry of Prostaglandins and Related Eicosanoids*. Edinburgh: Churchill-Livingstone.

Czervionke, R.L., Smith, J.B., Fry, G.L., Hoak, J.C. and Haycraft, D.L. 1979, Inhibition of prostacyclin by treatment of endothelium with aspirin. Correlation with platelet adherence. *Journal of Clinical Investigation*, **63:** 1089–1092.

Dalessio, D.J. 1972, *Wolff's Headache and Other Head Pain*. New York: Oxford University Press.

Dannhardt, G., Kiefer, W., Krämer, G., Maehrlein, S., Nowe, U. and Fiebeich, B. 2000, The pyrrole moiety as a template for COX-1/COX-2 inhibitors. *European Journal of Medicinal Chemistry*, **35:** 499–510.

Dannhardt, G. and Laufer, S. 2000, Structural approaches to explain the selectivity of COX-2 inhibitors: is there a common pharmacophore? *Current Medicinal Chemistry*, **7:** 1101–1112.

Dascombe, M.J. 1984, Effects of diflunisal on fever in the rabbit and the rat. *Journal of Pharmacy and Pharmacology*, **36:** 437–440.

Daun, J.M., Ball, R.W., Burger, H.R. and Cannon, J.G. 2000, Aspirin-induced increases in soluble IL-1 receptor type II concentrations in vitro and in vivo. *Journal of Leukocyte Biology*, **65:** 863–866.

David, M.J., Vignon, E., Peschard, M.J., Broquet, P., Louisot, P. and Richard, M. 1992, Effect of non-steroidal anti-inflammatory drugs (NSAIDs) on glycosyltransferase activity from human osteoarthritic cartilage. *British Journal of Rheumatology*, **31:** 13–17.

Davidson, E.M., Rae, S.A. and Smith, M.J.H. 1982, Leukotriene B4 in synovial fluid. *Journal of Pharmacy and Pharmacology*, **34:** 410.

Davies, D.T., Hughes, A. and Tonks, R.S. 1970, Glycolysis in human blood in the presence of sodium salicylate and the importance of the incubation medium. *Biochemical Pharmacology*, **19:** 1277–1285.

Davies, P. 1977, Lysosomal enzymes of phagocytic cells as humoral mediators of inflammation: modulation of their release by cyclic nucleotides. *In:* I.L. Bonta, J. Thompson and K. Brune (eds), *Inflammation: Mechanisms and their Impact on Therapy*, pp. 107–119. Basel: Birkhäuser.

Davies, P., Bonney, R.J., Humes, J.L. and Kuehl, F.A., Jr. 1979, Macrophages responding to inflammatory stimuli synthesise increased amounts of prostaglandins. *In:* K.D. Rainsford and A.W. Ford-Hutchinson (eds), *Prostaglandins and Inflammation*, pp. 143–149. Basel: Birkhäuser.

Davies, P., Bonney, R.J., Humes, J.L. and Kuehl, F.A. 1980, The role of macrophage secretory products in chronic inflammatory processes. *Journal of Investigative Dermatology*, **74:** 292–296.

Davies, P. and Macintyre, D.E. 1992, Prostaglandins and inflammation. *In:* J.I. Gallin, I.M. Goldstein and R. Snyderman (eds), *Inflammation: Basic Principles and Clinical Correlates*, pp. 123–128. New York: Raven Press.

Davis, R.M., Fisher, J.S. and McGowan, L.M. 1968, Local antiphlogistic activity of L-phenylalanine and L-tryptophane. *Journal of Endocrinology*, **41:** 603–604.

Dawkins, P.D., Gould, B.J. and Smith, M.J.H. 1966, Inhibitory effect of salicylate on the incorporation of L-[U-^{14}C]-leucine into the protein of rat tissue preparations in vitro. *Biochemical Journal*, **99:** 703–707.

Dawkins, P.D., McArthur, J.N. and Smith, M.J.H. 1971, Inhibition of protein biosynthesis in mouse liver by salicylate. *Biochemical Pharmacology*, **20:** 1303–1312.

Dawkins, P.D. and Smith, M.J.H. 1970, The effects of salicylate on the concentrations of amino acids in mouse tissues. *Journal of Pharmacy and Pharmacology*, **22:** 913–922.

Dawson, W. 1980, The comparative pharmacology of benoxaprofen. *Journal of Rheumatology*, **7 (Suppl. 6):** 5–11.

Day, R.O., McLachlan, A.J., Graham, G.G. and Williams, K.M. 1999, Pharmacokinetics of non-steroidal anti-inflammatory drugs in synovial fluid. *Clinical Pharmacokinetics*, **36:** 191–210.

De Alvare, L.R., Goda, K. and Kimura, T. 1976, Mechanism of superoxide anion scavenging reaction by bis-(salicylato)copper(II) complex. *Biochemical and Biophysical Research Communications*, **69:** 687–694.

De Maria, N., Colantoni, A., Zhu, Q.L., Joshibarve, S., Barve, S., McClain, C.J. and Van Thiel, D.H. 1997, Sodium salicylate reduces reperfusion injury in the liver by affecting NF-kappa B and inducible nitric oxide synthase activation. *Hepatology*, **26:** 904.

De Montis, M.G., Olianas, M.C., Mulas, G. and Tagliamonte, A. 1977, Evidence that only free tryptophan exchanges with the brain. *Pharmacological Research Communications*, **9:** 215–220.

De Vries, B.J., Van Den Berg, W.B. and Van De Putte, L.B.A. 1985, Salicylate-induced depletion of endogenous inorganic sulfate. Potential role in the suppression of sulfated glycosaminoglycan synthesis in murine articular cartilage. *Arthritis and Rheumatism*, **28:** 922–928.

De Vries, B.J., Van Den Berg, W.B., Vitters, E. and Van De Putte, L.B.A. 1988, Effects of NSAIDs on the metabolism of sulphated glycosaminoglycans in healthy and (post) arthritic murine articular cartilage. *Drugs*, **35:** 24–32.

De Vries, J. and Verboom, C.N. 1980, Effects of scavengers of superoxide radicals, hydrogen peroxide, singlet oxygen and hydroxyl radicals on malondialdehyde generation from arachidonic acid by bovine seminal vesicle microsomes. *Experientia (Basel)*, **36:** 1339–1340.

Dean, F.M. 1996, hP, the component of log P controlling structure–activity relationships amongst nonsteroidal anti-inflammatory drugs. *Journal of Pharmacy and Pharmacology*, **48:** 233–239.

Dearden, J.C. and George, E. 1977, Quantitative structure–activity study of inhibition of blood platelet aggregation by aspirin derivatives. *Journal of Pharmacy and Pharmacology*, **29:** 74P.

Dearden, J.C. and George, E. 1979, Anti-inflammatory potencies of some aspirin derivatives: a quantitative structure–activity study. *Journal of Pharmacy and Pharmacology* **(Suppl. 31):** 45P.

Deby, C. 1988, Metabolism of polyunsaturated fatty acids, precursors of eicosanoids. *In:* P.B. Curtis-Prior (ed.), *Prostaglandins. Biology and Chemistry of Prostaglandins and Related Eicosanoids*, pp. 11–36. Edinburgh: Churchill-Livingstone.

Dekel, S., Falconer, J. and Francis, M.J. 1980, The effect of anti-inflammatory drugs on glycosaminoglycan sulphation in pig cartilage. *Prostaglandins*, **4:** 133–140.

Delbarre, B., Delbarre, G. and Jobard, P. 1971, Action of acetylsalicylic acid on the autonomic nervous system. *Experientia (Basel)*, **27:** 922–924.

Delhon, A., Tarayre, J.P., Lauressergues, H. and Casadio, S. 1977, Study of the repercussions on blood of acute experimental inflammation in rats. IL blood changes in the course of carrageenan induced inflammation. *Arzneimittel Forschung*, **27(12):** 2308–2311.

Dewar, A.J. and Barron, G. 1977, The effects of excess vitamin A and acetylsalicylic acid on the stability of rat retinal lysosomes in vivo. *Experimental Eye Research*, **24:** 291–298.

DeWitt, D.L., El-Harith, E.A., Kraemer, S.A., Andrews, M.J., Yao, E.F., Armstrong, R.L. and Smith, W.L. 1990, The aspirin and heme-binding sites of ovine and murine prostaglandin endoperoxide synthases. *Journal of Biological Chemistry*, **265:** 5192–5198.

Dewse, C.D. 1977, Inhibitory action of aspirin on DNA synthesis in phytohaemagglutinin-stimulated human lymphocytes. *Journal of Pharmacy and Pharmacology*, **29:** 445–446.

Di Pasquale, G., Rassaert, C., Richter, R., Welaj, P., Gingold, J. and Singer, R. 1975, The anti-inflammatory properties of isoxicam (4-hydroxy-2-methyl-N-(5-methyl)-3-isoxolyl-2H-1,2-benzothiazine-3-carboxamide-1,1-dioxide). *Agents and Actions*, **5:** 256–263.

Di Perri, T. and Auteri, A. 1973, Anti-complement activity in vitro and in vivo of some non-steroidal anti-inflammatory drugs. *Bollettino della Societa Italiana di Biologia Sperimentale*, **49:** 220.

Di Rosa, M. 1974, Assessment of anti-inflammatory drugs in animals. *In:* G.-P. Velo, D.A. Willoughby and J.P. Giroud (eds), *Future Trends in Inflammation*, pp. 143–153. Padua: Piccin.

Di Rosa, M., Giroud, J.P. and Willoughby, D.A. 1971a, Studies on the mediators of the acute inflammatory response induced in rats in different sites by carrageenan and turpentine. *Journal of Pathology*, **104:** 15–29.

Di Rosa, M., Papadimitriou, J.M. and Willoughby, D.A. 1971b, A histopathological and pharmacological analysis of the mode of action of nonsteroidal anti-inflammatory drugs. *Journal of Pathology*, **105:** 239–256.

Diamantis, W., Kohlhepp, W.C., Haertlein, B., Melton, J. and Sofia, R.D. 1978, Meseclazone, 5-chlorosalicylic acid and acetylsalicylic acid. Comparison of their effects on in vitro and ex vivo platelet aggregation. *Thrombosis and Haemostasis*, **40:** 24–36.

Dick, W.C. 1972a, *An Introduction to Clinical Rheumatology*. Edinburgh: Churchill Livingstone.

Dick, W.C. 1972b, The use of radioisotopes in normal and diseased joints. *Seminars in Arthritis and Rheumatism*, **1:** 301–325.

Dieppe, P.A., Doherty, M., MacFarlane, D. and Maddison, P. 1985, *Rheumatological Medicine*. Edinburgh: Churchill Livingstone.

Dietze, G., Wicklmayr, M., Mayer, L., Bottger, I. and Von Funke, H. 1978, Bradykinin and human forearm metabolism: inhibition of endogenous prostaglandin synthesis. *Hoppe-Seyler's Zeitschrift fuer Physiologische Chemie*, **359:** 369–378.

Dinarello, C.A. and Wolff, S.M. 1978, Pathogenesis of fever in man. *New England Journal of Medicine*, **298:** 607–612.

Dingle, J.T. 1991, Cartilage maintenance in osteoarthritis: interaction of cytokines, NSAIDs and prostaglandins in articular cartilage damage and repair. *Journal of Rheumatology*, **18 (Suppl. 28):** 30–37.

Dingle, J.T. 1992, NSAIDs and human cartilage metabolism. *In:* K.D. Rainsford and G.P. Velo (eds), *Side-Effects of Anti-inflammatory Drugs 3, Inflammation and Drug Therapy Series*, Volume V, pp. 261–268. Dordrecht: Kluwer Academic Publishers.

Dingle, J.T. 1999, The effects of NSAID on the matrix of human articular cartilages. *Zeitschrift für Rheumatologie*, **58:** 125–129.

Dingle, J.T. and Shield, M.J. 1990, The interactions of cytokines, NSAIDs, and prostaglandins in cartilage destruction and repair. *Advances in Prostaglandin, Thromboxane, and Leukotriene Research*, **21:** 9–24.

Dinnendahl, V., Peters, H.D. and Schönhöfer, P.S. 1973, Effects of sodium salicylate and acetylsalicylic acid on cyclic $3',5'$-AMP-dependent protein kinase. *Biochemical Pharmacology*, **22:** 2223–2228.

Dinnermann, J.L., Lowenstein, C.J. and Snyder, S.H. 1993, Molecular mechanisms of nitric oxide regulation: potential relevance to cardiovascular disease. *Circulation Research*, **73:** 217–2222.

Dodel, R.C., Du, Y., Bales, K.R., Gao, F. and Paul, S.M. 1999, Sodium salicylate and 17β-estradiol attenuate nuclear transcription factor NF-κB translocation in cultured rat astroglial cultures following exposure to amyloid $A\beta_{1-40}$ and lipopolysaccharides, *Journal of Neurochemistry*, **73:** 1453–1460.

Doery, J.C.G., Hirsch, J. and De Gruchy, G.C. 1969, Aspirin: its effect on platelet glycolysis and release of adenosine diphosphate. *Science (Washington DC)*, **165:** 65–67.

Doherty, M. 1989, 'Chondroprotection' by non-steroidal anti-inflammatory drugs. *Annals of the Rheumatic Diseases*, **48:** 619–621.

Doig, M.V. and Ford-Hutchinson, A.W. 1980, The production and characterization of products of the lipoxygenase enzyme system released by rat peritoneal macrophages. *Prostaglandins*, **20:** 1007–1019.

Domenjoz, R. 1955, Pharmakotherapeutische Weiterentwicklung der Antipyretica-Antalgetica. *Archiv für Experimentelle Pathologie und Pharmakologie*, **225:** 14–42.

Done, A.K. 1983, Treatment of fever in 1982: a review. *American Journal of Medicine*, 27–35.

Dormandy, T.L. 1980, Free-radical reaction in biological systems. *Annals of the Royal College of Surgeons of England*, **62:** 188–194.

Dowes, F.R., Robin, M. and Steudle, R. 1975, *In:* G. Katona and J.R. Bleugno (eds), *Inflammation and Anti-Inflammatory Therapy*, pp. 319–325. New York: Wiley.

Dubas, T.C. and Parker, J.M. 1971, A central component in the analgesic action of sodium salicylate. *Archives Internationales de Pharmacodynamie et de Thérapie*, **194:** 117–122.

Dubois, R.N., Abramson, S.B., Crofford, L., Gupta, R.A., Simon, L.S., Van De Putte, B.A. and Lipsky, P.E. 1998, Cyclooxygenase in biology and disease. *FASEB Journal*, **12:** 1063–1073.

Dumonde, D.C. and Glynn, L.E. 1962, The production of arthritis in rabbits by an immunological reaction to fibrin. *British Journal of Experimental Pathology*, **43:** 373–383.

Duncan, M.W., Person, D.A., Rich, R.R. and Sharp, J.T. 1977, Aspirin and delayed type hypersensitivity. *Arthritis and Rheumatism*, **20:** 1174–1178.

Dykes, R.W. 1975, Nociception. *Brain Research*, **99:** 229–245.

Eakins, K.E., Higgs, G.A., Moncada, S. and Mugridge, K.G. 1980, Non-steroidal anti-inflammatory drugs which potentiate leukocyte migration in carrageenan-induced inflammation. *British Journal of Pharmacology*, **70:** 182.

Egan, R.W., Paxton, J. and Kuehl, F.A., Jr. 1976, Mechanism for irreversible self-deactivation of prostaglandin synthetase. *Journal of Biological Chemistry*, **251:** 7329–7335.

Egorin, M.J., Felsted, R.L. and Bachur, N.R. 1978, Salicylate effects the response of human lymphocytes to phytohemagglutinin isolectins. *Clinical Immunobiology and Immunopathology*, **10:** 1–10.

Einarsson, R., Eklund, H., Zeppezauer, E., Boiwe, T. and Branden, C.-I. 1974, Binding of salicylate in the adenosine-binding pocket of dehydrogenases. *European Journal of Biochemistry*, **49:** 41–47.

Emele, J.F. and Shanaman, J. 1963, Bradykinin writhing: a method for measuring analgesia. *Proceedings of the Society for Experimental Biology and Medicine*, **114:** 680–682.

Engelhart, G. 1978, Wirkung von nichtsteroidischen Antiphlogistika auf die Plasmaweißbindung des Corticosterons. *Arzneimittel Forschung*, **28:** 1714–1723.

Ennis, R.S., Granda, J.L. and Posner, A.S. 1968, Effect of gold salts and other drugs on the release and activity of lysosomal hydrolases. *Arthritis and Rheumatism*, **11:** 756–764.

Estes, D. and Kaplan, K. 1980, Lack of platelet effect with the aspirin analog, salsalate. *Arthritis and Rheumatism*, **23:** 1303–1307.

Evinger, M., Maeda, S. and Pesta, S. 1981, Recombinant human leukocyte interferon produced in bacteria has antiproliferative activity. *Journal of Biological Chemistry*, **256:** 2113–2114.

Exer, B., Krupp, P., Menassé, R. and Riesterer, L. 1976, Influence of adjuvant arthritis on connective-tissue metabolism. *Agents and Actions*, **6:** 651–656.

Failli, P., Cecchi, E., Tosti-Guerra, C., Mugelli, A., Laffi, G., Zilletti, L. and Giotti, A. 1998, Effect of some cyclooxygenase inhibitors on the increase in guanosine $3',5'$-cyclic monophosphate induced by NO-donors in human whole platelets. *British Journal of Pharmacology*, **123:** 1457–1463.

Famaey, J.-P. and Whitehouse, M.W. 1973, Interactions between non-steroidal anti-inflammatory drugs and biological membranes. II. Swelling and membrane permeability induced in some immunocompetent cells by various non-steroidal anti-inflammatory drugs. *Biochemical Pharmacology*, **22:** 2707–2717.

Famaey, J.-P. and Whitehouse, M.W. 1975, Interaction between nonsteroidal anti-inflammatory drugs and biological membranes – IV. Effects of nonsteroidal anti-inflammatory drugs and of various ions on the availability of sulfydryl groups on lymphoid cell and mitochondrial membranes. *Biochemical Pharmacology*, **24:** 1609–1615.

Farivar, R.S. and Brecher, P. 1996, Salicylate is a transcriptional inhibitor of the inducible nitric oxide synthase in cultured cardiac fibroblasts. *Journal of Biological Chemistry*, **271:** 31585–31592.

Farivar, R.S., Chobanian, A.V. and Brecher, P. 1996, Salicylate or aspirin inhibits the induction of the inducible nitric oxide synthase in rat cardiac fibroblasts. *Circulation Research*, **78:** 759–768.

Fassbender, H.G. 1976, Morphological aspects of rheumatoid arthritis and psoriatic arthritis. *In:* F.J. Wagenhaeuser (ed.), *Chronic Forms of Polyarthritis*, pp. 17–26. Berne: Hans Huber.

Ferreira, S.H. 1972, Prostaglandins, aspirin-like drugs and analgesia. *Nature New Biology*, **240:** 200–203.

Ferreira, S.H. 1979, Prostaglandins. *In:* J.C. Houch (ed.), *Chemical Messengers of the Inflammatory Process*, pp. 113–151. Amsterdam: Elsevier North Holland Biomedical Press.

Ferreira, S.H. 1980, Peripheral analgesia: mechanism of the analgesic action of aspirin-like drugs and opiate antagonists. *British Journal of Clinical Pharmacology*, **10:** 237S–245S.

Ferreira, S.H., Moncada, S., Parsons, M. and Vane, J.R. 1974, Proceedings: the concomitant release of bradykinin and prostaglandin in the inflammatory response to carrageenan. *British Journal of Pharmacology*, **52:** 108P–109P.

Ferreira, S.H., Moncada, S. and Vane, J.R. 1971, Indomethacin and aspirin abolish prostaglandin release from spleen. *Nature*, **231:** 237–239.

Ferreira, S.H., Moncada, S. and Vane, J.R. 1997, Prostaglandins and the mechanism of analgesia. *British Journal of Pharmacology*, **120:** 401–412.

Ferri, A., Calza, R., Pellegrini, A. and Cattani, L. 1994, Two distinct mechanisms of inhibition of platelet aggregation by acetylsalicylic acid. *Biochemical Molecular Biology International* **32:** 1101–1107.

Fiebich, B.L., Lieb, K., Hüll, M., Aicher, B., van Ryn, J., Pairet, M. and Engelhardt, G. 2000, Effects of caffeine and paracetamol alone or in combination with acetylsalicylic acid on prostaglandin E_2 synthesis in rat microglial cells. *Neuropharmacology*, **39:** 2201–2213.

Field, H.L. 1987, *Pain*. New York: McGraw-Hill Book Company.

Fierro, I.M., Colgan, S.P., Bernasconi, G., Petasis, N.A., Clish, C.B., Arita, M. and Serhan, C.N. 2003, Lipoxin A4 and aspirin-triggered 15-epi-lipoxin A4 inhibit human neutrophil migration: comparisions between synthetic 15 epimers in chemotaxis and transmigration with microvessel endothelial cells and epithelial cells. *Journal of Immunology*, **170:** 2688–2694.

Fink, M. and Irwin, P. 1982, Central nervous system effects of aspirin. *Clinical Pharmacology and Therapeutics*, **32:** 362–365.

Finlay, C., Davies, P. and Allison, A.C. 1975, Changes in cellular enzyme levels and the inhibition of selective release of lysosomal hydrolases from macrophages by indomethacin. *Agents and Actions*, **5:** 345–353.

Fjalland, B. 1974, Inhibition by non-steroidal anti-inflammatory agents of the release of rabbit aorta contracting substance and prostaglandins from chopped guinea-pig lungs. *Journal of Pharmacy and Pharmacology*, **26:** 448–451.

Flower, R.J. 1974, Drugs which inhibit prostaglandin biosynthesis. *Pharmacological Reviews*, **26:** 33–67.

Flower, R.J. and Vane, J.R. 1972, Inhibition of prostaglandin synthetase in brain explains the antipyretic activity of paracetamol (4-acetamidophenol). *Nature*, **240:** 410–411.

Flower, R.J., Gryglewski, R., Herbaczynska-Cedro, K. and Vane, J.R. 1972, The effects of anti-inflammatory drugs on prostaglandin biosynthesis. *Nature*, **238:** 104–106.

Ford-Hutchinson, A.W., Bray, M.A., Cunningham, F.M., Davidson, E.M. and Smith, M.J.H. 1981, Isomers of leukotriene B₄ possess different biological potencies. *Prostaglandins*, **21:** 143–152.

Ford-Hutchinson, A.W., Bray, M.A., Doig, M.V., Shipley, M.E. and Smith, M.J.H. 1980, Leukotriene B, a potent chemokinetic and aggregating substance released from polymorphonuclear leukocytes. *Nature*, **286:** 264–265.

Ford-Hutchinson, A.W., Walker, J.R., Connor, N.S. and Smith, M.J.H. 1977, Prostaglandin and leukocyte migration in inflammatory reactions. *Agents and Actions*, **7:** 469–472.

Ford-Hutchinson, A.W., Walker, J.R., Davidson, E.M. and Smith, M.J.H. 1978, PGI₂: a potential mediator of inflammation. *Prostaglandins*, **16:** 253–258.

Forrest, J. 1998, *Acute Pain: Pathophysiology and Treatment*. Grimsby (Ontario): Manticore.

Foye, W.O., Lai-Chen, Y.-L. and Patel, B.R. 1981, Inhibition of mucopolysaccharide sulfation. *Journal of Pharmaceutical Sciences*, **70:** 49–51.

Francesconi, R.P. and Mager, M. 1975, Salicylate, tryptophan, and tyrosine hypothermia. *American Journal of Physiology*, **228:** 1431–1435.

Frantz, B. and O'Neill, E.A. 1995, The effect of sodium salicylate and aspirin on NF-kappa B. *Science*, **270:** 2017–2019.

Fricchione, G.L., Bilfinger, T.V. and Stefano, G.B. 1998, Aspirin inhibits granulocyte adherence to saphenous vein endothelia in a process not mediated by nitric oxide. *International Journal of Cardiology*, **64 (Suppl. 1):** S29–S33.

Fridovich, I. 1975, Superoxide dismutases. *Annual Review of Biochemistry*, **44:** 147–159.

Fuccella, L.M. 1979, Clinical pharmacology of inhibitors of platelet aggregation. *Pharmacological Research Communications*, **11:** 825–852.

Fujihira, E., Mori, T. and Nakazawa, M. 1970, A simple method for evaluation of non-steroid anti-inflammatory drugs. *Pharmacometrics*, **4:** 903–910.

Fujii, K., Tajiri, K., Kajiwara, T., Tanaka, T. and Murota, K. 1989, Effects of NSAID on collagen and proteoglycan synthesis of cultures chrondrocytes. *Journal of Rheumatology*, **16 (Suppl. 18):** 28–31.

Gábor, M. 2000, *Mouse Ear Inflammation Models and their Pharmacological Applications*. Budapest: Akadémiai Kiadó.

Gabourel, J.D., Davies, G.H. and Rittenberg, M.B. 1977a, Effects of salicylate and phenobarbital on lymphocyte proliferation and function. *Clinical Immunobiology and Immunopathology*, **7:** 53–61.

Gabourel, J.D., Moore, M.A.S., Bagby, G.C., Jr. and Davies, G.H. 1977b, Effect of sodium salicylate on human and mouse granulopoiesis *in vitro*. *Arthritis and Rheumatism*, **20:** 59–64.

Gaginella, T.S. and Walsh, R.E. 1992, Sulfasalazine. Multiplicity of action. *Digestive Diseases and Sciences*, **37:** 801–812.

Galen, F.-X., Truchot, R. and Michel, R. 1974a, Influence of hydroxybenzoic acids and their iodized derivatives on the respiration of isolated mitochondria. *Biochemical Pharmacology*, **23:** 1367–1377.

Galen, F.-X., Truchot, R. and Michel, R. 1974b, Mechanism of action of iodo-4-salicylic acid on the respiration of isolated mitochondria. *Biochemical Pharmacology*, **23:** 1379–1385.

Gans, K.R., Keyner, S. and Orzechowski, R.F. 1980, Application of a radiometric ear assay for studies of adjuvant arthritis in rats. *Arthritis and Rheumatism*, **23:** 633–640.

Garcia Leme, J.G., Hamamura, L., Leite, M.P. and Rocha e Silva, M. 1973, Pharmacological analysis of the acute inflammatory process induced in the rat's paw by local injection of carrageenan and by heating. *British Journal of Pharmacology*, **48:** 88–96.

Garcia Rafenell, J., Planas, J.M. and Puig-Parellada, P. 1979, Comparison of the inhibitory effects of acetylsalicylic acid and trifusal on enzymes related to thrombosis. *Archives Internationales de Pharmacodynamie et de Thérapie*, **237:** 343–350.

Gautam, S.C., Pindolia, K.R., Noth, C.J., Janakiraman, N., Xu, Y.X. and Chapman, R.A. 1995, Chemokine gene expression in bone marrow stromal cells: downregulation with sodium salicylate. *Blood*, **86:** 2541–2550.

Gemmell, D.K., Cottney, J. and Lewis, A.J. 1979, Comparative effects of drugs on four paw oedema models in the rat. *Agents and Actions*, **9**: 107–116.

Genc, O., Turgut, S., Turgut, G. and Kortunay, S. 2003, Inhibition of spinal reflexes by acetylsalicylate and metamizol (dipyrone) in rats. *Pharmacology*, **69**: 123–126.

Ghosh, P., Wells, C., Smith, M. and Hutadilok, N. 1990, Chondroprotection, myth or reality: an experimental approach. *Seminars in Arthritis and Rheumatism*, **19**: 3–9.

Gierse, J.K., Hauser, S.D., Creely, D.P., Koboldt, C., Rangwala, S.H., Isakson, P.C. and Seibert, K. 1995, Expression and selective inhibition of the constitutive and inducible forms of human cyclo-oxygenase. *Journal of Biochemistry*, **305**: 479–484.

Gierse, J.K., Koboldt, C.M., Walker, M.C., Seibert, K. and Isakson, P.C. 1999, Kinetic basis for selective inhibition of cyclo-oxygenases. *Biochemistry Journal*, **339**: 607–614.

Gilchrest, H. and Watnick, A.S. 1977, Edema formation and neutrophil mobilization in the neutropenic rat. *Agents and Actions*, **7**: 529–532.

Gillard, J.W., Dixon, R., Ethier, D. and Evans, J. 1992, Therapeutic potential of 5-lipoxygenase inhibitors: the discovery and development of MK-886, a novel-mechanism leukotriene inhibitor. *In:* K.D. Rainsford and G.-P. Velo (eds), *Side Effects of Anti-inflammatory Drugs 3*, pp. 275–286. Dordrecht: Kluwer,

Gilroy, D.W., Colville-Nash, P.R., Willis, D., Chivers, J., Paul-Clark, M.J. and Willoughby, D.A. 1999, Inducible cyclooxygenase may have anti-inflammatory properties. *Nature Medicine*, **5**: 698–701.

Giuliano, F., Mitchell, J.A. and Warner, T.D. 2001, Sodium salicylate inhibits prostaglandin formation without affecting the induction of cyclooxygenase-2 by bacterial lipopolysaccharide *in vivo*. *Journal of Pharmacology and Experimental Therapeutics*, **299**: 894–900.

Glenn, E.M., Bowman, B.J. and Rohloff, N.A. 1979. Anomalous biological effects of salicylates and prostaglandins. *Agents and Actions*, **9**: 257–264.

Glenn, E.M., Bowman, B.J., Rohloff, N.A. and Seely, R.J. 1977, A major contributory cause of arthritis in adjuvant-inoculated rats: granulocytes. *Agents and Actions*, **7**: 265–282.

Glenn, E.M. and Gray, J. 1965a, Adjuvant-induced polyarthritis in rats: biologic and histologic background. *American Journal of Veterinary Research*, **26**: 1180–1194.

Glenn, E.M. and Gray, J. 1965b, Chemical changes in adjuvant-induced polyarthritis of rats. *American Journal of Veterinary Research*, **26**: 1195–1203.

Glenn, E.M. Rohloff, N., Bowman, B.J. and Lyster, S.C. 1973, The pharmacology of 2-(2-fluoro-4-biphenylyl)propionic acid (flurbiprofen). A potent non-steroidal anti-inflammatory drug. *Agents and Actions*, **3**: 210–216.

Goetzl, E.J. 1980, Mediators of immediate hypersensitivity derived from arachidonic acid. *New England Journal of Medicine*, **303**: 822–825.

Gold, E.W., Anderson, L.B., Miller, C.W. and Schwartz, E.R. 1976, Effect of salicylate on the surgical inducement of joint degeneration in rabbit knees. *Journal of Bone and Joint Surgery of America*, **58**: 1012–1015.

Gombar, V., Kapoor, V.K. and Singh, H. 1983, Quantitative structure–activity relationships. Anti-inflammatory activity of salicylic acid derivatives. *Arzneimittel Forschung*, **33**: 1226–1230.

Gorini, S., Valcavi, U. and Zonta-Bolego, N. 1963, Nuova derivati dell'acido O-cresotinico ad analgesia ed antiflogistica. *Bolletina della Societa Italiana de Biologia Spermentale*, **39**: 873–875.

Görög, P. and Kovaks, B. 1976, Chemical changes in adjuvant-induced polyarthritis of rats. *Agents and Actions*, **6**: 607–612.

Görög, P. and Kovaks, I.B. 1970, The inhibitory effect of non-steroidal anti-inflammatory agents on aggregation of red cells in vitro. *Journal of Pharmacy and Pharmacology*, **22**: 86–92.

Goto, K., Hisadome, M. and Imamura, H. 1977, Effects of anti-inflammatory agents on release of lysosomal enzymes from rabbit polymorphonuclear leucocytes induced by phospholipase C. *Yakugaku Zasshi*, **97**: 382–387.

Gotze, M. and Rossel, I. 1972, Lactic dehydrogenase and LDH-isoenzymes in synovia and synovial tissue. *Scandinavian Journal of Rheumatology*, **1**: 171–175.

Gould, B.J., Dawkins, P.D., Smith, M.J.H. and Laurence, A.J. 1966, The mechanism of the inhibition of aminotransferases by salicylate. *Molecular Pharmacology*, **2**: 526–533.

Gould, B.J., Huggins, A.K. and Smith, M.J.H. 1963, Effects of salicylate on glutamate dehydrogenase and glutamate decarboxylase. *Biochemical Journal*, **88**: 346–349.

Gould, B.J. and Smith, M.J.H. 1965, Inhibition of rat brain glutamate decarboxylase activity by salicylate in vitro. *Journal of Pharmacy and Pharmacology*, **17:** 83–88.

Graeme, M.L., Fabry, E. and Sigg, E.B. 1966, Mycobacterial adjuvant periarthritis in rodents and its modifications by anti-inflammatory agents. *Journal of Pharmacology and Experimental Therapeutics*, **153:** 373–380.

Green, A.Y., Green, D., Murray, P.A. and Wilson, A.B. 1971, Factors influencing the inhibitory action of anti-inflammatory drugs on carrageenan induced oedema. *British Journal of Pharmacology*, **41:** 132–139.

Greffrath, W., Kirschstein, T., Nawrath, H. and Treede, R.D. 2002, Acetylsalicylic acid reduces heat responses in rat nociceptive primary sensory neurons – evidence for a new mechanism of action. *Neuroscience Letters*, **320:** 61–64.

Grega, G.J., Svensjö, E. and Haddy, F.J. 1981, Macromolecular permeability of the microvascular membrane. *Microcirculation*, **1:** 325–341.

Greiling, H. and Kleesiek, K. 1982, *In:* E. Kaiser, F. Gabl, M.M. Müller and M. Bayer (eds), *XIth International Congress of Clinical Chemistry*, pp. 635–650. Berlin: de Gruyter.

Grilli, M., Pizzi, M., Memo, M. and Spano, P-F. 1996, Neuroprotection by aspirin and sodium salicylate through blockade of NF-κB activation. *Science*, **274:** 1383–1385.

Grisolia, S., Mendelson, J. and Diederich, D. 1969, Inactivation of enzymes by the therapeutic concentrations of aspirin and salicylate. *Nature*, **223:** 79–80.

Griswold, D.E., Marshall, P., Martin, L., Webb, E.F. and Zabko-Potapovich, B. 1991, Analgesic activity of SK&F 105809, a dual inhibitor of arachidonic acid. *Agents and Actions*, **32** (**Suppl.**): 113–117.

Griswold, D.E. and Ujeki, E.M. 1969, Immunosuppressant effect of salicylate and quinine on antibody forming cells. *European Journal of Pharmacology*, **6:** 56–60.

Grootveld, M. and Halliwell, B. 1986, Aromatic hydroxylation as a potential measure of hydroxyl-radical formation in vivo. *Biochemical Journal*, **237:** 499–504.

Gross, M. and Greenberg, L.A. 1948, *The Salicylates. A Critical Bibliographic Review*. New Haven; Hillhouse Press.

Grosser, N. and Schröder, H. 2000, A common pathway for nitric oxide release from NO-aspirin and glyceryl trinitrate. *Biochemical and Biophysical Research Communications*, **274:** 255–258.

Gryglewsky, R.J., Korbut, R., Ocetkiewicz, A. and Stachura, J. 1978, In vivo method for quantitation for anti-platelet potency of drugs. *Naunyn-Schmiedebergs Archives of Pharmacology*, **302:** 25–30.

Gryglewsky, R.J., Panczeko, B., Korbut, R., Grodzinska, L. and Ocetkiewicz, A. 1975, Corticosteroids inhibit prostaglandin release from perfused mesenteric blood vessels of rabbit and from perfused lungs of sensitized guinea pig. *Prostaglandins*, **10:** 343–355.

Guerinot, F., Poitou, P. and Bohuon, C. 1974, Serotonin synthesis in the rat brain after acetylsalicylic acid administration. *Journal of Neurochemistry*, **22:** 191–192.

Gugiyasu, K. 1976, Periosteal bone formation and arthritis in the pes of aspirin-treated Sprague–Dawley rats. *Medical Journal of the University of Kagoshima*, **28:** 1053–1061.

Guilbaud, G., Bonoist, J.M., Gautron, M. and Kayser, V. 1982, Aspirin clearly depresses responses of ventrobasal thalamus neurons to joint stimuli in arthritic rats. *Pain*, **13:** 153–163.

Gund, P. and Shen, T.Y. 1977, A model for the prostaglandin synthetase cyclooxygenation site and its inhibition by anti-inflammatory arylacetic acid. *Journal of Medicinal Chemistry*, **20:** 1145–1152.

Gutowska-Grzegorczyk, G. and Kałzak, M. 1968, Zmiany histochemiczne i. Enzymohistochemiczne w wątrobie i nerce królików w przewlekłym podawaniu kwasu acetylsalicylowego. *Rhgeumatologia*, VI: 181–185.

Guzman, F., Braun, C., Lim, R.K.S., Potter, G.D. and Rodgers, D.W. 1964, Narcotic and non-narcotic analgesics which block visceral pain by intra-arterial injection of bradykinin and other algesic agents. *Archives Internationales de Pharmacodynamie et de Thérapie*, **149:** 571–588.

Gyires, K. and Torma, Z. 1984, The use of the writhing test in mice for screening different types of analgesics. *Archives Internationales de Pharmacodynamie et de Therapie*, **267:** 131–140.

Habicht, J. and Brune, K. 1983, Inhibition of prostaglandin E_2 release by salicylates, benzoates and phenols: a quantitative structure–activity study. *Journal of Pharmacy and Pharmacology*, **35:** 718–723.

Haegerstam, G. and Edwall, L. 1977, Sodium acetylsalicylate and the role of prostaglandins in the mechanism of intradental pain. *Acta Odontologica Scandinavica*, **35:** 63–67.

Hägsstrom, J.Z. 1999, Leukotriene A$_4$ hydrolase and the committed step in leukotriene B$_4$ biosynthesis. *Clinical Reviews in Allergy and Immunology*, **17:** 111–131.

Haggag, A.A., Mohamed, H.F., Eldawy, M.A. and Elbahrawy, H. 1986, Biochemical studies on the anti-inflammatory activity of salicylates as superoxide radical scavengers. *IRCS Medical Science*, **14:** 1104–1105.

Hail, V., DeMello, M.C.F., Horakova, Z. and Beaven, M.A. 1977, Antiproliferative activity of anti-inflammatory drugs in two mammalian cell culture lines. *Journal of Pharmacology and Experimental Therapeutics*, **202:** 446–464.

Haluska, M.K., Walker, L.P. and Haluska, P.V. 2003, Genetic variation in cyclooxygenase 1: Effects on response to aspirin. *Clinical Pharmacology and Therapeutics*, **73:** 122–130.

Hamberg, M. and Samuelsson, B. 1974, Prostaglandin endoperoxides. VII. Novel transformations of arachidonic acid in guinea pig lung. *Biochemical and Biophysical Research Communications*, **61:** 942–949.

Han, P. and Ardlie, N.G. 1974, Platelet aggregation and release by ADP and thrombin: evidence for two separate effects of ADP on platelets, involvement of fibrinogen in release, and mechanism of inhibitory action of salicylic acid. *British Journal of Haematology*, **26:** 357–372.

Hannah, J., Ruyle, W.V., Jones, H., Matzuk, A.R., Kelly, K.W., Witzel, B.E., Holtz, W.J., Houser, R.W., Shen, T.Y. and Sarret, L.H. 1977, Discovery of diflunisal. *British Journal of Clinical Pharmacology*, **4:** 7S–13S.

Hannah, J., Ruyle, W.V., Jones, H., Matzuk, A.R., Kelly, K.W., Witzel, B.E. Holtz, W.J., Howser, R.A, Shen, T.Y, Sarett, L.H., Lotti, V.J., Ridley, E.A., Van Arman, C.G. and Winter, C.A. 1978, Novel analgesic–antiinflammatory salicylates. *Journal of Medicinal Chemistry*, **21:** 1093–1100.

Hansch, C. 1971, Quantitative structure–activity relationships in drug design. *In:* E.J. Ariens (ed.), *Drug Design*, pp. 271–341. New York: Academic Press.

Hansch, C. and Leo, A. 1979, *Substituted Constants for Correlation Analysis in Chemistry and Biology*. New York: Wiley-Interscience.

Hansch, C., Leo, A., Unger, S.H., Kim, K.H., Nikaitani, D. and Lien, E.J. 1973, 'Aromatic' substituent constants for structure–activity correlations. *Journal of Medicinal Chemistry*, **16:** 1207–1216.

Hansch, C. and Von Kaulla, K.N. 1970, Fibrinolytic congeners of benzoic and salicylic acid. A mathematical analysis of correlation between structure and activity. *Biochemical Pharmacology*, **19:** 2193–2200.

Hansen, H.S. 1974, Inhibition by indomethacin and aspirin of 15-hydroxyprostaglandin dehydrogenase in vitro. *Prostaglandins*, **8:** 95–105.

Hanzlik, P.J. 1927, *Actions and Uses of the Salicylates and Cinchophen in Medicine*. Baltimore: Williams & Wilkins Company.

Harford, D.J. and Smith, M.J.H. 1970, The effect of sodium salicylate on the release of acid phosphatase activity from rat liver lysosomes in vitro. *Journal of Pharmacy and Pharmacology*, **22:** 578–583.

Harris, E.D. 1981, *In:* W.N. Kelley, E.D. Harris, S. Ruddy and C.B. Sledge (eds), *Textbook of Rheumatology*, pp. 896–927. Philadelphia: Saunders.

Haslam, R.J. and McClenaghan, M.D. 1981, Measurement of circulating prostacyclin. *Nature*, **292:** 364–366.

Hatori, A., Shigematsu, A. and Tsuya, A. 1984, The metabolism of aspirin in rats: localization, absorption, distribution and excretion. *European Journal of Drug Metabolism and Pharmacokinetics*, **9:** 205–214.

Hattori, K. 1978a, Studies on anti-edematous effect of cataplasm (Report 2) 1. On the cataplasm contained steroidal hormone as its basis. 2. On effects of several cataplasm on guinea-pig intestinal contraction by extract collected from the rat paw edema induced by carrageenan. *Shika Gakuho*, **78:** 1731–1750.

Hattori, K. 1978b, Studies on anti-edematous effect of cataplasms (report 1), On the cataplasms contained methyl salicylate as its basis. *Shikwa Gakuho*, **78:** 771–784.

Hayasaka, S. and Sears, M. 1978, Effects of epinephrine, indomethacin, acetylsalicylic acid, dexamethasone, and cyclic AMP on the in vitro activity of lysosomal hyaluronidase from the rabbit iris. *Investigative Ophthalmology and Visual Sciences*, **17:** 1109–1113.

Haynes, D.R., Wright, P.F.A., Gadd, S.J., Whitehouse, M.W. and Vernon-Roberts, B. 1993, Is aspirin a prodrug for antioxidant cytokine-modulating oxymetabolites? *Agents and Actions*, **39:** 49–58.

He, T.-C., Cham, T.A., Bogelstein, B. and Kinzler, K.W. 1999, PPARδ is an APC-regulated target of nonsteroidal anti-inflammatory drugs. *Cell*, **99:** 335–345.

Hebborn, P. and Shaw, B. 1963, The action of sodium salicylate and aspirin on some kallikrein systems. *British Journal of Pharmacology*, **20**: 254–263.

Hedqvist, P., Gautam, N. and Lindbom, L. 2000, Interactions between leukotrienes and other inflammatory mediators/modulators in the microvasculature. *American Journal of Respiratory and Critical Care, Medicine*, **161 (Part 2)**: S117–S119.

Hellon, R., Townsend, Y., Laburn, H.P. and Mitchell, D. 1991, Mechanisms of fever. *In:* E. Schönbaum and P. Lomax, P (eds), *Thermoregulation: Pathology, Pharmacology and Therapy*, pp. 19–38. New York: Pergamon Press Inc.

Hemler, M.E., Crawford, C.G. and Lands, W.E.M. 1978, Lipoxygenation activity of purified prostaglandin-forming cyclooxygenase. *Biochemistry*, **17**: 1772–1779.

Henderson, B., Bitensky, L. and Chayen, J. 1978, Mitochondrial oxidative activity in human rheumatoid synovial lining cells. *Annals of the Rheumatic Diseases*, **37**: 548–551.

Henderson, B., Bitensky, L. and Chayen, J. 1979, Glycolytic activity in human synovial lining cells in rheumatoid arthritis. *Annals of the Rheumatic Diseases*, **38**: 63–67.

Henderson, B. and Glynn, L.E. 1981, Metabolic alterations in the synoviocytes in chronically inflamed knee joints in immune arthritis in the rabbit: comparison with rheumatoid arthritis. *British Journal of Experimental Pathology*, **62**: 27–33.

Henrotin, Y., Bassleer, C. and Franchimont, P. 1992, *In vitro* effects of etodolac and acetylsalicylic acid on human chondrocyte metabolism. *Agents and Actions*, **36**: 317–323.

Herman, J.H., Appel, A.M. and Hess, E.V. 1987, Modulation of cartilage destruction by select nonsteroidal antiinflammatory drugs. *Arthritis and Rheumatism*, **30**: 257–265.

Herman, J.H., Appel, A.M., Khosla, R.C. and Hess, E.V. 1989, *In vitro* effect of select nonsteroidal antiinflammatory drugs on the synthesis and activity of anabolic regulatory factors produced by osteoarthritic and rheumatoid synovial tissue. *Journal of Rheumatology*, **16**: 75–81.

Herman, J.H. and Hess, E.V. 1984, Nonsteroidal anti-inflammatory drugs and modulation of cartilaginous changes in osteoarthritis and rheumatoid arthritis. Clinical implications. *American Journal of Medicine*, **77(4B)**: 16–25.

Herman, J.H., Sowder, W.G. and Hess, E.V. 1994, Nonsteroidal antiinflammatory drug modulation of prosthesis pseudomembrane induced bone resorption. *Journal of Rheumatology*, **21**: 338–344.

Hermann, M., Kapiotis, S., Hofbauer, R., Exner, M., Seelos, C., Held, I. and Gmeiner, B. 1999a, Salicylate inhibits LDL oxidation initiated by superoxide/nitric oxide radicals. *FEBS Letters*, **445**: 212–214.

Hermann, M., Kapiotis, S., Hofbauer, R., Seelos, C., Held, I. and Gmeiner, B. 1999b, Salicylate promotes myeloperoxidase-initiated LDL oxidation: antagonization by its metabolite gentisic acid. *Free Radical Biology and Medicine*, **26**: 1253–1260.

Hernandez, M., de Arriba, A.F., Merlos, M., Fuentes, L., Respo, M.S. and Nieto, M.L. 2001, Effect of 4-trifluoromethyl derivatives of salicylate on nuclear factor kappB-dependent transcription in human astrocytoma cells. *British Journal of Pharmacology*, **132**: 547–555.

Hernandez, R., Alemany, M., Bozzo, J., Buchanan, M.R., Ordinas, A. and Bastida, E. 1993, Platelet adhesivity to subendothelium is influenced by polymorphonuclear leukocytes: studies with aspirin and salicylate. *Haemostasis*. **23**: 1–7.

Higgs, E.A., Moncada, S. and Vane, J.R. 1978, Inflammatory effects of prostacyclin (PGI_2) and 6-oxo-PGF_1 alpha in the rat paw. *Prostaglandins*, **16**,:153–162.

Higgs, G.A., Flower, R.J. and Vane, J.R. 1979, A new approach to anti-inflammatory drugs. *Biochemical Pharmacology*, **28**: 1959–1961.

Higgs, G.A., Salmon, J.A., Henderson, B. and Vane, J.R. 1987, Pharmacokinetics of aspirin and salicylate in relation to inhibition of arachidonate cyclooxygenase and antiinflammatory activity. *Proceedings of the National Academy of Science USA*, **84**: 1417–1420.

Higuchi, S., Osada, Y., Shioiri, Y., Tanaka, N., Otomo, S. and Aihara, H. 1985, The modes of antiinflammatory and analgesic actions of aspirin and salicylic acid. *Folia Pharmacologia Japonica*, **85**: 49–57.

Higuchi, S., Tanaka, N., Shioiri, Y., Otomo, S. and Aihara, H. 1986, Two modes of analgesic action of aspirin, and the site of analgesic action of salicylic acid. *International Journal of Tissue Reactions*, **8**: 327–321.

Hinz, B., Brune, K. and Pahl, A. 2000, Prostaglandin E_2 upregulates cyclooxygenase-2 expression in lipopolysaccharide-stimulated RAW 264.7 macrophages. *Biochemical and Biophysical Research Communications*, **272:** 744–748.

Hirschelman, R. and Bekemeier, H. 1977, On the mode of action of salicylic acid and its derivatives. *In:* H. Bekemeier (ed.), *100 Years of the Salicylic Acid as an Anti-rheumatic Drug*, pp. 114–136. Halle (Saale): Martin-Luther Universität Halle-Wittenberg.

Hladovec, J. 1979, Is the antithrombotic activity of 'antiplatelet' drugs based on protection of endothelium? *Thrombosis and Haemostasis*, **41:** 774–778.

Hochgesang, G.P., Rowlinson, S.W. and Marnett, L.J. 2000, Tyrosine-385 is critical for acetylation of cyclooxygenase-2 by aspirin. *Journal of the American Chemical Society*, **122:** 6514–6515.

Hoffmann, O., Klaushofer, K., Koller, K. and Peterlik, M. 1985, Prostaglandin-related bone resorption in cultured neonatal mouse calvaria: evaluation of biopotency of nonsteroidal anti-inflammatory drugs. *Prostaglandins*, **30:** 857–866.

Holman, R.T. 1951, *In:* R.T. Holman and S. Bergstrom (eds), *The Enzymes*, p. 559. New York: Academic Press.

Holtzman, M.J., Turk, J. and Shornick, L.P. 1992, Identification of a pharmacologically distinct prostaglandin H synthetase in cultured epithelial cells. *Journal of Biological Chemistry*, **267:** 21438–21445.

Hong, J., Ekdahl, N., Reynolds, H., Larsson, R. and Nilsson, B. 1999, A new in vitro model to study interaction between whole blood and biomaterials. Studies of platelet and coagulation activation and the effect of aspirin. *Biomaterials*, **20:** 603–611.

Hong, S.C.L. and Levine, L. 1976, Inhibition of arachidonic acid release from cells as the biochemical action of anti-inflammatory corticosteroids. *Proceedings of the National Academy of Sciences USA*, **73:** 1730–1734.

Horn, H., Preclik, G., Stange, E.F. and Ditschuneit, H. 1991, Modulation of arachidonic acid metabolism by olsalazine and other aminosalicylates in leukocytes. *Scandinavian Journal of Gastroenterology*, **26:** 867–879.

Horokova, Z., Bayer, B.M., Almeida, A.P. and Beavan, M.A. 1980, Evidence that histamine does not participate in carrageenan-induced pleurisy in the rat. *European Journal of Pharmacology*, **62:** 17–25.

Hoult, J.R.S. and Moore, P.K. 1978, Sulphasalazine is a potent inhibitor of prostaglandin 15-hydroxydehydrogenase: a possible basis for therapeutic action in ulcerative colitis. *British Journal of Pharmacology*, **64:** 6–8.

Hoult, J.R.S. and Moore, P.K. 1980, Effects of sulphasalazine and its metabolites on prostaglandin synthesis, inactivation and actions on smooth muscle. *British Journal of Pharmacology*, **68:** 719–730.

Hsia, J., Sarin, N., Oliver, J.H. and Goldstein, A.L. 1989a, Aspirin and thymosin increase interleukin-2 and interferon-gamma production by human peripheral blood lymphocytes. *Immunopharmacology*, **17:** 167–173.

Hsia, J., Simon, G.L., Higgins, N., Goldstein, A.L. and Hayden, F.G. 1989b, Immune modulation by aspirin during experimental rhinovirus colds. *Bulletin of the New York Academy of Medicine*, **65:** 45–56.

Hsia, J. and Tang, T. 1992, Aspirin as a biological response modifier. *In:* A.L. Goldstein and E. Garaci (eds), *Combination Therapies*, pp. 131–137. New York: Plenum Press.

Huang, C., Ma, Y.-Y., Hanenberger, D., Cleary, M.P., Bowden, G.T. and Dong, Z. 1997, Inhibition of ultraviolet B-induced activator protein-1 (AP-1) activity by aspirin in AP-1-luciferase transgenic mice. *Journal of Biological Chemistry*, **272:** 26325–26331.

Huang, W.T., Tsai, S.M. and Lin, M.T., 2001, Involvement of brain glutamate release in pyrogenic fever. *Neuropharmacology*, **41:** 811–818.

Huang, W.T., Wang, J.-J. and Lin, M.-T. 2004, Cyclooxygenase inhibitors attenuate augmented glutamate release in organum vasculosum laminae terminalis and fever induced by Staphylococcal enterotoxin A. *Journal of Pharmacological Sciences*, **94:** 192–196.

Hugenberg, S.T., Brandt, K.D. and Cole, C.A. 1993, Effect of sodium salicylate, aspirin and ibuprofen on enzymes required by the chondrocyte for synthesis of chondroitin sulfate. *Journal of Rheumatology*, **20:** 2128–2133.

Hulkower, K.I., Otis, E.R., Wernimont, A.K. and Bell, R.L. 1997, Stimulus dependence of nonsteroidal antiinflammatory drug potency in a cellular assay of prostaglandin H synthase-2. *European Journal of Pharmacology*, **331:** 79–85.

Humes, J.L., Winter, C.A., Sadowski, S.J. and Kuehl, F.A., Jr. 1981, Multiple sites on prostaglandin cyclooxygenase are determinants in the action of nonsteroidal antiinflammatory agents. *Proceedings of the National Academy of Sciences USA*, **78:** 2053–2056.

Hung, D.Y., Mellick, G.D., Masci, P.P., Whitaker, A.N., Whitehouse, M.W. and Roberts, M.S. 1998, Focused antithrombotic therapy: novel anti-platelet salicylates with reduced ulcerogenic potential and higher first-pass detoxification than aspirin in rats. *Journal of Laboratory and Clinical Medicine*, **132:** 469–477.

Hunskaar, S., Berge, O.-G. and Hole, K. 1986, A modified hot-plate test sensitive to mild analgesics. *Behaviour Brain Research*, **21:** 101–108.

Hunskaar, S., Fasmer, O.B. and Hole, K. 1985, Formalin test in mice, a useful technique for evaluating mild analgesics. *Journal of Neuroscience Methods*, **14:** 69–76.

Hurlbut, W.P. 1965, Salicylate: effects on ion transport and after potentials in frog sciatic nerve. *American Journal of Physiology*, **209:** 1295–1303.

Hurley, J.V., Ryan, G.B. and Friedman, A. 1968, The mononuclear response to intrapleural injection in the rat. *Journal of Pathology and Bacteriology*, **91:** 575–587.

Ignarro, L.J. 1972, Lysosome membrane stabilisation in vivo: effects of steroidal and nonsteroidal antiinflammatory drugs on the integrity of rat liver lysosomes. *Journal of Pharmacology and Experimental Therapeutics*, **182:** 179–188.

Ignarro, L.J. and Cech, S.Y. 1975, Lysosomal enzyme secretion from human neutrophils mediated by cyclic CMP: inhibition of cyclic GMP accumulation and neutrophil function by glucocorticosteroids. *Journal of Cyclic Nucleotide Research*, **1:** 283–292.

Ignarro, L.J. and Colombo, C. 1972, Enzyme release from guinea-pig polymorphonuclear leucocyte lysosomes inhibited in vitro by anti-inflammatory drugs. *Nature New Biology*, **239:** 155–157.

Ikeda, U., Kanabe, T., Kawahara, Y., Yokohama, M. and Shimada. K. 1996, Adrenomedullin augments inducible nitric oxide synthase expression in cytokine-stimulated cardic myocytes. *Circulation*, **94:** 2560–2565.

Inglot, A.D. and Wolna, E. 1968, Reactions of non-steroidal anti-inflammatory drugs with the erythrocyte membrane. *Biochemical Pharmacology*, **17:** 269–279.

Inoki, K., Hayashi, T., Kudo, T., Matsumoto, K., Oka, M. and Kotani, Y. 1977, Effects of morphine and acetylsalicylic acid on kinin forming enzyme in rat paw. *Archives Internationales de Pharmacodynamie et de Thérapie*, **228:** 126–135.

Inoki, R., Hayashi, T., Kudo, T. and Matsumoto, K. 1978, Effects of aspirin and morphine on the release of a bradykinin-like substance into the subcutaneous perfusate of the rat paw. *Pain*, **5:** 53–63.

Ishikawa, H., Mori, Y. and Tsurufugi, S. 1968, The inhibitory effects of some steroidal antiinflammatory agents on leukocyte emigration by the carboxymethyl cellulose pouch method. *Journal of the Pharmacological Society of Japan*, **88:** 1491–1493.

Iwata, H. 1987, Effect of anti-arthritic drugs for articular cartilage and synovial fluid. *In:* E.C. Huskisson and Y. Shiokawa (eds), *New Trends in Rheumatology*, pp. 35–46. Amsterdam: Excerpta Medica.

Iwata, H., Okamoto, H. and Koh, S. 1975, Effects of various drugs on serum free and total tryptophan levels and brain tryptophan metabolism in rats. *Japanese Journal of Pharmacology*, **25:** 303–310.

Jacobson, B. and Boström, H. 1964, Studies on the biochemistry of heart valves II. The effect of aging and antiinflammatory drugs on the synthesis of glucosamine 6-phosphate and phosphoadenosine phosphosulfate by bovine heart valves. *Biochimica Biophysica Acta*, **83:** 152–164.

Jaffe, E.A. and Weksler, B.B. 1979, Recovery of endothelial cell prostacyclin production after inhibition by low doses of aspirin. Journal of Clinical Investigation, **63:** 532–535.

Janakidevi, K. and Smith, M.J.H. 1970a, Effects of salicylate on RNA polymerase and on the incorporation of orotic acid and thymine into nucleic acids of rat foetuses in vitro. *Journal of Pharmacy and Pharmacology*, **22:** 249–252.

Janakidevi, K. and Smith, M.J.H. 1970b, Effects of salicylate on the incorporation of orotic acid into nucleic acids of mouse tissues in vivo. *Journal of Pharmacy and Pharmacology*, **22:** 51–55.

Jäpel, M., Lötzerich, H. and Rogalla, K. 1994, Role of acetylsalicylic acid in cytokine stimulation of macrophages in antibody-dependent cellular cytotoxicity (ADCC). *Mediators of Inflammation*, **3:** 419–424.

Jiang, M.-C., Liao, C.-F. and Lee, P.O. 2001, Aspirin inhibits matrix metalloproteinase-2 activity, increases E-cadherin production, and inhibits *in vitro* invasion of tumor cells. *Biochemical and Biophysical Research Communications*, **282:** 671–677.

Jilma, B., Blann, A., Pernerstorfer, T., Stohlawetz, P., Eichler, H.G., Vondrovec, B., Amiral, J., Richter, V. and Wagner, O.F. 1999, Regulation of adhesion molecules during human endotoxemia. No acute effects of aspirin. *American Journal of Respiratory and Critical Care Medicine*, **159**: 857–863.

Johnson, L.R. and Overholt, B.F. 1967, Release of histamine into gastric venous blood following injury by acetic or salicylic acid. *Gastroenterology*, **52**: 505–509.

Jones, H., Fordice, M.W., Greenwalt, R.B., Hannah, J., Jacobs, A., Ruyle, W.V., Walford, G.L. and Shen, T.Y. 1975, Synthesis and analgesic–antiinflammatory activity of some 4- and 5-substituted heteroarylsalicylic acids. *Journal of Medicinal Chemistry*, **21**: 1100–1104.

Jørgensen, O. 1964, Electron-microscopical studies of granulation formation in animals treated with anti-rheumatic drugs. *Acta Pathologica et Microbiologica Scandinavica*, **60**: 349–364.

Jørgensen, T.G., Weis-Fogh, U.S., Nielson, H.H. and Olesen, H.P. 1976, Salicylate- and aspirin-induced uncoupling of oxidative phosphorylation in mitochondria isolated from the mucosal membrane of the stomach. *Scandinavian Journal of Clinical and Laboratory Investigation*, **36**: 649–654.

Juan, H. and Lembeck, F. 1977, Prostaglandin F2alpha reduces the algesic effect of bradykinin by antagonizing the pain enhancing action of endogenously released prostaglandin E. *British Journal of Pharmacology*, **59**: 385–391.

Julius, D. and Basbaum, A.I. 2001, Molecular mechanisms of nociception. *Nature*, **413**: 203–210.

Jurivich, D.A., Sistonen, L., Kroes, R.A. and Morimoto, R.I. 1992, Effect of sodium salicylate on the human heat shock response. *Science*, **255**: 1243–1245.

Jurna, I., Spohrer, B. and Bock, R. 1992, Intrathecal injection of acetylsalicylic acid, salicylic acid and indomethacin depresses C fibre-evoked activity in the rat thalamus and spinal cord. *Pain*, **49**: 249–256.

Kahan, A., Perlik, F., Le Go, A., Delbarre, F. and Giroud, J.P. 1975, Adjuvant-induced arthritis in four inbred strains of rats. An in vitro study of peripheral T and B lymphocytes. *In:* J.P. Giroud, D.A. Willoughby and G.P. Velo (eds), *Future Trends in Inflammation II*, pp. 219–226. Basel: Birkhäuser.

Kalbhen, D.A., Karzel, K. and Domenjoz, R. 1967, The inhibitory effects of some antiphlogistic drugs on the glucosamine incorporation into mucopolysaccharides synthesized by fibroblast cultures. *Medicina et Pharmacologia Experimentalis*, **16**: 185–189.

Kalgutkar, A.S., Crews, B.C., Rowlinson, S.W., Garner, C., Seibert, K. and Marnett, L.W. 1998, Aspirin-like molecules that covalently inactivate cyclooxygenase-2. *Science*, **280**: 1268–1270.

Kam, C.M., Hudig, D. and Powers, J.C. 2000, Granzymes (lymphocyte serine proteases): characterization with natural and synthetic substrates and inhibitors. *Biochimica et Biophysica Acta*, **1477**: 307–323.

Kam, P.C. and See, A.U. 2000, Cyclooxygenase isoenzymes: physiological and pharmacological role. *Anaesthesia*, **55**: 442–449.

Kang, K., Bae, S.J., Kim, M., Lee, D.G., Cho, U., Lee, M.H., Lee, M.S., Nam, S., Kuettner, K.E. and Schwartz, D.E. 2000, Molecular characteristics of the inhibition of human neutrophil elastase by non-steroidal anti-inflammatory drugs. *Exp Mol Med*, **32**: 146–154.

Kaplan, D. 1973, Effect of sodium salicylate on the antibacterial activity of glucose oxidase. *Chemotherapy*, **19**: 235–242.

Kaplan-Harris, L. and Elsbach, P. 1980, The antiinflammatory activity of analogs of indomethacin correlates with their inhibitory effects on phospholipase A2 of rabbit polymorphonuclear leukocytes. *Biochimica et Biophysica Acta*, **618**: 318–326.

Kapusta, M.A. and Mendelson, J. 1967, The inhibition of adjuvant arthritis by statolon. *Arthritis and Rheumatism*, **10**: 288 (Abstracts).

Kargman, S., Wong, E., Greig, G.M., Falgueyret, J.-P., Cromlish, W., Ethier, D., Yergey, J.A., Riendeau, D., Evans, J.F., Kennedy, B., Tagari, P., Francis, D.A. and O'Neill, G.P. 1996, Mechanism of selective inhibition of human prostaglandin G/H synthase-1 and -2 in intact cells. *Biochemical Pharmacology*, **52**: 1113–1125.

Karler, R., Petty, W.C. and Sulkowski, T.S. 1968, Some comparative biochemical effects of 5-substituted congeners of salicylic acid. *Archives Internationales de Pharmacodynamie et de Thérapie*, **173**: 270–280.

Karzel, K., Aulepp, H. and Hack, G. 1973, Effects of recently developed antiphlogistic drugs on viability, reduplication, mean volume and volume distribution on mammalian cells cultured in vitro. *Pharmacology*, **10**: 272–290.

Kataoka, M., Tonooka, K., Ando, T., Imai, K. and Aimoto, T. 1997, Hydroxyl radical scavenging activity of nonsteroidal anti-inflammatory drugs. *Free Radical Research*, **27**: 419–427.

Katsuyama, K., Shichiri, M., Kato, H., Imai, T., Marumo, F. and Hirata, Y. 1999, Differential inhibitory actions by glucocorticoid and aspirin on cytokine-induced nitric oxide production in vascular smooth muscle cells. *Endocrinology*, **140:** 2183–2190.

Katz, J.M., Wilson, T.M., Skinner, S.J.M. and Gray, D.H. 1981, Bone resorption and prostaglandin production by mouse calvaria *in vitro*: response to exogenous prostaglandins and their precursor fatty acids. *Prostaglandins*, **22:** 537–551.

Kauke, Y., Bashey, R.I., Mori, Y. and Angrist, A.A. 1971, Effect of puromycin and salicylate on biosynthesis of mucopolysaccharides and glycoproteins in bovine heart valve. *Experimental and Molecular Pathology*, **15:** 336–344.

Kawashima, H., Kurozumi, S. and Hashimoto, Y. 1985, Aspirin inhibition of 1 alpha-hydroxyvitamin D3 or parathyroid hormone induced hypercalcemia *in vivo* in rats. A mechanism independent of prostaglandin biosynthesis inhibition. *Biochemical Pharmacology*, **34:** 1901–1906.

Kayashima, H., Koga, T. and Onoue, K. 1978, Role of T lymphocytes in adjuvant arthritis. II. Different subpopulations of T lymphocytes functioning in the development of the disease. *Journal of Immunology*, **120:** 1227–1131.

Kemp, A.S., Brown, S., Brooks, P.M. and Neoh, S.H. 1980, Migration of blood and synovial fluid neutrophils obtained from patients with rheumatoid arthritis. *Clinical and Experimental Immunology*, **39:** 240–246.

Kemp, A.S. and Smith, J. 1982, The effect of salicylate on human leucocyte migration. *Clinical and Experimental Immunology*, **49:** 233–238.

Kent, P.W. and Allen, A. 1968, The biosynthesis of intestinal mucins. The effect of salicylate on glycoprotein biosynthesis by sheep colonic and human gastric mucosal tissues in vitro. *Biochemical Journal*, **106:** 645–658.

Kepka-Lenhart, D., Chen, L.C. and Morris, S.M., Jr. 1996, Novel actions of aspirin and sodium salicylate: discordant effects on nitric oxide synthesis and induction of nitric oxide synthase mRNA in a murine macrophage cell line. *Journal of Leukocyte Biology*, **59:** 840–846.

Kim, H.M., Lee, E.H., Shin, T.K., Chung, C.K. and An, N.H. 1998, Inhibition of the induction of the inducible nitric oxide synthase in murine brain microglial cells by sodium salicylate. *Immunology*, **95:** 389–394.

Kimura, A., Roseto, J., Suh, K.Y., Cohen, A.M. and Bing, R.J. 1998, Effect of acetylsalicylic acid on nitric oxide production in infarcted heart in situ. *Biochemical and Biophysical Research Communications*, **251:** 874–878.

Kinsella, J.E., Hwang, D.H., Hu, P., Mai, J. and Shimp, J. 1979, Prostaglandins and their precursors in tissues from rats fed on trans, trans-linoleate. *Biochemical Journal*, **184:** 701–704.

Kirchner, T., Argientieri, D.C., Barbone, A.G., Singer, M., Steber, M., Ansell, J., Beers, S.A, Wachter, M.P., Malloy, E., Stewart, A. and Richie, D.M. 1997, Evaluation of the anti-inflammatory activity of a dual cyclo-oxygenase-2 selective/5-lipoxygenase inhibitor, RWJ 63556, in a canine model of inflammation. *Journal of Pharmacology and Experimental Therapeutics*, **282:** 1094–1101.

Kirkova, M., Ivancheva, E. and Russanov, E. 1995, Lipid-peroxidation and antioxidant enzyme-activity in aspirin-treated rats. *General Pharmacology*, **26:** 613–617.

Kirtikara, K., Swangkul, S. and Ballou, L.R. 2001, The analysis of nonsteroidal antiinflammatory drug selectivity in prostaglandin G/H synthase (PGHS)-null cells. *Inflammation Research*, **50:** 327–332.

Kleine, T.O. and Jungmann, U. 1977, Inhibitory and stimulating effects of prostaglandins A, E, F on the in vitro biosynthesis of glycosaminoglycans and protein from calf rib cartilage. *Pharmacological Research Communications*, **9:** 823–831.

Kleinerman, E.S., Louie, J.S., Wahl, L.M. and Muchmore, A.V. 1981, Pharmacology of human spontaneous monocyte-mediated cytotoxicity. *Arthritis and Rheumatism*, **24:** 774–780.

Klickstein, L.B., Shapleigh, C. and Goetzl, E.J. 1980, Lipoxygenation of arachidonic acid as a source of polymorphonuclear leukocyte chemotactic factors in synovial fluid and tissue in rheumatoid arthritis and spondyloarthritis. *Journal of Clinical Investigation*, **66:** 1166–1170.

Kluger, M.J., Bartfai, T. and Dinarello, C. (eds) 1998, *Molecular Mechanisms of Fever*. New York: Annals of New York Academy of Sciences, Volume 856.

Kofoed, J.A., Tocci, A.A. and Barcelo, A.C. 1973, The acidic glycosaminglycans of the synovial fluid in rheumatoid arthritis. *Experientia (Basel)*, **29:** 680–681.

Koga, T. and Pearson, C.M. 1973, Immunogenicity and arthritogenicity in the rat of an antigen from Mycobacterium tuberculosis wax D. *Journal of Immunology*, **111:** 599–608.

Koga, T., Kotani, S., Narita, T. and Pearson, C.M. 1976a, Induction of adjuvant arthritis in the rat by various bacterial cell walls and their water-soluble components. *International Archives of Allergy and Applied Immunology*, **51**: 206–213.

Koga, T., Van De Sande, B., Yeaton, R. and Pearson, C.M. 1976b, Reevaluation of inguinal lymph node injection for production of adjuvant arthritis in the rat. *International Archives of Allergy and Applied Immunology*, **51**: 359–367.

Koga, T., Tanaka, A. and Pearson, C.M. 1976c, Synergism of immunogenic and adjuvant-active components of mycobacterial wax D in the induction of adjuvant arthritis. *International Archives of Allergy and Applied Immunology*, **51**: 583–593.

Kohashi, O., Pearson, C.M., Watanabe, Y., Kotani, S. and Koga, T. 1976, Structural requirements for arthritogenicity of peptidoglycans from Staphylococcus aureus and Lactobacillus plan arum and analogous synthetic compounds. *Journal of Immunology*, **116**: 1635–1639.

Koo, H.N., Oh, S.Y., Kang, K.I., Moon, D.Y., Kim, H.D. and Kang, H.S. 2000, Modulation of HSP70 and HSP90 expression by sodium salicylate and aspirin in fish cell line CHSE-214. *Zoological Science*, **17**: 1275–1282.

Kopp, E. and Ghosh, S. 1994, Inhibition of NF-kappa B by sodium salicylate and aspirin. *Science*, **265**: 956–959.

Koppal, T., Petrova, T.V. and Van Eldik, L.J. 2000, Cyclopentenone prostaglandin 15-deoxy-$\Delta^{12,14}$-prostaglandin J_2 acts as a general inhibitor of inflammatory responses in activated BV-2 microglial cells. *Brain Research*, **867**: 115–121.

Korbut, R. and Moncada, S. 1978, Prostacyclin (PGI_2) and thromboxane A2 interaction in vivo. Regulation by aspirin and relationship with anti-thrombotic therapy. *Thrombosis Research*, **13**: 489–500.

Kourounakis, L. and Kapusta, M.A. 1974, Effect of Freund's adjuvant on the mitogenic responses of rat lymphocytes. *Annals of the Rheumatic Diseases*, **33**: 185–189.

Kourounakis, L. and Kapusta, M.A. 1976, Restoration of diminished T-cell function in adjuvant induced disease by methotrexate: evidence for two populations of splenic T-cell suppressors. *Journal of Rheumatology*, **3**: 346–354.

Krebs, H.A. and De Gasquet, P. 1964, Inhibition of gluconeogenesis by alpha-oxo acids. *Biochemical Journal*, **90**: 149–154.

Kuberasampath, T. and Rose, S.M. 1980, Effect of adjuvant arthritis on collagenase and certain lysosomal enzymes in relation to the catabolism of collagen. *Agents and Actions*, **10**: 78–84.

Kuehl, F.A., Jr. and Egan, R.W. 1978, Prostaglandins and related mediators in pain. *In:* K. Miehlke (ed.), *Diflunisal in Clinical Practice*, pp. 23–30. New York: Futura.

Kuehl, F.A., Jr., Humes, J.L., Egan, R.W., Ham, E.A., Beveridge, G.C. and Van Arman, C.G. 1977, Role of prostaglandin endoperoxide PGG_2, in inflammatory processes. *Nature*, **265**: 170–173.

Kuehl, F.A., Jr., Humes, J.L., Ham, E.A., Egan, R.W. and Dougherty, H.W. 1980, Inflammation: the role of peroxidase-derived products. *In:* B. Samuelsson, P. Ramwell and R. Paoletti (eds), *Advances in Prostaglandin and Thromboxane Research*, Vol. 6, pp. 77–86. New York: Raven Press.

Kuehl, F.A., Jr., Humes, J.L., Torchiana, M.L., Ham, E.A. and Egan, R.W. 1979, Oxygen-centred radicals in inflammatory processes. *In:* G. Weissman, B. Samuelsson and R. Paoletti (eds), *Advances in Inflammation Research*, Vol. 1, pp. 419–430. New York: Raven Press.

Kulkarni, S.K., Jain, N.K. and Singh, A. 2000, Cyclooxygenase isoenzymes and new therapeutic potential for selective COX-2 inhibitors. *Methods and Findings in Experimental and Clinical Pharmacology*, **22**: 291–298.

Kulonen, E. and Potila, M. 1975, Effect of the administration of antirheumatic drugs on experimental granuloma in rat. *Biochemical Pharmacology*, **24**: 219–225.

Kurose, I., Wolf, R., Miyasaka, M., Anderson, D.C. and Granger, D.N. 1996, Microvascular dysfunction induced by nonsteroidal anti-inflammatory drugs: role of leukocytes. *American Journal of Physiology*, **270**: G363–G369.

Kwon, G., Hill, J.R., Corbett, J.A. and McDaniel, M. 1997, Effects of aspirin on nitric oxide formation and de novo protein synthesis by RINm5F cells and rat islets. *Molecular Pharmacology*, **52**: 398–405.

Lafeber, F.P.G., Van Roy, H., Wilbrink, B., Huber-Bruning, O., Bijlsma, J.W.J. 1992, Human osteoarthritic cartilage is synthetically more active but in culture less vital than normal cartilage. *Journal of Rheumatology*, **19**: 123–129.

Lam, B.K. and Austen, K.F. 1992, Leukotrienes. Biosynthesis, release and actions. *In:* J.I. Gallin, I.M.

Goldstein and R. Snyderman (eds), *Inflammation: Basic Principles and Clinical Correlates*, pp. 139–147. New York: Raven Press.

Lands, W.E.M. 1979, The biosynthesis and metabolism of prostaglandins. *Annual Review of Physiology*, **41:** 633–652.

Lane, N.E., Bauer, D.C., Nevitt, M.C., Pressman, A.R. and Cummings, S.R. 1997, Aspirin and nonsteroidal antiinflammatory drug use in elderly women: effects on a marker of bone resorption. *The Study of Osteoporotic Fractures Research Group*, **24:** 1132–1136.

Laneuville, O., Breuer, D.K., DeWitt, D.L., Hla, T., Funk, C.D. and Smith, W.L. 1994, Differential inhibition of human prostaglandin endoperoxide H synthases-1 and –2 by nonsteroidal anti-inflammatory drugs. *Journal of Pharmacology and Experimental Therapeutics*, **271:** 927–934.

Lauritsen, K., Laursen, L.S., Bukhave, K. and Rask-Madsen, J. 1986, Effects of topical 5-aminosalicylic acid and prednisolone on prostaglandin E2 and leukotriene B4 levels determined by equilibrium in vivo dialysis of rectum in relapsing ulcerative colitis. *Gastroenterology*, **91:** 837–844.

Lauterburg, B.H., Grattagliano, I., Gmur, R., Stalder, M. and Hilderbrand, P. 1995, Noninvasive assessment of the effect of xenobiotics on mitochondrial-function in human beings – studies with acetylsalicylic-acid and ethanol with the use of the carbon 13-labelled ketoisocaproate breath test. *Journal of Laboratory and Clinical Medicine*, **125:** 378–383.

Law, E. and Lewis, A.J. 1977, The effect of systemically and topically applied drugs on the ultraviolet-induced erythema in the rat. *British Journal of Pharmacology*, **59:** 591–597.

Leader, J.E. and Whitehouse, M.W. 1966, Uncoupling of oxidative phosphorylation by some salicylamide derivatives. *Biochemical Pharmacology*, **15:** 1379–1387.

Lecomte, M., Laneuville, O., Ji, C., DeWitt, D.L. and Smith, W.L. 1994, Acetylation of human prostaglandin endoperoxide synthase-2 (cyclooxygenase-2) by aspirin. *Journal of Biological Chemistry*, **269:** 13207–13215.

Lee, J.I. and Burkart, G.J. 1998, Nuclear factor kappa B: important transcription factor and therapeutic target. *Journal of Clinical Pharmacology*, **38:** 981–983.

Lee, K.H. and Spencer, M.R. 1969, Studies on mechanism of action of salicylates. V. Effect of salicylic acid on enzymes involved in mucopolysaccharides synthesis. *Journal of Pharmaceutical Sciences*, **58:** 464–468.

Lehmeyer, J.E. and Johnston, R.B. 1978, Effect of anti-inflammatory drugs and agents that elevate intracellular cyclic AMP on the release of toxic oxygen metabolites by phagocytes: studies in a model of tissue-bound IgG. *Clinical Immunobiology and Immunopathology*, **9:** 482–490.

Lemay, S., Lebedeva, T.V. and Singh, A.K. 1999, Inhibition of cytokine gene expression by sodium salicylate in a macrophage cell line through an NF-κB-independent mechanism. *Clinical and Diagnostic Laboratory Immunology*, **6:** 567–572.

Leme, J.G., Bechara, G.H. and Sudo, L.S. 1977, The proinflammatory function of lymphocytes in non-immune inflammation: effect of steroidal and non-steroidal anti-inflammatory agents. *British Journal of Experimental Pathology*, **58:** 703–711.

Levitan, H. and Barker, J.L. 1972a, Effect of non-narcotic analgesics on membrane permeability of mulluscan neurones. *Nature New Biology*, **239:** 55–57.

Levitan, H. and Barker, J.L. 1972b, Membrane permeability: cation selectivity reversibly altered by salicylate. *Science (Washington DC)*, **178:** 63–64.

Levitan, H. and Barker, J.L. 1972c, Salicylate: a structure–activity study of its effects on membrane permeability. *Science (Washington DC)*, **176:** 1423–1425.

Lewis, D.A. 1970, The actions of some non-steroidal drugs on lysosomes. *Journal of Pharmacy and Pharmacology*, **22:** 909–912.

Lewis, G.P. 1977, The mediator concept as it developed: histamine and bradykinin. *In:* I.L. Bonta, J. Thompson and K. Brune (eds), *Inflammation: Mechanisms and their Impact on Therapy*, pp. 93–97. Basel: Birkhäuser.

Lewis, G.P. 1979, Introduction to prostaglandins and inflammation. *In:* K.D. Rainsford and A.W. Ford-Hutchinson (eds), *Prostaglandins and Inflammation*, pp. 7–15. Basel: Birkhäuser.

Lewis, G.P. and Piper, P.J. 1975, Inhibition of release of prostaglandins as an explanation of some of the actions of anti-inflammatory corticosteroids. *Nature*, **254:** 308–311.

Lewis, G.P. and Whittle, B.J.R. 1977, The inhibition of histamine release from rat peritoneal mast cells by non-steroid anti-inflammatory drugs and its reversal by calcium. *British Journal of Pharmacology*, **61:** 229–235.

Li, A., Seipelt, H., Muller, C., Shi, Y. and Artmann, M. 1999, Effects of salicylic acid derivatives on red blood cell membranes. *Pharmacology and Toxicology*, **85:** 206–211.

Li, S., Wang, Y., Matsumura, K., Ballou, L.R., Moreham, S.G. and Blatteis, C.M. 1999, The febrile response to lipopolysaccharide is blocked in cyclooxygenase-2 mice. *Brain Research*, **825:** 86–94.

Lim, R.K.S. 1966, Salicylate analgesia. *In:* M.J.H. Smith and P.K. Smith (eds), *Salicylates. A Critical Bibliographic Review*, pp. 155–202. New York: Wiley-Interscience.

Lim, R.K.S., Gutzman, F., Rodgers, D.W., Goto, K., Braun, C., Dickerson, G.D. and Engle, R.J. 1964, Site of action of narcotic and non-narcotic analgesics determined by blocking bradykinin-evoked visceral pain. *Archives Internationales de Pharmacodynamie et de Thérapie*, **152:** 25–58.

Lim, R.K.S., Kraautamer, G., Guzman, F. and Fulp, R.R. 1969, Central nervous system activity associated with the pain evoked by bradykinin and its alteration by morphine and aspirin. *Proceedings of the National Academy of Sciences USA*, **63:** 705–712.

Lindy, S., Uitto, J., Turto, H., Rokkanen, P. and Vainio, K. 1971, Lactate dehydrogenase in the synovial tissue in rheumatoid arthritis: total activity and isoenzyme composition. *Clinica Chimica Acta*, **31:** 19–23.

Lipiello, L., Hall, D. and Mankin, H.J. 1977, Collagen synthesis in normal and osteoarthritic human cartilage. *Journal of Clinical Investigation*, **59:** 593–600.

Liptay, S., Bachem, M., Häcker, G., Adler, G., Debatin, K.M. and Schmid, R.M. 1999, Inhibition of nuclear factor kappa B and induction of apoptosis in T-lymphocytes by sulfasalazine. *British Journal of Pharmacology*, **128:** 1361–1369.

Liu, L., Leech, J.A., Urch, R.B. and Silverman, F.S. 1997, In vivo salicylate hydroxylation: a potential biomarker for assessing acute ozone exposure and effects in humans. *American Journal of Respiratory and Critical Care Medicine*, **156:** 1405–1412.

Liyanage, S.P., Currey, H.L.F. and Vernon-Roberts, B. 1975, Influence of tubercle aggregate size on severity of adjuvant arthritis in the rat. *Annals of the Rheumatic Diseases*, **34:** 49–53.

Lo, C.J., Fu, F.R. and Cryer, H.G. 2000, Cyclooxygenase 2 (COX-2) gene activation is regulated by cyclic adenosine monophosphate. *Shock*, **13:** 41–45.

Lopez-Farre, A., Caramelo, C., Esteban, A., Alberola, M.L., Millas, I., Monton, M. and Casado, S. 1995, Effects of aspirin on platelet-neutrophil interactions. Role of nitric oxide and endothelin-1. *Circulation*, **91:** 2080–2088.

Lora, M., Denault, J.-B., Leduc, R. and de Brum-Fernandes, A.J. 1998, Systematic pharmacological approach to the characterization of NSAIDs. *Prostaglandins, Leukotrienes and Essential Fatty Acids*, **59:** 55–62.

Lora, M., Morisset, S., Menard, H.-A., Leduc, R. and de Brum-Fernandes, A.J. 1997, Expression of recombinant human cyclooxygenase isoenzymes in transfected COS-7 cells in vitro and inhibition by tenoxicam, indomethacin and aspirin. *Prostaglandins, Leukotrienes and Essential Fatty Acids*, **56:** 361–367.

Loveday, C. and Eisen, V. 1973, Suppression of lymphocyte transformation by salicylate. *Lancet*, **2:** 676.

Lukie, B.E. and Forstner, G.G. 1972, Synthesis of intestinal glycoproteins. Inhibition of (1–^{14}C) glucosamine incorporation by sodium salicylate in vitro. *Biochimica et Biophysica Acta*, **273:** 380–388.

Lund-Olesen, K. 1970, Oxygen tension in synovial fluids. *Arthritis and Rheumatism*, **13:** 769–776.

Lunec, J. and Dormandy, T.L. 1979, Fluorescent lipid-peroxidation products in synovial fluid. *Clinical Science*, **56:** 53–59.

Lunec, J., Halloran, S.P., White, A.G. and Dormandy, T.L. 1981, Free-radical oxidation (peroxidation) products in serum and synovial fluid in rheumatoid arthritis. *Journal of Rheumatology*, **8:** 233–245.

MacIntyre, D.E., Pearson, J.D. and Gordon, J.L. 1978, Localisation and stimulation of prostacyclin production in vascular cells. *Nature*, **271:** 549–551.

Maeda, M., Tanaka, Y., Suzuki, T. and Nakamura, K. 1977, Pharmacological studies on carprofen, a new non-steroidal anti-inflammatory drug, in animals. *Folia Pharmacology Japonica*, **73:** 757–777.

Majerus, P.W. and Stanford, N. 1977, Comparative effects of aspirin and diflunisal on prostaglandin synthetase from human platelets and sheep seminal vesicles. *British Journal of Clinical Pharmacology*, **4:** 15S–18S.

Maldofsky, H. and Warsh, J.J. 1978, Plasma tryptophan and musculoskeletal aspirin in non-articular rheumatism ('fibrositis syndrome'). *Pain*, **5:** 65–71.

Malemud, C.J., Papay, R.S., Hasler, P. and Kammer, G.M. 1994, The effect of nonsteroidal anti-inflammatory drugs on cAMP-dependent protein kinase-mediated phosphorylation by human chondrocytes in culture. *Clinical and Experimental Rheumatology*, **12:** 149–156.

Malmberg, A.B. and Yaksh, T.L. 1992a, Antinociceptive actions of spinal nonsteroidal anti-inflammatory agents on the formalin test in the rat. *Journal of Pharmacology and Experimental Therapeutics*, **263:** 136–146.

Malmberg, A.B. and Yaksh, T.L. 1992b, Hyperalgesia mediated by spinal glutamate or substance P receptor blocked by spinal cyclooxygenase inhibition. *Science*, **257:** 1276–1279.

Malmgren, R., Olssen, P. and Tornling, G. 1979, Uptake and release of serotonin in adhering platelets. Relationship to time and effect of acetylsalicylic acid. *Thrombosis Research*, **15:** 803–812.

Mancini, J.A., O'Neill, G.P., Bayly, C. and Vickers, P. 1994, Mutation of Ser-516 in human prostaglandin G/H synthase-2 to methionine or aspirin acetylation of this residue stimulates 15-R-HETE synthesis. *FEBS Letters*, **342:** 33–37.

Mancini, J.A., Vickers, P.J., O'Neill, G.P., Boily, C., Falgueyret, J.-P. and Reindeau, D. 1997, Altered sensitivity of aspirin-acetylated prostaglandin G/H synthase-2 to inhibition by nonsteroidal anti-inflammatory drugs. *Molecular Pharmacology*, **51:** 52–60.

Mangla, J.C., Kim, Y.M. and Rubulis, A.A. 1974, Adenylate cyclase stimulation by aspirin in rat gastric mucosa. *Nature*, **250:** 61–62.

Marcelon, G., Cros, J. and Guiraud, R. 1976, Activity of anti-inflammatory drugs on an experimental model of osteoarthritis. *Agents and Actions*, **6:** 191–194.

Martel, R.R., Klicius, J. and Herr, F. 1974, Determination of the therapeutic mean effective doses of several anti-inflammatory agents in adjuvant arthritic rats. *Canadian Journal of Physiology and Pharmacology*, **52:** 791–796.

Masferrer, J.L., Zweifel, B.S., Manning, P.T., Hauser, S.D., Leahy, K.M., Smith, W.G., Isakson, P.C. and Seibert, K. 1994, Selective inhibition of inducible cyclooxygenase 2 in vivo is antiinflammatory and nonulcerogenic. *Proceedings of the National Academy of Sciences USA*, **91:** 3228–3232.

Maskos, Z., Rush, J.D. and Koppenol, W.H. 1990, The hydroxylation of the salicylate anion by a fenton reaction and Γ-radiolysis: a consideration of the respective mechanisms. *Free Radical Biology and Medicine*, **8:** 153–162.

Mason, F.A., Hernandez, D.E., Nemeroff, C.B., Adcock, J.W., Hatley, O.L. and Prange, A.J., Jr. 1982, Interaction of neurotensin with prostaglandin E_2 and prostaglandin synthesis inhibitors: effects on colonic temperature in mice. *Regulatory Peptides*, **4:** 285–292.

Mathur, P.P., Smyth, R.D., Witmer, C.M. and Carr, G.S. 1978, Effects of a nonsteroidal anti-inflammatory agent and phenobarbital on hepatic microsomal mono-oxygenases in adjuvant disease in the rat. *Biochemical Pharmacology*, **27:** 1771–1774.

Matzner, Y., Drexler, R. and Levy, M. 1984, Effect of dipyrone, acetylsalicylic acid and acetaminophen on human neutrophil chemotaxis. *European Journal of Clinical Investigation*, **14:** 440–443.

Mayer, D.J. and Price, D.D. 1976, Central nervous system mechanisms of analgesia. *Pain*, **2:** 379–404.

McAdam, B.F., Catella-Lawson, F., Mardini, I.A., Kapoor, S., Lawson, J.A. and Fitzgerald, G.A. 1999, Systemic biosynthesis of prostacyclin by cyclooxygenase (COX)-2: the human pharmacology of a selective inhibitor of COX-2. *Proceedings of the National Academy of Sciences USA*, **96:** 272–277.

McArthur, J.N. and Dawkins, P.D. 1969, The effect of sodium salicylate on the binding of L-tryptophan to serum proteins. *Journal of Pharmacy and Pharmacology*, **21:** 744–750.

McArthur, J.N., Dawkins, P.D., Smith, M.J.H. and Hamilton, E.D.B. 1971a, Mode of action of antirheumatic drugs. *British Medical Journal*, **2:** 677–679.

McArthur, J.N. and Smith, M.J.H. 1969, Inhibition of glutamate decarboxylase by salicylate in vitro. *Journal of Pharmacy and Pharmacology*, **21:** 21–23.

McArthur, J.N. and Smith, M.J.H. 1972, Action of salicylates. *New England Journal of Medicine*, **287:** 361.

McArthur, J.N., Smith, M.J.H. and Dawkins, P.D. 1975, British Patent 1,415,506. *Chemical Abstracts*, **84:** 126765 f.

McArthur, J.N., Smith, M.J.H. and Hamilton, E.D.B. 1971b, Protein-bound peptides in human serum. *British Medical Journal*, **4:** 230.

McCord, J.M., Stokes, S.H. and Wong, K. 1979, Superoxide radical as a phagocyte-produced chemical mediator of inflammation. *In:* G. Weissman, B. Samuelsson and R. Paoletti (eds), *Advances in Inflammation Research*, Vol. 1, pp. 273–280. New York: Raven Press.

McCormack, K. and Brune, K. 1991, Dissociation between the antinociceptive and anti-inflammatory effects of the nonsteroidal anti-inflammatory drugs. A survey of their analgesic efficacy. *Drugs*, **41:** 533–547.

McCoubrey, A., Smith, M.H. and Lane, A.C. 1970, Inhibition of enzymes by alkylsalicylic acids. *Journal of Pharmacy and Pharmacology*, **22:** 333–337.

McKenzie, L.S. Horsburgh, B.A., Ghosh, P. and Taylor, T.K.F. 1976, Effect of anti-inflammatory drugs on sulphated glycosaminoglycan synthesis in aged human articular cartilage. *Annals of the Rheumatic Diseases*, **35:** 487–497.

McKenzie, L.S., Horsburgh, B.A., Ghosh, P. and Taylor, T.K.F. 1977, Sulphated glycosaminoglycan synthesis in normal and osteoarthritic hip cartilage. *Annals of the Rheumatic Diseases*, **36:** 369–373.

McLaughlin, S. 1973, Salicylates and phospholipid bilayer membranes. *Nature*, **243:** 234–236.

McNaughton, B.L. 1998, The neurophysiology of reminiscence. *Neurobiology of Learning and Memory*, **70:** 252–267.

Mehler, E.L. and Gerhards, J. 1987, Electronic determinants of the anti-inflammatory action of benzoic and salicylic acids. *Molecular Pharmacology*, **31:** 284–293.

Mehlman, M.A., Tobin, R.B. and Sporn, E.M. 1972, Oxidative phosphorylation and respiration by rat liver mitochondria from aspirin-treated rats. *Biochemical Pharmacology*, **21:** 3279–3285.

Meijer, C.J.M., Van De Putte, L.B.A., Lafeber, G.J.M., De Haas, E. and Cats, A. 1980, Membrane and transformation characteristics of lymphocytes isolated from the synovial membrane and paired peripheral blood of patients with rheumatoid arthritis. *Annals of the Rheumatic Diseases*, **39:** 75–81.

Mende, T.J. 1972, Enhancement of the antihemostatic effect of acetylsalicyic acid by ring iodination. *Pharmacology*, **7:** 249–254.

Mense, S. 1983, Basic neurobiologic mechanisms of pain and analgesia. *American Journal of Medicine*, **75:** 4–14.

Messing, R.B. and Lytle, L.D. 1977, Serotonin-containing neurons: their possible role in pain and analgesia. *Pain*, **4:** 1–21.

Metzke, H. 1977, Age dependence of adjuvant arthritis in rats. *Allergie und Immunologie*, **23:** 206–210.

Meyer, O.O. and Howard, B. 1943, Production of hypoprothrombinemia and hypercoagulability of the blood with salicylates. *Proceedings of the Society for Experimental Biology and Medicine*, **53:** 234–237.

Miehlke, K. (ed.) 1978, *Diflunisal in Clinical Practice*. New York: Futura.

Mifflin, R.C., Saada, J.I., Di Mari, J.F., Valentich, J.D., Adegboyega, P.A. and Powell, D.W. 2004, Aspirin-mediated COX-2 transcript stabilization via sustained p38 activation in human intestinal myofibroblasts. *Molecular Pharmacology*, **65:** 470–478.

Mikulikova, D. and Trnavsky, K. 1980, Influence of colchicine derivatives on lysosomal enzyme release from polymorphonuclear leukocytes and intracellular levels of cAMP after phagocytosis of monosodium urate crystals. *Biochemical Pharmacology*, **29:** 2146–2148.

Milanino, R. and Velo, G.P. 1981, Multiple actions of copper in control of inflammation: studies in copper-deficient rats. *In:* K.D. Rainsford, K. Brune and M.W. Whitehouse (eds), *Trace Elements in the Pathogenesis and Treatment of Inflammation*, pp. 209–230. Basel: Birkhäuser.

Militzer, K. 1975, Histometrical studies on induced paw inflammation by carrageenan and Freund's complete adjuvant in rats. *Arzneimittel Forschung*, **25:** 1884–1889.

Militzer, K. and Hirche, H. 1981, Prophylactic and therapeutic application of anti-inflammatory agents on the test method of carrageenan paw edema in the rat. *Arzneimittel Forschung*, **31:** 26–32.

Miller, W.S. and Smith, J.G. 1966, Effect of acetylsalicylic acid on lysosomes. *Proceedings of the Society for Experimental Biology and Medicine*, **122:** 634–636.

Mills, D.G., Hirst, M. and Philp, R.B. 1974a, The effects of some salicylate analogues on human blood platelets. 2. The role of platelet acetylation in the inhibition of platelet aggregation. *Life Sciences*, **14:** 673–684.

Mills, D.G., Philp, R.B. and Hirst, M. 1974b, The effects of some salicylate analogues on human blood platelets. 1. Structure activity relationships and the inhibition of platelet aggregation. *Life Sciences*, **14:** 659–672.

Milton, A.S. 1976, Modern views on the pathogenesis of fever and the mode of action of antipyretic drugs. *Journal of Pharmacy and Pharmacology*, **28 (Suppl. 4):** 393–399.

Milton, A.S. 1992, Antipyretic actions of aspirin. *In:* J.R. Vane and R. Botting (eds), *Aspirin and Other Salicylates*, pp. 213–244. London: Chapman Hall Medical.

Minta, J.O. and Williams, M.D. 1985, Some nonsteroidal antiinflammatory drugs inhibit the generation of superoxide anions by activated polymorphs by blocking ligand-receptor interactions. *Journal of Rheumatology*, **12:** 751–757.

Misawa, M., Nishimura, T., Yanaura, S., Ueno, K., Igarashi, T., Sati, T. and Kitagawa, H. 1985, Attenuation of aspirin analgesia by calcium loading in healthy subjects. *Journal of Pharmacobiodynamics*, **8:** 525–531.

Mitchell, J.A., Saunders, M., Barnes, P.J., Newton, R. and Belvisi, M.G. 1997, Sodium salicylate inhibits cyclo-oxygenase-2 activity independently of transcription factor (nuclear factor kappa B) activation: role of arachidonic acid. *Molecular Pharmacology*, **51:** 907–912.

Mitznegg, P., Estler, C.-J., Loew, F.W. and Van Seil, J. 1977, Effect of salicylates on cyclic AMP in isolated rat gastric mucosa. *Acta Hepato-Gastroenterologica*, **24:** 372–376.

Miyamoto, T., Ogino, N., Yamamoto, S. and Hayaishi, O. 1976, Purification of prostaglandin endoperoxide synthetase from bovine vesicular gland microsomes. *Journal of Biological Chemistry*, **251:** 2629–2636.

Mizushima, Y., Ishii, Y. and Masumoto, S. 1975, Physico-chemical properties of potent nonsteroidal anti-inflammatory drugs. *Biochemical Pharmacology*, **24:** 1589–1592.

Mobarok Ali, A.T.M. and Morley, J. 1980, Actions of aspirin and dipyridamole on lymphocyte activation. *Agents and Actions*, **10:** 509–512.

Mohr, W. and Wild, A. 1976, Adjuvant arthritis. *Arzneimittel Forschung*, **26:** 1860–1866.

Mohr, W. and Wild, A. 1977, Pannus development in adjuvant arthritis of the rat – an autoradiographic study. *Wiener Klinische Wochenschrift*, **89:** 757–765.

Moir, A.T.B. and Eccleston, D.J. 1968, The effects of precursor loading in the central metabolism of 5-hydroxyindoles. *Journal of Neurochemistry*, **15:** 1093–1108.

Moncada, S. 1999, Nitric oxide: discovery and impact on clinical medicine. *Journal of the Royal Society of Medicine*, **92:** 164–169.

Moncada, S., Palmer, R.M. and Higgs, E.A. 1991, Nitric oxide: physiology, pathophysiology, and pharmacology. *Pharmacological Reviews*, **43:** 109–142.

Moon, C., Ahn, M., Jee, Y., Heo, S., Kim, S., Kim, H., Sim, K.B., Koh, C.S., Shin, Y.G. and Shin, T. 2004, Sodium salicylate-induced amelioration of experimental autoimmune encephalomyelitis in Lewis rats is associated with the suppression of inducible nitric oxide synthase and cyclooxygenases. *Neuroscience Letters*, **356:** 123–126.

Moore, P.K. and Hoult, J.R. 1982, Selective actions of aspirin- and sulphasalazine-like drugs against prostaglandin synthesis and breakdown. *Biochemical Pharmacology*, **31:** 969–971.

Mori, T., Tsubura, Y., Fujihira, E., Otoma, S. and Nakazawa, M. 1970, Studies on rat adjuvant disease (III) Clinical and histologic changes of the disease. *Pharmacometrics*, **4:** 1051–1062.

Moroz, L.A. 1977, Increased blood fibrinolytic activity after aspirin ingestion. *New England Journal of Medicine*, **296:** 527–529.

Morton, D.M. and Chatfield, D.H. 1970, The effects of adjuvant-induced arthritis on the liver metabolism of drugs in rats. *Biochemical Pharmacology*, **19:** 473–481.

Moss, J., DeMello, M.C., Vaughan, M. and Beaven, M.A. 1976, Effect of salicylates on histamine and L-histidine metabolism. Inhibition of imidazoleacetate phosphoribosyl transferase. *Journal of Clinical Investigation*, **58:** 137–141.

Mossman, S.S., Coleman, J.M. and Gow, P.J. 1981, Synovial fluid lactic acid in septic arthritis. *New Zealand Medical Journal*, **93:** 115–117.

Movat, H.Z. 1979, Kinins and the kinin system as inflammatory mediators. *In:* J.C. Houch (ed.), *Chemical Mediators of the Inflammatory Process*, pp. 47–112. Amsterdam: Elsevier/North Holland Biomedical Press.

Muirden, K.D. and Peace, G. 1969, Light and electron microscopy studies in carrageenan, adjuvant, and tuberculin-induced arthritis. *Annals of the Rheumatic Diseases*, **28:** 392–401.

Mullnik, H. and Von Blomberg, M. 1980, Influence of anti-inflammatory drugs on the interaction of lymphocytes and macrophages. *Agents and Actions*, **10:** 512–515.

Murota, S.I., Abe, M., Otsuka, K. and Chang, W.-C. 1977, Stimulative effect of prostaglandins on production of hexosamine-containing substances by cultured fibroblasts (2). Early effect of various prostaglandins at various doses. *Prostaglandins*, **13:** 711–717.

Musil, J., Weissova, J., Adam, M. and Prokopec, J. 1968, The influence of anti-inflammatory drugs on the glycoprotein biosynthesis in vitro. *Pharmacology*, **1:** 295–302.

Mustard, J.F. and Packham, M.A. 1975, Platelets, thrombosis and drugs. *Drugs*, **9:** 19–76.

Muth-Selbach, U.S., Tegeder, I., Brune, K. and Geisslinger, G. 1999, Acetaminophen inhibits spinal prostaglandin E$_2$ release after peripheral noxious stimulation. *Anesthesiology*, **91:** 231–239.

Nakagawa, H. and Bentley, J.P. 1971, Salicylate-induced inhibition of collagen and mucopolysaccharide biosynthesis by a chick embryo cell-free system. *Journal of Pharmacy and Pharmacology*, **23:** 399–406.

Nakamura, H., Ishii, K., Imazu, C., Motoyoshi, S., Yokoyama, Y., Seto, Y. and Shimizu, M. 1982, Analgesic activity of a non-steroidal anti-inflammatory drug, zomepirac sodium, in experimental animals. *Folia Pharmacology Japonica*, **79:** 493–508.

Nakamura, H. and Shimizu, M. 1979, Accelerated granuloma formation in adjuvant-induced arthritic rats and its modification by antirheumatic drugs. *Journal of Pharmacology and Experimental Therapeutics*, **211:** 20–25.

Nakamura, H., Shimoda, A., Ishii, K. and Kadokawa, T. 1986, Central and peripheral analgesic action of non-acidic non-steroidal anti-inflammatory drugs in mice and rats. *Archives Internationale de Pharmacodynamie et de Therapie*, **282:** 16–25.

Nakao, S., Ogata, Y., Shimizu-Sasaki, E., Yamazaki, M., Furuyama, S. and Sugiya, H. 2000, Activation of NFκB is necessary for IL-1β-induced cyclooxygenase-2 (COX-2) expression in human gingival fibroblasts. *Molecular and Cellular Biochemistry*, **209:** 113–118.

Nakashima, T., Hori, T., Kiyohara, T. and Shibata, M. 1985, Effects of endotoxin and sodium salicylate on the preoptic thermosensitive neurons in tissue slices. *Brain Research Bulletin*, **15:** 459–463.

Narumiya, S., Salmon, J.A., Cottee, F.H., Weatherby, B.C. and Flower, R.J. 1981, Arachidonic acid 15-lipoxygenase from rabbit peritoneal polymorphonuclear leukocytes. Partial purification and properties. *Journal of Biological Chemistry*, **256:** 9583–9592.

Nathan, C. and Xie, Q.W. 1994, Regulation of biosynthesis of nitric oxide. *Journal of Biological Chemistry*, **269:** 13725–13728.

Neto, F.R. 1980, Further studies on the actions of salicylates on membranes. *European Journal of Pharmacology*, **68:** 155–162.

Neto, F.R. and Tarashi, T. 1976, Ionic mechanism of the salicylate block of nerve conduction. *Journal of Pharmacology and Experimental Therapeutics*, **199:** 454–463.

Neve, J., Parij, N. and Moguilevsky, N. 2001, Inhibition of the myeloperoxidase chlorinating activity by non-steroidal anti-inflammatory drugs investigated with a human recombinant enzyme. *European Journal of Pharmacology*, **417:** 37–43.

Newbold, B.B. 1963, Chemotherapy of arthritis induced in rats by mycobacterial adjuvant. *British Journal of Pharmacology*, **21:** 127–136.

Newbold, B.B. 1964, Lymphatic drainage and the adjuvant induced arthritis in rats. *British Journal of Experimental Pathology*, **45:** 375–383.

Newbold, B.B. 1974, Role of lymph nodes in adjuvant-induced arthritis in rats. *Annals of the Rheumatic Diseases*, **23:** 392–396.

Nicosia, S. 1999, Pharmacodynamic properties of leukotriene receptor antagonists. *Monaldi Archives of Chest Diseases*, **54:** 242–246.

Noordhoek, J., Nagy, M.R. and Bonta, I.L. 1977, Involvement of complement and kinins in some non-immunogenic paw inflammations in rats. *In:* I.L. Bonta (ed.), *Recent Developments in the Pharmacology of Inflammatory Mediators*, pp. 109–121. Basel: Birkhäuser.

Northover, B.J. 1977, Effect of indomethacin and related drugs on the calcium ion-dependent secretion of lysosomal and other enzymes by neutrophil polymorphonuclear leucocytes in vitro. *British Journal of Pharmacology*, **59:** 253–259.

Nose, M., Sasano, M. and Kawashima, Y. 1997, Salazosulfapyridine suppresses chondrocyte mediated degradation induced by interleukin 1β. *Journal of Rheumatology*, **24:** 550–554.

Nugteren, D.H. 1975, Arachidonate lipoxygenase in blood platelets. *Biochimica et Biophysica Acta*, **380:** 299–307.

Nunn, B. 1981, Evidence that platelet arachidonate metabolites play only a minor role in collagen-induced thrombocytopenia in mice. *British Journal of Pharmacology*, **74:** 915P.

Nusbickel, F.R. and Troyer, H. 1976, Histochemical investigation of adjuvant-induced arthritis. *Arthritis and Rheumatism*, **19:** 1339–1346.

Oates, J.A., Roberts, L.T., Sweatman, B.J., Maas, R.L., Gerkens, J.F. and Taber, D.F. 1980, Metabolism of the prostaglandins and thromboxanes. *In:* B. Samuelsson, P. Ramswell and R. Paoletti (eds), *Advances in Prostaglandin and Thromboxane Research*, pp. 35–41. New York: Raven Press.

O'Brien, J.R. 1968, Effects of salicylates on human platelets. *Lancet*, **1:** 779–783.

Oegema, T.R., Jr. and Thompson, R.C. 1986, Metabolism of chondrocytes derived from normal and osteoarthritic human cartilage. *In:* Kuettner, K. *et al.* (eds), *Articular Cartilage Biochemistry*, pp. 100–125. New York: Raven Press.

Oeth, P. and Mackman, N. 1995, Salicylates inhibit lipopolysaccharide-induced transcriptional activation of the tissue factor gene in human monocytic cells. *Blood*, **86:** 4144–4152.

Ohdo, S., Ogawa, N. and Song, J.G. 1995, Chronopharmalogical study of acetylsalicylic acid in mice. *European Journal of Pharmacology*, **293:** 151–157.

O'Kane, P.D., Queen, L.R., Ji, Y., Reebye, V., Stratton, P., Jackson, G. and Ferro, A. 2003, Aspirin modifies nitric oxide synthase activity in platelets: effects of acute versus chronic aspirin treatment. *Cardiovascular Research*, **59:** 152–159.

Oketani, K., Nagakura, N., Harada, K. and Inoue, T. 2001, In vitro effects of E3040, a dual inhibitor of 5-lipoxygenase and thromboxane A_2 synthetase, on eicosanoid production. *European Journal of Pharmacology*, **422:** 209–216.

O'Neill, E.A. 1998, A new target for aspirin. *Nature*, **396:** 15–16.

O'Neill, L.A., Barrett, M.L. and Lewis, G.P. 1987, Induction of cyclooxygenase by interleukin-1 in rheumatoid synovial cells. *FEBS Letters*, **212:** 35–39.

O'Neill, G.P., Mancini, J.A., Kargman, S., Yergey, J., Kwan, M.Y., Falgueyret, J.-P., Abramovitz, M., Kennedy, B.P., Ouellet, M., Cromlish, W., Culp, S., Evans, J.F., Ford-Hutchinson, A.W. and Vickers, P.J. 1994, Overexpression of prostaglandin G/H synthases-1 and -2 by recombinant vaccinia virus: inhibition by nonsteroidal anti-inflammatory drugs and biosynthesis of 15-hydroxyeicosatetraenoic acid. *Molecular Pharmacology*, **45:** 245–254.

Opelz, G., Terasaki, P.I. and Hirata, A.A. 1973, Suppression of lymphocyte transformation by aspirin. *Lancet*, **2:** 478–480.

Ortmann, R. and Perkins, J.P. 1977, Stimulation of adenosine $3',5'$-monophosphate formation by prostaglandins in human astrocytoma cells. Inhibition by nonsteroidal anti-inflammatory drugs. *Journal of Biological Chemistry*, **252:** 6018–6025.

O'Sullivan, M.G., Huggins, E.M. and McCall, C.E. 1993, Lipopolysaccharide-induced expression of prostaglandin H synthase-2 in alveolar macrophages is inhibited by dexamethasone but not by aspirin. *Biochemical and Biophysical Research Communications*, **191:** 1294–1300.

Otterness, I.G. and Bliven, M.L. 1985, Laboratory models for testing nonsteroidal anti-inflammatory drugs. *In:* J.G. Lombardino (ed.), *Non-steroidal-Anti-inflammatory Drugs*, pp. 113–251. New York: Wiley.

Otterness, I.G., Wiseman, E.H. and Gans, D.J. 1979, A comparison of the carrageenan edema test and ultraviolet light-induced erythema test as predictors of the clinical dose in rheumatoid arthritis. *Agents and Actions*, **9:** 177–183.

Oyanagui, Y. 1976, Participation of superoxide anions at the prostaglandin phase of carrageenan foot-oedema. *Biochemical Pharmacology*, **25:** 1465–1472.

Oyanagui, Y. 1978, Inhibition of superoxide anion production in non-stimulated guinea pig peritoneal exudate cells by anti-inflammatory drugs. *Biochemical Pharmacology*, **27:** 777–782.

Paalzow, L. 1973, Studies on the relationship between the analgesic activity of salicylic acid and the brain catecholamines in mice. *Acta Pharmacologica et Toxicologica*, **32:** 11–21.

Pablos, J.L., Santiago, B., Carreira, P.E., Galindo, M. and Gomez-Reino, J.J. 1999, Cyclooxygenase-1 and -2 are expressed by human T cells. *Clinical and Experimental Immunology*, **115:** 86–90.

Packham, M.A. 1982, Mode of action of acetylsalicylic acid. *In:* H.J.M. Barnett, J. Hirsh and J.F. Mustard (eds), *Acetylsalicylic Acid. New Uses for an Old Drug*, pp. 63–82. New York: Raven Press.

Paeile, C., Guiverneau, M. and Munoz, C. 1974, Influence of anti-inflammatory drugs on the electroencephalographic effect of morphine in rabbits. *Pharmacology*, **11:** 79–84.

Palmoski, M.J. and Brandt, K.D. 1979, Effect of salicylate on proteoglycan metabolism in normal canine articular cartilage in vitro. *Arthritis and Rheumatism*, **22:** 746–754.

Palmoski, M.J. and Brandt, K.D. 1982a, Partial reversal by beta-D-xyloside of salicylate-induced inhibition of glycosaminoglycan synthesis in articular cartilage. *Arthritis and Rheumatism*, **25:** 1084–1093.

Palmoski, M.J. and Brandt, K.D. 1982b, Aspirin aggravates the degeneration of canine joint cartilage caused by immobilization. *Arthritis and Rheumatism*, **25:** 1333–1342.

Palmoski, M.J. and Brandt, K.D. 1984, Effects of salicylate and indomethacin on glycosaminoglycan and prostaglandin E2 synthesis in intact canine knee cartilage *ex vivo*. *Arthritis and Rheumatism*, **27:** 398–403.

Palmoski, M.J. and Brandt, K.D. 1985, Proteoglycan depletion, rather than fibrillation, determines the effects of salicylate and indomethacin on osteoarthritic cartilage. *Arthritis and Rheumatism*, **28:** 548–553.

Palmoski, M.J., Coyler, R.A. and Brandt, K.D. 1980, Marked suppression by salicylate of the augmented proteoglycan synthesis in osteoarthritic cartilage. *Arthritis and Rheumatism*, **23:** 83–91.

Pan, Z., Camara, B., Gardner, H.W. and Backhaus, R.A. 1998, Aspirin inhibition and acetylation of the plant cytochrome P_{450}, allene oxide synthase, resembles that of animal prostaglandin endoperoxide H synthase. *Journal of Biological Chemistry*, **273:** 18139–18145.

Panush, R.S. and Anthony, C.R. 1976, Effects of acetylsalicylic acid on normal human peripheral blood lymphocytes. Inhibition of mitogen- and antigen-stimulated incorporation of tritiated thymidine. *Clinical and Experimental Immunology*, **23:** 114–125.

Park, B.K. and Leck, J.B. 1981, On the mechanism of salicylate-induced hypothrombinaemia. *Journal of Pharmacy and Pharmacology*, **33:** 25–28.

Parnham, M.J. 1980, The inflammatory response to lymph node cells from adjuvant-diseased rats: late changes in local and systemic leucocyte counts in the Wistar strain. *Journal of Pathology*, **132:** 11–21.

Parnham, M.J. 1998, Is there a COX-fight during inflammation? (Editorial). *Inflammation Research*, **47:** 43.

Parnham, M.J., Bonta, I.L. and Adolfs, M.J.P. 1978, Cyclic AMP and prostaglandin E in perfusates of rat hind paws during the development of adjuvant arthritis. *Annals of the Rheumatic Diseases*, **37:** 218–224.

Parrott, D.P. and Lewis, D.A. 1977, Protease and antiprotease levels in blood of arthritic rats. *Annals of the Rheumatic Diseases*, **36:** 166–169.

Pearson, C.M. 1966, Arthritis in animals. *In:* J.L. Hollander (ed.), *Arthritis and Allied Conditions*, pp. 119–129. London: Henry Kimpton.

Pearson, C.M. and Wood, F.D. 1963, Studies of arthritis and other lesions induced in rats by the injection of mycobacterial adjuvant VII. Pathologic details of the arthritis and spondylitis. *American Journal of Pathology*, **42:** 73–95.

Pearson, C.M. and Wood, F.D. 1964, Studies of polyarthritis and other lesions induced in rats by injection of mycobacterial adjuvant. I. General clinical and pathologic characteristics and some modifying factors. *Journal of Experimental Medicine*, **120:** 547–560.

Pecoud, A., Leimgruber, A. and Frei, P.C. 1980, Effect of one gold salt, of betamethasone, and of aspirin on the chemotaxis of human neutrophils measured in vitro. *Annals of the Rheumatic Diseases*, **39:** 25–30.

Pelletier, J.-P. 1999, The influence of tissue cross-talking in OA progression: role of nonsteroidal anti-inflammtory drugs. *Osteoarthritis and Cartilage*, **7:** 374–376.

Penniall, R. 1958, The effects of salicylic acid on the respiratory activity of mitochondra. *Biochimica Biophysica Acta*, **30:** 247–251.

Penrose, J.F. 1999, LTC4 synthase. Enzymology, biochemistry, and molecular characterization. *Clinical Reviews of Allergy and Immunology*, **17:** 133–152.

Perez, G.M., Melo, M., Keegan, A.D. and Zamorano, J. 2002, Aspirin and salicylates inhibit the IL-4 and IL-13 induced activation of STAT6. *Journal of Immunology*, **168:** 1428–1434.

Perlow, M., Dinarello, C.A. and Wolff, S.M. 1975, A primate model for the study of human fever. *Journal of Infectious Diseases*, **132:** 157–164.

Perper, R.J. and Orowsky, A.L. 1974, Enzyme release from human leukocytes and degradation of cartilage matrix. Effects of antirheumatic drugs. *Arthritis and Rheumatism*, **17:** 47–55.

Perry, K.H. 1968, Die Wirkung von Phenylbutazon und Natrium Salicylate auf die Glucosamin-6-Phosphat Synthese in der Magenschleimhaut von Ratten. *Archives Internationales de Pharmacodynamie et de Thérapie*, **176:** 337–359.

Peskar, B.M. 1985, Effects of sulphasalazine and 5-aminosalicylic acid on the human colonic prostaglandin system. *In:* L. Barbara, M. Miglioli and S.F. Phillips (eds), *New Trends in the Pathophysiology and Therapy of the Large Bowel*, pp. 185–196. Amsterdam: Elsevier.

Peters, H.D., Dinnendahl, V. and Schonhofer, P.S. 1975, Mode of action of anti-rheumatic drugs on the cyclic 3',5'-AMP regulated glycosaminoglycan secretion in fibroblasts. *Naunyn Schmiedebergs Archives of Pharmacology*, **289:** 29–40.

Peters-Golden, M. 1998, Molecular mechanisms of leukotriene synthesis: the changing paradigm. *Clinical and Experimental Allergy*, **28:** 1059–1065.

Peterson, D.A. and Gerrard, J.M. 1979, A hypothesis for the interaction of heme and arachidonic acid in the synthesis of prostaglandins. *Medical Hypotheses*, **5:** 683–694.

Petillo, J.J., Gulbenkian, A. and Tabachnik, I.I. 1969, Effects in vivo and in vitro of nonsteroidal anti-inflammatory drugs on (rat stomach) histidine decarboxylase. *Biochemical Pharmacology*, **18:** 1784–1788.

Petrescu, I. and Tarba, C. 1997, Uncoupling effects of diclofenac and aspirin in the perfused liver and isolated hepatic mitochondria of rat. *Biochimica et Biophysica Acta*, **1318:** 385–394.

Phadke, K., Nanda, S. and Lee, K. 1979, Release of proteases from cartilage cells as a result of activation by a macrophage factor – effects of some anti-inflammatory drugs. *Biochemical Pharmacology*, **28:** 3671–3673.

Pham Huy, D., Roch-Arveiller, M., Muntaner, O. and Giroud, J.P. 1985, Effect of some anti-inflammatory drugs on FMLP-induced chemotaxis and random migration of rat polymorphonuclear leucocytes. *European Journal of Pharmacology*, **111:** 251–256.

Philp, R.B., Paul, M.L., Killackey, J.J. and Killackey, B.A. 1983, The influence of dose, time of administration, body temperature and salicylate kinetics on the antithrombotic action of acetylsalicylic acid in male rats. *Haemostasis*, **13:** 42–52.

Picot, D., Loll, P.J. and Garavito, R.M. 1994, The X-ray crystal structure of the membrane prostaglandin H_2 synthase-1. *Nature*, **367:** 243–249.

Pierce, J.W., Read, M.A., Ding, H., Luscinskas, F.W. and Collins, T. 1996, Salicylates inhibit I kappa B-alpha phosphorylation, endothelial-leukocyte adhesion molecule expression, and neutrophil transmigration. *Journal of Immunology*, **156:** 3961–3969.

Piliero, S.J. and Colombo, C. 1969, Action of antiinflammatory drugs on the lysozyme activity and 'turbidity' of serum from rats with adjuvant arthritis or endocrine deficiency. *Journal of Pharmacology and Experimental Therapeutics*, **165:** 294–299.

Pillinger, M.H., Capodici, C., Rosenthal, P., Kheterpal, N., Hanft, S., Philips, M.R. and Weissmann, G. 1998, Modes of action of aspirin-like drugs: salicylates inhibit Erk activation and integrin-dependent neutrophil adhesion. *Proceedings of the National Academy of Sciences USA* **95:** 14540–14545.

Pini, L.A., Vitale, G. and Sandrini, M. 1997, Serotonin and opiate involvement in the antinociceptive effect of acetylsalicylic acid. *Pharmacology*, **54:** 84–91.

Poc̄wiardowska, E. 1976a, The effects of antipyretics on metabolism processes in rat liver mitochondria Part I. The action of sodium salicylate and pyrazolones on the reaction of respiratory chain. *Polish Journal of Pharmacology and Pharmacy*, **28:** 217–226.

Poc̄wiardowska, E. 1976b, The effects of antipyretics on metabolism processes in rat liver mitochondria Part II. The action of sodium salicylate, and pyrazolones on oxidation of α-ketoglutarate. *Polish Journal of Pharmacology and Pharmacy*, **28:** 227–231.

Podhaisky, H-P., Abate, A., Polte, T., Oberle, S. and Schroeder, H. 1997, Aspirin protects endothelial cells from oxidative stress – possible synergism with vitamin E. *FEBS Letters*, **417:** 349–351.

Pong, S.F., Demuth, S.M., Kineey, C.M. and Deegan, P. 1985, Prediction of human analgesic dosage of nonsteroidal anti-inflammatory drugs (NSAIDs) from analgesic ED_{50} values in mice. *Archives Internationales de Pharmacodynamie et de Therapie*, **273:** 212–220.

Porter, N.A., Wolf, R.A. and Weenen, H. 1980, The free radical oxidation of polyunsaturated lecithins. *Lipids*, **15:** 163–167.

Preston, F.E., Whipps, S., Jackson, C.A., French, A.J., Wyld, P.J. and Stoddard, C.J. 1981, Inhibition of prostacyclin and platelet thromboxane A2 after low-dose aspirin. *New England Journal of Medicine*, **304:** 76–79.

Provost, P., Samuelsson, B. and Radmark, O. 1999, Interaction of 5-lipoxygenase with cellular proteins. *Proceedings of the National Academy of Sciences USA*, **96:** 1881–1885.

Puig-Paradella, P. and Planas, J.M. 1978, Synovial fluid degradation induced by free radicals. In vitro action of several free radical scavengers and anti-inflammatory drugs. *Biochemical Pharmacology*, **27:** 535–537.

Radomski, M., Michalska, Z., Marcinkiewicz, E. and Gryglewski, R.J. 1986, Salicylates and 12-lipoxygenase activity in human washed platelets. *Pharmacological Research Communications*, **18:** 1015–1030.

Rådmark, O.P. 2000, The molecular biology and regulation of 5-lipoxygenase. *American Journal of Respiratory and Critical Care Medicine*, **161 (Part 2):** S11–S15.

Radwan, A.G. and West, G.B. 1968, The effect of non-steroidal anti-inflammatory drugs on histamine formation in the rat. *British Journal of Pharmacology*, **33:** 193–198.

Rae, S.A. and Smith, M.J. 1981, The stimulation of lysosomal enzyme secretion from human polymorphonuclear leucocytes by leukotriene B4. *Journal of Pharmacy and Pharmacology*, **33:** 616–617.

Rae, S.A., Davidson, E.M. and Smith, M.J.H. 1982, Leukotriene B4, an inflammatory mediator in gout. *Lancet*, **2:** 1122–1123.

Raghoebar, M., Van Den Berg, W.B. and Van Ginneken, C.A.M. 1988, Mechanisms of cell association of some non-steroidal anti-inflammatory drugs with isolated leucocytes. *Biochemical Pharmacology*, **37:** 1245–1250.

Rainsford, K.D. 1978, The effects of aspirin and other non-steroidal anti-inflammatory drugs on the gastrointestinal mucus glycoprotein biosynthesis in vivo. Relationship to ulcerogenic actions. *Biochemical Pharmacology*, **27:** 877–885.

Rainsford, K.D. 1984, *Aspirin and the Salicylates*. London: Butterworths.

Rainsford, K.D. 1987, Effects of antiinflammatory drugs on the release from porcine synovial tissue in vitro of interleukin 1 like cartilage degrading activity. *Agents and Actions*, **21:** 337–340.

Rainsford, K.D., Davies, A., Mundy, L. and Ginsburg, I. 1989, Comparative effects of azapropazone on cellular events at inflamed sites. Influence on joint pathology, leucocyte superoxide and eicosanoid production, platelet aggregation, synthesis of cartilage proteoglycans, synovial production and actions of interleukin 1-induced-cartilage resorption correlated with drug uptake into cartilage *in vitro*. *Journal of Pharmacy and Pharmacology*, **41:** 322–330.

Rainsford, K.D. and Ford-Hutchinson, A.W. (eds) 1979, *Prostaglandins and Inflammation*. Basel: Birkhäuser.

Rainsford, K.D., Rashad, S.Y., Revell, P.A., Low, F.M., Hemingway, A.P., Walker, F.S., Johnson, D., Stetsko, P., Ying, F.M. and Smith, F. 1992, Effects of NSAIDs on cartilage proteoglycan and synovial prostaglandin metabolism in relation to progression of joint deterioration in osteoarthritis. *In:* G. Balint, B. Gömör, and L. Hodinka (eds), *Rheumatology, State of the Art*, pp. 177–183. Amsterdam: Excerpta Medica.

Rainsford, K.D., Schweitzer, A. and Brune, K. 1981, Autoradiographic and biochemical observations on the distribution of non-steroidal anti-inflammatory drugs. *Archives Internationales de Pharmacodynamie et de Thérapie*, **250:** 180–194.

Rainsford, K.D., Schweitzer, A., Green, P., Whitehouse, M.W. and Brune, K. 1980a, Bio-distribution in rats of some salicylates with low gastric ulcerogenicity. *Agents and Actions*, **10:** 457–464.

Rainsford, K.D. and Swann, B.P. 1984, The biochemistry and pharmacology of oxygen radical involvement in eicosanoid production. *In:* J.V. Bannister and W.H. Bannister (eds), *The Biochemistry of Active Oxygen*, pp. 105–127. New York: Elsevier.

Rainsford, K.D., Tsang, S., Hunt, R.H. and Al-Jehani, N. 1995, Effects of non-steroidal anti-inflammatory drugs on prostaglandin H synthase isoenzyme 2 (cyclo-oxygenase 2). Production by porcine gastric mucosa in organ culture. *Inflammopharmacology*, **3:** 299–310.

Rainsford, K.D. and Whitehouse, M.W. 1980a, Anti-inflammatory/anti-pyretic salicylic acid esters with low gastric ulcerogenic activity. *Agents and Actions*, **10:** 451–456.

Rainsford, K.D. and Whitehouse, M.W. 1980b, Are all aspirins alike? A comparison of gastric ulcerogenicity with bioefficacy in rats. *Pharmacological Research Communications*, **12:** 85–95.

Ramstedt, U., Ng, J., Wigzell, H., Serhan, C.N. and Samuelsson, B. 1985, Action of novel eicosanoids lipoxin A and B on human natural killer cell cytotoxicity: effects on intracellular cAMP and target cell binding. *Journal of Immunology*, **135:** 3434–3438.

Reed, N.P., Balint, J.A. and Powers, S.R. 1968, Effect of aspirin on the incorporation of glucosamine into gastric mucosa. *Surgical Forum*, **19:** 286–288.

Reinisch, C.M., Dunzendorfer, S., Pechlaner, C., Ricevuti, G. and Wiedermann. 2000, The inhibition of oxygen radical release from human neutrophils by resting platelets is reversed by administration of acetylsalicylic acid or clopidogrel. *Free Radical Research*, **34:** 461–466.

Reunanen, M., Hanninen, O. and Hartiala, K. 1967, Inhibitory effect of salicylates and cincophen derivatives on amino acid incorporation. *Nature*, **218:** 918–919.

Revell, P.A., Mayston, V. and Davies, P.G. 1983, Fibronectin in the synovium of chronic inflammatory joint disease. *Annals of the Rheumatic Diseases*, **42:** 222.

Rinder, C.S., Student, L.A., Bonan, J.L., Rinder, H.M. and Smith, B.R. 1993, Aspirin does not inhibit adenosine diphosphate-induced platelet alpha-granule release. *Blood*, **82:** 505–512.

Rivers, J.P.W. and Frankel, T.L. 1981, Essential fatty acid deficiency. *British Medical Bulletin*, **37:** 59–64.

Rivkin, I., Rosenblatt, J. and Becker, E.L. 1975, The role of cyclic AMP in the chemotactic responsiveness and spontaneous motility of rabbit peritoneal neutrophils. The inhibition of neutrophil movement and the elevation of cyclic AMP levels by catecholamines, prostaglandins, theophylline and cholera toxin. *Journal of Immunology*, **115:** 1126–1134.

Robak, J. 1978, Adjuvant-induced and carrageenan-induced inflammation and lipid peroxidation in rat liver, spleen and lungs. *Biochemical Pharmacology*, **27:** 531–533.

Roberts, M.S. and Cross, S.E. 1999, Percutaneous absorption of topically applied NSAIDs and other compounds: Role of solute properties, skin physiology and delivery systems. *Inflammopharmacology*, **7:** 339–350.

Robinson, D. and Wilcox, P. 1969, Interaction of salicylates with rat liver lysosomes. *Biochemical Journal*, **115:** 54P.

Roch-Arveiller, M. and Giroud, J.P. 1979, Biological and pharmacological effects of carrageenan. *Pathologie et Biologie*, **27:** 615–626.

Roch-Arveiller, M., Huy, D.P., Maman, L., Giroud, J.P. and Sorenson, J.R. 1990, Non-steroidal anti-inflammatory drug–copper complex modulation of polymorphonuclear leukocyte migration. *Biochemical Pharmacology*, **39:** 569–574.

Roger, J. and Kalbhen, D.A. 1962, Adenosine triphosphate content of cartilage under the effect of different antirheumatic agents in vitro. *Arzneimittel Forschung*, **12:** 1512–1516.

Rohdenwald, P., Derendorf, H., Drehsen, G., Elger, C.E. and Knoll, O. 1982, Changes in cortical evoked potentials as correlates of the efficacy of weak analgesics. *Pain*, **12:** 329–341.

Rokosova-Cmuchalova, B. and Bentley, J.P. 1968, Relation of collagen synthesis to chondroitin sulfate synthesis in cartilage. *Biochemical Pharmacology*, **7 (Suppl.):** 315–328.

Rome, L.H. and Lands, W.E.M. 1975, Structural requirements for time-dependent inhibition of prostaglandin synthesis by anti-inflammatory drugs. *Proceedings of the National Academy of Sciences USA*, **72:** 4863–4865.

Römer, D. 1980, Pharmacological evaluation of mild analgesics. *British Journal of Clinical Pharmacology*, **10:** 247S–251S.

Rooney, P.J., Dick, W.C., Imrie, R.C., Turner, D., Buchanan, K.D. and Ardill, J. 1978, On the relationship between gastrin, gastric secretion, and adjuvant arthritis in rats. *Annals of the Rheumatic Diseases*, **37:** 432–435.

Rosenblum, W.I., El-Sabban, F. and Ellis, E.F. 1980, Aspirin and indomethacin, nonsteroidal anti-inflammatory agents alter the responses to microvascular injury in brain and mesentery. *Microvascular Research*, **20:** 374–378.

Rosenior, J.C. and Tonks, R.S. 1974, Salicylate inhibition of in vitro plasminogen activation by saline extracts of rat tissues. *Biochemical Pharmacology*, **23:** 2339–2341.

Rosenthale, M.E., Begany, A.J., Dervinis, A., Malis, J.L., Shriver, D.A., Datko, L.J. and Gluckman, M.T. 1974, Anti-inflammatory properties of 4,5-diphenyl-2-oxazolepropionic acid (oxaprozin). *Agents and Actions*, **4:** 151–159.

Rossi, A., Kapahl, P., Natoli, G., Takahashi, T., Chen, Y., Karin, M. and Santoro, M.G. 2000, Anti-inflammatory cyclopentenone prostaglandins are direct inhibitors of IκB kinase. *Nature*, **403:** 103–108.

Roth, G.J., Stanford, N. and Majerus, P.W. 1975, Acetylation of prostaglandin synthase by aspirin. *Proceedings of the National Academy of Sciences USA*, **72:** 3073–3076.

Rothschild, A.M., Cordiero, R.S. and Castania, A. 1975, Acute pulmonary edema and plasma kininogen consumption in the adrenaline-treated rat: inhibition by acetylsalicylic acid and resistance to salicylate and indomethacin. *Naunyn-Schmiedebergs Archiv für Pharmakologie*, **288:** 319–321.

Rotilio, D., Joseph, D., Hatmi, M. and Vargaftig B.B. 1984, Structural requirements for preventing the aspirin- and the arachidonate-induced inactivation of platelet cyclo-oxygenase: additional evidence for distinct enzymatic sites. *European Journal of Pharmacology*, **97:** 197–208.

Rouzer, C.A., Scott, W.A., Hamill, A.L. and Cohn, Z.A. 1980, Dynamics of leukotriene C production by macrophages. *Journal of Experimental Medicine*, **152:** 1236–1247.

Rubinstein, M., Giacomoni, D. and Packman, L.M. 1976, Effect of sodium salicylate on hamster cells in vitro. *Journal of Pharmaceutical Sciences*, **65:** 756–758.

Rumore, M.M., Aron, S.M. and Hiross, E.J. 1987, A review of mechanism of action of aspirin and its potential as an immunomodulating agent. *Medical Hypotheses*, **22:** 387–400.

Ryu, Y.S., Lee, J.H., Seok, J.H., Hong, J.H., Lee, Y.S., Lim, J.H., Kim, Y.M. and Hur, G.M. 2000, Acetaminophen inhibits iNOS gene expression in RAW 264.7 macrophages: differential regulation of NF-κB by acetaminophen and salicylates. *Biochemical and Biophysical Research Communications*, **272:** 758–764.

Saeed, S.A. and Cuthbert, J. 1977, On the mode of action and biochemical properties of anti-inflammatory drugs – II. *Prostaglandins*, **13:** 565–573.

Saeed, S.A. and Warren, B.J. 1973, On the mode of action and biochemical properties of anti-inflammatory drugs – I. *Biochemical Pharmacology*, **22:** 1965–1969.

Sagone, A.L. and Husney, R.M. 1987, Oxidation of salicylates by stimulated granulocytes: evidence that these drugs act as free radical scavengers in biological systems. *Journal of Immunology*, **138:** 2177–2183.

Sakitani, K., Kitade, H., Inoue, K., Kamiyama, Y., Nishizawa, M., Okumura, T. and Ito, S. 1997, The anti-inflammatory drug sodium salicylate inhibits nitric oxide formation induced by interleukin-1β at a translational step, but not at a transcriptional step, in hepatocytes. *Hepatology*, **25:** 416–420.

Samogyi, A., Berczi, I. and Selye, H. 1969, Inhibition by salicylates of various calcifying tissue reactions. *Archives Internationales de Pharmacodynamie et de Thérapie*, **177:** 211–223.

Sánchez de Miguel, L., Casado, S., Farre, J., Garcia-Duran, M., Rico, L.A., Monton, M., Romero, J., Bellver, T., Sierra, M.P., Guerra, J.I., Mata, P., Esteban, A. and Lopez-Farre, A. 1998, Comparison of in vitro effects of triflunisal and acetylsalicylic acid on nitric oxide synthesis by human neutrophils. *European Journal of Pharmacology*, **343:** 57–65.

Sánchez de Miguel, L., De Frutos, T., Gonzalez-Fernandez, F., Del Pozo, V., Lahoz, C., Jimenez, A., Rico, L., Garcia, R., Aceituno, E., Millas, I., Gomez, J., Farre, J., Casado, S. and Lopez-Farre, A. 1999, Aspirin inhibits inducible nitric oxide synthase expression and tumor necrosis factor-α release by cultured smooth muscle cells. *European Journal of Clinical Investigation*, **29:** 93–99.

Sancilio, L.F. 1969, Evans blue-carrageenan pleural effusion as a model for the assay of nonsteroidal antirheumatic drugs. *Journal of Pharmacology and Experimental Therapeutics*, **168:** 199–204.

Sandrini, M. 1999, Central effects of non-opioid analgesics. *CNS Drugs*, **12:** 337–345.

Sanduja, R., Loose-Mitchell, D. and Wu, K.K. 1990, Inhibition of de novo synthesis and message expression of prostaglandin H synthase by salicylates. *Advances in Prostaglandin, Thromboxane, and Leukotriene Research*, **21:** 149–152.

Sarraf, P., Mueller, E., Jones, D., King, F.J., DeAngelo, D.J., Partridge, J.B., Holden, S.A., Chen, L.B., Singer, S., Fletcher, C. and Spiegelman, B.M. 1998, Differentiation and reversal of malignant changes in colon cancer through PPAR gamma. *Nature Medicine*, **4:** 1046–1052.

Saunders, M.A., Sansores-Garcia, L., Gilroy, D.W. and Wu, K.K. 2001, Selective suppression of CCAAT/enhancer-binding protein beta binding and cyclooxygenase-2 promoter activity by sodium sallicylate in quiescent human fibroblasts. *Journal of Biological Chemistry*, **276:** 18897–18904.

Sawynok, J. and Liu, X.J. 2003, Adenosine in the spinal cord and periphery: release and regulation of pain. *Progress in Neurobiology*, **69:** 313–340.

Scadding, J.W. 1998, The neurophysiology of pain. *In:* P.J. Maddison, W.P. Isenberg and D.N. Glass (eds), *The Oxford Textbook of Rheumatology*, pp. 3936–3937. Oxford: Oxford University Press.

Schaible, H.G. 1998, The neurophysiology of pain. *In:* P.J. Maddison, W.P. Isenberg and D.N. Glass (eds), *The Oxford Textbook of Rheumatology*, pp. 487–498. Oxford: Oxford University Press.

Schaub, R.G., Rawlings, C.A. and Keith, J.C., Jr. 1979, Effect of long-term aspirin treatment on platelet adhesion to chronically damaged canine pulmonary arteries. *Thrombosis and Haemostasis*, **46:** 680–683.

Schiffmann, E. 1981, Molecular events in leukocyte chemotaxis; their possible roles in processing the chemical signal. *Bioscience Reports*, **1:** 89–99.

Schmelz, M., Weber, S. and Kress, M. 2000, Topical acetyl salicylate and dipyrone attenuate neurogenic protein extravasation in rat skin in vivo. *Neuroscience Letters*, **290:** 57–60.

■ CHAPTER 7 ■

Schmidt, C., Kosche, E., Baumeister, B. and Vetter, H. 1995, Arachidonic acid metabolism and intracellular calcium concentration in inflammatory bowel disease. *European Journal of Gastroenterology and Hepatology*, **7:** 865–869.

Schneider, C. and Brash, A.R. 2000, Stereospecificity of hydrogen abstraction in the conversion of arachidonic acid to 15*R*-HETE by aspirin-treated cyclooxygenase-2. *Journal of Biological Chemistry*, **275:** 4743–4746.

Schneider, W., Pappas, A. and Scheurlen, P.G. 1971, The effect of acetylsalicylic acid on the metabolism and transformability of human lymphocytes in vitro. *Klinische Wochenschrift*, **49:** 1187–1189.

Schneider, W., Pappas, A. and Scheurlen, P.G. 1972, Hexokinase inhibition and lymphocyte transformation. *Klinische Wochenschrift*, **50:** 261–263.

Schoener, E.P. and Wang, S.C. 1975, Leukocytic pyrogen and sodium acetylsalicylate on hypothalamic neurons in the cat. *American Journal of Physiology*, **229:** 185–190.

Schoener, E.P. and Wang, S.C. 1976, Observations on the central mechanisms of acetylsalicylate antipyresis. *Life Sciences*, **17:** 1063–1068.

Schönhöfer, P. 1967, Einige Untersuchungen über die enzymatische synthese von Glucosamin-6-Phosphat in Rattenleber-Homogenaten under ihre Beeinflussung in vivo. *Arzneimittel Forschung*, **17:** 602–605.

Schönhöfer, P. and Perry, K.H. 1967, The effect of a single large dose of salicylate and phenylbutazone on the glucosamine-6-phosphate synthesis *in vivo* by the liver and by the gastric mucosa of rats. *Medicina et Pharmacologie Experimentalis*, **17:** 175–182.

Schorderet, M. and Straub, R.W. 1991, Effect of non-narcotic analgesics and nonsteroid anti-inflammatory agents upon inorganic phosphates, intracellular potassium and impulse conduction in mammalian nerve fibres. *Biochemical Pharmacology*, **20:** 1355–1361.

Schror, K. 1997, Aspirin and platelets: the antiplatelet action of aspirin and its role in thrombosis treatment and prophylaxis. *Seminars in Thrombosis and Hemostasis*, **23:** 349–356.

Schwartz, J., Chan, C.C., Mukhopadhyay, S., McBride, K.J., Jones, T.M., Adcock, S., Moritz, C., Hedges, J., Grasing, K., Dobratz, D., Cohen, R.A., Davidson, M.H., Bachmann, K.A. and Gertz, B.J. 1999, Cyclooxygenase-2 inhibition by rofecoxib reverses naturally occurring fever in humans. *Clinical Pharmacology and Therapeutics*, **65:** 653–660.

Schwenger, P., Alpert, D., Skolnik, E.Y. and Vilcek, J. 1998, Activation of p38 mitogen-activated protein kinase by sodium salicylate leads to inhibition of tumor necrosis factor-induced IκBα phosphorylation and degradation. *Molecular and Cellular Biology*, **18:** 78–84.

Schwoch, G., Rudloff, V., Wood-Guth, I. and Passow, H. 1974, Effect of temperature on sulfate movements across chemically or enzymatically modified membranes of human red blood cells. *Biochimica and Biophysica Acta*, **339:** 126–138.

Scott, D.L., Almond, T.J., Walton, K.W. and Hunneyball, I.M. 1983, Involvement of fibronectin in fibrin and opsonisation in rheumatic diseases. *Annals of the Rheumatic Diseases*, **42:** 221.

Seed, J.C. 1965, A comparison of the antipyretic potency of aspirin and salicylic acid. *Clinical Pharmacology and Therapeutics*, **6:** 354–358.

Seibert, K., Zhang, Y., Leahy, K., Hauser, S., Masferrer, J., Perkins, W., Lee, L. and Isakson, P. 1994, Pharmacological and biochemical demonstration of the role of cyclooxygenase 2 in inflammation and pain. *Proceedings of the National Academy of Sciences USA* **91:** 12013–12017.

Selye, H. 1976, *Stress in Health and Disease*. Boston: Butterworths.

Serhan, C.N. 1997, Lipoxins and novel aspirin-triggered 15-epi-lipoxins (ATL): a jungle of cell-cell interactions or a therapeutic opportunity? *Prostaglandins*, **53:** 107–137.

Serhan, C.N., Fahlstadius, P., Dahlen, S.E., Hamberg, M. and Samuelsson, B. 1985, Biosynthesis and biological activities of lipoxins. *Advances in Prostaglandin Thromboxane and Leukotriene Research*, **15:** 163–166.

Serhan, C.N., Hamberg, M. and Samuelsson, B. 1984, Lipoxins: novel series of biologically active compounds formed from arachidonic acid in human leukocytes. *Proceedings of the National Academy of Sciences USA*, **81:** 5335–5339.

Serhan, C.N., Takano, T., Gronert, K., Chiang, N. and Clish, C.B. 1999, Lipoxin and aspirin-triggered 15-epi-lipoxin cellular interactions anti-inflammatory lipid mediators. *Clinical Chemistry and Laboratory Medicine*, **37:** 299–309.

Serup, J. and Ovesen, J.O. 1981, Salicylate-arthropathy. Accelerated coxarthrosis during long-term treatment with acetylsalicylic acid. *Schweizer Rundschau Medizinische Praxis*, **24:** 359–361.

Shackleford, R.E., Alford, P.B., Xue, Y., Thai, S-F., Adams, D.O. and Pizzo, S. 1997, Aspirin inhibits tumor necrosis factor-α gene expression in murine tissue macrophages. *Molecular Pharmacology*, **52:** 421–429.

Sharma, J.N., Zeitlin, I.J., Brooks, P.M., Buchanan, W.W. and Dick, W.C. 1980, The action of aspirin on plasma kininogen and other plasma proteins in rheumatoid patients: relationship to disease activity. *Clinical and Experimental Pharmacology and Physiology*, **7:** 347–354.

Sheppard, B.L. 1972, The effect of acetylsalicylic acid on platelet adhesion in the injured abdominal aorta. *Quarterly Journal of Experimental Physiology*, **57:** 319–323.

Shi, X., Ding, M., Dong, Z., Chen, F., Ye, J., Wang, S., Leonard, S.S., Castranova, V. and Vally-athan, V. 1999, Antioxidant properties of aspirin: characterization of the ability of aspirin to inhibit silica-induced lipid peroxidation, DNA damage, NF-κB activation, and TNF-α production. *Molecular and Cellular Biochemistry*, **199:** 93–102.

Shimanuki, T., Nakamura, R.M. and Dizerega, G.S. 1985, *In vivo* modulation of leukotaxis by non-steroidal anti-inflammatory drugs. *Agents and Actions*, **17:** 80–83.

Shimizu, M., Nakamura, H., Motoyoshi, S. and Yokoyama, Y. 1975, Pharmacological studies on 1-methyl-5-*p*-toluoylpyrrole-2-acetic acid (tolmetin), a new anti-inflammatory agent, in experimental animals I. Anti-inflammatory, analgesic and antipyretic activity. *Pharmacometrics*, **10:** 293–310.

Shimokawa, T. and Smith, W.L. 1992, Prostaglandin endoperoxide synthase. *Journal of Biological Chemistry*, **267:** 12387–12392.

Shimpo, M., Ikeda, U., Maeda, Y., Ohya, K-I., Murakami, Y. and Shimada, K. 2000, Effects of aspirin-like drugs on nitric oxide synthesis in rat vascular smooth muscle cells. *Hypertension*, **35:** 1085–1091.

Shukuya, R. 1954, On the kinetics of the human blood cholinesterase IV. The inhibition of cholinesterase by sodium salicylate. *Journal of Biochemistry*, **41:** 469–480.

Shyu, K.W., Lin, M.T. and Wu, T.C. 1984, Possible role of central serotoninergic neurons in the development of dental pain and aspirin-induced analgesia in the monkey. *Experimental Neurology*, **84:** 179–187.

Siegel, M.I., McConnell, R.T. and Cuatrecasas, P. 1979, Aspirin-like drugs interfere with arachidonate metabolism by inhibition of the 12-hydroperoxy-5,8,10,14-eicosatetranoic acid peroxidase activity of the lipoxygenase pathway. *Proceedings of the National Academy of Sciences USA*, **76:** 3774–3778.

Siegel, M.I., McConnell, R.T., Porter, N.A. and Cuatrecasas, P. 1980, Arachidonate metabolism via lipoxygenase and 12L-hydroperoxy-5,8,10,14-icosatetraenoic acid peroxidase sensitive to anti-inflammatory drugs. *Proceedings of the National Academy of Sciences USA*, **77:** 308–312.

Siegmund, E., Cadmus, R. and Lu, G. 1957, A method for evaluating both non-narcotic and narcotic analgesics. *Proceedings of the Society for Experimental Biology and Medicine*, **95:** 729–731.

Simchowitz, L., Mehta, J. and Spilberg, I. 1979, Chemotactic factor-induced generation of superoxide radicals by human neutrophils: effect of metabolic inhibitors and antiinflammatory drugs. *Arthritis and Rheumatism*, **22:** 755–763.

Simmons, D.L., Botting, R.M., Robertson, P.M., Madsen, M.L. and Vane, J.R. 1999, Induction of an acetaminophen-sensitive cyclooxygenase with reduced sensitivity to nonsteroid antiinflammatory drugs. *Proceedings of the National Academy of Sciences USA*, **96:** 3275–3280.

Skidmore, I.F. and Whitehouse, M.W. 1966, Biochemical properties of anti-inflammatory drugs. 8. Inhibition of histamine formation catalyzed by substrate specific mammalian histidine decarboxylases. Drug antagonism of aldehyde binding to protein amino groups. *Biochemical Pharmacology*, **15:** 1965–1983.

Skidmore, I.F. and Whitehouse, M.W. 1967, Biochemical properties of anti-inflammatory drugs. X. The inhibition of serotonin formation in vitro and inhibition of the esterase activity of alpha-chymyotrypsin. *Biochemical Pharmacology*, **16:** 737–751.

Sloboda, A.E., Oronsky, A.L. and Kerwar, S.S. 1988, Radiographic analysis of experimental rat arthritis. *In:* R.A. Greenwald and H.S. Diamond (eds), *CRC Handbook of Animal Models for the Rheumatic Diseases*, Vol. II, p. 159. Boca Raton: CRC Press.

Slowman-Kovacs, S.D., Albrecht, M.E. and Brandt, K.D. 1989, Effects of salicylate on chondrocytes from osteoarthritic and contralateral knees of dogs with unilateral anterior cruciate ligament transection. *Arthritis and Rheumatism*, **32:** 486–490.

Smith, J.B. and Willis, A.L. 1971, Aspirin selectively inhibits prostaglandin production in human platelets. *Nature New Biology*, **231:** 235–237.

CHAPTER 7

Smith, M.J.H. 1966, Metabolic effects of salicylates. *In:* M.J.H. Smith and P.K. Smith (eds), *The Salicylates, A Critical Bibliographic Review*, pp. 49–105. New York: Wiley-Interscience.

Smith, M.J.H. 1978a, Aspirin and prostaglandins: some recent developments. *Agents and Actions*, **8:** 427–429.

Smith, M.J.H. 1979, Prostaglandins and the polymorphonuclear leucocyte. *In:* K.D. Rainsford and A.W. Ford-Hutchinson (eds), *Prostaglandins and Inflammation*, pp. 91–103. Basel: Birkhäuser.

Smith, M.J.H. and Dawkins, P.D. 1971, Salicylate and enzymes. *Journal of Pharmacy and Pharmacology*, **23:** 729–744.

Smith, M.J.H., Ford-Hutchinson, A.W., Walker, J.R. and Slack, J.A. 1979, Aspirin, salicylate and prostaglandins. *Agents and Actions*, **9:** 483–487.

Smith, M.J.H., Gould, B.J. and Huggins, A.K. 1963, Inhibition of glutamate decarboxylase by salicylate congeners. *Biochemical Pharmacology*, **12:** 917–918.

Smith, M.J.H., Ford-Hutchinson, A.W. and Elliot, P.N. 1975a, Prostaglandins and the anti-inflammatory activities of aspirin and sodium salicylate. *Journal of Pharmacy and Pharmacology*, **27:** 473–478.

Smith, M.J.H., Hoth, M. and Davis, K. 1975b, Aspirin and lymphocyte transformation. *Annals of Internal Medicine*, **83:** 509–511.

Smith, M.J.H. and Jeffrey, S.W. 1956, The effects of salicylate on oxygen consumption and carbohydrate metabolism in the isolated rat diaphragm. *Biochemical Journal*, **63:** 524–528.

Smith, M.J.H. and Smith, P.K. (eds) 1966, *The Salicylates. A Critical Bibliographic Review.* New York: Wiley Interscience.

Smith, M.J.H. and Sturman, J.A. 1967, The mechanism of inhibition by salicylate of the pentose phosphate pathway in the human red cell. *Journal of Pharmacy and Pharmacology*, **19:** 108–113.

Smith, M.J.H. and Walker, J.R. 1980, The effects of some antirheumatic drugs on an *in vitro* model of human polymorphonuclear leucocyte chemokinesis. *British Journal of Pharmacology*, **69:** 473–478.

Smith, M.J.H., Walker, J.R., Ford-Hutchinson, A.W. and Penington, D.G. 1976, Platelets, prostaglandins and inflammation. *Agents and Actions*, **6:** 701–704.

Smith, R.J. 1977, Modulation of phagocytosis by lysosomal enzyme secretion from guinea-pig neutrophils: effect of non-steroidal anti-inflammatory agents and prostaglandins. *Journal of Pharmacology and Experimental Therapeutics*, **200:** 647–657.

Smith, R.J. 1978b, Nonsteroidal anti-inflammatory agents: regulators of the phagocytic secretion of lysosomal enzymes from guinea pig neutrophils. *Journal of Pharmacology and Experimental Therapeutics*, **207:** 618–629.

Snider, D.E., Jr. and Parker, C.W. 1975, Lower lymphocyte cyclic-AMP levels after aspirin. *New England Journal of Medicine*, **10:** 809–810, Letter.

Soekanto, A. 1994, Inhibition of osteoclast-like cell formation by sodium salicylate and indomethacin in mouse bone marrow culture. *Japanese Journal of Pharmacology*, **65:** 27–34.

Soekanto, A., Ohya, K. and Ogura, H. 1994, The effect of sodium salicylate on the osteoclast-like cell formation and bone resorption in a mouse bone marrow culture. *Calcified Tissue International*, **54:** 290–295.

Sofia, R.D. 1978, Comparative phlogistic activity of meseclazone, 5-chlorosalicylate acid, acetylsalicylate acid, phenylbutazone, indomethacin and hydrocortisone in various rat paw edema models. *Pharmacology*, **16:** 148–152.

Sofia, R.D., Danielson, L. and Vassar, H.B. 1979, Comparative effects of antiarthritic and other pharmacological agents in the 18-hour arthritis and carrageenan edema tests in rats. *Pharmacological Research Communications*, **11:** 179–193.

Sofia, R.D., Diamantis, W., Gordon, R., Kletzkin, M., Berger, F.M., Edelson, J., Singer, H. and Douglas, J.F. 1974, Pharmacology of a new nonsteroidal anti-inflammatory agent, 7-chloro-3,3a-dihydro-2-methyl-2H, 9H-isoxazole-(3,2–b)(1,3)-benzoxazin-9-one (W-2395). *European Journal of Pharmacology*, **26:** 51–62.

Sofia, R.D., Diamantis, W. and Ludwig, B.J. 1975, Comparative anti-inflammatory, analgesic, and antipyretic activities of 7-chloro-3,3a-dihydro-2-methyl-2H,9H-isoxazolo-(3,2-b)(1,3)-benzoxazin-9-one and 5-chlorosalicylic acid in rats. *Journal of Pharmaceutical Sciences*, **64:** 1321–1324.

Sofia, R.D., Vassar, H.B. and Nalepa, S.D. 1973, Systemic alterations in metal metabolism during inflammation as part of an integrated response to inflammation. *European Journal of Pharmacology*, **24:** 108–112.

Solheim, L.F., Ronningen, H. and Langeland, N. 1986a, Effects of acetylsalicylic acid and naproxen on the synthesis and mineralization of collagen in the rat femur. *Archives of Orthopaedic and Trauma Surgery*, **105:** 1–4.

Solheim, L.F., Ronningen, H., Barth, E. and Langeland, N. 1987, Effects of acetysalicylic acid on intramuscular bone matrix implants and composite grafts in rats. *Archives of Orthopaedic and Trauma Surgery*, **106:** 140–143.

Solheim, L.F., Ronningen, H. and Langeland, N. 1986b, Effects of acetylsalicylic acid and naproxen on bone resorption and formation in rats. *Archives of Orthopaedic and Trauma Surgery*, **105:** 137–141.

Solheim, L.F., Ronningen, H. and Langeland, N. 1986c, Effects of acetylsalicylic acid on heterotopic bone resorption and formation in rats. *Archives of Orthopaedic and Trauma Surgery*, **105:** 142–145.

Sorenson, J.R.J. 1982, The anti-inflammatory activities of copper complexes. *In:* H. Sigel (ed.), *Metal Ions in Biological Systems*, Vol. 14, pp. 77–124. New York: Marcel Dekker.

Spisani, S., Vanzini, G. and Traniello, S. 1979, Inhibition of human leucocytes locomotion by anti-inflammatory drugs. *Experientia*, **35:** 803–804.

Spławinski, J.A., Gorka, Z., Zacny, E. and Kaluza, J. 1977, Fever produced in the rat by intracerebral E-coli endotoxin. *Pflügers Archiv für die Gesampte Physiologie de Menschen und die Tiere*, **368:** 117–123.

Steer, K.A., Wallace, T.M., Bolton, C.H. and Hartog, M. 1997, Aspirin protects low density lipoprotein from oxidative modification. *Heart*, **77:** 333–337.

Sternberg, M., Peyroux, J., Grochulski, A., Engler, R., Feret, J., Moisy, M., Lagrue, G. and Jayle M.F. 1975, Biochemical criteria for the evaluation of drug efficiency on adjuvant arthritis and nephrotoxic serum nephritis in the rat: studies with phenylbutazone, L-asparaginase, colchicine, lysine acetylsalicylate, and pyridinol carbamate. *Canadian Journal of Physiology and Pharmacology*, **53:** 368–374.

Stevenson, M.N., Zhao, M.J., Asea, A., Coleman, C.N. and Calderwood, S.K. 1999, Salicylic acid and aspirin inhibit the activity of RSK2 kinase and repress RSK2-dependent transcription of cyclic AMP response element binding protein- and NF-kappa B- responsive genes. *Journal of Immunology*, **163:** 5608–5616.

Stone, C.A., Van Arman, C.G., Lotti, V.J., Minsker, D.H., Risley, E.A., Bagdon, W.J., Bokelman, D.L., Jensen, R.D., Mendlowski, B., Tate, C.L., Peck, H.M., Zwickley, R.E. and McKinney, S.E. 1977, Pharmacology and toxicology of diflunisal., *British Journal of Clinical Pharmacology*, **4 (Suppl. 1):** 19S–29S.

Stone, E. 1763, An account of the success of the bark of the willow in the cure of agues. In a letter to the Right Honourable George Earl of Macclesfield, President of R.S. from the Rev. Mr Edmund Stone, of Chipping-Norton in Oxfordshire. *Philosophical Transactions of the Royal Society London*, **53:** 195–200.

Strathy, G.M. and Gorski, J.P. 1987, Effects of anti-inflammatory drugs on cartilage recovery from catabolin-induced degradation. *Agents and Actions*, **21:** 149–159.

Strubelt, O. and Zetler, G. 1980, Anti-inflammatory effect of ethanol and other alcohols on the rat paw oedema and pleurisy. *Agents and Actions*, **10:** 279–286.

Stuart, J.J. and Pisko, E.J. 1981, Choline magnesium trisalicylate does not impair platelet aggregation. *Pharmacotherapeutica*, **2:** 547–551.

Sturman, J.A. and Smith, M.J.H. 1966, Effects of salicylate and γ-resorcylate (2:6-dihydroxybenzoate) on pathways of glucose metabolism in the human red cell. *Biochemical Pharmacology*, **15:** 1857–1865.

Sturman, J.A. and Smith, M.J.H. 1969, Effects of salicylates on glucose metabolism in the human red cell. *Biochemical Pharmacology*, **16:** 220–222.

Sugimoto, Y., Narumiya, S. and Ichikawa, A. 2000, Distribution and function of prostanoid receptors: studies from knockout mice. *Process in Lipid Research*, **39:** 289–314.

Suguro, T. 1977, A histochemical and ultrastructural study on articular cartilage lesions in allergic arthritis in the rabbit. *The Ryumachi*, **17:** 462–479.

Sulc, J. and Brozek, G. 1972, The effect of small doses of aspirin on some higher nervous functions. *Activitas Nervosa Superior (Praha)*, **14:** 111–113.

Sulc, J., Brozek, G. and Cmival, J. 1973, The effect of aspirin and a compound analgesic on EEG and performance. *Activitas Nervosa Superior*, **15:** 143.

Sunkel, C., Cillero, F., Armijo, M., Pina, M. and Alonso, S. 1978, Synthesis and pharmacological properties of eterylate, a new derivative of acetylsalicylic acid. *Arzneimittel Forschung*, **28:** 1692–1694.

■ CHAPTER 7 ■

Swedberg, M.D. 1994, The mouse grid-shock analgesia test: pharmacological characterization of latency to vocalization threshold as an index of antinociception. *Journal of Pharmacological and Experimental Therapy*, **269**: 1021–1028.

Swierkosz, T.A., Mitchell, J.A., Warner, T.D., Botting, R.M. and Vane, J.R. 1995, Co-induction of nitric oxide synthase and cyclo-oxygenase: interactions between nitric oxide and prostanoids. *British Journal of Pharmacology*, **114**: 1335–1342.

Swingle, K.F. 1974a, Evaluation for anti-inflammatory activity. *In:* R.A. Scherrer and M.W. Whitehouse (eds), *Anti-inflammatory Drugs, Chemistry and Pharmacology*, Vol. 2, pp. 33–122. New York: Academic Press.

Swingle, K.F. 1974b, Interaction of chloroquine and carrageenan. *Biochemical Pharmacology*, **15**: 1699–1774.

Tagliamonte, A., Biggio, G., Vargiu, L. and Gessa, G.L. 1973, Increase in brain tryptophan and stimulation of serotonin synthesis by salicylate. *Journal of Neurochemistry*, **20**: 909–912.

Takaoki, M., Yamashita, Y., Koike, K. and Matsuda, S. 1988, Effect of indomethacin, aspirin, and acetaminophen on *in vitro* antiviral and antiproliferative activities of recombinant human interferon-α_{2a}. *Journal of Interferon Research*, **8**: 727–733.

Takashiba, S., Van Dyke, T.E. and Amar, S. 1996, Inhibition of nuclear factor kappa B subunit p65 mRNA accumulation in lipopolysaccharide-stimulated human monocytic cells treated with sodium salicylate. *Oral Microbiology and Immunology*, **11**: 420–424.

Talbodec, A., Berkane, N., Blandin, V., Breittmayer, J.P., Ferrari, E., Frelin, C. and Vigne, P. 2000, Aspirin and sodium salicylate inhibit endothelin ETA receptors by an allosteric type of mechanism. *Molecular Pharmacology*, **57**: 797–804.

Tanaka, K. and Iizuka, Y. 1968, Suppression of enzyme release from isolated rat liver lysosomes by non-steroidal anti-inflammatory drugs. *Biochemical Pharmacology*, **17**: 2023–2032.

Tang, M.S., Copeland, R.A. and Penning, T.M. 1997, Detection of an Fe2+-protoporphyrin-IX intermediate during aspirin-treated prostaglandin H2 synthase II catalysis of arachidonic acid to 15-HETE. *Biochemistry*, **36**: 7527–7534.

Taylor, R.J. and Salata, J.J. 1976, Inhibition of prostaglandin synthetase by tolmetin (Tolectin, McN-2559), a new non-steroidal anti-inflammatory agent. *Biochemical Pharmacology*, **25**: 2479–2484.

Telias, I.D., Zvi, A.B. and Danon, A. 1985, Salicylate antagonizes the antiinflammatory action of aspirin in the rat. *Prostaglandins, Leukotrienes and Medicine*, **20**: 23–28.

Tempero, K.F., Cirillo, V.J. and Steelman, S.L. 1978, The clinical pharmacology of diflunisal. *In: Diflunisal in Clinical Practice. Proceedings of a Conference at the XIVth Congress of Rheumatology*, pp. 23–39. Chairman K. Hiehlke. New York: Futura.

Tenenbaum, J. 1984, Experimental models of osteoarthritis: a reappraisal. (Editorial). *Journal of Rheumatology*, **11**: 2.

The IUPHAR Compendium of Receptor Characterization and Classification 1998, IUPHAR Media Ltd., Burlington Press, Foxton, Cambridge (UK).

Thompkins, L. and Lee, K.H. 1969, Studies on the mechanism of action of salicylates. IV. Effect of salicylates on oxidative phosphorylation. *Journal of Pharmaceutical Sciences*, **58**: 102–105.

Thompkins, L. and Lee, K.H. 1975, Comparison of analgesic effects of isoteric variations of salicylic acid and aspirin (acetylsalicylic acid). *Journal of Pharmaceutical Sciences*, **64**: 760–763.

Tomoda, T., Takeda, K., Kurashige, T., Enzan, H. and Miyahara, M. 1994, Acetylsalicylate (ASA)-induced mitochondrial dysfunction and its potentiation by Ca^{2+}. *Liver*, **14**: 103–108.

Tordjman, C., Coge, F., Andre, N., Rique, H., Spedding, M. and Bonnet, J. 1995, Characterisation of cyclooxygenase 1 and 2 expression in mouse resident peritoneal macrophages in vitro; interactions of non-steroidal anti-inflammatory drugs with COX2. *Biochimica and Biophysica Acta*, **1256**: 249–256.

Travis, S. and Jewell, D.P. 1994, Salicylates for ulcerative colitis – their mode of action. *Pharmacology and Therapeutics*, **63**: 135–161.

Tremoli, E., Maderna, P., Eynard, A., Gregori, M. and Galli, G. 1986, *In vitro* effects of aspirin and non-steroidal anti-inflammatory drugs on the formation of 12-hydroxyecosatetraenoic acid by platelets. *Prostaglandins and Leukotrienes in Medicine*, **23**: 117–122.

Trnavska, Z. and Trnavsky, K. 1974, Influence of nonsteroidal anti-rheumatic drugs on collagen metabolism in rats with adjuvant-induced arthritis. *Pharmacology*, **12**: 110–116.

Trnavska, Z., Trnavsky, K. and Kuhn, K. 1968a, The influence of sodium salicylate on the metabolism of collagen. *Biochemical Pharmacology*, **17:** 1493–1500.

Trnavska, Z., Trnavsky, K. and Kuhn, K. 1968b, The influence of sodium salicylate on the metabolism of collagen in the lathyritic rat. *Biochemical Pharmacology*, **17:** 1501–1509.

Trnavsky, K. 1974, Some effects of antiinflammatory drugs on connective tissue metabolism. *In:* R.A. Scherrer and M.W. Whitehouse (eds), *Anti-inflammatory Agents*, Vol. 2, pp. 303–326. New York: Academic Press.

Tsai, A.L., Palmer, G., Wu, G., Peng, S., Okeley, N.M., van der Donk, W.A. and Kulmacz, R.J. 2002, Structural characterization of arachidonyl radicals formed by aspirin-treated prostaglandin H synthase 2. *Journal of Biological Chemistry*, **277:** 38311–38321.

Tschopp, T.B. 1979, Aspirin inhibits platelet aggregation on, but not adhesion to, collagen fibrils: an assessment of platelet adhesion and deposited platelet mass by morphometry and ^{51}Cr-labeling. *Thrombosis Research*, **11:** 619–632.

Tubaro, A., Dri, P., Delbello, G., Zilli, C. and Della Loggia, R. 1985, The croton oil ear test revisited. *Agents and Actions*, **17:** 347–349.

Turner, R.A., Johnson, J.A., Mountz, J.D. and Treadway, W.J. 1983, Neutrophil migration in response to chemotactic factors: effects of generation conditions and chemotherapeutic agents. *Inflammation*, **7:** 57–65.

Udassin, R., Ariel, I., Haskel, Y., Kitrosski, N. and Chevion, M. 1991, Salicylates as an in vivo free radical trap: Studies on ischemic insult to the rat intestine. *Free Radical Biology and Medicine*, **10:** 1–6.

Ueno, K., Kawashima, H., Ohnuma, N., Kurozumi, S., Hashimoto, Y. and Orimo, H. 1980, Aspirin inhibition of hypercalcemic effect of 1 alpha-hydroxyvitamin D_3 in rats. *Japanese Journal of Pharmacology*, **30:** 675–678.

Upmacis, R.K., Deeb, R.S. and Hajjar, D.P. 1999, Regulation of prostaglandin H_2 synthase activity by nitrogen oxides. *Biochemistry*, **38:** 12505–12513.

Valle, D. and Gaertner, J. 1993, Penetrating the peroxisome. *Nature*, **361:** 682.

Van Arman, C.G. 1976, Pathway to adjuvant arthritis. *Federation Proceedings*, **35:** 2442–2446.

Van Arman, C.G., Armstrong, D.A.J. and Kim, D.H. 1991, Antipyretics. *In:* E. Schönbaum and P. Lomax (eds), *Thermoregulation: Pathology, Pharmacology and Therapy*, pp. 55–104. New York: Pergamon Press.

Van Der Kraan, P.M., Vitters, E.L., De Vries, B.J., Van Den Berg, W.B. and Van De Putte, L.B.A. 1990, The effect of chronic paracetamol administration to rats on the glycosaminoglycan content of patellar cartilage. *Agents and Actions*, **29:** 218–223.

Van Dorp, D.A. 1971, Recent developments in the biosynthesis and the analysis of prostaglandins. *Annals of the New York Academy of Sciences*, **180:** 181–190.

Van Dorp, D.A. 1978, Isolation and properties of enzymes involved in prostaglandin biosynthesis. *Acta Biologica Medica Germanica*, **37:** 691–699.

Vane, J.R. 1971, Inhibition of prostaglandin synthesis as a mechanism of action for aspirin-like drugs. *Nature New Biology*, **231:** 232–235.

Vane, J.R. 1978, The mode of action of aspirin-like drugs. *Agents and Actions*, **8:** 430–431.

Vane, J.R. 1979, Summing up. *In:* K.D. Rainsford and A.W. Ford-Hutchinson (eds), *Prostaglandins and Inflammation*, pp. 113–118. Basel: Birkhäuser.

Vane, J.R., Backle, Y.S. and Botting, R.M. 1998, Cyclooxygenases 1 and 2. *Annual Reviews of Pharmacology and Therapy*, **38:** 97–110.

Vane, J.R. and Botting, R.M. 1995, New insights into the mode of action of anti-inflammatory drugs. *Inflammation Research*, **44:** 1–10.

Vane, J.R. and Botting, R.M. 1996, Overview – mechanisms of action of anti-inflammatory drugs. *In:* J. Vane, J. Botting and R. Botting (eds), *Improved Non-Steroidal Anti-Inflammatory Drugs. COX-2 Enzyme Inhibitors*, pp. 1–27. Dordrecht & London: William Harvey Press.

Vane, J.R. and Ferreira, S.H. (eds) 1979, *Inflammation and Anti-inflammatories*. Berlin: Springer Verlag.

Vanegas, H. and Schaible, H.G. 2001, Prostaglandins and cyclooxygenases in the spinal cord. *Progress in Neurobiology*, **64:** 327–363.

Van Oss, C.J., Friedmann, J.G. and Fontaine, M. 1961, Anticomplementary action of aspirin. *Nature*, **189:** 147.

Vargaftig, B.B. 1977, Carrageenan and thrombin trigger prostaglandin synthetase-independent aggregation of rabbit platelets: inhibition by phospholipase A2 inhibitors. *Journal of Pharmacy and Pharmacology*, **29:** 222–228.

Vargaftig, B.B. 1978a, Salicylic acid fails to inhibit generation of thromboxane A2 activity in platelets after in vivo administration to the rat. *Journal of Pharmacy and Pharmacology*, **30:** 101–104.

Vargaftig, B.B. 1978b, The inhibition of cyclooxygenase of rabbit platelets by aspirin is prevented by salicylic acid and phenanthrolines. *European Journal of Pharmacology*, **50:** 231–241.

Vargaftig, B.B., Clignard, M. and Benveniste, J. 1981, Present concepts on the mechanisms of platelet aggregation. *Biochemical Pharmacology*, **30:** 263–271.

Vargaftig, B.B. and Ferreira, S.H. 1981, Blockade of the inflammatory effects of platelet-activating factor by cyclo-oxygenase inhibitors. *Brazilian Journal of Medical and Biological Research*, **14:** 187–189.

Veihelmann, A., Landes, J., Hofbauer, A., Dorger, M., Refior, H.J., Messmer, K. and Krombach, F. 2001, Exacerbation of antigen-induced arthritis in inducible nitric oxide synthase-deficient mice. *Arthritis and Rheumatism*, **44:** 1420–1427.

Venerando, R., Motto, C., Pizzo, P., Rizzuto, R. and Siliprandi, N. 1996, Mitochondrial alterations induced by aspirin in rat hepatocytes expressing mitochondrially targeted green fluorescent protein (mtGFP). *FEBS Letters*, **382:** 256–260.

Vercesi, A.E. and Focesi, A., Jr. 1977, The effects of salicylate and aspirin on the activity of phosphorylase a in perfused hearts of rats. *Experientia (Basel)*, **33:** 157–158.

Vernon-Roberts, R., Liyanage, S.P. and Currey, H.L.F. 1976, Adjuvant arthritis in the rat. Distribution of fluorescent material after foodpad injection of rhodamine-labelled tubercle bacilli. *Annals of the Rheumatic Diseases*, **35:** 389–397.

Vigo, C., Lewis, G.P. and Piper, P.J. 1980, Mechanisms of inhibition of phospholipase A2. *Biochemical Pharmacology*, **29:** 623–627.

Vincent, J.E., Bonta, I.L. and Zijlstra, F.J. 1978, Accumulation of blood platelets in carrageenan rat paw oedema. Possible role in the inflammatory process. *Agents and Actions*, **8:** 291–295.

Vinegar, R., Schreiber, W. and Hugo, R. 1969, Biphasic development of carrageenan edema in rats. *Journal of Pharmacology and Experimental Therapeutics*, **166:** 96–103.

Vinegar, R., Truax, J.F. and Selph, J.L. 1976, Quantitative studies of the pathway to acute carrageenan inflammation. *Federation Proceedings*, **35:** 2447–2456.

Vinegar, R., Truax, J.F., Selph, J.L. and Johnston, P.R. 1990, New analgesic assay utilizing trypsin-induced hyperalgesia in the hind limb of the rat. *Journal of Pharmacological Methods*, **23:** 51–61.

Violi, F., Allessandri, C., Pratico, D., Guzzo, A., Ghiselli, A. and Balsano, F. 1989, Inhibition of cyclooxygenase-independent platelet aggregation by sodium salicylate. *Thrombosis Research*, **54:** 583–593.

Voigtländer, V., Hansch, G.M. and Rother, U. 1980, Effect of aspirin on complement in vivo. *International Archives of Allergy and Applied Immunology*, **61:** 145–149.

Von Kaulla, K.N. and Ens, G. 1967, On structure-related properties of synthetic organic clot-dissolving (thrombolytic) compounds. *Biochemical Pharmacology*, **16:** 1023–1034.

Wahn, H. and Hammerschmidt, S. 1998, Inhibition of PMN- and HOC1-induced vascular injury in isolated rabbit lungs by acetylsalicylic acid: a possible link between neutrophil-derived oxidative stress and eicosanoid metabolism? *Biochimica and Biophysica Acta*, **1408:** 55–66.

Waksman, B.H., Pearson, T.M. and Sharp, J.T. 1960, Studies of arthritis and other lesions induced in rats by injection of mycobacterial adjuvant II. Evidence that the disease is a disseminated immunologic response to exogenous antigen. *Journal of Immunology*, **85:** 403–417.

Walenga, R.W. and Bergström, W. 1985, Stimulation of calcium uptake in rat calvaria by prostacyclin. *Prostaglandins*, **29:** 191–202.

Walenga, R.W., Showell, H.J., Feinstein, M.B. and Becker, E.L. 1980, Parallel inhibition of neutrophil arachidonic acid metabolism and lysosomal enzyme secretion by nordihydroguaiaretic acid. *Life Sciences*, **27:** 1047–1053.

Walker, J.R. and Harvey, J. 1983, Actions of antiinflammatory drugs on leukotriene and prostaglandin metabolism: relationship to asthma and other hypersensitivity reactions. *In:* K.D. Rainsford and G.P. Velo, *Side-effects of Anti-inflammatory/Antirheumatic Drugs*, Vol. 6, pp. 227–236. New York: Raven Press.

Walker, J.R. and Smith, M.J.H. 1979, Adrenocortical stimulation and the anti-inflammatory actions of salicylates. *Journal of Pharmacy and Pharmacology*, **31:** 640–641.

Walker, J.R., Smith, M.J.H. and Ford-Hutchinson, A.W. 1976, Anti-inflammatory drugs, prostaglandins and leukocyte migration. *Agents and Actions*, **6:** 602–606.

Waller, S. 1973, Prostaglandins and the gastrointestinal tract. *Gut*, **14:** 402–417.

Wallis, W.J. and Simkin, P.A. 1993, Antirheumatic drug concentrations in human synovial fluid and synovial tissue. Observations on extravascular pharmacokinetics. *Clinical Pharmacokinetics*, **8:** 496–522.

Walz, D.T., Di Martino, M.J., Kuch, J.H. and Zuccarello, W. 1971, Adjuvant-induced arthritis in rats. I. Temporal relationship of physiological, biochemical, and hematological parameters. *Proceedings of the Society for Experimental Biology and Medicine*, **136:** 907–910.

Wang, Z.Y. and Brecher, P. 1999, Salicylate inhibition of extracellular signal-regulated kinases and inducible nitric oxide synthase. *Hypertension*, **34:** 1259–1264.

Wangel, A. and Klockars, M. 1977, Lymphocyte subpopulations in rheumatoid synovial tissue. *Annals of the Rheumatic Diseases*, **36:** 176–180.

Ward, P.A., Hugli, T.E. and Chenowith, D.A. 1979, Complement and chemotaxis. *In:* J.C. Houch (ed.), *Chemical Messengers of the Inflammatory Process*, pp. 153–178. Amsterdam: Elsevier/North Holland.

Warhurst, G., Lees, M., Higgs, N.B. and Turnberg, L.A. 1987, Pharmacokinetics of aspirin and salicylate in relation to inhibition of arachidonate cyclooxygenase and antiinflammatory activity. *Proceedings of the National Academy of Sciences USA*, 1987 **84:** 1417–1420.

Warne, P.J. and West, G.B. 1978, Inhibition of leukocyte migration by salicylates and indomethacin. *Journal of Pharmacy and Pharmacology*, **30:** 783–785.

Warnecke, P. and Seeber, S. 1969, Vergleich eineger Antiphlogistica in systemen neoplastiische Zellen. *Arzneimittel Forschung*, **19:** 975–977.

Warner, T.D. and Mitchell, J.A. 2003, Nonsteroidal antiinflammatory drugs inhibiting prostanoid efflux: as easy as ABC?. *Proceedings of the National Academy of Sciences USA*, **100:** 9108–9110.

Warner, T.D., Giuliano, F., Vojnovic, I., Bukasa, A., Mitchell, J.A. and Vane, J.R. 1999, Nonsteroid drug sensitivities for cyclo-oxygenase-1 rather than cyclo-oxygenase-2 are associated with human gastrointestinal toxicity: a full *in vitro* analysis. *Proceedings of the National Academy of Sciences USA* **96:** 7563–7568.

Waters, D.J., Caywood, D.D., Trachte, G.J., Turner, R.T. and Hodgson, S.F. 1991, Immobilization increases bone prostaglandin E, effect of acetylsalicylic acid on disuse osteoporosis studied in dogs. *Acta Orthopaedica Scandinavica*, **62:** 238–243.

Watnick, A.S. 1975, *In:* M.E. Rosenthale and H. Mansmann (eds), *Immunopharmacology*, pp. 155–171. New York: Spectrum.

Wax, J., Tessman, D.K., Winder, C.V. and Stephens, M.D. 1975, A sensitive method for the comparative bioassay of nonsteroidal anti-inflammatory compounds in adjuvant-induced primary inflammation in the rat. *Journal of Pharmacology and Experimental Therapeutics*, **192:** 166–171.

Weaver, L.C., Richards, A.B. and Martin, H.E. 1961, Analgesic and antipyretic properties of some aspirin derivatives. *Journal of Pharmacy and Pharmacology*, **13:** 105–110.

Weber, C., Erl, W., Pietsch, A. and Weber, P.C. 1995, Aspirin inhibits nuclear factor-kappa B mobilization and monocyte adhesion in stimulated human endothelial cells. *Circulation*, **91:** 1914–1917.

Wedmore, C.V. and Williams, T.J. 1981, Control of vascular permeability by polymorphonuclear leukocytes in inflammation. *Nature*, **289:** 645–650.

Weischer, C.H. and Anda, L.P. 1975, Activity of acid phosphatase in the carrageenan-induced edema: a report about the effects of antiphlogistics on this enzyme; biochemical and histochemical investigations. *International Journal of Clinical Pharmacology*, **12:** 258–260.

Weissman, G. 1968, Lysosomes. *In:* P.N. Campbell (ed.), *The Interaction of Drugs and Subcellular Components in Animal Cells*, pp. 203–212. London: Churchill.

Weissmann, G., Montesinos, M.C., Pillinger, M., Cronstein, B.M. 2002, Non-prostaglandin effects of aspirin III and salicylate: inhibition of NF kappa B (P105) – knockout mice. *Advances in Experimental Biology and Medical Sciences*, **507:** 571–577.

Weissmann, G., Smolen, J.E. and Hoffstein, S. 1978, Polymorphonuclear leukocytes as secretory organs of inflammation. *Journal of Investigative Dermatology*, **71:** 95–99.

Wennogle, L.P., Liang, H., Quintavalla, J.C., Bowen, B.R., Wasvart, J., Miller, D.B., Allentoff, A., Boyer, W., Kelly, M. and Marshall, P. 1995, Comparison of recombinant cyclooxygenase-2 to native isoforms: aspirin labeling of the active site. *FEBS Letters*, **371:** 315–320.

CHAPTER 7

Werb, Z. 1978, Biochemical actions of glucocorticoids on macrophages in culture. Specific inhibition of elastase, collagenase, and plasminogen activator secretion and effects on other metabolic functions. *Journal of Experimental Medicine*, **147:** 1695–1712.

Weser, U. and Schubotz, L.M. 1981, Catalytic reaction of copper complexes with superoxide. *In:* K.D. Rainsford, K. Brune and M.W. Whitehouse (eds), *Trace Elements in the Pathogenesis and Treatment of Inflammation*, pp. 103–120. Basel: Birkhäuser.

Westwick, J. 1979, Prostaglandins as mediators of inflammation-vascular aspects. *In:* K.D. Rainsford and A.W. Ford-Hutchinson (eds), *Prostaglandins and Inflammation*, pp. 59–63. Basel: Birkhäuser.

Whitehouse, D.J., Whitehouse, M.W. and Pearson, C.M. 1969, Passive transfer of adjuvant-induced arthritis and allergic encephalomyelitis in rats using thoracic duct lymphocytes. *Nature*, **224:** 1322.

Whitehouse, M.W. 1963, Uncoupling of oxidative phosphorylation by some arylacetic acids (anti-inflammatory or hypocholesterolaemic drugs). *Nature*, **201:** 629–630.

Whitehouse M.W. 1964, Biochemical properties of anti-inflammatory drugs – III. Uncoupling of oxidative phosphorylation in a connective tissue (cartilage) and liver mitochondria by salicylate analogues: relationship of structure to activity. *Biochemical Pharmacology*, **13:** 319–336.

Whitehouse, M.W. 1965, Some biochemical and pharmacological properties of anti-inflammatory drugs. *Progress in Drug Research*, **8:** 321–429.

Whitehouse, M.W. 1968, The molecular pharmacology of anti-inflammatory drugs: some possible mechanisms of action at the biochemical level. *Biochemical Pharmacology* **(Suppl):** 293–307.

Whitehouse, M.W. 1977, Some biochemical complexities of inflammatory disease affecting drug action. *In:* I.L. Bonta (ed.), *Recent Developments in the Pharmacology of Inflammatory Mediators*, pp. 135–147. Basel: Birkhäuser.

Whitehouse, M.W. 1978, The chemical nature of adjuvants. *In:* L.E. Glynn and M.W. Stewart (eds), *Immunochemistry: Advanced Textbook*, pp. 571–605. London: Wiley.

Whitehouse, M.W. and Beck, F.J. 1973, Impaired drug metabolism in rats with adjuvant-induced arthritis: a brief review. *Drug Metabolism and Distribution*, **1:** 251–255.

Whitehouse, M.W. and Beck, F.W.J. 1974, Standardisation of arthritogenic adjuvants for evaluating anti-inflammatory and immunosuppressant drugs. *Agents and Actions*, **4:** 227–229.

Whitehouse, M.W. and Boström, H. 1961, Studies on the action of some anti-inflammatory agents in inhibiting the biosynthesis of mucopolysaccharide sulphates. *Biochemical Pharmacology*, **7:** 135–150.

Whitehouse, M.W. and Boström, H. 1962, The effects of some anti-inflammatory (anti-rheumatic) drugs on the metabolism of connective tissues. *Biochemical Pharmacology*, **11:** 1175–1201.

Whitehouse, M.W. and Dean, P.G.D. 1965, Biochemical properties of anti-inflammatory drugs – V. Uncoupling of oxidative phosphorylation by some γ-resorcyl and other dihydroxybenzol compounds. *Biochemical Pharmacology*, **14:** 557–567.

Whitehouse, M.W., Orr, K.J., Beck, F.W.J. and Pearson, C.M. 1974, Freund's adjuvants: relationship of arthritogenicity and adjuvanticity in rats to vehicle composition. *Immunology*, **27:** 311–330.

Whitehouse, M.W., Rainsford, K.D., Ardlie, N.G., Young, I.G. and Brune, K. 1977, Alternatives to aspirin, derived from biological sources. *In:* K.D. Rainsford, K. Brune and M.W. Whitehouse (eds), *Aspirin and Related Drugs. Their Action and Uses*, pp. 43–57. Basel: Birkhäuser.

Whitehouse, M.W. and Skidmore, I.F. 1965, Concerning the regulation of some diverse biochemical reactions, underlying the inflammatory response, by salicylic acid, phenylbutazone and other acidic antirheumatic drugs. *Journal of Pharmacy and Pharmacology*, **17:** 668–670.

Wieth, J.O. 1970, Paradoxical temperature dependence of sodium and potassium fluxes in human red cells. *Journal of Physiology*, **207:** 563–580.

Wildfeuer, A. 1983, Effects of non-steroidal anti-inflammatory drugs on human leucocytes. *Zeitschrift für Rheumatology*, **42:** 16–20.

Wilhelmi, G. 1949, Ueber die pharmakologischen Eigenschaften von Irgapyrin, einem neuen Präparat aus der Pyrazolreihe. *Schweizer Medizinische Wochenschrift*, **79:** 577.

Wilhelmi, G. 1963, Salicylates and wound healing. *In:* A.St.J. Dixon, B.K. Martin, M.J.H. Smith and P.H.N. Wood (eds), *Salicylates. An International Symposium*, pp. 176–185. London: J. & A. Churchill.

Wilhelmi, G. 1978, Ferdernde und hemmende Einflüsse von Tribenosid und Acetylsalicylsäure auf die spontane Arthrose der Maus. *Arzneimittel Forschung*, **28:** 1724–1726.

Williams, D.M. and Johnson, N.W. 1976, Alteration in peripheral blood leukocyte distribution in response to local inflammatory stimuli in the rat. *Journal of Pathology*, **118:** 129–141.

Williams, T.J. 1979, Prostaglandin E$_2$, prostaglandin I$_2$ and the vascular changes of inflammation. *British Journal of Pharmacology*, **65:** 517–524.

Williams, T.J. and Peck, M.J. 1977, Role of prostaglandin-mediated vasodilation in inflammation. *Nature*, **270:** 530–532.

Willis, A.L. 1969, Release of histamine, kinin and prostaglandins during carrageenan induced inflammation in the rat. *In:* P. Mantagazza and E.W. Horton (eds), *Prostaglandins, Peptides and Amines*, p. 31. New York: Academic Press.

Willis, A.L. 1970, Identification of prostaglandin E$_2$ in inflammatory exudates. *Pharmacological Research Communications*, **2:** 279–304.

Willis, A.L., Davidson, P., Ramwell, P.W., Brocklehurst, W.E. and Smith, B. 1972, Release and actions of prostaglandins in inflammation and fever: inhibition by anti-inflammatory and anti-pyretic drugs. *In:* P.W. Ramwell and B.B. Pharis (eds), *Prostaglandins in Cellular Biology*, pp. 227–258. New York: Plenum Press.

Willoughby, D.A. and Dieppe, P. 1976, Prostaglandins in the inflammatory response – pro or anti? *In:* G.P. Lewis (ed.), *The Role of Prostaglandins in Inflammation*, pp. 14–25. Berne: Hans Huber.

Willoughby, D.A. and Flower, R.J. 1992, The anti-inflammatory action of the salicylates. *In:* J. Vane and R.M. Botting (eds), *Aspirin and the Other Salicylates*, pp. 141–165. London: Chapman and Hall Medical.

Willoughby, D.A., Moore, A. and Colville-Nash, P.R. 2000, COX-1, COX-2, and COX-3 and the future treatment of chronic inflammatory disease. *Lancet*, **355:** 646–648.

Wills, W.D. 1985, *The Pain System. The Neural Basis of Nociceptive Transmission in the Mammalian Nervous System*. Basel: Karger.

Winchurch, R.A., Foschi, G.V. and Walz, D.T. 1974, Inhibition of the lymphocyte-mediated cytotoxic reaction by anti-inflammatory drugs. *Journal of the Reticuloendothelial Society*, **15:** 112–117.

Winder, C.V., Wax, J., Burr, V., Been, M. and Rosiere, C.E. 1958, A study of pharmacological influences on the ultraviolet erythema in guinea pigs. *Archives Internationales de Pharmacodynamie et de Thérapie*, **116:** 261–292.

Winter, C.A. 1966, Nonsteroid anti-inflammatory agents. *Progress in Drug Research*, **10:** 139–203.

Winter, C.A., King, P.J., Tocco, D.J. and Tanabe, K. 1979, Analgesic activity of diflunisal [MK-647; 5-(2,4-difluorophenyl)salicylic acid] in rats with hyperalgesia induced by Freund's adjuvant. *Journal of Pharmacology and Experimental Therapeutics*, **211:** 678–685.

Winter, C.A., Risley, E.A. and Nuss, G.W. 1962, Carrageenan-induced oedema in the hind-paw of the rat as an assay for anti-inflammatory drugs. *Proceedings of the Society for Experimental Biology and Medicine*, **111:** 544–547.

Winter, C.A., Risley, E.A. and Silber, R.H. 1968, Anti-inflammatory activity of indomethacin and plasma corticosterone in rats. *Journal of Pharmacology and Experimental Therapeutics*, **162:** 196–201.

Wojtecka-Lukasik, E. and Dancewicz, A.M. 1974, Inhibition of human leucocyte collagenase by some drugs used in the therapy of rheumatic diseases. *Biochemical Pharmacology*, **23:** 2077–2081.

Wolna, E. and Inglot, A.D. 1973, Non-steroidal anti-inflammatory drugs: effects on the utilization of glucose and production of lactic acid in tissue culture. *Experientia (Basel)*, **29:** 69–71.

Woods, H.F., Stubbs, W.A., Johnson, G. and Alberti, K.G. 1974, Inhibition by salicylate of gluconeogenesis in the isolated perfused rat liver. *Clinical and Experimental Pharmacology and Physiology*, **1:** 535–540.

Worathumrong, N. and Grimes, A.J. 1975, Anaerobic glycolysis in normal human erythrocytes incubated in vitro with sodium salicylate. *Clinical Science and Molecular Medicine*, **49:** 375–384.

Wu, K.K., Chen, Y.E., Fordham, E., Ts'ao, C.H., Rayudu, G. and Matayachi, D. 1981, Differential effects of two doses of aspirin on platelets-vessel wall introduction in vivo. *Journal of Clinical Investigation*, **68:** 382–387.

Wu, K.K., Sanduja, R., Tsai, A-L., Ferhanoglu, B. and Loose-Mitchell, D.S. 1991, Aspirin inhibits interleukin 1-induced prostaglandin H synthase expression in cultured endothelial cells. *Proceedings of the National Academy of Sciences USA*, **88:** 2384–2387.

Xiao, G., Tsai, A.-H., Palmer, G., Boyar, W.C., Marshall, P.J. and Kulmacz, R.J. 1997, Analysis of the hydroperoxide-induced tyrosyl radicals and lipoxygenase activity in aspirin-treated human prostaglandin H synthase-2. *Biochemistry*, **36:** 1836–1845.

Xu, X.-M., Sansores-Garcia, L., Chen, X.-M., Matijevic-Aleksik, N., Du, M. and Wu, K.K. 1999,

Suppression of inducible cyclooxygenase 2 gene transcription by aspirin and sodium salicylate. *Proceedings of the National Academy of Sciences USA*, **96:** 5292–5297.

Yamada, K.M. and Olden, K. 1978, Fibronectins – adhesive glycoproteins of cell surface and blood. *Nature*, **275:** 179–184.

Yamamoto, S. 1988, Characterization of enzymes in prostanoid synthesis. *In:* P.B. Curtis-Prior (ed.), *Prostaglandins. Biology and Chemistry of Prostaglandins and Related Eicosanoids*, pp. 37–45. Edinburgh: Churchill-Livingstone.

Yamasaki, H. and Saeki, K. 1967, Inhibition of mast-cell degranulation by anti-inflammatory agents. *Archives Internationales de Pharmacodynamie et de Thérapie*, **168:** 166–179.

Yamazaki, R., Kusunoki, N., Matsuzai, T., Hashimoto, S., Kawai, S. 2002, Aspirin and sodium salicylate inhibit proliferation and induce apoptosis in rheumatoid synovial cells. *Journal of Pharmacy and Pharmacology*, **54:** 1675–1679.

Yazdi, M., Cheung, D.T., Cobble, S., Nimni, M.E. and Schonfeld, S.E. 1992, Effects of non-steroidal anti-inflammtory drugs on demineralized bone-induced bone formation. *Journal of Periodontal Research*, **27:** 28–33.

Yin, M.-J., Yamamoto, Y. and Gaynor, R.B. 1998, The anti-inflammatory agents aspirin and salicylate inhibit the activity of IκB kinase-β, *Nature*, **396:** 77–80.

Yin, M.-J., Yamamoto, Y. and Gaynor, R.B. 1999, Activation of nuclear factor κB as a target for anti-inflammatory therapy. *Gut*, **44:** 309–310.

Young, J.M., Panah, S., Satchawatcharaphong, C. and Cheung, P.S. 1996, Human whole blood assays for inhibition of prostaglandin G/H synthases-1 and -2 using A23187 and lipopolysaccharide stimulation of thromboxane B2 production. *Inflammation Research*, **45:** 246–253.

Yousefi, S., Chiu, J., Carandang, G., Archibeque, E.G., Vaziri, N. and Cesario, T.C. 1987, Effect of acetyl salicylic acid on production and action of leukocyte-derived interferons. *Antimicrobial Agents and Chemotherapy*, **31:** 114–116.

Yuan, C.-J., Mandal, A.K., Zhang, Z. and Mukherjee, A.B. 2000, Transcriptional regulation of cyclooxygenase-2 gene expression: novel effects of nonsteroidal anti-inflammatory drugs. *Cancer Research*, **60:** 1084–1091.

Yue, K.T.N. and Davis, J.W. 1978, Effect of oral aspirin on 'ecto-ATPase' activity of washed human platelets. *Clinical Pharmacology and Therapeutics*, **24:** 240–242.

Zatz, M.M., Skotnicki, A., Bailey, J.M., Oliver, J.H. and Goldstein, A.L. 1985, Mechanism of action of thymosin. II. Effects of aspirin and thymosin on enhancement of IL-2 production. *Immunopharmacology*, **9:** 189–198.

Zimmerman, G.A., Wiseman, G.A. and Hill, H.R. 1985, Human endothelial cells modulate granulocyte adherence and chemotaxis. *Journal of Immunology*, **134:** 1866–1874.

Zoete, V., Bailly, F., Maglia, F., Rougée, M. and Bennsasson, R.V. 1999, Molecular orbital therapy applied to the study of nonsteroidal anti-inflammatory drug efficiency. *Free Radical Biology and Medicine*, **26:** 1261–1268.

Zurier, R.B. and Sallas, M. 1973, Prostaglandin E_1 (PGE_1) suppression of adjuvant arthritis. *Histopathology. Arthritis and Rheumatism*, **16:** 251–258.

Zurier, R.B., Weissmann, G., Hoffstein, S., Kammerman, S. and Tai, H.H. 1974, Mechanisms of lysosomal enzyme release from human leukocytes II. Effects of cAMP and cGMP, autonomic agonists, and agents which affect microtubule function. *Journal of Clinical Investigation*, **53:** 297–309.

Metabolic and Related Effects of Salicylates

Increase ↑ or Decrease ↓ Effect (Conclusion)

1. Glucose and fatty acid metabolism a) Glucose and mitochondrial metabolism

Increase ↑ or Decrease ↓ Effect (Conclusion)	Drug	System	Reference (first author)
Na salicylate ↑ oxygen consumption in cerebral cortex and liver O_2 consumption is ↑ by acetylsalicylic acid	0.06–0.56mM Na salicylate, aspirin	*In vitro* rat liver and brain	Fishgold, J.T. 1951, *Am. J. Phys.*, **164:** 727–733
↓ Na salicylate ↓ cortisone-induced glycosuria and hyperglycemia in same animals after feeding	100mg salicylate	Adrenalectomised rats (*in vivo*)	Smith, M.J.H. 1952, *Nature*, **170:** 240
Salicylate ↑ blood glucose given orally but not intravenously	100mg salicylate	Male wistar rats (*in vivo*)	Smith, M.J.H. 1954, *Biochim. Bioplys. Acta*, **14:** 241–245
Salicylate ↑ glycogen breakdown and lactic acid accum. (aerobic and anaerobic), ↑ O_2 uptake and respiratory quotient and ↓ glucose uptake	5mM salicylate	Isolated rat diaphragm	Smith, M.J.H. 1956, *Biochem. J.*, **63:** 524–528
Salicylate ↓ creatine phosphate and ATP and ↑ inorganic phosphates	0.1–5mM salicylate	Isolated rat diaphragm	Smith, M.J.H. 1956, *Biochem. J.*, **64:** 589–592
↑ Conc of salicylic acid stimulated O_2 consumption	0.5 and 0.125mg/g body weight salicylate	Rat	Hsieh, A.C.L. 1959, *Br. J. Pharmac.*, **14:** 219–221
Inhibition of glutamic-pyruvic and glutamic-oxalo-acetic transaminase activities in all tissue types	resorcylic acid 5mM	Rat liver, kidney, brain and heart extracts	Huggins, A.K. 1961, *J. Pharm. Pharmac.*, **13:** 654–662
Salicylate caused an ↑ incorporation of radioactive succinate into fumarate and malate but ↓ formation of aspartate	5mM salicylate	Rat liver mitochondria	Bryant, C. 1962, *Biochem. J.*, **84:** 67
Correlation between blood glucose and plasma NEFA concs found with salicylic acids and congeners	4mg/kg salicylate	Rats (*in vivo*)	Fang, V.S. 1963, *Arch. Int. Pharmacodyn.*, **150:** 322–337
Salicylate congeners inhibited glycolytic reactions in mature erythrocytes	<20mM	Human red blood cell suspensions	Sturman, J.A. 1967, *Biochem. Pharm.*, **16:** 220–222
Na salicylate injected into vertebral artery caused marked hypoglycaemia	<25mg/kg	Cats (*in vivo*)	Gaitonde, B.B. 1967, *Br. J. Pharmac. Chemother.*, **30:** 554–560
↓ Release of fatty acids from adipose tissue in salicylate hypoglycaemic activity	<4mM/kg	Alloxan-diabetic rats (*in vivo*)	Fang, V. 1968, *J. Pharm. Sci.*, **57:** 2111–2116
Na salicylate ↑ glucose utilisation and lactate production	1–5mM Na salicylate	Heparinised human blood and washed human erthrocytes	Davies, D.T.P. 1970, *Biochem. Pharm.*, **19:** 1277–1285

Observation	Dose/Conc.	System	Reference
5-n-Alkysalicylates inhibited variety of enzymes, e.g. carbonic anhydrase etc.	10 mM	Isolated enzymes (that transform acidic substances)	McCoubrey, A. 1970, *J. Pharm. Pharmac.*, **22:** 333–337
Aspirin ↑ pyruvate utilisation, carbon dioxide incorporation and organic acid formation	0.2% aspirin in diet	Thiamine deficient rats (liver mitochondria)	Mehlman, M.A. 1971, *J. Biol. Chem.*, **246:** 1618–1622
↓ In erthrocyte 2,3-DPG conc related to dosage	400–600 mg/kg Na salicylate	Male Sprague–Dawley rats	Kravath, R.E. 1972, *Biochem. Pharm.*, **21:** 2656–2658
↓ Lactate and pyruvate by aspirin and adrenaline reversing effects of adrenaline	30 mg/kg aspirin	Rabbit (*in vivo*)	Laborit, H. 1973, *Agressologie*, **14:** 25–30
Salicylate ↑ activity of glucose oxidase at low concs	Na salicylate <20 mg	*Staphylococcus aureus* and *E. Coli*	Kaplan, D. 1973, *Chemotherapy*, **19:** 235–242
Biphasic lactic acid production; 2nd peak at high conc corresponds to toxic effects of drugs.	Aspirin, indomethacin, flufenamic acid (<10mM)	Primary monolayer chick embryo cells	Wolna, E. Englot, A.D. 1973, *Experientia*, **29:** 69–71
Production of lactate parallels glucose utilisation Chloroquine significantly ↓ heat production of erthrocytes, therefore suggesting inhibition of glycolysis	100 mM and 10mM (therapeutic doses)	Peripheral human blood erythrocytes	Sigidin, Ya A. 1975, *Byulleten Eksperimental Noi Biologii I Meditsiny*, **79:** 75–77
↑ Glucose consumption and ↑ lactate production, ↑ organic phosphate, erthrocytes unaffected by salicylate at concs that would stimulate glycolysis	<90 mM salicylate	Human erthrocytes	Worathumrong, N. 1975, *Clin. Sci. Mole. Med.*, **49:** 375–384
A sig. ↓ of TSH was seen after TRH injection on aspirin treatment with no changes noted for PRL	20 mg/dl salicylate	Human volunteers	Dussault, J.H. 1976, *J. Clin. Endocrinol. Metab.*, **43:** 232–235
Therapeutic drug (100–800 µM/kg) ↑ glucose utilisation, high doses (1–3 mM/kg) ↓ glucose utilisation. Benzoate caused no change	<3 mM/kg salicylate	Mice (*in vivo*)	Hull, E.W. *Clin. Biochem.*, 1977, **10:** 10
Salicylate competes with AMP moiety of NAD+	<20 mM salicylate	Isolated porcine lactate dehydrogenase	Cheshire, R.M. 1977, *Int. J. Biochem.*, **8:** 637–643
50% ↓ in mucosal ATP, t = 30min	Oral admin. of 2.6g aspirin	Human jejunal mucosal biopsies	Arvanitakis, C. 1977, *Gut*, **18:** 187–190
↑ Phosphorylase A levels with both drugs, t = 0–30mins	5 mM Na salicylate and aspirin	Perfused rat heart	Vercesi, A.F. 1977, *Experientia*, **33:** 157
↓ In basal glucose levels, ↑ basal insulin. ↑ Insulin with aspirin + Arg, ↓ glucose with aspirin + Arg. t = 3 days	<30 mg/100 ml aspirin	*In vitro* rat liver and brain	Giugliano, D. 1978, *Diabetiologia*, **14:** 359–362

1. Glucose and fatty acid metabolism b) Oxidative phosphorylation

Increase ↑ or Decrease ↓ Effect (Conclusion)	Drug	System	Reference
Anti-inflam. activity of salicylate is not as a result of the uncoupling of ox. phos.	500 mg/kg	Guinea pig (*in vivo*)	Marks, V. 1960, *Nature*, **187:** 610
Na salicylate and salicylic acid exhibited a greater inhibition effect than aspirin on the biosynthesis of citric acid	0–6 g doses	Human volunteers	Panagopoulos, K. 1961, *Med. Exp.*, **5:** 114–121
Various conclusions	40 mM salicylates NSAIDs	Rat liver mitochondria	Whitehouse, M.W. 1962, *Nature*, **196:** 1323–1324
Various effects on oxidase enzyme systems (salicylate, gamma-resorcylate and gentisate)	50 mM	Guinea pig liver mitochondria	Hines, W.J.H. 1963, *Biochem. Pharmac.*, **12:** 1109–1116
Salicylate caused an ↑ incorporation of carbon-14 in fumerate and malate in rat liver mitochondria incubated with labelled succinate	10 mM salicylate	Rat liver mitochondria	Smith, M.J.H. 1964, *Nature*, **202:** 96–97
Varying uncoupling activities of gamma-resorcyl and other dihydroxybenzol compounds	<25 mM	Rat liver mitochondria	Whitehouse, M.W. 1965, *Biochem. Pharmac.*, **14:** 557–567
Lysyl amino groups of certain proteins appear to be important binding sites for acidic anti-inflam. drugs	<5 mM	Rat liver mitochondria	Whitehouse, M.W. 1965, *J. Pharm. Pharmac.*, **17:** 668–671
Uncoupling properties of resorcinols, tropolones and diones	<50 mM	Rat liver mitochondria	Skidmore, I.F. 1965, *Biochem. Pharmac.*, **14:** 547–555
Thio analogues of salicylic acid uncoupled ox. phos. In varying degrees of potency	0.2–1.5 mM	Rat liver mitochondria	Burke, J.F. 1965, *Biochem. Pharmac.*, **14:** 1039–1048
Salicylamide derivatives had varying effects on uncoupling ox. phos.	<5 mM	Rat liver mitochondria	Leader, J.E. 1966, *Biochem. Pharmac.*, **15:** 1379–1387
Uncoupling activity of anthranilic acids correlated with affinity of drug for albumin amino groups	<20 mM	Rat liver mitochondria, bovine plasma albumin	Whitehouse, M.W. 1967, *Biochem. Pharmac.*, **16:** 753–760
I, Br and Cl derivatives of salicylic acid uncoupled ox. phos. They also stimulate respiration of liver homogenates. LD_{50} of salicylates dependent on potency as uncouplers	<10 mM	Rat liver mitochondria	Karler, R. 1968, *Arch. Int. Pharmacodyn.*, **173:** 270–280
5 pesticides did not uncouple or inhibit mitochondria in mice with severe symptoms of poisoning	<200 µM	Mitochondria preparations (mouse, housefly, honey bee)	Ilivicky, J. 1969, *Biochem. Pharmac.*, **18:** 1389–1401

Effect	Concentration	Preparation	Reference
RNA polymerases showed ↑ activity in presence of N6, O2'-dibutyryl cyclic AMP	1 mM salicylate	Rat liver nuclei	Jost, J.-P. 1971, *J. Bio. Chem.*, **246:** 1623–1629
Na salicylate and aspirin both inhibited the cyclic 3',5'-AMP-dependent protein kinase measured by phosphorylation of histone	<10 mM	Protein kinase from bovine diaphragm	Dinnendahl, V. 1973, *Biochem. Pharmac.*, **22:** 2223–2228
4-iodo-salicylic acid considered a uncoupling agent with similar properties to 2,4-dinitrophenol	<1 mM	Isolated rat liver mitochondria	Galen, F. 1974, *Biochem. Pharmac.*, **23:** 1379–1385
Salicylate binds in the hydrophobic adenosine-binding pocket, which is similar in all investigated dehydrogenases	0.34 mM	Isolated dehydrogenases	Einarsson, R. 1974, *Eur. J. Biochem.*, **49:** 41–47
Anionic form of uncoupler is located within the partly negatively charged phospholipid moeity of the membrane with its anionic group pointing towards the membrane		Rat liver mitochondria	Bakker, E.P. 1975, *Biochim. Biophys. Acta.*, **387:** 491–506.
Variety of agents demonstrated correlation between release of respiration and proton transport	various salicylates	Rat liver mitochondria	Cunarro, J. 1975, *Biochim. Biophys. Acta*, **387:** 234–240
Several NSAIDs found to inhibit mitochondrial ATPase, mitochondria ATPase Mg^{2+} and ATP conc dependent	<16 mM	Rat liver mitochondrial ATPase	Chatterjee, S.S. 1976, *Arzneim.-Forsch* (Drug Res.), **26:** 499–502
Na salicylate inhibited state 3 and stimulated state 4 of respiratory chain. Phenazone and aminophenazone only inhibited state 3	0–10 mM	Rat liver mitochondria	Pocwiarowska, E. 1976, *Pol. J. Pharmac. Pharm.*, **28:** 217–226
Salicylates inhibited both rat and chicken liver carboxylase activity	2.5 mM	Rat and chicken liver acetyl-CoA carboxylase	Dular, U., 1979, *Biochem. Pharmac* **28:** 715–718

1. Glucose and fatty acid metabolism c) Fatty acids (lipids)

Effect	Concentration	Preparation	Reference
Nicotinic acid and salicylic acid both ↓ plasma free fatty acids and prevented liver triglyceride accumulation	<300 mg/kg	Rat plasma (*in vivo*)	Bizzi, A. 1966, *Nature*, **209:** 1025–1026
Salicylic acid failed to counteract the accumulation of ketone bodies in plasma whereas nicotinic acid and 3,5-DMP both ↓ the ketone bodies	<100 mg/kg	Rat plasma (*in vivo*)	Bizzi, A. 1966, *Experientia*, **222:** 664
Changes in the composition of fatty acids in liver and brain after aspirin treatment	0.3 mg/g ASA	Rat liver and brain phospholipids (*in vivo*)	Krishnan, R.S. 1968, *Experientia*, **24:** 899–900
Both hypolipidemic agents inhibited acetyl CoA	<10 mM	Acetyl coenzyme A	Maragoudakis, M.E. 1969. *J. Bio. Chem.*, **244:**

Increase ↑ or Decrease ↓ Effect (Conclusion)	Drug	System	Reference
carboxylase		carboxylase isolated from chicken liver	5005–5013
Salicylate ↓ incorporation of acetate into fatty acids	1 mM	Cholesterol from rat liver homogenate	Klimov, A.N. 1969, Biochem. Pharmac., **18**: 1251–1252
Na salicylate and aspirin both caused dose-dependent inhibition of lipolysis in isolated fat cells	0–100 mM	Isolated rat fat cells	Schonhofer, P.S. 1973, Biochem. Pharmac., **22**: 629–637
Results indicated aspirin is not effective agent to reduce fat disposition in broilers	0–2 g/kg aspirin	Chickens	Brenes, A. 1982, Nut. Rep. Int., **26**: 501–508

2. Amino acid, protein and nucleic acid metabolism

Amino-aciduria was observed in all subjects with high blood salicylate levels	2000–4000 g ingested per patient	Human patients	Andrews, B.F. 1961, Am. J. Med. Sci., 67/411–70/414
Gamma-resorcylic acid inhibited glutamic-pyruvic and glutamic-oxaloacetic transaminases in rat liver, kidney, brain and heart extracts	5 mM	Rat tissue	Huggins, A.K. 1961, J. Pharm. Pharmac., **13**: 654–662
Salicylate uncouple ox. phos in connective tissue as well as muscle, liver, kidney etc.	0–5 mM	Bovine cartilage and rat liver	Whitehouse, M.W. 1964, Biochem. Pharmac., **13**: 319–336
Rat tissue aminotransferases with 20 amino acids. Activity was uneven in different tissues and salicylate was generally inhibitory	10 mM	Rat tissue	Gould, B.J. 1965, J. Pharm. Pharmac., **17**: 83–88
Salicylate inhibited rat brain glutamate decarboxylase activity *in vitro*	15–150 mM	Rat brain	Gould, B.J. 1965, J. Pharm. Pharmac., **17**: 15–18
Salicylate inhibited pig heart alanine and aspartate aminotransferase activities *in vitro*	0–50 mM	Aminotransferase preparations (commercial)	Gould, B.J. 1966, Mol. Pharmacol., **2**: 526–533
Salicylate 2.5 mM or greater inhibited creatine kinase activity *in vitro*	0.1–20 mM	Rat liver and diaphragm	Dawkins, P.D. 1966, Biochem. J., **99**: 703–707
Salicylate inhibited incorporation of leucine into proteins	0–3 mM	Rat liver	Reunanen, M. 1967, Nature, **218**: 918–919
Salicylate inhibiting thr and *chlorella* protein hydrolysate uptake was time and dose-dependent in both human and pig gastric-mucosal scrapings	3–60 mM	Gastric-mucosal scrapings of pig and man	Rainsford, K.D. 1968, Biochem. J., **111**: 37
Salicylate ↓ incorporation of ATP into RNA in rat liver. Salicylate inhibited DNA polymerase activity	3 mM+	Rat liver	Janakidevi, K. 1969, J. Pharm. Pharmac., **21**: 401
Salicylate inhibited activity of glutamate decarboxylase	0–250 mM	E. Coli and rat brain	McArthur, N. 1969, J. Pharm. Pharmac., **21**: 21–23

Observation	Concentration/Dose	Tissue/System	Reference
Salicylate significantly inhibited the incorporation of orotic acid at 30 min but not 6h	200 mg/kg	glutamate decarboxylase Mice (in vivo)	Janakidevi, K. 1970, J. Pharm. Pharmac., 22: 51–55
Salicylate inhibited incorporation of radiolabelled leucine into protein	0.3 mM+	Rat liver cell free protein	Burleigh, M. 1970, Biochem. J., 117: 68
Salicylate 10 mM+, ↑ asp, glut, tyr and ornithine and ↓ glutamate and gamma-aminobutyrate, ↓ amino acids in blood and kidney, ↑ in liver	0–20 mM and 75–600 mg/kg	Mouse liver prep and in vivo	Dawkins, P.D. 1970. J. Pharm. Pharmac., 22: 913–922
Salicylate inhibited glutamyl-transfer ribonucleic acid synthetase by comp inhib (gluamate) and comp-non-comp inhibitions with ATP and tRNA	0–12.5 mM	In vitro glutamyl-transfer ribonucleic acid synthetase	Burleigh, M. 1971. J. Pharm. Pharmac., 23: 590–594
Salicylate interfered with protein synthesis in mouse liver in vitro and in vivo	10–15 mM	Mouse liver	Dawkins, P.D. 1971, Biochem. Pharmac., 20: 1303–1312
Ro4-4608 (salicylate chalcone) = α-methyl 3.4 dopa ↓ the aromatic L-aminoacid decarboxylase activity	<10 mM	Guinea pig kidney	Lesieur-Demarkquilly, I. 1973, Annales Pharmaceutiques Francaises, 31(11): 705–708
Various effects of NSAIDs and immunosuppressive drugs on hepatic tryptophan pyrrolase	<340 mg/kg	Rat liver	Reinicke, C. 1973, Biochem. Pharmac., 22: 195–203
Guinea pig liver tryptophan pyrrolase activated by tryptophan but not salicylate	400 mg/kg salicylate	Guinea pig liver	Badawy, A.A.-B. 1974, Biochem. J., 138: 445–451
TA stimulation by NSAIDs is dependent on actions of endogenous corticosteroids	1 mM or <40 mg/kg	Rat liver tyrosine aminotransferase (TA) (isolated and in vivo)	Reinicke, C. 1975, Biochem. Pharmac., 24: 193–198

3. Transport and membranes

Observation	Concentration/Dose	Tissue/System	Reference
Human erythrocytes are protected or stabilised against hypotonic haemolysis with salicylates and other NSAIDs. At high concs, indomethacin, fenamates and chloroquine caused haemolysis	1 µM–10 mM	Erythrocyte membranes	Inglot, A.D. 1968, Biochem. Pharmac., 17: 269–279
NSAIDs inhibited macromolecule-induced red cell aggregation in vitro	100 µM	Rat red blood cells	Gorog, P. 1970, J. Pharm. Pharmac., 22: 86–92
NSAIDs induced cell swelling in vitro at high concs.	0–5 mM	Rat, rabbit and human lymphocytes	Famaey, J-P. 1973, Biochem. Pharmac., 22: 2707–2717
NSAIDs induced pseudo-energised high amplitude swelling. The amplitude of swelling is not dose-dependent but the rapidity is affected	0–5 mM	Rat liver	Famaey, J-P. 1973, Biochem. Pharmac., 22: 2693–2705

Increase ↑ or Decrease ↓ Effect (Conclusion)	Drug	System	Reference
↑ Conc of salicylate showed a saturation type decrease in SO_4^{2-}-permeability at both 4 and 37°C	0–25 mM	Human red blood cells	Schwoch, G. 1974, *Biochim. Biophys. Acta*, **339:** 126–138
A drug conc-dependent stimulation of a sulphide-disulphide interchange reaction in serum proteins	0–0.5 mM	Rat, rabbit and human lymphocyte membranes and rat liver mito	Famaey, J.-P. 1975, *Biochem. Pharmac.*, **24:** 1609–1615
NSAIDs inhibited heat-induced erythrocyte lysis, stabilised serum albumin and had high affinity for erythrocytes	3–300 µM	Erythrocytes and serum albumin	Mizushima, Y. 1975, *Biochem. Pharmac.*, **24:** 1589–1592
Salicylate, anoxia and 2,4-dinitrophenol stimulated xylose uptake	5 mM salicylate	Isolated rat soleus muscle	Korbl, G.P. 1977, *Biochim. Biophys. Acta*, **465:** 93–109
Acidic environment of cells may cause accumulation of acidic NSAIDs within cells	1–100 µM	Human erythrocytes and PMN	Brune, K. 1978, *Biochem. Pharmac.*, **27:** 525–530
Several NSAIDs inhibited α-amino-iso-butyric acid (AIB) and MeAIB uptake, which was conc dependent	0–15 mM	Rat hepatoma cells (HTC)	Bayer, B.M. 1980, *J. Biol. Chem.*, **255:** 8784–8790
Salicylate ↑ the unidirectional efflux of K^+ (Rb^+): dependent on extracellular Ca^{2+}. Incubating tissue with salicylate ↓ $[K^+]$ and ↑ $[Na^+]$	1–135 mM	Lactating rat mammary tissue	Shennan, D.B. 1992, *Biochem. Pharmac.*, **44:** 645–650
Various findings of the release of aspirin derivatives from human erthrocytes	1 mM	Human erythrocytes	Ohsako, M. 1995, *Biol. Pharm. Bull.*, **18:** 310–314

4. Metabolism and general pharmacology

Increase ↑ or Decrease ↓ Effect (Conclusion)	Drug	System	Reference
Na salicylate ↑ liver weights and thiamine concs in livers of adult but not young rats on normal but not thiamine-deficient diets. Thiamine content in kidney and brain not affected	0.5 of 10%/day p.o. in adults for 18 days, 0.3 ml of 5% p.o., 4 times a week	Rats (*in vivo*)	Cleland, J. 1946, *Aust. J. Exp. Biol. Med. Sci.*, **24:** 227–230
Na salicylate ↑ uric acid/creatine ration and slightly ↑ circulating eosinophils	4 or 6 g	Human volunteers	Roskam, J. 1951, *Lancet*, **11:** 375–376
Salicylate ↓ ascorbic acid and cholesterol conc in adrenals, reduced circulated eosinophils	500 mg/kg (*in vivo*)	Hypophysectomised rats injected with ACTH	Van Cauwenberge, H. 1951, *Lancet*, **11:** 374–375
Aspirin (p.o.) and Na salicylate (s.c.) ↓ adrenal ascorbate NA sali inhibited liver xanthine dehydrogenase and caused ↑ in activities in plasma of dehydrogenases and transaminases	300 mg/kg 0.75–1.5 mg/g body weight for 2 weeks	Rats (*in vivo*) Rabbits (*in vivo*)	Opitz, K. 1959, *Arzneim-forch. Drug Res.*, **9:** 618–620 Janota, I. 1960, *Nature*, **185:** 935–936

Effect	Dose	System	Reference
Dietary aspirin inhibited wound healing and granuloma tissue formation	75 or 150 mg/kg/day	Rats (*in vivo*)	Lee, K.H. *J. Pharmac. Sci.*, **57:** 1238–1240
Salicylate ↑ plasma free fatty acids. With fasted adult and suckling rats, those exposed to cold (2°C), adrenalectomised or hypothyroid animals free fatty acids declined with sali	100, 200 and 300 mg/kg	Rats (*in vivo*)	Bizzi, A. 1965, *Brit. J. Pharmacol*, **25:** 187–196
Salicylate inhibited long-chain fatty acids binding to Hu plasma proteins and bovine albumin	0.5–5 mM	Human and bovine plasma	Dawkins, P.D. 1970, *J. Pharm. Pharmac.*, **22:** 405–410
Sulfhydryl, albumin, aldehyde binding related to anti-inflammatory activities		Rodent systems	Grant, N. 1971, *Biochem. Pharmac.*, **20:** 2137–2140
Glucuronyl transferase activity. Salicylate inhibited steriod glucuronyl conjugation	10–250 mM	Rat liver homogenate	Sedlak, J. 1972, *Arch. Int. Pharmacodyn.*, **200:** 389–395
Na salicylate I.V. unchanged ascorbate in adrenal gland and stomach mucosa but ↑ plasma ACTH and ↓ plasma insulin but did not alter blood sugar. ↑ Blood levels of triglycerides and ↓ free fatty acid cholesterol and total lipids		Rabbits (*in vivo*)	Koyuncuoglu, H. 1976, *Pharmac. Res. Comm*, **8:** 267–278
Na salicylate inhibited growth of cells. Aspirin and Na salicylate showed equipotent conc related ↓ in cell growth and 3-H leucine incorporation into protein	100–200 µg/ml 1–5 mM	Hamster cell line BHK 21. Rat hepatoma cell line	Rubenstein, M. 1976, *J. Pharmaceut. Sci.*, **65:** 756–768 *and* Hial, V. 1977, *J. Pharmac. Exp. Ther.*, **202:** 446–454
Aspirin induced a serum immunoreactive insulin but not growth hormone in fasted male volunteers	1000 mg	Male volunteers	Syvalahti, E. *Acta Pharmacol. Toxicol.*, **37:** 336–344
Aspirin given in the diet for 24 days (1%) depressed adrenal weight, body weight and ovarian and uterine weight	1% in diet	Female rats (*in vivo*)	Chitwood, S.J. 1976, *Tox. App. Pharmac.*, **35:** 397–400 *and* Shaefer, J.M. 1975, *J. Reprod. Fert.*, **45:** 227–223
Salicylate inhibited net urea prodn., salicylate inhibited citrulline synthesis in isolated mito	5 mM	Adult male rat liver slices	Glasgow, A.M. 1977, *Proc. Soc. Exp. Biol. Med.*, **155:** 48–50
Aspirin, as well as paracetamol and indomethacin, inhibited lysolecitin acyltransferase	Ki (µM) 22–84 mM with asp, 8.9–20 mM para, 110–190 mM indo	Rat heart and liver, rabbit gastric mucosa	Shier, W.T. 1977, *Biochem. Biophys. Res. Comm*, **75:** 186–193

Increase ↑ Decrease ↓ Effect (Conclusion)	Drug	System	Reference
Aspirin taken by normal volunteers affected growth hormone response to hypoglycemia without affecting the plasma free fatty acids	3.2 g/d for 4 days	Human volunteers	Cavagnini, F. 1977, *Metabolism*, **26:** 193–200
Aspirin, diclofenac and ibuprofen showed drop in plasma ACTH, with no change in cortisol levels	50 mg/kg	Rats (*in vivo*)	Bruni, G., 1980, *Pharm. Res. Comm.*, **12:** 349–357
Na salicylate and aspirin inhibited 3-hydroxy steriod hydrogenase using 9,10 phenathrene-quinone (PQ) or androsterone (AN) as substrates	IC_{50} Na sal 256 μM (PQ), 77.8 μM (AN). ASP 2120 (PQ), 585 μM (AN)	Purified enzyme from male rat liver	Penning, T.M. 1983, *Proc. Nat. Acad. Sci. USA*, **80:** 4504–4508

5. Cytokine actions/production

Increase ↑ Decrease ↓ Effect (Conclusion)	Drug	System	Reference
Aspirin enhanced production of human IFN-alpha and IFN-gamma in conjunction with inducers	10 μg/ml	Human peripheral blood mononuclear cells	Yousefi, S. 1987, *Antimicrob. Agents Chemother.*, **31:** 114–16l
Aspirin, indomethacin and paracetamol did not inhibit antiviral and antiproliferative activities of rIFN-alpha2a	1 mM asp, 0.1 mM, indo 0.1 mM para	Human amnion FL cells, acute lymphoblastic leukaemia MOLT-4 cells and renal cell carcinoma NC 65 cells	Takaoki, M. 1988, *J. Interferon Res.*, **8:** 727–733
Aspirin had no influence on ADCC or the binding capacity of macrophages with regard to SW 948 tumour cells. Aspirin had no adverse effect on the capacity of macrophages for stimulation by IFN-gamma or IL-4	100 μg Aspirin	Mouse peritoneal macrophages	Japel, M. 1994, *Mediators of Inflam.*, **3:** 419–424
Aspirin suppressed LPS-inducible NF-kappa B binding to binding site in the TNF-alpha promoter, LPS-induced TNF-alpha mRNA accumulation and protein secretion	Therapeutic concs	Murine tissue macrophages	Shackelford, R.E. 1997, *Mol. Pharm.*, **52:** 421–429
Aspirin produced an ↑ in secretion of sIL-1RII, effect also seen in LPS treated cells (*ex vivo*). Cells *In vitro* asp also secreted significantly more sIL-1RII	81 mg every 2 days for 2 weeks or 10 mg/ml	Human blood mononuclear cells (*in vivo* and *in vitro*)	Daun, J.M. 1999, *J. Leukocyte Biol.*, **65:** 863–866
Na salicylate activated p38 MAPK but not ERK in 1h whereas IL-3, IL-5 and GM-CSF activated ERK but not MAPK in 1 h in EoL-1 cells	unknown	EoL-1 cells	Wong, C.K. 1999, *Immuno. Invest.*, **28:** 365–379

Effect	Concentration	System	Reference
Co-exposure to aspirin or salicylate + heat (24°C) enhanced and prolonged expression of HSP70 but inhibited HSP90. Indo had no effect	50 mM salicylate, 1 mM aspirin and 30 µM indo	Fish cell line CHSE-214	Koo, H.N. 2000, *Zoo. Sci.*, **17**: 1275–1282
Aspirin ↑ IL-4 secretion and RNA expression but did not affect IL-13, IL-2 or INF-gamma expression. Aspirin inhibited IL-4 but not IL-2, promoter driven chloramphenicol acetyltransferase expression	100 µM-1 mM	Human peripheral Blood T cells	Cianferoni, A. 2001, *Blood*, **97**: 1742–1749
Aspirin inhibited LPS-induced cell maturation and costimulatory molecule expression. Aspirin ↓ IL-12 and IL-10 production and ↑ TNF-alpha production	Therapeutic concs	Human dendritic cell line	Ling-Jun, H. 2001, *Scand. J. Rheum.*, **30**: 346–352
Salicylate activation of DNA binding by HSF was comparable to activation attained in heat shock. Na salicylate did not induce heat shock gene transcription	20-30 mM	Cultured human cells	Jurivich, D.A. 1992, *Science*, **255**: 1243–1245

6. Immune effects

Effect	Concentration	System	Reference
Aspirin showed strong anticomplementary action whereas salicylate did not. Quinone showed a slight inhibiting action whereas cortisone showed a strong action but only at concs above therapeutic doses	250–500 µM	Sheep erythrocytes	Van Oss, C.J. 1961, *Nature*, Jan 14th, p. 147
Salicylate and quinine both suppressed hemolysin plaque-forming cells. Serum hemolysin was not affected, salicylate reduced response to sheep erythrocyte antigen	300 µM	Sensitised mouse spleen cells	Griswold, D.E. 1969, *Eur. J. Pharm.*, **6**: 56–60
Benzyl and salicyl alcohol inhibited lymphocyte-mediated cytolysis when present in similar concs to anaesthetics known to block nerve impulse conduction	10–20 mM	Mouse lymphoid cells (*in vitro*)	Kemp, A.S. 1973, *Eur. J. Pharm.*, **3**: 674–677
Aspirin inhibited incorporation of 3H-thymidine in the lymphocyte transformation reaction in response to PHA. Na salicylate also suppressed lymphocyte transformation but salicin did not	10–40 mg/100 ml plasma	Human lymphocytes	Opelz, G. 1973, *Lancet*, Sept 1, p. 478–480
Aspirin, indomethacin and phenylbutazone inhibited lymphocyte activity	1–5 mM	Rat lymphocytes (*in vitro*)	Winchurch, R.A. 1974, *J. Reticuloendothelial Soc.*, **18**: 112–117
Na salicylate and caffeine ↑ serum interferon whereas morphine, hydromorphone, methadone, mescaline, trypan blue and Vit A ↓ serum interferon	0–300 mg/kg	Mice (*in vivo*)	Geber, W.F. 1974, *Arch. Int. Pharmacodyn. Ther.*, **214**: 322–327

Increase ↑ or Decrease ↓ Effect (Conclusion)	Drug	System	Reference
Aspirin failed to suppress lymphocytes responsiveness to PHA or allogenic lymphocytes	Asp, 600 mg 5 times a day for 14 days	Human lymphocytes	Smith, M.J. 1975, *Ann. Int. Med.*, **83:** 509–511
Aspirin decreased tumour incidence and severity of disease as opposed to ACTH, which increased both	325 mg/kg diet	CBA mice *in vivo* inoculated with a murine sarcoma virus	Seifter, E. 1975, *Life Sci.*, **16:** 629–636
Aspirin inhibited antigen-induced lymphocyte proliferation by anti-receptor alloantisera	Various doses of different % solutions of aspirin	Guinea pig (*in vivo*)	Geczy, A.F. 1975, *Eur. J. Immunol.*, **5:** 711–719
Aspirin inhibited lymphocyte blastogenesis to both mitogens and antigens at therapeutic concs. Inhibition was non-cytotoxic and partially reversible	0–40mg/100ml	Human peripheral blood lymphocytes (*in vitro*)	Panush, R.S. 1976, *Clin. Exp. Immunol.* **23:** 114–125
Na salicylate suppressed generation of mouse cytotoxic lymphocytes *in vitro*	0–50mg %	Human peripheral lymphocytes (*in vitro*)	Gabourel, J.D. 1977, *Clin. Immuno. Immunopath.*, **7:** 53–61
Aspirin in combination with DIPY showed an ↓ ability to produce leucocyte migration inhibitory activity when stimulated with concanavalin-A	50–60mg/kg	Human lymphocytes (*ex vivo*)	Coeugniet, E. 1977, *Acta Med. Scand.*, **201:** 197–201
Na salicylate inhibited generation of granulocyte and macrophage colonies when cultured with colony stimulating factor (CSF)	0–50mg %	Mouse or human bone marrow cultures (soft agar)	Gabourel, J.D. 1977, *Arth. Rheum.*, **20:** 59–64
Aspirin and placebo showed no significant difference on skin tests, lymphocyte proliferation and % of T cells (tested with three mitogens and four antigens)	4 g aspirin daily for 5 days	Human volunteers	Duncan, M.W. 1977, *Arth. Rheam.*, **20:** 1174–1178
Non-immune inflammation can be blocked by NSAIDs and steroidal agents	0–100 μg/ml	Rat paw lymphocytes	Leme, J.G. 1977, *Br. J. Exp. Path.*, **58:** 703–711
Aspirin inhibited C3 cleavage initiated by complement activators such as inulin or aggregated IgG. C3 cleavage is also suppressed at pH6.5 or below	50mM	Human serum (from volunteers)	Voigtländer, V. 1980, *Pharm.*, **8:** 613; *Int. Arch. Allergy Appl. Immunol.*, **61:** 145–149

Side Effects and Toxicology of the Salicylates

8

K.D. Rainsford

INTRODUCTION

The salicylates are, in comparison with other NSAIDs and analgesics, of relatively low toxicity when taken at therapeutic dosages (Lamont-Havers and Wagner, 1966; Smith, 1966; Cuthbert, 1974; Rainsford, 1975a; 1989a; 1997a; 1997b; Miller and Jick, 1977; Rainsford and Velo, 1984; 1987; 1992; Freie, 1996) (see Tables 8.1 and 8.2), and this is probably one of the main features accounting for the success of this group of drugs. Yet the discovery (or rediscovery), of some minor side effects at various times leading to concern about the safety of these drugs, usually leads to improved understanding about the relative importance of these effects.

During the therapeutic use of these salicylates the following, in some cases potentially serious, adverse reactions can occur (Freie, 1996):

1. Upper gastrointestinal haemorrhage and/or ulceration with generalised pain, dyspepsia, diarrhoea, constipation and other signs of abdominal discomfort.
2. Nephrotoxicity, principally from ingestion of aspirin *in combination with* other analgesic/anti-inflammatory drugs; also, rarely, renal cell carcinoma.
3. Hepatoxicity, more often manifest in certain disease states (e.g. systemic lupus erythematosus and rheumatoid arthritis).
4. Hypersensitivity reactions in the form of rashes, angioedema, urticarial weals or asthma, and rarely Stevens–Johnson and Lyell's syndromes.
5. Teratogenicity and reduced birth weight – the former especially from ingestion of large quantities of these drugs during the first trimester of pregnancy; this appears to be a general problem with many NSAIDs and analgesic drugs.
6. Central nervous system sensory reactions comprising tinnitus (known as 'salicylism' because of frequent occurrence at high doses of salicylates), loss of hearing, vertigo (with nausea) and myopia.
7. Blood dyscrasias (rarely), e.g. agranulocytosis, pancytopenia, aplastic anaemia, thromocytopaenia.
8. Reye's syndrome in children (see Hall and Glasgow, this volume, Chapter 9).

With the possible exception of Reye's syndrome, the pattern of adverse reactions with salicylates is, in general, similar to that of other NSAIDs (Biscarini, 1996). There is, however, marked variation in the frequency or relative risk of the occurrence of many of these adverse reactions among the NSAIDs and with aspirin (Rainsford and Velo, 1987; 1992; Rainsford, 1989a; 1997a; Walker and Rainsford, 1997;

TABLE 8.1

Overall pattern of adverse effects reported from aspirin in controlled clinical trials in clinical–epidemiology studies in patients with rheumatoid arthritis or osteoarthritis and adverse drug reaction (ADR) reports to regulatory agencies.

High frequency	Intermittent	Low frequency
Gastrointestinal: constipation, heartburn, abdominal pain, nausea	Gastrointestinal: dyspepsia, diarrhoea, stomatis	Gastrointestinal: GI bleeding/perforation, haematemesis, melaena, vomiting, abnormal liver function tests, jaundice
CNS: headache, dizziness, drowsiness	CNS: lightheadedness, vertigo	Renal: glomerular nephritis, haematuria, interstitial nephritis, nephrotic syndrome
Sensory: tinnitus (at high dose)	Sensory: hearing and visual disturbances	CNS: depression, dreams lack of concentration, insomnia, malaise, myalgia, muscle weakness
Dermatological: pruritus, skin eruptions, echymoses	Dermatological: sweating, purpura	Haematological: agranulocytosis, eosinopaenia, granulocytopenia, leucopenia, thrombocytopenia
	Cardiovascular: palpitations, oedema, dyspnoea	Dermatological: alopoecia, photosensitive dermatitis, skin rash
		Sensory: hearing impairment, thirst
		Cardiovascular: congestive heart failure.
		Reye's syndrome (in children).
		Other: eosinophilic pneumonitis, anaphylaxis, menstrual disorders, pyrexia

Rainsford and Powanda, 1998). The occurrence of the frequently observed adverse reactions is apparently greater in the elderly (more properly defined as those greater than 75 years of age rather than the conventional view of over 65 years) (Buchanan, 1990; Karsh, 1990; Griffin and D'Arcy, 1997; Beyth and Shorr, 1999; Fattinger *et al.*, 2000, Seinelä and Ahvenainen, 2000). With some reactions there may be pharmacological benefits evident with aspirin, e.g. the antiplatelet action that means this drug is at lower risk of producing the congestive heart failure in the elderly that is evident with other NSAIDs (Page and Henry, 2000).

In addition, salicylate poisoning (either accidental or deliberate) has been an important problem with the lay use of these drugs, although this problem has now been overtaken by more extensive use of paracetamol as an agent of suicide or accidental poisoning (Prescott, 1996; Rainsford and Powanda, 1998).

Of the adverse effects, the most widespread and clinically important is the development of gastrointestinal intolerance and mucosal damage (Tables 8.1 and 8.2). Hence these particular effects will receive prominent coverage in this chapter. The other effects, while certainly serious in their own right, are of somewhat lower frequency. Some side effects may be avoided or reversed either by cessation of the drug or, in the case of analgesic nephropathy and other symptoms of the 'analgesic abuse syndrome', by instituting appropriate legislative/educational procedures. With the regular ingestion of large quantities of salicylates such as that required for the treatment of arthritic conditions, gastrointestinal intolerance and mucosal damage is of very high frequency. This may be sufficiently serious to

TABLE 8.2

Principal adverse drug reactions from aspirin and related drugs.

Adverse reaction	Relative risk (RR)/incidence	
Peptic ulcer bleeds (PUBs) RRs (all NSAIDs)	NSAID or aspirin taken as single drug (cf. $1/10^5$ non-exposed, $4/10^5$ exposed)	4.1
	Upon switching	12.1
	Upon cessation	1.0
Acute renal failure	RR hospitalisation ($2/10^5$ exposed) Dose-dependent Higher (>8-fold) in first few months	2–4
Congestive heart failure	RR overall	1.5–2.4
Hepatic insufficiency	Rare	

Data from Pharmacoepidemicological studies derived from UK General Practice Database and other databases. (From Faust *et al.*, 1997; Henry, 1997; Garcia-Rodriguez, 2000 and Garcia-Rodriguez and Hernandez-Diaz, 2001; Griffin *et al.*, 2000).

require either: (a) discontinuation of therapy and resorting to more expensive and possibly higher risk therapy (i.e. by employing a more potent NSAID or use of ulcer-preventative agents; or (b) less satisfactory and often more toxic or risky analgesics (e.g. paracetamol, which may be hepatotoxic in the high doses used in the treatment of arthritis).

Many of the side effects associated with the use of salicylates (except teratogenicity) are not directly attributable to the effects of the drug alone but only become fully manifest in the presence of additional factors, which include:

1. The stress-associated or biochemical and pathological changes resulting from the disease itself or accompanying psychological state(s) (Juby and Davis, 1991; Peterson, 1995; Levenstein, 2000). Evidence of this is seen in gastrointestinal ulceration and haemorrhage, nephrotoxicity and hepatotoxicity generally attributed to the salicylates (Rainsford, 1975a; 1982a; Whitehouse, 1977; Whitehouse and Rainsford, 1977).
2. The influence of age and to some extent sex, which may lead to marked variations in pharmacokinetics of the salicylates.
3. Concurrent therapy or ingestion of other analgesic or anti-inflammatory drugs that may interact with salicylates or initiate biochemical or pharmacokinetic changes such as increasing the susceptibility towards salicylate-induced damage in some organ systems.
4. A primary effect of a salicylate in a one-organ system may, because of delicate inter-organ relationships, induce a secondary predisposition to potentially toxic effects in another. While there are very few clear examples of this, it is very likely to include the gastrointestinal mucosa–hepatic interrelationship(s), i.e. hepatic damage predisposing gastrointestinal ulceration and *vice versa* (Rainsford, 1982a).

OCCURRENCE OF ADVERSE EVENTS FROM NSAIDs, INCLUDING ASPIRIN

Among the most recent epidemiological studies is that by Garcia-Rodriguez (2000) and Garcia-Rodriguez and Hernandez-Diaz (2001), who analysed the occurrence of NSAID-associated adverse reactions in the GI tract, kidney and cardiovascular system using data from the UK General Practice Research Databases. These data showed the following general features:

CHAPTER 8

■ 369

1. The overall rate of hospitalisation for upper GI bleeding (UGIB) is 5 per 1000 person years in NSAID users compared with 1 per 1000 person years in non-users.
2. The pooled UGIB risk is about 4 with NSAID use (Table 8.2).
3. Risk factors for UGIB include sex, age and history of upper GI disease (peptic ulcer etc.) (Table 8.2).
4. The risk is maintained during treatment but returns to baseline following cessation of NSAIDs (Table 8.2), showing the importance of temporality in the relation to drug intake.
5. The rates of hospitalisation due to acute renal failure (ARF) are about 2 per 1 000 000 person years, thus representing about 1/50th the risk for UGIBs.

Incidence of adverse events

A comparison of all adverse reactions from aspirin with that from other frequently-used NSAIDs taken by rheumatic patients in a post-marketing study in German-speaking countries in Europe (Table 8.3) shows that aspirin accounts for the lowest number of adverse events per number of prescriptions, although the differences are not, overall, marked among the drugs. This may reflect patterns of use of aspirin compared with other NSAIDs – for instance, aspirin might be used more frequently in patients with relatively less serious disease and more potent NSAIDs in more severe arthritic states.

In other long-term studies in the USA and Europe, involving mostly post-marketing evaluations, the types of adverse reactions from aspirin compare with those of other frequently used NSAIDs and are principally evident in the GI tract and central nervous systems (Table 8.4). Similarly, GI and CNS adverse reactions predominate in adverse reactions in short-term trials (Table 8.5), indicating that although these trials may be less sensitive they are good predictors of major organ dysfunction.

GASTROINTESTINAL SIDE EFFECTS

Historical

Many clinicians in the early nineteenth century noted the symptoms of gastrointestinal distress and intolerance in patients who had ingested large quantities of the pure salicylic acid and its sodium salt, which had then become available for the treatment of fever and various rheumatic conditions. Maclagen (1879) of Dundee (Scotland) recorded that many of his patients being treated with salicylic acid developed irritation of the mouth and throat, so much so that he abandoned it for a while in

TABLE 8.3

Total adverse reactions in German-speaking countries from aspirin compared with other NSAIDs, observed in post-marketing surveillance studies. (From Brune et al., 1992.)

Drug	No. of adverse events	No. of prescriptions (total n = 36 147)	Percentage of total adverse events/no. of prescriptions
Aspirin	250	1211	20.6
Acemetacin	1553	3633	42.2
Diclofenac	4891	14477	33.8
Ibuprofen	1110	4037	27.5
Indomethacin	1693	3896	43.5
Ketoprofen	448	1183	37.9
Naproxen	282	1067	26.4
Piroxicam	488	1645	29.7
Tenoxicam	359	1075	33.4

TABLE 8.4

Most frequent adverse effects from aspirin and other NSAIDs in long-term clinical trial studies.

Adverse event	Drug/percentage adverse events			
	Aspirin	Diclofenac[a]	Ibuprofen	Naproxen[b]
GI system:				
Nausea and vomiting	9.4–14.8	1.0	7.4	3.9
Epigastric pain, abdominal pain and cramps	10.3–12.3	7.04	5.2	3–9
Diarrhoea and constipation	5.0	–	2.1	<3
Indigestion	2.5–3.8	–	0.7	–
Ulcers	1.7	0.1	–	<1
Gastritis	1.25	–	0.7	–
Stanatitis	0.9	–	–	–
CNS:				
Dizziness/vertigo	4.8	0.12	3.1	3–9
Headache	3.6	0.02	2.1	3–9
Tinnitus	14.0	–	2.1	3–9
Blurred vision	0.5	–	0.7	–
Other	6.0	0.14	–	–
Other:				
Peripheral oedema	1.7	–	–	3–9
Iron deficiency and decreased Hb	1.6	0.4	0.2	–
Rash	–	–	–	3–9

Data from:
[a]Brogden et al., 1980.
[b]Todd and Clissold, 1990.

favour of the less irritating derivative, salicin. It is clear from the studies reported at this time that very large doses of these salicylates were employed, and this probably accounted for the gastric intolerance. The aspects of a safe dosage were taken up critically by Dr Myers – a surgeon of the British Cold-stream Guards – in 1876 (Myers, 1876). He stated that a Dr Wolffberg had warned of the severe side effects to the alimentary tract from salicylic acid, following observations of massive haemorrhagic erosions of the mucous membranes of the stomach and intestines in several people who had died following ingestion of salicylic acid. Dr Wolffberg was also quoted as having observed haemorrhagic ulcerations in the large intestine of a dog given an enema containing 2 g of salicylic acid.

Apparently, unbeknown to these British clinicians, a year earlier (in 1875) the Swiss physician Buss had observed gastric irritancy in rabbits administered salicylic acid in order to study the antipyretic activity of this drug. Buss (1875) also observed that the gastric irritancy of salicylic acid could be reduced when it was given with sodium bicarbonate, a procedure that has subsequently been confirmed experimentally by others.

An increasing number of reports began to appear in the latter part of the nineteenth century of gastrointestinal haemorrhage and discomfort in patients taking salicylic acid or sodium salicylate (Editorial, 1876; Shaw, 1887, see Chapter 1), this no doubt being the impetus for the development of aspirin. Within a year of the introduction of aspirin there had been observations of its propensity to cause gastric disturbances, although curiously this was initially less than observed with salicylic acid. Chidichimo (1905) reported that oral dosing of aspirin 200 mg/kg to dogs caused vomiting. This observation was later confirmed by Chistoni and Lapresa (1909). It appears that this dose of aspirin was toxic to dogs since, in their studies, Chistoni and Lapresa reported that the dogs died within 30 hours. At autopsy they noted haemorrhage of the gastric mucosa and perforations of the intestinal mucosa.

TABLE 8.5

Adverse reactions occurring in short-term (usually 12-week) clinical trials (from Willkens, 1985).

	Percentage of patients with adverse reactions					
	Aspirin 2.4–4.8 g/d (n = 721)	Diclofenac 75–200 mg/d (n = 1227)	Ibuprofen 2.4 g/d (n = 74)	Indomethacin 75–125 mg/d (n = 130)	Naproxen 500 mg/d (n = 92)	Placebo (n = 359)
Total adverse effects	43	31	39	42	26	23
Severe adverse effects	9.2	4.6	5.4	10.0	3.3	4.5
Adverse effects by body system:						
Digestive	27	21	28	27	16	12
Nervous	8	6	8	21	8	7
Skin and appendages	3	3	5	1	3	4
Metabolic/nutritional	2	3	3	2	7	1
Special senses	20	3	5	2	0	2

Despite the earlier view, at the turn of the century, that aspirin was much less harmful to the gastrointestinal tract than salicylic acid or its sodium salt (Gazert, 1900) it soon became obvious that aspirin was in fact the more irritating drug in humans (Stockman, 1913; Sajous and Hundley, 1937). An enormous number of cases of gastrointestinal haemorrhage from aspirin were noted during the influenza epidemic just after the First World War (Sajous and Hundley, 1937). Roch (1912) appears to have been the first to observe that co-administration of sodium bicarbonate with aspirin markedly diminished the intolerance towards the latter.

Among the first definitive reports of the direct gastric irritant actions of aspirin were the gastroscopic observations by Douthwaite and Lintott (1938) on 16 patients given 15 grains (900 mg) of the drug. They noted that 13 of the 16 patients reacted to the drug by the appearance of hyperaemia and/or GI haemorrhage. They further observed that the calcium salt of aspirin 900 mg only caused a slight reaction in one out of five cases that were studied. Hurst and Lintott (1939) confirmed that aspirin particles were the major cause of damage in the gastroscopic observations on a man who had begun to bleed following ingestion of tablets of the drug. It is of interest that they also noted a reaction of the mucosa to the presence of aspirin particles, these being embedded in the mucosa as a result of what was presumed to be a gripping reaction by the muscularis mucosae.

There followed some reports challenging these observations that aspirin induced mucosal injury (see Chapter 1), although later research has certainly confirmed these original studies. However, in the ensuing debate about the role of tablets in the development of lesions to the gastric mucosa another interesting suggestion was made by Honingsberger (1943), who considered that the lesions developed from 'a capillary defect due to blood-borne aspirin or its compounds'. The observations of Hurst and Lintott (1939) on the embedding of aspirin particles into the gastric mucosa confirmed in a recent endoscopic photograph of a tablet of aspirin (Levy, 2000) and the suggestion of Honingsberger would appear to form the basis of the present-day concepts that aspirin and other salicylates have both local and systemic effects that are important in the genesis of gastric mucosal damage. Honingsberger's idea is of particular relevance today in view of the evidence that vascular injury combined with the platelet anti-aggregation effects and the ischaemic reactions in the mucosa to aspirin and related drugs are important early events in the development of mucosal damage.

The debate about aspirin being a major cause of upper gastrointestinal ulceration and haemorrhage, which had its beginnings in the 1920s and 1930s, appeared to subside during the Second World War. After the mid-1950s there was a steady increase in the number of publications devoted to this issue, and also the problem in general of salicylates and other NSAIDs (which thereafter become available) having gastrointestinal side effects.

Aspirin as a major cause of bleeding and ulceration

From the mid-1950s to the 1980s this was a hotly debated issue, with the value of epidemiological data in associating cause with effect being the centrepoint of the discussion. There were powerful industrial lobbies organised at the political, economic and academic levels mainly directed towards maintaining a defensive argument (e.g. the Aspirin Foundation – originally a group supported by Beechams Proprietaries, ICI Ltd, Miles Laboratories Ltd, Monsanto Ltd, Nicholas Laboratories Ltd, and Reckitt & Colman); on the other hand, medical researchers at all levels and non-aspirin commercial interests have continued to produce contrary evidence. The problem, as always in such debates, was to sort out the real issues and difficulties, since interests cloud scientific judgements.

The data from more recent epidemiological studies and drug regulatory agencies have now provided substantial evidence for association of aspirin intake with dose-related increase in bleeding and ulceration (Cuthbert, 1974; Rainsford and Velo, 1984; 1987; 1992; Freie, 1996; Rainsford, 1997a; Rainsford and Powanda, 1998; Rainsford et al., 2001). Many data have been derived from two major groups of aspirin users, namely: (a) arthritic patients who require large quantities of the drug over a long period for the relief of pain and control of inflammation; and (b) occasional use of the drug by the lay public who consume a whole range of different preparations, in differing doses, under a variety of conditions and for variable periods of time. Epidemiological studies attempt to embrace both these groups, and some critical factors that need to be considered from these studies include:

1. The need for reliable information relating overall drug dosage and the timing of this to effect(s).
2. The relative paucity of information from these studies on the stress factors (arising from the disease or stressful state being treated), or the use or ingestion of potentially co-ulcerogenic agents (e.g. alcohol or cigarettes) that are known to exacerbate aspirin injury.
3. The need for careful case-matched subjects and related controls in any study and reliable statistical methods.
4. Interaction with other analgesic and/or NSAIDs that are known to exacerbate gastrointestinal damage by aspirin.
5. The need to consider geographical, racial and sex status.

Early population studies

Some of the early literature highlighted the contrasting views that prevailed in the period of early investigations on the clinical impact of aspirin-associated GI bleeding and ulceration. Thus in a review of the early literature on the association between aspirin ingestion and massive haemorrhage published up until 1977, Shirley (1977) attempted to analyse the data reported by others according to the nine criteria suggested by Sir Austin Bradford Hill (Shirley, 1977) as guidelines for establishing the existence of a cause-and-effect relationship. In considering the strength of the association, it was concluded that in the studies (of the 1950s and 1960s) of patients who had consumed aspirin during the previous 24 hours, the association was statistically significant and thus proven. However, this author considered that the controls employed in these earlier studies may not have been entirely satisfactory. Other studies were likewise subject to methodological criticism, although the sheer number of independent reports by different workers did not appear to influence the conclusion by this author that an association was not obviously evident. Two rather surprising statements made by Shirley (1977) concerned experimental evidence for the association, and analogies with other drugs known to cause massive haemorrhage – both of which it was claimed were not available! Since these are two of the criteria required to meet the Hill guidelines, it is regrettable that this author did not consider the wealth of clinical and experimental literature available on these aspects at the time.

Piper and co-workers (1982) concluded that, according to the then acceptable epidemiological criteria, a causal relationship does exist between ingestion of aspirin and chronic gastric ulcer in respect of five out of seven criteria, the coherence of association and biological gradient being the two exceptions for which there is not clear evidence.

In contrast, Rees and Turnberg (1980) considered a selection of clinical and experimental evidence and concluded that aspirin ingestion *rarely* causes clinically significant gastric damage in normal subjects. This was a surprising conclusion, especially in view of the fact that many of the papers they quoted certainly failed to support their contention.

In another review, Rees and Turnberg (1981) qualified their conclusions by stating that:

> There is little evidence that aspirin predisposes to peptic ulceration or acute gastrointestinal bleeding. Frequent or heavy aspirin intake is associated with gastric ulcer or bleeding, although their incidence is rare. Whether these associations indicate a causal role for aspirin remains controversial.

There were some very interesting geographical trends in the associations of aspirin ingestion and upper gastrointestinal damage. One of the most significant of these concerns the epidemic of 'analgesic abuse' that developed in the post-Second World War period in the Eastern States of Australia (Douglas and Johnson, 1961; Senate Select Committee on Social Welfare, 1977; Duggan, 1980; Piper *et al.*, 1982). This syndrome is characterised by the development of both gastrointestinal and renal pathology, and is evident mostly in middle-aged females in the lower socio-economic groups. Aspirin ingestion has also been found (together with alcohol) to be a more prominent cause of upper gastrointestinal haemorrhage in low socio-economic classes, especially unskilled wage earners, in Nigeria (Ojiambio, 1965; Falaiye and Odutola, 1978). The epidemiological studies in Australia have shown that a relationship exists between chronic aspirin use and chronic gastric ulcer (Duggan, 1980).

One approach employed by some authors in comparing the effects of aspirin ingestion with a suitable control group has been to use data from subjects who have consumed paracetamol, which may be presumed to be non-toxic to the gastrointestinal tract (Piper *et al.*, 1977; Coggon and Langman, 1980). This may be a false assumption, since Hansen and Grossman (1978) have found that intravenous paracetamol (given with histamine) can produce gastric ulcers in cats. Also, recent studies by Garcia-Rodriguez (2000) and Garcia-Rodriguez and Hernandez-Diaz (2001) show that an appreciable number of cases of gastrointestinal haemorrhage can be attributed to paracetamol >2 g daily, and there is a direct dose–response relationship between paracetamol intake and upper GI bleeding and ulceration (Table 8.6).

Piper and co-workers (1977) in Sydney (Australia) found that there was a significantly higher proportion of patients with acute upper gastrointestinal haemorrhage or chronic ulcer who had consumed aspirin compared with those who had taken paracetamol. The evidence for ingestion of these drugs was derived not only from history, but also by testing for the presence of these drugs in the urine. The latter procedure must provide more conclusive evidence of recent drug ingestion (within 24 hours) than records of patient histories.

Lanas and co-workers (1992) provided an ingenious means of ascribing prior aspirin intake to its relationship to GI bleeding by measuring platelet cyclo-oxygenase inhibition in subjects admitted to hospital for upper or lower GI bleeding. Using this procedure, 21.5 per cent more patients with a history of aspirin use were uncovered. Moreover, there was an appreciable number of patients with colonic bleeding who had received aspirin/NSAIDs.

The magnitude of aspirin intake or abuse is a major factor in post-surgical cases (gastrectomy, vagotomy) and this is shown by data from Hirschowitz and Lanas (1998). Recurrence of ulcer and stenosis 'is the rule', and the authors consider that surreptitious aspirin, if discovered, should be a clear contraindication for elective surgery. Faivre and co-workers (1979) in France observed a significant relationship between aspirin ingestion established by the concentration of salicylates in plasma and acute gastrointestinal bleeding in patients admitted to hospital.

The studies of Piper and co-workers (1977) showed that a personality trait was evident in their chronic ulcer group: they found that heavy aspirin intake in women was strongly associated with neuroticism.

Pharmaco-epidemiological studies comparing GI ulceration and bleeding from aspirin with various NSAIDs and analgesics

In the past two decades a considerable number of studies have been performed to examine the factors accounting for peptic or gastric ulceration and bleeding following the intake of NSAIDs (Rainsford and Quadir, 1995; Henry *et al.*, 1998; Singh and Ramey, 1998). These data have been principally derived from large-scale databases, including retrospective and prospective studies undertaken using Medicaid, cohort or other regional health care maintenance organisations in the USA; the General Practice Research Database in the UK (Table 8.6); the Spanish Drug Monitoring System (Carvajal *et al.*, 1996); and other sources in Europe, Canada or Australia. They must represent data derived from populations principally from developed countries. Little is known about the risks or factors accounting for serious GI adverse reactions attributed to NSAIDs from less well-developed countries.

Analysis of the risks associated with intake of aspirin in some of these pharmaco-epidemiological studies present considerable methodological problems, largely because of the ubiquitous availability of this drug. It is important to define or eliminate some of the confounding issues that may influence interpretation of data from these studies. First, is the issue of the pattern of use of aspirin compared with other NSAIDs. The introduction of a wide range of NSAIDs in the past two to three decades has inevitably had a profound effect on the prescription of aspirin (Ahonen *et al.*, 1991). While probably

CHAPTER 8

TABLE 8.6

Epidemiological data from General Practice Database (UK) on peptic ulcer bleeding risks from aspirin, other NSAIDs and paracetamol.

Drug	Usage/factor		Relative risk
Aspirin	Overall use	Users	2.0
		cf Non-users	1.0
		Recent users	1.5
		Past users	1.1
	Dose	75–300 mg/d	2.1
		>400 mg/d	3.1
		<50 mg/d	0.7
	Period of use	1–60 d	4.5
		61–180 d	2.7
		181–730 d	1.9
		>730 d	1.6
	Concurrent NSAIDs	(low–med. dose)	2.4
		(high dose)	4.3
	Formulation	Plain *vs* coated, both (for GU and DU)	~2.0
Paracetamol		<1 g/d	1.0
		1–2 g/d	0.9
		2–4 g/d	3.4
		>4 g	6.5
		2 g with NSAID	4.2
		>2 g with NSAID	13.5
		cf NSAID alone	3.5
NSAIDs		Low dose	2.5
		High dose	5.0
	Duration	1–30 d	4.3
		>730 d	3.5
	Indication	OA	4.3 (adjusted); 3.7 (Crude)
		RA	4.7 (adjusted); 5.3 (Crude)
		Pain	4.2 (adjusted)
	Formulation	Plasma t$\frac{1}{2}$ <12 h	4.2 (high dose)
		≥12 h	5.4
		Slow release	6.2
		<12 h	2.4 (low dose)
		≥12 h	2.8
		Overall – slow release	3.9
Ibuprofen		Lowest risk (dose-dependent)	~1.0–2.0

Data from Henry, 1997; Garcia-Rodriguez, 2000, Garcia-Rodriguez and Hernandez-Diaz, 2001; Griffin, 2000.

used extensively for minor arthritic pain and inflammation on a non-prescription basis in those countries where it is widely available by over-the-counter and supermarket sales (e.g. in the UK, USA and Canada), this does not mean that aspirin is used widely by patients with more severe pain. Moreover, a complicating feature is that patients questioned about their use of aspirin as 'intake of other drugs' may not recall its use with accuracy because it is so ubiquitous and has associations as a supermarket commodity. The widespread use of low-dose aspirin as a prophylactic against coronary vascular diseases also raises complications concerning attribution of a particular NSAID as being a cause of GI injury.

Also, in some studies there are indications that several NSAIDs may be prescribed or taken by patients with rheumatic conditions (Berard *et al.*, 2000).

The second issue concerns the increasing use of NSAIDs in the elderly (Inman, 1987), who are also at risk of developing peptic ulcers and bleeding (PUBs) (Buchanan, 1990; MacDonald *et al.*, 1997; Menniti-Ippolito *et al.*, 1998; Beyth and Shorr, 1999). Given the increased use of NSAIDs in the elderly, it is difficult to establish precisely what age groups are particularly at risk and the reasons underlying this. Few of the earlier studies have attempted to define the risk of PUBs in relation to defined daily dosage or other denominator values that can be used for correction according to use.

A third issue is the coincident infection with *Helicobacter pylori*, which is also a major aetiologic agent in PUBs and gastritis (Taha *et al.*, 1993; Graham, 1996; Wilcox, 1997; Aalykke *et al.*, 1999; Santolaria *et al.*, 1999; Ng *et al.*, 2000) and may be significant in the development of dyspepsia (Veldhuyzen van Santen *et al.*, 2000).

Recent pharmaco-epidemiological studies have enabled comparisons to be made of the relative risks of PUBs or upper GI haemorrhage from aspirin compared with other NSAIDs (Tables 8.7 and 8.8) and with paracetamol (Table 8.6). Meta-analysis of published reports by Gabriel *et al.* (1991; see Table 8.7) has shown that aspirin is associated with an intermediate risk of serious upper GI complications, being slightly greater than with ibuprofen and naproxen and lower than with indomethacin and piroxicam. Henry and co-workers (1998) ranked NSAID-associated upper GI events with the plasma elimination half-life of each drug (Table 8.8). They observed some relation between these parameters, but the rank correlation was only moderately related even though the probability of the relationship was significant. Aspirin had an intermediate ranking. Diflunisal had a lower ranking and a relative risk closer to that of ibuprofen – a drug with lowest risk of upper GI events among the NSAIDs, even though it has a much longer plasma half-life than observed with drugs in the low–intermediate risk category.

The recent introduction of COX-2 selective drugs (the 'coxibs', e.g. rofecoxib, (Vioxx®) and celecoxib (Celebrex®)) as well as other established NSAIDs with high COX-2 selectivity (etodolate, meloxicam, nimesulide) have been developed to reduce the likelihood of serious upper GI ulceration and bleeding. Endoscopic and pharmaco-epidemiological studies have shown that etodolac, meloxicam and nimesulide have relatively low risks of causing GI injury compared with established NSAIDs (Traversa *et al.*, 1995; Kawai, 1998; Rainsford, 1998; 1999a; 2001; Degner and Richardson, 2001; Jones, 2001).

The newer coxibs show evidence of low endoscopically-observed upper GI injury and blood loss, along with a low incidence of serious GI adverse events (Hawkey, 1999a; Laine *et al.*, 1999; Langman *et al.*, 1999; Simon *et al.*, 1999; Hunt *et al.*, 2000). These drugs represent competitors for low ulcerogenic salicylates (e.g. salsalate) and for ibuprofen.

Oesophagitis and associated hiatus hernia is also a side effect noted with aspirin intake, although the frequency is lower than with peptic ulcer (Smith, 1978; Lanas and Hirschowitz, 1991).

TABLE 8.7

Relative risks of serious upper gastrointestinal complications from intake of aspirin and other NSAIDs (Derived from meta-analysis of Gabriel *et al.*, 1991.)

	RR	95% CI	Pos
Aspirin	3.38	2.26–5.01	3
Ibuprofen	2.27	1.85–2.80	1
Indomethacin	4.69	2.97–7.41	4
Naproxen	2.84	1.68–4.82	2
Overall	8.00	6.37–10.06	6
Overlap in groups rating*			1–3
Piroxicam	1.12	6.19–20.23	4

RR, relative risk or ratio; CI, confidence interval; Pos, position in ranking order (lowest to highest risk).
*Denotes groups where there is overlap based on position in ranking order.

TABLE 8.8

Ranking of individual NSAIDs by relative risk in comparison with plasma half-life. (From Henry *et al.*, 1998.)

Comparator	Rank by relative risk[a] (from lowest to highest RR)	Plasma half-lives (h)[b]		
		Main analysis	Sensitivity analysis I	Sensitivity analysis II
Ibuprofen	1	1.0	2.0	2.0
Diclofenac	2	1.5	1.5	1.5
Diflusinal	3	10.8	10.8	10.8
Fenoprofen	4	2.2	2.2	2.2
Aspirin	5	0.5	4.5	4.5
Sulindac	6	7.8	7.8	16.4
Naproxen	7	14.0	14.0	14.0
Indomethacin	8	3.8	3.8	3.8
Piroxicam	9	48.0	48.0	48.0
Ketoprofen	10	2.0	8.5	8.5
Tolmetin	11	6.8	6.8	6.8
Azapropazone	12	22.0	22.0	22.0
	Rank correlation	0.3692	0.5038	0.4733
	Kendall's tau (*p*-value)	(0.0947)	(0.0226)	(0.0322)

[a]The ranking by RR was obtained from the results of the summary ranking procedure, and the plasma half-life values were those published in reference texts.
[b]In the main analysis the $t_{\frac{1}{2}}$ for aspirin was used (rather than salicylic acid) and for ketoprofen in its conventional formulation. In sensitivity analysis I the $t_{\frac{1}{2}}$ for salicylic acid was used and the $t_{\frac{1}{2}}$ for ketoprofen in its sustained release formulation. Sensitivity analysis II was as for sensitivity analysis I except that the $t_{\frac{1}{2}}$ was substituted for the sulphide metabolite of sulindac with that of the parent drug.

Gastrointestinal side effects from salicylates in arthritic patients

Singh and colleagues (Singh *et al.*, 1994; Singh, 1998; Singh and Ramey, 1998) have devised a ranking of serious GI events as a GI Toxicity Index for NSAIDs based on data derived from the Arthritis, Rheumatism and Ageing Medical Information System (ARAMIS) prospective registry database (Table 8.9).

TABLE 8.9

GI toxicity index in patients with rheumatoid arthritis.

Drug	Number of patients	Toxicity index (mean ± SE)
Salsalate	187	0.81 ± 0.51
Ibuprofen	577	1.13 ± 0.29
Aspirin	1521	1.18 ± 0.18
Sulindac	562	1.68 ± 0.29
Diclofenac	415	1.81 ± 0.35
Naproxen	1062	1.91 ± 0.21
Tolmetin	243	2.02 ± 0.44
Piroxicam	814	2.03 ± 0.24
Fenoprofen	158	2.35 ± 0.55
Indomethacin	418	2.39 ± 0.34
Ketoprofen	259	2.65 ± 0.43
Meclofenamate	165	3.91 ± 0.54

Data derived from the Arthritis, Rheumatism and Ageing Medical Information System (ARAMIS) prospective observational database (Singh, 1998).

Salsalate has the lowest GI toxicity, while that of aspirin is only slightly greater than that of ibuprofen, which is a recognised 'bench standard' of low GI risk. The majority of conventional NSAIDs appear to have an appreciably greater toxicity. This may reflect differences in the patterns of use of these drugs in patients with rheumatoid arthritis. Aspirin, like salsalate and ibuprofen, may be used more frequently in subjects with mild disease, while those with more severe disease may require more potent NSAIDs and thus may have a higher risk for developing serious GI events.

The frequency of various gastrointestinal side effects attributed to aspirin or diflunisal in arthritic patients, as reported in controlled clinical trials of more than 200 patients (for reasonable statistical interpretation), is shown in Tables 8.4, 8.9 and 8.10. These data are compared with those from patients taking diflunisal, indomethacin or ibuprofen, which are standard drugs frequently employed as such in clinical trials. Overall, aspirin is comparable to indomethacin. As expected, all these drugs except diflunisal have a high incidence of gastrointestinal side effects. In comparison, diflunisal has an appreciably lower incidence of these side effects, comparable with ibuprofen.

When it comes to the examination of these data on the incidence of upper gastrointestinal ulceration, there are reasons to suspect that the incidence may be much higher than shown here since endoscopic observations were not performed in all patients in these studies (Tables 8.4, 8.9 and 8.10). The data in Table 8.10 probably represent the best available from upper GI endoscopy studies with various preparations and formulations of aspirin. They indicate that there is little difference in the propensity of the different formulations of aspirin for producing mucosal injury in the upper GI tract.

Assessment of gastrointestinal injury in humans

Gastroscopy and gastrointestinal blood loss methods

In the clinical assessment of gastrointestinal damage or bleeding, there are specific limitations concerning the reliability and accuracy of the techniques that are employed (Rainsford, 1975c; 1978a). Upper GI endoscopy is probably the most accurate and direct method for assessing both the extent and nature of the pathology in the stomach and duodenum (Holt, 1960; Edmar, 1971).

While used extensively on a routine basis, this technique only permits visualisation of mucosal changes at one point in time, under fasting conditions, and often with complications arising from the influences of pre- or concomitant medication, patient stress from anxiety, and occasional instrument-induced mucosal injury. Nonetheless, it is invaluable in giving insight by way of a macroscopic visualisation of the range of physio-pathological changes that occur in the mucosa. These include the subtle but important (micro-) vascular events underlying the appearance of erythema, blanching and ecchymosis ranging to the haemorrhagic or haemorrhagic lesions and frank ulcers.

In macroscopic appearance, aspirin tablets cause all of these changes in the stomach and duodenum of both normal volunteers (Table 8.10; Lanza et al., 1980) and patients with arthritic or other conditions (Gaucher et al., 1976). The radiochromium-labelled red blood cell technique which is frequently used to measure blood loss from the entire gastrointestinal tract from aspirin and related drugs has been used extensively for quantifying loss of blood in the entire GI tract (Watson and Pierson, 1960; (see also Table 8.11); Leonards, 1962; Goulston and Skyring, 1964; Atwater et al., 1965; Florkiewicz et al., 1967; Goulston and Cooke, 1968; Bouchier and Williams, 1969; Leonards and Levy, 1967; 1969; Croft et al., 1972; Ridolfo et al., 1973; Arvidsson et al., 1975; Vakil et al., 1977a; De Schepper et al., 1978a; 1978b; Lussier et al., 1978; Brandslund et al., 1979; Mielants et al., 1979; Dybdahl et al., 1980). With this method the patients red blood cells are labelled with ^{51}Cr and the amount of radiochromium in the faeces is determined following ingestion of drug or placebo. This faecal radioactivity is then related to that given quantity of blood, and the apparent blood loss thus calculated. Refinements of the original technique have been developed by a number of authors, including: (a) the use of bulk laxatives to regularise stool formation; (b) the use of markers to account for variability in GI transit; and (c) optimising the amount of radioactivity used in labelling red cells to enable measurement of blood loss over long periods of time (up to one month duration) and to reduce the error in estimation.

TABLE 8.10

Incidence and severity of gastric lesions or ulcers, observed gastroscopically, in volunteers and rheumatic patients taking salicylates compared with other NSAIDs or analgesic drugs.

Drug preparation	Daily dose (mg/d)	Percentage individuals with erosions or ulcers	Severity of damage	No. of patients and type	Reference (author(s), year)
Aspirin					
Various	2500+	85	14% GU	82 RA/OA + MA	Silvoso et al., 1979
	3000+	50	20% GU	26 RA/OA	Caruso and Bianchi Porro, 1980
	4000	46	NS	13 RA	Loebl et al., 1977
	3000	60	20% Haem	10 Vol	Axelson et al., 1977
	3900	100	NS	16 Vol	Chernish et al., 1979
	3000–3500	26	NS	43 RA	Atwater et al., 1965
Plain	4000	100	2.6 (score)	9 RA	Clarke et al., 1977
Slow release	4000	100	1.9 (score)		
Plain	3900	100	17 (G) 12 (D) (score)	20 Vol	Lanza et al., 1980
Buffered	3900	100	20 (G) 12 (D) (score)		
Enteric coated	3900	25	3 (G) 0 (D) (score)		
Plain	2400	100	10.7 (average erosions)	9 Vol	Hoftiezer et al., 1980
Enteric coated	2400	22	0.2		
	2600[a]	100	2.8 (score)	5 Vol	Lanza et al., 1986a
	2600[a]	100	3.6 (score)	10 Vol	Lanza et al., 1988
	3900[a]	100	3.5 (score)	30 Vol	Lanza et al., 1990
	2600[b]	100	3.6 (score)	31 Vol	Lanza et al., 1991
	3900[a]	100	3.4 (score)	14 Vol	Lanza et al., 1998
	3600[b]	100	3.1 (score)	30 Vol	Jiranek et al., 1989
	1200	100	3.47 (score)	17 Vol	Lanza et al., 1999
	2600	100	2.68 (score)	19 Vol	Stern et al., 1986
			2.88 (score)	8 Vol	Stern et al., 1987
			1.4	21 Vol	Berkowitz et al., 1986
Diflunisal	500–750	10	NS	20 OA	Caruso and Bianchi Porro, 1980

Drug	Dose				Reference
Ibuprofen	900–1700	17.7	NS	17 RA/OA	Caruso and Bianchi Porro, 1980
	2400[a]		1.3 (score)	15 Vol	Lanza et al., 1986b
			0.92 (score)	12 Vol	Lanza et al., 1987
			1.88 (score)	30 Vol	Lanza et al., 1989
	3200[a]		1.8 (score)	51 Vol	Lanza et al., 1999[c]
Indomethacin	100–150	30	NS	20 RA/OA	Caruso and Bianchi Porro, 1980
	150[a]		1.8 (score)	20 Vol	Lanza et al., 1983
	200[a]		2.25 (score)	12 Vol	Lanza et al., 1987
Nimesulide[d]	200[b]		0.7 (score)	35 Vol	Shah et al., 2001
Paracetamol	2600–5200	0	NS	14 RA	Caruso and Bianchi Porro, 1980
Rofecoxib[d]	250[a]		0.27 (score)	51 Vol	Lanza et al., 1999[c]

Abbreviations: RA, rheumatoid arthritis; OA, osteoarthritis; MA, mixed arthritic types (not RA or OA); Vol, volunteers; GU, gastric ulcers (diagnosed); Haem, showing haematemesis. Some patients in Silvoso et al. (1979) and Caruso and Bianchi Porro (1980) were on mixed medication including steroids; NS = not stated; G = gastric; D = duodenal mucosal scores..

Drugs taken orally for 6–7 (a) or 14–15 (b) days; (c) comparable data from same study; (d) COX-2 preferential (nimesulide) and selective (rofecoxib) inhibitors higher than indicated.

The evidence for this is from gastroscopy studies that have shown that gastroduodenal injury is more frequent in arthritic patients consuming all NSAIDs, not just aspirin alone.

Scores are mostly based on Mucosal Grading Scale as follows;

Score	Description
0	No visible lesions
1	1–10 petechiae
2	>10 petechiae
3	1–5 erosions
4	6–10 erosions
5	11–25 erosions
6	>25 erosions
7	Ulcer

From Lanza et al., 1975. In some studies scores 5–7 were combined to make 5.

TABLE 8.11

Comparison of the gastrointestinal blood loss (measured by the ^{51}Cr-labelled red-blood-cell technique) induced by various NSAIDs or analgesic preparations in humans.

Drug preparation	Dose (mg/d)	Mean 'blood loss' above control placebo (ml)	Number and type of patients	Reference (author(s), year)
Aspirin				
Various	1000	5.0	8 Prisoners	Ridolfo et al., 1973
	2100	2.3	24 Vol	Vakil et al., 1977a
	3000	6.5	10 Vol	De Schepper, 1978a; 1978b
	3600	1.3	6 RA	Lussier et al., 1978
	3800	4.3[a]	10 Vol	De Schepper et al., 1978a; 1978b
	4000	4.2[a]	14 RA	Loebl et al., 1977
	4800	2.6	10 Vol	Cohen, 1976
Plain	1000	7.8	18 Vol	Arvidsson et al., 1975
Buffered[b]	1000	3.3		
Plain	1500	3.0[c]	14 Vol ♂	Dybdahl et al., 1980
Microencapsulated	1500	1.2[c]		
Plain	4000	4.4	14 Vol ♂	Brandslund et al., 1979
Controlled release	4000	3.4		
Calcium salt[d]	5000	2.6	15 OA	Mielants et al., 1979
Enteric-coated	3000	1.29	17 OA	
Buffered	5000	2.5	15 OA	
Plain	2500	2.0	15 Vol	Leonards, 1962
Buffered (Alka Seltzer®)	2500	0.2	15 MA	Croft et al., 1972
Benorylate	4000	1.7	90 Vol	Watson and Pierson, 1960
Choline salicylate	2500	0.5	10 ♂ Vol	De Schepper et al., 1978a; 1978b
Diflunisal	500	0.06	12 Vol	Leonards and Levy, 1967
Salsalate (salicyl-salicyclic acid)	2000	0.1	8 Vol	Ridolfo et al., 1980
Ibuprofen	1600	0.25[a]	10 Vol	Ridolfo et al., 1980
Indomethacin	150	1.2[a]		Ridolfo et al., 1980
Paracetamol	4000	0	18	Goulston and Skyring, 1964

Abbreviations as in Table 8.10.

[a] Stool-marker techniques employed to correct for variations in daily blood loss due to varying time of gastrointestinal transit.

[b] Each 0.5 g tablet had 1.25 g $NaHCO_3$ + 0.5 g citric acid.

[c] Median values.

[d] No control apparent, enteric-coated given at dose required to achieve same serum salicylate levels as other 2 preparations.

The ^{51}Cr RBC technique does unfortunately have the following serious limitations:

1. False positive 'blood loss' may occur from the biliary excretion of radiochromium into the intestinal tract (Stephens and Lawrenson, 1969); biliary flow can also be stimulated by aspirin and related drugs (Schmidt et al., 1938; Bullock et al., 1970; Pugh and Rutishauser, 1978).
2. The inability to localise the source of bleeding to a particular region of the gastrointestinal tract (Rainsford, 1975c; 1978a).
3. The lack of correlation between the gastric lesions or blood observed in the stomach and the quantity of radiochromium therein following administration of aspirin to man or laboratory animals (Kuiper et al., 1969; Rainsford, 1978a).

Given these limitations, the data in Table 8.11 from ^{51}Cr blood-loss studies show the following points:

1. Plain aspirin tablets cause appreciable blood loss in normal volunteers and arthritic patients (when taken at doses for effective treatment of arthritis).
2. Certain buffered, controlled or sustained-release, and enteric-coated tablet preparations cause lower blood loss than plain aspirin, but the bleeding is still quite considerable.
3. Choline salicylate, diflunisal and salsalate (diplosal) cause virtually no blood loss, but benorylate is capable of causing some bleeding.

Comparing the data on gastroscopic observations (Table 8.10) with the radiochromium blood-loss studies (Table 8.11) shows that there is a relatively good overall correspondence between these two procedures for assessing gastrointestinal damage. This is remarkable, considering the aforementioned difficulties in assessing blood loss. It should be noted, however, that there is a somewhat weaker correlation between the results obtained with these two procedures when data from studies with other NSAIDs are compared (Rainsford, 1982a). Blood loss does, of course, represent a combination of damage to the mucosa and the antiplatelet effects of aspirin so the two measures are somewhat different (Hawkey et al., 1991; Hawkey, 1992; 1994).

In earlier studies, the appearance of occult blood in the faeces following aspirin ingestion was determined semi-qualitatively by the chemical test strip technique ('Occult test, Haematest, Haemastix'; Ross et al., 1964; Schwartz et al., 1983). Here the haeme from haemoglobin in the faeces is detected by oxidising benzidine (now discarded because of its carcinogenicity) or O-toluidine. This technique suffers from the disadvantage that peroxidases (present in ingested animal foods) also catalyse the oxidation of benzidine or O-toluidine and hence cause a large number of false positive results. Also, comparison of this method (using the radiochromium technique) with chemical analysis of the haemochromagen content of faeces (e.g. see Cohen et al., 1992) has revealed that the radiochromium technique is the most reliable of the two. Nonetheless, the early studies of the effects of various aspirin formulations and salicylate derivatives on the appearance of occult blood showed essentially similar results to those obtained with the radiochromium blood-loss technique (Table 8.11).

The lower incidence of gastrointestinal damage by various buffered, controlled or sustained-release, enteric-coated, or polymeric complexes (e.g. Aloxipirin®) of aspirin Tables 8.10 and 8.11 (see also Wood et al., 1962; Green, 1966; Thorsen et al., 1968; Martin, 1971; Clarke et al., 1977; Giroux et al., 1977; Meilants et al., 1979; Hoftiezer et al., 1980; Lanza et al., 1980), was attributed to delayed absorption of the aspirin from earlier types of these formulations compared with that of conventional tablets of aspirin (Martin, 1971; Rainsford, 1975c; 1975d; Day et al., 1976; Alpsten et al., 1982). The total bioavailability of most enteric-coated, sustained-release or polymeric complexes of aspirin available in recent years is comparable to aspirin itself (Martin, 1971; Day et al., 1976; Orozco-Alcala and Baum, 1979). Poor absorption was noted with some early enteric-coated preparations (Levy and Leonards, 1966), but this appears less of a problem with modern formulations of this kind than it was some years ago (Martin, 1971; Day et al., 1976).

The crystal form (Wheatley, 1964; see also Chapter 3) and mesh size of aspirin might be important in determining blood loss, although the results are debatable. Thus, Leonards and Levy (1967; 1969)

CHAPTER 8

found that fine particles of aspirin (of 80 mesh or finer) produced more gastrointestinal bleeding than that from coarse particles (<20 mesh), despite the fact that both preparations are absorbed at the same rate. However, the opposite results emerged from the studies of Györy and Stiel (1968). The aspirin preparations that these authors used were exactly defined (i.e. size 16 mesh, coarse; size 120, fine). In contrast, the preparations used by Leonards and Levy (1967; 1969) contained a whole spectrum of particles sized above or below the specified mesh size, so that it is difficult to draw satisfactory conclusions from the studies of these authors.

An important feature in contrasting the GI effects of aspirin versus that from non-acetylated salicylates is the finding that both sodium salicylate and salsalate (diplosal or salicylsalicylic acid) cause little, if any, blood loss (determined by the ^{51}Cr technique) when given orally (Table 8.11; Grossman et al., 1961; Thune, 1968; Leonards, 1969). Salsalate has also been shown to be relatively free from gastrotoxicity (see Rainsford and Buchanan, 1990) as observed in endoscopy studies (Edmar, 1971). The plasma half-life of salicylate following absorption and hydrolysis of salsalate is comparable to an equivalent of aspirin (i.e. with respect to salicylate content) (Leonards, 1969). However, it is possible that the relatively low gastric irritancy of salsalate is due to the slow rate of liberation of salicylate from this ester.

The dimeric form of aspirin, aspirin anhydride, which was once thought from the physicochemical evidence to have advantages over aspirin, and possibly low gastrointestinal irritancy (Garrett, 1959), has been shown to cause just as much bleeding in humans as aspirin (Kyriakopoulous et al., 1960; Wood et al., 1962) and also exhibits very poor bioavailability (Martin, 1971).

The paracetamol ester of aspirin, benorylate, was reported by Croft and co-workers (1972) to cause less gastrointestinal bleeding than aspirin itself. Paracetamol itself causes no significant bleeding compared with a placebo (Table 8.11). Benorylate undergoes hydrolysis following absorption to yield paracetamol, aspirin, salicylate and, presumably, metabolic congeners of salicylate (Robertson et al., 1972 – see also Chapter 3). While showing less gastrointestinal irritation than aspirin experimentally, benorylate has been associated with severe gastrointestinal haemorrhage, as shown by statistics of the UK Committee on the Safety of Medicines (Trewby, 1980). Furthermore, it can be seen from the data of Croft and co-workers (1972) that the bleeding caused by benorylate 4 g/d amounts to an average of 1.7 ml/d, a figure that is within the range of the blood-loss data caused by aspirin and reported by other authors (see Table 8.11). Also, benorylate has to be taken in much larger quantities by weight than aspirin to achieve comparable therapeutic benefits. It is therefore doubtful whether this drug is really a much safer alternative to aspirin in humans.

Following detailed studies on the biochemical effects of salicylates on the gastric mucosa of experimental animals (described later, p. 431) the concept was developed by Rainsford and Whitehouse (1978; 1980a; 1980b; 1980c; 1980d) that it is possible to fortify natural defences of the gastric mucosa against the irritant/ulcerogenic actions of aspirin and related drugs. Here certain nutrients (e.g. glucose with acetate or citrate salts) that have been found to overcome the biochemical 'deficiencies' induced by irritant salicylates have been added as adjuncts to aspirin to make specific formulations (AN-Sprin®).

In clinical studies of gastrointestinal blood loss (using the ^{51}Cr technique) one such formulation of AN-Sprin® was shown to produce no significant loss of blood over placebo (Whitehouse et al., 1981; 1984; Rainsford, 1984b). By comparison, using a buffered aspirin formulation (Bufferin®) showed significant blood loss of 6 ml/d. Pharmacokinetic studies have shown that the AN-Sprin® preparation has bioavailability equivalent with that of aspirin in man. Also, the AN-Sprin® formulation has the same potency (on the basis of aspirin mass) as that of aspirin itself. Thus this formulation shows considerable promise as being probably one of the safest, yet effective, formulations of aspirin that has been developed to date.

Intragastric cell exfoliation

Croft, (1963a; 1963b) adapted the cellular exfoliation technique originally developed for cytological studies to determine quantitatively the loss of cells caused by the irritant actions of aspirin. The quantity of exfoliated cells is estimated by determining the DNA content of gastric aspirates before and after ingestion of the drug. Aspirin has been clearly shown to enhance the normal rate of cellular exfoliation (Croft, 1963a; 1963b; Croft et al., 1966; Croft and Wood, 1967). The question arises whether or not the drug is causing damage to a presensitised mucosa produced as a consequence of the initial control washout used in the experimental design. The cell loss in the initial control period is often higher than in subsequent control periods or following the installation of aspirin (see Croft, 1963a). Repeated installation of control solutions after the initial period causes reproducible cell exfoliation, suggesting that there is a pronounced physical effect of instilling the solutions even during relatively short time periods (Croft, 1963a). The physical effect of the initial washing out of the gastric contents may be to remove effete cells, so that the irritant effects of aspirin may be greater on this presensitised mucosa than might be expected on an unwashed stomach.

Since the kinetics of cell turnover are affected by repeated intake of aspirin (Biasco et al., 1992a), this may influence the quantitative recovery of cells after repeated dosing compared with single doses of the drug. Also, it is difficult to determine the extent of gastric emptying that may take place during the instillation procedure; this should be corrected using the phenol red indicator method. Thus the volume of gastric aspirates collected may vary using Croft's method. Given these deficiencies, this technique does show that aspirin causes gastric cell exfoliation in normal and arthritic individuals (Croft, 1963a; 1963b; Croft and Wood, 1967). With some refinement it may be possible to extend the use of this valuable quantitative method, since there have been few studies since Croft's work in the mid-1960s. Faecal calprotectin (a protein on leucocytes) has been used as a method of assessing leucocyte loss from the GT tract (Melling et al., 1996).

Gastrointestinal protein loss

Beeken (1967) adapted the technique for studying protein enteropathy (Waldman, 1961) to study the effects of various aspirin mixtures. The procedure used is analogous to the ^{51}Cr-labelled red blood cell technique, except that ^{51}Cr-labelled albumin is given intravenously in place of ^{51}C-labelled red blood cells. He showed that oral ingestion of aspirin 600 mg three times daily for 7 days by 10 human volunteers (with no evident gastrointestinal pathology) caused an 87 per cent increase in the loss of ^{51}Cr-labelled albumin from the gastrointestinal tract (Beeken, 1967). Also, he found that four other aspirin preparations containing about the same quantity of aspirin but with added paracetamol, salicylamide, caffeine, citric acid or inorganic salt caused about the same amount of albumin loss from the gastro-intestinal tract.

The exact extent of this protein loss induced by long-term administration of salicylates has yet to be fully evaluated, but it could be related to the intestinal permeability changes from NSAIDs (see p. 386) – of particular significance in rheumatoid arthritis and related conditions where hypoalbuminaemia often occurs (Katz, 1977; Powanda, 1977). This is another technique that deserves further exploration for assessing the gastrointestinal effects of salicylates and related drugs.

Intragastric bleeding

The radiochromium red cell technique was essentially devised to measure total blood loss from the entire gastrointestinal tract, and thus gives no indication of the site of bleeding within the tract. Hunt (1979) and Hunt and Fisher (1980) devised a simple yet elegant and reliable method for measuring bleeding within the stomach. Their procedure involves measuring the haemoglobin content (based on the o-toluidine reaction with peroxidase) in gastric aspirates collected from fasted individuals. Variations in the rate of gastric emptying are accounted for by employing phenol red as a marker (Hunt and Fisher, 1980). These authors showed that instillation of 100 ml of a 1 g/1 solution of aspirin buffered to pH 3.25 caused an increase in the loss of blood into the gastric lumen, of 0.1 to 1.0 ml/d after about 80 minutes treatment with the drug. When aspirin solution buffered to pH 7.0 was introduced at this point, the bleeding continued to rise until about 120 minutes and then rapidly declined to become negligible at 150 minutes. The authors found that the plasma salicylate concentrations were the same over both periods following instillation of the acid and neutral-buffered aspirin solutions, respectively, showing that the neutral-buffered solution was absorbed at the same rate as the acid solution. The second neutral aspirin solution failed to produce blood loss, even though the drug was well absorbed. These data provide direct evidence supporting the value of buffering to neutral pH in preventing of gastric mucosal damage, as noted previously (p. 382). It also appears that buffering to neutral pH has an apparent protective effect on aspirin effects after the mucosa has been damaged by the drug. This technique has been employed by Wilson *et al.* (1985) and Hawkey (1992), who related the blood loss to endoscopic changes and measurement of prostaglandin production.

Permeability changes

Changes in the intestinal permeability of 'marker' molecules of various sizes have been employed by Bjarnason and co-workers (Bjarnason *et al.*, 1984; 1986a; 1986b; 1992; 1993; Bjarnason and Peters, 1987; Bjarnason and Macpherson, 1994; Sigthorsson *et al.*, 1998) and by others (Jenkins *et al.*, 1987; Jenkins, 1991; see also reviews by Bjarnason, 1988; 1994). These permeability changes induced by NSAIDs may have particular clinical significance, not only in relation to protein enteropathy but also in causing intestinal inflammation and serious (but rare), conditions such as diaphragm-like strictures, ulceration, peritonitis and colitis (Bjarnason *et al.*, 1986a, 1986b; 1993; 1997; Lang *et al.*, 1988; Colin *et al.*, 1992; Halter, 1993; Bjarnason and Macpherson, 1994; Halter *et al.*, 1995; Sigthorsson *et al.*, 1998). The selective permeability by different marker molecules is a key feature of the selectivity and specificity of this procedure (Bjarnason, 1988; 1994). Thus monosaccharides (e.g. α-L-ramnose (MWt 164)), being hydrophobic, exhibit transcellular movement through lipophobic aqueous pores in mucosal cell membranes; polyethylene glycol 400 (MWt 194–502) permeates through larger lipophilic pores on mucosal membranes; while ^{51}Cr-labelled ethylene diamine tetraacetic acid (^{51}Cr-EDTA or edetate) and disaccharides exhibit *inter*-cellular permeation. The urinary excretion of these markers is a convenient means of establishing the locus of permeability changes as well as serving as a quantitative means of determining permeability disruption by NSAIDs. The periodic excretion of ^{51}Cr-EDTA has been used extensively as a means of showing intestinal permeability of the intestine to different NSAIDs (Bjarnason *et al.*, 1993; Bjarnason, 1994; Sigthorsson *et al.*, 1998). There can be confounding factors, such as the presence of colitis, Crohn's disease, rheumatoid arthritis and intestinal parasitic infections, and the intake of pungent foods or irritant drugs or drinks, which all need to be controlled for any drug-related studies (Jenkins *et al.*, 1987; Bjarnason *et al.*, 1993; Bjarnason, 1994).

SUCROSE PERMEABILITY

A variant of the intestinal permeability model employed sucrose as a marker for determining effects of aspirin gastroduodenal permeability (Rabassa *et al.*, 1996). The method required intake of 100 g in 500 ml of sucrose followed by HPLC assay of urine samples collected over 5 hours for sucrose concentrations. Using balloon pyloric occlusion it was established that the site of increased sucrose permeability from aspirin was in the stomach.

Other markers of gastric permeability were investigated by Ivey and co-workers (Ivey and Schedl, 1970; Ivey and Parsons, 1975). Thus ^{51}CrCl$_3$ and phenol red (phenol-sulphonphthalein) were employed as non-absorbable markers, and the ratio of these used to establish breakdown of mucosal permeability (Ivey and Parsons, 1975).

In another study Ivey *et al.* (1974) showed that lithium was unsuitable as a marker, presumably because of ubiquitous transport by Na$^+$transporter systems.

Potential differences and ionic flux

Electropotential difference recordings have been shown to be of value for determining gastric mucosal integrity (Davenport, 1972; Murray *et al.*, 1974; Read and Fordtran, 1979; Pauranen, 1980; Cohen, 1981; Bruhn *et al.*, 1983; Nikolaides *et al.*, 1985). Tarnawski and Ivey (1978) showed that transmucosal potential difference measurements in humans were affected by aspirin and other mucosal barrier damaging agents in the same way as shown in rats. This technique, combined with measurement of ion flux, was shown to be of value for determining gastric mucosal injury in humans from aspirin and the effects of protective agents (Ivey *et al.*, 1975a; 1975b; 1980; Baskin *et al.*, 1976; Tarnawski *et al.*, 1980).

Factors affecting gastrointestinal damage

It is appropriate now to consider further some of the factors involved in the development of gastrointestinal damage by the salicylates in man. The following abbreviations will be employed to denote each technique:

CE = cellular exfoliation technique (gastric aspirates)
Clin = clinical observations
^{51}Cr = radiochromium-labelled red-blood-cell technique (in faeces)
G = gastroscopic observations
Hb = haemoglobin assays of gastric aspirates
ION = ionic flux measurements
OT = occult test for bleeding (in faeces)
PD = potential difference recordings
PE = permeability studies.

Acidity of the gastric contents

The acidity of the gastric contents appears to be a major factor in aspirin-induced gastrointestinal bleeding (Yeomans *et al.*, 1992). Two lines of evidence support this view. First, concurrent administration of buffers or antacids reduces the blood loss induced by aspirin (Matsumoto and Grossman, 1959; Pierson *et al.*, 1961; Leonards, 1962; Stubbé *et al.*, 1962; Wood *et al.*, 1962; Thorsen *et al.*, 1968; Leonards and Levey, 1969; St John and McDermott, 1970; Hunt and Fisher, 1980) (^{51}Cr, Hb, OT, G). Second, aspirin produces less gastrointestinal bleeding in achlorhydric patients than in normal individuals (Jabbari and Valberg, 1970; Tauxe *et al.*, 1975 (^{51}Cr, G). It should, however, be noted that achlorhydria may not confer resistance upon the gastric mucosa against ulcerogenic agents such as aspirin (Jabbari and Valberg, 1970; Tauxe *et al.*, 1975; Rafoth and Silvis, 1976). Curiously, no correlation exists between gastrointestinal bleeding induced by aspirin and the levels of maximal acid secretion induced in response to histamine (Jabbari and Valberg, 1970) (^{51}Cr). It appears, therefore, that although the presence of acid in the stomach contents is one factor affecting ulcerogenesis by aspirin, the acid-secreting capacity (except in the achlorhydric individual) may not be of particular relevance. Factors related to drug ionisation within the gastric lumen may also be relevant, and will be discussed later (p. 411).

Duodenal pH is markedly affected by the intake of food and cola drinks, but not cigarette smoking (McCloy *et al.*, 1984). Thus, after a meal the pH and H^+ concentration of the duodenal contents are reduced. Cola drinks increase the H^+ concentration and decrease pH as a consequence of their acidity. These factors may contribute to the development of gastroduodenal injury from aspirin.

The influence of foods

Ingestion of aspirin before meals produces the same blood loss as that when the drug has been taken with meals (Stephens *et al.*, 1968) (^{51}Cr), Significantly less blood loss occurs when aspirin is taken shortly after meals than when the drug is taken with meals (Lange, 1957) (OT). Thus for practical purposes the drug should be taken after meals to reduce the irritant action of raw drug particles.

Acute gastric injury by a single dose of 600 mg aspirin was reduced by 200 ml of 200 mg/ml chilli powder, the endoscopy scores being about $2\frac{1}{2}$ times less with chilli than with aspirin alone (Yeoh *et al.*, 1995) (G).

Milk phospholipids (2 g lyophilised preparation) taken with 2 g/d aspirin for 4 days reduced both antral and duodenal injury (Kivinen *et al.*, 1992) (G). This effect may be related to the influence of phospholipids on the mucosal protective barrier (see p. 446) Fish oil (6 capsules/day each containing 180 mg eicosapentenoic and 120 mg docosahexanoic acid) with aspirin 1.95 g/d for 21 days did not reduce the mucosal injury due to aspirin, even though there were no effects on mucosal PGE_2 or $PGF_{2\alpha}$ production (Faust *et al.*, 1990) (G). Similar results have been obtained by Feldman *et al.* (1990).

Dosage and timing of drug ingestion

Table 8.11 shows that there does not appear to be a dose–response effect of various aspirin prepara-tions in dose ranges of 1.0–4.0 g/d on gastrointestinal blood loss (^{51}Cr), and likewise with endoscopic observations of mucosal damage (Table 8.10). There is considerable variability in the data (Tables 8.10 and 8.11). There may, however, be difficulties with such a comparison of data from different studies conducted with differing methodologies. Also it is possible that as most studies were conducted at the high part of the dose–response curve, then relative differences may not be so apparent.

Much interest has been directed to establishing if low-dose aspirin such as that used for prevention of thrombo-embolic diseases may produce appreciable GI injury (Guslandi, 1997; Cole *et al.*, 1999; Lanas *et al.*, 2000). Aspirin 300 mg/d for 28 days produced gastric mucosal lesions in most human vol-unteers (Donnelly *et al.*, 2000) (G). In another study conducted over 8 days, 325 mg/d produced appre-ciable mucosal injury (Lanza median score 8.5; Fork *et al.*, 2000) (G).

In a long-term study that probably has greater significance for indicating the likelihood of gastric injury from aspirin at antithrombotic doses, 10, 81 and 325 mg/d aspirin for 1.5 and 3.0 months pro-duced gastric mucosal injury though the dose-related scores did not show marked dose-relationships (Cryer and Feldmann, 1999) (G). Injury was even evident in *Helicobacter pylori*-negative individuals. The dose of 325 mg/d aspirin also produced duodenal injury. The results from the 325 mg/d dosage are in accord with data from the Steering Committee of the Physician's Health Study Research Group (1989), in which subjects took 325 mg/d on alternate days for the prevention of myocardial infarction. These subjects had significantly more melaena and duodenal ulcers than those who took placebo. Overall these studies and a review of the literature (Guslandi, 1997) show that even low doses of aspirin are able to produce mucosal injury. The use of enteric-coated aspirin (300 mg/d) preparations would appear to result in somewhat less damage than observed with plain aspirin (Hawthorne *et al.*, 1991; Blondon *et al.*, 2000) (G). Higher doses of enteric-coated aspirin have been reported to produce almost negligible gastroduodenal injury (Lanza *et al.*, 1980; Petroski, 1993) (G), but other data (Table 8.10) show that this is not so clinically significant (see also Jaszewski *et al.*, 1989; Kelly *et al.*, 1996), and there is still blood loss with those preparations (^{51}Cr) (Table 8.11).

The studies of Hunt and Fisher (1980) suggest that blood loss can cease within 150 minutes follow-ing ingestion of aspirin despite continuing uptake of the drug into the mucosa – i.e. once the mucosa has been damaged (Hb). When aspirin is ingested repeatedly there is a progressive increase in blood loss, which peaks at about 3–5 days after the first doses of the drug (Dybdahl *et al.*, 1980; Konturek *et al.*, 1994c) (^{51}Cr). If drug intake ceases then it may take about 4–5 days for blood loss to return to normal, regardless of the aspirin preparation (^{51}Cr).

The blood loss after the first week of aspirin intake was found to be about the same as that following the second and subsequent weeks of drug ingestion (Leonards *et al.*, 1973; Arsenault *et al.*, 1975; Cohen, 1976) (^{51}Cr). Hence it appears that after a plateau of bleeding at about 3–5 days, followed by a decline in the bleeding thereafter.

The time course of blood loss (^{51}Cr) over a 14-day period of aspirin 2 g/d paralleled the gastric mucosal injury (Konturek, 1994c) (G, ^{51}Cr). These observations have given support to the suggestion (see pp. 409, 435) that there is adaptation to the mucosal injurious effects of aspirin following repeated ingestion of the drug.

Timing of aspirin intake may be a factor determining the occurrence and severity of mucosal injury. Moore and Goo (1987) found that volunteers who took aspirin 1.3 g in the morning (0800–1000 h) had about 40 to 50 per cent more lesions or percentage lesions than those who took the same dose in the evening (2000–2200 h) (G). In a related study the authors observed that there was a small increase in acid output around 0800–1000 h but a much greater peak at 2000–2200 h. However, the intragastric pH was increased at 0800 h to about pH 3.0 from a basal (overnight) level of about pH 1.5. Thus there is no apparent relationship between acid output and intragastric pH and the appreciably greater development of mucosal injury from aspirin when taken in the morning. Since many endoscopy studies involve dosing (either single or on the final day of multiple intake) of aspirin in the morning followed by endoscopy at 1–4 h thereafter, it would seem that the data from most of these studies would represent the peak ulcerogenic effects of the drug.

In relation to long-term exposure to aspirin it appears from pharmaco-epidemiological studies that the period of use does not confer an appreciable risk of PUBs (Table 8.6; Carvajal *et al.*, 1996; Garcia-Rodriguez, 2000), although it appears to decline slightly in the long term (0.5–1 year), perhaps as a consequence of 'drop-outs'.

Oral versus parenteral effects of salicylates

Grossman *et al.* (1961) and Mielants *et al.* (1979) showed that intravenous (i.v.) aspirin (in doses of 3 or 5 g of the sodium or lysine salts, respectively) caused significant blood loss (above placebos or controls) when given to hospitalised patients with peptic ulcer disease or those about to undergo orthopaedic surgery (^{51}Cr). In contrast, Cooke and Goulston (1969) failed to observe significant blood loss following i.v. administration of aspirin to normal volunteers under conditions in which oral aspirin did produce significant blood loss (^{51}Cr). It would therefore appear that the disease or stress states of those patients in the studies of Grossman *et al.* (1961) and Mielants *et al.* (1979) predisposed these individuals to the actions of parenterally administered aspirin. Clinical use of the injectable lysine salt of aspirin has also been implicated in the development of gastrointestinal damage (Aron *et al.*, 1970) (Clin).

Mielants *et al.* (1979) found that there was a positive correlation between the steady-state levels of serum salicylate (on day 7 of treatment) and the blood loss caused by either intravenous or enteric-coated, but not orally administered, aspirin (^{51}Cr). These authors argued that enteric-coated aspirin exerts a systemic action on the gastric mucosa resulting from intestinal absorption of the drug. Other authors have likewise assumed that enteric-coated aspirin has essentially a systemic mode of action on the gastric mucosa, and causes less mucosal damage (Hoftiezer *et al.*, 1980; Hawkey *et al.*, 1991) (G). Other studies indicate, however, that the pattern of metabolism of sustained-release preparations of aspirin is different to that observed with soluble aspirin in that much lower concentrations of aspirin are present after ingestion of the former compared to the latter preparations (McLeod *et al.*, 1986). If enteric-coated preparations have the same pattern of pharmacokinetics as sustained-release preparations, then part of the explanation as to why these preparations cause less blood loss than plain or soluble aspirin formulations may be the relatively small amount of aspirin in circulation (compared to its less ulcerogenic metabolite, salicylate) available to damage the gastric mucosa. Related to this is the observation that salicylate 3 g i.v. has not been found to cause significant blood loss above control values (Grossman *et al.*, 1961) (^{51}Cr). Thus the aspirin in enteric-coated or sustained-release preparations appears to undergo more extensive metabolism in the portal circulation to form the less irritant salicylate (McLeod *et al.*, 1986).

Cryer et al. (1999) found that application of a percutaneous preparation of aspirin 750 mg/d (in ethanol/propylene glycol) for 10 days produced gastro-duodenal injury in about one-half of the subjects (G). The bioavailability of aspirin from this preparation was about 4 to 8 per cent, showing that about 18 to 37 mg (2 µmol) aspirin per day delivered systemically is capable of causing moderate mucosal injury; this being paralleled by reduced gastroduodenal mucosal production of PGE_2 and $PGF_{2\alpha}$ (Cryer et al., 1999).

Although salicylate i.v. produces no appreciable histological signs of mucosal damage, some leucocyte infiltration into the gastric mucosa has been observed (Faggioli et al., 1970) (G). This suggests that some (albeit subtle) systemic effects are caused by salicylate on the gastric mucosa.

Blood group status, race, age and sex

Reports of gastrointestinal haemorrhage and ulceration have been obtained from Africa (Ojiambo, 1965; Sankale et al., 1977; Falaiye and Odutola, 1978), India (Hoon, 1969) and the Orient and Australia (Duggan, 1980) (Clin, G), although it is obvious that much of the published work has come from developed countries of the western world. There are no immediate indications of any racial factors that are suggestive of a predisposition or resistance to salicylate-induced gastrointestinal damage, even though quite pronounced geographical variations may be evident in the incidence of aspirin-related gastrointestinal ulceration (Douglas and Johnson, 1961; Lasagna, 1965; Piper et al., 1982).

Aspirin-induced gastrointestinal bleeding is not directly related to blood group status, age or possibly sex per se (^{51}Cr) (Holt, 1960; Stubbé et al., 1962). The high incidence of aspirin-related gastrointestinal haemorrhage and ulceration reported especially in middle-aged to old women (Stewart and Cluff, 1974; Miller and Jick, 1977) (Clin), may be a reflection of greater use of the drug by this group to ameliorate the symptoms of arthritis or a consequence of socio-economic factors underlying analgesic abuse (Piper et al., 1982).

In the cellular exfoliation studies of Horwick and Price-Evans (1966), a complex interaction has been observed between age, sex and blood group status and the effects of aspirin on gastric cell loss (CE). Men of blood group O secretor and non-secretor status were found to be more susceptible to aspirin-induced cellular exfoliation than women of the same blood group (CE). (Secretor status is a trait controlled by a Mendelian dominant gene, separate from the ABO(H) blood group gene. The ability to produce ABO(H) blood group substances in mucus secretions is defined as being of 'secretor status', inability to do so being termed a 'non-secretor'.) Also group O non-secretors were less susceptible to the effects of aspirin than secretors, whereas the converse was observed with individuals of group A blood status (CE). These observations can be contrasted with the evidence that individuals of blood group O status have an increased propensity to develop ulcers, especially in the duodenum (Aird et al., 1954; Langman, 1973; 1979) (CE). Also, individuals who are non-secretors of ABO(H) blood groups may be more liable to ulceration (Langman and Doll, 1965), although the evidence here is equivocal (Hoskins, 1967).

It is of interest that Horwick and Price-Evans (1966) found a weak correlation between age and sex, and the propensity towards aspirin-induced gastric cell loss. It was suggested by these authors that older men had a greater tendency to develop mucosal damage than women of the same age. When compared with the conflicting evidence from ^{51}Cr blood-loss studies (Holt, 1960; Stubbé et al., 1962), it would appear that there is a good case for a more thorough investigation into all genetic, age and sex factors likely to predispose individuals to salicylate-induced mucosal injury.

Exacerbation by alcohol (ethanol)

Epidemiological evidence indicates that there is strong synergism in the effects of combined aspirin plus alcohol ingestion and the development of haemorrhage (Needham et al., 1971) (Clin). In addition, alcohol is known to induce damage to the gastrointestinal mucosa experimentally in man, and its ingestion is frequently associated with the development of gastrointestinal haemorrhage and ulceration (Ivey *et al.*, 1980b). Repeated intake of alcohol with aspirin induces much greater blood loss than that caused by aspirin or alcohol *alone* (^{51}Cr) (Dobbing, 1967; Bouchier and Williams, 1969). In contrast, diflunisal taken with alcohol does not even cause any significant blood loss over placebo (^{51}Cr).

While these studies have emphasised the deleterious effects of combined alcohol *plus* aspirin ingestion, there are some possible beneficial effects of moderate alcohol ingestion that may be important in the context of the gastrointestinal tract. Thus, consumption of small quantities of alcohol may alleviate some of the symptoms of stress (e.g. tension). These stress factors can predispose the gastrointestinal mucosa to ulceration and haemorrhage either alone or in combination with NSAIDs (see p. 403). There is evidence that chronic alcohol ingestion may also induce an adaptive response towards deleterious effects on the gastric mucosa (Ivey *et al.*, 1980b). It is also possible that alcohol-induced stimulation of mucosal blood flow (Puurunen, 1980) may counteract the ischaemic reactions that are so important in initiating gastric mucosal damage (see pp. 426, 441).

When given in combination, alcohol may assist the absorption of lipophilic drugs such as salicylates by enhancing solubility of the drugs and causing breakdown of the mucosal barrier (see p. 424). Thus, separating alcohol intake from that of aspirin, it may be possible to reduce the incidence of the untoward effects of consumption of both these drugs.

Vitamin C (ascorbate) deficiency

The inability of man to synthesise ascorbate places him in a precarious position in that, apart from requiring dietary sources of this vitamin (Yew, 1973), there are drugs that depress levels of ascorbate, including aspirin and alcohol. There is considerable evidence that modern western diets are low in ascorbate, to the extent that a subclinical deficiency in this vitamin may exist with considerable consequences for body metabolism (Wilson, 1975) as well as for drug metabolism (Houston and Levy, 1975). Combination of this state with drugs that can induce a reduction in ascorbate levels may further exacerbate the situation (Dvorak, 1974; Wilson, 1975). This assumes considerable significance, since ascorbate is an essential cofactor for connective tissue (Barnes, 1975), and for other aspects of cellular metabolism (Wilson, 1975) necessary for maintenance of growth and regeneration of the gastrointestinal mucosa. Ascorbate deficiency is known to predispose the gastric mucosa to injury following exposure to stress, a state that results in depletion of ascorbate levels (Dvorak, 1974). Three days' treatment with vitamin C (480 mg b.i.d.) reduced the gastric damage from aspirin (400 mg b.i.d.) and restored the oxidant status (Pohle *et al.*, 2001).

Russell and co-workers (1968) observed that patients with gastrointestinal haemorrhage had depleted levels of ascorbate in their circulating leucocytes. They also found that aspirin depressed leucocyte ascorbate concentrations, and this has been confirmed by others (Wilson, 1975). Hence, Russell *et al.* (1968) have proposed that ascorbate deficiency may be a factor in the development of aspirin-induced gastrointestinal damage. Croft (1968) has also emphasised the significance of ascorbate deficiency in the aetiology of gastric haemorrhage. Co-ingestion of vitamin C has been found to reduce aspirin-induced gastric damage in humans by reducing oxyradicals that are produced during mucosal injury (Phole *et al.*, 2001).

Influence of Helicobacter pylori infection

Helicobacter pylori (HP), being a major cause of gastritis and gastric ulcers and exerting powerful mucosal inflammatory reactions during infection (Bodger and Crabtree, 1998), has presented a challenge to discovering associations between aspirin (or other NSAIDs) and the development of gastric ulcers, (Wilcox, 1997; Barkin, 1998; Hawkey, 1999b). Indeed the occurrence of gastritis from aspirin intake, possibly as a result of enhanced production of pro-inflammatory cytokines by the drug (Hamlet *et al.*, 1998), represents a similar pathological response to that from HP infection, even though the pattern of cytokine production and inflammatory reactions in the gastric mucosa is probably somewhat different with HP (Bodger and Crabtree, 1998). One of the central questions in view of the ubiquitous infection with HP is whether HP exacerbates aspirin injury. In contrast to this view, Hawkey (1999b) has proposed the view that HP-infected patients may be less prone to NSAID injury because of opposite effects of the organism compared with that of NSAIDs on prostaglandin production. Laine and co-workers (1995) found that PGE_2 production was increased by calcium ionophore stimulated gastric mucosal explants from HP-infected patients. However, after 1 or 4 weeks in the placebo arm of an endoscopy study comparing effects of etodolac and naproxen in HP-positive and HP-negative individuals, these differences in PGE_2 production became statistically non-significant. Thus, placebo treatment seemed slightly to increase PGE_2 production to nearer that of HP-positive individuals. Etodolac (a COX-2 inhibitory NSAID) had no effect on PGE_2 production in either HP-positive or HP-negative subjects, but naproxen had a pronounced (four-fold) inhibitory effect in HP-positive individuals whereas in HP-negative subjects it was reduced two-fold. Thus, COX-1/COX-2 NSAID treatment markedly affects PGE_2 production that is otherwise elevated in HP-infected subjects.

Direct effects of HP in stimulating PGE_2 have been shown in human gastric mucosal derived fibroblasts (Takahashi et al., 2000). This was shown to be related to induction of mRNA for COX-2 (Takahashi et al., 2000) and expression of immunoreactive COX-2 protein (Sawaoka et al., 1998), thus showing that the organism elicits transcriptional activation of COX-2 production as part of the mechanism of increased PGE_2 production. It therefore seems possible that HP infection could override the inhibitory effects of aspirin or other NSAIDs on PG production.

Using the sucrose permeability technique (PER), Rabassa et al. (1996) showed there were no differences in the permeability to this marker in HP-positive compared with HP-negative subjects. Konturek et al. (1998) found that 14 days' treatment with aspirin 2 g/d caused more prolonged bleeding (G, Cr) up to 14 days in HP-infected subjects, whereas in HP-negative individuals this was markedly reduced by day 14. The inference from this study was that HP impaired the adaptive responses to aspirin. Among the adaptive responses invoked in this study were increased mucosal DNA synthesis and luminal production of the growth factor TGFα (Konturek et al., 1997). Other factors could include acceleration of apoptosis by HP (Leung et al., 2000) and expression of the production of the protective heat shock protein (HSP)-70, which is increased during the adaptive response to aspirin (Konturek et al., 2001).

It is possible that the interactions between NSAIDs and HP may vary according to the NSAID. With non-aspirin NSAIDs, no interactions have been observed between HP and the ulcerogenic effects of these drugs (Thillainayagam et al., 1994) (G). This may be related to the variable effects of non-aspirin NSAIDs in inhibiting the HP aqueous extract-induced human neutrophil production of reactive oxygen species (Jones-Blackett et al., 1999).

Thus, although Hawkey (1999b) had accumulated considerable evidence for an exclusion of interaction between NSAIDs and HP (and their opposing effects on PG production), this effect may be of limited consequence, especially in the case of aspirin. Moreover, the irreversible inhibition of platelet COX-1 and prolonged effects on thromboxane A_2 production and the relatively long duration of mucosal production of PGE_2 (Feldman et al., 2000) may limit the potential for HP to stimulate PGE_2 production in aspirin-treated subjects and so overcome the inhibitory effects of the drug. With NSAIDs that are reversible or weak binding COX-1 inhibitors of PG production, there may be potential for HP to override their inhibitory effects on mucosal PG production and so confer benefits on mucosal protective mechanisms.

The stimulation of gastric and acid production by HP (Basso et al., 1999) suppression of protective heat shock proteins (Konturek et al., 2001) and pro-inflammatory reactions may also contribute to the complexities of HP–NSAID interactions.

Site and pathology of lesions

The sites of aspirin-induced gastrointestinal damage mostly appear to be confined to the fundic and antral mucosa and the duodenum (Paliard et al., 1971; Gaucher et al., 1976, Anseline, 1977; Rampon et al., 1978;

Silvoso *et al.*, 1979; Lockard *et al.*, 1980), although minor rectal irritation may also occur (Coldwell, 1966). Muir and Cossar (1955) investigated the effects of acute oral ingestion of aspirin by patients with peptic ulcers prior to elective gastrectomy. They observed mucosal irritation in the antrum and on the lesser curvature of the fundus. Microscopically, the lesions showed variable mucosal erosions, pronounced congestion in the mucosal capillaries, and polymorphonuclear and round cell infiltration. Such a picture is typical of acute erosive gastritis (Vickers and Stanley, 1963; Smith, 1966). Erosive and haemorrhagic gastritis are most frequently seen in rheumatoid arthritic patients on chronic aspirin therapy. A high proportion of these patients have gastric ulcers and erythema (Silvoso *et al.*, 1979). Also, duodenal mucosal lesions, erythema and ulcers are relatively frequent, although less so than damage to the stomach (Lockard *et al.*, 1980). In one limited study, healing of the gastric mucosal lesions was observed within 5 to 10 days of the cessation of aspirin therapy (Vickers and Stanley, 1963). However, it is difficult to place much reliability on these studies since it is not always possible to locate the original site of mucosal damage. Further studies are certainly needed to establish precisely the time and sequence of cellular responses involved in healing in humans.

Floate and Duggan (1978) have described the radiological appearance of an 'hour-glass' stomach in patients with benign chronic gastric ulcer who had clearly abused aspirin-containing analgesic preparations. Flattening of the antral region of the greater curvature is considered a radiological sign of NSAID gastropathy (Laveranstiebar *et al.*, 1994).

Smith (1978) considered that symptomatic oesophageal hiatus hernia, which occurs occasionally in patients on aspirin therapy, is due to direct injury of the oesophageal mucosa leading to oesophagitis during delayed passage of aspirin tablets from the mouth to the stomach.

Duodenitis may also be associated with aspirin intake (Gelzayd *et al.*, 1975). Diaphragm-like structures have been observed in the duodenum of patients that have taken aspirin (Ramaholimihaso *et al.*, 1998; Theifin *et al.*, 2000) and these resemble strictures seen in the intestine of patients receiving other NSAIDs long term (Bjarnason *et al.*, 1993).

Studies in laboratory mammals

Procedures for assessment of gastrointestinal damage

Studies on the side effects induced by aspirin and related drugs in laboratory species have been limited to investigations of the mechanisms of damage to the gastrointestinal mucosa, since these animals cannot give any clear indication of the side effects (e.g. nausea, indigestion, dyspepsia and epigastric pain). Drug-induced gastrointestinal irritation may be related in some way to the appearance of these subjective symptoms, but there may be other physiological responses to the drug – e.g. stimulation of the vagal–parasympathetic system – which could account for their development. As reviewed by Kauffman (1989), there have been very important lessons learned from studies of aspirin-induced gastric mucosal injury in laboratory animals. Much of the insight into the mechanisms of mucosal injury, which has necessarily involved invasive procedures that could not be employed ethically in humans, and the subsequent development of safer formulations or drug derivatives has come from studies in laboratory animals.

The most frequently employed method of assessing gastrointestinal damage is the determination of the number or severity of haemorrhagic lesions or ulcers observed at autopsy following drug administration (Otterness and Bliven, 1985; Rainsford, 1989b). The methods are simple, reproducible, and can be made quantitative (in a two-dimensional sense, i.e. the extent of the surface area of mucosa damaged). Confirmatory histological evidence of the extent of the lesions (i.e. in the mucosal submucosa) should be routinely employed to assist quantitative determination of the severity of damage, for example in differentiating ulcers *per se* from mucosal lesions. Also, histological assessment is important in order to discriminate the false negative results that can be obtained following treatments that cause the discharge of mucus from superficial mucous cells (e.g. copper complexes of salicylates; Rainsford and Whitehouse, 1976a; 1976b).

Some authors have employed a procedure to aid in visualising the lesions by giving an intravenous injection (before sacrifice) of a vital dye such as pontamine sky-blue (PSB). This dye is strongly bound to plasma proteins, and so these blue-stained proteins appear in sites of haemorrhagic erosions (Takagi and Kawashima, 1969; Brodie et al., 1970). However, this procedure causes a significant reduction in the number of lesions that develop in aspirin-treated rats compared with those who have not been injected with the dye (Rainsford, 1982a). It also appears that the injection of PSB is capable of altering the pharmacokinetics and bioefficacy of aspirin (Marquez and Roberts, 1972), which may explain the reduction in ulcerogenicity of aspirin evident in PSB-injected rats.

The conventional procedure of counting the number of gastric lesions seen in fasted animals yields results that can be quite variable, and this may require employment of larger numbers of animals per group to reduce the variance or error in the data obtained. One procedure has been developed where the animals are briefly exposed to mild stress – a procedure that specifically enhances the sensitivity of the mucosa to the effects of salicylates and related drugs (Rainsford, 1975b; 1977a; 1977b; 1989b). This procedure was developed on the basis that stress factors may predispose to the development of gastrointestinal damage by aspirin and related drugs. It involves exposing the animals to a brief period of cold or restraint stress following dosage of the drug (Rainsford, 1975b). This acute stressing is of such severity as to enhance the production of gastric lesions (compared with those in unstressed animals), but does not cause lesions or ulcers to develop in control animals (Rainsford, 1975b; 1977a). A synergistic interaction is evident with the combined effects of salicylates (or other NSAIDs) and stress (Rainsford, 1975b). The applied stress shifts the dose–response curve of most NSAID drugs to the left, and markedly reduces the intra- and inter-experimental variability (Rainsford, 1981a). Furthermore this assay is relevant to the human situation, since most individuals taking chronic aspirin are under some kind of stress (e.g. pain, psychosocial).

An increased susceptibility of the gastrointestinal mucosa to the ulcerogenic actions of NSAID has also been observed in rats in whom polyarthritis has been experimentally induced by the injection of Freund's adjuvant (Di Pasquale and Welaj, 1973; Shriver et al., 1977; Rainsford, 1981a; Whitehouse and Rainsford, 1982). There have been numerous suggestions that arthritic patients may be more prone to the ulcerogenic effects of NSAIDs (Siurala et al., 1965; Gibberd, 1966; Hollander, 1966; Hart, 1969; Olhagen, 1970; Haslock and Wright, 1974; Ivey and Clifton, 1974; Sun et al., 1974; Marcolongo et al., 1979), and hence this assay may be performed in adjuvant-arthritic rats. The combination of physical and pathological stresses (i.e. cold exposure of adjuvant-arthritic rats) enhances the gastric ulcerogenicity of NSAI drugs compared with that evident in cold-stressed normal rats or unstressed adjuvant-arthritic rats (Whitehouse and Rainsford, 1982). Adjuvant arthritic rats show abnormal drug metabolism (Beck and Whitehouse, 1973; Whiteshouse and Beck, 1973; Cawthorne et al., 1976) and this may account, in part, for the susceptibility of these diseased animals to ulcerogenic effects of NSAIDs. (Whitehouse, 1977).

There are marked variations in the incidence and severity of gastric lesions induced by aspirin and related drugs according to the species, sex, strain (of rats) and age of the laboratory animals (Anderson, 1958; Smith, 1966; Wilhelmi, 1974; Rainsford, 1975c; Wilhelmi and Menassé-Gydnia et al., 1972). Guinea pigs appear more sensitive to the gastric ulcerogenic effects of aspirin (Smith, 1966; Urushidani et al., 1978). This may be a reflection of the ease of mechanical disruption of the mucosa of this species compared with that of other laboratory animals (e.g. rats). For other NSAIDs, the guinea pig mucosa is also more sensitive than that of rats and mice (Wilhelmi, 1974). It was claimed by Anderson (1963) that the minimal dose for induction of gastric mucosal damage by salicylates in guinea pigs corresponds well with that required to cause bleeding in man (Roth and Valdes-Depena, 1963a). This finding can be corroborated in stressed rats, with a lower minimal effective dose for lesion production in this species than in unstressed guineas pigs.

Several authors have studied the effects of aspirin in cats (Roth and Valdes-Depena, 1963a; 1963b; Davis and Donnelly, 1968; Hansen et al., 1980). However the drug is usually toxic in this species (Davis and Donnelly, 1968), and this enhanced toxicity appears to be due to the extraordinarily long half-life of plasma salicylates (37.6 h after 44 mg/kg; Davis and Donnelly, 1968) and deficiency in the glucuronidation mechanism for detoxification of the drug (Williams, 1974). The possibility of producing unknown systemic toxicity with aspirin in cats strongly mitigates against the use of this species in

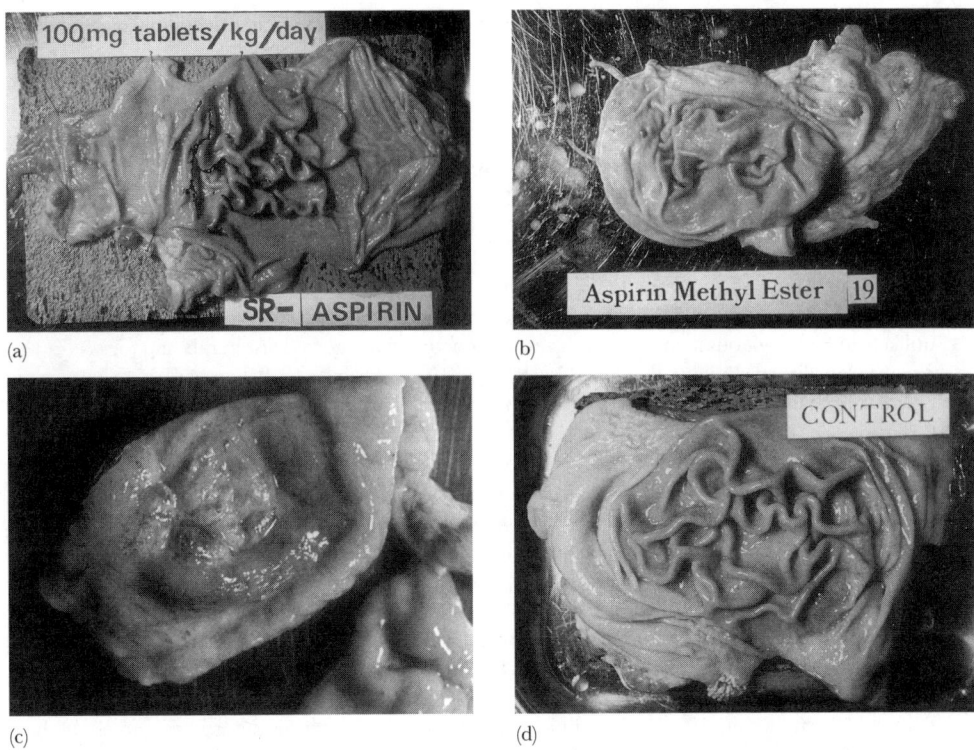

100mg tablets/kg/day

SR- | ASPIRIN

Aspirin Methyl Ester | 19

CONTROL

(a)   (b)   (c)   (d)

Figure 8.1 Gastric mucosal lesions in domestic pigs dosed orally for 10 days with aspirin and various analogues and formulations (a) Severe haemorrhagic lesions in a pig given aspirin 200 mg/kg per day as sustained-release tablets (given in 30 ml H_2O daily) showing haemorrhagic lesions and ulcers. (b) Absence of lesions in a pig given aspirin methyl ester 200 mg/kg per day (*cf* (a)). (c) Section of pylorus from pig given a sustained-release tablet formulation of aspirin 200 mg/kg per day showing a deep ulcer (about 2 cm in length). Such ulceration present in animals given tablets is not evident in animals given suspensions of the drug (d) Control given 30 ml/d H_2O daily. (From Rainsford, 1984a)

CHAPTER 8

laboratory investigations. Dogs and monkeys are occasionally employed in laboratory studies, but there are potential problems of strain differences or the presence of endemic parasites, particularly with monkeys (Rainsford, 1977b). Pigs have been successfully employed to study the effects of aspirin and related drugs (Figure 8.1). They are particularly valuable because their stomach and intestinal tract most closely resemble that of man of all the readily available laboratory species (Rainsford, 1975b; 1978c; Rainsford *et al.*, 1982d). The only disadvantage of using this species is the cost, and it is for this reason that rats are most commonly employed.

The Sprague–Dawley and the Japanese Donru strains of rat have been reported to be more sensitive to aspirin ulcerogenesis than the Wistar strain (Davis and Donnelly, 1968). The differences have been ascribed to intrinsic variations in mucosal susceptibility, acid and mucus production (Davis and Donnelly, 1968). No significant differences have been observed in the effects of aspirin in stressed Sprague–Dawley or Wistar rats (Rainsford, 1982a). In the early studies of Barbour and Dickerson (in 1938), marked seasonal variations were observed in the incidence of aspirin-induced gastric lesions in cats (Barbour and Dickerson, 1938). This has also been confirmed under modern conditions of standard temperature, relative humidity and lighting (Leng, 1978). Maximum aspirin-induced ulceration was found in the February to March period (in Germany), which differed from that observed with other ulcerogenic drugs. This suggests that there may be important seasonal factors in ulcerogenesis that may be relevant to aspirin gastrotoxicity in man (Leng, 1978; Olson, 1987).

Apart from measurement of mucosal lesions, other techniques for determining gastrointestinal damage include the following.

1. The ^{51}Cr-labelled red blood cell technique adapted from that used originally in man (Owen et al., 1954; Edelson and Douglas, 1973; Phillips, 1973; Menassé-Gydnia and Krupp, 1974; Smyth et al., 1976; Chernish et al., 1979; Menassé, 1979; Rainsford et al., 1978a) and dogs (Phillips and Palermo, 1977) and applied in animals suffers from the problems discussed previously (p. 383). Furthermore, in pigs it has been found that there is no relationship between the numbers of gastric mucosal lesions and the radiochromium content in gastro (Rainsford, 1978a). Thus this technique is unlikely to be an adequate method for assessing mucosal damage, at least in rats and pigs.

2. Another labelling technique, developed by Phillips (1973; Phillips and Palermo, 1977), shows some promise for following more accurate measurement of blood loss than the ^{51}Cr red blood cell technique (Chernish et al., 1979). With this method (Phillips, 1973) and using dogs (Rimbau et al., 1976 or pigs (Rainsford et al., 1995b; Phillips and Palermo, 1977; Ridolfo et al., 1980) ^{59}FeCl$_3$ is injected intravenously into the animal before the drug is given. The ^{59}FeCl$_3$ is presumably incorporated into haemoglobin in vivo, but it is possible that other iron-containing haemoproteins (e.g. cytochromes present in cells) could likewise be labelled. While dose-dependent appearance of the label has been observed in the faeces of dogs dosed with aspirin, the faecal ^{59}Fe could possibly be derived from cytochromes or bound to other non-haemoglobin-containing ferroproteins in cellular exfoliates. Also, it is possible for ^{59}Fe to appear in the intestine by transmucosal transfer or from breakdown of haemoproteins in the liver (Forth and Nell, 1981), but the quantitative significance of this has still to be determined. In principle, the Ridolfo et al. (1980) ^{59}Fe-labelling method is probably rather more specific than the Chernish et al. (1979) ^{59}Cr technique, but it still requires further investigation to establish its reliability for determining blood loss per se from the gastrointestinal tract.

3. Another marker of mucosal injury involving administering ^{3}H-inulin with the test drugs and subsequent plasma measurements of the label was found to relate well to gastric-duodenal injury in rats from aspirin 100 to 500 mg/kg (Flemström and Marsden, 1974; Wittmers et al., 1990).

4. The DNA-cellular exfoliation technique developed for use in man by Croft (1963a; 1963b) has been successfully employed in guinea pigs by Garner (1977a; 1977c) to study aspirin-induced gastric mucosal injury. This technique is relatively laborious for routine use, and can of course only be performed in fasted and anaesthetised animals.

5. Ezer and Szporny (1975) have developed an ingenious method for determining intestinal damage by measuring the reduction in tensile strength of the intestinal tract of rats caused by NSAIDs. The technique involves determining the manometric pressure required to induce bursting of the intestine following excision of the intestinal tract (Ezer and Szporny, 1975; Ezer et al., 1976). It does not appear to have been applied to the gastric region, but no doubt this could be so developed.

6. Occasionally the benzidine technique or the related Clinistix®-type methods have been applied to detect blood in gastric washings (Garner, 1977a) or faeces (Uhlenbrook, 1965). False positive reactions may be obtained from animals infested with nematodes (Uhlenbrook, 1965), although this is probably a remote possibility with modern animal laboratories. In any case, these procedures suffer from the same disadvantages as noted previously (p. 383) in man i.e. false positive reactions occurring from plant peroxidases.

7. Some investigators have studied the effects of aspirin and other salicylates on the gastric mucosa of pyloric-ligated rats (Shay et al., 1945; Lish et al., 1956; Johnson, 1966; Okabe et al., 1974a; 1974b). This method has obvious advantages in that changes in mucosal permeability and effects on gastric acid secretion caused by salicylates can be determined simultaneously with an assessment of gastric mucosal damage. There are, however, serious disadvantages with this technique. First, aspirin and other salicylates have been shown to cause fewer gastric lesions in pyloric-ligated rats and guinea pigs compared with those produced in non-ligated animals or even in pyloric-ligated control animals (Pauls et al., 1948; Lish et al., 1956; Anderson, 1965, Guth et al., 1976). This appar-

ent 'anti-ulcer' effect of salicylates has been attributed to the inhibitory effects of this drug on gastric acid secretion (Lish *et al.*, 1956) and biliary reflux (Guth *et al.*, 1976). Second, the Shay rat preparation is strictly an ulcer model induced by strong neurogenic stimuli emanating from pyloro-duodenal trauma induced by the ligation procedure and accompanied by enhanced gastric acid secretion (Anichov and Zavodskaya, 1968). It often results in death within 24 hours of ligation (Anichov and Zavodskaya, 1968), so there are obvious secondary pathological conditions developing during the experiment that complicate the interpretation of results obtained in the model.

8. Gastroscopy has been employed very successfully to observe the effects of salicylates or other NSAIDs *seriatum* on the gastric mucosa of dogs (Anichov and Zavodskaya, 1968; Sugawa *et al.*, 1971), pigs (Rainsford, 1978c; 1982a) and rats (Meyrat *et al.*, 1984; Fujii *et al.*, 1988; Alquorain *et al.*, 1993). It has the advantage in that the time course of the morphological effects of ulcerogenic drugs can be directly visualised by a minimally-invasive and non-terminal technique. With an accompanying biopsy of affected regions it allows both a quantitative and a qualitative evaluation of the ulcerogenic effects of NSAIDs. It should be noted, however, that complications may arise from physical trauma from introduction of the gastroscope and effects of the premedication. Nonetheless, a skilled operator will rarely (if ever) cause appreciable physical trauma, and possibly moderate premedication (i.e. without atropine) will have very little influence on the development of aspirin-induced lesions.

9. Many workers concerned with the physiological reactions (e.g. flux of Na^+, K^+, Cl^-) of the gastric mucosa to aspirin and other NSAI drugs have employed the Heidenhain procedure to develop a denervated pouch of the fundic (oxyntic) region of dog's stomach with a cannula inserted to facilitate sampling or installation of drug solutions. This procedure has many advantages, especially over the Shay rat preparation, since there is minimal complicating pathology and the experiments can be performed on conscious animals. It should be noted that the preparation has no nerve supply to the pouch, thus limiting interpretation to that of the drug effects on an *in vivo* fundic mucosal system devoid of normal neuroregulation. This can be an advantage if complicating neurogenic stimuli invoke pathological responses akin to those observed in the Shay rat preparation. In addition to measurements of ionic permeability (Davenport, 1964; 1965; 1969; Cooper *et al.*, 1966; Davidson *et al.*, 1966; Lin and Warrick, 1974) reflecting mucous membrane damage, several authors have determined blood loss (Davenport, 1965; 1969; Davidson *et al.*, 1966) by chemical assays or intraluminal appearance of radioactively-labelled albumin (Davenport, 1966)

10. Transmucosal potential difference recordings have been shown to give a reliable indication of the integrity of the gastric mucosa of laboratory animals as well as man (Davenport *et al.*, 1964; Smith *et al.*, 1971; Chvasta and Cooke, 1972; Ritchie and Fischer, 1973; Cooke and Kienzle, 1974; Baskin *et al.*, 1976; Caspary, 1978; Tarnawski and Ivey, 1978; Torchiana *et al.*, 1979; Rainsford and Willis, 1982; Cook *et al.*, 1996). A reduction in mucosal potential difference following ingestion of aspirin or other ulcerogenic drugs has been correlated to loss of Na^+ into the gastric lumen, bleeding and ultrastructural evidence of mucosal damage in man and laboratory animals (Smith *et al.*, 1971; Chvasta and Cooke, 1972; Ritchie and Fischer, 1973; Cooke and Kienzle, 1974; Baskin *et al.*, 1976; Rainsford and Willis, 1982). Quantitative assessment of the area and depth of mucosal lesions correlates with changes in transmucosal potential difference, especially in relation to the amount of deep mucosal damage (Cook *et al.*, 1996). This technique has advantages in that it is essentially a minimally-invasive procedure and it is a simple and very sensitive method for detecting the slightest changes in mucosal integrity. It does not, however, reveal the site of damage in the stomach, or disclose the severity of damage and underlying pathological changes in the gastric mucosa. Its use is probably confined to the determination of acute mucosal damage, since any chronic injury would probably have a more complex pathology and such changes as the formation of 'silent' or non-haemorrhaging ulcers may not be detected. Adaptive changes in the gastric mucosa leading to reduced mucosal damage (Konturek *et al.*, 1994a; 1994b; Wallace *et al.*, 1995) may also complicate the assessment of responses to the mucosa since the processes underlying adaptive responses may not reveal injury. One of the advantages with this technique is that acute changes in animals can be correlated with those in man using an identical experimental technique.

CHAPTER 8

11. An *ex vivo* and *in vitro* gastric chamber technique has been used to study the local effects of aspirin and other NSAIDs on the gastric mucosa (Alphin and Droppleman, 1971; Himal *et al.*, 1975; Labrid *et al.*, 1975; Harding and Morris, 1976; Kuo and Shanbour, 1976). With this technique sections of the mucosa of anaesthetised animals are drawn up into the base of a lucite chamber, the blood supply being left intact. Solutions are installed or collected for assay from the chamber, so enabling morphological and quantitative biochemical changes to be determined in particular regions of the gastrointestinal mucosa independent of influences on other parts of the tract. The technique must necessarily be performed in anaesthetised and somewhat traumatised animals, so there are some inevitable limitations arising from the operative technique.

12. The Ussing chamber has been used for studies of the effects of aspirin and other NSAIDs on acid/pepsin and ionic secretion, as well as electrophysiological and biochemical macroscopic or microscopic components of mucosal injury in amphibian gastric mucosa (Spenny and Bhown, 1977a; 1997b; Rowe *et al.*, 1985a; Soybel *et al.*, 1992; Nandi *et al.*, 1994; Takeuchi *et al.*, 1995). This system has particular advantages in that the low rate of metabolism in amphibia and the relatively thin mucosa enables quantitative separation of the time course of various morphological, physiological and biochemical events during mucosal injury under controlled conditions. The amphibian model can also be useful for microelectrode studies (Cheung *et al.*, 1985).

13. Measurement of neutrophil accumulation as a marker of inflammatory reactions accompanying the development of mucosal injury (Wallace *et al.*, 1990; Konturek *et al.*, 1993b; 1994a; 1994b; 1994d; Anthony *et al.*, 1996) is a useful quantitative measure of acute mucosal injury. The procedure involves assaying myeloperoxidase activity in post-mitochondrial/microsomal supernatants of homogenates. However, with adaptative changes upon repeated administration, neutrophil accumulation may not be such a predominant event in the entire gastric or intestinal mucosa (Rainsford *et al.*, 1995a) and so the utility of this method may be limited to use in chronic studies because of reduction in both mucosal injury and neutrophil penetration (Konturek *et al.*, 1994b).

14. Cultured primary or established cell lines have also been employed by many authors (e.g. Koelz *et al.*, 1977; 1978; Terano *et al.*, 1980; Allen *et al.*, 1991; Szabó *et al.*, 1997). Organ culture of pig or human fundic or antral mucosa has also proven very useful for studying effects of aspirin and other NSAIDs, especially in view of the advantages of studying drug effects in a multicellular system (Rainsford *et al.*, 1995a).

Comparison of the gastrointestinal damage by different salicylates

Clearly, with all the different techniques that have been employed by various authors, it is necessary to have some method of identifying the results obtained using particular techniques in order that important differences or correlations can be discriminated. For this purpose, the techniques will be identified in parenthesis by the following abbreviations.

CE = cellular exfoliation technique
^{51}Cr = radiochromium blood loss technique (in faeces)
EP = electrophysiological studies
EV = *ex vivo* chamber technique
^{59}Fe = radioactively-labelled iron technique
G = gastroscopy
Hb = chemical methods for detecting blood (as haemoglobin) in gastric aspirates
HP = Heidenhain pouch technique
IF = ionic flux measurements
LD = visual/microscopic assessment of lesions or ulcers
OT = occult test or chemical (Hb) methods for detecting blood in faeces
PD = transmucosal potential difference recordings
PL = pyloric-ligated (Shay) rat preparations

PSB = pontamine sky-blue technique
TS = measurement of tensile strength.

A comparison of the gastric ulcerogenic activity of various salicylates and aspirin preparations is shown in Tables 8.12 and 8.13 respectively (LD). In stressed rats there is a relatively good correspondence between the gastric ulcerogenic activity of the salicylates or their preparations (c.f. the appropriate standards) and the gastric irritancy (determined gastroscopically) or gastrointestinal bleeding from these drugs in man (*cf.* Tables 8.10, 8.11, 8.12 and 8.13). For example, benorylate, choline salicylate, diflunisal, diplosal, salicylamide and sodium salicylate are all relatively less ulcerogenic in the stomach than aspirin in man, stressed rats (LD – see Table 8.12) and unstressed rats, guinea pigs and dogs (Anderson, 1963; Fishler, 1964; Åberg, 1966; Rainsford and Whitehouse, 1976c; Rohrbach *et al.*, 1977; Torchiana *et al.*, 1979). The situation with aspirin anhydride is debatable. Gray and co-workers (Gray *et al.*, 1960) claimed that this drug was much less ulcerogenic than aspirin on the non-glandular mucosa of rats, a region that most others find is generally unaffected by aspirin (Rainsford, 1982a) (LD). While these authors employed a period of prolonged starvation (16–18 h) after oral dosing of these drugs to prior-fasted rats, damage to the non-glandular region has never been observed by this author in stressed or unstressed rats following such prolonged starvation (Rainsford, 1982a) (LD). Also, aspirin anhydride exhibits gastric ulcerogenic activity about equivalent to that of aspirin in stressed rats (LD – Table 8.12) and unstressed guinea pigs (Anderson, 1963) (LD), and causes gastrointestinal bleeding comparable to aspirin in man (Wood *et al.*, 1962).

TABLE 8.12

Acute gastric ulcerogenic activity of salicylates in stressed rats.

Drug[a]	Gastric lesion index	ED_{10} (mg/kg)
Salicylates		
Aspirin	63.0	5.04
Aspirin anhydride	49.0	–
Aspirin methyl ester	0	–
Benorylate	16.0	–
3,5-dibromo-aspirin	16.3	–
Difunisal	24.4	1.72
Difunisal methyl ester	0	–
Diplosal (salicylsalicylic acid)	6.5	–
Diplosal acetate	0	–
5-chlorosalicylic acid	19.3	0.34
Choline salicylate	0	–
Meseclazone	0	–
Methyl salicylate[b]	0	–
Salol (phenyl salicylate)	0	–
Salicylamide	0	–
Salicylic acid	16.0	60.70
Sodium salicylate	10.6	–
NSAIDs standards		
Flufenamic acid	–	20.50
Indomethacin	27.0	2.50
Paracetamol	0	–
Phenylbutazone	19.0	25.00

[a]All drugs tested in these studies were given orally (in H_2O) at 200 mg/kg, except for phenylbutazone 25 mg/kg and 3.5 dibromo-aspirin (equimolar with respect to aspirin at 282 mg/kg). Data from Rainsford 1975b; 1977b; 1978e; 1981a; Rainsford and White-house, 1976b; 1977; 1980b; 1980c; Whitehouse and Rainsford, 1980.

[b]Methyl salicylate is a potent irritant (non-haemorrhagic type) of the buccal, oesophageal and gastric mucosae

TABLE 8.13

Comparison of the gastric ulcerogenic activity of aspirin preparation in rats and pigs. (Data from Rainsford, 1978b; Rainsford and Whitehouse, 1980c.)

		Gastric mucosa	
Preparation	Rat lesion index	Pig lesion index	No. of ulcers (mean ± SE)
Commercial formulations			
Aspirin suspension (Monsanto)	63.0	44.0	0
Aspirin BP tablets (dispersed)	45.0	44.1	0
Alka-Seltzer® (Miles)[a]	21.0	ND	
Alixoprin® (Nicholas)[b]	29.2	ND	
Ascal® (ACF, Chemiefarma NV)[c]	31.8	ND	
Aspro-clear® (Nicholas)[d]	52.0	39.9	0.7 ± 0.3
Bayer, Timed-Release®[e]	47.6	ND	
Bufferin® (Bristol Meyers)	87.8	ND	
Ecotrin® (Smith Kline and French)	ND	34.8	1.8 ± 1.1
Lysine aspirin (Arthromedica)[h]	61.3	ND	
Disprin® = Solprin® (Reckitt and Colman)[i]	40.8	77.0	0
SRA® (Boots)	53.3	41.0	3.8 ± 1.5
Soluble preparations			
Aspirin + 1 equiv. NaOH	31.7	ND	
Aspirin + 1 equiv NaHCO₃	37.0	ND	
Aspirin + sodium acetate + glucose[j]	18.8	18.7	
Aspirin + sodium citrate + glucose[j]	24.6	ND	

[a]Alka-Seltzer® contains 12.5 equiv. $NaHCO_3$, 9 equiv. $Ca(H_2PO_4)_2$ per mole aspirin.

[b]Aloxiprin (Palaprin®) is a polymeric condensation product of aspirin with aluminium oxide containing 83 per cent aspirin by weight.

[c]Ascal® is a complex of calcium aspirin and urea in 1:1 equimolar proportions.

[d]Aspro-clear® contains 4.3 equiv. citric acid per mole aspirin.

[e]Bayer, Timed Release®, SRA® (= slow-release Aspirin®) are complexes of polymers and aspirin.

[f]Bufferin® contains 0.2 equiv. $Al(OH)_2$ glycine and 0.6 equiv. $MgCO_3$.

[g]Ecotrin® is an enteric-coated aspirin preparation.

[h]Lysine salt is unstable with prolonged storage.

[i]Disprin® = Solprin®. Contains 0.6 equiv. Ca_2CO_3 and 0.3 equiv. citric acid.

[j]Mixtures comprise 1:3:3 molar proportions with respect to aspirin = 1

ND = not determined.

In rat studies the drugs were given at 200 mg/kg for 2 hours and stressing at −15°C for 35 min. Pigs were dosed orally each day with aspirin 100 mg/kg preparation for 10 days. Aspirin preparations (a–e) given to rats were dissolved or dispersed in H_2O before administration; those tablet preparations dosed to pigs were given intact followed by 20–40 ml H_2O. No lesions were observed distal to the duodenum in the intestinal tract of pigs given aspirin, but some duodenal injury was observed. However, it was not possible to quantify this because of the diffuse appearance of this damage.

On comparing the gastric ulcerogenic activity of these commercially available salicylate derivatives with their relative therapeutic (anti-inflammatory or antipyretic) activities in rats (Rainsford and Whitehouse, 1976c; 1980c) and clinical assessment in man, the following points emerge:

1. While benorylate, diplosal, salicylamide and salol are less irritant than aspirin in man (see Broh-Kahn, 1960 and p. 386) and in laboratory animals (LD – Table 8.12), all of these drugs are less potent than aspirin as anti-inflammatory-antipyretic agents when compared on either a molar or a weight-for-weight basis.

2. Diflunisal, choline salicylate, sodium salicylate and salicylic acid are definitely less irritant to the gastric mucosa than aspirin (Stone *et al.*, 1977; Torchiana *et al.*, 1979 (LD – Table 8.14)). Diflunisal is some 8 to 13 times more potent than aspirin as an anti-inflammatory and analgesic agent, and although it is equipotent with aspirin as an antipyretic it does have a somewhat longer duration of

action in rats (Broh-Kahn, 1960; Rainsford and Whitehouse, 1976c; 1980c). In man, diflunisal taken at 500–750 mg daily (taken as two doses) is equivalent to aspirin 2–3 g daily (taken as three doses) for pain relief in osteoarthritis, with less gastroscopic signs of damage or bleeding (see Caruso *et al.*, 1978 and pp. 380, 382). Choline- and sodium salicylate are equipotent with aspirin as anti-inflammatory/analgesic/antipyretic agents in laboratory animals (Rainsford and Whitehouse, 1976c; 1980c), and both these drugs are as effective as aspirin in the treatment of rheumatoid and related arthropathies (Broh-Kahn, 1960; Essigman *et al.*, 1979. Since they, as well as diflunisal, are less gastrotoxic, it would appear that all these drugs have some distinct advantages over aspirin for the treatment of arthritic conditions, especially in patients at risk (see Chapter 11). The 'fishy' odour that develops in some patients taking choline salicylate could be a minor disadvantage with this salt.

3. Even the drugs mentioned in (2) above exhibit some degree of ulcerogenic activity, but this can (with the exception of methyl salicylate) be virtually eliminated by alkyl, aryl, glyceryl or 3-hydroxy-butyryl esterfication of carboxylic acid moiety (see Table 8.12; note point (1) above; see Åberg, 1966; Dittert *et al.*, 1968; Digenis and Swintosky, 1975; Rainsford and Whitehouse, 1976c; 1980b; Kumar and Billmora, 1978; Essigman *et al.*, 1979; Carter *et al.*, 1980; Diamantis *et al.*, 1980; Paris *et al.*, 1980; Whitehouse and Rainsford, 1980a) (LD). The carboxylic acid esters constitute examples of drug 'latentiation' (Digenis and Swintosky, 1975) or 'pro-drugs', and a critical factor in both their bioefficacy, as well as their freedom from gastrotoxicity, may be their rate of hydrolysis *in vivo* (Digenis and Swintosky, 1975; Rainsford *et al.*, 1980b). For instance, the ethyl carbonate ester of aspirin is hydrolysed at about the same rate as aspirin and is of comparable ulcerogenicity (Dittert *et al.*, 1968). Benorylate, the triglyceride esters of aspirin and cyclic aspirin triglycerides exhibit low ulcerogenicity, but also have poorer anti-inflammatory activity resulting from their slow absorption and presumably poor rate of generation of aspirin or salicylate *in vivo* (Kumar and Bellimora, 1978; Carter *et al.*, 1980; Paris *et al.*, 1980; Rainsford and Whitehouse, 1980a) (LD). Simple methyl esters of aspirin or diflunisal have low ulcerogenicity (Rainsford and Whitehouse, 1976c; 1980b; White-house and Rainsford, 1980) (LD). Aspirin and diflunisal methyl esters are absorbed at about the same rate as their parent acids, and both have bioefficacy comparable with their parent acids (Rainsford and Whitehouse, 1980).

Some phenolic esters of salicylates are also less ulcerogenic than their parent acids (Dittert *et al.*, 1968; Digenis and Swintosky, 1975; Inoue *et al.*, 1979) (LD). The O-phosphate ester of salicylic acid (salicylic acid dihydrogen phosphate or ϖ-phosphonoxybenzoic acid) may be well tolerated by the gastrointestinal tract of laboratory animals and man.

Inoue and co-workers synthesised a variety of saturated alkyl (C_2–C_8) and carboxyphenyl carbonate phenolic esters of salicylic acid, which they tested for hydrolysis by a human intestinal esterase preparation and for gastric ulcerogenicity in the rat (Inoue *et al.*, 1979) (LD). The *n*-heptanoyl and carboxyphenyl carbonate esters both exhibited much lower ulcerogenicity than aspirin (Inoue *et al.*, 1979) (LD). All the esters were hydrolysed by the enzyme preparation, and of these the *n*-heptanoyl ester proved to have superior analgesic activity to aspirin.

On comparing the ulcerogenic activities of various aspirin formulations, it can be seen that tableting is a major factor in ulcerogenesis (see Table 8.13; Phillips and Palermo, 1977; Ridolfo *et al.*, 1980) (LD). Under stringent test conditions in stressed rats or pigs, none of the conventional tablet preparations of aspirin shows any significant differences in gastric ulcerogenicity with one another (LD – Table 8.13) Many authors have shown the short-term protective effects to the gastric mucosa of having buffers or alkaline metal or alkaline earth ions present in the aspirin mixture (see Table 8.13; Büss and Balmer, 1962; Anderson, 1963; 1964a; Rainsford and Whitehouse, 1980c; 1980d; 1980e; Ridolfo *et al.*, 1980) (LD); see also Phillips and Palermo, 1977 (^{59}Fe); Garner, 1977b (CE); and Davenport, 1964; 1969 (HP-Hb)). Combining a buffer system with specific nutrients for the gastric tissues, e.g. glucose and citrate (which are added to overcome the breakdown in the biochemical defences induced by the drug), diminishes the ulcerogenic activity of aspirin and other salicylates much further than that attained by buffering alone (Rainsford and Whitehouse, 1980c; 1980d) (LD).

Copper complexes of aspirin, salicylic acid and other anti-inflammatory acids have been claimed to

CHAPTER 8

be less ulcerogenic than the parent drugs (Sorenson, 1976) (LD). Unfortunately, in the early studies of these copper complexes there were problems with the animal models employed (e.g. the Shay rat). Furthermore, a mucous effusion reaction induced by free cupric ions (liberated from the complexes by gastric H$^+$ ions) had the effect of masking the intrinsic irritancy of both the complexes and their dissociated anions) (Rainsford and Whitehouse, 1976a; 1976b; Rainsford, 1981b) (LD). It was only when careful histological examination revealed the disruption of mucosal cells associated with the mucous effusion response induced by the soluble or insoluble copper complexes of salicylates that a more realistic appraisal of the situation could be made (Rainsford and Whitehouse, 1976a; Rainsford, 1981b). Furthermore, species differences in the gastric ulcerogenicity of these copper complexes became obvious, with guinea pigs being more sensitive than rats (Lewis et al., 1981) (LD). Gastric potential difference was found to be correlated to gastric mucosal injury by copper aspirin (Alich et al., 1992) (LD). It was found that administering (p.o.) the copper salicylates in sunflower oil reduced their gastric irritancy and at the same time maintained their therapeutic activity (Rainsford, 1982c) (LD). The sunflower oil presumably diminishes the degree of direct contact between the irritant and gastric mucosa and may also provide a source of linoleate as a precursor for prostaglandin synthesis, which may be compromised by the salicylate moiety in the complex (Rainsford, 1982c) (LD). Comparison of the gastro-ulcerogenic effects of aspirin and other salicylates and NSAIDs with their anti-inflammatory effects in rats shows that there is a relationship between ulcerogenic and anti-inflammatory activity with most of these drugs (Rainsford, 1982c) (LD).

Factors influencing gastric ulcerogenicity

Alcohol consumption

The ubiquitous consumption of this drug, which in large quantities is gastrotoxic to man (Geall et al., 1970; Smith et al., 1971; Dinoso et al., 1972; Gazzard and Clark, 1978; Brassinne, 1979; Millan et al., 1980) and laboratory animals (Eastwood and Kirschner, 1974; Dinoso et al., 1976) and may also exacerbate aspirin-induced gastric injury in man (see p. 390), indicates that the interaction of ethanol with salicylates may be one of the most important factors to investigate in laboratory animals.

Acute oral administration of ethanol with aspirin increases the gastric mucosal injury in rats and dogs beyond that observed by giving aspirin alone (see Davenport, 1969 (HP-Hb); also Morris et al., 1972; Rainsford, 1982a; Lee et al., 1995 (LD)). However, repeated oral co-administration of ethanol (1–5 per cent) and aspirin for 5 days to rats caused no significant differences in the mucosal lesions or their severity compared with that from aspirin itself (Rainsford, 1982a) (LD). It is therefore possible that the gastric mucosa may adapt to the irritant effects of these drugs, a factor that was suggested by Hurley and Crandall (1963) to occur after chronic aspirin administration. Part of the effect of aspirin or other salicylates may be to inhibit ethanol-induced leukotriene C4 production (the effects of this leukotriene causes vasoconstriction and ischaemia in the gastric mucosa) (Trautmann et al., 1991).

Aspirin 200 mg/kg 4-hourly caused twice the damage in rats given 25 per cent v/v ethanol in drinking water for 21 days compared with animals that did not receive ethanol (Parmer et al., 1985) (LD). The injury from ethanol alone was no different than in controls, so the interaction from aspirin with ethanol was probably synergistic.

In relation to the hepatotoxic effects of ethanol and salicylates, it should be noted that hepatic injury will alter the capacity of the liver to form conjugates of salicylates with glycine or glucuronic acid. Furthermore, only reduction in albumin production by the liver as occurs in alcoholic cirrhotics (Brodie and Boobis, 1978) would be expected to reduce the plasma salicylate-binding capacity and aspirin esterase activity (Menguy et al., 1972; Juggi, 1975; Rainsford et al., 1982a) due to reduction of this important drug-carrying protein. The hypoalbuminaemia would lead to higher concentrations of the free (i.e. potentially gastrotoxic) form of the drug in circulation.

The effect of aspirin in causing an increase in post-prandial blood alcohol levels by inhibition of gastric alcohol dehydrogenase by alcohol does not affect gastric emptying (Gentry et al., 1999). Women have no appreciable first-pass effect with aspirin, and so have higher levels of aspirin when taken with alcohol or even without this drug.

Tobacco and nicotine

Ageel *et al.* (1983) and Parmar *et al.* (1985) showed that nicotine bitartrate 5–25 mg/ml in drinking water markedly enhanced the severity of gastric mucosal injury from aspirin 200 mg/kg compared with aspirin alone, while nicotine alone did not cause damage (LD). Thus, there appears to be a synergistic interaction between nicotine and aspirin.

In a model of cigarette smoke-induced effects on gastric functions in rats, Iwata *et al.* (1995) showed that this treatment attenuated the hyperaemia and lesions caused by hypertonic saline and speculated that cigarettes would also enhance the damage from aspirin (LD). In contrast to the results of Parmer *et al.* (1985), nicotine added to the drinking water for 50 days was not found to affect the development of gastric lesions induced in rats by aspirin although the hyperaemia induced by the latter was reduced (Battistel *et al.*, 1993) (LD). It is possible that long-term treatment with nicotine resulted in adaptive changes but did not lead to any aspirin–nicotine interactions in the gastric mucosa.

Disease/psychophysiological stress

As discussed previously, physical stresses and/or the stress of disease markedly enhances the susceptibility of the gastric mucosa to the ulcerogenic effects of salicylates. A synergy occurs between these drugs and the effects of exposure to physical (cold, restraint) stress (Brown *et al.*, 1975; 1978; Rainsford, 1975b; 1977a; 1977b; 1978b; Parmer and Hemmings, 1983; Robert *et al.*, 1989). A combination of factors may be involved in the manifestations of the synergistic interactions between physical stress and aspirin, including: (a) stimulation of vagal-parasympathetic activity and histamine release (from degranulating mast cells); (b) impairment of mucosal blood flow and development of local ischaemia; (c) specific sensitisation of acid-secreting parietal cells with accompanying stimulation of acid secretion; and (d) depression of energy metabolism (e.g. ATP production) in the gastric mucosa (Rainsford, 1978b; Rainsford and Whitehouse, 1980a; 1980e; Fiegler *et al.*, 1986). Mucosal prostaglandin production is not affected by exposure to cold (Robert *et al.*, 1989). Increased secretion of acid and decreased blood flow are, however, factors resulting from cold treatment in potentiating aspirin injury in this model (Robert *et al.*, 1989). It should be noted that both exposure to physical stress and administration of salicylates can produce these effects, so the combination of both treatments is truly synergistic. Psychoactive and cholinomimetic agents can produce synergistic effects with salicylates that resemble the effects of stressors (Rainsford, 1986; 1999b; Alshabanah *et al.*, 1993).

The increased susceptibility of the gastric mucosa of arthritic (compared to normal) rats towards the effects of aspirin and related drugs appears to be due to: (a) reduction in the oxidative metabolism of glucose and other metabolic functions in the liver (Barritt and Whitehouse, 1977; Whitehouse, 1977; Poon and Whitehouse, 1978) and possibly mucosa; (b) impairment of the liver (and possibly intestinal) metabolism and other drug-handling reactions of anti-inflammatory analgesic drugs (Morton and Chatfield, 1970; Whitehouse and Beck, 1973; Kato, 1977; Whitehouse, 1977; Mathew *et al.*, 1978); (c) reduction in the synthesis of the mucus barrier (Rainsford, 1982b); (d) accelerated gastric acid production (Rooney *et al.*, 1978), which may have consequences for gastric physiology beyond acid secretion; (e) increased plasma and organ levels of lysosomal hydrolases (Collins and Lewis, 1971; Yusibova *et al.*, 1978; Rao *et al.*, 1980); and (f) a possible reduction of the regenerative capacity of the mucosa resulting from (a).

That the stress of the arthritic disease is a major factor contributing to the evident gastrointestinal intolerance towards aspirin and other salicylates in humans has been strongly indicated (but not proved) in several studies (Kern *et al.*, 1957; Gibberd, 1966; Ivey and Clifton, 1974; Sun *et al.*, 1974). In many arthritic conditions there are indications that dyspepsia (Kern *et al.*, 1957; Gibberd, 1966) and bacterial infections (Olhagen, 1970; Haslock and Wright, 1974) may be prevalent, both of which can sensitise the gastrointestinal tract to the irritant actions of NSAIDs. Ivey and Clifton (1974) have reported that aspirin produced an abnormal high Na^+ and H^+ flux in the gastric mucosa of patients with rheumatoid arthritis compared with that in control subjects. The increased gastric permeability observed in rheumatoid arthritis was also seen in patients with benign gastric ulceration, suggesting an underlying gastric pathology in the rheumatoid group (Ivey and Clifton, 1974). These authors suggested that the persistent abnormal gastric mucosa in arthritics may predispose to gastric ulceration in these patients (Ivey and Clifton, 1974).

Siurrala and co-workers have reported histological evidence of gastritis and inflammatory cell infiltration with malabsorption in the intestinal tracts of patients with collagen diseases, including rheumatoid arthritics (Siurrala et al., 1965). These authors claimed that the histological changes they observed were unrelated to the ingestion of anti-inflammatory/analgesic drugs, especially since a number of their patients were not taking drugs at all (Siurrala et al., 1965). Further studies are needed firmly to establish the existence of a frail gastrointestinal mucosa predisposed to the irritant actions of aspirin and related drugs. It should be noted that arthritis is one of the diseases in which a stress adaptation does occur (Selye and Najusz, 1966), and this would be expected to have gastrointestinal manifestations. Subnormal mucosal prostaglandins in the elderly (Schlegel et al., 1977) may also contribute to susceptibility to aspirin in this group of patients.

Aspirin has been found to have greater gastro-ulcerogenic activity in portal hypertensive rats as a consequence of more profound effects on prostaglandin and nitric oxide production and microvascular injury (Sarfeh et al., 1988; Sarfeh and Tarnawski, 1991; Calatayad et al., 2001).

Haemorrhagic shock from bleeding, which produced gastric mucosal ischaemia, was found to increase the size of lesions and scores of mucosal lesions in anaesthetised dogs dosed intragastrically with 20 mmol aspirin (Casalnuovo et al., 1984). Another vascular factor that has been suggested as being important in potentiation of aspirin-induced gastric mucosal damage is portal hypertension. Sarfeh and Tarnawski (1991) showed that aspirin was more ulcerogenic in rats in whom portal hypertension had been induced by two-stage portal vein ligation compared with non-hypertensive rats. However, Espluges (2001) showed that there was an effect of anaesthetic and a complex interrelationship with the nitric oxide generating system in the gastric mucosa and smooth muscle that was important in determining the ulcerogenic effects of aspirin in portal hypertensive rats. In conscious hypertensive rats the effects of aspirin were less than in non-hypertensive animals, while in anaesthetised animals the reverse was observed and aspirin was clearly more ulcerogenic in hypertensive animals.

Nutritional status

More gastric erosions are observed in guinea pigs or rats following oral administration of aspirin to fasted than to replete or unstarved animals (Anderson, 1964b; 1964c; Rainsford, 1975d; Shriver et al., 1975) (LD). Anderson showed that the presence of a non-nutrient solid ('Hypercel') did not protect the mucosa against aspirin-induced gastric erosions in fasted guinea pigs (Anderson, 1964c) (LD). Likewise, co-administration of 15 to 30 per cent milk powder or a 10 per cent aqueous suspension of pellet food with aspirin to fasted guinea pigs did not cause any reduction in gastric mucosal damage compared with that with aspirin alone (Anderson, 1964c) (LD). Thus it appears there is some sensitising effect of the fasting procedure, independent of any effects of food in the stomach (Anderson, 1964c).

Considerable interest has been shown in the possible protective effect that certain dietary constituents, especially amino acids, glucose and certain tricarboxylate cycle intermediates, have against salicylate-induced gastric mucosal injury (Okabe et al., 1974b; 1975; 1976a; 1976b; 1976c; Takeuchi et al., 1976; MacDonald et al., 1977; Rainsford and Whitehouse, 1980a; 1980e). In a series of papers, Okabe and co-workers (Okabe et al., 1974b; 1975; 1976a; 1976c; Takeuchi et al., 1976) investigated the inhibitory effects of various amino acids on aspirin-induced gastric erosions and effects on acid and pepsin secretion. In the initial comparison of various amino acids, these authors showed that several basic amino acids were quite potent inhibitors of aspirin-induced mucosal injury and prevented hydrogen ion back-diffusion (see below) from the gastric lumen of Shay rats (Okabe *et al.*, 1976c) (LD). For some inexplicable reason these authors chose L-glutamine for a more detailed investigation (Okabe *et al.*, 1974b; 1975; 1976a; Takeuchi *et al.*, 1976), when it was quite clear from the comparative studies that glutamine was much less effective than many other amino acids in preventing aspirin injury and correcting certain physiopathological changes induced by this drug (Okabe *et al.*, 1976c) (LD-PL). The choice of the Shay rat preparation in which to perform the bulk of their studies (Okabe *et al.*, 1974a; 1975; 1976c) may have been unfortunate for Leeling and co-workers (1979), subsequently were unable to show any reduction in gastrointestinal blood loss in dogs by glutamine plus aspirin compared with aspirin alone (Ridolfo *et al.*, 1980) (Fe). While glutamine has some weak protective effects against aspirin injury in stressed rats, the timing of administration of this amino acid is critical to optimise its action (Rainsford and Whitehouse, 1980a) (LD).

The varying effects of different amino acids on acid and pepsin secretion and on hydrogen ion back diffusion in the mucosa (Okabe *et al.*, 1976c) raises the possibility that certain mixtures of amino acids may have a more beneficial effect against aspirin injury than individual amino acids. Indeed a casein hydrolysate mixture has been found to be effective in preventing aspirin-induced damage in stressed rats (Rainsford and Whitehouse, 1977) (LD), but this is not without some loss in anti-inflammatory efficacy of the drug (Rainsford and Whitehouse, 1977). Nonetheless, these results suggest that amino acids may have some (albeit weak) beneficial effects, and this may be of general nutritional significance.

Adding glucose to the drinking water of fasted rats has been shown to ameliorate aspirin-induced gastric mucosal damage (MacDonald *et al.*, 1977) (LD). However, oral co-administration of glucose with aspirin has no beneficial effects in stressed and starved rats (Rainsford and Whitehouse, 1980a) (LD).

The glycogenolytic hormone, glucagon, markedly reduces aspirin injury in dogs (see Chernish *et al.*, 1979 (HP); also Lin and Warrick, 1974 (^{51}Cr-transferrin)). These results suggest that physiological procedures for prolonged maintenance of glucose levels in blood and mucosa throughout the full length of starvation and aspirin injury may be important for determining more appropriately the susceptibility of the gastric mucosa to injury by this drug. Oral glucose with either amino or 2-oxoacids (especially those metabolised by the tricarboxylate cycle) has a more profound protective effect against aspirin injury in fasted and stressed rats and pigs than these constituents alone (Rainsford and Whitehouse, 1980a; Whitehouse *et al.*, 1984) (LD). The exact mechanism of this protective action of these nutrient mixtures involves a combination of correction of metabolic defects caused by the drug and by buffering both gastric acid and the acidic drug (Rainsford and Whitehouse, 1980a).

The metabolic effects of orally or systemically administered nutrient mixtures (such as glucose + acetate) is to restore the levels of ATP and lactate/glucose ratio, which are otherwise depressed by the drug, to near control levels (Rainsford and Whitehouse, 1980a). Co-dosing with glucose alone is sufficient to enhance energy (ATP) production in the gastric mucosa in aspirin + stressed animals, because it cannot promote its own metabolism. This is presumably due to depletion of intermediates of glycolysis and the tricarboxylate cycle. Salicylates inhibit enzymes of these metabolic pathways and mitochondrial oxidative phosphorylation (Fishgold *et al.*, 1951; Whitehouse, 1965) (see also p. 436). Also, stress reduces energy (ATP) production in the gastric mucosa. Thus the effect of administering precursors or intermediates of the tricarboxylate cycle is to override the inhibitory effects of salicylates upon this metabolic pathway, enabling glucose metabolism to proceed and restoring ATP production in the gastric mucosa. The mucosa is heavily dependent on the continual supply of nutrients for its metabolic demands from either the circulation and/or gastric lumen, so that any alteration of energy production due to inhibitory effects of salicylate on glucose metabolism will have effects on mucosal viability, resistance and functions. A further finding concerned with the lack of effects of glucose alone in protecting the mucosa from aspirin injury is that blood glucose levels are elevated in aspirin and aspirin + stressed rats, reflecting the block on glucose metabolism by the liver (Rainsford and Whitehouse, 1980a). Glucose is known to accelerate gastric acid production (Kowalewski and Kolodej, 1977) and so co-administration of this nutrient (alone) with aspirin may defeat the object of at least controlling gastric acid production, and so potentiate aspirin injury.

CHAPTER 8

More detailed biochemical mechanisms related to these effects will be discussed later. In the nutritional context, it is important to appreciate these results in recognising the susceptibility of the mucosa to aspirin injury. It may also help to resolve the observations made much earlier by Anderson regarding the susceptibility of the mucosa in fasted guinea pigs to aspirin (Anderson, 1964a; 1964b) (LD). Overall, it appears from all these studies with added nutrients that prolonged depression of the nutritional state during fasting is another factor in potentiating mucosal injury by aspirin and related drugs. The presence of bulk material in the gastric lumen confers little or no protective influence against injury.

Vitamin C status

Man, some non-human primates and guinea pigs are unable to produce ascorbate because of a lack of the hepatic enzyme 1-gluconolactone oxidase, which is required for ascorbate synthesis.

This inability, combined with the reported depressant effects of aspirin on leucocyte ascorbate levels (Russell et al., 1968) and the importance of this vitamin in tissue repair/protection, raises important questions about the role of ascorbate in aspirin-induced mucosal injury (Croft, 1968).

Aspirin given orally to scorbutic guinea pigs for 2 weeks caused more gastric mucosal erosions than that observed in normal animals (Russell and Goldberg, 1968) (LD). Ascorbate has been reported to ameliorate the gastric mucosal injury induced by the stress of starvation in rats (which can synthesise the vitamin; Cheney and Rudrud, 1974). However, inclusion of L-ascorbate 20–60 mg/kg in the rats' diet markedly increased the amount of gastric damage by aspirin 300 mg/kg compared with that in non-supplemented groups of animals (Lo and Konishi, 1978) (LD). Thus it appears that ascorbate can, in the rat at least, potentiate acute gastric mucosal injury by aspirin.

Curiously, a US patent filed by Mishikawa in 1990 (Mishikawa and Nakamura, 1992) claiming that ascorbic acid could reduce the gastric side effects of aspirin was granted, even though the positive evidence was available for over 15 years in the public domain!

The possibility exists that rats may not be the ideal species in which to perform these experiments vis-à-vis relevance to man. The results of these studies in rats (Russell and Goldberg, 1968; Lo and Konishi, 1978) do suggest, however, that the addition of vitamin C to overcome a deficiency induced by aspirin may be counterproductive, because the additional acidity from the vitamin may potentiate lesion development. Addition of the vitamin may also alter the metabolism of salicylates, so there may be added disadvantages, in a therapeutic sense, from megadose self-administration of this vitamin.

Dosage, route and timing of drug administration

Single oral doses of aspirin, salicylic acid and 5-chlorosalicylic acid cause an initial peak development of haemorrhagic gastric lesions in fasted (stressed) or unstressed adult rats at 1–2 hours, followed later (depending on the dose) after 16–24 hours of prolonged starvation by the appearance of deep craterous ulcers (Morris et al., 1967; Rainsford, 1975b; 1978b) (LD). With aspirin, peak erosions in pyloric-ligated (Shay) rats occurs at 3–7 hours (Okabe et al., 1975) (LD), further illustrating differences in the mucosal reactivity of this preparation. The initial peak of lesions from aspirin in rats follows (or coincides with) the peak plasma concentrations of salicylates. This lesion development clearly follows the gastric (fundic) mucosal concentrations of the drug (Morris et al., 1967; Brune et al., 1977a; Rainsford et al., 1981a) (LD). The more slowly absorbed drugs, meseclozone and diflunisal, have a much later initial peak of lesion development in stressed rats, at about 6 or more hours (Diamantis et al., 1980; Rainsford et al., 1981a; Rainsford, 1982a) (LD). The more complex metabolism of meseclozone accounts for more pronounced lesion development in unstressed rats at 24 hours compared with the earlier times (Diamantis et al., 1980; Rainsford, 1982a) (LD). In stressed rats, the full expression of ulcerogenic activity of diflunisal and meseclazone occurs at much earlier times (approximately 2 hours, depending on the dose; Rainsford, 1978b; 1982a) (LD).

Increasing doses of salicylates induce a characteristic bell-shaped response in the numbers of gastric mucosal lesions and their severity (Anderson, 1963; Horribin et al., 1974; Rainsford, 1978b) (LD). It is curious that in many laboratory species massive toxic doses of aspirin will cause little if any macroscopic signs of damage, although there is microscopic evidence of desquamation of superficial mucosal cells. A similar response has been observed in the stomach specimens taken from postmortems of individuals who died from salicylate poisoning (Rainsford, 1978b). These observations suggest that high doses of salicylates may limit the expression of ulcerogenesis, perhaps through suppression of acid and pepsin secretions, and physical effects of a bolus dose of an ingested acid causing mucosal denaturation.

In contrast to the situation in rats, peak lesion development from single oral doses of aspirin in guineas pigs appears only to occur at 2–3 hours, with no later signs of mucosal damage (Anderson, 1964c) (LD). Garner has shown that aspirin induces maximal cellular exfoliation in the guinea pig at 4 hours (Garner, 1977b; 1977c) (CE). This suggests that superficial erosion of mucosal cells occurs even after the appearance of haemorrhagic lesions. Prolonged intragastric infusion of aspirin over a 30-minute period to guinea pigs induces a greater rate of cellular exfoliation than a single infusion of the same dose of the drug (Garner, 1977a) (CE). This may reflect differences in the rate of uptake of the drug when given by these different procedures, even though the same total quantity of drug was given.

In dogs, intragastric installation of aspirin 12.5–20 mmol/l or salicylic acid 20 mmol/l in acidic media induces efflux of Na^+, K^+, Ca^{2+} and Cl^- ions, with loss of H^+ ions within 30 minutes of drug administration (Davenport, 1965; Chvasta and Cooke, 1972, Ivey et al., 1975b; Lin et al., 1975) (HP). Maximum changes in the concentrations of these physiological ions (used as indices of permeability changes in the gastric mucosa) are evident at 1 hour (Davenport, 1965; Chvasta and Cooke, 1972; Lin et al., 1975) (HP). Coincidentally, the intragastric potential (PD) which is another index of mucosal permeability, appears to decline rapidly – i.e. within the first 30 minutes following oral administration of salicylates to dogs (Torchiana et al., 1979) (HP-PD). The decline in PD persists beyond the maximal period of ionic permeability changes or bleeding (Torchiana et al., 1979) (HP-PD). Repeated administration of aspirin 20 mmol/l over a succession of 30-minute periods is capable of eliciting a progressive decline in mucosal potential (Ritchie and Fischer, 1973) (HP-PD). These results are comparable to observations of potential difference changes observed after oral aspirin in man (Caspary, 1978) (PD). Intragastric installation of salicylic acid 30 mmol/l in 160 mmol/l HCI to cats induces a statistically significant reduction in mucosal resistance at 5 minutes (Bruggeman et al., 1979), suggesting that breakdown in the physiological resistance of the mucosa can be a very rapid event.

There is variation in the effects of repeated (compared with single) oral doses of aspirin on the gastric mucosa of animals (but cf. man, see Graham et al., 1983). Hurley and Crandall (1963) originally suggested, following their gastroscopic observations, that the canine gastric mucosa may develop resistance to the erosive effects of repeated oral administration of aspirin (G). At about the same time, Fishler (1964) reported that aspirin 300 mg twice daily produced more gastric erosions and haemorrhage when dosed orally to dogs for 3 to 10 days then after 1 day's treatment (LD). Likewise, there was little difference between the gastric damage observed at 3, 6 or 10 days in dogs dosed orally with aspirin 15 g twice daily (Fishler, 1964) (LD). Since Fishler observed the gastric damage at autopsy whereas Hurley and Crandall performed repeated gastroscopies on anaesthetised dogs, it appears that either the anaesthetic or the gastroscopic procedures may have interfered with the development of gastric erosions induced by aspirin in the study by Hurley and Crandall.

In 1965, Anderson reported that repeated oral administration of aspirin 100 mg/kg per day to six guinea pigs for 11 days failed to produce any gastric mucosal damage (Anderson, 1965) (LD). The size of the experimental group was, however, insufficient to permit definite conclusions to be drawn.

The studies by Bolton and Cohen (1977) have shown that the content of haemoglobin, Na^+ ions and exfoliated cells (measured by the DNA content) was not significantly affected from days 1 to 11 following daily instillation of aspirin 20 mmol/l into gastric pouches of dogs (HP-CE, Hb). There was a slight increase in the rate of exfoliation of cells on days 13 and 15, but the authors did not consider this of significance, especially in relation to an adaptive effect by the mucosa (Bolton and Cohen, 1977).

■ CHAPTER 8 ■

TABLE 8.14

Comparison of the gastric ulcerogenicity of single compared with repeated oral doses of salicylates and other NSAIDs to rats.

Drug	Acute administration		Chronic administration	
	UD_{50}* (mg/kg p.o.)	Time of peak ulcerogenesis (h)	UD_{50}* (mg/kg p.o.)	Time of peak ulcerogenesis (h)
Salicylates				
Aspirin	65.2	2	54.2	2
5-chlorosalicylic acid	50.4	2	37.9	2
Diflunisal	169.6	6	115.7	4
Meseclazone	640.0	24	248.6	4
NSAIDs Standards				
Indomethacin	3.7	4	6.9	6
Phenylbutazone	154	4	115.9	6

*UD_{50} is the ulcerogenic dose for 50 per cent incidence of lesions determined at peak ulcerogenesis. Data from Diamantis *et al.*, 1980.

St John and co-workers (1973) also claimed that rats developed a refractory effect to repeated oral administration of aspirin (LD). However, it appears that these authors failed to fast their rats before the final dose of aspirin (in the repeat dose group), and they therefore compared their data on repeat dosing effects with the acute effects of the drug in fasted animals (St John *et al.*, 1973). More detailed studies by Diamantis and co-workers (1980) in unstressed rats (LD; see Table 8.14) and other work in stressed rats (Rainsford, 1982d) (LD) and unstressed mice (Tsodikov *et al.*, 1979) (LD) failed to confirm the notion of a refractory effect towards repeated oral dosing of aspirin in rats at least over a period of 1-week continuous administration. A slight enhancement is evident with repeated oral administration of 5-chlorosalicylic acid and diflunisal, and this is even more pronounced with meseclazone (LD; Table 8.14). In contrast to the effects of these salicylates, most arylalkanoic NSAIDS failed to show any differences in ulcerogenicity following repeated compared with single dosing (Diamantis *et al.*, 1980) (LD).

Orally administered aspirin and other salicylates generally induce more gastric mucosal damage than when these drugs are given parenterally, with the notable exception of the intravenous route. There is, as a consequence of the effect of parenterally administered aspirin, considerable evidence for a systemic component in salicylate-induced ulcerogenesis. In any discussion of the systemic component in drug-induced ulcerogenesis it is important to recognise the differences in drug distribution that occur following administration of the drug by different routes.

Relationship to drug kinetics; oral versus systemic effects

Pfeiffer and Lewandowski (1971) found that total plasma levels of salicylates obtained following subcutaneous (s.c.) or intrarectal (i.r.) administration of aspirin to rats were very much lower than when the drug was given by the oral (p.o.) or intraperitoneal (i.p.) routes. Frey and El-Sayed (1977) showed that the ratio of gastric mucosal to plasma levels of salicylates over the first hours, following s.c. administration of aspirin or salicylic acid were about one-ninth of that from p.o. administration of these drugs. This indicates that very low levels of the drug are present in the gastric mucosa following s.c. compared to p.o. administration. These differences in mucosal and plasma levels of salicylates are paralleled by the somewhat lower gastric mucosal damage achieved by the s.c. compared to the oral routes (Pfeiffer and Lewandowski, 1971) (PL-LD). Likewise, i.v. administered aspirin produces less damage than the oral or s.c. routes, even though the plasma salicylate levels are quite different when the drug is given by the s.c. and i.r. routes (Pfeiffer and Lewandowski, 1971) (PL-LD). Guth and Paulsen have reported that i.p. administration of sodium aspirin 15–20 mg/kg to 48-h fasted rats induced lesion development at 0.5–4.0 h (Guth and Paulsen, 1979a) (LD-PD). Higher i.p. doses of sodium aspirin (up to 400 mg/kg) have failed to elicit damage in 24-h fasted rats, but will produce damage in animals simultaneously exposed to cold stress (Rainsford, 1978b) (LD). Thus the prolonged period of fasting employed by Guth and Paulsen (1979a) appears to have constituted an additional stress component. Their studies are interesting in the context of mucosal ulcerogenesis by systemic compared to oral routes. Oral aspirin or salicylic acid usually induces breakdown in the mucosal permeability barrier that precedes lesion development, although Guth and Paulsen (1979a) showed the opposite effect following parenteral aspirin (LD, PD) It would appear that focal lesion development can be initiated, possibly deep in the mucosa by ischaemic reactions, independently of surface permeability changes. Thus, with parenteral aspirin, 'barrier' disruption is a consequence and not a cause of lesion development (Guth and Paulsen, 1979a).

A case for systemic aspirin being a major factor in gastric ulcerogenesis has been made by Hansen *et al.* (1978; 1980), based on their studies in cats. They observed ulcers in the fundus and antrum of conscious cats given either an i.v. infusion of sodium aspirin 0.06–4.0 mg/kg/h for 3–4 hours (which produced therapeutic levels of 0.06–1.4 mmol/l plasma salicylates, respectively) or a bolus i.v. injection of sodium aspirin 40 mg/kg, but only when histamine was given concurrently (Hansen *et al.*, 1978) (LD). The amount of mucosal damage following the bolus injection of aspirin was the same as that following an equivalent oral dose of the drug. Thus these authors concluded that aspirin has a similar mechanism for causing damage to the mucosa when given by the oral or the i.v route. Unfortunately, they overlooked in their interpretation some fundamental aspects about the pharmacokinetics of aspirin and salicylate distribution by oral compared to i.v routes.

Rowe *et al.* (1987) showed that i.v. salicylic acid produced the same degree of damage as i.v. aspirin in pyloric ligated and anaesthetised rats providing the gastric lumen was at acidic pH. Thus gastric pH may be a major determinant of salicylate-induced acute gastric mucosal injury when the drug is given parenterally.

Relationship to drug absorption in gastrointestinal mucosa

The pioneering theory of Martin (1963) stated that the uptake from the stomach of the non-ionised aspirin (R-COOH) form (according to the pH partition hypothesis) under acidic conditions results in ion trapping (of the R-COO$^-$ form) at neutral pH ($\sim$7.4) in the gastric cells, which then results in the cellular damage. This theory has been developed further by McCormack and Brune (1987) to explain the differences in GI mucosal injury by NSAIDs. Non-acidic drugs, being less likely to be absorbed at the rate of acidic drugs in the stomach, are considered less ulcerogenic for these reasons.

There is extensive evidence in support of the uptake of acidic drugs and salicylates in particular being related to ulcerogenicity (Brune *et al.*, 1977a; 1977b; 1979; 1984; Garner, 1977a; Rainsford and Brune, 1976; 1978; Kajii *et al.*, 1985; McCormack and Brune, 1987). The most direct evidence has come from autoradiographic studies in which the uptake of ^{3}H- or ^{14}C-labelled aspirin, salicylate or other NSAIDs has been studied with time (Brune *et al.*, 1977a; 1977b; 1979; 1984; Garner, 1977; Rainsford *et al.*, 1981b). Selective drug uptake into superficial mucous and parietal cells is a major feature of the early phase of mucosal injury, and is paralleled by fine structural changes and evidence of cellular damage in these cells (Rainsford, 1975d; Brune *et al.*, 1977a; 1977b; Figure 8.5; Rainsford and Brune, 1976; 1978; Rainsford *et al.*, 1981b).

The more lipophilic NSAID diflunisal, while having the same general property to accumulate in the gastric mucosal cells parallel with the development of mucosal damage (Rainsford *et al.*, 1981b; 1983), shows differing mucosal irritant responses to buffering of gastric contents (Nuernberg *et al.*, 1990) compared with that of aspirin; the acute mucosal irritancy from the latter being ameliorated when the gastric contents are buffered to neutral pH (Rainsford, 1975d). In contrast, Nuernberg *et al.* (1990) observed that there were no differences in ulcerogenicity when the gastric contents were buffered or un-buffered, even though the permeability of the serosal side of isolated rat stomachs to the drug was

■ CHAPTER 8 ■

markedly increased in the pH range of 4 to 5. The rate of absorption declined at higher and lower pH values. The peak absorption occurred at about the pK_a value of the drug showing the importance of pK_a in gastric absorption.

There are marked differences in the amount of aspirin in the gastric mucosa and blood circulation following oral compared with the parenterally administered drug.

First, the total concentrations of aspirin in the plasma (represented by the area under the plasma concentration curve in Figure 8.2 and mucosa are very much greater when the drug is given by the i.v. compared to the oral route (Lester *et al.*, 1946; Rowland *et al.*, 1967). Also, only 85 per cent of aspirin is bound to albumin and other plasma proteins, whereas by comparison greater than 95 per cent of salicylate is bound (Lester *et al.*, 1946; McArthur and Smith, 1969). Moreover, orally administered aspirin is more rapidly metabolised to salicylate than that from i.v. administration of the drug (Harris and Riegelman, 1969), there being appreciable hydrolysis by aspirin esterases in the gut and liver (Morgan and Truitt, 1965; Levy and Angelino, 1968; Harris and Riegelman, 1969; Juggi, 1975; Landecker *et al.*, 1977; Spenny and Nowell, 1979; Guth, 1982; Rainsford *et al.*, 1980a). Thus, when given intravenously much greater concentrations of the highly ulcerogenic parent drug are in the circulation relative to its less ulcerogenic metabolite, salicylate. This would be present in the gastric mucosa in appreciable quantities compared with that when an equal quantity of the drug is given orally. It should also be recalled that cats have a peculiar gastric sensitivity to aspirin, which might be due to their defective detoxification mechanisms (Davis and Donnelly, 1968; Williams, 1974). While there is no doubt that

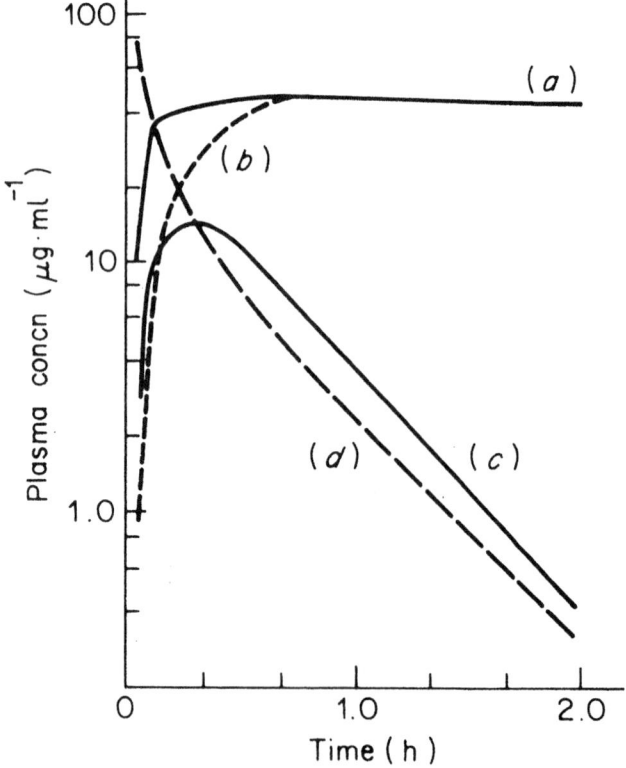

Figure 8.2 Comparison of the plasma levels in man of aspirin and salicylate following oral or intravenous (i.v.) administration of aspirin 650 mg. (a) i.v. salicylate, (b) oral salicylate, (c) oral aspirin, and (d) i.v. aspirin. (Based on Rowland *et al.*, 1967.) (From Rainsford (1984). Reproduced with permission of the publishers, Butterworths/ Heinemann.)

410 ■

these and other experiments (discussed above) do provide evidence for systemic as well as a local mode of action of aspirin on the gastric mucosa, the problem is specifically to ascribe quantitative significance to the effects when the drug is given by oral compared to i.v. routes.

Sex and age

There does not appear to be any difference in response to aspirin between male and female guinea pigs (Anderson, 1964a,b) (LD). However, aspirin induces significantly more lesions in pregnant rats (day 20) compared to non-pregnant rats (Takeuchi *et al.*, 1976) (PL-LD). This appears to be a consequence of hypersecretion of acid, but curiously there is hyposecretion of pepsin in both control and aspirin-treated pregnant rats compared with non-pregnant rats treated in the same way (Takeuchi *et al.*, 1976). Thus the mucosal sensitivity in the pregnant state may influence the pepsin-generated supply of protective amino acids necessary for maintenance of mucosal defence mechanisms.

Age appears to be a major factor in aspirin-induced ulcerogenesis. Young adult rats (100–200 g body weight) have been found to be more susceptible to aspirin-induced gastric damage than other older or younger animals (Wilhelmi and Menassé-Gdynia, 1972) (LD). These observations have been confirmed by Malcontenti-Wilson *et al.* (1998). They found that there is reduced adaptation of the gastric mucosa to the ulcerogenic effects of aspirin in mature and aged rats compared with younger animals, the area of mucosal damage being three-fold in aged over that in younger animals. Interestingly, diclofenac, ibuprofen, indomethacin and a COX-2 selective drug did not produce these effects related to age. The effect of ageing appears to be related to reduction in the resolution and regeneration of the mucosa, but not to expression of proliferating cell antigen or production of the growth factor TGF (Lee *et al.*, 1998) or age-related secretion of acid or other aggressive factors (Penny *et al.*, 1996).

Particle size of drug

Anderson observed that oral administration of fine particles (of size ⩽20 mm) of aspirin produced more lesions in guinea pigs than coarse particles (200–400 mm) of the drug (Anderson, 1964a) (LD). Furthermore, filtered suspensions of aspirin 14–56 mmol/l were found to produce the same number of gastric lesions as non-filtered solutions (Anderson, 1964c) (LD). Similar observations have been made in dogs (Hurley and Crandall, 1963) (G). The low concentration of aspirin (14 mmol/l) employed by Anderson is within the solubility range of the drug, even at the low pH present in the stomach. Thus damage can be initiated by the drug in solution, although it is clear that large particles of the drug can exacerbate damage. The observations in pigs (Rainsford, 1978c) that aspirin tablets are more ulcerogenic than suspensions or solutions of the drug (see Table 8.13) appears to reflect the additional effects of tableting for the development of mucosal damage. It should be recalled that Hurst and Lintott, in their earlier gastroscopic observations of the effects of aspirin tablets in man, noted that the drug particles appeared to be 'gripped' by the mucosa (Hurst and Lintott, 1939). It is possible that this mucosal reaction could ensure persistent contact of the corrosive drug particles with mucosa. Solutions of the drug will be distributed over a wider surface area of the stomach, and hence should induce less severe mucosal damage than caused by tablets.

Cholestasis

Salicylates are well known cholegogues (Geall *et al.*, 1970; Davenport, 1972; O'Brien and Silen, 1973; Cheung *et al.*, 1975; Semple and Russell, 1975; Guth *et al.*, 1976) (see p. 437) and it has been suggested from a considerable number of studies that bile salts may aggravate mucosal damage and potentiate the effects of aspirin. Cholestasis induced by bile duct ligation and resections 2 weeks prior to single doses of aspirin in rats resulted in a marked increase in gastric mucosal damage, the dose–response curve for ulcer-index being almost double in bile duct resected animals compared with sham-operated or controls (Dehpour *et al.*, 1998) (L).

Physicochemical and structural features of salicylates

One of the main features of the salicylates accounting for their gastric ulcerogenic activity is the presence of an unsubstituted carboxyl group. There are also complex physicochemical steric and other structural features that either enhance or diminish the ulcerogenicity of these benzoic acids. These can be summarised as follows (see also Figure 8.3):

1. Addition of a phenolic hydroxyl group at the 2-position enhances the ulcerogenicity inherent in benzoic acid (Anderson, 1963) (LD).
2. Most *O*-acetylsalicylic acids (i.e. aspirins) are more ulcerogenic than their respective non-acetylated phenols (i.e. salicylic acids) (Rainsford and Whitehouse, 1976c; 1980b; Rainsford, 1978e; Table 8.13). Possible biochemical reasons for this include the ability of a labile acetyl moiety to acetylate biomolecules (Rainsford *et al.*, 1983), and to be hydrolysed (enzymically and spontaneously) to liberate 2 moles of acid per mole of acetyl-salicylate, available to back-diffuse protons through the mucosal wall (see p. 427; Gibberd, 1966; Di Pasquale and Welaj, 1973). That the ulcerogenic activity (i.e. in addition to that of salicylate) is principally inherent in the acetyl moiety of aspirin is illustrated in the case of 2-methoxybenzoic acid (anisic acid), which in contrast to aspirin has only the same ulcerogenicity as salicylic acid (Rainsford, 1978e) (LD).
3. Addition of groups with increasing lipophilicity at the 3 position of salicylic acid results in a decrease in its ulcerogenicity (Rainsford, 1978e) (LD). In contrast, addition of these more lipophilic groups at a 4 or 5 position is, with the exception of diflunisal and flufenisal, associated with enhancement of the ulcerogenicity of salicylic acid (Rainsford and Whitehouse, 1976c; Rainsford, 1978e) (LD). The exceptions noted with the 5-(2′,4′-difluorophenyl) and 5′-(4′-fluorophenyl) salicylic acids, i.e. diflunisal and flufenisal, respectively, may be due to the pronounced electron-withdrawing effects of the fluoro groups added on the 5-phenyl moiety.
4. Curiously, the addition of bromo substituents at both the 3 and 5 positions decreases the ulcerogenicity of aspirin (Rainsford, 1978e) (LD). However, addition of a chloro-group at the 5 position alone makes little difference to the ulcerogenic activity of aspirin (Rainsford, 1978e) (LD). Thus it appears that the presence of disubstituted halogen groups at both the 3 and 5 positions is sufficient to elicit pronounced electron-withdrawing effects over the benzene nucleus, so reducing the ulcerogenicity inherent in aspirin.

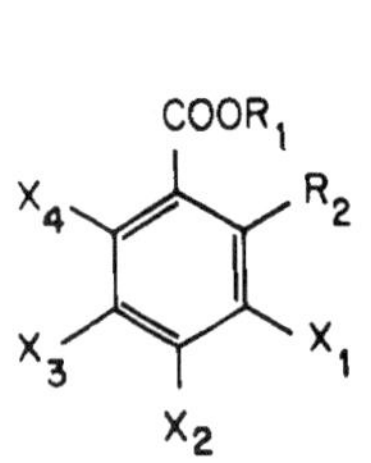

(1) *Enhanced* ulcerogenicity (of benzoic acid) in order by substitution with:

$R_1 = H$, $R_2 = -OCOCH_3 -OH> -H(X_1-X_4=-H)$,

and further when

$R_2 = -OCOCH_3$ or $-OH$

with X_2, X_3, X_4 = alkyl ($n = 1$–5) or halogen, or X_2, X_3 = phenyl.

(2) *Decreased* ulcerogenicity of aspirin, when $X_1 = C_6H_5-$, or alkyl ($n = 1$–5),

OR

$R_1 = -CH_3$, C_6H_5-, fatty acyl-triglyceryl, paracetamol etc.

Figure 8.3 Structural features of salicylates (benzoates) determining their ulcerogenicity.

5. In contrast to the situation with aspirin, a chloro group at the 5 position of salicylic acid almost doubles the ulcerogenicity of the latter (Rainsford, 1978e) (LD). However, similar substitution with a bromo group makes no difference to the ulcerogenicity of salicylic acid (Rainsford, 1978e) (LD). It would appear, therefore, that the ulcerogenicity of these halosalicylic acids can be markedly influenced by the relative inductive or electron-withdrawing activity of the individual halogen (i.e. chlorine being more active in this respect than bromine), as well as the relative bulkiness and lipophilicity of these groups (i.e. bromine being more so than chlorine).

6. Generally, the presence of non-halogen bulky lipophilic (i.e alkyl) groups at the 3, 4, and 5 positions reduces the ulcerogenicity of aspirin (Rainsford, 1978e) (LD). Again, this may be due to the intrinsic capacity of these compounds to acetylate biomolecules or to differences in their rate of hydrolysis following absorption from the gastric mucosa.

7. Detailed examination of the physiochemical properties, i.e. liposolubility (log P, measured by taking the logarithm of the solubility of the drug in an organic solvent, (e.g. n-octanol) compared with that in an aqueous medium (e.g. sodium phosphate buffer pH 7.4), or diffusion coefficient, log D) and dissociation constant (pK_a) in relation to the gastric ulcerogenicity data in Table 8.15 has revealed that none of these physiochemical features is alone important in determining the acute ulcerogenic activity of salicylates (Rainsford, 1981a) with the possible exception of pKa (Bjarnason and Rainsford, unpublished studies). However, combination of these properties is, within the limits of certain steric/electronic effects of individual substituents (as prime determinants of ulcerogenicity). These physicochemical properties are obviously important because they are known to influence the relative absorption of the salicylates across the gastric mucosa.

8. One common physiochemical property of these acids observed by X-ray crystallography (Wheatley, 1964) and laser Raman spectroscopy (Rainsford, 1978e) is that they all exist in the solid state as dimers, i.e. linked by alternating carbonyl and hydroxyl moieties of their carboxyl groups through hydrogen bonding. Laser Raman spectroscopy of solutions (prepared as salts) at pH 7.3 or as esters has shown that these derivatives exist as monomers (Rainsford, 1978e). It has been suggested that one of the reasons for the lower ulcerogenicity of neutral salts of salicylates compared with the free acids (dimers) could be due to the more energetic dimeric state favouring interaction and rupture of the hydrophobic bonds of the lipoproteins in mucosal cell surface membranes. In this scheme intramolecular interactions between the carbonyl and hydroxyl groups of the dimeric salicylic acids may be ruptured on passage of these molecules into the lipid phase of mucosal membranes. The hydrogen bonds between these moieties would then reform with suitable electron donor (or acceptor) groups within the hydrophobic zones of the membrane lipoproteins, thus causing disruption of the structural lipoproteins and consequent membrane labilisation. Monomers, being less lipophilic, would be less capable of these interactions.

Comparison of some of the factors influencing gastric mucosal damage by aspirin in man with those in laboratory animals

A summary of the main factors that certainly influence the development of gastric mucosal damage in man and laboratory animals is shown in Table 8.16. This enables a comparison to be made between the effects in laboratory animals and those in man. Some clinicians are very sceptical of studies in laboratory animals, and often overlook the value of such data. The converse is also true for some experimentalists in laboratory animals who regard data from human beings as being of limited value. It can be seen from this table that there is a relatively high degree of correspondence between the results obtained in man and laboratory animals. The three factors that do not correspond too well are the influence of sex and blood group status, and the findings concerning enteric-coated preparations of aspirin. In the case of the latter, this may be dependent on the dose employed and on the mucosal sensitivity of individuals. Arthritic patients who consume large quantities of aspirin (including the enteric-coated forms) may still be at risk with this formulation (see, for example, Table 8.10). An assessment of the contribution of blood group status is difficult to perform on the scale required for statistical

TABLE 8.15

Acute and chronic gastro-ulcerogenic activity of salicylates compared with other NSAIDs in different assay systems in rats.

Drugs	Acute ulcerogenic activities			ED_{10} cold treated rats		Chronic (7 day) ulcerogenicity in rats score			
	Minimum ulcerogenic dose mg/kg p.o. at		UD_{50} mg/kg p.o.			at dose	mortality	pK_a	Log D
	3h	24h		mg/kg p.o.	mmol/kg p.o.	(mg/kg/d) p.o.			
Aspirin	<15–30	<15–30	17.4 (8.7–34.8)	5.04	0.023	3.1 @ 2.15	0/10	3.5	−3.45
Azapropazone				240	0.80			6.5	−0.68
Diclofenac sodium	4–8	4–8		1.00	0.003				
Diflunisal				172	0.85			4.7	−1.57
Fenclofenac	400–800	200–800		184	0.62	0.8 @ 180	0/10	4.0	−2.73
Fenoprofen calcium	30–60	30–60		20.3	0.04			5.5	0.96
Flufenamic acid			174 (62.0–487)	20.5	0.073			4.9	−0.17
Ibuprofen	6–13	6–13	148 (62.0–355)	14.4	0.07	3.1 @ 244	2/10		
Indomethacin	1.3–2.5	2.5–5.0	6.6 (2.2–19.8)	2.5	0.0076	0.8 @ 5.7	0.10	5.3	−0.74
Ketoprofen	0.6–1.3	5–10	6.8 (2.4–19.0)						
Mefenamic acid			548 (62.0–355)	101	0.42			4.30	−0.9
Naproxen	2–4	2–4		1.6	0.007			5.30	−1.7
Oxaprozin				851	2.9			6.1	0.01
Phenylbutazone	40–80	38–75	96.0 (37.0–248)	25	0.081	1.1 @ 132	0/10	4.8	−1.92
Piroxicam				0.01	0.0003			4.8	−2.34
Salicylic acid				60.7	0.44			2.92	−5.54
Authors	Atkinson and Leach (1976)		Tsukada et al. (1978)	Rainsford (1981e)		Atkinson and Leach (1976)		Rainsford (1978c and unpublished data)	

Note
Correlations between acute ulcerogenicity in cold treated rats with pKa and log D have shown that best estimates are obtained with pKa although lipophilicity may play a minor role as well. (Bjarnason and Rainsford, unpublished studies).

TABLE 8.16

Summary of some factors implicated in the development of aspirin-induced gastric damage. (Modified from Rainsford, 1975c and 1984a)

Factor	Studies in man	Studies in experimental animals
Acidic nature of drug	+	+
Acidity of gastric contents	+	+
Particles of drug exacerbate damage	+[a]	+
Ameliorating effect of food	+	+
Sex	+	+[b]
Age	+	+
Blood group status	+ (−, ^{51}Cr)	ND
Exacerbation by alcohol	+	+
Development of tolerance ('refractory effect') on repeated oral administration	+	+
Reduced gastric damage by buffered/nutrient mixtures	+	+
Reduced gastric damage by enteric-coated preparations	+	−
Reduced gastric damage by esters	+	+
Vitamin C deficiency exacerbates damage	+	+
Exacerbation by disease or physical stress	+	+
Systemic component in ulcerogenesis	+	+

Where factor is positively implicated this is denoted by +. Negative implication is denoted by −, and ND = not determined.
[a]See also gastroscopic study of Thorsen et al., (1968.)
[b]Exacerbation in pregnant rats.

comparison in those few laboratory species with blood groups comparable to man (e.g. pigs and primates), but obviously such a study is required. Likewise, further studies are necessary on the exact role of sex status, and the influence of pregnancy and lactation on the development of gastric mucosal damage in both laboratory animals and man. It should also be noted that this information has largely been gathered from studies on aspirin, the most clinically used salicylate. Since this may, in fact, represent a special case, it is important to caution that this may not extend by way of comparison with other salicylates or NSAIDs in general. Nonetheless, available information on the other salicylates does indicate that most (if not all) of these factors are important in determining the capacity of these ulcerogenic drugs to exert their actions upon the gastric mucosa.

Intestinal damage

Salicylates probably induce less damage in the upper intestinal tract than in the stomach. However, there are indications that arthritic patients taking aspirin show considerable duodenal pathology (see p. 395).

Intestinal damage by aspirin in rats has been reported by Brodie and co-workers (Brodie et al., 1970) (PSB). However, in later studies from the same laboratory no intestinal lesions were observed following chronic oral administration of aspirin to rats (Stone et al., 1977) (LD). Likewise, no intestinal damage was found from acute or chronic administration of the drug to stressed or unstressed rats or guinea pigs (Rainsford, 1982a; 1982d; Horribin et al., 1974) (LD). Various tablet preparations of aspirin induce mucosal damage in the duodenum of pigs, whereas suspensions or solutions of this drug caused no such damage (Rainsford, 1978c) (LD). Thus, as with the stomach, tablet formulations of aspirin may be a major factor accounting for their irritancy or ulcerogenicity in the upper intestinal tract. Chronic oral administration of diflunisal to rats has been shown to induce intestinal damage (Stone et al., 1977) (LD), a factor that appears important for other NSAIDs causing similar damage to the intestinal tract, e.g. diclofenac, indomethacin, phenylbutazone, piroxicam and sulindac (Di Pasquale and Welaj, 1973; Cohen, 1976; Shriver *et al.*, 1977; Rainsford, 1982d) (LD). It should be noted that drugs that enter into the enterohepatic cycle in rats may not necessarily do so to the same extent in humans.

Light and/or electron-microscopic observations have shown that acute or chronic oral administration of aspirin to mice and dogs induces evidence of epithelial cell damage in the duodenum, jejunum and ileum, without there being macroscopic damage evident (Taylor and Crawford, 1968; Cassidy and Lightfoot, 1979; Djaldetti and Fishman, 1981). Subtle microscopic changes have also been observed in man following acute oral administration of aspirin (Hahn *et al.*, 1975; Ivey *et al.*, 1979). Rectal toxicity has been observed in rats following direct application of aspirin (Coldwell and Boyd, 1966).

Overall these studies suggest that minor microscopic changes may be induced in laboratory animals and man following aspirin administration. The damage can be exacerbated by tablet formulation of the drug.

Pathology of salicylate-induced gastrointestinal damage

The gastric mucosa

Extensive shedding and disruption of the superficial mucosal cells of the fundic and antral mucosa is evident within a few minutes of the oral administration of aspirin and other ulcerogenic salicylates to laboratory animals (Hingson and Ito, 1971; Pfeiffer and Weibel, 1973; Rainsford, 1975b; Harding and Morris, 1976; Rainsford and Brune, 1978; Robins, 1980a) as well as in man (Croft, 1963a; Baskin *et al.*, 1976). Selective damage occurs to parietal cells and the microvasculature (Figure 8.4) (Rainsford and Brune, 1978; Robins, 1980a). Subsequently erosions and haemorrhage develop, and the so-called haemorrhagic lesions of the glandular mucosa (which are the means of visually assessing damage) are evident within 15–30 minutes of administering aspirin and most other ulcerogenic salicylates (Anderson, 1963; 1964a; 1964b; Hingson and Ito, 1971; Hahn, 1973; Pfeiffer and Weibel, 1973; Rainsford, 1975d; Baskin *et al.*, 1976; Harding and Morris, 1976; Rainsford and Brune, 1976; 1978; Robins, 1980b) coincidental with rapid uptake of the drug into the gastric mucosa (Figures 8.5 and 8.6). Damage from oral administration of suspension of salicylates is usually confined to the fundic and, to a lesser extent, the antral regions of most species (Anderson, 1963; Hurley and Crandall, 1963; Roth and Valdes-Dapena, 1963a; Hingson and Ito, 1971; Pfeiffer and Weibel, 1975; Rainsford, 1975b; Shriver *et al.*, 1977; Rainsford and Brune, 1978). If tablets of aspirin are given to pigs, then more extensive damage and ulcers are observed in both the antrum and pylorus compared with those observed with suspensions or solutions of the drug (Rainsford, 1978c; Figure 8.1). Since gastric ulceration is generally observed in the antral region of humans (Menguy, 1972; MacDonald, 1973), it would appear that this is further evidence for tablets of aspirin being a predisposing factor in gastric pathology induced by this drug.

In addition to causing gross erosion of gastric mucous cells, electron microscopic evidence has shown that oral administration of aspirin or salicylic acid induces changes in these non-eroded mucous cells, including: (a) clumping of nuclear chromatin; (b) reduction in the density of the cytoplasmic ground substance; (c) intracellular vacuole formation ('vacuolation') and lysosomal disruption; (d) karyolysis; (e) redistribution and discharge of mucous granules; and (f) at later times swelling of mitochondria and endoplasmic reticulum (Hingson and Ito, 1971; Pfeiffer and Weibel, 1973; Rainsford, 1975d; Figure 8.4). Most of these changes probably reflect early events preceding cell death. Quantitatively it has been assessed by Cassidy and Lightfoot (1979), and 42.6 per cent of the gastric ridges exhibited abnormal ultrastructure following oral dosing of aspirin 10 mg/kg in phosphate buffer pH 7.1 to ether anaesthetised rats. This is surprising in view of the fact that much higher oral doses of aspirin + sodium bicarbonate mixtures given to unanaesthetised rats produced very little ultrastructural evidence of mucosal damage (Rainsford, 1975d) This could be a reflection of differences in the sensitivity of the mucosa towards the ether anaesthetic (a known mucosal irritant).

Since no apparent physical disruption of the intercellular tight junctions linking adjacent surface mucosal cells has been seen in either laboratory animals or man given aspirin (Frenning, 1971;

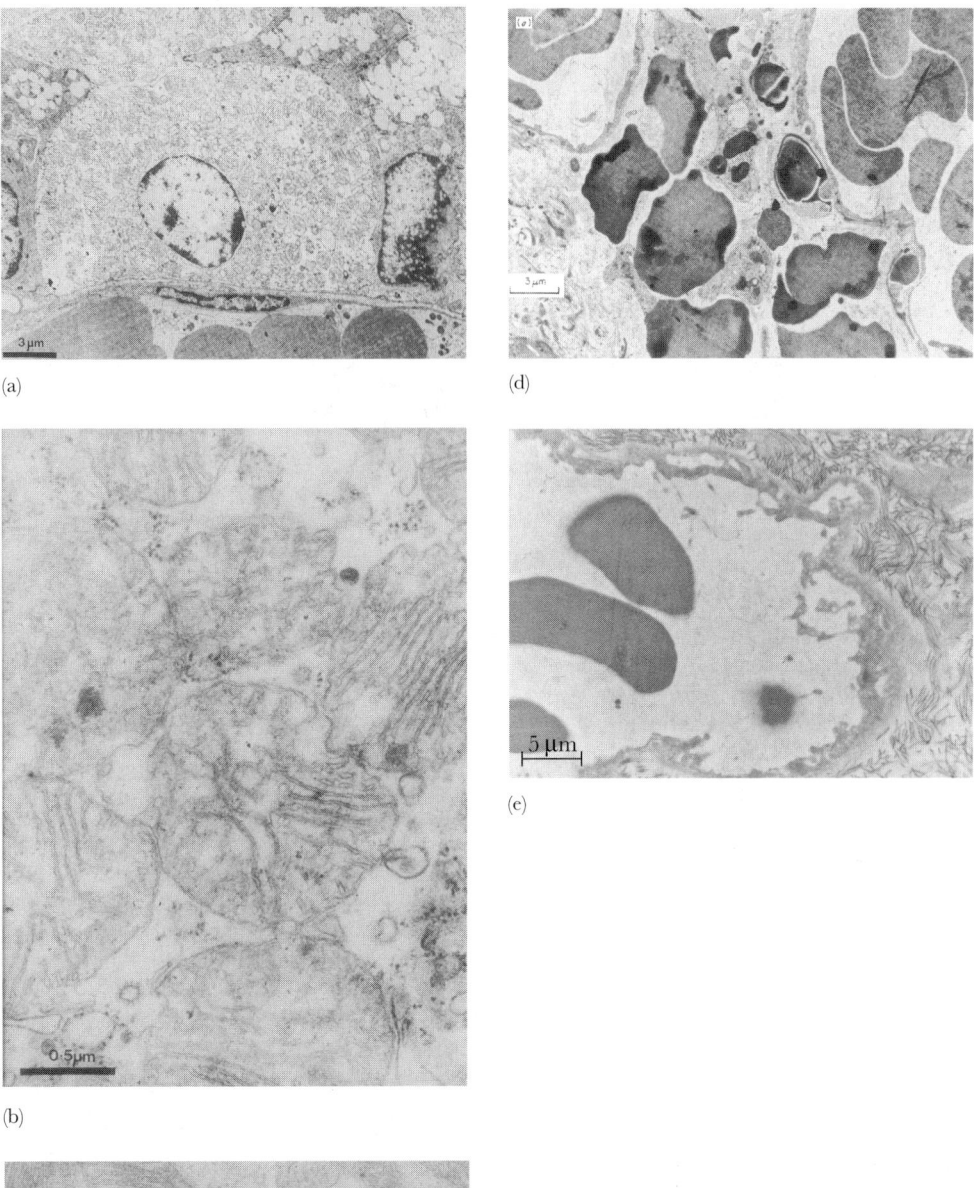

(a)

(d)

(b)

(e)

(c)

Figure 8.4 Electron micrographs of the fundic mucosa of fasted rats dosed orally for 2 hours with aspirin 200 mg/kg, showing selective damage to the acid secreting parietal cells (a), (b) and vascular damage with extravasation of red blood cells (d): (a) ×7488; (b) ×25 735. (c) Normal parietal cell mitochondria from a control rat given H₂O, for comparison with (a) and (b). (c) ×11 765. (e) Normal blood vessel for comparison with (d). (d) ×5000, (e) ×4000. (Rainsford, 1975d; 1977a; 1986; Rainsford *et al.*, 1984; 1985.)

■ 417

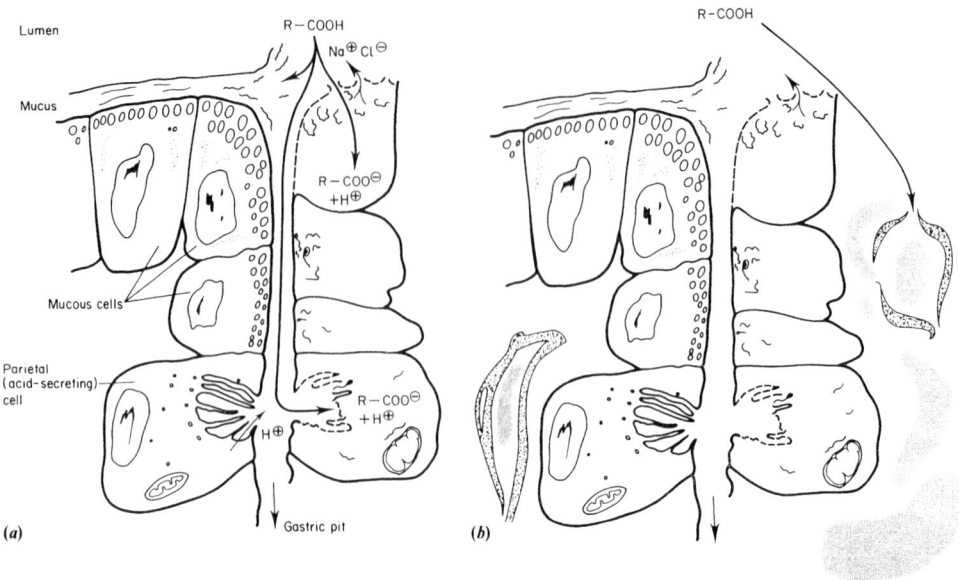

Figure 8.5 Relationship between aspirin accumulation (R—COOH) in mucous and parietal cells of the fundic mucosa and damage observed to their cells (maximal at 5–10 minutes following oral administration of the drug). Almost immediately there is selective damage to these cells. At the same time, there is disruption of endothelial cells of the microvasculature and associated loss of red blood cells into the interstitial space (see Figure 8.4). This may result from effects of the drug on endothelial cell and platelet thromboxane production, and could represent the focus for thrombohaemorrhagic phenomena and subsequent damage (e.g. from generation of free oxygen radicals from local tissue hypoxia). (From Rainsford (1984a). Reproduced with permission of the publishers, Butterworths/Heinemann).

Hingson and Ito, 1971; Pfeiffer and Weibel, 1973; Rainsford, 1975d; Baskin *et al.*, 1976; Rainsford and Brune, 1978), it appears that the permeability changes induced by aspirin or related drugs must occur in the surface membranes of the mucosal cells and not at the tight junctions between these cells (Baskin *et al.*, 1976; Frenning and Obrink, 1971). However, the bulk sloughing of superficial mucous cells that is evident during the first few minutes of salicylate administration could equally well account for these permeability changes.

Associated with these early changes in the superficial mucosa, there are also indications in experimental animal of salicylate-induced ischaemic reactions deeper within the mucosa adjacent to damaged cells (Frenning and Obrink, 1971; Rainsford, 1975d; 1983; Harding and Morris, 1976). The evidence for this comprises opening of endothelial junctions and damage to mucosal capillaries associated with loss of red blood cells into the interstitial space (Rainsford, 1975d; 1983; Robins, 1980b; Figure 8.4), and blanching of the surface epithelium (Harding and Morris, 1976). Extensive haemorrhage may infrequently occur in man (Roth, 1974; Lanas *et al.*, 1996; 2000) especially in arthritic patients (Gaucher *et al.*, 1976). Thus, Roth and Valdes-Dapena (1963a) have shown the presence of a fresh haemorrhagic extravasation in a patient who consumed only two aspirin tablets. The haemorrhage was so extensive that the patient subsequently required a subtotal gastrectomy.

Damage to the mucosal capillaries from salicylates may precipitate thrombohaemorrhagic phenomena (Selye, 1966) with hypoxia and ensuing autolytic and other degradative changes in these cells. There is both histo-enzymic (Ganter *et al.*, 1966; Jørgensen, 1976) and electron-microscopic evidence of lysosomal damage (Pfeiffer and Weibel, 1973; Rainsford, 1975b; Rainsford and Brune, 1978) in support of this suggestion. The pronounced antiplatelet effects of aspirin may account for prolongation of bleeding from injured sites (Hawkey *et al.*, 1991; Hawkey 1992; Lanas *et al.*, 1996).

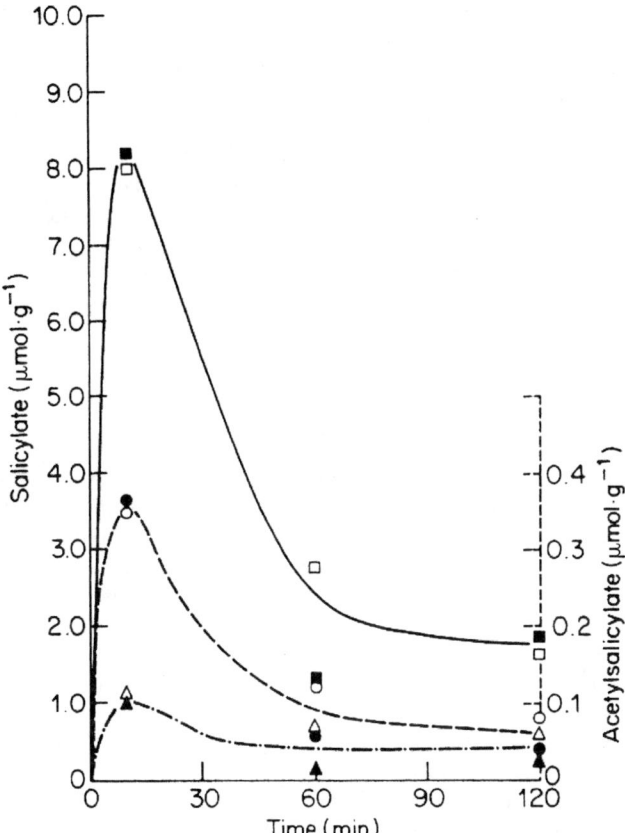

Figure 8.6 Gastric mucosal concentrations of salicylates following oral administration of aspirin 200 mg/kg to stressed (■ ● ▲) and non-stressed but fasted (□ ○ △) rats (from Rainsford, 1978b) (−) Total salicylates per g wet wt; (---) salicylates concentration per g wet wt; (−·−·−) aspirin concentration per g wet wt. (From Rainsford (1984). Reproduced with permission of the publishers, Butterworths/Heinemann.)

During the early stages of aspirin induced mucosal damage, in rats and pigs, there is selective damage to parietal cells located deep in the gastric mucosa (Pfeiffer and Weibel, 1973; Rainsford, 1975b; 1986; Rainsford and Brune, 1978; Robins, 1980b; Pfeiffer *et al.*, 1982; Rainsford and Willis, 1982; Rainsford *et al.*, 1982; 1984; Figure 8.4). The first ultrastructural signs of damage to these cells is the appearance of disrupted secretory canaliculi and dense bodies (presumably secondary lysosomes), followed 10 to 60 minutes later by the appearance of large vacuoles ('vacuolation') and disrupted mito-chondria (Pfeiffer and Weibel, 1973; Rainsford, 1975d; Rainsford and Brune, 1978; Robins, 1980b). Functional damage to parietal cell mitochondria can be seen histochemically (under the light micro-scope), where there is a reduction in the activity of the mitochondrial enzymes (e.g. succinic dehydro-genase and cytochrome oxidase), especially in the zone radiating away from a developing erosion (Ganter *et al.*, 1966; Rainsford, 1970; Jørgensen, 1976). Since relatively high concentrations of salicy-lates inhibit oxidoreductase enzymes of mitochondria *in vitro* (see Chapter 7, p. 280) this direct effect of the drugs could account for the reduction in histo-enzymic activity of these enzymes. There may also be some loss of mitochondrial enzymes by attack from proteases released from drug-damaged lyso-somes. The selective damage to large numbers of parietal cells by salicylates may cause an inherent weakening in deeper regions of the mucosa behind its normal defensive barrier (Rainsford and Brune, 1978).

Fine structural changes observed in human volunteers taking aspirin have mostly been investigated using biopsy material that only involves selection of the superficial mucosa and not the deeper regions that have been studied in laboratory animal models. Typical observations in these studies have shown that aspirin caused apical membrane rupture and loss or discharge of mucus granules (Hahn *et al.*, 1975; Baskin *et al.*, 1976; Pfeiffer *et al.*, 1982; Caselli *et al.*, 1995; 1997; McCarthy *et al.*, 1995). Although in one study the tight junctions between cells appear intact (Baskin *et al.*, 1976), in another there is clear opening of the basolateral elements of epithelial cells and pronounced oedema (McCarthy *et al.*, 1995). Scanning electron microscopy reveals a cobblestone-like appearance of the surface mucosa (Baskin *et al.*, 1976) and bag-like configurations, reflecting discharge of intracellular mucus and apical swelling (Harding and Morris, 1976; Pfeiffer *et al.*, 1982).

At 1 h after aspirin to HP-negative subjects there is evidence of capillaries having widening of endothelial cytoplasmic fenestrae and granular appearance of lumenal contents and sludging of red blood cells (McCarthy *et al.*, 1995). At 5 h there is evidence of swelling of mitochondria and their disruption with loss of cristae, although the tight junctions remain intact in the cells showing these changes. These mitochondrial changes are coincident with the known uncoupling effects of aspirin and salicylate (Jørgensen, 1976). The final structural observations of acute (single dose) effects of aspirin in superficial mucous cell injury in HP-negative subjects in the study by McCarthy *et al.* (1995) are in agreement with previous studies (Baskin *et al.*, 1976). The question of whether HP contributes to the development of gastritis associated with repeat dosing of aspirin is unresolved.

Thus, the principal cell types damaged during the early stages of gastric mucosal damage are the mucous, parietal and vascular-endothelial cells. In the case of the first two, the major factor accounting for their specific damage appears to be due to the rapid accumulation of drug anions (Martin, 1963; Rainsford and Brune, 1976; Brune *et al.*, 1977b). The passive diffusion of salicylic acids across the gastric mucosa is favoured by the acidic pH present in the stomach because, under these conditions of pH, the drug anions exist primarily in their non-ionic or protonated forms. This facilitates passage of the drugs into the lipophilic milieu of the mucosal surface membranes (Schanker *et al.*, 1957; Shore *et al.*, 1957). Having traversed the mucosal membranes, these drugs enter the cytosol and thus encounter a higher pH, dissociate protons from the carboxylic acid group, and then remain ionised and so pass more slowly into the circulation from the mucosal cells (Schanker *et al.*, 1957; Shore *et al.*, 1957; Rainsford, 1970; Rainsford and Brune, 1976). Thus the drugs become trapped as anions in the mucous and parietal cells (Figure 8.5). Autoradiographic evidence has been obtained to show the specific accumulation of orally administered ^{3}H-labelled salicylic acid into both these cell types (Brune *et al.*, 1977b), which supports this hypothesis (Martin, 1963; Rainsford and Brune, 1976). Most ulcerogenic salicylates are absorbed rapidly from the stomach and accumulate in the glandular mucosa prior to removal by the circulation (Schanker *et al.*, 1957; Shore *et al.*, 1957; Roth and Valdes-Dapena, 1963a; Brune *et al.*, 1977a; 1977b; Fromm and Kolis, 1982; Rainsford *et al.*, 1984; 1985), thus emphasising the massive transmucosal passage of these drugs.

Another cell that is affected by aspirin in rats is the mast cell. Acute oral administration of aspirin for 5 h to rats was observed by Räsänen and Taskinen (1973) to reduce the number of mast cells in the gastric mucosa and, at high doses, in the duodenum. An increase in the percentage of degranulated compared with non-degranulated mast cells 4 hours following oral administration of aspirin 100 mg/kg has also been shown by Mitra and Pal (1977), but not following the same dose of the drug given intravenously. The degranulation of mast cells by oral aspirin appears to be the cause of the increase in gastric mucosal histamine content induced by salicylates (Johnson, 1966; Johnson and Overholt, 1967).

The importance of mast cells in the development of mucosal injury is noted with other ulcerogens (Ursadi *et al.*, 1978). Its importance in aspirin injury is illustrated by the protective effects of the mast cell stabiliser, quazolast, against aspirin injury in rats (Fitzpatrick and Decktor, 1991) and the actions of a peptidoleukotriene antagonist with mast cell stabilising actions (Tabuchi and Kurebayashi, 1992).

The changes, including those in the endothelial cells following acute aspirin administration, also occur following chronic administration of the drug to rats (Rainsford *et al.*, 1982). However, the changes induced by chronic aspirin in parietal cells appear to be quite different to those seen following acute administration of this drug. In particular, a marked increase in autophagolysosomes has been observed in the cells of patients who have consumed aspirin 3 to 5 d prior to elective surgery for gastric or duodenal ulcers (Hahn *et al.*, 1975). Similar changes have also been observed in rats given twice-daily oral doses of aspirin 125 mg/kg for 14 d or pigs given a single daily dose of aspirin 100 mg/kg for 10 days (Rainsford *et al.*, 1982d). These changes indicate that there is a marked increase in the content and activity of lysosomes in parietal cells.

Robins (1980a) has claimed that aspirin and related drugs fail to induce the appearance of inflammatory cells (polymorphonuclears) 4 to 12 h after oral administration of these drugs to rats. However, aspirin has been shown to induce leucocyte infiltration including transient eosinophilia associated with oedema 2 to 3 h following oral administration of the drug to cats (Roth and Valdes-Dapena, 1963b) and up to 24 h following topical application of the drug onto the canine gastric mucosa (previously exteriorised with blood supply intact to the animals' abdominal wall; Stephens *et al.*, 1966). Transient eosinophilia has been observed in the gastric mucosa of rats given oral aspirin for 1 hr (Sugawa *et al.*, 1971), followed later at 4 h by a prolonged eosinophilia (Räsänen and Taskinen, 1973). While there may be species differences in the sequence of reactions by inflammatory cells to acute injury by aspirin, there is nonetheless infiltration of these cells into the gastric mucosa.

There is a growing body of evidence that implicates the interactions between polymorphonuclear neutrophil leucocytes (PMNs) and vascular changes in the pathogenesis of *acute* gastric mucosal injury from NSAIDs including aspirin. Early electronmicroscopic observations (Rainsford, 1977a; 1978b; Robins, 1980b) implicated the extravasation of blood cells into the interstitium in the pathogenesis of mucosal injury in laboratory animals. These changes have also been observed in humans (Frydman *et al.*, 1988; McCarthy *et al.*, 1995). The pioneering work of Wallace and colleagues (Wallace *et al.*, 1990; 1991; 1993; Asako *et al.*, 1992) in rats and of others in humans and rats (Konturek *et al.*, 1993a; Yoshida *et al.*, 1995) showed conclusively that neutrophils accumulate in the gastric mucosa and that their adhesion to endothelial cells was an important component of their accumulation and presumably their activation during the development of mucosal damage. The exact mechanisms by which adhesion of PMNs to the post-capillary venules induces endothelial cell injury and subsequent mucosal lesion development are not yet clear (Morise and Grisham, 1998) although mucosal contraction may be one factor (Anthony *et al.*, 1996). There have been indications of the variable expression of adhesion molecules on the surface of leucocytes and endothelia and implications that the pattern of expression of these adhesion molecules, which is controlled by pro-inflammatory cytokines and leukotriene B_4, may be significant in the accumulation and possible activation of leucocytes (Santucci *et al.*, 1995; Morise and Grisham, 1998; Morise *et al.*, 1998; Fiorucci *et al.*, 1999a; 1999b; 2000). Blockade of adhesion molecules with antibodies directed against endothelial cell adhesion molecules (ECAMS) such as P-selectin and ICAM-1 or E-selectin as well as CD18 or drug induced inhibition of leucocyte adhesion has been shown to attenuate the development of gastric mucosal injury in rats and rabbits (Wallace *et al.*, 1991; 1993; Low *et al.*, 1995). Variable expression of ICAM-1 or P-selectin has been shown using radio-iodinated monoclonal antibodies directed to these adhesion molecules following oral administration of 20 mg/kg of indomethacin p.o. (Morise *et al.*, 1998) and with other NSAIDs (Andrews *et al.*, 1994). These authors were not able to demonstrate extravascular leakage of leucocytes, although other studies have shown microvascular permeability changes with the same dose or lower of indomethacin (Gyömber *et al.*, 1996).

There appear to be important consequences in the development of mucosal injury from increases in tumour necrosis factor alpha (TNF-α) and leukotriene B_4 (LTB$_4$) both of which are enhanced during the development of mucosal injury following oral administration of NSAIDs (Rainsford *et al.*, 1984; Rainsford, 1986; Vaananen *et al.*, 1992; Kobayashi *et al.*, 1993; Appleyard *et al.*, 1996). An elegant model proposed by Morise and Grisham (1998) proposes that TNF-α, LTB$_4$ and oxygen radicals derived from the adhesion of leucocytes to endothelial cells elicits the intracellular production of reactive oxygen species and subsequent activation of the NFκB pathway (Figure 8.7). The subsequent release of active NFκB and consequent degradation of IκB via the 26 S-proteasome leads to the activation of genes that, among other things, control the transcription and translation of message required for the expression of ECAMS.

Earlier studies using conventional histological techniques have shown that repeated oral administration of aspirin and other salicylates induces extensive cellular infiltration into the gastric mucosa in

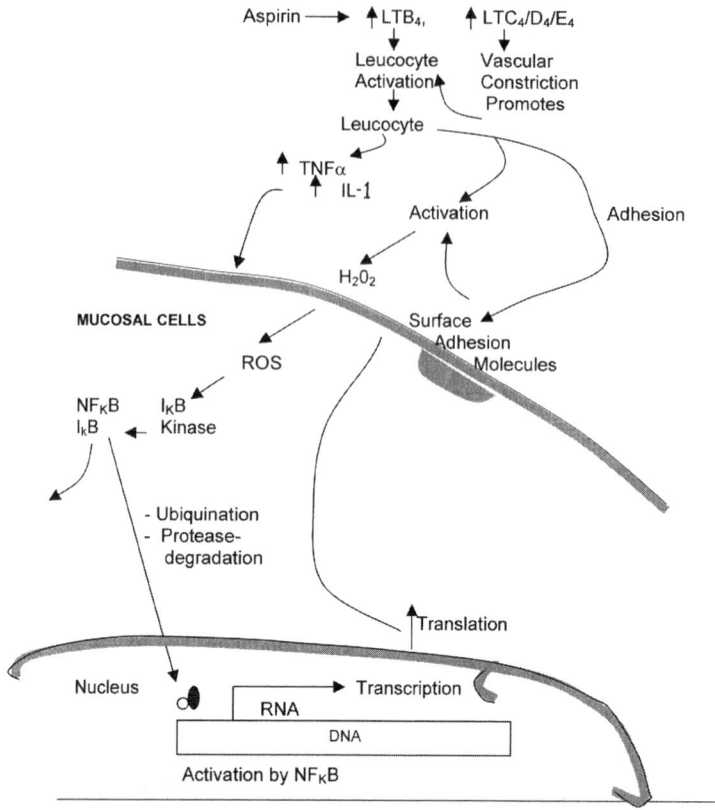

Figure 8.7 Postulated relationships between cytokine and leucocyte accumulation and activation and intracellular signalling events induced by aspirin. (Based on Morise and Grisham, 1998; Morise *et al.*, 1998; and others; see text.)

both humans (Sun *et al.*, 1974; Axelson *et al.*, 1977; Vakil *et al.*, 1977b; Marcolongo *et al.*, 1979) and laboratory animals (Hurley and Crandall, 1963; Rainsford, 1978b), this being principally around gastric lesions or ulcers. Using the myeloperoxidase (MPO) assay of neutrophil infiltration Konturek *et al.* (1994a; 1994c; 1994d) showed that repeated dosage of aspirin to rats and humans was accompanied by reduction in gastric lesions, which was paralleled by reduced MPO activity. In pigs, repeated daily dosage of aspirin 150 mg/kg per day produced no significant increase in gastric mucosal MPO in non-lesioned regions of the stomach, although there was bleeding from the GI tract and ulcers and lesions present at autopsy. There is, however, cellular infiltration around the ulcers, and this cellular infiltration may be of particular significance in chronic gastric ulceration, where a strong local immunological reactions may assist in maintaining the gastritis (Axelson *et al.*, 1977; Caselli *et al.*, 1995). The question of whether it is possible to distinguish particular histopathological changes attributable to aspirin or other NSAIDs has been examined (Stolte *et al.*, 1999). It would seem unlikely that there is a degree of specificity since the lesions from these drugs probably constitute a type of chemically-induced injury to distinguish it from that of infections, e.g. from *Helicobacter pylori* infection.

Both acute and chronic oral administration of aspirin and some other ulcerogenic salicylates cause a marked reduction in the histochemically determined content of muco-substances present on the apical surface and in superficial mucous cells (Roth and Valdes-Depena, 1963a; Menguy and Masters, 1965; Ganter *et al.*, 1966; Rainsford, 1970; Lindquist, 1971; Lev *et al.*, 1972 Jørgensen, 1976). Similar changes have been observed following treatment with other NSAIDs (Barroy *et al.*, 1970; Johansson and Lindquist, 1971). Chronic oral administration of aspirin to dogs was found by Häkkinen and co-workers (Häkkinen et al., 1968) to reduce the content of both immunologically and histochemically determined sulphated mucus in the superficial, but not the deeper, gastric pit mucous cells. It appears from these results that the effects of aspirin on mucus production which may in part be due to the inhibition of mucus synthesis (Rainsford and Smith, 1968; Rainsford, 1978f) may be manifest only in regions of the stomach exposed to high concentrations of the drug.

Repeated oral dosage of aspirin to rats (Yeomans et al., 1973) and mice (Tsodikov et al., 1979) increases the mitotic activity of the gastric mucosa, especially in regions where healing is evident (Yeomans et al., 1973), whereas repeated subcutaneous administration of aspirin is without effect (Max and Menguy, 1970). It has been suggested that the stimulation in mitotic activity by aspirin may be important in accelerating cell turnover (Max and Menguy, 1969; 1970). This may have consequences for the regenerative activity in the stomach (Lacy et al., 1993; 1996). However, it has been found that salicylates actually inhibit development of ulcers preformed by injection of acetic acid in rats and dogs (Sugawa et al., 1971; Takagi and Abe, 1974; Tanaka et al., 1982). Thus it appears that the gross demands of overall regeneration as in ulcer are blocked by salicylates, whereas the regeneration in the small area of an aspirin lesion may be enhanced.

Under in vitro culture conditions, the rat gastric mucosa appears remarkably resistant to the effects of aspirin. Focal erosions are only seen after the topical application of aspirin powder 1.0 mmol, whereas aspirin 1 to 20 mmol/kg normally causes damage in vivo (Terano et al., 1980). Several reasons could account for this difference in sensitivity. First, the neutral pH of the culture medium could be sufficient to protect the mucosa against deleterious effects of the drug, which in the in vivo studies, would obviously be established under acidic conditions. Second, it is possible that an intact blood supply may be necessary to have full expression of the ulcerogenic activity of this drug, e.g. through development of ischaemic reactions. Third, it is conceivable that the culture conditions may protect the mucosa by a superabundance of nutrients being available in the medium to overcome the metabolic effects of the drug, and by minimising the autolysis from liberation of cytotoxic degradation products.

THE INTESTINAL MUCOSA

Orally and rectally administered aspirin can cause microscopic damage to the surface epithelium of the upper or lower intestinal tract, respectively (Cassidy and Lightfoot, 1979; Djaldetti and Fishman, 1981). It is clear from early discussion (p. 415) that this is probably of only minor significance. Chronic oral administration of diflunisal 100 mg/kg per day to rats induces histopathological changes characteristic of ulcerative enteritis, but this does not appear to be evident in dogs (Stone et al., 1977). Installation into the colonic sacs of guinea pigs of aqueous suspensions of aspirin 6–30 mmol or salicylamide 6–30 mmol at acidic pH has been shown to result in extensive microscopic and ultrastructural changes in the surface epithelium (Stockinger, 1964). Much less damage has been found when the drug solutions are buffered to pH 7.4, illustrating that, as with the gastric mucosa, the damage to the intestinal mucosa is critically dependent on the pH of a drug solution. Ivey and co-workers (1979) have reported light- and electron-microscopic evidence of damage to the cells of villus tips, accompanied by an increase in the numbers of secondary lysosomes of these cells, 5 and 60 minutes after instillation of aspirin 40 mmol into the jejunal fistula of patients who had undergone a jejuno-ileal bypass operation. An increase in the numbers of secondary lysosomes, as well as multivesicular bodies, has been observed by Hahn and co-workers (1975) in the jejunal mucosa of patients who had previously taken aspirin prior to elective surgery. The increase in the numbers of secondary lysosomes seen in these acute dosage studies may be of particular relevance as a prelude to the development of a more complex pathology seen under repeated dosage conditions (Ivey et al., 1979; Lockard et al., 1980). Hahn has, however, suggested that these signs may represent repair processes (Hahn et al., 1975). Under in vitro conditions, aspirin 0.32 mmol/l causes only slight ultrastructural changes to the guinea pig ileum, comparing opening of intermediate junctions with slight enlargement of intercellular spaces (Callahan et al., 1980).

Biochemical effects

Effects on the mucosal barrier

The gastric mucosal barrier comprises: (a) the layer of mucus adherent to superficial mucous cells, as well as the mucus capable of being discharged in response to irritants from globules in the apical region surface mucous cells; (b) the membrane permeability barrier that exists between the interior of those mucosal cells bordering the gastric lumen and the milieu of the stomach contents; (c) the maintenance of adequate blood flow to ensure adequate supply of oxygen and nutrients for mucosal defences and removal of hydrogen ions and other, in some cases potentially noxious, waste products from the mucosa; and (d) the constant high rate of replacement of cells of the gastric mucosa, which ensures rapid regeneration of any damaged cells (Hollander, 1953; 1954; Lipkin, 1971; Skillman and Silen, 1972; Glass, 1976).

In terms of primary changes induced by salicylates, these involve effects on (a), (b) and (c). Event (d) is clearly a longer-term defence process that takes more time to be affected by these drugs; hence it is regarded as a secondary event, and will therefore, be considered later (pp. 434 and 442). All four components are in essence in a dynamic state in response to the irritant actions of the ulcerogenic salicylates. For instance, the discharge of mucus that occurs in response to irritation by salicylates is a dynamic protective response following damage to the mucous cell membranes by these drugs.

The layer of mucus, surface phospholipids and associated production of bicarbonate ions being constantly replaced in the normal physiological state by underlying mucous cells acts to protect the mucosa against peptic autodigestion and physical damage by drugs and noxious substances (Kodaira *et al.*, 1998) by:

1. Acting as a powerful viscous lubricant to protect the underlying mucosa from mechanical abrasion (Heatley, 1959; Davis, 1970; Allen, 1978; Allen and Garner, 1980).
2. Protecting the mucosa from direct contact by H^+ ions through physical diffusion (Heatley, 1950; Williams and Turnberg, 1980; Pfeiffer, 1981) and neutralisation of H^+ by bicarbonate ions secreted into the mucous layer from mucosal cells (Heatley, 1950; Allen and Garner, 1980; Williams and Turnberg, 1980; 1981; Pfeiffer, 1981).
3. Protection of the mucosa against pepsin autodigestion (Allen and Garner, 1980); this protective effect may occur by physical interaction of the acidic glycoproteins of mucus with pepsin (Prino *et al.*, 1972) and also by the creation of a less favourable pH gradient for peptic activity.
4. Acting as a physical barrier to undesirable macromolecules (which may include pepsin) or immunoglobulins) (Edwards, 1978).

DISRUPTION OF MUCUS AND MUCOSAL CELLS.

The disruption and sloughing away of the mucus-protective layer that occurs during the first few minutes of oral administration of aspirin and other ulcerogenic salicylates leaves the underlying mucosa exposed to irritant effects of these drugs and abrasive food particles, and also to attack by acid + pepsin (Lindquist, 1971). The mechanism of disruption of the mucus layer has been investigated by analytical ultracentrifugation using conventional velocity sedimentation analysis of both the solution and gel components of gastric mucus (Rainsford *et al.*, 1968; Rainsford, 1970; Johnson and Rainsford, 1972). It was found that aspirin and salicylic acid can induce a marked increase in the intermolecular interactions between adjacent mucus molecules of both the solution and gel-like components under the normal acidic conditions prevailing in the stomach (Rainsford *et al.*, 1968; Rainsford, 1970; Figure 8.8). However, at neutral pH (7.3) where no visible sloughing of the mucus layer occurs, the sedimentation behaviour of either porcine or human gastric mucus (Rainsford *et al.*, 1968; Rainsford, 1970) remains unchanged in the presence of neutral aspirin. Hence, it appears that the physical interactions that occur under acidic conditions arise when both the drugs and the acidic moieties (sulphate, glucuronate) of mucus glycoproteins are in their non-ionised or protonated forms. Based on this, it has been postulated (Rainsford, 1970; 1977; see Figure 8.8) that at acidic pH the non-ionised acidic groups present on both the drugs and the sulphate or glucuronic acid moieties of mucus glycoproteins interact with one another by hydrogen bonding. Under neutral conditions these groups are ionised and are mutually repulsed (Figure 8.8). Thus the interaction between the drugs and mucus molecules would appear to form the basis of clumping and subsequent sloughing away of the mucous layer such as seen when aspirin or salicylate are given orally. These would appear to be the first stages in the development of mucosal damage induced by the ulcerogenic salicylates. The disruption of the surface phospholipid protective and concomitant diminution of surface hydrophobicity is also a key initiating factor in the development of aspirin-induced mucosal damage (Lichtenberger *et al.*, 1985; 1995; 1996; Goddard *et al.*, 1987; Goddard and Lichtenberger, 1995).

CAPILLARY DAMAGE

The next site of damage to the mucosal barrier by ulcerogenic salicylates is to the capillary network in the mucosa. The damage observed ultrastructurally to the mucosal microvasculature (Figure 8.4) could be due to several drug effects. These include: opening of endothelial cell junctions following physiological changes induced by salicylates (e.g. histamine release, inhibition of vascular platelet prostaglandin production), and superoxide ($O_2^{\cdot}-$) or other radical-induced destruction of vascular or neighbouring cells following localised intestinal haemorrhage leading to a classical thrombohaemorrhagic reaction (Selye, 1966).

Salicylic and acetic acids (i.e. products of the mucosal hydrolysis of aspirin) both induce a marked increase of histamine into the gastric venous circulation (Johnson and Overholt, 1967; Johansson and Lindquist, 1971). This enhanced histamine production is due to increased synthesis of histamine from histidine, which has been shown to be initiated by aspirin (Petillo *et al.*, 1969). These effects would appear to be the basis for the observed increase in the total number of mast cells (as well as the proportion of degranulated cells) with release of amines (Daas *et al.*, 1977) that has been observed microscopically following oral aspirin (Räsänen and Taskinen, 1973).

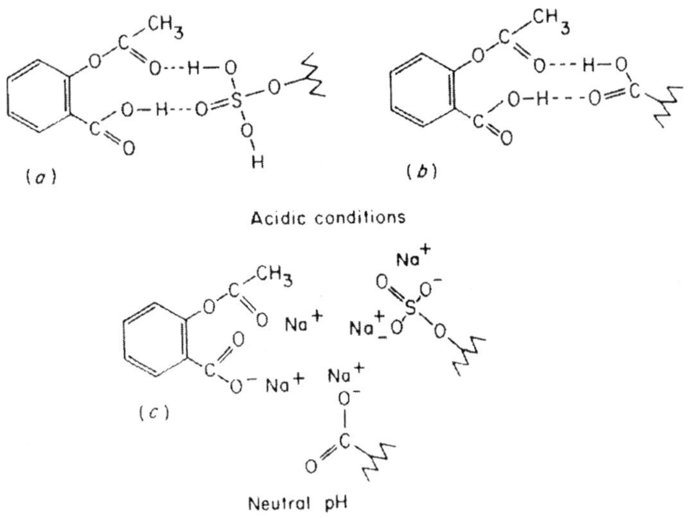

■ CHAPTER 8 ■

Figure 8.8 Postulated interaction between aspirin (or salicylic acid) and the acidic moieties (a) sulphate and (b) glucuronic acid of mucus glycoproteins under acidic but not neutral pH conditions (c). At low pH, hydrogen bonding occurs between the carbonyl and hydroxyl groups of the non-ionised (or protonated) drugs and the acidic groups on mucus glycoproteins. At neutral pH (7.3), acidic groups on both the drug and mucus glycoproteins are ionised with adjacent physiological counter-ions (e.g. Na^+, K^+ or Ca^{2+}) present to neutralise the negative charge of both the acidic groups. (Based on Rainsford, 1977.) (From Rainsford (1984). Reproduced with permission of the publishers, Butterworths/Heinemann.)

H$^+$ ION BACK-DIFFUSION, MUCOSAL BLOOD FLOW, PERMEABILITY CHANGES AND
ACID SECRETION

A variety of studies on experimentally induced gastric mucosal injury by salicylates and other agents strongly suggest that impairment of the blood flow through the stomach is a major factor accounting for mucosal injury, because of the inability to remove the hydrogen ions that back-diffuse into the mucosa (Kivilaasko and Silen, 1979). Since the actual concentration of H$^+$ in the gastric lumen is directly related to the net passage of these ions across the gastric mucosa (for subsequent removal by the mucosal circulation), then the effects of salicylates on acid secretion will determine the concentration of H$^+$ in the lumen available for this back-diffusion (Davenport, 1964; 1965; 1966; 1969; Davenport et al., 1964; Smith et al., 1971; Chvasta and Cooke, 1972; Ritchie and Fischer, 1973; Cooke and Kienzle, 1974; Lin and Warrick, 1974; Lin et al., 1975; Bruggeman et al., 1979; Kivilaasko and Silen, 1979; Flemström and Garner, 1980). Ritchie and Fischer (1973) suggested that the maintenance of a barrier against transmucosal transfer of H$^+$ ions would appear to depend on the presence of adequate mucosal production of ATP. The necessity of mucosal ATP synthesis for the functioning of the H$^+$/K$^+$-ATPase that controls H$^+$ production (Sachs et al., 1976) (via hormone influences on cyclic AMP synthesis) of H$^+$ ions shows further the importance that any drug effects on mucosal glycolytic and mitochondrial energy metabolism will have on the regulation of acid secretion (Hersey, 1974; 1977; 1980; 1981; Hersey et al., 1981; Fromm et al., 1975; Sarau et al., 1977; Thompson et al., 1977; Main and Pearce, 1978; Soll and Grossman, 1978; Hersey and Miller, 1981). Any effects of salicylates on the hormones involved in regulating both acid secretion (i.e. gastrin, acetylcholine and histamine) or calcium (Lodgson and Machen, 1981) and blood flow (i.e. histamine and PGE$_2$ and PGI$_2$) will also influence the total production, movement and toxicity of H$^+$ ions. Finally, there is the intrinsic ability of salicylic acids (when present in an acidic milieu in the stomach) to carry protons across into the mucosa, which contributes to the back-diffusion induced by the drug as well as that from breakdown in the mucosal membrane permeability barrier to H$^+$ ions. In the case of aspirin, the drug-mediated transmucosal transfer of protons could be of special significance since the liberation of salicylic and acetic acids by mucosal aspirin esterases would be expected to cause release of 2 moles of H$^+$ ions in the mucosa for each mole of aspirin absorbed (Figure 8.5).

In order to appreciate the significance of the effects of salicylates on those processes controlling H$^+$ ion dynamics and blood flow in the mucosa, it is necessary to examine in more detail their effects on the control of acid secretion and blood flow in the mammalian stomach (Figure 8.9). Also shown are the sites where aspirin is known to affect the biochemical process involved in acid secretion and/or mucosal blood flow; these processes may also be affected by other ulcerogenic salicylates.

It is important to appreciate the significance of the influences of blood flow on gastric mucosal damage induced by aspirin (Konturek et al., 1993b; 1994c). McGreevy and Moody showed that intra-arterial infusion of isoproterenol increased mucosal blood flow sufficiently to alleviate aspirin-induced damage in dogs (McGreevy and Moody, 1977). These and other studies (Moody et al., 1977) strongly suggest that increasing the mucosal blood flow has a 'cytoprotective' effect against aspirin injury. This could result from a variety of influences, in addition to the popular thesis that there is an enhancement in the removal of back-diffused H$^+$. These factors include an increase in the removal of the cytotoxic drug from the gastric mucosa, enhanced oxygenation of the tissue (restoring tissue respiration), and the removal of abnormally high concentrations of those metabolites that may have an untoward effect on mucosal metabolism. It is clear that mucosal blood flow has an important influence on the development of gastric injury by aspirin.

The question of how aspirin and related drugs influence acid secretion, blood flow and H$^+$ back-diffusion is complex, as any effect of aspirin or related drugs on one system can theoretically lead to a compensatory response on the others. Furthermore, there are indications that salicylates may influence these systems quite differently at various times, and especially when the drugs are given by different routes. Intragastric aspirin 20 mmol/l instilled into the unstimulated or prostaglandin-stimulated canine gastric pouch had no effect on gastric mucosal blood flow, but when the drug was given intravenously (3–50 mg/kg) into prostaglandin-stimulated pouch the mucosal blood flow was increased. In

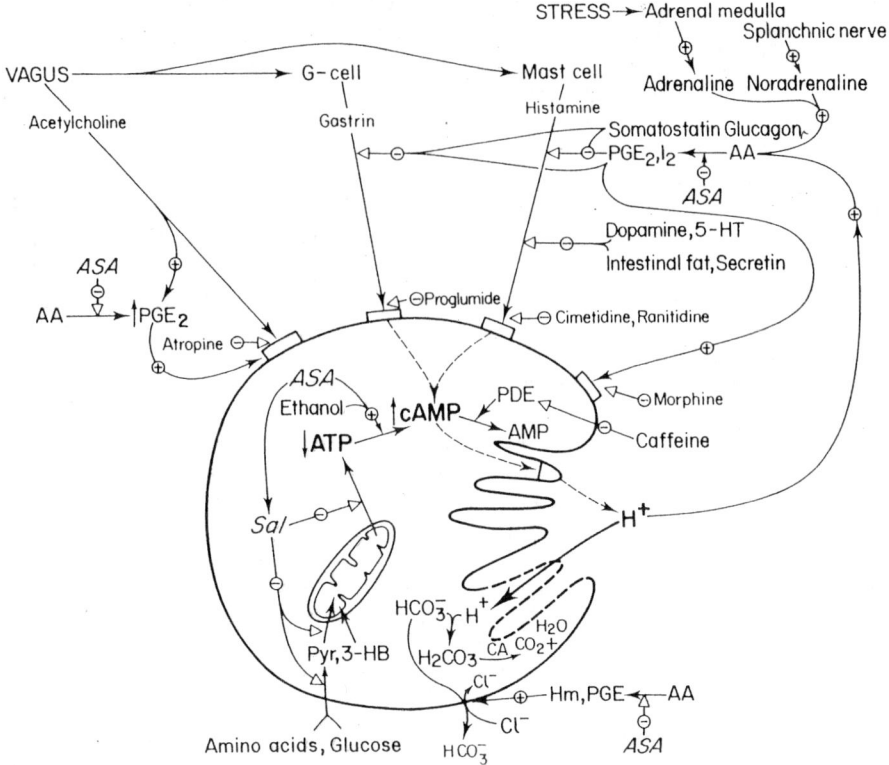

Figure 8.9 Actions of aspirin or salicylate on mediators and controls of acid secretory pathways. Positive effects are denoted ⊕, while negative effects are ⊖. Abbreviations: AA, amino acid; PDE, phosphodiesterase; 5-HT, 5-hydroxytryptamine or serontonin; Hm, histamine; 3-HB, 3-hydroxybutyrate. (From Rainsford (1984a). Reproduced with permission of the publishers, Butterworths/Heinemann.)

contrast, Gerkens *et al.* (1977) and Kauffman *et al.* (1980) found that intravenous aspirin 100 or 30 mg/kg, respectively, like indomethacin, reduced the basal or prostaglandin-stimulated blood flow in canine pouches. The difference between the results obtained by Bennett and Curwain (1977) and those of Gerkens *et al.* (1977) and Kauffman *et al.* (1980) cannot be ascribed to technique, since both Bennett and Curwain (1977) and Kaufmann *et al.* (1980) used radioactive aminopyrine to determine blood flow. While there are objections to this method (chiefly because the tracer, aminopyrine, a basic drug, can be variously distributed according to the pH status of the mucosa), the electromagnetic flow probe method used by Gerkens *et al.* (1977) appears to provide a more definitive measurement of blood flow. Thus, on balance it would appear that *parenterally* administered aspirin probably does reduce mucosal blood flow.

The differences in the results obtained from the workers who studied the effects of oral aspirin on mucosal blood flow (i.e. compare Augur, 1970; Lin and Warwick, 1974 and Cheung *et al.*, 1975 with Silen and O'Brien, 1973) would not also at first sight appear to be ascribed to technique, because all used the aminopyrine clearance method. However, in correspondence between Davenport and Monro (1973) and Silen and O'Brien (1973), one very important technicality emerged concerning the influence of pouch distension. It appears that the decrease in gastric mucosal blood flow by aspirin can be attenuated by increasing the pouch distension (this also enhances H$^+$ and Na$^+$ loss). Thus, in addition to the difficulties with the aminopyrine technique, there are problems with the maintenance of a suitable pouch pressure in these experiments, which confuses matters.

Two studies (Svanes *et al.*, 1979; McGreevy and Moody, 1981) using more direct measurements of blood flow with radioactively-labelled micropheres would appear to offer more precise evidence. McGreevy and Moody (1981) found no changes in mucosal blood flow in the unstimulated canine gastric pouches up to 10 minutes following installation of aspirin 20 mmol/l (in 140 mmol/l HCl), but by 20 minutes the blood flow had increased to damaged areas. Svanes and co-workers (1979) studied the distribution of radioactively-labelled microspheres in specific zones of the mucosa following the application of aspirin tablets 0.22 g to the antrum and fundus of anaesthetised cats, and observed a decrease in blood flow in a central zone of the fundic mucosa 4.5 h after aspirin administration, and an increase in blood flow to the muscle underlying the central and neighbouring zones of the fundic mucosa. From both these studies it appears that the increase in blood flow only occurs in the muscle layer underlying the areas damaged by aspirin, and only 20–30 min after damage to the surface mucosa has been initiated; this is well after the mucosal microcapillaries have been damaged (Rainsford, 1975c; 1983; Harding and Morris, 1976; Robins, 1980b).

A closer relationship of the microvascular injury (seen under the electron microscope) to the physio-pathological changes induced by salicylates can be seen in the studies of Bruggeman and co-workers (1979). They observed that instilling salicylic acid 20–30 mmol/l in 100–160 mmol/l HCl into canine gastric pouches induced a very rapid reduction (at 12 to 60 sec) in vascular resistance (measured by pressure-transducing methods under constant blood flow), reflecting vasodilatation (Bruggeman *et al.*, 1979). This effect was almost completely prevented by co-treatment with the H_1-receptor and H_2-receptor antagonists, mepyramine and cimetidine, suggesting that the vasodilatation was not evident with salicylic acid alone, the presence of added acid probably being linked to the back-diffusion of H^+ ions. This back-diffusion of acid was determined by the arteriovenous differences in H^+ content, and represents the first direct demonstration of this phenomena (Bruggeman *et al.*, 1979).

Bunce and co-workers (1982) showed that low concentrations of aspirin 5 mmol/l damaged the rat gastric mucosa without causing the H^+ back-diffusion seen at higher concentrations of the drug. These authors suggested that H^+ back-diffusion need not be a prerequisite for the development of damage. However, what these and other previous investigators had not been able to do was to determine the role played by the drug (anion) in buffering the H^+ – a factor that has been found appreciably to influence the H^+ status in the gastric mucosa (Rainsford *et al.*, 1982). Furthermore, there are direct drug-induced influences on H^+ production that should be considered.

In most reports, the reduction of H^+ concentration measured in vagal-denervated canine pouches following intragastric installation of salicylates was usually simply interpreted as reflecting a loss of H^+ ions (Davenport, 1964; 1965; 1966; 1969; Davenport *et al.*, 1964; Cooper *et al.*, 1966; Davidson *et al.*, 1966; Lin and Warrick, 1974), but completely ignored the effects of H^+ or HCO_3^- secretation by gastric parietal (*syn*-oxyntic) cells (Flemström and Garner, 1980). By the very nature of these experiments, it is not possible to ascertain the drug influences on acid secretion. Intragastric aspirin 20 mmol/l reduces the pentagastrin-stimulated acid output in vagally denervated canine gastric pouches (Bennett and Curwain, 1977). However, when intravenous aspirin is given with pentagastrin, the H^+ production in Heidenhain (vagally denervated) or intact canine pouches is increased (Bennett and Curwain, 1977; Gerkens *et al.*, 1977). Intraperitoneally administered aspirin 10 mg/kg reduces the basal (i.e. unstimulated) acid in ferrets with chronic fistula (Clark and Adamson, 1947). The low dose of aspirin employed in these later studies would not be sufficient to cause any increase in the permeability of the gastric mucosal barrier, so these results would appear to indicate direct inhibition of acid secretion. It is conceivable that the parenteral route is preferable for studying the influences of salicylates on gastric acid secretion *per se*, especially where there is evidence of concomitant lack of effects on the gastric permeability to H^+ ions. Thus it would appear from these collective studies of the effects of parenteral aspirin on acid secretion that aspirin inhibits *basal* secretion of acid, but also stimulates H^+ production when the parietal cell is sensitised by (penta)gastrin. Gastrin production does not appear to be affected by aspirin in normal subjects (Caldara *et al.*, 1978), so aspirin is unlikely to affect acid production via gastrin. Further evidence that aspirin inhibits basal acid secretion can be seen in electron micrographs of parietal cells from rats given oral aspirin (Rainsford *et al.*, 1982). The preponderance of tubovesicular structures and accompanying reduction in secretory canaliculi in the parietal cells of these animals are clear indications of reduced acid-secretory activity (Rainsford *et al.*, 1982). If orally administered aspirin reduces H^+ secretion, then the apparent loss of H^+ through the mucosa (observed with orally administered salicylates) could partly reflect this inhibition of acid-secretory activity.

Under conditions where the parietal cell is stimulated or sensitised (e.g. by gastrin, acetylcholine or histamine), the aspirin potentiation of acid secretion from disinhibition of the negative control by PGE_2 and PGI_2 following inhibition of PG production by aspirin (Figure 8.9) could be a major factor in ulcerogenesis. Thus, agents or conditions that stimulate the release of these secretagogues (e.g. alcohol, stress, histamine infusion and bethanechol chloride) all exacerbate mucosal damage induced by salicylates (Lin and Warwick, 1974; Rainsford, 1975b; 1977a; 1977b; 1978b; 1981a; 1981b; 1987; 1999b; Hansen *et al.*, 1978; Hansen and Grossman, 1980). The exacerbation of salicylate-induced mucosal damage by agents and conditions (e.g. physical stress, alcohol, histamine release) is due partly to stimulating release of secretagogues, which interact synergistically with aspirin to stimulate acid secretion. The release of these mediators could also have profound effects on vascular integrity, so exacerbating the direct influences of aspirin or other ulcerogenic salicylates and creating a localised tissue acidosis. Furthermore, aspirin-induced inhibition of secretion of the protective HCO_3^- anions (Garner, 1977b; Flemström and Garner, 1980) would prevent buffering of H^+ and contribute to local acidification in mucosal tissues.

The question of whether aspirin and related drugs stimulate the release of secretagogues has only received limited attention. While histamine production is stimulated by aspirin through direct enhancement of the enzyme, histidine decarboxylase, involved in histamine synthesis, (Petillo *et al.*, 1969), this drug has no influence on gastrin secretion in mice (Estler *et al.*, 1977) or man (Caldara *et al.*, 1978). Combinations of aspirin and salicylates do not affect erythrocyte acetylcholinesterase activity (Leeling *et al.*, 1978), suggesting that acetylcholine degradation may be unaffected. Nothing is known about the effects of these drugs on the enzyme choline acetyltransferase, which is involved in synthesis of acetylcholine. Since anticholinergics (e.g. atropine) and H_2-receptor antagonists (metiamide, cimetidine), as well as vagal denervation, all reduce aspirin-induced damage (in unstressed rats) and reduce acid secretion (Brodie and Chase, 1967a; 1967b; Brown *et al.*, 1978), it would appear highly probable that aspirin can indeed induce hyperproduction of both acetylcholine and histamine.

Since the actions of the secretagogues are mediators intracellularly, affecting the activities of the adenylate and guanylate cyclases (Batzri and Gardner, 1979; Becker and Ruoff, 1979; Soll and Wollin, 1979; Soumarin *et al.*, 1979; Thurston *et al.*, 1979; Wollin *et al.*, 1979), it is logical to ask what effect aspirin treatment has on the production of the respective cyclic nucleotides synthesised by these enzymes. Aspirin and salicylate have both been shown to stimulate cyclic AMP production in the rat fundic mucosa *in vitro* and *in vivo* (Mangla *et al.*, 1974a; Mitznegg *et al.*, 1977; Ueda *et al.*, 1991). Furthermore, the salicylate-induced elevation of cyclic AMP content *in vitro* is reversed by atropine and metiamide (Mitznegg *et al.*, 1977), both of which, upon acting on their respective receptors, block the activity of adenylate cyclase. Also, treatment with prostaglandin E_1 (PGE_1), which in common with PGE_2 and PGI_2 exerts powerful inhibitory effects on aspirin-induced mucosal injury (Carmichael *et al.*, 1978; Bommelaer and Guth, 1979; Miller and Tepperman, 1979; Cohen *et al.*, 1980; McGreevy and Moody, 1980; Tabata and Okabe, 1980; Hunt and Franz, 1981), has been found to inhibit the salicylate-induced rise in cyclic AMP levels (Mitznegg *et al.*, 1977). Thus the cyclic AMP regulation of acid secretion influenced by the secretagogues is one factor affected by aspirin.

The metabolism of pyruvate (arising from glucose) and of 3-hydroxybutyrate by the mitochondria is essential for: (a) providing energy in the form of ATP for the K^+/H^+-linked ATPase reaction involved in H^+ secretion from the parietal cells; (b) providing ATP to the cyclic AMP necessary to mediate responses from secretagogues; and (c) reducing equivalents (i.e. reduced pyridine nucleotides etc.) for the generation of protons from electron transport coupled to mitochondrial respiration. Aspirin and salicylates uncouple oxidative phosphorylation *in vitro* in the gastric mucosa and, at relatively high concentrations, inhibit the mitochondrial dehydrogenases (Whitehouse, 1965; Smith, 1966; Smith and Dawkins, 1971), so markedly reducing the mucosal content of ATP (Menguy *et al.*, 1973; Jørgensen *et al.*, 1976a; Spenny and Bhown, 1977b; Koelz *et al.*, 1978; Rainsford and Whitehouse, 1980a). Also, salicylates inhibit some of the enzymes involved in glucose metabolism (Whitehouse, 1965; Smith, 1966; Thompkins and Lee, 1969; Smith and Dawkins, 1971; Kravath *et al.*, 1972; Woods *et al.*, 1974), and aminotransferases (Gould *et al.*, 1966) thereby reducing the levels of intermediates entering the energy-yielding tricarboxylic acid cycle. So by directly depressing ATP production, aspirin and salicylate reduce the capacity of the mucosa to produce acid under basal conditions, even despite the strong stimulatory effects of both the secretagogues and aspirin on production of cyclic AMP. By forming cyclic AMP from ATP, this effect of secretagogues presumably also contributes further to reduction in ATP levels.

Inhibition of PGE_2/I_2 production by aspirin presumably affects secretagogue functions and regulation (Figure 8.9) by: (a) removing the inhibitory effects of these prostaglandins on the actions of histamine and gastrin; (b) stimulating the actions of acetylcholine (released on stimulation by the drug of vagal activity) on the cholinergic receptors on the parietal, G- and mast (enterochromaffin) cells; (c) attenuating or inhibiting the exchange of Cl^- ions with HCO_2^- so preventing the latter from intracellular buffering of H^+ back-diffused through the mucosa; and (d) facilitating the increased blood flow and vasodilatation (normally achieved through the synergistic interaction of PGE_2 with histamine), so reducing the effectiveness of blood flow in removing accumulated H^+.

The net effect of all these mechanisms on acid secretion will depend on the relative actions of aspirin or salicylate on the release and actions of secretagogues balanced against the availability of ATP for energy and the cyclic AMP production of the intracellular mediator. It is possible to envisage that, in the initial stages of gastric injury, there could be an overriding effect of secretagogue release in *stimulating* acid production, and only later, when sufficient salicylate has been generated to inhibit mitochondrial functions, will the effects of the secretagogues be effectively blocked.

With vagal stimulation by aspirin, stress or other stimuli, release of acetylcholine, gastrin and histamine occurs, leading to an initial stimulation in the production of H^+. However, when present in sufficiently high concentrations in the fundic cells aspirin/salicylate results in inhibition of mitochondrial and extra-mitochondrial ATP production or effects on ATPase so as to reduce the energy required for acid secretion, and in the disinhibition of acid secretion provided by the effects of aspirin/salicylate in reducing the negative control by PGE_2/I_2 that combine to yield the net reduction of acid secretion. The stimulation of acid production from cholinergics, gastrin and histamine and increased cAMP by aspirin (possibly as a result of phosphodiesterase inhibition) is potentially capable of being inhibited or modified by the reduction in ATP and PGs, the balance of these positive or negative effects yielding net increase or decrease in acid production.

The antagonists atropine (acetylcholine receptors), proglumide (gastrin receptors), cimetidine or ranitidine etc. (histamine receptors) and, H^+-pump blockers (e.g. omeprazole, pantoprazole), act on their respective receptors to block (partially or fully) the acid secretion.

One of the problems in trying to determine the biochemical effects of aspirin and salicylates (as they influence the secretagogue-mediated enhancement of acid secretion) is to understand how such an enhancement in acid production can occur despite the inhibitory effects of the drugs on production of ATP and reducing equivalents necessary for acid production. For instance, does the stimulation of histamine production increase the supply of blood-borne nutrients, which then overcomes the drug-induced depression of ATP production and supply of reducing equivalents?

Another phenomenon that is evident shortly after the oral administration of aspirin or salicylate is the rapid loss of intracellular ions (Na^+, K^+, Ca^{2+} and Cl^-) into the gastric lumen, with accompanying deterioration in the potential difference (or electrical resistance) between the gastric lumen and the mucosa and circulation (Davenport, 1964; 1965; 1966; 1969; Fishler, 1964; Cooper *et al.*, 1966; Davidson *et al.*, 1966; Smith *et al.*, 1971; Chvasta and Cooke, 1972; Chang *et al.*, 1973; Ritchie and Fischer, 1973; Cooke and Kienzle, 1974; Lin and Warwick, 1974; Lin *et al.*, 1975; Baskin *et al.*, 1976; Bolton and Cohen, 1977; Bruggeman *et al.*, 1979; Kivilaasko and Silen, 1979; Torchiana *et al.*, 1979). Lesser ulcerogenic salicylates (e.g. diflunisal) or salts of aspirin or salicylic acid buffered to pH greater than 5 cause little or no change in the mucosal permeability or transmucosal electropotential (Davenport, 1965; 1969; Bowen *et al.*, 1977; Torchiana *et al.*, 1979). Accompanying the permeability changes induced by ulcerogenic salicylates (Flemström and Marsden, 1974) is an apparent loss of hydrogen ions into the mucosa (Davenport, 1964; 1965; 1966; 1969; Davenport *et al.*, 1964; Smith *et al.*, 1971; Chvasta and Cooke, 1972; Chang *et al.*, 1973; Ritchie and Fischer, 1973; Cooke and Kienzle, 1974; Lin and Warwick, 1974; Bruggeman *et al.*, 1979; Kivilaasko, and Silen, 1979) and an increase in the permeability to high-molecular-weight substances (Beeken, 1967; Takagi and Kawashima, 1969; Flemström and Marsden, 1974; Davenport, 1976), which appears to be a prelude to bleeding (Davenport, 1965; 1969; Lin *et al.*, 1975). Ritchie and Fischer (1973) have provided evidence that the maintenance of mucosal integrity may depend on a fully operational glycolytic pathway for the subsequent production of ATP. Aspirin and salicylic acid affect mucosal ATP production (see p. 280), and may therefore, exert part of their actions (i.e. in addition to direct membrane effects) in breakdown of the mucosal permeability barrier by affecting the metabolic systems required to maintain the mucosal membranes.

That the gross permeability changes produced by oral aspirin require intimate contact of the drug is illustrated by comparing the effects of oral with parenteral aspirin (Guth and Paulsen, 1979a). When aspirin is given parenterally the formation of lesions (at 0.5 h) precedes the enhancement of ionic permeability and reduction in electropotential differences, whereas the reverse occurs when the drug is given orally (Guth and Paulsen, 1979a).

PEPSIN AND LYSOSOMAL/MUCOSAL AUTODIGESTION

Enhanced pepsin secretion has been observed 15 minutes following oral dosing of aspirin 250 mg/kg to rats (Ohe *et al.*, 1979) or on intragastric instillation of salicylic acid 20 mmol/l into the Heidenhain pouch of dogs (Johnson, 1972). The accelerated output of pepsin by these salicylates precedes the appearance of mucosal damage (Johnson, 1972; Dular and Dakshinamurti, 1979) and, in the case of aspirin, coincides with an increase in back-diffusion of acid into the mucosa (Ohe *et al.*, 1979). In view of the possibility that H^+ back-diffused into the mucosa might activate mucosal pepsinogen, Ohe and co-workers (1979; 1980a) examined the effects of aspirin on the activation of pepsinogen in rats. They found that oral administration of aspirin 250 mg/kg caused a transient increase in activated pepsinogen at 30 minutes after dosing (Johnson, 1972; Ohe *et al.*, 1979) which was prevented by the H_2-receptor antagonist cimetidine (Ohe *et al.*, 1980a). While the effects of cimetidine could be wide-ranging, it would appear that activation of mucosal pepsinogen is related to H^+ ion back-diffusion and is a factor involved in lesion development. In an earlier study, Mangla and co-workers (1974b) observed that intramuscular injection of aspirin 600 mg/kg failed to activate mucosal pepsinogen activity. This suggests that the drug has to be in contact with the luminal surface mucosa in order to elicit activation of pepsinogen. These authors also found that pepsinogen activation (as observed by Ohe and co-workers, 1979) could be due to direct effects of aspirin absorbed into the mucosa as well as to back-diffused H^+.

In addition to autodigestion by pepsin, it appears that aspirin and salicylate can induce the release of tissue-destructive hydrolytic enzymes by the direct effects of high concentrations of these drugs at low pH (as present in the stomach) in rupturing the lysosomal membrane (Lee and Spencer, 1969; Harford and Smith, 1970), and by the generalised effects of tissue injury and consequent free-radical-induced damage. Evidence for lysosomal enzyme release by these drugs has come from the histochemical and electron-microscopic observations (see p. 419), and from studies reported by Himal and co-workers (1975) on the release of these enzymes into lucite chambers containing explanted canine antral and duodenal mucosae. They found that aspirin 20 mmol/l in an acid-salt solution released acid phosphatase, β-galactosidase and β-*N*-acetylglucosaminidase (but not cathepsin D or arylsulphatase) in the antral mucosa coincident with an apparent loss of H^+ into the mucosa, and caused the appearance of mucosal lesions (Himal *et al.*, 1975). An almost opposite effect was observed in the duodenal mucosa, where, under the same conditions, no release of acid phosphatase, β-galactosidase or β-*N*-acetylglucosaminidase occurred and even the activities of cathepsin D, β-galactosidase and arylsulphatase were decreased (Himal *et al.*, 1975). Also, the H^+ loss was reduced and no mucosal damage was evident (Himal *et al.*, 1975). It appears from these studies that the duodenum has some intrinsic capacity to resist autolytic damage, and this may be related to the bicarbonate-secretory activity of this tissue, which has the effect of neutralising H^+ ions that would otherwise back-diffuse into the damaged mucosa.

PROSTAGLANDIN INHIBITION AND MUCOSAL GASTRO-PROTECTION

The role of prostaglandin synthase or cyclo-oxygenase isoform 1 (COX-1) in maintaining the physiological functions of the gastric mucosa and its inhibition by aspirin and other established NSAIDs has formed the mainstay of understanding the ulcerogenic effects of these drugs and the low GI irritancy of the newer COX-2 selective drugs (e.g. celecoxib, nimesulide, rofecoxib) that spare the COX-1 from inhibition in the stomach (Hawkey, 1999a; Whittle, 2001). While the newer COX-2 selective drugs have shown lower GI irritancy than COX-1/2 acting NSAIDs, the claim that this relates to selectivity of COX-2 inhibition may be a simplification (Rainsford, 2001). Moreover, there is a considerable body of evidence that COX-2 and its mRNA are present in the stomach (O'Neill and Ford-Hutchinson, 1993; Rainsford et al., 1995a; Romano et al., 1996; Zimmerman et al., 1998; Wallace, 1999), may be upregulated by cytokines (Mugridge et al., 1992; Romano et al., 1996) or during adaptive responses to aspirin (Davies et al., 1997; Brzozowski et al., 1998) or stress (Brzozowski et al., 2000), and are critically important in ulcer healing (Halter et al., 1997a; 1997b; Peskar and Maricic, 1998; Wallace, 1999). There are also indications that selectively inhibiting COX-1 or absence of the gene in COX-1 'knockout' mice may not relate to the development of gastric mucosal damage, while gastric mucosal damage can occur with COX-2 selective drugs without inhibition of mucosal PG production (Wallace, 1999; Wallace et al., 2000; Rainsford, 2001; Whittle, 2001). Combination of COX-1 with COX-2 selective drugs does lead to gastric erosions in rats, suggesting that inhibition of both isoforms is required for the development of mucosal damage (Wallace et al., 2000). Moreover, there are indications that COX-2 selective drugs may be associated with gastric injury (Kaplan-Machlis and Klostermeyer, 1999), so that the theory of COX-2 selectivity being related to fewer GI ulcers may not be entirely true (Rainsford, 2001).

Apart from being important in the control of vascular and acid-secretory functions, inhibition of prostaglandin biosynthesis may influence other processes underlying what has been described by Jacobson and popularised by Robert as a 'cytoprotective' effect (Pace-Asciak, 1972; Robert, 1975; 1981a; 1976; Mihas et al., 1976; Chadhury and Jacobson, 1978; Robert et al., 1979; Mózsik et al., 1987; Wilson, 1991; Peskar and Maricic, 1998). The concept of 'cyto-' or mucosal protection envisages ulcerogenic drugs or agents inducing a deficiency in endogenous protective prostaglandins. The bulk of the evidence in favour of this hypothesis has come from studies showing the very potent gastroprotective properties of E-type prostaglandins upon the development of gastric mucosal damage from aspirin (Carmichael et al., 1978; Bommelaer and Guth, 1979; Miller and Tepperman, 1979; Cohen et al., 1980; McGreevy and Moody, 1980; Tabata and Okabe, 1980; Waterbury et al., 1988; Johnston et al., 1995; Villar et al., 1998), as well as many other ulcerogenic NSAIDs or agents (Cohen et al., 1980; McGreevy and Moody, 1980; Tabata and Okabe, 1980; Rainsford et al., 1995b) although results are not always consistent (Ranta-Knuuttila et al., 2000). Unfortunately there are complications in interpreting results from studies on the protective effects of exogenous prostaglandins, because under these conditions prostaglandins may have specific pharmacological effects unrelated to their physiological functions (e.g. on acid secretion (Levine et al., 1982), haemodynamics, and metabolic rate of the mucosa (Larsen et al., 1992) and mucus effusion, see below), and also these studies do not specifically correlate the effects of drugs on endogenous prostaglandin production (Robert et al., 1979; Morris et al., 1984) in situ with either the ulcerogenic effects of the drugs or the suggested natural protective functions of the locally produced prostanoids.

The time sequence of effects of salicylates and dose-effects on gastric mucosal prostaglandin production have been extensively studied. Thus, reduction in the gastric venous content of immunoreactive prostaglandin E has been observed in dogs 15 to 30 minutes following oral administration of aspirin 20 mmol/l to dogs (Cheung et al., 1974), and in the gastro-epiploic circulation of the stomach after indomethacin administration to pigs (Rainsford, 1997). Also, inhibition of PGI_2 (prostacyclin) as well as PGE_2 production has been observed during the development of gastric lesions following acute or chronic administration of high doses of aspirin to laboratory animals (Whittle, 1978; Rainsford and Peskar, 1979; Konturek et al., 1981; Rainsford and Willis, 1982; Rainsford et al., 1984; 1985; Tepperman and Soper, 1986). The reduction of PGI_2 content is also evident after parenteral administration of the drug (Whittle, 1978; Rainsford and Peskar, 1979; Konturek *et al.*, 1981). In rats, the time course of inhibition by aspirin 200 mg/kg p.o. of mucosal PGE_2 begins at 5 minutes and is correlated with uptake of the drug into the gastric mucosa and the appearance of mucosal injury (Rainsford *et al.*, 1984; 1985). Shea-Donohue *et al.* (1990) observed that aspirin 25–150 mg/kg s.c. given to monkeys at −16 h and −45 min followed by endoscopy at 3 to 3.5 h reduced the mucosal production of prostaglandins, but doses of only 100 and 150 mg/kg s.c. produced gastric mucosal damage. Tepperman and Soper (1986) observed that oral administration of 100 mg/kg per day of aspirin resulted in gastric mucosal damage that was reduced over the period of 20 days of continuous daily dosing, and was accompanied by reduction in the mucosal concentrations of PGE_2, whereas the same dose of salicylic acid only produced mucosal damage slightly above that of placebo on day 10 but not at other times, and at no time inhibited the production of PGE_2. An important observation made by these authors was that a class of high-affinity binding sites for mucosal PGE_2 appeared on day 3, and low-affinity binding sites for this PG progressively increased over the period of 20 days after aspirin administration. Salicylic acid did not cause any changes in binding of PGE_2. This may represent an adaptive response to persistent irreversible inhibition by aspirin.

Lee *et al.* (1994) found that there was a relationship between time and the amount of the dose of aspirin given to human volunteers in causing reduction in gastric mucosal and platelet prostanoid production. Using gastric haemoglobin assays as a measure of gastric damage, they found that bleeding occurred at doses of aspirin, given for 2 d, that were higher than those required for inhibition of gastric mucosal PGs.

In human volunteers, several studies have shown that aspirin given for short periods of up to 3 days can exert long-lasting effects on the production of gastric mucosal prostaglandins and platelet TxB_2, and that the latter relates to the development of bleeding, including that from biopsy sites in the gastric mucosa (Faust *et al.*, 1990; Hawkey *et al.*, 1991; Hawkey, 1992; Lee *et al.*, 1994; Feldman *et al.*, 2000). Cryer and Feldman (1999) showed that low antithrombotic doses of aspirin (81 mg/d) or the standard therapeutic dose of 325 mg/d taken for 1.5 or 3 months resulted in a reduction in mucosal PG, reduced serum TxB_2, and endoscopically observed injury in the stomach, with there being little difference between the doses in the endoscopic scores of injury. The higher dose produced duodenal injury. These long-term studies with low doses of aspirin are interesting because they appear to show the absence of any adaptive changes to the mucosal injury, or the effects on prostaglandin production. Dermal application of 750 mg/d aspirin for 10 days (which, judged by plasma levels of salicylate, was absorbed by 4 to 8 per cent) reduced gastric and duodenal PGs and serum TxB_2 and resulted in significant gastric mucosal damage (Cryer *et al.*, 1999). Compared with other studies by the same group (see above; Feldman *et al.*, 2000), it appears that high(er) doses of aspirin may result in adaptation whereas lower doses do not, and this is paralleled by maintenance of the inhibitory effects of aspirin on prostanoid production.

In addition to aspirin irreversibly inhibiting the activity of COX-1 and converting the COX-2 form to produce 15-HETE and aspirin-triggered lipoxins (Serhan, 1999; Fiorucci *et al.*, 2003; see also Chapter 7, aspirin (but not indomethacin) has been shown to inhibit the gastric mucosal production of COX-2 in porcine fundic explants in organ culture (Rainsford *et al.*, 1995a), so that any protective effects from PGs or aspirin-triggered lipoxins derived from COX-2 would be blocked by aspirin. Thus, this inhibition of COX-2 expression by aspirin may also contribute to lesion development (Halter *et al.*, 1997a, 1997b). Alternatively, production of aspirin-triggered lipoxins may counter the leucocyte adhesion during aspirin injury (Fiorucci *et al.*, 2003).

In considering the actions of aspirin and related drugs on prostaglandin synthesis it is necessary to establish the following:

1. Does the inhibition of prostaglandin production *in vivo* correlate with the irritant or ulcerogenic actions of different salicylates?
2. What are the overall consequences of inhibiting prostaglandin production?
3. What are the respective effects on COX-1 and COX-2 in the gastric mucosa?

To answer the first question, it appears that aspirin given orally or pareterally does inhibit the mucosal production of PGE_2 and PGI_2, which is related to the degree of damage by the drug (Konturek *et al.*,

CHAPTER 8

1981; Rainsford et al., 1981; Rainsford and Willis, 1982). Inhibition of PG production in various in vitro systems has been shown to be correlated with gastric ulcerogenicity in rats, but physicochemical properties of the drugs play a major factor in the inter-relationship with inhibitory effects on COX-1 (Rainsford, 1988; Rainsford and Bjarnason, 2004; in press). Konturek and co-workers (1981) found, however, that an intragastric bolus dose of aspirin 0.6 mg/kg to rats followed by an infusion of the drug 0.4 mg/kg/h for 3 h produced a statistically significant reduction in PGE_2 and PGI_2 in the antrum and fundus, but no damage was observed to the mucosa in either of these areas. Low doses of aspirin were also found to inhibit PG production but did not cause gastric injury in rats (Ligumsky et al., 1983). Furthermore, the phenyl ester of aspirin (APE), as well as meseclazone, both of which are pro-drugs of low ulcerogenic activity (but capable of generating active PG synthesis inhibitors in vivo), failed to cause inhibition of either PGE_2 or PGI_2 production (with APE alone; Rainsford and Peskar, 1979) or a reduction in both these prostanoids (meseclazone; Rainsford et al., 1981; Rainsford and Willis, 1982) when given orally to pigs. Also, alterations in mucosal electropotential difference and ionic fluxes by aspirin do not correlate with changes in mucosal prostaglandin content (Ligumsky et al., 1982). Thus it would appear that there is not a good correlation between inhibition of PG production and the ulcerogenic activity of these drugs. However, a possible explanation of these results is that aspirin, like other NSAIDs, may prime the gastric mucosa by inhibiting mucosal prostaglandin production, which is followed by other biochemical effects, stress, H. pylori or other factors that then proceed to act to enable the expression of full ulcerogenic effects. In H. pylori-infected subjects, the recovery from inhibition of gastric PG production following 325 mg daily doses of aspirin for 72 h was slower than in non-infected subjects (Feldman et al., 2000).

Cryer et al. (1990) showed that salsalate 3.0 g/d for 7.5 d did not inhibit production of PGs in the gastric or duodenal mucosa of human volunteers and was without any injurious effects on the mucosae, while aspirin 3.8 g/d (which produced the same plasma levels of salicylates as that from salsalate) predictably inhibited mucosal PGs and caused mucosal damage. These results show that the effect of different salicylates on mucosal PG production does parallel their potential to cause mucosal damage.

While the relationship between inhibition of PG synthesis by salicylates in various cells of the gastric mucosa and the development of acute mucosal damage may not always be evident, it is worth considering further some of the processes that may be influenced by salicylate-induced reduction in PG production – an aspect that obviously invokes the second question posed above.

In addition to being involved in acid secretion and blood flow, the prostaglandins may exert their GI-protective effects (Baker et al., 1978a; 1978b; see also Wilson, 1991) by the following:

1. Increasing the physical outpouring of the protective mucus layer (Bolton et al., 1976; 1978; Bickel, 1981; Waterbury et al., 1988). Stimulation by PGE_2 of mucus synthesis is demonstrable in vitro and in vivo (Rainsford, 1980; Bickel, 1981). It is presumed that the enhanced output of mucus by prostaglandins is due to physical effects on the mucous cell surface membranes (Ligumsky et al., 1982; Allen et al., 1993).
2. Associated with the production of mucus is bicarbonate secretion, which contributes to the protective effects of PGs and is impaired by aspirin (Allen, 1978; Allen and Garner, 1980; Williams and Turnberg, 1981; Allen et al., 1993).
3. Direct membrane effects with influences on the process controlling ionic permeability (Bolton and Cohen, 1979; Dajani et al., 1978; 1979; Colton et al., 1979; Schiessel et al., 1980). Thus E-prostaglandins may tighten the mucosal barrier (Colton et al., 1979), so preventing the drop in mucosal potential difference that is caused by aspirin and related drugs (Bolton and Cohen, 1979; Colton et al., 1979; Dajani et al., 1979). The stimulation in the exchange of chloride for bicarbonate ions (Schiessel et al., 1980) may increase the intracellular pH, so resisting the pH reduction caused by H^+ back-diffusion. Shea-Donoghue et al. (1990) observed that aspirin 25–150 mg/kg inhibits K^+ uptake without affecting Na^+ transport or the Na^+/K^+ pump in isolated rat fundic and antral cells by a mechanism that is independent of the prostaglandin system (Koelz et al., 1978), but which may involve direct effects on cell membranes.
4. Effects on control of acid section (Pfeiffer and Lewandowski, 1972; Levine et al., 1982).

5. The regulation of circular muscle tone involved in gastrointestinal propulsion, bile flow and relaxation of the pyloric sphincter (Northover, 1967; Bennett et al., 1968; Bennett and Posner, 1971; Collier, 1974; Main and Whittle, 1974; Frankhuijen and Bonta, 1975; Beubler, 1978; Dajani et al., 1979; Nompleggi et al., 1980; Shea-Donohue et al., 1980; Anthony et al., 1996).

6. The regulation of prostaglandin synthesis by nitric oxide (Amin et al., 1995; Swierkosz et al., 1995; Sánchez de Miguel et al., 1999; Clancy et al., 2000) by increased production of eNOS (Fischer et al., 1999).

Regulation of these processes by prostaglandins is complex, and depends on the varying physiological responses of individual prostaglandins on different parts of the gastrointestinal tract. For example, the diarrhoaegenic effects of $PGF_{2\alpha}$ are due especially to stimulation of the duodenal circular smooth muscle (Beubler, 1978; Dajani et al., 1979). In contrast, the E-type prostaglandins cause inhibition of antral tone by either an increase or decrease in gastrointestinal motility (Bennett et al., 1968; Main and Whittle, 1974; Beubler, 1978; Nompleggi et al., 1980), depending on the species. Prostaglandin release occurs during vagal stimulation or after acetylcholine (Singh, 1980) and may be an important mediator of vagal functions (Coceani et al., 1967). PGI_2 (prostacyclin) has been reported to inhibit the basal and water-load-stimulated gastric emptying in monkeys (Shea-Donohue et al., 1980). Thus the whole aspect of the regulation of gastrointestinal motility depends on the type of prostaglandins, the species, and experimental conditions. That endogenous prostaglandins are per se important in gastric (and other physiological functions) has been illustrated by studies in rats deprived of the essential fatty acids necessary for prostaglandin production (Frankhuijen and Bonta, 1975).

Another consequence of aspirin blocking COX activities may involve the diversion of arachidonic acid to form peptido-leukotrienes and leukotriene B_4 (Rainsford, 1987; 1988; 1989a; 1999b). Inhibitors of leukotrienes prevent the development of gastric mucosal injury in rodents induced by aspirin and other NSAIDs (Rainsford, 1987; 1999b; Gyömber et al., 1996). This perturbation of leukotriene production may be responsible for altered mucosal defence leucocyte accumulation, and vascular injury observed during the early stages of mucosal damage by NSAIDs including aspirin (Rainsford, 1983; 1987; 1999b; Vananen et al., 1992; Gyömber et al., 1996).

The significance of salicylate-induced perturbations of prostaglandin-controlled gastrointestinal motility is profound, since the time of drug retention in the stomach and upper intestinal tract will influence the amount of the drug absorbed in these sites. Alterations in the prostaglandin-regulated contraction of the sphincter of Oddi (in the base of the bile duct) will influence the flow of bile, which (given other effects on the pyloric sphincter) will be available for reflux into the stomach (Main and Whittle, 1974) and consequent irritation of the mucosa therein (Guth, 1982). Also, the prostaglandin-regulated contraction of the pyloric sphincter with consequent effects upon fundic and antral tone will affect gastric distension, which in turn has influences on gastric acid secretion. There are several interesting observations regarding the effects of salicylates and prostaglandins upon gastric motility. First, there is the fundamental observation of Smith and Irving (1955) that sodium salicylate inhibits the passage of a barium meal in rats. Other investigators have observed gastric retention with aspirin and accompanying pylorospasm in man and in dogs (Schnedorf et al., 1936; Paul, 1943). Gastric mucosal distension is frequently observed in laboratory animals given salicylates, and is correlated with their irritant actions (Rainsford and Whitehouse, 1976c; Rainsford, 1981b; 1982b; 1982c).

Aspirin also enhances biliary secretion in laboratory animals (Schmidt et al., 1938; Bullock et al., 1970; Vaille et al., 1978; Ohe et al., 1980; Prigge and Gebhard, 1997) and, since bile salts (notably taurocholic and glycocholic acids) enhance the ulcerogenicity of aspirin (Semple and Russell, 1975; Guth, 1982), it is likely that enhanced bile secretion and reflux through the pylorus are major factors in the ulcerogenic actions of this and possibly other such drugs.

Aspirin 20 mmol/l has been found to inhibit the spontaneous movements of the rat ileum, which could be reversed by addition of an equimolar solution of sodium bicarbonate. This suggests that intestinal transit may also be slowed by aspirin. Some NSAIDs (e.g. fenamates) antagonise the contractions of the longitudinal ileal muscle of guinea pigs, induced by acetylcholine, histamine and transmural stimulation, which is reversed by addition of PGE_2 or $PGF_{2\alpha}$ (Famaey et al., 1977). This may be

CHAPTER 8

an *in vivo* function of aspirin that could account for the effects of this drug on intestinal motility (Barnes, 1965). Antagonism by aspirin and other NSAIDs of the prostaglandin-mediated increase in cyclic-AMP formation (Bali *et al.*, 1977) has been observed in human astrocytoma cells in culture (Ortmann and Perkins, 1977) and in canine parietal cells (Soll and Whittle, 1981). This may be an important action of these drugs in those prostaglandin-dependent functions mediated by actions on adenylate cyclase (Simon *et al.*, 1978) where a cytoprotective function has been ascribed to prostaglandins, e.g. in inhibiting or regulating acid secretion.

The mode of action of salicylates on the biosynthesis of the prostaglandins, hydroperoxyeicosate-taenoic acid and leukotrienes has already been discussed in Chapter 7. Of particular interest in the context of prostaglandin functions in the stomach are the attempts to overcome the inhibition of the so-called prostaglandin deficiency created by aspirin and other potent PG cyclo-oxygenase inhibitors by various pharmacological agents in order to protect against the ulceration induced by these drugs (see also pp. 445 *et seq.*). Sodium salicylate and diflunisal, which are relatively weak and/or reversible inhibitors of PG cyclo-oxygenase (Majerus and Stanford, 1977), have been found in rats to inhibit the ulcerogenic activity of aspirin and indomethacin, both of which are potent inhibitors of this enzyme (Ezer and Szporny, 1975; Ezer *et al.*, 1976; Correll and Jensen, 1979; Robert, 1981a). In human volunteers, diflunisal 8 mmol given 15 min prior to 2 mmol indomethacin prevented the drop in gastric mucosal potential difference and inhibition of mucosal PG production (Cohen, 1983). It is tempting to speculate that the gastroprotective functions of these salicylates may be due to their competitive combination with aspirin or indomethacin-sensitive sites on the prostaglandin cyclo-oxygenase (Humes *et al.*, 1981). However, in the case of the effects of the oral salicylate–aspirin mixture, Elliott (1979) has shown that the observed reduction in ulcerogenic activity is a consequence of an experimental artefact, since the high total dose of the drugs employed exceeds the optimal lesion-inducing dose. The protective effects of salicylate against indomethacin-induced gastric ulceration could also be due to an effect of salicylate in slowing the gastric absorption or in enhancing displacement of protein-bound indomethacin, since blood levels of the latter are significantly reduced in rats given the drug mixture as compared with indomethacin alone (Correll and Jensen, 1979). Similar pharmacokinetic interactions may arise with a paracetamol–aspirin mixture, which has been reported to exhibit lower gastric ulcerogenicity than that of aspirin and, coincidentally, a reduced inhibitory effect on PGE_2 and PGI_2 formation in rats (Van Kolfschoten *et al.*, 1980), while in human volunteers the protective effect of paracetamol against aspirin-induced gastric electropotential changes was blocked by prior treatment with indomethacin, implying that there is a PG mechanism underlying the protection by paracetamol (Stern *et al.*, 1984).

One possible way of overcoming the inhibitory effects of aspirin or related drugs on PG synthesis would be to provide a source of arachidonic acid or the fatty acid precursor of this substrate for the cyclo-oxygenase enzyme (Tarnawski *et al.*, 1985; 1989; Hawkey and Brown, 1986). Pritchard *et al.* (1988a) found that evening primrose oil (comprising 72 per cent linoleic acid and 9 per cent gamma-linoleic acid, both of which are precursors to substrates for COX activity) enhanced gastric PG production but did not protect against the gastric bleeding from aspirin 900 mg × 5 doses in human volunteers.

Mixtures of copper aspirinate or aspirin itself with sunflower oil (which is rich in linoleic acid esters) were found to have appreciably lower ulcerogenic activity than the same drugs suspended in water (Rainsford, 1982c; Alich *et al.*, 1992). However, linoleic acid could have effects on the distribution of the drug in the gastrointestinal tract such that not all the drug is absorbed from the oily mixture in the stomach at a rate comparable to that from an aqueous suspension. Also, co-oral administration of arachidonic acid with aspirin does not appear to affect the ulcerogenic activity of the latter (Rainsford and Whitehouse, 1980a), so the notion of providing a precursor substrate to overcome an aspirin-induced prostaglandin deficiency would not appear to have much to commend it. Although salicylate mixtures may present pharmacokinetic problems (noted above), it appears that the 16,16-dimethyl-prostaglandin E_2 does not affect aspirin absorption in rats (Guth and Paulsen, 1979b).

In addition to affecting COX activity in the stomach (with the consequent effects on prostaglandin-mediated processes), aspirin also inhibits the formation of membrane sterols (Russell and Goldberg, 1968; Perederii and Morozova, 1978) and lecithin (phosphatidylcholine) (Shier, 1977), the latter being the precursor phospholipid for arachidonic acid formation from lysolecithin. The consequences of this can be seen from the cycle of lecithin turnover. The inhibition of acyl-S-COA:lysolecithin acyltransferase was observed in rat gastric mucosa by Shier (1977) (with an inhibition constant $K_I = 84\,\mu mol/l$), the consequences of which would lead to: (a) a decrease in the substrate for phospholipase A_2 activity and subsequent prostaglandin formation; (b) a reduced content of one of the major stabilising constituents of membranes; and (c) an increase in the content of lysolecithin. Lysolecithin has been implicated in the pathogenesis of gastric ulceration in man (Johnson and McDermott, 1974; Bürgi et al., 1975) and in that experimentally induced in pigs (Bürgi et al., 1975; Kivilaasko et al., 1976). Furthermore, lysolecithin, a powerful detergent, destroys the canine gastric mucosa, with consequent change in the permeability to H^+ and Na^+ (Davenport, 1970), and induces powerful tonic contractions of all the principal regions of guinea pig stomach (Rogausch, 1978). Thus aspirin-induced stimulation of lysolecithin formation could have important consequences for gastric membrane stability, gastrointestinal propulsion, and the availability of membrane-derived substrates for prostaglandin formation.

Since part of the consequences of aspirin inhibiting PGI_2 (prostacyclin) in the gastric mucosa would be expected to block the preventative effects of PGI_2 on both platelet aggregation (MacIntyre et al., 1978; Czervione et al., 1979; Moncada and Vane, 1981; Hawkey, 1992) and vasoconstriction (Moncada and Vane, 1981) induced by endogenous thromboxane A_2 production (Whittle et al., 1981), it would appear that the aspirin-induced inhibition of PGI_2 production in the gastric microvasculature could be responsible for vasoconstriction. Also, the inhibition of platelet and vascular endothelial cell prostaglandin production would be expected to impair platelet aggregation and thus promote microthrombus formation. These combined events (i.e. vasoconstriction and impaired platelet aggregation) form the basis for the ischaemia and thrombohaemorrhagic phenomena evident in the early development of gastric injury by aspirin.

Platelets are likely to accumulate once the vascular tissue has been injured by aspirin. Thus Buchanan and co-workers (1981) have found that aspirin 10 mg/kg promotes platelet accumulation in the focus of an injured region of canine carotid arteries. This thrombogenic effect of aspirin persisted over 20 h, and was correlated with reduced PGI_2 production (Buchanan et al., 1981). It appears that as the irreversible effects of aspirin on platelets may persist for a long time, coincident with the inhibition by acetylation of the platelet prostaglandin COX-1 (Roth et al., 1975) and the 12 hydroperoxyeicosatetranoic acid (12-HPETE) peroxidase enzyme systems (Siegel et al., 1979), this could be a major factor in persistent bleeding following the intake of aspirin (Hawkey, 1992; 1994). The relatively low ulcerogenicity of salicylate and diflunisal compared with aspirin could be related to the weaker reversible effects of these drugs on platelet aggregation (Majerus and Stanford, 1977; Stone et al., 1977; Vargaftig, 1978). Likewise, the low ulcerogenicity of 2,3-diacetoxybenzoate (the 3-O-acetyl derivative of aspirin) could be related to the weaker activity of this compound as a platelet-aggregation inhibitor compared with aspirin (Mills et al., 1974; Whitehouse et al., 1977a). Thus there appears to be a relationship between the potency of salicylates as inhibitors of platelet aggregation and their ulcerogenic activity.

It must be noted, however, that there are other aspirin-sensitive responses to platelet aggregation (e.g. mediated by bradykinin; Vargaftig et al., 1981) and components of the coagulation cascade that combine to delay blood clotting (Gast, 1964). The non-platelet events in blood coagulation that are affected by aspirin or salicylate may include a reduction in circulating prothrombin (Meyer and Howard, 1943) and factor VII (Quick and Clesceri, 1960), though this may only be manifest following ingestion of large quantities of these drugs (Smith, 1966).

In contrast to the effects of aspirin and related drugs in causing inhibition of gastroduodenal prostaglandin production, sulphasalazine appears to increase production of colonic prostaglandins by inhibiting their breakdown (Peskar, 1985).

OXYRADICALS AND NITRIC OXIDE

Oxygen free-derived radicals (e.g. superoxide, $O_2^{\cdot-}$ and the hydroxyl radical $OH^{\cdot}$ generated during microhaemorrhage (Del Maestro et al., 1980; Granger et al., 1981; Greenwald, 1981; Tien et al., 1981), or the (as yet unidentified) $[O]^{\cdot}_x$ species produced from drug-induced perturbation of arachidonate metabolism (see p. 244), would be expected to induce localised tissue injury, for example by peroxidation of lipids and macromolecules in the focus of a vascular injury (Del Maestro *et al.*, 1980; Tien *et al.*, 1981). Studies by Granger and co-workers (1981) have notably provided evidence of superoxide production, independent of effects on histamine production or the inhibition of arachidonate metabolism, during the ischaemia induced by arterial haemorrhage in the ileal region. Similarly reperfusion injury in the stomach may be an oxyradical-mediated process involving leucocyte activation (Wada *et al.*, 1996). Thus the aspirin-induced microhaemorrhage observed from the gastric microvasculature is likely to be caused by direct effects on the endothelium (e.g. from accelerated histamine production, inhibition of platelet aggregation as well as membrane damage), so leading to production of tissue destructive free radicals.

Some evidence for the importance of free radicals in the development of aspirin-induced mucosal injury has come from studies using free radical scavengers, such as the antioxidants, butylated hydroxytoluene (BHT), butylated hydroxyanisole (BHA), and phenolic hydroxyl radical scavengers (e.g. MK-447 = 2-aminomethyl-4-*tert*-butyl-6-iodophenol; Rainsford, 1979; 1984; Van Kolfschoten *et al.*, 1983; Shriver *et al.*, 1980), dosmalfate (Le Kerneau *et al.*, 2000), vitamin E (Sugimoto *et al.*, 2000) and vitamin C (Pohle *et al.*, 2001). Co-administration of these agents with aspirin results in markedly less gastric mucosal damage compared with that induced by aspirin alone (Rainsford, 1979; 1984; Shriver *et al.*, 1980; Van Kolfschoten *et al.*, 1983). These findings imply that oxygen radicals are generated in the development of aspirin-induced damage to the mucous, parietal and endothelial cells, but it is not possible to discriminate whether these radicals arise in the microvascular component of tissue injury alone.

There do not appear to be any effects of aspirin on mucosal glutathione production (Ghanayem *et al.*, 1984) although other antioxidant defences may be affected by this drug (Mitobe *et al.*, 2000). Oxyradicals have also been implicated in the development of gastric mucosal damage from other NSAIDs (Vananen *et al.*, 1991).

The vascular component of aspirin-induced mucosal injury may also involve development of increased capillary fragility, perhaps after long-term administration of the drug (Frick, 1956). Frick (1956) has shown that aspirin induced a positive Rumpel–Meede test (which measures capillary fragility) in three patients who consumed the drug. The exact significance of this in relation to the drug-induced effects on microvascular injury is difficult to determine, but it is clearly a factor requiring further investigation.

Nitric oxide (NO) has been shown to be important in mucosal protection against indomethacin induced injury (Whittle *et al.*, 1990; 1995; Lamarque and Whittle, 1995; Mourad *et al.*, 2000). By employing arginine analogues that either inhibit or stimulate nitric oxide production it has been possible to show that NO is involved in acute stages of aspirin-induced gastric damage (Takeuchi *et al.*, 1998). The released NO may combine with oxyradicals (Morikawa *et al.*, 1995) to produce the very potent cell destructive peroxynitrite radical. Alternatively, when there is little production of oxyradicals that NO may be protective by its vascular effects (Whittle *et al.*, 1990; Moncada *et al.*, 1991) as well as stimulating bicarbonate secretion (Takeuchi *et al.*, 1994). The possible importance of NO in protecting the gastric mucosa from aspirin is shown by the low irritancy of the NO-releasing aspirin analogue NCX-4016 compared with that of aspirin (Tashima *et al.*, 2000).

Reduced production of mucus

Aspirin and other ulcerogenic drugs or agents all reduce the amount of the protective mucus layer (Roth, 1974; Rainsford, 1978f; 1982b; Azumi *et al.*, 1980; Moriga *et al.*, 1980). A reduction in the histochemically observed content of mucus following acute or chronic oral treatment with aspirin has already been mentioned. A quantitative reduction in the content of hexoses, hexosamines, sialic acid and sulphate moieties of mucus has also been observed following single or repeated doses of aspirin (Azumi *et al.*, 1980; Moriga *et al.*, 1980; Rainsford, 1982b).

The evidence showing that there is a relationship between the reduction in mucus and the development of gastric mucosal ulceration by aspirin and other NSAIDs has been reviewed (Rainsford, 1982b). This evidence can be summarised as follows:

1. Only ulcerogenic NSAIDs cause a reduction in mucus biosynthesis *in vivo*; NSAID or analgesic drugs with low or negligible ulcerogenic activity fail to show this effect (Rainsford, 1978f; 1982b).
2. The inhibition of mucus synthesis by aspirin *in vivo* occurs specifically at times when relatively high concentrations of the drug are present in the mucosa (Rainsford, 1978f).
3. Drugs or agents that increase mucus content or synthesis, e.g. carbenoxolone (Domsche *et al.*, 1972; Shillingford *et al.*, 1973), zolimidine (Murmann *et al.*, 1974a, 1974b), compound KL-11 (ε-*p*-chlorocarbonenzoxy-L-lysine-*O*-methyl-HCl) (Ezer and Szprony, 1970) and proglumide (Corinaldesi *et al.*, 1980), all markedly reduce the development of gastric mucosal erosions induced by aspirin or salicylate (Ezer and Szprony, 1970; Okabe *et al.*, 1976a; Corinaldesi *et al.*, 1980; Umetsu *et al.*, 1980).

The consequences of reducing the mucus protective layer have been detailed previously on p. 424, where it was pointed out that the reduction in mucus content arises from physical disruption of mucus and mucous globules. However, it has been shown that the longer-term effects following acute as well as chronic oral administration of aspirin and related drugs are principally due to direct inhibitory action of these drugs on the enzymes involved in the synthesis of mucus glycoproteins (Rainsford, 1978d). Aspirin or salicylate inhibit the following mucus glycoprotein synthesising enzymes: acetyl-ScoA synthetase (EC 6.2.1.1), L-glutamine-D-fructose-6-phosphate aminotransferase (EC 5.3.1.19), UDP-*N*-acetylglucosamine 4-epimeraes (EC 5.1.3.7), UDP-glucose-dehydrogenase (EC 1.1.1.22), UDP-glucuronosyltransferase (EC 2.4.1.17), and the transferase enzymes involved in the incorporation of *N*-acetylglucosamine, glucosamine, other hexoses and hexosamines and 'active' sulphate (3′-phosphoadenosyl-5′-phosphosulphate or PAPS) into mucosal glycoproteins (Rainsford, 1978f; 1982b). The reduction in mucus biosynthesis can also result from the *in vivo* reduction in mucosal ATP (Menguy *et al.*, 1973; Jørgensen *et al.*, 1976a, 1976b; Ohe *et al.*, 1980a; Rainsford and Whitehouse, 1980) necessary for the ATP-activation of sugars and sulphate moieties in mucus synthesis (Rainsford, 1982b). The reduction in mucus glycoprotein synthesis does not involve effects on the biosynthesis of the peptide core (Rainsford, 1978f). Intracellular release of autolytic lysosomal enzymes by salicylates also appears likely to induce degradation of mucus glycoproteins.

Cyclic AMP has been suggested as an important intracellular regulator of mucus biosynthesis, especially in view of the effect of some anti-ulcer drugs (e.g. carbenoxolone) as well as E-type prostaglandins in increasing the cellular content of cyclic AMP (Guslandi, 1980). Also, an increase in levels of cyclic AMP has been found to enhance the synthesis of sulphated mucopolysaccharides in fibroblasts in culture (Goggins *et al.*, 1972). The aspirin and salicylate-induced reduction in ATP content *in vivo* may lead to lower cyclic AMP levels, but paradoxically aspirin actually increases the total mucosal content of this second messenger (Mangla *et al.*, 1974b). Thus the reduction in mucus biosynthesis by salicylates cannot be simply reconciled with effects on cyclic AMP production, unless there are differences in the levels of this second messenger between mucous cells and the total cyclic AMP content of all mucosal cells as measured.

In view of the importance of stress states on the aetiology of aspirin-associated peptic ulceration (p. 403) and the fact that both stress and salicylate stimulate the production of ulcerogenic corticosteroids (Walker and Smith, 1979), it is conceivable that the accelerated release of these hormones could also reduce the content of mucus glycoproteins (Rainsford, 1982b). Thus in addition to direct effects of salicylates on enzymes controlling the synthesis and degradation of mucus glycoproteins, these drugs could have indirect effects mediated through adrenocortical stimulation (Rainsford, 1982b).

MUCOSAL DETOXIFICATION

Alterations in the formation of inactive (i.e. non-gastrotoxic) metabolites of salicylates may occur from chronic administration of salicylates or as a result of the effects of agents influencing these detoxifying reactions. Hietanen (1975) found that gastric and duodenal UDP-glucuronosyltransferase activity (which is responsible for the formation of glucuronic acid conjugates of salicylates) was reduced at 12 hours following a single oral dose of salicylic acid 138–276 mg/kg. This enzymic activity was restored to normal levels after 3 to 14 days repeated administration of the drug. Coincidentally, the gastric

CHAPTER 8

ulcerogenic activity of salicylic acid declined after the first day, suggesting that this detoxification developed upon repeated administration of the drug (Hietanen, 1975). Hepatic detoxification reactions were found by Hietanen to be quite stable, so that the adaptive detoxification reactions are specific to the gastroduodenal mucosae (Hietanen, 1975). The changes in UDP-glucuronosyltransferase activity could also be important in relation to the inhibitory effects of salicylate (or aspirin) on the synthesis of mucus glycoproteins. Thus, if both detoxification and glucuronic acid incorporation require the same enzyme, there could be preferential utilisation of the substrate for detoxification reactions when aspirin, salicylate or similar drugs are administered.

CELL TURNOVER AND MUCOSAL PROTEIN AND NUCLEIC ACID SYNTHESIS

The proliferative activity of the superficial gastric mucosa increases, especially in previously lesioned areas, after repeated oral or parenteral administration of aspirin (Max and Menguy, 1970; Yeomans et al., 1971; Tsodikov et al., 1979; Takeuchi and Johnson, 1982; Lacy et al., 1993; 1996; Ohning and Guth, 1995). The ability of aspirin to delay healing of severely damaged gastric tissue (Sugawa et al., 1971; Takagi and Abe, 1974) may be related to adrenocortical stimulation by the drug (Walker and Smith, 1979) or effects on growth factors (Stachura et al., 1994; 1996; Bamba et al., 1998) combined with the stress of the injury. However, it is possible that the direct inhibitory effects of aspirin or salicylate on protein and nucleic acid biosynthetic reactions that have been observed in vitro and in vivo in some systems (Dawkins et al., 1966; Rainsford and Smith, 1968; Janakidevi and Smith, 1970a; 1970b; 1970c; Burleigh and Smith, 1971; Dawkins et al., 1971; Rubinstein et al., 1976), may be relevant in this situation, i.e. where there are considerable demands upon the cellular regenerative capacity.

It has been claimed by Pichl and co-workers (1978) that gross protein biosynthesis in the gastric mucosa is inhibited in vivo by subcutaneous administration of aspirin 90–900 mg/kg (as the lysine salt) to rats. However, these authors made the mistake of concluding that they had observed an inhibitory effect of the drug in vivo when in fact they had observed an in vitro reduction in protein synthesis in tissues from rats dosed with aspirin. While aspirin and other salicylates do inhibit protein synthesis in the gastric mucosa of rats and pigs incubated in vitro (Rainsford and Smith, 1968; Rainsford, 1970; Pichl et al., 1978), and in the rat mucosa obtained following oral administration of the drug, no such effects have been observed on gross protein synthesis in vivo, despite exhaustive efforts to establish this point (Rainsford, 1970). It is still conceivable that there are local inhibitory effects of the salicylates on protein synthesis in the focus of regenerating tissue, but this aspect has yet to be established.

There is the possibility that a reduction of protein synthesis in the regenerating gastric mucosa could also reflect inhibition of messenger, transfer or ribosomal RNA synthesis, especially since in some other systems this site appears to be more sensitive to the inhibitory effects of salicylate (Janakidevi and Smith, 1970a; 1970b; 1970c; Rubinstein et al., 1976). Bali and co-workers (1970) found that a high oral dose of aspirin (800 mg/kg) stimulated the biosynthesis of RNA in the mouse gastric mucosa in vivo of 4-S RNA species. They concluded that the inhibition of protein synthesis was due to the reduction in the 'metabolic activity' of 28-S RNA. It is apparent that this and other interesting aspects of protein and RNA synthesis, especially in regenerating tissue, require further detailed investigation.

There is potential for drug-induced alterations in the production of cell-protective heat shock proteins (HSP) which in turn could influence all turnover. Thus, Jin et al. (1999) observed that HSP72 but not HSP6 or HSP90 was increased in the gastric mucosa in rats given 100 mg/kg/d aspirin for 20 days followed by a high dose of 250 mg/kg. This coincided with reduced PGE concentrations in the mucosa. The results may have relevance to adaptive changes caused by aspirin after long-term dosage.

IMMUNOLOGICAL CHANGES

At this level it is only possible to speculate about the possible non-Helicobacter pylori; (HP)-dependent immunological changes that have been observed histologically in man and laboratory animals following long-term oral administration of ulcerogenic salicylates (see pp. 391 and 416). Thus the extensive polymorph and round-cell infiltration that is so evident could be considered as the basis for a developing chronic disease pathology, together with any long-term effects of these drugs on cellular regeneration and turnover as previously mentioned. The unique immunological status of the gut in protecting the host from insult by a wide variety of environmental antigens could be compromised by long-term administration of ulcerogenic salicylates. Furthermore, the development in gastric ulcer disease of antibodies to parietal cell constituents (Fisher and Taylor, 1971) illustrates an autoimmune insult that could develop as a consequence of aspirin ulcerogenesis.

Mechanisms of gastric mucosal damage by salicylates

From the pathology of salicylate-induced gastric damage, it is apparent that salicylates have many effects that lead to the development of mucosal damage. The principal events and their time sequence are outlined in Table 8.17. There are several fundamental principles regarding the actions of the salicylates, including the following.

TABLE 8.17

Summary of principal physiopathological events that may occur in initiating gastric mucosal damage by aspirin. (Based on Rainsford, 1984a; 2001.)

Primary
1. Sloughing of mucus layer and mucous cells with loss of bicarbonate and disruption of protective phospholipid layer → ↓ permeability barrier
2. Selective damage to parietal cells
3. Tissue destruction from ↑O_2^-, O_x^-, OH·, ·NO + OH· → peroxynitrite and ↑ release of lysosomal enzymes
4. Damage to endothelial cells and impaired platelet aggregation, ↓ NO production → thrombohaemorrhagia
5. Histamine release from mast cells → vasoconstriction, ↑ acid production
6. Inhibition of prostaglandin E_2/I_2 from COX-1 inhibition → ↑ acid secretion, ↓ blood flow and ↓ cytoprotection
7. Increased leukotriene production (from diversion of arachidonate metabolism through the lipoxygenase pathway following COX-1/COX-2 inhibition, and leucocyte activation) → vasoconstriction (from peptido LTs), further recruitment and activation of leucocytes (by LTB_4) and ↓ NO production
8. H^+ back-diffusion → pepsinogen-autodigestion
9. ↑ cAMP production → ↑ H^+ production, then ↓ H^+ from ↓ ATP
10. Acetylation of biomolecules → altered metabolic regulation
11. Intracellular activation of signal transduction pathways
12. Reduction in sulphydryl content → altered redox state and activity of proteases

Secondary
1. ↓ Production mucus from ↓ ATP and ↓ activities of glycoprotein synthesis enzymes, ↓ PGs and ↑ cAMP; caspase induction from ↓ ATP → apoptosis
2. ↑ TNFα and IL-1 → mucosal inflammatory reactions
3. ↓ Repair regeneration of lesions and ulcers related to COX-2 inhibition, affects on growth factors and ↓ protein/nucleic acid biosynthesis
4. Exacerbation of injury by inflammatory reactions from concurrent *Helicobacter pylori* infection → gastritis
5. Stimulation of afferent (capsaicin-sensitive) sympathetic nerves → ↑ CGRP, substance P and other neurokinins → CNS responses (e.g. altered vagal, acid, vascular regulation)

Tertiary
1. Immunoreaction from gastric (leucocyte infiltration and production of parietal cell antibodies) → tissue self-destruction, gastritis and chronic ulceration
2. Variable adaptive changes from alterations in production of growth factors and apoptosis
3. Enhanced GI motility (amplitude) → alterations in gastric emptying and GI transit of food and drugs

CHAPTER 8

1. Salicylates exert systemic as well as local actions on the gastric mucosa. Local actions do, however, tend to predominate with oral administration of the drugs, simply because the gastric mucosa is presented with high concentrations of the drugs (Figures 8.5 and 8.6).
2. A corollary of (1) is that the drugs should be present in the gastric mucosal tissue to exert effects. There appear to be subtle effects mediated by these drugs on other organ systems e.g. vagal-parasympathetic/sympathetic nerve stimulation, which may contribute to the overall development of mucosal damage.
3. There are both concentration- and time-dependence of the effects of the salicylates on those systems involved in the development of mucosal damage. There is a strong element of interplay between the effects of the drugs on these systems, which contributes to the mechanism of gastric mucosal damage. These individual biochemical, cellular and physiological actions of salicylates are divided somewhat arbitrarily into categories of primary (or initial), secondary and tertiary events to indicate the likely time sequence of events in the development of mucosal damage (Table 8.17). There is, inevitably, some overlap between these events.

Summary

Salicylates vary in their propensity to induce gastric ulceration and haemorrhage in humans and in laboratory animals. Thus aspirin is amongst the most ulcerogenic of these drugs, while certain of its alkyl or aryl esters, salicylate (as the sodium salt or the free acid) and diflunisal are notably less irritant to the gastric mucosa. The gastric ulcerogenic activity of aspirin is enhanced by: (a) formulation as a tablet; (b) concurrent high alcohol intake; (c) vitamin C deficiency; (d) the stress resulting from inflammatory diseases, and exposure to physical and/or psychological stress; (e) gastric acidity; (f) certain bile salts refluxed into the stomach; (g) slowing of gastric emptying; and (h) the absence of certain nutrients necessary to protect the stomach against drug insult (e.g. glucose, amino acids). Blood group status, age and possibly sex also play an important part in predisposing individuals to the effects of aspirin, and the elderly arthritic female may be especially susceptible. Little is known about the factors underlying the development of intestinal damage, which may occur with some salicylates. While efflux of acid from the stomach, the presence of bile and the propensity of some salicylates (e.g. diflunisal) to undergo enterohepatic recirculation are obvious factors, coupled with the intrinsic irritancy of some of these salicylic acids, there are many other unresolved aspects of damage induced by salicylates in this region of the gastrointestinal tract.

The mechanism of gastric mucosal damage induced by aspirin and/or salicylate is summarised in Figures 8.5 to 8.8 and Table 8.17. It is clear that the mechanism(s) of gastric injury have essentially a multifactorial basis due to systemic as well as local actions of aspirin/salicylate on a number of biochemical cellular and integrated physiological systems. The differences inherent in the gastric ulcerogenicity of individual salicylates depend on the basic chemical structure of these compounds. Their relative actions upon these individual systems determine the profile of responses underlying the ulcerogenicity (or lack of it) of individual drugs.

Protection against aspirin-induced GI injury

Antacids and anti-acid secretory agents

A considerable number of studies in humans with other conventional NSAIDs have confirmed the demonstration of protection by control of acid secretion by antisecretory drugs (Huang and Hunt, 1996) thus giving, perforce, the generalisation that gastroduodenal injury by these drugs, like that from aspirin, is a gastric acid-related phenomena.

Antacids have been shown to have protective effects against acute and subacute gastric mucosal damage observed endoscopically in humans (Konturek *et al.*, 1991; Brzozowski *et al.*, 1993) and *in vitro* in amphibia (Müller *et al.*, 1985; Rowe *et al.*, 1985b). They are probably useful in mild to moderate injury, but have never achieved quite the recognition of H_2 receptor antagonists and H^+-pump blockers as preventative agents in aspirin- or NSAID-induced GI injury. Yet they are widely used especially as non-prescription agents by the public. In many cases their use is initiated to relieve symptomatic gastric side effects such as epigastric pain, heartburn and dyspepsia.

Evidence from studies in laboratory animal models indicates that the aluminium-antacids act as mild irritants by a non-prostaglandin, non-acid dependent system (Konturek *et al.*, 1991). The mechanism of mucosal protection by antacids against aspirin injury in the stomach appears to be more than simply neutralising the production of acid or gastric acidity, though these may play some role. Szelenyi (1984) showed that gastric mucosal PGE_2 is increased by pretreatment with $Al(OH)_3$ but not $CaCO_3$ antacids. Studies by Konturek *et al.* (1991) showed that acidified Maalox 70™ or $Al(OH)_3$ were more potent than their non-acidified forms in preventing acute gastric lesions induced in rats by acidified aspirin, as well as by ethanol, taurocholate and stress. These authors suggested that the action of these aluminium-containing antacids was probably due to their mild irritancy as shown by superficial exfoliation of the superficial mucosal cell layer in the stomach of these animals. There could also be some acceleration of cell death of cells in this layer that are already approaching the end of their life ('effete' cells). Konturek *et al.* (1991) claimed that protection by these antacids did not involve prostaglandins, since co-administered indomethacin did not affect their protection. However this is not definite proof, since reduction in gastric PG production by indomethacin is not the only action of this drug. They showed that reduction or elimination of acid production by administration of ranitidine did not affect the actions of the antacids, suggesting that their actions in mucosal protection do not depend on the effects on gastric acid secretion. Similar results were obtained by the same group with another aluminium-containing antacid, Supralox™ (Brzozowski *et al.*, 1993). These authors also found that the reduction in mucosal blood flow caused by acidified aspirin in rats was partly prevented by Supralox™. Several studies have also shown that these and other antacids may 'stabilise' or ensure longevity of growth factors in the stomach, so aiding (like sucralfate) the natural mucosal regeneration in lesioned/ulcerated areas by these endogenous repair agents (Konturek, 1993; Brzozowski *et al.*, 1994). Thus there appear to be multiple mechanisms of action of antacids in protection against aspirin or other NSAIDs. There may also be considerable variations between these agents (e.g. aluminium- compared with magnesium oxide-containing preparations) in their efficacy and modes of action.

The H_2-receptor antagonists were the first agents having single modes of action in suppressing or blocking acid production to be found to reduce aspirin-induced gastroduodenal injury in humans (Ivey *et al.*, 1975; MacKercher *et al.*, 1977; Pritchard *et al.*, 1988b) and laboratory animals (Brown *et al.*, 1975; 1978; Carmichael *et al.*, 1978; Konturek *et al.*, 1991; Ichikawa *et al.*, 1994). Studies have been performed showing protection against aspirin-induced gastroduodenal injury observed endoscopically and from intragastric blood loss assays in humans from H_2-receptor blockers (Pritchard *et al.*, 1988b). H_2-receptor antagonists and the H-pump blockers, which are probably even more effective protectants with specific actions, are considered to be proof of the concept that acid plays a major role in aspirin-induced upper GI injury (Yeomans *et al.*, 1992).

In endoscopy studies in male human volunteers, the H_2-receptor antagonist, ebrotidine (800 mg) given with aspirin (500 mg) twice daily for 3 days produced a mean number of lesions of 2.0 ± 0.3 (mean $\pm$ SEM) compared with that of aspirin alone being 3.7 ± 0.2 (Konturek *et al.*, 1992; 1993b). This is not a remarkable reduction in lesions considering the relatively low dose of aspirin employed. Transmucosal potential difference was reduced by 12 per cent while mucosal blood flow increased by 15 per cent in the corpus and 26 per cent in the antrum, and gastric microbleeding decreased with ebrotidine treatment, suggesting a moderate effect of acid control in these physiological parameters. Studies in rats by Konturek *et al.* (1992) showed that ebrotidine inhibits acid secretion, and increases mucosal blood flow and prostaglandin production in contrast to that from aspirin. Mucus secretion has been shown to be increased with an experimental H_2-receptor antagonist (Ichikawa *et al.*, 1994), and this may be a feature of H_2-antagonists.

Limitations of H_2-receptor antagonist therapy for long-term prevention of ulcers caused by aspirin or other NSAIDs was apparent in studies by O'Laughlin *et al.* (1982) and Jaszewski *et al.* (1989). This was evident in patients with rheumatic disease who had a history of aspirin-associated ulcers and received cimetidine and antacids in comparison with placebo and antacids while continuing aspirin therapy. There was no difference between the results with antacids alone compared with their combination with cimetidine. The similar study by Jaszewski *et al.* (1989) also showed that an 8-week course of cimetidine with antacids was ineffective in preventing gastric ulcers caused by enteric-coated aspirin. More potent H_2-receptor antagonists with longer half-lives in the circulation (e.g. famotidine) may be more effective in preventing aspirin-induced gastroduodenal injury (Pritchard *et al.*, 1988b).

The limitations to prophylaxis with H_2-receptor antagonists against gastroduodenal injury from aspirin and other NSAIDs may be related to the extent of acid-secretory control achieved by these drugs (Hunt *et al.*, 1995). Some newer H_2-receptor antagonists with added mucus/bicarbonate stimulating properties (e.g. see Morimoto *et al.*, 1994a; 1994b) may also prove more effective than traditional agents.

A more striking reduction in gastroduodenal injury by NSAIDs has been observed endoscopically with the H^+-pump blockers (Scheiman *et al.*, 1994; Huang and Hunt, 1996), which have more effective anti-acid secretory activity (Hunt *et al.*, 1995) although significantly mucosal injury is not completely eliminated compared with placebo.

Studies by Scheiman *et al.* (1994) showed that high doses of aspirin (650 mg q.i.d with placebo) taken for 2 weeks by healthy volunteers with omeprazole (20 mg once daily) resulted in fewer subjects (15 per cent) developing gastric lesions (grades 3–4) compared with those that received aspirin + placebo (70 per cent). Moreover, omeprazole-treated subjects did not develop duodenal injury, whereas half of those on aspirin alone developed erosions and 15 per cent had duodenal ulcers. An important observation in this study was that the serum salicylate levels were unaffected by the omeprazole treatment, thus eliminating a potentially important issue that the H^+-pump blocker might affect the gastric absorption of aspirin (Dal Negro, 1998) by raising the intragastric pH and so reducing the proportion of the non-ionised form of the drug – the species that is preferentially absorbed from the stomach.

Simon *et al.* (1995) found that 20 mg and 40 mg omeprazole (once daily for 14 days) reduced the gastric and duodenal lesion scores from that produced by a low dose of aspirin (300 mg/d × 14 d) of 12.4 ± 1.7 (mean $\pm$ SEM) to 2.9 ± 0.9 and 1.8 ± 0.5 respectively. There were no significant differences between the two doses of omeprazole. In a similarly designed study the same group showed reduction by the longer-acting H^+-pump blocker lansoprazole 15 mg/d × 14 d (lesion scores 3.6 ± 1.2) compared with aspirin 300 mg/d × 14 d (lesion scores 10.1 ± 1.4), a slightly less impressive reduction than achieved by the same group with omeprazole (Müller *et al.*, 1997). Interestingly, 300 mg/d ranitidine only reduced the damage score from aspirin by about half (lesion scores 5.8 ± 1.3), thus confirming the view that this H_2-receptor antagonist affords less protection than the two H^+-pump blockers. It is surprising that these anti-acid secretory agents did not completely eliminate mucosal injury from what is a relatively low dose (300 mg/d) of aspirin.

That protection by H^+-pump blockade of aspirin-induced gastric lesions in humans is a consequence of maintaining mucous membrane integrity by control of acid-related factors was shown in a study by Bergmann *et al.* (1992). They showed that 1 week prior treatment with lansoprazole 30 mg/d substantially restored the reduction in gastric mucosal potential difference from a single oral dose of aspirin 1 g over a 3-hour period of recording. Subsequent endoscopy showed that the Lanza scores (see p. 383) of mucosal injury were reduced from those produced by aspirin (2.25 ± 1.1; mean $\pm$ SEM) to 0.67 ± 0.98 on lansoprazole.

Extensive studies in a wide range of laboratory animal models have also shown that inhibition of acid-secretion by antisecretory agents prevents the development of acute and chronic gastric lesions or ulcers (Konturek *et al.*, 1991; Larsen *et al.*, 1992).

The mechanisms underlying the protective effects of these agents against aspirin or NSAIDs may involve more than the control of acid (and with it pepsin) secretion, and include restoration of mucosal permeability, mucosal blood flow, and partial reversal of the inhibition of prostaglandin production (Konturek *et al.*, 1991; Matsukura *et al.*, 1994). H^+-pump blockers may also reverse the vascular permeability and increase mucus production (Matsukawa *et al.*, 1994). The extent of these non-acid secretory effects may vary between H_2-receptor antagonists (Konturek *et al.*, 1991; Larsen *et al.*, 1992), and may even depend on the type of animal model employed.

A link between acid secretion and the activity of carbonic anhydrase in the stomach has been inferred from the studies of Puscas *et al.* (1989; 2000). In an endoscopy study in healthy male volunteers, Puscas *et al.* (1989) showed that treatment with the carbonic anhydrase inhibitor, acetazolamide (20 mg/d) caused a marked reduction in the Lanza scores of mucosal injury in subjects that received aspirin 1500 mg/d as well as from other NSAIDs. The dyspeptic symptoms were also reduced by acetazolamide.

These studies have not been without controversy. Cho and Pfeiffer (1984) showed that acetazolamide 100 or 200 mg/kg s.c. caused marked damage to the gastric mucosa of fasted rats. The logic of employing an agent that blocks secretion of the protective bicarbonate layer has been a limitation to acceptance of the view that carbonic anhydrase inhibitors could be useful in the prevention of gastric injury by aspirin or other drugs. There is a marked pharmacokinetic interaction between salicylate and acetazolamide, in which salicylate displaces acetazolamide from its binding sites on serum proteins, which could have toxic consequences (Sweeney *et al.*, 1986).

Prostaglandins

The earlier pioneering studies by the late André Robert at the Upjohn Laboratories in Kalamazoo (MI, USA), showing GI protection by prostaglandins, led to the concept that prostaglandins could induce 'cytoprotection' – a term arising from studies by Chaudhury and Jacobson (1978) to describe the mucosal protection against necrotising agents (also observed by Robert (1981a, 1981b) and extensively confirmed and extended by others) of the gastrointestinal tract by aspirin in laboratory animal models (Mózsik *et al.*, 1987).

Subsequently the use of the term 'cytoprotection' to describe this protective phenomenon by prostaglandins was critically evaluated, and the term is now properly confined to specific demonstration of cellular protection from injury by necrotising agents that occurs at the cellular level. There is evidence that prostaglandins do not exert cytoprotection from aspirin *in vitro* whereas they do exhibit mucosal protection *in vivo* (Henagen *et al.*, 1987; 1989). The effects of PGs on mucosal blood flow and the other non-'cytospecific' or indirect effects of these agents (e.g. on acid secretion; Mihas *et al.*, 1976; Wilson, 1991) mean that the observations of reduction in mucosal lesions, restoration of mucosal regeneration and ulcer repair mechanisms, which are integrally a major part of the actions of PGs. Such *in vivo* observations should be regarded as evidence of 'gastroprotection' when localised to a specific region of the GI tract or organ (in this case the gastric mucosa), or the more generalised form 'mucosal protection'. Extensive studies have shown that natural or synthetic prostaglandin derivatives protect against GI mucosal injury in humans from aspirin (Hawkey *et al.*, 1985; 1986; Wilson, 1991).

Present concepts of the effects of PGs now recognise that the direct effects of these agents on GI mucosal functions are receptor-mediated.

MISOPROSTOL

This PGE_1-diasteroisomeric analogue has probably been amongst the most successful of the PGs developed as anti-ulcer agents and applied for reducing GI mucosal injury from aspirin and related NSAIDs (Gullikson *et al.*, 1987; Jiranek *et al.*, 1989 Cryer, 1992; Jaszewski *et al.*, 1992; Walt, 1992; Silverstein *et al.*, 1995).

Recently, the recognition that even the low-dose aspirin used for prophylaxis of thromboembolic diseases is associated with a two- to three-fold increased risk of GI bleeding (Garcia-Rodriguez, 2000; Sorenson *et al.*, 2000; Garcia-Rodriguez and Hernandez-Diaz, 2001) has led to consideration of misoprostol for prophylactic use in low to moderate doses in subjects in need of aspirin prophylaxis. Donnelly *et al.* (2000) showed that 100 mg/d misoprostol was effective in reducing endoscopically-observed injury over 28 days of treatment with 300 mg/d aspirin; this is a somewhat high dose of the drug for prophylactic use in thromboembolic disease, but is one commonly used by the lay public. The odds ratio for developing injury at 28 days in subjects receiving misoprostol with aspirin was 0.18 (95 per

CHAPTER 8

cent C.I. 0.07–0.48), which is strikingly low. While this study was limited by the number of study subjects, it is an indication that this line of investigation should be extended in view of its significance for reducing the risks of aspirin injury. It is important, however, to establish what effects the combined treatment has on platelet aggregation and control of thrombotic reactions in subjects at risk from thromboembolic diseases. Fish oils containing omega-3 fatty acids may also have protection against aspirin-induced gastric injury (Pritchard *et al.*, 1988; Mandel *et al.*, 1994) and additionally may prevent thromboembolic diseases (Alharbi *et al.*, 1985; Faust *et al.*, 1990; Feldman *et al.*, 1990).

Phospholipids

The recognition that disruption of the phospholipid protective and concomitant diminution of surface hydrophobicity is probably one of the initiating factors in aspirin-induced gastric injury (Lichtenberger *et al.*, 1985; 1995; 1996; Goddard *et al.*, 1987; Dunjie *et al.*, 1993a; 1993b; Goddard and Lichtenberger, 1995; Giraud *et al.*, 1997; Anand *et al.*, 1999) has led to the development of a number of phospholipid preparations, including those associated with aspirin, as potential low-irritant formulations of aspirin or as prophylactics against aspirin or other NSAID injury. Among these developments is a phospatidylcholine (PC)–aspirin (PC–ASA) association complex developed by Lichtenberger and co-workers (1995). In rats, these authors found that the complex did not affect the capacity of aspirin to inhibit mucosal prostaglandin production. In a double-blind crossover study in humans, Anand *et al.* (1999) and Giraud *et al.* (1999) found that the PC–ASA formulation given three times daily with an aspirin dose of 1950 mg/d resulted in significantly fewer gastric erosions (mean, 2.9 ± 4.3 (SD)) compared with aspirin alone (8.7 ± 10.7 lesions). Duodenal lesions induced by aspirin were not significantly different with the PC formulation from those due to aspirin alone. Both ASA-PC and ASA reduced the content of 6-keto $PGF_{1\alpha}$ in the antral mucosa, showing that the mucosal protection afforded by PC is PG-independent.

Mucoprotectants

The strategy of stimulating mucus production, and probably associated bicarbonate secretion, has become a well-established approach to achieving gastric and intestinal protection against injury induced by aspirin and related drugs. Early studies suggested that carbenoxolone, a component of liquorice, protected against gastric mucosal damage in rats by the stimulation of mucus synthesis (Domsche *et al.*, 1972; 1977; Peskar, 1980) and lysosomal stabilisation (Symons *et al.*, 1978). Liquorice co-administered with aspirin was shown to cause less damage compared with aspirin alone to rats without affecting the absorption of the drug, as indicated by measurement of serum salicylate concentrations (Morgan *et al.*, 1983). Ibuprofen-induced gastric mucosal damage in rats was also reduced by deglycerrhised liquorice as well as carbenoxolone (Dehpour *et al.*, 1995). Part of the effects of these liquorice components may be related to their complexation with NSAIDs so acting as mucosal transport vehicles (Tolstijkov *et al.*, 1991).

Amidazopyridine (zolimidine) was shown to protect against gastric mucosal damage induced by aspirin, or the induction of experimental ulcers, by stimulation of mucus production (Murmann *et al.*, 1974). Zolimidine also inhibited acid production induced by secretagogues, but not antipeptic activity. Thus the protective effects of this drug may be attributed to a combination of effects on mucus production as well as the inhibition of acid secretion. In a study in human volunteers, 4 weeks prior administration of zolimidine 1200 mg/d t.i.d. protected against cellular exfoliation induced by a single dose of 1 g aspirin delivered intragastrically (Corinaldesi *et al.*, 1980).

Proglumide (DL-4 benzamido-N,N-dipropylglutaramic acid), an anti-ulcer agent that is an antagonist of gastrin receptors, was found to protect against aspirin induced mucosal injury as well as stimulating the otherwise inhibited production of gastric mucus.

A polyisoprenoid compound, geranyl geranyl acetone (GGA), protected against aspirin-induced mucosal injury in rats as well as that caused by other ulcerogenic agents, by restoration of mucus synthesis as well as blocking the hydrogen ion back-diffusion into the gastric mucosa and concomitant loss of sodium into the gastric lumen (Murakami et al., 1982a; 1982b).

Sucralfate and sofalcone have both been shown to have short, but not long-term, protective effects against gastric mucosal injury induced in humans by aspirin (Stern et al., 1986; 1987; 1989) and other NSAIDs, but their mechanisms of action may be different regarding the production of gastric mucus. Sofalcone reverses the inhibitory effects of aspirin on the synthesis of amino sugars and sulphated mucoproteins (Muramatsu et al., 1986), the major effect appearing to be due to the stimulation of sulphotransferase activity, which is otherwise inhibited in micromolar concentrations by aspirin (Slomiany et al., 1987). Curiously these authors also found that sucralfate was an inhibitor of sulphotransferase activity, so clearly the mucoprotectant effects of sucralfate must be different to those of sulfacone. Studies by Konturek and co-workers (1993) indicated that one of the principal effects of sucralfate is apparently to stabilise the basic fibroblast growth factor. Sucralfate has been shown to exhibit a protective effect in humans when administered before administration of aspirin (Stern et al., 1986; 1987). Sucralfate, in addition to protecting against aspirin-induced injury, also has a partially protective effect against injury induced by indomethacin and ethanol (Besner and Houle, 1985).

Another mucoprotectant agent, sulglycotide, a sulphated glycopeptide, was shown to have protective effects against aspirin injury in healthy volunteers by reversing the inhibition of cell turnover in the gastric mucosa induced by aspirin (Biasco et al., 1992b).

Zinc complexes had been shown in an extensive variety of animal model and human studies to have protective effects against aspirin-induced mucosal damage, as well as damage from other NSAIDs and ulcerogenic agents (Bravo et al., 1992; Cho and Pfeiffer, 1982; Rainsford, 1992; Rainsford and Whitehouse, 1992; Bulbena et al., 1993). Part of the protective effects of these zinc complexes may arise from the dissociation of zinc ions from the complexes and the formation of complexes with gastric mucus, so leading to stabilisation of the mucus (Rainsford, 1992). Also, potential for the stimulation of mucus glycoprotein secretion may arise from direct physical effects on the mucous cell membranes leading to release of mucus, as evidenced by observations under the scanning electron microscope (Bravo et al., 1992; Rainsford, 1992). This physical stabilisation of mucus and the stimulation of mucus outpouring may be important physical factors, but it is unlikely that there are effects on prostaglandin production leading to mucus secretion. Zinc may also stabilise lysosomal membranes and the membranes of mast cells, and may protect against the intragastric pressure increased by vagal stimulation (Cho and Pfeiffer, 1982). The apparent reduction in gastric acid secretion induced by zinc probably relates to the stabilisation of mast cells and protection against histamine release (Cho and Pfeiffer, 1982). Although there have been some attempts to employ zinc complexes in humans, to date this has not been an apparently successful procedure where patients are taking aspirin or other non-steroidals.

Miscellaneous agents

Bismuth subsalicylate, an old established anti-ulcer drug, has been shown to exert mucosal protection against oxyradical-induced cellular injury (Bagchi et al., 1999a; 1999b). Inter-related effects of oxyradicals on mucosal sulphydryls have been implicated from some studies (Ranta-Knuuttila et al., 1997) including those with various protective agents such as pyrazoles (Hauser and Szabo, 1991) and the calcium antagonist, verapamil (Gutierrez-Cabano, 1993). Paracetamol and β-carotene, both potential antioxidants, which protected against aspirin injury in rats, had little effect in human volunteers (Graham and Smith, 1985; Moses et al., 1988).

Capsiacin has been shown to protect against gastric injury induced by aspirin and other ulcerogens in rats by release of vasodilator neurokinins and CGRP from afferent sympathetic nerves leading to increased mucosal blood flow (Holzer et al., 1989; Brzozowski et al., 1996; Abdel-Salem et al., 1997; Mozsik et al., 1997). Chili (which contains capsiacin) also protects against aspirin-induced gastroduodenal lesions in humans (Yeoh et al., 1995). Another vascular mechanism for mucosal protection against aspirin involving endothelin receptors was highlighted by the studies of Duggan et al. (1999) who showed that the endothelin-1 antagonist, bosentan 700 mg, protected against aspirin 900 mg in human volunteers, but this was limited to 5 doses over a 12 h period possibly as a result of enzyme induction destroying the drug.

Some growth factors that have been implicated in growth repair, regeneration and adaptation (Brzozowski *et al.*, 1994; Scheiman *et al.*, 1997; Konturek *et al.*, 1998) have been found to have protective effects against aspirin-induced mucosal damage (Romano *et al.*, 1992; 1994). Irsogladine maleate has been shown to have anti-ulcer activity by increasing cAMP formation and the dibutyryl cAMP analogue has been found to have anti-ulcer activity as well (Ueda *et al.*, 1991). This is an interesting contrast to the effects of aspirin in increasing cAMP related stimulation of acid secretion (p. 430). Clearly there are aspects of cyclic nucleotide regulation in response to aspirin and anti-ulcer agents that require further investigation.

HEPATOTOXICITY

Aspirin is frequently associated with the elevation of liver transaminases and other indices of hepatotoxicity in patients with rheumatoid arthritis, juvenile rheumatoid arthritis, systemic lupus erythematosus Reiter's syndrome, and rheumatic fever (Manso *et al.*, 1956; Russell *et al.*, 1971; Iancu, 1972; Rich and Johnson, 1973; Iancu and Elian, 1974; Seaman *et al.*, 1974; Wolfe *et al.*, 1974; Zucker *et al.*, 1975; Ricks, 1976; Salzman *et al.*, 1976; Seaman and Plotz, 1976; Bernstein *et al.*, 1977; O'Gorman and Koff, 1977; Sbarbaro and Bennett, 1977; Carneskog *et al.*, 1980; Zimmerman, 1981; Singh *et al.*, 1992) as well as in Kawasaki's disease (Bertino, 1981). These effects are usually manifest when serum salicylate concentrations are higher than 25–30 mg/dl for 10 days or more (Rich and Johnson, 1973; Iancu and Elian, 1974; Seaman *et al.*, 1974; Wolfe *et al.*, 1974) and in patients with hypoalbuminaemia (Gitlin *et al.*, 1980). Epidemiological studies have shown that hepatotoxicity has been associated with intake of aspirin as well as of diclofenac, sulindac and some other NSAIDs, so for this reason hepatoxicity has been considered a class warning for this group of drugs (O'Brien and Bagby, 1985; Rainsford, 1989a; 1997; 1998; 1999a; Garcia Rodriguez *et al.*, 1994; Fry and Sreff, 1995; Farrell, 1997; Bjorkman, 1998; Tolman, 1998; Barbare *et al.*, 2001). Garcia Rodriguez *et al.* (1994) showed that there was a 10-fold increase in the risk of NSAIDs causing liver injury in patients with rheumatoid arthritis compared with those who had osteoarthritis. Furthermore, Reye's syndrome, which has been claimed to be associated with therapeutic doses of aspirin in children following viral infection (Starko *et al.*, 1980; Miller, 1981; Tansgard and Huttenlocher, 1981; Halpin *et al.*, 1982; Waldman *et al.*, 1982; Hurwitz *et al.*, 1985; Heubi *et al.*, 1987), is accompanied by microvesicular steatosis of the liver (Heubi *et al.*, 1987) or acute liver failure with encephalopathy (Sillanpää *et al.*, 1975) and is usually evident at serum concentrations of more than 1.0 mmol/l salicylate (Partin *et al.*, 1982; see also Glasgow, 1984; Hall, 1987). Much interest has therefore been shown in the mechanisms underlying these effects of salicylates at therapeutic doses, as well as the toxic manifestations of supratherapeutic doses. The frequent concurrent intake of many antibiotics with salicylates may complicate the attribution of cause of hepatic injury since antibiotics, especially amoxicillin with clavulanic acid and the cephalosporins, are well known to be hepatotoxic (Via *et al.*, 1997). Pre-existent liver damage and altered liver metabolism may have a profound effect on further developments of serious toxicity from salicylates and other drugs (Wilkinson and Schenker, 1976; Schenker *et al.*, 1999).

There are several sites of salicylate actions in the liver cell following administration of salicylic acid/salicylate or aspirin to laboratory animals that may be relevant to the hepatic reactions seen in patients. It should be noted, however, that the structure and function of liver systems in rodent models of hepatotoxicity may differ quantitatively or qualitatively in responsiveness to these and other drugs. This issue is of particular concern when it comes to interpretation of the peroxisomal proliferation effects of the salicylates as well as lipid lowering or antidiabetic agents.

Bullock *et al.* (1970) described fine structural changes in the livers of rats 2–4 hr after having been given a single i.p. dose of 400 mg/kg sodium salicylate, which notably included a marked increase in the number of peroxisomes, microbodies and multivesicular bodies in the Golgi region, while the ribosomal endoplasmic reticulum was normal. No other microscopic changes were observed. This suggests that the first changes induced by salicylate occur in the peroxisomal/Golgi system. An increase in the number of peroxisomes is evident after repeated daily feeding of 0.5–1.0 per cent of aspirin and 1 per cent salicylic acid in the diet for periods of 2–6 weeks (Hruban *et al.*, 1966; 1974). Moreover, those authors observed that chronic effects of these salicylates on the peroxisomal system resulted in marked changes in the appearance of the metrical plates, which comprise repeated folds of membranes, with catalase as an inclusion enzyme (Hruban *et al.*, 1974). These changes were also observed with the hypolipidaemic agent, clofibrate, and with dimethrin (Hruban *et al.*, 1974). Thus aspirin and salicylate would appear to have the characteristics of being classical hepatoperoxisomal proliferators. The extensive survey of the effects of other salicylates by Hruban *et al.* (1966) showed that this was also evident with salicylaldehyde, 2-methoxybenzoic acid and methyl salicylanilide. However, hexahydrosalicylic acid did not produce the same proliferation of peroxisomes but showed extrusion of crystalloid from these bodies. Thus the basic salicyl structure is a characteristic but not an absolute requirement for peroxisomal proliferation by this class of drugs.

Mice fed aspirin (dosage unstated) for 12 days also showed proliferation of peroxisomes but not the fibrillar inclusion structures (Hruban *et al.*, 1966).

Some ultrastructural investigations have been undertaken to determine the effects of aspirin on the liver of children who have been investigated for reactions to this drug. Iancu and Elian (1976) investigated a biopsy from a 9-year-old boy who had received aspirin 3.0–3.6 g/d q.i.d. (100 mg/kg) for 10 days for rheumatic fever and who had elevated serum transaminases. Marked dilatation of the rough endoplasmic reticulum, proliferation of the smooth endoplasmic reticulum, shrinking of the nuclei, and enlarged mitochondria with the appearance of crystalloid material were observed.

Partin *et al.* (1984) examined biopsies from children who had salicylate intoxication, and two cases of Reye's syndrome who had received salicylates. In the livers of patients with salicylate intoxication the light microscopic appearance was normal, as were the mitochondria, peroxisomes and glycogen granules observed by electron microscopy. Lipid accumulation was minimal and the succinic dehydrogenase activity was normal. The liver from the Reye's syndrome patients had extensive microvesicular fat, diminished glycogen and succinic dehydrogenase, and collapsed nuclei. Thus there appeared to be minimal changes in the patients with salicylate intoxication, but extensive liver injury in those who had Reye's syndrome.

When the biochemical and ultrastructural evidence is considered together, it is apparent that there is considerable variability in the nature of peroxisomal-related changes in the liver of animals given salicylates. There is an impression from the biochemical studies that abnormalities of mitochondrial fatty acid metabolism by salicylates can be differentiated from the effects of these drugs from those of lipid-lowering agents, whose primary mode of action is to influence lipid metabolism by peroxisomes. However, peroxisomal β-oxidation of fatty acids is induced by aspirin (Sakurai *et al.*, 1981). It is possible that there may be some inter-relationship between the mitochondrial effects of salicylates (including possibly their effects on oxidative phosphorylation) and metabolism by fatty acids on peroxisomes. Inflammation influences peroxisomal enzyme activity (Canonico *et al.*, 1977) and this may complicate the responses to salicylates.

The lack of evidence of structural changes in the peroxisomes of children who had taken aspirin compared with the effects seen in rodents given salicylates may be more of a reflection of species differences. It has been argued recently that the rat peroxisomal system may be substantially different from that in humans, and that this could present a major limitation in the extrapolation of findings in rats to prediction in humans.

The proliferation of peroxisomes caused by aspirin has been associated with lowering of both circulating and liver concentrations of triglycerides in rats given 1 per cent of the drug in the diet for 2 weeks (Ishii and Suga, 1979). This so-called 'clofibrate-like' effect of chronic aspirin treatment was ascribed to increased palmitoyl–CoA oxidising activity (Ishii and Suga, 1979). The peroxisomal proler-ator clofibrate or lipid-lowering-like activity has been associated with the activity of the Peroxisomal Proliferator-Activated Receptor-α (PPAR-α) transcription factor, whose activation has been implicated in the development of rodent hepatocellular carcinogenesis (Valle and Gartner, 1993; Kerston *et al.*, 2000; Roberts-Thomson, 2000). Among the ligands for PPAR-α are the clofibrate-like drugs, a range of eicosanoids, and unsaturated fatty acids including arachidonic acid, and this factor may induce COX-2 (Roberts-Thomson, 2000). The relevance of PPAR-α in the induction of liver proliferation, apoptosis and carcinogenesis seen in rodents has not been established in human cells, where its activation is associated with hypolipidaemia (Roberts, 1999). PPAR's may, however, protect against liver injury from hepatotoxicants (e.g. paracetamol) (Nicholls-Grzemski *et al.*, 2000).

The results of the studies by Ishii and Suga (1979) contrast with those of Deschamps *et al.* (1991), who found that a single high i.p. dose of aspirin or salicylic acid 3 mmol/kg to 48-h fasted mice resulted in reduced oxidation of radiolabelled palmitate and, with salicylic acid reduction, also in the oxidation of radiolabelled octanoic and butyric acids, the inhibition lasting for 9 hours. A marked decrease occurred in blood glucose with an increase in ketone bodies. This suggests that salicylates produced accelerated ketogenesis following extensive fasting. Following repeated administration (up to three times over 22 h) of salicylic acid 2 mmol/kg to fasted mice, the hepatic lipids and triglycerides were increased. From *in vitro* studies, Deschamps *et al.* (1991) concluded that salicylic acid decreased mitochondrial activation and consequent β-oxidation of long-chain fatty acids, possibly by sequestration of coenzyme A (used for formation of the glycine conjugate of salicylate; Tishler and Goldman, 1970) and of carnitine. This sequestration of coenzyme A is regarded as a major mitochondrial target of salicylates relevant in the development of hepatic reactions, including steatosis and steatohepatitis (Berson *et al.*, 1998; Pessayre *et al.*, 2001). The increase in ketone bodies might also be explained by inhibition of the tricarboxylic acid cycle by salicylate.

The differences in the effects of salicylate on the oxidation of palmitate observed by Deschamps *et al.* (1991) and Ishii and Suga (1979) might be due to timing of the dosing of the animals. Inhibition after a single dose of the drug might lead, upon its repeated administration, to adaptation, with consequent increase in enzymic activity. There may also be dose-dependent effects because the doses of salicylic acid used by Deschamps *et al.* (1991) were 540 mg/kg or 480 mg/kg i.p., which are very high and induced toxic levels of these drugs. The plasma concentration of salicylate at 0–1 h is approximately 5 mmol/l, which is very high by comparison with therapeutic doses of these drugs.

In vitro studies by Yoshida *et al.* (1998) have shown that the effects of salicylate on the oxidation of fatty acids are on the mitochondrial but not the peroxisomal enzymes. Thus salicylate up to 1 mmol/l concentration inhibits oxidation by mitochondria of radiolabelled linoleate, which is oxidised by peroxisomes (Yoshida *et al.*, 1988) and of octanoate (Graham and Parks, 1974).

Roberts and Knights (1992) and Knights (1998) have focused attention on the primary activation steps in the metabolism of fatty acids involving the formation of the ATP-dependent acid coenzyme A ligases (EC 6.2.1.1–2.1.3; AMP forming) in rat liver, which are differentially localised in mitochondria (for long-chain fatty acid metabolism) to those in peroxisomes and microsomal systems. Hypolipidaemic and peroxisomal proliferating agents (e.g. clofibrate) as well as the R(−) enantiomers of diastereo-isomeric 2-aryl propionic acids (e.g. ibuprofen) are metabolised by these long-chain fatty acid ligands by mitochondrial medium-chain fatty acid ligases (Vessey *et al.*, 1996; Knights, 1998). Mitochondrial medium-chain fatty acid ligases catalyse the formation of salicyl–CoA from acetyl CoA, which is then conjugated to glycine as part of the presumed detoxification pathway of the salicylate metabolism (Tishler and Goldman, 1970; Knights, 1998).

The results of these authors regarding the effects of salicylate on palmitate oxidation subcellular fractions are in agreement with those of Deschamps *et al.* (1991), who employed mouse liver preparations. However, the latter authors were unable to show any effects of even high concentrations of salicylate on the mitochondrial oxidation of octanoate. There may be species differences accounting for these varying drug effects *in vitro*.

Another locus for the actions of salicylates that involves the microsomal (in contrast to the peroxisomal) system is the cytochrome P_{450}-mediated ω-oxidation, which leads to formation of dicarboxylic acids (Bélanger and Attiste-Gbeassor, 1985). The cytochrome P_{450} isoforms 2C9 (CYP2C9), CYP3A4 and possibly others are involved in the oxidative metabolism of many NSAIDs, including the salicylates (Dupont *et al.*, 1999), and in the extensively studied cases of diclofenac CYP2C9 and CYP3A4 appear to be involved in adduct formation, which may underlie the immune-based hepatotoxicity from this drug (Hargues *et al.*, 1994; Miners and Birkett, 1998; Tang *et al.*, 1999).

Using the same dosing procedure as that employed by Hruban *et al.* (1966; 1974) and Ishii and Suga (1979), where 1 per cent aspirin was added to the diet of rats for 2 weeks, Okita (1986) found that these animals had a three-fold increase in laurate-ω-hydroxylation compared with that in control animals. There was no increase by aspirin treatment in the cytochrome P_{450} activities involving ethoxycoumarin or ethoxyresorufin O-de-ethylation, or benzphetamine N-demethylation or laurate 11-hydroxylation. However, the induction of the cytochrome P_{450}-mediated ω-hydroxylation, which led to the formation of the dicarboxylic acid derivative, occurred in post-mitochondrial supernatant fractions and was accompanied by increased cytosolic oxidation of 12-hydroxyl-lauric acid, an NAD^+-dependent enzyme system.

Neither aspirin 200 mg/kg or salicylic acid 300 mg/kg, either alone or with phenobarbital, were found to induce rat liver microsomal enzyme induction (as determined by the aminopyrine N-demethylase assay), whereas aminopyrine enhanced production of this enzyme (Reinicke, 1977). There was no interference with the tyrosine aminotransferase activity by these NSAIDs *in vitro*, although this activity was inhibited by indomethacin, ibuprofen, flufenamic and mefenamic acids (Reinicke and Klinger, 1971; Reinicke, 1975; 1977). However, aspirin did increase the activity of this enzyme in the livers of adrenalectomised and intact rats (Reinicke, 1977). The author postulated that there was an effect of aspirin independent of effects on corticosteroid regulation of the activity of this enzyme.

Benzo(a)pyrene metabolism has been found to be inhibited by aspirin in rat liver (Decloitre and Beurenkasse, 1980) thus raising the possibility that this could be another site of action on cytochrome P_{450} metabolism by this drug. Using sublines of the human hepatoma cell line HepG2 that differentially express cytochrome 2E1 (CYP 2E1), Wu and Cederbaum (2001) found that salicylate 1.0–10 mmol/l, while not being toxic to the E47 subline that expresses CYP 2E1, did enhance the cytotoxicity of arachidonic acid added to the cultures. An inhibitor of CYP 2E1 and an antioxidant inhibited the cytotoxicity and DNA fragmentation of salicylate and arachidonic acid. Salicylate also enhanced the cytotoxicity in pyrazole-treated rats. The drug also appeared to enhance CYP 2E1 levels and the production of its mRNA (Pankow *et al.*, 1994; Damme and Pankow, 1996). These observations may have clinical significance, especially when aspirin is taken with ethanol or other agents that are metabolised by CYP 2E1 and 3A4 (Dupont *et al.*, 1999).

These actions of aspirin in stimulating the ω-oxidation of lauric acid are similar to those of the perosisomal proliferators, clofibrate and di-(2-ethylhexyl) phthalate (Gibson *et al.*, 1982; Orton and Parker 1982; Lake *et al.*, 1984; Okita and Chance, 1984; Okita, 1986).

Viral infections can substantially reduce metabolism of salicylate as well as paracetamol (Jorup-Rönström *et al.*, 1986). Another factor linking the drug metabolism of salicylates to their hepatic effects concerns their metabolism to acyl-glucuronides; these may consequently lead to the formation by intramolecular rearrangement of drug-metabolite adducts that could become immunogenic as seen with other drugs (Lo *et al.*, 2001). The formation of acyl-glucuronides of diflunisal in rat livers and in humans appears to lead in rats to subsequent formation of glucuronyl adducts of proteins by nucleophilic substitution reactions (Dickinson *et al.*, 1989; 1991; 1993b; 1994; Dickinson and King, 1991; 1993; Wang and Dickinson, 1998), and these adducts may become immunogenic, so causing autoimmune-based injury in the liver and subsequent reactions in cirrhotics (MacDonald *et al.*, 1992). It is not known if the acyl-glucuronides of salicylate that form in humans (Tsuchiya and Levy, 1972) are capable of forming drug–protein conjugates as is seen with diflunisal, or whether these are so labile or subject to rapid hydrolysis by β-glucuronidases that their lifetime is limited. However, salicylate was found by Kyle and Kocsis (1986) to bind to liver as well as to kidney, the latter being more prominent. Radiolabelled salicylate was bound to proteins from kidney and liver mitochondria, with concomitant conversion to 2,3-dihydroxybenzoic acid. This and observations on the effects of ferrous ion and ascorbic acid suggested there was a mechanism involving the formation of oxyradicals possibly also involving cytochrome metabolism in this case (Kyle and Kocsis, 1986). Lipid peroxidation has been found to be increased and antioxidant enzymes decreased in the livers of aspirin-treated rats (Kirkova *et al.*, 1995) thus supporting the involvement of oxyradicals in the pathogenesis of liver injury by aspirin.

CHAPTER 8

A factor that could be important in the understanding of the effects of aspirin on the oxidation of lipids is the lipid peroxidation catalysed by mitochondria and microsomal systems. While it is understood that the metabolite of aspirin, salicylate, is an antioxidant and would be expected to scavenge oxyradicals (Colantoni *et al.*, 1998), it appears from the studies of Nakagawa *et al.* (1987) that a single oral dose of 50 mg/kg aspirin can cause a 30–50 per cent increase in lipid peroxidation (measured by the malondialdehyde reaction) in rat liver homogenates from fasted rats as well as in isolated mitochondria or microsomes from these animals. *In vitro* addition of 100 μmol/l salicylic acid to isolated mitochondria or microsomes did not cause increased lipid peroxidation, suggesting that there was some effect of aspirin administration *in vivo* mediating direct toxic effects on these subcellular fractions. The concomitant oral administration of clofibrate 100 mg/kg reversed the effect of aspirin in enhancing lipid peroxidation, which was shown to be related to the elevation of both glutathione peroxidase and non-protein thiol content by clofibrate. The hepatic triglycerides and phospholipid concentrations that were elevated by aspirin were also reduced by clofibrate. These authors also speculated that the effects of aspirin in mediating lipid peroxidation could be related to the effects of this drug on prostaglandin production. Perhaps this is linked to the effects of aspirin on the liver production of aspirin-triggered 15-epi-lipoxins (Titos *et al.*, 1999), which may confer some protective anti-inflammatory effects (Takaho *et al.*, 1997).

The well-established effects of salicylates on liver gluconeogenesis, mitochondrial uncoupling of oxidative phosphorylation, the inhibitory effects on mitochondrial dehydrogenases and the reduced production of ATP (Bryant *et al.*, 1963; Hines *et al.*, 1963; Whitehouse, 1965; Smith, 1966; Slater and Delaney, 1970; Smith and Dawkins, 1971; Mehlman *et al.*, 1972; Woods *et al.*, 1974; Chatterjee and Stefanovich, 1976; Pocwiardowska, 1976a; 1976b; Barritt and Whitehouse, 1977; Dawson, 1979; McDougall *et al.*, 1983; Baños and Reyes, 1989; Michalety *et al.*, 1989; Terada *et al.*, 1990; Petrescu and Tarba, 1997; Trost and Lemasters, 1997) may be important in initiating oxyradical injury, caspase activation and loss of cytochrome c, with consequent apoptosis of liver cells as seen with other mitochondrial toxins (Wallace and Starkhov, 2000). The role of direct cytotoxicity of salicylates on hepatic cells in culture (Tolman *et al.*, 1978; Chao *et al.*, 1988) in contrast to metabolic activation and metabolites of these drugs has not been determined.

NEPHROPATHY

The evidence of serious renal complications is comparatively low from normal therapeutic doses of salicylates, in relatively normal subjects including those with mild–moderate arthritic conditions (Smith, 1966; Nanra and Kincaid-Smith, 1972; Kimberley *et al.*, 1979; Prescott, 1979; 1984; Wortman *et al.*, 1980; Pommer *et al.*, 1989; Derby and Jick, 1991; Thuluvath *et al.*, 1994; Abbott and Fraser, 1998; Feinstein *et al.*, 2000; Koseki *et al.*, 2001). Nonetheless, the incidence of deaths due to nephropathy associated with aspirin intake (in the UK) appear relatively high (Table 8.18; Delzell and Shapiro, 1998). Dose, duration of intake and concomitant intake of other analgesics are clearly factors that affect the occurrence of chronic renal failure (Sandler *et al.*, 1991; McLaughlin *et al.*, 1998). Also, there has been much concern over the years about nephropathy associated with the abuse of analgesic–NSAID mixtures by the lay public (Irwin, 1976; Senate Select Committee on Social Welfare, 1977; Emkey and Mills, 1982; Hauser *et al.*, 1991; Fox, 1995; Gault and Barrett, 1998; Elserviers and De Broe, 1999), to the extent that legislative measures have even been employed in some countries to restrict the volume of analgesics on over-the-counter (OTC) non-prescription sale (Nordenfelt, 1972; Anonymous, 2001). The frequent association of intake of OTC (non-prescription) NSAIDs and analgesics with the development of renal syndromes (Derby and Jick, 1991; Whelton, 1995) has also led to concern about their widespread use by the lay public. Experimental and clinical evidence (reviewed later) shows that mixtures of analgesics and NSAIDs are more nephrotoxic than the individual drugs (Fox, 1995), and hence the move to limit the sale of these mixtures.

TABLE 8.18

Estimates of relative risks (and 95% confidence intervals) for analgesics associated with chronic renal disease. (From Delzell and Shapiro, 1998.)

Study (reference)	Definition of use	Aspirin Salicylates	Paracetamol acetaminophen	Any analgesic	Any combination Analgesic	Phenacetin combinations	Nonphenacetin combinations	Phenazones/ pyrazolones
Dubach, 1983								
Any regular use	Low or high NAPAP	0.8 (0.3–2.2)[b]	–			16 (3.9–66)[d]		–
Murray et al., 1983								
Any regular use	16+ doses/mo 30+ days	1.1 (0.8–1.6)	1.2 (0.7–2.1)	1.1 (0.8–1.5)	1.3 (0.8–2.1)	1.2 (0.7–1.9)	1.9 (0.8–4.6)	
Heavy use	3+ year	1.5 (0.8–2.7)	2.0 (0.6–6.7)	1.5 (0.9–2.6)	2.0 (0.9–4.3)	2.6 (1.1–6.1)	–	–
	3+ kg*	0.8 (0.3–2.1)	2.0 (0.1–30)	1.0 (0.5–2.3)	2.2 (0.5–10)	4.0 (0.4–37)	–	–
Sandler et al., 1989								
Any regular use	1+ dose/wk 1+ yr	2.0 (1.5–2.7)[c]	2.4 (1.7–3.6)[c]	1.5 (1.1–2.1)	–	3.2 (2.0–5.1)[c]		
Heavy use	1+ dose/day 1+ yr 2+ kg*	1.3 (0.7–2.5)	3.2 (1.1–9.8)	2.8 (1.9–4.2)	–	5.1 (1.8–15)		–
Pommer, 1989								
Any regular use	1+ yr 0.1+ kg*	2.3 (1.6–3.3)[c]	1.8 (1.2–2.9)[c]	–	1.9 (1.4–2.4)[c]	2.9 (2.1–4.0)[c]	1.2 (0.8–1.8)[c]	2.4 (1.7–3.3)[c]
Heavy use	1+ kg*	2.4 (1.4–4.2)	4.1 (1.3–12)	–	2.5 (1.8–3.5)	4.5 (2.6–7.7)	1.4 (0.7–2.7)[c]	3.6 (2.3–5.6)
	5+ kg*	4.4 (1.4–14)	–		3.0 (2.3–3.5)	9.2 (2.1–40)	4.0 (0.8–20)	18 (4.2–75)
Morians et al., 1990								
Any regular use	15+ doses/mo 30+ days	2.5 (1.2–5.2)	–	2.9 (1.8–4.7)	–	19 (2.3–157)	2.8 (1.1–7.3)	2.2 (0.9–5.3)
Perneger et al., 1994								
Any regular use	Avg annual use, >2 pills/wk	1.0 (0.8–1.4)[c]	1.4 (1.0–1.9)[c]	–	–	–	–	–
	1000+ pills	0.8 (0.6–1.0)[c]	1.2 (0.9–1.7)[c]	–	–	–	–	–
Heavy use	Avg annual use, >1 pill/day	1.1 (0.7–1.9)	2.1 (1.1–3.7)	–	–	–	–	–
	5000+ pills	1.0 (0.6–1.8)	2.4 (1.2–4.8)	–	–	–	–	–

*Cumulative dosage

[a] Relative Risks (RRs) for each category were adjusted for other types of analgesics only in the studies by Sandler et al., 1989; Morlans et al., 1990 and Perneger et al., 1994.

[b] A dash indicates that the study did not provide relevant data.

[c] Indicates crude relative risks and confidence intervals.

[d] RRs for urologic or renal disease mortality, 1968–1987.

Authors (year) of study reference as cited by Delzell and Shapiro (1998).

Aspirin alone does, in anti-arthritic doses (>2 g daily), induce a reversible decline in some components of renal function (i.e. reduced glomerular filtration with some water, sodium and potassium retention; Muther *et al.*, 1981; Plotz and Kimberley, 1981) but severe renal failure is very rare. Some patients with certain arthritic conditions (e.g. systemic lupus erythematosus, juvenile rheumatoid arthritis) may be particularly predisposed to renal failure (Huong *et al.*, 1997; 2001) especially from high-dose aspirin therapy (Muther *et al.*, 1981; Plotz and Kimberley, 1981). A particular drug–disease interaction may therefore occur in the development of aspirin-associated renal nephropathy.

The aspirin-associated nephropathy is a potential problem in two groups of individuals; those succumbing to socio-psychological factors underlying the syndrome of analgesic abuse (a select group), and those who consume drug mixtures (including aspirin) as in certain arthritic conditions.

Analgesic abuse syndrome

It is not possible to discuss the role of aspirin in the development of renal damage in the syndrome of analgesic abuse without considering the involvement of the other drugs present in analgesic mixtures that have also been implicated in this condition (Durbach *et al.*, 1991; Rahman *et al.*, 1993; Fox, 1995). There appear to be specific properties of these other analgesic/stimulant drugs that contribute to certain physical and psychological responses involved in the 'addition' to these drugs. For instance, caffeine is often present in analgesic mixtures, is known to have mild psychotropic effects (Nordenfelt, 1972), and also affects oxygen demand and renal oxygen supply in the medulla (Appel *et al.*, 1995). When combined with salicylamide or even aspirin or phenacetin, it is possible that the psychotropic effects of caffeine may be even more pronounced (Nordenfelt, 1972) as a result of the influences of these other analgesics upon the metabolism of prostaglandins and biogenic amines in the central nervous system. The social habit in some countries of ingesting analgesic mixtures with caffeine-containing beverages (e.g. coffee, tea) may also contribute to the abuse of analgesics.

The first association of analgesic nephropathy with a high intake of analgesics was revealed in Switzerland from the studies by Spühler and Zollinger (1953). These authors showed that various mixtures of agents containing particularly phenacetin and caffeine (in Saridon® and Kafa®) were implicated in the development of analgesic nephropathy. The abuse of these preparations was mostly by watchmakers working on assembly production lines. In the extreme situation, some individuals ingested sandwiches containing phenacetin preparations for lunch to overcome the tension and headaches associated with the close work involved in watchmaking (Dubach *et al.*, 1975). At that stage aspirin was not implicated, but it is well to note that a number of other analgesics were present in these mixtures. By 1964 it was generally suspected that phenacetin was the prime agent responsible for the development of renal damage, this conclusion being largely based on epidemiological evidence from Australia, the United Kingdom, the USA, Scandinavia and Switzerland (Nanra and Kincaid-Smith, 1972; Duggin, 1977). However, this was not then and nor has it since been without challenge (Nanra and Kincaid-Smith, 1972; Nordenfelt, 1972; Prescott, 1982). In fact, until the recognition of a possible synergism between the various drug mixtures, aspirin alone was also heavily implicated and such phrases in the literature as 'aspirin-containing analgesics' helped to embed the association in people's minds. As far as the involvement of phenacetin (present in drug mixtures) is concerned, it has frequently been associated with the development of cancer of the renal pelvis (Angervall *et al.*, 1969; Taylor, 1972; Rathert *et al.*, 1975), which is probably due to the formation of a paracetamol quinine-imine metabolite (de Vries, 1981; Nelson *et al.*, 1981; Prescott, 1996). Phenacetin has also been found to be carcinogenic in several assays (Farber, 1981; Weinstein *et al.*, 1981), and hence would appear to be a particularly noxious agent to include in analgesic preparations. It has been withdrawn from several countries, most recently in Japan.

Following the earlier reports in Spühler and Zollinger (1953), there were many studies published during the 1960s and 1970s regarding the nephropathy in patients attributed to analgesic abuse in Australia, Europe, New Zealand, the USA and South Africa (Abel, 1971; Nanra and Kincaid-Smith, 1972; Dubach *et al.*, 1975; Murray and Goldberg, 1970; 1975; Wiseman and Reinhert, 1975; Duggin,

1977; 1996; Muther *et al.*, 1981; Plotz and Kimberley, 1981). The reader is referred to earlier detailed reviews on this topic (Abel, 1971; Nanra and Kincaid-Smith, 1972; Murray and Goldberg, 1975; Prescott, 1979; 1982; 1984; 1996; Muther *et al.*, 1981; Plotz and Kimberley, 1981). In many of these countries there has been almost an epidemic of analgesic abuse with its associated kidney disease and this has been linked with the *per capita* intake of phenacetin (Duggin, 1977). The same *per capita* basis of analgesic intake could also be applied to aspirin. The highest incidence of analgesic nephropathy has been reported in Australia, with Switzerland, the Scandinavian countries, South Africa and Scotland following, in that order (Duggin, 1977). A relatively low incidence of analgesic nephropathy in the Western World appears to occur in the USA and Canada (Nanra and Kincaid-Smith, 1972; Duggin, 1977; Editorial, *British Medical Journal*, 1981). The high incidence in Australia has been attributed to socio-psychological factors (Dawborn, *et al.*, 1966; Abel, 1971); the hyper-concentration of drugs in the kidneys and urine resulting from the hot climate (Abel, 1971; Editorial, *British Medical Journal*, 1981); genetic predisposition (e.g. relationship to HLA haplotypes; Editorial, *British Medical Journal*, 1981); and/or the fact that medical practitioners in Australia have recognised analgesic nephropathy more often (Abel, 1971; Editorial, *British Medical Journal*, 1981).

Epidemiological studies on analgesic abuse in the eastern coastal region of Australia during the immediate post-Second World War period showed a clear pattern of predisposing factors. Those especially prone to abuse were women of low socio-economic status in heavy industrial areas and exposed to extremely potent advertising on the radio and television suggesting that a particular brand of analgesic would give a 'lift', alleviate tension and help them to cope with the problems of the day (Gillies and Skyring, 1968; Nanra and Kincaid-Smith, 1972; Senate Select Committee on Social Welfare, 1977). This combination of advertising and socio-economic circumstances was very successful in persuading many lonely housewives, often confined at home with small children, to take large quantities of these drugs that were freely available from many supermarkets or neighbourhood stores. The most extreme example of this analgesic abuse was seen in so-called 'analgesic tea parties' in some Australian suburbs, where analgesic powders or tablets were passed around on the tray together with the tea. A popular satirical play was even written about this in the 1960s, entitled *A Bex, a Cup of Tea and a Good Lie Down* (Duggin, 1977), 'Bex' being one of the popular brands of the analgesic mixtures.

Following publication of the epidemiological and experimental data in the 1960s and 1970s there was an ensuing public debate in Australia, hotly challenged by commercial interests. Nonetheless, it was clear that there was an analgesic abuse epidemic to the extent that several State and Federal Governments in Australia set up public enquiries. The major document that emanated from the Australian Federal Government's Senate Standing Committee on Social Welfare (Senate Select Committee on Social Welfare, 1977) is worth reading as a testimony to the deep social, economic and medical problems surrounding analgesic abuse and its relationship to other drug problems, not only in Australia, but also in other such Western societies. The report was uninhibited in recognising the major factors, its main conclusions (Senate Select Committee on Social Welfare, 1977) being summarised as follows:

1. Most compound (i.e. mixtures of) analgesics are habituating, and these account for the reported high incidence of 20 per cent of cases of kidney failure.
2. Women use more analgesics than men, and correspondingly they present with kidney disease five to six times more frequently than men.
3. Single analgesics (i.e. aspirin alone) are much safer.

Other reports at that time showed that there was a high incidence of analgesic abuse amongst high (secondary) school students in Australia (Irwin, 1976), and, paradoxically, in a group of 1456 individuals investigated in a town in Victoria where a high intake of analgesics containing aspirin (proven by urine testing) was *not* associated with any reduction in renal function (Christie *et al.*, 1976). The latter report is interesting because the study region was probably in an area of moderate climatic conditions away from industrial influences (though it should be noted that the authors did not provide adequate details of the group). Nonetheless, indications from this study reinforce the contention that the major problem in Australia was confined to hot industrial areas.

CHAPTER 8

■ 455

As a consequence of the reports of the Australian Governments, there followed a state-wide ban on the sale of analgesic mixtures to the lay public. There has probably also been a lowering of the incidence of analgesic nephropathy, as observed in Sweden following the ban of phenacetin-containing analgesics (Nordenfelt, 1972).

Studies by Murray and co-workers in Scotland showed that a large proportion of psychiatric patients who suffered from symptoms of analgesic abuse also had abnormal renal function and even chronic renal nephropathy (Murray *et al.*, 1970; Murray, 1973a; 1973b). Many of the patients were highly dependent on analgesic mixtures, which contained aspirin and/or phenacetin. Murray (1973a) defined the characteristics of this analgesic dependence as: a need to continue taking the analgesics; a tendency slowly to increase the dose, partly owing to tolerance and partly to treat symptoms of the analgesic; and a psychic dependence that results from appreciation of the psychotropic effects of the compound analgesic. Murray noted that severe withdrawal effects were rare in his patients (Murray, 1973a).

Schreiner and co-workers (1981) have reported the conditions under which analgesic nephropathy has developed in the USA. It was especially common in middle-aged women with histories of peptic ulcer, anaemia, psychiatric disease, headaches and arthralgias. Thus in many respects this situation is analogous to that observed in Australia, although the underlying social and economic factors may be rather different.

Gonwa and co-workers (1981) also reported a high incidence of analgesic nephropathy in a patient cohort in North Carolina. In some, the nephropathy progressed to 'end-stage' renal disease (Gonwa *et al.*, 1981; see also Murray *et al.*, 1983). Gonwa *et al.*, 1981 observed that if the analgesic abuse was diagnosed early enough, end-stage renal disease could be prevented.

In studies reported under the auspices of the Boston Collaborative Drugs Surveillance Program (BCDSP) it was found that daily analgesic use was evident in 7.2 per cent of patients, the total number of patients in the study being 6407 (Lawson, 1973). The incidence of renal damage was not significantly higher amongst these regular analgesic users than in those taking occasional quantities of these drugs or those who denied consumption of any analgesics. This overall trend would appear to reflect the underlying low incidence of analgesic abuse in the USA, although clearly there is a group of middle-aged women who are at risk in that nation (Schreiner *et al.*, 1981).

Regular analgesic use has been considered a risk factor in the development of renal cell carcinoma (McCreedie *et al.*, 1993; Gago-Dominguez *et al.*, 1999).

Studies in arthritic patients

There have been confusing reports in the literature concerning the possibility of salicylate ingestion being associated with renal pathology in arthritic patients (see also Chapter 10).

In a study in New Zealand of 763 patients with rheumatoid arthritis and 145 with osteoarthritis, no indications of renal injury (measured by standard biochemical and cellular techniques) were reported in those patients taking large quantities of aspirin (New Zealand Rheumatism Association Study, 1974). However, nephropathy was diagnosed in three patients who had taken analgesic mixtures comprising aspirin, phenacetin and caffeine or codeine, as well as in one patient who had taken aspirin and phenylbutazone (New Zealand Rheumatism Association Study, 1974). Likewise, Macklon and co-workers (1974) found no changes in renal function test (creatinine clearance, plasma urea, proteinuria, cellularuria) in rheumatoid arthritis patients who had consumed a cumulative dose of 5–20 kg aspirin. The same group also re-examined the effects of aspirin intake in rheumatoid patients using the sensitive ^{51}Cr-labelled EDTA clearance technique as a measure of renal function. Again no significant changes were observed (Akyol *et al.*, 1982).

Contrasting with these results, Burry and co-workers found in a retrospective study that high aspirin intake by rheumatoid arthritics was associated with an increase in urinary *N*-acetylglucosaminidase and a reduction in urine concentrating power, but no changes in ^{51}Cr-labelled EDTA clearance or proteinuria were observed (Burry *et al.*, 1976). In a prospective study, the same group found no significant changes in the glomerular or tubular function of 8 rheumatoid arthritis patients or 10 healthy volunteers who had taken a daily dose of aspirin 4 g for 10 days. A transient celluluria occurred at 3 days (Burry *et al.*, 1976), a phenomenon previously noted by Prescott (1965) in normal individuals, and this can be attributed to the sloughing of effete cells. Conversely, the urinary output of *N*-acetyl-glucosaminidase (NAG) – a marker enzyme for renal integrity – was increased over the 3–10 days of study (Burry *et al.*, 1976). A similar NAG enzymuria has been observed by Proctor and Kunin (1978) in rheumatoid arthritics consuming aspirin with other NSAIDs (especially the nephrotoxic drug D-penicillamine), but was not evident in those patients taking aspirin alone. In a study of the effects of graded doses of aspirin in human volunteers, no significant changes were observed in urinary excretion of NAG in individuals taking 650 or 1300 mg of the drug, but a dose of 1950 mg did cause a significant increase in the urinary output of this enzyme (Proctor and Kunin, 1978). Since this is within the dose range in which aspirin is taken for arthritis, it would appear that this enzymuria is indeed a real phenomenon in rheumatic patients. Increased urinary output of lactate dehydrogenase (another marker enzyme for cell damage) has also been shown following aspirin (Leatherwood and Plummer, 1969). The problem is knowing what these observations really mean, for in the study of Burry and co-workers the increased output of NAG was not associated with any appreciable deterioration in renal function (Burry *et al.*, 1976). It could be argued that NAG excretion may not reflect kidney damage. A further uncertainty is that Kimberley and Plotz (1977) found that creatinine clearance was depressed in rheumatoid subjects who had ingested aspirin. However, Berg (1977) did not find any reduction in creatinine clearance in 12 normal volunteers who had taken aspirin 4 g, although aspirin did reduce the urinary excretion of sodium. Also, aspirin has been found to antagonise the spironolactone-induced naturesis in man (Tweeddale and Ogilvie, 1973). In view of the observations of Muther and co-workers (1981) of an aspirin-induced depression of the renal clearance of creatinine, insulin and *p*-aminohippurate in 10 healthy volunteers under severe sodium restriction, it is possible that the sodium status and the consequences of aspirin ingestion on this may be an especially important factor in determining the actions of the drug on renal functions.

Some estimate of the risk of developing renal papillary necrosis from analgesic consumption by rheumatoid patients was derived from studies by Ferguson and co-workers (Ferguson *et al.*, 1977). These authors found that the overall risk was 10.6 per cent when a cumulative dose of more than 1 kg aspirin was taken in association with phenacetin, but was dramatically lower (0.3 per cent) when aspirin was taken without phenacetin (Ferguson *et al.*, 1977). These results further demonstrate the hazards of combinations of aspirin and phenacetin especially for the arthritic patient. Clearly the markedly lower incidence of papillary necrosis observed with aspirin alone (possibly within overall population limits) suggests that this drug has a relatively low toxicity in the kidney.

Of the other salicylates studied, diflunisal 250 mg twice daily has been reported to induce a lower renal excretion of NAG in osteoarthritic patients than that observed with aspirin 500 mg four times daily (Dieppe and Huskisson, 1978). In the earlier studies, large doses of sodium salicylate were found to induce albuminuria, celluluria and the appearance of casts in the urine of patients with rheumatic conditions (see Gross and Greenberg, 1948 for discussion). However, it is not possible in many of these early studies to determine the dose duration of treatment or even the disease status of the patients.

Analgesic nephropathy comprises a range of conditions, including acute and chronic renal failure, small bumpy kidneys, renal papillary necrosis, and chronic interstitial nephritis (Appel *et al.*, 1995). It is suggested that the spectrum of these conditions is changing because many of the drugs may also increase the risk of hypertension and cardiovascular disease (Faust *et al.*, 1997), especially in the elderly (Gurwitz *et al.*, 1994), as well as, glomerulonephritis and diabetes mellitus (Hopper *et al.*, 1989; Appel *et al.*, 1995). However, low dose aspirin may not be a risk in diabetic patients with microalbuminaemia (Hanson *et al.*, 2000).

In general, the risks of renal complications from OTC use of analgesics, including aspirin, in normal subjects is low (Whelton, 1995; Rainsford and Powanda, 1998), but may be more pronounced in the elderly (Ailabouni and Eknoyan, 1996; Caspi *et al.*, 2000).

CHAPTER 8

Pathology

The pathology of analgesic-induced kidney damage seen in humans varies considerably according to the severity of the condition (Dawborn *et al.*, 1966; Nanra and Kincaid-Smith, 1972). Characteristically the kidneys are small and shrunken, with the surface being raised into hypertrophied areas of tumour-like bars or in the form of small cortical cysts. The cortical nephrons may be hypertrophied and the papillae are necrotic and have a blackish-brown appearance (Nanra and Kincaid-Smith, 1972). Frank papillary necrosis is recognised microscopically by a total necrosis of all the medullary elements, i.e. the loops of Henle, vasa recta and collecting ducts (Spühler and Zollinger, 1953; Nanra and Kincaid-Smith, 1972). Kincaid-Smith considered that the papillary necrosis developed initially, followed by chronic interstitial nephritis and cortical damage (Dawborn *et al.*, 1966; Kincaid-Smith, 1967; Nanra and Kincaid-Smith, 1972). According to Kincaid-Smith, the disease develops in the papillae and results in a non-inflammatory necrosis in the cortex and interstitial damage (Dawborn *et al.*, 1966; Nanra and Kincaid-Smith, 1972). While this author has provided experimental evidence from studies in rats to support her views (Kincaid-Smith, 1967; Kincaid-Smith *et al.*, 1968), this concept has not been without challenge (Abrahams and Levin, 1968; Abrahams and Levinson, 1970; Abel, 1971). Again, other experimental evidence from studies in rats (Abrahams and Levin, 1968; Arnold *et al.*, 1976) and case reports (Viero and Cavallo, 1995) would appear to support Kincaid-Smith's concepts.

Gault and co-workers (1971) have performed light and electron-microscopic studies on the renal pathology of patients who had consumed a cumulative dose of at least 2 kg of both aspirin and phenacetin. The renal function of these patients varied from normal to 'end-stage' (Gault *et al.*, 1971). They observed that the earlier site for the development of the lesions was in the medulla, where there was an increase in intestitial collagen, focal thickening of the tubular basement membrane with degeneration atrophy of loss of the tubular epithelium, and the appearance of casts (Gault *et al.*, 1971). At this stage, renal function (based on creatinine clearance) was not demonstrably reduced. On progression, sclerotic or necrotic areas developed in the medullary elements and only at this stage was there any evidence of cortical damage. The cortical changes included: peritubular, interstitial and periglomerular fibrosis; tubular atrophy, dilation and thickening of the basement membrane; and in some cases round-cell infiltration. There was no evidence of any deposition of immunoglobulins, but bacterial infection and secondary pyelonephritis were observed (Gault *et al.*, 1971). The results of these authors support Kincaid-Smith's view that the first sign of damage is in the medulla (i.e. papillary necrosis).

Kimberley and co-workers (1979) reported light- and electron-microscopic observations in kidney biopsy from a patient with systemic lupus erythematosus who suffered acute renal failure after taking aspirin 3.6 g/d for 7 days and ibuprofen 1.6 g/d for 5 days. The patient was subsequently treated with prednisolone 60 mg/d for 3 days, after which the renal biopsy was taken (Kimberley *et al.*, 1979). Mild mesanglial hypercellularity was observed, but otherwise no fine structural evidence of damage was apparent in the glomeruli except for the appearance of small electron-dense deposits. The tubular epithelial cells showed patchy changes, i.e. disorientation or loss of the brush border, increased vasiculation, and irregular shaped mitochondria with distorted basal infoldings (Kimberley *et al.*, 1979). These authors considered that this was a case of reversible renal injury attributed to aspirin, but the patient was obviously on mixed medication and it is not possible to ascribe these pathological changes to one of these drugs. It is possible that mild injury is reversible, but, without sequential biopsies at the time of onset of renal failure it is difficult to draw further conclusions.

The main problems with most of the earlier histopathological studies performed on human material are (a) to determine the sequence of events with respect to time, and (b) discriminate the components of the analgesic mixtures mainly responsible for the observed changes. The alternative approach has been to try to reproduce the pathological changes seen in man by experiments in laboratory animals (Schnellmann, 1998). Here there has been much controversy over the experimental designs and animals employed by various workers. In the early studies (see Gross and Greenberg, 1948 for review), very high, mostly toxic, doses of aspirin or salicylate given orally produced albuminuria, celluluria, urinary casts and, occasionally, haematuria and histological signs of renal damage in a variety of laboratory animals. Papillary necrosis and lesions in the vasa recta were observed by Kincaid-Smith and co-workers in most rats dosed orally for 3–9 months with aspirin–phenacetin–caffeine (APC) mixtures, the daily dose of aspirin being 210 mg/kg (Kincaid-Smith et al., 1968; Saker and Kincaid-Smith,1969). Nanra and co-workers (1980) reported the development of renal papillary necrosis in most of the rats dosed orally for 72 weeks with one of: (a) aspirin 380 mg/kg per day + caffeine 140 mg/kg per day; (b) aspirin 504 mg/kg per day + salicylamide 168 mg/kg per day + caffeine 168 mg/kg per day; or (c) aspirin 500 mg/kg per day + caffeine 150 mg/kg per day. In contrast, Stone and co-workers (1977) only found renal papillary oedema in 1 of 30 and 4 of 30 rats dosed orally for 14 weeks with aspirin 100 or 200 mg/kg per day respectively, and in 2 of 30 rats dosed with diflunisal 100 mg/kg per day. No other signs of damage were observed in the other groups (30 rats each of both sexes dosed with aspirin 25 or 50 mg/kg per day (Stone et al., 1977). Likewise, mild papillary oedema was only observed in one of four dogs given aspirin 200 mg/kg per day and two of four dogs given diflunisal 50 mg/kg per day orally for 14 weeks (Stone et al., 1977). No other signs of damage were observed in the dogs given aspirin 25, 50 or 100 mg/kg per day or diflunisal 12.5, 25 or 50 mg/kg per day.

Oral feeding of aspirin, phenacetin and caffeine alone or in combination (up to their respective LD_{50} doses) to male and female mice for 73–80 weeks caused only mild changes, except in the groups given high doses of phenacetin (Macklin and Szot, 1980). Repeated oral administration of aspirin 100 or 1000 mg/kg per day to domestic pigs failed to cause any histologically observed changes to the kidneys (McIver and Hobbs, 1975; Rainsford and Challis, 1984).

Long-term (40–83 weeks) administration of aspirin 120–230 mg/kg per day to Fisher F344 rats resulted in renal papillary lesions and decreased urinary concentrating ability, whereas paracetamol 140–210 mg/kg per day under the same conditions did not cause any renal changes (Burrell et al., 1991a). Combination of aspirin and paracetamol led to the development of renal papillary lesions (Burrell et al., 1990; 1991b).

In a 6-month toxicity study performed under Good Laboratory Practice (GLP) conditions, Lehmann et al. (1996) compared the effects of 50–200 mg/kg per day of the mixture of aspirin, paracetamol and caffeine in the ratio of 5:4:1 by weight with those of the combinations without caffeine, or 200 mg/kg per day aspirin or paracetamol alone. Aside from decreased liver weights in male rats given aspirin, and gastrointestinal ulceration, the kidneys only showed 'initial' evidence of age-related changes comparable with those of the controls. Epithelial cell excretion was slightly increased in the 200 mg/kg per day aspirin group, but only at week 25, while protein and N-acetyl-glucosaminidase excretion were slightly increased in male rats that received 200 mg/kg per day paracetamol at the same time. Electron-microscopic observations revealed thickening of the basement membrane in rats that received 200 mg/kg per day aspirin.

These long-term studies in animal models are important in highlighting the differences in effects of the analgesics in relation to study design, dosage, and the minor time-dependent changes over the long term.

Overall, these chronic studies in laboratory species show that: (a) within the therapeutic dose ranges equivalent to those employed in man, neither aspirin nor diflunisal alone are significantly nephrotoxic; (b) massive toxic doses of analgesic mixtures containing aspirin induce renal papillary necrosis; and (c) high doses of aspirin or diflunisal *alone* can induce papillary oedema, which may be a prelude, when aspirin is present in mixtures at high doses, to the development of papillary necrosis. *In vitro* studies have shown that renal papillary cells from rats given NSAIDs are sensitive to the toxic action of aspirin and paracetamol (Whiting et al., 1999).

There still remains the question arising from the original suggestion by Kincaid-Smith that the damage first begins in the medulla (Dawborn et al., 1966; Nanra and Kincaid-Smith, 1972). The studies by Arnold and co-workers (1976) show that large oral doses of aspirin (300–600 mg/kg) induce cortical tubular necrosis in rats within 24 hours of drug administration. This suggests that damage to the tubules in the cortical region is a primary event. Interestingly, these authors found that tolerance may develop following long-term administration of aspirin with reduction in the tubular damage being evident (Arnold et al., 1973). The appearance of oedematous changes in the renal papillary zone of rats and dogs following high doses of aspirin or diflunisal (Stone et al., 1977) suggests that this may be a consequence of the earlier disturbances to the proximal tubular functions and general dysfunction of the distal tubular system (Porro et al., 1980).

It has been claimed that the rat is an unsuitable species in which to perform such studies of nephro-toxicity because this species has a different papillary structure and biochemical pathways in drug metabolism compared with those of humans (Duggin, 1977), while studies performed in other species would appear to support the main conclusions regarding the nephrotoxicity of aspirin in humans. Another aspect requiring more detailed analysis is the renal histology of the rat, which has a 'unipapil-lary' kidney – especially in the organisation of the tubular concentration mechanism, which differs appreciably from that of the multipapillary kidney in man. The pig appears to have a kidney structure of a multipapillary type and a renal drug detoxification mechanism resembling that in man, so the studies of the effects of aspirin in this species (McIver and Hobbs, 1975; Rainsford and Challis, 1982) appear to be more representative of the situation in man.

The possibility that the stress of a chronic inflammatory condition may predispose arthritic patients to renal damage by salicylates was investigated by studying the effects of chronic oral administration of aspirin 200 mg/kg per day to adjuvant-arthritic rats for 9 days (Rainsford and Challis, 1982). No changes were observed in the kidneys of these animals. Within the morphological limitations noted above, it appears that such a chronic inflammatory condition does not predispose to the development of renal damage by aspirin (Rainsford and Challis, 1982).

Biochemical effects of salicylates

From the above pathological evidence, it is necessary to consider the biochemical changes induced by the salicylates that contribute to the development of nephropathy in certain auto-immune conditions, e.g. systemic lupus erythematosus (Ohtomo et al., 1998) or diabetes mellitus where paradoxically it may be beneficial (Hopper et al., 1986; Moel et al., 1987; Bingulacpopovic et al., 1992; Guillausseau, 1994; Hamidou et al., 1995; Corsi et al., 1995; Contreras et al., 1997; Sawaki and Berger, 1998; Hanson et al., 2000) in renal transplantation (Murphy et al., 2001), or in immunoglobulin A nephropathy (Park et al., 1997) with complications of anti-phospholipid syndrome and in analgesic abuse. First it is necessary to consider the pharmacokinetics of the drugs and especially the influences of other analgesics or NSAIDs on the biodistribution of salicylates in the kidney.

Factors influencing drug and actions

Autoradiographic studies have shown that exceptionally high concentrations of radioactivity-labelled salicylate accumulate in the renal cortex (Rainsford et al., 1980a; 1981a). The reasons for the hypercon-centration of salicylate in this region are two-fold. First, salicylate is ultrafiltered in the glomeruli, then it undergoes back-diffusion in the proximal tubules and recycling via the blood back to the glomeruli (Beyer and Gelarden, 1978; Roch-Ramel et al., 1978). It seems reasonable to assume that this mechan-ism would also apply for all other salicylic acids but not to conjugates (e.g. salicylurate), which can accumulate in tubular cells (Cox et al., 1989) or that are excreted without recycling (Beyer and Gelar-den, 1978). The back-diffusion of salicylate is pH sensitive, so procedures that alkalinise the glomerular filtrate will reduce drug recycling. The second reason for the hyperconcentration is that some anal-gesics (e.g. aspirin, phenacetin and paracetamol) may covalently modify biomolecules in the kidney. For example, aspirin acetylates kidney proteins and other biomolecules, and the acetylation is espe-cially marked in the cortical zone (Caterson et al., 1978; Rainsford et al., 1981a; 1983). Thus aspirin, through its interaction with biomolecules (i.e. accompanying its metabolism to salicylate and acetyla-tion), causes the accumulation of both these products in the cortex (Whitehouse et al., 1977b). Forma-tion of acyl glucuronides of salicylate or diflunisal may lead to their reaction with biomolecules (Dickenson and King, 1991; 1993; 1996; Dickenson et al., 1993; King and Dickenson, 1993; Williams and Dickenson, 1994; Williams et al., 1995). Covalent binding to mitochondrial proteins has been observed in rats given [^{14}C]-salicylate; this binding being five times greater in aged animals (Kyle and Kocsis, 1985; 1986). It is possible that the paracetamol quinine-imine metabolite that forms from high doses of paracetamol or phenacetin and the alkyl products from phenacetin (de Vries, 1981; Nelson *et al.*, 1981) could also interact with biomolecules (e.g. RNA in the case of phenacetin; Nery, 1971) analogous to those acetylated by aspirin. The differential covalent modification of proteins by aspirin, paracetamol and phenacetin could enhance the nephrotoxicity of these drugs as combinations compared with the drugs alone.

A synergistic interaction between these drugs is seen with at least one important enzyme involved in regulating renal function – the prostaglandin (PG) endoperoxide synthetase system. These comprise COX-1, which is important in normal renal and associated vascular functions, and COX-2 in certain regions, e.g. the macula densa, whose production of PGs may be important in regulating glomerular filtration, proximal tubular ion reabsorption and renin production (Dubois *et al.*, 1998). COX-2 inhibition may cause impairment of renal functions (Dubois *et al.*, 1998; Heise *et al.*, 1998; Eras and Perazello, 2001) and death of renal medullary interstitial cells (Hao *et al.*, 1999). Inhibition of COX-2 causes upregulation of the renin–angiotensin system (Wolf *et al.*, 2000). In rabbit renal medullary interstitial cells, the selective COX-2 inhibitor SC-58236 but not the COX-1 inhibitor SC-58560 selectively blocked PGE_2 synthesis and caused cell death. Several unselective COX inhibitors produced similar effects to the COX-2 inhibitor, as did the antisense for COX-2. These results suggest that COX-2 inhibition is linked to cell death in these renal cells.

Using mouse inner medullary collecting duct cells, Rocha *et al.* (2001) showed that inhibition of COX-2 was not, alone, a factor in the death of these cells *in vitro*. Paracetamol 0.5 mmol/l arrested the cells in late G1 and S phases, the DNA synthesis being inhibited by effects on ribonucleotide reductase, producing a mixed form of cell death characterised by swollen cells and nuclei and apoptosis. The same concentration of aspirin or salicylic acid produced a reduction in the numbers of proliferating cells by about one-half.

Either paracetamol alone or that formed as the principal metabolite of phenacetin metabolism (Duggin and Mudge, 1976; Caterson *et al.*, 1978) is oxidised in the kidney by both the PG endoperoxide synthetase and the cytochrome P_{450} oxidase system (Mohandas *et al.*, 1981). The oxidation of paracetamol by the PG endoperoxide synthetase system, predominantly active in the medullary interstitial tissue (Limas *et al.*, 1976; Mohandas *et al.*, 1981), is inhibited by aspirin (Duggin and Mudge, 1976). This would force paracetamol to be oxidised by the cytochrome P_{450} system and, in the event of a glutathione (GSH) deficiency, could cause formation of appreciable quantities of its reactive quinine-imine metabolite (de Vries, 1981; Nelson *et al.*, 1981). The depletion of GSH levels could occur by: (a) salicylates inducing a leakage from cellular stores (Kaplowitz *et al.*, 1980); (b) fasting (Pessayre *et al.*, 1980); (c) chronic ethanol consumption (Peterson *et al.*, 1980); (d) certain chemical carcinogens (Guenthner and Oesch, 1981; Oesch, 1981); (e) excess paracetamol itself (Buttar *et al.*, 1977); and (f) a generalised reduction in ATP (Whitehouse, 1965; Smith and Dawkins, 1971). Reduced GSH levels could be one factor accounting for the enhanced renal toxicity of aspirin + paracetamol (or phenacetin) combinations (see p. 431), and these could also be a reflection of oxyradical injury. Also, the potential for ethanol to enhance the nephrotoxicity of these drug combinations should be noted, and this could contribute to reduction of GSH levels. This particular agent is known to enhance the general toxicity of paracetamol itself (Peterson *et al.*, 1980).

In addition to affecting formation of the mercapturate and cysteine conjugates of paracetamol (i.e. from glutathione), aspirin also influences the conjugation of paracetamol with sulphate, so increasing urinary excretion in rodents of its glucuronides (Wong *et al.*, 1976; Whitehouse *et al.*, 1977). Thus any condition or drug that affects the supply of precursors for conjugation could have a profound effect on the metabolic detoxification of salicylates or other NSAIDs or paracetamol (Bailey *et al.*, 1998). The pathological importance of the lack of capacity to form glucuronyl conjugates can be seen from the studies by Axelsen in the Gunn strain of rats; this is a mutant of the Wistar strain that has a deficiency of the enzyme UDP-glucuronosyltransferase (Axelson, 1975; 1980; Axelson and Burry, 1976). This author found that single or repeated doses of aspirin, paracetamol or phenacetin to homozygous Gunn rats readily induced renal papillary necrosis, whereas the same doses of these drugs failed to induce this lesion in either heterozygotes (which possess the UDP-glucuronosyltransferase activity) or in other normal rat strains (Axelson, 1975; 1980).

CHAPTER 8

The propensity of females to develop renal papillary necrosis could, in addition to socio-psychological influences, be related to their hormonal status. Thus Owen and Heywood (1980) found that oestrogen pretreatment of castrated male rats increased the development of necrosis and urinary γ-glutamyl transpeptidase output, an enzyme that these authors claimed is a useful indicator of renal damage. The value of enzymuria measurements in detecting renal damage by aspirin and other analgesics has been questioned following the studies of Plummer and co-workers (1975). These authors found that high doses of aspirin or phenacetin did induce the urinary output of lactate dehydrogenase (LDH – a usual marker of generalised cell injury) within 12–24 hours of dosage and, for phenacetin alone, of alkaline and acid phosphatase and glutamate dehydrogenase. However, the urinary excretion of these enzymes declined following repeated administration of these drugs, despite the development of extensive renal damage (Plummer et al., 1975). Thus enzymuria is only of value in determining acute changes in the rat, and this may only be a reflection of the transient celluluria attributed to the sloughing of effete cells (Prescott, 1965). Similar transient excretion of albumin and β_2 microglobulin following single doses of aspirin (Hustad et al., 1985) may lead to adaptive changes upon repeated intake of the drug. These results should be contrasted with the β-N-acetylglucosaminidase (NAG) enzymuria observed in rheumatoid subjects with renal damage following ingestion of nephrotoxic levels of aspirin and related drugs (Burry et al., 1976; Proctor and Kunin, 1978). It appears that the rat can somehow undergo certain type(s) of adaptive changes following repeated dosage of analgesics. Simple cell damage or sloughing may decline, but pathological changes proceed with autolytic changes. These adaptive changes may occur in man, so that 'silent' lesions may be evident following massive ingestion of analgesic/NSAIDs. Only in the presence of a disease pathology (e.g. in systemic lupus erythematosus), where there may be lysosomal fragility, will the enzymuria become evident. Clearly these are speculative concepts, but this aspect of potential drug–disease interaction(s) is of major importance in the context of the biochemical actions of the drugs.

It appears that high doses of salicylate itself can induce acute enzymuria in the rat (Trnavsky and Rovensky, 1976), so this phenomenon is not unique to acetylated salicylates. Differences in the metabolism of aspirin to salicylate would not be expected to influence the development of transient enzymuria. Indeed, this phenomenon may be a property of salicylate itself. The acetylation of enzymes and other biomolecules could produce other biochemical changes independent of the salicylate-induced transient celluluria/enzymuria. Whether or not these changes represent an inductive phase of renal damage is still debatable. There is considerable evidence that ischaemic reactions may be fundamental to the pathological changes induced by analgesic drugs that initiate renal damage and/or changes in renal function.

There is also a variety of drug effects on the prostaglandin–angiotensin, kinin–angiotensin, and renin–angiotensin systems that cause important physiopathological changes (Stygles et al., 1978; Fitzgerald et al., 1980; Keiser, 1980; Romero and Strong, 1997; Massie and Teerlink, 2000; Montini et al., 2000; Wolf et al., 2000). There is a complex physiological interplay between these systems, and Figure 8.10 is an attempt to illustrate these together with the effects of aspirin and salicylate upon components of these systems (based on Armstrong et al., 1976; Romero et al., 1976; Weber et al., 1976; Arnfelt-Rønne and Arrigoni-Martelli, 1977; Berl et al., 1977; Staszewska-Barczak, 1978; McGiff and Wong, 1979; Brooks et al., 1980; Plotz and Kimberley, 1981; Berg et al., 1990; Chun-Ming et al., 1999; Colleti et al., 1999).

The actions of prostaglandins E_2 and I_2 in acting as vasodilators contrast with those of $PGF_{2\alpha}$ and thomboxane A_2 (TXA_2) which are potent vasoconstrictors impairing blood flow (Gerber et al., 1978a; 1978b; Berg et al., 1990). The effects of PGE_2 and PGI_2 are to direct blood flow to the inner cortex at the expense of that in the outer region. Inhibition by aspirin of the synthesis of these PGs (although occurring at a lower rate in the cortex than in the medulla; Mohandas et al., 1981) would be expected to be more profound in the cortex because of the higher concentration of both the acetyl and salicyl moieties of aspirin present in the cortex (Rainsford et al., 1981b; Rainsford and Challis, 1984). Likewise, the aspirin-induced inhibition of nitric oxide production (Cheng et al., 2000) and kinin formation (Abe et al., 1978; Brooks et al., 1980; Berg et al., 1990) would be expected to impair vasodilatation and stimulation of PG production.

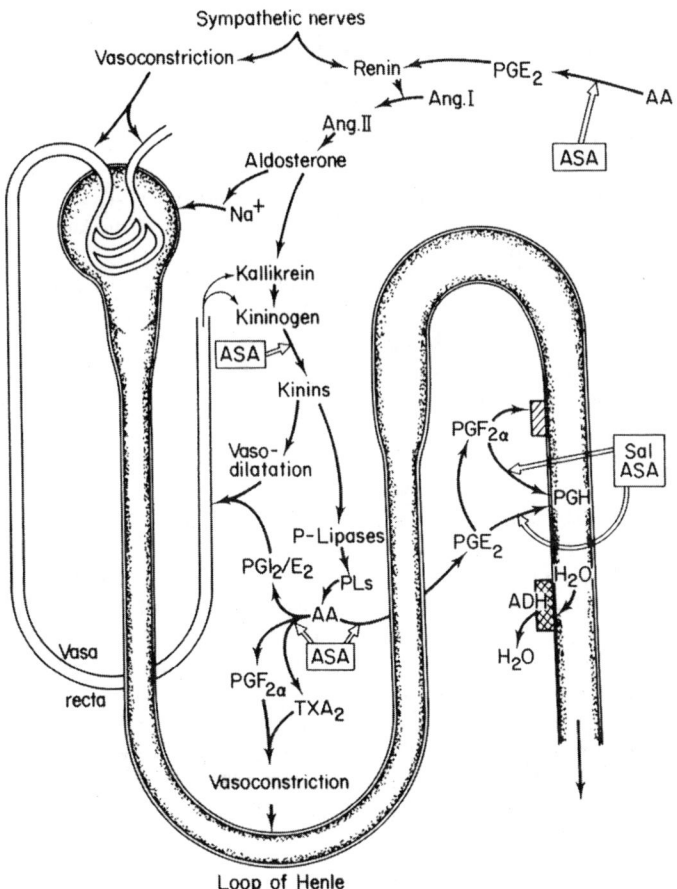

Figure 8.10 Action of aspirin and its metabolite salicylate upon kinin, renin and prostaglandin production in the kidney with consequences upon glomerular filtration of sodium and potassium ions, vascular functions of the vasa recta and renal tubular functions including water output. Aspirin inhibits renin production, which may in part be due to its inhibitory effects upon prostaglandin (PG) E_2 production. This causes reduced output of angiotensin II, aldosterone, sodium and water. The aldosterone-mediated production of kinins (with their consequent actions upon vasodilation and release of vasodilator prostaglandin E_2) is blocked by aspirin through inhibition of both renin activation and COX-2 derived prostaglandin synthesis. This results in vasoconstriction. The overall effects of aspirin upon these renal functions are more 'pharmacological' than 'pathological' and essentially comprises a 'blunting' of renal functions. Only with additive effects of high concentrations of other drugs, especially on oxidative metabolism of glucose and cytochrome P_{450} and cyclo-oxygenase-2 mediated renal metabolism, will pathological changes become evident. ADH = antidiuretic hormone; Ang I and Ang II = angiotensin I and II respectively; P-lipases = phospholipases and PLS = phospholipids. (From Rainsford (1984a). Reproduced with permission of the publishers, Butterworths/Heinemann.)

Sympathetic discharge, which induces vasoconstriction and ischaemia, could be especially important in disease–stress responses in the kidney. Sympathetic stimulation also enhances renin release, with consequent enhancement of renal prostaglandin production and stimulation of angiotensin II conversion, and subsequent aldosterone-mediated Na$^+$ secretion into the glomeruli. Aspirin inhibits renin release (Romero *et al.*, 1976; Brooks *et al.*, 1980) by impairing prostaglandin production (Romero *et al.*, 1976) probably by inhibiting COX-2 (Wolf *et al.*, 2000). The consequences of this would appear to be reflected in regulation of sodium excretion (Romero *et al.*, 1976), where reduced Na$^+$ excretion (Berg,

1977) could occur through aspirin-induced inhibition of prostaglandin (E_2) synthesis (Crowshaw, 1979). Reduction of K^+ and water output induced by aspirin (Berg, 1977) could also be caused by reduction in prostaglandin synthesis due to this drug (Crowshaw, 1979).

The involvement of renal prostaglandins in the regulation of renal blood flow may not be so much of a normal physiological function, as this only becomes evident in some altered physiological states, e.g. hormone or nerve-mediated vasoconstriction, renal artery or chronic inflammatory diseases (Crowshaw, 1979; Ergun et al., 2000). The overproduction of urinary PGE_2 in patients with systemic lupus erythematosus (SLE) may be a reflection of the inflammation of the kidneys of these patients (Kimberley et al., 1978; Crowshaw, 1979; Plotz and Kimberley, 1981). This inflammatory state may be a major factor accounting for the predisposition of aspirin-induced renal function, and cause renal damage in these patients (Kimberley and Plotz, 1977; Kimberley et al., 1978; Crowshaw, 1979; Plotz et al., 1981; Nakano et al., 1998). The inhibition of renal prostaglandin production by aspirin may have more profound effects in these patients because of the natural protective effects of prostaglandins on blood flow through the inflamed kidneys of those patients with systemic lupus erythematosus (SLE) and severe rheumatoid arthritis.

Normal subjects whose water/electrolyte status is affected by exercise may also be more susceptible to actions of aspirin on renal function (Zambraska et al., 1982).

In summary, the main pharmacological effects of aspirin on systems involved in regulating renal functions (Figure 8.10) are (a) decreases in blood flow, renin production, aldosterone levels, Na^+ excretion, water output and renal PGE_2 synthesis and nitric oxide production, and (b) increased blood pressure (Crowshaw, 1979). Thus, practically every step involved in maintenance of renal functions shown in Figure 8.10 is potentially capable of being interrupted by aspirin or salicylate. In the normal kidney, moderate to even high aspirin intake is obviously only going to cause minor blunting of renal functions, which is apparently readily reversed. Only where there is pre-existing pathology that is itself manifesting changes in kidney structure and function will there be untoward effects of aspirin mediated through disturbance of these physiological systems. The same comments are very likely to apply to other acidic NSAIDs (Crowshaw, 1979). Pharmacokinetic interactions and age exert profound effects on the components of kidney function and, where influencing renal prostaglandin production, influence the effects of salicylates and other NSAIDs (Table 8.18).

Renal metabolism and biosynthesis of macromolecules

Intravenous infusion of increasing doses of aspirin 7–200 mg/kg to anaesthetised dogs increased renal oxygen consumption (determined by measurements of the arteriovenous differences in oxyhaemoglobin concentration) and reduced sodium excretion and blood flow (Berg and Bergan, 1977). At plasma salicylate concentrations of 200–400 µg/ml (1.45–2.9 mmol/l) QO_2 increased by 36 per cent despite a 16 per cent decrease in renal blood flow, and renal lactate uptake was increased (Berg and Bergan, 1977). These results relate to the effects that aspirin has on renal mitochondrial metabolism (Mingatto et al., 1996; Corti et al., 1997; Pigoso et al., 1998; Al-Nasser, 1999) in which it can uncouple renal oxidative phosphorylation, thereby decreasing production of ATP. In isolated rat kidney tubules, aspirin 2 mmol/l inhibited the gluconeogenesis of 2-oxoglutarate, succinate and fructose, and also increased the respiratory rate and decreased intracellular concentrations of ATP (Dawson, 1975). The inhibition of gluconeogenesis was attributed to reduced ATP output (Dawson, 1975). Chronic feeding of 0.25 per cent aspirin to rats for 5 weeks was, however, found to reduce the renal phospho-enolpyruvate carboxylase activity (Madappally et al., 1972), so that inhibitory effects of aspirin on this enzyme could also account for the reduction by this drug in the gluconeogenic activity of the kidney. Prolonged feeding of aspirin 10 mg/kg to dogs caused uncoupling of oxidative phosphorylation in kidney mitochondria, and reduction in glycolytic activity in this tissue (Ratnikov, 1975). Thus reduction in renal ATP levels by aspirin could result from both reduction in the availability of oxidisable substrates as well as by uncoupling of oxidative phosphorylation. Reduced ATP levels could influence (a) the adenine–nucleotide regulation of xanthine dehydrogenase, leading to its conversion to the oxidase form capable of generating tissue-destructive superoxide radicals (Granger *et al.*, 1981), and (b) the availability of chemical energy for activation reactions to synthesise renal macromolecules, maintain tissue homeostasis and sustain tissue regeneration. The mitochondrial effects of salicylates may assume particular significance in those individuals with certain mitochondrial genome mutations where these have been associated with kidney disease (Singh *et al.*, 1996).

A reduction in glucosamine and, to a lesser extent, leucine turnover has been observed in microsomal and mitochondrial fractions from the kidneys of rats given aspirin in their drinking water, the consumption of drug being estimated at 800–100 mg/kg per day (Wheldrake, 1975). While this is indeed a high rate of consumption, it is possible that appreciable hydrolysis of the aspirin occurred by the drug being placed in drinking water. The inhibition of glycosaminoglycan and protein turnover could also result from direct inhibitory effects of the drug on enzymes involved in biosynthesis of these macromolecules (Whitehouse, 1965; Smith, 1966; Smith and Dawkins, 1971; Rainsford, 1980), as well as from general depression in ATP availability.

Phenacetin and indomethacin, like aspirin and salicylate, affect mitochondrial respiration, although both these drugs act by direct inhibition of mitochondrial respiratory enzymes (Cooney and Dawson, 1977; Dawson, 1979; Druery and Dawson, 1979). Salicylate, in contrast, uncouples oxidative phosphorylation at low concentrations and inhibits mitochondrial dehydrogenases and the cytochrome oxidase only at high concentrations (of salicylate) (Whitehouse, 1965; Smith, 1966; Thompkins and Lee, 1969; Smith and Dawkins, 1971). Thus combinations of aspirin with phenacetin could in part exert an enhanced nephrotoxicity compared with that of the drugs alone, because of the double attack of these drugs on components of the respiratory chain. Other NSAIDs (e.g. diclofenac, mefenamic acid) affect renal mitochondrial functions (Uyemura *et al.*, 1997). Thus combinations of NSAIDs including salicylates may have more pronounced mitochondrial effects than observed with the drugs alone. This combined effect may be a factor accounting for renal toxicity in patients taking mixtures of NSAIDs and analgesics.

In addition to affecting the energy metabolism of mitochondria, salicylates, like other NSAIDs, may affect caspases and the release of cytochrome C from mitochondria, leading to apoptosis (Wallace and Starkov, 2000).

HYPERSENSITIVITY REACTIONS AND ASTHMA

Aspirin and diflunisal, in common with other NSAIDs, induce a wide variety of hypersensitivity reactions in the skin (Giraldo *et al.*, 1969; Farr, 1970; Speer *et al.*, 1981). Upper respiratory tract reactions (asthma, etc.) are also particularly evident with aspirin (Giraldo *et al.*, 1969; Speer *et al.*, 1981). These conditions are manifest in symptoms such as skin rashes and eruptions, urticarial weals, angioedema, asthma, nasal polyps, rhinitis purpura and angina pectoris (Giraldo *et al.*, 1969; Moloney, 1977; Speer *et al.*, 1981). Overall the occurrence of all these conditions is not common, but the severity may vary depending on the condition (Szczeklik, 1980; Settipane, 1981).

The incidence of these anaphylactoid conditions has been variously estimated as occurring in between 0.004 and 0.3 per cent of the population consuming aspirin (Giraldo *et al.*, 1969; Kallos and Schlumberger, 1980; Settipane *et al.*, 1980). These conditions frequently occur in females, and in both sexes during the third to fifth decades of life.

Settipane (1981) defined the term 'aspirin intolerance' as broadly describing individuals presenting with acute urticaria–angioedema, bronchospasm, severe rhinitis or shock within 3 hours of aspirin ingestion. This author has discriminated two subgroups of individuals with aspirin intolerance; an urticarial group (which exhibits predominantly urticarial reactions upon challenge) and a bronchospastic type evident in individuals having a typical asthmatic reaction (Settipane, 1981). A similar subdivision has also been employed by Szczeklik (1980), and will be employed here. Aspirin intolerance occurs most frequently in the chronic urticarial type (23 per cent), whereas the bronchospastic type is evident in 4 per cent of individuals (Settipane, 1981).

CHAPTER 8

Aspirin-intolerant or aspirin sensitive asthma

Aspirin-intolerant asthma (AIA) was first recognised by Widal and co-workers (1922). This condition is an aggressive mucosal inflammatory disease combined with precipitation or eventual development of asthma and attacks of rhinitis, which occurs after ingestion of aspirin and most other NSAIDs (Szczeklik and Nizankowska, 2000). The clinical features of AIA are summarised in Table 8.19 (Stevenson, 1998). It is probably relatively undiagnosed in the population of asthmatics, in part because of the avoidance of NSAIDs by asthmatic patients who may be aware of the potential association between the intake of these drugs and asthma, and also owing to the lack of recognition in patients with mild NSAID-induced reactions because of the delayed onset (Szczeklik and Nizankowska, 2000; Szczeklik et al., 2000. Estimates of this under-reporting of aspirin-intolerant asthma (AIA) vary, but it is undoubtedly due to lack of routine challenge testing with aspirin and differences in responses to this with the route of administration of aspirin and the dose (Szczeklik and Nizankowska, 2000). Thus oral challenge with aspirin or its lysine salt (Carnimeo et al., 1981) remain the gold standard, but may precipitate asthmatic attacks, while nasal and bronchial routes for challenge are safer but less sensitive (Szczeklik and Nizankowska, 2000).

Surveys relying on patient history give a low prevalence of 3.8–4 per cent, whereas oral challenge (using aspirin) combined with spirometry measurements suggests that 8–28 per cent of asthmatics may be affected (Yunginger et al., 1973; Rachelefsky et al., 1975; Settipane et al., 1980).

The occurrence of AIA in the 'normal' population is estimated to be 0.3–0.62 per cent (Settipane et al., 1974; Hedman et al., 1999).

In adults AIA predominates in women, occurring approximately 2.5 times as often as in men, and the disease is probably more progressive and severe in females (Samter and Beers, 1968; Giraldo et al., 1969; Stevenson, 1984; Szczeklik and Nizankowska, 2000; Szczeklik et al., 2000). Sex hormones probably play a part in the disease, so underlying the predominance in women of other common autoimmune diseases, although the frequency may vary within these diseases (Whitacre et al., 1999). Recently, some insight into the natural history of the disease has come from a Pan-European study in 500 patients (Szczeklik et al., 2000). Symptoms start usually in the third decade, initially with persistent rhinitis and then progressing to asthma, aspirin intolerance and nasal polyposis. Atopy, which is present in about one-third of AIA patients, leads to earlier manifestations of rhinitis and asthma, but not of aspirin intolerance or nasal polyposis. While family history of aspirin intolerance was reported to be 6 per cent, this low rate might reflect poor recall or the underlying poor recognition of the condition, especially in the past. Whether the natural history of the disease in other populations in the world is similar to that in this report has still to be established.

Contrary to popular belief (Varner et al., 1998) aspirin-sensitive asthma has been considered relatively common amongst children (Settipane et al., 1980). Several reports suggest that the frequency (determined by oral challenge) in adults varies from 1.9 to 28 per cent of asthmatics (Falliers, 1974;

TABLE 8.19

Clinical features of asthmatic patients with aspirin disease. (From Stevenson, 1998.)

Category	Feature
Age of onset	>10–40 years
Rhinitis	Chronic congestion (>90%)
	Associated IgE-mediated (20%)
Nasal symptoms	Congestion, rhinorrhoea, anosmia, paranasal headache, sleep deprivation
Nasal examination	Pale, congested membrane, polyploid
Nasal smear	Eosinophils, mast cells, PMNs, bacteria (all variable)
Sinus radiographs	Abnormal with any pattern; parasinusitis most common
Asthma	Intermittent, often remitting when sinuses inactive
	Chronic, severe, especially when associated with chronic sinusitis
Sinusitis	Intermittent – chronic

Rachelefsky *et al.*, 1975; Settipane *et al.*, 1980). A familial coincidence has been suggested in some studies (Fisherman and Cohen, 1973), although the occurrence of this association is probably low (Settipane *et al.*, 1980).

Comparing aspirin-intolerant children with those exhibiting a tolerance to this drug, Rachelefsky and co-workers (1975) found the intolerant group was characterised by a greater number of females, an earlier onset of wheezing, and more sinusitis than was evident in the tolerant group. No differences were observed in the eosinophil count, mean IgE levels, or nasal eosinophilia. This suggests that there are no obvious immunological differences between these two groups of atopic individuals (Rachelefsky *et al.*, 1975). The suggestion that the increasing prevalence of asthma in children may be a consequence of declining use of the drug in this group (Varner *et al.*, 1998) requires careful evaluation in view of the above mentioned diagnostic issues.

Other hypersensitivity reactions may appear in conjunction with aspirin-sensitive asthma. Settipane and co-workers (1974) found, in a prospective study of 1372 atopic patients, that 2.7 per cent were intolerant to aspirin and had both asthma and rhinitis, whereas the frequency of aspirin intolerance amongst those with asthma was 3.8 per cent and with rhinitis alone was 1.4 per cent. In contrast, in a study of 205 patients, Speer and co-workers (1981) failed to confirm the widely-held view that aspirin-sensitive asthmatics have a strong tendency to develop nasal polyps.

The sequence of symptoms characterising the 'classic' case of aspirin sensitivity has been constructed as follows: during the third and fourth decades of life the patient starts to experience intense vasomotor rhinitis, accompanied over several months by chronic nasal congestion and the development of nasal polyps. Typical bronchial asthma and general symptoms of intolerance develop. After ingestion of aspirin or other NSAIDs by AIA subjects, an acute asthmatic attack can occur within minutes or up to 3 hours later (Szczeklik and Nizankowska, 2000). This is usually accompanied by profuse rhinorrhoea, infection or inflammation of the conjunctiva, periorbital oedema and sometimes a brilliant scarlet flushing appearance in the head and neck region (Szczeklik, 1980). The condition may be life-threatening in an appreciable proportion of patients, with death possibly occurring in highly-sensitive individuals following ingestion of even a single 300-mg tablet of aspirin.

While it is generally accepted that most patients who have AIA are also sensitive to other NSAIDs but not paracetamol, there are some reported exceptions. Recently there have been case reports of asthma being associated with the intake of paracetamol, and while these reports may be limited to a few individuals they do raise concerns about the accepted view that paracetamol can probably be taken safely by asthmatics, including those with AIA. There has been one study showing that the NSAID azapropazone does not elicit a response when given as a challenge to AIA subjects (Szczeklik, 1989). Tolerance has been shown to aspirin in AIA asthmatics (Stevenson *et al.*, 1980).

The sensitivity towards aspirin in asthma has also been shown to apply with some other agents, including benzoates, certain foods, cosmetics, colouring agents and especially the yellow azo dye, tartrazine (FD&C No 5) (Fisherman and Cohen, 1973; Yunginger *et al.*, 1973; Settipane *et al.*, 1974; Settipane and Pudapakkam, 1975; Rachelefsky *et al.*, 1975; Abrishami and Thomas, 1977; Szczeklik, 1980, Valverde *et al.*, 1980; Vargaftig *et al.*, 1980). The latter dyestuff is widely used for colouring foods, drinks, drugs and cosmetics (Settipane *et al.*, 1980) and was first shown by Lockey (1959), to be a cause of sensitivity. Some authors (Rachelefsky *et al.*, 1975; Abrishami and Thomas, 1977; Rudzki *et al.*, 1980; Szczeklik, 1980; Valverde *et al.*, 1980; Speer *et al.*, 1981), have reported a relatively high incidence (15–50 per cent) of tartrazine-induced bronchoconstriction in patients with aspirin-sensitive asthma, whereas others (Weltman *et al.*, 1978; Szczeklik, 1980; Speer *et al.*, 1981) have reported a low incidence. Psychological factors do not appear to be of any consequence in the determination of sensitivity to tartrazine or other drugs in patients with urticaria (Rudzki *et al.*, 1980), but psychological factors are known to precipitate asthmatic attacks (Berkow, 1977).

Clues in a patient's history that give rise to suspicions of AIA include: (a) an association of aspirin intake with the development of typical respiratory symptoms; (b) the development of severe asthma accompanied by chronic nasal congestion and rhinorrhoea; (c) the frequent appearance of nasal polyps; and (d) sudden attacks of asthma requiring emergency medical intervention (Szczeklik and Nizankowska, 2000).

CHAPTER 8

The familial history of AIA that has been suggested was found in one study of European patients to have an occurrence rate of 5.1 per cent (Lockley *et al.*, 1973) and, in a more recent one, of 6.0 per cent (also in European patients; Szczeklik *et al.*, 2000). There have been a few studies in siblings where there have been familial associations recognised and polymorphisms associated with variants in leukotriene C_4 (LTC_4) metabolism identified (Sanak and Szczeklik, 2000). The impression is that sex and atopy have a greater effect on the association with AIA than familial history (Szczeklik. *et al.*, 2000).

Immunological involvement

IgE levels have been variously reported as being normal (Rachelefsky *et al.*, 1975; Abrashami and Thomas, 1977) or above normal (Small *et al.*, 1981; De Weck, 1984), but there does not appear to be any consistency in this parameter in patients with aspirin-sensitive asthma (Small *et al.*, 1981). Many authors consider that these conditions do not have an immunological basis (Abrishami and Thomas, 1977; Kallos and Schlumberger, 1980; Szczeklik, 1980; Zhu *et al.*, 1992), but there are some puzzling features about these hypersensitivity states that do not completely remove this possibility. Eosinophilia is a common feature of aspirin intolerance (Abrishami and Thomas, 1977; Weltman *et al.*, 1978; Szczeklik, 1980).

The involvement of eosinophilia in chronic bronchial inflammation is a predominant feature of AIA (Bochenek *et al.*, 1996; Nasser *et al.*, 1996; Cowburn *et al.*, 1998). The eosinophilia that develops in AIA is accompanied by increased expression of IL-5, a key cytokine involved in the recruitment, maturation activation and rescue from apoptosis of eosinophils. In addition to increased production of pro-inflammatory cytokines, there is overproduction principally from eosinophils of cysteinyl-leukotrienes and an accompanied increase in LTC_4-synthetase that appears to correlate with aspirin sensitivity in AIA patients (Cowburn *et al.*, 1998).

The aetiology of AIA relates to observation of a frequent association with a chronic viral infection and auto-immunisation preceding the development of AIA (Szczeklik, 1988). In some respects AIA has the hallmarks of being an autoimmune syndrome, but there are indications of underlying genetic susceptibility, e.g. in LTC_4 synthetase (Sanak and Szczeklik, 2000).

Several studies appeared during the 1960s and 1970s suggesting that the salicyl or acetyl moieties of aspirin may be responsible for development of immuno-antigenicity in asthma (Schwartz and Amidon, 1966; Giraldo *et al.*, 1969; Christie *et al.*, 1976; Kallos and Schlumberger, 1978). Schwartz and Amidon (1966) showed that, under extreme alkaline conditions, aspirin (but not salicylic acid) could produce small quantities (0.01–0.1 per cent) of N-salicyl derivatives of glycine and ϵ-aminocaproic acid. These authors suggested that this could be a mechanism whereby aspirin combines with proteins to form antigens in hypersensitivity states.

Giraldo and co-workers (1969) showed that antigenicity and hypersensitivity reactions could be elicited in guinea pigs and rabbits given aspiryl chloride, but not by aspirin and salicylic acid. It is hard to envisage an activity resembling aspiryl chloride in aspirin itself.

De Weck (1971) provided evidence that aspirin anhydride (a frequent minor contaminant of aspirin preparations) could form protein conjugates *in vitro* and induce anti-aspiryl antibodies and contact sensitivity to aspirin anhydride in both guinea pigs and rabbits. The author also claimed that similar antibodies could develop in patients with aspirin intolerance after ingestion of the drug, which could cause skin sensitivity to aspiryl-polylysine (de Weck, 1971). In a later study, Bungaard and de Weck (1975) observed contact sensitivity in guinea pigs given aspirin anhydride, acetylsalicylic acid and salicylsalicylic acid (diplosal, salsalate). The clinical relevance of these studies was challenged by Kallos and Schlumberger (1978). In more recent studies no anti-aspirin antibodies have been found in aspirin-intolerant patients (Yurchak *et al.*, 1970; Schlumberger *et al.*, 1974). Diflunisal acyl glucuronide has been shown to form adducts with rat albumin leading to it being immunogenic (Worrall and Dickenson, 1995) so this may be a mechanism for the immunogenicity of this and possibly other salicylates.

The extensive acetylation of proteins, lipids and other biomolecules by aspirin has also been suggested as a basis of aspirin hypersensitivity (Minden and Farr, 1967; Flemström *et al.*, 1976). These authors have detected antibodies against aspirin-altered serum albumin in the sera from aspirin-intolerant patients and those with rheumatoid arthritis who have consumed aspirin. Unfortunately, no specific involvement can at present be ascribed to this reaction of aspirin-acetylated albumin in the pathogenesis of aspirin sensitivity. It is difficult to envisage cross-reactivity with other NSAIDs, tartrazine or benzoates in patients in whom the acetyl or salicyl moieties of aspirin might have initiated antigenic changes, unless there is a separate mechanism activated by these non-aspirin drugs (e.g. stimulation of SRS-A or cysteinyl-LT production). The possibility of chemical modification by aspirin of 'self' macromolecules in asthma still requires more detailed study, as it is conceivable that a form of 'silent' antigenic change may be elicited in certain individuals (e.g. those with a specific HLC haplotype or other genetic determinant).

An intriguing experiment was reported by Flemström and co-workers (1976), where they showed that a passive cutaneous anaphylaxis could be elicited in guinea pigs by prior injection with anti-dextran followed later by an oral dose of a mixture of dextran and aspirin 30–108 μmol. Since the pH of the gastric contents was sufficiently low (pH 2.6–2.9) to maintain the aspirin in the non-ionised form, it was suggested that the drug could, by breaking the mucosal barrier, aid the intragastric absorption of dextran (Flemström et al., 1976). These authors suggested that aspirin may contribute to sensitisation and allergic reactions to potentially antigenic food materials by facilitating their absorption from the stomach. The high incidence of sensitivity to foodstuffs in aspirin-sensitive individuals (Speer et al., 1981) would be one argument in support of this suggestion.

Some attention has been given regarding the possibility that activation of the complement pathway could be important in aspirin intolerance (Kallos and Schlumberger, 1980). Intolerant individuals lack plasma carboxypeptidase B, which normally inactivates C3a and C5a, so allowing overproduction of products of the complement pathway (Kallos and Schlumberger, 1980) The anti-complement activity of aspirin (Voigtländer et al., 1980), together with the effects of aspirin or salicylate on histamine production (Skidmore and Whitehouse, 1966; Conroy and de Weck, 1981) and release of even PGs or TXs, could explain the influence of these drugs in relieving bronchial asthma (Szczeklik, 1980). It is possible that in certain asthmatic conditions (e.g. where there is not a tendency to overproduce LTs and H(P)ETEs) aspirin may be beneficial because of the above-mentioned actions. In other states where a sensitivity to production or actions of LTs and/or H(P)ETE exists, there could be a profound effect on aspirin by diversion of arachidonate metabolism. Related to this are the observations of Szezeklik and co-workers (1977) that aspirin exerts greater inhibition of prostaglandin release from the nasal polyps of AIA compared with aspirin sensitive patients.

Since sodium cromoglycate and ketotifen, inhibitors of mast-cell-derived histamine, SRS-A and other anaphylatoxins, both protect against aspirin-induced asthma (Basomba et al., 1976; Szczeklik et al., 1980), it appears that mast-cell-derived products are produced in this type of aspirin intolerance. Aspirin 0.1–10 mmol/l as well as indomethacin. 0.01–1 mmol/l inhibit production of histamine from zymosan-activated serum-induced leucocytes of asthmatic/urticarial patients (Conroy and de Weck, 1981). Also salicylate inhibits histamine production by inhibiting histidine decarboxylase activity (Skidmore and Whitehouse, 1966), but it is not known if this is an important action of aspirin in vivo.

The concepts that α/β-adrenergic or cholinergic activities and histamine hyper-production may be major factors in aspirin-sensitive asthma have also received support from clinical and experimental studies (Basomba et al., 1976; Meilens and Rosenberg, 1976; Ito and Tajima, 1981). For example, aspirin or related drugs can sensitise the actions of propranolol and cholinergic stimulation in producing muscle contraction in the bronchus or trachea (Basomba et al., 1976; Ito and Tajima, 1981). Endogenous production of prostaglandins, which are known to cause feedback inhibition of the release of cholinergic and adrenergic transmitters, could also be important in these responses. This highlights a potential role of aspirin in inhibiting prostaglandin production, so causing overproduction of these mediators.

Early studies in the 1970s following the discovery of the role of PGs in inflammation indicated that the pathogenesis of aspirin-induced asthma was related to the effects of aspirin on inhibition of prostaglandin biosynthesis. In normal bronchial functions $PGF_{2\alpha}$ and TXA_2 induce bronchoconstriction, whereas PGE_2 induces bronchodilation, increased vasomotor tone, platelet aggregation and elevated cyclic AMP production (Speer et al., 1981). Influences of platelet accumulation and the release of broncho-effective mediators have recently attracted much attention as an aetiological factor in asthma

(Vargaftig, 1983). Inhibition of prostaglandin production might impair the functions of these mediators (Szczeklik, 1980; Settipane, 1981); however, inhibition of the production of all PGs by aspirin
would be expected to depress production of both bronchoconstrictor and bronchodilator
prostanoids. There is also evidence that aspirin down-regulates expression of COX-2 in nasal polps
of AIA patients (Picardo et al., 1999) suggesting that if there are any effects of aspirin on PGs this
will involve inhibiting both COX-1 and COX-2 enzyme production. A more appealing hypothesis
is that overproduction of leukotrienes and hydro(per)oxyeicosatetraenoic acids (H(P)ETEs) could
occur from inhibition of PG cyclo-oxygenase by aspirin and consequent diversion of arachidonic
acid through the lipoxygenase pathway (Chand and Altura, 1981; Speer et al., 1981). The well-
known hyperproduction of the slow-reacting substance in anaphylaxis (SRS-A), which is a mixture
of leukotrienes C_4 and D_4 in allergic states (Burka and Paterson, 1980; Rudzki et al., 1980; Yoshida
et al., 1998), together with its extraordinary potency as a bronchoconstrictor (Weiss et al., 1982),
make this mediator a strong candidate as one of the major effectors in asthma. Inhibitors of lipoxygenase activity or leukotrienes antagonists prevent bronchoconstriction, and this also supports the
idea of the SRS-A and related products being mediators of bronchoconstriction is asthma (Dawson
and Sweatman, 1980; Nijkamp and Ramakers, 1980; Drazen et al., 1999). Neither tartrazine nor its
metabolite sulphanilic acid inhibit prostaglandin synthesis in platelets (Vargaftig et al., 1980), guinea
pig lung microsomes (Gerber et al., 1979) or sheep seminal vesicles (Gerber et al., 1979). The lack of
inhibitory effects of tartrazine, including that of producing PGs and TXs in guinea pigs perfused
with arachidonic acid (Ceserani et al., 1978), has together with other evidence (Gerber et al., 1979;
Vargaftig et al., 1980) been used as an argument against tartrazine having any influence on
prostanoid productive in asthma. Yet nothing is known about the influences of this agent on activities of the lipoxygenases, especially in tissues from hypersensitivity states, so it is not possible to
eliminate the influences of this agent on arachidonate metabolism and any possible parallels with
aspirin.

Abnormalities of eicosanoid metabolism

There is evidence for a number of potential sites in the metabolism of eicosanoids being important in
the development of AIA. Among these are the cyclo-oxygenase(s) (Szczeklik, 1990), cysteinyl- or
peptido-leukotrienes including the LTC_4 synthetase (Isreal et al., 1993; Sanak and Szczeklik, 2000), and
the lipoxin pathway (Levy et al., 1993; Serhan, 1999; Sanak et al., 2000). Of these, the former two
essentially reflected amplification in the production of cysteinyl-LTs with concomitant reduction in the
production of PGE_2, which has anti-inflammatory activity (Sanak and Szczeklik, 2000). A recent refinement of the original COX hypothesis is that COX-2 is important in the pathogenesis of asthma and
atopy via the negative regulation by PGE_2 produced by this enzyme of Th_1-derived cytokines (Varner,
1999). The lipoxin hypothesis implies that there is reduced capacity in AIA patients to produce unique
aspirin-triggered 15-epi-lipoxins (Serhan et al., 1999; Sanak et al., 2000). Overall, these actions of
aspirin in AIA patients reflect an imbalance in the regulation of pro-inflammatory cystenyl LTs by
PGE_2 AND 15-epi-lipoxins. The ultimate actions in these events are probably directed towards the
antagonism of LTD_4 on its receptor and pathways of intracellular signal transduction.

Szczeklik (1990) advanced the theory about the role of COX in development of AIA, although this
was probably also envisaged by others at the time he proposed his view. This was of course developed
prior to the discovery of the COX isozymes. An updated version of this theory implicating COX-2 has
been developed by Varner (Varner, 1999; Picardo et al., 1999; Szczeklik et al., 2004). The basis of the
theory is that COX inhibition by aspirin and other NSAIDs is accompanied by excess production of
peptido-LTs, which can be shown to be increased in the urine of AIA and markedly increased upon
intake of aspirin (Nasser and Lee, 1998). Testing the increase of LTs in this hypothesis is possible in
part pharmacologically from the observations that potent and selective LTD_4/C_4 antagonists reduce
the respiratory symptoms in aspirin-intolerant subjects.

As indicated earlier there is a dual element to this COX hypothesis and the implication that there

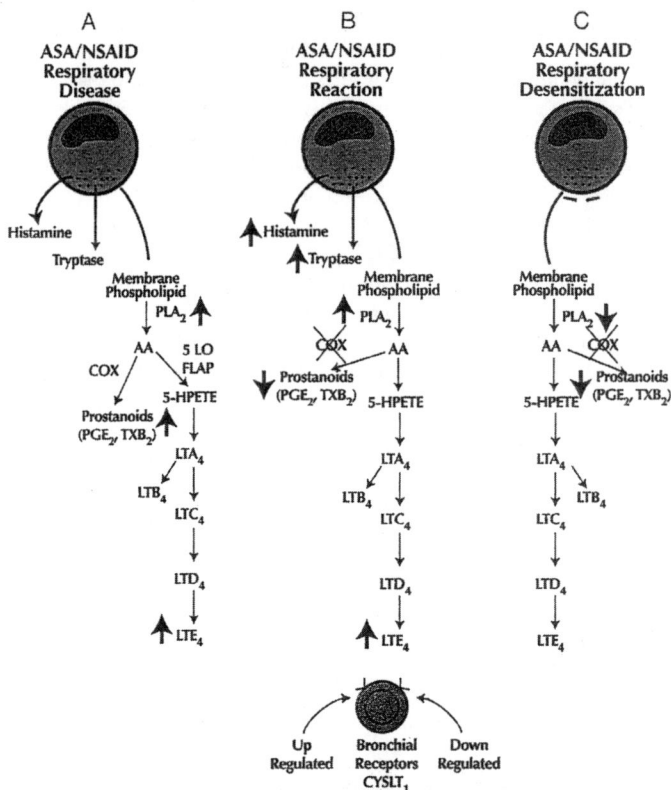

Figure 8.11 Pathogenesis of aspirin-associated respiratory reactions in respiratory disease reactions and desensiti-sation. A. Mast cell production of histamine and excess production of PGE_2, which negatively regulates the excessive production of 5-lipoxygenase (5-LO) products. B. During the aspirin reaction inhibition of COX blocks the negative control of PGE_2 on the 5-LO and enhances production of LTs and accompanying upregulation of peptido-LT receptors. C. Desensitisation follows from inhibition of histamine, with some reduction in LTs. (From Stevenson, 1998; reproduced with permission of the publishers of Immunology and Allergy Clinics of North America.)

are increased peptido-LTs associated with COX inhibition. This may cause shunting of arachidonate acid through the 5-lipoxygenese pathway following COX inhibition by aspirin and other NSAIDs.

However, the possibility that 'selective' reduction of PGE_2 following COX inhibition by these drugs and the infiltration and activation of eosinophils, which could then serve as a source of peptido-LTs, represents an additional component. COX inhibition by aspirin in the eosinophils may contribute further to the production of peptido-LTs, the latter having been shown to be increased or amplified in the airways of AIA patients who have taken aspirin (Nasser and Lee, 1998), As indicated above, COX-2 may be important in asthma and atopy (Varner, 1999), and the inhibition of this isoform may add to the effects of inhibition of COX-1 by aspirin.

An integrated view of the roles of eicosanoids along with the classical mediators of airway functions (histamine, tryptase) in manifestations of airway reactions, including that involved in airway reactivity and desensitisation is shown in Figure 8.11 (Stevenson, 1998).

The importance of the genetic regulation of a key enzyme, LTC_4-synthetase, in the pathway of peptido-LT production in AIA compared with that in aspirin-tolerant individuals has recently been investigated (Sanak and Szczeklik, 1998). This enzyme is expressed, along with 5-LO, in eosinophils macrophages and basophils. 5-LO is notably absent in epithelial and endothelial cells and in platelets, so in effect LTC_4-synthetase has no function in these cells. A common genetic variant of LTC_4 synthetase in which there is transversion of an adenine (A) to cytosine (C) at 444 bases upstream from the translation start was found at higher frequency in AIA patients (39 per cent) than in aspirin-tolerant (26 per cent) and normal subjects (25 per cent) (Sanak *et al.*, 1997; Sanak and Szczeklik, 1998).

Since 5-LO production appears to be unrelated to aspirin sensitivity (Sanak and Szczeklik, 2000) it would appear that the predominant variation in response in peptido-LT production from aspirin sensitivity in AIA subjects may reside in the -444; A→C promotor variant.

A secondary component to this effect of COX-inhibition in diverting arachidonate through the 5-lipoxygenase pathway may be related to a COX isoform which, when inhibited by an NSAID, contributes to the diversion of arachidonate to form lipoxygenase products. Also, COX-1 in platelets (which form an important component in the inflammatory reaction, especially in relation to PAF activation) is a likely candidate, but only 12-lipoxygenase activity is present in these cells and not 5-lipoxygenase. COX-2 induced by pro-inflammatory cytokines would be the other logical candidate, but COX-2 mRNA is down-expressed in nasal polyps of aspirin-sensitive asthmatics (Picardo *et al.*, 1999), raising the issue of how much inhibition of this enzyme would contribute to the diversion of arachidonate. It is possible that although the expression of message for COX-2 is reduced in AIA, there is sufficient enzyme present to be acetylated by aspirin and produce the unique COX-2-derived 15(R)-HETEs (Serhan *et al.*, 1999). Following transcellular movement from epithelial or endothelial calls to neutrophils and eosinophils, this 15(R)-HETE would be metabolised via 5-lipoxygenase and form 15-epi-lipoxins, which have anti-inflammatory, anti-bronchoconstrictor and antiproliferative activities (Tanaka *et al.*, 1997; Serhan *et al.*, 1999; see also Chapter 7, p. 270). Recent studies by Sanak *et al.* (2000) show that although asthmatics produce both lipoxins and 15-epi-lipoxins in whole blood *ex vivo*, those with AIA produce only about one-half the LXA_4 and 15-epi-LXA_4 found in the blood of aspirin-tolerant subjects. The reduced capacity to produce lipoxins is also evident when related to LTC_4 production under the same conditions, suggesting that reduction in biosynthetic capacity might not be related to the reduction in COX-2 expression that is responsible for producing 15(R)-HETE, the substrate for 5-LO (Sanak *et al.*, 2000)

It therefore appears that enhanced $LTC_4/D_4/E_4$ production by aspirin, coupled with reduced COX-1-derived PGE_2 and COX-2-derived lipoxins induced by this drug, accounts for the arachidonate component in acute AIA.

TERATOGENESIS AND PREGNANCY

The possibility that salicylates may produce fetal abnormalities, retardation of fetal growth and cause untoward effects on the maintenance of pregnancy (Silverman, 1974) has a controversial history. High doses of sodium salicylate, methyl salicylate and aspirin were reported in a considerable number of studies to induce congenital malformations and reduction in pathways in rodents (Warnaky and Takacs, 1959; Brown and West, 1964; Takacs and Warkany, 1968; Berstone and Monie, 1965). However, there were some negative studies, for example at doses of less than 3000 ppm of dietary methyl salicylate (Collins *et al.*, 1971). Serious interest has also been shown regarding whether this may occur in humans (Jackson, 1948; Goldman and Yukovak, 1964; Richards, 1969; 1971; 1972; Collins *et al.*, 1971; Nelson and Forfar, 1971; McNeil, 1973; Collins and Turner, 1975; Shapiro *et al.*, 1976; Slone *et al.*, 1977). Earlier pharmacological tests mentioned the danger of ingesting large quantities of salicylates during pregnancy and the reason given was that this drug may cause miscarriage or abortion (Jackson, 1948). No mention was made that salicylates could be teratogenic or affect growth. Jackson (1948) was probably the first to report a case observation of an 8-month pregnant woman who took aspirin 200 g with suicidal intent and later gave birth to a dead child. High concentrations of salicylate were found in the cord blood, and the author presumed that the child may have died *in utero* from salicylate poisoning. Following this observation, Jackson showed experimentally in rats and rabbits that: (a) salicylate is transported from mother to fetus; (b) the acute toxicity of sodium salicylate is the same for the fetus as for the mother; and (c) although liver glycogen was depleted by salicylate, there was no evidence of haemorrhage(s) or enhanced capillary permeability in animals that had died from this drug. Of course these studies were only concerned with the toxicity of the drug at or near term, and not with the development of congenital deformities *per se*. Factors relating to the timing of dosage of salicylate (Tagashira *et al.*, 1981; Fritz and Suter, 1985; Davis *et al.*, 1996; Foulon *et al.*, 1999; 2000; Cappon *et al.*, 2003) and the influence of factors controlling the transplacental movement of salicylate (Varma, 1988) have been established in rats. Maternal exposure to salicylate with measurement of the maternal to fetal transfer of salicylate following maternal infusion have been investigated in rabbits (Lukas *et al.*, 1987). The results showed that reduction in brain and liver weights were evident with a high-dose infusion of sodium salicylate on days 22–29 of pregnancy. However, recent studies indicate that aspirin may not be teratogenic in rabbits (Cook *et al.*, 2003).

Exposure to temperatures of 41–43°C increased the teratogenesis at a non-embryotoxic dose of aspirin. Aspirin 200 mg/kg caused a marked increase in the number of dead fetuses and skeletal abnormalities in mice (Tiboni *et al.*, 1998). Itami and Kanoh (1982; 1983; 1984) have shown that lipopolysaccharide enhances the embryotoxicity of aspirin and salicylic acid in pregnant rats. Thus there may be effects of both endotoxin and the body heat elevation produced as a consequence of its hypothalamic effects that influence the toxicity of salicylates to the fetus. Oxyradicals produced from lipopolysaccaride induction of cytokines by leucocytes may also contribute to development of these and other abnormalities (Karabulut *et al.*, 2000).

In vitro effects of salicylates have been studied in rat and mouse embryos at partial maturation (Greenaway *et al.*, 1985; Patierno *et al.*, 1989; Spézia *et al.*, 1992).

Comparisons of the effects of aspirin with other analgesic/NSAIDs have shown that paracetamol 250 mg/kg does not produce abnormalities or affect fetal growth in rats, in contrast to the same dose of aspirin, which did affect the development as indicated from previous studies (Lubawy and Burriss Garrett, 1977). Ibuprofen 25 mg/kg s.c. did not produce any teratogenic effects when given on day 10 of pregnancy to C37BL/6J mice, but significantly reduced the teratogenic effects of alcohol 5.8 g/kg, manifest by reduced fetal weight and dysmorphology. Similar protective effects were observed with aspirin 150 mg/kg p.o.

The first indications that salicylates might be teratogenic came from studies by Richards (1969b). This author reported an epidemiological study in South Wales (UK) of 833 pregnancies in which salicylates were implicated in causing fetal deformities during the first trimester of pregnancy. The deformities were chiefly in the central nervous system and alimentary tract. A previous survey conducted by the Royal College of General Practitioners (UK) of 1038 pregnancies concluded that there was no evidence of teratogenesis from salicylates (Slater, 1965). Nelson and Forfar (1971) reported a retrospective study of 1369 pregnancies, which implicated aspirin as well as other analgesics in causing congenital abnormalities. McNeil (1973) also reported eight suggestive cases of salicylate teratogenesis.

These studies have been followed by a considerable number of negative reports from large-scale epidemiological studies (Shapiro *et al.*, 1976; Slone *et al.*, 1976; CLASP Collaborative Group, 1995; Rai *et al.*, 2000) and case–control studies (Nielsen *et al.*, 2001). The whole question of teratogenicity of the salicylates in humans is still somewhat controversial. Evidence from the Boston University Drug Epidemiological Unit's Collaborative Perinatal Project on the incidence of malformations in 50 282 pregnancies suggested that aspirin is not teratogenic (Slone *et al.*, 1976). In a smaller study of 144 pregnancies in Sydney (Australia) by Turner and Collins (1975), salicylate ingestion determined by urinary salicylate assays was not associated with an increased incidence of congenital abnormalities. Østensen and Østensen (1996) showed that long-term intake of NSAIDs as part of antirheumatic therapy in patients with autoimmune rheumatic diseases was not associated with any teratogenic effects compared with those who did not receive these drugs. The patient numbers were relatively small (88), divided into 45 who took NSAIDs and 43 who did not, so there may be limitations to this study based on the small numbers. Østensen and Villiger (2001) have reviewed the safety of NSAIDs in lupus and have concluded the risks of teratogenicity are low, and this is supported by others (Huong *et al.*, 1997; 2001).

There have been concerns that NSAIDs may be associated with acute fatty liver of pregnancy,

CHAPTER 8

especially in individuals with hereditary defects in the trifunctional enzyme of long-chain fatty acid oxidation (Baldwin, 2000).

Experimental evidence in mice, rats, rabbits, dogs and ferrets shows that high doses of salicylates are teratogenic (Larsson *et al.*, 1963; 1964; Larsson and Boström, 1965; Kimmel *et al.*, 1971; Beall and Klein, 1977; Knight and Roe, 1978; Robertson *et al.*, 1979; Gulamhusein *et al.*, 1980; Depass and Weaver, 1982; Overman and White, 1983; Foulon *et al.*, 1999). The incidence and severity of these salicylate-induced deformities in rats and ferrets is dose-dependent (Kimmel *et al.*, 1971). There is a decrease in overall susceptibility to the drug with onset of gestation. The types of deformities produced in rats depend on the state of development. In one study the most frequent abnormalities (totalling 76 per cent) at day 9 of pregnancy were cranial abnormalities, skeletal dysplasia, umbilical and diaphragmatic hernia, and hydrocephalis (present in 11–31 per cent of surviving fetuses). By day 11 a different group of deformities was present, including clubfoot, polydactyly, abnormalities of the tail and kidneys, and cleft palate (present in 81 per cent of fetuses with abnormalities). These studies were performed with exceptionally high doses of salicylates, and it is obviously difficult to relate them to the situation in man. Gulamhusein and co-workers (1980) have shown that the ferret is a more susceptible species to salicylate-induced teratogenesis than the rat. They have suggested this may be a more suitable species in which to investigate teratogenesis by agents such as the salicylates.

Dietary restriction has been found markedly to enhance the teratogenicity of aspirin (Kimmel *et al.*, 1971). This may be a very important aspect, since individuals in lower socio-economic groups on poor nutrition may be more susceptible to aspirin-induced teratogenesis. This could explain some of the variations in the incidence of abnormalities reported in studies from different parts of the world. For instance, the positive reports of aspirin-associated teratogenesis in South Wales (a region with a relatively low socio-economic status) can be contrasted with the negative reports from the large-scale studies in the USA and others in Australia, where the populations may be nutritionally better off than in Wales (Turner and Collins, 1975). Hackman and Hurley (1984) showed that there is an interaction between dietary zinc deficiency and the mortality and numbers of malformations in Wistar and Sprague–Dawley pregnant rats. Variations in dietary zinc (0–1000 µg/g diet and normal chow) during gestation were induced with 250–750 mg/kg p.o. given daily on day 9. At levels of 0 and 4.5 µg/g zinc there were pronounced increases in the number of dead and malformed fetuses, although this varied with the strain of rat.

An interesting experimental complication arose from a well controlled study performed by Foulon *et al.* (1999) in which they observed that the food consumption and body weights of rats that received 180–300 mg/kg sodium salicylate p.o. declined between days 9 and 10. This influence of salicylate given by gavage on body weight and food consumption suggests that the drug may affect appetite. The same study also showed there was a dose-related increase in fetal loss, and a progressive increase in the number of supernumerary ribs.

Günther *et al.* (1988) found that 300 mg/kg salicylic acid given to magnesium-deficient pregnant rats on day 11 of gestation resulted in abnormalities of the thoracic and lumbar vertebrae, which were more abundant than in animals that received salicylic acid alone (these also had significant numbers of malformations). Vorhees *et al.* (1982) showed that 625 mg/kg aspirin produced maternal and offspring deficits when given to dams on day 11 but not day 12 of gestation. The reasons for these specific time-dependent changes are not known.

Guy and Sucheston (1986) showed that treatment with 250 mg/kg aspirin p.o. and ethanol (0.02 ml/g of 25 per cent v/v 95 per cent ethanol) i.p., but not aspirin or ethanol alone, to day 8 pregnant CD-1 mice caused a statistically significant increase in the number of animals with delayed development and malformations. Particularly noticeable were the numbers of urinogenital malformations.

The exposure of rats to 1000–3600 mg/m^3 toluene on day 10–13 of pregnancy was found markedly to increase the embryotoxicity and malformations and maternal toxicity of animals given 250 mg/kg aspirin p.o. (Ungvary *et al.*, 1983). The toluene treatment increased the maternal and fetal plasma salicylate concentrations, implying that the toxicity was related to the metabolism of aspirin to salicylate. Co-administration of glycine 2500–5000 mg/kg p.o. with aspirin and toluene reduced the numbers of dead or resorbed fetuses and malformations, especially those to the skeletal and internal organs. It is presumed from this study that the glycine conjugation to salicylate or other metabolites reduced the potential of the latter to cause fetal toxic effects.

Salicylate has been identified as the major metabolite present in rat fetal tissues following maternal ingestion of aspirin (Kimmel *et al.*, 1971). It is therefore presumed that salicylate is the offending component causing the development of fetal abnormalities. Yokoyama *et al.* (1984) found that treatment of day 11.5 rat embryos with 300 or 500 µg/ml sodium aspirin, or 230 or 385 µg/ml for 2 or 4 h respectively *in vitro*, resulted in more than 78 per cent malformations, including curly tail and cleft lip. The protein content and crown to rump length was more pronounced in the aspirin group, as were the numbers of oedematous facial and tail abnormalities. These authors concluded there were marked differences in the actions of aspirin compared with salicylate, implying that the latter is less toxic to the fetus.

Drug metabolism via the cytochrome P_{450} oxidative pathways leading to hydroxysalicylates with concomitant effects on calcium metabolism have been suggested from a brief study by Kitagawa *et al.* (1982). However, it is possible that alterations in the maternal vascular functions, especially those in the placenta, could be affected by aspirin, so causing placental arteriovenous anoxia. Vasoconstriction could ensue, so leading to the development of abnormalities. In support of this are the observations of Larsson and Boström (1965). They found that a single intramuscular injection of some salicylate to mice on the ninth or twelfth day of gestation induced skeletal and vascular abnormalities, which were correlated with the inhibition of the sulphation of chondroitin *in vitro* and *in vivo*. *P*-Hydroxybenzoic acid, given in the same way, failed to exhibit any such abnormalities or effects on chondroitin sulphate synthesis, showing that the effects are specific to salicylate.

Janakidevi and Smith (1970a; 1970b; 1970c) reported that salicylate inhibited RNA biosynthesis in mouse fetuses *in vitro* and *in vivo*. The *in vivo* effects of the drug were evident at very short time intervals after administration of salicylate 200 and 400 mg/kg i.p., but not at lower doses. The inhibitory actions of salicylate on RNA synthesis are due to inhibition of the Mn^{2+}-dependent RNA polymerase and not the Mg^{2+}-activated enzyme. Thus salicylate resembles α-amanitin but not those other inhibitors (e.g. actinomycin D and aflatoxin B_1) that influence the Mg^{2+} form of the enzyme system.

The possibility that salicylate-induced impairment of oxidative phosphorylation (Brody, 1956; Jeffrey and Smith, 1959; Bryant *et al.*, 1963; Hines *et al.*, 1963; Smith, 1966; 1968; Thompkins and Lee, 1969; Kalczak *et al.*, 1970) is involved in teratogenesis by this drug has received some support from observations that the uncoupling agent 2,4-dinitrophenol (6–10 mg/kg) induces similar abnormalities to those obtained by salicylates. It remains, however, for drug effects on ATP levels and mitochondrial oxidations to be demonstrated in fetal tissues following administration of aspirin or related drugs to the mother. Other drug effects include the inhibition of protein biosynthesis (Dawkins *et al.*, 1966; 1971) and prostaglandin production (Patrick and Challis, 1982; Zanagnolo *et al.*, 1996). Effects of aspirin on the patency of the ductus arteriosus have received particular attention, and have provided evidence for the importance of maintaining prostaglandin production in the fetus up to term. (Lewis and Schulman, 1973; Patrick and Challis, 1982).

Apart from producing abnormalities, it appears that salicylates might reduce birth weight and increase perinatal mortality. Such an association was shown by Collins and Turner (1975) in Sydney (Australia) with evidence of salicylate ingestion being obtained from urinary salicylate analysis. Negative reports that aspirin does not cause such effects have appeared from the Boston Collaborative Perinatal Project (Shapiro *et al.*, 1976), but the absence of data on urinary salicylates raises questions about the validity of this negative association.

Salicylates may also affect the fetus at or near term and on the course of the pregnancy as a consequence of effects on PG production (Levin, 1980; Patrick and Challis, 1982; Poyser, 1988). Reduction in preterm labour has been shown following intravenous infusion of aspirin (initial loading dose of 5.5–7.0 mmol/l per minute). This treatment also produced symptoms of vertigo, tinnitus, headache and hyperventilation, and, while not serious, it is possible that the hyperventilation may affect fetal oxygenation. Aspirin also attenuates mid-trimester abortions induced by combined urea and oxytocin treatment (Niebyl *et al.*, 1976), and thus it could be considered that aspirin has advantages in preventing a threatening preterm abortion and in low doses growth retardation and hypertension (Parazzini *et al.*, 1993). However, installation of aspirin 50–90 mg/kg into the rumen of fetal lambs causes an increase in pulmonary arterial pressure, which is directly related to contraction of the ductus arteriosus (Rudolph, 1981). It has been suggested that such an elevation of pulmonary pressure could stimulate muscle development in the small vessels of the lung, so interfering with the return to normal pulmonary vascular resistance that develops after birth (Rudolph, 1981). These effects could influence tissue oxygenation in the fetus, with long-term effects after birth, e.g. on brain development. Further effects of aspirin on PGE-related actions during closure of the persistent ductus in prematurely born infants have been explored by Coceani and Olley (1982). Cautious use of aspirin may be indicated, although attention to dosage is important.

The effects of aspirin intake during pregnancy on subsequent IQ of the developing child were studied in longitudinal studies, and one study indicated that there are no apparent effects (Klebanoff and Berendes, 1988) but in another, a negative relationship was shown, which was more evident in girls than boys (Pytkowicz Steissguth *et al.*, 1987). No effects of paracetamol intake on IQ were noted in this study. Further investigations are warranted to confirm the latter observations and to define the dose and other conditions accounting for any effects.

There may be complicating factors concerning effects on the newborn where there is maternal pulmonary hypertension diabetes mellitus and smoking in patients that have taken aspirin (van Marter *et al.*, 1996; Sibai *et al.*, 2000). Thus effects of maternal blood pressure and smoking may contribute to any untoward effects of aspirin on maternal-fetal PG-related effects in these patients.

The effect of aspirin and other analgesics/NSAIDs on male fertility has received some limited attention. Aspirin was found to reduce the concentrations of E and F prostaglandins in human semen samples, with no effect on the volume of semen (Collier and Flower, 1971). As a consequence of interest in the role of prostaglandins in spermatogenesis, studies have been performed on the effects of aspirin administration on sperm maturation in rats (Cenedella and Crouthamel, 1973; Abbatiello *et al.*, 1975; Scott and Persaud, 1978; Balasubramanian *et al.*, 1980). The results have been quite variable, and depend on the dose and timing of aspirin administration. No clear conclusions can be drawn at this stage to indicate the significance of these laboratory animal studies to man. Implantation of plastic rods containing 25 or 50 per cent aspirin in the region adjacent to the epididymus of rats caused temporary reduction in fertility, with no impairment of other reproductive functions (Ratnasooriya and Lionel, 1984).

MISCELLANEOUS SIDE EFFECTS

Cutaneous reactions that range from those frequently encountered (e.g. erythematous rash and urticaria) to the less frequent but more severe (e.g. bullous reactions, Stevens–Johnson and Lyell's syndromes) occur with the salicylates and are typical of those encountered with other NSAIDs (Rainsford, 1989a; Naldi *et al.*, 1999; Stevenson *et al.*, 2000; Wedi and Kapp, 2000; Kaufman and Kelly, 2001). The occurrence of minor skin reactions is relatively frequent among all the adverse drug reaction (ADRs) observed with aspirin and other NSAIDs (Table 8.1). The reporting rates per defined daily doses/1000 of nonserious reactions in a study in the regions of Friuli Venezia, Lombardy, Sicily and Vento in Italy are highest for aspirin with ibuprofen, paracetamol piroxicam and naproxen being among the lowest (Naldi *et al.*, 1999). The molecular basis for these reactions is not entirely understood (Stevenson *et al.*, 2000; Wedi and Kapp, 2000), although the concept of a reactive drug species being produced that elicits immune reactions may be a common feature with other NSAIDs (Rainsford, 1989a).

Salicylates have been implicated in the development of pulmonary oedema and respiratory failure (Heffner and Sahn, 1981; Seger, 1981; Zimmerman and Clemmer, 1981). Aspirin has potentially hazardous respiratory effects when taken during prolonged exercise in hot weather (Fred, 1981), and it appears that salicylate impairs the acclimatisation to work in high temperatures (Bass and Jacobson, 1965). It has been suggested that, during the acclimatisation process, salicylates impose a small endogenous heat load by uncoupling of oxidative phosphorylation in peripheral tissues (Bass and Jacobson, 1965).

There have been occasional reports of aspirin inducing blood dyscrasias (Beaver, 1965; Stare *et al.*, 1980). Effects of aspirin on platelet aggregation would be expected to have profound effects in bleeding states or in combination with drugs likely to promote this condition. It has been suggested that the effects of aspirin on the bone marrow could be a consequence of the acetylation by this drug of marrow macromolecules (Rainsford *et al.*, 1983). The reported depressant effects of salicylates on the eosinophil count (Smith, 1966) may be a consequence of adrenocortical stimulation by salicylates (Walker and Smith, 1979).

Feingold proposed that salicylates present in foods and drinks (notably the flavouring agent methyl salicylate) causes hyperactivity in children (Feingold, 1975a, 1975b; Stare *et al.*, 1980). It was claimed that elimination or reduction of salicylate-containing foods or drugs from the diet reduced the incidence of hyperactivity (Brenner, 1979; Stare *et al.*, 1980). Several authors have found that Feingold's low-salicylate diet did not produce any improvement, or only achieved rather inconsistent results (Cenedella and Crouthamel, 1973; Harner and Foyles, 1980; Stare *et al.*, 1980). Brenner (1979) noted that there was a significant increase in serum copper levels in children that responded to the restricted diet although the reasons for this and the consequences were not clear. Also, where there were violations of the diet it appears that children were unchanged in their behaviour (Adams, 1981). There is now much scepticism about the value of this diet (Cook and Woodhill, 1976) and there could be psychological and environmental factors peculiar to certain societies that contribute to the development of this syndrome (Cook and Woodhill, 1976).

Salicylates also induce loss of hearing, which is mostly reversible (Falbe-Hansen, 1941; McCabe and Dey, 1965; Perez de Moura and Hayden, 1968; Lucente, 1971; Morgan and Clark, 1998; Lue and Brownell, 1999; Cazals, 2000). This can be a major problem for patients on chronic salicylate therapy, but is most often reduced by lowering the quantity of drug ingested. Histological studies have failed to reveal any structural abnormalities induced by salicylates on the sensory epithelium stria vascularis, the spiral ganglion cells or the cochlea nerve in either man or laboratory animals (Myers and Bernstein, 1965; Perez de Moura, 1968). Also, no ultrastructural evidence of damage to the cochlea has been evident in rats given sodium salicylate 200 mg/kg per day for 5 days (Falk, 1974). No apparent changes are induced in the sodium, potassium or protein concentrations of the perilymph or endolymph, but hyperglycaemia in the inner-ear fluids and decreased malate dehydrogenase activity in the perilymph and endolymph have been observed (Silverstein *et al.*, 1967; Silverstein and Griffin, 1970). Electrophysiological studies in guinea pigs have shown that choline salicylate affects the hair cells of the cochlea: there is an increased threshold of sensitivity with a marked reduction in the range of dynamic microphonics (McPherson and Miller, 1974). Salicylates induce minor changes in cochlear function through activation of NMDA receptors (Guitton *et al.*, 2003).

Axonal neurodegeneration in the central nervous system has been reported in rats infected with *Trypanosoma brucei* and given 200 mg/kg sodium salicylate in their water supply (Quan *et al.*, 2000). There was extensive neuronal cell body damage in the cortex, hippocampus, striatum, thalamus and anterior olfactory nucleus. Accompanying this there was elevation of pro-inflammatory cytokines (IL-1β. TNFα), iNOS and COX-2 suggesting that there was extensive neural inflammation in these rats.

As reviewed in Chapter 13 there are indications that salicylates can have neuroprotective effects. However, in a rat model of temporal lobe epilepsy in which kainic acid is injected it was found that sodium salicylate caused focal haemorrhage and cell death in the hippocampus and entorhinal/piriform cortex (Najbauer *et al.*, 2000). The haemorrhage was not observed in rats given kainic acid or sodium salicylate alone showing that the drug exerted its effects only in animals with seizures. The cell injury in the hippocampus of animals given kainic acid and sodium salicylate comprised DNA fragmentation. There was damage to cells in the dorso-medial thalamic nuclei of animals given kainic acid alone or in combination with salicylate. There was evidence of cell proliferation around regions where there was injury and microglial activation. The authors suggested from these studies that there could be clinical implications for epileptic patients who take aspirin.

Of the CNS effects of salicylates, abnormal sleep patterns have been observed (Murphy *et al.*, 1994).

Aseptic meningitis has been reported with overdose of salicylates (Nair and Stacey, 1993) and this is probably a rare adverse reaction from these drugs.

Earlier reports that aspirin could induce chromosomal abnormalities (Jarvik *et al.*, 1971; Weinstein *et al.*, 1981) or like other analgesics may be mutagenic *in vitro* (Kuboyama and Fujii, 1992), have not been confirmed by others (Mauer *et al.*, 1970; Sankar and Geisler, 1971).

ACUTE SALICYLATE POISONING

Poisoning from ingestion of aspirin and, to a lesser extent, methyl salicylate is a problem confined to three main clinical situations (Smith, 1968; Temple, 1981):

1. Accidental ingestion of these salicylates in children of preschool age (Smith, 1968; Done, 1978; Temple, 1981; The National Poisons Information Service Monitoring Group, 1981).
2. Suicidal overdose in teenagers and adults, especially older women (Smith, 1968; Sellers *et al.*, 1981; Temple, 1981).
3. Therapeutic intoxication, which is evident in all age groups (Smith, 1968; Done, 1978; Temple, 1981).

Salicylates were a major cause of poisoning in the UK until the early 1970s, when tricylic antidepressants and alcohol replaced these drugs as the most common of the fatal poisons (White *et al.*, 1980). Similar trends have appeared in other countries (Fraser, 1980; Meridith *et al.*, 1981). Recently, paracetamol has also been recognised as a cause of death amongst the cases of analgesic poisoning admitted to hospitals in the UK (The National Poisons Information Service Monitoring Group, 1981), as well as in the USA and other Western countries (Sellers *et al.*, 1981; Prescott, 1996).

The introduction of safety packaging is claimed to have resulted in fewer deaths (Done, 1978; Clarke and Walton, 1979; Rodgers, 2002), but doubts have been expressed about this being of value in the UK (Fraser, 1980).

Toxic manifestations

The major toxic manifestations of salicylate poisoning have been extensively reviewed by Smith (1966; 1968), McQueen (1977), Done (1978), Atwood (1980) and Temple (1981). They may be grouped as indicated below.

Central nervous system (CNS) effects

Initially, tinnitus, deafness and vomiting occur, followed by delirium, convulsions and, ultimately, coma. The respiratory centre becomes directly stimulated by salicylate(s), and also becomes more sensitive to the pH changes from metabolic disturbances.

The respiratory centre is stimulated by massive entry of salicylate into the brain because of high free (i.e. unbound) concentrations of the drug in the plasma (Smith, 1968; Alvan *et al.*, 1981). This leads to increased alveolar ventilation, which, because of the consequent reduction in the partial pressure of CO_2 in the alveolar air and in blood, causes a rise in blood pH (Smith, 1968). A respiratory alkalosis ensues in adults, but seldom in young children (Smith, 1968; McQueen, 1977). This is limited by compensatory mechanisms, including buffering by the Hb–HbO systems (Hb = haemoglobin), exchange of intercellular cations, and the urinary excretion of HCO_3^- (Smith, 1968). This alteration in acid–base balance causes a respiratory alkalosis. Extreme hyperpnoea develops and contributes to dehydration, from vomiting (Smith, 1968; McQueen, 1977).

Initial agitation may be accompanied by a toxic psychosis with paranoid and hallucinatory behaviour and asterixis. When salicylates are ingested alone, unconsciousness is invariably accompanied by acidaemia (McQueen, 1977).

CNS effects will be exacerbated if depressant drugs (e.g. dextropropoxyphene or paracetamol) have been consumed with aspirin (McQueen, 1977).

Gastrointestinal effects

Vomiting may occur, with some blood present and accompanying severe substantial and epigastric pain (McQueen, 1977). One of the curious features of salicylate poisoning is that massive haemorrhage is rarely evident at death, despite the frequent blood loss observed at therapeutic doses of aspirin (Smith, 1968). In autopsies attended by this author, the stomach is often densely packed with tablets and there does not appear to be any way in which a haemorrhage can persist, even though bleeding may have occurred earlier. Even where gastric lavage has been employed, the amount of blood loss is often small (Atwood, 1980).

Metabolic effects and acid–base balance

As a consequence of the uncoupling effects of salicylate on mitochondrial oxidative phosphorylation there is an enhanced rate of catabolism in the body, leading to a dramatic increase in CO_2 production (Smith, 1968; McQueen, 1977). This produces a rise in PCO_2, and causes acidosis, and this is the opposite effect to that produced by stimulation of alveolar ventilation (Smith, 1968). The net result of these two effects depends on their relative magnitude, and usually the consequences of hyperventilation (i.e. respiratory alkalosis) will outstrip the peripheral respiratory acidosis (Smith, 1968). Prolonged exposure to salicylates leads ultimately to respiratory depression, whereupon the effects of acidosis in the peripheral tissues then ensue (Smith, 1968).

A third effect of salicylate on acid–base balance is the production of abnormally high concentrations of organic ions in the blood (Smith, 1968). This results from the derangement of carbohydrate, amino acid and lipid metabolism leading to hypersecretion of ketone bodies, as well as amino and organic acids (Smith, 1968; McQueen, 1977). Excretion of these acids depletes bicarbonate concentrations in the blood, and an acidaemia supervenes. If renal functions have been impaired, then the metabolic acidosis may be accelerated (McQueen, 1977). Young children are much more susceptible to the effects of a metabolic acidosis, while adults are notably more resistant (Smith, 1968).

The metabolic acidosis follows accumulation of organic acids, mostly 3-hydroxybutyrate and acetoacetate. This occurs because uncoupling of oxidative phosphorylation leads to an accumulation of metabolic intermediates (e.g. 2-oxoacids, amino acids, 3-hydroxybutyrate and acetoacetate), which cannot be interconverted because of the powerful inhibitory effects of salicylate on the activities of the transaminases, dehydrogenases and enzymes of gluconeogenesis (Smith, 1968; Smith and Dawkins, 1971; Madappally et al., 1972; Dawson, 1975; Millhorn et al., 1982). Also, the quantity of ketone bodies will be enhanced because of the inability of the acetate to be metabolised by the tricarboxylate cycle as a consequence of the inhibition by salicylate of transaminases and dehydrogenases (Smith, 1968).

Inhibition of erythrocyte 2,3-diphosphoglycerate (2,3-DPG) formation by salicylate will increase the affinity of oxygen for haemoglobin, as this glycolytic intermediate is necessary for regulating oxyhaemoglobin formation (Kravath et al., 1972). With the combined effects of lowering of the ability of HbO_2 to give up O_2 (because of the drug effects on 2,3-DPG levels), and of high oxygen utilisation because of uncoupling of oxidative phosphorylation effects, the CNS could ultimately become hypoxic. The salicylate-induced inhibition of glycolysis and the effects of uncoupling would reduce ATP levels so that this, combined with hypoxia, could account for the development of coma and ultimate loss of brain function and death. Direct poisoning of important brain functions (e.g. the respiratory centre) by salicylates would also be expected to contribute to death. The hyperthermia that develops as a con-

CHAPTER 8

sequence of uncoupling of oxidative phosphorylation by salicylate could also be especially important as a cause of death in infants (Smith, 1968). The long-term effects of hypoxia, impaired ATP production and hyperthermia could be especially important in the brains of the survivors. The prognosis for brain function, especially for the developing child, could be very bleak if the poisoning has not been treated quickly or the dose is so high that treatment is virtually ineffective.

Vomiting and electrolyte imbalance

Vomiting frequently develops, especially within 3–8 hours of salicylate poisoning in children (Smith, 1968). If severe and of long duration, this will produce a primary deficiency of sodium and potassium (Smith, 1968). Transport of Na^+ and K^+ across membranes will be inhibited because of the uncoupling of oxidative phosphorylation by salicylates (Smith, 1968), and thus electrolyte balance will often be severely affected in salicylate poisoning.

Pulmonary oedema

This may not always occur in salicylate poisoning (Smith, 1968). If evident, it could occur owing to the overwhelming inhibition of prostaglandin biosynthesis, as well as from over hydration (Smith, 1968).

Renal damage

Occasionally renal failure may present as a complication in salicylate poisoning, especially in patients who have previously abused analgesics (Smith, 1968; McQueen, 1977). Obviously this will impair elimination of both the products of acidosis as well as the drug, and could be a major factor in fatality in the absence of dialysis procedures.

Dosage

The fatal toxic dose of aspirin in adults has been estimated to be 500 mg/kg (Temple, 1981). Done (1978) estimated that in children, moderate poisoning from aspirin requires a dose of 240 mg/kg (rather than the oft-cited 120 mg/kg). Lethality in children does not usually occur below 480 mg/kg (Done, 1978), but this is very much age-dependent. This can be seen in the studies by Buchanec and co-workers (1981) on the pharmacokinetics (blood, urine levels) of salicylates following aspirin ingestion by 3–5-week-old children compared with 10–15 year olds. These authors found that much higher blood levels and a lower rate of elimination of salicylates were evident in the younger age group (Buchanec et al., 1981).

Normally, monitoring of plasma salicylates is mandatory during the management of salicylate intoxication (Done, 1960; 1978; Temple, 1981; The National Poisons Information Service Monitoring Group, 1981). There can, however, be problems with such determinations where the patient has been suspected of taking sustained-release or enteric-coated aspirin preparations, or especially an aspirin + paracetamol combination. This has been highlighted in several cases of poisoning that have been reported (Editorial, Lancet, 1981; Todd et al., 1981); here the salicylate determinations underestimated the amount of drug actually present because of the delayed absorption of the drug. Indeed, in one case salicylate was not detected at all in a single blood sample taken upon admission, and since the patient seemed well he was discharged home after gastric lavage only to die 15 hours later (Editorial, Lancet, 1981).

The combination of aspirin and paracetamol in an enteric-coated preparation is particularly hazardous, since paracetamol-induced hepatic failure may result from the initial release of paracetamol

TABLE 8.20

Nomogram relating serum salicylate concentration to symptoms.

S_o (mg/dl)	Grading	Symptoms
50	Not intoxicated	Asymptomatic
50–80	Mild	Hyperpnoea, marked lethargy and/or excitability; hypocapnia without acidosis
80–100	Moderate	Severe hyperpnoea, marked lethargy and/or excitability, vomiting in children, compensated metabolic acidosis
110	Severe	Coma, possible convulsions, uncompensated metabolic acidosis in children after 12 h
160	Usually lethal	

The approximate limits for estimating the levels of salicylate intoxication in this nomogram were calculated from the equation: log S_o + log S^* + 0.015T, where S_o = the initial serum salicylate concentration (mg/dl), S^* = measured salicylate concentration at time T (h). See Smith (1968) and Kravath *et al.* (1972) for gradings of severity.

followed later by salicylate poisoning from the delayed release of aspirin, largely in the intestine (Editorial, *Lancet*, 1981). To be sure of the possibility that paracetamol has not been taken with aspirin, it is obvious that serum paracetamol estimations should be performed routinely in any suspected case of salicylate poisoning.

Done (1960) devised a nomogram relating serum salicylate concentrations to the time from ingestion of aspirin (or salicylate) for a series of gradings from asymptomatic through to severe levels of intoxication from the drug (see Table 8.20). This nomogram is extensively used, and enables a ready and reliable estimate of the severity of intoxification (Done, 1960). Unfortunately, it appears that this nomogram cannot be applied to cases of poisoning from enteric-coated or sustained-release aspirin preparations because of the delayed release noted above.

The inevitable combination of vomiting, hyperapnoea and hyperthermia should always be considered as indicative of salicylate poisoning (Done, 1978).

Management

This aspect has been reviewed in depth by McQueen (1977), Done (1978), Atwood (1980), Temple (1981) and Brenner and Simon (1982), and may be found on the National Poisons Information Service (London) website (www.doh.gov.uk/npis.htm; accessed February, 2002) or, in the USA, the Center for Disease Control and Prevention, Atlanta, Georgia.

A summary of the essential clinical and laboratory findings and recommended management is listed below, and has been kindly provided by Dr Glyn Volans of the UK National Poisoning Advisory Services (Guy's and St. Thomas' Hospital, London; February, 2001).

Summary of clinical symptoms

Mild	Nausea, vomiting, epigastric pain, tinnitus, flushing
Moderate	Sweating, hyperventilation, dehydration, deafness, tremor, respiratory alkalosis with metabolic acidosis (acidosis predominant in children)
Severe	Hypo- or hyperglycaemia, hypokalaemia, hypo- or hypernatraemia, hypoprothrombinaemia, pyrexia (mostly in children), confusion, drowsiness, delirium, coma, convulsions (more common in children). CNS effects are usually relieved if the acidosis is corrected. Rarely, renal failure, pulmonary oedema or cardiovascular collapse. Death is usually due to cardiopulmonary arrest.

CHAPTER 8

Treatment (early management)

Adult   >120 mg/kg: give 50 g activated charcoal within 1 hour post-ingestion. For >250 mg/kg, consider gastric lavage followed by 50 g activated charcoal (Brenner and Simon, 1982; Mofenson *et al.*, 1985; Boldy and Vale, 1986; Barone *et al.*, 1988), again within 1 hour. In all cases, if >1 hour, give activated charcoal only.

Child   >120 mg/kg: give 1 g/1 kg activated charcoal within 1 hour of ingestion.

Treatment (maintenance management)

1. Check salicylate concentration 4 hours post-ingestion, then every 2–3 hours until peak concentration is achieved.
2. Monitor and correct urinary electrolytes arterial blood gases and pH, blood sugar and prothrombin time, and in moderate/severe cases monitor the central venous pressure.
3. Rehydrate with oral or i.v. fluids; large volumes may be necessary to counteract dehydration.
4. Repeat doses of activated charcoal (adult 25–50 g; child: 1 g/kg) every 4 hours until salicylate level has peaked.

Urinary alkalinisation

In the presence of alkaline urine (optimum pH 7.5–8.5), renal elimination of salicylate is enhanced 19- to 20-fold with an increase from a urine pH of 5 to 8. Hypokalaemia prevents urinary excretion of alkali, so must be corrected. Forced alkaline diuresis is not recommended.

Adult   1 litre of 1.26% sodium bicarbonate (isotonic) + 40 mmol potassium i.v. over 4 hours and/or 50 ml boluses of 8.4% sodium bicarbonate i.v., ideally via a central line.

Child   1 ml/kg 8.4% sodium bicarbonate (= 1 mmol/kg) + 20 mmol potassium diluted in 0.5 l dextrose or saline, infused at 2–3 ml/kg per hour.

Treatment – rehydration

Salicylate can produce marked dehydration from its various toxic actions. It is vital, particularly in the late-presenting patient, to rehydrate vigorously. Dehydration may cause an artificially high plasma concentration. In patients with severe dehydration, the central venous pressure should be monitored.

Gastric decontamination

Gastric decontamination may be useful in large aspirin overdoses. Early gastric lavage is preferred to ipecacuanha, as the latter is less likely to recover concreted aspirin and tends to cause protracted vomiting, which prevents retention of the activated charcoal (Curtis, 1984).

Correction of metabolic acidosis

In the presence of an acidosis, transfer of the salicylate ion into the CSF is enhanced. This results in CNS effects, including a rise in the body temperature, and indicates severe toxicity. This is particularly so in children, who quickly develop a metabolic acidosis. Correction of the metabolic acidosis usually resolves CNS effects.

Enhanced elimination – repeated dose-activated charcoal; alkalinisation of the urine

Activated charcoal adsorbs salicylates (Hillman and Prescott, 1985) but, as with most drugs, a degree of desorption may occur (Filippone, 1987). This can be minimised by giving repeated doses of activated charcoal. This regimen has been reported to enhance elimination of the salicylate ion from the blood back into the gut, a dialysis action, but its efficacy is doubtful and further investigation is required. However, it may still be effective in preventing delayed absorption, and is recommended for moderate to severe salicylate toxicity.

In the presence of alkaline urine, renal elimination of the salicylate ion is enhanced (19- to 20-fold with an increase from a pH of 5 to 8). Forced alkaline diuresis (FAD) used to be the standard method for producing alkaline urine. However, in recent years alkalinisation without forced diuresis has been shown to be as effective as FAD (Prescott *et al.*, 1982) and with fewer of the complications, such as electrolyte disturbances and fluid overload, to which the elderly and children are particularly susceptible. Therefore FAD is no longer recommended. Sodium bicarbonate should be administered to alkalinise the urine (optimum pH 7.5 to 8.5); the arterial pH should not rise above 7.6. It is very difficult to produce an alkaline urine if the patient is hypokalaemic, and therefore the serum potassium must be corrected. Alkalinisation of the urine is the recommended regime for moderate salicylate toxicity.

Haemodialysis

Haemodialysis is recommended for patients with severe salicylate intoxication. This includes those with coma, convulsions, acidaemia or hypoxia. Further indications include renal failure, congestive heart failure, non-cardiogenic pulmonary oedema, CNS effects not resolved by correction of acidosis, acid–base or electrolyte imbalance resistant to correction, or persistently high salicylate concentrations unresponsive to urinary alkalinisation.

Interpretation of peak salicylate concentration

Table 8.21 provides guidelines only. Clinical presentation is the most important factor in deciding management, particularly in the late presenting patient who may have a sub-toxic salicylate concentration but serious acid–base or CNS disturbances. More than one aspirin concentration must be determined to establish that the peak has been achieved. Laboratories using the Kodak analyser for measurement of aspirin may find a reduction in the true aspirin level (possible by as much as 40 per cent) when *N*-acetylcysteine (e.g. Parvolex™, Evans) has been administered.

TABLE 8.21

Management of salicylate poisoning based on information on peak salicylate concentrations.

Peak concentration interpretation		
Adult	Child/elderly	Management
<350 mg/l (2.52 mmol/l)	<250 mg/l (1.8 mmol/l)	Mild clinical effects, then further treatment unlikely to be necessary
<500 mg/l (3.6 mmol/l)	<350 mg/l (2.52 mmol/l)	Mild clinical effects, continue maintenance management
500–700 mg/l (3.6–5.04 mmol/l)	350–700 mg/l (3.24–5.04 mmol/l)	Moderate clinical effects, continue with maintenance management and institute urinary alkalinisation
>700 mg/l (5.04 mmol/l)	>700 mg/l >(5.04 mg/l)	Severe clinical effects, haemodialysis is recommended

CHAPTER 8

Available forms

Aspirin is commonly available as OTC preparations. It may have been taken alone as 75 mg, 300 mg, 325 mg, 500 mg or 530 mg tablets or capsules, or in combination with various other drugs, particularly paracetamol, codeine and caffeine. Formulations include tablets (including soluble and slow-release preparations), capsules, or powders.

Methyl salicylate is found in liniments and ointments, choline salicylate in teething gels, and salicylic acid in colloids and pastes.

Toxic doses (see also nomogram in Table 8.21)

The fatal dose of aspirin in adults is estimated at 500 mg/kg. Young children are most susceptible to salicylate toxicity. An acute fatal dose for children has not been established, but 4 g was fatal in a 4-year-old. Mild symptoms occur from plasma salicylate concentrations of 200 to 300 mg/l (Mongan *et al.*, 1973).

Metabolism (pharmacokinetics)

The peak plasma concentration of salicylate in therapy is reached 1 to 3 hours after ingestion (Levy and Leonards, 1966); however, it may take up to 24 hours to be reached in overdose. The elimination half-life is 2 to 4 hours in therapy, but in overdose may be up to 36 hours, and is typically 20 to 25 hours without treatment.

Biotransformation and elimination

Aspirin is rapidly converted to salicylic acid, which is then further metabolised to five main metabolites: salicyluric acid, salicyl phenol glucuronide, gentisic acid, acyl glucuronide and gentisuric acid (Levy and Leonards, 1966). The pathways to the first two of these involve saturable hepatic enzymes, so when a large amount of salicylate is ingested these enzymes quickly become saturated. This results in a change from first-order kinetics (where elimination is proportional to the plasma concentration) to zero-order kinetics (where only a certain amount is eliminated irrespective of the concentration). Thus salicylate may accumulate following mild therapeutic overdoses, particularly in children, and prior therapeutic use may increase the toxicity of an acute overdose.

Under zero-order kinetics the amount of salicylate excreted unchanged in the urine increases. With very high plasma concentrations, up to 80 per cent salicylate can be excreted unchanged in the urine (Snodgrass and Rumack, 1981). This pathway is sensitive to changes in urine pH; as the pH rises (i.e. the urine becomes more alkaline), excretion of salicylate is enhanced.

Concentration monitoring

Aspirin is insoluble in acid and may form concretions in the stomach wall. Absorption may continue for many hours, leading to delay in the peak salicylate concentration of up to 24 hours (McGuigan, 1986). It is advisable therefore to monitor the salicylate concentration every 2 to 3 h after the initial measurement to ensure the peak level has been reached.

Interpretation of salicylate concentrations

Interpretation of the salicylate concentration is not easy (compare with single ingestions of paracetamol), as many factors – peak levels, time since ingestion, co-ingestants, clinical condition including metabolic state, and age – need to be considered. A peak level of 450 mg/l (3.24 mmol/l) in a young adult at 6 hours post-ingestion is likely to give rise to far fewer effects than the same level on presentation at 24 hours, or in a child. The metabolic status is particularly important.

Alkalinisation of the urine

CHILDREN

Children quickly develop a metabolic acidosis following ingestion of salicylate in overdose, and respiratory alkalosis is rare under the age of 4 years. The likelihood of respiratory alkalosis increases with age until the age of 12 years, when the usual adult picture of respiratory alkalosis followed by metabolic acidosis is seen (Meredith and Vale, 1980). The underlying mechanisms for these age-dependent differences in acid–base balance are not understood. Since an acidosis enhances transfer of the salicylate ion across the blood–brain barrier, it is necessary to employ more vigorous therapy at lower salicylate concentrations in children than in adults.

THE ELDERLY

The elderly appear to be more susceptible to serious toxicity, and one reason for this may be that the renal capacity to deal with metabolic loads diminishes with age. Since urinary excretion is an important route of elimination for the salicylate ion, it is advisable to alkalinise the urine or haemodialyse at lower concentrations in the elderly than for healthy young adults.

Salicylate equivalent

All salicylates act by the same mechanisms, although different salicylates are not equivalent in efficacy or toxicity for weight. For example, 10 ml of a 35 per cent solution of methyl salicylate is equivalent to 4.89 g of aspirin, or 16.3×300-mg aspirin tablets (Vandenberg *et al.*, 1989). Five millilitres of Oil of Wintergreen is equivalent to about 23×300-mg aspirin tablets in content of salicylate (Johnson and Welch, 1984). One 15-g tube of Bonjela™ contains an aspirin equivalent of 970 mg (Johnson and Welch, 1984).

Conclusion

Salicylates are potentially very toxic, and an understanding of the mechanisms of toxicity and the reasoning behind various treatment regimes is essential in the management of intoxicated patients.

CONCLUSIONS

Of the side effects encountered with use of the salicylates, those affecting the gastrointestinal tract are most prevalent. They can be prevented largely by using formulations or derivatives that have proven lower ulcerogenicity, by avoidance of co-ulcerogens (e.g. alcohol, cigarettes) and exposure to stressful conditions, and by recognising the need to fortify the mucosal defences by way of nutrition.

■ CHAPTER 8 ■

Many of the other side effects could be prevented by: (a) recognising the presence of disease states that sensitise particular organs to the effects of salicylates (e.g. liver and kidney in SLE); (b) optimising dosage and drug monitoring in patients at risk (e.g. the elderly rheumatoid arthritic with hypoalbuminaemia); (c) avoiding the use of other acidic NSAIDs (especially to minimise untoward effects in the liver, renal and gastrointestinal tract); and (d) careful monitoring of patients on those toxic antirheumatic drugs (e.g. methotrexate, D-pencillamine, gold salts) that could predispose some organs (e.g. kidney) towards salicylate-induced damage.

A few side effects can only be avoided by withdrawing the drug (e.g. GI intolerance, reversible liver and renal injury, aspirin-sensitive asthma, tinnitus teratogenesis and difficult parturition).

Considering the widespread use of aspirin and other salicylates, the occurrence and severity of the side effects compared with other analgesics and/or NSAIDs, and the prospects of minimising the development of salicylate-induced side effects, this group of drugs can be regarded as being relatively safe. More careful use and application of these drugs through education of the professionals and public will also help to reduce the occurrence of side effects. More intensive research needs to be devoted towards understanding the development of salicylate-induced side effects and developing procedures to minimise their occurrence. The aim should be to make this relatively safe, cheap and immensely versatile group of drugs much safer.

REFERENCES

Aalykke, C., Lauritsen, J.M., Hallas, J., Reinholdt, S., Krogfelt, K. and Lauritsen, K. 1999, *Helicobacter pylori* and risk of ulcer bleeding among users of nonsteroidal anti-inflammatory drugs: a case–control study. *Gastroenterology*, **116:** 1305–1309.

Abbatiello, E.R., Kaminsky, M. and Weisbroth, S. 1975, The effect of prostaglandins and prostaglandin inhibitors on spermatogenesis. *International Journal of Fertility*, **20:** 177–182.

Abbott, F.V. and Fraser, M.I. 1998, Use and abuse of over-the-counter analgesic agents. *Journal of Psychiatry and Neuroscience*, **23:** 13–33.

Abdel-Salam, O.M.E., Mózsik, Gy. and Szolcsányl, J. 1997, The role of afferent sensory nerves in gastric mucosal protection. *In:* Gy. Mózsik, L. Nagy and A. Király (eds), *Twenty Five Years of Peptic Ulcer Research in Hungary (1971–1995)*, pp. 295–308. Budapest: Akadémiai Kiadó.

Abe, K., Irokawa, N., Yasujima, M., Seino, M., Chiba, S., Sakurai, Y., Yoshinaga, K. and Saito, T. 1978, The kallikrein–kinin system and prostaglandins in the kidney. Their relation to furosemide-induced diuresis and to the renin–angiotensin–aldosterone system in man. *Circulation Research*, **43:** 254–260.

Abel, J.A. 1971, Analgesic nephropathy – a review of the literature. *Clinical Pharmacology and Therapeutics*, **12:** 583–598.

Åberg, G. 1966, Der Effekt der Acetylsalicylsäure und der Disalicylsäure auf der Ventrikelschleimhaut bei Meerschweinchen. *Arzneimittel Forschung*, **16:** 898–899.

Abrahams, C. and Levin, N.W. 1968, Analgesic nephropathy. *Lancet*, **1:** 645.

Abrahams, C. and Levinson, C. 1970, Ultrastructure of the renal papilla in experimentally induced analgesic nephritis in rats. *South African Medical Journal*, **44:** 63–65.

Abrishami, M.A. and Thomas, J. 1977, Aspirin intolerance – A Review. *Annals of Allergy*, 39, 28–37.

Adams, W. 1981, Lack of behavioural effects from Feingold diet violations. *Perceptual and Motor Skills*, **52:** 307–313.

Ageel, A.M., Parmar, N.S. and Tariq, M. 1984, The effect of nicotine pre-treatment on the gastric mucosal damage induced by aspirin and reserpine in rats. *Life Sciences*, **34:** 751–756.

Ahonen, R., Enlund, H., Klaukka, T. and Martikainen, J. 1991, Consumption of analgesics and anti-inflammatory drugs in the Nordic countries between 1978–1988. *European Journal of Clinical Pharmacology*, **41:** 37–42.

Ailabouni, W. and Eknoyan, G. 1996, Nonsteroidal anti-inflammatory drugs and acute renal failure in the elderly. A risk–benefit assessment. *Drugs and Aging*, **9:** 341–351.

Aird, I., Bentall, H.H., Mehigan, J.A. and Roberts, J.A.F. 1954, The blood groups in relation to peptic ulceration and carcinoma of the colon, rectum, breast and bronchus. An association between the ABO blood groups and peptic ulceration. *British Medical Journal*, **2:** 315–321.

Akyol, S.M. Thompson, M. and Kerr, D.N.S. 1982, Renal function after prolonged consumption of aspirin. *British Medical Journal*, **284:** 631–632.

Alharbi, M.M., Islam, M.W., Alshabanah, O.A. and Algharably, N.M. 1995, Effect of acute administration of fish oil (omega-3 marine triglyceride) on gastric ulceration and secretion induced by various ulcerogenic and necrotizing agents in rats. *Food and Chemical Toxicology*, **33:** 553–558.

Alich, A.A., Wittmers, L.E., Anderson, L.A., Rieschl, E.M. and Peterson, P.L. 1992, Gastric mucosal damage due to aspirin and copper aspirinate assessed by gastric mucosal potential difference changes. *Journal of Pharmacological and Toxicological Methods*, **27:** 245–250.

Allen, A. 1978, Structure of gastrointestinal mucus glycoproteins and the viscous and gel forming properties of mucus. *British Medical Bulletin*, **34:** 28–33.

Allen, A., Flemström, G., Garner, A. and Kivilaakso, E. 1993, Gastroduodenal mucosal protection. *Physiological Reviews*, **73:** 823–857.

Allen, A. and Garner, A. 1980, Mucus and bicarbonate secretion in the stomach and their possible role in mucosal protection. *Gut*, **21:** 249–262.

Allen, C.N., Harpur, E.S., Gray, T.J.B. and Hirst, E.H. 1991, Toxic effects of non-steroidal anti-inflammatory drugs in a human intestinal epithelial cell line (HCT-8), as assessed by the MTT and neutral red assays. *Toxicology In Vitro*, **5:** 183–191.

Al-Nasser, I.A. 1999, Salicylate-induced kidney mitochondrial permeability transition is prevented by cyclosporin A. *Toxicology Letters*, **105:** 1–8.

Alphin, R.S. and Droppleman, D.A. 1971, New method for evaluating topical action of substances on gastric mucosa. *Journal of Pharmaceutical Sciences*, **60:** 1314–1316.

Alpsten, A., Bogentoft, C., Ekenved, G. and Sölvell, L. 1982, Gastric emptying and absorption of acetylsalicylic acid administered as enteric-coated micro granules. *European Journal of Clinical Pharmacology*, **33:** 57–61.

Alquorain, A.A., Satti, M.B., Marwah, S., Alnahdi, M. and Alhabdan, I. 1993, Non-steroidal anti-inflammatory drug-induced gastropathy – a comparative endoscopic and histopathological evaluation of the effects of tenoxicam and diclofenac. *Journal of International Medical Research*, **21:** 89–97.

Alshabanah, O.H., Islam, M.W., Algharably, N.M. and Alharbi, M.M. 1993, Effect of khatamines and their enantiomers on aspirin, indomethacin, phenylbutazone and reserpine induced gastric ulcers in rats. *Research Communications in Substances of Abuse*, **14:** 81–94.

Alvan, G., Bergman, U. and Gustafsson, L.L. 1981, High unbound fraction of salicylate in plasma during intoxication. *British Journal of Clinical Pharmacology*, **11:** 625–626.

Amin, A.R., Vyas, P., Attur, M., Leszczynska-Piziak, J., Patel, I.R., Weissmann, G. and Abramson, S.B. 1995, The mode of action of aspirin-like drugs: effect on inducible nitric oxide synthase. *Proceedings of the National Academy of Sciences USA*, **92:** 7926–7930.

Anand, B.S., Romero, J.J., Sanduja, S.K. and Lichtenberger, L.M. 1999, Phospholipid association reduces the gastric mucosal toxicity of aspirin in human subjects. *American Journal of Gastroenterology*, **94:** 1818–1822.

Anderson, K.W. 1958, XXXIIIth Meeting, Australia and New Zealand Association for the Advancement of Science.

Anderson, K.W. 1963, Some biochemical and physiological aspects of salicylate-induced gastric lesions in laboratory animals. *In:* A.St. J. Dixon, B.K. Martin, M.J.H. Smith and P.H.N. Wood (eds), *Salicylates: An International Symposium*, pp. 217–223. London: Churchill.

Anderson, K.W. 1964a, A study of the gastric lesions induced in laboratory animals by soluble and buffered aspirin. *Archives Internationales de Pharmacodynamie et de Thérapie*, **152:** 379–391.

Anderson, K.W. 1964b, A study of the gastric lesions induced by aspirin in laboratory animals. *Archives Internationales de Pharmacodynamic et de Thérapie*, **152:** 392–403.

Anderson, K.W. 1964c, Studies on the biochemical mechanism of the gastric erosion caused by aspirin. *Biochemical Pharmacology*, **13:** 1513–1517.

Anderson, K.W. 1965, Further studies on the aspirin-induced gastric lesions in the guinea pig. *Archives Internationales de Pharmacodynamie et de Thérapie*, **157:** 181–192.

Andrews, F.J., Malcontenti-Wilson, C. and O'Brien, P.E. 1994, Effect of non-steroidal anti-inflammatory drugs on LFA-1 and ICAM-1 expression in gastric mucosa. *American Journal of Physiology*, **266:** 657–664.

Angeel, A.M., Parmar, N.S. and Tariq, M. 1983, The effect of nicotine pre-treatment on the gastric mucosal damage induced by aspirin and reserpine in rats. *Life Sciences*, **34:** 751–756.

Angervall, L., Bengtsson, U., Zetterlund, C.G. and Zsigmund, M. 1969, Renal pelvic carcinoma in a Swedish district with abuse of a phenacetin-containing drug. *British Journal of Urology*, **41:** 401–405.

Anichov, S.V. and Zavodskaya, I.S. 1968, *The Experimental Basis of Gastric Ulcer Pharmacotherapy*. Oxford: Pergamon Press.

Anonymous. 2001, Safety of paracetamol packaging in the United Kingdom. *Prescrire International*, **10:** 189.

Anseline, P. 1977, Perforated peptic ulcer: an analysis of 246 cases. *Australian and New Zealand Journal of Medicine*, **47:** 81–85.

Anthony, A., Sim, R., Dhillon, A.P., Pounder, R.E. and Wakefield, A.J. 1996, Gastric mucosal contraction and vascular injury induced by indomethacin precede neutrophil infiltration in the rat. *Gut*, **39:** 363–368.

Appel, R.G., Bleyer, A.J. and McCabe, J.C. 1995, Case report: analgesic nephropathy: a soda and a powder. *American Journal of the Medical Sciences*, **310:** 161–166.

Appleyard, C.B., McCafferty, D.M., Tigley, A.W., Swain, M.G. and Wallace, J.L. 1996, Tumor necrosis factor mediation of NSAID-induced gastric damage: role of leukocyte adherence. *American Journal of Physiology*, **270:** G42–G48.

Armstrong, J.M., Blackwell, G.J., Flower, R.J., McGiff, J.C., Mullane, K.M. and Vane, J.R. 1976, Genetic hypertension in rats is accompanied by a defect in renal prostaglandin catabolism. *Nature*, **260:** 582–586.

Arnfelt-Rønne, I. and Arrigoni-Martelli, E. 1977, Renal prostaglandin metabolism in spontaneously hypertensive rats. *Biochemical Pharmacology*, **26:** 485–488.

Arnold, L., Collins, C. and Starmer, G.A. 1973, The short term effects of analgesics on the kidney with special reference to acetylsalicylic acid. *Pathology*, **5:** 123–134.

Arnold, L., and Collins, C., Starmer GA. 1976, Studies on the modification of renal lesions due to aspirin and oxyphenbutazone in the rat and the effects of 2:4 dinitrophenol. *Pathology*, **8:** 179–184.

Aron, E., Delbarre, B. and Jankowski, J.M. 1970, Soluble and injectable aspirin (lysine acetylsalicylate). Contribution to the study of digestive accidents with acetylsalicylic acid. *Archives Françaises des Maladies de l'Appereil Digestif*, **59:** 573–588.

Arsenault, A., Varady, J., Lebel, E. and Lussier, A. 1975, Effect of naproxen on gastrointestinal microbleeding following acetylsalicylate medication. *Journal of Clinical Pharmacology*, **15:** 340–346.

Arvidsson, B., Magnusson, B., Solvell, L., Magnusson, A. and Arnstad, L. 1975, Acetylsalicylic acid and gastrointestinal bleeding measurement of blood loss using a modified radioactive chromium method. *Scandinavian Journal of Gastroenterology*, **10:** 155–160.

Asako, H., Kubes, P., Wallace, J., Wolf, R.E. and Granger, D.N. 1992, Modulation of leukocyte adhesion in rat mesenteric venules by aspirin and salicylate. *Gastroenterology*, **103:** 146–152.

Atkinson, D.C. and Leach, E.C. 1976, Anti-inflammatory and related properties of 2-(2,4-dichlorophenoxy)phenylacetic acid (fenclofenac). *Agents and Actions*, **6:** 657–666.

Atwater, E.C., Mongan, E.S., Weiche, D.R. and Jacox, R.F. 1965, Peptic ulcer and rheumatoid arthritis. *Archives of Internal Medicine*, **115:** 184–189.

Atwood, S.J. 1980, The laboratory in the diagnosis and management of acetaminophen and salicylate intoxications. *Paediatric Clinics of North America*, **27:** 871–879.

Augur, N.A. 1970, Gastric mucosal blood flow following damage by ethanol, acetic acid, or aspirin. *Gastroenterology*, **58:** 311–320.

Axelsen, R.A. 1975, The induction of renal papillary necrosis in Gunn rats by analgesics and analgesic mixtures. *British Journal of Experimental Pathology*, **56:** 92–96.

Axelsen, R.A. 1980, Nephrotoxicity of mild analgesics in the Gunn strain of rat. *British Journal of Clinical Pharmacology*, **10**: 309S–312S.

Axelsen, R.A. and Burry, A.F. 1976, Bilirubin-associated renal papillary necrosis in the homozygous Gunn rat: light- and electron-microscopic observations. *Journal of Pathology*, **120**: 165–176.

Axelson, C.K., Christiansen, L.V., Johansen, A. and Poulsen, P.E. 1977, Comparative effects of tolferamic acid and acetylsalicylic acid on human gastric mucosa. A double blind crossover trial employing gastroscopy, external gastrocamera and multiple biopsies. *Scandinavian Journal of Rheumatology*, **6**: 23–27.

Azumi, Y., Ohara, S., Ishihara, K., Okabe, H. and Hotta, K. 1980, Correlation of quantitative changes of gastric mucosal glycoproteins with aspirin-induced gastric damage in rats. *Gut*, **21**: 533–536.

Bagchi, D., Carryl, O.R., Tran, M.X., Bagchi, M., Garg, A., Milnes, M.M., Williams, C.B., Balmoori, J., Bagchi, D.J., Mitra, S. and Stohs, S.J. 1999a, Acute and chronic stress-induced oxidative gastro-intestinal mucosal injury in rats and protection by bismuth subsalicylate. *Molecular and Cellular Biochemistry*, **196**: 109–116.

Bagchi, D., McGinn, T.R., Ye, X., Balmoori, J., Bagchi, M., Stohs, S.J., Kuszynski, C.A., Carryl, O.R. and Mitra, S. 1999b, Mechanism of gastroprotection by bismuth subsalicylate against chemically induced oxidative stress in cultured human gastric mucosal cells. *Digestive Diseases and Sciences*, **44**: 2419–2428.

Bailey, M.J., Worrall, S., De Jersey, J. and Dickinson, R.G. 1998, Zomepirac acyl glucuronide covalently modifies tubulin in vitro and in vivo and inhibits its assembly in an in vitro system. *Chemical and Biological Interactications*, **115**: 153–166.

Baker, R., Jaffe, B.M., Reed, J.D., Shaw, B. and Venables, C.W. 1978a, Endogenous prostaglandins and gastric secretion in the cat. *Journal of Physiology*, **278**: 451–460.

Baker, R., Jaffe, B.M., Reed, J.D., Shaw, B. and Venables, C.W. 1978b, Exogenous prostaglandins and gastric secretion in the rat. *Journal of Physiology*, **278**: 441–450.

Balasubramanian, A., Manimekalai, S., Singh, A.G. and Ramakrishnan, S. 1980, Short and long term effects of aspirin on testes of albino rats: a histological and biochemical study. *Indian Journal of Experimental Biology*, **18**: 1408–1410.

Baldwin, G.S. 2000, Do NSAIDs contribute to acute fatty liver of pregnancy? *Medical Hypotheses*, **54**: 846–849.

Bali, J.-P., Balmes, J.-L. and Baillat, X. 1970, Influence of aspirin on the metabolic activity of the ribonucleic acids of gastric and hepatic cells in the mouse. *Biologie et Gastro-Enterologie*, **1**: 5–12.

Bali, J.-P., Soumarmon, A., Lewin, M. and Bonfils, S. 1977, A gastric sensitive adenylate cyclase system in rat gastric mucosa. *In:* S. Bonfils (ed.), *1st International Symposium on Hormone Receptors in Digestive Tract Physiology, INSERM Symposium No. 3*, pp. 401–402. Amsterdam: Elsevier/North Holland.

Bamba, H., Ota, S., Kato, A. and Matsuzaki, F. 1998, Nonsteroidal anti-inflammatory drugs may delay the repair of gastric mucosa by suppressing prostaglandin-mediated increase of hepatocyte growth factor production. *Biochemical and Biophysical Research Communications*, **245**: 567–571.

Baños, G. and Reyes, P.A. 1989, A comparative study of the effect of ten non-steroidal anti-inflammatory drugs (NSAIDs) upon some mitochondrial and platelet functions. *International Journal of Biochemistry*, **21**: 1387–1394.

Barbare, J.C., Imbert, A. and Benkirane, A. 2001, Recent developments concerning drug-induced liver toxicity. *Presse Medicale*, **30**: 673–676.

Barbour, H.G. and Dickerson, V.C. 1938, Gastric ulceration produced in rats by oral and subcutaneous aspirin. *Archives Internationales de Pharmacodynamie et de Thérapie*, **58**: 78–87.

Barkin, J. 1998, The relation between *Helicobacter pylori* and non-steroidal anti-inflammatory drugs. *American Journal of Medicine*, **105**: 22S–27S.

Barnes, M.J. 1975, Function of ascorbic acid in collagen metabolism. *Annals of the New York Academy of Sciences*, **258**: 264–277.

Barnes, T.C. 1965, Effect of acetylsalicylic acid on movements of isolated intestine of the rat. *American Journal of Gastroenterology*, **44**: 476–480.

CHAPTER 8

Barone, J.A., Raia, J.J. and Huang, Y.C. 1988, Evaluation of the effects of multiple-dose activated charcoal on the absorption of orally administered salicylate in a simulated toxic ingestion model. *Annals of Emergency Medicine*, **17:** 34–37.

Barritt, G.J. and Whitehouse, M.W. 1977, Pathobiodynamics: effect of extrahepatic inflammation on calcium transport and drug metabolism by rat liver mitochondria in vitro. *Biochemical Medicine*, **17:** 99–115.

Barroy, J.P., Willems, G., Verbeustel, S.S. and Gerard, A. 1970, Histological changes of the gastric wall in dogs submitted to the action of phenylbutazone. *Acta Gastroenterologica Belgica*, **33:** 469–481.

Baskin, W.N., Ivey, K.J., Krause, W.J., Jeffrey, G.E. and Gemmell, R.T. 1976, Aspirin-induced ultrastructural changes in human gastric mucosa: correlation in the potential difference. *Annals of Internal Medicine*, **85:** 299–303.

Basomba, A., Romar, A., Pelaez, A., Villalmanzo, I.G. and Campos, A. 1976, The effect of sodium cromoglycate in preventing aspirin-induced bronchospasm. *Clinical Allergy*, **6:** 269–275.

Bass, D.E. and Jacobson, E.D. 1965, Effects of salicylate on acclimatization to work in the heat. *Journal of Applied Physiology*, **20:** 70–72.

Basso, D., Nagaglia, F., Brigato, L., DiMario, F., Rugge, M. and Plebani, M. 1999, *Helicobacter pylori* non-cytotoxic genotype enhances mucosal gastrin and mast cell tryptase. *Journal of Clinical Pathology*, **52:** 210–214.

Battistel, M., Plebani, M., DiMario, F., Jocic, M., Lippe, I.T. and Holzer, P. 1993, Chronic nicotine intake causes vascular dysregulation in the rat gastric mucosa. *Gut*, **34:** 1688–1692.

Batzri, S. and Gardner, J.D. 1979, Actions of histamine on cyclic AMP in guinea pig gastric cells: inhibition by H1 and H2 receptor antagonists. *Molecular Pharmacology*, **16:** 406–416.

Beall, J.R. and Klein, M.F. 1977, Enhancement of aspirin-induced teratogenicity by food restriction in rats. *Toxicology and Applied Pharmacology*, **39:** 489–495.

Beaver, W.T. 1965, Mild analgesics. A review of their clinical pharmacology. *American Journal of the Medical Sciences*, **250:** 577–604.

Beck, F.J. and Whitehouse, M.W. 1973, Effect of adjuvant disease in rats on cyclophosphamide and isophosphamide metabolism. *Biochemical Pharmacology*, **22:** 2453–2468.

Becker, M. and Ruoff, H.-J. 1979, Pentagastrin activation of adenytate cyclase in human gastric biopsy specimens. *Experientia (Basel)*, **35:** 781–782.

Beeken, W. 1967, Effects of salicylates on gastrointestinal protein loss in normal subject. *Gastroenterology*, **53:** 894–899.

Bélanger, P.M. and Attisté-Gbeassor, A. 1985, Effect of nonsteroidal anti-inflammatory drugs on the microsomal monooxygenase system of rat liver. *Canadian Journal of Physiology and Pharmacology*, **63:** 798–803.

Bennett, A. and Curwain, B.P. 1977, Effects of aspirin-like drugs on canine gastric mucosal blood flow and acid secretion. *British Journal of Pharmacology*, **60:** 499–504.

Bennett, A., Eley, K.G. and Scholes, G.B. 1968, Effect of prostaglandins E1 and E2 on intestinal motility in the guinea pig and rat. *British Journal of Pharmacology*, **34:** 639–647.

Bennett, A. and Posner, J. 1971, Studies on prostaglandin antagonists. *British Journal of Pharmacology*, **42:** 584–594.

Berard, A., Solomon, D.H. and Avorn, J. 2000, Patterns of drug use in rheumatoid arthritis. *Journal of Rheumatology*, **27:** 1648–1655.

Berg, K.J. 1977, Acute effects of acetylsalicylic acid on renal function in normal man. *European Journal of Clinical Pharmacology*, **11:** 117–123.

Berg, K.J. and Bergan, A. 1977, Effects of different doses of acetylsalicylic acid on renal oxygen consumption. *Scandinavian Journal of Clinical Laboratory Investigation*, **37:** 235–241.

Berg, K.J., Djøseland, O., Gjellan, A., Hundall, Ø., Knudsen, E.R., Rugstad, H.E. and Rønneberg, E. 1990, Acute effects of paracetamol on prostaglandin synthesis and renal function in normal man and in patients with renal failure. *Clinical Nephrology*, **34:** 255–262.

Bergmann, J.F., Chassany, O., Simoneau, G., Lemaire, M., Segrestaa, J.M. and Caulin, C. 1992, Protection against aspirin-induced gastric lesions by lansoprazole: simultaneous evaluation of functional and morphologic responses. *Clinical Pharmacology and Therapeutics*, **52:** 413–416.

Berkow, R. (ed.) 1977, *The Merck Manual of Diagnosis and Therapy*. Rahway, NJ: Merck, Sharp & Dohme Research Labs.

Berkowitz, J.M., Adler, S.N., Sharp, J.T. and Warner, C.W. 1986, Reduction of aspirin-induced gastroduodenal mucosal damage with ranitidine. *Journal of Clinical Gastroenterology*, **8:** 377–380.

Berl, T., Raz, A., Wald, H., Horowitz, J. and Czaczkes, W. 1977, Prostaglandin synthesis and the action of vasopressin: studies in man/rat. *American Journal of Physiology*, **232:** F529–F537.

Bernstein, B.H., Singsen, B.H., King, K.K. and Handon, V. 1977, Aspirin-induced hepatotoxicity and its effect on juvenile rheumatoid arthritis. *American Journal of Diseases of Children*, **131:** 659–663.

Bernstone, L.L. and Monie, I.W. 1965, Teratogenic effect of methyl salicylate and hypoxia in combination. *Anatomical Record*, **151:** 443.

Berson, A., Fromenty, B., Lettéron, P. and Pessayre, D. 1998, Rôle des mitochondries dans l'hépatotoxicité des médicaments. *Gastroenterologie Clinique et Biologie*, **22:** 59–72.

Bertino, J.S., Willis, E.D., Reed, M.D. and Speck, W.T. 1981, Salicylate hepatitis: a complication of the treatment of Kawasaki's disease. *American Journal of Hospital Pharmacy*, **38:** 1171–1172.

Besner, J.G. and Houle, J.M. 1985, Efficacy of sucralfate in protection of the gastric mucosa against lesions induced by acute administration of ethanol. *Clinical and Investigative Medicine*, **8:** 97.

Beubler, E. 1978, Prostaglandins and gastrointestinal functions. *Zeitschrift für Gastroenterologie*, **16:** 479–489.

Beyer, K.H. and Gelarden, R.T. 1978, Renal concentration gradients of salicylic acid and its metabolic congeners in the dog. *Archives Internationales de Pharmacodynamie et de Thérapie*, **231:** 180–195.

Beyth, R.J. and Shorr, R.I. 1999, Epidemiology of adverse drug reactions in the elderly by drug class. *Drugs and Ageing*, **14:** 231–239.

Biasco, G., Paganelli, G.M., Difebo, G., Siringo and S., Barbara, L. 1992a, Cell kinetic alterations induced by aspirin in human gastric mucosa and their prevention by a cytoprotective agent. *Digestion*, **51:** 146–151.

Biasco, G., Paganelli, G.M., Psilogenis, M., Nazzari, M., Difebo, G. and Siringo, S. 1992b, Effect of sulglycotide treatment on cell kinetics alterations induced by aspirin in humans. *Agents and Actions*, **(Special Issue):** C88–C89.

Bickel, M. 1981, Effect of 16,16-dimethyl prostaglandin E_2 on gastric mucus gel thickness. *Prostaglandins*, **21:** 63–66.

Bingulacpopovic, J., Juretic, D., Madeija, M., Cepelak, I., Papicfutac, D., Slijepcevic, M. and Lipovac, K. 1992, Comparison of protein glycation inhibitory and toxic effects of acetylsalicylic acid in experimental diabetes. *Acta Pharmaceutica*, **42:** 211–218.

Biscarini, L. 1996, Anti-inflammatory analgesics and drugs used in gout. *In:* M.N.G. Dukes (ed.), *Meyler's Side Effects of Drugs. An Encyclopedia of Adverse Reactions and Interactions*. 13th edn, pp. 204–264. Amsterdam: Elsevier.

Bjarnason, I. 1988, Non-steroidal anti-inflammatory drug-induced small intestinal inflammation in man. *In: Recent Advances in Gastroenterology*, No. 7. Melbourne and New York: Churchill Livingstone.

Bjarnason, I. 1994, Intestinal Permeability. *Gut*, **1 (Suppl.):** S18–S22.

Bjarnason, I., Hayllar, J., MacPherson, A.J. and Russell, A.S. 1993, Side effects of nonsteroidal anti-inflammatory drugs on the small and large intestine in humans. *Gastroenterology*, **104:** 1832–1847.

Bjarnason, I. and MacPherson, A.J.S. 1994, Intestinal toxicity of non-steroidal anti-inflammatory drugs. *Pharmacology and Therapeutics*, **62:** 145–157.

Bjarnason, I., MacPherson, A., Rotman, H., Schupp, J. and Hayllar, J. 1997, A randomized, double-blind, crossover comparative endoscopy study on the gastroduodenal tolerability of a highly specific cyclooxygenase-2 inhibitor, flosulide, and naproxen. *Scandinavian Journal of Gastroenterology*, **32:** 126–130.

Bjarnason, I. and Peters, J. 1987, Helping the mucosa make sense of macromolecules. *Gut*, **28:** 1057–1061.

Bjarnason, I., Peters, T.J. and Levi, A.J. 1986a, Intestinal permeability: clinical correlates. *Digestive Diseases*, **4:** 83–92.

Bjarnason, I., Smethurst, P., MacPherson, A., Walker, F., McElnay, J.C., Passmore, A.P. and Menzies, I.S. 1992, Glucose and citrate reduce the permeability changes caused by indomethacin in humans. *Gastroenterology*, **102:** 1546–1550.

Bjarnason, I., Williams, P., So, A., Zanelli, G.D., Levy, A.J., Gumpel, J.M., Peters, T.J. and Ansell, B. 1984, Intestinal permeability and inflammation in rheumatoid arthritis: effects of non-steroidal anti-inflammatory drugs. *Lancet*, **2:** 1171–1174.

Bjarnason, I., Zanelli, G., Prouse, P., Williams, P., Gumpel, M.J. and Levi, A.J. 1986b, Effect of non-steroidal anti-inflammatory drugs on the human small intestine. *Drugs*, **32:** 35–41.

Bjorkman, D. 1998, Nonsteriodal anti-inflammatory drug-associated toxicity of the liver, lower intestinal tract, and esophagus. *American Journal of Medicine*, **105 (5A):** 17S–21S.

Blondon, H., Barbier, J.P., Mahé, I., Deverly, A., Kolsky, H. and Bergmann, J.F. 2000, Gastroduodenal tolerability of medium dose enteric-coated aspirin: a placebo controlled endoscopic study of a new enteric-coated formulation versus regular formulation in healthy volunteers. *Fundamental and Clinical Pharmacology*, **14:** 155–157.

Bochenek, G. Nizankowska, E. and Szezeklik, A. 1996, Atropy trait in hypersensitivity to non-steroidal anti-inflammatory drugs. *Allergy*, **51:** 16–23.

Bodger, K. and Crabtree, J.E. 1998, *Helicobacter pylori* and gastric inflammation. *British Medical Bulletin*, **54 (1):** 139–150.

Boldy, D. and Vale, J.A. 1986, Treatment of salicylate poisoning with repeated oral charcoal. *British Medical Journal*, **292:** 136.

Bolton, J.P. and Cohen, M.M.1977, Effect of repeated aspirin administration on the gastric mucosal barrier and cell turnover. *Journal of Surgical Research*, **23:** 251–256.

Bolton, J.P. and Cohen, M.M. 1979, Effect of 16,16-dimethyl prostaglandin E_2 on the gastric mucosal barrier. *Gut*, **20:** 513–517.

Bolton, J.P., Palmer, D. and Cohen, M.M. 1976, Effect of the E2 prostaglandins on gastric mucus production in rats. *Surgical Forum*, **27:** 402–403.

Bolton, J.P., Palmer, D. and Cohen, M.M. 1978, Stimulation of mucus and nonparietal cell secretion by the E2 prostaglandins. *American Journal of Digestive Diseases*, **23:** 359–364.

Bommelaer, G. and Guth, P.H. 1979, Protection by histamine receptor antagonists and prostaglandin against gastric mucosal barrier disruption in the rat. *Gastroenterology*, **77:** 303–308.

Bouchier, I.A.D. and Williams, H.S. 1969, Determination of faecal blood-loss after combined alcohol and sodium acetylsalicylate intake. *Lancet*, **1:** 178–180.

Bowen, B.K., Krause, W.J. and Ivey, K.J. 1977, Effect of sodium bicarbonate on aspirin-induced damage and potential difference changes in human gastric mucosa. *British Medical Journal*, **2:** 1053–1055.

Brandslund, I., Rask, H. and Klitgaard, N.A. 1979, Gastrointestinal blood loss caused by controlled release and conventional acetylsalicylic acid tablets. *Scandinavian Journal of Rheumatology*, **8:** 209–213.

Brassinne, A. 1979, Effects of ethanol on plasma protein shedding in the human stomach. *Digestive Diseases and Sciences*, **24:** 44–47.

Bravo, M.L., Escolar, G., Navarro, C., Fontarnau, R. and Bulvena, O. 1992, Morphological study of gastric lesions developing in the rat under several damaging conditions – modifications induced by pretreatment with zinc acexamate. *Scanning Electron Microscopy*, **6:** 855–864.

Brenner, A. 1979, Trace mineral levels in hyperactive children responding to the Feingold diet. *Journal of Paediatrics*, **94:** 944–945.

Brenner, B.E. and Simon, R.R. 1982, Management of salicylate intoxication. *Drugs*, **24:** 335–340.

Brodie, D.A. and Chase, B.J. 1967a, Evaluation of gastric acid as a factor in drug induced gastric haemorrhage in the rat. *Gastroenterology*, **56:** 206–213.

Brodie, D.A. and Chase, B.J. 1967b, Role of gastric acid in aspirin-induced gastric irritation in the rat. *Gastroenterology*, **53:** 604–610.

Brodie, D.A., Tate, C.L. and Hooke, K.F. 1970, Aspirin: intestinal damage in rats. *Science*, **170:** 183–186.

Brodie, M.J. and Boobis, S. 1978, The effects of chronic alcohol ingestion and alcoholic liver disease on binding of drugs to serum proteins. *European Journal of Clinical Pharmacology*, **13:** 435–438.

Brody, T.M. 1956, Action of sodium salicylate and related compounds on tissue metabolism *in vitro*. *Journal of Pharmacology and Experimental Therapeutics*, **117:** 39–51.

Brogden, R.N., Heel, R.C., Pakes, G.E., Specht, T.M. and Avery, G.S. 1980, Diclofenac sodium: A review of its pharmacological properties and therapeutic use in rheumatic diseases and pain of varying origin. *Drugs*, **20:** 24–48.

Broh-Kahn, R.H. 1960, Choline salicylate: a new, effective, and well-tolerated analgesic, anti-inflammatory and anti-pyretic agent. *International Record of Medicine*, **173:** 217–233.

Brooks, P.M., Cossum, P.A. and Boyd, G.W 1980, Rebound rise in renin concentrations after cessation of salicylates. *New England Journal of Medicine*, **303:** 562–564.

Brown, P.A., Sawrey, J.M. and Vernikos-Danillis, J. 1975, Attenuation of salicylate and stress-produced gastric ulceration by metiamide. *Proceedings of the Western Pharmacology Society*, **18:** 123–127.

Brown, P.A., Sawrey, J.M. and Vernikos, J. 1978, Aspirin and indomethacin-induced ulcers and their antagonism by antihistamines. *European Journal of Pharmacology*, **51:** 275–283.

Brown, R.A. and West, G.B. 1964, Effect of acetylsalicylic acid on fetal rats. *Journal of Pharmacy and Pharmacology*, **16:** 563–565.

Bruggeman, T.M., Wood, J.G. and Davenport, H.W. 1979, Local control of blood flow in the dog's stomach: vasodilation caused by acid back-diffusion following topical application of salicylic acid. *Gastroenterology*, **77:** 736–744.

Brune, K., Fenner, H., Kurolski, M., Lanz, R. and members of the SPALA group. 1992, Adverse reactions to NSAIDs: Consecutive evaluation of 30,000 pateinets in rheumatology. *In:* K.D. Rainsford and G.P. Velo (eds), *Side Effects of Anti-inflammatory Drugs 3*, pp. 33–42, Dordrecht: Kluwer Academic Publishers.

Brune, K., Graf, P. and Rainsford, K.D. 1977a, A pharmacokinetic approach to the understanding of therapeutic effects and side effects of salicylates. *In:* K.D. Rainsford, K. Brune and M.W. Whitehouse (eds), *Aspirin and Related Drugs: Their Actions and Uses*, pp. 9–26. Basel: Birkhäuser.

Brune, K., Gubler, H. and Schweitzer, A. 1979, Authoradiographic methods for the evaluation of ulcerogenic effects of anti-inflammatory drugs. *Pharmacology and Therapeutics*, **5:** 199–207.

Bruhn, R., Lücker, P.W. and Penth, B. 1983, Untersuchungen zur Magenverträglichkeit von Tiaprofensäure im Vergleich zu Acetylsalicylsäure und Indometacin. *Therapiewoche*, **33:** 2213–2219.

Brune, K., Schweitzer, A. and Eckert, H. 1977b, Parietal cells of the stomach trap salicylates during absorption. *Biochemical Pharmacology*, **26:** 1735–1740.

Brune, K., Schweitzer, A. and Lanz, R. 1984, Importance of drug biodistribution and metabolism in the development of side-effects by anti-inflammatory/analgesic drugs. *Advances in Inflammation Research*, **6:** 9–15.

Bryant, C., Smith, M.J.H., Hines, W.J.W. 1963, Effects of salicylate and γ-resocylate on the metabolism of radioactive succinate and fumarate by rat liver mitochondria. *Biochemical Journal*, **86:** 391–396.

Brzozowski, T., Konturek, P.C., Konturek, S.J., Drozdowicz, D., Pajdo, R., Pawlik, M., Brzozowska, I. and Hahn, E.G. 2000, Expression of cyclooxygenase (COX)-1 and COX-2 in adaptive cytoprotection induced by mild stress. *Journal of Physiology (Paris)*, **94:** 83–91.

Brzozowski, T., Konturek, S.J., Majka, J., Drozdowicz, D., Phytopolonczyk, J. and Nauert, C. 1993, Supralox – a novel aluminium containing antacid with gastroprotective and ulcer healing activity. *European Journal of Gastroenterology and Hepatology*, **5:** 165–171.

Brzozowski, T., Konturek, S.J., Sliwowski, Z., Pytkopolonczyk, J., Szlachcic, A. and Drozdowicz, D. 1996, Role of capsaicin-sensitive sensory nerves in gastroprotection against acid-independent and acid-dependent ulcerogens. *Digestion*, **57:** 424–432.

Brzozowski, T., Majka, J., Konturek, S.J., Bielanski, W., Slomiany, B.L. and Garner, A. 1994, Gastroprotective activity and receptor expression of transforming growth factor alpha, epidermal growth factor and basic fibroblast growth factor in the rat stomach. *European Journal of Gastroenterology and Hepatology*, **6:** 337–343.

Brzozowski, T., Pierzchalski, P., Kwiecien, S., Pajdo, R., Hahn, E.G. and Konturek, S.J. 1998, Activation of genes for spasmolytic peptide, transforming growth factor alpha and for cyclooxygenase (COX)-1 and COX-2 during gastric adaptation to aspirin damage in rats. *Alimentary Pharmacology and Therapeutics*, **12:** 767–777.

Buchanan, M.R., Dejana, E., Gent, M., Mustard, J.F. and Hirsch, J. 1981, Enhanced platelet accumulation onto injured carotid arteries in rabbits after aspirin treatment. *Journal of Clinical Investigation*, **67:** 503–508.

Buchanan, W.W. 1990, Implications of NSAID therapy in elderly patients. *Journal of Rheumatology*, **20:** 29–32.

Buchanec, J., Galanda, V., Visnovsky, P. and Halakova, E. 1981, Effect of age on pharmacokinetics of salicylate. *Journal of Pediatrics*, **99:** 833–934.

Bulbena, O., Escolar, G., Navarro, C., Bravo, L. and Pfeiffer, C.J. 1993, Gastroprotective effect of zinc acexamate against damage induced by nonsteroidal anti-inflammatory drugs – a morphological study. *Digestive Diseases and Sciences*, **38:** 730–739.

Bullock, G.R., Delaney, V.B., Sawyer, B.C. and Slater, T.F. 1970, Biochemical and structural changes in rat liver resulting from the parietal administration of a large dose of sodium salicylate. *Biochemical Pharmacology*, **19:** 245–253.

Bunce, K.T., McCarthy, J.J., Spraggs, C.F. and Stables, R. 1982, Relationship between lesion formation and permeability of rat gastric mucosa to H^+ and other cations. *British Journal of Pharmacology*, **75:** 325–331.

Bungaard, H. and de Weck, A.L. 1975, The role of amino-reactive impurities in acetylsalicylic acid allergy. *International Archives of Allergy and Applied Immunology*, **49:** 119–124.

Bürgi, W., Kaufmann, H. and Clemencon, G. 1975, Isolation, identification and quantitative determination of lysolecithin in the human gastric juice. *Schweizerische Medizinische Wochenschrift*, **105:** 1814–1819.

Burka, J.F. and Paterson, N.A. 1980, Evidence for lipoxygenase pathway involvement in allergic tracheal contraction. *Prostaglandins*, **19:** 499–515.

Burleigh, M. and Smith, M.J.H. 1971, The mechanism of the inhibitory action of salicylate on glutamyl-transfer ribonucleic acid synthetase *in vitro*. *Journal of Pharmacy and Pharmacology*, **23:** 590–594.

Burrell, J.H., Yong, J.L.C. and MacDonald, G.J. 1990, Experimental analgesic nephropathy: changes in renal structure and urinary concentrating ability in Fischer 344 rats given continuous low doses of aspirin and paracetamol. *Pathology*, **22:** 33–44.

Burrell, J.H., Yong, J.L.C. and MacDonald, G.J. 1991a, Analgesic nephropathy in Fischer 344 rats: comparative effects of chronic treatment with either aspirin or paracetamol. *Pathology*, **23:** 107–114.

Burrell, J.H., Yong, J.L.C. and MacDonald, G.J. 1991b, Irreversible damage to the medullary interstitium in experimental analgesic nephropathy in F344 rats. *Journal of Pathology*, **164:** 329–338.

Burry, H.C., Dieppe, P.A., Breshnihan, F.B. and Brown, C. 1976, Salicylates and renal function in rheumatoid arthritis. *British Medical Journal*, **1:** 613–615.

Buss, C.E. 1875, Über die Anwendung der Salicylsäure als Antipyretikum. Doctoral Dissertation. Medical Faculty, University of Basel. Leipzig: J.H. Hirschfeld.

Büss, H. and Balmer, H. 1962, Carl Emil Buss (1849–1878) and the beginning of salicylic acid therapy. *Gesherus*, **19:** 130–154.

Buttar, H.S., Chow, A.Y.K. and Downie, R.H. 1977, Glutathione alterations in rat liver after acute and subacute oral administration of paracetamol. *Clinical and Experimental Pharmacology and Physiology*, **4:** 1–6.

Calatayud, S., Ramirez, M.C., Sanz, M.J., Moreno, L., Bosch, J., Pique, J.M. and Esplugues, J.V. 2001, Gastric mucosal resistance to acute injury in experimental portal hypertension. *British Journal of Pharmacology*, **132:** 309–317.

Caldara, R., Cutarelli, G., Ferrari, C. and Romussi, M. 1978, Failure of lysine acetylsalicylate and phenylbutazone to affect gastric secretion in healthy adults. *Acta Hepatogastroenterologica (Stuttgart)*, **25:** 219–221.

Callahan, R.F., Gebruers, E.M. and O'Regan, M.G. 1980, Ultrastructural effects of aspirin, prostaglandin E_1 and SITS on guinea pig ileum. *Irish Journal of Medical Science*, **149:** 452.

Cappon, G.D., Gupta, U., Cook, J.C., Tassinari, M.S. and Hurtt, M.E. 2003, Comparision of the developmental toxicology of aspirin (acetylsalicylic acid) in rats using selected dosing paradigms. *Birth Defects Research Part B, Developmental and Reproductive Toxicology*, **68:** 38–46.

Canonico, P.G., Rill, W. and Ayala, E. 1977, Effects of inflammation on peroxisomal enzyme activities, catalase synthesis, and lipid metabolism. *Laboratory Investigation*, **37:** 479–486.

Carmichael, H.A., Nelson, L.M. and Russell, R.I. 1978, Cimetidine and prostaglandin: evidence for different modes of action on the rat gastric mucosa. *Gastroenterology*, **74:** 1229–1232.

Carneskog, J., Florath-Ahlmen, M. and Olsson, R. 1980, Prevalence of liver disease in patients taking salicylates for arthropathy. *Hepato-gastroenterology*, **27:** 361–364.

Carnimeo, N., Resta, O., Foschino-Barbaro, M.P., Valerio, G. and Picca, V. 1981, Functional assessment of airways bronchoconstriction with nebulised acetyl salicylic acid. *Allergologia et Immunopathologia (Madrid)*, **9:** 1–8.

Carter, G.W., Young, P.R. and Swett, L.R. 1980, Pharmacological studies in the rat with [2-(1,3-dideconoyloxy)-propyl]2-acetyloxy-benzoate (A-45474): an aspirin pro-drug with negligible gastric irritation. *Agents and Actions*, **10:** 240–245.

Caruso, I. and Bianchi Porro, G. 1980, Gastroscopic evaluation of anti-inflammatory agents. *British Medical Journal*, **1:** 75–78.

Caruso, I., Fumagalli, M., Montrone, F., Vernazza, M., Porro, G.B. and Petrillo, M. 1978, Controlled, double-blind study comparing acetylsalicylic acid and diflunisal in the treatment of osteoarthritis of the hip and/or knee; long-term gastroscopic study. *In: Diflunisal in Clinical Practice. Proceedings of a Special Symposium held at the XIVth Congress of Rheumatology.* (Chairman, K. Miehlke), pp. 63–73. New York: Futura.

Carvajal, A., Pieto, J.R., Requejo, A.A. and Arias, L.H.M. 1996, Aspirin or acetaminophen? A comparison from data collected by the Spanish Drug Monitoring System. *Journal of Clinical Epidemiology*, **49:** 255–261.

Casalnuovo, C.A., Kraglund, K., Kristensen, I.B. and Amdrup, E. 1984, Canine gastric ulcer produced by hemorrhagic shock and aspirin. *Scandinavian Journal of Gastroenterology*, **19:** 487–491.

Caselli, M., LaCorte, R., DeCarlo, L., Aleotti, A., Trevisani, L., Ruina, M., Trotta, F. and Alvisi, V. 1995, Histological findings in gastric mucosa in patients treated with non-steroidal anti-inflammatory drugs. *Journal of Clinical Pathology*, **48:** 553–555.

Caselli, M., Ruina, M., LaCorte, R., Trevisani, L., Sartori, S., Dentale, A., Gaudenzi, P., Trotta, F. and Alvisi, V. 1997, Ultrastructural damage of gastric epithelium in patients taking NSAIDs. *Italian Journal of Gastroenterology and Hepatology*, **28:** 16–18.

Caspary, W.G. 1978, Transmurale elektrische Potentialdifferenz-Messung als Funktionsparameter der Magenmukosaschramke. *Zeitschrift für Gastroenterologie*, **16:** 126–136.

Caspi, D., Lubart, E., Graff, E., Habot, B., Yaron, M. and Segal, R. 2000, The effect of mini-dose aspirin on renal function and uric acid handling in elderly patients. *Arthritis and Rheumatism*, **43:** 103–108.

Cassidy, M.M. and Lightfoot, F.G. 1979, Electron microscopic study of gastrointestinal response to acetylsalicylic acid. *Journal of Submicroscopic Cytology*, **11:** 449–462.

Caterson, R.J., Duggin, G.G., Horvarth, J., Mohandas, J. and Tiller, D. 1978, Aspirin, protein transacetylation and inhibition of prostaglandin synthetase in the kidney. *British Journal of Pharmacology*, **64:** 353–358.

Cawthorne, M.A., Palmer, E.D. and Green, J. 1976, Adjuvant-induced arthritis and drug-metabolising enzymes. *Biochemical Pharmacology*, **25:** 2683–2688.

Cazals, Y. 2000, Auditory sensori-neural alterations induced by salicylate. *Progress in Neurobiology*, **62:** 583–631.

Cenedella, R.J. and Crouthamel, W.G. 1973, Effect of aspirin upon male mouse fertility. *Prostaglandins*, **4:** 285–290.

Ceserani, R., Colombo, M., Robuschi, M. and Bianco, S. 1978, Tartrazine and prostaglandin systems. *Prostaglandins and Medicine*, **1:** 499–505.

Chand, N. and Altura, B.M. 1981, Lipoxygenase pathway and hydroperoxy acids: possible relevance to aspirin-induced asthma and hyperirritability of airways in asthmatics. *Prostaglandins and Medicine*, **19:** 249–256.

Chang, R.S.K., Field, M. and Silen, W. 1973, Permeability of gastric mucosa to hydrogen and lithium. *Gastroenterology*, **64:** 593–598.

Chao, E.S., Dunbar, D. and Kaminsky, L.S. 1988, Intracellular lactate dehydrogenase concentration as an index of cytotoxicity in rat hepatocyte primary culture. *Cell Biology and Toxicology*, **4:** 1–11.

Chatterjee, S.S. and Stefanovich, V. 1976, Influence of anti-inflammatory agents on rat liver mitochondrial ATPase. *Arzneimittel Forschung*, **26:** 499–502.

Chaudhury, T.K. and Jacobson, E.D. 1978, Prostaglandin cytoprotection of gastric mucosa. *Gastroenterology*, **74:** 59–63.

Cheney, C.D. and Rudrud, E. 1974, Prophylaxis by vitamin C in starvation induced rat stomach ulceration. *Life Sciences*, **14:** 2209–2214.

Cheng, H.F., Wang, J.L., Zhang, M.Z., McKanna, J.A. and Harris, R.C. 2000, Nitric oxide regulates renal cortical cyclooxygenase-2 expression. *American Journal of Physiology – Renal Physiology*, **279:** 122–129.

Chernish, S.M., Rosenak, B.D., Brunelle, R.L. and Crabtree, R. 1979, Comparison of gastrointestinal effects of aspirin and fenoprofen. A double blind crossover study. *Arthritis and Rheumatism*, **22:** 376–383.

Cheung, L.Y., De, L. and Ashley, S.W. 1985, Intracellular microelectrode studies of necturus antral mucosa. *Gastroenterology*, **88:** 261–268.

Cheung, L.Y., Jubix, W. and Torma, M.J. 1974, Effects of aspirin on canine prostaglandin output and mucosal permeability. *Surgical Forum*, **25:** 407–409.

Cheung, L.Y., Moody, F.G. and Reese, R.S. 1975, Effect of aspirin, bile salt, and ethanol on canine gastric mucosal blood flow. *Surgery*, **77:** 786–792.

Chiba, T., Seino, Y., Goto, Y., Kadowaki, S., Taminato, T., Abe, H., Kato, Y., Matsukura, S., Nozawa, M. and Imura, H. 1978, Somatostatin release from isolated perfused rat stomach. *Biochemical and Biophysical Research Communications*, **82:** 731–737.

Chidichimo, F. 1905, Azione fisiologica e terapeutica dell' aspirin con speciale regardo all'utero. *Annalis di Ostetritia e Ginecologia*, **27:** 356–374.

Chistoni, A. and Lapresa, F. 1909, Richerche farmacologiche sull'aspirina. *Archivio di Farmacologia sperimentale e scienze affinini*, **8:** 63–80.

Cho, C.H. and Pfeiffer, C.J. 1982, The developing role of zinc as an antiulcer agent. *In:* C.J. Pfeiffer (ed.), *Drugs and Peptic Ulcer*, Vol. I, pp. 147–158. Boca Raton, FL: CRC Press.

Cho, C.H. and Pfeiffer, C.J. 1984, Study of the damaging effects of acetazolamide on gastric mucosa in rats. *In:* G. Mózsik, A. Pár and A. Bertelli (eds), *Recent Advances in Gastrointestinal Cytoprotection*, pp. 93–99. Budapest: Akademiai Kiado.

Christie, D., McPherson, L. and Kincaid-Smith, P. 1976, Analgesics and the kidney: a community-based study. *Medical Journal of Australia*, **2:** 527–529.

Chuan-Ming, H., Kōmhoff, M., Guan, YF, Redha, R. and Breyer, M.D. 1999, Selective targeting of cyclooxygenase-2 reveals its role in renal medullary interstitial survival. *American Journal of Physiology – Renal Physiology*, **277:** F352–F359.

Chvasta, T.E. and Cooke, A.R. 1972, The effect of several ulcerogenic drugs on the canine gastric mucosal barrier. *Journal of Laboratory and Clinical Medicine*, **79:** 302–315.

Cirstea, M., Suhaciu, G., Cirje, M. and Ciontescu, L. 1976, Immunological mechanisms in aspirin hypersensitivity. Studies on the immunogenicity of free aspirin. *Acta Allergologica*, **31:** 341–355.

Clancy, R., Varenika, B., Huang, W.Q., Ballou, L., Attur, M., Amin, A.R. and Abramson, S.B. 2000, Nitric oxide synthase/COX crosstalk: nitric oxide activates COX-1 but inhibits COX-2 derived prostaglandin production. *Journal of Immunology*, **165:** 1582–1587.

Clark, B.B. and Adamson, W.L. 1947, The effect of gastric antacids on gastric secretion as observed in the Cope pouch dog. *Gastroenterology*, **9:** 461–465.

Clarke, A. and Walton, W.W. 1979, Effect of safely packaging on aspirin ingestion by children. *Paediatrics*, **63:** 687–693.

Clarke, D.N., Mowat, N.A., Brunt, P.W. and Bain, L.S. 1977, A comparison of a new slow release aspirin ('slow aspirin') with plain aspirin in the treatment of rheumatoid disease. *Journal of International Medical Research*, **5:** 270–275.

Clasp Collaborative Group 1995, Low dose aspirin in pregnancy and early childhood development: follow-up of the collaborative low dose aspirin study in pregnancy. *British Journal of Obstetrics and Gynaecology*, **102:** 861–868.

Coceani, F. and Olley, P.M. 1982, Action of prostaglandin synthetase inhibitors on the ductus arteriosus: experimental and clinical aspects. *In:* H.J.M. Barnett, J. Hirsh and J.F. Mustard (eds), *Acetylsalicylic Acid: New Uses for an Old Drug*, pp. 109–122. New York: Raven Press.

Coceani, F., Pace-Asciak, C., Volta, F. and Wolfe, L.S. 1967, Effect of nerve stimulation on prostaglandin formation and release from the rat stomach. *American Journal of Physiology*, **213:** 1056–1064.

Coggon, D. and Langman, M.J.S. 1980, Aspirin, paracetamol, and haematemasis and malaena. *Gut*, **21:** A922 (Abstr. F12).

Cohen, A. 1976, Intestinal blood loss after a new anti-inflammatory drug, sulindac. *Clinical Pharmacology and Therapeutics*, **20:** 238–240.

Cohen, A., Boeijinga, J.K., Van Haard, P.M.M., Schoemaker, R.C. and Van Vliet-Verbeek, A. 1992, Gastrointestinal blood loss after non-steroidal anti-inflammatory drugs. Measurement by selective determintion of faecal porphyrins. *British Journal of Clinical Pharmacology*, **33:** 33–38.

Cohen, M.M. 1981, Prevention of aspirin-induced fall in gastric potential difference with prostaglandins. *Lancet*, **4:** 785.

Cohen, M.M. 1983, Diflunisal protects human gastric mucosa against damage by indomethacin. *Digestive Diseases and Sciences*, **28:** 1070–1077.

Cohen, M.M., Cheung, G. and Lyster, D.M. 1980, Prevention of aspirin-induced faecal blood loss by prostaglandin E_2. *Gut*, **21:** 602–606.

Colantoni, A., de Maria, N., Caraceni, P., Bernardi, M., Floyd, R.A. and Van Thiel, D.H. 1998, Prevention of reoxygenation injury by sodium salicylate in isolated-perfused rat livers. *Free Radicals in Biology and Medicine*, **25:** 87–94.

Coldwell, B.B. and Boyd, E.M. 1966, The acute rectal toxicity of acetylsalicylic acid. *Canadian Journal of Physiology and Pharmacology*, **44:** 909–918.

Cole, A.T., Hudson, N., Liew, L.C.W., Murray, F.E., Hawkey, C.J. and Heptinstall, S. 1999, Protection of human gastric mucosa against aspirin – enteric coating of dose reduction? *Alimentary Pharmacology and Therapeutics*, **13:** 187–193.

Colin, R., Hochain, P., Czernichow, P., Petit, A., Manchon, N.D. and Berkelmans, I. 1992, Nonsteroidal anti-inflammatory drugs and segmental non-gangrenous colitis: a case–control study. *European Journal of Gastroenterology and Hepatology*, **5:** 715–719.

Colletti, A.E., Vogl, H.W., Rahe, T. and Zambraski, E.J. 1999, Effects of acetaminophen and ibuprofen on renal function in anesthetized normal and sodium-depleted dogs. *Journal of Applied Physiology*, **86:** 592–597.

Collier, H.O.J. 1974, Inhibition of prostaglandins by aspirin and analgesics. *In:* H.J. Robinson and J.R. Vane (eds), *Prostaglandin Synthesis Inhibitors*, pp. 121–133. New York: Raven Press.

Collier, J.G. and Flower, R.J. 1971, Effect of aspirin on human seminal prostaglandin. *Lancet*, **2:** 852–853.

Collins, A.J. and Lewis, D.A. 1971, Lysosomal enzyme levels in the blood of arthritic rats. *Biochemical Pharmacology*, **20:** 251–253.

Collins, E. and Turner, G. 1975, Maternal effects of regular salicylate ingestion in pregnancy. *Lancet*, **2:** 335–338.

Collins, T.F.X., Hansen, W.H. and Keeler, H.V. 1971, Effect of methyl salicylate on rat reproduction. *Toxicology and Applied Pharmacology*, **18:** 755–765.

Colton, D.G., Callison, D.A. and Dajani, E.Z. 1979, Effects of a prostaglandin E1 derivative, SC-29333, and aspirin on gastric ionic fluxes and potential difference in dogs. *Journal of Pharmacology and Experimental Therapeutics*, **210:** 283–288.

Conroy, M.C. and De Weck, A.L. 1981, Effect of aspirin and indomethacin on histamine release from leukocytes of patients with suspected intolerance to aspirin. *International Archives of Allergy and Applied Immunology*, **66 (Suppl. 1):** 152–153.

Contreras, I., Reiser, K.M., Martinez, N., Giansante, E., Lopez, T., Suarez, N., Postalian, S., Molina, M., Gonzalez, F., Sanchez, M.R., Camejo, M. and Blanco, M.C. 1997, Effects of aspirin or basic amino acids on collagen cross-links and complications in NIDDM. *Diabetes Care*, **20:** 832–835.

Cook, G.A., Elliott, S.L., Skeljo, M.V., Giraud, A.S. and Yeomans, N.D. 1996, Correlation between transmucosal potential difference and morphological damage during aspirin injury of gastric mucosa in rats. *Journal of Gastroenterology and Hepatology*, **11:** 264–269.

Cook, J.C., Jacobson, C.F., Gao, F., Tasinari, M.S., Hurtt, M.E. and DeSesso, J.M. 2003, Analysis of the nonsteroidal anti-inflammatory drug literature for potential developmental toxicity in rats and rabbits. *Birth Defects Research Part B, Developmental and Reproductive Toxicology*, **68:** 5–26.

Cook, P.S. and Woodhill, J.M. 1976, The Feingold dietary treatment of the hyperkinetic syndrome. *Medical Journal of Australia*, **2:** 85–90.

Cooke, A.R. and Goulston, K. 1969, Failure of intravenous aspirin to increase gastrointestinal blood loss. *British Medical Journal*, **3:** 330–332.

Cooke, A.R. and Kienzle, M.G. 1974, Studies of anti-inflammatory drugs and aliphalic alcohols on antral mucosa. *Gastroenterology*, **66:** 56–62.

Cooney, G.J. and Dawson, A.G. 1977, Effects of indomethacin on the metabolism of glucose by isolated rat kidney tubules. *Biochemical Pharmacology*, **26:** 2463–2468.

Cooper, G.N., Meade, R.C. and Ellison, E.H. 1966, Heidenhain pouch bleeding due to oral salicylates. *Archives of Surgery*, **93:** 171–174.

Corinaldesi, R., Casadio, R., Sovera, A., Girotti, A., Pracio, A., Paparo, G.F. and Barbara, L. 1980, Zolimidine: protection against aspirin-induced gastric damage in man. *Drugs under Experimental Clinical Research*, **2:** 55–60.

Correll, T. and Jensen, K.M. 1979, Interaction of salicylates and other non-steroidal anti-inflammatory agents in rats as shown by gastro-ulcerogenic and anti-inflammatory activities, and plasma concentrations. *Acta Pharmacologica et Toxicologica*, **45:** 225–231.

Corsi, M., Lacerna, F., Laurora, G., Milani, M. and Borzone, A. 1995, Comparison of the effects of picotamide and aspirin on renal albumin excretion and cutaneous microcirculation in patients with type-II diabetes mellitus – a pilot study. *Current Therapeutic Research – Clinical and Experimental*, **56:** 1105–1110.

Corti, A., Degasperi, A., Colussi, S., Mazza, E., Amici, O., Cristalli, A., Prosperi, M., Scaiola, A., Vai, S., Notaro, P., Ceresa, F., Roselli, E., Settembre, A. and Santandrea, E. 1997, Evaluation of renal function during orthotopic liver transplantation. *Minerva Anestesiolgia*, **63:** 221–228.

Cowburn, A.S., Sladek, K., Soja, J., Adamek, L., Nizankowska, E., Szczeklik, A., Lam, B.K., Penrose, J.F., Austen, F.K., Holgate, S.T. and Sampson, A.P. 1998, Over expression of leukotriene C_4 synthase in bronchial biopsies from patients with aspirin – intolerant asthma. *Journal of Clinical Investigation*, **101:** 1–13.

Cox, P.G.F., Moons, W.M., Russel, F.G.M. and Van Ginneken, C.A.M. 1989, Renal handling of salicyluric acid in the isolated perfused rat kidney: evidence for accumulation in tubular cells. *Journal of Pharmacology and Experimental Therapeutics*, **251:** 750–755.

Croft, D.N. 1963a, Exfoliative cytology of the stomach after the administration of salicylates. *In:* A.St.J. Dixon, B.K. Martin, M.J.H. Smith and P.H.N. Wood (eds), *Salicylates. An International Symposium*, pp. 204–207. London: Churchill.

Croft, D.N. 1963b, Aspirin and the exfoliation of gastric epithelial cells. Cytological and biochemical observations. *British Medical Journal*, **2:** 897–901.

Croft, D.N. 1968, Aspirin, vitamin C deficiency and gastric haemorrhage. *Lancet*, **2:** 831–832.

Croft, D.N., Cuddigan, J.H.P. and Sweetland, C. 1972, Gastric bleeding and benorylate, a new aspirin. *British Medical Journal*, **3:** 545–547.

Croft, D.N., Pollock, D.J. and Coghill, N.F. 1966, Cell loss from human gastric mucosa measured by the estimation of deoxyribonucleic acid (DNA) in gastric washings. *Gut*, **7:** 333–343.

Croft, D.N. and Wood, P.H.N. 1967, Gastric mucosa and susceptibility to occult gastrointestinal bleeding caused by aspirin. *British Medical Journal*, **1:** 137–141.

Crowshaw, K. 1979, Prostaglandins and the side effects of anti-inflammatory drugs – the kidney. *In:* K.D. Rainsford and A.W. Ford-Hutchinson (eds), *Prostaglandins and Inflammation*, pp. 213–223. Basel: Birkhäuser.

Cryer, B. 1992, Effects of nonsteroidal anti-inflammatory drugs of endogenous gastrointestinal prostaglandins and therapeutic strategies for prevention and treatment of nonsteroidal anti-inflammatory drug-induced damage. *Archives of Internal Medicine*, **152:** 1145–1155.

Cryer, B. and Feldman, M. 1999, Effects of very low dose daily, long-term aspirin therapy on gastric, duodenal, and rectal prostaglandin levels and on mucosal injury in healthy humans. *Gastroenterology*, **117:** 17–25.

Cryer, B., Goldschmiedt, M., Redfern, J.S. and Feldman, M. 1990, Comparison of salsalate and aspirin on mucosal injury and on gastrointestinal prostaglandins. *Gastroenterology*, **99:** 1616–1621.

Cryer, B., Kliewer, D., Sie, H., McAllister, L. and Feldman, M. 1999, Effects of cutaneous aspirin on the human stomach and duodenum. *Proceedings of the Association of American Physicians*, **111:** 448–456.

Curtis, R.A. 1984, Efficacy of ipecac and activated charcoal/cathartic. Prevention of salicylate absorption in a simulated overdose. *Archives of Internal Medicine*, **144:** 48–52.

Cuthbert, M.F. 1974, Adverse reactions to non-steroidal anti-rheumatic drugs. *Current Medical Research and Opinion*, **2:** 600–609.

Czervione, R.L., Smith, J.B., Fry, G.L., Hoak, J.C. and Haycraft, D. 1979, Inhibition of prostacylin by treatment of endothelium with aspirin. Correlation with platelet adherence. *Journal of Clinical Investigation*, **63:** 1089–1092.

Daas, M., Gupta, M.B., Gupta, G.P. and Bhargawa, K.P. 1977, Biogenic amines in the pathogenesis of gastric ulceration induced by aspirin in rats. *Indian Journal of Medical Research*, **65:** 273–278.

Dajani, E.Z., Betermann, R.E., Roge, E.A.W., Schweingruber, F.L. and Woods, E.M. 1979, Canine gastrointestinal motility effects of prostaglandin F2 alpha in vivo. *Archives Internationales de Pharmacodynamie et de Thérapie*, **237:** 16–24.

Dajani, E.Z., Callison, D.A. and Betermann, R.E. 1978, Effects of E prostaglandins on canine gastric potential difference. *American Journal of Digestive Diseases*, **23:** 436–442.

Dal Negro, R. 1998, Pharmacokinetic drug interactions with anti-ulcer drugs. *Clinical Pharmacokinetics*, **35:** 135–150.

Damme, B. and Pankow, D. 1996, Induction of hepatic cytochrome P4502E1 in rats by acetylsalicylic acid or sodium salicylate. *Toxicology* **106:** 99–103.

Davenport, H.W. 1964, Gastric mucosal injury by fatty and acetylsalicylic acids. *Gastroenterology*, **46:** 245–253.

Davenport, H.W. 1965, Damage to gastric mucosa: effects of salicylates and stimulation. *Gastroenterology*, **49:** 189–196.

Davenport, H.W. 1966, Fluid produced by the gastric mucosa during damage by acetic and salicylic acids. *Gastroenterology*, **50:** 487–499.

Davenport, H.W. 1969, Gastric mucosal haemorrhage in dogs. Effects of acid, aspirin, and alcohol. *Gastroenterology*, **56:** 439–449.

Davenport, H.W. 1970, Effect of lysolecithin, digitonin, and phospholipase A upon the dog's gastric mucosal barrier. *Gastroenterology*, **59:** 505–509.

Davenport, H.W. 1972, The gastric mucosal barrier. *Digestion*, **5:** 162–165.

Davenport, H.W. 1976, Physiological parameters of the gastric mucosal barrier. *American Journal of Digestive Diseases*, **21:** 141–143.

Davenport, H.W. and Munro, D. 1973, Aminopyrine clearance in the damaged gastric mucosa reconciliation of conflicting data. *Gastroenterology*, **65:** 512–514.

Davenport, H.W., Warner, H.A. and Code, C.F. 1964, Functional significance of gastric mucosal barrier to sodium. *Gastroenterology*, **47:** 142–152.

Davidson, C., Hertig, D.H. and DeVine, R. 1966, Gastric haemorrhage in dogs. Effect of acid, aspirin and alcohol. *Clinical Pharmacology and Therapeutics*, **7:** 239–249.

Davies, N.M., Sharkey, K.A., Asfaha, S., MacNaughton, W.K. and Wallace, J.L. 1997, Aspirin causes rapid up-regulation of cyclo-oxygenase-2 expression in the stomach of rats. *Alimentary Pharmacology and Therapeutics*, **11:** 1101–1108.

Davis, D.P., Daston, G.P., Odio, M.R., York, R.G. and Kraus, A.L. 1996, Maternal reproductive effects of oral salicylic acid in Sprague–Dawley rats. *Toxicology Letters*, **84:** 135–141.

Davis, L.E. and Donnelly, E.J. 1968, Analgesic drugs in the cat. *Journal of the American Veterinary Medical Association*, **153:** 1161–1167.

Davis, S.S. 1970, Saliva is viscoelastic. *Experientia (Basel)*, **26:** 1298–1300.

Dawborn, J.F., Fairley, K.F., Kincaid-Smith, P. and King, W.E. 1966, The association of peptic ulceration, chronic renal disease, and analgesic abuse. *Quarterly Journal of Medicine*, **35:** 69–83.

Dawkins, P.D., Gould, B.J. and Smith, M.J.H. 1966, Inhibitory effect of salicyate on the incorporation of L-(U-^{14}C) leucine into the protein of rat tissue preparations in vitro. *Biochemical Journal*, **99:** 703–707.

Dawkins, P.D., McArthur, J.N. and Smith, M.J.H. 1971, Inhibition of protein biosynthesis in mouse liver by salicylate. *Biochemical Pharmacology*, **20:** 1303–1312.

Dawson, A.G. 1975, Effects of acetylsalicylate on gluconeogenesis in isolated rat kidney tubules. *Biochemical Pharmacology*, **24:** 1407–1411.

Dawson, A.G. 1979, Effects of phenacetin on respiration in mitochondria isolated from rat kidney, liver and brain. *Biochemical Pharmacology*, **28:** 3669–3671.

CHAPTER 8

Dawson, W. and Sweatman, W.J.F. 1980, The pharmacology of isamoxale. (2-methyl-n butyl-n (4-methyloxazol-2-yl) propanide LRCL 3950, a new anti-allergic compound. *British Journal of Pharmacology*, **71:** 387–398.

Day, R.O., Paull, P.D., Champion, G.D. and Graham, G.G. 1976, Evaluation of an enteric-coated aspirin preparation. *Australia and New Zealand Journal of Medicine*, **6:** 45–50.

Decloitre, F. and Beurenkasse, J. 1980, Inhibition of benzol(a)pyrene metabolic activation in the livers of rats treated with acetylsalicylic acid. *Chemico-Biological Interactions*, **30:** 367–371.

Degner, F. and Richardson, B. 2001, Review of gastrointestinal tolerability and safety of meloxicam. *Inflammopharmacology*, **9:** 71–80.

Dehpour, A.R., Mani, A.R., Alikhani, Z., Zeinoddini, M., Toor Savadkoohi, S., Ghaffari, K., Sharif, A., Sabbagh, B., Nowroozi, A. and Sadr, S. 1998, Enhancement of aspirin-induced gastric damage by cholestasis in rats. *Fundamental and Clinical Pharmacology*, **12:** 442–445.

Dehpour, A.R., Zolfaghari, M.E., Samadian, T., Kobarfard, F., Faizi, M. and Assari, M. 1995, Anti-ulcer activities of liquorice and its derivatives in experimental gastric lesion-induced by ibuprofen in rats. *International Journal of Pharmaceutics*, **119:** 133–138.

Del Maestro, R.L., Thaw, H.H., Björk, J., Planker, M. and Arfors, K.-E. 1980, Free radicals as mediators of tissue injury. *Acta Physiologica Scandinavica Supplementum*, **492:** 43–57.

Delzell, E. and Shapiro, S. 1998, A review of epidemiologic studies of nonnarcotic analgesics and chronic renal disease. *Medicine (Baltimore)*, **77:** 102–121.

Depass, L.R. and Weaver, E.V. 1982, Comparison of teratogenic effects of aspirin and hydroxyurea in the Fischer 344 and Wistar strains. *Journal of Toxicology and Environmental Health*, **10:** 297–305.

Derby, L.E. and Jick, H. 1991, Renal parenchymal disease related to over-the-counter analgesic use. *Pharmacotherapy*, **11:** 467–471.

Deschamps, D., Fisch, C., Fromenty, B., Berson, A., Degott, C. and Pessayre, D. 1991, Inhibition by salicylic acid of the activation and thus oxidation of long chain fatty acids. Possible role in the development of Reye's syndrome. *Journal of Pharmacology and Experimental Therapeutics*, **259:** 894–904.

De Schepper, P.J., Tjandramaga, T.B., De Roo, M., Verhaest, L., Daurio, C., Steeleman, S.L. and Tempero, K.F. 1978a, Gastrointestinal blood loss after diflunisal and after aspirin: effect of ethanol. *Clinical Pharmacology and Therapeutics*, **23:** 669–676.

De Schepper, P.J., Tjandramaga, T.B., Verhaerst, L., Daurio, C. and Steeleman, S.L. 1978b, Diflunisal versus aspirin: a comparative study of their effect of faecal blood loss, in the presence and absence of alcohol. *Current Medical Research and Opinion*, **5:** 520–524.

De Vries, J. 1981, Hepatotoxic metabolic activation of paracetamol and its derivatives phenacetin and benorylate: oxygenation or electron transfer. *Biochemical Pharmacology*, **30:** 399–402.

De Weck, A.L. 1971, Immunological effects of aspirin anhydride, a contaminant of commercial acetyl-salicylic acid preparations. *International Archives of Allergy and Applied Immunology*, **41:** 393–418.

De Weck, A.L. 1984, Pathophysiologic mechanisms of allergic and pseudo-allergic reactions to foods, food additives and drugs. *Annals of Allergy*, **53:** 583–586.

Diamantis, W., Melton, J., Sofia, R.D. and Ciofalo, V.B. 1980, Comparative gastric ulcerogenic effects of meseclazone, 5-chlorosalicylic acid and other nonsteroidal anti-inflammatory drugs following acute and repeated oral administration to rats. *Toxicology and Applied Pharmacology*, **52:** 454–461.

Dickinson, R.G. and King, A.R. 1991, Studies on the reactivity of acyl glucuronides – II. Interaction of diflunisal acyl glucuronide and its isomers with human serum albumin in vitro. *Biochemical Pharmacology*, **42:** 2301–2306.

Dickinson, R.G. and King, A.R. 1993, Studies on the reactivity of acyl glucuronides – V. Glucuronide-derived covalent binding of diflunisal to bladder tissue of rats and its modulation by urinary pH and beta-glucuronidase. *Biochemical Pharmacology*, **46:** 1175–1182.

Dickinson, R.G. and King, A.R. 1996, Vesico-hepato-renal cycling of acidic drugs via their reactive acyl glucuronide metabolites? Studies with diflunisal in rats. *Clinical and Experimental Pharmacology and Physiology*, **23:** 665–668.

Dickinson, R.G., Baker, P.V. and King, A.R. 1994, Studies on the reactivity of acyl glucuronides –

VII. Salicyl glucuronide reactivity in vitro and covalent binding of salicylic cid to plasma protein of humans taking aspirin. *Biochemical Pharmacology*, **47:** 469–476.

Dickinson, R.G., King, A.R., McKinnon, G.E., Hooper, W.D., Eadie, M.J. and Herkes, G.K. 1993, Studies on the renal excretion of the acyl glucuronide, phenolic glucuronides and sulphate conjugates of diflunisal. *British Journal of Clinical Pharmacology*, **35:** 609–613.

Dickinson, R.G., King, A.R. and Verbeeck, R.K. 1989, Elimination of diflunisal as its acyl glucuronide, phenolic glucuronide and sulfate conjugates in bile-exteriorized and intact rats. *Clinical and Experimental Pharmacology and Physiology*, **16:** 913–924.

Dieppe, P.A. and Huskisson, E.C. 1978, Diflunisal and acetylsalicylic acid: a comparison of efficacy in osteoarthritis; of nephrotoxicity, and of anti-inflammatory activity in the rat. *In: Diflusinal in Clinical Practice, Proceedings of a Conference, San Francisco 1977* (Chairman, K. Miehlke), pp. 57–61. New York: Futura Press.

Digenis, G.A. and Swintosky, J.V. 1975, Drug latentiation. *In: J.R. Gillette and J.R. Mitchell (eds), Concepts in Biochemical Pharmacology*, Part 3, pp. 86–112. Heidelberg: Springer-Verlag.

Dinoso, V.P., Chey, W.Y., Braverman, S.P., Rosen, A.P., Ottenberg, D. and Lorber, S.H. 1972, Gastric secretion and gastric mucosal morphology in chronic alcoholics. *Archives of Internal Medicine*, **130:** 715–719.

Dinoso, V.P., Ming, S.C. and McNiff, J. 1976, Ultrastructural changes of the canine gastric mucosa after topical application of graded concentrations of ethanol. *American Journal of Digestive Diseases*, **21:** 626–632.

Di Pasquale, G. and Welaj, P. 1973, Ulcerogenic potential of indomethacin in arthritic and non-arthritic rats. *Journal of Pharmacy and Pharmacology*, **25:** 831–832.

Dittert, L.W., Caldwell, H.C., Ellison, T., Irwin, G.M., Rivard, D.E. and Swintocrsky, J.G. 1968, Carbonate ester prodrugs of salicylic acid. Synthesis, solubility characteristics, in vitro enzymatic hydrolysis rates, and blood levels of total salicylate following oral administration to dogs. *Journal of Pharmaceutical Sciences*, **57:** 828–831.

Djaldetti, M. and Fishman, P. 1981, The effect of aspirin on mouse small intestinal mucosa. Ultrastructure studies. *Archives of Pathology and Laboratory Medicine*, **105:** 144–147.

Dobbing, J. 1967, Faecal blood-loss after sodium acetylsalicylate taken with alcohol. *Lancet*, **1:** 527–528.

Domsche, W., Domsche, S., Classen, M. and Demling, L. 1972, Some properties of mucus in patients with gastric ulcer. Effects of treatment with carbenoxolone sodium. *Scandinavian Journal of Gastroenterology*, **7:** 647–651.

Domsche, W., Domsche, S., Hagel, J., Demling, L. and Croft, D.N. 1977, Gastric epithelial cell turnover, mucus production, and healing of gastric ulcers with carbenoxolone. *Gut*, **18:** 817–820.

Done, A.K. 1960, Salicylate intoxication: significance of measurements in blood in cases of acute ingestion. *Pediatrics*, **26:** 800–807.

Done, A.K. 1978, Aspirin overdose: incidence, diagnosis, and management. *Paediatrics*, **62 (Suppl.):** 840–897.

Donnelly, M.T., Goddard, A.F., Filipowicz, B., Morant, S.V., Shield, M.J. and Hawkey, C.J. 2000, Low-dose misoprostol for the prevention of low-dose aspirin-induced gastroduodenal injury. *Alimentary Pharmacology and Therapeutics*, **14:** 529–534.

Douglas, R.A. and Johnson, E.D. 1961, Aspirin and chronic gastric ulcer. *Medical Journal of Australia*, **2:** 893–896.

Douthwaite, A.H. and Lintott, G.A.M. 1938, Gastroscopic observation of the effect of aspirin and certain other substances on the stomach. *Lancet*, **235:** 1222–1225.

Drazen, J.M., Isreal, E. and O'Byrne, P.M. 1999, Treatment of asthma with drugs modifying the leukotriene pathway. *New England Journal of Medicine*, **340:** 197–206.

Druery, C.J. and Dawson, A.G. 1979, Inhibition of respiration by phenacetin in isolated tubules and mitochondria of rat kidney. *Biochemical Pharmacology*, **28:** 57–61.

Dubach, U.C., Rosner, B. and Pfister, E. 1983, Epidemiologic study of abuse of analgesics containing phenacetin. Renal morbidity and mortality 1968–1979). *New England Journal of Medicine*, **308:** 357–362.

Dubach, U.C., Rosner, B. and Stürmer, T. 1975, Relation between regular intake of phenacetin-

CHAPTER 8

containing analgesics and laboratory evidence for urorenal disorders in a working female population of Switzerland. *Lancet*, **1:** 539–543.

Dubois, P., Degrave, E. and Vandenplas, O. 1998, Asthma and airway hyperresponsiveness among Belgian conscripts, 1978–1991. *Thorax*, **53:** 101–105.

Dubois, R.N., Abramson, S.B., Crofford, L., Gupta, R.A., Simon, L.S., Van De Putte, L.B. and Lipsky, P.E. 1998, Cyclooxygenase in biology and disease. *FASEB Journal*, **12:** 1063–1073.

Duggan, A.E., Stack, W., Hull, M., Filipowicz, B., Knifton, A., Crome, R., Weber, C., Bishop, A., Polak, J. and Hawkey, C.J. 1999, Protection against aspirin-induced human gastric mucosal injury by bosentan, a new endothelin-1 receptor antagonist. *Alimentary Pharmacology and Therapeutics*, **13:** 631–635.

Duggan, J.M. 1980, Gastrointestinal toxicity of minor analgesics. *British Journal of Clinical Pharmacology*, **10:** 407S–410S.

Duggin, G.G. 1977, Analgesic induced kidney disease. *Australian Journal of Pharmaceutical Sciences*, **6:** 44–48.

Duggin, G.G. 1996, Combination analgesic-induced kidney disease: the Australian experience. *American Journal of Kidney Diseases*, **28:** S39–S47.

Duggin, G.G. and Mudge, G.H. 1976, Phenacetin: renal tubular transport and intrarenal distribution in the dog. *Journal of Pharmacology and Experimental Therapeutics*, **199:** 10–16.

Dular, U. and Dakshinamurti, K. 1979, Effect of salicylates on acetyl coenzyme A carboxylase. *Biochemical Pharmacology*, **28:** 715–718.

Dunjic, B.S., Axelson, J., Arrajab, A., Larsson, K. and Bengmark, S. 1993a, Gastroprotective capability of exogenous phosphatidylcholine in experimentally induced chronic gastric ulcers in rats. *Scandinavian Journal of Gastroenterology*, **28:** 89–94.

Dunjic, B.S., Svenssion, I., Axelson, J., Adlercreutz, P., Arrajab, A. and Bengmark, S. 1993b, The rat gastric phospholipids – increased in ulcerated mucosa and decreased after healing. *European Surgical Research*, **25:** 376–382.

Dupont, I., Bodenez, B.F., Bardou, L., Guirriec, C., Stephan, N., Dreano, Y. and Lucas, D. 1999, Involvement of cytochromes P-450 2E1 and 3A4 in the 5-hydroxylation of salicylate in humans. *Drug Metabolism and Disposition*, **27:** 322–326.

Durbach, U.C., Rosner, B. and Stürmer, T. 1991, An epidemiologic study of abuse of analgesic drugs. Effects of phenacetin and salicylate on mortality and cardiovascular morbidity (1968 to 1987). *New England Journal of Medicine*, **324:** 155–160.

Dvorak, M. 1974, Effects of corticotrophin, starvation and glucose on ascorbic acid levels in the blood plasma and liver of piglets. *Nutrition and Metabolism*, **16:** 215–222.

Dybdahl, J.H., Daal, L.N., Larsen, S., Ekeli, H., Frislid, K., Wiik, I. and Arnstad, L. 1980, Acetylsalicylic acid-induced gastrointestinal bleeding determined by a ^{51}Cr method on a day to day basis. *Scandinavian Journal of Gastroenterology*, **15:** 887–895.

Eastwood, G.L. and Kirchner, J.P. 1974, Changes in the fine structure of mouse gastric epithelium produced by ethanol and urea. *Gastroenterology*, **67:** 71–84.

Edelson, J. and Douglas, J.F. 1973, Measurement of gastrointestinal blood loss in the rat: the effects of aspirin, phenylbutazone and seclazone. *Journal of Pharmacology and Experimental Therapeutics*, **184:** 449–452.

Editorial 1876, *Guy's Hospital Gazette*, **1 (NS):** 84–85.

Editorial 1980, *British Medical Journal*, **281:** 938–959.

Editorial 1981a, Analgesic nephropathy. *British Medical Journal*, **282:** 339–340.

Editorial 1981b, Analgesic poisoning. *Lancet*, **2:** 1210–1211.

Edmar, D. 1971, Effects of salicylates on the gastric mucosa as revealed by roentgen examination and the gastrocamera. *Acta Radiologica Supplementum*, **11:** 57–64.

Edwards, P.A.W. 1978, Is mucus a selective barrier to macromolecules? *British Medical Bulletin*, **34:** 55–56.

Elliott, P.N.C. 1979, The effect of combination of aspirin and sodium salicylate on the rat stomach. *British Journal of Pharmacology*, **67:** 484P–485P.

Elseviers, M.M. and De Broe, M. 1999, Analgesic nephropathy. Is it caused by multi-analgesic abuse or single substance use? *Drug Safety*, **20:** 15–24.

Emkey, R.D. and Mills, J.A. 1982, Aspirin and analgesic nephropathy. *Journal of the American Medical Association*, **247**: 55–57.

Eras, J. and Perazella, M.A. 2001, NSAIDs and the kidney revisted: are selective cyclooxygenase-2 inhibitors safe? *American Journal of the Medical Sciences*, **321**: 181–190.

Ergun, E.L., Caglar, M., Erdem, Y., Usalan, C., Ugur, O. and Duranay, M. 2000, Tc-99m DTPA acetylsalicylic acid (aspirin) renography in the detection of renovascular hypertension. *Clinical Nuclear Medicine*, **25**: 682–690.

Esplugues, J.V. 2001, Gastric mucosal resistance to acute injury in experimental portal hypertension. *British Journal of Pharmacology*, **132**: 309–317.

Essigman, W.M., Chamberlain, M.A. and Wright, V. 1979, Diflusinal in osteoarthrosis of the hip and knee. *Annals of the Rheumatic Diseases*, **38**: 148–151.

Estler, C.-J., Mitznegg, P., Domschke, W. and Demling, L. 1977, Effect of lysine-acetylsalicylate on serum gastrin levels. *Acta Hepatogastroenterologica*, **24**: 52–54.

Ezer, E., Paloisi, E., Hajos, Gy. and Szporny, L. 1976, Antagonism of the gastrointestinal ulcerogenic effect of some non-steroidal anti-inflammatory agents by sodium salicylate. *Journal of Pharmacy and Pharmacology*, **28**: 655–656.

Ezer, E. and Szprony, L. 1970, Prevention of experimental gastric ulcer in rats by a substance which increases biosynthesis of acid mucopolysaccharides. *Journal of Pharmacy and Pharmacology*, **22**: 143–144.

Ezer, E. and Szporny, L. 1975, A complementary method for the quantitative evaluation of rat stomach with Shay-ulcer. *Journal of Pharmacy and Pharmacology*, **27**: 866–867.

Faggioli, F., Gasbarrini, G., Scondotto, G.O.M., Mattei, M. and Zanetti, A. 1970, Effects of aspirin and salicylates on the gastric mucosa. *Gastroenterology and Clinical Medicine*, **51**: 298–319.

Faivre, J., Faivre, M., Lery, N., Ducluzeau, R., Moulinier, B. and Paliard, P. 1979, Aspirin and gastrointestinal bleeding. Interest of plasma salicylate determination. *Digestion*, **19**: 218–220.

Falaiye, J.M. and Odutola, T.A. 1978, The economic potential and the role of aspirin and alcohol ingestion in relation to haematemesis and melaena. *Nigerian Medical Journal*, **8**: 526–530.

Falbe-Hansen, J. 1941, Clinical and experimental studies on effects of salicylate and quinine on the ear. Acta Otolaryngologica Stockholm, **44 (Suppl.)**: 216.

Falk, S.H. 1974, Letter: Sodium salicylate. *Archives of Otolaryngology*, **99**: 393.

Falliers, C. 1974, Familial coincidence of asthma, aspirin intolerance and nasal polyposis. *Annals of Allergy*, **32**: 65–69.

Famaey, J.P., Fontaine, J. and Reuse, J. 1977, The effects of non-steroidal anti-inflammatory drugs on cholinergic and histamine-induced contractions of guinea pig isolated ileum. *British Journal of Pharmacology*, **60**: 165–171.

Farber, E. 1981, Chemical carcinogenesis. *New England Journal of Medicine*, **305**: 1379–1389.

Farr, R.S. 1970, Presidential message. The need to re-evaluate acetylosalicylic acid (aspirin). *Journal of Allergy*, **45**: 321–328.

Farrell, G.C. 1997, Drug-induced hepatic injury. *Journal of Gastroenterology and Hepatology*, **12**: S242–S250.

Fattinger, K., Roos, M., Vergeres, P., Holenstein, C., Kind, B., Masche, U., Stocker, D.N., Braunschweig, S., Kullak-Ublick, G.A., Galeazzi, R.L., Follath, F., Gasser, T. and Meier, P.J. 2000, Epidemiology of drug exposure and adverse drug reactions in two Swiss departments of internal medicine. *British Journal of Clinical Pharmacology*, **49**: 158–167.

Faust, T.W., Redfern, J.S., Podolsky, I., Lee, E., Grundy, S.M. and Feldman, M. 1990, Effects of aspirin on gastric mucosal prostaglandin E2 and F2 alpha content and on gastric mucosal injury in humans receiving fish oil or olive oil. *Gastroenterology*, **98**: 586–591.

Faust, T.W., Redfern, J.S., Podolsky, I., Lee, E., Grundy, S.M., Feldman, M. Feenstra, J., Grobbee, D.E., Mosterd, A. and Stricker, B.H.Ch. 1997, Adverse cardiovascular effects of NSAIDs in patients with congestive heart failure. *Drug Experience*, **17**: 166–180.

Feingold, B.F. 1975a, Hyperkinesis and learning disabilites linked to artificial food flavors and colors. *American Journal of Nursing*, **75**: 797–803.

Feingold, B.F. 1975b, Food additives in clinical medicine. *International Journal of Dermatology*, **14**: 112–114.

Feinstein, A.R., Heinemann, L.A.J., Curhan, G.C., Delzell, E., Deschepper, P.J., Fox, J.M., Graf, H., Luft, F.C., Michielsen, P., Mihatsch, M.J., Suissa, S., van der Woude, F. and Willich, S. 2000, Relationship between nonphenacetin combined analgesics and nephropathy: a review. Ad hoc Committee of the International Study Group on Analgesics and Nephropathy. *Kidney International*, **58:** 2259–2264.

Feldman, M., Shewmake, K. and Cryer, B. 1990, Effects of aspirin on gastric mucosal prostaglandin E_2 and $F_{2\alpha}$ content and on gastric mucosal injury in humans receiving fish oil or olive oil. *Gastroenterology*, **98:** 586–591.

Feldman, M., Shewmake, K. and Cryer, B. 2000, Time course inhibition of gastric and platelet COX activity by acetylsalicylic acid in humans. *American Journal of Physiology – Gastrointestinal, Liver Physiology*, **279:** G11113–G111120.

Ferguson, I., Johnson, R., Reay, B. and Wigley, R. 1977, Aspirin, phenacetin, and the kidney: a rheumatism clinic study. *Medical Journal of Australia*, **1:** 950–954.

Fiegler, M., Jávor, T., Nagy, L., Patty, I., Tárnok, F. and Mózsik, Gy. 1986, Biochemical background of the development of gastric mucosal damage in pylorus-ligated plus aspirin-treated rats. *International Journal of Tissue Reactions*, **VII:** 15–22.

Filippone, G.A. 1987, Reversible adsorption (desorption) of aspirin from activated charcoal. *Archives of Internal Medicine*, **147:** 1390–1392.

Fiorucci, S., Antonelli, E., Morelli, O., Federici, B. and Morelli, A. 1999a, Prevention of gastric damage by NO-releasing aspirin (NCX-4016) is mediated by inhibition of ICE-like proteases in gastric mucosa and endothelial cells. *Gastroenterology*, **114:** G0500.

Fiorucci, S., Antonelli, E., Santucci, L., Morelli, O., Miglietti, M., Federici, B., Mannucci, R., Del Soldato, P. and Morelli, A. 1999b, Gastrointestinal safety of nitric-oxide derived aspirin is related to inhibition of ICE-like cysteine proteases in rats. *Gastroenterology*, **116:** 1089–2206.

Fiorucci, S., Distrutti, E., Mencarelli, A., Morelli, A., Laufor, S.A., Cirino, G. and Wallace, J.L. 2003, Evidence that 5-lipoxygenase and acetylate cyclooxygenase 2-derived eicosanoids regulate leukocyte-endothelial adherence in response to aspirin. *British Journal of Pharmacology*, **139:** 1351–1359.

Fiorucci, S., Santucci, L., Cirino, G., Mencarelli, A., Familiari, L., Del Soldato, P. and Morelli, A. 2000, IL-1 beta converting enzyme is a target for nitric oxide-releasing aspirin: new insights in the anti-inflammatory mechanism of ntiric oxide-releasing nonsteroidal antiinflammatory drugs. *Journal of Immunology*, **165:** 5245–5254.

Fischer, H., Becker, J.C., Boknik, P., Huber, V., Luss, H., Neumann, J., Schmitz, W., Domschke, W. and Konturek, J.W. 1999, Expression of endoethelial cell-derived nitric oxide synthase (eNOS) is increased during gastric adaptation to chronic aspirin intake in humans. *Alimentary Pharmacology and Therapeutics*, **13:** 507–514.

Fisher, J.M. and Taylor, K.B. 1971, The significance of gastric antibodies. *British Journal of Haematology*, **20:** 1–7.

Fisherman, E.W. and Cohen, G.N. 1973, Aspirin disease, L-histidine and the sequential vascular response disease. *Annals of Allergy*, **31:** 555–560.

Fishgold, J.T., Field, J. and Hall, V.E. 1951, Effect of sodium salicylate and acetylsalicylate on metabolism of rat brain and liver *in vitro*. *American Journal of Physiology*, **164:** 727–733.

Fishler, J.J. 1964, Effects of aspirin, acetaminophen, and salicylamide on gastric mucosa of dogs. *American Journal of Digestive Diseases*, **9:** 465–470.

Fitzgerald, G.A., Hossmann, V., Hummerich, W. and Konrads, A. 1980, The renin–kallikrein–prostaglandin system: plasma active and inactive renin and urinary kallikrein during prostacyclin infusion in man. *Prostaglandins and Medicine*, **5:** 445–446.

Fitzpatrick, L.R. and Decktor, D.L. 1991, Gastroprotective and ulcer healing profile of the mast-cell stabilizer quazolast in rats. *Agents and Actions*, **33:** 330–336.

Flemström, G. and Garner, A. 1980, Stimulation of gastric acid and bicarbonate secretions by calcium in guinea pig stomach and amphibia isolated mucosa. *Acta Physiologica Scandinavica*, **110:** 419–426.

Flemström, G. and Marsden, N.V.B. 1974, Increased inulin absorption from the cat stomach exposed to acetylsalicylic acid. *Acta Physiologica Scandinavica*, **92:** 517–525.

Flemström, G., Marsden, N.V.B. and Richter, S. 1976, Passive cutaneous anaphylaxis in guinea pigs elicited by gastric absorption of dextran induced by acetylsalicylic acid. *International Archives of Allergy and Applied Immunology*, **51:** 627–636.

Floate, D.A. and Duggan, J.M. 1978, Hour-glass stomach: an explanation. *Medical Journal of Australia*, **2:** 674–676.

Florkiewicz, H., Kolber-Postepska, B. and Widomska-Czekajska, T. 1967, Treatment with aspirin and sodium salicylate and gastrointestinal haemorrhage. *Polski Tygodnik Lekarski*, **22:** 1283–1286.

Fork, F.T., Lafolie, P., Toth, E. and Lindgarde, F. 2000, Gastroduodenal tolerance of 75 mg clopidogrel versus 325 mg aspirin in healthy volunteers – a gastroscopic study. *Scandinavian Journal of Gastroenterology*, **35:** 464–469.

Forth, W. and Nell, G. 1981, Transport mechanisms involved in the absorption of heavy metals. In: K.D. Rainsford, K. Brune and M.W. Whitehouse (eds.) *Trace Metals in the Pathogenesis and Treatment of Inflammation*, pp. 23–36. Basel: Birkhäuser.

Foulon, O., Girard, H., Pallen, C., Urtizberea, M., Repetto-Larsay, M. and Blacker, A.M. 1999, Induction of supernumary ribs with sodium salicylate. *Reproductive Toxicology*, **13:** 369–374.

Foulon, O., Jaussely, C., Repetto, M., Urtizberea, M. and Blacker, A.M. 2000, Postnatal evolution of supernumerary ribs in rats after a single administration of sodium salicylate. *Journal of Applied Toxicology*, **20:** 205–209.

Fox, J.M. 1995, Kombinationsarzneimittel aus Paracetamol plus Acetylsalicylsäure: Nutzen und Risiken. *Der Schmerz*, **9:** 273–285.

Frankhuijen, A.L. and Bonta, I.L. 1975, Role of prostaglandins in tone and effector reactivity of the isolated rat stomach preparation. *European Journal of Pharmacology*, **31:** 44–52.

Fraser, N.C. 1980, Accidental poisoning deaths in British children. *British Medical Journal*, **280:** 1595–1598.

Fred, H.L. 1981, The 100 mile-run: preparation, performance, and recovery. A case report. *American Journal of Sports Medicine*, **9:** 258–261.

Freie, H.M.P. 1996, Antipyretic analgesics. *In:* M.N.G. Dukes (ed.), *Meyler's Side Effects of Drugs. An Encyclopedia of Adverse Reactions and Interactions*, 13th edn, pp. 191–203. Amsterdam: Elsevier.

Frenning, B. 1971, The effects of acetic and acetylsalicylic acids on the appearance of the gastric mucosal epithelia in the scanning electron microscope. *Scandinavian Journal of Gastroeneterology*, **6:** 605–612.

Frenning, B. and Obrink, K.J. 1971, The effects of acetic and acetylsalicylic acids on the appearance of the gastric mucosal surface epithelium in the scanning electronmicroscope. *Scandinavian Journal of Gastroenterology*, **6:** 605–612.

Frey, H.-H. and El-Sayed, M.A. 1977, Concentrations of acidic anti-inflammatory drugs in gastric mucosa. *Archives Internationales de Pharmacodynamie et de Thérapie*, **230:** 300–308.

Frick, P.G. 1956, Hemorrhagic diathesis with increased capillary fragility caused by salicylate therapy. *American Journal of the Medical Sciences*, **231:** 402–406.

Fritz, H. and Suter, H.P. 1985, Postnatal development of young rats following the treatment of the dams with sodium salicylate during late periods of pregnancy. *Drug Research*, **35:** 937–939.

Fromm, D. and Kolis, M. 1982, Effects of sodium salicylate and acetylsalicylic acid on intramural pH and ulceration of rabbit antral mucosa. *Surgery*, **91:** 438–447.

Fromm, D., Schwartz, J.H. and Quijaro, R. 1975, Effects of cyclic adenosime 3′5′-monophosphate and related agents on acid secretion by isolated rabbit gastric mucosa. *Gastroenterology*, **69:** 453–462.

Fry, S.W. and Seeff, L.B. 1995, Hepatotoxicity of analgesics and anti-inflammatory agents. *Gastroenterology Clinics of North America*, **24:** 875–905.

Frydman, G., O'Brien, P. and Phelan, D. 1988, Microvascular changes in the early damage of the gastric mucosa with ethanol and aspirin. *Digestive Diseases and Sciences*, **33:** 900.

Fujii, A., Kuboyama, N., Kobayashi, S., Namiki, Y. and Tamura, T. 1988, Time-course study of gastric damages in rats by anti-inflammatory drugs using a gastroscope and its quantifications. *Japanese Journal of Pharmacology*, **48:** 317–322.

Gabriel, S.E., Jaakkimainen, L. and Bombardier, C. 1991, Risk for serious gastrointestinal complications related to use of nonsteroidal anti-inflammatory drugs. A meta-analysis. *Annals of Internal Medicine*, **115:** 787–796.

Gago-Dominguez, M., Yuan, J.-M., Castelao, J.E., Ross, R.K. and Yu, M.C. 1999, Regular use of analgesics is a risk factor for renal cell carcinoma. *British Journal of Cancer*, **81:** 542–548.

Ganter, P., Julou, L. and Guyonnet, J.-C. 1966, Histochemical study of experimental gastric lesions induced by acetylsalicylic acid and other drugs in rats. *Laval Medical*, **37:** 416–434.

Garcia-Rodriguez, L.A. 2000, Gastrointestinal and renal complications of conventional NSAID: an epidemiological perspective. Abstracts of the VIIth World Conference on Clinical Pharmacology (CPT) and 4th Congress of the European Association for Clinical Pharmacology and Therapeutics. *British Journal of Pharmacology*, Abstr. No. 56: 15.

Garcia-Rodriguez, L.A. and Hernandez-Diaz, S. 2001, Relative risk of upper gastrointestinal complications among users of acetaminophen and nonsteroidal anti-inflammatory drugs. *Epidemiology*, **12:** 570–576.

Garcia-Rodriguez, L.A., Williams, R., Derby, L.E., Dean, A.D. and Jick, H. 1994, Acute liver injury associated with nonsteroidal anti-inflammatory drugs and the role of risk factors. *Archives of Internal Medicine*, **154:** 311–316.

Garner, A. 1977a, Assessment of gastric mucosal damage comparative effects of aspirin and fenclofenac on the gastric mucosa of the guinea pig. *Toxicology and Applied Pharmacology*, **42:** 477–486.

Garner, A. 1977b, Effects of acetylsalicylate on alkalinisation, acid secretion and ulcerogenic properties in the isolated gastric mucosa. *Acta Physiologica Scandinavica*, **99:** 281–291.

Garner, A. 1977c, Influence of salicylates on the rate of accumulation of deoxyribonucleic acid in gastric washings from the guinea pig. *Toxicology and Applied Pharmacology*, **42:** 119–128.

Garrett, E.R. 1959, The physical chemical evidence for aspirin anhydride as a superior form for the oral administration of aspirin. *Journal of the American Pharmaceutical Association*, **48:** 676–683.

Gast, L.G. 1964, Influence of aspirin on haemostatic parameters. *Annals of the Rheumatic Diseases*, **23:** 500–504.

Gaucher, A., Serot, J.-M., Gaucher, P., Rauber, G., Grignon, G., Pourel, J., Netter, P. and Faure, G. 1976, Gastric mucosa in patients with rheumatoid polyarthritis treated by anti-inflammatory agents. Clinical, radiological, endoscopic, anatomopathological and ultra structural study. *Nouvelle Presse Medicale*, **5:** 2781–2784.

Gault, M.H. and Barrett, B.J. 1998, Analgesic nephropathy. *American Journal of Kidney Diseases*, **32:** 351–360.

Gault, M.H., Blennerhassett, J. and Muehrcke, R.C. 1971, Analgesic nephropathy. A clinicopathologic study using electron microscopy. *American Journal of Medicine*, **51:** 740–756.

Gazert. 1900, Therapeutische Erfahrungen mit Aspirin. *Deutches Archiv für Klinische Medizin*, **68:** 142–154.

Gazzard, B.G. and Clark, M.C. 1978, Alcohol and the alimentary system. *Clinics in Endocrinology and Metabolism*, **7:** 429–446.

Geall, M.G., Phillips, S.F. and Summerskill, W.H.J. 1970, Profile of gastric potential difference in man. Effects of aspirin, alcohol, bile and endogenous acid. *Gastroenterology*, **58:** 437–443.

Gelzayd, E.A., Biederman, M.A. and Gelfand, D.W. 1975, Changing concepts of duodenitis. *American Journal of Gastroenterology*, **64:** 213–216.

Gentry, R.T., Baraona, E., Amir, I., Roine, R., Chayes, Z.W., Sharma, R. and Lieber, C.S. 1999, Mechanism of the aspirin-induced rise in blood alcohol levels. *Life Sciences*, **65:** 2505–2512.

Gerber, J.G., Branch, R.A., Nies, A.S., Gerkens, J.F., Shand, D.G., Hollifield, J. and Oates, J.A. 1978a, Prostaglandins and renin release: II. Assessment of renin secretion following infusion of PGI_2, E_2 and D_2 into the renal artery of anaesthetized dogs. *Prostaglandins*, **15:** 81–88.

Gerber, J.G., Nies, A.S., Friesinger, G.C., Gerkens, J.F., Branch, R.A. and Oates, J.A. 1978b, The effect of PGI_2 on canine renal function and hemodynamics. *Prostaglandins*, **16:** 519–528.

Gerber, J.G., Payne, N.A., Oelz, O., Nies, A.S. and Oates, J.A. 1979, Tartrazine and the prostaglandin system. *Journal of Allergy and Clinical Immunology*, **63:** 289–294.

Gerkens, J.F., Shand, D.G., Flexner, C., Nies, A.S., Oates, J.A. and Data, J.L. 1977, Effects of indomethacin and aspirin on gastric blood flow and acid secretion. *Journal of Pharmacology and Experimental Therapeutics*, **203:** 646–657.

Ghanayem, B.I., Ahmed, A.E. and Boor, P.J. 1984, Acetyl salicylic acid-induced gastric mucosal lesions: no role of gastric glutathione. *Research Communications in Chemical Pathology and Pharmacology*, **45:** 153–156.

Gibberd, F.B. 1966, Dyspepsia in patients with rheumatoid arthritis. *Acta Rheumatologica Scandinavica*, **12:** 112–121.

Gibson, G.G., Orton, T.C. and Tamburini, P.P. 1982, Cytochrome P-450 induction by clofibrate. *Biochemical Journal*, **203:** 161–168.

Gilles, M. and Skyring, A. 1968, Gastric ulcer, duodenal ulcer and gastric carcinoma: a case–control study of certain social and environmental factors. *Medical Journal of Australia*, **2:** 1132–1136.

Giraldo, B., Blumenthal, M.N. and Spink, W.W. 1969, Aspirin intolerance and asthma. A clinical and immunological study. *Annals of Internal Medicine*, **71:** 479–496.

Giraud, M.N., Sanduja, S.K., Felder, T.B., Illich, P.A., Dial, E.J. and Lichtenberger, L.M. 1997, Effect of omeprazole on the bioavailability of unmodified and phospholipid-complexed aspirin in rats. *Alimentary Pharmacology and Therapeutics*, **11:** 899–906.

Giroux, Y., Archambault, A., Farley, A., Gosselin, D., Orr, J.M., Marier, G. and Schipper, H.L. 1977, Effects on the gastric mucosa of coated acetylsalicylic acid: comparative study by endoscopy. *Union Medicale du Canada*, **106:** 841–847.

Gitlin, N. 1980, Salicylate hepatoxicity: the potential role of hypoalbumineania. *Journal of Clinical Gastroenterology*, **2:** 281–285.

Glasgow, J.F.T. 1984, Clinical features and prognosis of Reye's syndrome. *Archives of Disease in Childhood*, **59:** 230–235.

Glass, G.B.J. 1976, Gastric mucus and mucosal injury. *Materia Medica Polona*, **8:** 177–185.

Goddard, P.J., Hills, B.A. and Lichtenberger, L.M. 1987, Does aspirin damage canine gastric mucosa by reducing its surface hydrophobicity? *American Journal of Physiology*, **252:** 421–430.

Goddard, P.J. and Lichtenberger, L.M. 1995, In vitro recovery of canine gastric-mucosal surface hydrophobicity and potential difference after aspirin damage. *Digestive Diseases and Sciences*, **40:** 1357–1359.

Goggins, J.F., Johnson, G.S. and Pastan, I. 1972, The effect of dibutyryl cyclic adenosine monophosphate in synthesis of sulphated acid mucopolysaccharides by transformed fibroblasts. *Journal of Biological Chemistry*, **247:** 5759–5764.

Goldman, A.S. and Yakovak, W.C. 1964, Salicylate intoxication and congenital abnormalities. *Archives of Environmental Health*, **8:** 648–656.

Gonwa, T.A., Hamilton, R.W. and Buckalew, V.M. 1981, Chronic renal failure and end-stage renal disease in Northwest North Carolina. Importance of analgesic associated nephropathy. *Archives of Internal Medicine*, **141:** 462–465.

Gould, B.J., Dawkins, P.D., Smith, M.J.H. and Lawrence, A.J. 1966, The mechanisms of the inhibition of aminotransferases by salicylate. *Molecular Pharmacology*, **2:** 526–533.

Goulston, K. and Cooke, A.R. 1968, Alcohol, aspirin and gastrointestinal bleeding. *Lancet*, **4:** 664–665.

Goulston, K. and Skyring, A. 1964, Effect of paracetamol (N-acetyl-p-aminophenol) on gastrointestinal bleeding. *Gut*, **5:** 463–466.

Graham, A.D. and Park, M.V. 1974, Inhibition of the mitochondrial oxidation of octanoate by salicylic acid and related compounds. *Journal of Pharmacy and Pharmacology*, **26:** 531–534.

Graham, D.Y. 1996, Nonsteroidal anti-inflammatory drugs, *Helicobacter pylori*, and ulcers: where we stand. *American Journal of Gastroenterology*, **91:** 2080–2086.

Graham, D.Y., Lacey Smith, J. and Dobbs. 1983, Gastric adaption occurs with aspirin administration in man. *Digestive Diseases and Sciences*, **28:** 1–6.

Graham, D.Y. and Smith, J.L. 1985, Effects of aspirin and an aspirin–acetaminophen combination on the gastric mucosa in normal subjects – a double-blind endoscopic study. *Gastroenterology*, **88:** 1922–1925.

Granger, D.N., Rutili, G. and McCord, J.M. 1981, Superoxide radicals in feline intestinal ischemia. *Gastroenterology*, **81:** 22–29.

Gray, J.E., Jones, P.M. and Feenstra, E.S. 1960, Comparable effects of acetylsalicylic acid and acetylsalicylic anhydride on the non-glandular portion of the rat stomach. *Toxicology and Applied Pharmacology*, **2:** 514–522.

Green, D.M. 1966, Tablets of coated aspirin microspherules – a new dosage form. *Journal of New Drugs*, **6:** 294–304.

CHAPTER 8

Greenaway, J.C., Mirkes, P.E., Walker, E.A., Juchau, M.R., Shepart, T.H. and Fantel, A.G. 1985, The effect of oxygen concentration on the teratogenicity of salicylate, niridazole, cyclophosphamide, and phosphoramide mustard in rat embryos in vitro. *Teratology*, **32:** 287–295.

Greenwald, R.A. 1981, Oxyradicals and connective tissue. *Journal of Rheumatology*, **8:** 185–187.

Griffin, J.P. and D'Arcy, P.F. 1997, *A Manual of Adverse Drug Reactions*. Amsterdam: Elsevier.

Griffin, M.R., Yared, A. and Ray, W.A. 2000, Nonsteroidal anti-inflammatory drugs and acute renal failure in elderly persons. *American Journal of Epidemiology*, **151:** 488–496.

Gross, M. and Greenberg, L.A. 1948, *The Salicylates. A Critical Biblographic Review*. New Haven: Hillhouse Press.

Grossman, M.I., Matsumoto, K.K. and Lichter, R.J. 1961, Fecal blood loss produced by oral and intravenous administration of various salicylates. *Gastroenterology*, **40:** 383–388.

Guenthner, J.M. and Oesch, F. 1981, Modulation of epoxide hydrolase activity; effect on the activation of benzo[a] pyrene and its covalent binding to DNA in the nucleus. *Trends in Pharmacological Sciences*, **2:** 129–132.

Guillausseau, P.J. 1994, Pharmacological prevention of diabetic microangiopathy. *Diabetes and Metabolism*, **20:** 219–228.

Guitton, M.J., Caston, J., Ruel, J., Johnson, R.M., Pujol R. and Puel, J.L. 2003, Salicylate induces tinnitus through activation of cochlear NMDA receptors. *J Neurosci*, **23:** 3944–3952.

Gulamhusein, A.P., Harrison-Sage, C., Beck, F. and Al-Alousi, A. 1980, Salicylate-induced teratogenesis in the ferret. *Life Sciences*, **27:** 1799–1805.

Gullikson, G.W., Anglin, C.P., Kessler, L.K., Smeach, S., Bauer, R.F. and Dajani, E.Z. 1987, Misoprostol attenuates aspirin-induced changes in potential difference and associated damage in canine gastric mucosa. *Clinical and Investigative Medicine*, **10:** 145–151.

Günther, T., Rebentisch, E., Vormann, J., Konig, M. and Ising, H. 1988, Enhanced ototoxicity of gentamicin and salicylate caused by Mg deficiency and Zn deficiency. *Biological Trace Element Research*, **16:** 43–50.

Gurwitz, J.H., Avorn, J., Bohn, R.L., Glynn, R.J., Monane, M. and Mogun, H. 1994, Initiation of antihypertensive treatment during nonsteroidal anti-inflammatory drug therapy. *Journal of the American Medical Association*, **272:** 781–786.

Guslandi, M. 1980, Review: effect of anti-ulcer drugs on gastric mucosa barrier and possible cyclic AMP involvement. *International Journal of Clinical Pharmacology*, **18:** 140–143.

Guslandi, M. 1997, Gastric toxicity of antiplatelet therapy with low-dose aspirin. *Drugs*, **53:** 1–5.

Guth, P.H. 1982, Pathogenesis of gastric mucosal injury. *Annual Review of Medicine*, **33:** 183–196.

Guth, P.H. and Paulsen, G. 1979a, Effect of parental aspirin on the gastric mucosal barrier in the rat. *Digestion*, **19:** 93–98.

Guth, P.H. and Paulsen, G. 1979b, Prostaglandin cytoprotection does not involve interference with aspirin absorption. *Proceedings of the Society for Experimental Biology and Medicine*, **162:** 128–130.

Guth, P.H., Paulson, G., Lynn, D. and Aures, D. 1976, Mechanism of prevention of aspirin-induced gastric lesions by bile duct ligation in the rat. *Gastroenterology*, **71:** 750–753.

Gutierrez-Cabanco, C.A. 1993, Prostaglandins and sulfhydryls may mediate gastric protection induced by verapamil in rats. *Digestive Diseases and Sciences*, **38:** 2043–2048.

Guy, J.F. and Sucheston, M.E. 1986, Teratogenic effects on the CD-1 mouse embryo exposed to concurrent doses of ethanol and aspirin. *Teratology*, **34:** 249–261.

Gyömber, E., Vattay, P., Szabo, S. and Rainsford, K.D. 1996, Effect of lipoxygenase inhibitors and leukotriene antagonists on acute gastric mucosal lesions and chronic gastritis in rats. *Journal of Gastroenterology and Hepatology*, **11:** 922–927.

Györy, A.Z. and Stiel, J.N. 1968, Effect of particle size on aspirin induced gastrointestinal bleeding. *Lancet*, **2:** 300–302.

Hackman, R.M. and Hurley, L.S. 1984, Interactions of salicylate, dietary zinc and genetic strain in teratogenesis in rats. *Teratology*, **30:** 225–236.

Hahn, K.-J. 1973, Cause of injuring effect of acetylsalicylic acid and phenylbutazone on stomach mucosa. *Verhandlungen der Deutschen Gesellschaft für Innere Medizin*, **79:** 828–830.

Hahn, K.-J., Krischofski, D., Weber, E. and Morgenstern, E. 1975, Morphology of gastrointestinal effects of aspirin. *Clinical Pharmacology and Therapeutics*, **17:** 330–338.

Häkkinen, I.P.T., Johansson, R. and Pantio, M. 1968, An immunological and histoimmunological study of gastric sulphogylcoproteins in healthy and aspirin treated dogs. *Gut*, **9:** 712–716.

Hall, R.W. 1987, Aspirin and Reye's syndrome – do parents know? *Journal of the Royal College of General Practitioners*, **37:** 459–460.

Halpin, T.J., Holtzhauer, F.J., Campbell, R.J., Hall, L.J., Correa-Villasenor, A., Lanase, R., Rice, J. and Hurwitz, E.S. 1982, Reye's syndrome and medication use. *Journal of the American Medical Association*, **248:** 687–691.

Halter, F. 1993, Diaphragm disease of the ascending colon – association with sustained-release diclofenac. *Journal of Clinical Gastroenterology*, **16:** 74–80.

Halter, F., Kaufmann, M., Schweizer, W. and Ruchti, C. 1995, Diaphragmes (diaphragm disease) de l'intstin grêle et du côlon induits par les AINS. *Acta Endoscopia*, **25:** No. 2.

Halter, F., Peskar, B., Rainsford, K.D. and Schmassmann, A. 1997a, Cytoprotection and healing: two unequal brethren. *Inflammopharmacology*, **5:** 407–414.

Halter, F., Rainsford, K.D., Sirko, S.P. and Schmassmann, A. 1997b, NSAID-induced mucosal injury: Analysis of gastric toxicity of new generation NSAIDs; ulcerogenicity compared with ulcer healing. *Yale Journal of Biology and Medicine*, **70:** 33–43.

Hamidou, M.A., Moreau, A., Jego, P., Testa, A., Banisadr, F., Buzelin, F. and Grolleau, J.Y. 1995, Captopril and aspirin in treatment of renal microangiopathy in primary antiphospholipid syndrome. *American Journal of Kidney Diseases*, **25:** 486–488.

Hamlet, A., Lindholm, C., Nilsson, O. and Olbe, L. 1998, Aspirin-induced gastritis, like *Helicobacter pylori*-induced gastritis, disinhibits acid secretion in humans: relation to cytokine expression. *Scandinavian Journal of Gastroenterology*, **33:** 346–356.

Hansen, D., Aures, D. and Grossman, M.I. 1980, Comparison of intravenous and intragastric aspirin in production of antral gastric ulcers in cats. *Proceedings of the Society for Experimental Biology and Medicine*, **164:** 589–592.

Hansen, D.G., Aures, P. and Grossman, M.I. 1978, Histamine augments gastric ulceration produced by intravenous aspirin in cats. *Gastroenterology*, **74:** 540–543.

Hansen, D.G. and Grossman, M.I. 1978, Production of gastric ulcers in cats by intravenous acetaminophen plus histamine. *Clinical Research*, **26:** 110A.

Hanson, H.P., Gaede, P.H., Jensen, B.R. and Parving, H.H. 2000, Lack of impact of low-dose acetylsalicylic acid on kidney function in type 1 diabetic patients with microalbuminuria. *Diabetes Care*, **23:** 1742–1745.

Hao, C.M., Komhoff, M., Guan, Y.F., Redha, R. and Breyer, M.D. 1999, Selective targeting of cyclooxygenase-2 reveals its role in renal medullary interstitial cell survival. *American Journal of Physiology – Renal Physiology*, **277:** 352–359.

Harding, R.K. and Morris, G.P. 1976, Pathological effects of aspirin and of haemorrhagic shock on the gastric mucosa of the rat. *In:* O. Johari and R.P. Becker (eds), *Scanning Electron Microscopy, Part V. Proceedings of a Workshop on Advances in the Biomedical Applications of the SEM*, pp. 253–262. Chicago: ITT Research Institute.

Harford, D.J. and Smith, M.J.H. 1970, The effects of sodium salicylate on the release of acid phosphatase activity from rat liver lysosomes in vitro. *Journal of Pharmacy and Pharmacology*, **22:** 578–583.

Hargues, S.J., Amouzedeh, H.R., Pumford, N.R., Myers, T.G., McCoy, S.C. and Pohl, L.R. 1994, Metabolic activation and immunochemical localization of liver protein adducts of the nonsteroidal anti-inflammatory drug diclofenac. *Chemical Research in Toxicology*, **7:** 575–582.

Harner, I.C. and Foyles, R.A. 1980, Effect of Feingold's K-P diet on a residential mentally handicapped population. *Journal of the American Dietetic Association*, **76:** 575–578.

Harris, P.A. and Riegelman, S. 1969, Influence of the route of administration on the area under the plasma concentration time curve. *Journal of Pharmaceutical Sciences*, **58:** 71–75.

Hart, F.D. 1969, Rheumatoid arthritis; extra-articular manifestations. *British Medical Journal*, **3:** 131–136.

Haslock, I. and Wright, V. 1974, Arthritis and intestinal disease. *Journal of the Royal College of Physicians of London*, **8:** 154–162.

Hauser, A.C., Derfler, K. and Balcke, P. 1991, Progression of renal insufficiency in analgesic nephropathy: impact of continuous drug abuse. *Journal of Clinical Epidemiology*, **44:** 53–56.

Hauser, J. and Szabo, S. 1991, Extremely long protection by pyrazole derivatives against chemically-induced gastric mucosal injury. *Journal of Pharmacology and Experimental Pharmaceutics*, **256:** 592–598.

Hawkey, C.J. 1992, The ulcerogenic and anti-haemostatic effects of NSAIDs in the gut. *In:* K.D. Rainsford and G.P. Velo (eds), *Side Effects of Anti-inflammatory Drugs 3*, pp. 54–58. Dordrecht: Kluwer Academic Publishers.

Hawkey, C.J. 1994, Aspirin and gastrointestinal bleeding. *Alimentary Pharmacology and Therapeutics*, **8:** 141–146.

Hawkey, C.J. 1999a, COX-2 inhibitors. *Lancet*, **353:** 307–314.

Hawkey, C.J. 1999b, Personal Review: *Helicobacter pylori*, NSAIDs and cognitive dissonance. *Alimentary Pharmacology and Therapeutics*, **13:** 695–702.

Hawkey, C.J. and Brown, G. 1986, Essential fatty acids and aspirin induced damage to human gastric mucosa. *Digestive Diseases and Sciences*, **31:** 41.

Hawkey, C.J., Hawthorne, A.B., Hudson, N., Cole, A.T., Mahida, Y.R. and Daneshmend, T.K. 1991, Separation of the impairment of hemostasis by aspirin from mucosal injury in the human stomach. *Clinical Science*, **81:** 565–573.

Hawkey, C.J., Simpson, G. and Somerville, K.W. 1985, Reduction by enprostil of aspirin induced bleeding from human gastric mucosa. *Gut*, **26:** 560–564.

Hawkey, C.J., Simpson, G. and Somerville, K.W. 1986, Reduction by enprostil of aspirin induced blood loss from human gastric mucosa. *American Journal of Medicine*, **81:** 50–53.

Hawthorne, A.B., Mahida, Y.R., Cole, A.T. and Hawkey, C.J. 1991, Aspirin-induced gastric mucosal damage: prevention by enteric-coating and relation to prostaglandin synthesis. *British Journal of Clinical Pharmacology*, **32:** 77–83.

Heatley, N.G. 1950, Some experiments on partially purified gastrointestinal mucosubstance. *Gastroenterology*, **37:** 304–312.

Heatley, N.G. 1959, Mucosubstance as a barrier to diffusion. *Gastroenterology*, **37:** 313–318.

Hedman, J., Kaprio, J., Poussa, T. and Nieminen, M.M. 1999, Prevalence of asthma, aspirin intolerance, nasal polyposis and chronic obstructive pulmonary disease in a population-based study. *International Journal of Epidemiology*, **28:** 717–722.

Heffner, J.E. and Sahn, S.A. 1981, Salicylate-induced pulmonary oedema. Clinical features and prognosis. *Archives of Internal Medicine*, **95:** 405–409.

Heise, G., Grabensee, B., Schrör, K. and Heering, P. 1998, Different actions of the cyclooxygenase 2 selective inhibitor flosulide in rats with passive Heymann nephritis. *Nephron*, **80:** 220–226.

Henagan, J.M., Schmidt, K.L. and Miller, T.A. 1987, Failure of prostaglandin to prevent aspirin damage in canine gastric mucosa in vitro. *Gastroenterology*, **92:** 1432.

Henagan, J.M., Schmidt, K.L. and Miller, T.A. 1989, Prostaglandin prevents aspirin injury in the canine stomach under in vivo but not in vitro conditions. *Gastroenterology*, **97:** 649–659.

Henbi, J.E., Partin, J.C., Partin, J.S. and Schubert, W.K. 1987, Reye's syndrome: current concepts. *Hepatology*, **7:** 155–164.

Henry, D. 1997, Variability in risk of major upper gastro-intestinal complications with individual NSAIDs. Importance of drug dose and half-life; results of meta analysis. *In:* K.D. Rainsford (ed.), *Side Effects of Anti-inflammatory Drugs IV*, p. 327. Dordrecht: Kluwer Academic Publishers, 327.

Henry, D., Drew, A. and Beuzeville, S. 1998, Adverse drug reactions in the gastrointestinal system attributed to ibuprofen. *In:* K.D. Rainsford and M.C. Powanda (eds), *Safety and Efficacy of Non-Prescription (OTC) Analgesics and NSAIDs*, pp. 19–45. Dordrecht: Kluwer Academic Publishers.

Hersey, S.J. 1974, Interactions between oxidative metabolism and acid secretion in gastric mucosa. *Biochimica et Biophysica Acta*, **344:** 157–203.

Hersey, S.J. 1977, Metabolic changes associated with gastric stimulation. *Gastroenterology*, **73:** 914–919.

Hersey, S.J. 1980, ATP metabolism in isolated gastric glands. *Annals of the New York Academy of Sciences*, **341:** 274–282.

Hersey, S.J. 1981, Energy source for secretion in gastric glands. *Federation Proceedings*, **40:** 2511–2518.

Hersey, S.J., Chew, C.S., Campbell, L. and Hopkins, E. 1981, Mechanisms of action of SCN in isolated gastric glands. *American Journal of Physiology*, **240:** G232–G238.

Hersey, S.J. and Miller, M. 1981, Metabolic properties of gastric glands. *In: Membrane Biophysics: Structure and Function in Epithelia*, pp. 175–187. New York: Alan R. Liss Inc.

Hersey, S.J., Miller, M. and Owirodu, A. 1982, Role of glucose metabolism in acid production by isolated gastric glands. *Biochimica et Biophysica Acta*, **714:** 143–151.

Heubi, J.E., Partin, J.C., Partin, J.S. and Schubert, W.K. 1987, Reye's syndrome: current concepts. *Hepatology*, **7:** 155–164.

Hietanen, E. 1975, Mucosal and hepatic metabolism during the spontaneous disappearance of salicylate-induced gastric erosions. *American Journal of Digestive Diseases*, **20:** 31–41.

Hillman, R.J., and Prescott, L.F. 1985, Treatment of salicylate poisoning with repeated oral charcoal. *British Medical Journal*, **291:** 1472.

Himal, H.S., Greenberg, L., Boutros, M.I.R. and Waldron-Edward, D. 1975, Effects of aspirin on ionic movement and acid hydrolase activity of explants of canine anural and duodenal mucosae. *Gastroenterology*, **69:** 439–447.

Hines, W.J.W., Bryant, C. and Smith, M.J.H. 1963, Effects of salicylate, γ-resorcylate and gentisate on oxidase systems from guinea pig mitochondria. *Biochemical Pharmacology*, **12:** 1109–1116.

Hingson, D.J. and Ito, S. 1971, Effect of aspirin and related compounds on the fine structure of mouse gastric mucosa. *Gastroenterology*, **61:** 156–177.

Hirschowitz, B.I. and Lanas, A. 1998, Intractable upper gastrointestinal ulceration due to aspirin in patients who have undergone surgery for peptic ulcer. *Gastroenterology*, **114:** 883–892.

Hoftiezer, J.W., Burks, M., Silvoso, G.R. and Ivey, K.J. 1980, Comparison of the effects of regular and enteric-coated aspirin on gastroduodenal mucosa of man. *Lancet*, **2:** 609–612.

Hollander, F. 1953, Some recent contributions to the physiology of gastric mucus secretion. *Journal of the National Cancer Institute*, **13:** 989–1005.

Hollander, F. 1954, The two component mucous barrier. Its activity in protecting the gastroduodenal mucosa against peptic ulceration. *Archives of Internal Medicine*, **93:** 107–119.

Hollander, J.E. (ed.) 1966, *Arthritis and Allied Conditions: A Textbook of Rheumatology*. Philadelphia: Lea and Fabiger.

Holt, P.R. 1960, Measurement of gastrointestinal blood loss in subjects taking aspirin. *Journal of Laboratory and Clinical Medical*, **56:** 717–726.

Holzer, P., Pabst, M.A. and Lippe, I.T. 1989, Intragastric capsaicin protects against aspirin-induced lesion formation and bleeding in the rat gastric mucosa. *Gastroenterology*, **96:** 1425–1433.

Honingsberger, M. 1943, Aspirin and gastric haemorrhage. *British Medical Journal*, **2:** 57.

Hoon, J.R. 1969, Aspirin gastritis examined with intragastric photography, *Industrial Medicine and Surgery*, **38:** 262–272.

Hopper, A.H., Tindall, H. and Davies, J.A. 1986, Aspirin/dipyridamole treatment reduces proteinuria in diabetic nephropathy. *Transplantation Proceedings*, **18:** 1644.

Hopper, A.H., Tindall, H. and Davies, J.A. 1989, Administration of aspirin–dipyridamole reduces proteinuria in diabetic nephropathy. *Nephrology, Diathesis and Transplantion*, **4:** 140–143.

Horribin, D.F., Amnku, M.S. and Nassar, B.A. 1974, Aspirin and arteriolar responses to noradrenaline. *Lancet*, **1:** 567–568.

Horwich, L. and Price-Evans, D.A. 1966, Influence of the ABO blood group and salivary ABH secretor status on the cell-removing effect of aspirin on human gastric mucosa. *Gut*, **7:** 525–530.

Hoskins, L.C. 1967, The ABO blood group antigens and their secretion by healthy and diseased gastric mucosa. *Annals of the New York Academy of Science*, **140:** 848–860.

Houston, J.B. and Levy, G 1975, Modification of drug biotransformation by vitamin C in man. *Nature*, **255:** 78–79.

Hruban, Z., Gotoh, M., Slesers, A. and Chou, S.-F. 1974, Structure of hepatic microbodies in rats treated with acetylsalicylic acid, clofibrate and dimethrin. *Laboratory Investigation*, **30:** 64–75.

Hruban, Z., Swift, H. and Slesers, A. 1966, Ultrastructural alterations of hepatic microbodies. *Laboratory Investigation*, **15:** 1884–1901.

CHAPTER 8

Huang, J. and Hunt, R.H. 1996, A clinician's view of strategies for preventing NSAID-induced gastrointestinal ulcers. *Inflammopharmacology*, **4:** 17–30.

Humes, J.L., Winter, C.A., Sadowski, S.J. and Kuehl, F.A. 1981, Multiple sites on prostaglandin cyclooxygenase are determinants in the action of nonsteroidal antiinflammatory agents. *Proceedings of the National Academy of Sciences of the USA*, **78:** 2053–2056.

Hunt, J.N. 1979, A procedure for measuring gastric bleeding caused by drugs. *Digestive Diseases and Sciences*, **24:** 525–528.

Hunt, J.N. and Fisher, M.A. 1980, Aspirin-induced gastric bleeding stops despite rising plasma salicylate. *Digestive Diseases and Sciences*, **25:** 135–139.

Hunt, J.N. and Franz, D.R. 1981, Effect of prostaglandin E$_2$ on gastric mucosal bleeding caused by aspirin. *Digestive Diseases and Sciences*, **26:** 301–305.

Hunt, R.H., Bowen, B., Mortensen, E.R., Simon, T.J., James, C., Cagliola, A., Quan, H. and Bolognese, J.A. 2000, A randomized trial measuring fecal blood loss after treatment with rofecoxib, ibuprofen, or placebo in healthy subjects. *American Journal of Medicine*, **109:** 201–226.

Hunt, R.H., Cederberg, C., Dent, J., Halter, F., Howden, C., Marks, I.N.S., Rune, S. and Walt, R.P. 1995, Optimizing acid suppression for treatment of acid-related diseases. *Digestive Diseases and Sciences*, **40:** 24S–49S.

Hurwitz, E.S., Barrett, M.J., Bregman, D., Gunn, W.K., Schonberger, L.B., Fairweather, W.R., Drage, J.S., Lamontagne, J.R., Kaslow, R.A., Burlington, D.B., Quinnann, G.F., Parker, R.A., Phillips, K., Pinsky, P., Dayton, D. and Dowdle, W.R. 1985, Public Health Service Study on Reye's Syndrome and Medication. *New England Journal of Medicine*, **313:** 849–857.

Huong, D.L.T., Wechsler, B., Vauthier-Brouzes, D., Beaufils, H., Lefebvre, G. and Piette, J.C. 2001, Pregnancy in past or present lupus nephritis: a study of 32 pregnancies from a single centre. *Annals of the Rheumatic Diseases*, **60:** 599–604.

Huong, D.L.T., Wechsler, B., Vauthier-Brouzes, D., Seebacher, J., Lefebvre, G., Bletry, O., Darbois, Y., Godeau, P. and Piette, J.C. 1997, Outcome of planned pregnancies in systemic lupus erythematosus: a prospective study of 62 pregnancies. *British Journal of Rheumatology*, **36:** 772–777.

Hurley, J.W. and Crandall, L.A. 1963, The effects of various salicylates upon the dog's stomach: a gastroscopic photographic evaluation. *In:* A.St.-J. Dixon, B.K. Martin, M.J.H. Smith and P.H.N. Wood (eds), S*alicylates; An International Symposium*, pp. 213–216. London: Churchill.

Hurst, A. and Lintott, G.A.M. 1939, Aspirin as a cause of haematemasis: a clinical and gastroscopic study. *Guy's Hospital Reports*, **89:** 173–176.

Hurwitz, E.S., Barrett, M.J., Bregman, D., Gunn, W.J., Schonberger, L.B., Fairweather, W.R., Drage, J.S., Lamontagne, J.R., Kaslow, R.A. and Burlington, D.B. 1985, Public Health Service study on Reye's syndrome and medications. Report of the pilot phase. *New England Journal of Medicine*, **313:** 849–857.

Husted, S.E., Kremmer Nielsen, H., Petersen, T., Mogensen, C.E. and Geday, E. 1985, Acute effects of acetylsalicylic acid on renal and hepatic function in normal humans. *International Journal of Clinical Pharmacology, Therapy and Toxicology*, **23:** 141–144.

Iancu, T. 1972, Serum transaminases and salicylate therapy. *British Journal of Medicine*, **2:** 167.

Iancu, T. and Elian, E. 1974, Letter: Aspirin-induced abnormalities of liver function. *American Journal of Diseases of Children*, **128:** 116–117.

Iancu, T. and Elian, E. 1976, Ultrastructural changes in aspirin hepatoxicity. *American Journal of Clinical Pathology*, **66:** 570–575.

Ichikawa, T., Ishihara, K., Ogata, Y., Ohara, S., Saigenji, K. and Hotta, K. 1994, Effects of Z-300, a new histamine H$_2$-receptor antagonist, on mucin biosynthesis in rat gastric mucosa. *Japanese Journal of Pharmacology*, **65:** 63–66.

Inman, W.H. 1987, Non-steroidal anti-inflammatory drugs: assessment of risks. *European Journal of Rheumatology and Inflammation*, **8:** 71–85.

Inoue, M., Morikawa, M., Tsuboi, M. and Sugiura, M. 1979, Species difference and characterization of intestinal esterase activity of ester-type drugs. *Japanese Journal of Pharmacology*, **29:** 9–16.

Irwin, R.P. 1976, Minor analgesic use among high-school students: indications for renal morbidity and mortality. *Medical Journal of Urology*, **113:** 653–657.

Ishii, H. and Suga, T. 1979, Clofibrate-like effects of acetylsalicylic acid on peroxisomes and on hepatic and serum triglyceride levels. *Biochemical Pharmacology*, **28:** 2829–2833.

Isreal, E. Fischer, A.R. and Rosenberg, M.A. 1993, The pivotal role of 5-lipoxygenase products in the reaction of aspirin – sensitive asthmatics to aspirin. *American Journal of Respiratory and Critical Care Medicine*, 148; 1447–1451.

Itami, T. and Kanoh, S. 1982, Studies on the pharmacological bases of fetal toxicity of drugs. (I). Relation of fetal toxicity and tissue concentration of acetylsalicylic acid with pyrogen in pregnant rats. *Folia Pharmacologica Japonica*, **79:** 357–867.

Itami, T. and Kanoh, S. 1983, Studies on the pharmacological bases of fetal toxicity of drugs. (IV). Effect of endotoxin and starvation on serum protein binding of salicylic acid in pregnant rats. *Japanese Journal of Pharmacology*, **33:** 1199–1204.

Itami, T. and Kanoh, S. 1984, Studies on the pharmacological bases of fetal toxicity of drugs. (VII). Enhancement effect of bacterial pyrogen on the fetal toxicity of salicylic acid. *Folia Pharmacologica Japonica*, **84:** 411–416.

Ito, Y. and Tajima, K. 1981, Spontaneous activity in the trachea of dogs treated with indomethacin: an experimental model for aspirin-related asthma. *British Journal of Pharmacology*, **73:** 563–571.

Ivey, K.J., Baskin, W. and Jeffrey, G. 1975a, Effect of cimetidine on gastric potential difference in man. *Lancet*, **2:** 1072–1073.

Ivey, K.J., Baskin, W.N., Krause, W.J. and Terry, B. 1979, Effect of aspirin and acid on human jejunal mucosa. An ultrastructure study. *Gastroenterology*, **76:** 50–56.

Ivey, K.J. and Clifton, J.A. 1974, Back-diffusion of hydrogen ions across gastric mucosa of patients with gastric ulcer and rheumatoid arthritis. *British Medical Journal*, **I (896):** 16–19.

Ivey, K.J., Paone, D.B. and Krause, W.J. 1980a, Acute effect of systemic aspirin on gastric mucosa in man. *Digestive Diseases and Sciences*, **25:** 97–99.

Ivey, K.J. and Parsons, C. 1975, Are non-absorbable indicators of comparable value in the human stomach made abnormal by taurocholic acid? *Clinical and Experimental Pharmacology and Physiology*, **2:** 151–157.

Ivey, K.J., Parsons, C. and Gray, C. 1974, Failure of lithium to provide a marker for gastric hydrogen ion back-diffusion in man. *Gastroenterology*, **66:** 69–72.

Ivey, K.J., Parsons, C. and Weatherby, R. 1975b, Effect of prednisolone and salicylic acid on ionic fluxes across the human stomach. *New Zealand Journal of Medicine*, **5:** 408–412.

Ivey, K.J. and Schedl, H.P. 1970, Gastric on absorbable indicators for studies in man. *Gastroenterology*, **59:** 234–239.

Ivey, K.J., Tarnawski, A., Stachura, J., Werners, H., Macht, T. and Burks, M. 1980b, The induction of gastric mucosal tolerance to alcohol by chronic administration. *Journal of Laboratory and Clinical Medicine*, **96:** 922–932.

Iwata, F., Zhang, X.Y. and Leung, F.W. 1995, Aggravation of gastric mucosal lesions in rat stomach by tobacco cigarette smoke. *Digestive Diseases and Sciences*, **40:** 1118–1124.

Jabbari, M. and Valberg, L.S. 1970, Role of acid secretion in aspirin-induced gastric mucosal injury. *Canadian Medical Association Journal*, **102:** 178–181.

Jackson, A.V. 1948, Toxic effects of salicylate on the fetus and mother. *Journal of Pathology*, **60:** 587–593.

Janakidevi, K. and Smith, M.J.H. 1970a, Differential inhibition of RNA polymerase activities by salicylate in vitro. *Journal of Pharmacy and Pharmacology*, **22:** 58–59.

Janakidevi, K. and Smith, M.J.H. 1970b, Effects of salicylate on the incorporation of orotic acid into nucleic acids of mouse tissues in vivo. *Journal of Pharmacy and Pharmacology*, **22:** 51–55.

Janakidevi, K. and Smith, M.J.H. 1970c, Effects of salicylate on RNA polymerase activity and on the incorporation of orotic acid and thymidine into the nucleic acids of rat fetuses in vitro. *Journal of Pharmacy and Pharmacology*, **22:** 249–252.

Jarvik, L.F., Fleiss, J.L., Mauer, I. and Weinstein, D. 1971, Acetyl salicylic acid and chromosome damage. *Sciences (Washington DC)*, **171:** 829–830.

Jaszewski, R., Calzada, R. and Dhar, R. 1989, Persistence of gastric ulcers caused by plain aspirin or nonsteroidal anti-inflammatory agents in patients treated with a combination of cimetidine, antacids, and enteric-coated aspirin. *Digestive Diseases and Sciences*, **34:** 1361–1364.

Jaszewski, R., Graham, D.Y. and Stromatt, S.C. 1992, Treatment of nonsteroidal anti-inflammatory drug-induced gastric-ulcers with misoprostol – a double-blind multicenter study. *Digestive Diseases and Sciences*, **37:** 1820–1824.

Jeffrey, S.W. and Smith, M.J.H. 1959, The effects of salicylates on oxygen consumption and carbohydrate metabolism in the isolated rat diaphragm. *Biochemical Journal*, **72:** 462–465.

Jenkins, A. 1991, Do non-steroidal anti-inflammatory drugs increase colonic permeability? *Gut*, **32:** 66–69.

Jenkins, R.T., Rooney, P.J., Jones, D.B., Bienenstock, J. and Goodacre, R.L. 1987, Increased intestinal permeability in patients with rheumatoid arthritis: a side-effect of oral nonsteroidal anti-inflammatory drug therapy? *British Journal of Rheumatology*, **26:** 103–107.

Jin, M., Otaka, M., Okuyama, A., Itoh, S., Otani, S., Odashima, M., Iwabuchi, A., Konishi, N., Wada, I., Pacheco, I., Itoh, H., Tashima, Y., Masamune, O. and Watanabe, S. 1999, Association of 72-kDa heat shock protein expression with adaptation to aspirin in rat gastric mucosa. *Digestive Diseases and Sciences*, **44:** 1401–1407.

Jiranek, G.C., Kimmey, M.B., Saunders, D.R., Willson, R.A., Shanahan, W. and Silverstein, F.E. 1989, Misoprostol reduces gastroduodenal injury from one week of aspirin. An endoscopic study. *Gastroenterology*, **96:** 656–661.

Johansson, H. and Lindquist, B. 1971, Anti-inflammatory drugs and gastric mucus. *Scandinavian Journal of Gastroenterology*, **6:** 48–54.

Johnson, A.G. and McDermott, S.J. 1974, Lysolecethin: a factor in the pathogenesis of gastric ulceration. *Gut*, **15:** 710–713.

Johnson, L.R. 1966, Histamine liberation by gastric mucosa of pylorus-ligated rats damaged by acetic or salicylic acids. *Proceedings of the Society for Experimental Biology and Medicine*, **121:** 384–390.

Johnson, L.R. 1972, Pepsin secretion during damage by ethanol and salicylic acid. *Gastroenterology*, **62:** 412–416.

Johnson, L.R. and Overholt, B.F. 1967, Release of histamine into gastric venous blood following injury by acetic or salicylic acid. *Gastroenterology*, **52:** 505–509.

Johnson, P. and Rainsford, K.D. 1972, The physical properties of mucus: preliminary observations on the sedimentation behaviour of porcine gastric mucus. *Biochimica et Biophysica Acta*, **286:** 72–78.

Johnson, P.N. and Welch, D.W. 1984, Methyl salicylate/aspirin (salicylate) equivalence: who do you trust? *Veterinary and Human Toxicology*, **26:** 317–318.

Johnston, S.A., Leib, M.S., Firrester, S.D. and Marini, M. 1995, The effect of misoprostol on aspirin-induced gastroduodenal lesions in dogs. *Journal of Veterinary Internal Medicine*, **9:** 32–38.

Jones, R.A. 2001, Etodolac (Lodine®) profile of an established selective COX-2 inhibitor. *Inflammopharmacology*, **9:** 63–70.

Jones-Blackett, S., Hull, M.A., Davies, G.R. and Crabtree, J.E. 1999, Non-steroidal anti-inflammatory drugs inhibit *Helicobacter pylori*-induced human neutrophil reactive oxygen metabolite production In vitro. *Alimentary Pharmacology and Therapeutics*, **13:** 1653–1661.

Jørgensen, T.G. 1976, Ulcer formation and histochemical changes in rat-stomach mucosa induced by acetylsalicylic acid. *Acta Pathologica et Microbiologica Scandinavica*, **84:** 64–72.

Jørgensen, T.G., Kaplan, E.L. and Peskin, G.W. 1974, Salicylate effects on gastric acid secretion. *Scandinavian Journal of Clinical and Laboratory Investigation*, **33:** 31–38.

Jørgensen, T.G., Weis-Fogh, U.S., Nielson, H.H. and Olsen, H.P. 1976a, Salicylate and aspirin-induced uncoupling of oxidative phosphorylation in mitochondria isolated from the mucosal membrane of the stomach. *Scandinavian Journal of Clinical and Laboratory Investigation*, **36:** 649–654.

Jørgensen, T.G., Weis-Fogh, U.S. and Olesen, H.P. 1976b, The influence of acetylsalicylic acid (aspirin) on gastric mucosal content of energy-rich phosphate bond. *Scandinavian Journal of Clinical and Laboratory Investigations*, **36:** 771–777.

Jorup-Rönström, C., Beermann, B., Wåhlin-Boll, E., Melander, A. and Britton, S. 1986, Reduction of paracetamol and aspirin metabolism during viral hepatitis. *Clinical Pharmacokinetics*, **11:** 250–256.

Juby, A. and Davis, P. 1991, Psychological profiles of patients with upper gastrointestinal symptomatology induced by non-steroidal anti-inflammatory drugs. *Annals of the Rheumatic Diseases*, **50:** 211–213.

Juggi, J.S. 1975, Tissue acetylsalicylic acid esterase activity in rats with acute and chronic liver damage from carbon-tetrachloride and ethanol. *Enzyme*, **20:** 183–187.

Kajii, H., Horie, T., Hayashi, M. and Awazu, S. 1985, Fluorescence study on the interaction of salicylate with rat small intestinal epithelial cells: possible mechanism for the promoting effects of salicylate in drug absorption in vivo. *Life Sciences*, **37:** 523–530.

Kalczak, M., Gutowska-Grzegorczyk, G. and Maldyke, E. 1970, The effect of chronic administration of acetylsalicylic acid on the rabbits liver. *Polish Medical Journal*, **9:** 128–134.

Kallos, P. and Schlumberger, H.D. 1978, 'Immunogenic impurities' in acetylsalicylic acid. *Journal of Pharmacy and Pharmacology*, **30:** 67–68.

Kallos, P. and Schlumberger, H.D. 1980, The pathomechanism of acetylsalicylic acid intolerance. A hypothesis. *Medical Hypotheses*, **6:** 487–490.

Kaplan-Machlis, B. and Klostermeyer, B.S. 1999, The cyclooxygenase-2 inhibitors: safety and effectiveness. *Annals of Pharmacotherapy*, **33:** 979–988.

Kaplowitz, N., Kuhlenkamp, G., Godstein, L. and Reeve, J. 1980, Effects of salicylates and phenobarbital on hepatic glutathione in the rat. *Journal of Pharmacology and Experimental Therapeutics*, **212:** 240–245.

Karabulut, A.K., Ülger, H. and Pratten, M.K. 2000, Protection by free oxygen radical scavening enzymes against salicylate-induced embryonic malformations in vitro. *Toxicology in Vitro*, **14:** 297–307.

Karsh, J. 1990, Adverse reactions and interactions with aspirin. Considerations in the treatment of the elderly patient. *Drug Safety*, **5:** 317–327.

Kato, R. 1977, Drug metabolism under pathological and abnormal physiological states in animals and man. *Xenobiotica*, **7:** 25–92.

Katz, W.A. 1977, *Rheumatic Diseases: Diagnosis and Management*. Philadelphia: Lippincott.

Kauffman, G. 1989, Aspirin-induced gastric mucosal injury: lessons learned from animal models. *Gastroenterology*, **96:** 606–614.

Kauffman, G.L., Aures, D. and Grossman, M.I. 1980, Intravenous indomethacin and aspirin reduce basal gastric mucosal blood flow in dogs. *American Journal of Physiology*, **238:** G131–G134.

Kaufman, D.W. and Kelly, J.P. 2001, Acetylsalicylic acid and other salicylates in relation to Stevens–Johnson syndrome and toxic epidermal necrolysis. *British Journal of Clinical Pharmacology*, **51:** 174–176.

Kawai, S. 1998, Cycloxygenase selectivity and the risk of gastro-intestinal complications of various non-steroidal anti-inflammatory drugs: a clinical consideration. *Inflammation Research*, **47 (Suppl. 2):** S102–S106.

Keiser, H.R. 1980, The kallikrein–kinin system in essential hypertension. *Clinical and Experimental Hypertension*, **2:** 675–691.

Kelly, J.P., Kaufman, D.W., Jurgelon, J.M., Sheehan, J., Koff, R.S. and Shapiro, S. 1996, Risk of aspirin-associated major upper-gastrointestinal bleeding with enteric-coated or buffered product. *Lancet*, **348:** 1413–1416.

Kern, F., Clark, G.M. and Lukens, J.G. 1957, Peptic ulceration occurring during therapy for rheumatoid arthritis. *Gastroenterology*, **33:** 25–33.

Kerston, S., Desvergne, B. and Wahli, W. 2000, Roles of PPARs in health and disease. *Nature*, **405:** 421–424.

Kimberley, R.P., Gill, J.R., Bowden, R.E., Keiser, H.R. and Plotz, P.H. 1978, Elevated urinary prostaglandins and the effects of aspirin on renal functions in lupus erythematosus. *Annals of Internal Medicine*, **89:** 336–341.

Kimberley, R.P. and Plotz, P.H. 1977, Aspirin-induced depression of renal function. *New England Journal of Medicine*, **296:** 418–424.

Kimberley, R.P., Sherman, R.L., Mouradian, J. and Lockshin, M.D. 1979, Apparent acute renal failure associated with therapeutic aspirin and ibuprofen administration. *Arthritis and Rheumatism*, **22:** 281–285.

Kimmel, C.A., Wilson, J.G. and Schumacher, H.J. 1971, Studies on metabolism and identification of the causative agent in aspirin tetragenesis in rats. *Teratology*, **4:** 15–24.

Kincaid-Smith, P. 1967, Pathogenesis of the renal lesion associated with the abuse of analgesics. Lancet, **1:** 859–862.

Kincaid-Smith, P., Saker, B.M. and McKenzie, I.F.C. 1968, Lesions in the vasa recta in experimental analgesic nephropathy. *Lancet*, **1:** 24.

King, A.R. and Dickinson, R.G. 1993, Studies on the reactivity of acyl glucuronides – IV. Covalent binding of diflunisal to tissues of the rat. *Biochemical Pharmacology*, **45:** 1043–1047.

Kirkova, M., Ivancheva, E. and Russanov, E. 1995, Lipid peroxidation and antioxidant enzyme activity in aspirin-treated rats. *General Pharmacology*, **26:** 613–617.

Kitagawa, H., Saito, H., Ueno, K., Naminohira, S., Igarashi, T., Satoh, T. and Sakai, T. 1982, Acetylsalicylic acid induced fetal toxicity and drug metabolism. *Journal of Pharmaco-Dynamics*, **5:** 39.

Kivilaasko, E., Ehnholm, C., Kalima, T.V. and Lempinen, M. 1976, Duodenogastric reflux of lysolecithin in the pathogenesis of experimental porcine stress ulceration. *Surgery*, **79:** 65–69.

Kivilaasko, E. and Silen, W. 1979, Pathogenesis of experimental gastric-mucosal injury. *New England Journal of Medicine*, **301:** 364–369.

Kivinen, A., Tarpila, S. Salminen, S. and Vapaatalo, H. 1992, Gastroprotection with milk phospholipids – a 1st human study. Milchwissenschaft. *Science International*, **47:** 694–696.

Klebanoff, M.A. and Berendes, H.W. 1988, Aspirin exposure during the first 20 weeks of gestation and IQ at four years of age. *Teratology*, **37:** 249–255.

Knight, E. and Roe, D.A. 1978, Effects of salicylamide and protein restriction on the skeletal development of the rat fetus. *Teratology*, **18:** 17–22.

Knights, K.M. 1998, Role of hepatic fatty acid: coenzyme A ligases in the metabolism of xenobiotic carboxylic acids. *Clinical and Experimental Pharmacology and Physiology*, **25:** 776–782.

Kobayashi, K. Fukuda, T., Higuchi, K., Nakamura, H. and Arkawa, T. 1993, Role of leukotrienes in indomethacin-induced mucosal damage in rats. *Journal of Clinical Gastroenterology*, **17 (Suppl. 1):** S11–S14.

Kodaira, H., Kagoshima, M. and Ishii, K. 1998, Role of sialomucin in the protective effect of aspirin on the gastric mucosa of rat. *Naunyn-Schmiedeberg's Archives of Pharmacology*, **358:** 4060.

Koelz, H.R., Sachs, G., Fischer, J.A. and Blum, A.L. 1977, Inhibition of cation transport of isolated rat gastric cells by salicylate. *In:* S. Bonfils (ed.), *1st International Symposium on Hormone Receptors in Digestive Tract Physiology, INSERM Symposium No. 3*, p. 403. Amsterdam: Elsevier/North Holland.

Koelz, H.R., Fischer, J.A., Sachs, G. and Blum, A.L. 1978, Specific effect of acetylsalicylic acid on cation transport of isolated gastric mucosal cells. *American Journal of Physiology*, **235:** E16–E21.

Konturek, J.W., Dembinski, A., Konturek, S.J. and Domschke, W. 1997, *Helicobacter pylori* and gastric adaptation to repeated aspirin administration in humans. *Journal of Physiology and Pharmacology*, **48:** 383–391.

Konturek, J.W., Dembinski, A., Stoll, R., Domschke, W. and Konturek, S.J. 1994c, Mucosal adaptation to aspirin induced gastric damage in humans. Studies on blood flow, gastric mucosal growth, and neutrophil activation. *Gut*, **35:** 1197–1204.

Konturek, J.W., Dembinski, A., Stoll, R., Konturek, M. and Domschke, W. 1993a, Gastric mucosal blood flow and neutrophil activation in aspirin-induced gastric mucosal damage in man. *Scandinavian Journal of Gastroenterology*, **28:** 767–771.

Konturek, J.W, Fischer, H., Konturek, P.C., Huber, V., Boknik, P., Luess, H., Neumann, J., Brzozowski, T., Schmitz, W., Hahn, E.G., Domschke, W. and Konturek, S.J. 2001, Heat Shock Protein 70 (HSP70) in gastric adaptation to aspirin in *Helicobacter pylori* infection. *Journal of Physiology and Pharmacology*, **52:** 153–164.

Konturek, J.W., Konturek, S.J., Stachura, J. and Domschke, W. 1998, *Helicobacter pylori*-positive peptic ulcer patients do not adapt to aspirin. *Alimentary Pharmacology and Therapeutics*, **12:** 857–864.

Konturek, J.W., Stachura, J., Dembinski, A., Stoll, R. and Domschke, W. 1994c, Do infiltrating leukocytes contribute to the adaptation of human gastric mucosa to continued aspirin (ASA) administration? *Gastroenterology*, **106:** 110.

Konturek, S.J. (1993) New aspects of clinical pharmacology of antacids. *Journal of Physiology and Pharmacology*, **44:** 5–21.

Konturek, S.J., Brzozowski, T., Drozdowicz, D. and Majka, J. 1992, Ebrotidine, a novel H-2-receptor antagonist with local gastroprotective activity. *European Journal of Gastroenterology and Hepatology*, **3:** 941–947.

Konturek, S.J., Brzozowski, T., Garlici, J., Majka, J., Stachura, J. and Nauert, C. 1991, Intragastric pH in the gastroprotective and ulcer-healing activity of aluminum-containing antacids. *Digestion*, **49:** 140–150.

Konturek, S.J., Brzozowski, T., Pkerzchalski, P., Kwiecien, S., Pajdo, R., Hahn, E.G. and Konturek, S.J. 1998, Activation of genes for spasmolytic peptide, transforming growth factor alpha and for cyclooxygenase (COX)-1 and COX-2 during gastric adaptation to aspirin damage in rats. *Alimentary Pharmacology and Therapeutics*, **12:** 767–777.

Konturek, S.J., Brzozowski, T., Stachura, J., Dembinski, A. and Majka, J. 1994a, Role of gastric blood-flow, neutrophil infiltration, and mucosal cell-proliferation in gastric adaptation to aspirin in the rat. *Gut*, **35:** 1189–1196.

Konturek, S.J., Brzozowski, T., Stachura, J. and Majka, J. 1994b, Role of neutrophils and mucosal blood-flow in gastric adaptation to aspirin. *European Journal of Pharmacology*, **253:** 107–114.

Konturek, S.J., Kwiecien, N., Sito, E., Obtulowicz, W., Kaminski, K. and Oleksy, J. 1993b, Effects of ebrotidine on aspirin-induced gastric-mucosal damage and blood-flow in humans. *Scandinavian Journal of Gastroenterology*, **28:** 1047–1050.

Konturek, S.J., Piastucki, I., Brzozowski, T., Radecki, T., Dembinska-Kiec, A. and Zmuda, A. 1981, Role of prostaglandins in the formation of aspirin-induced gastric ulcers. *Gastroenterology*, **80:** 4–9.

Koseki, Y., Terai, C., Moriguchi, M., Uesato, M. and Kamatani, N. 2001, A prospective study of renal disease in patients with early rheumatoid arthritis. *Annals of the Rheumatic Diseases*, **60:** 327–331.

Kowaleski, K. and Kolodej, A. 1977, Effect of intra-arterial infusion of glucose on secretory activity of isolated canine stomach. *Pharmacology*, **15:** 324–330.

Kravath, R.E., Abel, G., Colli, A., McNamara, H. and Cohen, M.I. 1972, Salicylate poisoning effect on 2,3-diphosphoglycerate levels in the rat. *Biochemical Pharmacology*, **21:** 2656–2658.

Kuboyama, N. and Fujii, A. 1992, Mutagenicity of analgesics, their derivatives, and anti-inflammatory drugs with S-9 mix of several animal species. *Journal of the Nikon University School of Dentistry*, **34:** 183–195.

Kuiper, D.H., Overholt, B.F., Fall, D.J. and Pollard, H.M. 1969, Gastroscopic findings and faecal blood loss following aspirin administration. *American Journal of Digestive Diseases*, **14:** 761–769.

Kumar, R. and Billimora, J.D. 1978, Gastric ulceration and the concentration of salicylate in plasma in rats after the administration of ^{14}C-labelled aspirin and its synthetic triglyceride, 1,3-dipalmitoyl-2(2′-acetoxy-[^{14}C] carboxylenzoyl) glycerol. *Journal of Pharmacy and Pharmacology*, **30:** 754–758.

Kuo, Y.-J. and Shanbour, L.L. 1976, Mechanism of action of aspirin on canine gastric mucosa. *American Journal of Physiology*, **230:** 762–767.

Kyle, M.E. and Kocsis, J.J. 1985, The effect of age on salicylate-induced nephrotoxicity in male rats. *Toxicology and Applied Pharmacology*, **81:** 337–347.

Kyle, M.E. and Kocsis, J.J. 1986, Metabolism of salicylate by isolated kidney and liver mitochondria. *Chemico-Biological Interactions*, **59:** 325–335.

Kyriakopoulous, A.A., Clark, M.L., Mock, D.C. and Hagans, J.A. 1960, A comparative study of gastrointestinal bleeding incident to the administration of aspirin, aspirin anhydride and placebo. *Clinical Research*, **8:** 202.

Labrid, C., Dureng, G. and Boero, C. 1975, Value of the 'gastric chamber', a new technique performed ex vivo, for the study of the gastric mucosa of the rat. *Comptes Rendu des Séances de la Societé de Biologie*, **169:** 566–573.

Lacy, E.R., Cowart, K.S., King, J.S., Delvalle, J. and Smolka, A.J. 1996, Epithelial response of the rat gastric mucosa to chronic superficial injury. *Yale Journal of Biology and Medicine*, **69:** 105–118.

Lacy, E.R., Morris, G.P. and Cohen, M.M. 1993, Rapid repair of the surface epithelium in human gastric mucosa after acute superficial injury. *Journal of Clinical Gastroenterology*, **17:** 125–135.

Laine, L., Cominelli, F., Sloane, R., Vasini-Raggi, V., Marin-Sorenson, M. and Weinstein, W.M. 1995, Interaction of NSAIDs and *Helicobacter pylori* on gastrointestinal injury and prostaglandin production: a controlled double-blind trial. *Alimentary Pharmacology and Therapeutics*, **9:** 127–135.

Laine, L., Harper, S., Simon, T., Bath, R., Johanson, J., Schwartz, H., Stern, S., Quan, H. and Bolognese, J. 1999, A randomised trial comparing the effect of rofecoxib, a cyclooxygenase 2-specific inhibitor, with that of ibuprofen on the gastroduodenal mucosa of patients with osteoarthritis. Rofecoxib Osteoarthritis Endoscopy Study Group. *Gastroenterology*, **177:** 776–783.

Lake, B.G., Tredger, J.M., Gray, T.J.B., Stubberfield, C.R., Hodder, K.D., Gangolli, S.D. and Williams, R. 1984, The effect of peroxisome proliferators on the metabolism and spectral interaction of endogenous substrates of cytochrome P-450 in rat heptic microscomes. *Life Sciences*, **35:** 2621–2626.

Lamarque, D. and Whittle, B.J.R. 1995, Involvement of superoxide and xanthine oxidase in neutrophil-independent rat gastric damage induced by NO donors. *British Journal of Pharmacology*, **116:** 1843–1848.

Lamont-Havers, R.W. and Wagner, B.M. 1966, *Effects of Chronic Salicylate Administration (Conference). National Institute of Arthritis and Metabolic Diseases*. Washington, DC: US Government Printing Office.

Lanas, A., Arroyo, M.T., Esteva, F., Cornudella, R., Hirschowitz, B.I. and Sainz, R. 1996, Aspirin related gastrointestinal bleeders have an exaggerated bleeding time response due to aspirin use. *Gut*, **39:** 654–660.

Lanas, A., Bajador, E., Serrano, P., Fuentes, J., Carreno, S., Guardia, J., Sanz, M., Montoro, M. and Sainz, R. 2000, Nitrovasodilators, low-dose aspirin, other nonsteroidal anti-inflammatory drugs, and the risk of upper gastrointestinal bleeding. *New England Journal of Medicine*, **343:** 834–839.

Lanas, A. and Hirschowitz, B.I. 1991, Significant role of aspirin use in patients with esophagitis. *Journal of Clinical Gastroenterology*, **13:** 622–667.

Lanas, A., Sekar, M.C. and Hirschowitz, B.I. 1992, Objective evidence of aspirin use in both ulcer and nonulcer upper and lower gastrointestinal bleeding. *Gastroenterology*, **103:** 862–869.

Landecker, K.D., Wellington, J.E., Thomas, J.H. and Piper, D.W. 1977, Gastric ulcer, aspirin esterase and aspirin. *In:* K.D. Rainsford, K. Brune and M.W. Whitehouse (eds), *Aspirin and Related Drugs: Their Actions and Uses*, pp. 71–79. Basel: Birkhäuser.

Lang, J., Price, A.B., Levi, A.J., Burke, M., Gumpel, J.M. and Bjarnason, I. 1988, Diaphragm disease: pathology of disease of the small intestine induced by non-steroidal anti-inflammatory drugs. *Journal of Clinical Pathology*, **41:** 516–526.

Lange, H.F. 1957, Salicylates and gastric haemorrhage I. Occult bleeding. *Gastroenterology*, **33:** 770–777.

Langman, M.J.S. 1973, Blood groups and alimentary disorders. *Clinics in Gastroenterology*, **2:** 497–506.

Langman, M.J.S. 1979, *The Epidemiology of Chronic Digestive Disease*. London: Edward Arnold.

Langman, M.J.S. and Doll, R. 1965, ABO blood group and secretor status in relation to clinical characteristics of peptic ulcers. *Gut*, **6:** 270–273.

Langman, M.J.S., Jensen, D.M., Watson, D.J., Harper, S.E., Zhao, P.L., Quan, H., Bolognese, J.A. and Simon, T.J. 1999, Adverse upper gastrointestinal effects of rofecoxib compared with NSAIDs. *Journal of the American Medical Association*, **282:** 1929–1933.

Lanza, F.L. 1984, Endoscopic studies of gastric and duodenal injury after the use of ibuprofen, aspirin, and other nonsteroidal anti-inflammatory agents. *American Journal of Medicine*, **77:** 19–24.

Lanza, F.L., Evans, D.G. and Graham, D.Y. 1991, Effect of *Helicobacter pylori* infection on severity of gastroduodenal mucosal injury after the acute administration of naproxen or aspirin to normal volunteers. *American Journal of Gastroenterology*, **86:** 735–737.

Lanza, F.L., Fakouhi, D., Rubin, A., Davis, R.E., Rack, M.F., Nissen, C. and Geis, S. 1989, A double-blind placebo-controlled comparison of the efficacy and safety of 50, 100 and 200 micrograms of misoprostol QID in the prevention of ibuprofen-induced gastric and duodenal mucosal lesions and symptoms. *American Journal of Gastroenterology*, **84:** 633–636.

Lanza, F.L., Graham, D.Y., Davis, R.E. and Rack, M.F. 1990, Endoscopic evaluation of cimetidine and sucralfate for prevention of naproxen-induced gastroduodenal injury: effects of scoring method. *Digestive Diseases and Sciences*, **35:** 1494–1499.

Lanza, F.L., Nelson, R.S. and Greenberg, B.P. 1983, Effects of fenbufen, indomethacin, naproxen, and placebo on gastric mucosa of normal volunteers. *American Journal of Gastroenterology*, **75 (4B):** 75–79.

Lanza, F.L., Nelson, R.S. and Rack, M.F. 1984, A controlled endoscopic study comparing the toxic effects of sulindac, naproxen, aspirin and placebo on the gastric mucosa of healthy volunteers. *Journal of Clinical Pharmacology*, **24:** 89–95.

Lanza, F.L., Peace, K., Gustitus, L., Rack, M.F. and Dickson, B. 1988, A blinded endoscopic comparative study of misoprostol versus sucralfate and placebo in the prevention of aspirin-induced gastric and duodenal ulceration. *American Journal of Gastroenterology*, **83:** 143–146.

Lanza, F.L., Rack, M.F., Lynn, M., Wolf, B.S. and Sanda, M. 1987, An endoscopic comparison of the effects of etodolac, indomethacin, ibuprofen, naproxen and placebo on the gastrointestinal mucosa. *Journal of Rheumatology*, **14:** 338–341.

Lanza, F.L., Rack, M.F., Simon, T.J., Lombardi, A., Reyes, R. and Suryawanshi, S. 1998, Effects of alendronate on gastric and duodenal mucosa. *American Journal of Gastroenterology*, **93:** 753–757.

Lanza, F.L., Rack, M.F., Simon, T.J., Quan, H., Bolognese, J.A., Hoover, M.E., Wilson, F.R. and Harper, S.E. 1999, Specific inhibition of cyclooxygenase-2 with MK-0966 is associated with less gastroduodenal damage than either aspirin or ibuprofen. *Alimentary Pharmacology and Therapeutics*, **13:** 761–767.

Lanza, F.L., Royer, G. and Nelson, R. 1975, An endoscopic evaluation of the effects of nonsteroidal anti-inflammatory drugs on the gastric mucosa. *Gastrointestinal Endoscopy*, **21:** 103–105.

Lanza, F.L., Royer, G.L. and Nelson, R.S. 1980, Endoscopic evaluation of the effects of aspirin, buffered aspirin, and enteric-coated aspirin on gastric and duodenal mucosa. *New England Journal of Medicine*, **303:** 136–138.

Lanza, F.L., Royer, G.L., Nelson, R.S., Rack, M.F., Seckman, C.E. and Schwartz, J.H. 1986b, Effects of acetaminophen on human gastric mucosal injury caused by ibuprofen. *Gut*, **27:** 440–443.

Lanza, F.L., Royer, G.L., Nelson, R.S., Seckman, C.E., Schwartz, J.H., Rack, M.F. and Gernaat, C.M. 1986a, Effects of flurbiprofen and aspirin on the gastric and duodenal mucosa. *American Journal of Medicine*, **80 (Suppl. 3A):** 31–35.

Larsen, K.R., Dajani, E.Z. and Ives, M.M. 1992, Anti-ulcer drugs and gastric-mucosal integrity – effects of misoprostol, 16,16-dimethyl-PGE2, and cimetidine on hemodynamics and metabolic rate in canine gastric-mucosa. *Digestive Diseases and Sciences*, **37:** 1029–1038.

Larsson, K.S. and Boström, H. 1965, Teratogenic action of salicylates related to the inhibition of mucopolysaccharide synthesis. *Acta Paediatrica Scandinavica*, **54:** 43–48.

Larsson, K.S., Boström, H. and Ericson, B. 1963, Salicylate-induced skeletal and vessel malformations in mouse embryos. *Acta Paediatrica Scandinavica*, **52:** 36–40.

Larsson, K.S., Ericson, B. and Böstrom, H. 1964, Salicylate-induced malformations in mouse embryos. *Acta Morphologica Neerlando-Scandinavica*, **6:** 35–44.

Lasagna, L.A. 1965, Drug interaction in the field of analgesic drugs. *Proceedings of the Royal Society of Medicine*, **58:** 978–983.

Laveranstiebar, R.L., Laufer, I. and Levine, M.S. 1994, Greater curvature antral flattening – a radiologic sign of NSAID-related gastropathy. *Abdominal Imaging*, **19:** 295–297.

Lawson, D.H. 1973, Analgesic consumption and impaired renal function. *Journal of Chronic Diseases*, **26:** 39–45.

Leader, M.A. and Neuwirth, E. 1978. Clinical research and the noninstitutional elderly: a model for subject recruitment. *Journal of the American Geriatric Society*, **26:** 27–31.

Leatherwood, P.D. and Plummer, D.T. 1969, The excretion of lactate dehydrogenase in human urine after the ingestion of aspirin. *Biochemical Journal*, **114:** 197–202.

Lee, K.H. and Spencer, M.R. 1969, Studies on mechanism of action of salicylates v. effect of salicylic acid on enzymes involved in mucopolysaccherides synthesis. *Journal of Pharmaceutical Sciences*, **58:** 464–468.

Le Kerneau, J., Gompel, H., Sorkine, M., Gilton, A., Gonzalez-Dunia, J. and Valiente, R. 2000, A comparative study of the gastroprotective action of three doses of dosmalfate vs. placebo in the prevention of acute aspirin-induced gastric lesions. *Drugs of Today*, **36:** 67–72.

Lee, M., Cryer, B. and Feldman, M. 1994, Dose effects of aspirin on gastric prostaglandins and stomach mucosal injury. *Annals of Internal Medicine*, **120:** 184–189.

Lee, M., Barnes, J., Devi, G., Yang, Y., Henderson, G. and Schenker, S. 1995, Chronic ethanol exposure renders the gastric-mucosa more susceptible to acute aspirin-induced injury in rats. *Gastroenterology*, **108:** A146.

Lee, M., Hardman, W.E. and Cameron, I. 1998, Age-related changes in gastric mucosal repair and proliferative activities in rats exposed acutely to aspirin. *Gerontology*, **44:** 198–203.

Leeling, J., Bare, J. and Medon, P. 1978, The effect of acetylsalicylic acid/salicylic acid mixtures on acetylcholinesterase activity in vitro. *Arzneimittel Forschung*, **28:** 1727–1728.

Leeling, J.L., Johnson, N. and Helms, R.J. 1979, Influence of L-glutamine on aspirin-induced gastrointestinal microbleeding in dogs. *Journal of Pharmacy and Pharmacology*, **31:** 63.

Lehmann, H., Hirsch, U., Bauer, M., Greischel, A., Schmid, J. and Schneider, P. 1996, Studies on the chronic oral toxicity of an analgesic drug combination consisting of acetylsalicylic acid, paracetamol and caffeine in rats including an electronmicroscopic evaluation of kidneys. *Arzneimittel Forschung*, **46:** 895–905.

Leng, E. 1978, Monthly variations in drug-induced gastric ulcers in rats. *International Journal of Biometerology*, **22:** 190–196.

Leonards, J.R. 1962, Aspirin and blood loss from the gastrointestinal tract. *Federation Proceedings*, **21:** 452.

Leonards, J.R. 1969, Absence of gastrointestinal bleeding following administration of salicylsalicylic acid. *Journal of Laboratory and Clinical Medicine*, **74:** 911–914.

Leonards, J.R. and Levy, G. 1967, The role of dosage in aspirin-induced gastrointestinal bleeding. *Clinical Pharmacology and Therapeutics*, **8:** 400–408.

Leonards, J.R. and Levy, G. 1969, Biopharmaceutical aspects of aspirin-induced gastrointestinal blood loss in man. *Journal of Pharmaceutical Sciences*, **58:** 1277–1279.

Leonards, J.R., Levy, G. and Niemczura, R. 1973, Gastrointestinal blood loss during prolonged aspirin administration. *New England Journal of Medicine*, **289:** 1020–1022.

Lester, D., Lolli, G. and Greenberg, L.A. 1946, The fate of acetylsalicylic acid. *Journal of Pharmacology*, **87:** 329–342.

Leung, W.K., To, K.F., Chan, F.K.L., Lee, T.L., Chung, S.C.S. and Sung, J.J.Y. 2000, Interaction of *Helicobacter pylori* eradication and non-steroidal anti-inflammatory drugs on gastric epithelial apoptosis and proliferation: implications on ulcerogenesis. *Alimentary Pharmacology and Therapeutics*, **14:** 879–885.

Lev, R., Siegel, H.I. and Glass, G.B. 1972, Effects of salicylates on the canine stomach: a morphological and histochemical study. *Gastroenterology*, **62:** 970–980.

Levenstein, S. 2000, The very model of a modern etiology: a biopsychosocial view of peptic ulcer. *Psychosomatic Medicine*, **62:** 176–185.

Levi, G. and Angelino, N.J. 1968, Hydrolysis of aspirin by rat small intestine. *Journal of Pharmaceutical Sciences*, **57:** 1449–1450.

Levin, D.L. 1980, Effects of inhibition of prostaglandin synthesis on fetal development, oxygenation, and the fetal circulation. *Seminars in Perinatology*, **4:** 35–44.

Levine, R.A., Kohen, K.R., Schwartzer, E.H. and Ramsey, C.E. 1982, Prostaglandin E2-histamine in interactions on cAMP, cGMP and acid production in isolated pendic glands. *American Journal of Physiology*, **242:** G21–G26.

Levy, B.D., Bertram, S., Tai, H.H., Israel, E., Fischer, A., Drazen, J.M. and Serhan, C.N. 1993, Agonist-induced lipoxin A4 generation: detection by a novel lipoxin A4-ELISA. *Lipids*, **28:** 1047–1053.

Levy, D.J. 2000, An aspirin tablet and a gastric ulcer. *New England Journal of Medicine*, **343:** 863.

Levy, G. and Angelino, N.J. 1968, Hydrolysis of aspirin by rat small intestine. *Journal of Pharmaceutical Sciences*, **57:** 1449.

Levy, G. and Leonards, J.R. 1966, Absorption, metabolism, and excretion of salicylates. *In:* M.J.H. Smith and P.K. Smith (eds), *The Salicylates: A Critical Bibliographic Review*, pp. 5–48. New York: Wiley-Interscience.

Lewis, A.J., Smith, W.E. and Brown, D.H. 1981, A comparison of the anti-inflammatory effects of copper complexes in different modes of inflammation. *In:* K.D. Rainsford, K. Brune and M.W. Whitehouse (eds), *Trace Elements in the Pathogenesis and Treatment of Inflammation*, pp. 327–338. Basel: Birkhäuser.

Lewis, R.B. and Schulman, J.D. 1973, Influence of acetylsalicylic acid, an inhibitor of prostaglandin synthesis, on the duration of human gestation and labour. *Lancet*, **2:** 1159–1161.

Lichtenberger, L.M., Richards, J.E. and Hills, B.A. 1985, Effect of 16,16-dimethyl prostaglandin-E2 on the surface hydrophobicity of aspirin-treated canine gastric mucosa. *Gastroenterology*, **88:** 308–314.

Lichtenberger, L.M. Ulloa, C., Vanous, A.L., Romero, J.J., Dial, E.J., Illich., P.A. and Walters, E.T. 1996, Zwitterionic phospholipids enhance aspirin's therapeutic activity, as demonstrated in rodent model systems. *Journal of Pharmacology and Experimental Therapeutics*, **277:** 1221–1227.

Lichtenberger, L.M., Wang, Z.-M., Romero, J.J., Ulloa, C., Perez, J.C., Giraud, M.-N. and Barreto, J.C. 1995, Non-steroidal anti-inflammatory drugs (NSAIDs) associate with Zwitterionic phospholipids – insight into the mechanism and reversal of NSAID-induced gastrointestinal injury. *Nature Medicine*, **1:** 154–158.

Ligumsky, M., Grossman, M.I. and Kauffman, G.L. 1982, Endogenous gastric mucosal prostaglandins: their role in mucosal integrity. *American Journal of Physiology*, **242:** G337–G341.

Ligumsky, M., Golanska, E.M., Hansen, D.G. and Kauffman, G.L. 1983, Aspirin can inhibit gastric mucosal cyclo-oxygenase without causing lesions in rat. *Gastroenterology*, **84:** 756–761.

Limas, C., Limas, C.J. and Gesell, M.S. 1976, Effects of indomethacin on renal interstitial cells. *Laboratory Investigation*, **34:** 522–528.

Lin, T.-M. and Warrick, M.W. 1974, Action of acetylsalicylic acid and glucagon on acid secretion mucosal blood flow, bleeding and net ion efflux in the fundic pouch of pentagastrin-simulated dogs. *Archives Internationales de Pharmacodynamie et de Thérapie*, **210:** 279–287.

Lin, T.-M., Warrick, M.W., Evans, D.C. and Nash, J.F. 1975, Action of the anti-inflammatory agents, acetylsalicylic acid, indomethacin and feroprofen on the gastric mucosa of dogs. *Research Communications in Chemical Pathology and Pharmacology*, **11:** 1–14.

Lindquist, B. 1971, Effekten av acetylsalicylat på ventrikelslemhinnan. En histokemisk studie av ventrikelslemmet hos råtta. *Nordisk Medicin*, **21:** 74–77.

Lipkin, M. 1971, In 'defence' of the gastric mucosa. *Gut*, **12:** 599–603.

Lish, P.M., Dungan, K.W. and Robbins, S.R. 1956, Gastrointestinal pharmacology of antipyretic-analgesic agents. I. Effects on acid secretion and ulcer formation in the Shay rat. *Archives Internationales de Pharmacodynamie et de Thérapie*, **119:** 389–397.

Lo, A., Addison, R.S., Hooper, W.D. and Dickinson, R.G. 2001, Disposition of naproxen, naproxen acyl glucuronide and its rearrangement isomers in the isolated perfused rat liver. *Xeniobiotica*, **31:** 309–319.

Lo, G.Y. and Konishi, F. 1978, Synergistic effect of vitamin C and aspirin on gastric lesions in the rat. *American Journal of Clinical Nutrition*, **31:** 1397–1399.

Lockard, O.O. Jr, Ivey, K.J., Butt, J.H., Silvoso, G.R., Sisk, C. and Holt, S. 1980, The prevalence of duodenal lesions in patients with rheumatic diseases on chronic aspirin therapy. *Gastrointestinal Endoscopy*, **26:** 5–7.

Lockey, S.D. 1959, Allergic reactions due to FD & C yellow No. 5 tartrazine, an aniline dye used as a coloring and identifying agent in various steroids. *Annals of Allergy*, **17:** 719–721.

Lockley, R.F., Rucknagel, D.L. and Vansselow, N.A. 1973, Familial occurrence of asthma, nasal polyps and aspirin in tolerance. *Annals of Internal Medicine*, **78:** 56–63.

Loebl, D.H., Craig, R.M., Culic, D.D., Ridolfo, A.S., Falk, J. and Schmid, F.R. 1977, Gastrointestinal blood loss. Effect of aspirin, fenoprofen and acetaminophen in rheumatoid arthritis as determined by sequential gastroscopy and radioactive fecal markers. *Journal of the American Medical Association*, **237:** 976–981.

Logsdon, G.D. and Machen, T.E. 1981, Involvement of extracellular calcium in gastric stimulation. *American Journal of Physiology*, **241:** G365–G375.

Low, J., Grabow, D., Sommers, C., Wallace, J., Lesch, M., Finkel, M., Schrier, D., Metz, A. and Conroy, M.C. 1995, Cytoprotective effects of C10959 in the rat gastric mucosa: modulation of leukocyte adhesion. *Gastroenterology*, **109:** 1224–1233.

Lubawy, W.C. and Burriss Garrett, R.J. 1977, Effects of aspirin and acetaminophen on fetal and placental growth in rats. *Journal of the Pharmaceutical Sciences*, **66:** 111–113.

Lucente, G. 1971, Aspirin and the otolaryngologist. *Archives of Otolaryngology*, **94:** 443–444.

Lue, A.J.-C. and Brownell, W.E. 1999, Salicylate induced changes in outer hair cell lateral wall stiffness. *Hearing Research*, **135:** 163–168.

Lukas, J.C., Rosenkrantz, T.S., Raye, J.R., Porte, P.J. and Philipps, A.F. 1987, Intrauterine growth retardation after long-term maternal salicylate administration in the rabbit. *American Journal of Obstetrics and Gynecology*, **156:** 245–249.

Lussier, A., Arsenault, A. and Varaday, J. 1978, Gastrointestinal microbleeding after aspirin and naproxen. *Clinical Pharmacology and Therapeutics*, **23:** 402–407.

MacDonald, A., Dekanski, J.B., Gottfried, S., Parke, D.G. and Sacra, P. 1977, Effects of blood glucose levels on aspirin-induced gastric mucosal damage. *American Journal of Digestive Diseases*, **22:** 909–914.

MacDonald, J.I., Wallace, S.M., Mahachai, V. and Verbeek, R.K. 1992, Both phenolic and acyl glucuronidation pathways of diflunisal are impaired in liver cirrhosis. *European Journal of Clinical Pharmacology*, **42:** 471–474.

MacDonald, T.M., Morant, S.V., Robinson, G.C., Shield, M.J., McGilchrist, M.M., Murray, F.E. and McDevitt, D.G. 1997, Association of upper gastrointestinal toxicity of non-steroidal anti-inflammatory drugs with continued exposure: cohort study. *British Medical Journal*, **315:** 1333–1337.

MacDonald, W.C. 1973, Correlation of mucosal history and aspirin intake in chronic gastric ulcer. *Gastroenterology*, **65:** 381–389.

MacIntyre, D.E., Pearson, J.D. and Gordon, J.L. 1978, Localisation and stimulation of prostacyclin production in vascular cells. *Nature*, **271:** 549–551.

MacKercher, P.A., Ivey, K.J., Baskin, W.N. and Krause, W.J. 1977, Protective effect of cimetidine on aspirin-induced gastric mucosal damage. *Annals of Internal Medicine*, **87:** 676–679.

Macklin, A.W. and Szot, R.J. 1980, Eighteen-month oral study of aspirin, phenacetin and caffeine, in CS7B1/6 mice. *Drug and Chemical Toxicology*, **3:** 135–163.

Macklon, A.F., Craft, A.W., Thompson, M. and Kerr, D.N.S. 1974, Aspirin and analgesic nephropathy. *British Medical Journal*, **1:** 597–600.

Maclagan, T.J. 1879, The treatment of acute rheumatism by salicin and salicylic acid. *Lancet*, **1:** 875–877.

Madappally, M.M., Mackerer, C.R. and Mehlman, M.A. 1972, The inhibitory effects of acetylsalicylic acid feeding on gluconeogenic enzymes in rat liver and kidney. *Life Sciences*, **11:** 77–85.

Main, I.H.M. and Pearce, J.B. 1978, Effects of calcium on acid secretion from the rat isolated gastric mucosa during stimulation with histamine, pentagastrin, methacholine and dibutyrylcyclic adenosine-3′,5′-monophospate. *British Journal of Pharmacology*, **64:** 359–368.

Main, I.H.M. and Whittle, B.J.R. 1974, Regulation by prostaglandins of gastric secretion. *In:* H.J. Robinson and J.R. Vane (eds), *Prostaglandin Synthesis Inhibitors*, pp. 363–372. New York: Raven Press.

Majerus, P.W. and Stanford, N. 1977, Comparative effects of aspirin and diflunisal on prostaglandin synthetase from human platelets and sheep seminal vesicles. *British Journal of Clinical Pharmacology*, **4:** 15S–18S.

Malcontenti-Wilson, C., Schulz, S., Andrews, F.J. and O'Brien, P.E. 1998, Aged gastric mucosa: mechanisms of vulnerability. *Journal of Gastroenterology and Hepatology*, **13:** S204–S208.

Mandel, K.G., Bertram, T.A., Eichold, M.K., Pepple, S.C. and Doyle, M.J. 1994, Fatty acid mediated gastroprotection does not correlate with prostaglandin elevation in rats exposed to various chemical insults. *Veterinary Pathology*, **31:** 679–688.

Mangla, J.C., Kim, Y.M. and Rubulis, A.A. 1974a, Adenyl cyclase stimulation by aspirin in rat gastric mucosa. *Nature*, **250:** 61–62.

Mangla, J.C., Kim, Y.M. and Turner, M.D. 1974b, Are pepsinogens activated in gastric mucosa after aspirin-induced injury? *Experientia (Basel)*, **30:** 727–729.

Manso, C., Taranta, A. and Nydick, I. 1956, Effect of aspirin administration on serum oxaloacetic and glutamic pyruvic transaminases in children. *Proceedings of the Society for Experimental Biology and Medicine*, **93:** 84–88.

Marcolongo, R., Bayeli, P.F. and Montaguani, M. 1979, Gastrointestinal involvement in rheumatoid arthritis: a biopsy study. *Journal of Rheumatology*, **6:** 163–173.

Marquez, M. and Roberts, D.J. 1972, Changes in the potency of aspirin in the presence of protein-bound dye. *Journal of Pharmacy and Pharmacology*, **24:** 658–660.

Martin, B.K. 1963, Accumulation of drug anions in gastric mucosal cells. *Nature*, **198:** 896–897.

Martin, B.K. 1971, The formulation of aspirin. *Advances in Pharmaceutical Sciences*, **3:** 107–171.

Massie, B.M. and Teerlink, J.R. 2000, Interaction between aspirin and angiotensin-converting enzyme inhibitors: real or imagined. *American Journal of Medicine*, **109:** 431–433.

Mathew, P.P., Smyth, R.D., Witmer, C.M. and Carr, G.S. 1978, Effects of a nonsteroidal anti-inflammatory agent and phenobarbital on hepatic microsomal mono-oxygenases in adjuvant disease in the rat. *Biochemical Pharmacology*, **27:** 1771–1774.

Matsukura, H., Masuda, M., Kawaguchi, K., Uchida, A. and Kamishiro, T. 1994, Cytoprotective effect of NC-1300-0-3 against gastric lesions induced by necrotizing agents in rats. *Japanese Journal of Pharmacology*, **65:** 9–18.

Matsumoto, K.K. and Grossman, M.I. 1959, Quantitative measurement of gastrointestinal blood loss during ingestion of aspirin. *Proceedings of the Society for Experimental Biology and Medicine*, **102:** 517–519.

Mauer, I., Weinstein, D. and Solomon, H.M. 1970, Acetylsalicylic acid: no chromosome damage in human leukocytes. *Science*, **169:** 198–201.

Max, M. and Menguy, R. 1969, Influence of aspirin and phenylbutazone on rate of turnover of gastric mucosal cells. *Digestion*, **2:** 67–72.

Max, M. and Menguy, R. 1970, Influence of adrenocorticotropin, cortisone, aspirin, and phenylbutazone on the rate of exfoliation and the rate of renewal of gastric mucosal cells. *Gastroenterology*, **58:** 329–336.

McArthur, J.N. and Smith, M.J.H. 1969, The determination of the binding of salicylate to serum proteins. *Journal of Pharmacy and Pharmacology*, **21:** 589–594.

McCabe, P. and Dey, F. 1965, The effect of aspirin upon auditory sensitivity. *Annals of Otology, Rhinology and Laryngology*, **74:** 312–325.

McCarthy, C.J., Sweeney, E. and Omorain, C. 1995, Early ultrastructural changes of antral mucosa with aspirin in the absence of *Helicobacter pylori. Journal of Clinical Pathology*, **48:** 994–997.

McCloy, R.F., Greenberg, G.R. and Baron, J.H. 1984, Duodenal pH in health and duodenal ulcer disease: effect of a meal, Coca-Cola, smoking, and cimetidine. *Gut*, **25:** 386–392.

McCormack, K. and Brune, K. 1987, Classical absorption theory and the development of gastric mucosal damage associated with the non-steroidal anti-inflammatory drugs. *Archives of Toxicology*, **60:** 261–269.

McCreedie, M., Stewart, J.H. and Day, N.E. 1993, Different roles for phenacetin and paracetamol in cancer of the kidney and renal pelvis. *International Journal of Cancer*, **53:** 245–249.

McDougall, P., Markham, A., Cameron, I. and Sweetman, A.J. 1983, The mechanism of inhibition of mitochondrial oxidative phosphorylation by the non-steroidal anti-inflammatory agent diflunisal. *Biochemical Pharmacology*, **32:** 2595–2598.

McGiff, J.C. and Wong, P.Y. 1979, Compartmentalisation of prostaglandins and prostacyclin within the kidney: implications for renal function. *Federation Proceedings*, **38:** 89–93.

McGreevy, J.M. and Moody, F.G. 1977, Protection of gastric mucosa against-induced erosions by enhanced blood flow. *Surgical Forum*, **28:** 357–359.

McGreevy, J.M. and Moody, F.G. 1980, A mechanism for prostaglandin cytoprotection. *British Journal of Surgery*, **67:** 873–876.

McGreevy, J.K. and Moody, F.G. 1981, Focal microcirculatory changes during the production of aspirin-induced gastric mucosal erosions. *Surgery*, **89:** 337–341.

McGuigan, M.A. 1986, Death due to salicylate poisoning in Ontario. *Canadian Medical Association Journal*, **135:** 891–894.

McIver, M.A. and Hobbs, J.B. 1975, The failure of high doses of aspirin to produce renal lesions in pigs. *Medical Journal of Australia*, **1:** 197–199.

McLaughlin, J.K., Lipworth, L., Chow, W.-H. and Blot, W.J. 1998, Analgesic use and chronic renal failure: a critical review of the epidemiologic literature. *Kidney International*, **54:** 679–686.

McLeod, L.J., Roberts, M.S., Cossum, P.A. and Vial, J.H. 1986, The effects of different doses of some acetylsalicylic acid formulations on platelet function and bleeding times in healthy subjects. *Scandinavian Journal of Haematology*, **36:** 379–384.

McNeil, J.R. 1973, The possible teratogenic effect of salicylates on the developing fetus. *Clinical Paediatrics*, **12:** 347–350.

McPherson, D.L. and Miller, J.M. 1974, Choline salicylate. Effects on cochlear function. *Archives of Otolaryngology*, **99:** 302–308.

McQueen, E.G. 1977, Salicylate toxicology. *In:* K.D. Rainsford, K. Brune and M.W. Whitehouse (eds), *Aspirin and Related Drugs: Their Actions and Uses*, pp. 97–108. Basel: Birkhäuser.

Michaletz, P.A., Cap, L., Alpert, E. and Lauterburg, B.H. 1989, Assessment of mitochondrial function in vivo with a breath test utilizing alpha-ketoisacaproic acid. *Hepatology*, **10:** 829–832.

Mehlman, M.A., Tobin, R.B. and Sporn, E.M. 1972, Oxidative phosphorylation and respiration by rat liver mitochondria from aspirin-treated rats. *Biochemical Pharmacology*, **21:** 3279–3285.

Melling, T.R., Aabakken, L., Røseth, A. and Osnes, M. 1996, Faecal calprotectin shedding after short-term treatment with non-steroidal anti-inflammatory drugs. *Scandinavian Journal of Gastroenterology*, **31:** 339–344.

Menassé, R. 1979, Evaluation of ulcerogenic effects. *Pharmacology and Therapeutics*, **5:** 191–197.

Menassé-Gydnia, R. and Krupp, P. 1974, Quantitative measurement of gastrointestinal bleeding in rats: the effect of non-steroidal anti-inflammatory drugs. *Toxicology and Applied Pharmacology and Experimental Therapeutics*, **184:** 389–396.

Menguy, R. 1972, Acute gastric mucosal bleeding. *Annual Review of Medicine*, **23:** 297–312.

Menguy, R., Desbaillets, L. and Masters, Y.F. 1973, Influence of aspirin and ethanol on energy metabolism in the gastric mucosa. *Gastroenterology*, **64:** 772.

Menguy, R., Desbaillets, L., Okabe, S. and Masters, Y.F. 1972, Abnormal aspirin metabolites in patients with cirrhosis and its possible relationship to bleeding in cirrhotics. *Annals of Surgery*, **172:** 412–417.

Menguy, R. and Masters, Y.F. 1965, Effects of aspirin on gastric mucous secretion. *Surgery, Gynaecology and Obstetrics*, **120:** 92–98.

Menniti-Ippolito, F., Maggini, M., Raschetti, R., Da Ca, R., Traversa, G. and Walker, A.M. 1998, Ketorolac use in outpatients and gastrointestinal hospitalization: a comparison with other non-steroidal anti-inflammatory drugs in Italy. *European Journal of Clinical Pharmacology*, **54:** 393–397.

Meredith, T.M., Vale, J.A. and Goulding, R. 1981, The epidermiology of acute acetaminophen poisoning in England and Wales. *Archives of Internal Medicine*, **141:** 397–400.

Meyer, O.O. and Howard, B. 1943, Production of hypoprothrombinemia and hypocoagulability of the blood with salicylates. *Proceedings of the Society for Experimental Biology and Medicine*, **53:** 234–237.

Meyrat, P., Baumgartner, A., Kappeler, M. and Halter, F. 1984, Simple gastroscopy technique in the rat. *Digestive Diseases and Sciences*, **29:** 327–329.

Mielants, H., Veys, E.M., Verbruggen, G. and Shelstraete, K. 1979, Salicylate-induced gastrointestinal bleeding: comparison between soluble buffered, enteric-coated and intravenous administration. *Journal of Rheumatology*, **6:** 210–218.

Mielens, E. and Rosenberg, F.J. 1976, Dual effects of aspirin in guinea pig lungs. *British Journal of Pharmacology*, **57:** 495–500.

Mihas, A.A., Gibson, R.G. and Hirschowitz, B.I. 1976, Inhibition of gastric secretion in the dog by 16,15-dimethyl prostaglandin E2. *American Journal of Physiology*, **230:** 351–356.

Millan, M.S., Morris, G.P., Beck, I.T. and Henson, J.T. 1980, Villous damage induced by suction biopsy and by acute ethanol intake in normal human small intestine. *Digestive Diseases and Sciences*, **25:** 513–525.

Miller, P. 1981, Reye's syndrome. *Infection Control*, **2:** 150–152.

Miller, R.R. and Jick, H. 1977, Acute toxicity of aspirin in hospitalised medical patients. *American Journal of Medical Sciences*, **274:** 271–279.

Miller, T.A. and Tepperman, B.L. 1979, Effect of prostaglandin E2 on aspirin-induced gastric mucosal injury. *Journal of Surgical Research*, **26:** 10–17.

Millhorn, D.E., Eldridge, F.L and Waldrop, T.G. 1982, Effects of salicylate and 2-4 dinitrophenol on respiration and metabolism. *Journal of Applied Physiology*, **53:** 925–929.

Mills, D.G., Philp, R.B. and Hirst, M. 1974, The effects of some salicylate analogues on human platelets. 1. Structure activity relationships and the inhibition of platelet aggregation. *Life Sciences*, **14:** 659–672.

Minden, P. and Farr, R.S. 1967, Human antibodies against acetylsalicylic acid altered human serum albumin. *Arthritis and Rheumatism*, **10:** 399.

Miners, J.O. and Birkett, D.J. 1998, Cytochrome P4502C9: an enzyme of major importance in human drug metabolism. *British Journal of Clinical Pharmacology*, **45:** 525–538.

Mingatto, F.E., Santos, A.C., Uyemura, S.A., Jordani, M.C. and Curti, C. 1996, *In vitro* interaction of nonsteroidal anti-inflammatory drugs on oxidative phosphorylation of rat kidney mitochondria: respiration and ATP synthesis. *Archives of Biochemistry and Biophysics*, **334:** 303–308.

Mishikawa, Y. and Nakamura, K. 1992, Analgesic and Anti-inflammatory Medicine. US Patent No. 5,128,334, dated 7 July 1992.

Mitobe, Y., Hiraishi, H., Sasai, T., Shimada, T. and Terano, A. 2000, The effects of aspirin in antioxidant defences of cultured rat gastric mucosal cells. *Alimentary Pharmacology and Therapeutics*, **14:** 10–17.

Mitra, R. and Pal, S.P. 1977, Inhibition of mast cell production by L-glutamine in aspirin-induced ulceration in rat stomach. *Indian Journal of Physiology and Pharmacology*, **21:** 374–378.

Mitznegg, P., Estler, C.-J., Loew, F.W. and van Seil. J. 1977, Effect of salicylates on cyclic AMP in isolated rat gastric mucosa. *Gastroenterologica*, **24:** 372–376.

Moel, D.I., Safirstein, R.L., McEvoy, R.C. and Hsueh, W. 1987, Effect of aspirin on experimental diabetic nephropathy. *Journal of Laboratory and Clinical Medicine*, **110:** 300–307.

Mofenson, H.C., Carracio, T.R., Greensher, J., D'Agostino R. and Rossi, A. 1985, Gastrointestinal dialysis with activated charcoal and cathartic in the treatment of adolescent intoxications. *Clinical Pediatrics*, **24:** 678–684.

Mohandas, J., Duggin, G.G., Horvath, J.S. and Tiller, D.J. 1981, Metabolic oxidation of acetaminophen (paracetamol) mediated by cytochrome P-450 mixed-function oxidase and prostaglandin endoperoxide sytherease in the rabbit kidney. *Toxicology and Applied Pharmacology*, **61:** 252–259.

Moloney, J.R. 1977, Nasal polyps, nasal polypectomy, asthma, and aspirin sensitivity. Their association in 445 cases of nasal polyps. *Journal of Laryngology and Otology*, **91:** 837–846.

Moncada, S., Palmer, R.M.J. and Higgs, E.A. 1991, Nitric oxide; physiology, pathophysiology and pharmacology. *Pharmacological Reviews*, **43:** 109–142.

Moncada, S. and Vane, J.R. 1981, Prostacyclin and blood coagulation. *Drugs*, **21:** 430–437.

Mongan, E., Kelly, P., Nies, K., Porter, W.W. and Paulus, H.E. 1973, Tinnitus as an indication of therapeutic serum salicylate levels. *Journal of the American Medical Association*, **226:** 142–145.

Montini, G., Sacchetto, E., Murer, L., Dall'Amico, R., Masiero, M., Passerini-Glazel, G. and Zacchello, G. 2000, Renal glomerular response to the inhibition of prostaglandin E2 synthesis and protein loading after the relief of unilateral ureteropelvic junction obstruction. *Journal of Urology*, **163:** 556–560.

Moody, F., McGreevy, J., Zalewsky, C., Cheung, L. and Simons, M. 1977, The cytoprotective effect of mucosal blood flow in experimental erosive gastritis *Upsala Journal of Medical Sciences*, **82:** 264.

Moore, J.G. and Goo, R.H. 1987, Day and night aspirin-induced gastric mucosal damage and protection by ranitidine in man. *Chronobiology International*, **4:** 111–116.

Morgan, A. and Clark, D. 1998, CNS adverse effects of nonsteroidal anti-inflammatory drugs. Therapeutic implications. *CNS Drugs*, **9:** 281–290.

Morgan, A.M. and Truitt, E.B. 1965, Evaluation of acetylsalicylic acid esterase in aspirin metabolism. *Journal of Pharmaceutical Sciences*, **54:** 1640–1646.

Morgan, R.J., Nelson, L.M., Russell, R.I. and Docherty, C. 1983, The protective effect of deglycyrrhinized liquorice against aspirin and aspirin plus bile acid-induced gastric mucosal damage, and its influence on aspirin absorption in rats. *Journal of Pharmacy and Pharmacology*, **35:** 605–607.

Moriga, M., Aono, M., Murakami, M. and Uchino, H. 1980, The role of gastric mucosal hexosomine in aspirin-induced ulcers. *Gastroenterologia Japonica*, **15:** 1–13.

Morikawa, M., Inoue, M., Tokumaru, S. and Kogo, H. 1995, Enhancing and inhibitory effects of nitric oxide on superoxide anion generation in human polymorphonuclear leukocytes. *British Journal of Pharmacology*, **115:** 1302–1306.

Morimoto, Y., Oshima, S., Hara, H. and Sukamoto, T. 1994a, Effects of KB-5492, a new anti-ulcer agent, on ethanol- and acidified aspirin-induced gastric mucosal damage *in vivo* and *in vitro*. *Japanese Journal of Pharmacology*, **64**: 41–47.

Morimoto, Y., Shimohara, K., Oshima, S., Hara, H. and Sukamoto, T. 1994b, Effects of KB-5492, a new anti-ulcer agent with a selective affinity for the sigma-receptor, on aspirin-induced disruption of the rat gastric mucosal barrier. *Japanese Journal of Pharmacology*, **64**: 49–55.

Morise, Z. and Grisham, M.B. 1998, Molecular mechanisms involved in NSAID-induced gastropathy. *Journal of Clinical Gastroenterology*, **27**: S87–S90.

Morise, Z., Komatsu, S., Fuseler, J.W., Granger, D.N., Perry, M., Issekutz, A.C. and Grisham, M.B. 1998, ICAM-1 and P-selectin expression in a model of NSAID-induced gastropathy. *American Journal of Physiology*, **274**: G246–G252.

Morlans, M., Laporte, J.R., Vidal, X., Cabeza, D. and Stolley, P.D. 1990, End-stage renal disease and non-narcotic analgesics: a case-control study. *British Journal of Clinical Pharmacology*, **30**: 717–723.

Morris, C.H., Christian, J.E., Miya, T.S. and Hansen, W.G. 1972, Effect of aspirin and ethanol on the gastric mucosa of the rat. *Journal of Pharmaceutical Sciences*, **61**: 815.

Morris, E., Noujan, A.A., Miya, T.S. and Christain, J.E.E. 1967, Relationship between lesion production, absorption, and distribution of ^{14}C-labelled acetylsalicylic acid. *Journal of Pharmaceutical Sciences*, **56**: 896–899.

Morris, G.P., Wallace, J.L. and Harding, P.L. 1984, Effects of prostaglandin E_2 on salicylate-induced damage to the rat gastric mucosa: cytoprotection is not associated with preservation of the gastric mucosal barrier. *Canadian Journal of Physiology and Pharmacology*, **62**: 1065–1069.

Morton, D.M. and Chatfield, D.H. 1970, The effects of adjuvant-induced arthritis on the liner metabolism of drugs in rats. *Biochemical Pharmacology*, **19**: 473–481.

Moses, F., Kikendall, J.W., Bowen, P. and Young, T.R. 1988, Chronic beta-carotene supplementation does not alter the acute endoscopic response of gastric mucosa to aspirin. *American Journal of Gastroenterology*, **83**: 1037.

Mourad, F.H., Khuri, M., Shouaib, F. and Nassar, C.F. 2000, Protective effect of the nitric oxide donor molsidomine on indomethacin and aspirin-induced gastric injury in rats. *European Journal of Gastroenterology and Hepatology*, **12**: 81–84.

Mózsik, Gy., Abdel-Salem, O.M.E. and Szolcsányi, J. 1997, *Capsaicin-sensitive Afferent Nerves in Gastric Mucosal Damage and Protection*. Budapest: Akadémiai Kiadó.

Mózsik, Gy., Jávor, T., Kitajima, M., Pfeiffer, C.J., Rainsford, K.D., Simon, L. and Szabo, S. (eds) 1987, *Advances in Gastrointestinal Cytoprotection: Topics 1987*. Budapest: Akademiai Kiado.

Mugridge, K.G., Perretti, M., Wallace, J.L. and Parente, L. 1992, Possible factors involved in the protective effects of interleukin-1 in aspirin- and indomethacin-induced gastric damage. *In:* K.D. Rainsford and G.P. Velo (eds), *Side-effects of Anti-inflammatory Drugs 3*, pp. 135–144. Dordrecht: Kluwer.

Muir, A. and Cossar, T.A. 1955, Aspirin and ulcer. *British Medical Journal*, **2**: 7–12.

Müller, P., Dammann, H.-G. and Simon, B. 1985, Schutzwirkung von zwei Antazida bei acuter Acetylsalicyl-säure-Schädigung der menschlichen Magenschleimhaut. *Arzneimittel Forschung*, **35**: 1862–1864.

Müller, P., Fuchs, W. and Simon, B. 1997, Studies on the protective effect of lansoprazole on human gastric mucosa against low-dose acetylsalicylic acid – an endoscopic controlled double-blind study. *Arzneimittel-Forschung*, **47**: 758–760.

Murakami, M., Oketani, K., Fujisaki, H. and Wakabayashi, T. 1982a, Effect of geranylgeranylacetone (GGA) on gastric lesions induced by topical aspirin plus HCl. *Japanese Journal of Pharmacology*, **32**: 921–924.

Murakami, M., Oketani, K., Fujisaki, H., Wakabayashi, T., Ohgo, T. and Okabe, S. 1982b, Effects of the antiulcer drug geranylgeranylacetone on aspirin-induced gastric ulcers in rats. *Japanese Journal of Pharmacology*, **32**: 299–306.

Muramatsu, M., Murakami, S., Arai, I., Isobe, Y., Hirose, H., Usuki, C. and Aihara, H. 1986, Effect of sofalcone on acute gastric mucosal lesions induced by aspirin and ethanol in reference to the biosynthesis of gastric mucosal glycoprotein. *Research Communications in Chemical Pathology and Pharmacology*, **54**: 321–337.

Murmann, W., Carminatti, G.M. and Cattaneo, R. 1974a, Pharmacology of zolimidine (2-(p-methyl-sulfonylphenyl)-imidazo (1,2-a) pyridine) a new non-anticholinergic gastroprotective agent. II). Protective activity against drug-induced experimental ulcers. *Panminerva Medica*, **16:** 321–334.

Murmann, W., Carminati, G.M. and Cattaneo, R. 1974b, Pharmacology of zolimidine (2-(p-methylsul-fonylphenyl)-imidazo (1,2-a) pyridine) a new non-anticholinergic gastro protective agent. IV. Enhancement of gastric mucus production in different experimental models. *Panminerva Medicine*, **16:** 347–359.

Murphy, G.J., Taha, R., Windmill, D.C., Metcalfe, M. and Nicholson, M.L. 2001, Influence of aspirin on early allograft thrombosis and chronic allograft nephropathy following renal transplantation. *British Journal of Surgery*, **88:** 261–266.

Murphy, P.J., Badia, P., Myers, B.L., Boecker, M.R. and Wright, K.P. 1994, Nonsteroidal anti-inflammatory drugs affect normal sleep patterns in humans. *Physiology and Behaviour*, **55:** 1063–1066.

Murray, H.S., Storottman, M.P. and Cooke, A.R. 1974, Effect of several drugs on gastric potential difference in man. *British Medical Journal*, **1:** 19–21.

Murray, R.M. 1973a, Dependence on analgesic nephropathy (to alcohol and other drugs). *British Journal of Addiction*, **68:** 265–272.

Murray, R.M. 1973b, The origins of analgesic nephropathy. *British Journal of Psychiatry*, **123:** 99–106.

Murray, R.M., Timbury, G.C. and Linton, A.L. 1970, Analgesic abuse in psychiatric patients. *Lancet*, **1:** 1303–1305.

Murray, T. and Goldberg, M. 1975, Analgesic abuse and renal disease. *Annual Review of Medicine*, **26:** 537–550.

Murray, T.G., Stolley, P.D., Anthony, J.C., Schinnar, R., Hepler-Smith, E. and Jeffreys, J.L. 1983, Epidemiologic study of regular analgesic use and end-stage renal disease. *Archives of Internal Medicine*, **143(9):** 1687–1693.

Muther, R.S., Potter, D.M. and Bennett, W.M. 1981, Aspirin-induced depression of glomerular filtration rate in normal humans: role of sodium balance. *Annals of Internal Medicine*, **94:** 317–321.

Myers, A.B.R 1876, Salicin in acute rheumatism. *Lancet*, **2:** 676–677.

Myers, E. and Bernstein, J. 1965, Salicylate ototoxicity. A clinical and experimental study. *Archives of Otolaryngology*, **82:** 483–493.

Nair, J. and Stacy, M. 1993, Aseptic meningitis associated with salicylate overdose. *Psychosomatics*, **34:** 372.

Najbauer, J., Schuman, E.M. and Mamelak, A.N. 2000, The aspirin metabolite sodium salicylate causes focal cerebral hemorrhage and cell death in rats with kainic acid-induced seizures. *Neuroscience*, **99:** 107–117.

Nakagawa, M., Ishihara, N., Shimokawa, T. and Kojima, S. 1987, Effect of clofibrate on lipid peroxidation in rats treated with aspirin and 4-pentenoic acid. *Journal of Biochemistry*, **101:** 81–88.

Nakano, M., Ueno, M., Hasegawa, H., Watanabe, T., Kuroda, T., Ito, S. and Arakawa, M. 1998, Renal haemodynamic characteristics in patients with lupus nephritis. *Annals of the Rheumatic Diseases*, **57:** 226–230.

Naldi, L., Conforti, A., Venegoni, M., Troncon, M.G., Caputi, A., Ghiotti, E., Cocci, A., Moretti, U., Velo, G.P. and Leone, R. 1999, Cutaneous reactions to drugs. An analysis of spontaneous reports in four Italian regions. *British Journal of Clinical Pharmacology*, **48:** 839–846.

Nandi, J., Crockett, J. and Levine, R.A. 1994, A possible role of protein kinase C in augmenting H+ secretion by nonsteroidal anti-inflammatory drugs. *Journal of Pharmacology and Experimental Therapeutics*, **269:** 932–940.

Nanra, R.S., Daniel, V. and Howard, M. 1980, Analgesic nephropathy induced by common propriety mixtures. *Medical Journal of Australia*, **1:** 486–487.

Nanra, R.S. and Kincaid-Smith, P. 1972, Chronic effect of analgesics on the kidney. *Progress in Biochemical Pharmacology*, **7:** 285–323.

Nasser, S.M.S., Christie, P.E., Pfister, R., Sousa, A.R., Walls, A., Schmitz-Schumann, M. and Lee, T.H. 1996, Inflammatory cell populations in bronchial biopsies from aspirin-sensitive asthmatic subjects. *American Journal of Respiratory Critical Care Medicine*, **153:** 90–96.

Nasser, S.M.S. and Lee, T.H. 1998, Leukotrienes in aspirin – sensitive asthma. *In:* A. Szezeklik, R. Gregglewski and J. Vane (eds), *Eicosanoids, Aspirin and Asthma*, pp. 317–335. New York: Marcel Dekker.

Needham, C.D., Kyle, J., Jones, P.F., Johnson, S.J. and Kerridge, D.F. 1971, Aspirin and alcohol in gastrointestinal haemorrhage. *Gut*, **12:** 819–821.

Nelson, M.M. and Forfar, J.O. 1971, Associations between drugs administered during pregnancy and congenital abnormalities of the fetus. *British Medical Journal*, **1:** 523–527.

Nelson, S.D., Forte, A.J., Vaishnav, Y., Mitchell, J.R., Gillette, J.R. and Hinson, J.A. 1981, The formation of arylating and alkylating metabolites of phenacetin in hamsters and hamster liver microsomes. *Molecular Pharmacology*, **19:** 140–145.

Nevi, R. 1971, The binding of radioactive label from labelled phenacetin and related compounds to rat tissues in vivo and to nucleic acids and bovine plasma albumin in vitro. *Biochemical Journal*, **122:** 311–315.

New Zealand Rheumatism Association Study 1974, Aspirin and the kidney. *British Medical Journal*, **1:** 593–596.

Ng, T.M., Fock, K.M., Khor, J.L., Teo, E.K., Sim, C.S., Tan, A.L. and Machins, D. 2000, Non-steroidal anti-inflammatory drugs. *Helicobacter pylori* and bleeding gastric ulcer. *Alimentary Pharmacology and Therapeutics*, **14:** 203–209.

Nicholls-Grzemski, F.A., Calder, I.C., Priestly, B.G. and Burcham, P.C. 2000, Clofibrate-induced *in vitro* hepatoprotection against acetaminophen is not due to altered glutathione homeostasis. *Toxicological Sciences*, **56:** 220–228.

Niebl, J.R., Blake, D.A., Burnett, L.S. and King, T.M. 1976, The influence of aspirin on the course of induced midtrimester abortion. *American Journal of Obstetrics and Gynaecology*, **124:** 607–610.

Nielsen, G.L., Sorensen, H.T., Larsen, H. and Pedersen, L. 2001, Risk of adverse birth outcome and miscarriage in pregnancy users of non-steroidal anti-inflammatory drugs: population based observational study and case-control study. *British Medical Journal*, **322:** 266–270.

Nijkamp, F.P. and Ramakers, A.G.M. 1980, Prevention of anaphylatic bronchoconstriction by a lipoxygenase inhibitor. *European Journal of Pharmacology*, **62:** 121–122.

Nikolaides, N., König, A., Ballke, E.H., Griefahn, B. and Jährig, K. 1985, The influence of acetylsalicylic acid on the transepithelial potential difference of gastric mucosa in children. *Klinische Wochenschrift*, **63:** 184–187.

Nompleggi, D., Meyers, L., Ramwell, P., Castell, D. and Dubois, A. 1980, PGE_2 involvement in the regulation of gastric emptying. *In:* B. Samuelsson, P. Ramwell and R. Paoletti (eds), *Advances in Prostaglandin and Thromoxane Research*, Vol. 8, pp. 1587–1588. New York: Raven Press.

Nordenfelt, O. 1972, Deaths from renal failure in abusers of phenacetin-containing drugs. *Acta Medica Scandinavica*, **191:** 11–16.

Northover, B.J. 1967, The effect of anti-inflammatory drugs on vascular smooth muscle. *British Journal of Pharmacology*, **31:** 483–493.

Nuernberg, B., Szelenyi, I., Schneider, T. and Brune, K. 1990, Diflunisal disposition: role of gastric absorption in the development of mucosal damage and anti-inflammatory potency in rodents. *Drug Metabolism and Disposition*, **18:** 937–942.

O'Brien, P. and Silen, W. 1973, Effect of bile salts and aspirin on the gastric mucosal blood flow. *Gastroenterology*, **64:** 246–253. (reply: *Gastroenterology*, **64:** 514, 1973).

O'Brien, W.M. and Bagby, G.F. 1985, Rare adverse reactions to nonsteroidal antiinflammatory drugs. *Journal of Rheumatology*, **12:** 13–20.

Oesch, F. 1981, Fate of epoxides. *Advances in Experimental Medicine and Biology*, **136:** 39–52.

O'Gorman, T. and Koff, R.S. 1977, Salicylate hepatitis. *Gastroenterology*, **72:** 726–728.

Ohe, K., Hayashi, K., Shirakawa, T., Yamada, K., Kawasaki, T. and Miyoshi, A. 1980b, Aspirin and taurocholate-induced metabolic damage in mammalian gastric mucosa in vitro. *American Journal of Physiology*, **239:** G457–G462.

Ohe, K., Ueno, N., Yokoya, H. and Miyoshi, A. 1979, The activation of pepsinogen inside the gastric mucosa caused by the hydrogen ion back-diffusion during the course of aspirin ulceration. *Hiroshima Journal of Medical Science*, **28:** 1–11.

Ohe, K., Yokoya, H., Kitaura, T. and Miyoshi, A. 1980a, Increase in pepsin content in gastric mucosa during the course of aspirin and taurocholate-induced gastric ulceration in rats. *Digestive Diseases and Sciences*, **25:** 849–856.

Ohning, B.V. and Guth, P.H. 1995, Time-course of mucosal cell proliferation following acute aspirin injury in rat stomach. *Digestive Diseases and Sciences*, **40:** 1351–1356.

Ohtomo, Y., Matsubara, T., Nishizawa, K., Unno, A., Motohashi, T. and Yamashiro, Y. 1998, Nephropathy and hypertension as manifestations in a 13-y-old girl with primary antiphospholipid syndrome. *Acta Paediatrica*, **87:** 903–907.

Ojiambio, H.P. 1965, The pattern of duodenal ulceration in Nairobi. *East African Medical Journal*, **42:** 629–633.

Okabe, S., Honda, K., Takeuchi, K. and Takagi, K. 1975, Inhibitory effect of L-glutamine on gastric irritation and back-diffusion of gastric acid in response to aspirin in the rat. *American Journal of Digestive Diseases*, **20:** 626–631.

Okabe, S., Hung, C.R., Takeuchi, K. and Takagi, K. 1976a, Effects of L-glutamine of acetyl/salicylic acid or taurocholic acid-induced gastric lesions and secretory changes in pylorus-ligated rats under normal or stress conditions. *Japanese Journal of Pharmacology*, **26:** 455–460.

Okabe, S., Takata, Y., Takeuchi, K., Naganuma, T. and Takagi, K. 1976b, Effects of carbenoxolone Na on acute and chronic gastric ulcer models in experimental animals. *American Journal of Digestive Diseases*, **21:** 618–625.

Okabe, S., Takeuchi, K., Honda, K. and Takagi, K. 1976c, Effects of various amino acids on gastric lesions induced by acetylsalicylic acid (ASA) and gastric secretion in pylorus-ligated rats. *Arzneimittel Forschung*, **26:** 534–537.

Okabe, S., Takeuchi, K., Nakamura, K. and Takagi, K. 1974a, Pathogenesis of gastric lesions induced by aspirin in the pylorus-ligated rat. *Japanese Journal of Pharmacology*, **24:** 363–371.

Okabe, S., Takeuchi, K., Nakamura, K. and Takagi, K. 1974b, Inhibitory effects of L-glutamine on the aspirin-induced gastric lesions in the rat. *Journal of Pharmacy and Pharmacology*, **26:** 605–611.

Okita, R. 1986, Effect of acetylsalicylic acid on fatty acid omega-hydroxylation in rat liver. *Pediatric Research*, **20:** 1221.

Okita, R. and Chance, C. 1984, Induction of laurate ω-hydroxylase by di(2-ethylhexyl)phthalate in rat liver microsomes. *Biochemical Biophysical Research Communications*, **121:** 304–309.

O'Laughlin, J.C., Silvoso, G.K. and Ivey, K.J. 1982, Resistance to medical therapy of gastric ulcers in rheumatic disease patients taking aspirin. A double-blind study with cimetidine and follow-up. *Digestive Diseases and Sciences*, **27:** 976–980.

Olhagen, B. 1970, The intestine and rheumatism. *Acta Rheumatologica Scandinavica*, **16:** 177–183.

Olson, C.E. 1987, A chronobiologic approach to ethanol and acidified aspirin injury of the gastric-mucosa in the rat. *Chronobiology International*, **4:** 19–29.

O'Neill, G.P. and Ford-Hutchinson, A.W. 1993, Expression of mRNA for cyclooxygenase-1 and cyclooxygenase-2 in human tissues. *FEBS Letters*, **330:** 156–160.

Orozco-Alcala, J.J. and Baum, J. 1979, Regular and enteric-coated aspirin: a re-evaluation. *Arthritis and Rheumatism*, **22:** 1034–1037.

Ortmann, R. and Perkins, J.P. 1977, Stimulation of adenosine 3'5'-monophosphase formation by prostaglandins in human astrocytema cells. Inhibition by non-steroidal anti-inflammatory agents. *Journal of Biological Chemistry*, **252:** 6018–6025.

Orton, T.C. and Parker, G.L. 1982, The effect of hypolipidemic agents on the hepatic microsomal drug-metabolising enzyme system of the rat. Induction of cytochrome P-450 with specificity toward terminal hydroxylation of lauric acid. *Drug Metabolism Disposition*, **10:** 110–115.

Østensen, M. and Østensen, H. 1996, Safety of non-steroidal anti-inflammatory drugs in pregnancy patients with rheumatic disease. *Journal of Rheumatology*, **23:** 1045–1049.

Østensen, M. and Villiger, P.M. 2001, Nonsteroidal anti-inflammatory drugs in systemic lupus erythematosus. *Lupus*, **10:** 135–139.

Otterness, I.G. and Bliven, M.L. 1985, Laboratory methods for testing nonsteroidal anti-inflammatory drugs. *In:* J.G. Lombardino (ed.), *Nonsteroidal Antiinflammatory Drugs*, pp. 113–251. New York: Raven Press.

Overman, D.O. and White, J.A. 1983, Comparative teratogenic effects of methyl salicylate applied orally or topically to hamsters. *Teratology*, **28**: 421–426.

Owen, C.A., Bollman, J.L. and Grindlay, J.H. 1954, Radiochromium-labelled erythrocytes for the detection of gastrointestinal haemorrhage. *Journal of Laboratory and Clinical Medicine*, **4**: 238–245.

Owen, R.A. and Heywood, R. 1980, Renal toxicity of aspirin to rats pre-treated with ethinyl oestradiol. *Toxicology Letters*, **5**: 269–274.

Pace-Asciak, C. 1972, Prostaglandin synthetase activity in the rat stomach fundus. Activation by L-norepinephrine and related compounds. *Biochimica et Biophysica Acta*, **280**: 161–171.

Page, J. and Henry, D. 2000, Consumption of NSAIDs and the development of congestive heart failure in elderly patients: an underrecognized public health problem. *Archives of Internal Medicine*, **160**: 777–784.

Paliard, P.M., Moulinier, B., Eyraud, J., Bruhiere, J. and Vedrienne, J. 1971, Stenosing esophagitis, postbulbar duodenal ulcer and aspirin addiction. *Lyon Medical*, **225**: 995–996.

Pankow, D., Damme, B. and Schror, K. 1994, Acetylsalicylic acid-induction of cytochrome P-450 2E1. *Archives of Toxicology*, **68**: 261–265.

Parazzini, F., Benedetto, C., Frusca, T., Gregorini, G., Bocciolone, L., Marozio, L., Romero, M., Danesino, V., Degaetano, G., Gastaldi, A., Massobrio, M., Remuzzi, G., Tognoni, G., Benedetto, C., Frusca, T., Gregorini, G., Guaschino, S., Romero, M., Bianchi, C., Bocciolone, L., Valcamonico, A. and Giambuzzi, M. 1993, Low-dose aspirin in prevention and treatment of intrauterine growth retardation and pregnancy-induced hypertension. *Lancet*, **341**: 396–400.

Paris, G.Y., Garmaise, D., Cimon, D.G., Swett, L., Carter, G.W. and Young, P. 1980, Glycerides as prodrugs 2. 11,3-Dialkonyl-2-(2 methyl-4-OXO-1,3-benzodioxan-2-y1)glycerides (cyclic aspirin triglycerides as anti-inflammatory agents. *Journal of Medicinal Chemistry*, **23**: 79–82.

Park, C.W., Kim, Y.O., Shin, Y.W., Song, H.C., Jin, D.C., Kim, S.Y., Choi, E.J., Chang, Y.S. and Bang, B.K. 1997, Long-term benefits of aspirin plus dipyridamol and angiotensin-converting enzyme inhibitor therapy in patients with immunoglobulin A nephropathy. *Kidney International*, **51**: 969.

Parmar, N.S. and Hemmings, G. 1983, The effect of 4-methoxy-5,7,3',4'-tetrahydroxyflavan on the restraint of induced gastric ulceration augmented by aspirin, a gastric mucosal barrier breaker. *Research Communications in Chemical Pathology and Pharmacology*, **41**: 337–340.

Parmer, N.S., Tariq, M. and Ageel, A.M. 1985, Effect of nicotine, alcohol and caffeine pre-treatment on the gastric mucosal damage induced by aspirin, phenylbutazone and reserpine in rats. *Japanese Journal of Pharmacology*, **39**: 1–6.

Partin, J.S., Partin, J.C., Schubert, W.K. and Hammond, J.G. 1982, Serum salicylate concentrations in Reye's disease. A study of 130 biopsy-proven cases. *Lancet*, **1**: 191–194.

Partin, J.S., Daugherty, C.C., McAdams, A.J., Partin, J.C. and Schubert, W.K. 1984, A comparison of liver ultrastructure in salicylate intoxication and Reye's syndrome. *Hepatology*, **4**: 687–690.

Patierno, S.R., Lehman, N.L., Henderson, B.E. and Landolph, J.R. 1989, Study of the ability of phenacetin, acetaminophen, and aspirin to induce cytotoxicity, mutation, and morphological transformation in C3H/1OT½ clone 8 mouse embryo cells. *Cancer Research*, **49**: 1038–1044.

Patrick, J.E. and Challis, J.R.G. 1982, The role of prostaglandins and their inhibitors of reproduction. *In:* H.J.M. Burnett, J. Hirsh and J.F. Mustard (eds), *Acetylsalicylic Acid: New Uses for an Old Drug*, pp. 123–136. New York: Raven Press.

Paul, W.D. 1943, The effct of acetylsalicylic acid (aspirin) on the gastric mucosa. *Journal of the Iowa State Medical Society*, **33**: 155–158.

Pauls, F., Wick, A.N. and McKay, E.M. 1948, Inhibition of gastric ulceration in the rat by *O*-hydroxy (salicylic) acid. *Science (Washington DC)*, **107**: 19–20.

Pauranen, J. 1980, Effect of prostaglandin E$_2$, cimetidine, and atropin on ethanol-induced gastric mucosal damage in the rat. *Scandinavian Journal of Gastroenterology*, **15**: 458–488.

Penney, A.G., Andres, F.J. and O'Brien, P.E. 1996, Influence of age on natural and delayed healing of experimentally-induced gastric ulcers in rats. *Digestive Diseases and Sciences*, **41**: 1838–1844.

Perederii, O.F. and Morozova, R.P. 1978, Squalene and sterols of rat stomach tissue and the effect of acetylsalicylic acid on their content. *Ukrayin'skyi Biokhimichnyl Zhurnal*, **50**: 777–782.

Perez De Moura, L.F. and Hayden, R.C. Jr 1968, Salicylate ototoxicity. A human temporal bone report. *Archives of Otolaryngology*, **87:** 368–372.

Perneger, T.V., Whelton, P.K. and Klag, M.J. 1994, Risk of kidney failure associated with the use of acetaminophen, aspirin and nonsteroidal anti-inflammatory drugs. *New England Journal of Medicine*, **331:** 1675–1679.

Peskar, B.M. 1980, Effect of carbenoxolone on prostaglandin synthesizing and metabolizing enzymes and correlation with gastric mucosal carbenoxolone concentrations. *Scandinavian Journal of Gastroenterology*, **65 (Suppl.):** 109–112.

Peskar, B.M. 1985, Effects of sulphasalazine and 5-aminosalicylic acid on the human colonic prostaglandin system. *In:* L. Barbara, M. Miglioli and S.F. Phillips (eds), *New Trends in the Pathophysioogy and Therapy of the Large Bowel*, pp. 185–196. Amsterdam: Elsevier.

Peskar, B.M. and Maricic, N. 1998, Role of prostaglandins in gastroprotection. *Digestive Diseases and Sciences*, **9 (Suppl.):** 23S–29S.

Pessayre, D., Fromenty, B. and Mansoouri, A. 2001, Drug-induced microvesicular steatosis and steatohepatitis. *In:* J.L. Lemasters and A.-L. Nieminen (eds), *Mitochondria in Pathogenesis*, pp. 489–517. New York: Kluwer/Plenum.

Pessayre, D., Wandscheer, J.-C., Cobert, B., Level, R., Degott, C., Batt, A.M., Martin, N. and Benhamou, J.P. 1980, Additive effects of inducers and fasting on acetaminophen hepatoxicity. *Biochemical Pharmacology*, **29:** 2219–2223.

Peterson, F.J., Holloway, D.E., Erickson, R.R., Duquette, P.H., McLain, C.J. and Holtzman, J.L. 1980, Ethanol induction of acetaminophen toxicity and metabolism. *Life Sciences*, **27:** 1705–1711.

Peterson, W.L. 1995, The role of acid in upper gastrointestinal haemorrhage due to ulcer and stress-related mucosal damage. *Alimentary Pharmacology and Therapeutics*, **9:** 43–46.

Petillo, J.J., Gulbenkian, A. and Tabachnick, I.I.A. 1969, Effects in vivo and in vitro of nonsteroidal anti-inflammatory drugs on (rat stomach) histidine decarboxylase. *Biochemical Pharmacology*, **18:** 1784–1788.

Petrescu, I. and Tarba, C. 1997, Uncoupling effects of diclofenac and aspirin in the perfused liver and isolated hepatic mitochondria of rat. *Biochimica et Biophysica Acta*, **1318:** 385–394.

Petroski, D. 1993, Endoscopic comparison of 3 aspirin preparations and placebo. *Clinical Therapeutics*, **15:** 314–320.

Pfeiffer, C.J. 1981, Experimental analysis of hydrogen ion diffusion in gastrointestinal mucus glycoprotein. *American Journal of Physiology*, **240:** G176–G182.

Pfeiffer, C.J., Harding, R.K. and Morris, G.P. 1982, Ultrastructural aspects of salicylate-induced damage to the gastric mucosa. *In:* C.J. Pfeiffer (ed.), *Drugs and Peptic Ulcer*, Vol. II, pp. 109–126. Boca Raton, FL: CRC Press.

Pfeiffer, C.J. and Lewandowski, L.G. 1971, Comparison of gastric toxicity of acetylsalicylic acid with rate of administration in the rat. *Archives Internationales de Pharmacodynamie et de Thérapie*, **190:** 5–13.

Pfeiffer, C.J. and Lewandowski, L.G. 1972, Wirkung von Prostaglandin und Aspirin auf die Magensekretion des Laberfrettehaus. *Leber Magen Darm*, **2:** 142–144.

Pfeiffer, C.J. and Weibel, J. 1973, The gastric mucosal response to acetylsalicylic acid in the ferret. An ultrastructure study. *American Journal of Digestive Diseases*, **18:** 834–846.

Phillips, B.M. 1973, Aspirin-induced gastrointestinal microbleeding in dogs. *Toxicology and Applied Pharmacology*, **24:** 182–189.

Phillips, B.M. and Palermo, B.T. 1977, Physical form as a determinant of effect of buffered acetylsalicylate formulations of GI microbleeding. *Journal of Pharmaceutical Sciences*, **66:** 124–126.

Phole, T., Brzozowski, T., Becker, J.C., Van Der Voort, I.R., Markmann, A., Konturek, S.J., Moniczewski, A. and Domschke, W. 2001, Role of reactive oxygen metabolites in aspirin-induced gastric damage in humans: gastroprotection by vitamin C. *Alimentary Pharmacology and Therapeutics*, **15:** 677–687.

Picardo, C. Fernandez-Morata, J.C., Juan, M., Roca-Ferrer, J., Fuentes, M., Xaubet, A. and Mullol, J. 1999, Cyclooxygenase-2 mRNA is down expressed in nasal polyps from aspirin-sensitive asthmatics. *American Journal of Respiratory Critical Care Medicine*, **160:** 291–296.

■

CHAPTER 8

■

Pichl, J., Mitznegg, P., Subramanian, N., Sprügel, W., Domschke, W. and Estler, C.-J. 1978, Lysine acetylisalicylate inhibits the protein synthesis in the rat gastric mucosa in vivo and in vitro. *Acta Hepato-Gastroenterologica*, **25**: 52–54.

Pierson, R.N., Holt, P.R., Watson, R.M. and Keating, R.P. 1961, Aspirin and gastrointestinal bleeding. Chromate[51] blood loss studies. *American Journal of Medicine*, **31**: 259–265.

Pigoso, A.A., Uyemura, S.A., Santos, A.C., Todrigues, T., Mingatto, F.E. and Curti, C. 1998, Influence of nonsteroidal anti-inflammatory drugs on calcium efflux in isolated rat renal cortex mitochondria and aspects of the mechanisms involved. *International Journal of Biochemistry and Cell Biology*, **30**: 961–965.

Piper, D.W. Gellaty, R. and McIntosh, J. 1982, Analgesic drugs and peptic ulcer: Human studies. *In:* C.J. Pfeiffer (ed.), *Drugs and Peptic Ulcer Disease*, pp. 75–93. Boca Raton, FL: CRC Press.

Piper, D.W., Grieg, M., Landecker, K.D., Shinners, J., Waller, S. and Canalese, J. 1977, Analgesic intake and chronic gastric ulcer. Acute upper gastrointestinal haemorrhage, personality traits and social class. *Proceedings of the Royal Society of Medicine*, **70 (Suppl. 7)**: 11–15.

Plotz, P.H. and Kimberly, R.P. 1981, Acute effects of aspirin and acetaminophen on renal function. *Archives of Internal Medicine*, **141**: 343–348.

Plummer, D.T., Leatherwood, P.D. and Blake, M.E. 1975, Urinary enzymes and kidney damage by aspirin and phenacetin. *Chemico-Biological Interactions*, **10**: 277–284.

Pocwiardowska, E. 1976a, The effects of antipyretics on metabolism processes in rat liver mitochondria. Part I. The action of sodium salicylate, and pyrazolones on the reaction of respiratory chain. *Polish Journal of Pharmacology and Pharmacy*, **28**: 217–226.

Pocwiardowska, E. 1976b, The effects of antipyretics on metabolic processes in rat liver mitochondria. Part II. The action of sodium salicylate, and pyrazolones on oxidation of alpha-keto-glutarate. *Polish Journal of Pharmacology and Pharmacy*, **28**: 227–231.

Pohle, T., Herbst, H., Konturek, J.W. and Domschke, W. 2001, Functional and structural studies on gastric mucosa and ulcer healing. *Microscopy Research and Technique*, **53**: 323–324.

Pommer, W., Bronder, E., Greiser, E., Helmert, U., Jesdinsky, H.J., Klimpel, A., Borner, K. and Molzahn, M. 1989, Regular analgesic intake and the risk of end-stage renal failure. *American Journal of Nephrology*, **9**: 403–412.

Poon, P.Y.W. and Whitehouse, M.W. 1978, Pathobiodynamics: changes in ascorbate metabolism in rats with peripheral inflammation. *Biochemical Medicine*, **20**: 81–86.

Porro, G.B., Caruso, I., Ciabattoni, G., Patrignani, P., Patrono, C. and Pugliese, F. 1980, Effects of acetylsalicylic acid, phenacetin, paracetamol, and caffeine on renal tubular epithelium. *British Journal of Pharmacology*, **72**: 3771–3778.

Powanda, M.C. 1977, Changes in body balances of nitrogen and other key nutrients: description and underlying mechanisms. *American Journal of Clinical Nutrition*, **30**: 1254–1268.

Poyser, N.L. 1988, Parturition and the termination of pregnancy. *In:* P.B. Curtis-Prior (ed.), *Prostaglandins: Biology and Chemistry*, pp. 463–475. London and Edinburgh: Churchill Livingstone.

Prescott, L.F. 1965, Effects of acetylsalicylic acid, phenacetin, paracetamol, and caffeine on renal tubular epithelium. *Lancet*, **2**: 91–96.

Prescott, L.F. 1979, The nephrotoxicity and hepatotoxicity of antipyretic analgesics. *British Journal of Clinical Pharmacology*, **7**: 453–462.

Prescott, L.F. 1980, Hepatotoxicity of mild analgesics. *British Journal of Clinical Pharmacology*, **10**: 373S–379S.

Prescott, L.F. 1982, Analgesic nephropathy: a reassessment of the role of phenacetin and other analgesics. *Drugs*, **23**: 75–149.

Prescott, L.F. 1984, Renal damage in man from ingestion of antiinflammatory and analgesic drugs. *In:* K.D. Rainsford and G.P. Velo (eds), *Side-effects of Anti-inflammatory/Analgesic Drugs*, pp. 109–118. New York: Raven Press.

Prescott, L.F. 1996, Paracetamol (Acetaminophen). *A Critical Bibliographic Review*. London: Taylor & Francis.

Prescott, L.F., Balali-Mood, M., Critchley, J.A.J.H., Johnstone, A.F. and Proudfoot, A.T. 1982, Diuresis or urinary alkalinisation for salicylate poisoning. *British Medical Journal*, **285**: 1383–1386.

Prichard, P.J., Daneshmend, T.K., Millns, P.J., Edwards, T.J., Bhaskar, N.K. and Hawkey, C.J. 1988b, Use of endoscopy and blood measurement to show dose dependent protection of human gastric mucosa by famotidine against aspirin. *Gut*, **29:** A729.

Prigge, W.F. and Gebhard, R.L. 1997, Oral acetylsalicylic acid induces biliary cholesterol secretion in the rat. *Lipids* **32:** 753–758.

Prino, G., Lietti, A. and Allegra, G. 1972, Antipeptic activity of a sulphate glycopeptide different behaviour with purified human pepsin or human gastric juice. *American Journal of Digestive Diseases*, **17:** 863–867.

Pritchard, P., Brown, G., Bhaskar, N. and Hawkey, C. 1988, The effect of dietary fatty acids on the gastric production of prostaglandins and aspirin-induced injury. *Alimentary Pharmacology and Therapeutics*, **2:** 179–184.

Proctor, R.A. and Kunin, C.M. 1978, Salicylate-induced ensymuria: comparison with other anti-inflammatory agents. *American Journal of Medicine*, **65:** 987–993.

Pugh, P.M. and Rutishauser, S.C.B. 1978, Comparative effects of 2,4-dinitrophenol and sodium salicylate on bile section in the dog, cat, rabbit and guinea-pig. *General Pharmacology*, **9:** 119–121.

Puscas, I., Coltau, M., Puscas, I.I., Maghiar, A., Herea, I., Hecht, A. and Puscas, C. 2000, Proton pump inhibitors protect gastric mucosa from aspirin-induced injuries in parallel with carbonic anhydrase inhibition. *Gastroenterology*, **118:** 5920.

Puscas, I., Hajdu, A., Buzas, G. and Bernath, Z. 1989, Prevention of non-steroidal anti-inflammatory agents induced acute gastric mucosal lesions by carbonic anhydrase inhibitors. An endoscopic study. *Acta Physiologica Hungarica*, **73:** 279–283.

Puurunen, J. 1980, Effect of prostaglandin E_2, cimetidine, and atropine on ethanol-induced gastric mucosal damage in the rat. *Scandinavian Journal of Gastroenterology*, **15:** 485–488.

Pytkowicz Steissguth, A., Treder, R.P., Barr, H.M., Shepard, T.H., Archie Bleyer, W., Sampson, P.D. and Martin, D. 1987, Aspirin and acetaminophen use by pregnant women and subsequent child IQ and attention decrements. *Teratology*, **35:** 211–219.

Quan, N., Mhlanga, J.D.M., Whiteside, M.B., Kristensson, K. and Herkenham, M. 2000, Chronic sodium salicylate treatment exacerbates brain neurodegeneration in rats infected with *Trypanosoma brucei*. *Neuroscience*, **96:** 181–194.

Quick, A.J. and Clesceri, L. 1960, Influence of acetylsalicylic acid and salicylamide on the coagulation of blood. *Journal of Pharmacology*, **128:** 500–504.

Rabassa, A.A., Goodgame, R., Sutton, F.M., Ou, C.N., Rognerud, C. and Graham, D.Y. 1996, Effects of aspirin and *Helicobacter pylori* on the gastroduodenal mucosal permeability to sucrose. *Gut*, **39:** 159–163.

Rachelefsky, G.S., Coulson, A., Siegel, S.C. and Stiehm, E.R. 1975, Aspirin intolerance in chronic childhood asthma: detected by oral challenge. *Paediatrics*, **56:** 443–448.

Rafoth, R.J. and Silvis, S.E. 1976, Gastric ulceration associated with aspirin ingestion in an achlorhydric patient: a case report. *American Journal of Digestive Diseases*, **21:** 279–280.

Rahman, A., Segasothy, M., Samad, S.A., Zulfiqar, A. and Rani, M. 1993, Analgesic use and chronic renal disease in patients with headache. *Headache*, **33:** 442–445.

Rai, R., Backos, M., Baxter, N., Chilcott, I. and Regan, L. 2000, Recurrent miscarriage – an aspirin a day? *Human Reproduction*, **15:** 2220–2223.

Rainsford, K.D. 1970, *Effects of Salicylates on Gastric Tissues*, PhD Thesis, University of London.

Rainsford, K.D. 1975a, Aspirin. Actions and uses. *Australian Journal of Pharmacy*, **56:** 373–382.

Rainsford, K.D. 1975b, A synergistic interaction between aspirin, or other non-steroidal anti-inflammatory drugs, and stress which produces severe gastric mucosal damage in rats and pigs. *Agents and Actions*, **5:** 553–558.

Rainsford, K.D. 1975c, The biochemical pathology of aspirin-induced gastric damage. *Agents and Actions*, **5:** 326–344.

Rainsford, K.D. 1975d, Electronmicroscopic observations on the effects of orally administered aspirin and aspirin bicarbonate mixtures on the development of gastric mucosal damage in the rat. *Gut*, **16:** 514–527, 1010.

Rainsford, K.D. 1977a, Aspirin and gastric ulceration. Light and electronmicroscopic observations in a model of aspirin plus stress ulceration. *British Journal of Experimental Pathology*, **58:** 215–219.

Rainsford, K.D. 1977b, The comparative gastric ulcerogenic activities of non-steroidal anti-inflammatory drugs. *Agents and Actions*, **7:** 573–577.

Rainsford, K.D. 1978a, Is the radiochromium-red cell technique a valid method of measuring gastrointestinal blood loss or damage following aspirin administration. *Drugs under Experimental and Clinical Research*, **4:** 183–189.

Rainsford, K.D. 1978b, The role of aspirin in gastric ulceration. Some factors involved in the development of gastric mucosal damage induced by aspirin in rats exposed to various stress conditions. *American Journal of the Digestive Diseases*, **23:** 521–530.

Rainsford, K.D. 1978c, Gastric mucosal ulceration induced in pigs by tablets but not suspensions or solutions of aspirin. *Journal of Pharmacy and Pharmacology*, **30:** 129–131.

Rainsford, K.D. 1978d, Effects of anti-inflammatory drugs on mucus production: relationship to ulcerogenesis. *In:* C.J. Pfeiffer (ed.), *Drugs and Peptic Ulcer Disease*, Vol. 2, pp. 227–236. Boca Raton, FL: CRC Press.

Rainsford, K.D. 1978e, Structure–activity relationships of non-steroid anti-inflammatory drugs. I. Gastric ulcerogenic activity. *Agents and Actions*, **8:** 587–605.

Rainsford, K.D. 1978f, The effects of aspirin and other non-steroid anti-inflammatory/analgesic drugs on gastro-intestinal mucus glycoprotein biosynthesis in vivo: relationship to ulcerogenic actions. *Biochemical Pharmacology*, **27:** 877–885.

Rainsford, K.D. 1979, Prostaglandins and the development of gastric mucosal damage by anti-inflammatory drugs. *In:* K.D. Rainsford and A.W. Ford-Hutchinson (eds), *Prostaglandins and Inflammation*, pp. 193–210. Basel: Birkhäuser.

Rainsford, K.D. 1980, Aspirin, prostaglandins and mucopolysaccharide/glycoprotein secretion. *Agents and Actions*, **19:** 520–521.

Rainsford, K.D. 1981a, Comparison of the gastric ulcerogenic activity of new non-steroidal anti-inflammatory drugs in stressed rats. *British Journal of Pharmacology*, **73:** 226P–227P.

Rainsford, K.D. 1981b, Reactions of the gastric mucosa to orally administered copper and related metal complexes. *In:* K.D. Rainsford, K. Brune and M.W. Whitehouse (eds), *Trace Elements in the Pathogenesis and Treatment of Inflammation*, pp. 369–387. Basel: Birkhäuser.

Rainsford, K.D. 1982a, An analysis of the gastro-intestinal side-effects of non-steroidal anti-inflammatory drugs, with particular reference to comparative studies in man and laboratory species. *Rheumatology International*, **2:** 1–10.

Rainsford, K.D. 1982b, Effects of anti-inflammatory drugs on mucus production: relationship to ulcerogenesis. *In:* C.J. Pfeiffer (ed.), *Drugs and Peptic Ulcer Disease*, pp. 227–236. Boca Raton, FL: CRC Press.

Rainsford, K.D. 1982c, Development and therapeutic actions of oral copper complexes of anti-inflammatory drugs. *In:* J.R.J. Sorenson (ed.), *Inflammatory Diseases and Copper*, pp. 375–390. Clifton, NJ: Human Press.

Rainsford, K.D. 1982d, A profile of the gastric ulcerogenic activity of benoxaprofen compared with other non-steroidal anti-inflammatory drugs in rats, stressed or given alcohol, and in pigs. *European Journal of Rheumatology and Inflammation*, **5:** 148–164.

Rainsford, K.D. 1983, Microvascular injury during gastric mucosal damage by anti-inflammatory drugs in pigs and rats. *Agents and Actions*, **13:** 457–460.

Rainsford, K.D. 1984a, *Aspirin and the Salicylates*. London: Butterworths.

Rainsford, K.D. 1984b, Mechanisms of gastrointestinal ulceration by non-steroidal anti-inflammatory/analgesic drugs. *In:* K.D. Rainsford and G.P. Velo (eds), *Side-Effects of Anti-inflammatory/Analgesic Drugs, Advances in Inflammation Research*, Vol. 6, pp. 51–64. New York: Raven Press.

Rainsford, K.D. 1986, Structural damage and changes in eicosanoid metabolites in the gastric mucosa of rats and pigs induced by anti-inflammatory drugs of varying ulcerogenicity. *International Journal of Tissue Reactions*, **VIII:** 1–14.

Rainsford, K.D. 1987, Effects of lipoxygenase inhibitors and leukotriene antagonists on the development of gastric mucosal lesions induced by non-steroidal anti-inflammatory drugs in cholinomimetic treated mice. *Agents and Actions*, **21:** 316–319.

Rainsford, K.D. 1988, Interplay between anti-inflammatory drugs and eicosanoids in gastrointestinal damage. *In:* K. Hillier (ed.), *Eicosanoids and the Gastrointestinal Tract*, pp. 111–128. Lancaster: MTP Press.

Rainsford, K.D. 1989a, Concepts of the mode of action and toxicity of anti-inflammatory Drugs. A basis for safer and more selective therapy and for future drug development. *In:* K.D. Rainsford and G.-P. Velo (eds), *New Developments in Antirheumatic Therapy*, pp. 37–92. Lancaster: Kluwer Academic Publishers.

Rainsford, K.D. 1989b, Gastrointestinal side-effects. *In:* J.Y. Chang and A.J. Lewis (eds), *Pharmacological Methods in the Control of Inflammation*, pp. 343–362. New York: A.R. Liss.

Rainsford, K.D. 1992, Protective effects of a slow-release zinc complex, zinc monoglycerolate [Glyzinc™] on the gastro-intestinal mucosa of rodents. *Experimental and Clinical Gastroenterology*, **1:** 349–360.

Rainsford, K.D. (ed.) 1997a, *Side Effects of Anti-inflammatory Drugs IV*. Dordrecht: Kluwer Academic Publishers.

Rainsford, K.D. 1997b, Gastrointestinal adaptation, regulation of eicosanoids, and mucosal protection from NSAIDs. *In:* K.D. Rainsford (ed.), *Side Effects of Anti-inflammatory Drugs 4*, pp. 197–205. Dordrecht: Kluwer Academic Publishers.

Rainsford, K.D. 1998, An analysis from clinico-epidemiological data of the principal adverse events from the COX-2 selective NSAID, nimesulide, with particular reference to hepatic injury. *Inflammopharmacology*, 6, 203–221.

Rainsford, K.D. 1999a, Relationship of nimesulide safety to its pharmacokinetics: assessment of adverse reactions. *Rheumatology*, **38 (Suppl. 1):** 4–10.

Rainsford, K.D. 1999b, Inhibition by leukotriene inhibitors, calcium and platelet activating factor antagonists of acute gastric and intestinal damage in arthritic rats and cholinomimetic treated mice. *Journal of Pharmacy and Pharmacology*, **51:** 331–339.

Rainsford, K.D. 2001, The ever-emerging anti-inflammatories. Have there been any real advances? *Journal of Physiology*, **95:** 11–19.

Rainsford, K.D., Adesioye, J. and Dawson, S. 2001, Relative safety of NSAIDs and analgesics for non-prescription use or in equivalent doses. *Inflammopharmacology*, **8:** 351–359.

Rainsford, K.D. and Bjarnason, I. 2003, In press.

Rainsford, K.D. and Brune, K. 1976, Role of the parietal cell in gastric damage induced by aspirin and related drugs: implications for safer therapy. *Medical Journal of Australia*, **1:** 881–883.

Rainsford, K.D. and Brune, K. 1978, Selective cytotoxic actions of aspirin on parietal cells: a principal factor in the early stages of aspirin-induced gastric damage. *Archive für Toxikologie*, **40:** 143–150.

Rainsford, K.D. and Buchanan, W.W. 1990, Aspirin versus the non-acetylated salicylates. *Baillieres Clinical Rheumatology*, **4:** 247–268.

Rainsford, K.D. and Challis, D.R. 1984, Effects of aspirin and related drugs on the kidneys of rats exposed to various stress states, and in pigs. *In:* K.D. Rainsford and G.P. Velo (eds), *Side Effects of Anti-inflammatory/Analgesic Drugs*, pp. 159–164. New York: Raven Press.

Rainsford, K.D., Ford, N.L.G., Brooks, P.M. and Watson, H.M. 1980a, Plasma aspirin esterases in normal individual patients with alcoholic liver disease and rheumatoid arthritis: characterisation and importance of the enzymic components. *European Journal of Clinical Investigation*, **10:** 413–420.

Rainsford, K.D., Fox, S.A. and Osborne, D.J. 1984, Comparative effects of some non-steroidal anti-inflammatory drugs on the ultrastructural integrity and prostaglandin levels in the rat gastric mucosa: relationship to drug uptake. *Scandinavian Journal of Gastroenterology*, **19 (Suppl. 101):** 55–68.

Rainsford, K.D., Fox, S.A. and Osborne, D.J. 1985, Relationship between drug absorption, inhibition of cyclooxygenase and lipoxygenase pathways, and development of gastric mucosal damage by non-steroidal anti-inflammatory drugs in rats and pigs. *In:* M.J. Bailey (ed.), *Advances in Prostaglandins, Leukotrienes and Lipoxins*, pp. 639–653. New York: Plenum Press.

Rainsford, K.D., Perkins, W.E. and Stetsko, P.I. 1995b, Chronic effects of misoprostol in combination with the NSAID, diclofenac, on gastrointestinal tract of pigs. Relation to diarrheagenic activity, leukocyte infiltration, and mucosal leukotrienes. *Digestive Diseases and Sciences*, **40:** 1435–1444.

CHAPTER 8

Rainsford, K.D. and Peskar, B.M. 1979, Relationship between gastric ulceration and mucosal prostaglandin concentrations following chronic oral administration of aspirin preparations to pigs. *In:* K. Brune and M. Bagglioni (eds), *Arachidonic Acid Metabolism in Inflammation and Thrombosis*, pp. 293–297. Basel: Birkhäuser.

Rainsford, K.D., Peskar, B.M. and Brune, K. 1981a, Relationship between inhibition of prostaglandin production and gastric mucosal damage induced by anti-inflammatory drugs may depend on type of drugs and species. *Journal of Pharmacy and Pharmacology*, **33:** 127–128.

Rainsford, K.D. and Powanda, M.C. (eds) 1998, *Safety and Efficacy of Non-Prescription (OTC) Analgesics and NSAIDs*. Dordrecht: Kluwer Academic Publishers.

Rainsford, K.D. and Quadir, M. 1995, Gastrointestinal damage and bleeding from non-steroidal anti-inflammatory drugs. I. Clinical and epidemiological aspects. *Inflammopharmacology*, **3:** 169–190.

Rainsford, K.D., Schweitzer, A. and Brune, K. 1981b, Autoradiographic and biochemical observations on the distribution of non-steroid anti-inflammatory drugs. *Archives Internationales de Pharmacodynamie et de Thérapie*, **250:** 180–193.

Rainsford, K.D., Schweitzer, A. and Brune, K. 1983, Distribution of the acetyl compared with the salicyl moiety of acetylsalicylic acid. Acetylation of macromolecules in organs wherein side-effects are manifest. *Biochemical Pharmacology*, **32:** 1301–1308.

Rainsford, K.D., Schweitzer, A., Green, P., Whitehouse, M.W. and Brune, K. 1980b, Bio-distribution in rats of some salicylates with low gastric ulcerogenicity. *Agents and Actions*, **10:** 457–464.

Rainsford, K.D. and Smith, M.J.H. 1968, Inhibition of protein biosynthesis by salicylate in gastric-mucosal scrapings of pig and man in vitro. *Biochemical Journal*, **111:** 37.

Rainsford, K.D., Tsang, S., Hunt, R.H. and Al-Jehani, N. 1995a, Effects of non-steroidal anti-inflammatory drugs on prostaglandin H synthase isoform 2 (cyclo-oxygenase 2) production by porcine gastric mucosa in organ culture. *Inflammopharmacology*, **3:** 299–310.

Rainsford, K. and Velo, G.P. (eds) 1984, Side-Effects of Antiinflammatory/Analgesic Drugs. *Advances in Inflammation Research*, Vol. 6. New York: Raven Press.

Rainsford, K.D. and Velo, G.P (eds) 1987, S*ide Effects of Anti-inflammatory Drugs. Part I: Clinical and Epidemiological Aspects*. Lancaster: MTP Press Ltd.

Rainsford, K.D. and Velo, G.P (eds) 1992, *Side Effects of Anti-Inflammatory Drugs 3*. Dordrecht: Kluwer Academic Publishers.

Rainsford, K.D., Watkins, J. and Smith, M.J.H. 1968, Aspirin and mucus. *Journal of Pharmacy and Pharmacology*, **20:** 941–943.

Rainsford, K.D. and Whitehouse, M.W. 1976a, Gastric mucus effusion elicited by oral copper compounds: Potential anti-ulcer activity. *Experentia (Basel)*, **32:** 1172–1173.

Rainsford, K.D. and Whitehouse, M.W. 1976b, Concerning the merits of copper aspirin as a potential anti-inflammatory drug. *Journal of Pharmacy and Pharmacology*, **28:** 83–86.

Rainsford, K.D. and Whitehouse, M.W. 1976c, Gastric irritancy of aspirin and its congeners: anti-inflammatory activity without this side-effect. *Journal of Pharmacy and Pharmacology*, **28:** 599–601.

Rainsford, K.D. and Whitehouse, M.W. 1977, Non-steroid anti-inflammatory drugs: combined assay for anti-edemic potency and gastric ulcerogenesis in the same animal. *Life Sciences*, **21:** 371–378.

Rainsford, K.D. and Whitehouse, M.W. 1978, Australian Patent 511852 (to Australian National University, Canberra, Australia.) *Chemical Abstracts*, **95:** 49410.

Rainsford, K.D. and Whitehouse, M.W. 1980a, Biochemical gastroprotection from acute ulceration induced by aspirin and related drugs. *Biochemical Pharmacology*, **29:** 1281–1289.

Rainsford, K.D. and Whitehouse, M.W. 1980b, Anti-inflammatory/anti-pyretic salicylic acid esters with low gastric ulcerogenic activity. *Agents and Actions*, **10:** 451–456.

Rainsford, K.D. and Whitehouse, M.W. 1980c, Are all aspirins really alike? A comparison of gastric ulcerogenicity with bio-efficacy in rats. *Pharmacological Research and Communications*, **12:** 85–95.

Rainsford, K.D. and Whitehouse, M.W. 1980d, Gastroprotective and anti-inflammatory properties of green-lipped mussel (*Perna canaliculus*) preparation. *Arzneimittel Forschung*, **30:** 2128–2132.

Rainsford, K.D. and Whitehouse, M.W. 1980e, Protection of the gastric mucosal lining from damage by aspirin and related drugs. European patent Application 79300053.0 Pending (to Australian National University, Canberra, Australia).

Rainsford, K.D. and Whitehouse, M.W. 1992, Anti-ulcer activity of a slow release zinc complex, zinc monoglycerolate [Glyzinc™]. *Journal of Pharmacy and Pharmacology*, **44:** 476–482.

Rainsford, K.D. and Willis, C.M. 1982, Relationship of gastric mucosal damage induced in pigs by antiinflammatory drugs to their effects on prostaglandin production. *Digestive Diseases and Sciences*, **27:** 624–635.

Rainsford, K.D., Willis, C.M., Walker, S.A. and Robins, P.G. 1982, Electron microscopic observations comparing the gastric mucosal damage induced in rats and pigs by benoxaprofen and aspirin, reflecting their differing actions as prostaglandin-synthesis inhibitors. *British Journal of Experimental Pathology*, **63:** 25–34.

Ramaholimihaso, F., Diebold, M.D. and Thiefin, G. 1998, Aspirin-induced duodenal diaphragm-like stricture. *Gastroenterologie Clinique et Biologique*, **22:** 650–651.

Rampon, S., Bussiere, J.L., Lopitaux, R., Sauvezie, B., Fermaud, H., Rivoal, A. and Bergeron, A. 1978, Gastric mucosa and anti-inflammatory agents. A comparative study by fibroscopy. *Revue du Rhumatisme et des Maladies Osteo-articulaires*, **45:** 153–160.

Ranta-Knuuttila, T., Kiviluoto, T., Karkkainen, P. and Kivilaakso, E. 1997, The novel thiol modulating agent, OR-1384, protects rat gastric mucosa against ethanol and aspirin (ASA) induced injury. *Gastroenterology*, **112:** A266.

Ranta-Knuuttila, T., Mustonen, H. and Kivilaakso, E. 2000, Topical prostaglandin E-2 protects isolated gastric mucosa against acidified taurocholate – but not ethanol – or aspirin-induced injury. *Digestive Diseases and Sciences*, **45:** 99–104.

Rao, C.N., Rao, V.H., Verbruggen, L. and Orloff, S. 1980, Effect of bioflavanoids on lysosomal acid hydrolases and lysosomal stability in adjuvant induced arthritis. *Scandinavian Journal of Rheumatology*, **9:** 280–284.

Räsänen, T. and Taskinen, E. 1973, The count of mast cells, tissue eosinophils, and epithelial mitoses of rat gastrointestinal mucosa after aspirin treatment. *Acta Physiologica Scandinavica*, **89:** 182–186.

Rascu, A., Manger, K., Rubbert, A., Manger, B. and Kalden, J.R. 1996, SLE and pregnancy: own experiences and literature review of the last 15 years. *Aktuelle Rheumatologie*, **21:** 98–105.

Rathert, P., Melchior, H. and Lutzeyer, W. 1975, Phenacetin: a carcinogen for the urinary tract. *Journal of Urology*, **113:** 653–657.

Ratnasooriya, W.D. and Lionel, N.D.W. 1984, Effect of aspirin-containing silastic implants placed adjacent to epididymis on fertility of rats. *Indian Journal of Experimental Biology*, **22:** 75–77.

Ratnikov, V.I. 1975, Characteristics of the energy metabolism in the kidneys in analgesic nephropathy. *Voprosy Meditsinskoi Khimii*, **21:** 154–158.

Read, N.W. and Fordtran, J.S. 1979, The role of intraluminal junction potentials in the generation of the gastric potential difference in man. *Gastroenterology*, **5:** 932–938.

Rees, W.D.N. and Turnberg, L.A. 1980, Reappraisal of the effects of aspirin on the stomach. *Lancet*, **2:** 410–413.

Rees, W.D.N. and Turnberg, L.A. 1981, A reappraisal of the effects of aspirin on the stomach. *In:* R. Hallam, L. Goldman and G.R. Fryers (eds), *Aspirin Symposium 1980. Royal Society of Medicine International Congress and Symposium. Series No. 39*, pp. 15–19. London, New York: Royal Society of Medicine/Academic Press/Grune & Stratton.

Reinicke, C. 1975, Influence of non-steroid anti-inflammatory drugs (NSAIDs) on hepatic tyrosine aminotransferase (TA) activity in rats in vitro and in vivo. *Biochemical Pharmacology*, **24:** 193–198.

Reinicke, C. 1977, Influence of non-steroidal amti-inflammatory drugs (NSAIDS) on liver enzymes. *Drugs under Experimental and Clinical Research*, **2:** 139–153.

Reinicke, C. and Klinger, W. 1971, Effects of non-steroid antirheumatic agents on microsomal drug-metabolizing enzymes of rat liver. *Biochemical Pharmacology*, **20:** 1405–1412.

Rich, R.R. and Johnson, J.S. 1973, Salicylate hepatotoxicity in patients with juvenile rheumatoid arthritis. *Arthritis and Rheumatism*, **16:** 1–9.

Richards, I.D. 1969, Congenital malformations and environmental influences in pregnancy. *British Journal of Preventative and Social Medicine*, **23:** 218–225.

CHAPTER 8

Richards, I.D. 1971, Incidence of congenital defects in South Wales, 1964–6. *British Journal of Preventive and Social Medicine*, **25:** 59–64.

Richards, I.D. 1972, A retrospective enquiry into possible teratogenic effects of drugs in pregnancy. *Advances in Experimental Medicine and Biology*, **27:** 441–455.

Ricks, W.B. 1976, Letter: Salicylate hepatotoxicity in Reiter's syndrome. *Annals of Internal Medicine*, **84:** 52–53.

Ridolfo, A.S., Crabtree, R.E., Johnson, D.W. and Rockhold, F.W. 1980, Gastrointestinal microbleeding: comparison between benoxaprofen and other non-steroidal anti-inflammatory agents. *Journal of Rheumatology*, **7 (Suppl. 6):** 36–47.

Ridolfo, A.S., Rubin, A., Crabtree, R.E. and Gruber, C.M. 1973, Effects of fenoprofen and aspirin on gastrointestinal microbleeding in man. *Clinical Pharmacology and Therapeutics*, **14:** 226–230.

Rimbau, V., Lopez, R., Forn, J. and Torralba, A. 1976, Neuva técnica en rata para la determinación de tolerancias gastro intestinales mediante radioisótopas. *Archivos de Farmacologia y Toxicologica*, **5:** 203–209.

Ritchie, W.P. and Fischer, R.P. 1973, Effect of metabolic inhibitors on the gastric mucosal barrier. *Surgery*, **73:** 614–622.

Robert, A. 1975, An intestinal disease produced experimentally by a prostaglandin deficiency. *Gastroenterology*, **69:** 1045–1047.

Robert, A. 1976, Antisecretory, anti-ulcer, cytoprotective and diarrheogenic properties of prostaglandins. *In:* B. Samuelsson and R. Paoletti (eds), *Advances in Prostaglandin and Thromboxane Research*, Vol. 2, pp. 507–520. New York: Raven Press.

Robert, A. 1981a, Current history of cytoprotection. *Prostaglandins*, **21:** 89–96.

Robert, A. 1981b, Gastric cytoprotection by sodium salicylate. *Prostaglandins*, **21:** 139–146.

Robert, A., Leung, F.W., Kaise, D.G. and Guth, P.H. 1989, Potentiation of aspirin-induced gastric lesions by exposure to cold in rats. *Gastroenterology*, **97:** 1147–1158.

Robert, A., Nezamio, J.E., Lancaster, C. and Hanchar, A.J. 1979, Cytoprotection by prostaglandins in rats. Prevention of gastric necrosis produced by alcohol, HC1, NaOH, hypertonic NaCl, and thermal injury. *Gastroenterology*, **77:** 433–434.

Roberts, B.J. and Knights, K.M. 1992, Inhibition of rat peroxisomal palmitoyl-CoA ligase by xenobiotic carboxylic acids. *Biochemical Pharmacology*, **44:** 261–267.

Roberts, R.A. 1999, Peroxisome proliferators: mechanisms of adverse events in rodents and molecular basis for species differences. *Archives of Toxicology*, **73:** 413–418.

Robertson, A., Glynn, J.P. and Watson, A.K. 1972, The absorption and metabolism in man of 4-acetamidophenyl-2-acetoxybenzoate (benorylate). *Xenobiotica*, **2:** 339–347.

Robertson, R.T., Allen, H.L. and Bokelman, D.L. 1979, Aspirin: teratogenic evaluation in the dog. *Teratology*, **20:** 313–320.

Roberts-Thomson, S.J. 2000, Peroxisome proliferator-activated recptors in tumorigenesis: targets for tumour promotion and treatment. *Immunology and Cell Biology*, **78:** 436–441.

Robins, P.G. 1980a, Gastrointestinal erosions and the lack of inflammatory response. *Journal of Pharmacy and Pharmacology*, **32:** 307–308.

Robins, P.G. 1980b, Ultrastructural observations on the pathogenesis of aspirin induced gastric erosions. *British Journal of Experimental Pathology*, **61:** 497–504.

Roch, M. 1912, Acide acetyl-salicylique et salicylate de soude. *Bulletin Général de Thérapeutique Médicale, Chirurgical, Obstétricale et Pharmaceutique*, **163:** 218–223.

Rocha, G.M., Michea, L.F., Peters, E.M., Kirby, M., Xu, Y., Ferguson, D.R. and Burg, M.B. 2001, Direct toxicity of nonsteroidal anti-inflammatory drugs for renal medullary cells. *Proceedings of the National Academy of Sciences of the USA*, **98:** 5317–5322.

Roch-Ramel, F., Roth, L., Arnow, J. and Weiner, I.M. 1978, Salicylate excretion in the rat: free flow micropuncture experiments. *Journal of Pharmacology and Experimental Therapeutics*, **207:** 737–747.

Rodgers, G.B. 2002, The effectiveness of child-resistant packaging for aspirin. *Archives of Pediatric and Adolescent Medicine*, **156:** 929–933.

Rogausch, H. 1978, The effect of lysolecithin on contractile force of isolated gastric smooth muscle. *Research in Experimental Medicine (Berlin)*, **173:** 9–15.

Rohrbach, P., Laroche, M.J. and Teche, G. 1977, Benorylate toxicity in animals and renal excretion in man. *Therapie*, **32:** 89–98.

Romano, M., Kraus, E.R., Boland, C.R. and Coffey, R.J. 1994, Comparison between transforming growth factor alpha and epidermal growth factor in the protection of rat gastric mucosa against drug induced injury. *Italian Journal of Gastroenterology*, **26:** 223–228.

Romano, M., Lesch, C.A., Meise, K.S., Veljaca, M., Sanchez, B., Kraus, E.R., Boland, C.R., Guglietta, A. and Coffey, R.J. 1996, Increased gastroduodenal concentrations of transforming growth factor alpha in adaptation to aspirin in monkeys and rats. *Gastroenterology*, **110:** 1446–1455.

Romano, M., Polk, W.H., Awad, J.A., Arteaga, C.L., Nanney, L.B., Wargovich, M.J., Kraus, E.R., Boland, C.R. and Coffey, R.J. 1992, Transforming growth factor-alpha protection against drug-induced injury to the rat gastric-mucosa *in vivo*. *Journal of Clinical Investigation*, **90:** 2409–2421.

Romero, J.C., Dunlap, C.L. and Strong, C.G. 1976, The effect of indomethacin and other anti-inflammatory drugs on the renin–angiotensin system. *Journal of Clinical Investigation*, **58:** 282–288.

Romero, J.C. and Strong, C.G. 1977, Hypertension and the interrelated renal circulatory effects of prostaglandins and the renin–angiotensin system. *Mayo Clinic Proceedings*, **52:** 462–464.

Rooney, P.J., Dick, W.C., Imrie, R.C., Turner, D., Buchanan, K.D. and Ardill, J. 1978, On the relationship between gastrin, gastric secretion, and adjuvant arthritis in rats. *Annals of the Rheumatic Diseases*, **37:** 432–435.

Ross, G., Gray, C.H., De Silva, S. and Newman, J. 1964, Assessment of routine tests for occult blood in faeces. *British Medical Journal*, **1:** 1351–1354.

Roth, G.J., Stanford, N. and Majerus, P.W. 1975, Acetylation of prostaglandin synthase by aspirin. *Proceedings of the National Academy of Sciences of the USA*, **72:** 3073–3076.

Roth, J.L.A. 1974, Drug-induced lesions. *In:* H.L. Bockus (ed.), *Gastroenterology*, Vol. 1, pp. 487–514. Philadelphia: W.B. Saunders.

Roth, J.L.A. and Valdes-Dapena, A. 1963a, Mechanisms of salicylate gastro-intestinal erosion and hemorrhage. *In:* A.St.J. Dixon, B.K. Martin, M.J.H. Smith and P.H.N. Wood (eds), *Salicylates: An International Symposium*, pp. 224–225. London: Churchill.

Roth, J.L.A. and Valdes-Dapena, A. 1963b, Topical action of salicylates on the buccal mucosa of man and on the stomach of the cat. *In:* S.C. Skoryna (ed.), *Pathophysiology of Peptic Ulcer*, pp. 245–251. Montreal: McGill University Press.

Rowe, P.H., Kasdon, E., Starlinger, M.J., Hollands, M.J. and Silen, W. 1985b, Prostaglandins prevent red streaks (RS) in rat gastric mucosa caused by intravenous (IV) aspirin but not by salicylate. *Digestive Diseases and Sciences*, **30:** 398.

Rowe, P.H., Lange, R., Marrone, G., Matthews, J.B., Kasdon, E. and Silen, W. 1985a, In vitro protection of amphibian gastric mucosa by nutrient HCO_3-against aspirin injury. *Gastroenterology*, **89:** 767–778.

Rowe, P.H., Starlinger, M.J., Kasdon, E., Hollands, M.J. and Silen, W. 1987, Parenteral aspirin and sodium salicylate are equally injurious to the rat gastric mucosa. *Gastroenterology*, **93:** 863–871.

Rowland, M., Riegelman, S., Harris, P.A., Sholkoff, S.D. and Erying, E.J. 1967, Kinetics of acetylsalicylic acid disposition in man. *Nature*, 215, 413–414.

Rubinstein, M., Ciacomoni, D. and Pachman, L.M. 1976, Effect of sodium salicylate on hamster cells in vitro. *Journal of Pharmaceutical Sciences*, **65:** 756–758.

Rudolph, A.M. 1981, Effects of aspirin and acetaminophen in pregnancy and in the newborn. *Archives of Internal Medicine*, **141 (3 Special Number):** 358–363.

Rudzki, E., Czubalski, K. and Grzywa, Z. 1980, Detection of urticaria with food additives intolerance by means of diet. *Dermatologia*, **161:** 57–62.

Russell, A.J., Sturge, R.A. and Smith, M.A. 1971, Serum transaminases during salicylate therapy. *British Medical Journal*, **2:** 428–429.

Russell, R.I. and Goldberg, A. 1968, Effect of aspirin on the gastric mucosa of guinea-pigs on a scorbutogenic diet. *Lancet*, **2:** 606–608.

Russell, R.I., Goldberg, A., Williamson, J.M. and Wares, E. 1968, Ascorbic acid levels in leukocytes of patients with gastrointestinal haemorrhage. *Lancet*, **2:** 603–606.

CHAPTER 8

Sachs, G., Chang, H.H., Rabone, E., Schackman, R., Lewin, M. and Saccomani, G. 1976, A nonelectrogenic H$^+$. *Journal of Biological Chemistry*, **251**: 7690–7698.

Sajous, C.E. and Hundley, P. 1937, *In:* G.M. Piersol and E.L. Bortz (eds), *Cyclopaedia of Medicine*, Vol. II. Philadelphia: Davis.

Saker, B.M. and Kincaid-Smith, P. 1969, Papillary necrosis in experimental analgesic nephropathy. *British Medical Journal*, **1**: 161–162.

Sakurai, T., Miyazawa, S., Osumi, T., Furuta, S. and Hashimoto, T. 1981, Induction of peroxisomal β-oxidation by the administration of acetylsalicylic acid. *Toxicology and Applied Pharmacology*, **59**: 8–14.

Salzmann, D.A., Gall, E.P. and Robinson, S.F. 1976, Aspirin-induced hepatic dysfunction in a patient with adult rheumatoid arthritis. *American Journal of Digestive Diseases*, **21**: 815–820.

Samter, M. and Beers, R.F. Jr 1968, Intolerance to aspirin. Clinical studies and consideration of its pathogenesis. *Annals of Internal Medicine*, **68**: 975–983.

Sanak, M., Levy, B.D., Clish, C.B., Chiang, N., Gronot, K., Mastalerz, L., Serhan, C.N. and Szczeklik, A. 2000, Aspirin-intolerant asthmatics generate more lipoxins than aspirin-intolerant asthmatics. *European Respiratory Journal*, **16**: 44–49.

Sanak, M., Simon, H.U. and Szczeklik, A. 1997, Leukotriene C$_4$ synthase promoter polymorphism and risk of aspirin-induced asthma. *Lancet*, **350**: 1599–1600.

Sanak, M. and Szczeklik, A. 1998, Leukotriene C$_4$ synthase promotor polymorphism in bronchial asthma. *Allergy*, **53**: 112–118.

Sanak, M. and Szczeklik, A. 2000, Genetics of aspirin induced asthma. *Thorax*, **55 (Suppl. 2)**: 50–52.

Sánchez de Miguel, L., De Frutos, T., Gonzalez-Fernandez, F., Del Pozo, V., Lahoz, C., Jimenez, A., Rico, L., Garcia, R., Aceituno, E., Millas, I., Gomez, J., Farre, J., Casado, S. and Lopez-Farre, A. 1999, Aspirin inhibits inducible nitric oxide synthase expression and tumor necrosis factor-α release by cultured smooth muscle cells. *European Journal of Clinical Investigation*, **29**: 93–99.

Sandler, D.P., Burr, R. and Weinberg, C.R. 1991, Nonsteroidal anti-inflammatory drugs and the risk for chronic renal disease. *Annals of Internal Medicine*, **115**: 165–172.

Sandler, D.P., Smith, J.C., Weinberg, C.R., Buckalew, V.M. Jr, Dennis, V.W., Blythe, W.B. and Burgess, W.P. 1989, Analgesic use and chronic renal disease. *New England Journal of Medicine*, **320**: 1238–1243.

Sankale, M., Sou, A.M., Coly, D., Agbetra, M. and Dupuy-Dourreau, R. 1977, Drug-induced haemorrhagic gastritis in African blacks (preliminary study, apropos of 8 cases observed in Dakar). *Bulletin de la Societé, Medicale d'Afrique Noire de Langue Française*, **22**: 384–389.

Sankar, D.V. and Geisler, A. 1971, Mouse leukocyte chromosome system. Effect of chlorpromazine and aspirin. *Research Communications in Chemical Pathology and Pharmacology*, **2**: 477–482.

Santolaria, S., Lanas, A., Benito, R., Pérez-Aisa, M.A., Montoro, M. and Sainz, R. 1999, *Helicobacter pylori* infection is a protective factor for bleeding gastric ulcers but not for bleeding duodenal ulcers in NSAID users. *Ailmentary Pharmacology and Therapeutics*, **13**: 1511–1518.

Santucci, L., Fiorucci, S., Di Matteo, F.M. and Morelli, A. 1995, Role of tumor necrosis factor-α release and leukocyte margination in indomethacin-induced gastric injury in rats. *Gastroenterology*, **108**: 393–401.

Sarau, H.M., Foley, J.J., Moonsammy, G. and Sachs, G. 1977, Metabolism of dog gastric mucosa. Levels of glycolytic, citric acid cycle and other intermediates. *Journal of Biological Chemistry*, **252**: 8572–8581.

Sarfeh, I.J. and Tarnawski, A. 1991, Increased susceptibility of the portal hypertensive gastric mucosa to damage. *Journal of Clinical Gastroenterology*, **13**: S18–S21.

Sarfeh, I.J., Tarnawski, A., Hajduczek, A., Stachura, J., Bui, H.X. and Krause, W.J. 1988, The portal hypertensive gastric mucosa – histologic, ultrastructural and functional analysis after aspirin-induced damage. *Surgery*, **104**: 79–85.

Sawacki, P.T. and Berger, M. 1998, Pharmacological treatment of diabetic patients with cardiovascular complications. *Journal of Internal Medicine*, **243**: 181–189.

Sawaoka, H., Kawano, S., Tsuji, S., Tsujii, M., Sun, W., Gunawan, E.S. and Hori, M. 1998, *Helicobacter pylori* infection induces cyclooxygenase-2 expression in human gastric mucosa. *Prostaglandins, Leukotrienes and Essential Fatty Acids*, **59 (5)**: 313–316.

Sbarbaro, J.A. and Bennett, R.M. 1977, Aspirin hepatotoxicity and disseminated intravascular coagulation. *Annals of Internal Medicine*, **86:** 183–185.

Schanker, L.S., Shore, P.A., Brodie, B.B. and Hogben, C.A.M.1957, Absorption of drugs from the stomach I. The rat. *Journal of Pharmacology and Experimental Therapeutics*, **120:** 528–529.

Scheiman, J.M., Behler, E.M., Loeffler, K.M. and Elta, G. 1994, Omeprazole ameliorates aspirin-induced gastroduodenal injury. *Digestive Diseases and Sciences*, **39:** 97–103.

Scheiman, J.M., Meise, K.S., Greenson, J.K. and Coffey, R.J. 1997, Transforming growth factor-alpha (TGF-alpha) levels in human proximal gastrointestinal epithelium – effect of mucosal injury and acid inhibition. *Digestive Diseases and Sciences*, **42:** 333–341.

Schenker, S., Martin, R.R. and Hoyumpa, A.M. 1999, Antecedent liver disease and drug toxicity. *Journal of Hepatology*, **31:** 1098–1105.

Schiessel, R., Matthews, J., Barzilai, A., Merhav, A. and Silen, W. 1980, PGE$_2$ stimulates gastric chloride transport possible key to cytoprotection. *Nature*, **283:** 671–673.

Schlegel, W., Wenk, K., Dollinger, H.C. and Raptis, S. 1977, Concentrations of prostaglandin A-, E- and F-like substances in gastric mucosa or normal subjects and of patients with various gastric diseases. *Clinical Science and Molecular Medicine*, **52:** 255–258.

Schlumberger, H.D., Löbbecke, E.-A. and Kallos, P. 1974, Acetylsalicylic acid intolerance. *Acta Medica Scandinavica*, **196:** 451–458.

Schmidt, C.R., Beazell, J.M., Atkinson, A.J. and Ivy, A.C. 1938, The effect of therapeutic agents on the volume and constituents of the bile. *American Journal of Digestive Diseases*, **5:** 613–617.

Schnedorf, J.G., Bradley, W.B. and Ivey, A.C. 1936, Effect of acetylsalicylic acid upon gastric activity and the modifying action of calcium gluconate and sodium bicarbonate. *American Journal of Digestive Diseases and Nutrition*, **3:** 239–245.

Schnellmann, R.G. 1998, Analgesic nephropathy in rodents. *Journal of Toxicology and Environmental Health*, Part B, **1:** 81–90.

Schreiner, G.E., McAnally, J.F. and Winchester, J.F. 1981, Clinical analgesic nephropathy. *Archives of Internal Medicine*, **141:** 349–357.

Schwartz, M.A. and Amidon, G. 1966, Reaction of aspirin with amines. Potential mechanism. *Journal of Pharmaceutical Sciences*, **55:** 1464–1465.

Schwartz, S., Dahl, J., Ellefson, M. and Ahlquist, D. 1983, The 'HemoQuant' Test: a specific and quantitative determination of heme (hemoglobin) in feces and other materials. *Clinical Chemistry*, **29:** 2061–2067.

Scott, J.E. and Persaud, T.V.N. 1978, A quantitative study of the effects of acetylsalicylic acid on spermatogenesis and organs of the rat. *International Journal of Fertility*, **23:** 282–287.

Seaman, W.E., Ishak, K.G. and Plotz, P.H. 1974, Aspirin-induced hepatotoxicity in patients with systemic lupus erytheratosus. *Annals of Internal Medicine*, **80:** 1–8.

Seaman, W.E. and Plotz, P.H. 1976, Effect of aspirin on liver tests in patients with RA or SLE in normal volunteers. *Arthritis and Rheumatism*, **19:** 155–160.

Seger, D. 1981, Pulmonary oedema complication of salicylate intoxication. *Annals of Emergency Medicine*, **10:** 549–551.

Seinela, L. and Ahvenainen, J. 2000, Peptic ulcer in the very old patients. *Gerontology*, **46:** 271–275.

Sekine, T., Cha, S.H., Tsuda, M., Apiwattanakul, N., Nakajima, N., Kanai, Y. and Endou, H. 1998, Identification of multispecific organic anion transporter 2 expressed predominantly in the liver. *FEBS Letters*, **429:** 179–182.

Sellers, E.M., Marshman, J.A., Kaplan, H.L., Giles, H.G., Kapur, B.M., Busto, U., MacLeod, S.M., Stapleton, C. and Sealey, F. 1981, Acute and chronic drug abuse emergencies in Metropolitan Toronto. *International Journal of the Addictions*, **16:** 283–303.

Selye, H. 1966, *Thrombohemorrhagic Phenomena*. Springfield, IL: C.C. Thomas.

Selye, H. and Najusz, E. 1966, Stress and rheumatic diseases. *In:* J.L. Hollander (ed.), *Arthritis and Allied Conditions*, pp. 618–630. New York: Henry Kimpton.

Semple, P.F. and Russell, R.I. 1975, Role of bile acids in the pathogenesis of aspirin-induced gastric mucosal hemorrhage in rats. *Gastroenterology*, **68:** 67–70.

Senate Select Committee On Social Welfare 1977, *Drug Problems in Australia – An Intoxicated Society*. Report to The Parliament of the Commonwealth of Australia. Canberra: Australian Government Publishing Service.

Serhan, C.N. 1999, Lipoxins and aspirin-triggered 15-epi-lipoxins. *In:* J.I. Gallin and R. Snyderman (ed.), *Information: Basic Principles and Clinical Correlates*, pp. 373–385. Philadelphia: Lippincott, Williams & Wilkins.

Serhan, C.N., Tanako, T., Gronert, K., Chiang, N. and Chish, C.B. 1999, Lipoxin and aspirin-triggered 15-epi-lipoxin cellular interactions: anti-inflammatory lipid mediators. *Clinical Chemistry and Laboratory Medicine*, **37:** 299–309.

Settipane, G.A. 1981, Adverse reactions of aspirin and related drugs. *Archives of Internal Medicine*, **141:** 328–332.

Settipane, G.A., Chafee, F.H. and Klein, D.E. 1974, Aspirin intolerance. II. A prospective study in an atopic and normal population. *Journal of Allergy and Clinical Immunology*, **53:** 200–204.

Settipane, G.A. and Pudupakkam, R.K. 1975, Aspirin intolerance. III. Subtypes, familial occurrence, and cross-reactivity with tartrazine. *Journal of Allergy and Clinical Immunology*, **56:** 215–221.

Settipane, R.A., Constantine, H.P. and Settipane, G.A. 1980, Aspirin intolerance and recurrent urticaria in normal adults and children. Epidemiology and review. *Allergy*, **35:** 149–154.

Shah, A.A., Thjodleifsson, B., Murray, F.E., Kay, E., Barry, M., Sigthorsson, G., Gudjonsson, H., Oddsson, E., Price, A.B., Fitzgerald, D.J. and Bjarnason, I. 2001, Selective inhibition of COX-2 in humans is associated with less gastrointestinal injury: a comparison of nimesulide and naproxen. *Gut*, **48:** 339-346.

Shapiro, S., Monson, R.R., Kaufman, D.W., Siskind, V., Heinonen, O.P. and Slone, D. 1976, Perinatal mortality and birth weight: relation to aspirin taken during pregnancy. *Lancet*, **1:** 1375–1380.

Shaw, L.E. 1887, Cases of haemorrhage occurring during treatment by salicylate of soda. *Guy's Hospital Reports*, **XLIV:** 125–135.

Shay, H., Komarov, S.A., Fels, S.S., Meranze, D., Grunstein, M. and Siplet, H. 1945, A simple method for the uniform production of gastric ulceration in the rat. *Gastroenterology*, **5:** 43–61.

Shea-Donohue, P.T., Meyers, L., Castell, D.O. and Dubois, A. 1980, Effect of prostacyclin on gastric emptying and secretion in rhesus monkeys. *In:* B. Samuelsson, P.W. Ramwell and R. Paoletti (eds), *Advances in Prostaglandin and Thromboxane Research*, Vol. 8, pp. 1557–1558. New York: Raven Press.

Shea-Donohue, T., Steel, L., Montcalm-Mazzilli, E. and Dubois, A. 1990, Aspirin-induced changes in gastric function: role of endogenous prostaglandins and mucosal damage. *Gastroenterology*, **98:** 284–292.

Shier, W.T. 1977, Inhibition of acyl coenzyme A: lysolecithin acetyltransferases by local anaesthetics, detergents and inhibitors of cyclic nucleotide phosphodiesterases. *Biochemical and Biophysical Research Communications*, **75:** 186–193.

Shillingford, J., Lindup, W.E. and Parke, D.G. 1973, Effect of carbenoxolone on some aspects of the gastrointestinal mucosa. *Transactions of the Biochemical Society*, **1:** 966–968.

Shirley, E. 1977, A review of papers purporting to show a cause and effect relationship between aspirin ingestion and massive gastrointestinal haemorrhage. *Proceedings of the Royal Society of Medicine*, **70:** 4–10.

Shore, P.A., Brodie, B.B. and Hogben, C.A.M. 1957, The gastric secretion of drugs: A pH partition hypothesis. *Journal of Pharmacology and Experimental Therapeutics*, **119:** 361–369.

Shriver, D.A., Dove, P.A., White, C.B., Sandor, A. and Rosenthale, M.E. 1977, A profile of the gastrointestinal toxicity of aspirin. *Toxicology and Applied Pharmacology*, **42:** 75–83.

Shriver, D.A., Thompson, J.W., Scott, C.K., Moore, L.E., Woolf, G. and Mack, V.J. 1980, The gastric antisecretory and antiulcer activity of MK-447, an enhancer of prostaglandin synthesis. *Life Sciences*, **27:** 2483–2487.

Shriver, D.A., White, C.B., Sandor, A. and Rosenthale, M.E. 1975, A profile of the rat gastrointestinal toxicity of drugs used to treat inflammatory diseases. *Toxicology and Applied Pharmacology*, **32:** 73–83.

Sibai, B.M., Caritis, S., Hauth, J., Lindheimer, M., Vandorsten, J.P., MacPherson, C., Klebanoff, M., Landon, M., Miodovnik, M., Paul, R., Meis, P., Dombrowski, M., Thurnau, G., Roberts, J. and McNellis, D. 2000, Risks of pre-eclampsia and adverse neonatal outcomes among women with pregestational diabetes mellitus. *American Journal of Obstetrics and Gynecology*, **182:** 364–369.

Siegel, M.I., McConnell, R.T. and Cuatrecasas, P. 1979, Aspirin-like drugs interfere with arachidonate metabolism by inhibition of the 12-hydroperoxy 5,8,10,14-eicosatetraenoic acid peroxidase. *Proceedings of the National Academy of Sciences of the USA*, **76:** 3771–3778.

Sigthorsson, G., Tibble, J., Hayllar, J., Menzies, I., MacPherson, A., Moots, R., Scott, D., Gumpel, M.J. and Bjarnason, I. 1998, Intestinal Permeability and Inflammation in Patients on NSAIDs. *Gut*, **43:** 506–511.

Sillanpää, M., Mäkelä, A.-L. and Koivikko, A. 1975, Acute liver failure and encephalopathy (Reye's syndrome) during salicylate therapy. *Acta Paediatrica Scandinavica*, **64:** 877–880.

Silverman, H.M. 1974, Fetal and newborn adverse drug reactions. A survey of the recent literature. *Drug Intelligence and Clinical Pharmacy*, **8:** 690–693.

Silverstein, F.E., Graham, D.Y., Senior, J.R., Wyn Davies, H., Struthers, B.J., Bittman, R.M. and Geis, S.G. 1995, Misoprostol reduces serious gastrointestinal complications in patients with rheumatoid arthritis receiving nonsteroidal anti-inflammatory drugs. *Annals of Internal Medicine*, **123:** 241–249.

Silverstein, H., Bernstein, J. and Davies, D. 1967, Salicylate ototoxicity: A biochemical and electrophysiological study. *Annals of Otology, Rhinology and Laryngology*, **76:** 118–128.

Silverstein, H. and Griffen, W. 1970, Salicylate ototoxicity. *In:* Parapells (ed.), *Biochemical Mechanisms in Hearing and Deafness*, pp. 223–246. Springfield, IL: C.C. Thomas.

Silvoso, G.R., Ivey, K.J., Butt, H.H., Lockard, O.O., Holt, S.D., Sisk, C., Bastkin, W.N., Mackercher, P.A. and Hewett, J. 1979, Incidence of gastric lesions in patients with rheumatic disease on chronic aspirin therapy. *Annals of Internal Medicine*, **91:** 517–520.

Simon, B., Elsner, H. and Muller, P. 1995, Protective effects of omeprazole against low-dose acetylsalicylic-acid – an endoscopic controlled double-blind study in healthy volunteers. *Arzneimittel-Forschung*, **45:** 701–703.

Simon, B., Kather, H.J. and Kommerell, B. 1978, Effects of prostaglandins and their methylated analogues upon human adenylate cyclase in the upper gastrointestinal tract. *Digestion*, **17:** 547–533.

Simon, L.S., Weaver, A.L., Graham, D.Y., Kivitz, A.J., Lipskey, P.E., Hubbard, R.C., Isakson, P.C., Verburg, K.M., Yu, S.S., Zhao, W.W. and Geis, G.S. 1999, Anti-inflammatory and upper gastrointestinal effects of celecoxib in rheumatoid arthritis: a randomised controlled trial. *Journal of the American Medical Association*, **282:** 1921–1928.

Singh, G. 1998, Recent considerations in nonsteroidal anti-inflammatory drug gastropathy. *America Journal of Medicine*, **105 (1B):** 31S–38S.

Singh, G. and Ramey, D.R. 1998, NSAID-induced gastrointestinal complications: the ARAMIS perspective – 1997. *Journal of Rheumatology*, **26 (suppl. 51):** 8–16.

Singh, G., Ramey, D.R., Morfeld, D. and Fries, J.F. 1994, Comparative toxicity of non-steroidal anti-inflammatory agents. *Pharmacology and Therapeutics*, **62:** 175–191.

Singh, H., Chugh, J.C., Shembesh, A.H., Ben-Musa, A.A. and Mehta, H.C. 1992, Hepatotoxicity of high dose salicylate therapy in acute rheumatic fever. *Annals of Tropical Paediatrics*, **12:** 37–40.

Singh, J. 1980, Prostaglandin release from rat stomach following vagal stimulation or administration of acetylcholine. *European Journal of Pharmacology*, **65:** 39–48.

Singh, P.R., Santella, R.N. and Zawada, E.T. 1996, Mitochondrial genome mutations and kidney disease. *American Journal of Kidney Diseases*, **28:** 140–146.

Siurala, M., Julkunen, H., Toivonen, S., Pelkonen, R., Saxen, E. and Pitkänen, E. 1965, Digestive tract in collagen diseases. *Acta Medica Scandinavica*, **178:** 13–25.

Skidmore, I.F. and Whitehouse, M.W. 1966, Biochemical properties of anti-inflammatory drugs VIII. Inhibition of histamine formation catalyzed by substrate-specific mammalian histidine decarboxylases. Drug antagonism of aldehyde binding to protein amine groups. *Biochemical Pharmacology*, **15:** 1965–1983.

Skillman, J.J. and Silen, W. 1972, Gastric mucosal barriers. *Surgical Annals*, **4:** 213–237.

Slater, B.C.S. 1965, Teratogenic effects of salicylates. *In:* J.M. Robson, F.M. Sullivan and R.L. Smith (eds), *Symposium on Embryopathic Activity of Drugs*, p. 241. London: Churchill.

Slater, T.F. and Delaney, V.B. 1970, Liver adenosine triphosphate content and bile flow rate in the rat. *Biochemistry Journal.*, **116:** 303–308.

Slomiany, A., Liau, Y.H., Sarosiek, J. and Slomiany, B.L. 1987, Enzymatic sulfation of mucin in gastric mucosa – effect of aspirin and sofalcone. *Gastroenterology*, **92:** 1645.

Slone, D., Siskund, V., Heinonen, O.P., Monson, R.R., Kaufman, D.W. and Shapiro, S. 1976, Aspirin and congenital malformations. *Lancet*, **1:** 1373–1375.

Small, P., Frenkiel, S. and Black, M. 1981, Multifactorial aetiology of nasal polyps. *Annals of Allergy*, **46:** 123–126.

Smith, B.M., Skillman, J.J., Edwards, B.G. and Silen, W. 1971, Permeability of the human gastric mucosa. Alteration by acetylsalicylic acid and ethanol. *New England Journal of Medicine*, **285:** 716–721.

Smith, M.J.H. 1966, Toxicology. *In:* M.J.H. Smith and P.K. Smith (eds), *The Salicylates. A Critical Bibliographic Review*, pp. 233–306. New York: Wiley-Interscience.

Smith, M.J.H. 1968, The metabolic basis of the major symptoms in acute salicylate intoxication. *Clinical Toxicology*, **1:** 387–407.

Smith, M.J.H. and Dawkins, P.D. 1971, Salicylate and enzymes. *Journal of Pharmacy and Pharmacology*, **23:** 729–744.

Smith, M.J.H. and Irving, J.D. 1955, The effect of salicylate on the passage of a barium meal in the rat. *British Journal of Radiology*, **28:** 39–41.

Smith, V.M. 1978, Association of aspirin ingestion with symptomatic esophageal hiatus hernia. *Southern Medical Journal*, **71:** 45–47.

Smyth, R.D., Mathew, P.P., Procacci, R.L., Carr, G.S. and Reavey-Cantwell, N.H. 1976, Comparative effects of fenclorac and indomethacin on gastrointestinal blood loss in the rat. *Toxicology and Applied Pharmacology*, **38:** 507–515.

Snodgrass, W. and Rumack, B.I.I. 1981, Salicylate toxicity following therapeutic doses in young children. *Clinical Toxicology* **18:** 247–259.

Soll, A.H. and Grossman, M.I. 1978, Cellular mechanisms in acid secretion. *Annual Review of Medicine*, **28:** 495–507.

Soll, A.H. and Whittle, B.J.R. 1981, Prostacyclin analogues inhibit canine parietal cell activity and cyclic AMP formation. *Prostaglandins*, **21:** 353–365.

Soll, A.H. and Wollin, A. 1979, Histamine and cyclic AMP isolated canine parietal cells. *American Journal of Physiology*, **237:** E444–E445.

Sorensen, H.T., Mellemkjaer, L., Blot, W.J., Nielsen, G.L., Steffensen, F.H., McLaughlin, J.K. and Olsen, J.H. 2000, Risk of upper gastrointestinal bleeding associated with use of low-dose aspirin. *American Journal of Gastroenterology*, **95:** 2218–2224.

Sorenson, J.R.J. 1976, Copper chelates as possible active forms of the antiarthritic agents. *Journal of Medicinal Chemistry*, **19:** 135–148.

Soumarin, A., Ghesquier, D. and Lewin, M.J.M. 1979, Regulation of gastric secretion. *In:* G. Rosselion, P. Fromageot and S. Bonfils (eds), *Hormone Receptors and Nutrition*, pp. 349–354. Amsterdam: Elsevier/North Holland.

Soybel, D.I., Davis, M.B. and West, A.B. 1992, Effects of aspirin on pathways of ion permeation in Necturus antrum: role of nutrient HCO_3. *Gastroenterology*, **103:** 1475–1485.

Speer, F., Denison, T.R. and Baptiste, J.E. 1981, Aspirin allergy. *Annals of Allergy*, **46:** 317–320.

Spenny, J.G. and Bhown, M. 1977a, Effect of acetylsalicylic acid on the amphibian gastric mucosa. I. Electrophysiological and permeability changes. *Gastroenterology*, **73:** 785–789.

Spenny, J.G. and Bhown, M. 1977b, Effect of acetylsalicylic acid on the gastric mucosa. II. Mucosal ATP and phosphocreatine content, and salicylate effects on mitochondrial metabolism. *Gastroenterology*, **73:** 995–999.

Spenny, J.G. and Nowell, R.M. 1979, Acetylsalicylate hydrolase of rabbit gastric mucosa. Isolation and purification. *Drug Metabolism and Disposition*, **7:** 215–219.

Spézia, F., Fournex, R. and Vannier, B. 1992, Action of allopurinol and aspirin on rat whole-embryo cultures. *Toxicology*, **72:** 239–250.

Spühler, O. and Zollinger, H. 1953, Die chronische interstitielle Nephritis. *Zeitschrift für Klinische Medizin*, **151:** 1–9.

St John, D.J.B. and McDermott, F.T. 1970, Influence of achlorhydria on aspirin-induced occult gastrointestinal blood loss: studies in Addisonian pernicious anaemia. *British Medical Journal*, **2:** 450–452.

St John, D.J.B., Yeomans, N.D., McDermott, F.T. and de Boer, W.G.R.M. 1973, Adaption of the gastric mucosa to repeated administration of aspirin in the rat. *American Journal of Digestive Diseases*, **18:** 881–886.

Stachura, J., Konturek, J.W., Dembinski, A. and Domschke, W. 1994, Do infiltrating leukocytes contribute to the adaptation of human gastric mucosa to continued aspirin administration? *Scandinavian Journal of Gastroenterology*, **29:** 966–972.

Stachura, J., Konturek, J.W., Dembinski, A. and Domschke, W. 1996, Growth markers in the human gastric mucosa during adaptation to continued aspirin administration. *Journal of Clinical Gastroenterology*, **22:** 282–287.

Stare, F.J., Whelan, E.M. and Sheridan, M. 1980, Diet and hyperactivity: is there a relationship? *Paediatrics*, **66:** 521–525.

Starko, K.M., Ray, C.G., Dominguez, L.B., Stromberg, W.L., Woodall, D.F. 1980, Reye's syndrome and salicylate use. *Paediatrics*, **66:** 859–864.

Staszewska-Barczak, J. 1978, Role of renal prostaglandins in circulatory homeostasis. *Contributions in Nephrology*, **11:** 179–188.

Steering Committee of the Physicians' Study Research Group 1989, Final report on the aspirin component of the ongoing physicians' health study. *New England Journal of Medicine*, **321:** 129–135.

Stephens, F.O. and Lawrenson, K.B. 1969, ^{51}Cr excretion in bile. *Lancet*, **1:** 158–159.

Stephens, F.O., Milton, G.W. and Lowenthal, J. 1966, Effect of aspirin on explanted gastric mucosa. *Gut*, **7:** 223–227.

Stephens, F.O., Milverton, E.J., Hambley, C.K. and Van der Ven, E.K. 1968, The effect of food on aspirin induced gastro intestinal blood loss. *Digestion*, **1:** 267–276.

Stern, A.I., Hogan, D.L., Kahn, L.H. and Isenberg, J.I. 1984, Protective effect of acetaminophen against aspirin- and ethanol-induced damage to the human gastric mucosa. *Gastroenterology*, **86:** 728–733.

Stern, A.I., Ward, F. and Hartley, G. 1986, Gastric mucosa and aspirin – protective effect of sucralfate pretreatment in man. *Australian and New Zealand Journal of Medicine*, **16:** 612.

Stern, A.I., Ward, F. and Hartley, G. 1987, Protective effect of sucralfate against aspirin-induced damage to the human gastric mucosa. *American Journal of Medicine*, **83:** 83–85.

Stern, A.I., Ward, F. and Sievert, W. 1989, Lack of gastric mucosal protection by sucralfate during long-term aspirin ingestion in humans. *American Journal of Medicine*, **86 (Suppl. 6A):** 66–69.

Stevenson, D.D. 1984, Diagnosis, prevention and treatment of adverse reactions to aspirin and non-steroidal anti-inflammatory drugs. *Journal of Allergy and Clinical Immunology*, **74:** 617–622.

Stevenson, D.D. 1998, Adverse reactions to nonsteroidal anti-inflammatory drugs. *Immunology and Allergy Clinics of North America*, **18:** 773–798.

Stevenson, D.D., Simon, R.A. and Mathison, D.A. 1980, Aspirin-sensitive asthma: tolerance to aspirin after positive oral aspirin challenges. *Journal of Allergy and Clinical Immunology*, **66:** 82–88.

Stevenson, D.D., Simon, R.A., Mathison, D.A. and Christiansen, S.C. 2000, Montelukast is only partially effective in inhibiting aspirin responses in aspirin-sensitive asthmatics. *Annals of Allergy, Asthma and Immunology*, **85:** 477–482.

Stewart, R.B. and Cluff, L.E. 1974, Gastrointestinal manifestations of adverse drug reactions. *American Journal of Digestive Diseases*, **19:** 1–7.

Stockinger, L. 1964, Licht- und Elektronmikroscopische untersuchungen der lokelen Einwirkung einiger Salicylate auf di Magen-Darm-Schleimhaut. *Arzneimittel Forschung*, **14:** 360–363.

Stockman, R. 1913, The action of salicylic acid and chemically allied bodies in rheumatic fever. *British Medical Journal*, **1:** 597–600.

Stolte, M., Panayiotou, S. and Schmitz, J. 1999, Can NSAID/ASA-induced erosions of the gastric mucosa be identified at histology? *Pathology Research and Practice*, **195:** 137–142.

Stone, C.A., Van Arman, G.G., Lotti, V.J., Minsker, D.H. Risley, E.A., Bagdon, W.J., Bokelman, D.L., Jensen, R.D., Mendlowski, B., Tate, C.L., Peck, H.M., Zwickley, R.E. and McKinney, S.E. 1977, Pharmacology and toxicology of diflunisal. *British Journal of Clinical Pharmacology*, **4:** 19S–29S.

Stubbé, L.Th.Fl., Pietersen, J.H. and Van Heulen, C. 1962, Aspirin preparations and their noxious effects on the gastrointestinal tract. *British Medical Journal*, **1:** 675–680.

Stygles, V.G., Smith, W.L., Reinke, D.A. and Hook, J.B. 1978, Prostaglandin-forming cyclooxygenase in renal medulla of spontaneously hypertensive rats during development. *Biology of the Neonate*, **33:** 309–313.

Sugawa, C., Lucas, C.E. and Walt, A.J. 1971, Serial endoscopic observations of the effects of histamine, aspirin, steroids, and alcohol on standardised gastric ulcers in dogs. *Gastrointestinal Endoscopy*, **18:** 56–58.

Sugimoto, N., Yoshida, N., Yoshikawa, T., Nakamaura, Y., Ithikawa, H., Naito, T. and Kondo, M. 2000, Effect of vitamin E on aspirin-induced gastric mucosal injury in rats. *Digestive Diseases and Sciences*, **45:** 599–605.

Sun, D.C.H., Roth, S.H., Mitchell, C.S. and Englund, D.W. 1974, Upper gastrointestinal disease in rheumatoid arthritis. *American Journal of Digestive Diseases*, **19:** 405–410.

Svanes, K., Leiknes, K.A., Verhang, J.E. and Soreide, O. 1979, Aspirin damage to ischemic gastric mucosa in shocked cats. *Scandinavian Journal of Gastroenterology*, **14:** 633–639.

Sweeney, K.R., Chapron, D.J., Brandt, L.J., Gomolin, I.H., Feig, P.U. and Kramer, P.A. 1986, Toxic interaction between acetazolamide and salicylate: case reports and a pharmacokinetic explanation. *Clinical Pharmacology and Therapeutics*, **40:** 518–524.

Swierkosz, T.A., Mitchell, J.A., Warner, T.D., Botting, R.M. and Vane, J.R. 1995, Co-induction of nitric oxide synthase and cyclo-oxygenase: interactions between nitric oxide and prostanoids. *British Journal of Pharmacology*, **114:** 1335–1342.

Symons, A.M., Johnston, B. and Parke, D.V. 1978, Effect of sodium carbenoxolone on lysosomal enzyme release. *Biochemical Pharmacology*, **27:** 2461–2463.

Szabó, I., Bodis, B., Nemeth, P. and Mózsik, Gy. 1997, Comparative viability studies on isolated gastric mucosal mixed cells and hepatoma and myeloma cell lines with ethanol, indomethacin and their combination. *In: Biochemical Pharmacology as an Approach to Gastrointestinal Disorders*, pp. 387–394. Dordrecht: Kluwer Academic Publishers.

Szczeklik, A. 1980, Analgesics, allergy and asthma. *British Journal of Clinical Pharmacology*, **10:** 401S–405S.

Szczeklik, A. 1988, Aspirin-induced asthma as a viral disease. *Clinical Allergy*, **18:** 15–20.

Szczeklik, A. 1989, Aspirin-induced asthma: new insights into pathogenesis and clinical presentation of drug intolerance. *International Archives of Allergy and Applied Immunology*, **1:** 70–75.

Szczeklik, A. 1990, The cyclooxygenase theory of aspirin-induced asthma. *European Respiratory Journal*, **3:** 588–593.

Szczeklik, A., Czerniawska-Mysik, G., Serwonska, M. and Kuklinski, P. 1980, Inhibition by ketotifen of idiosyncratic reactions to aspirin. *Allergy*, **35:** 421–424.

Szczeklik, A., Gryglewski, R.J., Olszewski, E., Dembinska-Kiec, A., and Szerniawska-Mysik, G. 1977, Aspirin-sensitive asthma: the effect of aspirin on the release of prostaglandins from nasal polyps. *Pharmacology Research Communications*, **9:** 415–425.

Szczeklik, A. and Nizankowska, E. 2000, Clinical features and diagnosis of aspirin induced asthma. *Thorax*, **55 (Suppl. 2):** 50–52.

Szczeklik, A., Nizankowska, E. and Duplaga, M., on behalf of the AIANE Investigators. 2000, Natural history of aspirin-induced asthma. *European Respiratory Journal*, **16:** 432–436.

Szczeklik, A., Sanak, M., Nizankowska-Mogilnicka, E. and Kielbasa, B. 2004, Aspirin intolerance and the cyclooxygenase-leukotriene pathways. *Current Opinion in Pulmonary Medicine*, **10:** 51–56.

Szelényi, I. 1984, Functional cytoprotection by certain antacids. *In:* Gy. Mózsik, A. Pár and A. Bertelli (eds), *Recent Advances in Gastrointestinal Cytoprotection*, pp. 73–82. Budapest: Akademiai Kiado.

Tabata, K. and Okabe, S. 1980, Effects of 16,16-dimethyl prostaglandin E2 methyl ester on aspirin and indomethacin induced gastrointestinal lesions in dogs. *Digestive Diseases and Sciences*, 25, 439–448.

Tabuchi, Y. and Kurebayashi, Y. 1992, Effect of DS-4574, a novel peptidoleukotriene antagonist with mast cell stabilizing action, on acute gastric lesions and gastric secretion in rats. *Japanese Journal of Pharmacology*, **60:** 335–340.

Tagashira, E., Nakao, K., Urano, T., Ishikawa, S., Hiramori, T. and Yanaura, S. 1981, Correlation of teratogenicity of aspirin to the stage-specific distribution of salicylic acid in rats. *Japanese Journal of Pharmacology*, **31:** 563–571.

Taha, A.S., Dahill, S., Nakshabendi, I., Lee, F.D., Sturrock, R.D. and Russell, R.I. 1993, Duodenal histology, ulceration, and *Helicobacter pylori* in the presence or absence of non-steroidal anti-inflammatory drugs. *Gut*, **34:** 1162–1166.

Takacs, E. and Warkany, J. 1968, Experimental production of congenital cardiovascular malformations in rats by salicylate poisoning. *Teratology*, **1:** 109–118.

Takagi, K. and Abe, Y. 1974, Studies on the healing of experimental ulcer in the rat II: influence of anti-inflammatory drugs on the healing of the acetic ulcer and the components in gastric tissue. *Japanese Journal of Pharmacology*, **24:** 345–356.

Takagi, K. and Kawashima, K. 1969, Effects of some anti-inflammatory drugs on capillary permeability of the gastric mucosa of the rat. *Japanese Journal of Pharmacology*, **19:** 431–439.

Takahashi, M., Katayama, Y., Takada, H., Kuwayama, H. and Terano, A. 2000, The effect of NSAIDs and a COX-2 specific inhibitor on Helicobacter pylori-induced PGE_2 and HGF in human gastric fibroblasts. *Alimentary Pharmacology and Therapeutics*, **14 (Suppl. 1):** 44–49.

Takaho, T., Fiore, S., Maddox, J.F., Brady, H.R., Petasis, N.A. and Serhan, C.N. 1997, Aspirin-triggered 15-epi-lipoxin A_4 and LXA_4 stable analogs are potent inhibitors of acute inflammation: evidence for anti-inflammatory receptors. *Journal of Experimental Medicine*, **185:** 1693–1704.

Takeuchi, K. and Johnson, L.R. 1982, Effect of cell proliferation and loss on aspirin-induced gastric damage in the rat. *American Journal of Physiology*, **243:** G463–G468.

Takeuchi, K., Okabe, S. and Takagi, K. 1976, Effect of L-glutamine on acetylsalicylic acid-induced gastric lesions in pregnant and non-pregnant rats. *Japanese Journal of Pharmacology*, **26:** 267–269.

Takeuchi, K., Ohuchi, T. and Okabe, S. 1994, Endogenous nitric oxide in gastric alkaline response in the rat stomach after damage. *Gastroenterology*, **106:** 367–374.

Takeuchi, K., Miyake, H. and Okabe, S. 1995, Mucosal ulceration in isolated amphibia stomachs in vitro. Roles of nutrient HCO_3^- and endogenous prostaglandins. *Digestion*, **56:** 357–363.

Takeuchi, K., Yasuhiro, T., Asada, Y. and Sugawa, Y. 1998, Role of nitric oxide in pathogenesis of aspirin-induced gastric mucosal damage in rats. *Digestion*, **59:** 298–307.

Tanaka, H., Shuto, K. and Muromo, H. 1982, Effect of N-acetyl-L-glutamine aluminum complex (KW-110), an antiulcer agent, on the non-steroidal anti-inflammatory drug-induced exacerbation of gastric ulcer in rats. *Japanese Journal of Pharmacology*, 32, 307–313.

Tanako, T., Fiore, S., Maddox, J.F., Brady, H.R., Petasis, N.A. and Serhan, C.N. 1997, Aspirin-triggered 15-epi-lipoxin A_4 (LX A_4) and LX A_4 stable analogues are potent inhibitors of acute inflammation: Evidence for anti-inflammatory receptors. *Journal of Experimental Medicine*, **185:** 1693–1704.

Tang, W., Stearns, R.A., Wang, R.W., Chiu, S.-H.L. and Baillie, T.A. 1999, Roles of human hepatic cytochrome P450s 2C9 and 3A4 in metabolic activation of diclofenac. *Chemical Research in Toxicology*, **12:** 192–199.

Tarnawski, A. and Ivey, K.J. 1978, Transmucosal potential-difference profile in rat upper gastrointestinal tract. A simple model for testing gastric effects of pharmacologic agents. *Canadian Journal of Physiology and Pharmacology*, **56:** 471–473.

Tarnawski, A., Ivey, K.J., Krause, W.J., Sherman, D., Burks, M. and Hewett, J. 1980, Quantitative analysis of human parietal cells after pentagastrin: correlation with gastric potential difference. *Laboratory Investigation*, **42:** 420–426.

Tarnawski, A., Hollander, D., Stachura, J., Krause, W.J. and Gergely, H. 1989, Protection of the rat gastric mucosa against aspirin injury by arachidonic acid – a dietary prostaglandin precursor of fatty acid. *European Journal of Clinical Investigation*, **19:** 278–290.

Tarnawski, A., Hollander, D., Stachura, J., Krause, W.J., Dadyfakza, V., Gergely, H. and Zipser, R.D. 1985, Can arachidonic acid (prostaglandin precursor) protect the gastric mucosa against aspirin injury? Macroscopic, histologic, ultrastructural and functional time sequence. *Gastroenterology*, **88:** 1610.

Tashima, K., Fujita, A., Umeda, M. and Takeuchi, K. 2000, Lack of gastric toxicity of nitric oxide-releasing aspirin, NCX-4016 in the stomach of diabetic rats. *Life Sciences*, **67:** 1639–1652.

Tauxe, R.V., Wright, J.F. and Hirschowitz, B.I. 1975, Marginal ulcer in achlorhydric patients. *Annals of Surgery*, **181:** 455–457.

Taylor, J.S. 1972, Carcinoma of the urinary tract and analgesic abuse. *Medical Journal of Australia*, **1:** 407–409.

Taylor, L.A. and Crawford, L.J. 1968, Aspirin-induced gastrointestinal lesions in dogs. *Journal of the American Veterinary Medical Association*, **152:** 617–619.

Temple, A.R. 1981, Acute and chronic effects of aspirin toxicity and their treatment. *Annals of Internal Medicine*, **141:** 364–369.

Tepperman, B.L. and Soper, B.D. 1986, Prostaglandin E_2 binding sites in porcine oxyntic mucosa: effects of salicylates. *Canadian Journal of Physiology and Pharmacology*, **64:** 515–520.

Terada, N., Kizaki, Z., Inoue, F., Kodo, N., Ochi, M., Furukawa, N., Kinugasa, A. and Sawada, T. 1990, The effects of salicylate on ketogenesis, gluconeogenesis and urea production in rat liver perfusion. *Acta Paediatrica Japonica*, **32:** 456–461.

Terano, A., Krause, W.J., McKenzie W., Mahoney, J. and Ivey, K.J. 1980, Effects of aspirin on rat gastric mucosa in organ culture: histological study. *Gastroenterology*, **78:** 1277.

The National Poisons Information Service Monitoring Group 1981, Analgesic poisoning: a multicentre, prospective survey. *Human Toxicology*, **1:** 7–23.

Thiefin, G., Barraya, R., Ramaholimihasso, F. and Diebold, M.D. 2000, Aspirin-induced duodenal diaphragm. *Presse Medicale*, **2:** 1351–1352.

Thillainayagam, A.V, Tabaqchali, S., Warrington, S.J. and Farthing, M.J.G. 1994, Interrelationships between *Helicobacter pylori* infection, non-steroidal anti-inflammatory drugs and gastro-duodenal disease – a prospective study in healthy volunteers. *Digestive Diseases and Sciences*, **39 (5):** 1085–1089.

Thompkins, L. and Lee, K.H. 1969, Studies on the mechanism of action of salicylates IV: effect of salicylates on oxidative phosphorylation. *Journal of Pharmaceutical Science*, **58:** 102–105.

Thompson, W.J., Chang, L.K., Rosenfeld, G.C. and Jacobson, E.D. 1977, Activation of rat gastric mucosal adenylyl cyclase by secretory inhibitors. *Gastroenterology*, **72:** 251–254.

Thorsen, W.B., Western, D., Tanaka, Y. and Morrissey, J.F. 1968, Aspirin injury to the gastric mucosa. Gastrocamera observations of the effect of pH. *Archives of Internal Medicine*, **121:** 499–506.

Thuluvath, P.J., Ninkovic, M., Calam, J. and Anderson, M. 1994, Mesalazine induced interstitial nephritis. *Gut*, **35:** 1493–1496.

Thune, S. 1968, Gastrointestinal blödning och salicylika. En jämförande undersökning mellan acetylsalicylsyra och salicylsalicylsyra. *Nordisk Medicin*, **79:** 312–314.

Thurston, D., Aao, P. and Wilson, D.E. 1979, Cyclic nucleotides and the regulation of canine gastric acid secretion. *Digestive Diseases and Sciences*, **24:** 257–264.

Tiboni, G.M., Iammarrone, E. and Piccirillo, G. 1998, Aspirin pretreatment potentiates hyperthermia-induced teratogenesis in the mouse. *American Journal of Obstetrics and Gynecology*, **178:** 270–279.

Tien, M., Svingen, B.A. and Aust, S.D. 1981, Superoxide dependent lipid peroxidation. *Federation Proceedings*, **40:** 179–182.

Tishler, S.L. and Goldman, P. 1970, Properties and reactions of salicyl-coenzyme A. *Biochemical Pharmacology*, **19:** 143–150.

Titos, E., Chiang, N., Serhan, C.N., Romano, M., Gaya, J., Pueyo, G. and Clària, J. 1999, Hepatocytes are a rich source of aspirin-triggered 15-epi-lipoxin A_4. *American Journal of Physiology, Cell Physiology*, **277:** C870–C877.

Todd, P.A. and Clissold, S.F. 1990, Naproxen. A reappraisal of its pharmacology, and therapeutic use in rheumatic diseases and pain states. *Drugs*, **40:** 91–137.

Todd, P.J., Sills, J.A., Harris, F. and Cohen, J.M. 1981, Problems with overdoses of sustained-release aspirin. *Lancet*, **1:** 777.

Tolman, K.G. 1998, Hepatotoxicity of non-narcotic analgesics. *American Journal of Medicine*, **105:** 13S–19S.

Tolman, K.G., Peterson, P., Gray, P. and Hammar, S.P. 1978, Hepatotoxicity of salicylates in monolayer cell cultures. *Gastroenterology*, **74:** 205–208.

Tolstijkov, G.A., Waltina, L.A., Murinov, Y.I., Davydova, V.A., Tolstikova, T.G., Bondarev, A.I.,

Zarudyi, F.S. and Lazareva, D.N. 1991, Beta-glycyrrhizic acid-nonsteroidal inflammatory agent complexes as new transport formulations. *Khimiko-Farmatsevticheskii Zhurnal*, **25:** 29–32.

Tonsgard, J.H. and Huttenlocher, P.R. 1981, Salicylates and Reye's syndrome. *Pediatrics*, **68:** 747–748.

Torchiana, M.L., Wiese, S.R. and Westrick, B.L. 1979, Comparison of the effects of diflusinal and other salicylates on the intragastric electropotential. *Journal of Pharmacy and Pharmacology*, **31:** 112–113.

Trautmann, M., Peskar, B.M. and Peskar, B.A. 1991, Aspirin-like drugs, ethanol-induced rat gastric injury and mucosal eicosanoid release. *European Journal of Pharmacology*, **201:** 53–58.

Traversa, G., Walker, A.M., Ippolito, F.M., Caffari, B., Capurso, L., Dezi, A., Koch, M., Maggini, M., Alegiani, S.S. and Raschetti, R. 1995, Gastroduodenal toxicity of different nonsteroidal anti-inflammatory drugs. *Epidemiology*, **6:** 49–54.

Trewby, P.N. 1980, Drug-induced peptic ulcer and gastrointestinal bleeding. *British Journal of Hospital Medicine*, **23:** 185–188.

Trnavsky, K. and Rovensky, J. 1976, Influence of some antirheumatic drugs and cytostatics on urinary enzyme level. *Pharmacology*, **14:** 378–384.

Trost, L.C. and Lemasters, J.J. 1997, Role of the mitochondrial permeability transition in salicylate toxicity to cultured rat hepatocytes: implications for the pathogenesis of Reye's syndrome. *Toxicology and Applied Pharmacology*, **147:** 431–441.

Tsodikov, G.V., Klimenko, V.C. and Laz'kova, S.N. 1979, Proliferative activity of epithelium of the surface and pits of the gastric mucosa after aspirin injury. *Bulletin of Experimental Biology and Medicine*, **88:** 1492–1496.

Tsuchiya, T. and Levy, G. 1972, Biotransformation of salicylic acid to its acyl and phenolic glucuronides in man. *Journal of Pharmaceutical Sciences*, **61:** 800–801.

Tsukada, W., Tsubokawa, M., Masukawa, T., Kojima, H., Kasahara, A. 1978, Pharmacological study of 6,11-dihydro-11-oxodibenz[b,e]oxepin-3-acetic acid (oxepinac): a new antiinflammatory drug. *Arzneimittel Forschung*, **28:** 428–438.

Turner, G. and Collins, E. 1975, Foetal effects of regular salicylate ingestion in pregnancy. *Lancet*, **2:** 338–339.

Tweeddale, M.G. and Ogilvie, R.I. 1973, Antagonism of spironolactone-induced natricresis by aspirin in man. *New England Journal of Medicine*, **289:** 198–200.

Ueda, F., Watanabe, M., Hirata, Y., Kyoi, T. and Kimura, K. 1991, Changes in cyclic-AMP content of rat gastric mucosa induced by ulcerogenic stimuli – in relation to the antiulcer activity of irsogladine maleate. *Japanese Journal of Pharmacology*, **55:** 493–499.

Uhlenbrook, K. 1965, Zur Frage der Entstehung okkulter Blutungen im Gefolge der Therapie mit Salicylaten. *Arzneimittel Forschung*, **16:** 174–177.

Umetsu, T., Kimura, K., Sanai, K. and Iwaki, K. 1980, Effect of proglumide on glycoprotein synthesis in aspirin-induced gastric erosions in rats. *European Journal of Pharmacology*, **64:** 69–77.

Ungvary, G.Y., Tatrai, E., Lorincz, M. and Barcza, G.Y. 1983, Combined embryotoxic action of toluene, a widely used industrial chemical, and acetylsalicylic acid (aspirin). *Teratology*, **27:** 261–269.

Ursadi, M.M., Franceshini, J. and Mizzotti, B. 1978, Preliminary investigation on mast degranulation and prostaglandin involvement in experimental gastric ulceration. *Prostaglandins*, **15:** 507–512.

Urushidani, T., Okabe, S., Takuchi, K. and Takagi, K. 1978, Strain differences in aspirin-induced gastric ulceration in rats. *Japanese Journal of Pharmacology*, **28:** 569–578.

Uyemura, S.A., Santos, A.C., Mingatto, F.E., Jordani, M.C. and Curti, C. 1997, Diclofenac sodium and mefenamic acid: potent inducers of the membrane permeability transition in renal cortex mitochondria. *Archives of Biochemistry and Biophysics*, **342:** 231–235.

Vaananen, P.M., Keenan, C.M., Grisham, M.B. and Wallace, J.L. 1992, Pharmacological investigation of the role of leukotrienes in the pathogenesis of experimental NSAID gastropathy. *Inflammation*, **16:** 227–240.

Vaille, C., Roze, C., de la Tour, J., Souchard, M., Chariot, J. and Debray, C. 1978, Action de l'aspirine sur les sécrétions bilare et pancreatique du rat. *Pharmaceutiques Françaises*, **36:** 207–212.

Vakil, B.J., Kulkarni, R.D., Kulkarni, V.N., Mehta, D.J., Gharpure, M.B. and Pispati, P.K. 1977a, Estimation of gastro-intestinal blood loss in volunteers treated with non-steroidal anti-inflammatory agents. *Current Medical Research and Opinion*, **5:** 32–37.

CHAPTER 8

Vakil, B.J., Shah, P.N., Dalal, N.J., Wagholikar, U.N. and Pispati, P.K. 1977b, Endoscopic study of gastrointestinal injury with non-steroidal anti-inflammatory drugs. *Current Medical Research and Opinion*, **5:** 38–42.

Valle, D. and Gartner, J. 1993, Penetrating the peroxisome. *Nature*, **361:** 682–684.

Valverde, E., Vich, J.M., Garcia-Calderon, J.V. and Garcia-Calderon, P.A. 1980, In vitro stimulation of lymphocytes in patients with chronic urticaria induced by additives and food. *Clinical Allergy*, **10:** 691–698.

Vananen, P.M., Keenan, C.M., Grisham, M.B. and Wallace, J.L. 1992, Pharmacological investigation of the role of leukotrienes in the pathogenesis of experimental NSAID gastropathy. *Inflammation*, **16:** 227–240.

Vananen, P.M., Meddings, J.B. and Wallace, J.L. 1991, Role of oxygen-derived radicals in indomethacin-induced gastric injury. *American Journal of Physiology*, **261:** G470–G475.

Vandenberg, S.A., Smolinske, S.C., Spoerke, D.G. and Rumack, B.H. 1989, Non-aspirin salicylates: conversion factors for estimating aspirin equivalency. *Veterinary and Human Toxicology*, **31:** 49–50.

Van Kolfschoten, A.A., Dembinska-Kiec, A. and Basista, M. 1980, Interaction between aspirin and paracetamol on the production of prostaglandins in the rat gastric mucosa. *Journal of Pharmacy and Pharmacology*, **33:** 462–463.

Van Kolfschoten, A.A., Zandberg, P., Jager, L.P. and Van Noordwijk, J. 1983, Protection by paracetamol against various gastric irritants in the rat. *Toxicology and Applied Pharmacology*, **69:** 37–42.

Van Marter, L.J., Leviton, A., Allred, E.N., Pagano, M., Sullivan, K.F., Cohen, A. and Epstein, M.F. 1996, Persistent pulmonary hypertension of the newborn and smoking and aspirin and nonsteroidal anti-inflammatory drug consumption during pregancy. *Pediatrics*, **97:** 658–663.

Vargaftig, B.B. 1978, Salicylic acid fails to inhibit generation of thromboxane A_2 in platelets after in vivo administration to the rat. *Journal of Pharmacy and Pharmacology*, **30:** 101–104.

Vargaftig, B.B. 1983, Detection of urticaria with food additives intolerance by means of diet. *In:* J. Morley and K.D. Rainsford (eds), *The Pharmacology of Asthma, Proceedings of an ARC Workshop*, pp. 151–179. Basel: Birkhäuser.

Vargaftig, B.B., Bessot, J.C. and Pauli, G. 1980, Is tartrazine-induced asthma related to inhibition of prostaglandin biosynthesis? *Respiration*, **39:** 276–282.

Vargaftig, B.B., Chignard, M. and Benveniste, J. 1981, Present concepts on the mechanisms of platelet aggregation. *Biochemical Pharmacology*, **30:** 263–271.

Varma, D.R. 1988, Modification of transplacental distribution of salicylate in rats by acidosis and alkalosis. *British Journal of Pharmacology*, **93:** 978–984.

Varner, A.E. 1999, The cyclooxygenase theory of atopy and asthma. *Pediatric Asthma, Allergy and Immunology*, **13:** 43–50.

Varner, A.E., Busse, W.W. and Lemanske, R.F. 1998, Hypothesis: decreased use of pediatric aspirin has contributed to the increasing prevalence of childhood asthma. *Annals of Allergy, Asthma and Immunology*, **81:** 347–351.

Veldhuyzen van Santen, S.J.O., Flook, N., Chiba, N., Armstrong, D., Barkun, A., Bradette, M., Thomson, A., Bursey, F., Blackshaw, P., Frail, D. and Sinclair, P. 2000, An evidence-based approach to the management of uninvestigated dyspepsia in the era of *Helicobacter pylori*. *Canadian Medical Association*, **(Suppl. 12):** 162.

Vessey, D.A., Hu, J. and Kelley, M. 1996, Interaction of salicylate and ibuprofen with the carboxylic acid: CoA ligases from bovine liver mitochondria. *Journal of Biochemical Toxicology*, **11:** 73–78.

Vial, T., Biour, M., Descotes, J. and Trep, C. 1997, Antibiotic-associated hepatitis: update from 1990. *Annals of Pharmacotherapy*, **31:** 204–220.

Vickers, F.N. and Stanley, M.M. 1963, Aspirin gastritis: gastroduodenoscopic observations. Four case reports. *Gastroenterology*, **44:** 419–423.

Viero, R.M. and Cavallo, T. 1995, Granulomatous interstitial nephritis. *Human Pathology*, **26:** 1347–1353.

Villar, D., Buck, W.B. and Gonzalez, J.M. 1998, Ibuprofen, aspirin and acetaminophen toxicosis and treatment in dogs and cats. *Veterinary and Human Toxicology*, **4:** 156–163.

Voigtländer, V., Hänsch, G.M. and Rother, U. 1980, Anticomplement activity of aspirin. *International Research Communications System. Medical Sciences*, **8:** 613.

Vorhees, C.V., Klein, K.L. and Scott, W.J. 1982, Aspirin-induced psychoteratogenesis in rats as a function of embryonic age. *Teratogenesis, Carcinogenesis and Mutagenesis*, **2:** 77–84.

Wada, K., Kamisaki, Y., Kitano, M., Kishimoto, Y., Nakamoto, K. and Itoh, T. 1996, A new gastric ulcer model induced by ischemia-reperfusion in the rat: role of leukocytes on ulceration in rat stomach. *Life Sciences*, **59:** P295–P301.

Waldman, R.J., Hall, W.N., McGee, H. and Amburg, G.V. 1982, Aspirin as a risk factor in Reye's syndrome. *Journal of the American Medical Association*, **247:** 3089–3094.

Waldman, T.A. 1961, Gastrointestinal protein loss demonstrated by ^{51}Cr-labelled albumin. *Lancet*, **2:** 121–123.

Walker, F.S. and Rainsford, K.D. 1997, Do NSAIDs adversely affect joint pathology in osteoarthritis? *In: Side Effects of Anti-inflammatory Drugs IV*, pp. 43–53. Dordrecht: Kluwer Academic Publishers.

Walker, J.R. and Smith, M.J.H. 1979, Adrenocortical stimulation and the anti-inflammatory actions of salicylates. *Journal of Pharmach and Pharmacology*, **31:** 640–641.

Wallace, J., Arfors, K.E. and McKnight, G.W. 1991, A monoclonal antibody against the CD18 leukocyte adhesion molecule prevents indomethacin-induced gastric damage in the rabbit. *Gastroenterology*, **100:** 878–883.

Wallace, J.L. 1999, Selective COX-2 inhibitors: is the water becoming muddy? *Trends in Pharmacological Sciences*, **20:** 4–6.

Wallace, J.L., Keenan, C.M. and Granger, D.N. 1990, Gastric ulceration induced by non-steroidal anti-inflammatory drugs is a neutrophil-dependent process. *Americal Journal of Physiology*, **259:** G462–G467.

Wallace, J.L., McKnight, G.W. and Bell, C.J. 1995, Adaptation of rat gastric mucosa to aspirin requires mucosal contact. *American Journal of Physiology – Gastrointestinal and Liver Physiology*, **31:** G134–G138.

Wallace, J.L., McKnight, G.W., Miyasaka, M., Tamatani, T., Paulson, J., Anderson, D.C., Granger, D.N. and Kubes, P. 1993, Role of endothelial adhesion molecules in NSAID-induced gastric mucosal injury. *American Journal of Physiology*, **265:** G993–G998.

Wallace, J.L., McKnight, G.W., Reuter, B.K. and Vergnolle, N. 2000, NSAID-induced gastric damage in rats: requirement for inhibition of both cyclooxygeanse 1 and 2. *Gastroenterology*, **119:** 706–714.

Wallace, K.B. and Starkov, A.A. 2000, Mitochondrial targets of drug toxicity. *Annual Reviews of Pharmacology and Toxicology*, **40:** 353–388.

Walt, W.O. 1992, Misoprostol for the treatment of peptic ulcer and anti-inflammatory drug-induced gastroduodenal ulceration. *New England Journal of Medicine*, **327:** 1575–1580.

Wang, M. and Dickinson, R.G. 1998, Hepatobiliary transport of diflunisal conjugates and taurocholate by the perfused rat liver: The effect of chronic exposure of rats to diflunisal. *Life Sciences*, **62:** 751–762.

Warkany, J. and Takacs, E. 1959, Experimental production of congenital malformations in rats by salicylate poisoning. *American Journal of Pathology*, **35:** 315–331.

Waterbury, L.D., Fisher, P.E., Garay, G.L., Petty, T. and Rosenkranz, R.P. 1988, Stimulation of mucus production and prevention of aspirin induced ulcerogenesis by enprostil in the rat. *Proceedings of the Western Pharmacology Society*, **31:** 21–23.

Watson, R.M. and Pierson, R.N. 1960, Absence of gastrointestinal bleeding following administration of salicylic acid. *Federation Proceedings*, **19:** 191.

Webber, P.E., Larsson, C., Anggard, E., Hamberg, M., Corey, E.J., Nicolaou, K.C. and Samuelsson B. 1976, Stimulation of renin release from rabbit renal cortex by arachidonic acid and prostaglandin endoperoxides. *Circulation Research*, **39:** 868–874.

Wedi, B. and Kapp, A. 2000, Aspirin-induced adverse skin reactions: new pathophysiological aspects. *Thorax*, **55 (Suppl. 2):** S70–S71.

Weinstein, D., Katz, M. and Kazmer, S. 1981, Use of rat/hamster 5–9 mixture in the Ames mutagenicity assay. *Environmental Mutagens*, **3:** 1–9.

CHAPTER 8

Weiss, J.W., Drazen, J.M., Coles, N., McFadden, E.R., Weller, P.F., Corey, E.J., Lewis, R.A. and Austen, K.F. 1982, Bronchoconstrictor effects of leukotriene (in humans). *Science (Washington DC)*, **216:** 196–198.

Weltman, J.K., Szaro, R.P. and Settipane, G.A. 1978, An analysis of the role of lgE in intolerance to aspirin and tartrazine. *Allergy*, **34:** 273–281.

Wheatley, P.J. 1964, The crystal and molecular structure of aspirin. *Journal of the Chemical Society*, **1964:** 6036–6048.

Wheldrake, J.F. 1975, The effect of aspirin (acetyl-salicylate) on macromolecule turnover in rat kidney and liver. *Experientia (Basel)*, **31:** 559–560.

Whelton, A. 1995, Renal effects of over-the-counter analgesics. *Journal of Clinical Pharmacology*, **35:** 454–463.

Whitacre, C.C., Reingold, S.C., O'Looney, P.A. and The Task Force on Gender, Multiple Sclerosis and Autoimmunity. 1999, A gender gap in autoimmunity. *Science*, **283:** 1277–1278.

White, L.E., Driggers, D.A. and Wardinsky, T.D. 1980, Poisoning in childhood and adolescence: A study of 111 cases admitted to a military hospital. *Journal of Family Practice*, **11:** 27–31.

Whitehouse, L.W., Paul, C.J., Wong, L.G. and Thomas, B.H. 1977b, Effect of aspirin on a subtoxic dose of 14C-acetaminophen in mice. *Journal of Pharmaceutical Sciences*, **66:** 1399–1403.

Whitehouse, M.W. 1965, Some biochemical and pharmacological properties of anti-inflammatory drugs. *Progress in Drug Research*, **8:** 321–429.

Whitehouse, M.W. 1977, Some biochemical complexities of inflammatory disease affecting drug action. *In:* I.L. Bonta (ed.), *Recent Developments in the Pharmacology of Inflammatory Mediators, Agents and Actions Supplement No. 2*, pp. 135-147. Basel: Birkhäuser.

Whitehouse, M.W. and Beck, F.J. 1973, Impaired drug metabolism in rats with adjuvant-induced arthritis: a brief review. *Drug Metabolism and Disposition*, **1:** 251–255.

Whitehouse, M.W. and Rainsford, K.D. 1977, Side-effects of anti-inflammatory drugs. *In:* I.L. Bonta, J. Thompson and K. Brune (eds), *Inflammation: Mechanisms and their Impact on Therapy*, pp. 171–187. Basel: Birkhäuser.

Whitehouse, M.W. and Rainsford, K.D. 1980, Esterification of acidic anti-inflammatory drugs suppresses, their gastrotoxicity without adversely affecting their anti-inflammatory activity in rats. *Journal of Pharmacy and Pharmacology*, **32:** 795–796.

Whitehouse, M.W. and Rainsford, K.D. 1982, Comparison of the gastric ulcerogenic activities of different salicylates. *In:* C.J. Pfeiffer (ed.), *CRC Handbook – Drugs and Peptic Ulcer Disease*, Vol. 2, pp. 127–141. Boca Raton, FL: CRC Press.

Whitehouse, M.W., Rainsford, K.D., Ardlie, N.G., Young, I.G. and Brune, K. 1977, Alternatives to aspirin, derived from biological sources. *In:* K.D. Rainsford, K. Brune and M.W. Whitehouse (eds), *Aspirin and Related Drugs: Their Actions and Uses*, pp. 43–57. Basel: Birkhäuser.

Whitehouse, M.W., Rainsford, K.D., Mearrick, P., Percy, N. and Osborn, P.G. 1981, Development of a less gastro-irritant soluble aspirin formulation (Preface to Clinical Study). *Proceedings of the Australian Rheumatic Association*.

Whitehouse, M.W., Rainsford-Koechli, V. and Rainsford, K.D. 1984, Aspirin gastrotoxicity: protection by various strategens. *In:* K.D. Rainsford and G.P. Velo (eds), *Side-Effects of Anti-inflammatory/Analgesic Drugs, Advances in Inflammation Research*, Vol. 6, pp. 77–87. New York: Raven Press.

Whiting, P.H., Tisocki, K. and Hawksworth, G.M. 1999, Human renal medullary interstitial cells and analgesic nephropathy. *Renal Failure*, **21:** 387–392.

Whittle, B.J.R. 2001, Basis of gastrointestinal toxicity of non-steroid anti-inflammatory drugs. *In:* J.R. Vane and R.M. Botting (eds), *Therapeutic Roles of Selective COX-2 Inhibitors*, pp. 329–354. London: William Harvey Press.

Whittle, B.J.R., Broughton-Smith, N.K., Moncada, S. and Vane, J.R. 1978, Actions of prostacyclin (PGI_2) and its product, 6-oxo-$PGF_{1 alpha}$ on the rat gastric mucosa in vivo and in vitro. *Prostaglandins*, **15:** 955–967.

Whittle, B.J.R., Kauffman, G. and Moncada, S. 1981, Vasoconstriction with thromboxane A2 induces ulceration of the gastric mucosa. *Nature*, **292:** 472–474.

Whittle, B.J.R., László, F., Evans, S.M. and Moncada, S. 1995, Induction of nitric oxide synthase and microvascular injury in the rat jejunum provoked by indomethacin. *British Journal of Pharmacology*, **116:** 2286–2290.

Whittle, B.J.R., Lopes-Bermonte, J. and Moncada, S. 1990, Regulation of gastric mucosal integrity by endogenous nitric oxide: interactions with prostanoids and sensory neuropeptides in the rat. *British Journal of Pharmacology*, **99:** 607–611.

Widal, M.F., Abrami, P. and Lermoyez, J. 1922, Anaphylaie et idiosyncrasie. *Presse Medicale*, **30:** 189–192.

Wilcox, C.M. 1997, Relationship between nonsteroidal anti-inflammatory drug use, *Helicobacter pylori*, and gastroduodenal mucosal injury. *Gastroenterology*, **113:** S85–S89.

Wilhelmi, G. 1974, Species differences in susceptibility to the gastro-ulcerogenic action of anti-inflammatory agents. *Pharmacology*, **11:** 220–230.

Wilhelmi, G. and Menassé-Gydnia, R. 1972, Gastric mucosal damage induced by non-steroid anti-inflammatory agents in rats of different ages. *Pharmacology*, **8:** 321–328.

Wilkens, R.F. 1985, Worldwide clinical safety experience with diclofenac. *Seminars in Arthritis and Rheumatism*, **15 (Suppl. 1):** 105–110.

Wilkinson, G.R. and Schenker, S. 1976, Effects of liver disease on drug disposition in man. *Biochemical Pharmacology*, **25:** 2675–2681.

Williams, A.M. and Dickinson, R.G. 1994, Studies on the reactivity of acyl glucuronides – VI. Modulation of reversible and covalent interaction of diflunisal acyl glucuronide and its isomers with human plasma protein in vitro. *Biochemical Pharmacology*, **47:** 457–467.

Williams, A.M., Worrall, S., De Jersey, J. and Dickinson, R.G. 1995, Studies on the reactivity of acyl glucuronides – VIII. Generation of an antiserum for the detection of diflunisal-modified proteins in diflunisal-dosed rats. *Biochemical Pharmacology*, **49:** 209–217.

Williams, R.T. 1974, Inter-species in the metabolism of xenobiotics. *Transactions of the Biochemical Society*, **2:** 359–377.

Williams, S.E. and Turnberg, L.A. 1980, Retardation of acid diffusion by pig gastric mucosa: a potential role in mucosal protection. *Gastroenterology*, **79:** 299–304.

Williams, S.E. and Turnberg, L.A. 1981, Demonstration of a pH gradient across mucus adherent to rabbit gastric mucosa: evidence for a 'mucus–bicarbonate' barrier. *Gut*, **22:** 94–96.

Wilson, C.W.M. 1975, Clinical pharmacological aspects of ascorbic acid. *Annals of the New York Academy of Sciences*, **258:** 355–376.

Wilson, D.E. 1991, Role of prostaglandins in gastroduodenal mucosal protection. *Journal of Clinical Gastroenterology*, **13 (Suppl. 1):** S65–S71.

Wilson, J.H.P., Koole-Lesuis, H., Edixhoven-Bosdijk, A., Van Den Berg, J.W.O. and Von Essen, H.A. 1985, Determination of haemoglobin in gastric aspirates. *Journal of Clinical Chemistry and Clinical Biochemistry*, **23:** 841–844.

Wiseman, E.H. and Rheinert, T. 1975, Anti-inflammatory drugs and renal papillary necrosis. *Agents and Actions*, **5:** 322–325.

Wittmers, L.E., Anderson, L.A., Fall, M.M. and Alich, A.A. 1990, Intragastric insulin as a measure of mucosal damage cause by aspirin. *Journal of Pharmacological Methods*, **24:** 229–239.

Wolf, K., Castrop, H., Hartner, A., Goppelstrube, M., Hilgers, K.F. and Kurtz, A. 2000, Inhibition of the renin–angiotensin system upregulates cyclooxygenase-2 expression in the macula densa. *Hypertension*, **34:** 503–507.

Wolfe, J.D., Metzger, A.L. and Goldstein, R.C. 1974, Aspirin hepatitis. *Annals of Internal Medicine*, **30:** 74–76.

Wollin, A., Soll, A.H. and Samloff, I.M. 1979, Actions of histamine, secretin, and PGE_2 on cyclic AMP production by isolated canine fundic mucosal cells. *American Journal of Physiology*, **237:** E437–E443.

Wong, L.T., Solomanaj, G. and Thomas, B.H. 1976, Metabolism of [^{14}C] paracetamol and its interactions with aspirin in hamsters. *Xenobiotica*, **6:** 575–584.

Wood, P.H.N., Harvey-Smith, E.A. and Dixon, A.St.-J. 1962, Salicylates and gastrointestinal bleeding. Acetylsalicylic acid and aspirin derivatives. *British Medical Journal*, **1:** 669–675.

Woods, H.F., Stubbs, W.A., Johnson, G. and Alberti, K.G.M.M. 1974, Inhibition by salicylate of gluco-neogenesis in the isolated perfused rat liver. *Clinical and Experimental Pharmacology and Physiology*, **1:** 535–540.

Worrall, S. and Dickinson, R.G. 1995, Rat serum albumin modified by diflunisal acyl glucuronide is immunogenic in rats. *Life Sciences*, **56:** 1921–1930.

Wortman, D.W., Kelsch, R.C., Kuhns, L., Sullivan, D.B. and Caddidy, J.J. 1980, Renal papillary necrosis in juvenile rheumatoid arthritis. *Journal of Paediatrics*, **97:** 37–40.

Wu, D. and Cederbaum, A.L. 2001, Sodium salicylate increases CYP2E1 levels and enhances arachidonic acid toxicity in HepG2 cells and cultured rat hepatocytes. *Molecular Pharmacology*, **59:** 795–805.

Yeoh, K.G., Kang, J.Y., Yap, I., Guan, R., Tan, C.C., Wee, A. and Teng, C.H. 1995, Chili protects against aspirin-induced gastroduodenal mucosal injury in humans. *Digestive Diseases and Sciences*, **40:** 580–583.

Yeomans, N.D., Saint John, D.J. and de Boer, W.G. 1973, Regeneration of gastric mucosa after aspirin-induced injury in the rat. *American Journal of Digestive Diseases*, **18:** 773–780.

Yeomans, N.D., Skeljo, M.V. and Giraud, A.S. 1992, The role of acid regulation in the treatment of NSAID-induced mucosal damage. *Digestion*, **51:** 3–10.

Yew, M. 1973, 'Recommended daily allowances' for vitamin C. *Proceedings of the National Academy of Sciences of the USA*, **70:** 969–972.

Yokoyama, A., Takakubo, F., Eto, K., Ueno, K., Igarashi, T., Satoh, T. and Kitagawa, H. 1984, Teratogenicity of aspirin and its metabolite, salicylic acid, in cultured rat embryos. *Research Communications in Chemical Pathology and Pharmacology*, **46:** 77–91.

Yoshida, S., Nakagawa, H., Yamawaki, Y., Sakamoto, H., Akahori, K., Nakabayashi, M., Sakamoto, M., Hasegawa, H., Shoji, T., Tajima, T. and Amayasu, H. 1998, Bronchial hyperresponsiveness, hypersensitivity to analgesics and urinary leukotriene E$_4$ excretion in patients with aspirin-intolerant asthma. *International Archives of Allergy and Immunology*, **117:** 146–151.

Yoshida, N., Yoshikawa, T., Nakamura, Y., Arai, M., Matsuyama, K., Iinuma, S., Yagi, N., Naito, Y., Miyasaka, M. and Kondo, M. 1995, Role of neutrophil-mediated inflammation in aspirin-induced gastric mucosal injury. *Digestive Diseases and Sciences*, **40:** 2300–2304.

Yoshida, Y., Fujii, M., Brown, F.R. III and Singh, I. 1988, Effect of salicylic acid on mitochondrial-peroxisomal fatty acid catabolism. *Pediatric Research*, **23:** 338–341.

Yoshida, Y., Wang, S. and Osame, M. 1998, Aspirin induces short-chain free fatty acid accumulation in rats. *European Journal of Pharmacology*, **349:** 49–52.

Yunginger, J.W., O'Connell, E.J. and Logan, G.B. 1973, Aspirin-induced asthma in children. *Journal of Paediatrics*, **82:** 218–221.

Yurchak, A.M., Wicher, K. and Arbesman, K.E. 1970, Immunologic studies on aspirin. Clinical studies with aspiryl–protein conjugates. *Journal of Allergy*, **46:** 245–253.

Yusibova, N.A.Y., Bezkrovnaya, V.G. and Chekhovskaga, T.A. 1978, *Vestsi Akademii Nauk Belaruskai SSR*, **1:** 128–129.

Zambraski, E.J., Rofrano, T.A. and Ciccone, C.D. 1982, Effects of aspirin treatment on kidney function in exercising man. *Medicine and Science in Sports and Exercise*, **14:** 419–423.

Zanagnolo, V., Dharmarajan, A.M., Endo, K. and Wallach, E.E. 1996, Effects of acetylsalicylic acid (aspirin) and naproxen sodium (naproxen) on ovulation, prostaglandin, and progesterone production in the rabbit. *Fertility and Sterility*, **65:** 1036–1943.

Zhu, D., Becker, W.M., Schulz, K.H., Schubele, K. and Schlaak, M. 1992, The presence of specific IgE to salicyloyl and O-methylsalicyloyl in aspirin-sensitive patients. *Asian and Pacific Journal of Allergy and Immunology*, **10:** 25

Zimmerman, G.A. and Clemmer, T.P. 1981, Acute respiratory failure during therapy for salicylate intoxication. *Annals of Emergency Medicine*, **10:** 104–106.

Zimmerman, H.J. 1981, Effects of aspirin and acetaminophen on the liver. *Annals of Internal Medicine*, **141:** 333–342.

Zimmerman, K.C., Sarbia, M., Schror, K. and Weber, A.A. 1998, Constitutive cyclooxygenase-2 expression in healthy human and rabbit gastric mucosa. *Molecular Pharmacology*, **54:** 536–540.

Zucker, P., Daum, F. and Cohen, M.I. 1975, Aspirin hepatitis. *American Journal of Diseases of Children*, **129:** 1433–1434.

Reye's Syndrome and Aspirin

J.F.T. Glasgow and S.M. Hall

INTRODUCTION

Reye's syndrome (RS) is a serious, acute encephalopathy, principally of childhood, which is associated with selective hepatic dysfunction; occasionally adults are affected. The primary insult is to mitochondria, and evidence for this relates both to structural integrity and biochemical function. Aetiopathogenesis is likely to be multifactorial and although still unclear, some clues have emerged. There is a sizeable literature from the USA since the early 1970s, much of which is based on data from a national reporting scheme that began in 1973. Epidemiological surveillance of RS in the UK and Ireland took place between 1981 and 2001. No other countries have ongoing surveillance schemes, but individual case reports and time-limited series, usually based on hospital referrals, have been published from many parts of the world.

Following recognition in the early 1980s of a possible association with aspirin medication and subsequent public health intervention, there has been a striking reduction in the number of cases reported both in the UK and the USA – a decline described recently in the *New England Journal of Medicine* as a 'public health triumph' (Monto, 1999). The intervening years have seen several related developments; one was the description of a group of novel inherited metabolic disorders (IMD) with features that mimic RS. Simultaneously, less expensive and commercially available laboratory equipment simplified diagnosis of these conditions. Together these innovations have tended to alter not only the way clinicians regard a diagnosis of RS (if indeed it may still be considered a single entity), but also the efforts directed towards a more specific diagnosis. RS needs to be recognised early and aggressive management initiated in a paediatric intensive care unit. Even with optimum management, mortality and morbidity are significant, and those who apparently have made a full recovery and are attending normal school tend to show subtle cognitive and academic impairments.

The purpose of this chapter is to describe clinical, epidemiological and laboratory findings and discuss differential diagnosis and relevant inherited metabolic disorders (IMD), followed by principles of management, prognosis and outcome. Efforts to elucidate causation are described, and in particular the place of aspirin usage is reviewed in detail. Current understanding of pathogenesis is discussed and recent experimental data summarised – data showing that salicylate and its metabolites interact with cell biochemistry and may plausibly be involved in pathogenesis. Finally, the importance of present and future vigilance and the types of action required to prevent resurgence of this devastating encephalopathy are emphasised.

BACKGROUND

RS was putatively first described in 1929 by the late Lord Brain (Brain *et al.*, 1929), long before the definitive publication(s) in the autumn of 1963 (Johnson *et al.*, 1963; Reye *et al.*, 1963). These papers reported a total of 36 children seen since the early 1950s with acute encephalopathy and fatty infiltration of the liver and other organs. Reye reported 21 seriously ill children, all but 2 of whom were comatose at the time of admission; 17 died. Since 1973 the current authors have dealt with 79 patients in Northern Ireland, 56 of whom have been prospectively studied, and were reviewed for a recent National Workshop (NWS) on RS and Inherited Metabolic Disorders; 23 of these have been reported previously (Glasgow, 1984). In the UK as a whole, 450 cases were reported between 1981 and 2001 (Hall and Lynn, 2002), details of which are given in Table 9.1.

DEFINITION

The British Paediatric Surveillance Unit defines RS in children less than 16 years of age as an unexplained, non-inflammatory encephalopathy, associated with serum aspartate, or alanine aminotransferases, or plasma ammonia more than three times the upper normal laboratory range, or hepatic

TABLE 9.1

Reye's syndrome in the UK and Ireland 1981/82–1996/97.

Reporting period (August–July)	Total reports from the UK	Revised diagnosis (inherited metabolic disorder in brackets)	*Cases of Reye's syndrome	Number of deaths (of cases)
1981/82	47	7 (3)	40	26
1982/83	69	10 (6)	59	34
1983/84	93	12 (3)	81	36
1984/85	64	8 (2)	56	32
1985/86	53[1]	13 (4)	39	22
1986/87	47	21 (11)	26	13
1987/88	44	12 (3)	32	19
1988/89	31	13 (6)	18	9
1989/90	24[1]	8 (5)	15	7
1990/91	25	13 (8)	12	5
1991/92	23[2]	6 (5)	15	6
1992/93	21[3]	10 (6)	5	4
1993/94	20[4]	13 (7)	3	3
1994/95	17[2]	3 (2)	12	3
1995/96	18[1]	2 (1)	15	7
1996/97	7	2 (2)	5	4
1997/98	11	4 (2)	7	5
1998/99	11	4 (3)	7	2
1999/00	4	1 (1)	3	2
2000/01	3[†]	2 (1)	0	0
Total	632	164 (81)	450	239

*Compatible with the diagnosis.
[†]To April 2001.
[1]Follow-up not received for one case; [2]Follow-up not received for two cases; [3]Follow-up not received for five cases and one case did not meet the case definition; [4]Follow-up not received for four cases.

fatty infiltration that is microvesicular in appearance and panlobular in distribution. Diagnostic criteria are thus very non-specific.

CLINICAL FEATURES

RS is widely considered to be a biphasic illness. First there is a viral prodrome, most frequently with upper respiratory or 'flu-like symptoms, varicella or, less frequently, gastroenteritis. The interval between the prodrome and the second (neurological) phase tends to be 1–7 (median 3) days. The neurological phase marks an abrupt deterioration of the patient's condition. It begins with vomiting, which tends to be repeated, profuse and often contains altered blood. In one study, symptomatology of RS patients showed differences when compared with those who had similar viral illnesses uncomplicated by RS. In the RS group, headache, anorexia, vomiting, abdominal pains, altered behaviour, drowsiness or loss of consciousness were significantly more likely, probably reflecting the onset of rising intracranial pressure (Hall *et al.*, 1988). The mechanism of vomiting remains unclear and is likely multifactorial. Elevated polyamines have been documented in RS, and spermine, derived from ornithine, induces vomiting when given systemically to volunteers (Campbell and Brown, 1985). Appearance of vomiting is thought to relate to the onset of encephalopathy the pathogenesis of which has been discussed by DeLong and Glick (1982). De Vivo (1988) makes the point, however, that children questioned carefully after recovery can recall the early period of vomiting – an observation suggesting that cerebral function and memory appear to be preserved.

Neurological signs

Changes in consciousness and demeanour are most apparent, beginning with lethargy and disinterest in surroundings. Early features can be subtle, and parents often feel a child to be more ill than doctors may discern (J.C. Partin, 1988). Drowsiness, or combativeness with irrational behaviour (agitated delirium), commonly follows, and/or the child becomes progressively less responsive and lapses into coma. Papilloedema is almost invariably absent. Hyperventilation and pupillary abnormalities, evidence of pyramidal tract dysfunction and abnormal responses to external stimuli (see below) tend to occur over the course of a few hours to several days. RS is an unpredictable disorder, and there is considerable variation in the rate of neurological progression: some patients do not progress to unconsciousness, others recover well in spite of a poor outlook (see below), and others that receive scant treatment make a virtual spontaneous recovery.

The neurological features of RS can be graded according to clinical course and expression of illness. Various groups categorise the features in slightly differing ways. That of Lovejoy and colleagues (1974) focuses on the degree of neurological involvement. It is a modification of the system of progressive cerebral dysfunction set forth in the classic monograph of Plum and Posner (1966): Lovejoy grade I consists of vomiting with lethargy and sleepiness (although not mentioned in this classification, by implication grade 0 is a child that is awake and alert); grade II, of disorientation, delirium and combativeness, neurogenic hyperventilation, brisk reflexes and appropriate responses to noxious stimuli. Grade III consists of coma, hyperventilation and decorticate posturing on stimulation, with preservation of pupillary and oculocephalic (doll's eye) reflexes. In grade IV there is deepening coma, decerebrate posturing, loss of the oculocephalic reflexes, large fixed pupils, and dysconjugate eye movements in response to caloric stimulation of the oculovestibular reflex. Grade V is marked by features of brainstem death with respiratory arrest, flaccidity and areflexia. (Another grade, not part of Lovejoy, is grade VI; this unclassifiable stage refers to those in whom therapeutic neuromuscular paralysis has been commenced). With the provisos mentioned, grades I–V are identical to those used by the Centers for Disease Control and Prevention at Atlanta, USA (CDC) (see Belay *et al.*, 1999).

Alerting features

To summarise therefore, any child that, after a brief, acute, febrile (presumed viral) illness, develops unexplained vomiting and/or neurological or behavioural symptoms or signs of any type or degree, especially accompanied by clouding of consciousness (although this need not be present), should be investigated promptly for RS.

INVESTIGATIONS

Electroencephalography

The EEG in children with RS or metabolic encephalopathy tends to be non-specific. From the outset, even before clinical symptoms arise, slow waves of the delta frequency (1–3/s) appear, more often posteriorly, and may persist after consciousness has returned. The pattern tends to correlate with CSF glutamine levels. If raised ICP develops, cerebral hypoxia/ischaemia results in flattening of the fast activity, until with brain death the record becomes isoelectric (Brown and Imam, 1991).

Tissue pathology

Liver

Hepatic changes represent a continuum, appearing soon after the onset of vomiting – the point from which neurological involvement is thought to arise. The severity of the encephalopathy seems to correlate with their evolution and magnitude. Earliest abnormalities are loss of glycogen and cytoplasmic swelling, most apparent at the periphery of the lobule. These changes become panlobular as the disorder progresses. Another early feature is the appearance of panlobular, microvesicular fat droplets, largely of triglyceride. In mild cases lipid tends to clear in 2 to 5 days, but in the more severe cases this can persist for up to 9 days. Such changes are also seen in other organs, principally the proximal renal tubule and loop of Henle, myocardium, pancreas and lymph nodes. Inflammation and cell necrosis are inconspicuous, and biliary stasis or duct system abnormalities absent (Brown and Imam, 1991). However, fatty infiltration of the liver is a non-specific finding (Bonnell and Beckwith, 1986) and is reported in some β-oxidation disorders (Treem et al., 1986), glycogen storage diseases (Leonard, 1988), numerous other nutritional and metabolic disorders, and certain drug and toxin ingestions (see summary in Mowat, 1994).

Ultrastructural and histochemical changes are essential for the diagnosis of RS. Electron microscopic changes early on show swollen pleomorphic mitochondria, loss of dense bodies with proliferation of the smooth endoplasmic reticulum and peroxisomes (Brown and Madge, 1972). Later, the mitochondria become large with progressively disrupted matrices (Partin et al., 1971; Bove et al., 1975). Like hepatic fatty infiltration, peroxisomal proliferation is not specific to RS – in contradistinction to the mitochondrial changes, which are unique and are absent in other acute intoxications (Partin et al., 1984), metabolic encephalopathies (Treem et al., 1986) or disorders of mitochondrial DNA (Poulton and Brown, 1995).

Histochemical stains show marked reduction in activities of mitochondrial enzymes such as succinic acid dehydrogenase and cytochrome oxidase. Biochemical assays of these and other mitochondrial enzymes (ornithine transcarbamylase, carbamyl phosphate synthetase, glutamate dehydrogenase, pyruvate dehydrogenase, pyruvate carboxylase, monoamine oxidase) are temporarily reduced; this reduction is maximal on the first day of illness, but can persist in survivors to the seventh day. By contrast, activities of cytosolic enzymes are maintained throughout (Mowat, 1994).

Brain

Macroscopically, brain oedema is present, perhaps leading to secondary brainstem compression and herniation at the tentorium (with consequent III nerve injury) and/or the foramen magnum; hence the range and sequence of neurological signs. Laminar neuronal necrosis is consistent with a degree of cerebral hypoxia. Much of the brain swelling is caused by large fluid-filled blebs within myelin sheaths. Endothelial cells show tubuloreticular inclusions (said to indicate interferon production). Astrocytes tend to be swollen and depleted in glycogen (J.S. Partin, 1988). As in other hyperammonaemic conditions, throughout the brain protoplasmic astrocytes are enlarged as type II Alzheimer cells; they are thought to convert ammonia to glutamine (Cavanagh and Kyu, 1971). Electronmicroscopy (performed on material from a few severely ill patients undergoing craniectomy) shows that the mitochondria of neurones, but not of other cells, have changes similar to those seen in hepatocytes (J.S. Partin, 1988). However, it is well documented that mitochondrial enzyme activities are relatively unaffected in brain (and muscle) compared to liver (Robinson *et al.*, 1978; Table 9.2).

Skeletal muscle

Skeletal muscle cells show abnormal lipid accumulation, glycogen depletion and mitochondrial injury, especially in Type I fibres.

Laboratory diagnosis

Liver function tests

Striking (though selective) abnormalities in liver function are universal. Serum aspartate aminotransferase (AST) and alanine aminotransferase (ALT) are elevated from twice to greater than 100 times normal; the magnitude of the increase *per se* bears no relationship to illness severity or prognosis. Hyperammonaemia is present in most cases, although it may be transient. Most authorities agree that the height of the blood ammonia is predictive of disease severity and outcome; our findings in 56 cases treated over a 7-year period confirm this association (unpublished data). The comatose patient with an initial blood ammonia greater than 5 times normal will likely have a poor outcome (Huttenlocher, 1972; Roe *et al.*, 1975). Prothrombin time is also consistently prolonged. On the other hand, serum bilirubin tends to be normal. Slight elevations are seen in shocked patients; 18 per cent in our series. Hypoglycaemia is a feature, especially in babies and in the more severely ill; in two-thirds of our patients, initial blood glucose was less than 2.6 mmol/l.

TABLE 9.2

Sites of reduced mitochondrial enzyme activity in Reye's syndrome (Robinson *et al.*, 1978).

Enzyme	Location		
	Liver	Muscle	Brain
Citrate synthetase	Low	Normal	Normal
Glutamate dehydrogenase	Low	–	Normal
Pyruvate carboxylase	Low	–	–
Pyruvate dehydrogenase	Low	Normal	Normal
Succinate dehydrogenase	Low	Normal	Normal

Other biochemistry

Electrolyte values tend to be normal while blood urea and creatinine are mildly elevated, indicating a degree of dehydration. A mixed metabolic acidosis/respiratory alkalosis may be present in comatose patients.

CSF examination

Although the criterion of a non-inflammatory encephalopathy is that there be less than 8 leucocytes/μl, lumbar puncture as part of the initial assessment should not be done unless the diagnosis, especially in babies, is in doubt and any increase in intracranial pressure has been adequately controlled. Low levels of CSF glucose are found in relation to subnormal blood glucose levels; profoundly low values were present in 6 critically ill babies out of 28 reported earlier (Glasgow, 1984).

DIFFERENTIAL DIAGNOSIS

Differential diagnosis depends upon the age of the child and whether coma is present. The sequence of events in which a mild viral illness is followed by vomiting and selective hepatic dysfunction, agitated delirium and coma is characteristic of RS (see above). In the conscious patient, consideration must be given to anicteric hepatitis A or B, Epstein–Barr viral hepatitis, cytomegalovirus hepatitis, or α_1-antitrypsin deficiency. Varicella hepatitis is less likely in the absence of generalised varicella-zoster infection (Lichtenstein *et al.*, 1983). Margosa oil and hypoglycins (unripe ackee fruit) are known to cause similar illnesses, and isopropyl alcohol (Glasgow and Ferris, 1969) has been reported in association with a Reye-like illness. Drug intoxication with salicylates or paracetamol, or anticonvulsant therapy with sodium valproate, are much more likely possibilities in Western society. However, a careful history and appropriate laboratory tests should settle the matter. Occasionally, a primary intra-abdominal emergency – intussusception or volvulus – may confuse the unwary, as vomiting, confusion and agitation are to the fore.

In comatose patients it is essential to consider subarachnoid bleeding or intracranial infections (various types), though abnormal liver biochemistry would not normally be expected; these and other more common differentials are shown in Table 9.3.

Inherited metabolic disorders (IMD)

There are a number of 'Reye-like' IMDs whose clinical and pathological manifestations may be identical to those of RS. Where an initial report of RS to the British surveillance scheme has been revised

TABLE 9.3

Differential diagnosis of the comatose Reye's syndrome patient.*

Widespread hypoxic/ischaemic insult – including liver and brain	Intracranial bleeding (e.g. subarachnoid)
Septicaemic shock	Fulminant hepatitis – especially in early infancy
Severe generalised viral infection – e.g. adenovirus, varicella	Haemorrhagic shock encephalopathy syndrome
	Severe dehydration
Salmonellosis, shigellosis	Apparent life-threatening event
Intracranial infection – e.g. meningitis, encephalitis	Metabolic disorders (see Table 9.4)
	Drugs/toxins (see Table 9.5)

*Implies abnormal liver biochemistry.

TABLE 9.4

Reye's syndrome can be confused with inherited metabolic disorders.*

***

Urea cycle – any of the six enzymes can be deficient; ornithine transcarbamylase is the most common.

Organic acidurias – branched chain amino acid disorders – propionic, methylmalonic, or isovaleric acidaemia; glutaric aciduria type II.

Ketogenesis – e.g. deficiencies of various acyl CoA dehydrogenases have been reported, of which medium chain CoA dehydrogenase is the most common; hydroxymethyl glutaric aciduria.

Primary disorders of carnitine metabolism – e.g. carnitine deficiency, carnitine palmitoyl transferase deficiency.

Disorders of carbohydrate metabolism – e.g. fructose 1,6-diphosphatase deficiency, fructosaemia, galactosaemia.

Other genetic disorders – e.g. cystic fibrosis, α_1-antitrypsin deficiency, familial erythrophagocytic reticulosis.

***

*De Vivo (1988) reviews these and other disorders, which are grouped according to predominant clinical and biochemical features.

subsequently, the final diagnosis has most often been of one of these IMDs (Table 9.1). Since reporting began in the UK and Ireland in 1981, 164 cases (26 per cent of total reports) were so revised, half of which were shown to have an IMD. This proportion will increase as other, as yet unrecognised, defects are elucidated. Groups of IMDs likely to be confused with RS are shown in Table 9.4, and have been reviewed (Robinson, 1987; Roe, 1988; De Vivo, 1988). It is in this group of patients that urgent investigations are particularly important (see below).

As will be outlined below there is currently no evidence that 'classic (N. American) RS', with which there is an epidemiological association with recent aspirin use, is an IMD as we presently understand that term. The fact that there may be an as yet to be determined, innate susceptibility to the disorder, possibly of genetic origin, should not alter this view.

AETIOLOGY AND PATHOGENESIS

In spite of intensive research into RS, especially during the 1970s and 1980s, its aetiology and pathogenesis remain poorly understood, though some pointers have emerged. In an individual with an as yet undefined innate susceptibility, a cascade of interrelated exogenous and endogenous events occurs – viral invasion and replication, viral spread and cytokine response, possibly linked to nutritional factors and/or the effects of drugs, toxins or poisons – all leading to a profound catabolic reaction and mitochondrial failure. The evidence supporting this hypothesis, which will now be reviewed, is derived from clinical, laboratory and epidemiological research. Two types of epidemiological study have elucidated the principal risk factors for development of RS: descriptive and analytical. The former collect data on the incidence of the disease in defined populations, as well as on its distribution by age and sex, ethnic and socio-economic group, geography and trends over time. The latter tests specific hypotheses about causation, and may take the form of cohort or of case–control studies in which patients with the disease of interest are compared with those who do not develop it.

Exogenous factors

Exogenous factors include viral infection, nutritional factors, drugs and toxins.

Viral infection

Viral infection appears to be a *sine qua non* in RS, although reports of viral isolation from liver, brain or muscle are rare (Partin *et al.*, 1976). Much of the preliminary information that led to the hypothesis outlined above was derived from descriptive epidemiological data from the national surveillance scheme for RS run in the USA since 1973 by the CDC (Centers for Disease Control, 1985; Belay *et al.*, 1999). The CDC data indicate that there were clear winter peaks in incidence of RS in the 1970s and early 1980s, as well as years of higher incidence coinciding with influenza epidemics. They also show that most cases are associated with a viral prodrome, typically and most frequently confirmed virologically as influenza B, but also influenza A and varicella. It has been estimated that during epidemics there may be RS attack rates of 50, 3 and 0.3/100000 children aged less than 17 years infected with influenza B, A and varicella, respectively (summarised by De Vivo, 1988).

At least 16 other viral infections, involving both DNA and RNA viruses, have been associated with RS, but most are isolated case reports and there is little epidemiological evidence to support their having a significant role (Linnemann *et al.*, 1975). There are still many unanswered questions about the precise role of the viral prodrome in the pathogenesis of RS (Mowat, 1988). Davis, using large intravenous doses of influenza B (a virus known to secrete the toxin neuraminidase), has studied experimental infection in mice. Although lethargy, seizures, coma and death resulted and hepatic fatty infiltration developed in association with disturbed liver function, mitochondrial disruption, as recognised in RS, was not induced (Davis *et al.*, 1983); addition of aspirin did not enhance toxicity (Davis *et al.*, 1985).

Information on the occurrence of RS in the UK is derived from the British Reye's Syndrome Surveillance Scheme (BRSSS), which began in August 1981 and was continued via the British Paediatric Surveillance Unit from 1986 until April 2001. This survey had two principal objectives: first, to describe the characteristics of RS in the UK and Ireland and to monitor long-term trends, and second, to provide a centralised case register for more detailed clinical, epidemiological and laboratory research (Hall and Lynn, 2002).

In order to facilitate comparisons with RS in the USA, a similar case definition was used in the UK to that employed by the CDC. There should be *unexplained* non-inflammatory encephalopathy and at least one of the following: serum hepatic transaminases (see above) elevated at least three times above normal, blood ammonia raised at least three times, and panlobular, microvesicular fatty infiltration of the liver. This case definition is very sensitive, but during the 1980s it became apparent that it is rather non-specific because there are many conditions (see Tables 9.3–9.5) that also fulfil the criteria (Green and Hall, 1992). Some of these mimic RS not only clinically and biochemically, but also on light microscopy of liver cells.

Case ascertainment of RS for the BRSSS was principally through clinical reporting by paediatricians, but death certificates and microbiology reports were also monitored. When it began in 1981, reporting was on a 'passive' basis via a pre-circulated case report form sent to all hospital paediatricians throughout the UK and Ireland. In 1986, however, the reporting became 'active' with the introduction of the British Paediatric Surveillance Unit (BPSU) scheme (Hall and Glickman, 1988), in which paediatricians are sent a card each month reminding them to report cases of a range of rare disorders or, if none have been seen, make a nil return.

TABLE 9.5

Exogenous toxins and drugs implicated in causation, or which can mimic Reye's syndrome.

Aspirin	Latex paints
Aflatoxins	Isopropyl alcohol
Ackee fruit products – hypoglycin-A or 4-pentanoic acid	Margosa oil
Emulsifiers (industrial surfactants)	Pteridine
Endotoxin tumour necrosis factor, IL-1, IL-6	Outdated tetracycline
Hornet stings/bee venom	Sodium valproate
Insecticides – mosquito repellents	Warfarin

The findings of the first 5 years of surveillance of RS in the UK suggested that, while there were some similarities with the USA, there were also major differences (Hall and Bellman, 1985). The overall incidence was similar (0.3–0.6 per 100 000 children <16 years of age). The most striking difference, however, was in age: although cases up to age 17 years were reported, the median age of the British cases was 14 months, compared to 8–9 years in the American patients at that time. Another difference was the absence of the marked February–March peak of RS observed each year in the USA. Moreover, while nearly 60 per cent of the British cases, like the American cases, were associated with a respiratory prodromal illness, a much smaller proportion was associated with varicella, although a much larger proportion had a gastrointestinal prodrome. Furthermore, although only about 50 per cent of the cases were fully investigated virologically, a wide variety of viral infections were identified; these included influenza and varicella, but also a range of other viruses. The strong association with influenza B, influenza A and varicella reported in the USA was not observed here, although a possible explanation was that there has not been a major influenza epidemic in the UK since 1976.

RS has been reported in many countries other than the USA and the UK. Apart from the original publication by Reye et al. (1963), a further series from Australia of cases occurring between 1972 and 1986 was described by Orlowski et al. (1990). Bellman and Hall (1985) reviewed reports of RS from current and previous Commonwealth countries and found published series from New Zealand, Zimbabwe, Canada, India, Sri Lanka, Malaysia, Singapore, Jamaica, Nigeria and South Africa. Yamashita et al. (1984) reviewed published and unpublished data on the epidemiological and clinical features of RS in Asian countries. They described series from Japan, Korea, Taiwan, Thailand, Malaysia, Singapore and Indonesia. Gauthier et al. (1989) described 12 cases of certain or probable RS admitted over an 18-year period to a paediatric hospital in Montreal.

The literature on RS from Europe other than the UK is relatively sparse: de Villemeur et al. (1993) described nine cases of 'idiopathic' RS admitted to a hospital in Paris over an unspecified period. A population-based study in France conducted over a 12-month period between 1995 and 1996 found an estimated annual incidence of RS of 0.079 per 100 000 children under 15 (Autret-Leca, 2001). Thabet et al. (2002) reported on 14 patients with 'idiopathic' RS (in whom inherited metabolic disorders had been excluded) admitted to their hospital in Paris between 1991 and 2001. Larsen (1997) refers to a retrospective study of the incidence of RS in Denmark in 1979, which was found to be 0.09 per 100 000 children. A retrospective, questionnaire-based survey of paediatric hospitals in Norway yielded three cases, giving an incidence of 0.05 per 100 000 children between 1977 and 1981 (Aagenes, 1984). Another such survey in Spain between 1980 and 1984 found 57 cases, an incidence of 0.12 per 100 000 children under 15 (Palomeque et al., 1986). A retrospective enquiry of paediatric hospitals in West Germany ascertained 15 cases of RS, giving an incidence of 0.04–0.05 per 100 000 children under 18 (Gladtke and Schauseil-Zipf, 1987).

No countries, however, apart from the USA and the UK, have long-term surveillance schemes involving reporting of cases of RS to a national epidemiological centre, using a standard case definition. All the surveys listed above are time-limited series, often referral-centre based and frequently only involving parts of countries, so that denominators for calculating incidence are unclear. They are mostly retrospective studies involving hospital record reviews and the validity of some, given the non-specific case definition of RS, is questionable. Larsen (1997), in commenting on the low incidence found in Denmark, notes that no surveillance system has ever existed in that country and that diagnostic awareness of RS is probably low, resulting in underascertainment. Bellman and Hall (1985), in comparing reported incidences from other countries with the much higher rates in the USA and UK, remark that differences may reflect ascertainment efforts. These in turn are determined both by diagnostic awareness coupled with aggressive investigation of suspected cases (see also Lichtenstein et al., 1983), and by the intensity of the surveillance methodology. For example, one study found six times more RS hospitalisations, using hospital inpatient data, than cases reported during the same period (1991–1994) to the National Surveillance System in the USA (Sullivan et al., 2000).

Epidemiological inferences drawn from retrospective, time-limited series involving relatively small numbers of cases need to be treated with caution. However, some observations are consistent when comparing reports of RS from other countries with those from the USA; the average age of cases is

CHAPTER 9

much younger in the former, and the seasonal peaks in association with influenza epidemics are not observed. The reason for this difference is unknown, but it may reflect more valid data on 'classic' RS (see page 567) in the USA with its larger population base under surveillance.

Nutritional factors

Ferrets rendered arginine deficient and infected with influenza B virus show elevated free fatty acids, hepatic microvesicular fat and mitochondrial swelling. The model is of limited usefulness, as the deficiency itself causes such profound metabolic disturbances (Deshmukh *et al.*, 1982; 1983). This was also a concern in the Davis mouse model, where some of the changes may have been related to starvation. These and other (as yet unsuccessful) attempts to create an animal model of RS have been reviewed (Crocker, 1988; Kilpatrick-Smith *et al.*, 1989).

Exogenous toxins and drugs

A number of drugs, toxins and poisons have been associated with, or can mimic, RS (Table 9.5). The strongest and most consistent association has been with the use of aspirin given during the viral prodrome of RS. The evidence will be discussed in detail below. Aflatoxins, produced by many fungal species that contaminate foods, have been implicated, especially in Thailand (Olsen *et al.*, 1971). A study of blood and urine in the USA, however, found aflatoxin levels in cases and controls did not differ significantly (Nelson *et al.*, 1980). Perhaps a concept more appropriate to RS has been the idea that exogenous agents (viral infection plus toxin) act synergistically. In Nova Scotia, it was observed that cases tended to appear after local areas of forest had been heavily sprayed with insecticide. Taking up this idea, Crocker's group inoculated suckling mice with encephalomyocarditis virus in combination with insecticide-containing polychlorinated biphenyls (industrial emulsifiers/surfactants). The combination of virus and toxin resulted in greater mortality, development of fatty liver with structural and functional mitochondrial abnormalities, and depressed interferon production (also found in RS), than those treated with virus alone (summarised by Crocker, 1988). In subsequent experiments influenza B virus was used, which produced similar though not identical findings (Crocker *et al.*, 1986). However, a subsequent epidemiological study in Maine found no evidence to support a link with agricultural spraying (Wood and Bogdan, 1986).

Reye's syndrome and aspirin

The background to the association between aspirin and RS is as follows: in the early 1960s and 1970s several authors observed that there were clinical and biochemical similarities between RS and salicylism. Furthermore, a high proportion of patients in individual case reports gave a history of aspirin treatment for the viral prodrome (Rodgers, 1985). In one later case–control study a dose–response relationship was noted between stage of coma and total dose of aspirin received (Starko *et al.*, 1980). A study in the early 1980s found higher mean salicylate levels in RS patients on admission compared to controls (Partin *et al.*, 1982) and another, a correlation between salicylate level at presentation and the maximum level of coma achieved (Tonsgard and Huttenlocher, 1981). In 1983 a study demonstrating histopathological similarities between RS and salicylate toxicity was published (Starko and Mullick, 1983). These observations, together with experimental work suggesting that the association between RS and aspirin could be biologically plausible, prompted a number of analytic epidemiological studies.

The US case–control studies

The most controversial and widely publicised studies linking RS and aspirin were the six epidemiological case–control studies conducted in the USA between 1978 and 1987 (Starko *et al.*, 1980; Halpin *et al.*, 1982; Waldman *et al.*, 1982; Hurwitz *et al.*, 1985; 1987; Forsyth *et al.*, 1989). All six demonstrated a significant association with salicylate usage (specified as aspirin in four) by RS patients during their prodromal illnesses.

The findings of the first three studies, conducted between 1978 and 1980 (Starko *et al.*, 1980; Halpin *et al.*, 1982; Waldman *et al.*, 1982) led the United States Surgeon General in June 1982 to issue a public and professional warning about a possible RS–aspirin association (Centers for Disease Control, 1982). An earlier caution had appeared in 1980 when the results of the studies first became known, although all were not yet published (Centers for Disease Control, 1980). Parents and doctors were advised to avoid using aspirin in children and teenagers with symptoms of influenza and chickenpox. The American Academy of Pediatrics initially endorsed this warning but later that year reversed its decision, as did the Secretary of the Department of Health and Human Services who decided at the end of 1982 not to order the warning labels for all aspirin-containing products that had been considered earlier in the year. The reason for these reversals was the considerable controversy that had arisen over the conduct of the epidemiological studies. This was so intense that the United States Public Health Service (PHS) set up a task force to undertake yet another study that would attempt to overcome the various potential biases and criticisms that had been associated with these three studies. These criticisms have been reviewed in detail (RS Working Group, 1982; Daniels *et al.*, 1983; Rodgers, 1985) but can be summarised as follows:

1. *Selection bias.* Did the cases really have RS? Was it certain, for example, that they did not include patients with aspirin toxicity? Were the controls equally at risk of developing RS in respect of all other parameters except aspirin usage (they were not always matched for type, severity and viral cause of the prodrome).
2. *Recall bias.* Memory is inevitably less precise for parents of controls than for those of the cases, because the event tends to be less significant; furthermore, the control interviews took place after a longer interval than those for cases, thus exacerbating this less satisfactory recall. (Nevertheless, the fact that more controls than cases reported the use of acetaminophen (paracetamol) – containing medications, suggests that the recall of parents of cases was not greater than that of parents of controls for fever medications.)
3. *Data collection bias.* The interviewers were aware of the research hypothesis and of the designation of the subjects (case or control).
4. *Categorisation bias.* Product identification was more accurate for controls than for cases because the former were interviewed at home. (However, in some studies the investigators made efforts to inspect the actual containers of medication provided to the case children.)
5. *Temporal precedence bias.* Exposure to the risk factor should precede onset of disease. The clinical marker for onset of RS is ambiguous, but is normally taken to be the onset of vomiting; however, onset might be earlier. It was only possible to assess exposure timing in one part of a single study, because of the methods of data collection.

Re-analyses of the raw data from some of the earlier case–control studies concluded that the biases were so significant that none of them had satisfactorily demonstrated an association between aspirin and RS (Daniels *et al.*, 1983). The Task Force Study, which had an initial pilot phase in 1984, attempted to overcome these and other criticisms, but it still retained potential for biases that were again pointed out (White, 1986). The most important of these in the USA was the widespread knowledge in 1984 of the RS–aspirin association, which meant that there could have been preferential diagnosis and/or reporting of patients who fulfilled diagnostic criteria whenever there was a history of aspirin ingestion. Furthermore, parents may consciously (perhaps with litigation in mind) or unconsciously have reported aspirin use once they knew the diagnosis.

Although the findings of the first stage of the PHS study were not published until October 1985 (Hurwitz *et al.*, 1985), the results became widely known at the end of 1984. This had only been intended as a pilot phase to establish the appropriate methodology, but the United States Department of Health considered the findings of such significance that voluntary product labelling was requested in early 1985 and a public service education campaign was immediately initiated. Legislation requiring product labelling was passed in 1986.

Although action had been taken as a result of its pilot phase, the main PHS study proceeded in 1985 and 1986 and the results were published in 1987 (Hurwitz *et al.*, 1987). This confirmed the earlier findings in showing a significant association between RS and aspirin. Twenty-six of 27 cases (96 per cent) versus 53/140 controls (38 per cent) had been exposed, for an odds ratio of 40 with a lower 95 per cent confidence limit of 5.8. In an accompanying editorial, Dr Edward Mortimer summarised the issues that had led to the controversy over the earlier studies. He asked why it had taken so long for package labelling to be mandated and suggested two reasons: first, that case control studies are susceptible to challenge on the basis of bias; and second, that 'some aspirin companies have exploited the inherent weaknesses of retrospective studies to exert pressure on the federal government, the American Academy of Pediatrics and other organisations to prevent warnings about the risk of RS attendant upon aspirin use' (Mortimer, 1987).

Although no major criticisms of the main study subsequently appeared in the literature, there remained some concern about methodology, mainly relating to certain biases that, it was felt, still had not been satisfactorily overcome. A further study (the so-called 'Yale' Study which was industry funded) was therefore conducted by investigators who had not been involved in any previous study, and every attempt was made to overcome all earlier biases (Forsyth *et al.*, 1989).

The findings of the Yale Study, published in May 1989, confirmed the results of those conducted previously: 21/24 cases (88 per cent) versus 8/48 controls (17 per cent) had been exposed to aspirin, for an odds ratio of 35, lower 95 per cent confidence limit 4.2 (Forsyth *et al.*, 1989). Sub-analyses of the data, addressing the various biases, continued to show significant association with aspirin. The authors of the Yale Study quoted Sir Austin Bradford Hill: 'The more anxious we are to prove that a difference between groups is the result of some particular action we have observed, the more exhaustive should be our search for an alternative explanation of how that difference has arisen'. They concluded that they had satisfactorily addressed alternative explanations and that the association between aspirin and RS was a real one.

The British Risk Factor Study

At the beginning of 1984 a 2-year British risk factor study was initiated (Hall *et al.*, 1988). Because of the differences between RS epidemiology in the UK and that in the USA, it was considered that a descriptive study to generate aetiological hypotheses that might warrant further more detailed research should be undertaken. Cases were ascertained through the BRSSS, and parents interviewed. In addition to satisfying the surveillance scheme diagnostic criteria, cases were also allotted a 'Reye score' – a measure of how closely the illness fitted the clinical, biochemical and pathological characteristics of the North American cases. An individual who was blinded to risk factor exposure carried out scoring.

At the end of 1984, because of the publicity about the RS–aspirin association in the USA, the medication data for the first year of the study were analysed. As this revealed that more than 50 per cent of the cases had been exposed to aspirin, a comparison group was recruited in the second year of the study. The findings of the British RS risk factor study can be summarised as follows.

A significant case–comparison excess of aspirin exposure (63/106, 59 per cent versus 48/185, 26 per cent; $p < 0.0000001$) was found; there was a significant relationship between Reye score and aspirin exposure. RS cases were more likely than the comparisons to have had more than one aspirin preparation, and to have been given adult formulations. Although this study had important methodological limitations, the data suggested an association between RS and aspirin exposure. No major findings emerged to contradict those of the US studies.

During the first half of 1986 the Committee on Safety of Medicines in the UK, in reviewing the evidence for the aspirin–RS association, concluded that public health action should be taken. Accordingly, in mid-June all UK doctors were sent a letter warning of the association and advising them not to prescribe aspirin for children under 12 years of age, except for certain chronic illnesses such as juvenile chronic arthritis or other connective tissue disorders. A notice appeared in the British Medical Journal (Anon., 1986), there was considerable publicity in the media, and warning leaflets were placed on pharmacy counters and in GP surgeries. Warning labelling was required to be placed on all aspirin-containing preparations; this stated 'DO NOT GIVE TO CHILDREN UNDER 12 UNLESS YOUR DOCTOR TELLS YOU TO'. The pharmaceutical industry voluntarily withdrew paediatric aspirin products. From April 1998, all aspirin-containing medications sold in the UK have also been required by the Medicines Control Agency to state in the patient information leaflet: 'there is a possible association between aspirin and Reye's syndrome when given to children with a fever.' In April 2002, the Committee on Safety of Medicines issued new advice that aspirin should also be avoided in children aged 12–15 years if feverish (Anon., 2002). This was because 10 of 17 aspirin-associated cases reported after 1986 were aged over 12 (Hall and Lynn, 1999). The advice was further modified in April 2003 to state 'Do not give to children under 16 years unless on the advice of a doctor.' From October 2003 all products containing aspirin were required by statute to include this label warning (Anon., 2003).

RS and aspirin – evidence for a causal relationship

Since epidemiological studies can only show associations between diseases and risk factors and not prove causality, this section summarises the evidence linking aspirin and RS in the light of standard criteria for judging causality.

First, the association should be consistent on replication. All six US case–control studies and the British Risk Factor Study demonstrated a significant association between RS and taking aspirin for prodromal symptoms. Two other studies have, however, not demonstrated an association (Orlowski *et al.*, 1987; 1990; Committee on Reye's Syndrome Research, 1983).

The second study by Orlowski *et al.*, was an expansion of the first, with more cases and with a control group. Subjects were children hospitalised at three centres in Australia between 1972 and 1986 and data were obtained by retrospective case-note review. Aspirin exposure was reported in only 8 per cent of cases and 3 per cent of controls, and it was concluded that there was no association between aspirin and RS. These studies were criticised on a number of grounds (Baral, 1988; Hall, 1988; McGee and Sienko, 1988; Hurwitz and Mortimer, 1990). The rigour of the methodology was certainly not comparable to the later United States studies, especially the Yale study, for example.

Baral (1988), one of Reye's co-authors in the original description of RS in Sydney, Australia (Reye *et al.*, 1963), noted that, in contrast to Orlowski's findings, 11 of the 21 patients in that study had aspirin exposure recorded in the case notes and suspected that many more had in fact received it. Morgan, another original co-author, stated that 'Australia almost certainly led the world in aspirin consumption in the 1950s and 1960s' (Morgan, 1985). Orlowski *et al.* (1990) presented data showing that paracetamol rather than aspirin had subsequently dominated the analgesic/antipyretic market in Australia from 1970 onwards. McGee and Sienko (1988), Hall (1988), and Hurwitz and Mortimer (1990) all highlighted the epidemiological differences between Orlowski's cases and those studied in the USA, in particular the young age. They surmised that the Australian cases were mostly *not* 'classic' RS (see below and Hardie *et al.*, 1996), and this was confirmed when the cases were subsequently reviewed, using more precise diagnostic criteria for RS (Orlowski, 1999). Orlowski concluded that none of the original 49 patients could be diagnosed as having certain RS. Hurwitz and Mortimer (1990) suggested that the absence of 'classic' RS reflected the lack of aspirin use in Australia (in contrast to the situation in the USA in the 1970s).

The second case–control study, which was said not to demonstrate an association between aspirin and RS, was conducted in Japan, but details of the methodology were never published in a peer-review journal – thus precluding proper evaluation.

CHAPTER 9

There have been no other analytic studies of the RS–aspirin association elsewhere in the world. Some of the descriptive case series cited earlier noted whether aspirin exposure was recorded in case notes. For example, Aagenes (1984) found none in his series of three cases; Gladtke and Schauseil-Zipf (1987) found 3 among their 15 cases; 23 of 57 cases recorded in Spain had had aspirin (Palomeque *et al.*, 1986); 52 of 73 patients admitted to Bangkok Children's Hospital between 1979 and 1982 had taken 'salicylates' (Yamashita *et al.*, 1985). These data do not contribute usefully to the evidence linking aspirin and RS because they were not collected from caregivers in a systematic fashion, it is often not clear that the cases really did have 'classic' RS, and there are no comparison groups. However, it is of interest that de Villemeur *et al.* (1993), in their series of patients referred with a preliminary diagnosis of RS, observed that 7 of 9 patients who were finally designated as having 'idiopathic' RS had apparently had aspirin, compared to only 4 of 21 who were eventually found to have a Reye-like IMD. Autret-Leca *et al.* (2001) found that 8 (89 per cent) of 9 cases of RS in France in 1995 and 1996 in whom IMDs had been excluded had been exposed to aspirin compared to 44 per cent of French children in general in 1992. Until the results of that study were available, there was no mandated warning on the use of aspirin in children in France. Thabet *et al.* (2002) noted that 12 of 14 patients admitted to their hospital in Paris between 1991 and 2001 with idiopathic RS had been given aspirin. They commented: 'unlike widespread opinion, severe RS without identified metabolic disorder seems not to disappear in our country.'

A further piece of evidence supporting the consistency of the aspirin–RS association was the finding of an increased risk of RS among children taking long-term salicylate therapy for connective tissue disorders (Remington *et al.*, 1985; Rennebohm *et al.*, 1985).

Second, the association should be strong: all of the analytic epidemiological studies showed highly significant associations between aspirin exposure and RS. Although this was reduced on re-analysis of the early US studies, the rigorous method of addressing biases in the most recent, the Yale study, yielded a convincing strength of association.

Third, the association should be specific: other exogenous agents have been linked with Reye-like illness (see above and Table 9.5), for example the anti-epileptic drug sodium valproate, aflatoxin in Thailand, bongkrekate in Indonesia, Margosa oil in southern Asia, unripe ackee fruits in Jamaica, hopentanate in Japan, and insecticides in Canada (Stumpf, 1995). However, none has been convincingly demonstrated, in a case–control study where RS cases fully satisfied diagnostic criteria, to have a significant association with this disorder. The issue has been raised as to whether use of ibuprofen, also a non-steroidal anti-inflammatory drug, in children with chickenpox or influenza, might pose an increased risk of RS (Prior *et al.*, 2000). Studies on animal models suggested that this was unlikely (Mukhopadhyay *et al.*, 1992) and this lack of association was supported by an evaluation of ibuprofen sales data (as a surrogate for its use) and surveillance data for RS in the USA. The introduction of prescription paediatric ibuprofen in 1989 and over-the-counter products in 1995 has not been associated with an increase in reporting of RS (Prior *et al.*, 2000). There is no evidence that RS is associated with paracetamol. In fact the USA case control studies described above, demonstrated an inverse relationship between RS and use of acetaminophen for the viral prodrome.

It appears from the findings of several of the analytic studies, particularly that by Hall and colleagues (1988), that a proportion of 'Reye's syndrome' patients develop the condition without exposure to aspirin. It is possible, however, that the proportion who were not exposed did not in fact have RS but one of the IMDs or some other condition that can mimic the syndrome. This hypothesis was strengthened by the observation that the higher the 'Reye score' in the British RS cases, the more likely they were to be have been exposed to aspirin. It seems likely, therefore, that Reye's *syndrome* is just that – a heterogeneity of disorders that might consist of two groups. The first, 'idiopathic' RS (the 'classic North American type' of older cases with a respiratory or varicella prodrome, profuse vomiting and encephalopathy), may almost invariably be associated with aspirin, although feasibly occasional cases may occur without it, depending on the nature and extent of the subject's innate susceptibility (see below) to the initial mitochondrial insult. The second, a mix of conditions including the RS-mimicking IMDs, tends to present in younger patients and aspirin is unlikely to have a role in precipitating encephalopathic episodes. This hypothesis is supported by the findings of a further study based on the BRSSS descriptive data (Hardie *et al.*, 1996), described below.

Fourth, there should be a dose–response relationship. The early clinical studies of RS in which the 'response', measured by severity of the encephalopathy, was apparently related to blood salicylate levels were later criticised and the results considered unreliable on the grounds that the salicylate methodology was insufficiently precise (Clark *et al.*, 1985). Among the US epidemiological studies (in which aspirin exposure data were obtained by parental interview) one, the PHS 'Pilot Study', reported that there was no relationship between the daily dose of aspirin taken per unit body weight and the deepest grade of coma attained. Furthermore, there was no difference between the average daily doses taken by cases and controls. However, the PHS 'Main Study' and the subsequent Yale Study both demonstrated a significant excess aspirin dosage among cases compared to controls that took aspirin. Neither study looked at dosage in relation to clinical severity.

It appears, however, that there is a substantially increased risk of RS even at low dosage. The Yale Study demonstrated that even a low total dose (defined as <45 mg/kg) was strongly associated with a risk of RS, with an odds ratio of 20.0. The PHS Main Study showed that, although cases who were given aspirin were more likely to have received higher dosages than controls also given the drug, nevertheless 3 of the 26 RS cases had received <10 mg/kg per day. Cases had a median daily dosage of 26.4 mg/kg, the lowest values being 4.1 mg/kg; figures for the controls were 11.1 and 2.4 mg/kg, respectively. A further analysis of these data concluded that there was a dose–response effect; the authors commented, however, that 'no safe dose (of aspirin) exists' (Pinsky *et al.*, 1988).

Although the British Risk Factor Study did not formally study dosage, it was noted that RS patients who received aspirin were more likely to have been given adult preparations than comparison cases who had taken the medication. Nevertheless, at least one of the cases that developed classic, influenza-associated RS, had had only one tablet (300 mg) before onset. However, the absence of a consistent dose–response relationship is not incompatible with the concept of aspirin as a causative factor in RS, if its role is idiosyncratic or as an adjuvant to an innate susceptibility. As is the case with penicillin hypersensitivity, dosage may be irrelevant.

Fifth, there should be a temporal association between the risk factor and the disease. The US case–control studies went to great lengths to demonstrate in each case that aspirin was given before the onset of encephalopathy (defined as the onset of vomiting and/or certain neurological symptoms). Although epidemiologically rigorous this may not have been necessary, because it is possible that aspirin might have an exacerbatory role in aetiology rather than acting as a prime mover. The validity of the case–control studies has, however, been challenged by one author on the basis that medication exposure after the onset of vomiting was not taken into account and that anti-emetics may have a causal role in RS (Casteels-Van Daele, 1991; Casteels-Van Daele *et al.*, 2000). These views have been criticised on a number of grounds, and are not widely supported (Khan *et al.*, 1993; Hall, 1994).

Another aspect of the temporal association is the surveillance of trends and other changes in the incidence of RS, and comparison of these with trends in aspirin usage. Only the USA and the UK have appropriate data because of their long-term national RS surveillance schemes. Various studies in the USA have shown that physician prescribing, drug store purchasing and parental use of children's aspirin for fever have declined considerably since 1981 (Barrett *et al.*, 1986; Remington *et al.*, 1986). In parallel with this, the annual incidence of RS in the USA has also strikingly declined. For example, in 1980 there were 555 cases reported to the national surveillance scheme compared to no more than 2 per year between 1994 and 1997 (Belay *et al.*, 1999). One elegant study demonstrated that this decline paralleled exactly the upsurge in professional and lay publications mentioning the link with aspirin, which began in the USA in 1980 (Soumerai *et al.*, 1992).

Trends in RS in the UK are more difficult to interpret (Table 9.1). This is in part because the reporting scheme was initiated more recently than that in the USA, and hence longer-term trends are not available to facilitate comparison. Numbers also are smaller. Moreover, the method of case ascertainment was changed from July 1986, the month after the action by the Committee on Safety of Medicines, which had the potential to exert a confounding effect.

The introduction of an active reporting scheme via the BPSU in 1986, which would have been expected to enhance RS ascertainment, did not have this effect (Hall and Lynn, 1999). The most that can be said is that the existing decline levelled out, to be followed by a further substantial decline in case reports (Hall and Lynn, 2000). In parallel with these trends, a survey of aspirin usage by British parents showed that children with febrile illnesses were 17 times less likely to have received aspirin in 1988/89 than in 1985/86 (Porter *et al.*, 1990). It is notable also that, in spite of an influenza epidemic in the winter of 1989/90 – the largest since 1976 – no accompanying upsurge in the incidence of RS materialised.

In addition to the decline in incidence since 1986, certain clinical and epidemiological character-istics of reported RS in the UK and Ireland also changed; most notably, the median age almost halved. One study explored the relationship between these trends and the 1986 aspirin warning (Hardie *et al.*, 1996). The hypotheses were that patients reported to have 'RS' are a heterogeneous group having a number of disorders, and that these disorders include not only the Reye-like IMDs but also a separate group in which aspirin is a risk factor. The latter would be clinically and epidemiologically distinct, and resemble the North American cases. All cases reported to the BRSSS between 1982 and 1990 were allotted a score in which patients showing the typical clinical and pathological features of 'classical' (North American) RS scored highly. The findings showed that high-scoring cases were significantly more likely to have occurred in the period before June 1986 than in the same period afterwards. They were also more likely to have received aspirin, and to be older, than the lower-scoring cases. The study therefore provided further evidence to support the association between aspirin and RS. It also provided a counter-argument to those critics of the evidence linking aspirin and RS, who suggested that the decline in RS could be explained by increased recognition of metabolic, viral or toxic diseases (Casteels-Van Daele and Egermont, 1994; Orlowski, 1999; Casteels-Van Daele, 2000). If this were the case all score categories should have declined at equal rates, whereas the decline was significantly greater in the subset of patients who most resembled 'classic' RS.

Gauthier *et al.* (1989) demonstrated a decline in cases of RS at their hospital in Montreal between 1970 and 1987, but unfortunately only included subjects in whom a *discharge* diagnosis of RS had been made. There was a decline after 1980, both in cases in whom the authors' retrospective appraisal led to a revision of the original diagnosis (presumably because such cases were now receiving an alternative discharge diagnosis because of heightened diagnostic awareness of 'Reye-like' conditions) *and* in cases considered to have 'certain' RS. It would have been more illuminating if the authors had included all cases *admitted* with Reye-like features, because the real issue is whether there has been a decline in all patients presenting to hospital with the clinical and pathological features of RS whatever the final diag-nosis. If all RS can be explained by a mix of inherited metabolic disorders and other conditions that mimic it, then there should have been no change in the admission rate of such cases. Unfortunately no study has been undertaken to document this; however, anecdotal experience at referral centres is that there has been a dramatic decline in patients with these manifestations (Sarnaik, 1999; JFT Glasgow, unpublished data; J. Orlowski, discussion at International Workshop 'Reye's Syndrome Revisited', Leuven, May, 1996; J. Partin, personal communication).

Finally, supportive experimental evidence demonstrating that there is *biological plausibility* to the hypothesis that aspirin has a causal role in RS, is outlined below in the sections on endogenous factors and innate susceptibility.

Endogenous factors

In addition to exogenous agents such as viruses, drugs and/or toxins, it has been suggested that endogenous factors, for example an unusual metabolite, or some component of the immune response, might play a role in the aetiopathogenesis of RS.

Immune response

Many metabolic effects observed in RS may be mediated via the immune response. Active in this regard might be monokines, such as tumour necrosis factor (TNF), interleukin-1 (IL-1) and endotoxin, released from bowel flora (Larrick and Kunkel, 1986). Kirn and colleagues (1982) have shown that Kupffer cell function is impaired and endotoxaemia commonly results after infection with a number of viruses. Endotoxin provokes macrophages to release TNF, a known mediator of metabolic toxicity. TNF contributes to the inflammatory response by stimulating various cell types (including fat cells); this results in catabolism characterised by hypertriglyceridaemia (Beutler *et al.*, 1985). In rat hepatocytes TNF inhibits fatty acid oxidation, an effect enhanced by addition of IL-1 or IL-6 (Nachiappen *et al.*, 1994). Experimental, sub-lethal doses of endotoxin to rats cause an increase in plasma ammonia, free fatty acids and lactate, fatty infiltration of the liver, and mitochondrial damage (Yoder, 1985). Young animals are known to be more sensitive than matures to the effects of TNF and endotoxin. Endotoxin-like activity has been found in plasma and CSF of children with RS, and has been linked early in the course of the illness to clinical and encephalographic markers of neurological disability (Cooperstock *et al.*, 1975; Kirn *et al.*, 1982). Moreover, aspirin enhances the release of TNF by mouse macrophages (Larrick and Kunkel, 1986). Thus immune mediators provide possible links between viral infection, aspirin and metabolic failure.

The metabolic disturbance

Initially, massive catabolism, possibly of muscle (DeLong and Glick, 1982) and fat, is a predominant early feature of RS, generated perhaps by elevated levels of catacholamines found in blood and CSF; the degree of elevation correlates with the neurological age at the time of admission (see page 557; De Vivo, 1988). This elevation is likely to be consequent upon the period of hyperexcitability ('toxic delirium') associated with sympathetic overactivity, and manifest as fever, tachypnoea, tachycardia and pupillary dilatation (De Vivo, 1988). Most of the other disturbed biochemistry is attributable to primary mitochondrial failure (see review by Osterloh *et al.*, 1989): Levels of alanine, aspartate, alpha-amino-n-butyrate, glutamine and lysine are characteristically elevated in plasma and urine; by contrast, citrulline and arginine are low (Hilty *et al.*, 1974). Citrulline and its precursors (such as carbamyl phosphate) accumulate and leak from mitochondria into the cytosol, where transformation occurs to orotic acid. This metabolite interferes with the liver's limited ability to oxidise fatty acid, synthesise lipoprotein, and secrete triglyceride as very low density lipoprotein particles; accordingly, this decreases in the early days of illness. Inability to further oxidise fatty acids and 3-carbon fragments (alanine, pyruvate, and lactate) necessitates an increase in the rate of glycolysis to meet energy demand. Hypoglycaemia is the result of both glycogen depletion and decreased gluconeogenesis.

Under the influence of elevated growth hormone, glucocorticoids and glucagon (fasting stress), there is massive lipolysis and elevation of glycerol and free fatty acids. A huge concentration of fatty acids (>3mmol/l) is presented for β-oxidation, but typically ketosis is disproportionately low for the degree of starvation (Haymond *et al.*, 1978). Short-chain fatty acids are also disproportionately elevated (60–100 μmol/l; Trauner *et al.*, 1975). Up to 55 per cent of serum free fatty acids is in the form of dicarboxylic acids, which are formed by microsomal (P_{450}) ω-oxidation when β-oxidation is impaired. Serum levels of these acids correlate with blood ammonia concentrations. Although medium- (C_6–C_{12}) and long-chain (C_{14}–C_{18}) dicarboxylic acids are elevated, the latter are not normally present. Dicarboxylic acids, especially the long chains, are toxic to mitochondria *in vitro* (Tonsgard and Getz, 1985). In comatose RS patients, 86 per cent (mean) of free fatty acids are in this form; 31 per cent in the less severely affected (Tonsgard, 1985).

Aprille (1988) considered several possibilities to account for mitochondrial failure of RS, and concluded that neither multiple mitochondrial enzyme deficiencies nor uncoupling of oxidative phosphorylation (aspirin is a weak uncoupler – see below) could account for the laboratory and experimental evidence. She thought the most likely candidate was inhibition of oxidative phosphorylation at any of its intermediate steps – electron transport, nicotinamide adenine dinucleotide ($NADH_2$/NAD) translocase, phosphate transport, or ATP synthetase. This is a view shared by Corkey and colleagues (1988; Figure 9.1), although others appear to favour the alternative hypothesis – suppression of mitochondrial biogenesis (De Vivo, 1988). Whatever the precise mechanism, β-oxidation is blocked at multiple sites.

CHAPTER 9

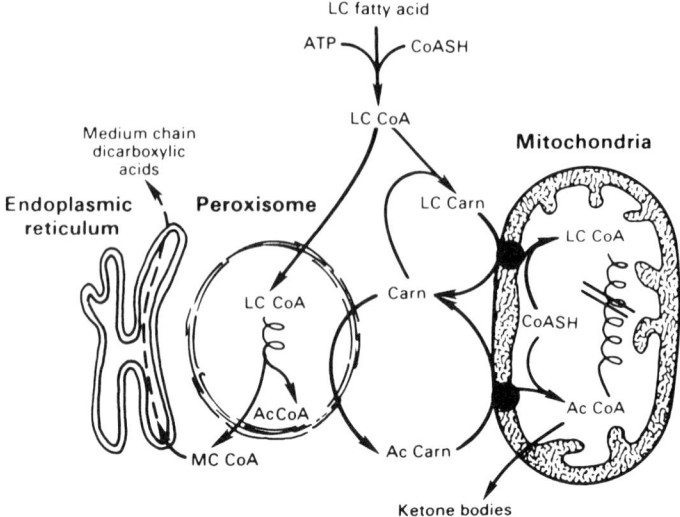

Figure 9.1 Model illustrating interactions between mitochondrial β-oxidations, peroxisomal β-oxidation, and ω-oxidation in the endoplasmic reticulum. The model shows oxidation of activated long-chain (LC) fatty acids by both mitochondria and peroxisomes. The end product of mitochondrial β-oxidation is primarily ketone bodies, whereas peroxisomal β-oxidation yields acetyl carnitine (Ac carn) and medium chain (MC) acyl CoA esters. ω-Oxidation converts these MC intermediates to dicarboxylic acids. A putative block in mitochondrial β-oxidation would increase fatty acid oxidation by peroxisomes and endoplasmic reticular pathways. (From Corkey *et al.*, 1988.)

The current authors have shown, using tritiated palmitate in fibroblast cultures, that β-oxidation measured by standard means is blocked by salicylates. Since inhibition was not demonstrable in cell lines lacking the long-chain 3-hydroxyacyl-CoA dehydrogenase (LCHAD), it was deduced that it was this enzyme, one component of the mitochondrial trifunctional enzyme (MTE), that was being inhibited by salicylate. Moreover, in cell extracts, salicylate and one of its principal metabolites – hydroxy-hippurate (HHA or salicylurate) – were shown reversibly to inhibit HAD activities. Kinetic studies using pure short-chain HAD and long-chain HAD (using MTE) showed that HHA and salicylate were competitive inhibitors of the former but mixed (non-competitive) inhibitors of the latter; importantly, both compounds inhibited the combined three-step, MTE reaction in a physiological direction (Glasgow *et al.*, 1999).

Mowat (1988) has summarised the evidence for mitochondrial susceptibility. One can envisage RS occurring as a result of a viral infection or viral-derived host material acting directly on mitochondria to impair production of some component of the cell membrane; or acting on ribosomes causing secondary loss of mitochondrial proteins; or alternatively acting on the ADP-requiring protease activity essential for the assembly of ribosomally derived protein. Regarding the second suggestion, it should be remembered that in viral infection viral RNA frequently diverts the biosynthetic pathways of the host endoplasmic reticulum into the production of viral proteins (Meert *et al.*, 1990). Thus it is at least possible that viral infection might adversely affect organelle function and be aggravated by proteolysis and lipolysis mediated by fatty acid and dicarboxylic acid accumulation impairing *inter alia* oxidative phosphorylation. Short- and medium-chain acyl CoA esters, which sequester free coenzyme A, accumulate within mitochondria, blocking energy production and thereby account for the occurrence of hyperammonaemia, hypoglycaemia, and many other metabolic intermediates present in RS (Corkey *et al.*, 1988).

A majority of mitochondrial proteins are synthesised outwith mitochondria in the cytosol and are taken up by the organelle. In RS, however, ATP/energy production within mitochondria is significantly reduced. Proper processing of imported protoenzymes requires an ATP-dependent protease to cut these down to final size prior to assembly of the holoenzyme complexes on the inner membrane (Van Coster *et al.*, 1991). As a consequence, catalytic activity fails.

Innate (genetic) susceptibility

That there must be some innate host susceptibility or idiosyncrasy in the aetiopathogenesis of RS has been deduced from the observation that the antecedent viral infections are extremely common, whereas RS is rare. Nevertheless, both recurrence and a family history of RS (which might be expected in a genetically determined disorder) are most unusual (Belay *et al.*, 1999). In fact, clinicians generally agree that if either of these two features is present then an alternative diagnosis, such as one of the IMDs that mimic RS, is a much more likely probability (see Tables 9.4 and 9.6).

Recent work using cultured fibroblasts from recovered RS patients and controls provides some evidence of innate susceptibility (Glasgow *et al.*, 1999), and of a biological difference between Reye's syndrome and control cells. Fibroblasts from RS patients were significantly more sensitive to inhibition of β-oxidation by salicylate than control cells. The differences were demonstrable at levels well within the therapeutic range of plasma salicylate in the 56 patients (serum range on admission: zero, 2.5; median 0.9 mmol/l), and those reported in the literature (Partin *et al.*, 1982). In control cells, 1 mmol/l salicylate significantly *stimulated* palmitate oxidation, an effect quite opposite to that noted in RS cells (Figure 9.2); higher concentrations caused increasing inhibition in each group of cells. Stimulatory then inhibitory effects of salicylate on β-oxidation were reported at similar concentrations in rat liver slices (Maddaiah and Miller, 1989). Thus stimulatory and inhibitory effects of salicylate were independent; only the latter was due to blocking of LCHAD activity.

How might this stimulation of β-oxidation, at therapeutic salicylate levels, be induced? Salicylate, although not HHA or gentisate, is known to uncouple mitochondrial β-oxidation from phosphorylation. Thus the stimulatory effects in control cells might be due to the presence of an uncoupling protein. In RS cells, it is possible that salicylate does not uncouple oxidative phosphorylation because a specific target protein is absent. It is suggested that this target protein is likely one of the family of mitochondrial uncoupler proteins (UCPs) (Ricquier and Bouillaud, 2000) whose role is still unclear. Their expression is increased under conditions in which long chain fatty acids (LCFA) accumulate, prompting the suggestion that by stimulating mitochondrial respiration (Boss *et al.*, 2000) they enhance β-oxidation, thus protecting against cell apoptosis. Lack of a specific UCP (howsoever caused) might explain the difference in β-oxidation response to low concentrations of salicylate between controls (and LCHAD-deficient cells) on the one hand, and RS cells on the other – a difference that warrants further investigation. This sensitivity of RS cells, if widely expressed in body tissues, could explain the reaction of certain individuals to the action of aspirin on the MTE–LCHAD enzyme system that might contribute to the pathogenesis of RS (Glasgow and Middleton, 2001).

This is a subtler concept than the single gene:enzyme defect, a number of which can mimic RS (Table 9.4). Finally, skin fibroblasts from RS patients do not show differences from controls in the three

TABLE 9.6

Features suggesting an inherited metabolic disorder as the cause of acute encephalopathy (modified from Green and Hall, 1992).

Presence of at least one of the following should trigger consideration/investigation of an IMD:
Age at onset <3 years.
Absence of a history of viral prodrome clearly separated from onset of the encephalopathy.
Past history of encephalopathic episodes: vomiting in association with viral infections, unexplained
 failure to thrive, neurodevelopmental disorder, or apparent life-threatening event(s).
Family history (immediate/extended) of RS or R-LS, unexplained encephalopathy, or SIDS.

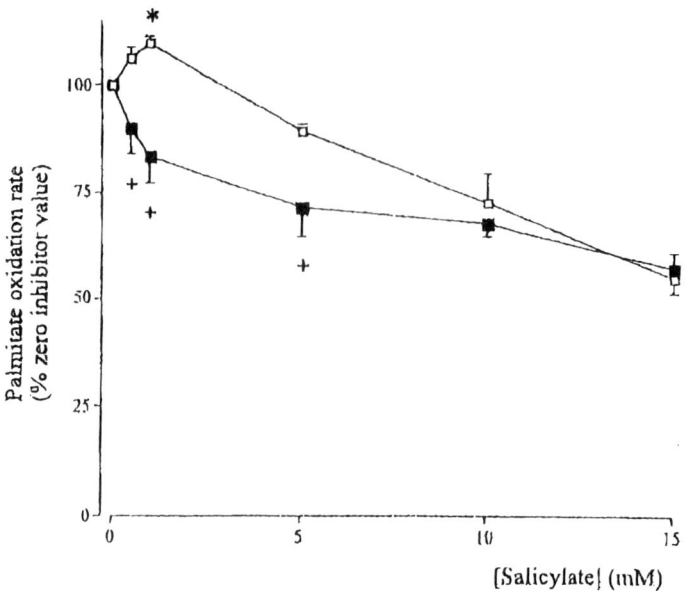

Figure 9.2 Salicylate inhibits palmitate β-oxidation differentially in fibroblasts from controls ($n = 6$) and RS patients ($n = 6$) (□ and ■, respectively). Rates are a percentage of the zero value within each independent experiment, and are presented as means ± SEM. These zero salicylate rates were 68 ± 12 and 72 ± 16 pmol/min per mg ± SEM, respectively, for controls or RS cells. Where the error bars are not visible they fall within the symbol. *$P < 0.01$ (paired t-test) for control rates at 1 mM salicylate (74 ± 12 pmol/min per mg ± SEM) compared to rates at zero salicylate. +$P < 0.01$ (ANOVA) for RS compared to controls at the same concentration of salicylate. (From Glasgow *et al.*, 1999, with permission from Elsevier Science.)

component activities of the mitochondrial trifunctional enzyme (MTE) or in rates of palmitate oxidation, that would suggest an absence of a specific (inherited) enzyme in this part of the β-oxidation pathway.

NEUROLOGICAL INVOLVEMENT AND ENDOGENOUS FACTORS

The pathogenesis of the encephalopathy, which follows hepatopathy by several days, is also unclear, although a large number of neurotoxins are present in liver failure (Brown and Imam, 1991). This is most readily explained also as an acute energy crisis secondary to increased levels in fatty acid (and dicarboxylic acids), a low CSF glucose (see above), and hyperammonaemia – likely in combination. The latter, however, does not appear to initiate the encephalopathy, as most children do not have an elevated ammonia at onset of vomiting, when only lethargy tends to be present (Heubi *et al.*, 1984); this is likely to be a compounding factor. There is no inflammation or infection of the nervous system. It is undetermined whether damage to neuronal mitochondria is the cause of the encephalopathy, or if this is secondary to perturbations initiated by events elsewhere (e.g. in the hepatocyte). An alternative is that the catabolic response described above, modulated by viral infection and/or the immune response, is the prime mover, or that some combination of these mechanisms play a part (see De Vivo, 1988).

However, since raised intracranial pressure (ICP) does not always correlate with the degree of encephalopathy, there is probably injury to other elements such as endothelial cells. These cells of brain capillaries are unique. Sealed together by tight junctions, they lack the fenestrations seen in other capillary beds, and hence substances being transported from blood to brain interstitium must of necessity be transcellular rather than intercellular. Endothelial cells are surrounded by pericytes with phagocytic properties, and these, with astrocytes' foot processes, constitute the blood–brain barrier. In metabolic terms this is a highly active area; endothelial cells have up to five times more mitochondria than hepatocytes or muscle cells (Oldendorf and Brown, 1975). Impaired fatty acid oxidation in endothelial cells may in part contribute to cytotoxic cerebral oedema, and will be exacerbated by defective oxidative phosphorylation that interferes with glucose transport from blood to brain. Cerebral hypoxia, severe acid–base disturbance and/or systemic hypotension compounds such insults. Progression and duration of the cerebral disturbance ultimately determines outcome.

MANAGEMENT PRINCIPLES

Management is founded upon the twin aims of preventing, minimising and correcting metabolic abnormalities, and of controlling increased or increasing intracranial pressure (ICP). Success will depend on the patient being managed in a centre that has a paediatric intensive care unit, and is equipped for optimum laboratory monitoring and ideally for the investigation of IMDs. In this regard it is crucial that the diagnosis of RS (or Reye-like syndrome) be considered early, and definitive therapy instituted before brain damage occurs.

Urgent investigations

When the diagnosis of RS is suspected, confirmatory laboratory tests (serum AST/ALT, blood ammonia, and prothrombin time) are mandatory, as is the immediate determination of plasma glucose. Liver biopsy requires consideration, depending upon the precise clinical features; ideally, it is desirable to obtain electronmicroscopy of liver tissue to confirm (or exclude) the characteristic ultra-structural changes of RS. Additional tests likely to be abnormal in RS are serum 3-OH butyrate, free fatty acids, glycerol, pyruvate, lactate, uric acid and creatinine phosphokinase. Serum electrolytes, urea, creatinine, calcium, phosphate and osmolality determinations are also useful. However, because a number of IMDs can mimic RS, studies to confirm or exclude such must also be undertaken as soon as possible; details of these investigations have been outlined (Green and Hall, 1992). These investigations include supporting biochemical markers such as urine orotic acid, urine and plasma amino acids and carnitine, including acyl-carnitine fractions. An immediate urine sample should be collected for detection of an organic aciduria by gas chromatography–mass spectrometry (GC-MS) prior to the commencement of intravenous glucose, or within 2 hours; thereafter metabolism tends to revert to normal. The novel technique using tandem MS that is presently available only in some supraregional centres will become more widely used in the next few years. The methodology has been modified to use dried blood spots on Guthrie cards (Millington et al., 1997), and feasibility studies have recently been carried out to assess this approach in neonatal screening for medium-chain acyl-CoA dehydrogenase deficiency (Clayton et al., 1998). Precise diagnostic tests and the way in which samples should be obtained and stored are important (Green and Hall, 1992) both for the disorders listed (Table 9.4) and for some of the more recently described mitochondrial disorders (mtDNA; see review by Poulton and Brown, 1995).

Essential components of treatment are as follows

Given our ignorance of aetiopathogenesis, management is empirical and is based upon a general understanding of the pathophysiological disturbances. There are no reports of randomised comparisons of therapeutic regimens. As in the case of any seriously ill patient, resuscitation and stabilisation are essential prior to any other consideration. The structured approach is used, with correction of any airway, breathing, or cardio-circulatory abnormalities. High-flow oxygen is given. The presence of hypoxaemia or shock will exaggerate development of cerebral oedema and increase the risk of raised ICP, and must be corrected. Dehydration should be corrected, but care taken not to overload the circulation; the current authors think that maintenance fluids ought to be restricted to about 65–70 per cent of normal; on the other side of the Atlantic, however, this view has been challenged (Partin, 1988a).

If RS is suspected, an i.v. infusion of glucose should be started immediately. Once the diagnosis is confirmed, this should never be interrupted. As well as correcting hypoglycaemia (two-thirds of the authors' cases had plasma glucose <2.6 mmol/l), a high concentration of i.v. glucose importantly minimises catabolism of protein and fat (see above). While some groups recommend a 10–15 per cent glucose solution, others prefer higher concentrations (20–30 per cent) in order to maintain blood glucose of 11–22 mmol/l, and have a direct effect in diminishing cerebral oedema (J.C. Partin, 1988).

All patients should be transferred to units that can provide intensive nursing and medical care. Thereafter the authors' practice has been to monitor those less severely affected (grades I–II), whereas the child in Lovejoy grade III and beyond is managed in a more aggressive fashion. Following general anaesthesia, some centres use hyperventilation (HV) under neuromuscular blockade ($PaCO_2 < 3.5 kP_a$) to control ICP. It is essential also to insert central venous and arterial cannulae in order to facilitate frequent blood sampling (thus avoiding unnecessary discomfort, which can increase ICP), provide access for hypertonic infusion, and monitor systemic blood pressure (BP).

Management of raised intracranial pressure

This cannot be inferred from physical signs – as mentioned, papilloedema is almost invariably absent, and it is essential to monitor ICP using an in situ measuring device. The cerebral ventricular catheter, although invasive, is the 'gold standard', whereas a subdural or subarachnoid 'bolt', and especially an extradural device, are much less so (Brown and Imam, 1991; Sharples, 1991). This is attached to a pressure transducer and display/recording equipment. The aim is to maintain ICP at less than 20–25 mmHg in order to prevent cerebral perfusion pressure (the difference between mean arterial BP and ICP) falling to less than 40–50 mmHg (Jenkins et al., 1987). Other factors that tend to increase ICP include constriction of the neck veins, head-down tilt, or poor management – inadequately controlled ventilation, poor maintenance of sedation/neuromuscular blockade, excessive fluid infusion, recurrence of hypoglycaemia or seizures – as well as factors that cause arousal, such as pharyngeal suction, coughing, or painful procedures. Use of lumbar puncture has been mentioned. The nuances of how the medications are used have been reviewed (Glasgow et al., 1985).

Should ICP be critically raised greater than 35 mmHg, or be elevated in spite of the above measures, control may be attempted using diuretic therapy and CSF drainage. Intravenous mannitol, possibly supplemented with frusemide, is used provided serum osmolality is less than 320 mol/kg, otherwise renal failure and acidosis may ensue; therapy may need to be repeated. Should osmolality be greater than 320 mol/kg, management is more challenging and controversial. Some centres will continue the use of HV, perhaps commenced at an earlier stage (see above). This acts by reducing the cerebral vascular volume; however if cerebral blood flow is already reduced (this is often uncertain), ischaemic damage may be induced or exacerbated. In this difficult situation, others recommend i.v. thiopentone perhaps with phenytoin, the latter to control possible latent seizure activity. Intravenous barbiturate acts in a similar fashion to HV but beneficially reduces cerebral metabolic rate also, and hence oxygen requirements; however, arterial BP may need to be supported using a dopamine infusion. Finally, if in spite of these measures ICP remains unacceptably high, i.v. dexamethazone is given and hypothermia induced (30–31°C) using a servo-controlled cooling blanket further to reduce cerebral metabolism. Cerebral decompression (bifrontal craniectomy) is a desperate procedure, though it is claimed that a few patients have recovered well thereafter. There is no consensus of evidence or opinion, and management approaches in relation to ICP are likely to evolve further.

Other measures

As in chronic hepatic encephalopathy, enemas, neomycin and lactulose have been used to combat ammonia production; however, evidence of efficacy is lacking. Nor is there evidence that use of systemic L-carnitine is beneficial. On the contrary, it may increase mitochondrial uptake of long-chain fatty acids, enhance accumulation of acyl CoA intermediates, and further burden β-oxidation (Walter, 1996).

The future

No trials of any form of therapy have as yet been carried out. If RS were to appear again in significant numbers, it would be essential to study the brain disorder using modern methodologies so as to formulate more specific hypotheses that could then be tested appropriately.

PROGNOSIS

The neurological phase tends to last about 1–4 days, and complete recovery of organ function is expected in those who survive. Quality of survival in determined *inter alia* by the degree and duration of the intracranial hypertension, and the presence during the acute illness of systemic hypotension, hyperpyrexia and various metabolic disturbances including hypoglycaemia, cerebral hypoxia, high levels of free fatty acids and hyperammonaemia (De Vivo, 1988). The latter was especially highlighted in the recent review of 1207 cases of RS (1981–1997) from the USA. Cases with an elevated blood ammonia (i.e. $>26\,\mu$mol/l, which appears to be the upper normal limit in the laboratories concerned) had a significantly higher risk of neurological morbidity and mortality than those with normal levels – relative risks 4.1 and 3.4, respectively (Belay *et al.*, 1999). Efforts in the mid-1980s in Northern Ireland to acquaint colleagues working in Accident and Emergency departments and paediatrics with the earliest features of RS may have helped the prognosis, although it remains unproven whether early diagnosis can reduce the number of comatose cases, brain-damaged patients and/or deaths (Partin, 1988a). Among the 56 cases seen by the authors between 1979 and 1986, a poor outcome occurred in 27 per cent and mortality was 12 per cent (Glasgow *et al.*, 1986). Overall, UK mortality remains at 40–60 per cent. This must be compared to 31 per cent in the USA. Neurological morbidity in survivors is 10–60 per cent. Those who reach the stage of coma have a significantly reduced chance of making a satisfactory recovery. In the Partins' large series of 243 patients (1963–1986), the proportion that died or were brain damaged in grades I–II was zero, compared to 21 per cent and 50 per cent in grades III and IV, respectively. Factors associated with a poor outcome are shown in Table 9.8.

The findings in Northern Ireland are similar (Table 9.7): those with a shorter than average interval between prodrome and neurological deterioration tend to have a worse outlook (grades III, IV) than those less severely affected (grades I, II) (Glasgow, unpublished observations). The worst prognosis is seen in infants with a very short prodrome and rapid neurological progression. On the other hand, the literature contains examples of patients with a very mild illness consisting of vomiting (and hepatic dysfunction) but a paucity of neurological signs – a stage I illness (Lichtenstein *et al.*, 1983; Heubi *et al.*, 1984).

Follow-up of survivors

Several descriptive studies have been published (see, for example, Bruner *et al.*, 1979; Shaywitz *et al.*, 1982). The current authors have conducted two case–control follow-up studies (the sibling nearest in age acting as control) of intellectual and academic performance in 22 former RS patients (Duffy *et al.*, 1991). Damage to the developing brain that result in neuropsychological deficits may only become

TABLE 9.7

Outcome of 56 cases of Reye's syndrome* according to neurological grade on admission.

Grade	Number patients	Full recovery	% recovery
I	7	7	100
II	19	17	90
III	24	16	67
IV	6	2	33

*Glasgow (unpubl. data); no patients in grade V (brainstem death) on admission.

apparent much later, when maturation deviates from normal and abilities may be simply less well acquired. Moreover, mild impairments, perhaps clinically undetectable in themselves, may exert a cumulative influence on the development of other functions, leading to more significant disabilities (Baron *et al.*, 1995). Hence the need for outcome studies during the school years and beyond. Accordingly, the period of follow-up has been extended and 18 of these patient–sibling pairs studied. These adolescents had recovered well and had no obvious neurological impairment. On the basis of formal testing, general intellectual functioning and verbal and visuospatial skills were significantly reduced if RS occurred during infancy (Meekin *et al.*, 1999). A similar profile was evident on the measure of self-esteem. Differences for measures of behavioural problems failed to reach significance. Another significant variable was the development of unconsciousness (i.e. progression to grade III or more) during encephalopathy. Although this was not so closely associated with less good outcomes, it was linked to relative deficits on the scale of cognitive ability. However, only those differences on the Wechsler scale of verbal abilities reached statistical significance. This emphasises again the need for careful, long-term follow-up and assessment.

Parent support groups have played a key role in such campaigns in the past (Monto, 1999); one – The National Reye's Syndrome Foundation of the UK – is continuing this work by producing educational material for both parents and professionals. A recent National Workshop highlighted the need for ongoing education for professionals, many of whom have never seen a case of RS.

TABLE 9.8

Factors indicating a poor prognosis in Reye's syndrome.

Infants <12 months of age*.
EEG shows marked slowing.
Minimum cerebral perfusion pressure <40–50 mmHg (see Jenkins *et al.*, 1987).
Rapid neurological progression to Grade IV encephalopathy.
Blood ammonia > five times normal; or ammonia > twice upper normal limit and prothrombin time >3 seconds above normal (predicts progression to deeper coma grades in those who are not comatose initially (see Heubi *et al.*, 1984).
AST/ALT ratio < unity.
Creatinine phosphokinase >10 times normal.
Free fatty acids excessively elevated and/or marked increase in long-chain dicarboxylic acids (Tonsgard, 1985).

*Recent review of 1207 USA cases found age <5 years (see Belay *et al.*, 1999).

CONCLUSION

There is now a substantial body of evidence that there is an association between aspirin and 'classic' RS. There has been a dramatic decline in this condition both in the UK and Ireland and in the USA since these countries took public health measures to reduce the use of aspirin in children. In fact the current extreme rarity, coupled with the low specificity of case definition, means that any patient in whom RS is considered a diagnostic possibility *must* be investigated for the Reye-like IMDs. These are now a more likely diagnosis, especially in children less than 3 years of age (Belay *et al.*, 1999).

It will, however, be important to continue monitoring the incidence of RS because, as time passes since the initial publicity, there may be loss of parental awareness of the link. The importance of not giving children aspirin (except where therapeutically indicated, as for example in Kawasaki disease or stroke prevention) needs to be continually emphasised, even though there are now warnings both on the exterior of packaging and on the enclosed information sheets. Any upsurge, which is most likely to occur in the setting of an influenza epidemic or pandemic, should trigger a renewed public education campaign. In the USA the authors of the Yale Study (Forsyth *et al.*, 1989) observed that they were surprised still to be able to ascertain cases of RS in 1987, after the public health campaigns of the early 1980s. It has also been shown in the UK that parental awareness, although high in 1986 (Hall, 1987), diminished rapidly in 1987 (Fairhurst, 1988), and another study showed the same decline in 1988/89 (Porter *et al.*, 1990). In 1998, a survey conducted by The Royal Pharmaceutical Society of Great Britain showed that 70 per cent of parents questioned, were unaware of the warning not to give aspirin to children under 12 years of age (Anonymous, 1998). Individual aspirin-associated cases continue to be seen (McGovern *et al.*, 2001; Da Silveira *et al.*, 2002; Bhutta *et al.*, 2003) and reported (Hall and Lynn, 2000). Clearly there is a continuing need for warning labels, and there may be a need for a 'booster' education campaign if current trends in RS reporting show signs of reversing. It is hoped, however, that RS will remain extremely rare, and never become the important cause of childhood mortality and morbidity that it was in earlier decades.

ACKNOWLEDGEMENTS

The BRSSS and the research programme at The Queen's University of Belfast into the effects of aspirin on β-oxidation was generously supported by The National Reye's Syndrome Foundation of the UK. We also thank Dr Bruce Middleton, University of Nottingham, for helpful comments on the manuscript, and Mr Raymond Moore, Dr Alison Gray and Ms Joanne Hill for expert laboratory assistance.

REFERENCES

Aagenes, O. 1984, Severe Reye's syndrome in Norway, 1977–1981. *Tidsskr Nor Loegeforen*, **104:** 36–37.

Anon. 1986, CSM update: Reye's syndrome and aspirin. *British Medical Journal*, **292:** 1590.

Anon. 1998, News from the Royal Pharmaceutical Society of Great Britain: 'Ask before you buy painkillers' say pharmacists. London: RPS.

Anon. 2002, Aspirin and Reye's syndrome in children up to and including 15 years of age. *Current Problems in Pharmacovigilance*, **28:** 4.

Anon. 2003, www.mhra.gov.uk

Aprille, J.R. 1988, Metabolic and physiological consequences of mitochondrial dysfunction in Reye's syndrome. *In:* C. Wood (ed.), *Reye's Syndrome*, pp. 20–44. London: Royal Society of Medicine.

Autret-Leca, E., Jonville-Bera, A., Llau, M., Bavoux, F., Saudubray, J., Laugier, J., Devictor, D. and Barbier, P. 2001, Incidents of Reye's syndrome in France: a hospital-based survey. *Journal of Clinical Epidemiology*, **54:** 857–862.

Baral, J. 1988, Aspirin and Reye syndrome. *Pediatrics*, **82:** 135.

Baron, I.S., Fennell, E.B. and Voeller, K. 1995, *Pediatric Neuropsychology in the Medical Setting*. New York: Oxford University Press.

Barrett, M.J., Hurwitz, E.S., Schonberger, L.B. and Rogers, M.F. 1986, Changing epidemiology of Reye syndrome in the United States. *Pediatrics*, **77:** 598–602.

Belay, E.D., Bresee, J.S., Holman, R.C., Khan, A.S., Shahriari, A. and Schonberger, L.B. 1999, Reye's syndrome in the United States from 1981 through 1997. *New England Journal of Medicine*, **340:** 1377–1382.

Bellman, M.H. and Hall, S.M. 1985, Reye's syndrome in the British Commonwealth. *In:* J.D. Pollack (ed.), *Reye's Syndrome*, Vol. IV, pp. 15–31. Bryan, Ohio: National Reye's Syndrome Foundation.

Beutler, B., Mahoney, J., LeTrange, N., Pekala, P. and Cerami, A. 1985, Purification of cachectin: a lipoprotein lipase-suppressing hormone secreted by endotoxin-induced RAW 264.7 cells. *Journal of Experimental Medicine*, **161:** 984–992.

Bhutta, A.T., Van Savell, H. and Schexnayder, S.M. 2003, Reye's syndrome: down but not out. *Southern Medical Journal*, **96:** 43–45.

Bonnell, H.J. and Beckwith, J.B. 1986, Fatty liver in sudden childhood death. Implications for Reye's syndrome? *American Journal of Diseases of Children*, **140:** 30–33.

Boss, O., Hagen, T. and Lowell, B.B. 2000, Uncoupling proteins 2 and 3: potential regulators of mitochondrial energy metabolism. *Diabetes*, **49:** 143–156.

Bove, K.E., McAdams, J.A., Partin, J.C., Partin, J.S., Hug, G. and Schubert, W.K. 1975, The hepatic lesion in Reye's syndrome. *Gastroenterology*, **69:** 685–687.

Brain, W.R., Hunter, D. and Turnbull, H.M. 1929, Acute meningoencephalomyelitis of childhood: report of six cases. *Lancet*, **1:** 221–227.

Brown, J.K. and Imam, H. 1991, Interrelationships of the liver and brain with special reference to Reye syndrome. *Journal of Inherited and Metabolic Diseases*, **14:** 436–458.

Brown, R.E. and Madge, G.E. 1972, Fatty acids and mitochondrial injury. *New England Journal of Medicine*, **268:** 287–288.

Bruner, R.L., O'Grady, D.J., Partin, J.C., Partin, J.S. and Schubert, W.K. 1979, Neuropsychological consequences of Reye's syndrome. *Journal of Pediatrics*, **95:** 706–711.

Campbell, R.A. and Brown, R.E. 1985, Polyamines and the Reye–Johnson syndrome (abstract). *In:* J.D. Pollack (ed.), *Reye's Syndrome*, Vol. IV, pp. 306. Bryan, Ohio: National Reye's Syndrome Foundation.

Casteels-Van Daele, M. 1991, Reye syndrome or side-effects of anti-emetics? *European Journal of Pediatrics*, **150:** 456–459.

Casteels-Van Daele, M. and Eggermont, E. 1994, Reye's syndrome. *British Medical Journal*, **308:** 919–920.

Casteels-Van Daele, M., Van Geet, C., Wouters, C. and Eggermont, E. 2000, Reye's syndrome revisited: a descriptive term covering a group of heterogeneous disorders. *European Journal of Paediatrics*, **159:** 641–648.

Cavanagh, J.B. and Kyu, M.H. 1971, Type II Alzheimer changes experimentally produced in astrocytes in rats. *Journal of Neurological Science*, **12:** 63–75.

Centers for Disease Control 1980, Reye's syndrome – Ohio, Michigan. *CDC Morbidity and Mortality Weekly Report*, **29:** 532–539.

Centers for Disease Control 1982, Surgeon-General's Advisory on the use of salicylates and Reye's syndrome. *CDC Morbidity and Mortality Weekly Report*, **31:** 289–290.

Centers for Disease Control 1985, Reye's syndrome – United States, 1984. *CDC Morbidity and Mortality Weekly Report*, **34:** 13–16.

Clark, J.H., Nagamori, K. and Fitzgerald, J.F. 1985, Confirmation of serum salicylate levels in Reye's syndrome: a comparison between the Natelson colorimetric method and high performance liquid chromatography. *Clinica Chimica Acta*, **145:** 243–247.

Clayton, P.T., Doig, M., Ghafari, S., Meaney, C., Taylor, C., Leonard, J.V., Morris, M. and Johnson, A.W. 1998, Screening for medium chain acyl-CoA dehydrogenase deficiency using electrospray ionisation tandem mass spectrometry. *Archives of Disease in Childhood*, **79:** 109–115.

Committee on Reye's Syndrome Research 1983, Japanese Ministry of Health and Welfare Official Report. Unpublished.

Cooperstock, M.S., Tucker, R.P. and Baublis, J.V. 1975, Possible pathogenetic role of endotoxin in Reye's syndrome. *Lancet*, **1**: 1272–1274.

Corkey, B.E., Hale, D.E., Glennon, M.C., Kelly, R.J., Coates, P.M., Kilpatrick, L. and Stanley, C.A. 1988, Relationship between unusual acyl CoA profiles and the pathogenesis of Reye's syndrome. *Journal of Clinical Investigation*, **88**: 782–788.

Crocker, J.F.S. 1988, The role of exogenous agents in Reye's syndrome. *In:* C. Wood (ed.), *Reye's Syndrome*. pp. 45–63. London: Royal Society of Medicine.

Crocker, J.F.S., Renton, K.W., Lee, S.H., Rozee, K.R., Digout, S.C. and Malatjalian, D.A. 1986, Biochemical and morphological characteristics of a mouse model of Reye's syndrome induced by the interaction of an influenza B virus and a chemical emulsifier. *Laboratory Investigation*, **54**: 32–40.

Daniels, S.R., Greenberg, R.S. and Ibrahim, M.A. 1983, Scientific uncertainties in the studies of salicylate use and Reye's syndrome. *Journal of the American Medical Association*, **249**: 1311–1316.

Da Silveira, E.B., Young, K., Rodriguez, M. and Ameen, N. 2002, Reye's syndrome in a 17 year old male, is this disease really disappearing? *Digestive Diseases and Sciences*, **47**: 1959–1961.

Davis, G.E., Cole, L.L., Lockwood, S.T. and Kornfeld, M. 1983, Experimental influenza B virus toxicity in mice. *Laboratory Investigation*, **48**: 140–147.

Davis, G.E., Green, C.L. and Wallace, J.M. 1985, Influenza B virus model of Reye's syndrome in mice: the effect of aspirin. *Annals of Neurology*, **18**: 556–559.

De Villemeur, T.B., Nuttin, C., Poggi, F., Carre, M., Bonnefont, J., Rabier, D., Charpentier, C., Brivet, M. and Saudubray, J.M. 1993, Metabolic investigations in Reye's syndrome and Reye-like syndrome. *In:* J.P. Buts and E.M. Sokal (eds), *Management of Digestive and Liver Disorders in Infants and Children*, pp. 615–622. Amsterdam: Elsevier.

De Vivo, D.C. 1988, Treatment of Reye's syndrome: concensus and controversy. *In:* C. Wood (ed.), *Reye's Syndrome*, pp. 133–155. London: Royal Society of Medicine.

DeLong, G.R. and Glick, T.H. 1982, Encephalopathy of Reye's syndrome: a review of pathogenetic hypotheses. *Pediatrics*, **69**: 53–63.

Deshmukh, D.R., Deshmukh, G.D., Shope, T.C. and Radin, N.S. 1983, Free fatty acids in an animal model of Reye's syndrome. *Biochemica Biophysica Acta*, **753**: 153–158.

Deshmukh, D.R., Maassab, H.F. and Mason, M. 1982, Interactions of aspirin and other potential etiological factors in an animal model of Reye syndrome. *Proceedings of the National Academy of Science (USA)*, **79**: 7557–7560.

Duffy, J., Glasgow, J.F.T., Patterson, C.C., Clarke, M.J. and Turner, I.F. 1991, A sibling-controlled study of intelligence and academic performance following Reye's syndrome. *Developmental Medicine and Child Neurology*, **33**: 811–815.

Fairhurst, H. 1988, Survey of analgesic use (letter; comment). *Journal of the Royal College of General Practitioners*, **39**: 76.

Forsyth, B.W., Horwitz, R.I., Acampora, D., Shapiro, E.D., Viscoli, C.M., Feinstein, A.R., Henner, R., Holabird, N.B., Jones, B.A., Karabelas, A.D., Kramer, M.S., Miclette, M. and Wells, J. 1989, New epidemiologic evidence confirming that bias does not explain the aspirin/Reye's syndrome association. *Journal of the American Medical Association*, **261**: 2517–2524.

Gauthier, M., Guay, J., Lacroix, J. and Lortie, A. 1989, Reye's syndrome: a reappraisal of diagnosis in 49 presumptive cases. *American Journal of Diseases in Childhood*, **143**: 1181–1185.

Gladtke, E. and Schauseil-Zipf, U. 1987, Reye's syndrome. *Monatsschr Kinderheilkd*, **135**: 699–704.

Glasgow, J.F.T. 1984, Clinical features and prognosis of Reye's syndrome. *Archives of Disease in Childhood*, **59**: 230–235.

Glasgow, J.F.T. and Ferris, J.A.J. 1969, Encephalopathy and visceral fatty infiltration of probable toxic aetiology. *Lancet*, **1**: 451–453.

Glasgow, J.F.T., Hicks, E.M., Jenkins, J.G., Keilty, S.R., Black, G.W. and Fannin, T.F. 1985, Reye's syndrome. *British Journal of Hospital Medicine*, **34**: 42–46.

Glasgow, J.F.T., Jenkins, J.G., Hicks, E.M., Keilty, S.R., Crean, P.M., Black, G.W. and Fannin. T.F. 1986, The prognosis for Reye's syndrome in Ireland can be improved. *Irish Journal of Medical Science*, **155**: 111–116.

Glasgow, J.F.T. and Middleton, B. 2001, Reye syndrome: insights on causation and prognosis. *Archives of Disease in Childhood*, **85:** 351–353.

Glasgow, J.F.T., Middleton, B., Moore, R., Gray, A. and Hill, J. 1999, The mechanism of inhibition of β-oxidation by aspirin metabolites in skin fibroblasts from Reye's syndrome patients and controls. *Biochimica et Biophysica Acta*, **1454:** 115–125.

Glick, T.H., Lakosky, W.H., Levitt, L.P., Mellin, H. and Reynolds, D.W. 1970, Reye's syndrome: an epidemiological approach. *Pediatrics*, **46:** 371–377.

Green, A. and Hall, S.M. 1992, Investigation of metabolic disorders resembling Reye's syndrome. *Archives of Disease in Childhood*, **67:** 1313–1317.

Hall, R.W. 1987, Aspirin and Reye's syndrome – do parents know? (see comments). *Journal of the Royal College of General Practitioners*, **37:** 459–460.

Hall, S.M. 1988, Reye study criticized (letter). *Pediatrics*, **82:** 391–394.

Hall, S.M. 1994, Reye's syndrome (letter; comment). *British Medical Journal*, **309:** 411.

Hall, S.M. and Bellman, M.H. 1985, Reye's syndrome in the British Isles: the British Paediatric Association/PHLS Communicable Diseases Surveillance Centre Joint Surveillance Scheme. *In:* J.D. Pollack (ed.), *Reye's Syndrome*, Vol. IV, pp. 32–46. Bryan, Ohio: National Reye's Syndrome Foundation.

Hall, S.M. and Glickman, M. 1988, The British Paediatric Surveillance Unit. *Archives of Disease in Childhood*, **63:** 344–346.

Hall, S.M. and Lynn, R. 1998, Reye Syndrome. *In:* M. Guy, A. Nicoll and R. Lynn (eds), *British Paediatric Surveillance Unit, 12th Annual Report*. London: RCPCH.

Hall, S.M. and Lynn, R. 1999, Reye's syndrome. *New England Journal of Medicine*, **341:** 845.

Hall, S.M. and Lynn, R. 2002, Reye syndrome. *In:* R. Lynn, A. Nicoll, J. Rahi and C. Verity (eds), *British Paediatric Surveillance Unit, 16th Annual Report*, London: RCPCH.

Hall, S.M., Plaster, P.A., Glasgow, J.F. and Hancock, P. 1988, Preadmission antipyretics in Reye's syndrome. *Archives of Disease in Childhood*, **63:** 857–866.

Halpin, T.J., Holtzhauer, F.J., Campbell, R.J., Hall, L.J., Correa-Villasenor, A., Lanese, R., Rice, J. and Hurwitz, E.S. 1982, Reye's syndrome and medication use. *Journal of the American Medical Association*, **248:** 687–691.

Hardie, R.M., Newton, L.H., Bruce, J.C., Glasgow, J.F.T., Mowat, A.P., Stephenson, J.P.B. and Hall, S.M. 1996, The changing clinical pattern of Reye's syndrome 1982–1990. *Archives of Disease in Childhood*, **74:** 400–405.

Haymond, M.W., Karl, I.E., Keating, J.P. and De Vivo, D.C. 1978, Metabolic response to hypertonic glucose administration in Reye's syndrome. *Annals of Neurology*, **3:** 207–215.

Heubi, J.E., Daugherty, C.C., Partin, J.S., Partin, J.C. and Schubert, W.K. 1984, Grade I Reye's syndrome – outcome and predictors of progression to deeper coma grades. *New England Journal of Medicine*, **311:** 1539–1542.

Hilty, M.D., Romshe, C.A. and Dalamater, P.V. 1974, Reye's syndrome and hyperaminoacidemia. *Journal of Pediatrics*, **84:** 362–365.

Hurwitz, E.S., Barrett, M.J., Bregman, D., Gunn, W.J., Pinsky, P., Schonberger, L.B., Drage, J.S., Kaslow, R.A., Burlington, D.B., Quinnan, G.V., LaMontagne, J.R., Fairweather, W.R., Dayton, D. and Dowdle, W.R. 1987, Public Health Service study of Reye's syndrome and medications. Report of the main study (published erratum appears in *Journal of the American Medical Association*, 1987 June 26; **257 (24):** 3366). *Journal of the American Medical Association*, **257:** 1905–1911.

Hurwitz, E.S., Barrett, M.J., Bregman, D., Gunn, W.J., Schonberger, L.B., Fairweather, W.R., Drage, J.S., LaMontagne, J.R., Kaslow, R.A., Burlington, D.B., Quinnan, G.V., Parker, R.A., Phillips, K., Pinsky, P., Dayton, D. and Dowdle, W.R. 1985, Public Health Service study on Reye's syndrome and medications. Report of the pilot phase. *New England Journal of Medicine*, **313:** 849–857.

Hurwitz, E.S. and Mortimer, E.A. 1990, A catch in the Reye is awry. *Cleveland Clinic Journal of Medicine*, **57:** 318–319.

Huttenlocher, P.R. 1972, Reye's syndrome: relationship of outcome to therapy. *Journal of Pediatrics*, **80:** 845–850.

Jenkins, J.G., Glasgow, J.F.T., Black, G.W., Fannin, T.F., Hicks, E.M. and Keilty, S.R. 1987, Reye's syndrome: assessment of intracranial monitoring. *British Medical Journal*, **294:** 337–338.

Johnson, G.M., Scurletis, T.D. and Carroll, N.B. 1963, A study of 16 fatal cases of encephalitis-like disease in North Carolina children. *North Carolina Medical Journal*, **24:** 464–473.

Khan, A.S., Kent, J. and Schonberger, L.B. 1993, Aspirin and Reye's syndrome. *Lancet*, **341:** 968.

Kirkham, F.J. 2001, Non-traumatic coma in children. *Archives of Disease in Childhood*, **85:** 303–312

Kilpatrick-Smith, L., Hale, D.E. and Douglas S.D. 1989, Progress in Reye syndrome: epidemiology, biochemical mechanisms and animal models. *Digestive Diseases*, **7:** 135–146.

Kirn, A., Gut, J. and Gendroult, J. 1982, Interaction of viruses with sinusoidal cells. *In:* H. Popper and F. Schaffner (eds), pp. 377–392. *Progress in Liver Diseases*, Vol. 7. Orlando: Grune and Stratton Inc.

Larrick, J.W. and Kunkel, S.L. 1986, Is Reye's syndrome caused by augmented release of tumour necrosis factor? *Lancet*, **2:** 132–133.

Larsen, S.U. 1997, Reye's syndrome. *Medical Science Law*, **37:** 235–239.

Leonard, J.V. 1988, *In:* C. Wood (ed.), *Reye's Syndrome*, p. 128. London: Royal Society of Medicine London (comments in a discussion).

Lichtenstein, P.K., Heubi, J.E., Daugherty, C.C., Farrell, M.K., Sokol, R.J., Rothbaum, R.J., Suchy, F.J. and Balistreri, W.F. 1983, Grade I Reye's syndrome: a frequent cause of vomiting and liver dysfunction after varicella and upper respiratory tract infection. *New England Journal of Medicine*, **309:** 133–139.

Linnemann, C.C., Shea, L., Partin, J.C., Schubert, W.K. and Schiff, G.M. 1975, Reye's syndrome: epidemiologic and viral studies, 1963–1974. *American Journal of Epidemiology*, **101:** 517–526.

Lovejoy, F.H., Smith, A.L., Bresnan, M.J., Wood, J.N., Victor, D.I. and Adams, P.C. 1974, Clinical staging in Reye syndrome. *American Journal of Diseases of Children*, **128:** 36–41.

Maddaiah, V.T. and Miller, P.S. 1989, Effects of ammonium chloride, salicylate and carnitine on palmitic acid oxidation in rat liver slices. *Pediatric Research*, **25:** 119–123.

McGee, H.B. and Sienko, D.G. 1988, A catch in the Reye. *Pediatrics*, **82:** 390–399.

McGovern, M.-C., Glasgow, J.F.T. and Stewart, M.C. 2001, Reye's syndrome and aspirin: lest we forget. *British Medical Journal*, **322:** 1591–1592.

Meekin, S., McCusker, C., Glasgow, J.F.T. and Rooney, N. 1999, A long-term follow-up of cognitive, emotional and behavioural sequelae to Reye's syndrome. *Developmental Medicine and Child Neurology*, **41:** 549–553.

Meert, K.L., Kauffman, R.E., Deshmukh, D.R. and Sarnaik, A.P. 1990, Impaired oxidative metabolism of salicylate in Reye's syndrome. *Developmental Pharmacology and Therapeutics*, **15:** 57–60.

Millington, D., Kodo, N., Terada, N., Roe, C.R. and Chace, D. 1997, The analysis of diagnostic markers of genetic disorders in human blood using mass spectrometry. *International Journal of Mass Spectrometry and Ion Processes*, **111:** 211–218.

Monto, A.S. 1999, The disappearance of Reye's syndrome – a public health triumph. *New England Journal of Medicine*, **340:** 1423–1424.

Morgan, G. 1985, Reye's syndrome: the beginning. *In:* J.D. Pollack (ed.), *Reye's Syndrome*, Vol. IV, pp. 2–8. Bryan, Ohio: National Reye's Syndrome Foundation.

Mortimer, E.A. 1987, Reye's syndrome, salicylates, epidemiology and public health policy. *Journal of the American Medical Association*, **257:** 1941.

Mowat, A.P. 1988, Endogenous factors which may contribute to Reye's syndrome. *In:* C. Wood (ed.), *Reye's Syndrome*, pp. 64–79. London: Royal Society of Medicine.

Mowat, A.P. 1994, Reye's syndrome. *In: Liver Disorders in Childhood*, 3rd edn. London: Butterworth.

Mukhopadhyay, A., Sarnaik, A.P. and Deshmukh, D.R. 1992, Interactions of ibuprofen with influenza infection and hyperammonaemia in an animal model of Reye's syndrome. *Pediatric Research*, **31:** 258–260.

Nachiappen, V., Curtiss, D., Corkey, B.E. and Kilpatrick-Smith, L. 1994, Cytokines inhibit fatty acid oxidation in isolated rat hepatocytes: Synergy among TNF, IL-6 and IL-1. *Shock*, **1:** 123–129.

Nelson, D.B., Kimbrough, R. and Landrigan, P.S. 1980, Aflatoxin and Reye's syndrome: a case–control study. *Pediatrics*, **66:** 865–869.

Oldendorf, W.H. and Brown, J.W. 1975, Greater number of capillary endothelial cell mitochondria in brain than muscle. *Proceedings of the Society for Experimental Biology and Medicine*, **149:** 736–738.

Olsen, L.C., Bourgeois, C.H., Cotton, R.B., Harikul, S., Grossman, R.A. and Smith, T.J. 1971, Encephalopathy and fatty degeneration of the viscera in Northeastern Thailand. Clinical syndrome and epidemiology. *Pediatrics*, **47:** 707–716.

Orlowski, J.P. 1999, Whatever happened to Reye's syndrome? Did it ever really exist? *Critical Care Medicine*, **27:** 1582–1587.

Orlowski, J.P., Campbell, P. and Goldstein, S. 1990, Reye's syndrome: a case–control study of medication use and associated viruses in Australia. *Cleveland Clinic Journal of Medicine*, **57:** 323–329.

Orlowski, J.P., Gillis, J. and Kilham, H.A. 1987, A catch in the Reye. *Pediatrics*, **80:** 638–642.

Osterloh, J., Cunningham, W., Dixon, A. and Combest, D. 1989, Biochemical relationships between Reye's syndrome and Reye-like metabolic and toxicological syndromes. *Medical Toxicological and Adverse Drug Exposure*, **4:** 272–294.

Palomeque, A., Domenech, P., Martinez-Guttierez, A., Asencio, M.J., Belmonte, J.A., Calvo, C., Delgado, A., Escudero, G., Gomez Calzado, A. and Labay, M. 1986, Reye's syndrome in Spain 1980–1984. *An Esp Pediatr*, **24 (5):** 285–289.

Partin, J.C. 1979, Reye's syndrome (encephalopathy and fatty liver): diagnosis and treatment. *Gastroenterology*, **69:** 511–518.

Partin, J.C. 1988, General management of Reye's syndrome. *In:* C. Wood (ed.), *Reye's Syndrome*, pp. 156–172. London: Royal Society of Medicine.

Partin, J.C., Partin, J.S., Saalfield, K., Schubert, W.K. and Jacobs, R. 1976, Isolation of influenza virus from liver and muscle biopsy specimens from a surviving case of Reye's syndrome. *Lancet*, **2:** 599–602.

Partin, J.C., Schubert, W.K. and Partin, J.S. 1971, Mitochondrial ultrastructure in Reye's syndrome (encephalopathy and fatty degeneration of the viscera). *New England Journal of Medicine*, **185:** 1339–1343.

Partin, J.S. 1988, Ultrastructural changes in liver, muscle and brain in Reye's syndrome. *In:* C. Wood (ed.), *Reye's Syndrome*, pp. 1–19. London: Royal Society of Medicine.

Partin, J.S., Daugherty, C.C., McAdams, A.J., Partin, J.C. and Schubert, W.K. 1984, A comparison of liver ultra-structure in salicylate intoxication and Reye's syndrome. *Hepatology*, **4:** 687–690.

Partin, J.S., Partin, J.C., Schubert, W.K. and Hammond, J.G. 1982, Serum salicylate concentrations in Reye's disease. A study of 130 biopsy-proven cases. *Lancet*, **1:** 191–194.

Pinsky, P.F., Hurwitz, E.S., Schonberger, L.B. and Gunn, W.J. 1988, Reye's syndrome and aspirin. Evidence for a dose–response effect. *Journal of the American Medical Association*, **260:** 657–661.

Plum, F. and Posner, J.B. 1966, *Diagnosis of Stupor and Coma*, pp. 56–67. Philadelphia: F.A. Davis Co.

Porter, J.D., Robinson, P.H., Glasgow, J.F.T., Banks, J.H. and Hall, S.M. 1990, Trends in the incidence of Reye's syndrome and the use of aspirin. *Archives of Disease in Childhood*, **65:** 826–829.

Poulton, J. and Brown G.K. 1995, Investigation of mitochondrial disease (review). *Archives of Disease in Childhood*, **73:** 94–97.

Prior, M.J., Nelson, E.B. and Temple, A.R. 2000, Pediatric ibuprofen use increases while incidence of Reye's syndrome continues to decline. *Clinical Pediatrics*, **39:** 245–247.

Remington, P.L., Rowley, D., McGee, H., Hall, W.N. and Monto, A.S. 1986, Decreasing trends in Reye's syndrome and aspirin use in Michigan, 1979 to 1984. *Pediatrics*, **77:** 93–98.

Rennebohm, R.M., Heubi, J.E., Daugherty, C.C. and Daniels, S.R. 1985, Reye syndrome in children receiving salicylate therapy for connective tissue disease. *Journal of Pediatrics*, **107:** 877–880.

Reye, R.D.K., Morgan, G. and Baral, J. 1963, Encephalopathy and fatty degeneration of the viscera. A disease entity in childhood. *Lancet*, **2:** 749–752.

Ricquier, D. and Bouillaud, F. 2000, The uncoupling protein homologues: UCP1, UCP2, UCP3, StUCP and AtUCP. *Biochemistry Journal*, **345:** 161–179.

Robinson, B.H., Taylor, J., Cutz, E. and Gall, D.G. 1978, Reye's syndrome: preservation of mitochondria in brain and muscle compared with liver. *Pediatric Research*, **12:** 1045–1047.

Robinson, R.O. 1987, Differential diagnosis of Reye's syndrome. *Developmental Medicine and Child Neurology*, **29:** 110–120 (and subsequent correspondence).

Rodgers, G.C. Jr. 1985, Analgesics and Reye syndrome: fact or fiction? *In:* J.D. Pollack (ed.), *Reye's Syndrome*, Vol. IV, pp. 117–134. Bryan, Ohio: National Reye's Syndrome Foundation.

Roe, C.R. 1988, Metabolic disorders producing a Reye-like syndrome. *In:* C. Wood (ed.), *Reye's Syndrome*, pp. 85–107. London: Royal Society of Medicine.

Roe, C.R., Schonberger, L.B., Gelbach, S.H., Weis, L.A. and Siudbury, J.B. 1975, Enzymatic alterations in Reye's syndrome: prognostic implications. *Pediatrics*, **55**: 119–126.

RS Working Group 1982, Reye's syndrome and salicylates: a spurious association. *Pediatrics*, **70**: 158–160.

Sarnaik, A.P. 1999, Reye's syndrome: hold the obituary. *Critical Care Medicine*, **27**: 1674–1676.

Sharples, P.M. 1991, Intracranial pressure monitoring in comatose children: concepts and controversies. *Current Paediatrics*, **1**: 46–48.

Shaywitz, S.E., Cohen, P.M., Cohen, D.J., Mikkelson, E., Morowitz, G. and Shaywitz, B.A. 1982, Long-term consequences of Reye's syndrome. *Journal of Pediatrics*, **100**: 41–46.

Soumerai, S.B., Ross-Degnan, D. and Kahn, J.S. 1992, Effects of professional and media warnings about the association between aspirin use in children and Reye's syndrome (review). *Milbank Quarterly*, **70**: 155–182.

Starko, K.M. and Mullick, F.G. 1983, Hepatic and cerebral pathology findings in children with fatal salicylate intoxication: further evidence for a causal relation between salicylate and Reye's syndrome. *Lancet*, **1**: 326–329.

Starko, K.M., Ray, C.G., Dominguez, L.B., Stromberg, W.L. and Woodall, D.F. 1980, Reye's syndrome and salicylate use. *Pediatrics*, **66**: 859–864.

Stumpf, D.A. 1995, Reye syndrome: an international perspective. *Brain and Development*, **17 (Suppl.)**: 77–78

Sullivan, K.M., Belay, E.D., Durbin, R.E., Foster, D.A. and Nordenberg, D.F. 2000, Epidemiology of Reye's syndrome, United States, 1991–1994. Comparison of CDC Surveillance and Hospital Admission Data. *Neuroepidemiology*, **19**: 338–344.

Sullivan, K.M., Remington, P.L., Halpin, T.J. and Hurwitz, E.S. 1988, Reye's syndrome among patients with juvenile rheumatoid arthritis. *Journal of the American Medical Association*, **260**: 3434–3435.

Thabet, F., Durand, P., Chevret, L., Fabre, M., Debray, D., Brivet, M. and Devictor, D. 2002, Syndrome de reye severe: a propos de 14 cas pris en charge dans une unite de reanimation pediatrique pendant 11 ans. *Archives Pediatriques*, **9**: 581–586.

Tonsgard, J.H. 1985, Serum dicarboxylic acids in patients with Reye syndrome. *Journal of Pediatrics*, **109**: 440–445.

Tonsgard, J.H. and Getz, G.S. 1985, The effect of Reye's syndrome serum on isolated chinchilla liver mitochondria. *Journal of Clinical Investigation*, **76**: 816–825.

Tonsgard, J.H. and Huttenlocher, P.R. 1981, Salicylates and Reye's syndrome (letter). *Pediatrics*, **68**: 747–748.

Trauner, D.A., Nyhan, W.L. and Sweetman, L. 1975, Short chain organic acidaemia in Reye's syndrome. *Neurology*, **25**: 296–298.

Treem, W.R., Witzleben, C.A., Piccoli, D., Stanley, C.A., Hale, D.A., Coates, P.M. and Watkins, J.B. 1986, Medium chain and long chain acyl CoA dehydrogenase deficiency: clinical, pathological and ultrastructural differentiation from Reye's syndrome. *Hepatology*, **6**: 1270–1278.

Van Coster, R.N., De Vivo, D.C., Blake, D., Lombes, A., Barrett, R. and DiMauro, S. 1991, Adult Reye's syndrome: a review of new evidence for a generalised defect in intramitochondrial enzyme processing. *Neurology*, **41**: 1815–1820.

Waldman, R.J., Hall, W.N., McGee, H. and Van Amburg, G. 1982, Aspirin as a risk factor in Reye's syndrome. *Journal of the American Medical Association*, **247**: 3089–3094.

Walter, J.H. 1996, L-Carnitine. *Archives of Disease in Childhood*, **74**: 475–478.

White, J.M. 1986, Reye's syndrome and salicylates. *New England Journal of Medicine*, **314**: 920.

Wood, R.B. Jr. and Bogdan, G.F. 1986, Reye's syndrome and spruce budworm insecticide spraying in Maine, 1978–1982. *American Journal of Epidemiology*, **124**: 671–677.

Yamashita, F., Ono, E., Kimura, A. and Yoshida, I. 1985, Reye's syndrome in Asian countries. *In:* J.D. Pollack (ed.), *Reye's Syndrome*, Vol. IV, pp. 47–59. Bryan, Ohio: National Reye's Syndrome Foundation.

Yoder, M.C., Egler, J.M., Yudkoff, M., Chatten, J., Douglas, S.D. and Polin, R.A. 1985, Metabolic and morphologic changes that mimic Reye syndrome after endotoxin administration in rats. *Infection and Immunology*, **47**: 329–331.

CHAPTER 9

Salicylates in the Treatment of Acute Pain

K.D. Rainsford

INTRODUCTION

A definition of pain is that which occurs as 'an unpleasant sensory and emotional experience associated with actual or potential tissue damage, or described in terms of such damage' (The International Association for the Study of Pain (IASP) as quoted by Forrest, 1998). Though pain is mediated by biochemical algogens that are mediators released from injured tissues and activate nociceptors, it is actually a psychobehavioural and sensory experience achieved by this process of activation of nociceptive pathways (Forrest, 1998).

The concept of acute pain is the 'normal predicted physiological response to an adverse chemical, thermal or mechanical stimulus' that is 'associated with surgery, trauma or acute illness' (Carr and Goudas, 1999). Acute pain essentially arises from substantial injury of body tissue with activation of nociceptive transducers at the site of local tissue damage (Loeser and Melzack, 1999). The activation of nociceptive pathways occurs from components of the tissue-derived 'inflammatory soup', and leads to transduction of the signals to Aδ and C fibres in the dorsal horn (Loeser and Melzack, 1999). Drugs such as aspirin and other NSAIDs lead to changes in the 'inflammatory soup', and produce pain relief mainly by restoration of nociceptive sensitivity to the resting state (Loeser and Melzack, 1999). Paracetamol may act on components of the nociceptive pathways mediated by the inflammatory soup in a different manner to aspirin and the NSAIDs. Recent evidence, for instance, implicates paracetamol inhibiting leucocyte-derived radicals produced by myeloperoxidase and superoxide pathways (Graham *et al.*, 1999). Moreover, the previously held belief that paracetamol does not inhibit prostaglandin production has recently been refuted (Graham *et al.*, 1999). Some suggestions have also implicated paracetamol affecting the activity of a new isoform of the prostaglandin G/H synthase (cyclo-oxygenase)-3 as a likely site of action of the drug in the nervous system. (Chandrasekharan *et al.*, 2003).

Patients' attitudes, beliefs and personalities strongly affect their immediate experience of acute pain (Carr and Goudas, 1999). Recent definitions of acute pain and the transition to chronic pain have essentially been revised, since acute pain from tissue injury may last upwards of 1 month (Carr and Goudas, 1999). However, there is evidence that acute pain may rapidly evolve into chronic states (Carr and Goudas, 1999) and thus there are differences in the nature of the transition from acute to chronic pain as well as the precision of the dichotomy between these two. For the purposes of the discussion in this section concerning the nature of the responses to acute pain, these will be defined by the clinical models and states that are defined as being of acute character. The discussion of the various types of headache will be considered as a special case of acute pain, with migraine and related chronic forms of headache being considered separately.

A century of use of aspirin for the relief of a variety of acute painful conditions is proof positive of its widespread utility, efficacy and acceptance worldwide. This has not, of course, been without controversy. Among the issues has been the question of safety. Adverse reactions in the form of gastrointestinal ulceration and bleeding, acute renal failure, asthma and the association in children with Reye's syndrome have been among the major concerns. Competition from other analgesics and NSAIDs, including paracetamol, ibuprofen, ketoprofen, naproxen and dipyrone, employed for minor painful conditions (as well as for control of fever) appears to have had relatively little impact on the use of aspirin for these states. No doubt these other analgesics have blunted the potential sales of aspirin, but the drug still continues to be sold successfully for pain relief.

While this section will largely focus of the role of aspirin and paracetamol in pain relief, the relative effects of NSAIDs other than aspirin will also be considered by way of comparison of efficacy and safety.

CONCEPTS OF THE MODE OF ACTION IN ACUTE PAIN

The basic mechanisms of nociception and pain pathways are considered in detail elsewhere in this book (Chapter 7; Figure 7.10, p. 292). Here it is useful briefly to summarise the relevant mechanisms underlying the clinical manifestations of acute pain (see also Besson, 1999).

Tissue destruction activates nociceptors and initiates local reactions in which a variety of mediators are released sequentially. Initially this involves release of histamine and 5-hydroxytryptamine (5-HT or serotonin) from mast cells and production of kinins from plasma proteins. Then activation of kinin receptors and increased intracellular calcium results in arachidonic acid release by the activation of phospholipase A_2, and this results in the stimulation of prostaglandin production. Neurotransmitters (including the excitatory amino acids glutamate and aspartate, and the peptides (substance P and calcitonin gene-related peptide) that are released in the dorsal horn are modulated by encephalins and dynophin. Nitric oxide is released, and protein kinase C activation is also among the early events. Expression of the oncogenes *c-fos* and *c-jun* in the dorsal horn neurones accompanies the induction of inducible cyclo-oxygenase-2 (COX-2), that is responsible for the amplified production of prostaglandins. The induction of COX-2 in the spinal cord, especially the dorsal horn, is of particular significance in relation to the actions of the NSAIDs and paracetamol (Buritova *et al.*, 1996; Pitcher and Henry, 1999), since this may represent a major site of their actions in controlling spinal transmission of pain responses (Malmberg and Yaksh, 1992) along with supraspinal effects (Bannwarth *et al.*, 1995; Björkman, 1995; Sandrini, 1999), aside from their actions arising from mediators produced in the peripheral 'inflammatory soup'.

EARLY OBSERVATIONS IN VARIOUS PAIN STATES

Gross and Greenberg (1948) reviewed the early clinical trials on the efficacy of aspirin and other analgesics in a mixture of acute and chronic pain states. Simon (1931) found that aspirin had analgesic activity when given to 45 patients who had 'constant pain from arthritis, neuritis and headaches' in doses of 10–20 grains (648–1296 mg); moderate to complete relief was achieved in about half the patients. Curiously, the combination of aspirin with magnesium oxide was sometimes found more effective than with aspirin alone.

Lewy (1942) observed pain relief from repeated intake of doses of 0.15 g aspirin in several hundred patients with inoperable cancer or during puerperium (the 6-week period following child delivery, which was presumably normal); pain relief was achieved in 75 per cent of patients with cancer and 84.5 per cent of those with post-partum pain. Unfortunately, half of the group of 233 patients who received placebo exhibited pain relief that he did not consider in comparison with the groups that received aspirin. An appreciable placebo effect in 60 per cent of subjects was also observed by Jellinek (1946) in comparing the efficacy of three unspecified analgesic preparations for relief of frequent headaches.

Bywaters (1963) considered the effectiveness of salicylate compared with cortisone in treating rheumatic fever and judged that, apart from severe cases of rheumatic carditis, both were no better than bed rest alone 'despite the dramatic effects of both on fever and joint inflammation'. In more severe cases he considered that 'salicylates are potentially dangerous as they may occasionally produce heart failure and precipitate pulmonary oedema'.

Beaver (1965) undertook a comprehensive review of trials in which aspirin had been compared with placebo, other salicylates, non-narcotic analgesics (phenacetin = acetophenetidin; acetaminophen = paracetamol; antipyrine = phenazone; dipyrone = metamizole = methampyrone; aminopyrine = amidopyrine) and various NSAIDs available at the time. These were considered effective, but by this time few controlled trials had been performed to what would be regarded as acceptable criteria by today's standards. Table 10.1, from Beaver (1965), summarises the trials reviewed by this author in which aspirin had been shown to have superiority over placebo in acute or mixed acute/chronic pain states. It is clear from these studies that by the mid-1960s aspirin had been shown to be more effective than placebo in doses of 650 mg or greater, and in some studies at the lower dose of 325 mg.

EFFICACY IN VARIOUS PAIN CONDITIONS

Several reviews (Gold and Cattell, 1965; Moertel *et al.*, 1972; 1974; Vandam, 1972; Wallenstein, 1975; Beaver, 1981; Cooper, 1981; Amadio, 1984; Rainsford, 1984; Vane and Botting, 1992; Moote, 1996; Brasseur, 1997; Edwards *et al.*, 1999; McQuay and Moore, 1999) have considered the efficacy of aspirin and other analgesics or NSAIDs in various pain models. The general conclusions from these reviews are that aspirin does show analgesic activity that is comparable with other analgesics/NSAIDs, although the relative dose effects may vary somewhat between different studies.

Laska and co-workers (1982) undertook an extensive series of studies (10) in 4234 patients in three models of clinical pain – post-episiotomy, uterine cramping and postsurgical (about one-third whom had undergone Caesarean section or other gynaecological surgery) – investigating the efficacy of 325, 650 and 1300 mg aspirin compared with placebo in relief of pain. An important (and indeed fundamental) outcome from these studies was the development of a mathematical model that characterised the probability that an analgesic provides complete pain relief as a function of its dose, the severity of the underlying pain intensity, and the aetiology and type of painful state. The development of this model and non-parametric statistical analysis of the data to fit the model enabled comparisons of observed with expected pain relief, for which there was good correspondence.

Figure 10.1 shows the complete pain relief achieved with the three dose levels of aspirin compared with the estimation of initial pain intensities, showing the exponential relationship between these parameters; the data were from eight of the studies in Laska and co-workers' report.

The overall results showed that the numerical score quantifying surgical pain is 1.4 times that greater than for severe episiostomy pain, and that is 3.2 times greater than from uterine cramping. These quantitative estimations of differences in initial pain intensity are important for giving a basis to what is clinically obvious but intrinsically difficult to quantify. Mathematical modelling that was developed by these authors and the quantal responses obtained with aspirin as a standard. Recently, a systematic review of pain and analgesic response in a range of acute pain models showed that the event rates for placebo were lower for dental than post-surgical pain (Barden *et al.*, 2004).

A recent careful and detailed systematic review of 72 published studies of randomised single-dose

TABLE 10.1

Early controlled human analgesic assays that have demonstrated the superiority of aspirin compared with placebo. (From Beaver, 1965; references to individual studies cited in this paper.)

Investigator(s)	Pain state	Aspirin dose (mg)	Administration
Beecher et al., 1953	Postoperative	600	Single dose, p.r.n. crossover
Brennan, 1963	Postoperative dental, outpatients	650	Multiple dose, p.r.n.
Bruni and Holt, 1965	Postpartum	650	4 doses, p.r.n.
Carlsson and Magnusson, 1955	Headache, outpatients	1000	Single dose, p.r.n. crossover
Currier and Westerberg, 1958	Headache, outpatients	650	Single dose, p.r.n. partial crossover
De Kornfeld and Lasagna, 1959	Postpartum	600	Single dose, p.r.n.
De Kornfeld et al., 1962	Postpartum	650	Single and multiple dose, p.r.n.
Feinberg et al., 1962	Mixed musculoskeletal, outpatients	325 or 650	q.i.d. × 1 week
Forrest, pers. comm.	Mixed acute and chronic	300 and 900	Single dose, p.r.n. crossover
Frey, 1961	Headache, inpatients and outpatients	650	Single dose, p.r.n. crossover, non-crossover
Kantor et al. (see Beaver, 1965)	Postoperative and fracture	600	Single dose, p.r.n.
Kantor et al., 1964	Postpartum	600 and 1200	Single dose, p.r.n.
Lasagna et al., 1958	Postpartum	600 (?)	Single dose, p.r.n.
Magee and De Jong, 1959	Headache, outpatients	600 or 1200	Multiple dose, p.r.n.
Marrs et al., 1959	Mixed chronic and acute	325	q.4h
Murray, 1964	Headache, outpatients	163, 325 and 650	Single dose, p.r.n.
Orkin et al., 1957	Postpartum	600	Multiple dose, p.r.n.
Sevelius and Colmore, 1964	Postpartum	325 (?)	2 doses, p.r.n.
Sunshire et al., 1964	Mixed acute	600	Single dose, p.r.n.
Uhland, 1962	Postpartum and mixed	325 or 650	Multiple dose, p.r.n.
Valentine and Martin, 1959	Postoperative	325	Single or multiple dose, p.r.n. (?)
Zelvelder, 1961	Mixed chronic and acute	500	Multiple dose, p.r.n.

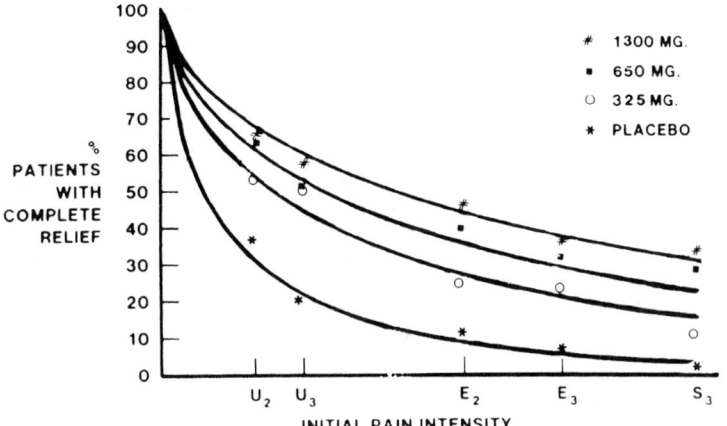

Figure 10.1 Influence of initial pain intensity in three different pain models comprising moderate and severe uterine cramping (U_2 and U_3 respectively), moderate and severe episiostomy (E_2 and E_3 respectively) and severe post-surgical pain (S_3) on the pain relief achieved by doses of 325 to 1300 mg aspirin. Based on data from 4,234 patients performed in clinical experiments conducted at two hospitals in San Juan, Puerto Rico and Caracas, Venezuela by Laska and co-workers (1982). These authors derived a quantal response model to model the binary response to achieving complete pain relief thus: The mathematical probability, **P**, that a patient derives pain relief from **d** mg of aspirin when she has a pain of aetiology, **t**, and reports the initial pain severity, s_t is given by the expression

$$P\;(Complete\;Relief\mid\,t,\,s_t,\,d)=1-F\,(\lambda s_t{}^{\alpha}\,d^{-\beta})$$

where **F** is the distribution of the random error and **λ**, **α** and **β** are constants unknown parameters derived from the data. Values for **s** are none = 0, mild =1, moderate = 2, and severe = 3. These numbers are employed in the subscripts for U = uterine, E = episiostomy and S = surgical pain. (Figure reproduced from Laska *et al.* (1982) with permission of the publisher of the Journal of Clinical Pharmacology.)

trials in 3253 patients given aspirin compared with 3297 that received placebo showed that in the dose range of 600 to 1200 mg aspirin there was a clear dose–response effect for pain relief with this drug over that from placebo (Edwards *et al.*, 1999). Also, most interestingly, this review showed that the type of pain model and pain measurement, the sample size, quality of study design and study duration had no significant impact on the results. The lack of relationship of the latter three parameters is curious, since methodological aspects would be expected to play a considerable part in determination of the efficacy of analgesics (Moore *et al.*, 1996). However, it must be concluded that the selection of the studies based on these authors' rigorous criteria must have been such as to remove any major sources of experimental error so that the effects of aspirin were clearly defined above that of placebo.

Edwards and co-workers (1999) also found in their systematic review that the pain relief with aspirin was equivalent to that of paracetamol on a milligram dose basis, and in both these the dose–response curves were similar. The single lowest dose (600–650 mg) of aspirin produced significantly greater drowsiness and symptoms of gastric irritation compared with placebo, but other adverse events recorded (headache, nausea, dizziness and vomiting) were the same in the two groups. The relative risks (presumably determined from the odds ratios) of all adverse reactions in 53 of the trials was 1.2 (95 per cent confidence interval being 1.03–1.4), showing that the risks of these minor adverse reactions were low.

McQuay and Moore (1999) performed systematic reviews of the published literature on published randomised controlled trials to test the evidence for differences in the analgesic efficacy and adverse reactions of NSAIDs and dipyrone given by different routes in treatment of acute and chronic pain conditions. They concluded that only in treating renal colic did intravenously administered NSAIDs or dipyrone have faster onset of action over that given by other routes. Otherwise, there was no difference in pain relief whether the NSAIDs were given orally or by injection, thus suggesting that the former is by far the most suitable route. None of the studies reviewed by these authors included any salicylate preparations.

In a similar systematic review of published randomised controlled clinical trials of topically applied NSAIDs that included salicylate-containing preparations in treatment of acute and chronic painful conditions, it was found that the relative benefits were 1.7 (1.5–1.9) and 2.0 (1.5–2.7) respectively for these states (McQuay and Moore, 1999). Thus the benefits of these topical treatments with NSAIDs are, overall, relatively small. In acute painful conditions diethyl salicylate and salicylic acid exhibited relative benefits that were within the range of approximately 1 to 4, and this was similar to that of most other NSAIDs. The data from trials with ketoprofen, felbinac, ibuprofen and piroxicam were somewhat more impressive than those from the salicylates. This may reflect the quality of the trials undertaken. In 2-week treatments of chronic painful conditions triethylolamine salicylate had a variable relative benefit that was at the lower end (below 1.0), while diethylamine salicylate was above that and ranged to about 8.0. In one study flufenamate + salicylate was among the most potent treatment and its effect was about equivalent to that of the best, which was the diclofenac plaster (although in another study it was slightly less effective).

COMPARATIVE EFFECTS IN ACUTE PAIN STATES

Dental surgery

The determination of acute pain relief from analgesics and NSAIDs in acute dental surgery (e.g. removal of impacted wisdom teeth/molars) has been a valuable model to determine the relative efficacy of these drugs (Cooper, 1981; Dionne, 1998; Dionne and Cooper, 1999; Barden *et al.*, 2004). Among the earlier studies, Cooper (1981) found that aspirin was equipotent with paracetamol in relieving dental surgical pain at oral doses of 650 mg each. Seymour and Rawlins (1982) found that a higher dose of aspirin 1200 mg taken orally was superior to paracetamol 1000 mg given intravenously or dihydrocodeine 25 mg orally. Aspirin 650 mg was found equally effective with zomepirac 25–100 mg (a drug which is now discontinued) in relieving the pain secondary to the surgical removal of impacted third molars (Cooper, 1980; Cooper *et al.*, 1980). Similar results were obtained when zomepirac was compared with an APC (aspirin, paracetamol, codeine) formulation (Baird and Turek, 1980; Mehlish, 1981). While these APC formulations are not normally considered for the general use by the lay public because of the risks of analgesic nephropathy (see Chapter 8), there are indications from other studies that these particular combinations do produce superior analgesia in some conditions, compared with that achieved by doubling the dose of one of the constituents (Beaver, 1981; Dionne and Cooper, 1999). APC preparations may be specifically indicated in relief of acute painful conditions, such as in dental surgery, where the quantity of these drug mixtures will be small and their use limited to a short period of postoperative treatment. The rationale of using a mixture of aspirin with codeine could be related to the individual peripheral and central modes of actions respectively (Lim, 1966). Aspirin has often been used as a comparator drug for investigating the effects of new NSAIDs in the dental pain model. Thus Ruedy (1973) observed that naproxen 400 mg was superior to aspirin 325 mg + codeine 30 mg in 62 patients who were suffering moderate to severe pain following dental surgery. It could be argued that the dose required for anti-inflammatory activity from aspirin was suboptimal in this study; a feature that may be important in view of the acute surgery leading to appreciable local inflammation and the importance of this in generating pain.

In a double-blind trial in 600 patients, the NSAID, fenbufen 500 mg, was found to be slightly better than aspirin 750 mg and both drugs were superior to placebo in relieving pain from the removal of wisdom teeth (Henrickson *et al.*, 1979). The difference between the two drugs, although statistically significant, was marginal.

Several studies have shown that diflunisal is an effective analgesic for the treatment of postoperative pain arising from dental surgery (Ackerman and Braun, 1978; Peterson, 1979; Ihalainen, 1980; Forbes *et al.*, 1982). This drug is superior to placebo, and at doses of 250 and 1000 mg has been found to be superior to aspirin 650 mg for total peak pain relief after 12 hours of self-medication (Forbes *et al.*, 1982) and to 1000 mg ibuprofen (Nyström *et al.*, 1988). Aspirin 650 mg was found to be equieffective with ibuprofen 400 mg and bromfenac 100 mg in patients who had third molar extractions (Forbes *et al.*, 1991; 1992). Aspirin 650 mg was also equiactive with 100–400 mg ibuprofen (Cooper *et al.*, 1980; Jain *et al.*, 1986) and zomepirac (now discontinued) 25 mg, but higher doses of zomepirac (50 and 100 mg) gave superior pain relief to that of aspirin 650 mg (Cooper *et al.*, 1980; Cooper, 1981).

A study reported by Hepsö and co-workers (1976) highlights a major problem with aspirin, in that treatment with this drug before the extraction of wisdom teeth increases bleeding time and blood loss from the tooth socket. Also, there is a marked increase in the occurrence of ecchymosis and haematoma in these patients. It appears that aspirin should not be taken *before* such surgery but only *after*, for pain relief following operation. Alternatively, if such prior treatment is required then it would be more appropriate to choose diflunisal or ibuprofen, since these drugs do not appreciably inhibit platelet aggregation or prolonged bleeding time (Majerus and Stanford, 1977; Rainsford, 1999), and furthermore would be expected to have a more prolonged effect than aspirin as a consequence of their potency and longer plasma half-life (Rainsford, 1996).

In postoperative dental pain, aspirin 900 mg has been shown to be equally effective to pirprofen 300 mg (Hutlin and Olander, 1981). Likewise, standard analgesic doses of aspirin (e.g. 300–1200 mg) are equally effective in relieving surgical pain compared with naproxen (Mahler *et al.*, 1976), indoprofen 100–200 mg (Okun *et al.*, 1979), proquazone 75–300 mg (Kantor *et al.*, 1977; Forbes *et al.*, 1980) and tiaprofenic acid (Cutting and Thornton, 1981).

However, aspirin 650 mg was less effective than 200 or 400 mg suprofen (now discontinued) (Cooper *et al.*, 1983; Sunshine *et al.*, 1983). A longer duration of analgesia has been observed with 1000 mg diflunisal than with aspirin 650 mg (Beaver *et al.*, 1983). However, this may be an unfair comparison since the plasma elimination half-life is much greater with diflunisal (5–15 h) compared with that of aspirin (2–4 h) as salicylate (Rainsford, 1996). The proper comparison would have probably involved using two doses of aspirin, the second being 3–4 h after the first.

Markowitz *et al.* (1985) observed that meclofenamate sodium 100 or 200 mg was more effective than 650 mg buffered aspirin in treatment of post-surgical dental pain. Forbes *et al.* (1990) found that ketolorac tromethamine 10 mg was superior to aspirin 650 mg in pain relief following third molar surgery.

The use of aspirin as a standard for comparison with other analgesics or NSAIDs has now been replaced in recent years by ibuprofen (Dionne and Cooper, 1999), probably because of its efficacy and greater gastrointestinal tolerability than aspirin (Rainsford *et al.*, 1997). A combination of optimal doses of ibuprofen 400 mg with codeine 30 mg was shown to be superior to aspirin in third molar surgery (Frame *et al.*, 1986).

Seymour and co-workers (Seymour and Rawlins, 1982; Seymour *et al.*, 1992) have compared the effects of soluble aspirin with aspirin tablets in pain relief following dental surgery, and related these effects to pharmacokinetics of the formulations. Thus, soluble aspirin 600 mg produced significant analgesia only at 45 minutes, whereas the higher dose of 1200 mg produced more sustained analgesia from 45 to 240 minutes after ingestion. These authors found there was a significant correlation between analgesia and plasma salicylate but not aspirin, concentrations with the 1200 mg dose. In a follow-on of this study the authors showed that doses of sodium salicylate 537 mg and 1074 mg (these being equimolar with those of the 600 mg and 1200 mg doses of aspirin used in the first study) failed to produce analgesia in postoperative dental pain, even though the peak plasma levels of salicylate were comparable at the lower dose and about 2.5 times greater than aspirin at the higher dose of the drug (Seymour *et al.*, 1984). These results suggest that both the acetyl and salicyl moieties of aspirin are required for acute pain relief in this model.

Aspirin 1.2 g was more effective for pain relief by female patients undergoing third molar surgery when taken as a soluble formulation than as plain aspirin tablets (Seymour *et al.*, 1986). The time of onset of pain relief and duration was earlier with the soluble formulation (Seymour *et al.*, 1986). These authors found that there was a significant correlation between pain scores following intake of either formulation and plasma esterase activity, and they suggested that the latter could be regarded as a determinant of patient analgesic response. No rationale for this determinant was proposed by the authors in relation to the mechanism(s) of analgesia by aspirin, so it is presumed that the pharmacokinetic effect of aspirin in inhibiting aspirin esterase activity is the underlying premise to this determinant.

CHAPTER 10

Skjelbred (1984) examined the longer-term analgesia and anti-inflammatory effects from aspirin 2 g and 4 g daily doses when taken 3 h postoperatively and continued for 3 days. This author found that the lower dose of aspirin tended to increase the postoperative swelling, which on days 3 and 6 averaged 109 per cent and 133 per cent respectively compared with placebo. The higher dose of aspirin produced significant reduction in swelling. Pain (assessed by the visual analogue scale technique) was significantly reduced with both doses of aspirin over the period of 4 to 24 hours following intake, but there was no significant difference between the analgesic effects of the two doses. Subjective postoperative bleeding was evident with the 2 g dosage but not 4 g aspirin, and the latter was responsible for appreciable tinnitus. This author has claimed from previous studies that paracetamol reduces swelling and that this drug is more effective than aspirin (Skjelbred, 1984), which is surprising in view of the accepted view that paracetamol has no anti-inflammatory activity.

Collectively, these findings on the effects of aspirin in producing both pain relief and anti-oedemic or anti-inflammatory effects suggest that to optimise these benefits and reduce the risk of bleeding it is preferable to take doses of at least 1.2 g singularly 2–3 hours after the surgery and to maintain intake t.i.d or q.i.d. for at least 1–2 days thereafter. Soluble aspirin would appear more effective than plain aspirin, especially in view of the faster onset of action and lower risk of gastrointestinal adverse reactions from the former than the latter.

Combinations of aspirin with opioids or other analgesics have attracted considerable interest in treatment of post-surgical dental pain (Moore et al., 1998; Moore, 1999; McQuay and Moore, 1999). The aspirin combinations with codeine and other opioids in studies of comparative analgesia in dental surgery (Moore and McQuay, 1997; Zhang and Po, 1997; Moore et al., 1998; Moore, 1999) make these aspirin drug combinations of considerable interest for comparative studies because of the belief that a higher 'ceiling' of analgesia may be achieved by aspirin–codeine and other mixtures than with the drugs taken alone (Beaver, 1988).

In a double-blind trial in patients with moderate to severe pain from dental surgery, the compound of aspirin 375 mg with codeine phosphate 30 mg and caffeine citrate 30 mg was less effective than 400 mg ibuprofen (Squires and Masson, 1981). It could be argued from the studies of the effects of aspirin alone (Seymour and Rawlins, 1982; Seymour et al., 1992) that the dose of aspirin employed by Squires and Masson (1981) was insufficient to give adequate analgesia, such that the ceiling of analgesia would be increased by the codeine plus caffeine.

Forbes et al. (1981) examined the pain relief achieved by a combination of aspirin, codeine, phenacetin and caffeine (APC–codeine) in patients who had undergone surgery for removal of impacted third molars. The APC–codeine mixture was found to be equivalent to hydrocodone–paracetamol. The results suggested that each of the components in the analgesic mixture had a contribution towards the total and peak analgesia.

The combination of aspirin 650 mg plus codeine 60 mg was significantly more effective than placebo but only slightly more effective than aspirin 650 mg, and then only at 1–4 hours after treatment of moderate to severe pain resulting from surgical removal of third molars (Sunshine et al., 1983). The overall pain relief was slightly greater with the combinations than with aspirin alone, while the time to peak analgesia was 79 minutes with the former and 100 minutes with aspirin (Sunshine et al., 1983). Thus the impression is that with adjustment of dosage aspirin alone affords the predominant analgesic effectiveness in acute dental surgery. Similar impressions are apparent from the data comparing ibuprofen plus codeine with that of codeine alone (Dionne, 1999). The major differences in pain relief with this combination are at 2–3 hours after dosage, but the major part of analgesic activity is due to ibuprofen itself (Dionne and Cooper, 1999).

Dahl et al. (1985) found that aspirin 500 mg plus codeine phosphate 30 mg provided superior pain relief to aspirin 500 mg alone following third molar surgery. Moreover, the number of tablets of the combination used by these subjects was fewer than in those who received aspirin alone.

Various combinations of ibuprofen with codeine were found by Frame et al. (1986) to be superior to aspirin alone in relief of pain from third molar surgery.

Sex differences in analgesic efficacy of an aspirin–codeine–caffeine–butalbital as well as a paracetamol–codeine combination were found in the pain relief following surgical removal of third molars (Forbes *et al.*, 1986). It appeared from this study that female patients experienced more effective pain relief than male subjects.

In a later study from the same group, paracetamol 600 mg plus codeine 60 mg was found to be superior in pain relief from third molar surgery to aspirin 650 mg (Forbes *et al.*, 1990). Morse *et al.* (1987) found that diflunisal was superior to an aspirin plus codeine combination in post-treatment pain following surgical treatment for periodontal disease.

A meta-analysis of 18 randomised double-blind trials of the effectiveness of oral tramidol 50–150 mg compared with aspirin 650 mg plus codeine 60 mg, or paracetamol 650 mg plus dextropropoxyphene, showed that the aspirin–codeine combination was equivalent to tramadol and the paracetamol plus dextropropoxyphene in moderate to severe post-surgical pain, including that from dental surgery (Moore and McQuay, 1997).

A later randomised double-blind controlled study in third molar surgical patients by Moore *et al.* (1998) showed that the total pain relief scores (TOTPAR), maximum pain relief scores, sum of pain intensity difference scores, peak pain intensity difference and global evaluations were all found to be statistically significant for the aspirin 650 mg plus codeine 60 mg combination compared with placebo. The TOTPAR scores for the aspirin–codeine combination were greater in the first 3-hour period than with the weak narcotic analgesic, tramadol, but the 6-hour scores were not significantly different between these two drugs. In a later review by the senior author it was concluded that tramadol had maximal analgesic efficacy in oral surgery equivalent to that of 60 mg codeine, but less than aspirin plus codeine or other combinations of NSAIDs or paracetamol with codeine (Moore, 1999).

In conclusion, it appears that: (a) good analgesia can be achieved with anti-inflammatory activity from aspirin at >650-mg single doses; (b) codeine combinations with aspirin are probably more effective than aspirin alone; (c) these aspirin–codeine combinations are superior to weak opioid analgesics such as codeine or tramadol; (d) a single dose of diflunisal has a longer duration of action than of a single dose of aspirin; and (e) some more potent NSAIDs, especially those with pronounced analgesic activity (e.g. ketorolac, meclofenamate sodium), may have more potent analgesic effects than aspirin.

General postoperative pain

Aspirin has been used extensively for relief of other non-dental postoperative pain, although its use has probably been supplanted by other NSAIDs, paracetamol and ibuprofen, the former two groups on the presumed basis of safety probably more than efficacy. The use of aspirin in control of both acute inflammatory events underlying the injury from surgery and consequent pain has its roots firmly established on the basis of peripheral anti-inflammatory effects combined with central mechanisms of analgesia (McCormack and Brune, 1991; Cashman and McAnulty, 1995; Walker, 1995; Cashman, 1996).

Comparison of aspirin with the other NSAIDs in pain has, generally speaking, yielded data that show these drugs are equivalent in pain relief (Cashman and McAnulty, 1995) providing the doses and timing are such that they are pharmacologically comparable (Mahler *et al.*, 1976; Okun *et al.*, 1979; Cushman and McAnulty, 1995; Barden *et al.*, 2004). The effectiveness of intravenously administered aspirin has been investigated in a number of surgical conditions. This route of administration may have a number of therapeutic benefits arising from the pharmacokinetic aspects, since the first-pass metabolism of aspirin is lower than if it were taken orally and the analgesia might be greater because of the potency of aspirin being greater than that of salicylate in the control of inflammation and pain and the production of inflammatory mediators, especially prostaglandin E_2.

In gynaecological surgery, Kweekel-de Vries *et al.* (1974), found that 1.8 g lysine acetylsalicylate (= 1 g aspirin) i.v. gave pain relief in four patients comparable with that of 10 mg morphine hydrochloride i.m. Korttila *et al.* (1980) found that intravenously administered lysine acetylsalicylate 12.5 or 25 mg/kg was comparable regarding pain relief with intramuscular oxycodone chloride 0.15 mg/kg in patients undergoing varicose vein surgery. Blendinger and Eberlein (1980) found that intravenous lysine acetylsalicylate 1.8 g i.v. given postoperatively to patients who had undergone abdominal hysterectomy, removal of ovarian cysts or surgery for an ectopic pregnancy had mean scores for pain relief

CHAPTER 10

and pain intensity difference that reached a maximum at 30 minutes and remained at this level for the next 90 minutes. In contrast, these scores of pain relief only peaked at 60 minutes after intravenous administration of 1 g dipyrone and then declined over the next hour. It would appear from these observations that aspirin was superior to an equivalent dose of dipyrone. The relatively small numbers of subjects (17 in all) in this study limit the conclusions that can be drawn from it. The effectiveness of intravenous lysine acetylsalicylate 1.8 g for superficial surgery, including that from varicose vein surgery, was confirmed by Tigerstedt *et al.* (1981). These authors observed, however, that oxycodone 4 mg had the best peak effect, and that combination of this opioid with other analgesics probably provides greater comfort. For pain arising from major surgery, combinations of aspirin or other NSAIDs and opioid analgesics has a mechanistic basis in the recognition of differing analgesia afforded by opioid compared with non-opioid analgesics, with the consequent potential for additive or synergistic effects in achieving analgesia (Fields, 1987).

Combinations of aspirin with other opioids are often employed, but the data to support their use in different conditions, especially the optimal doses for effects, are relatively few. Thus a combination of aspirin 650 mg with pentazocine 50 mg has been found to be more effective than aspirin 650 mg alone in the treatment of postoperative pain, and the pain rating with this combination was 'very good to excellent' (Calimlin *et al.*, 1997). Half dosage of the combined preparation was equi-analgesic with aspirin 650 mg, and both these were superior to placebo (Calimlin *et al.*, 1997). The combination of a centrally acting analgesic such as pentazocine with aspirin as a peripheral analgesic could be especially valuable for abdominal and other severe forms of deep pain that occur following major surgical operations.

The question as to whether codeine or caffeine enhance the analgesic efficacy of aspirin in postoperative pain has been examined by systemic review of published literature and meta-analysis of randomised controlled trials (Zhang and Po, 1997). The results of this analysis have shown that codeine 60 mg leads to a small but not clinically significant increase in TOTPAR scores compared to that of aspirin 650 mg, as well as in the proportion of patients exhibiting moderate to excellent pain relief. The addition of caffeine conferred no benefit compared with that of aspirin. The question as to whether the same dose of codeine would enhance the analgesic effects of higher-dose (e.g. 1.0–1.2 g) aspirin requires resolution, since it may be argued that lower doses of aspirin for relief of surgical pain may not be sufficient to produce the full anti-inflammatory component of peripheral analgesia, such that there would be added benefit to achieve the 'analgesic ceiling' (Beaver, 1988).

Corpataux *et al.* (1997) found that the addition of a low dose of lysine acetylsalicylate (LAS) (90 mg) with prilocaine 0.5 per cent mixed and given i.v. produced significant analgesia following application of a forearm tourniquet and reduced requirements for morphine. These authors suggested that the combination of LAS with prilocaine may be useful as a postoperative analgesic.

Postoperative pain relief from rectal administration of aspirin has been shown by Spansberg *et al.* (1996) to be achieved in patients undergoing surgery for femoral neck fracture with spinal anaesthesia.

These studies show that aspirin is effective in relief of operative pain when given i.v. or rectally. There may be advantages in reducing the requirements for morphine and other opioids (Pang *et al.*, 1999) by added treatment with parenteral or rectal aspirin.

Postpartum and episiotomy

The use of salicylates for the treatment of postpartum pain is historical, going back to the use of willow bark extracts in Roman times. Aspirin with or without codeine has been employed extensively in treating this condition and compares favourably with other NSAI/analgesic drugs (Bloomfield *et al.*, 1974; 1976; 1977; 1981; 1983; 1986; Messer *et al.*, 1980; Cooper, 1981; Windle *et al.*, 1989). In the treatment of postpartum pain, aspirin–phenacetin–caffeine mixtures are no more effective than aspirin alone (Lasagna, 1965). However, a single oral dose of 800 mg aspirin and 65 mg caffeine provided more pain relief than 1000 mg paracetamol or the combination of 648 mg paracetamol and 648 mg aspirin (Rubin and Winter, 1984).

The 4-fluoro-(5-phenyl) derivative of aspirin, flufenisal (MK-835) (now withdrawn), which was the progenitor to diflunisal, was equally effective as aspirin in the treatment of episiotomy pain (Bloomfield *et al.*, 1970). Diflunisal has also been shown to be effective in this condition (Melzack *et al.*, 1983). Intravenous LAS 1.8 g was found more effective than dipyrone 1 g i.v. in the treatment of pain following episiotomy (Blendinger and Eberlein, 1980).

In a brief report, Dundee and McAteer (1981) claimed that LAS 1.8 g given i.v. to 30 patients undergoing gynaecological and orthopaedic operations produced very poor analgesia, especially by comparison with that from morphine 10 mg treatment in the parallel series of 20 patients. In the treatment of the pain following abdominal surgery, LAS 12.5 mg/kg was found to be equi-analgesic to oxycodone (Tammisto *et al.*, 1980). The latter authors considered 1.8 g LAS equi-analgesic to about 6 mg oxycodone.

As with general surgical pain, an appreciable placebo effect is evident in pain intensity difference scores and other indices of pain relief at periods up to 3 hours postpartum (Bloomfield *et al.*, 1981) or following episiotomy (Bloomfield *et al.*, 1970), but the placebo effect then declines, so enabling substantial clinical and statistically significant differences to be apparent up to 6–8 hours following a single dose of aspirin or other NSAID.

Dysmenorrhoea

This is a syndrome characterised by the presence of crampy pain in the lower abdominal region, thighs and back, just before and after the onset of menstruation (Berkow, 1977; Dawood, 1981; 1988). Headaches, pelvic soreness, increased frequency of defecation and urination are accompanying symptoms. Primary dysmenorrhoea occurring shortly after the menarche is probably related to a variety of developmental controls and psychological factors influencing the production and action of the sex steroids, as well as to intrauterine devices (Berkow, 1977; Dawood, 1988). Secondary dysmenorrhea, which occurs later in life, may often have a more complex pathological basis – e.g. endometriosis, pelvic lesions, pelvic tumours etc. (Berkow, 1977; Dawood, 1988). Primary dysmenorrhea is the most frequent cause of school absenteeism amongst adolescent females in the USA, one study reporting a 14 per cent incidence in that country (Klein *et al.*, 1981). Abnormalities of prostaglandin production have been implicated in the pathogenesis of dysmenorrhoea (Halbert *et al.*, 1976; 1981). Several NSAI drugs that inhibit prostaglandin production have been shown to reduce the symptoms of this condition (Budoff, 1979; 1980; Chan *et al.*, 1979; Dawood, 1981; 1988; Rosenwaks *et al.*, 1981) coincident with reduction in menstrual prostaglandins (Dawood, 1981; 1988). In one study it was claimed that aspirin was less effective in both pain relief and reducing blood levels of the $PGF_{2\alpha}$ metabolite, 13,14-dihydro-15-keto-$PGF_{2\alpha}$, than naproxen (Rosenwaks *et al.*, 1981). It was quite apparent that the authors of this study had employed sub-optimal doses of aspirin, thus making their comparison of dubious value. In another study, Klein and co-workers (1981) showed that aspirin 600 mg four times daily taken for the full month did reduce pain and improved behavioural features compared with placebo in a group of 29 adolescents (who were carefully selected to match population psychological characteristics, but were not in themselves an aberrant group). Benorylate has also been used to treat primary dysmenorrhea (Prasad, 1980). Diflunisal 1000 mg has been shown to give pain relief in dysmenorrhoea, and significant but not complete relief of uterine contractile amplitude, maximal pressure, basal tone and frequency of contractions (Forman *et al.*, 1982). The uterine activity was abolished in those subjects that received nifedepine 30 mg in addition to diflunisal (Forman *et al.*, 1982). This interesting observation implies that there are added benefits of employing a calcium antagonist to block calcium ion-mediated smooth muscle contractions.

Other NSAIDs, notably ibuprofen and the fenamates, are widely used for the relief of pain in primary dysmenorrhea (Owen, 1984; Rainsford, 1999), where it is considered that these drugs (especially the fenamates (Owen, 1984) may be more effective than ibuprofen, indomethacin and naproxen.

A recent systematic review by Zhang and Po (1998) compared the efficacy and safety of aspirin, ibuprofen, mefenamic acid, naproxen and paracetamol in treatment of primary dysmenorrhea reported in 56 randomised controlled trials. In 55 of these there were comparisons with placebo, while in 12 there were direct comparisons. All the NSAIDs proved more effective than placebo, but paracetamol did not prove to be so. The requirement for additional 'rescue' analgesia was less with ibuprofen and naproxen than with aspirin or paracetamol. There were no differences between ibuprofen and naproxen, but side effects, especially nausea, were more prevalent with naproxen. The total of all adverse events, or those comprising nausea, dizziness or headache, were not statistically different among the other drugs. Ibuprofen was considered to have the most favourable risk–benefit ratio.

Recent studies showing the effectiveness of the selective COX-2 inhibitor rofecoxib, initially at 50 mg followed by 25 mg once daily, being comparable with that of naproxen 550 mg q. 12 h (Brown et al., 1999) underlies the importance of COX-2 derived prostaglandins in pain in primary dysmenorrhea. Whether rofecoxib and other COX-2 selective drugs will supplant aspirin, the fenamates, ibuprofen or naproxen in treating pain in primary dysmenorrhea will depend on factors such as cost, acceptance and symptomatic adverse reactions, the importance of which have not yet been determined.

Migraine and headache

In an earlier review of the use of salicylates in the treatment of headaches and migraine (collectively cephalagia), Lim (1966) emphasised the categories of headache and the appropriateness of salicylates in the treatment of these various types of headache. In summary, headaches may be defined as and to originate as follows:

1. Vascular headaches of the migrainous and associated types are unilateral in focus, with throbbing, which begin in and about an eye and are accompanied by nausea, vomiting and anorexia (Lim, 1966; Berkow, 1977; Blau, 1992). Migraine appears to originate from a brief vasoconstriction followed by prolonged vasodilatation of cranial vessels, with local perivascular oedema and tenderness (Lim, 1966; Lance et al., 1983). The prodromal symptoms (e.g. flashes of light, hemianophia, paraesthesias) are probably due to intracerebral vasoconstriction (Berkow, 1977; Lance et al., 1983; Blau, 1992). The head pain that follows may result from vasodilatation of the scalp arteries (Lance et al., 1983). Other vascular headaches may in part resemble migraine include: (a) the premenstrual or menstrual types; (b) cluster headaches; (c) those from toxic manifestations of infections or actual poisoning, alcoholism etc.; and (d) hypertension headaches (Lim, 1966; Blau, 1992).
2. Family history of migraine is common, but there do not appear to be differences between the sexes except where oral contraceptives are taken, in which case there is an increased frequency of migraine attacks (Blau, 1992; Tietjen, 2000). Interest in the possible genetic associations with migraine has yielded some conflicting data (Buzzi et al., 2000; Chabriat et al., 2000). In young women, cigarette smoking and association of anti-cardiolipin antibodies in those taking oral contraceptives may potentiate migraine-associated aura in those with migraine stroke (Tietjen, 2000). Associations have been reported of stroke with migraine (Tietjen, 2000), sudden hypertension (paroxysmal), and the transient hypertension that can occur during migraine or cluster headache (Mathew, 1999). Migraine or headache from ocular vascular changes in normal pressure glaucoma patients (Cursiefen et al., 2000) underlies the importance of vascular–ischaemic factors in precipitating migraine (Blau, 1992; Tietjen, 2000).
3. Theories regarding the pathogenesis of migraine have been considered by Blau (1992) to be unpersuasive because they fail to account for the complex symptoms in this condition. As pointed out in a review by Mascia et al. (1998) of the role of dopamine in migraine, many of the hypotheses have highlighted only one aspect of migraine pathophysiology without offering a comprehensive model for the cause. These authors highlighted the potential for reduced release of 5-HT (serotonin) between migraine attacks. They also reviewed the literature on the evidence in favour of an intrinsic hypersensitivity of dopamine receptors in migraineurs. While dopamine D_2 receptor antagonists may be useful for prophylaxis and may abort migraine attacks, many of these agents lack selectivity and a considerable number of the trials have been poorly controlled. The concept that there may be

enhanced platelet aggregation and subsequent increased production of the thrombogenic prostanoid thromboxane A_2 with accompanying formation of platelet aggregates has received considerable support (Deshmukh *et al.*, 1976; Couch and Hassanein, 1977; Horrobin, 1977; Malmgren *et al.*, 1978; Masel *et al.*, 1980; Oxman *et al.*, 1982; Blau, 1992). The utility of aspirin in prevention or treatment of migraine probably relates in part to its antithrombotic effects (Deshmukh *et al.*, 1976; D'Andrea *et al.*, 1984), but can only be considered a partial factor (Masel *et al.*, 1980). From a pharmacological viewpoint, the effectiveness of 5-hydroxytryptamine (5HT-1) agonists such as sumatriptan (Tansey *et al.*, 1993; Pini *et al.*, 1999) gives support to the role of 5-HT in the development and, possibly, some of the initiating events in migraine and headache. Since 5-HT stimulates platelet aggregation, the activation of platelets would seem a useful target for therapy.

4. Psychogenic, i.e. tension or nervous headaches of the musculocontraction type, headaches may develop in states of hysteria and anxiety. They will generally be bitemporal and are made worse by emotional disturbances. Muscle tension headaches are intermittent and focus in the fronto-occipital region, accompanied by a feeling of local stiffness or tightness in the neck.

5. Headaches from intra- and extracranial organic diseases may vary in severity and focal development. The intracranial conditions may include CNS tumours, meningitis, syphilis, tuberculosis, subarachnoid haemorrhage and Paget's disease. Extracranial causes may include eyestrain, earache, sinusitis and oral (including dental) abnormalities. Amongst the latter should be included headaches arising from traumatic conditions.

Obviously, the correct diagnosis of the type and origin of the headache is mandatory before instituting therapy. For conditions in categories (1), (3) and (4) above, it is well known that salicylates (aspirin, diflunisal) give effective symptomatic relief. In fact, the widespread use of aspirin by the lay public for the self-treatment of 'common' headaches of the psychogenic (group 4) and non-migrainous vascular types (group 1) is testimony in itself to the value of this drug.

Efficacy in migraine

The treatment of migraine involves a variety of non-pharmacological procedures (biofeedback, trigger factor control, lifestyle interventions and cognitive therapy) and pharmacological agents (Grazzi *et al.*, 1995; Leone *et al.*, 1995; Anon., 1998). Among the latter are agents to *abort* migraines (analgesic mixtures, sumatriptan and congeners, and dihydroergotamine as tablets or nasal spray), or to *prevent* migraines (avoidance of trigger foods, or prophylaxis with propanolol, methysergide or timolol; Anon., 1998). Evidence suggests that in acute therapy, ergot derivatives and sumatriptan are not superior to simple analgesics and NSAIDs (Higelin and Annoni, 1998). 'Triple therapy', combining an NSAID such as aspirin with metoclopramide and ergotamine, may improve the tolerance and effectiveness of the ergot (Von Seggern and Adelman, 1996). Although aspirin and other minor analgesics and NSAIDs have long been used in treating established migraine attacks, their efficacy varies according to the nature of the migraine or headache.

Of much interest is the demonstration that migraine patients have increased platelet aggregability, and that reversal of this by inhibitors of platelet aggregation (such as aspirin) may be useful in the prophylaxis or treatment of migraine (Hilton and Cummings, 1971; Deshmukh *et al.*, 1976; Couch and Hassanein, 1977; Masel *et al.*, 1980; Oxman *et al.*, 1982; D'Andrea *et al.*, 1984). Those migraine patients with excessive platelet aggregation respond well to treatment with aspirin during attacks (Hilton and Cummings, 1971; D'Andrea *et al.*, 1984).

Some of the variability in response may be due to (a) the necessity to optimise critically the dose of aspirin that inhibits platelet aggregation but not vascular prostacyclin production (Fucella, 1979), and (b) the sex-related differences in anti-thrombotic effects of aspirin arising from higher rate of drug absorption in females compared with males (Kelton *et al.*, 1981).

The possibility that a defect in platelet aggregability is a major factor in migraine is indeed very tantalising. Alterations in 5-HT (serotonin) uptake (Malmgren *et al.*, 1978) combined with accumulation of platelets in the cerebral vasculature and their hyperproduction of vasoactive prostaglandins and thromboxane A_2 could represent the important link between serotonin and prostaglandins in this condition. Clearly, one important aspect requiring research is how platelets are involved in the cycle of initial vasoconstriction followed by vasodilatation during the early development of a migraine attack. This has major significance for therapy. If cerebral platelet accumulation and differential production of vasoconstrictive thromboxane A_2 occurs in platelets followed by vasodilatory PGI_2, then it may be very important carefully to optimise the dose of aspirin or like salicylates in order to achieve specific effects on the control of vasoactive prostanoids as certain stages in the development of an attack.

The strategy of focusing on the platelet as a target for prevention of migraine has been extended in other studies. Thus the combination of dipyridamole (an inhibitor of adenosine uptake which affects the primary phase of platelet aggregation) with aspirin (a secondary phase inhibitor) has been shown, in a double-blind crossover study, to be a very effective prophylactic (Masel et al., 1980).

Temporary sensitisation of central trigeminal neurons has been found in patients with migraine headache which is relieved by 1000 mg i.v. LAS (Kaube et al., 2002).

Analgesic combinations that include aspirin have been explored in several studies for treatment of migraine. Thus Lipton et al. (1998), Goldstein et al. (1999) and Silberstein et al. (1999) studied the effects of a combination of aspirin 250 mg, paracetamol 250 mg and caffeine 65 mg (Excedrin™, a preparation that is commonly used in the USA) for the relief of pain and disability in a series of trials in migraine sufferers who met the International Headache Society (IHS) criteria for migraine with or without aura, and for whom over-the-counter medications would be indicated. They observed that the combination analgesic was superior to placebo in all parameters measured in the 1- or 2- to 6-hour post-treatment period. This combination therapy, Excedrin™, has been approved by the FDA as a new indication for the 'temporary relief of the complete range of symptoms associated with migraine headache' (Hussar, 2000).

Concerns have been expressed about the limitations of this combination therapy (Maizels, 1998). Indeed, in a letter to the Editor, Strong (1998), as a migraine-sufferer who responds to analgesics (including Excedrin™), described a self-experiment in which he took 100 mg caffeine and claimed relief of migraine induced by 250 mg monosodium glutamate and 90 g ricotta cheese; the lower dose of 50 mg caffeine was ineffective. While not being an objective experiment, this report highlights the possibility that caffeine may have important pain-relieving effect in the treatment of migraine, even though studies in experimental animal models indicate the addition of caffeine confers little if any benefit (Engelhardt et al., 1997). This aspect certainly deserves further examination.

The combination of centrally acting analgesics (such as dextropropoxyphene) with aspirin may reduce the severity and duration of migraine attacks comparable with the effect achieved by ergotamine; the particular benefit is that the aspirin–dextropropoxyphene mixture causes much less nausea and vomiting than is observed with ergotamine (Hakkarainen et al., 1978; 1980).

Boureau et al. (1994) compared the effects of the combination of aspirin 1 g, codeine 25 mg and paracetamol 400 mg with that of aspirin 1 g alone and placebo in a randomised double-blind, double-dummy, multicentre crossover study in 198 patients with acute migraine attacks. Approximately half of the patients in each of the two drug groups had almost complete disappearance of pain after 2 hours, whereas 29.8 per cent experienced this level of relief on placebo. The surprising thing in these studies was the relatively small differences between the drug treatments and placebo – i.e. about 20 per cent. Also, there was no difference between the aspirin–codeine–paracetamol combination and aspirin alone.

Two studies compared the effectiveness of LAS containing 900 mg aspirin with metoclopramide 10 mg as an effervescent formulation with that of placebo in randomised double-blind, multicentre trial designs in the treatment of patients fulfilling IHS criteria for acute migraine (Chabriat et al., 1994; Henry et al., 1995). There was about 20 per cent difference in both studies of the headache relief between control and placebo. There was a similar level of placebo responsiveness as in the above-mentioned study by Boureau et al. (1994). There were smaller but statistically significant differences between the drug and placebo treatments in relief of nausea and vomiting, as well as in global outcomes.

Tfelt-Hansen and co-workers (1995) compared the effects of the combination of LAS (equivalent to 900 mg aspirin, presumably given orally although this was not stated in the paper) and metoclopramide 10 mg with that of oral sumatriptan 100 mg and placebo in a double-blind randomised, three-parallel group, multicentre study in 421 patients with migraine. Both treatments were superior to placebo, and were approximately equivalent in achieving relief of headache. However, the aspirin–metoclopramide combination was superior to sumatriptan in the relief of nausea and the occurrence of side effects.

While orally-administered aspirin has some benefit in migraine (O'Neill and Mann, 1978; Ross-Lee *et al.*, 1982a; 1982b; Lange *et al.*, 2000), it is suggested that parenterally administered aspirin may have advantages in the treatment of migraine owing to faster absorption and shorter onset of action than when the drug is taken orally (Noda *et al.*, 1985). Noda *et al.* (1985) observed benefit to three patients with migraine when given i.v. lysine acetyl-salicylate (LAS) containing 450 mg aspirin. Diener (1999) reported studies on behalf of a multicentre study group in which i.v. LAS 1.8 g (equivalent to 1 g aspirin) was compared with sumatriptan 6 mg s.c. and placebo in a double-blind, double-dummy, randomised study in 275 patients fulfilling IHS diagnostic criteria for migraine with or without aura. Both treatments were significantly superior to placebo in decreasing headache. However, sumatriptan was more effective (in 91 per cent patients) than LAS (in 73.9 per cent patients). On the other hand, LAS was better tolerated, with 7.6 per cent of patients having adverse events compared with 37.8 per cent from sumatriptan, there being a wide range of adverse reactions from the latter, ranging from nausea to injection site reactions, and tiredness/weariness/fatigue.

Limmroth *et al.* (1999a) compared the efficacy of LAS 1000 mg i.v. with ergotamine 0.5 mg s.c. in a randomised double-blind crossover study in 56 patients with acute migraine. While LAS was equi-effective with ergotamine in controlling pain, it had faster onset of action and fewer side effects than the latter. Interestingly, LAS had no effects on the blood flow through the extra- and intracranial arteries; ergotamine only affected blood flow in the mid-cerebral artery. The results suggested there was no relation between the relief of headache and blood flow through the key blood vessels implicated in the pathogenesis of migraine (Lance *et al.*, 1983).

While there is some variability between the results obtained in the different studies, the results overall suggest that by far the greatest degree of relief of acute migraine is obtained with aspirin at about 1 g alone. The addition of either centrally acting analgesics or anti-migraine drugs or metoclopramide, though rather limited, may represent a small added benefit. It is possible that the nature of the measurements in the reported trials may be such as to limit the accuracy of determining the pain relief of the added anti-migraine or analgesic agents because the pain relief from aspirin alone is at the upper end of the dose–response curve, as well as the inherent inter-subject variability.

Non-migraine headache

The use of non-migrainous muscle-contraction type or 'simple' headache as a model for determining pain relief from analgesics, especially over-the-counter drugs, has been applied in several investigations (van Graffenried *et al.*, 1980; Schachtel *et al.*, 1991a). Aspirin is a well-established treatment for this condition when taken in various formulations (Ross-Lee *et al.*, 1982b; Limmroth *et al.*, 1999b). It is also effective in the treatment of high altitude headache (Burtscher *et al.*, 1998; 2001).

Von Graffenried *et al.* (1980) showed that there was a dose–response effect from aspirin when taken at 250–1000 mg compared with that of placebo during a 3-hour period of evaluation using a visual analogue scale and verbal rating. Peters and co-workers (1983) showed that aspirin 650 mg was superior to paracetamol 1 g and placebo at 2 hours for the relief of tension–vascular type headaches. However, there were no differences between the two drug treatments in subjects that had moderate headache in general, or in a sub-group of those patients with tension-type headaches over a 6-hour period. The two drug treatments were superior to placebo, although there was an appreciable placebo response in all groups.

Caffeine has been considered as an additive for the treatment of headache with a number of other analgesics (Von Graffenried *et al.*, 1980; Migliardi *et al.*, 1994). A combination of aspirin 1 g with caffeine 64 mg showed greater pain relief than 1 g paracetamol when assessed over a 4-hour period using a variety of pain measures (Schachtel *et al.*, 1991b).

A major issue in the studies of the effects of caffeine is the influence of caffeine-containing beverages that may be consumed during self-administration of analgesic combinations on the outcome of the pain assessments. Migliardi *et al.* (1994) considered this aspect in their investigations of the effects of caffeine on tension-type headache as defined by the IHS criteria. In six studies of a total of 2811 patients comparing the effects of caffeine 130 mg in addition to paracetamol 1 g or aspirin 500 mg plus paracetamol 500 mg, they observed that caffeine significantly enhanced the pain relief over 4 hours compared with that of the analgesics alone, independent of any influence of intake of caffeine-containing beverages. It was evident in these studies that there was a high placebo effect in the measurements of total pain relief and pain intensity differences from the pooled data in all the trials. The placebo response was affected by the 'usual' caffeine consumption, with increasing pain intensity difference scores being observed with increase in the consumption of caffeine-containing beverages. The overall incidence of the side effects in patients was not different in the four studies in which the effects of paracetamol/aspirin/ caffeine (17 per cent) were examined compared with the two studies in which paracetamol/caffeine (21 per cent) was examined. However, the 'stomach discomfort' experienced by those taking paracetamol/aspirin/caffeine (9 per cent) or paracetamol/caffeine (9 per cent) was higher than those taking paracetamol alone (5 per cent) or placebo (5 per cent). Nervousness and dizziness was higher (4–7 per cent each) in the caffeine combination groups than in the paracetamol (1–2 per cent) or placebo (1 per cent) groups. There was no difference in the incidence of 'stomach discomfort' in the two analgesic groups, suggesting that aspirin did not contribute to the discomfort from caffeine. This study is important, for it is an extensive comparison of the influence of caffeine in relation to both efficacy and frequently observed side effects from both the analgesics and caffeine. While there may be added analgesic benefits from what is a high dose (130 mg) of caffeine, this is more than offset by appreciable side effects involving the gastric and central nervous systems.

Comparisons of the effects of other NSAIDs (especially those available over-the-counter) on the relief of symptoms of headache do not appear to have received much attention. Diamond (1983) compared the effect of 400 and 800 mg ibuprofen with that of aspirin 650 mg and placebo in subjects with muscular contraction-type headache. All drug treatments produced pain intensity difference scores that were lesser than placebo, although there was appreciable response from the latter. The physicians' global assessment indicated that the two dose levels of ibuprofen produced greater pain relief than placebo, but not compared with that from aspirin. It could be argued that the dose of aspirin in this study was lower than that which would be regarded as equivalent for pain relief to the low dose of 400 mg ibuprofen. There were an appreciable number of drop-outs (33 from 141 subjects in total) in this study, so it is difficult to attach much statistical credence to comparisons across the drug treatment groups.

Nebe *et al.* (1995) compared the effect of 200 mg ibuprofen with that of aspirin 500 mg and placebo in a double-blind, double-dummy, three-way crossover study in patients who regularly experience headache. After 1 hour the response to ibuprofen was greater than that from aspirin, but by the end of the observation period of 150 minutes the differences were no longer significant but both drugs were superior to placebo. It would therefore appear that the initial pain relief from ibuprofen is greater than that from aspirin, but this will obviously depend of the dose of both drugs.

Overall, the studies in headache indicate that there is an appreciable placebo effect whose basis is not apparently understood. Aspirin alone would appear to be as effective as paracetamol in the relief of tension-type headaches, and might be more effective in patients with vascular-tension type headaches, perhaps as a response to antiplatelet effects. Aspirin might be less effective than ibuprofen, depending on the dose taken, but this aspect deserves further consideration in patients with headaches of various types. Though caffeine produces added pain relief, this potential benefit is offset by the appreciable occurrence of minor but discomforting side effects in the gastric and central nervous systems.

Minor aches and pains

The use of aspirin and other salicylates for the relief of tonsillopharyngeal pain, low back pain, upper respiratory tract infections, minor aches and pains is legendary (Muckle, 1974; 1977; 1980; O'Grady *et al.*, 1975; Dyment, 1986; Buchanan and Rainsford, 1990; Hall, 1990; Jones, 1997; Eccles *et al.*, 2003).

Throat pain from tonsillopharyngitis has been employed as a model for evaluating the comparative effects of mild analgesics (Schachtel *et al.*, 1988; Boureau, 1998; Eccles *et al.*, 2003). This is an interesting model, as the appreciable local inflammation might be expected to lead to the discrimination of drugs with an appreciable anti-inflammatory effect from those with simple analgesic activity and little or no anti-inflammatory activity. Using their throat pain model, Schachtel *et al.* (1988) showed that aspirin 800 mg relieved the symptoms and gave greater pain relief than placebo. Moreover, combination of caffeine 64 mg with aspirin 800 mg gave superior relief of symptoms to aspirin alone over the period of 0.5–2 h. Those patients with fever also showed antipyresis from both treatments.

Time-dependent relief of pain parameters associated with sore throat has been found with two 400 mg effervescent aspirin tablets (Eccles *et al.*, 2003).

A syrup formulation of aspirin 100 mg/5 ml gave less pain relief following tonsillectomy than the same dose of ibuprofen (Parker *et al.*, 1986). While there was some pain relief from the aspirin treatment, approaching that from ibuprofen, this was not significantly different from placebo (Parker *et al.*, 1986). However, this study may not have enabled adequate discrimination of the effects of aspirin compared with placebo because there were only 33 patients in the aspirin group compared with 44 who received ibuprofen; in this respect the study was unbalanced. Moreover, ibuprofen is about two to three times more potent as an analgesic and anti-inflammatory agent and, since the same doses of both drugs were given, this adds further to the comparisons of these drugs being unfair. The salicylate gel formulation comprising 8.7 per cent choline salicylate with 0.01 per cent cetalkonium chloride was significantly more effective than placebo in a double-blind trial in 49 patients with ulcerative pharyngitis (Daniel *et al.*, 1983). These studies deserve further investigation to establish if gel formulation of aspirin and various combinations with caffeine and other analgesics are effective in pain relief in throat pain models.

Salicylates have been extensively employed in back pain (Jones, 1997). Depending on dose, aspirin has equivalent pain relief to paracetamol with or with dextroproxyphene, but under some conditions may be less effective than ketorolac (Catapano, 1996).

In a randomised double-blind study in 30 patients with low back pain, diflunisal 1000 mg daily b.i.d. was more effective in pain relief than paracetamol 4000 mg daily b.i.d. (Hickey, 1982).

In sports injuries, soft-tissue injuries and musculoskeletal pain, oral aspirin has been used extensively (Muckle, 1974; 1980; Dyment, 1986). Topical formulations of salicylates are often employed for these conditions, and several studies have shown that they work, in part, by a counter-irritant mechanism (Collins *et al.*, 1984; Green, 1991; Taniguchi *et al.*, 1994). An aerosol formulation of two alkyl ester–salicylate formulations applied to the forearm of volunteers produced erythema and elevation of blood salicylate levels at 20–30 minutes after application (Collins *et al.*, 1984). Methyl salicylate was found to be absorbed at earlier times than the ethyl ester (Collins *et al.*, 1984). In acute sprains, aspirin 3.6 mg/d was as effective as diclofenac sodium 150 mg/d (Duncan and Farr, 1988). In 1290 patients with acute sprains and tendinitis, aspirin 4 g/d, while giving pain relief, was found to be less effective than piroxicam 40 mg/d (Heere, 1988). In other comparative studies in generalised sports injuries, aspirin has been shown to be as effective as naproxen (Andersen and Gotzsche, 1984). However, in relief of pain in soft-tissue injuries in professional footballers, Muckle (1977) observed that aspirin 3.6 g/d, while effective, was less so than flurbiprofen 150 mg/d.

In a variety of acute sports injuries and other musculoskeletal conditions, aspirin 650 mg was equal to or slightly less effective than etodolac 25–400 mg, depending on the dose of the latter (Pena, 1990).

The variations in effectiveness of aspirin in the treatment of soft-tissue injuries compared with that of other NSAIDs are clearly a function of the relative doses of those drugs.

The use of non-acetylated salicylates (e.g. salsalate (disalcid), choline or other salts of salicylate) for soft-tissue injuries may confer benefits by having less gastrointestinal irritancy than is observed with aspirin or more potent NSAIDs (Buchanan and Rainsford, 1990).

The tolerability of the three most commonly used analgesics, aspirin, ibuprofen and paracetamol, was recently compared in a large-scale randomised, blinded, multicentre study in 8677 adults who received these drugs for acute painful conditions by prescription from 1108 general practitioners (Moore *et al.*, 1999). A high proportion of patients (99.5 per cent) were evaluable, of whom 95 per cent had adhered to the study protocol. The treatment groups were well balanced with respect to demographic features, treatment indications, concurrent conditions and treatments. Most of the subjects received the drugs for musculoskeletal conditions (approximately one-third of the total), colds/flu (about one-fifth), backache (one-eighth), sore throat and headache (about one-tenth each). The percentage of subjects that withdrew from the study because of lack of efficacy was about the same across the treatment groups (aspirin, 7.0 per cent; ibuprofen 6.1 per cent; paracetamol 6.9 per cent). This indicates that at the doses employed (aspirin and paracetamol up to 3 g daily and ibuprofen up to 1.2 g daily), these drugs produced equivalent responses. Overall the adverse reactions were minor in severity, but were more prevalent in the aspirin group (18.7 per cent) than in the ibuprofen (13.7 per cent) or paracetamol (14.5 per cent) groups. The rates of withdrawal because of side effects were about the same across the treatment groups (5–7 per cent). Abdominal pain and dyspepsia were the most frequent side effects in all the groups. The occurrence of these symptoms was greater in the aspirin group (9.9 per cent) than in those that received ibuprofen (4.2 per cent) or paracetamol (6.1 per cent). This study is important for showing that, under conditions where the drugs all presumably show equivalent pain relief, gastric symptoms are about twice those from aspirin as from the other two drugs.

Topical aspirin has been found to be as effective as lidocaine for the treatment of post-herpetic neuralgia, with over 70 per cent of patients responding to both treatments (Tajti *et al.*, 1999; Balakrishnan *et al.*, 2001). Bareggi *et al.* (1998) found that aspirin 750 mg applied topically in diethyl ether produced significant pain relief in 82.6 per cent of patients with moderate to severe pain from acute and post-herpetic neuralgia compared with 15.4 per cent of those who received oral aspirin 500 mg. Overall pain relief was achieved in 95 per cent of those patients who received the topical aspirin compared with 21 per cent of those that received oral aspirin. The local levels of acetylsalicylate in the applied skin area were approximately 80–100-fold higher than after oral aspirin, while the circulating levels of acetylsalicylate and salicylate were much lower. Topically-applied aspirin 75 mg in a Vaseline-based moisturiser was found to be as effective as 375–750 mg oral aspirin for treatment of acute herpetic neuralgia but had a longer time of onset of pain relief (Balakrishnan *et al.*, 2001). These results show the potential benefits of local application of aspirin in a suitable solvent for a particularly difficult painful condition, and that the benefits may be related to the local high concentration of aspirin in the skin.

Aspirin administered topically in chloroform has been found to be an effective adjuvant therapy in the treatment of chronic neurogenic pain (Tharion and Bhattacharji, 1997).

Cancer pain

The World Health Organisation (WHO) guidelines provide for drugs to be administered immediately when there is pain, to be given 'by the clock' rather than on demand, in a graded series – the so-called 'three-step ladder', starting with non-opioid weaker analgesics (aspirin or paracetamol), then weak opioids (codeine, tramadol), and finally strong opioids (morphine, fentanyl) (Stjernsward, 1988; Skaer, 1993; Dahiwal *et al.*, 1995; Hagen *et al.*, 1997; Mortimer and Bartlett, 1997). NSAIDs, in general, are regarded as part of the adjunctive group of medications, along with tricyclic antidepressants, benzodiazepines, palliative radiation therapy, and psychological techniques that are employed in the overall management of pain in the cancer patient (Meyers and Meyers, 1987; Skaer, 1993; Mortimer and Bartlett, 1997). Psychological factors, especially the ability to cope with the stress of chronic pain, beliefs, and adjustment are major factors affecting the patient (Jensen *et al.*, 1991) that doubtless also influence the response to analgesic therapy. Aspirin, like other NSAIDs and paracetamol, enhances the effectiveness of weak oral narcotics (e.g. codeine) in achieving pain relief in patients with mild to moderate cancer pain (Moertel *et al.*, 1971; Creagan and Wilkinson, 1989; Minotti *et al.*, 1989).

Algorithms for assisting in the choice of therapies have been developed (Du Pen *et al.*, 1999), but the focus in employing these has been the use of chemotherapeutic agents and strong opioids (e.g. morphine). The potential for NSAIDs, including aspirin, to reduce tumour mass, proliferation and invasion linked to relief of chronic pain has not formed a consideration in the therapeutic approaches, although (as noted in Chapters 7 and 13) the salicylates and other NSAIDs have potential prophylactic effects in controlling tumour growth and proliferation.

Wallenstein (1975) emphasised the importance of controlling for the multiplicity of variables inherent in determining measurements of analgesic responses in the clinical setting in cancer patients. In the studies performed over 25 years by this author at the Sloane Kettering Cancer Center, objective measurement of pain responses was achieved by employment of full-time nurse observers to visit the patients at their bedside at hourly intervals and question them about their pain responses. This author and colleagues employed double-blind crossover, randomised, placebo-controlled study designs. The observations of this author are described in detail, since they probably represent among the first comparative studies of drug efficacy with aspirin, various other salicylates, paracetamol and combinations with caffeine in cancer pain under carefully controlled conditions.

In 294 patients with moderate to severe pain Wallenstein (1975), in 10 separate studies, found that a single dose of 600 mg aspirin gave statistically significant pain relief above that of a lactose placebo by measuring the area under the curve over a 6-hour period of the change in pain intensity. The mean ± s.d. of the scores in these 10 studies from aspirin was 4.24 ± 1.08, and from placebo was 2.21 ± 0.97. These observations show that about half the pain relief was attributable to a placebo effect. The total change in pain intensity from aspirin 400–900 mg exhibited a linear dose relationship, but the peak values appeared to plateau after 600 mg. The lowest dose of aspirin (400 mg) showed statistically significant difference in pain intensity difference scores from placebo. The highest dose of aspirin (900 mg) showed an increased duration of effect compared with the two lower doses.

In 27 patients studied using the same design, Wallenstein (1975) showed that the effects of 600 mg aspirin were equivalent to those of the same quantity of paracetamol, and both were superior to placebo. However, 600 mg salicylamide only exhibited transient pain relief over that of placebo at 1 hour after drug administration.

In a further study of 41 patients, Wallenstein (1975) observed that combinations of 210 mg aspirin with 150 mg paracetamol alone or with 30 mg caffeine gave 'somewhat superior' pain relief compared with either 360 mg aspirin or paracetamol, but the differences did not achieve statistical significance although all treatments were superior to placebo. A linear log dose–response relationship was observed for all the aspirin and paracetamol treatments, either alone or in combination, with the curves overlapping. However, the log dose–response curve obtained from combination with caffeine was shifted to the left, showing that caffeine, even at the low dose of 30 mg, has an added analgesic effect.

That aspirin with paracetamol was equiactive, with no added or synergistic activity being evident, is interesting in reflecting the lack of differences in mechanisms of pain relief with these two drugs.

In another group of 20 patients it was found that a single dose of 32 mg codeine gave pain intensity difference scores over a 6-hour period of observation that were between aspirin 600 mg and placebo, although the difference between the codeine and placebo treatments was statistically significant. The combination of aspirin 600 mg with codeine 32 mg produced additive effects compared with the drugs alone. In a similar factorial study 10 mg morphine i.m. gave greater pain relief than 600 mg aspirin, but the combination only produced slightly greater differences in pain intensity scores than the morphine alone.

Both the codeine–aspirin and morphine–aspirin studies indicate that these drugs show no interactions and have differing mechanisms of pain relief, as would be expected from knowledge of their pharmacological mechanisms.

Aspirin was compared with seven other NSAIDs and paracetamol for treatment of cancer pain in a double-blind randomised study (Ventafridda et al., 1990). The protocol provided for each drug to be given for 1 week to eight patients and for another week to a further eight patients. It is questionable whether a 1-week treatment period is sufficient to allow for full development of pain relief, and this may have in part accounted for only 48 of 65 subjects having completed the first week of treatment and 41 the second week. The authors concluded that diclofenac, indomethacin and naproxen were the most effective drugs, and were 'well tolerated'. While this study confirms the use of NSAIDs (including aspirin) as a first step in pain relief, the problem of drop-outs and influence on the power of the study limits conclusions regarding the relative efficacy of individual drugs.

Aspirin has been found to be effective in the treatment of tumours of the pancreas (Moertel et al., 1971), osteoid osteoma (Giannikas et al., 1977; Orlowski and Mercer, 1977; Cohen et al., 1983; Bednar et al., 1993; Hasegawa et al., 1993), osteoblastoma of the spine (Nemoto et al., 1990), and head and neck cancers (Saxena et al., 1994).

Many studies examining the efficacy of NSAIDs in treatment of cancer pain have been performed in relatively small numbers of patients, and often these comprise patients with various tumours with varying states of disease progression. While not controlling for the latter, the problem of small numbers of patients in trials has been addressed by Eisenberg et al. (1994) in a meta-analysis of 25 published randomised-controlled trials that met inclusion criteria for 1545 patients. In 13 single-dose studies, aspirin gave equivalent pain relief to three other NSAIDs, and all were superior to placebo and roughly equivalent in pain relief to 5–10 mg i.m. morphine. Recommended dose levels and supramaximal single doses of NSAIDs achieved the same pain relief. The authors suggested that this indicates a ceiling of analgesic relief that is effective at normal recommended doses. Addition of weak opioids to NSAIDs did not produce superior analgesia over that of the NSAIDs alone when taken either singularly or as multiple doses. This observation contrasts with other studies, including those from single-dose double-blind studies in 37 patients with advanced cancer pain who received APC (224 mg aspirin, 32 mg caffeine, 160 mg phenacetin) with 4.5 mg oxycodone hydrochloride and 0.38 mg oxycodone terphthalate, compared with those given 100 mg zomepirac (a drug that is now discontinued). Equivalent pain relief was achieved with the latter NSAID compared with that of the APC–oxycodone mixture.

Minotti et al. (1989) performed a randomised double-blind study, 99 patients were planned to receive aspirin 640 mg plus codeine 40 mg, diclofenac sodium 50 mg and the anti-depressant analgesic nefopam 60 mg, all q.i.d. for 10 days. While all the treatments produced significant pain relief, only 26.3 per cent completed the planned treatment period, the mean period for being on treatments being 4.65 days. The drop-outs were for lack of efficacy and adverse reactions. The patients that received nefopam had a significantly shorter period of treatment than those on the other NSAIDs. The aspirin–codeine combination was, in this study, equivalent to that of diclofenac.

Intravenous and intrathecal lysine acetylsalicylate have been shown to be useful in treating chronic pain in cancer patients (Devoghel, 1983; Sacchetti et al., 1984; Pellerin et al., 1987). However, in a study of pain from bone involvement from tumours ketoprofen 400 mg i.v. was superior to aspirin 1 g i.v. as single doses, no adverse reaction being observed with either drug. Flurbiprofen has been found effective in treating bone pain from metastasis of breast cancer (Lomen et al., 1986) indicating the potential for other NSAIDs to be used in this condition.

Lumbar puncture administration of 120–720 mg aspirin produced relief of pain in 78 per cent of cases of cancer patients; those with bone metastases arising from breast or lung cancers or myelomas showed the greatest response, with those suffering from visceral (e.g. pancreatic) tumours the least. An important observation made in this study was that the immediate analgesia from a single injection often lasted as long as 3–4 weeks. The authors of this study considered that the main indication for this lumbar cannula-administered aspirin was pain associated with osteolytic metastases from adenocarcinomas and myelomas. The only major complication was fatigue. The lack of respiratory depression observed with strong opioids offers considerable advantages for this treatment, especially where the pain can be very severe. These encouraging studies warrant further investigation, especially to confirm the long-term benefits of this treatment in bone cancer patients, in whom pain control without adverse reactions is a major issue.

In their study of the mode of action of NSAIDs (including aspirin) in pain relief in the benign tumour osteoid osteoma, Hasagawa et al. (1993) observed that immunoreactivity to the S-100 and protein gene product 9.5, which are nerve-fibre specific proteins, was particularly prevalent in lesioned areas near tumours, while that for PGE_2 was less so. Since the five cases that were examined received NSAIDs, it is not surprising that the immunoreactive PGE_2 was less abundant. Thus the conclusion of these authors that the nerves adjacent to osteoid osteomas mediate the pain rather than the effects of PGE_2 from osteoblasts is in a sense a self-fulfilling observation, since it would not have been possible to discriminate the effect on PGE_2 production because of the lack of a placebo control in this study.

Overall, it would appear that there are particular advantages for using i.v. or other parenteral routes for administering aspirin in relieving pain from bone cancers. The pain relief from a single i.v. dose of the drug compared with that from oral aspirin or other NSAIDs given i.v. or orally in this group of patients deserves further investigation. Intravenous aspirin may also confer benefits in patient-controlled morphine analgesia, by reducing the amount of morphine that is required (Pang et al., 1999).

The potential for long-term pain relief from locally applied aspirin (or that which is administered to a blood supply localised to the tumour(s)) is particularly important, and may relate to the long-term effects of aspirin in acetylating cyclo-oxygenases, so leading to irreversible inactivation of these enzymes.

Another potential benefit of aspirin related to its effects in irreversibly blocking cyclo-oxygenases, and thus production of prostaglandins, is in the treatment of diarrhoea during radiotherapy of visceral organs. Radiation therapy leads to production of prostaglandins that cause diarrhoea, as well as smooth muscle contractions (Mennie *et al.*, 1975). In a randomised double-blind trial in 28 women who were receiving radiotherapy for uterine cancer, Mennie *et al.* (1975) observed that buffered aspirin reduced the number of bowel motions, abdominal pain and flatulence that is attributable to radiotherapy.

While it could be argued that NSAIDs exist with greater potency and efficacy than aspirin, the significant potential for long-term benefits arising from irreversible blockade of cyclo-oxygenase-derived prostaglandins from aspirin formulations specifically delivered to sites of tumour invasion or where radio- or other therapies are applied may give aspirin some specific advantages over the other drugs.

REFERENCES

Ackerman, K. and Braun, H.D. 1978, Diflunisal in the treatment of postoperative pain in oral surgery. *In:* K. Miehkle (Chairman), *Diflunisal in Clinical Practice. Proceedings of a Conference at the XIVth International Congress of Rheumatology, San Francisco*, pp. 143–148. New York: Futura.

Amadio, P. 1984, Peripherally acting analgesics. *American Journal of Medicine*, **76:** 17–25.

Andersen, L.A. and Gotzsche, P.C. 1984, Naproxen and aspirin in acute musculoskeletal disorders: a double-blind parallel study in patients with sports injuries. *Pharmatherapeutica*, **3:** 531–537.

Anon. 1998, Current treatment strategies for migraine disorders. *American Journal of Managed Care*, **4:** S618–S629.

Baird, W.M. and Turek, T. 1980, Comparison of zomepirac, APC with codeine, codeine and placebo in the treatment of moderate and severe postoperative pain. *Journal of Clinical Pharmacology*, **20:** 243–249.

Balakrishnan, S., Bhushan, K., Bhargava, V.K. and Pandhi, P. 2001, A randomized parallel trial of topical aspirin-moisturizer solution vs. oral aspirin for acute herpetic neuralgia. *International Journal of Dermatolology*, **40:** 535–538.

Bannwarth, B., Demotes-Maynard, F., Schaeverbeke, T., Labat, L. and Dehais, J. 1995, Central analgesic effects of aspirin-like drugs. *Fundamental and Clinical Pharmacology*, **9:** 1–7.

Barden, J., Edwards, J.E., McQuay, H.J. and Moore, A.R. 2004, Pain and analgesic response after third molar extraction and other postsurgical pain. *Pain*, **107:** 86–90.

Bareggi, S.R., Pirola, R., De Benedittis, G. 1998, Skin and plasma levels of acetylsalicylic acid: a comparison between topical aspirin/diethyl ether mixture and oral aspirin in acute herpes zoster and post-herpetic neuralgia. *European Journal of Clinical Pharmacology*, **54:** 231–235.

Beaver, W.T. 1965, Mild analgesics. A review of their clinical pharmacology. *American Journal of Medical Science*, **250:** 577–604.

Beaver, W.T. 1981, Aspirin and acetaminophen as constituents of analgesic combinations. *Archives of Internal Medicine*, **141:** 293–300.

Beaver, W.T. 1988, Impact of non-narcotic oral analgesics on pain management. *American Journal of Medicine*, **84 (5A):** 3–15.

Beaver, W.T., Forbes, J.A. and Shackleford, R.W. 1983, A method for the 12-hour evaluation of analgesic efficacy in outpatients with postoperative oral surgery pain. Three studies of diflunisal. *Pharmacotherapy*, **3:** 23S–37S.

Bednar, M.S., McCormack, R.R., Glasser, D. and Weiland, A.J. 1993, Osteoid osteoma of the upper extremity. *Journal of Hand Surgery*, **18A:** 1019–1025.

Beecher, H.K., Keats, A.S., Mosteller, F. and Lasagna, L. 1953, The effectiveness of oral analgesics (morphine, codeine, acetylsalicylic acid) and the problem of placebo 'reactors' and 'non-reactors'. *Journal of Pharmacology and Experimental Therapeutics*, **109:** 393–400.

Berkow, R. (ed.) 1977, Primary Dysmenorrhea. *In: The Merck Manual*, 14th edn, pp. 1679–1680. Rahway, NJ: Merck Sharp & Dohme.

Besson, J.M. 1999, The neurobiology of pain. *Lancet*, **353:** 1610–1615.

Björkman, R. 1995, Central antinociceptive effects of non-steroidal anti-inflammatory drugs and paracetamol. *Acta Anaesthesiologica Scandinavica*, **39 (Suppl. 103):** 9–44.

Blau, J.N. 1992, Migraine: theories of pathogenesis. *Lancet*, **339:** 1202–1209.

Blendinger, I. and Eberlein, H.J. 1980, Comparison of intravenous acetylsalicylic acid and dipyrone in postoperative pain: an interim report. *British Journal of Clinical Pharmacology*, **10:** 339S–341S.

Bloomfield, S.S., Barden, T.P. and Hille, R. 1970, Clinical evaluation of flufenisal, a long acting analgesic. *Clinical Pharmacology and Therapeutics*, **11:** 747–754.

Bloomfield, S.S., Barden, T.P. and Mitchell, J. 1974, Comparative efficacy of ibuprofen and aspirin in episiotomy pain. *Clinical Pharmacology and Therapeutics*, **15:** 565–570.

Bloomfield, S.S., Barden, T.P. and Mitchell, J. 1976, Aspirin and codeine in two postpartum pain models. *Clinical Pharmacology and Therapeutics*, **20:** 499–503.

Bloomfield, S.S., Barden, T.P. and Mitchell, J. 1977, Naproxen, aspirin, and codeine in postpartum uterine pain. *Clinical Pharmacology and Therapeutics*, **21:** 414–421.

Bloomfield, S.S., Barden, T.P., Mitchell, J. and Bichlmeir, G. 1981, A comparison of flurbiprofen, aspirin and placebo in postpartum uterine pain. *Current Therapeutic Research*, **30:** 670–679.

Bloomfield, S.S., Cissell, G.B., Mitchell, J. and Barden, T.P. 1983, Codeine and aspirin analgesia in postpartum uterine cramps: qualitative aspects of quantitative assessment. *Clinical Pharmacology and Therapeutics*, **34:** 488–495.

Bloomfield, S.S., Mitchell, J., Cissell, G. and Barden, T.P. 1986, Flurbiprofen, aspirin, codeine, and placebo for postpartum uterine pain. *American Journal of Medicine*, **80 (3A):** 65–70.

Boureau, F. 1998, Multicenter study of the efficacy of ibuprofen compared with paracetamol in throat pain associated with tonsilitis. *In: K.D. Rainsford and M.C. Powanda (eds), Safety and Efficacy of Non-Prescription (OTC) Analgesics and NSAIDs*, pp. 119–121. Dordrecht: Kluwer Academic Publishers.

Boureau, F., Joubert, J.M., Lasserre, V., Prum, B. and Delecoeuillerie, G. 1994, Double-blind comparison of an acetaminophen 400 mg-codeine 25 mg combination versus aspirin 1000 mg and placebo in acute migraine attack. *Cephalalgia*, **14:** 156–161.

Brasseur, L. 1997, A review of current pharmacological treatment of pain. *Drugs*, **53 (Suppl. 2):** 10–17.

Brenman, H.S. 1963, Oral analgesics. Preoperative vs postoperative use. *Journal of the American Dental Association*, **67:** 23–27.

Brown, J., Morrison, B.W., Bitner, M., Woosley, C., Sandler, M., Dunkley, V.C., Earl, H., Johnson, L., Kotey, P. and Seidenberg, B. 1999, COX-2 specific inhibitor, VIOXX™, is effective in the treatment of primary dysmenorrhea. *In: Proceedings of the International Conference on Inflammopharmacology and 6th Symposium on Side-Effects of Anti-inflammatory Drugs. Chateau Elan, Braselton, Georgia (USA) 23–26 May 1999*, Abstract No. 15.

Bruni, J.R. and Holt, R.E. 1965, Controlled double-blind evaluation of three analgesic medications for post-partum discomfort. *Obstetrics and Gynecology*, **25:** 76–81.

Buchanan, W.W. and Rainsford, K.D. 1990, Aspirin and nonacetylated salicylates: use in inflammatory injuries incurred during sporting activities. *In: W.B. Leadbetter, J.A. Buckwalter and S.L. Gordon (eds), Sports-Induced Inflammation*, pp. 431–441. Park Ridge, IL: American Academy of Orthopaedic Surgeons.

Budoff, P.W. 1979, Use of mefenamic acid in the treatment of primary dysmenorrhea. *Journal of the American Medical Association*, **241:** 2713–2716.

Budoff, P.W. 1980. Mefenamic acid therapy in dysmenorrhea. *In: B. Samuelsson, P.W. Ramwell and R. Paoletti (eds), Advances in Prostaglandin and Thromboxane Research*, Vol. 8, pp. 1449–1453.

Buritova, J., Chapman, V., Honoré, P. and Besson, J.-M. 1996, Selective cyclooxygenase-2 inhibition reduces carrageenan oedema and associated spinal c-Fos expression in the rat. *Brain Research*, **715:** 217–220.

Burtscher, M., Likar, R., Nachbauer, W. and Philadelphy, M. 1998, Aspirin for prophylaxis against headache at high altitudes: randomised, double-blind, placebo controlled trial. *British Medical Journal*, **316:** 1057–1058.

Burtscher, M., Likar, R., Nachbauer, W., Philadelphy, M., Puhringer, R. and Lammle, T. 2001, Effects of aspirin during exercise on the incidence of high-altitude headache: a randomized, double-blind, placebo-controlled trial. *Headache*, **41:** 542–545.

Buzzi, M.G., Di Gennaro, G., D'Onofrio, M., Ciccarelli, O., Santorelli, F.M., Fortin, D., Nappi, G., Nicoletti, F. and Casali, C. 2000, mtDNA A3243G MELAS mutation is not associated with multigenerational female migraine. *Neurology*, **54:** 1005–1007.

Bywaters, E.G.L. 1963, Clinical implications of the anti-inflammatory effects of salicylate. *In:* A.St.J. Dixon, B.K. Martin, M.J.H. Smith and P.H.N. Wood (eds), *Salicylates. An International Symposium*, pp. 154–160. London: J.A. Churchill Ltd.

Calimlim, J.F., Wardell, W.M., Davies, H.T., Lasagna, L. and Gillies, A.J. 1977, Analgesic efficacy of an orally administered combination of pentazocine and aspirin, with observations on the use and statistical efficacy of GLOBAL subjective efficacy ratings. *Clinical Pharmacology Therapeutics*, **21:** 34–43.

Carlsson, A. and Magnusson, T. 1955, Failure to demonstrate antipyretic and analgesic properties of salicylamides. *Acta Phrmacologica et Toxicologica*, **11:** 248–253.

Carr, D.B. and Goudas, L.C. 1999, Acute pain. *Lancet*, **353:** 2051–2058.

Cashman, J.C. and McAnulty, G. 1995, Nonsteroidal anti-inflammatory drugs in perisurgical pain management. *Drugs*, **49:** 51–70.

Cashman, J.N. 1996, The mechanisms of action on NSAIDs in analgesia. *Drugs*, **52 (Suppl. 5):** 13–23.

Catapano, M.S. 1996, The analgesic efficacy of ketolorac for acute pain. *Journal of Emergency Medicine*, **14:** 67–75.

Chabriat, H., Joire, J.E., Danchot, J., Grippon, P. and Bousser, M.G. 1994, Combined oral lysine acetylsalicylate and metoclopramide in the acute treatment of migraine: a multicentre double-blind placebo-controlled study. *Caphalalgia*, **14:** 297–300.

Chabriat, H., Vahedi, K., Clark, C.A., Poupon, C., Ducros, A., Denier, C., Le Bihan, D. and Bousser, M.G. 2000, Decreased hemispheric water mobility in migraine related to mutation of CACNA1A gene. *Neurology*, **54:** 510–512.

Chan, W.Y., Dawood, M.Y. and Fuchs, F. 1979, Relief of dysmenorrhea with the prostaglandin synthetase inhibitor ibuprofen: effects on prostaglandin levels in menstrual fluid. *American Journal of Gynecology* **135:** 102–108.

Chandrasekharan, N.V., Dai, H., Roos, K.L., Evanson, N.K., Tomsik, J., Elton, T.S. and Simmons, D.L. 2003, COX-3, a cyclooxygenase-1 variant inhibited by acetaminophen and other analgesic/antipyretic drugs: cloning, structure, and expression. *Proceedings of the National Academy of Sciences, USA*, **99:** 13926-13931.

Cohen, M.D., Harrington, T.M. and Ginsburg, W.W. 1983, Osteoid osteoma: 95 cases and a review of the literature. *Seminars Arthritis and Rheumatism*, **12:** 265–281.

Collins, A.J., Notarianni, L.J., Ring, E.F. and Seed, M.P. 1984, Some observations on the pharmacology of 'deep-heat', a topical rubifacient. *Annals of Rheumatic Diseases*, **43:** 411–415.

Cooper, S.A. 1980, Efficacy of zomepirac in oral surgery pain. *Journal of Clinical Pharmacology*, **20:** 230–242.

Cooper, S.A. 1981, Comparative analgesic efficacies of aspirin and acetaminophen. *Archives of Internal Medicine*, **141:** 282–285.

Cooper, S.A., Reynolds, D.C., Kruger, G.O. and Gottlieb, S. 1980, An analgesic relative potency assay comparing zomepirac sodium and aspirin. *Journal of Clinical Pharmacology*, **20:** 98–105.

Cooper, S.A., Wagenberg, B., Eskow, R. and Zissu, J. 1983, Double-blind evaluation of suprofen and aspirin in the treatment of periodontal pain. *Pharmacology*, **27 (Suppl. 1):** 23–30.

Corpataux, J.B., Van Gessel, E.F., Donald, F.A., Forster, A. and Gamulin, Z. 1997, Effects of postoperative analgesia of small-dose lysine acetylsalicylate added to prilocaine during intravenous regional anaesthesia. *Anesthesia and Analgesia*, **84**: 1081–1085.

Couch, J.R. and Hassanein, R.S. 1977, Platelet aggregability in migraine. *Neurology*, **27**: 843–848.

Creagan, E.T. and Wilkinson, J.M. 1989, Pain relief in terminally ill patients. *American Family Physician*, **40**: 133–140.

Currier, R.D. and Westernberg, M.R. 1958, Evaluation of a salicylamide compound in the treatment of headache. *University of Michigan Medical Bulletin*, **24**: 415–418.

Cursiefen, C., Wisse, M., Cursiefen, S., Junemann, A., Martus, P. and Korth, M. 2000, Migraine and tension in high-pressure and normal-pressure glaucoma. *American Journal of Opthalmology*, **129**: 102–104.

Cutting, C.J. and Thornton, E.D. 1981, A comparative trial of tiaprofenic acid ('Surgam') versus aspirin in the control of pain following injury. *Pharmatherapeutica*, **2**: 509–512.

Dahl, E., Feldman, G. and Jonsson, E. 1985, Acetylsalicylic acid compared with acetylsalicylic acid plus codeine as postoperative analgesics after removal of impacted mandibular third molars. *Swedish Dental Journal*, **9**: 207–212.

D'Andrea, G., Toldo, M., Cananzi, A. and Ferro-Milone, F. 1984, Study of platelet activation in migraine: control by low doses of aspirin. *Stroke*, **15**: 271–275.

Daniel, A., Hartmann, P., Giumelli, B., Poitou, P. and Vezin, J.C. 1983, A study of the antalgic efficacy of a choline salicylate + cetalkonium chloride gel on various mouth ulcerative lesions. *Therapie*, **38**: 81–84.

Dawood, M.Y. 1981, Dysmenorrhea and prostaglandins: pharmacological and therapeutic considerations. *Drugs*, **22**: 42–56.

Dawood, M.Y. 1988, Nonsteroidal anti-inflammatory drugs and changing attitudes towards dysmenorrhea. *American Journal of Medicine*, **84 (Suppl. 5A)**: 23–29.

De Kornfeld, T.J. and Lasagna, L. 1959, Clinical trial of two analgesics: WIN 14,098 and ethoheptazine. *Federation Proceedings*, **18**: 382 (abstract) 1509.

De Kornfeld, T.J., Lasagna, L. and Frazier, T.M. 1962, A comparative study of five proprietary analgesic compounds. *Journal of the American Medical Association*, **182**: 1315–1318.

Deshmukh, S.V., Meyer, J.S. and Mouche, R.J. 1976, Platelet dysfunction in migraine: effect of self-medication with aspirin. *Thrombosis and Haemostasis*, **36**: 319–324.

Devoghel, J.C. 1983, Small intrathecal doses of lysine-acetylsalicylate relieve intractable pain in man. *Journal of International Medical Research*, **11**: 90–91.

Dhaliwal, H.S., Sloan, P., Arkinstall, W.W., Thirlwell, M.P., Babul, N., Harsanyi, Z. and Darke, A.C. 1995, Randomized evaluation of controlled-release codeine and placebo in chronic cancer pain. *Journal of Pain Symptom Management*, **10**: 612–623.

Diamond, S. 1983, Ibuprofen versus aspirin and placebo in the treatment of muscle contraction headache. *Headache*, **23**: 206–210.

Diener, H.C., for the ASASUMAMIG Study Group, 1999, Efficacy and safety of intravenous acetylsalicylic acid lysinate compared to subcutaneous sumatriptan and parenteral placebo in the acute treatment of migraine. A double-blind, double-dummy, randomized, multicentre, parallel group study. *Cephalalgia*, **19**: 581–588.

Dionne, A. and Cooper, A. 1999, Use of ibuprofen in dentistry. *In:* K.D. Rainsford (ed.), *Ibuprofen. A Critical Bibliographic Review*, pp. 407–430. London, Taylor and Francis.

Dionne, R.A. 1998, Evaluation of analgesic mechanisms and NSAIDs for acute pain using the oral surgery model. *In:* K.D. Rainsford and M.C. Powanda (eds), *Safety and Efficacy of Non-Prescription (OTC) Analgesics and NSAIDs*, pp. 105–117. Dordrecht: Kluwer Academic Publishers.

Duncan, J.J. and Farr, J.E. 1988, Comparison of diclofenac sodium and aspirin in the treatment of acute sprain injuries. *American Journal of Sports Medicine*, **16**: 656–659.

Dundee, J.W. and McAteer, E. 1981, Intravenous salicylate for postoperative pain. *Lancet*, **1**: 154.

Du Pen, S.L., Du Pen, A.R., Polissar, N., Hansberry, J., Miller Kraybill, B., Stillman, M., Panke, J., Everly, R. and Syrjala, K. 1999, Implemented guidelines for cancer pain management: results of a randomized controlled clinical trial. *Journal of Clinical Oncology*, **17**: 361–370.

Dyment, P.G. 1986, Management of minor soft tissue trauma in adolescent athletes. *Journal of Adolescent Health Care*, **7 (Suppl. 6)**: 133S–135S.

Eccles, R., Loose, I., Jawad, M. and Nyman, L. 2003, Effects of acetylsalicylic acid on sore throat pain and other pain symptoms associated with acute upper respiratory tract infection. *Pain Medicine*, **4:** 118–124.

Edwards, J.E., Oldman, A.D., Smith, L.A., Carroll, D., Wiffen, P.J., McQuay, H.J. and Moore, R.A. 1999, Oral aspirin in postoperative pain: a quantitative systematic review. *Pain*, **81:** 289–297.

Eisenberg, E., Berkey, C.S., Carr, D.B., Mosteller, F. and Chalmers, T.C. 1994, Efficacy and safety of nonsteroidal antiinflammatory drugs for cancer pain: a meta-analysis. *Journal of Clinical Oncology*, **12:** 2756–2765.

Engelhardt, G., Mauz, A.B. and Pairet, M. 1997, Role of caffeine in combined analgesics from the point of view of experimental pharmacology. *Arnzeimittel Forschung*, **47:** 917–927.

Feinberg, I., Carey, J., Hussussian, J. and Arias, B. 1962, Treatment of painful skeletal muscle disorders. A report of a double blind study of methocarbamol, aspirin and placebo. *American Journal of Orthopedics*, **4:** 280–286.

Fields, H.L. 1987, *Pain*. New York: McGraw-Hill Book Company.

Forbes, J.A., Beaver, W.T., Jones, K.F., Edquist, I.A., Gongloff, C.M., Smith, W.K., Smith, F.G. and Schwartz, M.K. 1992, Analgesic efficacy of bromfenac, ibuprofen and aspirin in postoperative oral surgery pain. *Clinical Pharmacology and Therapeutics*, **51:** 343–352.

Forbes, J.A., Bowser, M.W., Calderazzo, J.P. and Foor, V.M. 1981, An evaluation of the analgesic efficacy of three opioid–analgesic combinations in postoperative oral surgery pain. *Journal of Oral Surgery*, **39:** 108–112.

Forbes, J.A., Butterworth, G.A., Burchfield, W.H. and Beaver, W.T. 1990, Evaluation of ketorolac, aspirin and an acetaminophen-codeine combination in postoperative oral surgery pain. *Pharmacotherapy*, **10:** 77S–93S.

Forbes, J.A., Calderazzo, J.P., Bowser, M.W., Foor, V.M., Shackleford, R.W. and Beaver, W.T. 1982, A 12–hour evaluation of the analgesic efficacy of diflunisal, aspirin and placebo in post-operative dental pain. *Journal of Clinical Pharmacology*, **22:** 89–96.

Forbes, J.A., Edquist, I.A., Smith, F.G., Schwartz, M.K. and Beaver, W.T. 1991, Evaluation of bromfenac, aspirin, and ibuprofen in postoperative oral surgery pain. *Pharmacotherapy*, **11:** 64–70.

Forbes, J.A., Jones, K.F., Smith, W.K. and Gongloff, C.M. 1986, Analgesic effect of an aspirin–codeine–butalbital–caffeine combination and an acetaminophen–codeine combination in postoperative oral surgery pain. *Pharmacotherapy*, **6:** 240–247.

Forbes, J.A., White, R.W., White, E.H. and Hughes, M.K. 1980, An evaluation of the analgesic efficacy of proquazone and aspirin in postoperative dental pain. *Journal of Clinical Pharmacology*, **20:** 465–474.

Forman, A., Andersson, K.-E. and Ulmsten, U. 1982, Combined effects of diflunisal and nifedipine on uterine contractility in dysmenorrhoeic patients. *Prostaglandins*, **23:** 237–246.

Forrest, J. 1998, *Acute Pain: Pathophysiology and Treatment*. Grimsby (Ontario): Manticore Publishers.

Forrest, W.J. 1966, Personal communications as to W.T. Beaver; See Beaver (1965), Chairman of the Veterans Administration Co-operative Analgesic Study Group.

Frame, J.W., Fisher, S.E., Pickvance, N.J. and Skene, A.M. 1986, A double-blind placebo-controlled comparison of three ibuprofen/codeine combinations and aspirin. *British Journal of Oral Maxillofacial Surgery*, **24:** 122–129.

Frey, G.H. 1961, The role of placebo response in clinical headache evaluations. *Headache*, **1:** 31–38.

Fuccella, L.M. 1979, Clinical pharmacology of inhibitors of platelet aggregation. *Pharmacy Research and Communications*, **11:** 825–852.

Giannikas, A., Papachristou, G., Tiniakos, G., Chrysafidiss and G. and Hartofilakidis-Garofallidis, G. 1977, Osteoid osteoma of the termina phalanges. *Hand*, **9:** 295–300.

Gold, H. and Cattell, McK. 1965, Mild analgesics. A review of their clinical pharmacology. *American Journal of Medicine Science*, **250:** 577–604.

Goldstein, J., Hoffman, H.D., Armellino, J.J., Hamelsky, S.W., Couch, J., Blumenthal, H. and Lipton, R.B. 1999, Treatment of severe, disabling migraine attacks in an over-the-counter population of migraine sufferers: results from three randomized, placebo-controlled studies of the combination of acetaminophen, aspirin, and caffiene. *Cephalagia*, **19:** 684–691.

CHAPTER 10

Graham, G.G., Day, R.O., Milligan, M.K., Ziegler, J.B. and Kettle, A.J. 1999, Current concepts of the actions of paracetamol (acetaminophen) and NSAIDs. *Inflammopharmacology*, **7:** 255–263.

Grazzi, L.M., D'Amico, D., Moschiano, F. and Bussone, G. 1995, A review of the treatment of primary headaches. Part 1: Migraine. *Italian Journal of Neurological Science*, **16:** 577–586.

Green, B.G. 1991, Interactions between chemical and thermal cutaneous stimuli: inhibition (counter-irritation) and integration. *Somatosensory and Motor Research*, **8:** 301–312.

Gross, M. and Greenberg, L.A. (eds), 1948, *The Salicylates: A Critical Bibliographic Review*. New Haven, USA: Hillhouse Press.

Hagen, N.A., Elwood, T. and Ernst, S. 1997, Cancer pain emergencies: a protocol for management. *Journal of Pain and Symptom Management*, **14:** 45–50.

Hakkarainen, H., Gustafsson, B. and Stockman, O. 1978, A comparative trial of ergotamine tartrate, acetyl salicylic acid and a dextropropoxyphene compound in acute migraine attacks. *Headache*, **18:** 35–39.

Hakkarainen, H., Quiding, H. and Stockman, O. 1980, Mild analgesics as an alternative to ergotamine in migraine. A comparative trial with acetylsalicylic acid, ergotamine tartrate and dextropropoxyphene compound. *Journal of Clinical Pharmacology*, **20:** 590–595.

Halbert, D.R., Demers, L.M. and Darnell Jones, D.E. 1976, Dysmenorrhea and prostaglandins. *Obstetric and Gynecological Surveys*, **31:** 77–81.

Hall, S.M. 1990, Treatment of viral pharyngitis or flu. *British Medical Journal*, **301:** 1165.

Hasegawa, T., Hirose, T., Sakamoto, R., Seki, K., Ikata, T. and Hizawa, K. 1993, Mechanism of pain in osteoid osteomas: an immunohistochemical study. *Histopathology*, **22:** 487–491.

Heere, L.P. 1988, Piroxicam in acute musculoskeletal disorders and sports injuries. *American Journal of Medicine*, **84 (5A):** 50–55.

Henrickson, P.-A., Tjernberg, Aa., Ahlstrom, U. and Peterson, L.-E. 1979, Analgesic efficacy and safety of fenbufen following surgical removal of a lower wisdom tooth: a comparison with acetylsalicylic acid and placebo. *Journal of International Medical Research*, **7:** 107–116.

Henry, P., Hiesse-Provost, O., Dillenschneider, A., Ganry, H. and Insuasty, J. 1995, Efficacy and tolerance of an effervescent aspirin-metoclopramide combination in the treatment of a migraine attack. Randomized double-blind study using a placebo. *Presse Medicale*, **24:** 254–258.

Hepsö, H.U., Loekken, P., Björnson, J. and Godal, H.C. 1976, Double-blind crossover study of the effect of acetylsalicylic acid on bleeding and post-operative course after bilateral oral surgery. *European Journal of Clinical Pharmacology*, **10:** 217–225.

Hickey, R.F.J. 1982, Chronic low back pain; a comparison of diflunisal with paracetamol. *New Zealand Medical Journal*, **95:** 312–314.

Higelin, F. and Annoni, J.M. 1998, Medical treatment of migraine: from mechanisms of action to contraindications. *Schweizer Medizinische Wochenschrift*, **128:** 374–383.

Hilton, B.P. and Cummings, J.N. 1971, An assessment of platelet aggregation induced by 5-hydroxy-tryptamine. *Journal of Clinical Pathology*, **24:** 250–258.

Horrobin, D.F. 1977, Hypothesis: prostaglandins and migraine. *Headache*, **17:** 113–117.

Hussar, D.A. 2000, New drugs of 1999. *Journal of the America Pharmacological Association*, **40:** 181–221.

Hutlin, M. and Olander, K.-J. 1981, Comparison of pirprofen, acetylsalicylic acid, and placebo in post-operative pain after oral surgery. *In:* J.K. van der Korst (Chairman), ®*Rengasil as an Antirheumatic and Analgesic Drug*. ®Rengasil Symposium, Paris. Ciba-Geigy, Basel.

Ihalainen, U. 1980, The efficacy and tolerability of diflunisal and ASA in the relief of postoperative pain in oral surgery. *Proceedings of the Finnish Dental Society*, **76:** 262–266.

Jain, A.K., Ryan, J.R., McMahon, G., Kuebel, J.O., Walters, P.J. and Noveck, C. 1986, Analgesic efficacy of low-dose ibuprofen in dental extraction. *Pharmacotherapy*, **6:** 318–322.

Jellinek, E.M. 1946, Clinical tests on comparative effectiveness of analgesic drugs. *Biometrics Bulletin*, **2:** 87–91.

Jensen, M.P., Turner, J.A., Romano, J.M. and Karoly, P. 1991, Coping with chronic pain: a critial review of the literature. *Pain*, **47:** 249–283.

Jones, A.K. 1997, Primary care management of acute low back pain. *Nurse Practitioner*, **22:** 50–52.

Kantor, T.G. 1966, Personal communication to W. T. Beaver, see Beaver (1965).

Kantor, T.G., Laska, E., Meisner, M., Sunshine, A. 1964. *In: Drug Addiction and Narcotics Bulletin*, Appendix 20, p. 3921. See Beaver (1965).

Kantor, T.G., Meisner, M., Laska, E. and Sunshine, A. 1964, Analgesic effects of aspirin in postpartum pain. *Federation Proceedings*, **23:** 176.

Kantor, T.G., Streem, A. and Laska, E. 1977, Estimates of doses of antiinflammatory drugs in man by testing analgesic potency. I. 1-isopropyl-4-phenyl-7-methyl-2 (IH) quinazolone versus aspirin. *Arthritis and Rheumatism*, **20:** 1381–1387.

Kantor, T.G., Sunshine, A., Laska, E., Meisner, M. and Hopper, M. 1966, Oral analgesic studies: pentazocine hydrochloride, codeine, aspirin, and placebo and their influence on response to placebo. *Clinical Pharmacology and Therapeutics*, **7:** 447–454.

Kaube, H., Katsarava, Z., Przywara, S., Drepper, J., Ellrich, J. and Diener, H.C. 2002, Acute migraine headache: possible sensitization of neurons in the spinal trigeminal nucleus? *Neurology*, **58:** 1234–1238.

Kelton, J.G., Carter, C.J., Rosenfeld, J., Massicotte-Nolan, M.P. and Hirsch, J. 1981, Sex-related differences in the efficacy of acetylsalicylic acid (ASA): the absorption of ASA and its effect on collagen-induced thromboxane B_2 generation. *Thrombosis Research*, **24:** 163–168.

Klein, J.R., Litt, I.F. and Udall L. 1981, The effects of aspirin on dysmenorrhea in adolescents. *Journal of Pediatrics*, **98:** 987–990.

Korttila, K., Pentti, O.M. and Auvinen, J. 1980, Comparison of i.m. lysine acetylsalicylate and oxycodone in the treatment of pain after operation. *British Journal of Anaesthiology*, **52:** 613–617.

Kweekel-deVries, W.J., Spierdijk, J., Mattie, H. and Hermans, J.M.H. 1974, A new soluble acetylsalicylic acid derivative in the treatment of postoperative pain. *British Journal of Anaesthiology*, **46:** 133–135.

Lance, J.W., Lambert, G.A., Goadsby, P.J. and Duckworth, J.W. 1983, Brainstem influences on the cephalic circulation: experimental data from cat and monkey of relevance to the mechanism of migraine. *Headache*, **23:** 258–265.

Lange, R., Schwarz, J.A. and Hohn, M. 2000, Acetylsalicylic acid effervescent 1000 mg (Aspirin) in acute migraine attacks; a multicentre, randomized, double-blind, single dose, placebo-controlled parallel group study. *Cephalalgia*, **20:** 663–667.

Lasagna, L. 1965, Drug interactions in the field of analgesic drugs. *Proceedings of the Royal Society of Medicine*, **58 (11):** 978–983.

Lasagna, L., Laties, V.G. and Dohan, J.L. 1958, Further studies on the 'pharmacology' of placebo administration. *Journal of Clinical Investigation*, **37:** 533–537.

Laska, E.M., Sunshine, A., Wanderling, J.A. and Meisner, M.J. 1982, Quantitative differences in aspirin analgesia in three models of clinical pain. *Journal of Clinical Pharmacology*, **22:** 531–542.

Leone, M., Grazzi, L., D'Amico, D., Moschiano, F. and Bussone, G. 1995, A review of the treatment of primary headaches. Part 1: Migraine. *Italian Journal of Neurological Science*, **16:** 577–586.

Lewy, R.B. 1942, Comparative studies in pain control. *Anesthesia and Analgesia*, **21:** 255–272.

Lim, R.K.S. 1966, Salicylate analgesia. *In:* M.J.H. Smith and P.K. Smith (eds), *Salicylates. A Critical Bibliographical Review*, pp. 155–202. New York: Wiley-Interscience.

Limmroth, V., Katsarava, Z. and Diener, H.C. 1999b, Acetylsalicylic acid in the treatment of headache. *Cephalalgia*, **19:** 545–551.

Limmroth, V., May, A. and Diener, H. 1999a, Lysine-acetylsalicylic acid in acute migraine attacks. *European Neurology*, **41:** 88–93.

Lipton, R.B., Stewart, W.F., Ryan, R.E. Jr., Saper, J., Silberstein, S. and Sheftell, F. 1998, Efficacy and safety of acetaminophen, aspirin, and caffeine in alleviating migraine headache pain: three double-blind, randomized, placebo-controlled trials. *Archives of Neurology*, **55:** 210–217.

Loeser, J.D. and Melzack, R. 1999, Pain: an overview. *Lancet*, **353:** 1607–1609.

Lomen, P.L., Samal, B.A., Lamborn, K.R., Sattler, L.P. and Crampton, S.L. 1986, Flurbiprofen for the treatment of bone pain in patients with metastatic breast cancer. *American Journal of Medicine*, **80:** 83–87.

Magee, K.R. and De Jong, R.N. 1959. *University of Michigan Medical Bulletin*, **25:** 74.

Mahler, D.L., Forrest, W.H., Brown, C.R., Shroff, P.F., Gordon, H.E., Brown, B.W. and James, K.E. 1976, Assay of aspirin and naproxen analgesia. *Clinical Pharmacology and Therapeutics*, **19:** 18–23.

Maizels, M. 1998, Limitations of acetaminophen, aspirin, and caffeine in alleviating migraine. *Archives of Neurology*, **55:** 1590.

Majerus, P.W. and Stanford, N. 1977, Comparative effects of aspirin and diflunisal on prostaglandin synthetase from human platelets and sheep seminal vesicles. *British Journal of Clinical Pharmacology*, **4 (Suppl. 1):** 15S–18S.

Malmberg, A.B. and Yaksh, T.L. 1992, Hyperalgesia mediated by spinal glutamate or substance P receptor blocked by cyclooxygenase inhibition. *Science*, **257:** 1276–1279.

Malmgren, K., Ollsson, P., Thornling, G. and Unge, G. 1978, Acetylsalicylic asthma and migraine – a defect in serotonin (5-HT) uptake in platelets. *Thrombosis Research*, **13:** 1137–1139.

Markowitz, N.R., Young, S.K., Rohrer, M.D. and Turner, J.L. 1985, Comparison of meclofenamate sodium with buffered aspirin and placebo in the treatment of postsurgical dental pain. *Journal of Oral Maxillofacial Surgery*, **43:** 517–522.

Marrs, J.W., Glas, W.W. and Silvani, J. 1959, D-Propoxyphene hydrochloride. *American Journal of Pharmacy*, **131:** 271–276.

Mascia, A., Afra, J. and Schoenen, J. 1998, Dopamine and migraine: a review of pharmacological, biochemical, neurophysiological, and therapeutic data. *Cephalalgia*, **18:** 174–182.

Masel, B.E., Chesson, A.L., Peters, B.H., Levin, H.S. and Alperin, J.B. 1980, Platelet antagonists in migraine prophylaxis. A clinical trial using aspirin and dipyridamole. *Headache*, **20:** 13–18.

Mathew, N.T. 1999, Migraine and hypertension. *Cephalalgia*, **19 (Suppl. 25):** 17–19.

McCormack, K. and Brune, K. 1991, Dissociation between the antinociceptive and anti-inflammatory effects of the nonsteroidal anti-inflammatory drugs. A survey of their analgesic efficacy. *Drugs*, **41:** 533–547.

McQuay, H. and Moore, A. 1999, *An Evidence-Based Resource for Pain Relief.* Oxford: Oxford University Press.

Mehlish, D.R. and Joy, E.D. 1981, Zomepirac sodium vs APC with codeine for oral surgery. *Journal of Oral Surgery*, **39:** 426–429.

Melzack, R., Jeans, M.E., Kinch, R.A. and Katz, J. 1983, Diflunisal (1000 mg single dose) versus acetaminophen (650 mg) and placebo for the relief of post-episiotomy pain. *Current Therapeutic Research*, **34:** 929–939.

Mennie, A.T., Dalley, V.M., Dinneen, L.C. and Collier, H.O. 1975, Treatment of radiation induced gastrointestinal distress with acetylsalicylate. *Lancet*, **2:** 942–943.

Messer, R.H., Vaughn, T. and Harbert, G. 1980, Clinical evaluation of zomepirac and APC with codeine in the treatment of postpartum episiotomy pain. *Journal of Clinical Pharmacology*, **20:** 279–284.

Meyers, F.J. and Meyers, F.H. 1987, Management of chronic pain. *American Family Physician*, **36:** 139–146.

Migliardi, J.R., Armellino, J.J., Friedman, M., Gillings, D.B. and Beaver, W.T. 1994, Caffeine as an analgesic adjuvant in tension headache. *Clinical Pharmacology and Therapy*, **56:** 576–586.

Minotti, V., Patoia, L., Roila, F., Basurto, C., Tonato, M., Pasqualucci, V., Maresca, V. and Del Favero, A. 1989, Double-blind evaluation of analgesic efficacy of orally administered diclofenac, nefopam, and acetylsalicylic acid (ASA) plus codeine in chronic cancer pain. *Pain*, **36:** 177–183.

Moertel, C.G., Ahmann, D.L., Taylor, W.F. and Schwartau, N. 1971, Aspirin and pancreatic cancer pain. *Gastroenterology*, **60:** 552–553.

Moertel, C.G., Ahmann, D.L., Taylor, W.F. and Schwartau, N. 1972, A comparative evaluation of marketed analgesic drugs. *New England Journal of Medicine*, **286:** 813–815.

Moertel, C.G., Ahmann, D.L., Taylor, W.F. and Schwartau, N. 1974, Relief of pain by oral medications. A controlled evaluation of analgesic combinations. *Journal of the American Medical Association*, **229:** 55–59.

Moore, A., McQuay, H. and Gavaghan, D. 1996, Deriving dichotomous outcome measures from continuous data in randomised controlled trials of analgesics. *Pain*, **66:** 229–237.

Moore, N., Van Ganse, E., Le Parc, J.-M., Wall, R., Schneid, H., Farhan, M., Verriere, F. and Pelen, F. 1999, The PAIN study: Paracetamol, Aspirin and Ibuprofen New Tolerability Study. A large-scale, randomised clinical trial comparing the tolerability of aspirin, ibuprofen and paracetamol for short-term analgesia. *Clinical Drug Investigations*, **18:** 89–98.

Moore, P.A. 1999, Pain management in dental practice: tramadol vs. codeine combinations. *Journal of the American Dental Association*, **130:** 1075–1079.

Moore, P.A., Crout, R.J., Jackson, D.L., Schneider, L.G., Graves, R.W. and Bakos, L. 1998, Tramadol hydrochloride: analgesic efficacy compared with codeine, aspirin with codeine, and placebo after dental extraction. *Journal of Clinical Pharmacology*, **38:** 554–560.

Moore, R.A. and McQuay, H.J. 1997, Single-patient data meta-analysis of 3453 postoperative patients: oral tramadol versus placebo, codeine and combination analgesics. *Pain*, **69:** 287–294.

Moote, C.A. 1996, Ibuprofen arginine in the management of pain – a review. *Clinical Drug Investigation*, **11:** 1–7.

Morse, D.R., Furst, M.L., Koren, L.Z., Bolanos, O.R., Esposito, J.V. and Yesilsoy, C. 1987, Comparison of diflunisal and an aspirin-codeine combination in the management of patients having one-visit endodontic therapy. *Clinical Therapy*, **9:** 500–511.

Mortimer, J.E. and Bartlett, N.L. 1997, Assessment of knowledge about cancer pain management by physicians in training. *Journal of Pain and Symptom Management*, **14:** 21–28.

Muckle, D.S. 1974, Comparative study of ibuprofen and aspirin in soft tissue injuries. *Rheumatology and Rehabilitation*, **13:** 141–147.

Muckle, D.S. 1977, A double-blind trial of flurbiprofen and aspirin in soft-tissue trauma. *Rheumatology and Rehabilitation*, **16:** 58–61.

Muckle, D.S. 1980, A double-blind trial of ibuprofen and aspirin in the treatment of soft-tissue injuries sustained in professional football. *British Journal of Sports Medicine*, **14:** 46–47.

Murray, W.J. 1964, Evaluation of aspirin treatment of headache. *Clinical Pharmacology and Therapeutics*, **5:** 21–25.

Nebe, J., Heier, M. and Diener, H.C. 1995, Low-dose ibuprofen in self-medication of mild to moderate headache: a comparison with acetylsalicylic acid and placebo. *Cephalalgia*, **15:** 531–535.

Nemoto, O., Moser, R.P., Vandam, B.E., Aoki, J. and Gilkey, F.W. 1990, Osteoblastoma of the spine – a review of 75 cases. *Spine*, **15:** 1272–1280.

Noda, S., Itoh, H., Umezaki, H. and Fukuda, Y. 1985, Successful treatment of migraine attacks with intravenous injection of aspirin. *Journal of Neurology, Neurosurgery and Psychology*, **48:** 1187.

Nyström, E., Gustafsson, I. and Quiding, H. 1988, The pain intensity at analgesic intake, and the efficacy of diflunisal in single doses and effervescent acetaminophen in single and repeated doses. *Pharmacotherapy*, **8:** 201–209.

O'Grady, F.W., Lambert, H.P. and Begent, R.H. 1975, Sensible prescribing. III. Acute otitis media and sore throat. *Practitioner*, **214:** 421–427.

Okun, R., Green, J.W. and Shackleford, R.W. 1979, An analgesic comparison study of indoprofen versus aspirin and placebo in surgical pain. *Journal of Clinical Pharmacology*, **19:** 487–492.

O'Neill, B.P. and Mann, J.D. 1978, Aspirin prophylaxis in migraine. *Lancet*, **ii:** 1179–1181.

Orkin, L.R., Joseph, S.I. and Helrich, M. 1957, Effects of mild analgesics on post-partum pain: a method for evaluating analgesics. *New York Journal of Medicine*, **57:** 71–78.

Orlowski, J.P. and Mercer, R.D. 1977, Osteoid osteoma in children and young adults. *Pediatrics*, **59:** 526–532.

Owen, P.R. 1984, Prostaglandin synthetase inhibitors in the treatment of primary dysmenorrhea. Outcome trials reviewed. *American Journal of Obstetrics and Gynecology*, **148:** 96–103.

Oxman, T.E., Hitzemann, R.J. and Smith, R. 1982, Platelet membrane lipid composition and the frequency of migraine. *Headache*, **22:** 261–267.

Pang, W., Mok, M.S., Ku, M.C. and Huang, M.H. 1999, Patient-controlled analgesia with morphine plus lysine acetyl salicylate. *Anesthesia and Analgesia*, **89:** 995–998.

Parker, D.A., Gibbin, K.P. and Noyelle, R.M. 1986, Syrup formulations for post-tonsillectomy analgesia: a double-blind study comparing ibuprofen, aspirin and placebo. *Journal of Laryngology Otolaryngology*, **100:** 1055–1060.

Pellerin, M., Hardy, F., Abergel, A., Boule, D., Palacci, J.H., Babinet, P., Ng Wingtin, L., Glowinski, J., Amiot, J.-F., Mechali, D., Colbert, N. and Starkman, M. 1987, Douleur chronique rebelle des cancéreaux. Intérât de l'injection intrarachidienne d'acetylsalicylate de lysine. Soixante observations. *La Presse Médicale*, **16:** 1465–1468.

Pena, M. 1990, Etodolac: analgesic effects in musculoskeletal and postoperative pain. *Rheumatology International*, **10 (Suppl.):** 9–16.

Peters, B.H., Fraim, C.J. and Masel, B.E. 1983, Comparison of 650 mg aspirin and 1000 mg acetaminophen with each other, and with placebo in moderately severe headache. *American Journal of Medicine*, **74:** 36–42.

Peterson, J.K. 1979, Diflunisal, a new analgesic, in the treatment of postoperative pain following removal of impacted third molars. *International Journal of Oral Surgery*, **8:** 102–113.

Pini, L.A., Fabbri, L. and Cavazzuti, L. 1999, Efficacy and safety of sumatriptan 50 mg in patients not responding to standard care, in the treatment of mild to moderate migraine. The sumatriptan 50 mg Italian Study Group. *International Journal of Clinical Pharmacology and Research*, **19:** 57–64.

Pitcher, G.M. and Henry, J.L. 1999, NSAID-induced cyclooxygenase inhibition differentially depresses long-lasting versus brief synaptically-elicited responses of rat spinal dorsal horn neurons in vivo. *Pain*, **82:** 173–186.

Prasad, R. 1980, The treatment of primary dysmenorrhoea with benorylate. *Practitioner*, **224:** 325–327.

Rainsford, K.D. 1984, *Aspirin and the Salicylates*. London: Butterworth.

Rainsford, K.D. 1996, Mode of action, uses, and side effects of anti-inflammatory drugs. *In:* K.D. Rainsford (ed.), *Advances in Anti-Rheumatic Therapy*, pp. 59–111. Boca Raton, FL: CRC Press.

Rainsford, K.D. 1999, Pharmacology and toxicology of ibuprofen. *In:* K.D. Rainsford (ed.), *Ibuprofen: A Critical Bibliographic Review*, pp. 143–275. London: Taylor & Francis.

Rainsford, K.D., Roberts, S.C. and Brown, S. 1997, Ibuprofen and paracetamol: relative safety in non-prescription dosages. *Journal of Pharmacy and Pharmacology*, **49:** 345–376.

Rosenwaks, Z., Jones, G.S., Henzl, M.R., Dubin, N.H., Rhodgaonka, R.B. and Hoffmann, S. 1981, Naproxen sodium, aspirin and placebo in primary dysmenorrhea. Reduction of pain and blood levels of prostaglandin $F_{2\alpha}$ metabolite. *American Journal of Obstetrics and Gynecology*, **140:** 592–598.

Ross-Lee, L., Eradie, M.J. and Tyrer, J.H. 1982a, Aspirin treatment of migraine attacks: clinical observations. *Cephalagia*, **2:** 71–76.

Ross-Lee, L., Eradie, M.J. and Tyrer, J.H. 1982b, Aspirin treatment of migraine attacks: blood level data. *Cephalagia*, **2:** 9–14.

Rubin, A. and Winter, L. 1984, A double-blind randomized study of an aspirin/caffeine combination versus acetaminophen/aspirin combination versus acetaminophen versus placebo in patients with moderate to severe post-partum pain. *Journal of International Medical Research*, **12:** 338–345.

Ruedy, J. 1973, A comparison of the analgesic efficacy of naproxen and acetylsalicylic acid-codeine in patients with pain after dental surgery. *Scandinavian Journal of Rheumatology*, **Suppl. 2:** 60–63.

Sacchetti, G., Camera, P., Rossi, A.P., Martoni, A., Bruni, G. and Pannuti, F. 1984, Injectable ketoprofen vs. acetylsalicylic acid for the relief of severe cancer pain: a double-blind, crossover trial. *Drug Intelligence and Clinical Pharmacy*, **18:** 403–406.

Sandrini, M. 1999, Central effects of non-opioid analgesics – review of actions and their clinical implications. *CNS Drugs*, **12:** 337–345.

Saxena, A., Andley, M. and Gnanasekaran, N. 1994, Comparison of piroxicam and acetylsalicylic acid for pain in head and neck cancers: a double-blind study. *Palliative Medicine*, **8:** 223–229.

Schachtel, B.P., Fillingim, J.M., Lane, A.C., Thoden, W.R. and Baybutt, R.I. 1991b, Caffeine as an analgesic adjuvant. A double-blind study comparing aspirin with caffeine to aspirin and placebo in patients with sore throat. *Archives of Internal Medicine*, **151:** 733–737.

Schachtel, B.P., Fillingim, J.M., Thoden, W.R., Lane, A.C. and Baybutt, R.I. 1988, Sore throat pain in the evaluation of mild analgesics. *Clinical Pharmacology and Therapeutics*, **44:** 704–711.

Schachtel, B.P., Thoden, W.R., Konerman, J.P., Brown, A. and Chaing, D.S. 1991a, Headache pain model for assessing and comparing the efficacy of over-the-counter analgesic agents. *Clinical Pharmacology and Therapeutics*, **50:** 322–329.

Sevelius, H. and Colmore, J.P. 1964, Experimental design for evaluation of non-narcotic agents. A comparison of aspirin, buffered aspirin, sustained-release salicylate and placebo. *Journal of New Drugs*, **4:** 337–346.

Seymour, R.A. and Rawlins, M.D. 1982, Efficacy and pharmacokinetics of aspirin in post-operative dental pain. *British Journal of Clinical Pharmacology*, **13:** 807–810.

Seymour, R.A., Rawlins, M.D. and Clothier, A. 1984, The efficacy and pharmacokinetics of sodium salicylate in post-operative dental pain. *British Journal of Clinical Pharmacology*, **17:** 161–163.

Seymour, R.A., Weldon, M., Kelly, P., Nicholson, E. and Hawkesford, J.E. 1992, An evaluation of buffered aspirin and aspirin tablets in postoperative pain after third molar surgery. *British Journal of Clinical Pharmacology*, **33:** 395–399.

Seymour, R.A., Williams, F.M., Luyk, N.M., Boyle, M.A., Nicholson, E., Ward Booth, P. and Rawlins, M.D. 1986 Comparative efficacy of soluble aspirin and aspirin tablets in postoperative dental pain. *European Journal of Clinical Pharmacology*, **30:** 495–498.

Silberstein, S.D., Armellino, J.J., Hoffman, H.D., Battikha, J.P., Hamelsky, S.W., Stewart, W.F. and Lipton, R.B. 1999, Treatment of menstruation-associated migraine with the nonprescription combination of acetaminophen, aspirin, and caffeine: results from three randomized, placebo-controlled studies. *Clinical Therapeutics*, **21:** 475–491.

Simon, F.A. 1931, A clinical comparison of acetylsalicylic acid with and without magnesium oxide. *Journal of Laboratory and Clinical Medicine*, **16:** 1064–1066.

Skaer, T.L. 1993, Management of pain in the cancer patient. *Clinical Therapeutics*, **15:** 638–649.

Skjelbred, P. 1984, The effects of acetylsalicylic acid on swelling, pain and other events after surgery. *British Journal of Clinical Pharmaceutics*, **17:** 379–384.

Spansberg, N.L., Ankerr-Molleer, E., Dahl, J.B., Schultz, P. and Christensen, E.F. 1996, The value of continuous blockade of the lumbar plexus as an adjunct to acetylsalicylic acid for pain relief after surgery for femoral neck fractures. *European Journal of Anaesthesiology*, **13:** 410–412.

Squires, D.J. and Masson, E.L. 1981, A double-blind comparison of ibuprofen, ASA–codeine–caffeine compound and placebo in the treatment of dental surgery pain. *Journal of International Medicine Research*, **9:** 257–260.

Stjernsward, J. 1988, WHO cancer pain relief programme. *Cancer Survival*, **7:** 195–208.

Strong, F.C. III 1998, It may be the caffeine in Extra Strength Excedrin that is effective for migraine. *Journal of Pharmacy and Pharmacology*, **49:** 1260.

Sunshine, A., Laska, E., Meisner, M. and Morgan, S. 1964, Analgesic studies of indomethacin as analysed by computer technique. *Clinical Pharmacology and Therapeutics*, **5:** 699–707.

Sunshine, A., Marrero, I., Olson, N.Z., Laska, E.M. and McCormick, N. 1983, Oral analgesic efficacy of suprofen compared to aspirin, aspirin plus codeine, and placebo in patients with postoperative dental pain. *Pharmacology*, **27 (Suppl. 1):** 31–40.

Tajti, J., Szok, D. and Vecsei, L. 1999, Topical acetylsalicylic acid versus lidocaine for post-herpetic neuralgia: results of a double-blind comparative trial. *Neurobiology*, **7:** 103–108.

Tammisto, T., Tigerstedt, I. and Korttila, K. 1980, Comparison of lysine acetylsalicylate and oxycodone in postoperative pain following abdominal surgery. *Annales Chirurgiae et Gynecologiae (Helsinki)*, **69:** 287–292.

Taniguchi, Y., Deguchi, Y., Saita, M. and Noda, K. 1994, Antinociceptive effects of counter-irritants. *Nippon Yakurigaku Zasshi*, **104:** 433–446.

Tansey, M.J., Pilgrim, A.J. and Lloyd, K. 1993, Sumatriptan in the treatment of migraine. *Journal of Neurological Science*, **114:** 109–116.

Tfelt-Hansen, P., Henry, P., Mulder, L.J., Scheldewaert, R.G., Schoenen, J. and Chazot, G. 1995, The effectiveness of combined oral lysine acetylsalicylate and metoclopramide compared with oral sumatriptan for migraine. *Lancet*, **346:** 923–926.

Tharion, G. and Bhattacharji, S. 1997, Aspirin in chloroform as an effective adjuvant in the management of chronic neurogenic pain. *Archives of Physical Medicine and Rehabilitation*, **78:** 437–439.

Tietjen, G.E. 2000, The relationship of migraine and stroke. *Neuroepidemiology*, **19:** 13–19.

Tigerstedt, I., Leander, P. and Tammisto, T. 1981, Postoperative analgesics for superficial surgery. Comparison of four analgesics. *Acta Anaesthesiologica Scandinavica*, **25:** 543–547.

Uhland, H. 1962, A double blind study of a new analgesic combination. *Northwest Medicine*, **61:** 843–845.

Valentine, G. and Martin, S.J. 1959, Evaluation of a new analgesic agent D-propoxyphene hydrochloride (Darvon®) in preanesthetic and postanesthetic management. *Anesthesia and Analgesia*, **38:** 55–58.

Vandam, L.D. 1972, Analgetic drugs – the mild analgetics. *New England Journal of Medicine*, **286:** 20–23.

Vane, J.R. and Botting, R.M. 1992, Analgesic actions of aspirin. *In: Aspirin and Other Salicylates*, pp. 166–212. London: Chapman & Hall Medical.

Ventafridda, V., De Conno, F., Panerai, A.E., Maresca, V., Monza, G.C. and Ripamonti, C. 1990, Non-steroidal anti-inflammatory drugs as the first step in cancer pain therapy: double-blind, within-patient study comparing nine drugs. *Journal of International Medical Research*, **18:** 21–29.

Von Graffenried, B., Hill, R.C. and Nuesch, E. 1980, Headache as a model for assessing mild analgesic drugs. *Journal of Clinical Pharmacology*, **20:** 131–144.

Von Seggern, R.L. and Adelman, J.U. 1996, Cost considerations in headache treatment. Part 2: Acute migraine treatment. *Headache*, **36:** 493–502.

Walker, J.S. 1995, NSAIDs: an update on their analgesic effects. *Clinical and Experimental Pharmacology and Physiology*, **22:** 855–860.

Wallenstein, S.L. 1975, Analgesic studies of aspirin in cancer patients. *In:* T.L.C. Dale (ed.), *Proceedings of the Aspirin Symposium. Royal College of Surgeons, London, May 30th 1975*, pp. 5–18. London: Royal Society of Medicine.

Windle, M.L., Booker, L.A. and Rayburn, W.F. 1989, Postpartum pain after vaginal delivery. A review of comparative analgesic trials. *Journal of Reproductive Medicine*, **34:** 891–895.

Zelvelder, W.G. 1961, Mededelingen van de afdeling klinisch geneesmiddelenonderzoek tno. De beoordeling van een pijnstillend middel. Verslag van een dubbel-blind onderzoek naar de doeltreffendheid van acetosal, acetylaminofenol en den placebo van pain bij de behandeling van ein bij chronische, in een verpleeghuis verzorgde patienten. *Nederlandisch Tijschrift voor Geneeskunde*, **1005:** 1697–1702.

Zhang, W.Y. and Po, A.L. 1997, Do codeine and caffeine enhance the analgesic effect of aspirin? A systematic overview. *Journal of Clinical Pharmacology and Therapeutics*, **22:** 79–97.

Zhang, W.Y. and Po, A.L.W. 1998, Efficacy of minor analgesics in primary dysmenorrhoea: a systematic review. *British Journal of Obstetrics and Gynaecology*, **105:** 780–789.

Acetylsalicylic Acid for the Prevention and Treatment of Thromboembolic Diseases

K.E. Webert and J.G. Kelton

THE ROLE OF PLATELETS IN HAEMOSTASIS AND THROMBOSIS

Platelets play a pivotal role in haemostasis, interacting with the coagulation cascade on both a macro and a molecular level. Platelets circulate as tiny anuclear cells, which are produced by the megakaryocytes in the bone marrow. Platelets are released from the bone marrow and circulate for 7–10 days before being cleared by the reticuloendothelial system. The normal platelet count ranges from $150–450 \times 10^9/l$, and very low platelet counts are associated with bleeding, with very high platelet counts being associated with thrombosis. Platelets bind to damaged endothelium through a number of receptors. The most important receptor is glycoprotein Ib (GP Ib), which is complexed with glycoprotein V and IX. GP Ib binds to the subendothelium through a large fibrillary protein called von Willebrand factor (vWF) (Figure 11.1). The complex of GP Ib, V and IX is present in high concentrations on the platelet surface at approximately 20 000–40 000 receptors per platelet. Other less abundant platelet receptors also participate in adhesion. In particular, glycoprotein Ia/IIa binds to collagen exposed on the damaged blood vessel wall, glycoprotein IIb/IIIa binds to fibrinogen and glycoprotein Ic/IIa binds to fibronectin.

The binding of the platelets to the vessel wall, in a process called adhesion, results in activation of the platelets, initiation of shape change, and extension of the platelets in a process called pseudopod formation. These energy-dependent processes occur through mobilisation of ionised calcium within the platelet membrane.

Once the platelet has adhered to the vessel wall, the platelet becomes activated. The activation steps are complex, but, in general, most of the previously described receptors can bind internal platelet G proteins that initiate signal transduction. There are two major signal transduction pathways. The first is the activation of an enzyme, phospholipase C, which cleaves phosphatidylinositol-biphosphate and ultimately results in the release of ionised calcium. The other pathway of signal transduction is the prostaglandin one: arachidonic acid is released from membrane phospholipids and is sequentially converted into a number of prostaglandins, the most important being thromboxane A_2. This pathway will be described in greater detail subsequently. Additionally, activation results in the release of platelet-granular contents. Dense granules, named because they appear dense under electron microscopic investigation, release adenosine diphosphate (ADP) and serotonin. Alpha granules release a variety of substances, including clotting factors such as fibrinogen, von Willebrand factor, and factor V; platelet-specific components including platelet factor 4, beta thromboglobulin and platelet-derived growth factor; and, finally, plasma proteins including albumin, immunoglobulins, and high-molecular-weight kininogen (Figure 11.2).

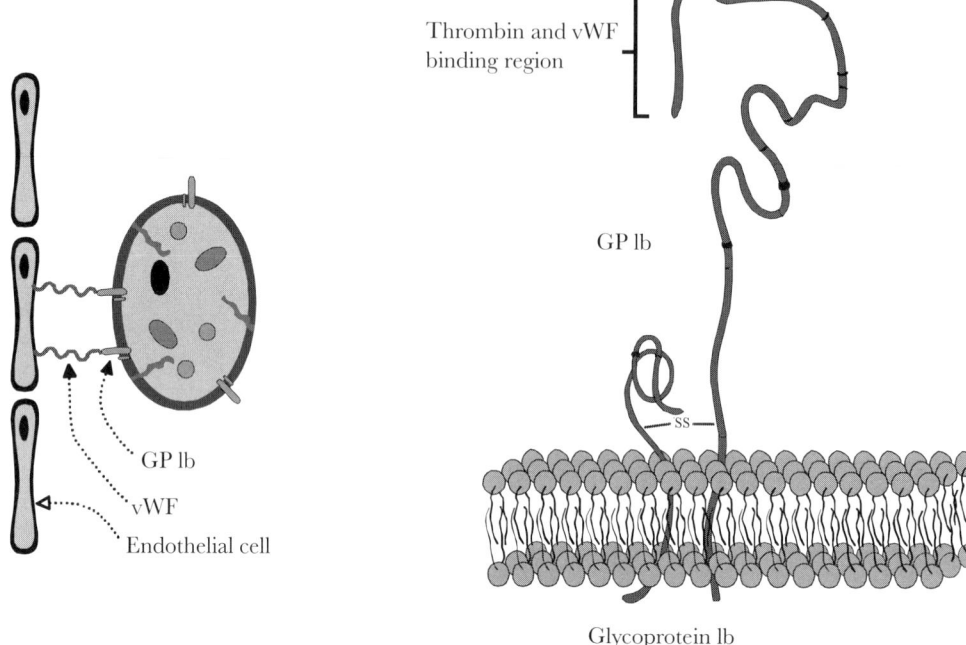

Thrombin and vWF
binding region

GP lb

GP lb

vWF

Endothelial cell

ss

Glycoprotein lb

Figure 11.1 Platelet adhesion. Glycoprotein 1b (GP1b) is a receptor found in high concentrations on the platelet surface. *In vivo*, GP1b binds to the subendothelium through a large fibrillary protein called von Willebrand factor (vWF). This results in platelet adhesion to the subendothelium. GP1b also serves as a binding site for thrombin.

Activation of platelets through adhesion, as well as through the binding of a variety of agonists, including ADP, serotonin and thromboxane A_2, results in a conformational change in a platelet glycoprotein complex termed glycoprotein IIb/IIIa (GP IIb/IIIa). GP IIb/IIIa is the most abundant platelet glycoprotein, with unactivated platelets having approximately 25 000–50 000 copies and activated platelets having an additional 25 000–30 000 copies of this glycoprotein. This receptor recognises the arginine–glycine–aspartate (RGD) sequence of fibrinogen and similar molecules. The cross-linking of platelets by fibrinogen results in platelet aggregation (Figure 11.3). Although the prostaglandin pathway is important for activation of GP IIb/IIIa, prostaglandin-independent pathways can also activate platelets and glycoprotein IIb/IIIa. This explains why aspirin is not entirely effective in preventing platelet aggregation, and why aspirin does not result in serious platelet-mediated bleeding in most individuals.

Platelet function can be down-regulated by the levels of cyclic adenosine monophosphate (cAMP) and cyclic guanosine monophosphate (cGMP). For example, nitric oxide, also known as endothelial relaxation factor, inhibits platelet activation by increasing the levels of cGMP. Prostaglandin (PGI$_2$), produced by the vasculature, also inhibits platelet function by raising platelet cAMP. Consequently, aspirin, through inhibition of the production of PGI$_2$, theoretically could have the net effect of increasing platelet function. This prothrombotic effect of aspirin has been demonstrated in experimental animals but not in humans.

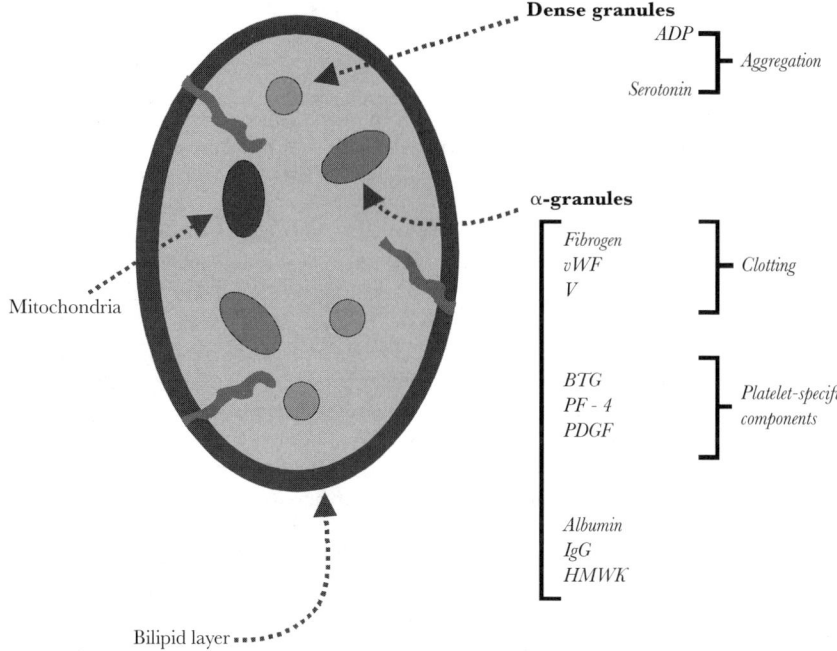

Figure 11.2 Anatomy of a platelet. Platelets contain two types of granules: dense granules and alpha granules. Dense granules contain adenosine diphosphate (ADP) and serotonin. Alpha granules contain various substances including clotting factors, platelet specific components, and plasma proteins. Note that platelets lack a nucleus; therefore, irreversibly activated enzymes are not regenerated during the platelet's seven to ten day lifespan. Abbreviations: von Willebrand's factor, vWF; beta-thromboglobulin, βTG; platelet factor 4, PF4; platelet derived growth factor, PDGF; high molecular weight kininogen, HMWK.

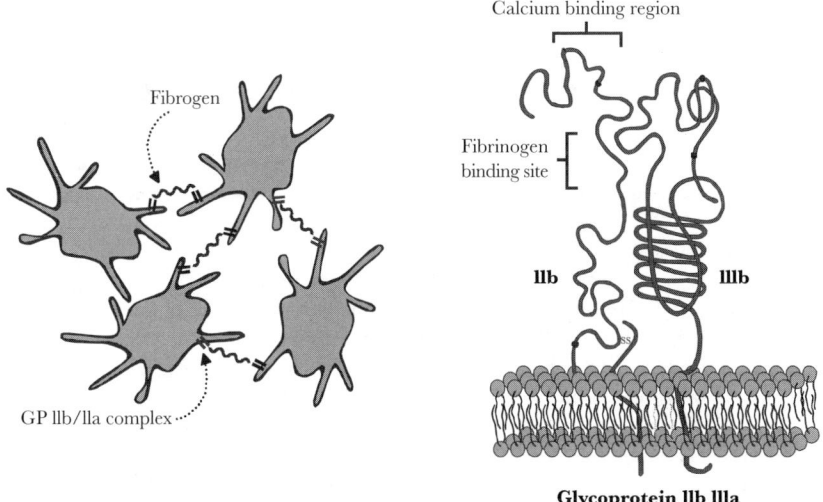

Figure 11.3 Platelet aggregation. Activation of platelets results in a conformational change in platelet GP IIb/IIIa. The receptor binds fibrinogen and similar molecules. A single fibrinogen is able to bind to and bridge two GP IIb/IIIa molecules and it is hypothesised that the cross-linking of platelets by fibrinogen results in platelet aggregation.

THE ROLE OF PLATELETS IN ACUTE AND CHRONIC VASCULAR INJURY

As noted previously, platelets play a pivotal role in haemostasis; consequently, they play a key role in thrombosis. Platelets bind to sites of vascular damage. These sites of damage are consistent and predictable. For example, because of local disturbances and turbulence in blood flow, injury typically occurs in the vessels at areas of bifurcation or branch points. Consequently, platelet thrombi at the site of vessel damage occur in anatomically predictable areas. The degree of vessel damage is related to both intrinsic and extrinsic factors such as the individual's blood pressure, cholesterol levels and the use of cigarettes. As platelets adhere to the damaged vessel wall, they release components from their granules (see Figure 11.2) that attract other platelets. Some of these components include platelet-derived growth factor, which leads to smooth muscle proliferation and vessel repair. With time, repeated damage to the vessel wall induces a chronic inflammatory process. Phagocytic cells, lymphocytes and neutrophils accumulate in the vessel wall, forming an atherosclerotic plaque. The plaque has a distinct pathology, with a fibrous cap formed of connective tissue and a centre rich in free and phagocytic cell-laden lipids. The fast flow rate within the arterial vessel makes the protruding plaque prone to rupture. Damage to the plaque is frequently spontaneous; however, it can also happen with sudden changes in blood flow, as can happen with exertion. The coagulation cascade is triggered by the release of tissue factor from the ruptured endothelial and phagocytic cells. The tissue factor binds to activated factor VII, which circulates in small amounts. Together, these factors initiate the coagulation cascade, and this triggers the formation of additional platelet thrombi. The platelet aggregates, which form on the surface and at the edges of the atherosclerotic plaque, can break off and be carried distally, where they can result in thrombotic events affecting the areas supplied by the distal vessels. For example, plaques within the carotid artery typically lead to platelet thrombi within the cerebral vasculature, especially the middle cerebral arteries. Plaques within the coronary arteries lead to platelet thrombi within the coronary circulation, which is manifested as angina, unstable angina, or myocardial infarction. The simultaneous activation of platelets, as well as the activation of the coagulation cascade, explains why the recent introduction of a combination antiplatelet plus anticoagulant therapy has led to significant therapeutic advances in the treatment of patients with a number of thrombotic vascular disorders.

EFFECTS OF ACETYLSALICYLIC ACID ON PLATELET FUNCTION

Arachidonate is released by the action of phospholipase on the cell membrane phospholipids. The free arachidonic acid serves as the precursor for the biosynthesis of eicosanoids. Free arachidonic acid is metabolised by two different enzyme pathways. The lipoxygenase pathway produces the leukotrienes, lipoxins, and 15-hydroxyeicosatetraenoic acids (Figure 11.4). The cyclo-oxygenase pathway synthesises the thromboxanes and several prostaglandins that have platelet reactivity. Cyclo-oxygenase (COX) or prostaglandin endoperoxide H synthase changes arachidonate to prostaglandin G_2 and further catalyses peroxidation to form prostaglandin H_2. Prostaglandin H_2 serves as the substrate for the synthesis of a variety of prostaglandins, which is tissue specific. These include prostaglandin E_2, prostaglandin D_2, prostaglandin $F_{2\alpha}$, prostaglandin I_2 (prostacyclin), and thromboxane A_2. Some, such as PGI_2, are synthesised in the vessel wall and have platelet inhibitory effects. In contrast, thromboxane A_2 is synthesised in platelets and has a proaggregating effect. Thromboxane A_2 is a vasoconstrictor and also a platelet-aggregating agent.

One of the key and rate-limiting enzymes in the prostaglandin pathway is cyclo-oxygenase, also known as endoperoxide H synthase. Cyclo-oxygenase exists as two isoforms; cyclo-oxygenase-1 (COX-1) and cyclo-oxygenase-2 (COX-2). The isoforms are encoded by different genes. The two enzymes are similar in size (71 kD) and in enzyme kinetics, and have 75 per cent amino acid homology (Patrono, 1994; Williams and DuBois, 1996). Although both enzymes have similar activities, their regulation is different because of differences in the gene promoter sites. The difference in gene regulation

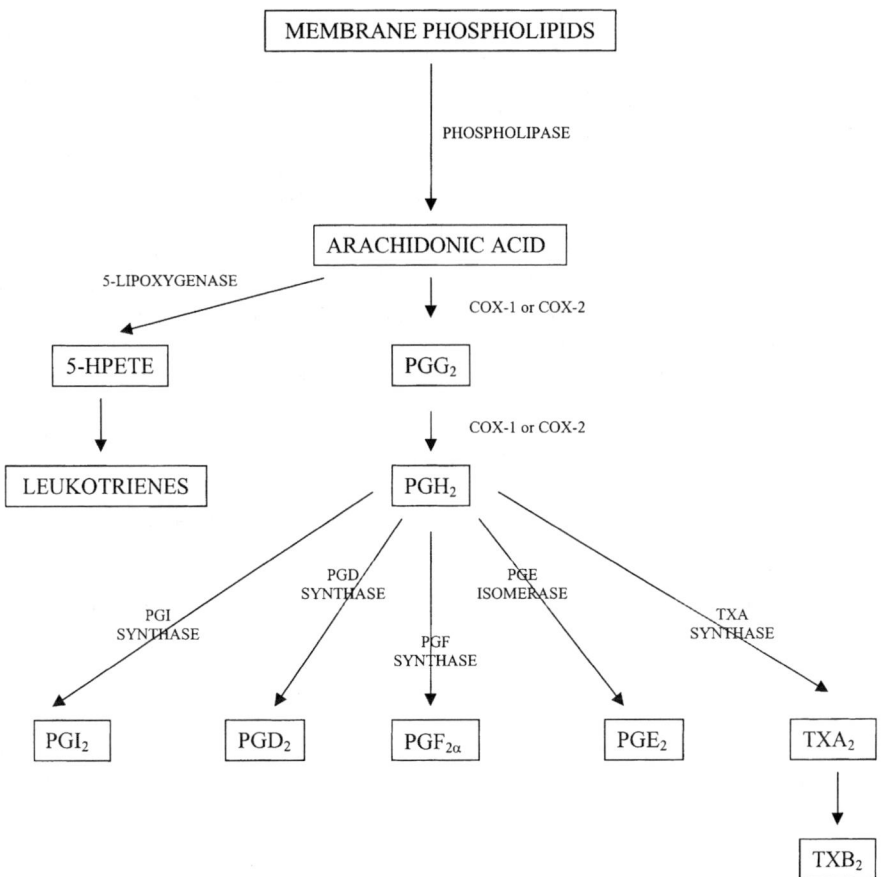

Figure 11.4 Metabolism of arachidonic acid. Arachidonic acid is metabolised by the lipogenase and cyclooxygenase pathways. Metabolism by the lipoxygenase pathway produces leukotriense, lipoxins and 15-hydroxyeicosatetraeonic acids. The cyclooxygenase pathway results in the synthesis of thromboxanes and prostaglandins. Abbreviations: cyclooxygenase, COX; prostaglandin G_2, PGG_2; prostaglandin H_2, PGH_2; prostacyclin synthase, PGI synthase; prostacyclin I_2, PGI_2; prostaglandin F synthase, PGF synthase; prostaglandin $F_{2\alpha}$, $PFG_{2\alpha}$; prostaglandin E isomerase, PGE isomerase; thromboxane A synthase, TXA synthase; thromboxane A_2, TXA_2; thromboxane B_2, TXB_2

may be due to the distinct roles of the COX-1 and COX-2 proteins. COX-1 is constitutively expressed on the endoplasmic reticulum membrane of all cells. High concentrations of COX-1 are found in platelets, vascular endothelial cells, renal collecting tubule cells, and gastric mucosa cells. It is likely that COX-1 has a regulating function by altering local prostaglandin concentrations to control local tissue perfusion, participating in haemostasis and protection of the gastrointestinal tract mucosa (Schrör, 1997; Kaplan-Machlis and Klostermeyer, 1999).

COX-2 is not normally expressed in cells, and it is expressed in response to various stimuli including growth factors, cytokines and bacteria lipopolysaccharide (Patrono, 1994). When induced, the levels of COX-2 increase within hours after a single stimulus (Vane *et al.*, 1998). COX-2 plays a role in defence functions of the organism (Schrör, 1997), and is involved in inflammation, mitogenesis and specialised signal transduction. Consequently, COX-2 is the mechanism by which cellular prostaglandin levels are elevated in inflammation (Kaplan-Machlis and Klostermeyer, 1999).

The antithrombotic action of aspirin, or acetylsalicylic acid (ASA), is due to inhibition of platelet function by inhibition of thromboxane A_2 synthesis, which is mediated by irreversible acetylation of cyclo-oxygenase. Aspirin inhibits COX-1 approximately 200-fold more potently than it inhibits COX-2 (Schrör, 1997), and this explains the different dosage requirements of aspirin as an anti-thrombotic (COX-1) and an anti-inflammatory drug (COX-2). Much smaller doses of aspirin are used to inhibit platelet function than are required when aspirin is used as an anti-inflammatory agent.

Acetylation of COX-1 renders it completely inactive. However, acetylated COX-2 still has some activity and can convert arachidonate to 15-R-hydroxyeicosatetraenoic acid (Williams and DuBois, 1996). Neither acetylated COX-1 nor COX-2 is capable of producing prostaglandin H_2, which is the precursor of prostaglandin and thromboxane synthesis. The result is the inhibition of production of downstream metabolites, including thromboxane A_2, and the subsequent inhibition of platelet aggregation. This inhibition of COX is especially effective in platelets. Because platelets do not have nuclei, they lack the ability to synthesise damaged platelet proteins such as acetylated COX-1 and consequently aspirin inhibits platelet function for the lifespan of that platelet (7–10 days).

Aspirin does not prevent initial platelet adhesion. Similarly, many agonists that initiate platelet aggregation are not inhibited by aspirin. Thus the inhibition of COX by aspirin only eliminates the thromboxane A_2-dependent mechanism. Overall, the antithrombotic effect of aspirin is relatively modest (Schrör, 1997).

The inhibition of cyclo-oxygenase by aspirin is accomplished by selective acetylation of the cyclo-oxygenase. Aspirin acetylates the hydroxyl group of a serine residue at position 530 in COX-1 and position 516 in COX-2 (Patrono, 1994; Bjorkman, 1998). Acetylation of COX prevents its substrate, arachidonic acid, from accessing the catalytic site of the enzyme. Recent experiments have demonstrated that acetylation at Ser_{530} does not explain the inhibition of the catalytic function of COX-1 by aspirin. When Ser_{530} was replaced with alanine, which does not have an acetylation site, the catalytic activity of COX was not affected; however, aspirin no longer had an inhibitory effect (Shimokawa and Smith, 1992). We now recognise that the key aminoacid required for the COX enzyme function is the tyrosine at position 385. Arachidonic acid must travel through a tunnel formed by the COX enzyme to reach the tyrosine at the catalytic centre. The Ser_{530} is located in a narrow part of the tunnel, and its acetylation sterically inhibits the access of arachidonic acid to the catalytic site (Loll *et al.*, 1995).

PHARMACOKINETICS OF ASA

Aspirin is quickly absorbed from the stomach and small intestine. Peak plasma salicylate levels occur within 1–2 hours after ingestion. Enteric-coated forms of aspirin have a slower absorption, taking up to 3–4 hours for peak plasma levels to be reached. The half-life of aspirin in the plasma is approximately 15 minutes before it is deacetylated into salicylic acid. Salicylic acid does not inhibit COX. The half-life of salicylic acid in the serum is dependent on the dosage. For example, at doses less than 600 mg, the elimination of salicylic acid is by first-order kinetics within a half-life of approximately 5 hours, while at doses greater than 4 g per day, elimination is by zero-order kinetics and the half-life may be 15 hours or longer. Despite the short half-life of aspirin in the serum, the duration of the effects of aspirin on platelets is much longer. This is due to the fact that aspirin (but not salicylate) irreversibly inhibits platelet cyclo-oxygenase (COX-1), and thus the inhibition of platelet aggregation is seen for up to 10 days, the life-span of the platelet. Since approximately 10 per cent of circulating platelets are replaced every 24 hours, 5 to 6 days following the ingestion of aspirin, approximately 50 per cent of the platelets will function normally (Patrono *et al.*, 1998). There is evidence that platelet–platelet 'cross-talk' means that the newly released platelets can recruit and participate in normal function with previously aspirinised platelets.

The clinical pharmacology of aspirin has been investigated *in vivo* by measuring the urinary excretion of thromboxane metabolites and serum levels of thromboxane B_2 (Patrignani *et al.*, 1982). Doses of 100 mg of aspirin almost entirely inhibited the synthesis of thromboxane A_2, and the inhibition occurred before the appearance of aspirin in the systemic circulation (Patrignani *et al.*, 1982). This suggests that the rapid inhibition of COX may be due to the acetylation of the enzymes within the portal circulation.

The inhibition of platelet COX by aspirin is cumulative. For example, the administration of less than 50 mg of aspirin per day will progressively inhibit platelet function, but the effect takes days. Higher doses of ASA have a more rapid onset of action (Patrignani *et al.*, 1982).

DOSE–EFFECT RELATIONSHIPS OF ASA

Clinically, aspirin's antithrombotic activity is mainly due to its ability to inhibit platelet cyclo-oxygenase (COX-1). As noted, maximal inhibition of platelet thromboxane synthesis occurs at low doses of aspirin, typically 100–200 mg per day in an adult. Since the inhibition of COX is a saturable effect, increasing the dose further has very little clinical effect. This feature of maximal effect at low doses with an absence of dose effect has been observed in clinical trials, described subsequently (Patrono *et al.*, 1998).

The minimum effective dose of aspirin required for antithrombotic efficacy continues to be uncertain. Various randomised trials have shown that aspirin has some efficacy at doses of 50–100 mg per day. Doses as low as 10–20 mg per day have been shown to result in a 61 per cent inhibition of serum thromboxane formation. However, it was felt that this inhibition of thromboxane formation was unlikely to be clinically relevant, as it was not associated with a prolongation of bleeding time (Schrör, 1997). In healthy volunteers, doses of ASA as low as 30 mg per day had some antiplatelet activity (Schrör, 1997; Patrono *et al.*, 1998).

The generalisability of the data from experiments determining the optimal antithrombotic dose of aspirin in healthy volunteers to patients with atherosclerotic disease has been questioned. Even in healthy subjects, there are differences in the efficacy of aspirin in different vascular regions. In healthy volunteers it is assumed that the endothelium is functionally intact, and therefore platelet adhesion to the vessel wall should not occur. In contrast, patients with atherosclerotic disease have platelet activation and adhesion to areas of injured endothelium and plaques. Platelets adhering to the subendothelium might transfer prostaglandin precursors to the endothelial cells of the vasculature, which might result in enhanced prostacyclin synthesis by the vascular endothelium (Schrör, 1997).

Several studies have demonstrated variability in the inhibition of platelet function by aspirin among individuals (Schrör, 1997). A possible explanation for the 'aspirin resistance' in some subjects may involve the induction of COX-2. The antiplatelet effects of aspirin at doses less than 300 mg are due to the inhibition of COX-1, with the enzymatic action of COX-2 being essentially unaffected. Therefore, if expression of COX-2 is increased, the antiplatelet action of low-dose aspirin might be less effective (Schrör, 1997). It is possible that COX-2 is induced in the vessel wall of patients with atherosclerotic disease under the influence of cytokines or growth factors. The increased expression of COX-2 may contribute to an overall increase in thromboxane biosynthesis and apparent aspirin resistance, and this might explain why some studies suggest that in patients with cerebrovascular disease, high-dose aspirin may be more effective than lower doses (Schrör, 1997).

Studies have indicated that the mechanism of platelet activation may affect the efficacy of inhibition of platelet aggregation by aspirin (Schrör, 1997). For instance, aspirin does not block the potentiation of proaggregatory factors by catecholamines. Platelet activation by sheer stress does not involve stimulation of platelet-dependent thromboxane formation and therefore is resistant to the effects of aspirin.

CHAPTER 11

THE EFFECTS OF ASA ON COAGULATION FACTORS, OTHER THAN PLATELETS

ASA also has other effects upon the haemostatic system as well as those on platelets. These effects include enhancement of fibrinolysis and inhibition of coagulation factors (Patrono, 1998). Enhanced fibrinolysis is seen *in vivo* with high doses of aspirin, 650 mg twice a day (Bjornsson *et al.*, 1988), and is due to the N-acetylation of lysyl residues of fibrinogen (Patrono *et al.*, 1998).

Aspirin also suppresses the coagulation cascade by at least two mechanisms. First, the salicylate component of aspirin has an anti-vitamin K effect at high doses of more than 1500 mg per day (Quick and Cleasceri, 1960). Second, aspirin at doses greater than 500 mg causes a decrease in the rate of thrombin generation in whole blood. The effects of aspirin on thrombin generation may be due to an interaction with platelet phospholipids (Szczeklik *et al.*, 1996). It has also been suggested that this effect may occur as a consequence of impaired platelet coagulant activity secondary to inhibition of thromboxane-dependent platelet aggregation (Patrono *et al.*, 1998). Inhibition of thrombin formation has been seen with single doses of 500 mg and daily doses of 300 mg.

ASA FOR THE PREVENTION OF VASCULAR THROMBOEMBOLIC DISEASE: EPIDEMIOLOGICAL STUDIES

Vascular thrombosis, including cardiovascular, neurovascular and peripheral vascular disorders, is the leading cause of morbidity and mortality in the western world. Platelets play a pivotal role in the ongoing pathogenesis and acute events in all of these disorders. Aspirin, the most widely used antiplatelet agent, has been evaluated in many thousands of patients in many different methodologically sound clinical trials. In general, aspirin has proved useful (20–25 per cent reduction) in preventing recurrent cardiovascular events. The studies are generally divided into two groups: primary prevention studies investigate the ability of drugs to prevent the first vascular event, and secondary prevention studies are designed to prevent the subsequent occurrence of a vascular event after an initial event. In general, secondary prevention studies represent a high risk population.

Primary prevention of vascular disease

Cardiovascular disease

Two large trials have studied the administration of aspirin to males without a history of myocardial infarction or stroke (Peto *et al.*, 1988; Steering Committee of the Physicians' Health Study Research Group, 1989). The intention was to determine if an antiplatelet agent, aspirin, could prevent cardiovascular endpoints. The Physicians' Health Study was a double-blind, placebo-controlled, randomised trial designed to determine if aspirin administered at a dose of 325 mg every other day reduced mortality from cardiovascular disease. The effect of β-carotene supplementation on the risk of cancer was also evaluated. The study enrolled 22 071 male physicians aged 40 to 84 years of age living in the USA. The principal outcome of the study was cardiovascular death. After 5 years, there was no difference in the rates of cardiovascular death in the aspirin-treated group versus the placebo group (0.23 per cent versus 0.24 per cent per year). All-cause deaths were also not different between the groups. However, there was a significant difference in the rates of myocardial infarction in patients treated with aspirin compared to those treated with placebo (relative risk 0.26 per cent versus 0.44 per cent per year), which represented a 44 per cent reduction in risk. The benefits of aspirin were significant for both fatal and non-fatal myocardial infarction. The combined outcome of non-fatal myocardial infarction, non-fatal stroke and death from cardiovascular causes was also significantly reduced in the group treated with aspirin when compared with placebo (0.56 per cent versus 0.68 per cent per year, relative risk reduction 18 per cent, $P = 0.01$). The effects of aspirin on the risk of myocardial infarction were modified by two risk factors; age and cholesterol level. The reduction in the risk of myocardial infarction associated with aspirin use was seen only in patients older than 49 years. The beneficial effects on aspirin on myocardial infarction were apparent at all levels of serum cholesterol, but were greatest at lower levels. The men treated with aspirin had a slight increase in the frequency of haemorrhagic stroke.

The British Doctors' Study (Peto *et al.*, 1988) was a much smaller study, which could explain its somewhat different results. This study enrolled 5139 British male physicians aged 50 to 78 years who did not have a history of stroke, myocardial infarction or peptic ulcer disease. Patients were randomly assigned to aspirin 500 mg daily or aspirin avoidance. After 6 years of follow-up, there was no significant difference between groups in the combined end point of important cardiovascular events, myocardial infarction, stroke and total cardiovascular mortality. There was a trend toward a reduced incidence of myocardial infarction (0.9 per cent per year versus 0.93 per cent per year, relative risk reduction 3 per cent, $P = $NS). Total mortality was 10 per cent lower in the treated group versus the control group; however, this was not statistically significant. An overview of the Physicians' Health Study and the British Doctors' Study (Hennekens *et al.*, 1989) also found a reduced frequency of acute myocardial infarction in the aspirin-treated group (risk reduction 32 per cent, $P < 0.001$).

In 1994, the Antiplatelet Trialists' Collaboration published a collaborative overview of randomised trials of antiplatelet therapy (Antiplatelet Trialists' Collaboration, 1994). These trials included approximately 28 000 low-risk or primary prevention subjects. Treatment with an average of 5 years of antiplatelet therapy resulted in a significant reduction of 5 per 1000 in non-fatal myocardial infarction.

The Thrombosis Prevention Trial (Medical Research Council's General Practice Research Framework, 1998) also studied men who had not experienced episodes of ischaemic heart disease. However, this trial differed from the trials described previously in that it selected men at increased risk of cardiovascular events. The study included 5499 men aged 45 to 69 years. These men scored in the top 20 per cent of a risk score that included characteristics such as smoking, family history, body mass index, blood pressure, serum cholesterol, plasma fibrinogen, and factor VII activity, and were therefore considered to be higher risk for ischaemic heart disease. Men were assigned to aspirin (75 mg, controlled-release formulation) or placebo, and to warfarin to maintain an international normalised ratio of 1.5 or placebo. The design was multifactorial, and the groups included: warfarin and aspirin; warfarin and placebo aspirin; placebo warfarin and aspirin; and placebo warfarin and placebo aspirin. The aspirin therapy resulted in a significant reduction in ischaemic cardiac events by 20 per cent. Much of this effect was due to a 32 per cent ($P = 0.004$) reduction in non-fatal events.

In summary, because of the very low absolute risk of cardiovascular events in the healthy men enrolled in these studies, the absolute benefit of aspirin therapy is less dramatic than the effect observed in higher risk populations. Typically, aspirin will prevent four important cardiovascular events per 1000 subjects treated for 5 years (Patrono, 1994). The recently published recommendations of the Fifth ACCP Consensus Conference (Cairns *et al.*, 1998) do not recommend the routine use of aspirin for primary prevention of coronary artery disease in low-risk individuals, such as those younger than 50 years of age without a history of a vascular event. However, aspirin at daily doses of 80–325 mg should be considered for individuals greater than 50 years of age who have at least one major risk factor for coronary artery disease and who do not have a contraindication to aspirin.

Cerebrovascular disease

In the Physicians' Health Study (Steering Committee of the Physicians' Health Study Research Group, 1989), 22 071 males aged 40 to 84 years were randomised to placebo or aspirin. There were 119 events of stroke in the aspirin group and 98 events in the placebo group. This represented a relative risk of 1.22, which was not statistically significant. In the subgroup analysis, aspirin use was associated with an increased risk of haemorrhagic stroke that was of borderline statistical significance (relative risk 2.14, $P = 0.06$).

In the British Doctors' Study (Peto *et al.*, 1988), there were significantly fewer confirmed transient ischaemic attacks in the aspirin group (0.16 per cent per year versus 0.28 per cent per year, $P < 0.05$). However, there was no difference in strokes (0.32 per cent per year versus 0.29 per cent per year, NS), but more disabling strokes in the aspirin group (0.19 per cent per year versus 0.07 per cent per year, $P < 0.05$).

As discussed previously, the Antiplatelet Trialists' Collaboration has published a collaborative overview of randomised trials of antiplatelet therapy (Antiplatelet Trialists' Collaboration, 1994). Among these trials were approximately 28 000 low-risk or primary prevention subjects. Treatment with prolonged antiplatelet therapy did not result in any decrease in non-fatal strokes among these low-risk subjects.

In summary, there is little evidence that aspirin treatment will prevent ischaemic neurological events.

Secondary prevention of vascular disease

Cardiovascular disease

Secondary prevention means that a drug is evaluated for its ability to prevent an event after an initial event. In 1988, the Antiplatelet Trialists' Collaboration (Antiplatelet Trialists' Collaboration, 1988) analysed 25 randomised trials of antiplatelet treatment of patients with a history of transient ischaemic attack, occlusive stroke, unstable angina or myocardial infarction. The analysis included approximately 29 000 patients. Overall, aspirin (and other antiplatelet agents) gave a 15 per cent reduction in vascular mortality and a 30 per cent reduction in non-fatal vascular events (stroke or myocardial infarction). Additionally, there were no differences between the effects seen in patients with histories of cerebral or cardiac disease. Therefore, it was concluded that antiplatelet agents can reduce the incidence of serious vascular events by approximately 25 per cent among patients at risk for occlusive vascular disease.

In 1994, the Antiplatelet Trialists' Collaboration published a second collaborative overview of randomised trials of antiplatelet therapy (Antiplatelet Trialists' Collaboration, 1994). The overview included about 20 000 patients with a prior history of myocardial infarction. The mean duration of antiplatelet therapy was 27 months. Aspirin therapy resulted in significant reductions in non-fatal reinfarction (18 per 1000 treated; $P < 0.001$). Additionally, treatment with antiplatelet therapy caused a highly significant reduction of approximately 36 per 1000 in the risk of another vascular event ($P < 0.001$).

Cerebrovascular disease

The Antiplatelet Trialists' Collaboration (Antiplatelet Trialists' Collaboration, 1988) analysis of 25 randomised trials of antiplatelet treatment in patients with a history of transient ischaemic attack, occlusive stroke, unstable angina or myocardial infarction documented a 25 per cent reduction in the incidence of serious vascular events.

The Swedish Aspirin Low-Dose Trial (SALT Collaborators Group, 1991) enrolled patients who had experienced either a transient ischaemic attack or a minor stroke. The study included 1360 patients aged 50 to 79 years who were treated with either aspirin 75 mg daily, or placebo. The median duration of follow-up was 32 months. The aspirin-treated group had an 18 per cent reduction in the risk of stroke or death (relative risk 0.82, $P = 0.02$) and a 17 per cent reduction in the risk of stroke, myocardial infarction or vascular death (relative risk 0.83, $P = 0.03$).

The Dutch TIA trial (Dutch TIA Trial Study Group, 1991) also investigated the effectiveness of aspirin in patients with a mild stroke or a transient ischaemic attack. The 3131 patients were treated with 30 or 283 mg of aspirin daily. In the group assigned to receive 30 mg of aspirin, the frequency of death from vascular causes, non-fatal stroke, or non-fatal myocardial infarction was 14.7 per cent compared to 15.2 per cent in the group treated with 283 mg of aspirin. The daily dose of 30 mg of aspirin was therefore not less effective than 283 mg of aspirin in preventing the combined outcome of death from vascular causes, non-fatal stroke, or non-fatal myocardial infarction (age- and sex-adjusted hazard ratio 0.91, NS). In addition, there were fewer bleeding events in the group taking the lower dose of aspirin.

The efficacy of aspirin, at doses ranging from 50–1500 mg daily, at reducing the risk of recurrent vascular complications in cerebrovascular patients has been studied in at least 10 trials (Patrono, 1994). A collaborative overview of these randomised trials (Antiplatelet Trialists' Collaboration, 1994) involved more than 10 000 patients with a history of stroke or transient ischaemic attack. The patients were treated with antiplatelet therapy for a mean duration of 33 months, and this treatment reduced the risk of another vascular event by 37 per 1000 ($P < 0.001$). There was also a large and significant reduction in the occurrence of non-fatal stroke (20 prevented per 1000; $P < 0.001$). The reduction in vascular deaths was 11 per 1000 ($P < 0.05$). Therefore, the results of the overview support the conclusion that antiplatelet therapy reduces both the vascular and all-cause mortality in patients with a history of stroke or transient ischaemic attack.

The published recommendations from the ACCP Consensus Conference on Antithrombotic Therapy (Albers *et al.*, 1998) suggest that every patient who has experienced an atherothrombotic stroke or transient ischaemic attack and has no contraindication should receive an antiplatelet agent such as aspirin to reduce the risk of recurrent stroke and other vascular events. The optimal dose of aspirin is not known. Although there are theoretical reasons why higher doses of aspirin may be more effective in patients with established atherosclerotic disease, no good evidence exists that proves one dose is more effective than another. Furthermore, few side effects occur at the lower doses. Therefore, the current recommendation is that patients receive aspirin as the initial therapy at a dose of 50–325 mg daily.

ASA FOR THE TREATMENT OF ACUTE VASCULAR THROMBOEMBOLIC EVENTS

Treatment of acute myocardial infarction

The ISIS-2 pilot study (ISIS Pilot Study Investigators, 1987) included 619 patients with suspected acute myocardial infarction. The patients were randomised in a multifactorial design to treatment with streptokinase or placebo, intravenous heparin or no heparin, and aspirin (325 mg alternate days) or placebo. Aspirin therapy resulted in a decrease of early mortality that was statistically significant. A non-significant decrease in re-infarction and stroke was also seen.

The ISIS-2 trial (ISIS Collaborative Group, 1998) studied 17 187 patients with suspected acute myocardial infarction randomised to treatment with intravenous streptokinase and/or oral aspirin, 160 mg daily for 1 month, or neither. Patients allocated to aspirin alone had a significant reduction in vascular mortality (9.4 per cent versus 11.8 per cent, odds reduction 23 per cent, $P < 0.001$). The reduction in mortality associated with aspirin therapy continued during the 15-month follow-up period. Aspirin was also associated with a significant reduction in non-fatal re-infarction (1.0 per cent versus 2.0 per cent) and non-fatal stroke (0.3 per cent versus 0.6 per cent). The administration of aspirin was not associated with a significant increase in cerebral haemorrhage or serious bleeds. The ISIS-2 study demonstrated that short-term aspirin therapy in patients with acute myocardial infarction decreased both mortality and the risk of re-infarction. Furthermore, the combination of streptokinase and aspirin was found to be more beneficial than either agent alone. In all groups of patients studied, the beneficial effects of aspirin were additive to those of streptokinase. The incidence of vascular deaths was 8.0 per cent among patients on both agents compared with 13.2 per cent among those on neither agent (odds reduction 42 per cent, 95 per cent confidence limit 34–50 per cent).

The Antiplatelet Trialists' Collaboration's overview analysis of randomised trials of antiplatelet therapy (Antiplatelet Trialists' Collaboration, 1994) included 145 randomised trials of antiplatelet therapy and 29 randomised comparisons among antiplatelet regimens. Together, over 20 000 patients with suspected or definite acute myocardial infarction were available for analysis. Treatment with antiplatelet therapy produced a reduction of 38 per 1000 in the risk of a vascular event. Most of this reduction was in vascular deaths. There was also a significant reduction in non-fatal re-infarction during the month of treatment, and a significant reduction in non-fatal stroke.

In summary, it is currently recommended that all patients with acute myocardial infarction be treated with aspirin at a dose of 160–325 mg orally as soon as possible. The patients should be continued on a daily dose of aspirin indefinitely (Cairns et al., 1998).

Treatment of unstable angina

The Veterans' Administration Cooperative Study (Lewis et al., 1983) studied 1338 men with a diagnosis of unstable angina. Patients were randomly assigned to daily treatment with 324 mg of aspirin in an effervescent buffered powder or placebo for 12 weeks. Over the follow-up period of 12 weeks, treatment with aspirin significantly decreased the incidence of death or myocardial infarction by 51 per cent. In the treatment group, 5 per cent of patients died or experienced an acute myocardial infarction compared to 10 per cent in the placebo group. Aspirin treatment also resulted in a 55 per cent reduction in all-cause mortality ($P = 0.001$) in fatal or non-fatal myocardial infarction.

The Research Group on Instability in Coronary Artery Disease (RISC) studied men with unstable angina or non-Q-wave myocardial infarction (RISC Group, 1990). The study was a prospective, randomised, double-blind, placebo-controlled, multicentre study conducted in Sweden. A total of 796 men were randomised to either aspirin 75 mg daily or placebo and 5 days of intermittent intravenous heparin or placebo. Initially, it was planned that all patients would be followed for at least 1 year prior to the completion of the study. However, based on the published results of the ISIS-2 study (ISIS Collaborative Group, 1988), the Safety Committee recommended an early ending of the trial. The minimum period of treatment was 3 months. The men treated with aspirin had a significant reduction in the risk of myocardial infarction and a death. After 5 days the risk ratio was 0.43 (95 per cent confidence interval 0.21–0.91), at 1 month it was 0.31 (0.18–0.53), and at 3 months it was 0.36 (0.23–0.57).

In summary, aspirin reduces the incidence of myocardial infarction and death in patients with coronary instability. It is recommended that patients with unstable angina be treated with aspirin 160–325 mg daily, as soon as possible after the diagnosis of unstable angina is made. The aspirin should be continued indefinitely (Cairns et al., 1998).

Treatment of acute ischaemic stroke

The use of aspirin in patients with acute ischaemic stroke had been investigated in two large trials; the Chinese Acute Stroke Trial and the International Stroke Trial (CAST Collaborative Group, 1997; International Stroke Trial Collaborative Group, 1997).

In the International Stroke Trial (IST), patients received up to 14 days of antithrombotic therapy that was started as soon as possible after the stroke. The trial was a multifactorial design, with the patients being assigned aspirin 300 mg daily or aspirin avoidance and unfractionated heparin 5000 units or 12 500 units twice daily. The primary outcomes of the study were death within 14 days and death or dependency at 6 months. The trial enrolled 19 435 patients with acute ischaemic stroke. At 14 days, the patients treated with aspirin had fewer deaths (9.0 versus 9.4 per cent). At 6 months, there was a trend to a smaller percentage of patients in the aspirin-treated group being dead or dependent (62.2 versus 63.5 per cent, $P = 0.07$). Patients treated with aspirin also had fewer recurrent ischaemic strokes (2.8 versus 3.9 per cent, $P < 0.001$). There was no excess of haemorrhagic strokes in the patients treated with aspirin. The patients treated with aspirin had a significant reduction in the combined outcome of death or non-fatal recurrent stroke (11.3 versus 12.4 per cent, $P = 0.02$).

The Chinese Acute Stroke Trial (CAST) studied 21 106 patients with acute ischaemic stroke randomised to aspirin 160 mg per day or placebo within 48 hours of onset of ischaemic stroke. The therapy was continued in hospital for up to 4 weeks. There was a reduction in mortality during the treatment period in the patient group treated with aspirin (3.3 versus 3.9 per cent, $P = 0.04$). Additionally, there were fewer recurrent ischaemic strokes in the group treated with aspirin (1.6 versus 2.1 per cent, $P = 0.01$). However, the patients treated with aspirin had a trend towards increased incidence of haemorrhagic stroke (1.1 per cent versus 0.9 per cent, NS). Aspirin caused a 12 per cent proportional risk reduction for the combined in-hospital endpoint of death or non-fatal stroke at 4 weeks (5.3 per cent versus 5.9 per cent, $P = 0.03$).

When the results of the CAST and IST are combined, they show that aspirin therapy started early in hospital prevents death and recurrent ischaemic stroke. Treatment with aspirin results in 9 to 10 fewer deaths or non-fatal recurrences of stroke per 1000 in the first few weeks after the acute stroke ($P = 0.001$) and prevents approximately 13 deaths or dependent patients after several weeks or months of follow-up (CAST Collaborative Group, 1997; International Stroke Trial Collaborative Group, 1997).

Recommendations from the American College of Chest Physicians' (ACCP) Conference on Anthithrombotic Therapy in 1998 indicate that all patients with ischaemic stroke who are not receiving tPA, intravenous heparin or oral anticoagulation should be administered aspirin 160–325 mg daily, within 48 hours of stroke onset (Albers *et al.*, 1998).

NOVEL ANALOGUES OF SALICYLATES AND FUTURE DEVELOPMENTS

Based on the current understanding of the different functions of COX-1 and COX-2, the development of novel non-steroidal anti-inflammatory medications has been possible. Theoretically, a more effective and safer medication would be selective for COX-2. This would inhibit inflammatory processes and could avoid many of the side effects of non-steroidal anti-inflammatories such as gastric mucosal irritation and bleeding. Selective COX-2 inhibitors have been developed and are being used to treat patients with arthritis. Short-term studies with durations up to 7 days using therapeutic and 'supertherapeutic' doses of celecoxib and rofecoxib, two COX-2 inhibitors approved for use in the USA, have shown no effect on platelet activity or bleeding time (Kaplan-Machlis and Klostermeyer, 1999; Schachna and Ryan, 1999). However, recent studies have suggested that the expression of COX-2 by inflammatory and vascular cells in atherosclerotic arteries may be important in the pathogenesis of atherosclerosis (Stemme *et al.*, 2000). Therefore, the selective COX-2 inhibitors are not currently thought to have a role in the prevention and treatment of thromboembolic conditions because of the lack of an antiplatelet effect. However, future studies may demonstrate their anti-inflammatory effect to be of benefit in certain thromboembolic conditions.

REFERENCES

Albers, G.W., Easton, J.D., Sacco, R.I. *et al.* 1998, Antithrombotic and thrombolytic therapy for ischemic stroke. *Chest*, **114**: 683S–698S.

Antiplatelet Trialists' Collaboration 1988, Anti-platelet Trialists Collaboration: secondary prevention of vascular disease by prolonged anti-platelet therapy. *British Medical Journal*, **296**: 320–331.

Antiplatelet Trialists' Collaboration 1994, Collaborative overview of randomised trials of antiplatelet therapy – I: prevention of death, myocardial infarction, and stroke by prolonged antiplatelet therapy in various categories of patients. *British Medical Journal*, **308:** 81–106.

Bjorkman, D.J. 1998, The effect of aspirin and nonsteroidal anti-inflammatory drugs on prostaglandins. *American Journal of Medicine*, **105 (1B):** 8S–12S.

Bjorsson, T.D., Schneider, D.E. and Berger, H. 1988, Aspirin acetylates fibrinogen and enhances fibrinolysis: fibrinolytic effect is independent of changes in plasminogen activator levels. *Journal of Pharmacology and Experimental Therapy*, **250:** 154–160.

Cairns, J.A., Théroux, P., Lewis, H.D. *et al.* 1998, Antithrombotic agents in coronary artery disease. *Chest*, **114:** 611S–633S.

Chinese Acute Stroke Trial (CAST) Collaborative Group 1997, CAST: A randomized placebo-controlled trial of early aspirin use in 20 000 patients with acute ischemic stroke. *Lancet*, **349:** 1641–1649.

Dutch TIA Trial Study Group 1991, A comparison of two doses of aspirin (30 mg vs. 283 mg a day) in patients after a transient ischemic attack or minor ischemic stroke. *New England Journal of Medicine*, **325:** 1261–1266.

Hennekens, C.H., Buring, J.E., Sandercock, P. *et al.* 1989, Aspirin and other antiplatelet agents in the secondary and primary prevention of cardiovascular disease. *Circulation*, **80:** 749–756.

International Stroke Trial Collaborative Group 1997, The International Stroke Trial (IST): a randomized trial of aspirin, subcutaneous heparin, both, or neither among 19 435 patients with acute ischemic stroke. *Lancet*, **349:** 1569–1581.

ISIS (International Studies of Infarct Survival) Pilot Study Investigators 1987, Randomised factorial trial of high-dose intravenous streptokinase, of oral aspirin and of intravenous heparin in acute myocardial infarction. *European Heart Journal*, **8:** 634–642.

ISIS-2 Collaborative Group 1988, Randomised trial of intravenous streptokinase, oral aspirin, both, or neither among 17 187 cases of suspected acute myocardial infarction: ISIS-2. *Lancet*, **ii:** 349–360.

Kaplan-Machlis, B. and Klostermeyer, B.S. 1999, The cyclooxygenase-2 inhibitors: safety and effectiveness. *Annals of Pharmacotherapy*, **33:** 979–988.

Lewis, H.D. Jr., Davis, J.W., Archibald, D.G. *et al.* 1983, Protective effects of aspirin against acute myocardial infarction and death in men with unstable angina: results of a Veterans' Administration Cooperative Study. *New England Journal of Medicine*, **309:** 396–403.

Loll, P.J., Picot, D. and Garavito, R.M. 1995, The structural basis of aspirin activity inferred from the crystal structure of inactivated prostaglandin H2-synthase. *Nature (Structural Biology)*, **2:** 637–643.

Medical Research Council's General Practice Research Framework 1998, Thrombosis Prevention Trial: randomized trial of low intensity oral anticoagulation with warfarin and low-dose aspirin in the primary prevention of ischemic heart disease in men at increased risk. *Lancet*, **351:** 233–241.

Patrignani, P., Filabozzi, P. and Patrono, C. 1982, Selective cumulative inhibition of platelet thromboxane production by low-dose aspirin in healthy subjects. *Journal of Clinical Investigation*, **69:** 1366–1372.

Patrono, C. 1994, Aspirin as an antiplatelet drug. *New England Journal of Medicine*, **330 (18):** 1287–1294.

Patrono, C., Coller, B., Dalan, J. *et al.* 1998, Platelet-active drugs. The relationships among dose, effectiveness, and side effects. *Chest*, **114:** 470S–488S.

Peto, R., Gray, R., Collins, R., *et al.* 1988, Randomised trial of prophylactic daily aspirin in British male doctors. *British Medical Journal*, **296:** 313–316.

Quick, A.J. and Cleasceri, L. 1960, Influence of acetylsalicylic acid and salicylamide on the coagulation of blood. *Journal of Pharmacology and Experimental Therapeutics*, **128:** 95–99.

RISC Group 1990, Risk of myocardial infarction and death during treatment with low dose aspirin and intravenous heparin in men with unstable coronary artery disease. *Lancet*, **336:** 827–830.

SALT Collaborators Group 1991, Swedish Aspirin Low-Dose Trial of 75 mg aspirin as secondary prophylaxis after cerebrovascular ischemic events. *Lancet*, **338:** 1345–1349.

Schachna, L. and Ryan, P.F.J. 1999, COX-2 inhibitors: the next generation of nonsteroidal anti-inflammatory drugs. *Medical Journal of Australia*, **171:** 175–176.

Schrör, K. 1997, Aspirin and platelets: the antiplatelet action of aspirin and its role in thrombosis treatment and prophylaxis. *Seminars in Thrombosis and Hemostasis*, **23 (4):** 349–356.

Shimokawa, T. and Smith, W.L. 1992, Prostaglandin endoperoxide synthase: the aspirin acetylation region. *Journal of Biology and Chemistry*, **267:** 12387–12392.

Smith, W.L. and DeWitt, D.L. 1995, Biochemistry of prostaglandin endoperoxide H synthase-1 and synthase-2 and their differential susceptibility to nonsteroidal anti-inflammatory drugs. *Seminars in Nephrology*, **15:** 179–194.

Steering Committee of the Physicians' Health Study Research Group 1989, Final report on the aspirin component of the ongoing Physicians' Health Study. *New England Journal of Medicine*, **321 (3):** 129–135.

Stemme, V., Swedenborg, J., Claesson, H., *et al.* 2000, Expression of cyclo-oxygenase-2 in human atherosclerotic carotid arteries, *European Journal of Vascular and Endovascular Surgery*, **20 (2):** 146–152.

Szczeklik, A., Musial, J., Undas, A. *et al.* 1996, Inhibition of thrombin generation by aspirin is blunted in hypercholesterolemia. *Atherosclerosis, Thrombosis and Vascular Biology*, **16:** 948–954.

Vane, J.R., Bakhle, Y.S. and Botting, R.M. 1998, Cyclooxygenases 1 and 2. *Annual Review of Pharmacology and Toxicology*, **38:** 97–120.

Williams, C.S. and DuBois, R.N. 1996, Prostaglandin endoperoxide synthase: why two isoforms?. *American Journal of Physiology*, **270:** G393–G400.

Use of Salicylates in Rheumatic and Related Conditions

W.W. Buchanan, W.F. Kean and K.D. Rainsford

INTRODUCTION

The salicylates, in whatever form – plant extracts or concoctions, tablets or other oral formulations, or topical preparations – have probably been extensively used throughout time for the treatment of pain, joint inflammation and fever in rheumatic and other painful conditions. Yet aside from the observations of the Reverend Edward Stone (1763) and the clinical observations in the nineteenth century (see Chapter 1), there were no formal clinical trials on the use of salicylates in treating rheumatic diseases until the introduction of corticosteroids in the late 1940s and 1950s, when studies were performed to compare the efficacy of these drugs to what was apparently a 'recognised standard', aspirin (Copeman, 1964; Roth, 1988a; Rainsford, 1999). Indeed it was not until the pioneering double-blind trials of aspirin compared with cortisone carried out by the Empire Rheumatism Council (1955; 1957) and the joint Medical Research Council/Nuffield Foundation (1954; 1955; 1957a; 1957b; 1959) and the Medical Research Council (1954; 1957) was there any real basis to the efficacy of aspirin in the treatment of rheumatic disease or the use of aspirin as a standard in clinical trials of these conditions. For the best part of three decades after these initial studies, aspirin was recognised as a standard for comparison in most clinical trials of the newer NSAIDs and other antirheumatic agents. These earlier studies of the effects of aspirin in rheumatic fever contrast to the paucity of reports of clinical trials with aspirin in osteoarthritis until the 1960s, when again it became a basis for comparison in trials with newer NSAIDs (Von Rechenberg, 1961; Lewis and Furst, 1987; 1994; Kean *et al.*, 1999). It is only since the substantial numbers of clinical trials with ibuprofen during the 1980s and 1990s that this drug has supplanted aspirin as the accepted standard for comparison (Kean *et al.*, 1999; Rainsford, 1999a).

The salicylates' pre-eminence in the treatment of rheumatic diseases for almost 100 years is reflected in the quotation by Duthie in 1963:

> Like everyone else, we have been co-operating with the drug industry for a long time, trying to find a substitute for aspirin. I would not care to say how many promising drugs we have screened, that the industry had already assayed using the standard tests for anti-inflammatory activity. In clinical trials none of them have come up to the level of effectiveness of aspirin.

In overall efficacy in the relief of rheumatic symptoms only corticosteroids were of greater value until the past 40 years, when an increasing number of alternative non-steroidal anti-inflammatory drugs has emerged to challenge this position.

The original doses and dose frequency of salicylates were determined from trial and error applications. No properly controlled therapeutic trials of aspirin and the salicylates in the treatment of musculoskeletal disease were performed until the late 1960s, when new NSAIDs were introduced. The primary interest of most investigators was the therapeutic efficacy and toxicity of the newer NSAIDs for the treatment of rheumatoid disease, osteoarthritis and other musculoskeletal problems. Today, salicylates are extensively used in the treatment of a wide range of clinical conditions such as musculoskeletal disease, mechanical pain or injury, thromboembolic and cardiovascular disease, and as possible prevention agents in some forms of cancer. There is a wide range of formulations including tablets, liquids, injectables, ointments, and numerous mixed or compound products. Like most pharmaceutical agents aspirin and the salicylates have a toxicity profile, such that the therapeutic benefits must be weighed against the risk of adverse outcome.

HISTORY

While the formal history of aspirin and the salicylates has been previously discussed in Chapters 1 and 2, there are a few points germane to their clinical therapeutic use that may be useful here. Although it is frequently stated in many books and papers that various plant and tree extracts containing salicylates have been used since ancient times, there are some notable exceptions (such as the 1684 *London Practice of Physick* by T. Willis (1621–1675; Willis, 1992) and the 1659 *School of Physick* by Nicholas Culpeper (1616–1654; Culpeper, 1993). There is an overall opinion, not wisdom, about these ancient remedies. Most of the extracts used in the past had insufficient salicylate to have any effect, and there were no critical observations and no experiments – i.e. no evidence of efficacy (Manjo, 1975).

It was a clergyman, not a doctor, who must take credit for the first clinical trial of salicylates. The Reverend Edward Stone (1732–1770), of Chipping Norton, Oxfordshire, performed clinical trials of extracts of willow bark in 50 patients. Each received 20–60 g of the powdered bark orally, and the results in terms of reducing fever were uniformly successful. The Reverend Stone's observations were made over a period of 6 years (Stone, 1763). The outcome of his report to the Philosophical Society of London occasioned a burst of clinical studies corroborating his findings and, more importantly, the beginning of efforts by chemists to identify the active ingredient. Stone's studies were observations, but were nevertheless sound and have stood the severest test of all, time. Recent studies showing the therapeutic effectiveness of standardised preparations of willow bark extract (Assalix™, Bionorica AG, Neumarkt/Opf., Germany) in the treatment of osteo- and rheumatoid arthritis and in low back pain (Chrubasik and Roufogalis, 2000; Chrubasik *et al.*, 2000; Schmid *et al.*, 2001) have produced clear proof of the antirheumatic efficacy of this natural salicylate therapy. It was Thomas J. Maclagan (1863–1903), of Dundee, Scotland, who first observed in his studies on salicin in acute rheumatic fever that the greater the patient's pain, the greater was the relief (Maclagan, 1876a; 1876b). NSAIDs work so effectively in acute gouty arthritis and less so in patients with mild pain with osteoarthritis. In contrast, most patients with rheumatoid arthritis have moderate to severe pain. The response to analgesics may, therefore, depend on the degree of pain; the greater the intensity, the greater the response to analgesics. Nearly a century later, Lee and colleagues confirmed that the response to aspirin was greater in patients with more severe rheumatoid arthritis (Lee *et al.*, 1973).

Unlike the present day practice of extensive pharmacological studies in experimental animals, few studies were performed with aspirin and other salicylates, and even those that were done were of doubtful relevance. The studies by Professor Doctor Heinrich Dreser (1860–1929), Director of the Bayer Company at Elberfeld, Germany who oversaw the discovery of acetylsalicylic acid, compared its effects with those of salicylic acid on the tails of fish (Dreser, 1907) – hardly a relevant model for studying gastric irritancy! Only acetylsalicylic acid did not cause opacification, so it was presumed that the acetyl derivative was not irritant, in contrast to salicylic acid, which did cause opacification. Buss (1875) had previously reported the effects of salicylate on the gastric mucosa in experimental animals, which showed evidence of mucosal damage. In 1877, Bälz reported that salicylates produced nausea, vomiting and intestinal bleeding. Gazert (1900) claimed from clinical observations that gastrointestinal disturbances were less frequent with acetylsalicylic acid than with salicylic acid, and this was subsequently confirmed by Görges (1902). Several studies thereafter showed that aspirin was irritant to the gastric mucosa of laboratory animals (Gross and Greenberg, 1948). However, it was not until the late 1930s in England that the damaging effects of aspirin on the gastric mucosa were observed (Douthwaite and Lintott, 1938), although dyspepsia had been well documented by several early workers (Wohlgemuth, 1899). Finlay and Lucus (1879) came to the conclusion that sodium salicylate had no effect on the heart lesions of acute rheumatic fever, which promoted Sir William Osler (1849–1919) to conclude that 'the salicylates are useless' in his landmark textbook, *The Principles and Practice of Medicine* (Osler, 1892). It was Wohlgemuth (1899) who also stated 'dass die Heikraft des Aspirin natürlich dieselbe ist, wie die der Salicylsäure selbst' (that the healing power of aspirin is naturally the same as salicylic acid itself), a fact we now know today. The acetyl moiety confers no advantage in terms of either analgesia or anti-inflammatory action, at least as tested in rheumatoid arthritis (Preston *et al.*, 1989).

RHEUMATOID ARTHRITIS

It may be thought that aspirin and other salicylate preparations would have quickly proved to be the physician's Excalibur in the treatment of rheumatoid arthritis. This, however, was not to be the case. Maclagan (1876a; 1876b) was not particularly impressed with salicin in chronic rheumatics, although he admitted that it sometimes did 'good'. Likewise, Stricker (1876) considered the efficacy of salicylic acid in chronic joint rheumatism doubtful. These early workers had previously used salicylates in acute rheumatic fever, where the effects were more dramatic. In his textbook *The Principles and Practice of Medicine*, Sir William Osler (1849–1919) recommended high doses of salicylates for acute rheumatic fever but did not even mention the use of salicylates in rheumatoid arthritis. As Goodwin and Goodwin (1981) noted, even as late as 1941 the *Rheumatism Reviews*, written by the leaders in American rheumatology, failed to mention the use of salicylates in the treatment of rheumatoid arthritis (Hench *et al.*, 1949). In the mid-1930s, Cecil sought the opinion of 16 of his American colleagues interested in rheumatoid arthritis. Salicylates did not rank in the first three treatments by any of these 16 arthritis experts (Cecil, 1934), and Cecil concluded that 'All agreed that salicylates were of value only for their analgesic effect'. However, by the late 1940s aspirin in large doses, taken indefinitely, had become the treatment of choice (Hench *et al.*, 1949).

Goodwin and Goodwin (1981) have argued that the final acceptance of salicylates for the treatment of rheumatoid arthritis was stimulated by the 'wonder drug' cortisone, introduced by the late Nobel Prize winner Philip S. Hench (1896–1965) in 1949 (Hench *et al.*, 1949). Cortisone was quickly proved to be associated with numerous adverse reactions, and in clinical therapeutic trials proved no better than high-dose salicylate therapy (Medical Research Council and Nuffield Joint Committee 1954; 1955; 1957a; 1957b; 1959). Thus it was that aspirin and other salicylates were finally established in the treatment of rheumatoid arthritis, remained the dominant therapy throughout the 1970s and 1980s, and were the principal agents used as standard comparators to new NSAIDs. Surprisingly, it was not until 1965 that an anti-inflammatory action of aspirin was demonstrated in rheumatoid arthritis by the reduction of joint swelling and associated relief of pain with doses of aspirin of between 3.6 and 7.5 g daily (Fremont-Smith and Boyles, 1965). The study was not blinded but compliance was probably high, as the study was conducted in hospital. A significant finding was that when aspirin was discontinued and replaced by pure analgesic agents, swelling of joints recurred within 72 hours. This report was followed by one by Mainland and Sutcliffe (1965), who reported a 7-day double-blind placebo-controlled trial in 492 patients with rheumatoid arthritis, in which a statistically significant reduction in the number of swollen joints, grip strength and walking time, but not in the duration of morning stiffness, occurred with daily doses of 3 to 9 g aspirin compared to placebo. Boardman and Hart (1967)

CHAPTER 12

TABLE 12.1

Mean differences in clinical and laboratory parameters between placebo and aspirin, indomethacin, ibuprofen and prednisone in patients with rheumatoid arthritis.

Clinical parameter	Aspirin	Indomethacin	Ibuprofen	Prednisolone
Pain index	−6	−10	−6	−7
Articular index	−6	−10	−6	−7
Grip strength mmHg				
Right	+12.3	+14.5	+17.5	+16.5
Left	+10.6	+11.1	+12.7	+11.3
Digital joint circumference (mm)				
Right	−1	−4	0	−4
Left	−3	−3	−3	−1
^{99mc}Tc knew joint uptake ($\% \times 10^{-7}$)				
Right	−9.3	−12.0	−8.3	−11.7
Left	−7.5	−10.9	−10.9	−12.1

Based on data in Buchanan *et al.* (1985). N = 37 patients $P < 0.05$.

were next to describe a double-blind study of prednisone 7.5 mg versus placebo, paracetamol (acetaminophen) 6 g versus placebo, and high (3.9 g) and low (2.6 g) daily doses of salicylate versus placebo in 15 patients with rheumatoid arthritis. Each treatment period lasted a week, with a 14-day period free of drugs. Anti-inflammatory effects were determined by measurement of the circumference of the proximal interphalangeal joints, and only prednisone (see Table 12.1) and the high dose of salicylate were shown to reduce joint size. This has led to the belief that 3 g or more daily of salicylate is required for an anti-inflammatory effect. However, the method of measurement is relatively insensitive, and so an anti-inflammatory effect with doses less than 3 g daily may occur but not be demonstrated with the ring-size measurement. Multz *et al.* (1974) also demonstrated that 3.6 g daily of salicylate had an anti-inflammatory effect compared to placebo, and that only 3 days of treatment were required to demonstrate this effect. Those who organise clinical therapeutic trials of non-steroidal anti-inflammatory analgesics in rheumatoid arthritis might take note of the fact that pain and joint inflammation responses are evident with 1 week of therapy, and do not improve over subsequent weeks (Rosenbloom *et al.*, 1985).

Ritchie *et al.* (1968) and Calabro and Paulus (1970) confirmed the superiority of aspirin (5.0 g daily) over placebo, as did Mainland and Sutcliffe (1965) with 3.9 g per day. The superiority of sodium salicylate, 4.0 g daily, over placebo was shown by Deodhar *et al.* (1973), while other authors proved the same with other salicylate preparations, including Aloxiprin (the aluminium formulation of aspirin; Huskisson and Hart, 1972), benorylate (Bain and Burt, 1970; Sperryn *et al.*, 1972; Maneshka, 1973; Sasisikhar *et al.*, 1973; Champion and Graham, 1978; Beales *et al.*, 1982), diflunisal (Palmer *et al.*, 1981), and salicylamide (Calabro and Paulus, 1970). Innumerable studies have shown that aspirin and other non-acetylated salicylates, prescribed in therapeutic doses, are equivalent in both analgesic and anti-inflammatory effects to other established non-steroidal anti-inflammatory analgesics, including ibuprofen (Jasani *et al.*, 1968; Chalmers, 1969; Dick-Smith, 1969; Brooks *et al.*, 1970; 1973; Dick *et al.*, 1970a; 1970b; Huskisson *et al.*, 1970; Sasaki, 1970; Phillips and Rogerson, 1972; Deodhar *et al.*, 1973; Hingorani, 1973; Dornan and Reynolds, 1974; Blechman *et al.*, 1975; Darlington and Coomes, 1975; Mena *et al.*, 1977; Arendt-Racine *et al.*, 1978; Blechman and Lechner, 1979; Cardoe and Fowler, 1979; Australasia Multicentre Trial Group, 1980; Meinicke and Danneskiold-Samsøe, 1980; Palmer *et al.*, 1981); indomethacin (Pinals and Frank, 1967; Franke and Manz, 1972; Huskisson and Hart, 1972; Deodhar *et al.*, 1973; 1977; Haslock *et al.*, 1975; Mavrikakis *et al.*, 1977; Bijlsma, 1978; Mutru *et al.*, 1978; Barnes *et al.*, 1979; McMahon and Cash, 1979); azapropazone (Grennan *et al.*, 1974); phenylbutazone (Haslock *et al.*, 1971; Stephens *et al.*, 1979); indoprofen (Loizzi *et al.*, 1980); sublindac (Bower *et al.*, 1979); ketoprofen (Lussier *et al.*, 1976; Capell *et al.*, 1979); naproxen (Diamond *et al.*, 1973; 1975;

Bowers *et al.*, 1975; Hill *et al.*, 1975; Alexander *et al.*, 1978; Arendt-Racine *et al.*, 1978; Kajandev and Martio, 1978; Melton *et al.*, 1978; Capell *et al.*, 1979; Australasia Multicentre Trial Group, 1980; Gall *et al.*, 1982); flurbiprofen (Barraclough *et al.*, 1974; Stephens *et al.*, 1979. Dequeker and Mardjuardi, 1981); piroxicam (Ward *et al.*, 1978; Balogh *et al.*, 1979; Wilkens *et al.*, 1982); benoxaprofen (Gum, 1980); diclofenac (Soloman and Abrams, 1974; Huskisson *et al.*, 1982); fenoprofen (Gum, 1976; Signler *et al.*, 1976); pirprofen (Davis *et al.*, 1979; Singleton and Wild, 1980; Roth *et al.*, 1981); tiaprofenic acid (Reynolds *et al.*, 1977; Borrachero del Campo, 1978; Caruso *et al.*, 1978; Ciocci, 1978; Huskisson and Scott, 1978; Stephens *et al.*, 1979; Australasia Multicentre Trial Group, 1980); fenclofenac (Hill and Hill, 1977); oxaprozin (Hubsher *et al.*, 1983); tolmetin (Schneyer, 1975); the fenamates (Barnado *et al.*, 1966; Multz *et al.*, 1978; Cardoe and Fowler, 1979; Gowans, 1981) and other analgesics (Lee *et al.*, 1974b; 1975; Hardin and Kirk, 1979). These studies confirm the general phenomenon that can be described by the equation:

$$\text{Aspirin or salicylate} = \text{NSAID-1} = \text{NSAID-2} = \ldots \text{NSAIDn}$$

in terms of both analgesia and anti-inflammatory effects.

Confirmation of the basis of the equation was elegantly established by Lee and colleagues (1976) in patients with rheumatoid arthritis treated with aspirin and 13 other non-steroidal anti-inflammatory analgesics for 2 weeks. Moreover, this equivalence in therapeutic response of aspirin and other NSAIDs has been shown in sports and other acute injuries (Muckle, 1974a; 1974b; 1980; Krishnan, 1977; Anderson and Gotzsche, 1984; Aghabubian, 1986; Brown *et al.*, 1986; Indelicato, 1986; Stull and Joki, 1986; Jenner, 1987; Duncan, 1988; Duncan and Farr, 1988; Sloan *et al.*, 1989; Weiler, 1992; Leadbetter, 1995).

Many of the therapeutic trials of salicylates in rheumatoid arthritis have been performed in the short term. Indeed, as mentioned above, this is all that is required to confirm analgesic and anti-inflammatory effect and equivalence with other non-steroidal anti-inflammatory/analgesics. Although these studies confirm the *pharmacological* action of aspirin and other salicylates, they allow no assessment of the *therapeutic* value of the drug with continuous use. To this end, Capell and her colleagues (1979) reported their experience with enteric-coated sodium salicylate, 3.9 g daily, in 95 patients with rheumatoid arthritis over a 12-month period. Eighty-four per cent of the patients failed to complete 1 year of treatment, nearly 50 per cent as a result of adverse effects, mostly due to dyspepsia. In comparison to other non-steroidal anti-inflammatory analgesics, the enteric-coated sodium salicylate group had the highest number of drop-outs, only exceeded by those patients prescribed placebo.

Other long-term studies of aspirin over 6 to 12 months have confirmed the higher drop-out rate with aspirin compared to other non-steroidal anti-inflammatory analgesics, again largely due to dyspepsia (Gowans, 1981; Roth *et al.*, 1982; Appelrouth *et al.*, 1987; Bernhard *et al.*, 1987; Mullen, 1987). This type of study gives a much better appreciation of how a non-steroidal anti-inflammatory/analgesic will perform in the 'real world', in comparison to the artificial situation of a double-blind controlled trial performed on 'squeaky-clean' patients. Thus in essence all NSAIDs, including the salicylates, will give equivalent therapeutic effects in terms of pain relief and control of joint swelling when used at their optimal dosages in rheumatic diseases. However, aspirin will cause a greater number of 'drop-outs' in long-term therapy over several months, principally owing to dyspepsia. It is not known if the non-acetylated salicylate salsalate (diplosal), which is notably less likely to cause gastrointestinal bleeding or ulceration, also produces drop-outs from dyspepsia as seen with aspirin.

Aspirin versus non-acetylated salicylates

In the past, aspirin was considered to be a better analgesic (Beaver, 1965; Lim, 1966) and anti-inflammatory agent (Adams and Cobb, 1967) than the non-acetylated salicylic acid. Also, Seed (1965) claimed that aspirin had greater antipyretic potency than sodium salicylate. Recognition of the rapid hydrolysis of aspirin to salicylate led to the view that the major part of the clinical effects of aspirin is due to its metabolite, salicylate (Beaver, 1965). However, there are consequences of this conversion. Aspirin binds to serum albumin, causing permanent alteration (Hawkins *et al.*, 1968; 1969; Pinckard *et al.*, 1968), and acetylates a number of other tissue components, including DNA (Pinckard *et al.*, 1968; 1970; Hawkins *et al.*, 1969; Buchanan *et al.*, 1979a). Aspirin, unlike non-acetylated salicylates, also has an irreversible effect of thromboxane biosynthesis in platelets (Dromgoole *et al.*, 1982; Burris *et al.*, 1987; Danesh *et al.*, 1989) as a result of cyclo-oxygenase (COX-1) inhibition (Ferreira *et al.*, 1971; Vane, 1971; 1987; Estes and Kaplan, 1980; Rosenkranz *et al.*, 1986; Higgs *et al.*, 1988). The original work of Vane (1971) showed sodium salicylate to be 50 times less potent than aspirin in inhibiting cyclo-oxygenase *in vitro*. Unlike acetyl-salicylic acid, both sodium salicylate and paracetamol have no effect on inducible nitric oxide synthetase (Amin *et al.*, 1995), although sodium salicylate at suprapharmacological doses does inhibit nitrite production.

With evidence from these earlier studies of such greater 'potency', it might be expected that acetylsalicylic acid would be more powerful in relieving pain and reducing inflammation in rheumatoid arthritis than any of the non-acetylated salicylate preparations (salicylic acid, sodium salicylate, salsalate, magnesium salicylate, choline salicylate, benorylate and diflunisal) but this has proven not to be the case either in animals (Collier, 1969) or in patients with rheumatoid arthritis (Blechman and Lechner, 1979; Singleton, 1980; Rothwell, 1983; McPherson, 1984; Roth, 1985; Albengres *et al.*, 1988; Multicentre Salsalate/Aspirin Comparison Study Group, 1989; Preston *et al.*, 1989; April *et al.*, 1990).

The fact that acetylsalicylic acid and non-acetylated salicylates are equivalent in both analgesic and anti-inflammatory effects in rheumatoid arthritis may simply be due to the fact that acetylsalicylic acid is rapidly hydrolysed by gastrointestinal esterases to salicylic acid, so that only a relatively small amount is absorbed (Graham, 1977a; 1977b; see also Chapter 4). In inflammatory exudates salicylic acid is the predominant agent, exceeding acetylsalicylic acid by some 30 to 50 times (Graham, 1977a; 1977b), since the plasma half-life ($t\frac{1}{2}$), of acetylsalicylic acid is only 10 to 15 minutes. The evidence from recent molecular biological studies (Chapter 7) suggests that acetylsalicylate and salicylate are equipotent inhibitors of the induction of COX-2 in submicromolar concentrations, which may be related to differences in tissue sensitivities to inhibition of cyclo-oxygenase. Thus, for example, platelets that only have COX-1 may be more sensitive to acetylsalicylic acid than synovial tissue (Estes and Kaplan, 1980). Moreover, Mitchell *et al.* (1997) have shown that non-acetylated salicylates are poor inhibitors of cyclo-oxygenase in pure, cell-free enzyme cell membrane systems, but are potent inhibitions of both COX-1 and COX-2 in whole cell systems, especially of the latter. Vane (1987) noted that a daily dose of 3 g of salicylate reduces the urinary output of prostaglandin metabolites by some 88 to 95 per cent, which would certainly explain the anti-inflammatory effect and why it may be equipotent to acetyl salicylic acid *in vivo*. This confirms the view that aspirin and salicylic acid have the capacity to inhibit prostaglandin synthesis to an equivalent extent *in vivo* in humans. Higgs *et al.* (1988) also showed that reduction in cyclo-oxygenase in rats occurred with amounts of salicylate equivalent to their dosage in humans. Crook and Collins (1975) and Robinson (1978) showed that a small daily dose of aspirin (600 mg) completely inhibited prostaglandin synthetase in the synovium of rheumatoid joints, thus suggesting that a much lower dose than the 3 g daily recommended by Boardman and Hart (1967) was required to achieve an anti-inflammatory effect. It must be remembered that Boardman and Hart (1967) used a relatively crude method of measuring reduction in inflammation, namely the circumference of the proximal interphalangeal joints of the fingers by jewellers' rings. There are, as yet, no *in vivo* studies on the degree of cyclo-oxygenase inhibition and anti-inflammatory effects of different salicylates in joints or sites of inflammation, or whether anti-inflammatory effects might be influenced by changes in the lipoxygenase pathway (Brooks and Day, 1991).

Bonta *et al.* (1977) showed that aspirin would reduce carrageenan-induced paw oedema in rats raised on diets deficient in normal precursors of prostaglandin, suggesting that effects other than those on prostaglandin synthesis may be responsible for the anti-inflammatory effect. Acetylsalicylic acid and the salicylates are known to have diverse biological effects, so that inhibition of prostaglandins alone may represent but a single, albeit defining property of the drug. There is growing evidence from the studies of Weissman and his colleagues (Abramson and Weissman, 1989; Abramson *et al.*, 1990; Weissman, 1991) and others (Altman, 1990; Diaz-Gonzalez *et al.*, 1995) that effects on polymorphonuclear leucocytes may be an alternative explanation for the anti-inflammatory action of aspirin and non-acetylated salicylates.

Non-acetylated salicylates (e.g. salsalate, salicylate) have the advantage over acetylsalicylic acid in that they produce less gastroduodenal damage (Leonards, 1969; Cohen, 1979; Simon and Mills, 1980a; 1980b; Buchanan and Rainsford, 1990) and fewer renal effects (Antillon *et al.*, 1989; Bergamo *et al.*, 1989) and are less likely to induce asthma (Dromgoole *et al.*, 1982; Slepian *et al.*, 1985). Paulus (1989) has challenged the continued use of acetylsalicylic acid in the treatment of rheumatoid arthritis, suggesting that non-acetylated preparations should be prescribed in preference.

Recent rediscovery of the antirheumatic properties of willow bark extract (WBE; Reamalex™, Gerard House Ltd, Luton, UK) has been shown in a double-blind, placebo-controlled trial in patients with rheumatoid and osteoarthritis (Mills *et al.*, 1996). While the pain relief from this preparation was modest, other trials with WBE preparations in the treatment of osteoarthritis and low back pain have yielded more striking effects (Chrubasik and Eisenberg, 2000; Chrubasik *et al.*, 2000).

Clinical use

Paulus (1989) has stated that aspirin should not be generally used in the drug treatment of rheumatoid arthritis. Yet Fries *et al.* (1993) found a low gastrointestinal toxicity score with aspirin, especially enteric-coated aspirin, compared with other non-steroidal anti-inflammatory analgesics. In this study the dose was comparatively low, i.e. 1 to 3 g daily – certainly less than that generally recommended for achieving anti-inflammatory effect (Boardman and Hart, 1967a), in contrast with the potent inhibitory effect on gastric mucosal prostaglandin production by low-dose aspirin (Cryer and Feldman, 1999).

Although aspirin and non-acetylated salicylates have been shown to be equivalent in both analgesic and anti-inflammatory action to other non-steroidal anti-inflammatory analgesics, the incidence of side effects, especially in the gastrointestinal tract (see also Chapter 8), has often been higher (Deodhar *et al.*, 1973; Furst *et al.*, 1987a; 1987b). Csuka and McCarty (1989) pointed out, however, that the newer non-steroidal anti-inflammatory analgesics have been compared with fixed doses of aspirin that were often too small to produce optimal effect, and they considered that the ready availability of aspirin as an over-the-counter preparation and its cheapness contributed to its low esteem among general physicians. The current authors agree with the opinion expressed recently by Wollheim (1996), that aspirin may no longer be considered the first choice in drug therapy in rheumatoid arthritis because of the recent popularity of the COX-2 selective NSAIDs; however, this may not apply to non-acetylated salicylates.

It is frequently not appreciated that the difference (or '*delta*') in the response to drugs like acetylsalicylic acid and the non-acetylated salicylates prescribed in appropriate anti-inflammatory dosage compared with placebo (Boardman and Hart, 1967a) to patients with rheumatoid arthritis is relatively small. This was amply demonstrated by Deodhar and colleagues (1973), who studied 37 patients in a double-blind, crossover trial of sodium salicylate, indomethacin, ibuprofen and prednisone. All the active compounds were significantly better than placebo, but they were no different from each other. The patients were treated as inpatients throughout the study, which may have blunted the response, but ensured compliance. The means of the differences in the outcome measures were small.

Aspirin and the salicylate preparations are frequently prescribed on a four-times-a-day basis. This is unnecessary, since two of the biotransformation pathways – salicylic acid to salicyluric acid, and salicylic acid to salicyl phenolic glucuronide – are limited by Michaelis–Menten kinetics (Furst *et al.*, 1979). This means that as the dose increases so these enzymes are saturated, resulting in an increase of the plasma salicylate half-life of salicylate with dosage (Levy and Tsuchiya, 1972; Pedersen and Fitzgerald, 1984). With a low dose of acetylsalicylic acid, e.g. 300 to 600 mg, the plasma $t_\frac{1}{2}$ is $1\frac{1}{2}$ to 2 hours, while with 3 g or more daily it is of the order of 18 hours or more. Therefore, aspirin need not be prescribed more than twice daily with anti-inflammatory doses (Levy and Giacomini, 1978; Cassell *et al.*, 1979; Keystone *et al.*, 1982). Less frequent dosing might have the advantage of improving patient compliance

(Gatley, 1968; Hussar, 1975; Clinite and Kabat, 1976; Hulka *et al.*, 1976), but pain is usually the 'driving force' persuading patients to take their drugs. Comparable plasma levels of salicylate have been noted with different strengths of plain and enteric-coated aspirin (Paton and Little, 1980).

Steady-state plasma salicylate levels vary widely in patients with rheumatoid arthritis (Paulus *et al.*, 1971; Champion *et al.*, 1975; Graham *et al.*, 1977a; 1977b). Salicylate is bound to plasma albumin, so that hypoalbuminaemia will result in lower total serum concentrations (Borga *et al.*, 1976). Serum salicylate concentrations do not correlate either with the dose or with body size, except at the extremes, but body movement tends to decrease the level of this drug (Paulus *et al.*, 1971; Graham *et al.*, 1977a; 1977b). The serum concentration of salicylate is, however, almost entirely controlled by the activity of the enzymes involved in hepatic biotransformation (Levy *et al.*, 1965; Levy and Tsuchiya, 1972; Tsuchiya and Levy, 1972; Gupta *et al.*, 1975; Furst *et al.*, 1977; Graham *et al.*, 1977a; 1977b). Pharmacokinetic parameters of the salicylates are essentially unchanged by old age (Roberts *et al.*, 1983), but, as previously mentioned, lower plasma concentrations would be expected if hypoalbuminaemia was present. Gender differences in salicylate absorption and clearance are probably influenced by hormonal factors (Kelton *et al.*, 1981; Miners *et al.*, 1986), but are not of a magnitude to have any significant clinical effect. There have been few chronobiological studies, but a shorter plasma half life ($t\frac{1}{2}$) has been observed when a salicylate dose is given at 10 p.m. than at 6 a.m. (Markiewicz and Semenowicz, 1979), which may be due to circadian changes in urinary pH (Ayers *et al.*, 1977). A number of factors other than ingestion of antacids and exercise can alter urinary pH and influence renal excretion of salicylates (Schachter and Manis, 1958). Salicylates also induce their own biotransformation, with a decrease in total salicylate concentrations occurring after 5 to 6 weeks (Rumble *et al.*, 1980; Olsson, 1983; Owen *et al.*, 1989). Unbound plasma salicylate also probably falls, since salivary salicylate concentrations have been noted to fall (Rumble *et al.*, 1980). Plasma binding of salicylate is unchanged during chronic aspirin administration (Owen *et al.*, 1989). The decline in concentrations of plasma salicylates over time has been explained by the induction of the enzyme responsible for formation of salicylurate (Furst *et al.*, 1977; Rumble *et al.*, 1980; Day *et al.*, 1983). It is not known whether the decline in serum salicylate concentrations over time has clinical consequences.

Because of the considerable variation in plasma salicylate concentrations (Paulus *et al.*, 1971; Champion *et al.*, 1975; Graham *et al.*, 1977a; 1977b), there is no value in monitoring plasma salicylate levels at any one point in time (Mandelli and Tognoni, 1980; Tugwell *et al.*, 1984). The generally accepted concentration of 150 µg/ml is considered necessary for an anti-inflammatory effect (Koch-Weser, 1972). The preferred range of steady-state concentrations of 150 to 300 µg/ml in rheumatoid arthritis (Dromgoole *et al.*, 1981) now need to be challenged. Monitoring free salicylate levels might seem rational (Levy and Moreland, 1984), but in clinical practice this has not proven to be the case (Gurwich *et al.*, 1984). More free salicylate is present at higher total salicylate concentrations (Furst *et al.*, 1979) and in the elderly because of lower serum albumin concentrations (Wallace and Verbeeck, 1987; Kean and Buchanan, 1990). The clinical adage of 'push to tinnitus, then back off slightly' cannot be recommended now, especially in the elderly with pre-existing hearing loss (Dromgoole *et al.*, 1981). Although adverse reactions to aspirin and non-acetylated salicylates are considered to be concentration-related effects, side effects may occur at concentrations generally not considered to be toxic (Gurwich *et al.*, 1984; Owen *et al.*, 1989).

Juvenile rheumatoid arthritis

Clinical experience (Bywaters, 1977; Schaller, 1978; Holister, 1985) and therapeutic trials (Mäkelä *et al.*, 1979; Brewer, 1982; Brewer and Giannini, 1982; Brewer *et al.*, 1982) have proven the efficacy of salicylate therapy in juvenile rheumatoid arthritis. Only minor differences in efficacy have been demonstrated between salicylates and other NSAIDs (Brewer, 1982), but salicylates have in general proven more toxic (Barron *et al.*, 1982). Although no significant differences were found by Mäkelä *et al.* (1979) among three salicylate preparations (micro-encapsulated aspirin, enteric-coated aspirin and standard aspirin), the enteric-coated form resulted in a lower incidence of gastric bleeding (12 per cent) than the other two preparations (22–28 per cent). Oroyco-Alcala and Baum (1979) showed that the enteric-coated aspirin produced similar salicylate plasma concentrations to regular and buffered aspirin, and had a higher degree of acceptance in children because of ease of administration. Wide variation in plasma salicylate concentrations can be expected (Bardare *et al.*, 1978), and it may require doses of greater than 100 mg/kg per day to achieve salicylate concentrations of greater than 20 g/dl (Doughty *et al.*, 1980). In general, plasma salicylate concentrations of 0.2 μg/ml can be achieved after 5 days with doses of between 70 and 80 mg/kg per day, but higher doses of 100 mg/kg per day may be required to achieve this, and even higher concentrations in one-third of patients (Doughty *et al.*, 1980).

Elevation of hepatic enzymes can be expected in at least one-third of patients, usually in those receiving high doses of salicylate and who have hypoalbuminaemia (Mäkelä *et al.*, 1979). The incidence of side effects can be reduced by monitoring plasma salicylate concentrations (Bardare *et al.*, 1978; Mäkelä *et al.*, 1979; Pachman *et al.*, 1979) (but see earlier comments concerning the value of monitoring blood levels). It has been estimated that 50 per cent of patients respond to NSAID therapy within 2 weeks, while some 25 per cent take 12 or more weeks (Lovell *et al.*, 1984). The mean time of response is approximately 30 days, which is certainly much longer than that in adult rheumatoid arthritis, where efficacy can be demonstrated in a week or less (Lovell *et al.*, 1984). This would suggest a poorer response in children than in adults with rheumatoid arthritis, which may be due to less severe pain and inflammation, since these determine the 'delta' of the response. Children appear to experience less pain in arthritic joints than adults (Laaksomen and Laine, 1961). If pain and inflammation persist despite adequate salicylate therapy, another NSAID should be prescribed (Schlegel and Paulus, 1986). However, multiple NSAID therapy, which is frequent (Lee *et al.*, 1974a), should be avoided (Roth, 1988a; 1988b). Liquid preparations, such as that available with ibuprofen, are less toxic than other forms of salicylates (Giannini *et al.*, 1990).

Mention has already been made of hepatic complications with salicylate therapy. It is perhaps less well known that chronic salicylism (Everson and Krenzelok, 1986) and renal papillary necrosis have been reported in children treated for long periods with salicylates (Wortmann *et al.*, 1980; Allen *et al.*, 1986). The most serious complication is Reye's syndrome, which has been reported among patients with juvenile rheumatoid arthritis (Remington *et al.*, 1985; Sullivan *et al.*, 1988) and adult Still's disease (Tumiati *et al.*, 1989). As discussed in Chapter 9, there is still controversy regarding the role of aspirin in causing Reye's syndrome. However, while doubt remains it seems prudent to avoid aspirin therapy in juveniles. It is not clear whether Reye's syndrome occurs with non-acetylated salicylate therapy.

Salicylazosulphapyridine (sulphasalazine)

Sulphasalazine has been used since the early 1940s for the treatment of inflammatory joint disease including rheumatoid arthritis (Svartz, 1948).

In the early 1940s, Dr Nana Svartz sanctioned the introduction of sulphasalazine for the treatment of inflammatory joint disease of the rheumatoid type (Svartz, 1948). By the late 1940s the drug was being criticised for its side effect profile. However, the dosage utilised may have been unnecessarily high. Over the succeeding years the drug was extensively used and investigated in the management of ulcerative colitis, Crohn's disease, and other types of inflammatory bowel disease (Klotz *et al.*, 1980), and it was subsequently shown that the 5-aminosalicylate moiety was the active metabolite in inflammatory bowel disease (Schroder and Campbell, 1972), and that the sulphapyridine moiety was also possibly an active species in the treatment of inflammatory rheumatoid disease and other inflammatory joint disorders (Das and Eastwood, 1974; Molin and Stendahl, 1979; Hoult and Moore, 1980; Klotz *et al.*, 1980; McConkey *et al.*, 1980). Sulphasalazine has also been shown to have antibacterial properties on the gut flora (Olhagen, 1970; Haslock and Wright, 1974). The thiol reactivity and the 'antibacterial' action of sulphasalazine are the most likely properties relating to its influence on the reactive arthritides.

Sulphasalazine differs from other salicylates in having slow-acting or disease-modifying activity, being classified as a DMARD (Bax, 1992). A summary of the concepts of action of sulphasalazine in

CHAPTER 12

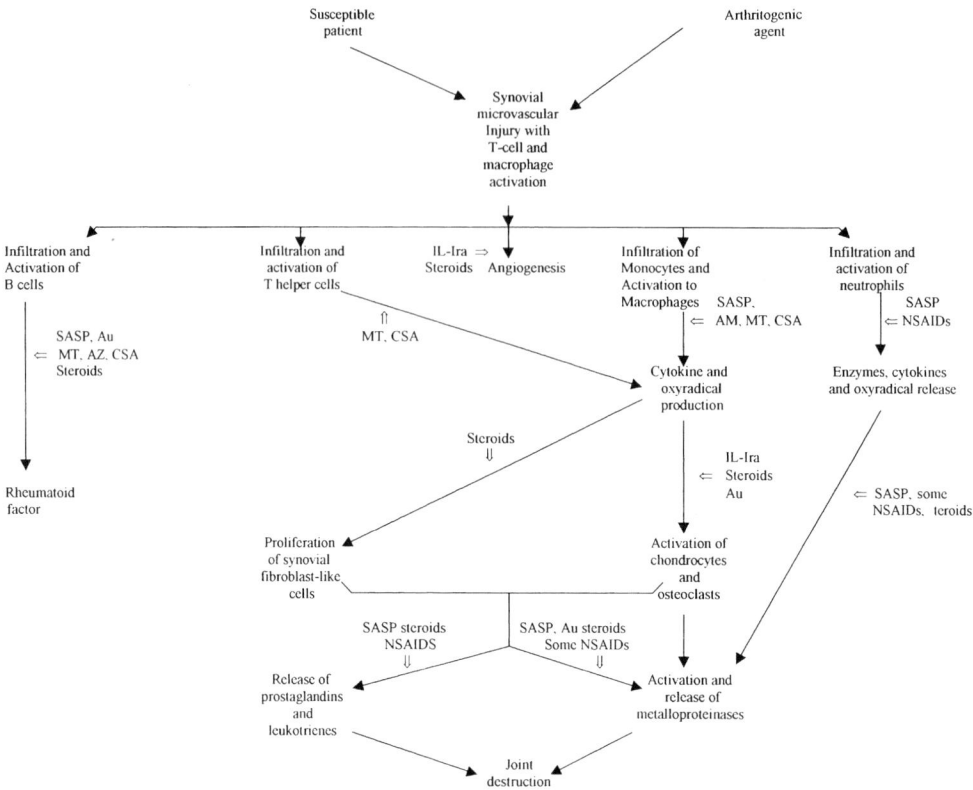

Figure 12.1 Actions of sulphasalazine (salicylazosulphapyridine; SASP) compared with other DMARDs and NSAIDs on components of the joint pathogenesis of rheumatoid arthritis. Abbreviations: Au, gold salts (thiomalate or auranofin); SASP, salicyl-azosulphapyridine or sulphasalazine; CSA, cyclosporin A; AZ, azathioprine; MT, methotrexate; IL-1ra, interleukin-1 receptor antagonist. Inhibitory sites of action shown by cross-slash. It should be noted that most of these drug effects have been derived from *in vitro* data, and only in some cases has information been derived from *ex vivo* studies in cells derived from rheumatic patients. (Modified from Rainsford, 1996.)

the major pathological pathways in rheumatoid arthritis as compared with other DMARDs and NSAIDs is shown in Figure 12.1 (from Rainsford, 1996).

McConkey and colleagues (1976; 1978; 1980) and Bird and colleagues (1982) were amongst the first in the early 1980s to re-establish the effectiveness of sulphasalazine in the management of rheumatic diseases, using clinically effective doses with an acceptable toxicity profile. Aspects of the history of the development of sulphasalazine have been reviewed by McConkey (1986). Plasma levels of the 5-aminosalicylate metabolite are generally low, and suggest that synovial levels of this salicylate derivative will also be low. It is unlikely that the 5-aminosalicylate component alone is a major influence in the management of rheumatoid disease and the other inflammatory arthritides (Bax, 1992). It is more likely that the thiol complex sulphapyridine is the active species, with properties similar to the other thiols such as the gold complexes and D-penicillamine, and drugs with thiol activity such as the chloroquine, Studies on the use of the 5-aminosalicylate moiety in the management of inflammatory bowel disease suggest that the majority of the toxicity from the sulphasalazine compound is related to the sulphapyridine group (Berry *et al.*, 1983; Das, 1983; Sack and Peppercorn, 1983). Individuals with abnormal N-acetylation may not be able to detoxify the sulphasalazine and may be susceptible to toxicity in this regard. Other side effects observed from sulphasalazine (see Bax, 1992) include male infertility

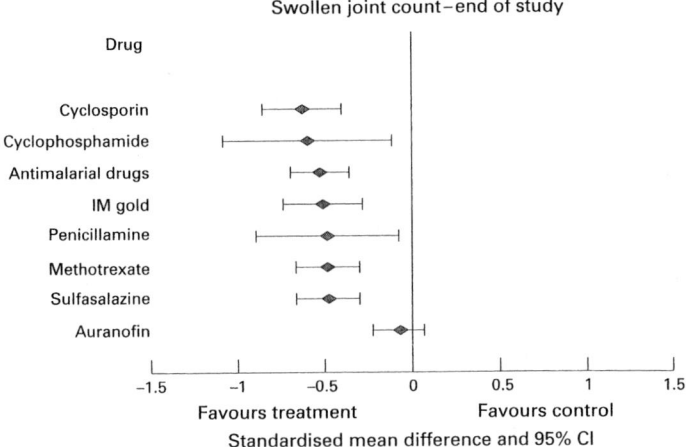

Figure 12.2 Summary of the effects of sulphasalazine compared with the other 'established' DMARDs for the treatment of rheumatoid arthritis based on meta-analysis by the Cochrane Musculoskeletal Group of the Cochrane Collaboration (from Tugwell *et al.*, 2000). The results show that sulphasalazine has similar benefits to other DMARDs in comparison with placebo, with the exception of auranofin, which was no better than placebo.

(possibly reversible) (Birnie *et al.*, 1981; Toovey *et al.*, 1981), hepatotoxicity (Kanner *et al.*, 1978; Matek *et al.*, 1980), lung injury (Rosenhall and Uddenfeldt, 1979) and embryotoxicity (Craxi and Pagliarello, 1980; Mogadam *et al.*, 1981).

Tugwell *et al.* (2000) conducted a meta-analysis of randomised controlled trials performed with sulphasalazine and compared these with seven other 'established' DMARDs in the treatment of rheumatoid arthritis. The data formed part of the evidence-based analysis in the Cochrane Database of Systematic Records. The analysis of data of swollen joint counts from seven trials with sulphazalazine showed there was a clear benefit from this drug, with a standardised mean difference of −0.43 with confidence intervals (−0.61 to −0.24) that do not cross the line from that favouring treatment to that favouring control, showing there is a clear difference between sulphasalazine and placebo (Figure 12.2). Sulphasalazine showed no differences compared with cyclosporin, cyclophosphamide antimalarial drugs, intramuscular gold salts and methotrexate, while all these were superior to auranofin, which did not show any benefit.

In analysing the drop-outs due to toxicity in related efficacy (as measured by Fender joint count) there were no differences between sulphasalazine, intramuscular gold, penicillamine and leflunomide, but these had drop-outs rates (ranging from about 12 to 27 per cent that were higher than those from auranofin and anti-malarial drugs.

A summary of the treatment benefits in therapy of rheumatoid arthritis (see Tugwell *et al.*, 2000) based on the Cochrane data analysis stated that it was found that sulphazalazine 2–3 g/d for 6 months reduced pain for 66 per cent of people and reduced overall pain for 51 per cent compared with placebo. Gastric side effects were the most frequent of all adverse reactions, and comprised stomach upset and digestion problems in 10 per cent of patients. Skin rashes were present in 7 per cent of patients, while 2 per cent developed abnormal white blood cell counts and 4 per cent had liver problems. Up to 22 per cent of patients stopped treatment because of side effects.

A meta-analysis of five randomised controlled trials of sulphasalazine treatment for ankylosing spondylitis, which met acceptable criteria for methodologic quality, showed consistent clinical benefit compared with placebo for duration and severity of morning stiffness, general well-being, ESR and IgA (Ferraz *et al.*, 1990).

Other salicylates

Diflunisal was selected as the best of a group of salicylate derivatives that was developed over a course of 15 years in the 1970s (Miehlke, 1977). Diflunisal is a more potent anti-inflammatory agent in animal models than aspirin, and has a unique effect on the reduction of leukotriene-B_4 production from poly-morphonuclear leucocytes. The drug has a low intrinsic gastrotoxicity. It is of interest that diflunisal was not further promoted for use in rheumatoid arthritis.

A trial of magnesium dithiosalicylate 3 g/day in the treatment of rheumatoid arthritis (Dequeker *et al.*, 1980) has shown that this drug compared favourably with aspirin 3 g/day. The thiol component of the magnesium thiosalicylate was of interest to the investigators because of the possible disease modifying activity. There was a high incidence of gastrointestinal side effects and hypersensitivity reactions in a large number of patients. The drug had to be withdrawn in a number of patients.

OSTEOARTHRITIS

Osteoarthritis, the oldest known and most prevalent arthritic disorder, is a major cause of disability, especially in the elderly (Lawrence, 1977). It is found in all races (Valkenburg, 1981), and is evident radiologically in more than 80 per cent of persons over the age of 55 (Lawrence, 1977). However, only some 15 per cent of persons with radiological evidence of the disease are symptomatic. After the age of 50, the increase in severe forms of the disease is not linear but exponential (Lawrence *et al.*, 1966). When all ages are considered, males and females are equally affected. The role of obesity has long been debated, but there is sound evidence that obesity is associated with knee osteoarthritis, especially in women (Felson *et al.*, 1988). There is a positive correlation between bone density and the development of osteoarthritis (Solomon *et al.*, 1982), and with prolonged overuse of a joint or a group of joints associated with certain occupations, e.g. elbow disease in foundry workers (Mintz and Fraga, 1973).

Osteoarthritis is generally considered to be primary, i.e. idiopathic, or secondary to trauma or underlying disease of the joint (Bullough, 1998; see also Kean and Buchanan, 2002 for a review of the aetiology and pathology of this disease). The distinction is arbitrary and of no great relevance to medical drug therapy, as the same drugs are prescribed in both. The disease is generally considered to commence with cartilage failure. The surface of the cartilage undergoes fibrillation, with the formation of deep clefts resulting in ultimate total destruction of the cartilage. This results in loss of cartilage water and release into the joint cavity of proteoglycans, pieces of cartilage and collagen type II fibres, which cause a chemical synovitis and a localised immune reaction, presumably to these degraded macromolecules (Moskowitz *et al.*, 1970; Goldenberg *et al.*, 1982; Jasin, 1985; Vetto *et al.*, 1990; Helbig *et al.*, 1998). The adjacent bone remodels, with increase in density (eburnation) and osteophyte formation. Bone cysts or geodes with well demarcated margins develop in subchondral bone, and contain synovial fluid and pieces of cartilage. They result from fluid being pressed into channels in the denuded articular surface (Landells, 1953). Release of hydroxyapatite crystals in the shards of fragmented cartilage also plays a role in the development of synovitis (Myers *et al.*, 1992; Bullough, 1998).

Osteoarthritis was previously considered to be a degenerative process, the result of wear and tear on the articular cartilage. However, the disease process is clearly an inflammatory condition with associated synovitis (Bullough, 1998). This can histologically resemble rheumatoid arthritis, with the formation of pannus, especially in hip joint disease. The sedimentation rate and other acute phase reactants may be moderately increased in early stages of the disease (Spector *et al.*, 1997). The synovial fluid contains destructive enzymes, and the fluid and synovial lining contain both COX-2 as well as COX-1 derived prostanoids (Crofford, 1999).

Salicylates in osteoarthritis

Aspirin has been studied in a number of controlled double-blind clinical therapeutic trials in osteoarthritis, and found to be effective against placebo and equivalent in analgesic effect to phenylbutazone (Scharff, 1974), tolmetin (April *et al.*, 1975; Muller *et al.*, 1977), naproxen (Melton *et al.*, 1978; Martinez *et al.*, 1980), indomethacin (Martinez *et al.*, 1980), fenoprofen (Brooke, 1976; Diamond, 1976; McMahon *et al.*, 1976), sulindac (Bower, 1979); diclofenac (Joubert *et al.*, 1974), tiaprofenic acid (Chahade and Josef, 1978; Huskisson and Scott, 1978; Valtonen, 1979a; Peyron, 1982), fenbufen (Chalem *et al.*, 1977; Valtonen, 1979b) and ibuprofen (Peyron and Doury, 1982). Likewise, non-acetylated salicylates have proven effective in osteoarthritis (Goldenberg *et al.*, 1974) and a liquid formulation of salicylate has been found mostly for elderly patients (Nevinny and Gowans, 1960). Diflunisal has been found equivalent to aspirin (Bresniham *et al.*, 1978; Dieppe and Huskisson, 1978; 1979; Tait *et al.*, 1978; Wojtulewski *et al.*, 1978; Essigman *et al.*, 1979), ibuprofen (Brooks *et al.*, 1982a) and naproxen (Wojtulewski *et al.*, 1978). Aspirin has also been compared with benoxaprofen (Alarcon-Segovia, 1980), phenylbutazone (Scharff, 1974; Rothstein, 1975), tolmetin (Muller *et al.*, 1977), diflunisal (Grayson, 1978a; 1978b; Dieppe and Huskisson, 1979), naproxen (Melton *et al.*, 1978; Martinez *et al.*, 1980), fenoprofen (Brooke, 1976; Diamond, 1976; McMahon *et al.*, 1976), indomethacin (Martinez *et al.*, 1980), diclofenac (Joubert *et al.*, 1974; Chahade and Josef, 1978), sulindac (Chahade and Josef, 1978; Huskisson and Scott, 1978) and fenbufen (Valtonen, 1979b).

Intra-articular aspirin like that of corticosteroids was found to be no better than saline in controlling joint inflammation (Rylance *et al.*, 1980) suggesting that the intra-articular delivery of drugs may not be of additional value.

Salicylates were used in the treatment of several forms of arthritis, including osteoarthritis, prior to the development of formalised trials (Rosenbloom *et al.*, 1985). By the time alternative anti-inflammatory agents became available aspirin was already established as an effective form of therapy and became the standard against which other agents were compared (Rosenbloom *et al.*, 1985). In spite of the proliferation of non-steroidal drugs, and while not monopolising the field, salicylates remain a common comparison drug. Since patients requiring therapy are often in middle and late life, research has been directed at developing agents that are more efficacious, less toxic, or more convenient to take. Data on the comparative effects of salicylates and placebo come from trials in which placebo washout periods have been followed by the administration of aspirin. In these trials, aspirin is invariably more efficacious than placebo. More side effects are seen in subjects on aspirin, but placebo controls are not exempt from this phenomenon, in spite of the pharmacological inactivity of the compounds.

In a review of clinical therapeutic trials of NSAIDs in osteoarthritis up to 1985, Rosenbloom *et al.* (1985) found that aspirin remained a common comparison drug (18 of 65 studies). In 17 studies a double-blind randomised, controlled design was used, while eight employed a crossover. The number of subjects varied from 20 to 140, and total study duration from 2 to 28 weeks. Aspirin dosage ranged from 1.3 to 4.5 g/day in divided dosage. In seven studies the method of assessing compliance to the therapeutic regimen was clearly defined, while in the remaining studies it was not clear whether compliance was quantified or assumed to be adequate. Long-term studies in osteoarthritis have been few (Appelrouth *et al.*, 1987; Mullen, 1987; Anderson *et al.*, 1989). Compared with nabumetone, drop-outs were much more common with aspirin, being 20 and 60 per cent respectively (Mullen, 1987). More recently, ibuprofen has replaced aspirin as the drug of choice for comparison with a standard in clinical trials with newer NSAIDs (Kean *et al.*, 1999).

Current management of osteoarthritis involves stepwise introduction of non-pharmacological treatment (exercises, weight loss, joint protection, support) and paracetamol (up to 4 g daily) for pain control, and capsaicin or topical anti-inflammatory creams (March, 1997). Should pain relief be inadequate, then ibuprofen or other low GI-irritant NSAIDs should be used at low analgesic doses (*cf.* higher doses where anti-inflammatory effects are evident) progressing to higher doses and where necessary gastroprophylaxis (with e.g. misoprostol, anti-acid secretory agents, mucoprotectants, etc.) in subjects having GI side effects or those at risk of bleeding.

There has, however, been much debate about whether it is beneficial to employ NSAIDs in comparison with paracetamol as first-time treatment for pain in osteoarthritis. The American College of Rheumatology (ACR) Subcommittee on Osteoarthritis Guidelines (2000) has made recommendations for medical management of osteoarthritis of the knee and hip and they suggest that while paracetamol is appropriate for NSAID-naïve patients, there are concerns about the hepatic safety of paracetamol. This appears to have attracted much discussion in the form of Letters to the Editor of *Arthritis and Rheumatism* (Brandt, 2001; Dart and Kuffner, 2001; Dieppe, 2001), including one from the drug manufacturer (Lynch and Fox, 2001). The response by Dr Hochberg for the ACR Subcommittee pointed out that although the risks of paracetamol toxicity (including that in patients with a history of liver disease) was low, reports and the manufacturers' guidelines about use of this drug with alcohol indicated that the risk could not be avoided.

A critical review by Felson (2001) of the case for use of paracetamol or NSAIDs in treatment of osteoarthritis included a detailed analysis of three trials where paracetamol was compared with NSAIDs. The results showed that the differences between the two sets of treatments (different NSAIDs were employed in the three trials) were relatively small, but favoured use of NSAIDs. Since the number of patients in these trials ranged from 61 to 112 and the studies were performed for 6 weeks, it can hardly be said that these results were definitive indications for preferences one way or the other.

Where aspirin fits into the place of therapy in osteoarthritis is now unclear. Clearly GI safety will be an issue, although the recent pharmaco-epidemiological analysis of Garcia-Rodriguez and Hernandez-Diaz (2001) has shown that there is a marked increase in risk of upper gastrointestinal complications when doses of greater than 2 g/d paracetamol are taken (see also Chapter 8). Thus it may be advantageous to re-evaluate the efficacy of aspirin and particularly the low ulcerogenic drugs, e.g. salsalate and ibuprofen, in comparison with paracetamol in large-scale randomised, controlled trials especially for long-term (e.g. 1–2 years) therapy of osteoarthritis, focusing on adverse reactions in the GI tract, liver and kidney as well as efficacy.

Side effects

The proportion of patients experiencing drug-related side effects while on salicylates ranged from 0 to 97 per cent, being most often in the order of 24–64 per cent. Between 6 and 50 per cent of patients (average 25 per cent) had to discontinue therapy because of side effects. Nausea and dyspepsia are the most frequent side effects (Buchanan and Brooks, 1991).

MISCELLANEOUS MUSCULOSKELETAL DISORDERS

Aspirin has been used in the management of several musculoskeletal disorders, including juvenile rheumatism, the reactive arthritides, soft tissue lesions and injury. Aspirin has also been of value in the treatment of gout, provided that it is used in high dosage. However, lack of patient compliance and poor patient understanding of the chemical mechanisms suggests that aspirin should not be used in the treatment of gout, since low doses can exacerbate the problem.

Gout

The effect of large doses of salicylates (especially salicylic acid or sodium salicylate) on enhancing the excretion of uric acid has been a well-established effect known for over two centuries (Gross and Greenberg, 1948; Smith, 1966). Among the earliest reports were those by Fawcett (1896), who employed the hyperbromite assay to analyse urea in four cases of patients with gout who received 600 mg of sodium salicylate four times daily and compared the results with the effects of colchicine (Colchicum). The pronounced uricosuric effect was evident and comparable with that of Colchicum.

The effects of salicylates on the excretion of urate are dose-dependent. Aspirin or salicylate in doses greater than 3.5 g per day have a uricosuric effect on the kidney (Smith, 1966; Fanelli and Weiner, 1979). However, at doses of less than 3.5 g per day uric acid retention occurs, and thus these salicylates are contraindicated in the treatment of acute gout. In addition, it is possible that salicylates may have an antagonistic effect to some uricosuric drugs and therefore should not be given concomitantly with these types of agents.

It is possible that salicylate may mobilise gout tophi (Smith, 1966). Several studies have indicated that one of the major effects of salicylate is to displace urate from its binding sites on circulating plasma proteins (Schlosstein et al., 1973; Whitehouse et al., 1970; 1973; Postlethwaite et al., 1974; Kippen et al., 1974). Small doses of salicylate depress tubular secretion of uric acid in the kidney, but at high doses the drug inhibits the tubular reabsorption of urate, with a net effect that there is near complete glomerular filtration of the urate metabolite (Smith, 1966). A study by Postlethwaite and colleagues (1974) indicated that aspirin displaces urate from protein binding in 40 per cent of patients with chronic renal insufficiency. Salicylate blunts the effects of probenecid in excretion of urate (Fanelli and Weiner, 1979), suggesting there is a common effect of both drugs on the renal excretion of urate.

Low doses of diflunisal, 250–375 mg per day have been shown to increase uric acid excretion (Dresse et al., 1975; Tempero et al., 1977; van Loenhout et al., 1981). Diflunisal antagonises the urate retention observed during hydrochlorothiazide therapy (Tempero et al., 1977). This suggests that diflunisal has a uricosuric action, but this has not yet been determined. However, it is possible that the drug could be useful for the management of the treatment of gout. It would be expected that this low dose of diflunisal would also have a low incidence of gastrointestinal side effects.

Systemic lupus erythematosus

Systemic lupus erythematosus (SLE) is a complex of disease expression profiles, and it is therefore difficult to evaluate the general clinical outcome in therapeutic studies. In one study aspirin 3.6 g per day was found to be superior to ibuprofen taken in high dosage, up to 2.4 g per day (Karsh et al., 1980). It is important to note that patients with hepatic or renal dysfunction and systemic lupus should be treated with caution with aspirin, the salicylates and other NSAIDs.

DRUG KINETICS AND MONITORING

Blood or plasma salicylate levels have been readily available for drug monitoring over the course of many years. Although this a useful tool in attempting to monitor therapeutic benefits and minimise toxicity, it has been identified that there is a significant intersubject variation with respect to plasma levels of salicylates during aspirin therapy (Graham et al., 1977a). There is also a high variability in salicylate determination in different laboratories (Mutchie et al., 1980). The time of blood sampling is important (Gupta et al., 1975; Day et al., 1979). It is important to observe that single doses of aspirin undergo first-order kinetics, whereas long-term (daily use) of aspirin results in Michaelis–Menten or zero-order kinetics. The high variability in patient response to aspirin and salicylate also reflects a significant variability in the rate of absorption of aspirin and salicylates, the metabolism, distribution, and rate of absorption of these compounds. Day and colleagues (Graham *et al.*, 1977a; 1977b; Day *et al.*, 1983; 1987a; 1987b; 1988a; 1988b; Day, 1985) observed an increased capacity of the liver to metabolise salicylate to salicylurate, associated with a reduction in steady-state serum concentrations of salicylate within 4 weeks of chronic salicylate treatment in patients with rheumatoid arthritis. This induction of salicylurate formation could account for the variability in the relationship between blood levels of the active drug (aspirin, salicylate) and the therapeutic response as discussed by Gupta *et al.* (1975). In addition, low albumin concentrations (as identified in elderly patients and/or in rheumatoid arthritis patients) and abnormal drug binding characteristics of plasma protein may contribute to the variability in response to aspirin and the salicylates (Ballantyne *et al.*, 1971; Piafsky, 1980). It has been suggested by O'Malley and colleagues (1971) that altered drug metabolism in the elderly may contribute to the variable response to aspirin and the salicylates.

A total plasma salicylate value of approximately 1 mmol/l (150 µg/ml) is considered the plasma level necessary to achieve therapeutic activity in the management of inflammatory arthritis. Adverse effects are identified above this level. Mongan and colleagues (1973) have reported that tinnitus occurred in rheumatoid arthritis patients with serum salicylate levels greater than 1.5 mmol/l (284 µg/ml). Recovery from tinnitus usually occurs 24 hours after the drug is stopped (Miller, 1978). The serum salicylate levels are of value in monitoring toxicity, especially in the avoidance of hepatotoxicity in patients with abnormal liver function and inflammatory joint disease.

Salivary salicylate concentrations correlated with total and free plasma salicylate levels in rheumatoid arthritis (Caruso *et al.*, 1978). Where measurement is available, saliva sample monitoring may be acceptable to some patients.

In general, high-dose aspirin correlates with clinical effectiveness.

Interactions with anti-inflammatory drugs

The larger the number of concomitant drugs taken, the greater the incidence of adverse drug effects and possible effects on efficacy. Multi-drug use increases the potential for drug–drug interactions, which result in altered pharmacokinetics and pharmacodynamics (Grennan and Aarons, 1984; Brater 1986; Brooks and Buchanan, 1991). Corticosteroids decrease the levels of salicylate during long term treatment with aspirin (Klinenberg and Miller, 1965). There is an increase in urinary secretion of salicylate. Aspirin diminishes the plasma half-life and renal clearance of dexamethasone, but enhances hepatic clearance and the C-6 hydroxylation of the steroid in rats (Kassem and Schulte, 1981). It is postulated that aspirin may influence corticosteroid metabolism and distribution in a similar manner in patients with arthritis. Aspirin has been shown to antagonise the anti-inflammatory activity of indomethacin in chronic and acute models of inflammation induced in rats (Swingle *et al.*, 1970; Van Arman *et al.*, 1973; Khalili-Varasteh *et al.*, 1976), but no such effect has been identified with benorylate (Khalili-Varasteh *et al.*, 1976). In rats salicylic acid affects the pharmacokinetics of indomethacin (Yesair *et al.*, 1970). In human studies, aspirin has had a variable effect on the pharmacokinetics of indomethacin. Some investigators identified no differences in plasma levels or binding of indomethacin given orally along with aspirin (Dresse, 1975; Garnham *et al.*, 1975; Tempero *et al.*, 1977; Van Loenhout *et al.*, 1981), but buffered aspirin increased the absorption of indomethacin and this was associated with increase in side effects (Garnham *et al.*, 1975). However, in other studies the total serum concentration and half-life of indomethacin were reduced by concurrent oral aspirin. In addition, the level of free indomethacin increased (Kaldestad *et al.*, 1975; Day *et al.*, 1988c). Studies in rats have shown that concurrent oral administration of aspirin or salicylate with indomethacin reduces the plasma concentrations of indomethacin (Correll and Jensen, 1979). However, indomethacin reduces the plasma salicylate concentration when given concurrently with aspirin. These effects are consistent with the reduced anti-inflammatory activity of combinations compared with indomethacin alone (Swingle *et al.*, 1970; van Arman *et al.*, 1973; Khalili-Varasteh *et al.*, 1976). The benefits of reduced gastrointestinal toxicity of salicylate or diflunisal (but not aspirin combined to indomethacin) compared with indomethacin alone has been considered for commercial exploitation. One product, Pelsonin®, which is an indomethacin/salicylate combination, was investigated in Hungary (Ezer *et al.*, 1979). Favourable therapeutic benefits were identified in patients with rheumatoid arthritis, and there was lower gastro-intestinal blood loss as determined by the ^{51}Cr-labelled red blood cell technique compared with that observed for indomethacin alone. Later development of this formulation was terminated for unknown reasons.

While the plasma levels of many NSAIDs (including indomethacin) are reduced by aspirin in humans and in experimental animals, leading to reduced efficacy for some NSAIDs (Rubin *et al.*, 1973; Mason and McQueen, 1974; Warrick *et al.*, 1974; Chaplin *et al.*, 1975; Wiseman *et al.*, 1975; Brooks and Khong, 1977; Thompson *et al.*, 1979; Willis *et al.*, 1980; Miller, 1981; Williams *et al.*, 1981), no such differences have been observed with sudoxicam and flurbiprofen (Wiseman *et al.*, 1975; Brooks and Khong, 1977). Studies by Kean *et al.*, 1981 and by others (Miller, 1981) have shown that use of a single drug is better than drugs taken in combination in comparable dosages. However, combined oral aspirin and rectal indomethacin have been shown to have additive therapeutic benefits when compared to either drug alone (Ekstrand *et al.*, 1981). However, the serum half-life of indomethacin was lower. Thus while it may be advantageous to combine some NSAIDs, it may be necessary to adjust dosages. The overall value of combined salicylate/NSAID compounds has to be weighed against the potential for the induction of gastrointestinal toxicity with the combination therapy.

ADVERSE EVENTS IN RHEUMATIC PATIENTS

The evidence for the development of adverse reactions from salicylates compared with other NSAIDs and analgesics and their mechanisms has been considered in Chapter 8. Here we consider the clinical implications of adverse reactions as they are relevant to patients with rheumatic conditions, as well as practical issues concerning the management of these conditions. Many of the details concerning key points mentioned here will be found in the comprehensive review of Brooks and Buchanan (1991), as well as in Chapter 8.

Gastrointestinal toxicity

One of the major problems with aspirin and other non-acetylated salicylates is toxicity in the gastro-intestinal tract (Walt *et al.*, 1986; Freis *et al.*, 1989; 1993; Brooks and Buchanan, 1991; Freis, 1995; Singh, 1996). This can be broadly classified as:

1. Acute gastrointestinal haemorrhage and ulceration
2. Microbleeding
3. Gastric and duodenal erosions and ulceration
4. Dyspepsia
5. Small and large bowel effects.

Gastrointestinal haemorrhage and ulceration

It was Buss who in 1875 probably first reported the effects of salicylic acid on the stomachs of experimental animals (i.e. rabbits). Bälz (1877) was probably the first to observe that salicylic acid caused intestinal bleeding, vomiting and nausea. It was not until the initial gastroscopic studies by Douthwaite and Lintott (1938) showing mucosal injury in volunteers after aspirin ingestion, and reports of aspirin possibly causing haematemesis (Hurst and Lintott, 1939; Hurst, 1943; Parry and Wood, 1967), that a causal relationship was first taken seriously.

Up until the late 1980s many still doubted whether aspirin was a major cause of haematemesis, largely because of the low incidence and inadequacy of the literature (Rooney and Kean, 1987) and also because rechallenge did not cause recurrence of bleeding. Nearly half a century after Douthwaite and Lintott's article in 1938, it was finally accepted that aspirin did cause bleeding. Aspirin was considered among the worst offenders (Rainsford, 1982; Graham and Smith, 1986; Semble and Wu, 1987), but other NSAIDs introduced in the 1980s can also cause appreciable ulcers and bleeding (Langman, 1988). The evidence that non-acetylated salicylates lead to a lower incidence of the complication is based on some endoscopic observations and clinical, but endoscopic evidence (Cohen, 1979; Silvoso *et al.*, 1979; Caruso and Porro, 1980; Hoftiezer *et al.*, 1980; Lanza *et al.*, 1980; 1981; Dromgoole *et al.*, 1981; Kilander and Doterall, 1983; Roth, 1988a; 1988b; Buchanan and Rainsford, 1990).

Several risk factors have been identified that lead to an increased likelihood of bleeding: age over 60 years, female sex, greater severity of disease, high dosage and chronic use of aspirin, smoking and ethanol use, and past history of peptic ulcer disease (Goulston and Cooke, 1978; Willoughby *et al.*, 1986; Faulkner *et al.*, 1988; Llewellyn and Pritchard, 1988; Rosen and Fleischer, 1989) It should be noted that not all authors have noticed the elderly to be more vulnerable (Jick *et al.*, 1985; 1987a; 1987b; Kafetz, 1988), and this may be related to individual definitions of 'elderly'. (Buchanan and Brooks, 1991). However, there is now extensive literature indicating a predisposition of elderly subjects to aspirin or other NSAID-induced gastrointestinal bleeding (Alexander *et al.*, 1983; Booker, 1983; Caradoc-Davies, 1984), in whom mortality is high. This may be related to lower gastric mucosal PGE_2 production in elderly subjects (Cryer *et al.*, 1992a). Why elderly females are more prone to bleeding from aspirin has not been explained, but it may relate to the reduction in salivary flow and the consequent protective effect of mucus, bicarbonate and epidermal growth factor in the gastric mucosa, reduction of the mucosal prostaglandin concentrations, or low aspirin esterase activity. The higher pH of the stomach contents from reduced acid production and chronic atrophic gastritis might, on the other hand, be expected to be protective. Severity of disease, in particular rheumatoid arthritis, has been identified as a risk factor.

Malone *et al.*, 1986 and Voskuyl *et al.* (1993) have suggested that the 'rheumatoid' stomach may be more prone to bleeding. Higher doses of aspirin certainly are more likely to lead to bleeding (Kaufman *et al.*, 1993), but even small over-the-counter doses can cause haematemesis. The chronic use of aspirin has been reckoned to increase the likelihood of a haematemesis three-fold, but the risk is highest within 6 months of starting treatment rather than later. Smoking probably predisposes to bleeding by reducing mucosal prostaglandin concentrations (Cryer *et al.*, 1992b) and impairing the healing of aspirin-induced erosions and ulcers (Willoughby *et al.*, 1986). Alcohol, especially in the form of spirits, affects the mucosal surface, but does not by itself lead to erosions, ulcers or bleeding, but there is increasing evidence that bleeding is more likely in patients who take both aspirin and alcohol (Quick, 1966; 1971; Needham *et al.*, 1971; Goulston and Cooke, 1978; Burbridge *et al.*, 1984). Similarly, oral corticosteroids may (especially at low dosages) not cause erosions or ulcers (Bianchi Porro and Pace, 1988), but when taken with aspirin certainly increase the incidence of bleeding or ulcers (Fries *et al.*, 1989). It is uncertain whether a pre-existing peptic ulcer is more prone to bleed with aspirin as evidenced from promotion by aspirin of bleeding from biopsy sites (O'Laughlin *et al.*, 1981a), although there is no evidence that adrenocorticosteroid administration leads to peptic ulceration (Conn and Poynard, 1985). It can only be presumed that if bleeding does not occur from a pre-existing ulcer, then the gastric mucosa of such a patient may be more prone to developing ulcers and bleeding with aspirin.

Bleeding is certainly much more likely if the patient has a bleeding disorder or is being treated with heparin or oral anticoagulants. Low doses of aspirin with anticoagulant therapy also causes an increase in bleeding. Although a meta-analysis of four studies demonstrated an increase in bleeding when aspirin in low doses (100 to 500 mg per day) was combined with oral anticoagulant therapy, the mortality from such haemorrhage appeared to be less (Fiore *et al.*, 1993). Turpie *et al.* (1993) concluded that the risk of bleeding with low-dose aspirin, 100 mg per day, for patients treated with warfarin was offset by an overall reduction in mortality. Several groups have reported no serious bleeds and no deaths with low aspirin daily doses of 100 mg and warfarin (Fagan *et al.*, 1994; Younossi *et al.*, 1997). Large doses of aspirin can induce hypoprothrombinaemia and increase bleeding time.

Topical administration of methylsalicylate in patients on warfarin is a generally unrecognised potential hazard (Littleton, 1989; 1990; Yip *et al.*, 1990). Warfarin interactions have also been reported with traditional medicines including, Kwan Loong Medicated Oil, which contains methyl salicylate, and Danshen, the root of *Salvia miltiorrhiza* (Tam *et al.*, 1995). Diflunisal displaces warfarin from its protein binding sites, causing a fall in blood levels, which must be kept in mind in patients receiving anticoagulant therapy (Tempero *et al.*, 1977; Serlin *et al.*, 1980a; 1980b; Verbeeck *et al.*, 1980; Furst, 1991). The potential danger of haemorrhage when aspirin is given with heparin is well documented, and this combination is generally not recommended although there might be a place for a non-acetylated salicylate or one of the new COX-2 inhibitors, or paracetamol.

It should be noted that analysis of risk factors in acute gastrointestinal haemorrhage are invariate. The variables being independent, and so it is not possible to determine the relative importance of the various risk factors (Fries *et al.*, 1989; 1995).

Acute gastrointestinal haemorrhage has been shown to cost more than the price of medication (Zeidler, 1992). Fries (1991) has suggested it is the second mostly deadly rheumatic disease, with an especially high mortality (Griffith *et al.*, 1988), particularly in the elderly (Guess *et al.*, 1983). The Committee of the Safety of Medicines Update (1986) of the UK (now superseded by the Medicines Control Agency) has recommended not giving NSAIDs to patients who have had ulcers. The risk factors that have been identified in non-steroidal anti-inflammatory analgesic-induced acute gastrointestinal haemorrhage have already been discussed. However, it is still not possible in an individual patient to predict the complication. Patients usually have no symptoms prior to a bleed (Carson *et al.*, 1987; Graham, 1988; Levey *et al.*, 1988), which Skander and Ryan (1988) have suggested may in part be because of relief of pain by NSAID therapy. Moreover, acute gastrointestinal haemorrhage is not associated with dyspepsia (Lakai *et al.*, 1987; Bijlsma *et al.*, 1988a; 1988b; Upadhyay *et al.*, 1988; 1990), gastric erosions or micro-bleeding (Ivey, 1986), or chronic iron deficiency anaemia (Upadhyay *et al.*, 1990). Infection with *H. pylori* is certainly associated with peptic ulceration (Soll *et al.*, 1991), but the role of this organism in predisposing to NSAID-induced gastric and duodenal erosions and haemorrhage (Thillainayagam *et al.*, 1994; Chan *et al.*, 1997) remains controversial, although Wolfe *et al.* (1999) consider it to be a greater risk and that the infection should be eradicated. Infection with *H. pylori* can lead to symptoms of dyspepsia (Upadhyay *et al.*, 1988).

As previously discussed, there is a large corpus of evidence to support the view that non-acetylated salicylates are less injurious to the gastric mucosa than aspirin. Roth (1988a; 1988b); Rainsford and Buchanan (1990). Greene and Winickoff (1992) have recommended a non-acetylated salicylate in preference to aspirin, but others, notably Koff and Shapiro (1996), maintain that there is equal risk of bleeding with the two preparations. The problem is that the incidence of acute gastrointestinal bleeding from NSAIDs with low ulcerogenicity is not high (Jaszewski *et al.*, 1989; Langman, 1989; Henry *et al.*, 1996). The evidence in favour of the use of non-acetylated salicylates over aspirin is based largely on gastroscopic studies (Roth, 1988a; 1988b) and low GI blood loss (Chapter 8; Table 8.11). It is certainly our recommendation that if salicylate therapy is to be used, it should preferably be the non-acetylated form (e.g. salsalate).

It should also be noted that peptic ulcers in patients with rheumatoid arthritis may heal (Gerber *et al.*, 1981; Graham *et al.*, 1988). Highly selective COX-2 inhibitors, such as celecoxib and rofecoxib (= coxibs) have a lower incidence of acute gastrointestinal ulcers and haemorrhage than reference NSAIDs (Yeomans *et al.*, 1998a; Cannon *et al.*, 1999; Emery *et al.*, 1999; Simon *et al.*, 1999; Feldman and McMahon, 2000; Oviedo and Wolfe, 2001). Recent studies on the long-term GI effects of NSAIDs and coxibs have shown that combination of aspirin at prophylactic antithrombotic doses in combination with rofecoxib results in a two-fold increase in serious GI complications, whereas combination with celecoxib does not result in an increase in these complications (Silverstein *et al.*, 2000).

What is also lacking is data on risk factors, e.g. smoking. Enteric-coated aspirin has been noted to reduce both gastric (Silvoso *et al.*, 1979; Lanza *et al.*, 1980) and duodenal ulcers (Lockard *et al.*, 1980). However, no differences were noted by Jaszewski *et al.* (1989). It is worth noting that drugs such as slow-release ketoprofen, which are largely absorbed from the small and large intestine, may also cause acute gastrointestinal haemorrhage (Collins *et al.*, 1988), as pro-drugs, such as sulindac (Carson *et al.*,

CHAPTER 12

1987). So-called buffered aspirin (the only truly buffered one is Alka-Seltzer®) has been found to be less gastrotoxic in some studies (Lanza *et al.*, 1980; Hoftiezer *et al.*, 1982) but no different in others (Lanza, 1984) compared to plain aspirin.

Although Fries *et al.* (1989; 1993) have identified NSAIDs as a major potential hazard in the treatment of patients with arthritis, especially in relation to acute gastrointestinal haemorrhage studies on the relative toxicity of NSAIDs showed that aspirin was found to be the least toxic of this class of drugs. This, however, was probably due to the widespread use of enteric-coated aspirin by the patients in the study, and certainly by the low doses, which varied from 1000 to 3000 mg/day (Fries *et al.*, 1993). The claim by these workers that aspirin was more toxic than DMARDs would not be unquestioningly accepted by the majority of clinical rheumatologists.

There is no evidence that antacids prevent acute gastrointestinal haemorrhage (Singh and Fries, 1996). They only have a weak, if any, protective effect (Caldwell *et al.*, 1987; Lanza *et al.*, 1988; Agrawal and Saggioro, 1991; Wolfe *et al.*, 1999).

There is evidence that H_2 antagonists partly prevent the damaging effects not only of aspirin (Lolu-dice *et al.*, 1981; O'Laughlin *et al.*, 1982) on the gastroduodenal mucosa but also of other NSAIDs. (Ehsanallah *et al.*, 1988; Page *et al.*, 1988; Robinson *et al.*, 1989; Agrawal *et al.*, 1991; Bank *et al.*, 1991). In some studies their protective effects were only found in the duodenum (Ehsanallah *et al.*, 1988; Robinson *et al.*, 1989; Agrawal *et al.*, 1991), while healing of ulcers in some patients with continued use of NSAIDs has been reported (Walan *et al.*, 1989). Hudson *et al.* (1997a) found that the H_2 receptor antagonist famotidine was effective in healing NSAID ulcers and in maintaining healing. However, Singh and Fries (1996) found H_2 antagonists to be ineffective in preventing gastric and duodenal lesions. Armstrong and Blower (1987) have reported acute gastrointestinal bleeding in patients receiving NSAIDs and H_2 antagonists both together.

Omeprazole and other proton-pump inhibitors have been shown in short-term studies to protect the gastric and duodenal mucosa from injury from aspirin (Daneshmend, 1990; Scheiman, 1996) and other NSAIDs (Oddson *et al.*, 1990; Ekstrom *et al.*, 1996; Taha *et al.*, 1996; Hudson *et al.*, 1997b; Cullen *et al.*, 1998). It should be noted that ulcer healing with omeprazole occurs in a proportion of patients with the continuous use of NSAID therapy (Walan *et al.*, 1989), but healing has been observed in patients with peptic ulcers despite continuous intake of aspirin (O'Laughlin *et al.*, 1981b). Omeprazole has been noted to be better than ranitidine in healing gastric and duodenal ulcers (Yeomans *et al.*, 1998a; Walan *et al.*, 1989). Singh and Fries (1996), however, found that omeprazole did not protect either the stomach or the duodenal mucosa from NSAID damage. No large-scale long-term prospective study has been done to examine the effects of H-pump blockers or H_2-receptor antagonists for the prevention of acute gastrointestinal bleeds from NSAIDs, especially in at risk subjects. The studies quoted refer to endoscopic findings of erosions and ulcers, and not to bleeding.

There is considerable evidence that the prostaglandin analogue misoprostol, both in short-term (Graham *et al.*, 1988; Eular *et al.*, 1990; Agrawal *et al.*, 1991; Geis *et al.*, 1991; Saggioro *et al.*, 1991; Jazewski *et al.*, 1992; Grazioli *et al.*, 1993) and in long-term use, i.e. up to 6 months (Silverstein *et al.*, 1986; Roth *et al.*, 1989; Graham *et al.*, 1993; Furst, 1994; Hollander, 1994; Raskin *et al.*, 1996) does protect both the gastric and the duodenal mucosa from the damaging effects of NSAID therapy in about half the patients. Roth *et al.* (1989) clearly demonstrated the healing of gastroduodenal injury by misoprostol in patients with rheumatoid arthritis who were receiving aspirin, as did Katz and Shriver (1989) and Euler *et al.* (1990). Misoprostol has been found more effective than sulcralfate (Agrawal *et al.*, 1991), H_2 receptor antagonists (Roth *et al.*, 1989; Agrawal *et al.*, 1991; Lancaster-Smith *et al.*, 1991; Robinson *et al.*, 1991; Tildesley *et al.*, 1993) and proton pump inhibitors (Hawkey *et al.*, 1998) in preventing both gastric and duodenal mucosal damage. Recent large-scale trials (OMNIUM and ASTRONAUT) and other studies evaluated using evidence-based criteria (Rostrom *et al.*, 1999) indicate that omeprazole 20 and 40 mg daily is probably about the most effective in achieving ulcer-healing and dyspepsia control in patients receiving NSAIDs for 6 months, or in secondary prophylaxis thereafter for 6 months, than misoprostol 200 mg twice daily. Both these drugs were more effective than ranitidine 150 mg twice daily. The differences in ulcer healing or secondary prophylaxis between omeprazole and misoprostol, though statistically significant, are relatively small. Moreover, there are still some 10–30 per cent of patients who do not respond in the long term. But does misoprostol prevent acute gastrointestinal haemorrhage? The only controlled trial to answer this important question was conducted by Silverstein *et al.* (1995), who showed that this and other serious complications such as perforation were reduced by approximately 50 per cent. Misoprostol has therefore been recommended for patients at risk from acute gastrointestinal haemorrhage, especially females over 60 years of age. The question, however, remains whether misoprostol should be prescribed for all patients receiving salicylate or NSAID therapy (Cullen and Hawkey, 1993), especially in view of the frequent occurrence of diarrhoea with misoprostol.

Dyspepsia

Dyspepsia is by far the most common problem encountered with salicylate therapy and, as previously discussed, at least 5 to 15 per cent of patients will be required to discontinue salicylate therapy within 6 months of treatment (Capell *et al.*, 1979). Some 60 per cent of patients will have dyspepsia, but the incidence depends on how symptoms are elicited (Huskisson *et al.*, 1974). Dyspepsia as a result of salicylate therapy is dose-related (Pullar *et al.*, 1985), but does not correlate with plasma salicylate levels (Dromgoole *et al.*, 1981). Dyspepsia is more common in patients with a past history of duodenal ulcer (Muir, 1963). There is little evidence that food or antacids reduce dyspepsia (Rainsford, 1982; 1984b), but food will reduce the T_{max} and salicylate levels (Spiers and Malone, 1967; Wood, 1967; Volans, 1974). There is some evidence that cimetidine (Bijlsma, 1988a; 1988b; Taha *et al.*, 1996), famotidine or sucralfate (Caldwell *et al.*, 1987) and misoprostol (Graham *et al.*, 1988) reduce dyspepsia, although the latter not infrequently causes diarrhoea (Walt, 1992). Dyspepsia was not associated in one study with *H. pylori* infection (Upadhyay *et al.*, 1988), or with gastric and/or duodenal erosions and microbleeding (Muir, 1963; Welch *et al.*, 1978; Silvoso *et al.*, 1979; Lanza, 1984; Collins *et al.*, 1986; Graham and Smith, 1986; Armstrong and Blower, 1987; Jorde and Burhol, 1987; Shallcross and Heatley, 1990). Apart from those patients who have had peptic ulcers and are more prone to develop dyspepsia, it is not possible to predict which patients will get dyspepsia (Muir, 1963). Although dyspepsia is common with aspirin (Hoftiezer *et al.*, 1982; O'Laughlin *et al.*, 1982) the incidence with buffered aspirin is the same (Batterman, 1958; Cronk, 1958) and is only slightly reduced by choline magnesium trisalicylate and choline salicylate (Goldenberg *et al.*, 1978), although the taste of the latter troubles many patients. The studies of Capell *et al.* (1979) showed that even with enteric-coated aspirin, 39 out of 95 patients with rheumatoid arthritis (41 per cent) had to stop treatment because of dyspepsia, the duration of the study being 1 year.

Oesophageal, small and large bowel complications

Salicylates and NSAIDs cause damage to the gastrointestinal tract, other than in the stomach and duodenum (Somerville and Hawkey, 1986). Oesophageal damage has been reported as a result of salicylate and NSAID irritation (Agdal *et al.*, 1970; Bataille *et al.*, 1982; Semble *et al.*, 1989; Eng and Sabanathan, 1991; Minocha and Greenbaum, 1991; Ecker and Karsh, 1992), especially in patients with oesophageal reflux (Eng and Sabanathan, 1991). Oesophageal complications are, in our experience, likely to occur in frail elderly female patients who have decreased salivary flow. Occasionally, serious complications such as haemorrhage, ulceration and strictures may occur (Heller *et al.*, 1982; De Caestecker and Heading, 1988).

Inflammation and ulceration in the small intestine due to long-term NSAID therapy is now well established (Mielants and Veys, 1985; Rooney and Bjarnason, 1991; Sigthorsson *et al.*, 1998). However, the study of Sigthorsson *et al.* (1998) showed that long-term use of aspirin, like that of the pro-drug nabumetone, was not associated with intestinal inflammation as observed with various NSAIDs. Inflammation of the ileum and caecum may result (Bjarnason *et al.*, 1987) in an appearance not dissimilar from that of Crohn's disease (Bannerjee, 1989). Blood and protein loss may occur (Bjarnason *et al.*,

1987) as a result of ulceration (Madhok *et al.*, 1986; Morris *et al.*, 1991), which may lead to severe haemorrhage and perforation (Langman *et al.*, 1985). Band-like strictures may occur from NSAIDs causing intestinal obstruction (Bjarnason *et al.*, 1986a; 1986b; 1987; 1988; 1993; Lang *et al.*, 1988). Increased intestinal permeability has been demonstrated as a result of NSAID therapy using ^{51}Cr EDTA (Bjarnason *et al.*, 1984; 1986a; 1986b; Jenkins *et al.*, 1987; Mielants *et al.*, 1991) and inert saccharides (Juby *et al.*, 1986). Oral corticosteroids do not have this effect (Mielants *et al.*, 1991). Glucose–citrate formulations of indomethacin (Bjarnason *et al.*, 1990) and metronidazole (Bjarnason *et al.*, 1992) reduce the increase in permeability caused by NSAIDs.

Faecal calprotectin (Tibble *et al.*, 1999; 2001) and indium-labelled polymorphonuclear scans (Rooney *et al.*, 1986) may be helpful in identifying the problem. It must always be remembered that secondary amyloidosis may affect the ileum and ascending colon, leading to ulceration and refractory iron deficiency anaemia and intestinal haemosiderosis (Lee *et al.*, 1978).

Perforation and haemorrhage from NSAID-induced damage to the large bowel have been reported (Langman *et al.*, 1985; Holt *et al.*, 1993), as well as strictures (Huber *et al.*, 1991). There are reports of colitis resulting from NSAID use in the elderly (Bjarnason *et al.*, 1992; Gibson *et al.*, 1992). In an autopsy study of deaths in patients with rheumatoid arthritis, Brooks *et al.* (1975) noted that perforation of colonic diverticulae was the most common iatrogenic cause of death, but as a consequence of oral corticosteroid therapy. Leach and Callum (1981) reported benign colonic fistula formation in patients receiving aspirin and oral corticosteroids. Aspirin suppositories can cause proctocolitis and rectal bleeding (Levy and Gaspar, 1975; Pearson *et al.*, 1983; Holt *et al.*, 1993). Inflammatory bowel disease may be worsened with NSAID therapy (Gilat, 1979; Rampton and Sladen, 1981). However, aspirin may reduce diarrhoea from bacterial infection (Atkinson and Collier, 1980).

Renal adverse events

Salicylates, like other NSAIDs, are potentially nephrotoxic (Scott *et al.*, 1964; Kincaid-Smith *et al.*, 1968; Murray and Goldberg, 1975a; 1975b; Nanra and Kincaid-Smith, 1975; Kimberley *et al.*, 1978b; Nanra *et al.*, 1978; Plotz and Kimberley, 1981; Nanra, 1983; Unsworth *et al.*, 1987). Non-acetylated salicylates, being weak prostaglandin inhibitors, appear less toxic than aspirin. There are five renal syndromes now recognised:

1. Haemodynamic or functional acute renal failure (Plotz and Kimberley, 1981; Clive and Stoff, 1984a; 1984b; Dunn, 1987; Blackshear *et al.*, 1983; 1985; Perneger *et al.*, 1994)
2. Hyponatraemia
3. Hyperkalaemia
4. Interstitial nephritis
5. Analgesic nephropathy.

The first three are predictable, being related to inhibition of renal prostaglandins (Levenson *et al.*, 1982; Patrono, 1986; Zambraski *et al.*, 1988). The first three reactions are seen especially in elderly patients.

The pharmacological effects of aspirin and non-acetylated salicylates on renal function (Beeley and Kendall, 1971; Kimberley and Plotz, 1977; Kimberley *et al.*, 1978; Shelley, 1978; Munther and Bennett, 1980; Levenson *et al.*, 1982; Patrono, 1986) include:

1. Reduction in glomerular filtration and renal blood flow
2. Transient loss of tubular cells on starting aspirin therapy
3. Interstitial nephritis
4. Analgesic nephropathy
5. Nephrotic syndrome.

Sites for drug interactions at the level of tubular functions

Reduced glomerular filtration and renal blood flow are related to inhibition of renal prostaglandins (Beeley and Kendall, 1971; Kimberley and Plotz, 1977; Brooks and Cossum, 1978; Kimberley *et al.*, 1979; Garella and Matarese, 1984; Arroyo *et al.*, 1986; Patrono, 1986). This, however, only occurs in patients with impaired renal function, and not in normals (Blackshear *et al.*, 1985; Brater, 1990a; Prescott and Martin, 1992). In some rheumatic conditions e.g. severe rheumatoid arthritis or ankylosing spondylitis there may be intrinsically abnormal functions (Lourie *et al.*, 1977) that are exacerbated by the intake of aspirin or other analgesics. Reduction in renal function occurs in patients with both aspirin and non-acetylated salicylates, but impaired renal function quickly returns to normal when the aspirin is discontinued (Kimberley *et al.*, 1978). Reduction in renal function has been reported with salsalate (Abraham and Stillman, 1987), although less so than with aspirin (Stillman and Schlesinger, 1990). Renal function quickly returns to normal with discontinuation of salicylate therapy (Kimberley *et al.*, 1978a). There are indications of interactions with drug-induced hepatic dysfunction that influence the development of renal effects from NSAIDs (Zipser, 1986) as well as the influence of other drugs (Zipser and Henrich, 1986).

Patients who are hypovolaemic owing to diminished fluid intake or excessive loss, who are elderly and receiving treatment for congestive cardiac failure, or who have hepatic cirrhosis, may develop acute renal failure (Clive and Stoff, 1984a; 1984b; Blackshear *et al.*, 1985; Arroyo *et al.*, 1986; Brater, 1990a; 1990b; Buchanan and Brooks, 1991). No detrimental effect has been noted in patients with normal renal function, even when sodium is depleted (Staessen *et al.*, 1983). It is clear that these complications are preventable.

Acute interstitial nephritis is a rare complication of NSAID therapy, usually occurring 6 months after starting therapy (Bender *et al.*, 1984).

A small number of patients have features of a hypersensitivity reaction, with eosinophilia (Blackshear *et al.*, 1985). The condition is usually reversible following discontinuation of the drug (Blackshear *et al.*, 1985). Acute interstitial nephritis has been reported with diflunisal (Chan *et al.*, 1980), but to our knowledge it has not been described with aspirin or other non-acetylated salicylates.

Although phenacetin was the initial component in analgesic mixtures implicated as a cause of analgesic nephropathy, it is now clear that the syndrome can occur with aspirin and other NSAIDs (Buchanan *et al.*, 1979a; Muhalwas *et al.*, 1981). However, several long-term studies of aspirin have failed to identify even a single case (Macklon *et al.*, 1974; Akyol *et al.*, 1982; Emkey and Mills, 1982; New Zealand Rheumatism Association Study, 1974). With regard to the New Zealand Study, comprising 908 patients on high anti-inflammatory doses of aspirin, Murray and Goldberg (1975a) did find, subsequent to the report, five cases of renal papillary necrosis. Cancer of the bladder is a well-recognised complication of analgesic nephropathy (McCready, 1982), and has been reported in elderly patients taking aspirin (Prescott, 1982).

Nephrotic syndrome is an uncommon complication of NSAID therapy, and usually resolves on discontinuation of the drug. A mild case has been reported with salsalate (Valles and Tovar, 1987).

As discussed previously, aspirin and other non-acetylated salicylates blunt the effect of thiazide and loop diuretics, and angiotensin-converting enzyme inhibitors (Buchanan and Brooks, 1991). Reimann *et al.* (1985) recorded a slight fall in serum concentration of lithium and a slight increase in renal excretion following administration of 4 g of aspirin. However, other workers have not found any differences with co-administration of the two drugs, either with anti-inflammatory or small analgesic doses of aspirin or sodium salicylate (Ragheb, 1987; Langlois and Paquette, 1994). No differences were noted with lysine acetylsalicylic acid (Singer *et al.*, 1981).

The nephrotoxicity of cyclosporin may be exacerbated by cyclo-oxygenase inhibition by NSAIDs (Buchanan, 1992), although low dose aspirin has been shown to alleviate cyclosporin nephrotoxicity in the rat (Adams *et al.*, 1993). Kovarik *et al.* (1993) found no interaction of cyclosporin with aspirin in a daily dose of 2880 mg in normal subjects. The pharmacokinetics of salsalate have been studied in patients undergoing renal dialysis (Williams *et al.*, 1986). Because elimination of salicylic acid is impaired in dialysis, the dosage of salsalate should be reduced.

CHAPTER 12

Aspirin has been found to increase the risk of haemorrhage after lithotrypsy (Ruiz and Saltzman, 1990) and prostatectomy (Watson *et al.*, 1990). Aspirin has proved successful in inhibiting platelet aggregation in elderly uraemic patients with heparin-induced thrombocytopaenia (Matsuo *et al.*, 1989).

Hepatic toxicity

Increased serum transaminase concentrations were noted in the mid 1950s in children with rheumatic fever treated with aspirin (Mydick *et al.*, 1955; Manso *et al.*, 1956), but this was largely ignored until the 1970s, when it became apparent that the same also occurred in a number of conditions including juvenile rheumatoid arthritis (Rich and Johnson, 1973; Athreya *et al.*, 1975), adult rheumatic disease (Seaman and Plotz, 1976; Wilson, 1976) and systemic lupus erythematosus (Seaman *et al.*, 1974; Seaman and Plotz, 1976), Sjögren's syndrome (Webb *et al.*, 1973) and Reiter's syndrome (Ricks, 1976). The complication occurs in approximately 5 per cent of patients with adult rheumatoid disease (Paulus *et al.*, 1982) and although it occurs with other NSAIDs the rate appears much lower (Paulus, 1982). However, with paracetamol (acetaminophen), diclofenac, sulindac and phenylbutazone the rate of occurrence is higher (Katz and Love, 1991). It is not known whether non-acetylated salicylates produce less hepatotoxicity than aspirin, but a severe case of eosinophilic hepatitis as a result of hypersensitivity with eosinophilia has been reported with choline magnesium trisalicylate (Nadkarmi *et al.*, 1992). The explanation for the disease non-specificity (Zimmerman, 1974; 1981; Zucker *et al.*, 1975; Buchanan and Brooks, 1991; Schiff and Maddrey, 1994; Lee, 1995) is probably that it is associated with hypoalbuminaemia, since there is an inverse correlation (Gitlin, 1980), and the damaging effect on monolayer cultures of hepatocytes can be reduced by the addition of albumin (Tolman *et al.*, 1978). Increased serum transaminase concentrations have, however, been observed in normal subjects receiving aspirin (Garber *et al.*, 1975). The phenomenon appears to be dose-dependent, usually occurring when serum concentrations of salicylate exceed 25 mg/dl (O'Gorman and Koff, 1977). Patients with increased serum transaminase levels are usually asymptomatic (O'Gorman and Koff, 1977) and quickly return to normal on stopping aspirin, or even if aspirin is continued (Russell *et al.*, 1971). Occasionally patients may be symptomatic, with anorexia and nausea (O'Gorman and Koff, 1977); rarely, the patient may become severely ill with hepatomegaly, hyperbilirubinaemia and an increased prothrombin time (Benson, 1983). This, however, is extremely rare. Generally it is recommended that high doses of aspirin be stopped in patients with elevated serum transaminase levels when the prothrombin time is increased (Dromgoole *et al.*, 1981; Katz and Love, 1991). The histological findings in the liver include centrilobular hepatocellular degeneration with periportal inflammatory infiltrates of polymorphonuclear leucocytes (Seaman and Plotz, 1976), but similar changes are found in patients with primary Sjögren's syndrome who have mitochondrial antibodies (Webb *et al.*, 1975; Whaley *et al.*, 1975a; 1975b; Whaley and Webb, 1977). Biopsy findings do not include cholestasis or any feature of other types of iatrogenic liver disease (Dawes and Symmons, 1992; Lee, 1995).

Aspirin allergy

Aspirin allergy can be subdivided into three groups:

1. Bronchial asthma
2. Urticaria and angio-oedema; and
3. Anaphylactic shock (Buchanan *et al.*, 1979a).

Lamson and Thomas (1932) appear to have been the first to describe asthma as an untoward effect of aspirin therapy. Aspirin hypersensitivity is relatively rare in healthy subjects, but one in five patients with bronchial asthma can be expected to have worsening of their asthma when taking aspirin (McDonald *et al.*, 1972; Morassut *et al.*, 1989). Aspirin-induced asthma appears to be peculiarly common in patients with nasal polyps, the triad of nasal polyps, aspirin and asthma being known eponymously as Samter's syndrome (Samter and Beers, 1967; 1968; Samter, 1969a; 1970; Zeitz and Jarmoszuk, 1985; Zeitz, 1988). The pathogenesis of aspirin-induced asthma is not entirely certain, but most evidence points to inhibition of cyclo-oxygenase (Szczeklik et al., 1979; 1986; Szczeklik, 1983; 1990) with perhaps release of anaphylactic mediators such as the leukotrienes (i.e. SRS-As), as a result of a switch to the leukotriene pathway (Dromgoole et al., 1982; Szczeklik, 1983a). In support of inhibition of cyclo-oxygenase being the cause of aspirin-induced asthma is the fact that the more potent NSAID inhibitors of cyclo-oxygenase are more likely to cause asthma (Szczeklik, 1980; 1986; Zeitz, 1988), while paracetamol (acetaminophen) (Szczeklik, 1986) does not, and non-acetylated salicylates have a much lower incidence (Dromgoole et al., 1982; Settipane, 1988; Szczeklik et al., 1990). However, the latter have been noted occasionally to cause an exacerbation of asthma (Chudwin et al., 1986), and aspirin has been found, in a small proportion of patients with bronchial asthma, to cause improvement (Szczeklik and Nizankowska, 1983). Intermittent use of aspirin has been noted by Strom et al. (1987) to result in an increase in exacerbation of asthma. In contrast to these studies on aspirin-related hypersensitivity, there have been some reports indicating that patients with asthma may have their symptoms relieved by aspirin (Szczeklik et al., 1978).

In the Samter syndrome, eosinophilia is common. Aspirin reactions leading to asthma have been blocked by ketoprofen (Delaney, 1983) and modified by H_1 antihistamines (Szczeklik and Serwonska, 1979) and sodium cromoglycate (Basomba et al., 1976) although the latter is not very effective (Dahl, 1981). There was evidence in the past that the allergic reaction was the result of tartrazine, which does not inhibit prostaglandins, but if it does occur it would seem to be rare (Szczeklik, 1986). Likewise, there is little current evidence that it is due to a contaminant aspirin anhydride that produces aspiryl conjugates in vitro and to which antibodies develop (De Weck, 1971).

Pulmonary oedema, not due to cardiac disease, can rarely occur in patients receiving aspirin therapy, and is probably due to an increase in pulmonary lymph flow or an allergic-type pneumonia (Davis and Burch, 1974; Bowers et al., 1977).

Urticaria and angio-oedema are more common than asthma as a result of aspirin therapy (Speer, 1975). In patients with urticaria it is prudent to avoid aspirin or salicylates, since exacerbations can persist for up to 3 weeks after ingestion (Monroe and Jones, 1977). Aspirin will also cause an exacerbation of urticaria in patients already suffering from the disease (Warin and Smith, 1982; Schlegel and Paulus, 1986). Like asthma the pathogenesis may be related to inhibition of prostaglandins (Szczeklik et al., 1977), since other NSAIDs can provoke similar reactions while acetaminophen does not (Buchanan et al., 1979a). Although aspirin intolerance remains for life, Speer (1975) considered in a study extending over 25 years that the prognosis was good. Death from anaphylactic shock is extremely rare (Speer, 1975). Aspirin is however more likely than non-acetylated salicylates to cause such reactions (Samter and Beers, 1968; Smith, 1968), although a severe anaphylactic reaction has been reported after choline magnesium trisalicylate (Nadkarni et al., 1992). Severe anaphylactic reactions are, however, extremely rare. The commonest complication is acute bronchospasm, which usually occurs in patients with pre-existing bronchial asthma and nasal polyps (Samter's syndrome) (Zeitz, 1988). Up to 20 per cent of patients with bronchial asthma cannot take aspirin (Morassut et al., 1989), and a dose as low as 40 mg may precipitate an asthma attack (Oates et al., 1988). Bronchospasm can be of varying severity (Samter, 1970; 1977), and is probably brought about by inhibition of prostaglandin biosynthesis (Szczeklik et al., 1975) rather than a true allergy, although increases in IgE concentrations and eosinophilia may be found (Buchanan et al., 1979a). Inhibition of cyclo-oxygenase may result in diversion to lipoxygenase products, but this awaits confirmation (Strom et al., 1987). Desensitisation to aspirin in aspirin-sensitive asthmatic patients is more effective in upper respiratory symptoms than in bronchospasm (Pleskow et al., 1982; Chudwin et al., 1986). Patients who take aspirin intermittently rather than on a continuous basis are more likely to develop allergic reactions (Strom et al., 1987), as are patients with pre-existing allergies to foods etc. (Settipane et al., 1974; Buchanan et al., 1979a). Urticaria, unlike asthma, does not appear to be due to prostaglandin inhibition (Settipane, 1983).

Ototoxicity

Ototoxicity is a common and troublesome adverse effect of salicylate therapy, especially with high dosage (Myers *et al.*, 1963). With reduction in dosage, both tinnitus and deafness rapidly clear up (Silverstein *et al.*, 1967). No morphological changes have been described with the development of auditory changes (Bernstein and Weiss, 1967). Hearing loss is sensory neural, but vestibular changes may also occur (Silverstein *et al.*, 1967). Tinnitus in normal subjects occurs when the serum salicylate concentration exceeds 25 mg/dl, but patients with pre-existing hearing loss, such as the elderly, may not experience tinnitus even with extremely high serum levels (Morgan *et al.*, 1973; Halla and Hardin, 1988). Indeed, tinnitus and deafness may not occur in elderly patients who otherwise manifest symptoms of salicylate intoxication (Taggart *et al.*, 1991; Taggart, 1992). Although minimum thresholds for both tinnitus (6 mg/100 ml) and deafness (22 mg/dl) have been suggested (Myers *et al.*, 1963), the studies of Halla and Hardin (1988) showed that both these symptoms could occur below these levels. Day *et al.* (1989) studied the relationship between unbound 'free' plasma salicylate concentrations and ototoxicity in normal subjects treated with aspirin. Hearing loss and tinnitus intensity increased with aspirin dosage, and with both unbound and total salicylate concentrations. These authors could not conclude from their studies that unbound salicylate concentration was a better predictor of salicylate-induced ototoxicity than total salicylate concentration.

Pulmonary oedema

Pulmonary oedema was first reported in patients with acute rheumatic fever who were being treated with large doses of salicylates (Reid *et al.*, 1950). However, by the mid-1970s it became apparent that patients with no cardiac disease could develop pulmonary oedema from salicylates (Granville-Grossman and Sergeant, 1960; Davis and Burch, 1974; Hrnicek *et al.*, 1974; Tashima and Rose, 1974; Broderick *et al.*, 1976). Increased pulmonary lymph flow, which has been recorded in sheep infused with aspirin, may be responsible (Bowers *et al.*, 1977). Pulmonary oedema occurs in patients with serum salicylate concentrations in the toxic range, i.e. greater than 3040 µg/dl.

In a study of 111 patients with salicylate intoxication, Walters *et al.* (1983) found that cigarette smoking, a component of metabolic acidosis along with salicylate ingestion were risk factors for development of pulmonary oedema. The authors considered the aetiology to be multifactorial, but primarily due to altered vascular permeability in the lungs.

Cardiovascular adverse effects

Fluid retention occurs with both aspirin and non-acetylated salicylates, and may lead to pedal oedema and even precipitate congestive heart failure in the elderly (Hanzlik *et al.*, 1917; Berg, 1977; Feenstra *et al.*, 1997; Van den Ouweland *et al.*, 1988; Page and Henry, 2000). The relative risks in case–controlled studies from non-aspirin NSAIDs (odds ratios, adjusted 2.8 (CI 1.3–5.2) were similar to those from all NSAIDs (2.8; CI 1.5–5.1) when taken in the week prior to hospitalisation (Page and Henry, 2000). There was, however, a marked increase in risk with use of long $t_{\frac{1}{2}}$ NSAIDs (Page and Henry, 2000). An unusual case of fluid retention simulating congestive heart failure after ingestion of 1500 mg of aspirin was recently reported in a 29-year-old woman whose only significant medical history was that of migraine. A challenge test proved positive and produced symptoms of congestive heart failure (Manfredini *et al.*, 2000). Aspirin does lead to an increase in blood pressure (Walter *et al.*, 1981; Mills *et al.*, 1982), but perhaps less so than other NSAIDs such as indomethacin and naproxen (Chalmers *et al.*, 1984). Non-prescription aspirin has been noted to cause an increase in blood pressure, but only in those patients who are hypertensive (Bradley, 1991). There have been reports that aspirin may induce angina (Al-Abassi, 1983; Miwa *et al.*, 1983) and cause coronary artery spasm in aspirin-induced asthma (Habbab *et al.*, 1986), but this has yet to be confirmed. There is evidence that NSAIDs, including salicylates, may cause an increase in cardiac sensitivity to ouabain (Wilkerson and Glenn, 1977). Analgesic doses of aspirin increase the plasma concentration of glyceryl trinitrate (Rey *et al.*, 1992), leading to an increased effect of the latter. On the other hand, antiplatelet doses have only a slight effect on the plasma glyceryl trinitrate levels, which does not appear to be clinically significant (Rey *et al.*, 1992). These pharmacokinetic effects of aspirin may also explain the influences on the haemodynamic effects of nitroglycerine (Weber *et al.*, 1983). After coronary artery bypass surgery, patients who had taken small doses of aspirin for at least 3 months required more glyceryl trinitrate than those who had not taken aspirin (Rey *et al.*, 1992). Glyceryl trinitrate has an effect on platelet aggregation, and when aspirin and glyceryl trinitrate are taken together the effect is additive (Karlberg *et al.*, 1993). The mechanism of the interaction between aspirin and glyceryl trinitrate is not understood, but may relate to the inhibition of prostaglandins by aspirin (which suppress the vasodilator effects of glyceryl tri-nitrate) and the reduction of hepatic blood flow through the liver by aspirin, thus diminishing the bio-transformation of glyceryl trinitrate.

Salicylate cerebral effects

Tachypnoea, confusion, ataxia oliguria and insomnia occurring in adult patients receiving high doses of salicylate are frequently not recognised as being due to salicylate intoxication (Anderson *et al.*, 1976). Salicylates readily penetrate the blood–brain barrier, although the concentration in cerebrospinal fluid is much lower than in plasma (Brem *et al.*, 1973; Buchanan and Rabinowitz, 1974; Bannwarth *et al.*, 1989).

Effects on joint destruction

It has been suggested that NSAIDs may accelerate cartilage destruction, especially in the hip (Solomon, 1973; Ronningen and Langeland, 1979; Newman and Ling, 1985). However, other authors (Dieppe *et al.*, 1993; Williams *et al.*, 1993) have claimed that there is no convincing evidence that NSAIDs adversely affect cartilage metabolism in osteoarthritis, while there is one report that aspirin may have a protective effect (Chrisma *et al.*, 1972). This and the issue regarding the purported chondroprotective effects of some NSAIDs has been the subject of much debate. Concepts of the actions of NSAIDs in the osteoarthritic joint (see Figure 12.3) show that there is a complex series of interactions between the drugs and the disease state. Factors such as biomechanical responses to varying degrees of joint destruction, and potential overuse of joints that would otherwise be painful, as a consequence of the analgesic effects of the drugs, so contributing to the complexity of the issue, whether or not NSAIDs accelerate cartilage destruction. Clinical trials to examine this issue have been undertaken by Rashad *et al.* (1989; 1992). In the first of these (Rashad *et al.*, 1989) compared indomethacin with azapropazone in the course of hip joint osteoarthritis. The findings suggested that indomethacin cause more rapid destruction of cartilage compared with azapropazone when loss of joint space was employed as the criterion, and this was corroborated with *in vitro* studies showing that this indomethacin caused inhibition of proteoglycan synthesis in both normal porcine and human osteoarthritic cartilage (Rainsford *et al.*, 1989; 1992). Other studies identified some other NSAIDs (Rainsford *et al.*, 1992), which are potent prostaglandin synthesis inhibitors, as potential causes of joint destruction in osteoarthritis (Rashad *et al.*, 1992). Indomethacin has also been shown to stimulate release of pro-inflammatory cytokines from human osteoarthritic and normal porcine synovium *in vitro* (Rainsford *et al.*, 1997). Indomethacin is the only NSAID to affect synovial perfusion as tested by ^{133}Xe clearance, and this may be relevant to its effects on joint destruction (Dick *et al.*, 1970; 1971a; 1971b; 1971c).

Earlier studies showed that in the concentrations that are found in serum and synovial fluid, both aspirin and sodium salicylate suppress proteoglycan synthesis in normal cartilage in *in vitro* studies (Pal-moski *et al.*, 1980; Ghosh, 1988; Brandt, 1993; see also Chapter 7). The inhibitory effects are greater in

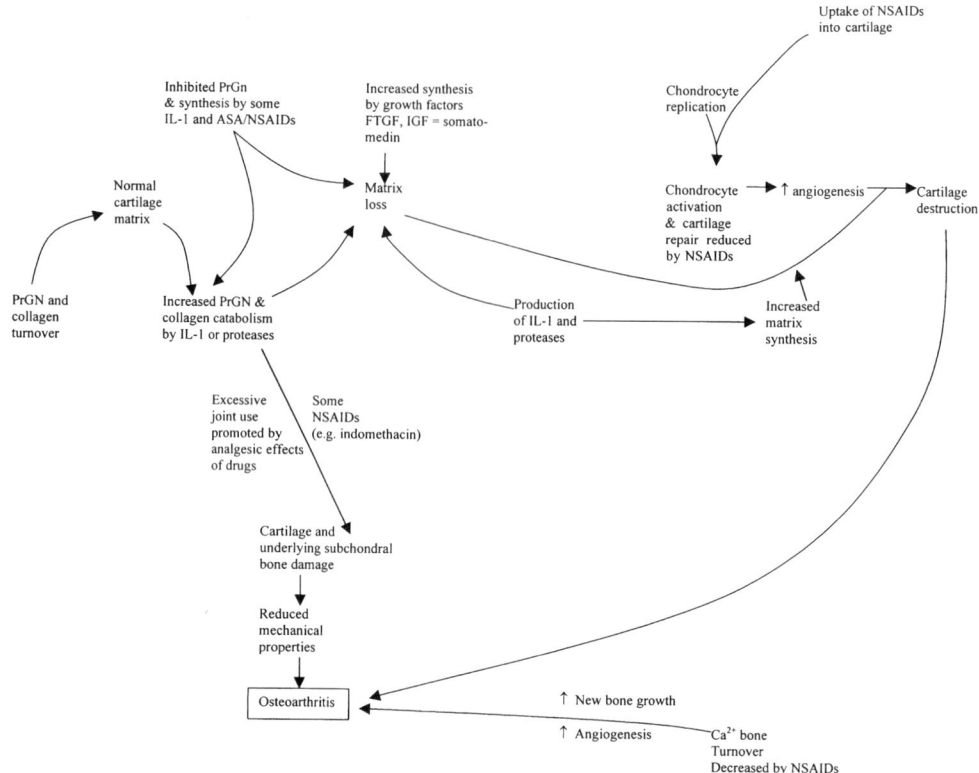

Figure 12.3 Factors influencing the joint destruction in osteoarthritis and the variable effects of NSAIDs. These drugs vary considerably in their propensity to affect different biochemical and cellular events underlying cartilage and bone metabolism and responses to inflammatory mediators and destructive enzymes.

cartilage slices from dogs with experimental osteoarthritis, and *in vivo* studies in this canine model show an increase in degenerative changes with aspirin, when serum concentrations of 20 to 25 µg/dl are present, while in un-operated joints and in healthy control dogs cartilage remains histologically and biochemically normal (Jobanputra and Nuki, 1994). However, it has to be noted that to date no long-term clinical study of aspirin and salicylate therapy has been performed in patients with osteoarthritis.

COX-2 enzyme has been identified as increased in osteoarthritic cartilage, which raises the question as to the effect of COX-2 inhibitors on cartilage integrity (Amin *et al.*, 1997). Some recent studies have shown that cartilage matrix synthesis is affected differentially by NSAIDS with varying COX-2 selectivity, some stimulating synthesis by indirectly affecting release of interleukin 1, others directly inhibiting synthesis by effects on synthetic enzymes (including the COX-2 inhibitor, nimesulide), and others having no affect (Dingle, 1999; Pelletier, 1999). Aspirin falls into the latter group, in having no effect on cartilage matrix synthesis in human arthritic and normal cartilage (Dingle, 1999; Pelletier, 1999), although other studies show that salicylate does inhibit proteoglycan synthesis (Palmoski *et al.*, 1980; Rainsford *et al.*, 1992).

Thyroid function

Aspirin has been used to treat subacute (de Quervain's) thyroiditis but if symptoms are severe oral corticosteroids are usually prescribed. Salicylates in large doses have been noted to depress serum cholesterol concentrations in patients with coronary artery disease (Alexander *et al.*, 1959), hypothyroidism and other conditions associated with hypercholesterolaemia (Alexander and Johnson, 1956). Aspirin does affect thyroid function tests and changes include competitive inhibition of thyroid hormone binding to serum carrier proteins (Wolff and Austen, 1958; Larsen, 1972). Thus serum protein-bound iodine concentrations fall, although free thyroxine levels remain unaltered (Wolff and Austen, 1958) or only slightly increased (Larsen, 1972). The thyroid uptake of iodine is decreased (Wolff and Austen, 1958), as is the clearance of iodine (Wolff and Austen, 1958). Aspirin also affects the peripheral metabolism of thyroxine (Wolff and Austen, 1958; Woeber and Ingbar, 1964) and blunts the response of the release of thyroid-stimulating hormone (Dussault *et al.*, 1976; Ramay *et al.*, 1976). Aspirin also inhibits the hepatic conversion of thyroxine to tri-iodothyronine (Chopra *et al.*, 1980). Similar abnormalities in thyroid function have also been observed with salsalate (McConnell, 1989; 1992). Despite these changes, patients remain euthyroid.

Reye's syndrome

This syndrome has been reviewed in Chapter 9 and by Porter *et al.* (1990). It is also now known as the 'macrophage activation' syndrome (Prieur and Stephen, 1994), and carries the eponym of the Australian physician R.D.K. Reye, who first described this condition in the *Lancet* of 1963 (Reye *et al.*, 1963), although it was also possibly noted earlier in 1929 by Brain *et al.* (1929). It is discussed here in the context of treating children who have juvenile rheumatoid arthritis (or other rheumatic disease) with salicylates. The clinical features include an encephalopathy, varying in severity from drowsiness to deep coma, associated with severe vomiting and hepatomegaly. Serum transaminase, lactic dehydrogenase and ammonia concentrations are significantly elevated, and acid–base disturbances, hypoglycaemia and coagulation abnormalities may ensue in severe cases. Surprisingly, jaundice is absent. Most cases have been reported in children between the ages of 5 and 15 years, with a mortality of between 20 and 40 per cent (Corey *et al.*, 1976; Buchanan and Brooks, 1991). The syndrome has most frequently followed a viral infection, such as influenza B or varicella, for which the patient received aspirin (Johnson *et al.*, 1963). Microvascular fatty changes in hepatocytes (Prescott, 1986) were originally considered pathognomic of the condition, but the same is also found in salicylate intoxication (Starko and Mullick, 1983). Both influenza B virus and salicylic acid have been noted in mice to inhibit fatty acid beta oxidation (Trauner *et al.*, 1988), which may explain the increase in free fatty acids and ammonia. It has also been suggested that children who develop Reye's syndrome may have lower aspirin esterases (Tomasova *et al.*, 1984), or that the syndrome may result from an aspirin-induced aberration in the immune response to a viral infection (Bailey *et al.*, 1982).

Clinical reports and case–control studies in the 1980s established an association with treatment with aspirin (Starko *et al.*, 1980; Halpin *et al.*, 1982; Waldman *et al.*, 1982; Hurwitz *et al.*, 1985; 1987a; 1987b; Glen-Bott, 1987; Hall *et al.*, 1988), which has largely been borne out by a fall in mortality after public health follow-up in 1980, the Surgeon General's warnings in the USA (Surgeon General, 1982), and labelling of aspirin bottles cautioning the risk of Reye's syndrome (Arrowsmith *et al.*, 1987; Hurwitz, 1989; Davis and Buffler, 1992). Forsyth *et al.* (1989) stated that bias accounted for the association of aspirin and Reye's syndrome. The use of aspirin in the USA certainly fell after the warnings about the risk (Taylor *et al.*, 1985; Remington *et al.*, 1986; Arrowsmith *et al.*, 1987; Belay *et al.*, 1999), but there could be other factors underlying this reduced incidence. A dose–response relationship has been reported in some studies (Pinsky *et al.*, 1988; Hurwitz, 1989), but cases occurred in which other patients have received doses well below the antipyretic level of 80 mg/kg per day (Buchanan and Brooks, 1991).

No evidence was found to relate the onset of Reye's syndrome to whether aspirin was taken fasting or after meals (Williams *et al.*, 1990). Although the evidence is strong that aspirin plays a role in the pathogenesis of the syndrome, patients have been reported who had not taken the drug (Buchanan and Brooks, 1991), leading several workers to doubt the association (Orlowski, 1984; Anon., 1987; Barel and Orlowski, 1988; Editorial, 1990). An Australian study failed to demonstrate a link between aspirin and Reye's syndrome, but in this study aspirin use was based on chart reviews only, the patients were much younger than those who usually develop Reye's syndrome and the use of aspirin was uncommon (Orlowski *et al.*, 1990). The possibility that Reye's syndrome may just be disappearing and not be associated with the decline in the use of aspirin has been proposed but is difficult to substantiate (Orlowski *et al.*, 1987).

It is particularly noteworthy that Reye's syndrome has been observed on long-term treatment with aspirin in a few patients with juvenile rheumatoid arthritis (Daum *et al.*, 1976; Christoffersen *et al.*, 1980; Young *et al.*, 1984; Remington *et al.*, 1985; Rom *et al.*, 1987), systemic lupus erythematosus (Hansen *et al.*, 1985) and connective tissue disease (Rennebohm *et al.*, 1985; Barrett *et al.*, 1986). Reye's syndrome must, however, be differentiated from chronic salicylism, which can occur in juvenile rheumatoid arthritis with long-term high-dose salicylate therapy (Everson and Krenzelok, 1986). The syndrome has also been reported in adult patients with Still's disease (Stillman *et al.*, 1983; Tumiati *et al.*, 1989).

Since no safe level of aspirin use can be recommended (Pinsky *et al.*, 1988) and it is impossible to predict which child may develop Reye's syndrome, it seems prudent not to prescribe aspirin to children for whatever reason. Nevertheless, aspirin is still recommended for the treatment of acute rheumatic fever (Marshall, 1990; Bisno, 1997; Cassidy, 1997; Gibofsky and Zabriskie, 1997; 1998; Alsaeid and Majeed, 1998; Williams, 1998) and juvenile rheumatoid arthritis (Cassidy, 1997; Pachman and Poznanski, 1997; Prieur, 1998; Sherry *et al.*, 1998). The exception to this view would appear to be that of Ansell (1998), who does not mention salicylates in the treatment of rheumatoid factor positive juvenile rheumatoid polyarthritis but does state that good responses are obtained from aggressive therapy with sulphasalazine or methotrexate. Pachman and Poznanski (1997) are the only authors who warn against using salicylates if influenza and varicella infection are present. Although there has been a tendency to use alternative NSAIDs, in particular ibuprofen and naproxen (both in liquid form) (Goldenstein-Schauberg, 1997), it is surprising that so many authorities still recommend aspirin despite the warnings regarding Reye's syndrome (Rahwan and Rahwan, 1986). There are no reports, to our knowledge, on non-acetylated salicylates causing Reye's syndrome. For infants, i.e. children less than 2 years of age, acetaminophen and ibuprofen should be the drugs of choice for the treatment of fever (Lesko and Mitchell, 1999). Note, however, should be made of a syndrome resembling Reye's syndrome in mice infected with the influenza virus and treated with ibuprofen (Mukhopadhyay *et al.*, 1992).

A number of inborn errors of metabolism can mimic Reye's syndrome, and clinical differentiation may be difficult. However, the majority of these occur in children under 3 years of age, and are characterised by recurrent episodes, hypoglycaemia, muscle weakness and cardiac enlargement, and are present in siblings (Roe *et al.*, 1986; Anon., 1987; Rowe *et al.*, 1988; Forsyth *et al.*, 1991; Green and Hall, 1992). Ultrastructural changes occurring in the liver are quite different from those found in Reye's syndrome, especially regarding changes in the mitochondria. In Reye's syndrome the mitochondria show pleomorphic changes with loss of dense granules, while in inborn errors of metabolism the mitochondria are normal in both size and appearance (Partin *et al.*, 1984; Green and Hall, 1992).

FACTORS AFFECTING SIDE EFFECTS AND THERAPEUTIC EFFECTS

Age and sex

Age *per se* does not appear markedly to affect plasma concentrations of salicylates (Castleden *et al.*, 1977; Salem and Stevenson, 1977; Cuny *et al.*, 1979; Roberts *et al.*, 1983; Ho *et al.*, 1985). However, minor differences between the young and the elderly have been observed, although there do not appear to have any clinical significance. They include increase in the plasma half-life, $t_{\frac{1}{2}}$ (Curry *et al.*,

1979); increase in the area under the curve, AUC (Salem and Stevenson, 1977); increase in the apparent volume of distribution (Curry *et al.*, 1979; Larkin, 1979; Roberts *et al.*, 1983; Ho *et al.*, 1985); and increase in the level of the metabolite salicyluric acid (Montgomery and Sitar, 1981; Ho *et al.*, 1985). The increase in the apparent volume of distribution can be attributed to low serum concentrations of albumin, which are known to be low in the elderly, especially those poorly nourished and with advanced illness and severe debility (Greenblatt *et al.*, 1982). There is a correlation between serum concentrations of salicyluric acid and serum creatinine, and higher concentrations in elderly patients can be explained by diminished renal elimination (Gunsberg *et al.*, 1984).

Sex differences have been noted in the plasma concentrations of both acetyl salicylic acid and salicylate, with greater values noted in females (Coppe *et al.*, 1981; Kelton *et al.*, 1981; Buchanan *et al.*, 1983; Ho *et al.*, 1985). Mena *et al.* (1982) noted there were sex differences in the absorption of sodium salicylate. Kelton *et al.* (1981) ascribed the increased plasma levels of salicylates to the increased absorption in females, but did not prove that it was not the result of a volume effect, i.e. a small 'pool' of distribution due to smaller lean body mass. It seems more likely to be the result of a lower formation of salicyluric acid in females (Sechserova *et al.*, 1979). It is, however, strange that if this is the case the T_{max} was found to be shorter in males in the study of Miaskiewicz *et al.* (1982). Higher concentrations of plasma acetyl salicylic acid in females may be due to lower amounts of aspirin esterase in this sex (Menguy *et al.*, 1972; Windorfer *et al.*, 1974).

Sex-related differences in absorption of acetylsalicylic acid have been observed both in humans (Kelton *et al.*, 1981; Miaskiewicz *et al.*, 1982; Trnavska and Trnavsky, 1983) and in animals (Buchanan *et al.*, 1983). However, these differences may be the effect of distribution, i.e. a pool effect, since the lean body mass or total body water is reduced in females. Graham *et al.* (1977a; 1977b) have reported differences in plasma clearance of salicylates, T_{max}, in males and females. These differences would seem unlikely to have any clinical consequences in patients with osteoarthritis or other rheumatic disease treated with aspirin, but may be of importance in prevention of thrombosis.

More important than the pharmacokinetics of aspirin and salicylates in the elderly are the pharmacodynamic effects. Adverse reactions are not only increased in elderly persons (Alexander *et al.*, 1983; Morgan and Furst, 1986), but also occur at lower plasma concentrations (Grigor *et al.*, 1987). The elderly often have multiple illnesses and as a result take more medications (Vestel, 1978; Schmucker, 1984), including over-the-counter medications (Cupit, 1982; Lamy, 1982). Thus, this is of greater concern than any minor differences in pharmacokinetics (Kean and Buchanan, 1985; Buchanan, 1990). Despite the need for caution in the use of aspirin in the elderly and the requirement for careful adjustment of dosage and frequent monitoring, this drug may well have specific advantages in providing both antithrombotic activity as well as pain relief in some rheumatic patients who are at risk of developing thromboembolic disease.

Many patients with osteoarthritis are elderly, and there is evidence that they suffer from more toxicity than young patients (Schlegel and Paulus, 1986). However, many elderly patients take multiple medications and have multiple pathologies (Royal College of Physicians, 1984), and the relationship between the number of drugs and adverse side effects is not linear but exponential (Nolan and O'Malley, 1988). It seems possible that the elderly are no more prone to adverse drug reactions than the young if the number of drugs they are taking and the number of diseases they have are taken into account (Steel *et al.*, 1981). In a study of flurbiprofen, Sheldrake *et al.* (1977) found the incidence of adverse effects to be lower in the elderly than in younger patients. However, in this study the elderly patients were 'squeaky clean' and not the frail elderly with multiple organ failure found in nursing homes. It thus depends on the definition of elderly. Otto von Bismarck (1815–1898), the first to define the aged, simply recorded it as 65 years or older. Most patients up to age 75 years remain in reasonably good health, and it is only over this age that multiple organ failure begins. Elsewhere we have noted that the elderly do not comprise a homogeneous group, and that assessment and definition should not be made on chronology but on the basis of organ failure (Kean and Buchanan, 1987a; 1987b; Buchanan and Kean, 1992).

Schlegel and Paulus (1986) opined that the elderly may be more prone to side effects because of forgetfulness and lack of compliance. Forgetfulness may lead to taking an increased dose of salicylate, but it could be argued that lack of compliance, which is especially notorious with salicylates (Beck *et al.*, 1988), might lower the incidence of side effects!

Physiological changes that occur with ageing have the potential for altering drug disposition (Ouslander, 1981; Buchanan, 1990). Aspirin and salicylates are passively absorbed, and so the changes in the stomach and small intestine have no demonstrable effect (Bender, 1968). Reduction in lean body mass (Novack, 1972) and hypoalbuminia (Woodford-Williams, 1964) occur with ageing, but have little clinical consequence (Larking, 1979). The effects of the displacement of salicylate from albumin-binding sites by other drugs are potentially of significance (Wallace and Whiting, 1976). Biotransformation is potentially compromised in non-synthetic pathways, with diminution of the P_{450} mixed oxidases, but salicylic acid is essentially conjugated by a synthetic reaction so the rate of biotransformation is not affected (Curry *et al.*, 1979; Roberts *et al.*, 1983; Ho *et al.*, 1985). In the elderly, the plasma clearance of aspirin and salicylate, half-lives of their elimination (Curry *et al.*, 1979; Salem and Stephenson, 1979), and volume of distribution (Curry *et al.*, 1979; Salem and Stevenson, 1979; Roberts *et al.*, 1983) have been noted to be increased although not confirmed by Ho *et al.* (1985). Renal clearance gradually falls with age, and salicyluric acid has been noted to correlate with renal clearance; hence this may be why elderly patients have lower salicyluric acid clearance with increase in serum salicyluric acid levels (Montgomery and Sitar, 1981; Ho *et al.*, 1985). However, none of the changes reported in elderly patients are sufficient to lead to clinical consequences, and it is probable that they are of no clinical relevance (Montgomery *et al.*, 1986).

TOXICITY

Salicylate toxicity has been claimed to be higher in elderly patients (Grigor *et al.*, 1987). Tinnitus cannot be relied upon to warn of toxicity, since many elderly patients are deaf (Halla and Hardin, 1988). More care is required, especially in monitoring plasma salicylate levels, since these vary five- to six-fold after a single dose (Preston *et al.*, 1989). There is, however, growing evidence that acute gastrointestinal haemorrhage may be more common in elderly females (Buchanan and Brooks, 1991). Stern *et al.* (1984) provided evidence for a protective effect of paracetamol (acetaminophen) against damage to the human gastric mucosa, perhaps by causing an increase in COX-2 isoenzyme (Gretzer, 1998) but recent epidemiological data suggest there may be an increase in risks of upper GI bleeding and ulceration from coadministered paracetamol and NSAIDs including aspirin (Garcia-Rodriguez and Hernandez-Diaz, 2001).

Salicylate overdose and death can occur in elderly subjects with less than twice the recommended therapeutic serum levels of 25 to $30 \mu g/100 ml$ (Done, 1968). Whether this can be explained by the lower serum albumin concentrations (Hill, 1973; Wallace and Whiting, 1976; Morgan and Furst, 1986) and reduced capacity of salicylate binding to albumin (Netter *et al.*, 1985) in the elderly is not known. Frequently unrecognised in the adult, and especially the elderly, are cognitive decline (Saag *et al.*, 1995) and subacute and acute confusional states, ataxia, oliguria and insomnia (Anderson *et al.*, 1976). Tachypnoea is almost always present in salicylism, and indeed may be the first symptom. Elderly patients are also prone to pulmonary oedema (Heffner and Sahn, 1981; Walters *et al.*, 1983). Salicylates should never be 'pushed to the limits of toxicity, then backed off', and patients should be warned of the dangers of supplementation with over-the-counter salicylate preparations.

CONCLUSIONS

Aspirin has had a long history of effective use in treating pain and soft tissue inflammation in rheumatic conditions. Its use has been justified from results of clinical trials developed with a variety of methodologies. The current trend towards using low gastrotoxic NSAIDs (ibuprofen, COX-2 selective drugs) clearly presents a challenge to aspirin. It is evident, however, that salsalate, salicin and Willow Bark Extract (*Assalix*) may have a special place in treating certain rheumatic conditions, although long-term randomised controlled trials in large patient groups would be necessary before the place of these non-acetylated salicylate preparations in the therapy of different rheumatic states can be conclusively identified.

REFERENCES

Abraham, P.A. and Stillman, M.T. 1987, Salsalate exacerbation of chronic renal insufficiency. Relation to inhibition of prostaglandin synthesis. *Archives of Internal Medicine*, **147:** 1674–1676.

Abramson, S. and Weissman, G. 1989, The mechanisms of action of nonsteroidal antiinflammatory drugs. *Clinical and Experimental Rheumatology*, **7 (Suppl. 3):** S163–S170.

Abramson, S.B., Cherksey, B., Gude, D., Leszcynska-Piziak, J., Philips, M.R., Blau, L. and Weissman, G. 1990, Nonsteroidal anti-inflammatory drugs exert differential effects on neutrophil function and plasma membrane viscosity. Studies in human neutrophils and liposomes. *Inflammation*, **14:** 11–30.

Adams, K.E., Brown, P.A., Heys, S.D. and Whiting, P.H. 1993, Alleviation of experimental cyclosporin A nephrotoxicity by low dose aspirin in the rat. *Biochemical Pharmacology*, **46:** 2104–2108.

Adams, S.S. and Cobb, R. 1967, Non-steroidal anti-inflammatory drugs. *Progress in Medicinal Chemistry*, **5:** 59–138.

Agdal, N. 1979, Drug-induced oesophageal damage: a review and report of a fatal case of indomethacin-induced ulceration. *Ugesler Laeger*, **141:** 3019–3021.

Aghabubian, R.V. 1986, Comparison of diflunisal and acetaminophen in the management of Grade 2 ankle sprain. *Clinical Therapeutics*, **8:** 520–526.

Aghabubian, R.V., Volturo, G.A. and Heifetz, I.N. 1986, Comparison of diflunisal and naproxen in management of acute low back strain. *Clinical Therapeutics*, **9:** 47–51.

Agrawal, N.M., Roth, S., Graham, D.Y., White, R.H., Germain, B., Brown, J.A. and Stomatt, S.C. 1991, Misoprostol compared with sucrafiltrate in the prevention of nonsteroidal anti-inflammatory drug-induced gastric. A randomized, controlled trial. *Annals of Internal Medicine*, **115 (3):** 195–200.

Agrawal, N.M. and Saggioro, A. 1991, Treatment and prevention of NSAID induced gastroduodenal mucosal damage. *Journal of Rheumatology*, **28:** 15–18.

Akyol, S.M., Thompson, M. and Kerr, D.N.S. 1982, Renal function after prolonged consumption of aspirin. *British Medical Journal*, **284:** 631–632.

Alarcon-Segovia, T. 1980, Long-term treatment of symptomatic osteoarthritis with benoxaprofen. Double blind comparison with aspirin and ibuprofen. *Journal of Rheumatology*, **6 (Suppl.):** 89–99.

Albengres, E., Pinquier, J.L., Riant, P., Bree, F., Urian, S., Barre, J. and Tillement, J.P. 1988, Pharmacologic criteria for risk – benefit evaluation of NSAIDs. *Scandinavian Journal of Rheumatology*, **73 (Suppl.):** 3–15.

Alexander, A.M., Veitch, G.B. and Wood, J.B. 1975, Anti-rheumatic and analgesic drug usage and acute gastrointestinal bleeding in elderly patients. *Journal of Clinical and Hospital Pharmacy*, **10:** 89–93.

Alexander, W.D. and Johnson, K.W.M. 1956, Comparison of effects of acetylsalicylic acid with DL tri-iodothyronine in patients with myxoedema. *Clinical Science*, **15:** 593–601.

Alexander, W.D., MacDougall, A.I., Oliver, M.F. and Boyd, G.S. 1959, The effect of salicylate on the serum lipids and lipoproteins in coronary artery disease. *Clinical Science*, **18:** 195–203.

Allen, R.C., Petty, R.E., Lirenman, D.S., Malleson, P.N. and Lazer, R.M. 1986, Renal papillary necrosis in children with chronic arthritis. *Journal of Pediatrics*, **140:** 16–20.

Alsaeid, K. and Majeed, H.A. 1998, Acute rheumatic fever: diagnosis and treatment. *Pediatric Annals*, **27:** 295–300.

Altman, R.D. 1990, Neutrophil activation; an alternative to prostaglandin inhibition as the mechanism of action for NSAIDs. *Seminars in Arthritis and Rheumatism*, **(Suppl. 2):** 1–54.

CHAPTER 12

American College of Rheumatology Subcommittee on Osteoarthritis Guidelines 2000, Recommendations for the medical management of osteoarthritis of the knee and hip: 2000 update. *Arthritis and Rheumatism*, **44:** 1905–1915.

Amin, A.R., Attur, M., Patel, R.N., Thakker, G.D., Marshall, P.J., Rediske, J., Stuchin, S.A., Patel, I.R. and Abramson, S.B. 1997, Superinduction of cycloxygenase-2 activity in human osteoarthritis-affected cartilage. Influence of nitric oxide. *Journal of Clinical Investigation*, **99:** 1231–1237.

Amin, A.R., Vyas, P., Attur, M., Leszczynska-Piziak, J., Patel, I.R., Weissmann, G. and Abramson, S.B. 1995, The mode of action of aspirin-like drugs: effect on inducible nitric oxide synthase. *Proceedings of the National Academy of Sciences, USA*, **92:** 7926–7930.

Anderson, J.J., Firschein, H.E. and Meeman, R.F. 1989, Sensitivity of a health status measure to short-term clinical changes in arthritis. *Arthritis and Rheumatism*, **32:** 844–850.

Anderson, L.A. and Gotzsche, P.C. 1984, Naproxen and aspirin in acute musculoskeletal disorders. A double-blind, parallel study in patients with sports injuries. *Pharmatherapeutica*, **3:** 531–537.

Anderson, L.G. and Bina, P.R.C. 1980, Double-blind crossover trial comparing fenbufen and acetyl salicylic acid in rheumatoid arthritis. *Arzneimittal-Forschüng*, **30:** 735–739.

Anderson, R.J., Potts, D.E. and Gabow, P.A. 1976, Unrecognized adult salicylate intoxication. *Annals of Internal Medicine*, **85:** 745–760.

Anon. 1987, Reye's syndrome and aspirin: epidemiological associations and inborn errors of metabolism (Editorial). *Lancet*, **2:** 429–431.

Ansell, B. 1998, Juvenile rheumatoid arthritis (rheumatoid factor positive polyarthritis). *In:* P.J. Maddison (ed.), *Oxford Textbook of Rheumatology*, pp. 1629–1638.

Antillon, M., Cominelli, F., Reynolds, T.B. and Zipser, R.D. 1989, Comparative acute effects of diflunisal and indomethacin on renal function in patients with cirrhosis and ascites. *American Journal of Gastroenterology*, **84:** 386–391.

Appelrouth, D.J., Baim, S., Chang, R.W., Cohen, M.H., Englund, D.W., Germain, B.F., Hartman, S.S., Jaffer, A., Mullen, B.J. and Smith, F.E. 1987, Comparison of the safety and efficacy of nabumetone and aspirin in the treatment of osteoarthritis in adults. *American Journal of Medicine*, **83:** 78–81.

April, P., Abeles, M., Baraf, H., Cohen, S., Curran, N., Doucette, M., Eklholm, B., Goldlust, B., Knee, C.M., Lee, E. *et al.* 1990, Does the acetyl group of aspirin contribute to the antiinflammatory efficacy of salicylic acid in the treatment of rheumatoid arthritis? *Seminars in Arthritis and Rheumatism*, **19 (4 Suppl. 2):** 20–28.

April, P., Deighton, M.N. and Termulo, C. 1975, Comparison of tolmetin with aspirin in the treatment of rheumatoid arthritis. *In:* J. Ward (ed.), *Tolmetin: A New Non-steroidal Anti-inflammatory Agent*, pp. 47–56. Princeton: Excerpta Medica.

Arendt-Racine, E.C., Atkinson, M.H., Decoteau, W.E., Flatt, V.L. and Varady, J. 1978, Drug trial in rheumatoid arthritis. A new design. *Clinical Pharmacology and Therapeutics*, **23:** 233–240.

Armstrong, C.P. and Blower, A.L. 1987, Non-steroidal anti-inflammatory drugs and life threatening complications of peptic ulceration. *Gut*, **28:** 527–532.

Arrowsmith, J.B., Kennedy, D.L., Kuritzsky, J.N. and Faich, G.A. 1987, Maternal patterns of aspirin use and Reye syndrome reporting, United States, 1980–1985. *Pediatrics*, **79:** 858–863.

Arroyo, V., Gines, P., Rimolo, A. and Gaha, J. 1986, Renal function abnormalities, prostaglandins and effects of nonsteroidal anti-inflammatory drugs in cirrhosis and ascites. An overview with emphasis on pathogenesis. *American Journal of Medicine*, **81 (Suppl. 2B):** 104–122.

Athreya, B.H., Moser, G., Cecil, H.S. and Myers, A.R. 1975, Aspirin-induced hepatotoxicity in juvenile rheumatoid arthritis. *Arthritis and Rheumatism*, **18:** 347–353.

Atkinson, D.C. and Collier, H.O.J. 1980, Salicylates: molecular mechanism of therapeutic action. *In: Advances in Pharmacology and Chemotherapy*, Vol. 17, pp. 233–287. Amsterdam: Academic Press Inc.

Australasia Multicentre Trial Group 1980, The simultaneous assessment of four non-steroidal anti-inflammatory drugs in rheumatoid arthritis using a simple and rapid time design. *Journal of Rheumatology*, **7:** 857–864.

Ayers, J.W., Weidler, D.J., MacKichan, J. and Wagner, J.G. 1977, Circadian rhythm of urinary pH in man with and without chronic antacid adiministration. *European Journal of Clinical Pharmacology*, **12:** 415–420.

Bailey, J.M., Low, C.E. and Papillo, M.B. 1982, Reye's syndrome and aspirin use: a possible immuno-logical relationship. *Prostaglandins, Leukotrienes and Medicine*, **8:** 211–218.

Ballantyne, F.C., Fleck, A. and Dick, W.C. 1971, Albumin metabolism in rheumatoid arthritis. *Annals of the Rheumatic Diseases* **30:** 265–270.

Balogh, Z., Papazoglou, S.N., Macleod, M. and Buchanan, W.W. 1979, A crossover trial of piroxicam, indomethacin and ibuprofen in rheumatoid arthritis. *Current Medical Research and Opinion*, **6:** 148–153.

Bälz, E. 1877, Salicylsäure, Salicylsaures Natron und Thymol in ihrem Einfluss auf Krankheiten. *Archiv Heilkund*, **18:** 60–82.

Banerjee, A.K. 1989, Enteropathy induced by non-steroidal anti-inflammatory drugs. *British Medical Journal*, **298:** 1539–1540.

Bank, S., Greenberg, R.E., Magier, D. and Lavin, P.T. 1991, The efficacy and tolerability of famoti-dine and ranitidine on the healing of active duodenal ulcer and during six-month maintenance treat-ment, with special reference to NSAID/aspirin-related ulcers. *Clinical Therapeutics*, **13:** 304–318.

Bannwarth, B., Netter, P., Pourel, J., Royer, R.J. and Gaucher, A. 1989, Clinical pharmacokinetics of nonsteroidal anti-inflammatory drugs in the cerebrospinal fluid. *Biomedicine and Pharmacotherapy*, **43:** 121–126.

Bardare, M., Cislaghi, G.U., Mandelli, M. and Sereni, F. 1978, Value of monitoring plasma salicylate levels in treating juvenile rheumatoid arthritis. *Archives of Diseases in Childhood*, **53:** 381–385.

Bardhan, K.D., Bjarnason, I., Scott, D.L., Griffin, W.M., Fenn, G.C., Shield, M.J. and Morant, S.V. 1993, The prevention and healing of acute non-steroidal anti-inflammatory drug-associated gastroduo-denal mucosal damage by misoprostol. *British Journal of Rheumatology*, **32:** 990–995.

Barel, J. and Orlowski, J.P. 1988, Aspirin and Reye's syndrome. *Pediatrics*, **82:** 135–136.

Barnardo, D.E., Currey, H.L.F., Mason, R.M., Fox, W.R. and Weatherall, M. 1966, Mefenamic acid and flufenamic acid compared with aspirin and phenylbutazone in rheumatoid arthritis. *British Medical Journal*, **2:** 342–343.

Barnes, C.G., Berry, H., Carter, M.E., Downie, W.W., Fowler, P.D., Moll, J.M., Perry, J.D., Sawaf, M.S. and Wright, V. 1979, Diclofenac sodium (Voltarol) and indomethacin: a multicentre comparative study in rheumatoid arthritis and osteoarthritis. *Rheumatology and Rehabilitation*, **2:** 135–143.

Barraclough, R.E., Lenaghan, E. and Muirden, K.D. 1974, A comparison of flurbiprofen and aspirin in the treatment of rheumatoid arthritis. *Medical Journal of Australia*, **2:** 925–927.

Barrett, M.J., Hurwitz, E.S., Patriarca, P.A., Schonberger, L.B., Micheals, R., Jaffee, R. and Lehberger, L. 1986, Reye syndrome in connective tissue disease. Letter. *Journal of Pediatrics*, **108:** 1043–1044.

Barron, K.S., Person, D.A. and Brewer, E.J. 1982, The toxicity of non-steroidal anti-inflammatory drugs (NSAIDs) in juvenile rheumatoid arthritis (JRA). *Journal of Rheumatology*, **9:** 149–155.

Basomba, A., Romar, A., Pelaez, A., Villalmanzo, I.G. and Campos, A. 1976, The effect of sodium chromoglycate in preventing aspirin induced bronchospasm. *Clinical Allergy*, **6:** 269–275.

Bataille, C., Soumagne, D., Loly, J. and Brassine, A. 1982, Oesophageal ulceration due to indomethacin. *Digestion*, **24:** 66–68.

Batterman, R.D. 1958, Comparison of buffered and unbuffered acetylsalicylic acid. *New England Journal of Medicine*, **258:** 213–219.

Bax, D.E. 1992, Sulfonamides. *In:* J.S. Dixon and D.E. Furst (eds), *Second-Line Agents in the Treatment of Rheumatic Diseases*, pp. 267–286. New York: Marcel Dekker.

Beales, D.L., Burry, H.C. and Grahame, R. 1972, Comparison of aspirin and benorylate in the treat-ment of rheumatoid arthritis. *British Medical Journal*, **2:** 483–485.

Beall, S., Gardner, J. and Coxley, D. 1983, Anterolateral compartment syndrome related to drug-induced bleeding: a case report. *American Journal of Sports Medicine*, **11:** 454–455.

Beaver, W.T. 1965, Mild analgesics. A review of their clinical pharmacology. *American Journal of the Medical Sciences*, **250:** 577–604.

Beck, N.C., Parker, J.C., Frank, R.G., Geden, E.A., Kay, D.R., Gamache, M., Shivvers, N., Smith, E. and Anderson, S. 1988, Patients with rheumatoid arthritis at high risk for noncompliance with salicy-late treatment regimens. *Journal of Rheumatology*, **15 (7):** 1081–1084.

CHAPTER 12

Beeley, L. and Kendall, M.J. 1971, Effect of aspirin on renal clearance of 125-diatrizoate. *British Medical Journal*, **1:** 707–714.

Belay, E.D., Bresel, J.S., Holman, R.C., Khan, A.S., Shahriari, A. and Schonberger, L.B. 1999, Reye's syndrome in the United States from 1981 through 1997. *New England Journal of Medicine*, **340:** 1377–1382.

Bender, A.D. 1968, Effect of age on intestinal absorption: implications for drug absorption in the elderly. *Journal American Geriatric Society*, **16 (12):** 1331–1339.

Bender, W.L., Wheldon, A., Beschorner, W.E., Darwish, M.O., Hall-Craggs, M. and Soley, K. 1984, Interstitial nephritis, proteinuria and renal failure caused by non-steroidal anti-inflammatory drugs. *American Journal of Medicine*, **76:** 1006–1012.

Benson, G.D. 1983, Hepatotoxicity following the therapeutic use of antipyretic analgesics. *American Journal of Medicine*, **75 (Suppl.):** 85–93.

Berg, K.J., 1977, Acute effects of acetylsalicylic acid on renal function in normal man. *European Journal of Clinical Pharmacology*, **11:** 117–123

Berg, K.J., Forre, O., Djoseland, O., Mikkelson, M., Narverud, J. and Rugstad, H.E. 1989, Renal side effects of high and low cyclosporin A doses in patients with rheumatoid arthritis. *Clinical Nephrology*, **31:** 232–238.

Bergamo, R.R., Cominelli, F., Kopple, J.D. and Zipser, R.D. 1989, Comparative acute effects of aspirin, diflunisal, ibuprofen and indomethacin on renal function in healthy man. *American Journal of Nephrology*, **9:** 460–463.

Bernhard, G.C., Appelrouth, D.J., Bankhurst, A.D., Biundo, J., Bockow, B.I., Brobym, R.D., Brodsky, A.L., Burch, F.X., Chang, R.W. and Cohen, M.H. 1987, Long-term treatment of rheumatoid arthritis comparing nabumetone with aspirin. *American Journal of Medicine*, **83:** 44–49.

Bernstein, J.M. and Weiss, A.D. 1967, Further observations on salicylate ototoxicity. *Journal of Laryngology and Otology*, **81:** 915–920.

Berry, H., Bloom, B., Fernandes, L. and Morris, M. 1983, Comparison of timegadine and naproxen in rheumatoid arthritis. A placebo controlled trial. *Clinical Rheumatology*, **2:** 357–361.

Bianchi Porro, G. and Pace, F. 1988, Ulcerogenic drugs and upper gastrointestinal bleeding. *Bailière's Clinical Gastroenterology*, **2:** 309–327.

Bijlsma, A. 1978, The long-term efficacy and tolerability of Voltaren (diclofenac sodium) and indomethacin in rheumatoid arthritis. *Scandinavian Journal of Rheumatology*, **22 (Suppl):** 74–80.

Bijlsma, J.W.J. 1988a, Treatment of endoscopy-negative NSAID-induced upper gastrointestinal symptoms with cimetidine: an international multicentre collaborative study. *Alimentary Pharmacology and Therapeutics*, **2:** 75–83.

Bijlsma, J.W.J. 1988b, Treatment of NSAID-induced gastrointestinal lesions with cimetidine: an international multicentre collaborative study. *Alimentary Pharmacology and Therapeutics*, **2:** 85–96.

Bird, H., Dixon, J.S., Pickup, M.E, Rhind, V.M., Lowe, J.R., Lee, M.R. and Wright, V. 1982, A biochemical assessment of sulphasalazine in rheumatoid arthritis. *Journal of Rheumatology*, **9:** 36–45.

Birnie, G.G., McLeod, T.I.F. and Watkinson, G. 1981, Incidence of sulphasalazine-induced male infertility. *Gut*, **22:** 452–455.

Bisno, A.L. 1997, Rheumatic fever. *In:* W.N. Kelly, E.D. Harris, S. Ruddy and C.B. Sledge (eds), *Textbook of Rheumatology*. 5th edn, Vol. 2, pp. 1225–1240. Philadelphia: W.B. Saunders.

Bjarnason, I. 1988, Non-steroidal anti-inflammatory drug-induced small intestine inflammation in man. *In:* R. Pounder (ed.), *Recent Advances in Gastroenterology*, pp. 23–46. Edinburgh: Churchill Livingstone.

Bjarnason, I., Hayllar, J., MacPherson, A.J. and Russell, A.S. 1993, Side effects of nonsteroidal anti-inflammatory drugs on the small and large intestine in humans. *Gastroenterology*, **104:** 1832–1847.

Bjarnason, I., Peters, T.J. and Levi, A.J. 1986a, Intestinal permeability: clinical correlates. *Digestive Diseases*, **4:** 83–92.

Bjarnason, I., Price, A.B., Zanelli, G., Smethurst, P., Burke, M., Gumpel, J.M. and Levi, A.J. 1988, Clinicopathological features of nonsteroidal antiinflammatory drug-induced small intestinal strictures. *Gastroenterology*, **94:** 1070–1074.

Bjarnason, I., Williams, P., Smethurst, P., Peters, T.J. and Levi, A.J. 1986b, The effect of non-steroidal anti-inflammatory drugs and prostaglandins on the permeability of the human small bowel. *Gut*, **27:** 1292–1297.

Bjarnason, I., Williams, P., So, A., Zanelli, G.D., Levi, A.J., Gumpel, J.M., Peters, T.J. and Ansell, B. 1984, Intestinal permeability and inflammation in rheumatoid arthritis: effects of non-steroidal anti-inflammatory drugs. *Lancet*, **2:** 1171–1174.

Bjarnason, I., Zanelli, G., Prouse, P., Williams, P., Gumpel, M.J. and Levi, A.J. 1986c, Effect of non-steroidal anti-inflammatory drugs on the human small intestine. *Drugs*, **32:** 35–41.

Bjarnason, I., Zanelli, G., Smith, T., Prouse, P., Williams, P., Smethurst, P., Delacey, G., Gumpel, M.J. and Levi, A.J. 1987, Non-steroidal anti-inflammatory drug-induced intestinal inflammation in humans. *Gastroenterology*, **93:** 480–489.

Blackshear, J.L., Davidman, M. and Stillman, T. 1983, Identification of risk for renal insufficiency from non-steroidal anti-inflammatory drugs. *Archives of Internal Medicine*, **143:** 1130–1134.

Blackshear, J.L., Napier, J.S., Davidman, M. and Stillman, M.T. 1985, Renal complications of non-steroidal anti-inflammatory drugs: Identification and monitoring of those at risk. *Seminars in Arthritis and Rheumatism*, **14:** 163–175.

Blechman, W.J. and Lechner, B.L. 1979, Clinical comparative evaluation of choline magnesium trisalicylate and acetylsalicylic acid in rheumatoid arthritis. *Rheumatology and Rehabilitation*, **18:** 119–124.

Blechman, W.J., Schmid, F.R., April, P.A., Wilson, C.H. and Brooks, C.D. 1975, Ibuprofen or aspirin in rheumatoid arthritis therapy. *Journal of the American Medical Asssociation*, **233:** 336–339.

Boardman, P.L. and Hart, F.D. 1967, Clinical measurement of the anti-inflammatory effects of salicylates in rheumatoid arthritis. *British Medical Journal*, **4:** 264–268.

Bonta, I.L., Bult, H., Vincent, J.E. and Zijlstra, F.J. 1977, Acute anti-inflammatory effects of aspirin and dexamethasone in rats deprived of endogenous prostaglandin precursors. *Journal of Pharmacy and Pharmacology*, **29:** 1–7.

Booker, J.A. 1983, Acute upper gastrointestinal bleeding in the aged. *Medical Journal of Australia*, **2:** 422.

Borga, O., Cederlof, I.O., Ringberger, V.A. and Norlin, A. 1976, Protein binding of salicylate in uremic and normal plasma. *Clinical Pharmacology and Therapeutics*, **20:** 464–475.

Borrachero Del Campo, J. 1978, Comparative double-blind and open trials of sulindac and acetyl salicylic acid in the treatment of rheumatoid arthritis. *European Journal of Rheumatology and Inflammation*, **1:** 16–17.

Bower, R.J., Umbenhauer, E.R. and Hereus, V. 1979, Clinical evaluation of sulindac. A new non-steroidal anti-inflammatory drug. *In:* G. Weissmann, B. Samuelsson and R. Paoletti (eds), *Advances in Inflammation Research*, Vol. 1, pp. 559–567. New York, Raven Press.

Bowers, R.E., Brigham, K.L. and Owen, P.J. 1975, Salicylate pulmonary edema: the mechanism in sheep and review of the clinical literature. *Annual Review of Respiratory Diseases*, **114:** 261–270.

Bowers, R.E., Brigham, K.L. and Owen, P.J. 1977, Salicylate pulmonary edema: the mechanism in sheep and review of the clinical literature. *American Review of Respiratory Diseases*, **115:** 261–268.

Bradley, J.G. 1991, Nonprescription drugs and hypertension. Which one affects blood pressure? *Postgraduate Medicine*, **89:** 195–197.

Brain, W.R., Hunter, D. and Turnbull, H.M. 1929, Acute meningo-encephalomyelitis of childhood: report of 6 cases. *Lancet*, **1:** 221–227.

Brandt, K.D. 1993, NSAIDs in the treatment of osteoarthritis. Friends or foes? *Bulletin of Rheumatic Diseases*, **42:** 1–4.

Brandt, K.D. 2001, A critique of the 2000 update of the American College of Rheumatology recommendations for the management of hip and knee osteoarthritis. *Arthritis and Rheumatism*, **44:** 2451–2455.

Brater, D.C. 1986, Drug–drug and drug–disease interactions with nonsteroidal anti-inflammatory drugs. *American Journal of Medicine*, **80 (Suppl. 1A):** 62.

Brater, D.C. 1990a, Adverse effects of non-steroidal anti-inflammatory drugs on renal function. *Annals of Internal Medicine*, **112:** 559–560.

Brater, D.C. 1990b, Eicosanoids and hypertension. *Western Journal of Medicine*, **153 (2):** 194–195.

Brem, J., Pereli, E., Gopalen, S.K. and Miller, T.B. 1973, Salicylism, hyperventilation and the central nervous system. *Journal of Pediatrics*, **83:** 264–268.

CHAPTER 12

Bresnihan, B., Hughes, G. and Essigman, W.K. 1978, Diflunisal in the treatment of osteoarthritis: a double-blind study comparing diflunisal with ibuprofen. *Current Medical Research and Opinion*, **5:** 556–561.

Brewer, E.J. 1982, The Pediatric Rheumatology Collaborative Study Group – the first nine years. *Journal of Rheumatology*, **9:** 1–2.

Brewer, E.J. and Giannini, E.H. 1982, Methodology and studies of children with juvenile rheumatoid arthritis. *Journal of Rheumatology*, **9:** 107–139.

Brewer, E.J., Giannini, E.H. and Person, D.A. 1982, *Juvenile Rheumatoid Arthritis*, 2nd edn, pp. 205–208. Philadelphia: W.B. Saunders.

British Journal of Clinical Pharmacology, **19 (5):** 675–684.

Broderick, T.W., Rienke, R.T. and Goldman, E. 1976, Salicylate-induced pulmonary edema. *American Journal of Rheumatology*, **127:** 865–866.

Brooke, J.W. 1976, Fenoprofen therapy in large joint osteoarthritis: double-blind comparison with aspirin and long-term experience. *Journal of Rheumatology*, **3 (Suppl.):** 71–75.

Brooks, C.D., Schlagel, C.A., Sekjar, N.C. and Sobota, J.T. 1973, Tolerance and pharmacology of ibuprofen. *Current Therapeutic Research*, **15:** 180–190.

Brooks, C.D., Schmid, F.R., Biudo, J., Blau, S., Gonzalez-Alcover, R., Gowans, J.D.C., Hurd, E., Partridge, R.E.H. and Tarpley, E.L. 1970, Ibuprofen and aspirin in the treatment of rheumatoid arthritis: a cooperative double-blind trial. *Rheumatolology and Physical Medicine*, **11 (Suppl.):** 48–63.

Brooks, P.M. and Cossum, P. 1978, Salicylates and creatinine clearance reevaluated. *Australian and New Zealand Journal of Medicine*, **8:** 660.

Brooks, P.M. and Day, R.O. 1991, Nonsteroidal anti-inflammatory drugs – differences and similarities. *New England Journal of Medicine*, **324:** 1716–1725.

Brooks, P.M., Dougan, M.A., Mugford, S. and Meffin, E. 1982, Comparative effectiveness of 5 analgesics in patients with rheumatoid arthritis and osteoarthritis. *Journal of Rheumatology*, **9:** 723.

Brooks, P.M. and Khong, T.K. 1977, Flurbiprofen-aspirin interaction: a double-blind crossover study. *Current Medical Research and Opinion*, **5:** 53–57.

Brooks, P.M., Stephens, W.H., Stephens, M.E.D. and Buchanan, W.W. 1975, How safe are antirheumatic drugs? A study of possible iatrogenic deaths in patients with rheumatoid arthritis. *Health Bulletin*, **33:** 108–111.

Brown, F.L., Bodison, S., Dixon, J., Davis, W. and Mowoshawski, Y. 1986, Comparison of diflunisal and acetaminophen with codeine in the treatment of initial or recurrent low back strain. *Clinical Therapeutics*, **9:** 52–58.

Buchanan, M.R., Blajchman, M.A., Fejana, E., Mustard, J.F., Senyi, A.F. and Hirsh, J. 1979, Shortening of the bleeding time in thrombocytopenic rabbits after exposure of jugular vein to high aspirin concentration. *Prostaglandins in Medicine*, **3:** 333–342.

Buchanan, M.R., Vazquez, M.J. and Gimbrone, M.A. Jr. 1983, Arachidonic acid metabolism and the adhesion of human polymorphonuclear leukocytes to cultured vascular endothelial cells. *Blood*, **62 (4):** 889–895

Buchanan, N. and Rabinowitz, L. 1974, Infantile salicylism – a reappraisal. *Journal of Pediatrics*, **84:** 391–400.

Buchanan, W.W. 1990, Implications of NSAID therapy in elderly patients. *Journal of Rheumatology*, **17 (Suppl. 20):** 29–32

Buchanan, W.W. 1992, Cyclosporin: clinical efficacy and toxicity in patients with rheumatoid arthritis. *In:* K.D. Rainsford and G.-P. Velo (eds), *Side Effects of Anti-inflammatory Drugs*, Vol. III, pp. 301–310. Lancaster: Kluwer Academic Publishers.

Buchanan, W.W. and Brooks, P.M. 1991, Prediction of organ system toxicity with antirheumatic drug therapy. *In:* N. Bellamy (ed.), *Prognosis in the Rheumatic Diseases*, pp. 403–450. Dordrecht: Kluwer Academic Publishers.

Buchanan, W.W. and Kean, W.F. 1992, Antirheumatic drug therapy in the elderly. *In:* K.D. Rainsford and G.P. Velo (eds), *Side Effects of Anti-Inflammatory Drugs*, Vol. 3, pp. 248–260. Lancaster: Kluwer Academic Publishers.

Buchanan, W.W. and Kean, W.F. 2001, Rheumatoid arthritis as seen through long distance spectacles. *Inflammopharmacology*, **9:** 3–22.

Buchanan, W.W. and Kean, W.F. 2002a, Osteoarthritis I: epidemiological risk factors and historical considerations. *Inflammopharmacology*, **10:** 1–21.

Buchanan, W.W. and Kean, W.F. 2002b, Osteoarthritis II: Pathology and pathogenesis. *Inflammopharmacology*, **10:** 23–52.

Buchanan, W.W. and Rainsford, K.D. 1990, Aspirin and nonacetylated salicylates: use in inflammatory injuries incurred during sporting activities. *In:* W.B. Leadbetter, J.A. Buckwalter and S.L. Gordon (eds), *Sports Induced Inflammation: Clinical and Basic Science Concepts*, pp. 431–441. Park Ridge: American Academy of Orthopaedic Surgeons.

Buchanan, W.W., Rooney, P.J. and Rennie, J.A.M. 1979, Aspirin and the salicylates. *Clinical Rheumatological Diseases*, **5:** 499–539.

Bullough, P.G. 1998, Osteoarthritis and related disorders. Pathology. *In:* J.H. Klippel and P.A. Dieppe (eds), *Rheumatology*, 2nd edn, pp. 1–8. London: Mosby International.

Burbidge, E.J., Lewis, R.D. and Halsted, C.H. 1984, Alcohol and the gastrointestinal tract. *Medical Clinics of North America*, **68:** 77–89.

Burris, S.M., Smith, C.M. II, Rao, G.H. and White, J.G. 1987, Aspirin treatment reduces platelet resistance to deformation. *Arteriosclerosis*, **7:** 385–388.

Buss, C.E. 1875, *Über die Anwendung der Salicylsäure als Antipyreticum*. Leipzig: J.B. Hirschfeld. (Inaugural Dissertation zur Erlangung der Doktowürde der Hohen Medizinischen Fakultät der Universität Basel.)

Bywaters, E.G.L. 1977, The history of pediatric rheumatology. *Arthritis and Rheumatism*, **20 (Suppl.):** 145–152.

Calabro, J.J. and Paulus, H.E. 1970, Anti-inflammatory effect of acetyl salicylic acid in rheumatoid arthritis. *Clinical Orthopedics and Related Research*, **71:** 124–131.

Caldwell, J.R., Roth, S.H., Wu, W.C., Semble, E.L., Castell, D.O., Heller, M.D. and Marsh, W.H. 1987, Sucralfate treatment of non-steroidal anti-inflammatory drug-induced gastrointestinal symptoms and mucosal damage. *American Journal of Medicine*, **83 (Suppl. 3B):** 74–82.

Cannon, G.W., Caldwell, J.R., Holt, P., McLean, B., Seidenberg, B., Bolognese, J., Ehrich, E., Mukhopadhyay, S. and Daniels, B. 2000, Rofecoxib, a specific inhibitor of cyclooxygenase 2, with clinical efficacy comparable with that of diclofenac sodium: results of a one-year, randomized, clinical trial in patients with osteoarthritis of the knee and hip. Rofecoxib Phase III Protocol 035 Study Group. *Arthritis and Rheumatism*, **43:** 978–987.

Capell, H.A., Rennie, J.A.N., Rooney, P.J., Murdock, R.M., Hole, D.J., Dick, W.C. and Buchanan, W.W. 1979, Patient compliance: a novel method of testing non-steroidal anti-inflammatory analgesics in rheumatoid arthritis. *Journal of Rheumatology*, **6:** 584–593.

Caradoc-Davies, T.H. 1984, Nonsteroidal anti-inflammatory drugs, arthritis, and gastrointestinal bleeding in elderly in-patients. *Age and Ageing*, **13:** 295–298.

Cardoe, N. and Fowler, P.D. 1979, Diclofenac sodium (Voltarol): a double-blind comparative study with ibuprofen in patients with rheumatoid arthritis. *Rheumatology and Rehabilitation*, **2 (Suppl.):** 89–99.

Carson, D.A., Chen, P.P., Kipps, D.J., Radoux, V., Jirik, F.R., Goldfien, R.D., Fox, R.I., Silverman, G.J. and Fong, S. 1987, Idiotypic and genetic studies of human rheumatoid factors. *Arthritis and Rheumatism*, **30:** 1321–1325.

Carson, J.L., Strom, B.L., Morse, M.L., West, S.L., Soper, K.A., Stolley, P.D. and Jones, J.K. 1987, The relative gastrointestinal toxicity of nonsteroidal anti-inflammatory drugs. *Archives of Internal Medicine*, **147:** 1054–1059.

Caruso, I. and Bianchi Porro, G. 1980, Gastroscopic evaluation of anti-inflammatory agents. *British Medical Journal*, **280:** 75–78.

Caruso, I., Fumagalli, M., Monterone, F., Vernazza, M., Bianchi Porro, G. and Petrillo, M. 1978, Controlled, double-blind study comparing acetylsalicylic acid and diflunisal in the treatment of osteoarthritis of the hip and/or the knee; long-term gastroscopic study. *In:* K. Miehlke (ed.), *Diflunisal in Clinical Practice. Proceedings of a Conference at the XIVth International Congress of Rheumatology, San Francisco*, pp. 63–74. New York, Futura.

Cassell, S., Furst, D., Dromgoole, S. and Paulus, H. 1979, Steady-state serum salicylate levels in hospitalised patients with rheumatoid arthritis. Comparison of two dosage schedules of choline magnesium trisalicylate. *Arthritis and Rheumatism*, **22:** 384–388.

Cassidy, J.T. 1997, Juvenile rheumatoid arthritis. *In:* W.N. Kelly, E.D. Harris, S. Ruddy and C.B. Sledge (eds), *Textbook of Rheumatology*, 5th edn, Vol. 2, pp. 1207–1224. Philadelphia: W.B. Saunders.

Castleden, C.M., Volans, C.N. and Raymond, K. 1977, The effect of ageing on drug absorption from the gut. *Age and Ageing*, **6 (3):** 138–143

Cecil, R.L. 1934, The medical treatment of rheumatoid arthritis. *Journal of the American Medical Association*, **103:** 1583–1589.

Chahade, W.H. and Josef, H. 1978, Clinical evaluation of the efficacy and tolerance of sulindac in patients with osteoarthritis of the hip and/or knee during 144 weeks: comparative study with aspirin during the first 96 weeks. *European Journal of Rheumatology and Inflammation*, **1:** 41–44.

Chalem, F., Pena, M. Lizarazo, H. and Farias, P. 1977, Comparison of fenbufen and aspirin in the treatment of rheumatoid arthritis. *Current Therapeutic Research*, **22:** 769–783.

Chalmers, J.P., West, M.J., Weng, L.M., Bune, A.J. and Graham, J.R. 1984, Effects of indomethacin, sulindac, naproxen, aspirin and paracetamol in treated hypertensive patients. *Clinical Experimental Hypertension*, **6:** 1077–1093.

Chalmers, T.M. 1969, Clinical experience with ibuprofen in the treatment of rheumatoid arthritis. *Annals of the Rheumatic Diseases*, **28:** 513–517.

Champion, G.D., Day, R.O. and Graham, G.G. 1975, Salicylates in rheumatoid arthritis. *Clinical Rheumatology*, **1:** 245–265.

Champion, G.D. and Graham, G.G. 1978, Pharmacokinetics of non-steroidal anti-inflammatory agents. *Australian and New Zealand Journal of Medicine* **8 (Suppl. 1):** 94–100.

Chan, K., Sung, J., Chung, S., To, K.F., Yung, M.Y., Leung, V.K.S., Lee, Y.T., Chan, C.S., Li, E.K. and Woo, J. 1997, Randomised trial of eradication of *Helicobacter pylori* before nonsteroidal anti-inflammatory drug therapy to prevent peptic ulcers. *Lancet*, **350:** 975–979.

Chan, L.K., Winearls, C.G., Oliver, D.O. and Dunnill, M.S. 1980, Acute interstitial nephritis and erythroderma associated with diflunisal. *British Medical Journal*, **1:** 84–85.

Chaplin, M.D., Chu, N.I., Rice, B.G. and Hama, K.M. 1975, Effect of repeated dosing with aspirin on plasma levels of naproxen (D-2-(6-methyl-2-naphtyl)-propionic acid) in rats. *Proceedings of the Western Pharmacological Society*, **18:** 62–66.

Chopra, I.J., Van Herle, A.J., Teco, G.N. and Nguyen, A.H. 1980, Serum-free thyroxine in thyroidal and nonthyroidal illnesses: a comparison of measurements by radioimmunoassay, equilibrium dialysis, and free thyroxine index. *Journal of Clinical Endocrinology and Metabolism*, **51:** 135–143.

Chrisman, O.D., Snook, G.A. and Wilson, T.C. 1972, The protective effect of aspirin against degeneration of human articular cartilage. *Clinical Orthopaedic and Related Research*, **84:** 193–196.

Christoffersen, P., Faarup, P., Geertinger, P. and Krogh, P. 1980, Reye's syndrome in a child on long-term salicylate medication. *Forensic Science International*, **15:** 129–133.

Chrubasik, S. and Eisenberg, E. 2000, Efficacy and safety of *Salix* extract preparations. *In:* S. Chrubasik and B.D. Roufogalis (eds), *Herbal Medicinal Products for the Treatment of Pain*, pp. 74–84. Lismore: Southern Cross University Press.

Chrubasik, S., Eisenberg, E., Balan, E., Weinberger, T., Luzzati, R. and Conradt, C. 2000, Treatment of low back pain exacerbations with willow bark extract: a randomised double-blind study. *American Journal of Medicine*, **109:** 9–14.

Chrubasik, S. and Roufogalis (eds) 2000, Efficacy and safety of salix extract preparations. *In: Herbal Medicinal Products for the Treatment of Pain*, pp. 74–78. Lismore, NSW: Southern Cross University Press.

Ciocci, A. 1978, Sulindac in the treatment of rheumatoid arthritis. *European Journal of Rheumatology and Inflammation*, **1:** 55–57.

Clinite, J.C. and Kabat, H.F. 1976, Improving patient compliance. *Journal of the American Pharmacy Association*, **16:** 74–76, 85.

Clive, D.M. and Stoff, J.S. 1984a, Renal syndromes associated with nonsteroidal anti-inflammatory drugs. *New England Journal of Medicne*, **310:** 563–572.

Clive, D.M. and Stoff, J.S. 1984b, Renal syndromes associated with nonsteroidal anti-inflammatory drugs in the kidney. *Medical Clinics of North America*, **68**: 371–572.

Cohen, A. 1979, Fecal blood loss and plasma salicylate study of salicylsalicylic acid and aspirin. *Journal of Clinical Pharmacology*, **19**: 242–247.

Collier, H.O.J. 1969, A pharmacological analysis of aspirin. *Advances in Pharmacology and Chemotherapy*, **1**: 333–405.

Collins, A.J., Davies, J. and Dixon, A.St.-J. 1986, Contrasting presentation and findings between patients with rheumatic complaints taking non-steroidal anti-inflammatory drugs and a general population referred for endoscopy. *British Journal of Rheumatology*, **25**: 50–53.

Collins, A.J., Davies, J. and Dixon, A.S. 1988, A prospective endoscopic study of the effect of Orudis and Oruvail on the upper gastrointestinal tract, in patients with osteoarthritis. *British Journal of Rheumatology*, **27 (2)**: 106–109.

Committee On Safety Of Medicines Update 1986, Non-steroidal anti-inflammatory drugs and serious gastrointestinal reactions. *British Medical Journal*, **292**: 1190–1191.

Conn, H.O. and Poynard, T. 1985, Adrenocorticosteroid administration and peptic ulcer: a critical analysis. *Journal of Chronic Diseases*, **38**: 457–468.

Copeman, W.S.C. 1964, *Textbook of the Rheumatic Diseases*, 3rd edn. Edinburgh and London: Livingstone.

Coppe, D., Wessinger, S.J., Ransil, B.J., Harris, W. and Salzman, E. 1981, Sex differences in the platelet response to aspirin. *Thrombosis Research*, **23**: 1–21.

Corell, T. and Jensen, K.M. 1979, Interaction of salicylates and other non-steroidal anti-inflammatory agents in rats as shown by gastro-ulcerogenic and anti-inflammatory activities, and plasma concentrations. *Acta Pharmacologica et Toxicologica*, **45**: 225–231.

Corey, L., Rubin, R.J., Hattwick, M.A.W., Moble, G.R. and Cassidy, E. 1976, A nation-wide outbreak of Reye's syndrome: its epidemiologic relationship to influenza B. *American Journal of Medicine*, **61**: 615–625.

Craxi, A. and Pagliarello, G. 1980, Possible embryotoxicity of sulfasalazine. *Archives of Internal Medicine*, **140**: 1674.

Croffort, L.J. 1999, COX-2 in synovial tissues. *Osteoarthritis and Cartilage*, **7**: 406–408.

Cronk, G.A. 1958, Laboratory and clinical studies with buffered and non-buffered acetylsalicylic acid. *New England Journal of Medicine*, **258**: 219–222.

Crook, D. and Collins, A. 1975, Prostaglandin synthetase activity from human rheumatoid synovial tissue and its inhibition by non-steroidal anti-inflammatory drugs. *Prostaglandins*, **9**: 857–865.

Cryer, B. and Feldman, M. 1999, Effects of very low-dose daily, long-term aspirin therapy on gastric, duodenal, and rectal prostaglandin levels and on mucosal injury in healthy humans. *Gastroenterology*, **117**: 17–25.

Cryer, B., Lee, E. and Feldman, M. 1992b, Factors influencing gastroduodenal mucosal prostaglandin concentrations: Roles of smoking and aging. *Arrivals of Internal Medicine*, **116**: 636.

Cryer, B., Redfern, J.S. and Goldsmiedt, M. 1992a, Effect of aging on gastric duodenal mucosal prostaglandin concentrations in humans. *Gastroenterology*, **102**: 1118–1123.

Csuka, M.E. and McCarty, D.J. 1989, Aspirin and the treatment of rheumatoid arthritis. *Rheumatic Disease Clinics of North America*, **15**: 439–454.

Cullen, D. and Hawkey, C.J. 1993, Arthrotec for all? *Annals of the Rheumatic Diseases*, **52**: 841–842.

Cullen, L., Kelly, L., Connor, S.O. and Fitzgerald, D.J. 1998, Selective cyclooxygenase-2 inhibition by nimesulide in man. *Journal of Pharmacology and Experimental Therapeutics*, **287 (2)**: 578–582

Culpeper, N. 1993, *Culpeper's School of Physick or the Experimental Practice of the Whole Art*. London. Printed for N. Brook at the Angel in Cornhill 1659. Reprinted by the Classics of Medicine Library, New York.

Cuny, G., Royer, R.J., Mur, J.M., Sero, J.M., Fauer, G., Netter, P., Maillard, A. and Penn, F. 1979, Pharmacokinetics of salicylates in the elderly. *Gerontology*, **25**: 49–55.

Cupit, G.C. 1982, The use of non-prescription analgesics in an older population. *Journal of the American Geriatic Society*, **30 (Suppl. 11)**: 76–80.

Curry, S.H., Whelpton, R., De Schepper, P.J., Vranckx, S. and Schiff, A.A. 1979, Kinetics of fluphenazine after fluphenazine dihydrochloride, enanthate and decanoate administration to man. *British Journal of Clinical Pharmacology*, **7**: 325–331.

Dahl, R. 1981, Oral and inhaled sodium cromoglycate in challenge test with food allergens or acetyl-salicylic acid. *Allergy*, **36:** 161–165.

Danesh, B.J., McLaren, M., Russell, R.I., Lowe, G.D. and Forbes, C.D. 1989, Comparison of the effect of aspirin and choline magnesium trisalicylate on thromboxane biosynthesis in human platelets: role of the acetyl moiety. *Haemostasis*, **19:** 169–173.

Daneshmend, T.K. 1990, Diseases and drugs but not food decrease ketoconazole 'bioavailability'. *British Journal of Clinical Pharmacology*, **29 (6):** 783–784.

Darlington, L.G. and Coomes, E.M. 1975, Comparison of benorylate and ibuprofen in the treatment of established rheumatoid arthritis. *Rheumatology and Rehabilitation*, **14:** 76–80.

Dart, R.C. and Kuffner, E.K. 2001, Use of acetaminophen in alcoholic patients: comment on the 2000 update of the American College of Rheumatology recommendations for the management of hip and knee osteoarthritis. *Arthritis and Rheumatism*, **44:** 2449.

Das, K.M. 1983, Pharmacotherapy of inflammatory bowel disease. Part 1. Sulfasalazine. *Postgraduate Medicine*, **74:** 141–148.

Das, K.M. and Eastwood, M.A. 1974, The role of the colon in the metabolism of saliculazosulphapyridine. *Scandinavian Journal of Gastroenterology*, **9:** 137–141.

Daum, F., Zucker, P. and Cohen, M.I. 1976, Acute liver failure and encephalopathy (Reye's syndrome) during salicylate therapy. *Acta Paediatrica Scandinavica*, **65:** 747.

Davis, D.L. and Buffler, P. 1992, Reduction of deaths after drug labelling for risk of Reye's syndrome. *Lancet*, **340:** 1042.

Davis, J.D., Struth, A.G., Turner, R.A., Pisko, E.J. and Ruchte, L.R. 1979, Pirprofen and aspirin in the treatment of rheumatoid arthritis. *Clinical Pharmacology and Therapeutics*, **25:** 618–623.

Davis, P., Tannenbaum, H., Kraag, G., Brandwein, S., Boate, B., Grace, M. and Auclair, C. 1983, Efficacy and tolerance of Sergam (tiaprofunic acid) compared with enteric-coated ASA (ECA) in rheumatoid arthritis. *Third International Seminar on the Treatment of Rheumatic Diseases, Israel, 13–20 Nov, 1983*. Also quoted in Sorkin, E.M. and Brodie, R.N. 1985, Tiaprofenic acid. A review of its pharmacological properties and therapeutic efficacy in rheumatic diseases and pain states. *Drugs*, **29:** 208–235.

Davis, P.R. and Burch, R.E. 1974, Pulmonary edema and salicylate intoxication. *Annals of Internal Medicine*, **80:** 553–566.

Dawes, P.T. and Symmons, D.P.M. 1992, Short-term effects of antirheumatic drugs. *Balliere's Clinical Rheumatology*, **6:** 117–140.

Day, R., Shen, D. and Azarnoff, D. 1979, *Clinical Pharmacology and Therapeutics*, **25:** 220.

Day, R.O., Shen, D.D. and Azarnoff, D.L. 1983, Induction of salicyluric acid formation in rheumatoid arthritis patients treated with salicylates. *Clinical Pharmacokinetics*, **8:** 263–271.

Day, R.O. and Brooks, P.M. 1987b, Variations in response to non-steroidal anti-inflammatory drugs. *British Journal of Clinical Pharmacology*, **23:** 655–658.

Day, R.O., Furst, D.E., Dromgoole, S.H. and Paulus, H.E. 1988a, Changes in Salicylate serum concentration and metabolism during chronic dosing in normal volunteers. *Biopharmacy and Drug Disposition*, **9:** 273–283.

Day, R.O., Graham, G.G., Bieri, D., Brown, M., Cawns, D., Harris, G., Hounsell, J., Platt-Hepworth, S., Reeve, R., Sambrook, P.N. and Smith, J. 1989, Concentration–response relationships for salicylate-induced ototoxicity in normal volunteers. *British Journal of Clinical Pharmacology*, **28:** 695–702.

Day, R.O., Graham, G.G., Williams, K.M. and Brooks, P.M. 1988b, Variability in response to NSAIDs. Fact or fiction? *Drugs*, **36:** 643–651.

Day, R.O., Graham, G.G., Williams, K.M., Champion, G.D. and De Jager, J. 1987a, Clinical pharmacology of non-steroidal anti-inflammatory drugs. *Pharmacology and Therapeutics*, **33:** 383–433.

Day, R.O., Harris, G., Brown, M., Graham, G.G. and Champion, G.D. 1988c, Interaction of salicylate and corticosteroids in man. *British Journal of Clinical Pharmacology*, **26:** 334–337.

De Caestecker, J.S. and Heading, R.C. 1988, Iatrogenic oesphageal ulceration with massive hemorrhage and stricture formation. *British Journal of Clinical Practice*, **42:** 212–214.

Delaney, J.C. 1983, The effect of ketotifen on aspirin-induced asthmatic reactions. *Clinical Allergy*, **13:** 247–251.

Deodhar, S.D., Dick, W.C., Hodgkinson, R. and Buchanan, W.W. 1973, Measurement of clinical response to anti-inflammatory drugs in rheumatoid arthritis. *Quarterly Journal of Medicine*, **42:** 287–401.

Deodhar, S.D., McLeod, M.M., Dick, W.C. and Buchanan, W.W. 1977, A short-term comparative trial of salsalate and indomethacin in rheumatoid arthritis. *Current Medical Research and Opinion*, **5:** 185–188.

Dequeker, J. and Mardjuardi, A. 1981, Treatment of rheumatoid arthritis with flurbiprofen: a comparison with enteric-coated aspirin. *Current Medical Research and Opinion*, **7:** 418–422.

Dequeker, J., Stevens, E. and Wuyts, L. 1980, A controlled trial of magnesium dithiosalicylate compared with aspirin in rheumatoid arthritis. *Current Medical Research and Opinion*, **6:** 589–592.

De Silva, M., Hazleman, B.L. and Dippy, J.E. 1980, Diflunisal and aspirin: a comparative study in rheumatoid arthritis. *Rheumatology and Rehabilitation*, **19:** 126–130.

De Weck, A.L. 1971, Immunological effects of aspirin anhydride, a contaminant of commercial acetylsalicylic acid preparations. *International Archives of Allergy and Applied Immunology*, **41:** 393–418.

Diamond, H. 1976, Double-blind crossover study of fenoprofen and aspirin in osteoarthritis. *Journal of Rheumatology*, **3 (Suppl.):** 67–70.

Diamond, H., Alexander, S., Kerzell, W., Lussier, A., Odone, D. and Tompkin, S.R. 1973, A multicentre double-blind crossover comparison study of naproxen and aspirin in patients with rheumatoid arthritis. *Scandinavian Journal of Rheumatology*, **(Suppl. 2):** 171–175.

Diamond, H., Alexander, S., Kuzell, W., Lussier, A., Odone, D. and Tompkins, R. 1975, Naproxen and aspirin in rheumatoid arthritis: a multicentre double-blind crossover comparison study. *Journal of Clinical Pharmacology*, **15:** 335–339.

Diaz-Gonzalez, F., Conzalez-Alvarao, I., Campanero, M.R., Mollinedo, F., del Pozo, M.A., Munoz, C. Pivel, J.P. and Sanchez-Madrid, F. 1995, Prevention of in vitro neutrophil-endothelial attachment through shedding of L-selectin by nonsteroidal antiinflammatory drugs. *Journal of Clinical Investigation*, **95:** 1756–1765.

Dick, W.C., Deodhow, S.D., Provan, C.J., Nuki, G. and Buchanan, W.W. 1971, Isotope studies in normal and diseased knee joints. ^{99mm}Tc uptake related to clinical assessment and to synovial perfusion measured by the ^{133}Xe clearance technique. *Clinical Science*, **40:** 327–336.

Dick, W.C., Grayson, M.F., Woodburn, A., Nuki, G. and Buchanan, W.W. 1970c, Indices of inflammation activity. Relationship between isotope studies and clinical methods. *Annals of the Rheumatic Diseases*, **29:** 643–648.

Dick, W.C., Neufeld, R.R., Prentice, A.G., Woodburn, A., Whaley, K., Nuki, G. and Buchanan, W.W. 1970a, Measurement of joint inflammation. A radioisotopic method. *Annals of the Rheumatic Diseases*, **39:** 135–137.

Dick, W.C., Nuki, G., Deodhar, S. and Buchanan, W.W. 1970b, Some aspects in the quantitation of inflammation in joints of patients suffering from rheumatoid arthritis. *Rheumatology and Physical Medicine*, **11 (Suppl.):** 40–47.

Dick, W.C., Onge, R.A., Whaley, K., Gillespie, F., Boyle, J.A., Jasani, M.K. and Buchanan, W.W. 1969, Measurement of synovial blood flow in normal and diseased joints. *Annals of the Rheumatic Diseases*, **28:** 197–198.

Dick-Smith, J.B. 1969, Ibuprofen, aspirin and placebo in treatment of rheumatoid arthritis. A double-blind clinical trial. *Medical Journal of Australia*, 853–859.

Dieppe, P.A. 2001, Concerns about the methodology used in developing the 2000 update of the American College of Rheumatology recommendations for the management of hip and knee osteoarthritis. *Arthritis and Rheumatism*, **44:** 2450–2451.

Dieppe, P.A., Frankel, S.J. and Toth, B. 1993, Is research into the treatment of osteoarthritis with nonsteroidal anti-inflammatory drugs misdirected? *Lancet*, **341:** 353–354.

Dieppe, P.A. and Huskisson, E.C. 1978, Diflunisal and acetylsalicylic acid: a comparison of efficacy in osteoarthritis, of nephrotoxicity, and of anti-inflammatory activity in the rat. *In:* K. Miehlke (ed.), *Diflunisal in Clinical Practice. Proceedings of a Conference at the XIVth International Congress of Rheumatology, San Francisco*, pp. 57–61. New York: Futura.

Dieppe, P.A. and Huskisson, E.C. 1979, Diflunisal and aspirin: a comparison of efficacy and nephrotoxicity in osteoarthritis. *Rheumatology and Rehabilitation*, **18:** 53–56.

Dingle, J.T. 1999, The effect of nonsteroidal anti-inflammatory drugs on human articular cartilage glycosaminoglycan synthesis. *Osteoarthritis and Cartilage*, **7:** 313–314.

Done, A.K. 1968, Treatment of salicylate poisoning: review of personal and published experiences. *Clinical Toxicology*, **1:** 451.

Dornan, J. and Reynolds, W.J. 1974, Comparison of ibuprofen and acetylsalicylic acid in the treatment of rheumatoid arthritis. *British Medical Journal*, **4:** 82–84.

Doughty, R.A., Giesecke, L. and Athreye, B. 1980, Salicylate therapy in juvenile rheumatoid arthritis. *American Journal of Diseases in Childhood*, **134:** 461–463.

Douthwaite, A.H. and Lintott, G.A.M. 1938, Gastroscopic observation of the effect of aspirin and certain other substances on the stomach. *Lancet*, **ii:** 1222–1225.

Dreser, H. 1907, Über modifizierte Salicylsäuren. *Medizinische Klinik*, **3:** 390–393.

Dresse, A., Fischer, P., Gerard, M.A., Tempero, K.F. and Verhaest, L. 1975, Hypouricemic acid derivative, in normal humans. *Proceedings of the VIth International Congress in Pharmacology, Helsinki, Finland.* Abstract Number 611.

Dromgoole, S.H., Furst, D.E., Desiraju, R.K., Nayak, R.K., Kirschenbaum, M.A. and Paulus, H.E. 1982, Tolmetin kinetics and synovial fluid prostaglandin E levels in rheumatoid arthritis. *Clinical Pharmacology and Therapeutics*, **32:** 371–377.

Dromgoole, S.H., Furst, D.E. and Paulus, H.E. 1981, Rational approach to the use of salicylates in the treatment of rheumatoid arthritis. *Seminars in Arthritis and Rheumatism*, **11:** 257–283.

Duncan, J.J. 1988, Comparison of diclofenac sodium and aspirin in the treatment of acute sports injuries. *American Journal of Sports Medicine*, **16 (6):** 656–659.

Duncan, J.J. and Farr, J.E. 1988, Comparison of diclofenac sodium and aspirin in the treatment of acute sports injuries. *American Journal of Sports Medicine*, **16:** 656–659.

Dunn, M. 1984, Nonsteroidal anti-inflammatory drugs and renal function. *Annual Review of Medicine*, **35:** 411–428.

Dunn, M. 1987, The role of arachidonic acid metabolites in renal homeostasis non-steroidal anti-inflammatory drugs renal function and biochemical, histological and clinical effects and drug interactions. *Drugs*, **33 (Suppl. 1):** 56–66.

Dussault, J.H., Parlow, A., Letarte, J., Guyda, H. and Laberge, C. 1976, TSH measurements from blood spots on filter paper: a confirmatory screening test for neonatal hypothyroidism. *Journal of Pediatrics*, **89:** 550–552.

Duthie, J.J.R. 1963, Concluding Remarks. *In:* A.St.-J. Dixon, B.K. Martin, M.J.H. Smith, P.H.N. Wood (eds), *Salicylates. An International Symposium*, pp. 288–292. London: J. & A. Churchill.

Ecker, G.A. and Karsh, J. 1992, Naproxen induced ulcerative esophagitis. *Journal of Rheumatology*, **19:** 646–647.

Editorial 1990, A catch in the Reye is awry. *Cleveland Clinics Journal of Medicine*, **57:** 318–320.

Ehsanallah, R.S.B., Page, M.C., Tildesley, G. and Wood, J.R. 1988, Prevention of gastroduodenal damage by nonsteroidal anti-inflammatory drugs: controlled trial with ranitidine. *British Medical Journal*, **297:** 1017–1021.

Ekstrom, P., Carling, L., Wetterhus, S., Wingren, P.E., Anker-Hansen, O., Lundegardh, G., Thorhallsson, E. and Unge, P. 1996, Prevention of peptic ulcer and dyspeptic symptoms with omeprazole in patients receiving continuous non-steroidal anti-inflammatory drug therapy. A Nordic multicentre study. *Scandinavian Journal of Gastroenterology*, **31 (8):** 753–758.

Elkstrand, R., Alvan, G., Magnusson, A., Oliw, E., Palmer, L. and Rane, A. 1981, Additive clinical effect of indomethacin suppositories during salicylate therapy in rheumatoid patients. *Scandinavian Journal of Rheumatology*, **10:** 69–75.

Emery, P., Zeidler, H., Kvien, T.K., Guslandi, M., Naudin, R., Stead, H., Verburg, K.M., Isakson, P.C., Hubbard, R.C. and Geis, G.S. 1999, Celecoxib versus diclofenac in long-term management of rheumatoid arthritis: randomised double-blind comparison. *Lancet*, **354 (9196):** 2106–2111.

Emkey, R.D. and Mills, J.A. 1982, Aspirin and analgesic nephropathy. *Journal of the American Medical Association*, **247:** 55.

Empire Rheumatism Council 1955, Multi-centre controlled trial comparing cortisone acetate and acetyl salicylic acid in the long-term treatment of rheumatoid arthritis. *Annals of the Rheumatic Diseases*, **14:** 353–368.

Empire Rheumatism Council 1957, Multi-centre controlled trial comparing cortisone acetate and acetyl salicylic acid in the long-term treatment of rheumatoid arthritis. *Annals of the Rheumatic Diseases*, **16:** 277–289.

Eng, J. and Sabanathan, S.1991, Drug-induced esophagitis. *American Journal of Gastroenterology*, **86 (9):** 1127–1133.

Essigman, W.K., Chamberlain, M.A. and Wright, V. 1979, Diflunisal in osteoarthrosis of the hip and knee. *Annals of the Rheumatic Diseases*, **38:** 148–151.

Estes, D. and Kaplan, K. 1980, Lack of platelet effect with the aspirin analog, salsalate. *Arthritis and Rheumatism*, **23:** 1303–1307.

Eular, A.R., Sofdi, M., Jaszewski, R., Welsh, J., Le, V., Raskin, J., Fleischmann, R., Razzaque, M. and Champion, C. 1990, A report of three multiclinic trials evaluating arbaprostil in arthritic patients with ASA/NSAID gastric mucosal damage. The Upjohn Company Arbaprostil ASA/NSAID Gastric Mucosal Damage Treatment Study Groups. *Gastroenterology*, **98:** 1549–1557.

Everson, G.W. and Krenzelok, E.P. 1986, Chronic salicylism in a patient with juvenile rheumatoid arthritis. *Clinical Pharmacy*, **5:** 334–341.

Ezer, E., Palosi, E., Hajos, G., Rosdy, B. and Szporny, L. 1979, Comparative pharmacology of a 1:10 combination of indomethacin-sodium salicylate. *Agents and Actions*, **9:** 117–123.

Fagan, S.C., Kertland, H.R. and Tietjen, G.E. 1994, Safety of combination aspirin and anticoagulation in acute ischemic stroke. *Annals of Pharmacology*, **28:** 441–443.

Fanelli, G.M. Jr and Weiner, I.M. 1979, Urate excretion: drug interactions. *Journal of Pharmacology and Experimental Therapeutics*, **210:** 186–195.

Faulkner, G., Prichard, P., Somerville, K. and Langman, M.J. 1988, Aspirin and bleeding peptic ulcers in the elderly. *British Medical Journal*, **297:** 1311–1313.

Feenstra, J., Grobbe, D.E., Mosterd, A. and Stricker, B.H.Ch. 1997, Adverse cardiovascular effects of NSAIDs in patients with congestive heart failure. *Drug Safety*, **17:** 166–180.

Feldman, M. and McMahon, A.T. 2000, Do cyclooxygenase-2 inhibitors provide benefits similar to those of traditional nonsteroidal anti-inflammatory drugs, with less gastrointestinal toxicity? *Annals of Internal Medicine*, **132 (2):** 134–143.

Felson, D.T. 2001, The verdict favours nonsteroidal antiinflammatory drugs for treatment of osteoarthritis, and a plea for more evidence on other treatments. *Arthritis and Rheumatism*, **44:** 1477–1480.

Felson, D.T., Anderson, J.J., Naimark, A., Walker, A.M. and Meenan, R.F. 1988, Obesity and knee osteoarthritis. The Framingham study. *Annals of Internal Medicine*, **109:** 18–24.

Ferraz, M.B., Tugwell, P., Goldsmith, C.H. and Atra, E. 1990, Meta-analysis of sulfasalazine in ankylosing spondylitis. *Journal of Rheumatology*, **17:** 1482–1486.

Ferriera, S.H., Moncada, S. and Vane, J.R. 1971, Indomethacin and aspirin abolish prostaglandin release from the spleen. *Nature New Biology*, **231:** 237–239.

Finlay, D.W. and Lucus, R.H. 1879, Salicylate and alkaline treatment of acute rheumatism with an analysis of 158 cases. *Lancet*, **2:** 420–421.

Fiore, L., Brophy, M., Deykin, D., Cappelleri, J. and Lau, J. 1993, The efficacy and safety of the addition of aspirin in patients treated with oral anticoagulants after heart valve replacement. *Blood*, **82 (10 Suppl. 1):** 409a.

Forsyth, B.W., Horwitz, R.I., Acampora, D., Shapiro, E.D., Viscoli, C.M., Feinstein, A.R., Henner, R., Halabird, N.B., Jones, B.A. and Karabelas, A.D. 1989, New epidemiologic evidence confirming that bias does not explain the aspirin/Reye's syndrome association. *Journal of the American Medical Association*, **261:** 2517–2524.

Forsyth, B.W., Shapiro, E.D., Horwitz, R.I., Viscoli, C.M. and Acampora, D. 1991, Misdiagnosis of Reye's-like illness. *American Journal of Diseases in Childhood*, **145:** 964–966.

Franke, M. and Manz, G. 1972, Benorylate and indomethacin in the treatment of rheumatoid disease. A double-blind clinical trial. *Current Therapeutic Research*, **14:** 113–122.

CHAPTER 12

Fremont-Smith, K. and Boyles, T.B. 1965, Salicylate therapy in rheumatoid arthritis. *Journal of the American Medical Association*, **192:** 113–116.

Fries, J.F. 1991, The hierarchy of quality-of-life assessment, the Health Assessment Questionnaire (HAQ), and issues mandating development of a toxicity index. *Controlled Clinical Trials*, **12:** 106S–117S.

Fries, J.F. 1995, ARAMIS and toxicity measurement. *Journal of Rheumatology*, **22:** 995–997.

Fries, J.F., Miller, S.R., Spitz, P.W., Williams, C.A., Hubert, H.B. and Bloch, D.A. 1989, Toward an epidemiology of gastropathy associated with non-steroidal anti-inflammatory drugs and hospitalization for upper gastrointestinal bleeding. *Gastroenterology*, **96:** 647–655.

Fries, J.F., Williams, C.A., Ramey, D.R. and Bloch, D.A. 1993, The relative toxicity of alternative therapies for rheumatoid arthritis: implications for the therapeutic progression. *Seminars in Arthritis and Rheumatism*, **23 (Suppl.):** 68–73.

Furst, D.E., 1991, Pharmacokinetic interactions and adverse drug experiences in rheumatoid arthritis. *Inflammopharmacology*, **1:** 69–77.

Furst, D.E., Blocka, K., Cassell, S., Harries, E.R., Hirschberg, J.M., Josephson, N., Lachenbruch, P.A., Trimble, R.B. and Paulus, H.E. 1987b, A controlled study of concurrent therapy with a nonacetylated salicylate and naproxen in rheumatoid arthritis. *Arthritis and Rheumatism*, **30:** 146–154.

Furst, D.E., Gupta, N. and Paulus, H.E. 1977, Salicylate metabolism in twins. Evidence suggesting a genetic influence and induction of salicylate formation. *Journal of Clinical Investigation*, **60:** 32–42.

Furst, D.E., Sarkissian, E., Blocka, K., Cassell, S., Dromgoole, S., Harries, E.R., Hirschberg, J.M., Josephson, N. and Paulus, H.E. 1987a, Serum concentrations of salicylate and naproxen during concurrent therapy in patients with rheumatoid arthritis. *Arthritis and Rheumatism*, **30:** 1157–1161.

Furst, D.E., Tozer, T.N. and Melmon, K.L. 1979, Salicylate clearance, the resultant of protein-binding and metabolism. *Clinical Pharmacology Therapeutics*. **26:** 380–389.

Gall, E.P., Caperton, E.M., McComb, J.E., Messner, R., Multz, C.V., O'Hanlan, M. and Willkens, R.F. 1982, Clinical comparison of ibuprofen, fenoprofen calcium, naproxen and tolmetin sodium in rheumatoid arthritis. *Journal of Rheumatology*, **9:** 402–407.

Garber, E., Craig, R.M. and Bahu, R.M. 1975, Aspirin hepatitis. *Annals of Internal Medicine*, **82:** 592–593.

Garcia-Rodriguez, L.A. and Hernandez-Diaz, S. 2001, Relative risk of upper gastrointestinal complications among users of acetaminophen and nonsteroidal anti-inflammatory drugs. *Epidemiology*, **12:** 570–576.

Garella, S. and Matarese, R.A. 1984, Renal effects of prostaglandins and clinical adverse effects of nonsteroidal anti-inflammatory agents. *Medicine*, **63:** 165–181.

Garnham, J.G., Raymond, K., Shotton, E. and Turner, P. 1975, The effect of buffered aspirin on plasma indomethacin. *European Journal of Clinical Pharmacology*, **8:** 107–113.

Gatley, M.S 1968, To be taken as directed. *Journal of the Royal College of General Practice*, **16:** 39–44.

Gazert, H. 1900, Therapeutische Erfahrungen mit Aspirin. *Deutches Archiv fur Klinische Medizin*, **68:** 142–154.

Geis, G.S., Stead, H., Wallenmark, C.-B. and Nicholson, P.A. 1991, Prevalance of mucosal lesions in the stomach and duodenum due to chronic use of NSAIDs in patients with rheumatoid arthritis or osteoarthritis, an interim report on prevention by Misoprostil of diclofenc-associated lesions. *Journal of Rheumatology*, **18 (Suppl. 28):** 11–14.

Gerber, L.J., Rooney, P.J. and McCarthy, D.M. 1981, Healing of peptic ulcers during continuing anti-inflammatory drug therapy in rheumatoid arthritis. *Journal of Clinical Gastroenterology*, **3:** 7–11.

Ghosh, P. 1988, Anti-rheumatic drugs and cartilage. *Baillieré's Clinical Rheumatology*, **2:** 309–338.

Giannini, E.H., Brewer, E.J., Miller, M.L., Gibbas, D., Passo, M.H., Bernstein, B., Person, D.A., Fink, C.W. and Sawyer, L.A. 1990, Liquid ibuprofen in the treatment of juvenile rheumatoid arthritis. Results of the double–blind and open studies conducted by the Pediatric Rheumatology Collaborative Study Group. *Journal of Paediatrics*, **117:** 645–652.

Gibofsky, A. and Zabriskie, J.B. 1997, Rheumatic fever: etiology, diagnosis, and treatment. *In:* W.J. Koopman (ed.), *Arthritis and Allied Conditions*, 13th edn, Vol. 2, pp. 1581–1594. Baltimore: William and Williams.

Gibofsky, A. and Zabriskie, J.B. 1998, Rheumatic fever. *In:* P.J. Maddison, D.A. Isenberg, P. Woo and D.M. Glass (eds), *Oxford Textbook of Rheumatology*, 2nd edn, Vol. 2, pp. 972–982. Oxford: Oxford University Press.

Gibson, G.R., Whitacre, E.B. and Ricotti, C.A. 1992, Colitis induced by nonsteroidal anti-inflammatory drugs. Report of four cases and review of the literature. *Archives of Internal Medicine*, **152:** 625–632.

Gilat, T. 1979, Etiology of inflammatory bowel disease. *Journal of Clinical Gastroenterology*, **1:** 299–300.

Gitlin, N. 1980, Salicylate hepatotoxicity: the potential role of hypoalbuminaemia. *Journal of Clinical Gastroenterology*, **2:** 281–285.

Glen-Bott, A.M. 1987, Aspirin and Reye's syndrome: a reappraisal. *Medical Toxicology*, **2:** 161–165.

Goldenberg, A., Rudnicki, R.D. and Koonce, M.L. 1978, Clinical comparison of efficacy and safety of choline magnesium trisalicylate and indomethacin in treating osteoarthritis. *Current Therapeutic Research*, **24:** 245–259.

Goldenberg, D.L. 1974, Aspirin hepatotoxicity. *Annals of Internal Medicine*, **80:** 773.

Goldenberg, D.L., Egan, M.S. and Cohen, A.S. 1982, Inflammatory synovitis in degenerative joint disease. *Journal of Rheumatology*, **9:** 204–209.

Goldenstein-Schaunberg, C. and Yoshinari, N.H. 1997, Drug treatment in juvenile chronic arthritis (in Portuguese). *Review Hospital Clinical Faculty of Medicine, Sao Paulo*, **52:** 90–95.

Goodwin, J.S. and Goodwin, J.M 1981, Failure to recognise efficacious treatments: a history of salicylate therapy in rheumatoid arthritis. *Perspectives in Biology and Medicine*, **25:** 78–92.

Görges. 1902, Über neure Arzneimittel: Aspirin und Digitalis-Dialysat. *Berliner Klinische Wochenschrift*, **30:** 753–755.

Goulston, K. and Cooke, A.R. 1978, Alcohol, aspirin and gastrointestinal bleeding. *British Medical Journal*, **4:** 664–665.

Gowans, J.D.C. 1981, Six-month, double-blind comparison of sodium meclofenamate (Meclomen) with buffered aspirin in the treatment of rheumatoid arthritis. *Current Medical Research and Opinion*, **7:** 384–391.

Graham, D.J. and Smith, J.L. 1986, Aspirin and the stomach. *Annals of Internal Medicine*, **104:** 390–398.

Graham, D.Y. 1988, Misoprostol and ulcer prophylaxis. *Lancet*, **2:** 1484–1485.

Graham, D.Y., Agrawal, N.M. and Roth, S.H. 1988, Prevention of NSAID-induced gastric ulcer with misoprostol: multicentre, double-blind, placebo-controlled trial. *Lancet*, **2:** 1277–1280.

Graham, D.Y. and Smith, J.L. 1988, Gastroduodenal complications of chronic NSAID therapy. *American Journal of Gastroenterology*, **83:** 1081–1084.

Graham, D.Y., White, R.J., Moreland, L.W., Schubert, T.T., Katz, R., Jaszewski, R., Tindall, E., Triadafilopoulos, G., Stromatt, S.C. and Teoh, L.S. 1993, Duodenal and gastric ulcer prevention with misoprostol in arthritis patients taking NSAIDs. Misoprostol Study Group. *Annals of Internal Medicine*, **119:** 257–262.

Graham, G.G., Champion, G.D., Day, R.O., Kaski, A.L., Hills, L.G. and Paull, P.D. 1977b, *In:* K.D. Rainsford, K. Brune and M.W. Whitehouse (eds), *Aspirin and Related Drugs. Their Actions and Uses*, pp. 37–42. Basel: Birkhäuser.

Graham, G.G., Champion, G.D., Day, R.O. and Paull, P.D. 1977a, Patterns of plasma concentrations and urinary excretion of salicylate in rheumatoid arthritis. *Clinical Pharmacology and Therapeutics*, **22:** 410–420.

Granville-Grossman, K.L. and Sergeant, H.G.S. 1960, Pulmonary oedema due to salicylate intoxication. *Lancet*, **i:** 575–577.

Grayson, M.F. 1978a, A clinical trial of diflunisal against aspirin in osteoarthritis. *Rheumatology and Rehabilitation*, **17:** 265–269.

Grayson, M.F. 1978b, Two trials of diflunisal in osteoarthritis. *Current Medical Research and Opinion*, **5:** 567–571.

Grazioli, I., Avoss, M., Bogliolo, A., Broggini, M., Carcassi, A., Carcassi, U., Cecconami, L., Ligniere, G.C., Colombo, B., Consoli, G., Di Matteo, L., Fioravanti, A., Frizziero, L., Lamontagna, G., Lapadula, G., Macarri, G., Mangiameli, A., Marcolongo, R., Perpignano, G., Pipitone, V., Quattrocchi, G., Saggioro, A., Tirri, G. and Todesco, S. 1993, Multicenter study of the safety/efficacy of misoprostol in the prevention and treatment of NSAID-induced gastroduodenal lesions. *Clinical and Experimental Rheumatology*, **11**: 289–294.

Green, A. and Hall, S.M. 1992, Investigation of metabolic disorders resembling Reye's syndrome. *Archives of Diseases in Children*, **67**: 1313–1317.

Greenblatt, D.J., Divoll, M., Abernethy, D.R. and Shader, R.I. 1982, Physiologic changes in old age: relation to altered drug disposition. *Journal of the American Geriatricians Society*, **30 (11 Suppl.)**: S6–S10.

Greene, J.M. and Winickoff, R.N. 1992, Cost-conscious prescribing of nonsteroidal anti-inflammatory drugs for adults with arthritis. A review and suggestions. *Archives of Internal Medicine*, **152** (10): 1995–2002.

Grennan, D.M. and Aarons, L. 1984, Salicylate–NSAID interactions. *Annals of the Rheumatic Diseases*, **43**: 351–360

Grennan, D.M., MacLeod, M. and Kennedy, A.C. 1974, Preliminary clinical evaluation of azapropazone in rheumatoid arthritis. *Current Medical Research and Opinion*, **2**: 67–71.

Gretzer, B., Ehrlich, K., Maricic, N., Lambrecht, N., Respondek, M. and Peskar, B.M. 1998, Selective cyclo-oxygenase-2 inhibitors and their influence on the protective effect of a mild irritant in the rat stomach. *British Journal of Pharmacology*, **123** (5): 927–935.

Griffin, M.R., Ray, W.A. and Schaffner, W. 1988, Non-steroidal anti-inflammatory drug use and death from peptic ulcer in elderly persons. *Annals of Internal Medicine*, **109**: 359–363.

Grigor, R.R., Spitz, P.W. and Furst, D.E. 1987, Salicylate toxicity in elderly patients with rheumatoid arthritis. *Journal of Rheumatology*, **14 (1)**: 60–66.

Gross, M. and Greenberg, L.A. 1948, *The Salicylates. A Critical Bibliographic Review*. New Haven: Hillhouse Press.

Guess, H.A., West, R., Strand, L.M., Helston, D., Lydick, E.G., Bergman, U. and Wolski, K. 1988, Fatal upper gastrointestinal hemorrhage or perforation among users and nonusers of non-steroidal anti-inflammatory drugs in Saskatchewan, Canada 1983. *Journal of Clinical Epidemiology*, **41**: 35–45.

Gum, O.B. 1976, Fenprofen in rheumatoid arthritis: a controlled crossover multicentre study. *Journal of Rheumatology*, **3 (Suppl. 2)**: 26–31.

Gum, O.R. 1980, Long term efficacy and safety of benoxaprofen: comparison with aspirin and ibuprofen in patients with active rheumatoid arthritis. *Journal of Rheumatology*, **7 (Suppl. 6)**: 76–88.

Gunsberg, M., Bochner, F., Graham, G., Imhoff, D., Parsons, G. and Cham, B. 1984, Disposition of and clinical response to salicylates in patients with rheumatoid disease. *Clinical Pharmacology and Therapeutics*, **35 (5)**: 585–593.

Gupta, N., Sarkissian, E. and Paulus, H. 1975, Correlation of plateau serum salicylate level with rate of salicylate metabolism. *Clinical Pharmacology and Therapeutics*, **18**: 350–355.

Gurwich, E.L., Raees, S.M., Skosey, J. and Niazi, S. 1984, Unbound plasma salicylate concentration in rheumatoid arthritis patients. *British Journal of Rheumatology*, **23**: 66–73.

Habbab, M.A., Szwed, S.A. and Haft, J.J. 1986, Is coronary artery spasm part of aspirin-induced asthma syndrome? *Chest*, **90**: 141–143.

Hall, S.M., Plaster, P.A., Glasgow, J.F. and Hancock, P. 1988, Pre-admission antipyretics in Reye's syndrome. *Archives of Diseases in Childhood*, **63**: 857–866.

Halla, J.T. and Hardin, J.G. 1988, Salicylate ototoxicity in patients with rheumatoid arthritis; a controlled study. *Annals of Rheumatic Diseases*, **47**: 134–137.

Halpin, T.J., Holtzhauer, F.J., Campbell, R.J., Hall, L.J., Correa-Villasenor, A., Lanese, R., Rice, J. and Hurwitz, E.S. 1982, Reye's syndrome and medication use. *Journal of the American Medical Association*, **248**: 687–691.

Hansen, J.R., McCray, P.B., Bale, J.F. Jr, Corbett, A.J. and Flanders, D.J. 1985, Reye syndrome associated with aspirin therapy for systemic lupus erythematosus. *Pediatrics*, **76**: 202–205.

Hanzlik, P.J., Scott, R.W. and Reycraft, J.L. 1917, The salicylates VIII. Salicyl edema. *Archives of Internal Medicine*, **20**: 329–340.

Hardin, J.G. Jr and Kirk, K.A. 1979, Comparative effectiveness of five analgesics for the pain of rheumatoid synovitis. *Journal of Rheumatology*, **6**: 405.

Haslock, D.I., Nicholson, P.A. and Wright, V. 1971, The treatment of rheumatoid arthritis. A comparison of phenylbutazone and benorylate. *Clinical Trials Journal*, **8**: 43–50.

Haslock, I., Omar, A.S. and Wright, V. 1975, A comparison of micro-encapsulated aspirin and indomethacin in the treatment of rheumatoid arthritis. *Journal of Clinical Practice*, **29:** 311–314.

Haslock, L. and Wright, V. 1974, Arthritis and intestinal disease. *Journal of the Royal College of Physicians of London*, **8:** 154–161.

Hawkey, C.J., Karrasch, J.A., Szczepanski, L., Walker, D.G., Barkun, A., Swannell, A.J. and Yeomans, N.D. 1998, Omeprazole compared with misoprostol for ulcers associated with nonsteroidal anti-inflammatory drugs. Omeprazole versus Misoprostol for NSAID-induced ulcer management (OMNIUM) Study Group. *New England Journal of Medicine*, **338:** 727–734.

Hawkins, D., Pinckard, R.N., Crawford, I.P. and Farr, R.S. 1969, Structural changes in human serum albumin induced by ingestion of acetylsalicylic acid. *Journal of Clinical Investigation*, **48:** 536–542.

Hawkins, D., Pinckard, R.N. and Farr, R.S. 1968, Acetylation of human serum albumin by acetylsalicylic acid. *Science*, **160:** 780–781.

Heffner, J.E. and Sahn, S.A. 1981, Salicylate-induced pulmonary edema. *Annals of Internal Medicine*, **95:** 405–409.

Helbig, B., Gross, W.L., Borisch, B., Starz, H. and Muller-Hermelink, H.K. 1998, Characterization of synovial macrophages by monoclonal antibodies in rheumatoid arthritis and osteoarthritis. *Scandinavian Journal of Rheumatology*, **76 (Suppl.):** 61–66.

Heller, S.R., Fellows, I.W., Ogilvie, A.L. and Atkinson, M. 1982, Non-steroidal anti-inflammatory drugs and benign oesophageal stricture. *British Medical Journal*, **285:** 167–168.

Hench, P.S., Kendall, E.C., Slocomb, C.H. and Polley, H.F. 1949, The effect of a hormone of the adrenal cortex (17-hydroxy-11-dehydro-cortisone: compound E) and of pituitary adrenocorticotrophic hormone on rheumatoid arthritis. *Proceedings of the Mayo Clinic*, **24:** 181–197.

Henry, D.A., Johnson, A., Dobson, A. and Duggan, J. 1987, Fatal peptic ulcer complications and the use of non-steroidal anti-inflammatory drugs, aspirin and corticosteroids. *British Medical Journal*, **295:** 1227–1229.

Henry, D.A., Lim, L.L., Garcia-Rodriguez, L.A., Perez Gutthann, S., Carson, J.L., Griffin, M., Savage, R., Logan, R., Moride, Y., Hawkey, C., Hill, S. and Fries, J.T. 1996, Variability in risk of gastrointestinal complications with individual non-steroidal anti-inflammatory drugs: results of a collaborative meta-analysis. *British Medical Journal*, **312 (7046):** 1563–1566.

Higgs, G.A., Follenfant, R.L. and Garland, L.G. 1988, Selective inhibition of arachidonate 5–lipoxygenase by novel acetohydroxamic acids: effects on acute inflammatory responses. *British Journal of Pharmacology*, **94:** 547–551.

Hill, H.F., Hill, A.F.S., Mowat, A.G., Ansell, B.M., Nathews, J.A., Seifert, M.H., Gumpel, J.M. and Christie, G.A. 1975, Multicentre double-blind crossover trial comparing naproxen and aspirin in rheumatoid arthritis. *Scandinavian Journal of Rheumatology*, **2 (Suppl.):** 176–181.

Hill, M.F.M. and Hill, A.G.S. 1977, Fenclofenac and soluble aspirin in rheumatoid arthritis: a comparative trial. *Proceedings of the Royal Society of Medicine*, **70 (Suppl. 6):** 27–30.

Hingorani, K. 1973, Double-blind study of benorylate and ibuprofen in rheumatoid arthritis. *Rheumatology and Rehabiltation*, **11 (Suppl.):** 39–47.

Ho, P.C., Triggs, E.J., Bourne, D.W. and Heazlewood, V.J. 1985, The effects of age and sex on the disposition of acetylsalicylic acid and its metabolites. *British Journal of Clinical Pharmacology*, **19 (5):** 675–684.

Hoftiezer, J.W., Burks, M., Suilvoso, G.F. and Ivey, K.J. 1980, Comparison of the effects of regular and enteric-coated aspirin on gastroduodenal mucosa in man. *Lancet*, **2:** 609–612.

Hoftiezer, J.W., O'Laughlin, J.C. and Ivey, K.J. 1982, Effects of 24 hours of aspirin, bufferin, paracetamol and placebo on normal human gastroduodenal mucosa. *Gut*, **23 (8):** 692–697.

Hollister, J.-R. 1985, Aspirin in juvenile rheumatoid arthritis. *American Journal of Diseases in Childhood*, **139:** 866–867.

Holt, S., Rigoglioso, V., Sidhu, M., Irshad, M., Howden, C.W. and Mainero, M. 1993, Nonsteroidal anti-inflammatory drugs and lower gastrointestinal bleeding. *Digestive Diseases and Sciences*, **38:** 1619–1623.

Hoult, J.R.S. and Moore, P.K. 1980, Effects of sulfasalazine and its metabolites on prostaglandin synthesis, inactivation and actions on smooth muscle. *British Journal of Pharmacology*, **68:** 719–730.

CHAPTER 12

Hrnicek, G., Skelton, J. and Miller, W.C. 1974, Pulmonary edema and salicylate intoxication. *Journal of the American Medical Association*, **230:** 866–870.

Huber, T., Ruchti, C. and Halter, F. 1991, Nonsteroidal anti-inflammatory drug-induced strictures: a case report. *Gastroenterology*, **100:** 1119–1122.

Hubsher, J.A., Walker, B.R. and Appelrouth, D.J. 1983, Oxaprozin once daily and ibuprofen Q.I.D. in the treatment of rheumatoid arthritis: a multicentre study. *Clinical Pharmacology and Therapeutics*, **33:** 267–270.

Hudson, N., Murray, F.E., Cole, A.T., Filipowicz, B. and Hawkey, C.J. 1997b, Effect of sucralfate on aspirin-induced mucosal injury and impaired haemostasis in humans. *Gut*, **41 (1):** 19–23.

Hudson, N., Taha, A.S., Russell, R.I., Trye, P., Cottrell, J., Mann, S.G., Swanell, A.J., Sturrock, R.D. and Hawkey, C.J. 1997a, Famotidine for healing and maintenance in nonsteroidal anti-inflammatory drug-associated gastroduodenal ulceration. *Gastroenterology*, **112 (6):** 1817–1822.

Hulka, B.S., Cassel, J.C., Kupper, L.L. and Burdette, J.A. 1976, Communication, compliance, and concordance between physicians and patients with prescribed medications. *American Journal of Public Health*, **66:** 847–853.

Hurst, A. 1943, Aspirin and gastric haemorrhage. *British Medical Journal*, **1:** 768.

Hurst, A. and Lintott, G.A.M. 1939, Aspirin as a cause of haematemesis: a clinical and gastroscopic study. *Guy's Hospital Reports*, **89:** 173–176.

Hurwitz, E.S. 1989, Reye's syndrome. *Epidemiology Reviews*, **11:** 249–253.

Hurwitz, E.S., Barnett, M.J., Bregman, D., Gunn, W.J., Pinsky, P., Schonberger, L.B., Drage, J.S., Kaslow, R.A., Burlington, D.B. and Quinnan, G.V. 1987a, Public Health Service study of Reye's syndrome and medications. Report of the main study. *Journal of the American Medical Association*, **257:** 1905–1911.

Hurwitz, E.S., Barrett, M.J. and Bregman, D. 1985, Public Health Service study on Reye's syndrome and medications: report of the pilot phase. *New England Journal of Medicine*, **313:** 849–857.

Hurwitz, E.S., Barrett, M.J. and Bregman, D. 1987b, Public Health Service study on Reye's syndrome and medication. *Journal of the American Medical Association*, **257:** 1905–1987.

Huskisson, E.C., Dieppe, P.A., Scott, J. and Jones, H. 1982, Diclofenac sodium, diflunisal and naproxen: patient preference for anti-inflammatory drugs in rheumatoid arthritis. *Rheumatology and Rehabilitation*, **21:** 238–242.

Huskisson, E.C. and Hart, F.D. 1972, The use of indomethacin and aloxiprin at night. *Practitioner*, **208:** 248–251.

Huskisson, E.C. and Scott, J. 1978, Sulindac. Trials of a new anti-inflammatory drug. *Annals of the Rheumatic Diseases*, **37:** 89–92.

Huskisson, E.C., Shenfield, G.M., Taylor, R.T. and Hart, F.D. 1970, A new look at ibuprofen. *Rheumatology and Physical Medicine*, **11 (Suppl.):** 88–92.

Huskisson, E.C., Wojtulewski, J.A., Berry, H., Scott, J., Hart, F.D. and Balme, J.J. 1974, Treatment of rheumatoid arthritis with fenoprofen: comparison with aspirin. *British Medical Journal*, **1:** 176–180.

Hussar, D.A. 1975, Patient noncompliance. *Journal of the American Pharmacology Association*, **15:** 183–190, 201.

Indelicato, P.A. 1986, Comparison of diflunisal and acetaminophen with codeine in the treatment of mild to moderate pain due to strains and sprains. *Clinical Therapeutics*, **8:** 269–274.

Ivey, K.J. 1986, Gastrointestinal intolerance and bleeding with non-narcotic analgesics. *Drugs*, **32 (Suppl. 4):** 71–89.

Jasani, M.K., Downie, W.W., Samuels, B.M. and Buchanan, W.W. 1968, Ibuprofen in rheumatoid arthritis. Clinical study of analgesic and anti-inflammatory activity. *Annals of the Rheumatic Diseases*, **27:** 457–462.

Jasin, H.E. 1985, Autoantibody specificities of immune complexes sequestered in articular cartilage of patients with rheumatoid arthritis and osteoarthritis. *Arthritis and Rheumatism*, **28:** 341–348.

Jaszewski, R., Calzada, R. and Dhar, R. 1989, Persistence of gastric ulcers caused by plain aspirin or nonsteroidal antiinflammatory agents in patients treated with a combination of cimetidine, antacids, and enteric-coated aspirin. *Digestive Diseases and Sciences*, **34 (9):** 1361–1364.

Jaszewski, R., Graham, D.Y. and Stromatt, S.C. 1992, Treatment of nonsteroidal anti-inflammatory drug-induced gastric ulcers with misoprostol. A double-blind multicenter study. *Digestive Diseases and Sciences*, **37:** 1820–1824.

Jenkins, R.T., Rooney, P.J., Jones, D.B., Bienenstock, J. and Goodacre, R.L. 1987, Increased permeability in patients with rheumatoid arthritis: a side effect of oral nonsteroidal anti-inflammatory drug therapy? *British Journal of Rheumatology*, **26:** 103–107.

Jenner, P.N. 1987, Nabumetone in the treatment of skin and soft tissue injury. *American Journal of Medicine*, **83 (Suppl. 4B):** 101–106.

Jick, H., Field, A.D. and Perera, D.R. 1985, Certain nonsteroidal anti-inflammatory drugs and hospitalisation for upper gastrointestinal bleeding. *Pharmacotherapy*, **5:** 280–284.

Jick, H. and Porter, J. 1978, Drug-induced gastrointestinal bleeding. *Lancet*, **2:** 87–89.

Jick, S.S., Perera, D.R., Walker, A.M. and Jick, H. 1987a, Non-steroidal anti-inflammatory drugs and perforated peptic ulcers. *Lancet*, **2:** 1398.

Jick, S.S., Perera, D.R., Walker, A.M. and Jick, H. 1987b, Non-steroidal anti-inflammatory drugs and hospital admission for perforated peptic ulcer. *Lancet*, **2:** 380–382.

Jobanputra, P. and Nuki, G. 1994, Nonsteroidal anti-inflammatory drugs in the treatment of osteoarthritis. *Current Opinion in Rheumatology*, **6:** 433–439.

Johnson, G.M., Scurletis, T.D. and Carroll, N.B. 1963, A study of sixteen fatal cases of encephalitis-like disease in North Carolina children. *North Carolina Medical Journal*, **24:** 464–473.

Jones, M.P., Schubert, M.L. and Smith, J.L. 1991, Controversies, dilemmas, and dialogues. What do you recommend for prophylaxis in an elderly woman with arthritis requiring NSAIDs for control? *American Journal of Gastroenterology*, **86:** 264–268.

Jorde, R. and Burhol, P.G. 1987, Asymptomatic peptic ulcer disease. *Scandinavian Journal of Gastroenterology*, **22:** 129–134.

Joubert, P.H., Kushlick, A.R., McNeill, W.G., Sheard, E.S. and Muller, F.O. 1974, South African multicentre trial with Voltaren in osteoarthritis of the knee. *South African Medical Journal*, **48:** 1973–1978.

Juby, L.D., Axon, A.T., Wright, V., Winstanley, P. and Rothwell, J. 1986, Intestinal permeability and inflammation in rheumatoid arthritis. *British Journal of Rheumatology*, **25 (2):** 226–227.

Kafetz, K. 1988, Gastrointestinal hemorrhage in elderly people. *British Journal of Hospital Medicine*, **40:** 207–209.

Kajandev, A. and Martio, J. 1978, Diclofenac sodium (Voltaren) and naproxen in the treatment of rheumatoid arthritis; a comparative double-blind study. *Scandinavian Journal of Rheumatology*, **(Suppl. 22):** 57–62.

Kaldestad, E., Hansen, T. and Brath, H.K. 1975, Interaction of indomethacin and acetylsalicylic acid as shown by the serum concentrations of indomethacin and salicylate. *European Journal of Clinical Pharmacology*, **9:** 199–207.

Kanner, R.S., Tedesco, F.J. and Kalser, M.H. 1978, Azulfidine-(sulfasalazine)-induced hepatic injury. *American Journal of Digestive Diseases*, **23:** 956–958.

Karlberg, K.-E., Ahlner, J., Henriksson, P., Torgård, K., Sylvén, C. 1993, Effects of nitroglycerin on platelet aggregation beyond the effects of acetylsalicylic acid in healthy subjects. *American Journal of Cardiology*, **71:** 361–364.

Karsh, J., Kimberly, R.A., Stahl, N.I., Platz, P.H. and Decker, J.L. 1980, Comparative effects of aspirin and ibuprofen in the management of systemic lupus erythematosus. *Arthritis and Rheumatism*, **23:** 1401–1404.

Kassem, M.A. and Schulte, K.E. 1981, Influence of phenylbutazone, mofebutazone and aspirin on the pharmacokinetics of dexamethasone in the rat. *European Journal of Drug Metabolism and Pharmacokinetics*, **6:** 11–20.

Katz, L.M. and Love, P.Y. 1991, Hepatic dysfunction in association with NSAIDs. *In: Nonsteroidal Anti-inflammatory Drugs: Subpopulation. Therapy and Drug Delivery Systems.* New York: Marcel Dekker.

Kaufman, D.W., Kelly, J.P., Sheehan, J.E., Laszlo, A., Wiholm, B.E., Alfredsson, L., Koff, R.S. and Shapiro, S. 1993, Nonsteroidal anti-inflammatory drugs use in relation to major upper gastrointestinal bleeding. *Clinical Pharmacology and Therapy*, **53:** 485–494.

Kean, W.F. and Buchanan, W.W. 1985, Drug therapy in the elderly. *In: Rheumatic Therapeutics*, pp. 457–476. New York: McGraw Hill.

Kean, W.F. and Buchanan, W.W. 1987a, Pharmacokinetics of NSAID with special reference to the elderly. *Singapore Medical Journal*, **28**: 383–389.

Kean, W.F. and Buchanan, W.W. 1987b, Antirheumatic drug therapy in the elderly: a case of failure to identify the correct issues? *Journal of the American Geriatric Society*, **35 (4)**: 363–364.

Kean, W.F. and Buchanan, W.W. 1990, Pharmacological and pharmacodynamic implications of non-steroidal anti-inflammatory drug therapy in the elderly. *Canadian Journal of Gastroenterology*, **4**: 126–130.

Kean, W.F., Buchanan, W.W. and Rainsford, K.D. 1999, Therapeutics of ibuprofen in rheumatic and other chronic and painful diseases. Chapter 6. *In:* K.D. Rainsford (ed.), *Ibuprofen. A Critical Bibliographic Review*, pp. 277–353. London: Taylor & Francis.

Kean, W.F., Kraag, G.R., Rooney, P.J. and Capell, H.A. 1981, Clinical therapeutic trial of aspirin and azapropazone in rheumatoid arthritis when prescribed singly and in combination. *Current Medical Research and Opinion*, **7**: 164–167.

Kelly, J.P., Kaufman, D.W., Jurgelon, J.M., Sheehan, J., Koff, R.S. and Shapiro, S. 1996, Risk of aspirin-associated major upper-gastrointestinal bleeding with enteric-coated or buffered product. *Lancet*, **23 (348)**: 1413–1416.

Kelton, J.G., Carter, C.J., Rosenfeld, J., Massicotte-Nolan, M.P. and Hirsch, J. 1981, Sex-related differences in the efficacy of acetylsalicylic acid (ASA): the absorption of ASA and its effect on collagen-induced thromboxane B_2 generation. *Thrombosis Research*, **24**: 163–168.

Keystone, E.C., Paton, T.W., Littlejohn, G., Verdejo, A., Piper, S., Wright, L.A. and Goldsmith, C.H. 1982, Steady-state plasma levels of salicylate in patients with rheumatoid arthritis: effects of dosing interval and tablet strength. *Canadian Medical Association Journal*, **127**: 283–286.

Khalili-Varasteh, H., Rosner, I. and Legros, J. 1976, Benorylate interaction with indomethacin and phenylbutazone. *Archives Internationales de Pharmacodynamie et de Therapie*, **219**: 149–159.

Kilander, A. and Doterall, G. 1983, Endoscopic evaluation of the comparative effects of acetylsalicylic acid and choline magnesium trisalicylate on human and gastric duodenal mucosa. *British Journal of Rheumatology*, **22**: 36–40.

Kimberley, R.P., Bowden, R.E. and Keiser, H.R. 1978, Reduction of renal function by newer non-steroidal anti-inflammatory drugs. *American Journal of Medicine*, **64**: 804–807.

Kimberley R.P., Gill, J.R. and Bowden, R.E. 1978, Elevated urinary prostaglandins and the effects of aspirin on renal function in lupus erythematosus. *Annals of Internal Medicine* **89**: 336–341.

Kimberley, R.P. and Plotz, P.H. 1977, Aspirin-induced depression of renal fundus. *New England Journal of Medicine*, **296**: 418–424.

Kimberley, R.P., Sherman, R.L., Mouradian, J. and Lockshin, M.D. 1979, Apparent acute renal failure associated with therapeutic aspirin and ibuprofen administration. *Arthritis and Rheumatism*, **22**: 281–285.

Kincaid-Smith, P., Saker, B.M., McKenzie, I.F.C. and Muirden, K.D. 1968, Lesions in the blood supply of the pipilla in experimental analgesic nephropathy. *Medical Journal of Australia*, **1**: 203–206.

Kippen, I., Whitehouse, M.W. and Klinenberg, J.R. 1974, Occasional Survey. Pharmacology of uricosuric drugs. *Annals of the Rheumatic Diseases*, **33**: 391–396.

Klinenberg, J.R. and Miller, F. 1965, Effects of corticosteroids on blood salicylate concentrations. *Journal of the American Medical Association*, **194**: 131–134.

Klotz, U., Maier, K., Fischer, C. and Heinkel, K. 1980, Therapeutic efficacy of sulfasalazine and its metabolites in patients with ulcerative colitis and Crohn's disease. *New England Journal of Medicine*, **303**: 1499–1502.

Koch-Weser, J. 1972, Drug therapy. Serum drug concentrations as therapeutic guides. *New England Journal of Medicine*, **287**: 227–231.

Koff, R.S. and Shapiro, S. 1996, Risk of aspirin-associated major upper gastrointestinal bleeding with enteric coated or buffered product. *Lancet*, **248**: 1413–1416.

Kovarik, J.M., Mueller, E.A., Gaber, M., Johnston, A. and Jaehnchen, E. 1993, Pharmacokinetics of cyclosporin and steady-state aspirin during co-administration. *Journal of Clinical Pharmacology*, **33**: 513–521.

Laaksomen, A.-L. and Laine, V. 1961, A comparative study of joint pain on adult and juvenile rheumatoid arthritis. *Annals of the Rheumatic Diseases*, **20**: 386–390.

Lakai, E.N., Smith, J.L., Lidsky, M.D. and Graham, D.Y. 1987, Gastroduodenal mucosa and dyspeptic symptoms in arthritic patients during chronic non-steroidal anti-inflammatory drug use. *American Journal of Gastroenterology*, **82:** 1153–1158.

Lamson, R.W. and Thomas, R. 1932, Some untoward effects of acetylsalicylic acid. *Journal of the American Medical Association*, **99:** 107.

Lamy, P.P. 1982, Over-the-counter medication: the drug interactions we overlook. *Journal of the American Geriatrics Society*, **30 (Suppl. 11):** 69–75.

Lamy, P.P. 1986, Renal effects of nonsteroidal anti-inflammatory drugs. Heightened risk to the elderly. *Journal of the American Geriatric Society*, **34:** 361–367.

Lancaster-Smith, M.J., Jaderberg, M.E. and Jackson, D.A. 1991, Ranitidine in the treatment of nonsteroidal anti-inflammatory drug associated gastric and duodenal ulcers. *Gut*, **32:** 252–255.

Landells, J.W. 1953, The bone cysts of osteoarthritis. *Journal of Bone and Joint Surgery*, **35B:** 643–650.

Lang, J., Price, A.B., Levi, A.J., Burke, M., Gumpel, J.M. and Bjarnason, I. 1988, Diaphragm disease: pathology of disease of the small intestine induced by non-steroidal anti-inflammatory drugs. *Journal of Clinical Pathology*, **41 (5):** 516–526.

Langlois, R. and Paquette, D. 1995, Increased serum lithium levels due to ketorolac therapy. *Canadian Medical Association Journal*, **152 (2):** 152–153.

Langman, M.J. 1988, Ulcer complications and non-steroidal anti-inflammatory drugs. *American Journal of Medicine*, **84:** 15–19.

Langman, M.J. 1989, Epidemiologic evidence on the association between peptic ulceration and anti-inflammatory drug use. *Gastroenterology*, **96:** 640–646.

Langman, M.J.S., Morgan, L. and Worrall, A. 1985, Use of anti-inflammatory drugs by patients admitted with small or large bowel perforations and hemorrhage. *British Medical Journal*, **290:** 347–349.

Lanza, F., Peace, K., Gustitus, L., Rack, M.F. and Dickson, B. 1988, A blinded endoscopic comparative study of misoprostol versus sucralfate and placebo in the prevention of aspirin-induced gastric and duodenal ulceration. *American Journal of Gastroenterology*, **83:** 143–146.

Lanza, F.-L. 1984, Endoscopic studies of gastric and duodenal injury after the use of ibuprofen, aspirin and other non-steroidal anti-inflammatory agents. *American Journal of Medicine*, **77:** 19–24.

Lanza, F.-L., Royer, G.L. and Nelson, R.S. 1980, Endoscopic evaluation of the effects of aspirin, buffered aspirin and enteric-coated aspirin on gastric and duodenal mucosa. *New England Journal of Medicine*, **303:** 136–138.

Lanza, F.-L., Royer, G.L., Nelson, R.S., Chen, T.T., Seckman, C.E. and Rack, M.F. 1981, A comparative endoscopic evaluation of the damaging effects on non-steroidal anti-inflammatory agents on the gastric and duodenal mucosa. *American Journal of Gastroenterology*, **175:** 17–21.

Larkin, P.W. 1979, Salicylate binding: effect of age and regular aspirin ingestion. *Australian Journal of Pharmaceutical Science*, **8:** 123–124.

Larsen, P.R. 1972, Salicylate-induced increases in free triiodothyronine in human serum. Evidence of inhibition of triiodothyronine binding to thyroxine-binding globulin and thyroxine-binding prealbumin. *Journal of Clinical Investigation*, **51:** 1125–1134.

Lawrence, J.S. 1977, *Rheumatism in Populations*. London: William Heinemann Medical Books.

Lawrence, J.S., Bremner, J.M. and Bier, F. 1966, Osteoarthrosis prevalence in the population and relationship between symptoms and X-ray changes. *Annals of the Rheumatic Diseases*, **25:** 1–24.

Leach, R.D. and Callum, K.C. 1981, Salicylates, steroids and benign gastrocolic fistula. *British Journal of Clinical Practice*, **35:** 338.

Leadbetter, W.B. 1995, Anti-inflammatory therapy in sports injury: the role of nonsteroidal drugs and corticosteroid injection. *Clinics in Sports Medicine*, **14:** 353–410.

Lee, F.D., El-Ghobarey, A.F., Buchanan, W.W. and Browne, M.K. 1978, Rheumatoid arthritis, amyloidosis, intestinal haemosiderosis: a new clinical syndrome? *Hematology*, **7:** 121–124.

Lee, P., Ahola, J.J., Grennan, D., Brooks, P.P. and Buchanan, W.W. 1974a, Observations on drug prescribing in rheumatoid arthritis. *British Medical Journal*, **1:** 424–426.

Lee, P., Andersen, J.A., Miller, J., Webb, J. and Buchanan, W.W. 1976, Evaluation of analgesic action and efficacy of antirheumatic drugs. Study of 10 drugs in 684 patients with rheumatoid arthritis. *Journal of Rheumatology*, **3**: 283–294.

Lee, P., Baxter, A., Dick, W.C. and Webb, J. 1974b, An assessment of grip strength measurement in rheumatoid arthritis. *Scandinavian Journal of Rheumatology*, **3**: 17–23.

Lee, P., Kennedy, A.C., Anderson, J. and Buchanan, W.W. 1974c, Benefits of hospitalisation in rheumatoid arthritis. *Quarterly Journal of Medicine*, **43**: 205–214.

Lee, P., Watson, M., Webb, J., Anderson, S. and Buchanan, W.W. 1975, Therapeutic effectiveness of paracetamol in rheumatoid arthritis. *International Journal of Clinical Pharmacology and Biopharmacy*, **2**: 168–175.

Lee, P., Webb, J., Anderson, J. and Buchanan, W.W. 1973, Method of assessing therapeutic potential of anti-inflammatory drugs in rheumatoid arthritis. *British Medical Journal*, **2**: 685–688.

Lee, W.M. 1995, Drug-induced hepatotoxicity. *New England Journal of Medicine*, **333**: 1118–1146.

Leonards, J.R. 1969, Absence of gastrointestinal bleeding following administration of salicylsalicylic acid. *Journal of Laboratory and Clinical Medicine*, **74**: 911–914.

Lesko, S.M. and Mitchell, A.A. 1999, The safety of acetaminophen and ibuprofen among children younger than two years old. *Pediatrics*, **104**: 39.

Levenson, D.J., Simmons, C.E. and Brenner, B.M. 1982, Arachidonic acid metabolism, prostaglandins and the kidney. *American Journal of Medicine*, **72**: 354–374.

Levey, M., Killer, D.R., Kaufman, D.W., Siskind, V., Schwingl, P., Rosenberg, L., Strom, B. and Shapiro, S. 1988, Major upper gastrointestinal tract bleeding. Relation to the use of aspirin and other non-narcotic analgesics. *Archives of Internal Medicine*, **148**: 281–285.

Levy, G. and Giacomini, K.M. 1978, Rational aspirin dosage regimens. *Clinical Pharmacology and Therapeutics*. **23**: 247–252.

Levy, G. and Tsuchiya, T. 1972, Salicylate accumulation kinetics in man. *New England Journal of Medicine*, **287**: 430–432.

Levy, G., Tsuchiya, T. and Ansel, L.P. 1965, Salicylurate formation demonstration of Michaelis–Menten kinetics in man. *Journal of Pharmaceutical Sciences*, **54**: 496.

Levy, N. and Gaspar, E. 1975, Rectal bleeding and indomethacin suppositories. *Lancet*, **1**: 577.

Levy, R.H. and Moreland, T.A. 1984, Rationale for monitoring free drug levels. *Clinical Pharmacokinetics*, **1**: 1–9.

Lewis, A.J. and Furst, D.E. (eds) 1987, *Nonsteroidal Anti-Inflammatory Drugs. Mechanisms and Clinical Uses*. New York: Marcel Dekker.

Lewis, A.J. and Furst, D.E. (eds) 1994, *Nonsteroidal Anti-Inflammatory Drugs. Mechanisms and Clinical Uses*. 2nd edn. New York; Marcel Dekker.

Lim, R.K.S. 1966, Salicylate analgesia. *In:* M.J.H. Smith and P.K. Smith (eds), *Salicylates. A critical bibliographic review*, pp. 155–202. New York: Wiley-Interscience.

Littleton, F. Jr 1990, Warfarin and topical salicylates. *Journal of the American Medical Association*, **263**: 2888.

Littleton, M.T. 1989, Complications of multiple trauma. *Critical Care and Nursing Clinics of North America*, **1**: 75–84.

Llewellyn, J.G. and Pritchard, M.H. 1988, Influence of age and disease state in nonsteroidal anti-inflammatory drug associated gastric bleeding. *Journal of Rheumatology*, **15**: 691–694.

Lockard, O.O. Jr, Ivey, K.J., Butt, J.H., Silvoso, G.R., Sisk, C. and Holt, S. 1980, The prevalence of duodenal lesions in patients with rheumatic diseases on chronic aspirin therapy. *Gastrointestinal Endoscopy*, **26 (1):** 5–7.

Lo Iudice, T.A., Saleem, T. and Lang, J.A. 1981, Cimetidine in the treatment of gastric ulcer induced by steroidal and non-steroidal anti-inflammatory agents. *American Journal of Gastroenterology*, **75**: 104–110.

Loizzi, P.P., Muratori, M., Bruni, G. and Sachetti, G. 1980, A double-blind crossover comparison between indoprofen and aspirin in rheumatoid arthritis. *Current Medical Research and Opinion*, **6**: 598–605.

Lourie, S.H., Denman, S.J. and Schroeder, E.T. 1977, Association of renal papillary necrosis and ankylosing spondylitis. *Arthritis and Rheumatism*, **20**: 917–920.

Lovell, D.J., Giannini, E.H. and Brewer, E.J. 1984, Time course of response to non-steroidal anti-inflammatory drugs in juvenile rheumatoid arthritis. *Arthritis and Rheumatism*, **27**: 1433–1437.

Ludwin, D. and Alexopoulou, I. 1993, Cyclosporin A nephropathy in patients with rheumatoid arthritis. *British Journal of Rheumatology*, **32 (Suppl. 1):** 60–64.

Lussier, A., Camerlain, N., Menard, H., Myal, D. and Wehner, S. 1976, A double-blind crossover evaluation of keoptofen and aspirin in rheumatoid arthritis. *Scandinavian Journal of Rheumatology*, **14 (Suppl.):** 99–104.

Lynch, J.M. and Fox, T.M. 2001, Use of acetaminophen in the treatment of osteoarthritis in patients with liver disease: comment on the 2000 update of the American College of Rheumatology recommendations for the management of hip and knee osteoarthritis. *Arthritis and Rheumatism*, **44:** 2448–2449.

Macklon, A.F., Craft, A.W., Thompson, M. and Kerr, D.N.S. 1974, Aspirin and analgesic nephropathy. *British Medical Journal*, **1:** 597–600.

MacLagan, T.J. 1876a, Treatment of acute rheumatism by salicin. *Lancet*, **i:** 342–343.

MacLagan, T.J. 1876b, The treatment of acute rheumatism by salicin and salicylic acid. *Lancet*, **i:** 627.

Madhok, R., MacKenzie, J.A., Lee, F.D., Bruckner, F.E., Terry, T.R. and Sturrock, R.D. 1986, Small bowel ulceration in patients receiving non-steroidal anti-inflammatory drugs for rheumatoid arthritis. *Quarterly Journal of Medicine*, **58 (225):** 53–58.

Mainland, D. and Sutcliffe, M.I. 1965, Aspirin in rheumatoid arthritis, a seven day double-blind trial – preliminary report. *Bulletin of Rheumatic Diseases*, **16:** 264–268.

Mäkelä, A.L., Yrjana, T. and Mattila, M. 1979, Dosage of salicylates for children with juvenile rheumatoid arthritis. A prospective clinical trial with three different preparations of acetyl salicylic acid. *Acta Paediatrica Scandinavica*, **68:** 423–430.

Malone, D.E., McCormack, P.A., Daly, I., Jones, B., Long, A., Bresnihan, B., Molony, I. and O'Donoghue, D.P. 1986, Peptic ulcer in rheumatoid arthritis – intrinsic or related to drug therapy? *British Journal of Rheumatology*, **25:** 342–344.

Mandelli, M. and Tognoni, G. 1980, Monitoring plasma concentrations of salicylate. *Clinical Pharmacokinetics*, **5:** 424–440.

Maneksha, S. 1973, 'Safaprin' and benorylate – a comparative trial of two new preparations of aspirin and paracetamol in the treatment of rheumatoid arthritis and osteoarthritis. *Current Medical Research and Opinion*, **1:** 563–569.

Manfredini, R., Ricci, L., Giganti, M., La Cecilia, O., Kuwornu, H., Chierici, F. and Gallerani, M. 2000, An uncommon case of fluid retention simulating congestive heart failure after aspirin consumption. *American Journal of the Medical Sciences*, **320:** 72–74.

Manjo, G. 1975, *The Healing Hand, Man and Wound in the Ancient World*. Cambridge, MA: Harvard University Press. (Reprinted by the Classics of Medicine Library. Division of Gryphon Editions. New York, NY, USA, 1991.)

Manso, C., Taranta, A. and Nydick, I. 1956, Effect of aspirin administration on serum glutamic oxalacetic and glutamic pyruvic transaminases in children. *Proceedings of the Society for Experimental Biology and Medicine*, **93:** 84–88.

March, L.M. 1997, Osteoarthritis. Start with education, exercise, joint protection and effective analgesia. *Medical Journal of Australia*, **166:** 98–103.

Markiewicz, A. and Semenowicz, K. 1979, Time dependent changes in the pharmacokinetics of aspirin. *International Journal of Clinical Pharmacology*, **17:** 409–411.

Marshall, R.L. 1990, Ibuprofen and aspirin in acute rheumatic fever. Letter. *Journal of the American Medical Association*, **263:** 1633.

Martinez, L.M., Homan, K.I., Smyth, C.J. and Vaughan, J.H. 1980, A comparison of naproxen, indomethacin and aspirin in osteoarthritis. *Journal of Rheumatology*, **7:** 711–716.

Mason, R.W. and McQueen, E.G. 1974, Protein binding of indomethacin: binding of indomethacin to human plasma albumin and its displacement from binding by ibuprofen, phenylbutazone and salicylate, in vitro. *Pharmacology*, **12:** 12–19.

Matek, W., Rösch, W. and Becker, V. 1980, Pathological changes in lungs and liver after therapy of ulcerative colitis. *Medizinischer Fortschritt*, **98:** 491–496.

Matsuo, T., Yamada, T., Chikahira, Y. and Kadowaki, S. 1989, Effect of aspirin on heparin-induced thrombocytopenia in a patient requiring haemodialysis. *Blut*, **59:** 393–395.

Mavrikakis, M.E., MacLeod, M., Buchanan, W.W. *et al.* 1977, Mefenamic acid: an under-rated antirheumatic? *Current Medical Research and Opinion*, **4:** 535–539.

McConkey, B., Amos, R.S., Crockson, R.A., Crockson, A.P. and Walsh, L. 1978, Salazopyrin in rheumatoid arthritis. *Agents and Actions*, **8:** 438–441.

McConkey, B., Amos, R.S., Durham, S., Forster, P.J.G., Hubball, S. and Walsh, L. 1980, Sulphasalazine in rheumatoid arthritis. *British Medical Journal*, **280:** 442–444.

McConkey, B., Davies, P., Crockson, R.A., Crockson, A.P., Butler, M. and Constable, D.J. 1976, Dapsone in rheumatoid arthritis. *Rheumatology and Rehabilitation*, **15:** 230–234.

McConnell, R.J. 1989, Salsalate alters thyroid function test results. *Arthritis and Rheumatism*, **32:** 1344.

McConnell, R.J. 1992, Abnormal thyroid function test results in patients taking analgesics. *Journal of the American Medical Association*, **267:** 1242–1243.

McCreadie, M. 1982, Analgesics in cancer of the renal pelvis in New South Wales. *Cancer*, **49:** 2617–2625.

McDonald, J.R., Methison, D.A. and Stevenson, D.D. 1972, Aspirin intolerance in asthma. Detection by oral challenge. *Journal of Allergy and Clinical Immunology*, **50:** 198–207.

McMahon, M.F. and Cash, H.C. 1979, An open assessment of the efficacy and tolerability of diclofenac sodium (Voltarol) in patients with rheumatic disease and a comparative study of diclofenac sodium (Voltarol) with indomethacin in patients with osteoarthritis and rheumatoid arthritis. *Rheumatology and Rehabilitation*, **Suppl. 2:** 81–88.

McMahon, F.G., Jain, A. and Orel, A. 1976, Controlled evaluation of fenoprofen in geriatric patients with osteoarthritis. *Journal of Rheumatology*, **3 (Suppl.):** 76–82.

McPherson, T.C. 1984, Salsalate for arthritis: a clinical evaluation. *Clinical Therapeutics*, **6:** 388–403.

Medical Research Council and Nuffield Foundation Joint Committee on Clinical Trials of Cortisone, ACTH and Other Therapeutic Measures in Chronic Rheumatic Diseases 1959, *Annals of the Rheumatic Diseases*, **18:** 173.

Medical Research Council and Nuffield Foundation Joint Committee on Clinical Trials of Cortisone. ACTH and other Therapeutic Measures in Chronic Rheumatic Diseases 1954, *British Medical Journal*, **1:** 1223.

Medical Research Council and Nuffield Foundation Joint Committee on Clinical Trials of Cortisone, ACTH and other Therapeutic Measures in Chronic Rheumatic Diseases 1955, *British Medical Journal*, **2:** 695.

Medical Research Council and Nuffield Foundation Joint Committee on Clinical Trials of Cortisone, ACTH and other Therapeutic Measures in Chronic Rheumatic Diseases 1957a, *British Medical Journal*, **1:** 837.

Medical Research Council and Nuffield Foundation Joint Committee on Clinical Trials of Cortisone, ACTH and Other Therapeutic Measures in Chronic Rheumatic Diseases 1957b, *British Medical Journal*, **2:** 199.

Meinicke, J. and Danneskiold-Samsøe, B. 1980, Diclofenac sodium (Voltaren) and ibuprofen in rheumatoid arthritis. *Scandinavian Journal of Rheumatology*, **(Suppl. 35):** 2–8.

Melton, J.W. III, Lussier, A., Ward, J.R., Neustadt, D. and Multz, C. 1978, Naproxen vs. aspirin in osteoarthritis of the hip and knee. *Journal of Rheumatology*, **5:** 338–346.

Mena, H.R., Caperton, E.M., Liebl, D.H. and Zuckner, J. 1977, High doses of ibuprofen in rheumatoid arthritis: a comparison with aspirin. *Journal of the Louisiana State Medical Society*, **129:** 263–267.

Mena, H.R., Ward, J.R., Zuckner, J., Wolski, K.P., Briney, W.G., Miaskiewicz, S.L., Shively, C.A. and Vesell, E.S. 1982, Sex differences in absorption kinetics of sodium salicylate. *Clinical Pharmacology and Therapeutics*, **31 (1):** 30–37.

Menguy, R., Desbaillets, L., Masters, Y.F. and Okabe, S. 1972, Evidence for a sex-linked difference in aspirin metabolism. *Nature*, **239:** 102–103.

Mhajo, G. 1975, *The Healing Hand. Man and Wound in the Ancient World.* Cambridge, MA: Harvard Press. (Reprinted by The Classics of Medicine Library, New York, 1991).

Miaskiewicz, S.L., Shively, C.A. and Vessell, E.S. 1982, Sex differences in absorption kinetics of sodium salicylate. *Clinical Pharmacology and Therapeutics*, **31:** 30–37.

Miehlke, K. 1977, Diflunisal in clinical practice. *Proceedings of a special symposium held at the XIV Congress of Rheumatology, San Francisco, California, USA.* Chairman: Professor K. Miehlke, Futura Publishing Company Inc., New York.

Mielants, H., Goemaere, S., De Vos, M., Schelstraete, K., Goethals, K., Maertens, M., Ackerman, C. and Veys, E.M. 1991, Intestinal mucosal permeability in inflammatory rheumatic diseases. I. Role of antiinflammatory drugs. *Journal of Rheumatology*, **18 (3):** 389–393.

Mielants, H. and Veys, E.M. 1985, NSAIDs and the leaky gut. *Lancet*, **i:** 218.

Miller, D.R. 1981, Combination use of nonsteroidal anti-inflammatory drugs. *Drug Intelligence and Clinical Pharmacology*, **15:** 3–7.

Miller, R.R. 1978, Deafness due to plain and long-acting aspirin tablets. *Journal of Clinical Pharmacology*, **18:** 468–471.

Mills, E.H., Whitwort, J.A., Andrews, J. *et al.* 1982, Non steroidal anti-inflammatory drugs and blood pressure. *Australian and New Zealand Journal of Medicine*, **12:** 478–482.

Mills, S.Y., Jacoby, R.K., Chacksfield, M. and Willoughby, M. 1996, Effect of a proprietary herbal medicine on the relief of chronic arthritic pain: a double-blind study. *British Journal of Rheumatology*, **35:** 874–878.

Miners, J.O., Grgurinovich, N., Whitehead, A.G., Robson, R.A. and Birkett, D.J. 1986, Influence of gender and oral contraceptive steroids on the metabolism of salicylic acid and acetylsalicylic acid. *British Journal of Clinical Pharmacology*, **22:** 135–142.

Minocha, A. and Greenbaum, D.S. 1991, Pill-esophagitis caused by nonsteroidal antiinflammatory drugs. *American Journal of Gastroenterology*, **86:** 1086–1089.

Mintz, G. and Fraga, A. 1973, Severe osteoarthritis of the elbow in foundry workers. *Archives of Environmental Health*, **27:** 78–80.

Misra, S.P., Kumar, N. and Anand, B.S. 1990, Aspirin concurrently administered with ranitidine does not delay healing of duodenal ulcer. *Australian and New Zealand Journal of Medicine*, **20:** 201.

Mitchell, J.A., Saunders, M., Barnes, P.J., Newton, R. and Belvisi, M.G. 1997, Sodium salicylate inhibits cyclo-oxygenase-2 activity independently of transcription factor (nuclear factor κ) activation: role of arachidonic acid. *Molecular Pharmacology*, **51:** 907–912.

Miwa, K., Kambara, H. and Kawai, C. 1979, Variant angina aggravated by aspirin. *Lancet*, **2:** 1382.

Mogadam, M., Dobbins, W.O., Korelitz, B.I. and Ahmed, S.W. 1981, Pregnancy in inflammatory bowel disease: effect of sulfasalazine and corticosteroids on fetal outcome. *Gastroenterology*, **80:** 72–76.

Molin, L. and Stendahl, O. 1979, The effect of sulfasalazine and its active components on human polymorphonuclear leukocyte function in relation to ulcerative colitis. *Acta Medica Scandinavica*, **206:** 451–457.

Mongan, E., Kelly, P., Nies, K., Porter, W.W. and Paulus, H.E. 1973, Tinnitus as an indication of therapeutic serum salicylate levels. *Journal of the American Medical Association*, **226:** 142–145.

Monroe, E.W. and Jones, H.E. 1977, Urticaria. An updated review. *Archives of Dermatology*, **113:** 80–90.

Montgomery, P.R. and Sitar, D.S. 1981, Increased serum salicylate metabolites with age in patients receiving chronic acetylsalicylic acid therapy. *Gerontology*, **27:** 329–333.

Montgomery, P.R., Berger, L.G., Mitenko, P.A. and Sitar, D.S. 1986, Salicylate metabolism: effects of age and sex in adults. *Clinical Pharmacolgy and Therapeutics*, **39:** 571–576.

Morassut, P., Yang, W. and Karsh, J. 1989, Aspirin intolerance. *Seminars in Arthritis and Rheumatism*, **19:** 22–30.

Morgan, E., Kelly, P., Mies, K., Porter, W.W. *et al.* 1973, Tinnitus as an indication of therapeutic serum levels. *Journal of the American Medical Association*, **226:** 142–145.

Morgan, J. and Furst, D.E. 1986, Implications of drug therapy in the elderly. *Clinical Rheumatic Diseases*, **12:** 227–244.

Morris, A.J., Madhok, R., Sturrock, R.D., Capell, H.A. and MacKenzie, J.F. 1991, Enteroscopic diagnosis of small bowel ulceration in patients receiving non-steroidal anti-inflammatory drugs. *Lancet*, **337:** 520.

Moskowitz, R.W., Schwartz, H.J., Michel, B., Ratnoff, O.D. and Astrup, T. 1970, Generation of kinin-like agents by chondroitin sulfate, heparin, and human articular cartilage. Possible pathophysiologic implications. *Journal of Laboratory and Clinical Medicine*, **76:** 790–794.

Muckle, D.S. 1974a, A double-blind trial of ibuprofen and aspirin in the treatment of soft tissue injuries sustained in professional football. *British Journal of Sports Medicine*, **7:** 46–47.

Muckle, D.S. 1974b, Comparative study of ibuprofen and aspirin in soft-tissue injuries. *Rheumatology and Rehabilitation*, **13:** 142–145.

Muir, A. 1963, Salicylates, dyspepsia and peptic ulceration. *In:* A. Dixon, B.K. Martin, M.J.H. Smith and P.H.N. Wood (eds), *Salicylates. An International Symposium*, p. 230. Boston: Brown.

Mukhopadhyay, A., Sarnaik, A.P. and Deshmukh, D.R. 1992, Interactions of ibuprofen with influenza infection and hyperammonemia in an animal model of Reye's syndrome. *Pediatric Research*, **31:** 258–260.

Mulhawas, K.K., Shah, G.M. and Wimer, R.L. 1981, Renal papillary necrosis caused by long-term ingestion of pentanocine and aspirin. *Journal of the American Medical Association*, **246:** 867–868.

Mullen, B.J. 1987, Results of a six-month study comparing the safety and efficacy of nabumetone and aspirin in the treatment of osteoarthritis. *American Journal of Medicine*, **83:** 74–77.

Muller, F.O., Gosling, J.A. and Erdmann, G.H. 1977, A comparison of tolmetin with aspirin in the treatment of osteo-arthritis of the knee. *South African Medical Journal*, **51:** 794–796.

Multicentre Salsalate/Aspirin Comparison Study Group 1989, Does the acetyl group of aspirin contribute to the antiinflammatory efficacy of salicylic acid in the treatment of rheumatoid arthritis? *Journal of Rheumatology*, **16:** 321–327.

Multz, C.V., Bernhard, G.C., Blechman, W.C., Zane, S., Restifo, R.A. and Varady, J.G. 1974, A comparison of intermediate dose aspirin and placebo in rheumatoid arthritis. *Clinical Pharmacology and Therapeutics*, **15:** 310–315.

Multz, C.V., Bobyn, R.D. and Caldwell, J.R. 1978, Sodium meclofenamate (Meclomen) vs aspirin for rheumatoid arthritis. *Current Therapeutic Research*, **23:** S72–S80.

Munther, R.S. and Bennett, W.M. 1980, Effects of aspirin on glomerular filtration rate in normal humans. *Annals of Internal Medicine*, **92:** 386–387.

Murray, T. and Goldberg, M. 1975, Analgesic abuse and renal disease. *Annual Review of Medicine*, **26:** 537–550.

Murray, T. and Goldberg, M. 1975, Chronic interstitial nephritis: etiologic factors. *Annals of Internal Medicine*, **82(4):** 453–459.

Mutchie, K.D., Saunders, G.H., Hanissian, A.S. and Poe, T.E. 1980, Interlaboratory salicylate variability. *Journal of Rheumatology*, **7:** 737–740.

Mutru, O., Penttila, M., Pesonen, J., Salmela, P., Suhonen, O. and Sonck, T. 1978, Diclofenac sodium (Voltaren) and indomethacin in the ambulatory treatment of rheumatoid arthritis: a double-blind multicentre study. *Scandinavian Journal of Rheumatology*, **22:** 51–56.

Mydick, I., Yang, J., Stollerman, G.H., Wroblewski, F. and La Due, J.S. 1955, The influence of rheumatic fever on serum concentrations of the enzyme glutamic oxalacetic transaminase. *Circulation*, **12:** 795.

Myers, E.N., Bernstein, J.M. and Fostiropoulos, G. 1963, Salicylate ototoxicity. A clinical study. *New England Journal of Medicine*, **273:** 587–590.

Myers, S.L., Flusser, D., Brandt, K.D. *et al.* 1992, Prevalence of cartilage shards in synovium and their association with synovitis in patients with early and end-stage osteoarthritis. *Journal of Rheumatology*, **19:** 1247–1251.

Nadkarni, M.M., Peller, C.A. and Retig, J. 1992, Eosinophilic hepatitis after ingestion of choline magnesium trisalicylate. *American Journal of Gastroenterology*, **87:** 151–153.

Nanra, R.S. 1983, Renal effects of antipyretic analgesics. *American Journal of Medicine*, **75:** 70–81.

Nanra, R.S. and Kincaid-Smith, P. 1975, Renal papillary necrosis in rheumatoid arthritis. *Medical Journal of Australia*, **1:** 194–197.

Nanra, R.S., Stuart-Taylor, J., De Leon, A.H. and White, K. 1978, Analgesic nephropathy: etiology, clinical syndrome and clinicopathologic correlations in Australia. *Kidney International*, **13:** 79–92.

Needham, C.D., Kyle, J., Jones, P.F., Johnson, S.J. and Kerridge, D.F. 1971, Aspirin and alcohol in gastrointestinal hemorrhage. *Gut*, **12:** 819–821.

Netter, P., Faure, G., Regent, M.C., Procknal, J.A. and Levy, G. 1985, Salicylate kinetics in old age. *Clinical Pharmacology and Therapeutics*, **38:** 6–11.

Nevinny, D. and Gowans, J.C.D. 1960, Observations in the usefulness of a new liquid salicylte in arthritis. *Internal Records in Medicine*, **173:** 242–247.

Newman, N.M. and Ling, R.S. 1985, Acetabular bone destruction related to non-steroidal anti-inflammatory drugs. *Lancet*, **2:** 11–14.

New Zealand Rheumatism Association, 1974, Aspirin and the kidney. *British Medical Journal*, **1:** 593–596.

Nolan, L. and O'Malley, K. 1988, Prescribing for the elderly. Part 1. Sensitivity of the elderly to adverse drug reactions. *Journal of the American Geriatric Society*, **36:** 142–149.

Novak, L.P. 1972, Aging, total body potassium, fat free mass, and cell mass in males and females between the ages of 18 and 35 years. *Journal of Gerontology*, **27:** 438–443.

Nydick, J., Tang, J., Stollerman, G.H. *et al.* 1955, The influence of rheumatic fever on serum concentrations of the enzyme, glutamic oxalacetic transaminase. *Circulation*, **12:** 795–800.

Oates, J.A., Fitzgerald, G.A., Branch, R.A., Jackson, E.K., Knapp, H.R. and Roberts, L.J. II 1988, Clinical implications of prostaglandin and thromboxane A2 formation (2). *New England Journal of Medicine*, **319:** 761–767.

Oddson, E., Gudjonsson, H. and Thjodleifsson, B. 1990, Protective effect of omeprazole or ranitidine against naproxen induced damage to the human gastroduodenal muosa. *Scandinavian Journal of Gastroenterology*, **25 (Suppl. 176):** 25.

O'Gorman, T. and Koff, R.S. 1977, Salicylate hepatitis. *Gastroenterology*, **72:** 726–728.

O'Laughlin, J.C., Hoftiezier, J.W. and Ivey, K.J. 1981a, Healing of aspirin-associated peptic ulcer disease despite continued salicylate ingestion. *Archives of Internal Medicine*, **141:** 781–783.

O'Laughlin, J.C., Hoftiezier, J.W., Mahoney, J. and Ivey, K.J. 1981b, Does aspirin prolong bleeding from gastric biopsies in man? *Gastrointestinal Endoscopy*, **27:** 1–5.

O'Laughlin, J.C., Silvoso, G.K. and Ivey, K.J. 1982, Resistance to medical therapy of gastric ulcers in rheumatic disease patients taking aspirin: a double-blind study with cimetidine and follow-up. *Digestive Diseases and Sciences*, **27:** 926–980.

Olhagen, B. 1970, The intestine and rheumatism. *Acta Rheumatica Scandinavica* **16:** 177–183.

Olssen, B. 1983, Decreasing serum salicylate concentrations during long-term administration of acetyl-salicylic acid in healthy volunteers. Discussion of possible clinical implications. *Scandinavian Journal of Rheumatology*, **12:** 81–84.

O'Malley, K., Crooks, J., Duke, E. and Stevenson, I.H. 1971, Effect of age and sex on human drug metabolism. *British Medical Journal*, **3:** 607–609.

Orlowski, J.P. 1984, Reye's syndrome. How strong is the association? *Postgraduate Medicine*, **75:** 47–54.

Orlowski, J.P., Campbell, P. and Goldstein, S. 1990, Reye's syndrome: a case control study of medication use and associated viruses in Australia. *Cleveland Clinical Journal of Medicine*, **57:** 323–329.

Orlowski, J.P., Gillis, J. and Kilham, H.A. 1987, A catch in the Reye. *Pediatrics*, **80:** 638–642.

Oroyco-Alcala, J.J. and Baum, J. 1979, Regular and enteric-coated aspirin: a re-evaluation. *Arthritis and Rheumatism*, **22:** 1034–1037.

Osler, W. 1892, *The Principles and Practice of Medicine*. New York: D. Appleton and Company. (Reprinted, by the Classics of Medicine Library, Birmingham, Alabama, USA, 1978.)

Ouslander, J.G. 1981, Drug therapy in the elderly. *Annals of Internal Medicine*, **95 (6):** 711–722.

Oviedo, J.A. and Wolfe, M.M. 2001, Clinical potential of cyclo-oxygenase-2 inhibitors. *BioDrugs*, **15:** 563–572.

Owen, S.G., Roberts, M.S., Friesen, W.T. and Francis, H.W. 1989, Salicylate pharmacokinetics in patients with rheumatoid arthritis. *British Journal of Clinical Pharmacology*, **28:** 449–461.

Pachman, L.M., Olufs, R., Procknal, J.A. and Levy, G. 1979, Pharmacokinetic monitoring of salicylate therapy in children with juvenile rheumatoid arthritis. *Arthritis and Rheumatism*, **22:** 826–831.

Pachman, L.M. and Poznanski, A.K. 1997, Juvenile (rheumatoid arthritis). *In:* W.J. Koopman (ed.), *Arthritis and Allied Conditions*, 13th edn, Vol. 1, pp. 1155–1196. Baltimore, MD: Williams and Wilkins.

Page, J. and Henry, D. 2000, Consumption of NSAIDs and the development of congestive heart failure in elderly patients. An unrecognized public health problem. *Achives of Internal Medicine*, **160:** 777–784.

Page, M.C., Tildesley, G. and Wood, J.R. 1988, Prevention of gastroduodenal damage induced by non-steroidal anti-inflammatory drugs: controlled trials of ranitidine. *British Medical Journal*, **297:** 1017–1021.

Palmer, D.G., Ferry, D.G., Gibbins, B.L., Hall, S.M., Grennan, D.M., Lum, J. and Myers, D.B. 1981, Ibuprofen and diflunisal in rheumatoid arthritis: a double-blind comparative trial. *New Zealand Journal of Medicine*, **94:** 45–47.

Palmoski, M.J., Colyer, R.A. and Brandt, K.D. 1980, Marked suppression by salicylate of the augmented proteoglycan synthesis in osteoarthritic cartilage. *Arthritis and Rheumatism*, **23:** 83–91.

Parry, D.J. and Wood, P.H.N. 1967, Relationship between aspirin taking and gastroduodenal hemorrhage. *Gut*, **8:** 301–305.

Partin, J.S., Daugherty, C.C., McAdams, A.J., Partin, J.C. and Schubert, W.K. 1984, A comparison of liver ultrastructure in salicylate intoxication and Reye's syndrome. *Hepatology*, **4:** 687–690.

Paton, J.W. and Little, A.H. 1980, A note on enteric-coated ASA. *Canadian Journal of Hospital Pharmacy*, **33:** 21.

Patrono, C. 1986, Inhibition of renal prostaglandin synthesis in man: methodological and clinical implications. *Scandinavian Journal of Rheumatology*, **62:** 14–25.

Paulus, H.E. 1982a, Government affairs: FDA Arthritis Advisory Committee Meeting. *Arthritis and Rheumatism*, **25:** 1124–1125.

Paulus, H.E. 1982b, An overview of benefit/risk of disease modifying treatment of rheumatoid arthritis as of today. *Annals of the Rheumatic Diseases*, **1:** 26–29.

Paulus, H.E. 1989, Aspirin versus nonacetylated salicylate. *Journal of Rheumatology*, **16:** 264–265.

Paulus, H.E., Siegel, M., Mongan, E., Okun, R. and Calabro, J.J. 1971, Variations of serum concentrations and half-life and salicylate in patients with rheumatoid arthritis. *Arthritis and Rheumatism*, **14:** 527–531.

Pearson, D.J., Stones, N.H. and Bentley, S.J. 1983, Proctocolitis induced by salicylate and associated with asthma and recurrent nasal polyps. *British Medical Journal*, **287:** 1675.

Pederson, A.K. and Fitzgerald, G.A. 1984, Dose-related kinetics of aspirin. Presystemic acetylation of platelet cyclooxygenase. *New England Journal of Medicine*, **311:** 1206–1211.

Pelletier, J.P. 1999, The influence of tissue cross-talking on OA progression: role of nonsteroidal anti-inflammatory drugs. *Osteoarthritis and Cartilage*, **7:** 374–376.

Perneger, T.V., Whelton, P.K. and Klag, M.J. 1994, Risk of kidney failure associated with the use of acetaminophen, aspirin and nonsteroidal anti-inflammatory drugs. *New England Journal of Medicine*, **331:** 1675–1679.

Peyron, J. and Doury, P. 1982, Comparative study of pirprofen and ketoprofen in the treatment of osteoarthritis of the hip and of the knee. *Nouvelle Presse Medicale*, **11:** 2497–2499.

Peyron, J.G. 1982, Double-blind comparative studies of tiaprofenic acid in degenerative joint diseases. *Journal of Rheumatology*, **7:** 151–158.

Phillips, J. and Rogerson, R. 1972, A double-blind comparative trial of aspirin preparation and ibuprofen in general practice. *British Journal of Clinical Practice*, **26:** 113–115.

Piafsky, K.M., 1980, Disease-induced changes in the plasma binding of basic drugs. *Clinical Pharmacokinetics*, **5:** 246–262.

Pinals, R.S. and Frank, S. 1967, Relative efficacy of indomethacin and acetylsalicylic acid in rheumatoid arthritis. *New England Journal of Medicine*, **276:** 512–514.

Pinckard, R.N., Hawkins, D. and Farr, R.S. 1968, In vitro acetylation of plasma proteins, enzymes and DNA by aspirin. *Nature*, **219:** 68–69.

Pinckard, R.N., Hawkins, D. and Farr, R.S. 1970, The inhibitory effect of salicylate on the acetylation of human albumin by acetylsalicylic acid. *Arthritis and Rheumatism*, **13:** 361–368.

Pinsky, P.F., Hurwitz, E.S., Schonberger, L.B. and Gunn, W.J. 1988, Reye's syndrome and aspirin. Evidence for a dose-response effect. *Journal of the American Medical Association*, **260:** 657–661.

Pleskow, W.W., Stevenson, D.D., Mathison, D.A., Simon, R.A., Schatz, M. and Zeiger, R.S. 1982, Aspirin desensitization in aspirin-sensitive asthmatic patients: clinical manifestations and characterization of the refractory period. *Journal of Allergy and Clinical Immunology*, **69:** 11–19.

Plotz, P.H. and Kimberley, R.P. 1981, Acute effects of aspirin and acetaminophen on renal function. *Archives of Internal Medicine*, **3, 141:** 343–348.

Porter, J.D., Robinson, P.H., Glasgow, J.F., Banks, J.H. and Hall, S.M. 1990, Trends in the incidence of Reye's syndrome and the use of aspirin. *Archives of Diseases in Childhood*, **65:** 826–829.

Postlethwaite, A.E., Gutman, R.A. and Kelley, W.N. 1974, Salicylate-mediated increase in urate removal during hemodialysis: evidence for urate binding to protein *in vivo*. *Metabolism*, **23:** 771–777.

Prescott, L.F. 1982, Analgesic nephropathy: a reassessment of the role of phenacetin and other analgesic drugs. *Drugs*, **23:** 75–149.

Prescott, L.F. 1986, Liver damage with non-narcotic analgesics. *Medical Toxicology*, **1 (Suppl. 1):** 44–56.

Prescott, L.F. and Martin, U. 1992, Current status of nephrotoxicity by non-steroidal anti-inflammatory drugs. *In:* K.D. Rainsford and G.-P. Velo (eds), *Side Effects of Anti-inflammatory Drugs*, Vol. 3, pp. 223–235. Dordrecht: Kluwer Academic Publishers.

Preston, S.J., Arnold, M.H., Beller, E.M., Brooks, P.M. and Buchanan, W.W. 1989, Comparative analgesic and anti-inflammatory properties of sodium salicylate and acetylsalicylic acid (aspirin) in rheumatoid arthritis. *British Journal of Clinical Pharmacology*, **27:** 607–611.

Prieur, A.M. 1998, Juvenile chronic arthritis. *In:* J.H. Kleppel and O.A. Dieppe (eds), *Rheumatology*, 2nd edn, Vol. 1 **(5):** 21.1–21.10. London: Mosby.

Prieur, A.M. and Stephen, J.L. 1994, Macrophage activation syndrome in pediatric rheumatic diseases. *Review of Rheumatology*, **61B:** 385–388.

Pringle, J.A., Byers, P.D. and Brown, M.E.A. 1978, Immunofluorescence in osteoarthritis. *Nature*, **274:** 94–96.

Pullar, T., Zoma, A., Madhok, R., Hunter, J.A. and Capell, H.A. 1985, Have the newer NSAIDs contributed to the management of rheumatoid arthritis? *Scottish Medical Journal*, **3:** 161–163.

Quick, A.J. 1966, Salicylates and bleeding: the aspirin tolerance test. *American Journal of Medical Science*, **252 (3):** 265–269.

Quick, A.J. 1971, Aspirin, platelets and bleeding. *Journal of Pediatrics*, **79:** 873–874.

Ragheb, M.A. 1987, Aspirin does not significantly affect patients' serum lithium levels. *Journal of Clinical Psychiatry*, **48:** 425.

Rahwan, G.L. and Rahwan, R.G. 1986, Aspirin and Reye's syndrome: the change in prescribing habits of health professionals. *Drug Intelligence and Clinical Pharmacy*, **20:** 143–145.

Rainsford, K.D. 1982, An analysis of the gastrointestinal side-effects of non-steroidal anti-inflammatory drugs, with particular reference to comparative studies in man and laboratory species. *Rheumatology International*, **2:** 1–10.

Rainsford, K.D. (ed.) 1984a, *Anti-inflammatory and Anti-rheumatic Drugs*. Boca Raton: CRC Press.

Rainsford, K.D. 1984b, Biochemical effects of anti-inflammatory drugs in inflammatory diseases. *In:* W.R.N. Williamson (ed.), *Anti-Inflammatory Compounds*, pp. 132–152. New York: Marcel Dekker.

Rainsford, K.D. 1996, Mode of action, uses, and side-effects of anti-inflammatory drugs. *In:* K.D. Rainsford (ed.), *Advances in Anti-Rheumatic Therapy*, pp. 59–111. Boca Raton, FL: CRC Press.

Rainsford, K.D. 1999, History and development of ibuprofen. *In:* K.D. Rainsford (ed.), *Ibuprofen. A Critical Bibliographic Review*, pp. 3–24. London: Taylor & Francis.

Rainsford, K.D. and Buchanan, W.W. 1990, Aspirin versus the non-acetylated salicylates. *In: Baillière's Clinical Rheumatology*, pp. 247–268. London: Baillière Tindall.

Rainsford, K.D., Davies, A., Mundy, L. and Ginsburg, I. 1989, Comparative effects of azapropazone on cellular events at inflamed sites. Influence on joint pathology in arthritic rats, leucocyte superoxide and eicosanoid production, platelet aggregation, synthesis of cartilage proteoglycans, synovial production and actions of interleukin-1 in cartilage resorption correlated with drug uptake into cartilage *in vitro*. *Journal of Pharmacy and Pharmacology*, **41:** 322–330.

Rainsford, K.D., Rashad, S.Y., Revell, P.A., Low, F.M., Hemingway, A.P., Walker, F.S., Johnson, D., Stetsko, P., Ying, C. and Smith, F. 1992, Effects of NSAIDs on cartilage proteoglycan and synovial prostaglandin metabolism in relation to progression of joint deterioration in osteoarthritis. *In:* A. Bálint, B. Gömör and L. Hodinka (eds), *Rheumatology, State of the Art*, pp. 177–183. Amsterdam, London, New York and Tokyo: Excerpta Medica.

Rainsford, K.D., Ying, C. and Smith, F.C. 1997, Effects of meloxicam compared with other NSAIDs, on cartilage proteoglycan metabolism, synovial prostaglandin E2, and production of interleukins 1, 6 and 8, in human and porcine explants in organ culture. *Journal of Pharmacy and Pharmacology*, **49:** 991–998.

Ramay, J.N., Burrow, G.N., Spaulding, S.W., Donabedian, R.K, Speroff, L. and Frantz, A.G. 1976, The effect of aspirin and indomethacin on the TRH response in man. *Journal of Clinical Endocrinology and Metabolism*, **43:** 107–114.

Rampton, D.S. and Sladen, G.E. 1981, Relapse of ulcerative proctocolitis during treatment with non-steroidal anti-inflammatory drugs. *Postgraduate Medicine*, **57:** 297–299.

Rashad, S., Low, F., Revell, P., Hemingway, A., Rainsford, K. and Walker, F. 1989, Effect of non-steroidal anti-inflammatory drugs on course of osteoarthritis. *Lancet*, **2:** 519–522.

Rashad, S., Rainsford, K., Revell, P., Low, F., Hemingway, A. and Walker, F. 1992, The effects of NSAIDs on the course of osteoarthritis. *In:* A. Bálint, B. Gömör and L. Hodinka (eds), *Rheumatology, State of the Art*, pp. 184–188. Amsterdam, London, New York and Tokyo: Excerpta Medica, .

Raskin, J.B., White, R.H. and Jackson, J.E. 1996, Misoprostol dosage in the prevention of nonsteriodal anti-inflammatroy drug-induced gastric and duodenal ulcers: a comparison of three regimens. *Annals of Internal Medicine*, **123:** 344–350.

Reid, J., Walson, R.D. and Sproull, D.H. 1950, The mode of action of salicylate in acute rheumatic fever. *Quarterly Journal of Medicine*, **19:** 1–12.

Reimann, I.W., Galbs, E., Fischer, C. and Froehlich, J.C. 1985, Influence of intravenous acetylsalicylic acid and sodium salicylate on human renal function and lithium clearance. *European Journal of Clinical Pharmacology*, **29:** 435–441.

Remington, P.L., Rowley, D., McGee, H., Hall, W.N. and Monto, A.S. 1986, Decreasing trends in Reye syndrome and aspirin use in Michigan, 1979 to 1984. *Pediatrics*, **77:** 93–98.

Remington, P.L., Shabino, C.L., McGee, H., Preston, G., Sarniak, A.P.P. and Hall, W.M. 1985, Reye syndrome and juvenile rheumatoid arthritis in Michigan. *American Journal of Diseases in Children*, **139:** 870–872.

Rennebohm, R.M., Heubi, J.E., Dougherty, C.C. and Daniels, S.R. 1985, Reye's syndrome in children receiving salicylate therapy for connective tissue disease. *Journal of Pediatrics*, **107:** 877–880.

Rey, E., El-Assaf, H.D., Richard, M.O., Weber, S., Bourdon, A., Picard, G. and Olive, G. 1983, Pharmacological interaction between nitroglycerin and aspirin after acute and chronic aspirin treatment of healthy subjects. *European Journal of Clinical Pharmacology*, **25:** 779–782.

Reye, R.D.K., Morgan, G. and Baral, J. 1963, Encephalopathy and fatty degeneration of the viscera: a disease entity in childhood. *Lancet*, **2:** 749–752.

Reynolds, P.M.G., Rhymer, A.R., MacLeod, M. and Buchanan, W.W. 1977, Comparison of sulindac and aspirin and rheumatoid arthritis. *Current Medical Research and Opinion*, **4:** 485–491.

Rich, R.R. and Johnson, J.S. 1973, Salicylate hepatotoxicity in patients with juvenile rheumatoid arthritis. *Arthritis and Rheumatism*, **16:** 1.

Ricks, W.B. 1976, Salicylate hepatotoxicity in Reitner's syndrome (Letter). *Annals of Internal Medicine*, **84:** 52–53.

Ritchie, D.M., Boyle, J.A., McInnes, J.M., Jasani, M.K., Dalakos, T.G., Grieveson, P. and Buchanan, W.W. 1968, Clinical studies with an articular index for the assessment of joint tenderness in patients with rheumatoid arthritis. *Quarterly Journal of Medicine*, **37:** 393–406.

Roberts, M.S., Rumble, R.H., Wanwimolruk, S., Thomas, D. and Brooks, P.M. 1983, Pharmacokinetics of aspirin and salicylate in elderly subjects and in patients with alcoholic liver disease. *European Journal of Clinical Pharmacology*, **25 (2):** 253–261.

Robinson, D.R., McGuire, M.B., Bastian, D., Kantrowitz, F. and Levine, L. 1978, The effects of anti-inflammatory drugs on prostaglandin production by rheumatoid synovial tissue. *Prostaglandins in Medicine*, **1:** 461–477.

Robinson, H., Abruzzo, J.L., Miyara, A., and Ward, J.R. 1989, Concomitant tometin and aspirin therapy for rheumatoid arthritis. *In:* J.R. Ward (ed), *Tometin: A New Non-steroidal Anti-inflammatory Agent*, pp. 88–97. Princeton: Excerpta Medica.

Robinson, M., Mills, R.J. and Euler, A.R. 1991, Ranitidine prevents duodenal ulcers associated with non-steroidal anti-inflammatory drug therapy. *Alimentary Pharmacology and Therapeutics*, **5:** 143–150.

Roe, C.R., Millington, D.S., Maltby, D.A. and Kennbrew, P. 1986, Recognition of medium-chain acyl-CoA dehydrogenase deficiency in asymptomatic siblings of children dying of sudden infant death or Reye-like syndromes. *Journal of Paediatrics*, **108**: 13–18.

Rom, G.C., Munden, P.M., Waagner, D. and Ledbetter, E.O. 1987, Reye's syndrome and juvenile rheumatoid arthritis: a case report in Texas. *Texas Medicine*, **83**: 46–47.

Ronningen, H. and Langeland, N. 1979, Indomethacin treatment in osteoarthritis of the hip joint. *Acta Orthopedia Scandinavia*, **50**: 169–174.

Rooney, P.J. and Bjarnason, I. 1991, NSAID gastropathy: not just a pain in the gut! Editorial. *Journal of Rheumatology*, **18**: 796–798.

Rooney, P.J., Jenkins, R.T., Smith, K.M. and Cotes, G. 1986, Indium-labelled polymorphonuclear leucocyte scans in rheumatoid arthritis – an important clinical cause of false positive results. *British Journal of Rheumatology*, **25**: 167–170.

Rooney, P.J. and Kean, W.F. 1987, Reinforcement of bias in the medical literature: nonsteroidal anti-inflammatory drugs and the stomach – a case in point. *British Journal of Rheumatology*, **26**: 231–233.

Rosen, A.M. and Fleischer, D.E. 1989, Upper GI bleeding in the elderly: diagnosis and management. *Geriatrics*, **44**: 25–40.

Rosenbloom, D., Brooks, P., Bellamy, N. and Buchanan, W.W. 1985, *Clinical Trials in the Rheumatic Diseases. A Selected Critical Review*. New York: Praeger.

Rosenhall, L. and Uddenfeldt, M. 1979, Lung manifestations after salazopyridine intake. *Lakartidningen*, **76**: 3923–3924.

Rosenkranz, B., Fischer, C., Meese, C.O. and Frölich, J.C. 1986, Effects of salicylic and acetylsalicylic acid alone and in combination on platelet aggregation and prostanoid synthesis in man. *British Journal of Clinical Pharmacology*, **21**: 309–317.

Rostrom, A., Maetzel, A., Tugwell, P. and Wells, G. 1999, Ulcer disease and non-steroidal anti-inflammatory drugs: etiology and treatment. *In: Evidence Based Gastroenterology and Hepatology*, pp. 97–117. London: British Medical Journal Books.

Roth, H., Levassear, Y.J. and Ryan, R. 1981, A controlled, multicentre, twelve month study of pirprofen and acetylsalicylic acid in rheumatoid arthritis. *In:* J.K. van der Korst (ed.), *A New Antirheumatic Analgesic Agent Pirprofen (Rengasil®)*, pp. 58–78. Bern: Hans Huber.

Roth, S., Agrawal, N., Mahowald, M., Montoya, H., Robbins, D., Miller, S., Nutting, E., Woods, E., Crager, M. and Nissen, C. 1989, Misoprostol heals gastroduodenal injury in patients with rheumatoid-arthritis receiving aspirin. *Archives of Internal Medicine*, **149**: 775–779.

Roth, S.H. 1982, Treatment of rheumatoid arthritis. *Journal of American Medical Association*, **248**: 546.

Roth, S.H. 1985, Special studies of diclofenac and safety: gastrointestinal, renal, hepatic, and other consequences of therapy. *Seminars in Arthritis and Rheumatism*, **15 (2 Suppl. 1)**: 99–104.

Roth, S.H. 1988a, NSAID and gastropathy: a rheumatologist's review. *Journal of Rheumatolology*, **15**: 912–919.

Roth, S.H. 1988b, Salicylates revisited. Are they still the hallmark of anti-inflammatory therapy? *Drugs*, **36**: 1–6.

Rothstein, J. 1975, Phenylbutazone and aspirin in osteoarthritis – a controlled study. *Current Therapeutic Research*, **17**: 444–451.

Rothwell, K.G. 1983, Efficacy and safety of a non-acetylated salicylate, choline magnesium trisalicylate, in the treatment of rheumatoid arthritis. *Journal of International Medical Research*, **11**: 343–348.

Rowe, P.C., Valla, D. and Brusilow, S.W. 1988, Inborn errors of metabolism in children referred with Reye's syndrome: a changing pattern. *Journal of the American Medical Association*, **260**: 3167–3170.

Royal College of Physicians 1984, Medication for the elderly. *Journal of the Royal College of Physicians (London)*, **18**: 7–9.

Rubin, A., Rodda, B.E., Warrick, P., Gruber, C.M. and Ridolfo, A.S. 1973, Interactions of aspirin with nonsteroidal anti-inflammatory drugs in man. *Arthritis and Rheumatism*, **16**: 635–645.

Rugstad, H.E., Hundal, O., Holme, I., Herland, O.B., Husby, G. and Giercksky, K.E. 1986, Piroxicam and naproxen plasma concentrations in patients with osteoarthritis: relation to age, sex, efficacy and adverse events. *Clinical Rheumatology*, **5 (3)**: 389–398.

CHAPTER 12

Rumble, R.H., Brooks, P.M. and Roberts, M.S. 1980, Metabolism of salicylate during chronic aspirin therapy. *British Journal of Clinical Pharmacology*, **9**: 41–45.

Russell, A.S., Sturge, R.A. and Smith, M.A. 1971, Serum transaminases during salicylate therapy. *British Medical Journal*, **2**: 428–432.

Rylance, H,J., Chalmers, T.M. and Elton, R.A. 1980, Clinical trials of intra-articular aspirin in rheumatoid arthritis. *Lancet*, **2**: 1099–1102.

Saag, K.G., Rubenstein, L.M., Chrischilles, E.A. and Wallace, R.B. 1995, Nonsteroidal antiinflammatory drugs and cognitive decline in the elderly. *Journal of Rheumatology*, **22**: 2142–2147.

Sack, D.M. and Peppercorn, M.A. 1983, Drug therapy of inflammatory bowel disease. *Pharmacotherapy*, **3**: 158–176.

Saggioro, A., Alvisi, V., Blasi, A., Dobrilla, G., Marcolongo, R. and Fioravanti, A., 1991, Misoprostol prevents NSAID-induced gastroduodenal lesions in patients with osteoarthritis and rheumatoid arthritis. *Italian Journal of Gastroenterology*, **23**: 119–123.

Salem, S.A. and Stevenson, I.H. 1977, Absorption kinetics of aspirin and quinine in elderly subjects (proceedings). *British Journal of Clinical Pharmacology*, **4 (3)**: 397P

Samter, M. and Beers, R.F. Jr. 1968, Intolerance to aspirin. Clinical studies and consideration of its pathogenesis. *Annals of Internal Medicine*, **68**: 975–983.

Samter, M. 1969a, Reactions to drugs. *Illinois Medical Journal*, **136**: 159–165.

Samter, M. 1969b, The acetyl- in aspirin. *Annals of Internal Medicine*, **71**: 208–209.

Samter, M. 1970, The pathogenesis of reactions to drugs. *Folia Allergologia (Roma)*, **17**: 414.

Samter, M. 1971, Allergies of the skin: expected versus actual incidence possible explanations for the difference, and therapeutic approaches. *Advances in Biology of the Skin*, **11**: 337–343.

Samter, M. 1977, Aspirin, salicylates, and the magic of diets. *Cutis*, **20**: 18, 24, 52.

Sasaki, S. 1970, Clinical trial of ibuprofen in Japan. *Rheumatology and Physical Medicine*, **11 (Suppl.)**: 32–39.

Sasisekhar, P.R., Penn, R.G., Haslock, I. and Wright, V. 1973, A comparison of benorylate and aspirin in the treatment of rheumatoid arthritis. *Rheumatology and Rehabilitation*, **12 (Suppl.)**: 31–38.

Schachter, D. and Manis, J.G. 1958, Salicylate and salicyl conjugates: fluorometric estimation, biosynthesis and renal excretion in man. *Journal of Clinical Investigation*, **10**: 484–489.

Schaller, J.G. 1978, Chronic salicylate administration in juvenile rheumatoid arthritis; aspirin 'hepatitis' and its clinical significance. *Pediatrics*, **62 (Suppl.)**: 916–925.

Scharff, U. 1974, A double-blind comparison of phenylbutazone and aspirin in osteoarthritis. *Current Therapeutic Research*, **16**: 1264–1269.

Scheiman, J. 1996, NSAIDS, gastrontestinal injury, and cytoprotection. *Gastroenterology Clinics of North America*, **25**: 279–298.

Schiff, E.R. and Maddrey, W.C. 1994, Can we prevent nonsteroidal anti-inflammatory drug-induced hepatic failure? *Gastrointestinal Diseases Today*, **3**: 7–13.

Schlegel, S.I. and Paulus, H.E. 1986, NSAIDs, use in rheumatic disease, side-effects and interactions. *Bulletin of Rheumatic Diseases*, **6**: 1–7.

Schlosstein, L.H., Kippen, I., Whitehouse, M.W., Bluestone, R., Paulus, H.B. and Klinenberg, J.R. 1973, Studies with some novel uricosuric agents and their metabolites: correlation between clinical activity and drug-induced displacement of urate from its albumin binding sites. *Journal of Laboratory and Clinical Medicine*, **82**: 412–418.

Schmid, B., Lüdtke, R., Selbmann, H.-K., Kötter, I., Tschirdewahn, B., Schaffner, W. and Heide, L. 2001, Efficacy and tolerability of a standardized willow bark extract in patients with osteoarthritis: randomised placebo-controlled, double-blind clinical trial. *Phytotherapy Research*, **15**: 344–350.

Schmucker, D.L. 1984, Drug disposition in the elderly: a review of the critical factors. *Journal of the American Geriatric Society*, **32**: 144–149.

Schneyer, J.J. 1975, Tolmetin versus aspirin in the treatment of rheumatoid arthritis: a controlled double-blind study. *In:* J.R. Ward (ed.), *Tolmetin: A New Non-steroidal Anti-inflammatory Agent*, pp. 34–46. Princeton: Excerpta Medica.

Schröder, H. and Campbell, D.E.S. 1972, Absorption, metabolism, and excretion of salicylazosulfapyridine in man. *Clinical Pharmacology and Therapeutics*, **4**: 539–551.

Scott, J.T., Denman, A.M. and Dorling, J. 1964, Renal irritation caused by salicylates. *Lancet*, **1:** 344–348.

Seaman, W.E., Ishak, K.G. and Plotz, P.H. 1974, Aspirin-induced hepatotoxicity in patients with systemic lupus erythematosus. *Annals of Internal Medicine*, **80:** 1–12.

Seaman, W.E. and Plotz, P.H. 1976, Effect of aspirin on liver tests in patients with rheumatoid arthritis or SLE and in normal volunteers. *Arthritis and Rheumatism*, **19:** 155–160.

Sechserova, M., Sechser, T., Raskova, H., Elis, J., Vanecek, J. and Polak, L. 1979, Ontogenic drug studies in calves. I. Age-dependent salicylate levels and metabolism. *Arzneimittel Forschung*, **29 (4):** 650–651.

Seed, J.C. 1965, Antipyretic activity of aspirin and salicylic acid. *Clinical Pharmacology and Therapeutics*, **6:** 354.

Semble, E.L. and Wu, W.C. 1987, Anti-inflammatory drugs and gastric mucosal damage. *Seminars of Arthritis and Rheumatism*, **16:** 271–286.

Semble, E.L., Wu, W.C. and Castell, D.O. 1989, Nonsteroidal anti-inflammatory drugs and esophageal injury. *Seminars in Arthritis and Rheumatism*, **19:** 99–109.

Serlin, M.J., Mossman, S., Sibeon, R.G. and Breckenridge, A.M. 1980a, The effect of diflunisal on the steady state pharmacodynamics and pharmacokinetics of warfarin (proceedings). *British Journal of Clinical Pharmacology*, **9:** 287P–288P.

Serlin, M.J., Mossman, S., Sibeon, R.G., Tempero, K.F. and Breckenridge, A.M. 1980b, Interaction between diflunisal and warfarin. *Clinical Pharmacology and Therapeutics*, **28:** 493–498.

Settipane, G.A. 1983, Aspirin and allergic diseases: a review. *American Journal of Medicine*, **74:** 102–109.

Settipane, G.A. 1988, Aspirin sensitivity and allergy. *Biomedicine and Pharmacotheraphy*, **42:** 493–498.

Settipane, G.A., Chafee, F.H. and Klein, D.E. 1974, Aspirin intolerance. II. A prospective study in an atopic and normal population. *Journal of Allergy and Clinical Immunology*, **53:** 200–204.

Shallcross, T.M. and Heatley, R.V. 1990, Effect of nonsteroidal anti-inflammatory drugs on dyspeptic symptoms. *British Medical Journal*, **300:** 568–569.

Sheldrake, F.E., Webber, J.M. and Marsh, B.D. 1977, A long-term assessment of flurbiprofen. *Current Medical Research and Opinion*, **5:** 106–116.

Shelley, J.H. 1978, Pharmacological mechanisms of analgesic nephropathy. *Kidney International*, **13:** 15–26.

Sherry, D.D., Mellins, E.D. and Nepom, B.S. 1998, Pauciarticular onset juvenile chronic arthritis. *In:* P.J. Maddison, D.A. Isenberg, P. Wood and D.N. Glaso (eds), *Oxford Textbook of Rheumatology*, 2nd edn, Vol. 2, 5.6.1, pp. 1099–1114, Oxford: Oxford University Press.

Signler, J.W., Ridolofo, A.S. and Bluhm, G.B. 1976, Comparison of benefit to risk ratios of aspirin and fenoprofen: controlled multi-centre study in rheumatoid arthritis. *Journal of Rheumatology*, **2 (Suppl.):** 49–60.

Sigthorsson, G., Tibble, J., Hayllar, J., Menzies, I., MacPherson, A., Moots, R., Scott, D., Gumpel, M.J. and Bjarnason, I. 1998, Intestinal permeability and inflammation in patients on NSAIDs. *Gut*, **43 (4):** 506–511.

Silverstein, H., Bernstein, J.M. and Davies, D.G. 1967, Salicylate ototoxicity: a biochemical and electrophysiological study. *Annals of Otology*, **76:** 118–125.

Silverstein, F.E., Kimmey, M.B., Saunders, D.R. and Levine, D.S. 1986, Gastric protection by misoprostol against 1300 mg of aspirin. An endoscopic study. *Digestive Diseases and Sciences*, **31:** 137S–141S.

Silverstein, F.E., Graham, D.Y., Senior, J.R., Davies, H.W., Struthers, B.J., Bittman, R.M. and Beis, G.S. 1995, Misoprostol reduces serious gastrointestinal complications in patients with rheumatoid arthritis receiving non-steroidal anti-inflammatory drugs. A randomized, double-blind, placebo-controlled trial. *Annals of Internal Medicine*, **123:** 241–249.

Silverstein, F.E., Faich, G., Goldstein, J.L., Simon, L.S., Pincus, T., Whelton, A., Makuch, R., Eisen, G., Agrawal, N.M., Stenson, W.F., Burr, A.M., Zhao, W.W., Kent, J.D., Lefkowith, J.B., Vergurg, K.M. and Geis, G.S. 2000, Gastrointestinal toxicity with celecoxib vs non-steroidal anti-inflammatory drugs for osteoarthritis and rheumatoid arthritis: the CLASS study: a randomized controlled trial. Celecoxib Long-term Arthritis Safety Study. *Journal of the American Medical Association*, **284:** 1247.

CHAPTER 12

Silvoso, G.R., Ivey, K.J., Butt, J.H., Lockard, O.O., Holt, S.D., Sisk, C., Baskin, W.N., MacKercher, P.A. and Hewett, J. 1979, Incidence of gastric lesions in patients with rheumatic disease on chronic aspirin therapy. *Annals of Internal Medicine*, **91:** 517–520.

Simon, L.S. and Mills, J.A. 1980, Drug therapy with non-steroidal anti-inflammatory drugs (first of two parts). *New England Journal of Medicine*, **302:** 1179–1185.

Simon, L.S. and Mills, J.A. 1980, Nonsteroidal anti-inflammatory drugs (second of two parts). *New England Journal of Medicine*, **302:** 1237–1243.

Simon, L.S., Weaver, A.L., Graham, D.Y., Kivitz, A.J., Lipsky, P.E., Hubbard, R.C., Isakson, P.C., Verburg, K.M., Yu, S.S., Zhao, W.W. and Geis, G.S. 1999, Anti-inflammatory and upper gastrointestinal effects of celecoxib in rheumatoid arthritis: a randomized controlled trial. *Journal of the American Medical Association*, **282:** 1921–1928.

Singer, L., Imbs, J.L., Darian, J.M., Singer, P., Krieger-Finance, F., Schmidt, M. and Schwartz, J. 1981, Risque d'intoxication par le lithim en cas de traitement associe par les anti-inflammatoires non-steroidens. *Therapie*, **36:** 323–326.

Singh, G. and Fries, J.F. 1996, Famotidine to prevent peptic ulcer caused by NSAIDs. *New England Journal of Medicine*, **335 (17):** 1322.

Singh, G., Ramey, D.R., Morfeld, D., Shi, H., Hatoum, H.T. and Fries, J.F. 1996, Gastrointestinal tract complications of nonsteroidal anti-inflammatory drug treatment in rheumatoid arthritis. A prospective observational cohort study. *Archives of Internal Medicine*, **156 (14):** 1530–1536.

Singleton, C.M. and Wild, J.H. 1980, A double-blind comparison of aspirin and pirprofen in the treatment of rheumatoid arthritis. *Journal of Rheumatology*, **7:** 865–870.

Singleton, P.T. Jr. 1980, Salsalate; its role in the management of rheumatic disease. *Clinical Therapeutics* **3:** 80–102.

Skander, M.P. and Ryan, F.P. 1988, Non-steroidal anti-inflammatory drugs and pain free peptic ulceration in the elderly. *British Medical Journal*, **297:** 833–834.

Slepian, I.K., Mathews, K.P. and McLean, J.A. 1985, Aspirin-sensitive asthma. *Chest*, **87:** 386–391.

Sloan, J.P., Hain, R. and Pownall, R. 1989, Benefits of early anti-inflammatory medication following acute ankle injury. *Injury*, **20:** 81–83.

Smith, M.J.H. 1966, Metabolic effects of salicylates. *In:* M.J.H. Smith and P.K. Smith (eds), *Salicylates. A Critical Bibliographic Review*, pp. 49–105. New York: Wiley-Interscience.

Smith, M.J.H. 1966, Toxicology. *In:* M.J.H. Smith and P.K. Smith (eds), *Salicylates. A Critical Bibliographic Review*, p. 223. New York: Wiley-Interscience.

Smith, M.J.H. 1968, The metabolic basis of the major symptoms in acute salicylate poisoning. Clinical Toxicology, **1:** 387–407.

Soll, A.H., Weinstein, W.M., Kurata, J. and McCarthy, D. 1991, Non-steroidal anti-inflammatory drugs and peptic ulcer disease. *Annals of Internal Medicine*, **114:** 307–319.

Solomon, L. and Abrams, G. 1974, Voltaren in the treatment of rheumatoid arthritis. *South African Medical Journal*, **11:** 949–952.

Solomon, L., Schnitzler, C.M. and Browett, J.P. 1982, Osteoarthritis of the hip: the patient behind the disease. *Annals of the Rheumatic Diseases*, **41:** 118–125.

Somerville, K.W. and Hawkey, C.J. 1986, Non-steroidal anti-inflammatory agents and the gastrointestinal tract. *Postgraduate Medical Journal*, **62:** 23–28.

Spector, T.D., Hart, D.J. and Mandra, D. 1997, Low level increases in serum C-reactive protein are present in early osteoarthritis of the knee and predict progressive disease. *Arthritis and Rheumatism*, **40:** 723–727.

Speer, F. 1975, Aspirin allergy: a clinical study. *Southern Medical Journal*, **68:** 314–318.

Sperryn, P.N., Hamilton, E.B.D. and Parsons, V. 1972, Double-blind comparison of aspirin and 4-(acetamido)phenyl-2-acetoxy-benzoate (benorylate) in rheumatoid arthritis. *Annals of the Rheumatic Diseases*, **32:** 157–161.

Spiers, A.S.D. and Malone, H.F. 1967, Effect of food on aspirin absorption. *Lancet*, **i:** 440.

Staessen, J., Fagard, R., Lijmem, P., Moerman, E., De Schaepdryver, A. and Amery, A. 1983, Effects of prostaglandin synthesis inhibition on blood pressure and humoral factors in exercising, sodium-deplete normal man. *Journal of Hypertension*, **1:** 123–130.

Starko, K.M. and Mullick, F.G. 1983, Hepatic and cerebral pathology findings in children with fatal salicylate intoxication: further evidence for a causal relation between salicylate and Reye's syndrome. *Lancet*, **1:** 326–329.

Starko, K.M., Ray, C.G., Dominguez, L.B., Stomberg, W.L. and Woodall, D.E. 1980, Reye's syndrome and salicylate use. *Pediatrics*, **66:** 859–864.

Steel, K., Gertman, P.M. and Crescenzi, C. 1981, Iatrogenic illness on a general medical service at a university hospital. *New England Journal of Medicine*, **304:** 638–642.

Stephens, W.H., El-Ghobarey, A.F., MacLeod, M.M. and Buchanan, W.W. 1979, A double-blind crossover trial of mefenamic acid, sulindac and flurbiprofen in rheumatoid arthritis. *Current Medical Research and Opinion*, **5:** 754–758.

Stern, A.I., Hogan, D.L., Kahn, L.H. and Isenberg, J.I. 1984, Protective effect of acetaminophen against aspirin- and ethanol-induced damage to the human gastric mucosa. *Gastroenterology*, **86 (4):** 728–733.

Stillman, A., Giller, H., Shillington, D., Sobonya, R., Payne, C.M., Ettinger, D. and Lee, S.M. 1983, Reye's syndrome in the adult: case report and review of the literature. *American Journal of Gastroenterology*, **78:** 365–368.

Stillman, M.J. and Schlesinger, P.A. 1990, Nonsteroidal anti-inflammatory drug nephrotoxicity. Should we be concerned? *Archives of Internal Medicine*, **150:** 268–270.

Stone, E. 1763, An account of the success of the bark of the willow in the cure of agues. (A letter to the Right Honourable George Earl of Macclesfield, President of the Royal Society.) *Philosophical Transactions of the Royal Society of London*, **53:** 195–200.

Stricker, S. 1876, Ueber die Resultate der Behandlung der Polyarthritic rheumatica mit Salicylsäure. *Berliner Klinische Wochenschrift*, **13:** 1–2, 15–16, 99–103.

Strom, B.L., Carson, J.L., Morse, M.L., West, S.L. and Soper, K.A. 1987, The effect of indication on hypersensitivity reactions associated with zomepirac sodium and other nonsteroidal anti-inflammatory drugs. *Arthritis and Rheumatism*, **30:** 1142–1148.

Stull, P.A. and Joki, P. 1986, Comparison of diflunisal and naproxen in the treatment of tennis elbow. *Clinical Therapeutics*, **9 (Suppl. C):** 62–66.

Sullivan, K.M., Remington, P.L., Hurwitz, E.S. and Halpin, T.J. 1988, Reye's syndrome among patients with juvenile rheumatoid arthritis. *Journal of the American Medical Association*, **260:** 3434–3435.

Surgeon General Advice on the Use of Salicylates on Reye's Syndrome 1982, *Morbidity and Mortality Weekly Report*, **31:** 289–290.

Svartz, N. 1941, Ett nytt sulfonaamidpreparat. Förelöpande meddelande. *Nodesk Medicin*, **9:** 554.

Svartz, N. 1948, The treatment of rheumatic polyarthritis with acid azo compounds. *Rheumatism*, **4:** 180–185.

Swingle, K.F., Grant, T.J., Jaques, L.W. and Kvam, D.C. 1970, Interactions of anti-inflammatory drugs in carrageenan-induced foot edema of the rat. *Journal of Pharmacology and Experimental Therapeutics*, **172:** 423–425.

Szczeklik, A. 1980, Analgesics, allergy and asthma. *British Journal of Clinical Pharmacology*, **2:** 401S–405S.

Szczeklik, A. 1983, Asthma, aspirin and leukotrienes. *Bulletin of European Physiopathology and Respirology*, **19:** 531–538.

Szczeklik, A. 1983b, Anti-cyclo-oxygenase agents and asthma. *Journal of Asthma*, **1:** 23–29.

Szczeklik, A. 1986, Analgesics, allergy and asthma. *Drugs*, **4:** 148–163.

Szczeklik, A. 1990, The cyclooxygenase theory of aspirin-induced asthma. *European Respiratory Journal*, **3:** 588–593.

Szczeklik, A., Gryglewski, R.J. and Czerniawska-Mysik, G. 1975, Relationship of inhibition of prostaglandin biosynthesis by analgesics to asthma attacks in aspirin-sensitive patients. *British Medical Journal*, **1:** 67–69.

Szczeklik, A., Gryglewski, R.J. and Czerniawska-Mysik, G. 1977, Clinical patterns of hypersensitivity to nonsteroidal anti-inflammatory drugs and their pathogenesis. *Journal of Allergy and Clinical Immunology*, **60:** 276–284.

Szczeklik, A., Gryglewski, R.J., Czerniawska-Mysik, G. and Nizankowska, E. 1986, Aspirin sensitive asthma and arachidonic acid transformation. *New England Respiratory and Allergy Proceedings*, **7:** 21–25.

■ CHAPTER 12 ■

Szczeklik, A., Gryglewski, R.J. and Nizankowska, E. 1978, Asthma relieved by aspirin and by other cyclo-oxygenase inhibitors. *Thorax*, **33:** 664–665.

Szczeklik, A., Gryglewski, R.J. and Nizankowska, E. 1979, Asthma and nonsteroidal anti-inflammatory drugs. *Annals of Internal Medicine*, **90:** 126–127.

Szczeklik, A. and Nizankowska, E. 1983, Asthma improved by aspirin-like drugs. *British Journal of Diseases of the Chest*, **77:** 153–158.

Szczeklik, A. and Serwonska, M. 1979, Inhibition of idiosyncratic reactions to aspirin in asthmatic patients by clemastine. *Thorax*, **34:** 654–657.

Taggart, A.J. 1992, Symptomatic salicylate ototoxicity: a useful indicator of serum salicylate concentrations. *Annals of the Rheumatic Diseases*, **51:** 704.

Taggart, H., McDermott, B.J., Beringer, T.R.O., Wilson, G., Roberts, S.D. and Taggart, A.J. 1991, The pharmacokinetics of benorylate in young volunteers and elderly patients with osteoarthritis and rheumatoid arthritis. *Drug Investigation*, **3:** 39–44.

Taha, A.S., Hudson, N., Hawkey, C.J., Swannell, A.J., Trye, P.N., Cottrell, J., Mann, S.G., Simon, T.J., Sturrock, R.D. and Russell, R.I. 1996, Famotidine for the prevention of gastric and duodenal ulcers caused by nonsteroidal antiinflammatory drugs. *New England Journal of Medicine*, **224:** 1435–1439.

Tait, G.B., Lim, C.M., Highton, T.C., Keary, P.J. and Laurent, M.R. 1978, *In: Diflunisal in Clinical Practice. Proceedings of a Conference at the XIVth International Congress of Rheumatology, San Francisco*, pp. 43–56. New York: Futura.

Tam, L.S., Chan, T.Y.L., Leung, W.K. and Critchley, J.A.J.H. 1995, Warfarin interactions with traditional medicines: danshen and methyl salicylate oil. *Australian and New Zealand Journal of Medicine*, **25:** 258.

Tashima, C.K. and Rose, M. 1974, Pulmonary edema and salicylates. *Annals of Internal Medicine*, **81:** 274–280.

Taylor, J.P., Gustafson, T.L., Johnson, C.C., Brandenburg, N. and Gleyen, W.P. 1985, Antipyretic use among children during the 1983 influenza season. *American Journal of Diseases in Children*, **139:** 486–488.

Tempero, K.F., Cirillo, V.J. and Steelman, S.L. 1977, Diflunisal: a review of pharmacodynamic properties, drug interactions, and special tolerability studies in humans. *British Journal of Clinical Pharmacology*, **4:** 315.

Thillainayagam, A.V., Tabaqchali, S., Warrington, S.J. and Farthing, M.J.G. 1994, Interrelationships between *Helicobactor pylori* infection, nonsteroidal anti-inflammatory drugs and gastroduodenal disease. *Digestive Diseases and Sciences*, **39:** 1085–1089.

Thompson, T.A., Borman, C.H., Goodblatt, R.S. and Roth, W.J. 1979, Effects of aspirin on ^{14}C-pirprofen disposition in rats. *Journal of Pharmaceutical Sciences*, **68:** 996–998.

Tibble, J., Sigthorsson, G., Foster, R., Sherwood, R., Fagerhol, M. and Bjarnason, I. 2001, Faecal calprotectin and faecal occult blood tests in the diagnosis of colorectal carcinoma and adenoma. *Gut*, **49(3):** 402–408.

Tibble, J.A., Sigthorseson, G., Foster, R., Scott, D., Fagerhol, M.L., Roseth, A. and Bjarnason, I. 1999, High prevalence of NSAID enteropathy as shown by a simple faecal test. *Gut*, **45:** 362–366.

Tildsley, G., Ehsanullah, R.S. and Wood, J.R. 1993, Ranitidine in the treatment of gastric and duodenal ulcers associated with non-steroidal anti-inflammatory drugs. *British Journal of Rheumatology*, **32:** 474–478.

Tolman, K.G., Petersen, P., Gray, P. and Hammar, S.P. 1978, Hepatotoxicity of salicylate in monolayer cell cultures. *Gastroenterology*, **74:** 205–208.

Tomasova, H., Nevoral, J., Pachl, J. and Kind, V. 1984, Aspirin esterase and Reye's syndrome. *Lancet*, **2:** 43.

Toovey, S., Hudson, E., Hendry, W.F. and Levi, A.J. 1981, Sulphasalazine and male infertility: reversibility and possible mechanism. *Gut*, **22:** 445–451.

Trauner, D.A., Horvath, E. and Davis, L.E. 1988, Inhibition of fatty acid beta oxidations by influenza B virus and salicylic acid in mice: implications for Reye's syndrome. *Neurology*, **38:** 239–241.

Trnavska, Z. and Trnavsky, K. 1983, Sex differences in the pharmacokinetics of salicylates. *European Journal of Clinical Pharmacology*, **25:** 679–682.

Tsuchiya, T. and Levy, G. 1972, Biotransformation of salicylic acid to its acyl and phenolic glucuronides in man. *Journal of Pharmaceutical Sciences*, **61:** 800–801.

Tugwell, P., Hart, L., Kraag, G., Park, A., Dok, C., Bianchi, F., Goldsmith, C. and Buchanan, W.W. 1984, Controlled trial of clinical utility of serum salicylate monitoring in rheumatoid arthritis. *Journal of Rheumatology*, **11:** 457–461.

Tugwell, P., Wells, G., Strand, V., Maetzel, A., Bombardier, C., Crawford, B., Dorrier, C. and Thompson, A. 2000, Clinical improvement as reflected in measures of function and health-related quality of life following treatment with leflunomide compared with methotrexate in patients with rheumatoid arthritis: sensitivity and relative efficiency to detect a treatment effect in a twelve-month, placebo-controlled trial. Leflunomide Rheumatoid Arthritis Investigators Group. *Arthritis and Rheumatism*, **43:** 506–514.

Tumiati, B., Azzalito, C. and Veneziani, M. 1989, Reye's syndrome in an adult following therapy with salicylates in Still's disease. *Italian Medicina (Florence)*, **9:** 64–65.

Turpie, A.G., Gent, M., Laupacis, A., Latour, Y., Gunstensen, J., Basile, F., Klimek, M. and Hirsh, J. 1993, A comparison of aspirin with placebo in patients treated with warfarin after heart-valve replacement. *New England Medical Journal*, **329:** 524–529.

Unsworth, J., Sturman, S., Lumec, J. and Blake, D.R. 1987, Renal impairment associated with non-steroidal anti-inflammatory drugs. *Annals of the Rheumatic Diseases*, **46:** 233–236.

Upadhyay, R., Howatson, A., McKinlay, A., Danesh, B.J.Z., Sturrock, R.D. and Russell, R.I. 1988, *Campylobacter pylori*-associated gastritis in patients with rheumatoid arthritis taking non-steroidal anti-inflammatory drugs. *British Journal of Rheumatology*, **27:** 113–116.

Upadhyay, R., Torley, H.I., McKinlay, A.W., Sturrock, R.D. and Russell, R.I. 1990, Iron deficiency anaemia in patients with rheumatic disease receiving non-steroidal anti-inflammatory drugs: the role of upper gastrointestinal lesions. *Annals of the Rheumatic Diseases*, **49:** 359–362.

Valkenburg, H.A. 1981, Clinical versus radiological osteoarthritis in the general population. *In:* J.G. Peyron (ed.), *Epidemiology of Osteoarthritis*, pp 53–58. Paris: Geigy.

Valles, M. and Tovar, J.L. 1987, Salsalate and minimal-change nephritic syndrome (Letter). *Annals of Internal Medicine*, **107 (1):** 116.

Valtonen, E.J. 1979a, Symptomatic response of osteoarthritis to benorylate. A dose discrimination study. *Scandinavian Journal of Rheumatology*, **25 (Suppl.):** 9–14.

Valtonen, E.J. 1979b, Clinical comparison of fenbufen and aspirin in osteoarthritis. *Scandinavian Journal of Rheumatology*, **27 (Suppl.):** 1–7.

Van Arman, C.G., Nuss, G.W. and Risley, E.A. 1973, Interactions of aspirin, indomethacin and other drugs in the adjuvant-induced arthritis in the rat. *Journal of Pharmacology and Experimental Therapeutics*, **187:** 400–414.

Van Den Ouweland, F.A., Gribnau, F.W. and Meybloom, R.H. 1988, Congestive heart failure due to nonsteroidal anti-inflammatory drugs in the elderly. *Age and Ageing*, **17:** 8–16.

Van Loenhout, J.W., Van De Putte, L.B.A., Gribnaue, F.W.J. and Van Ginneken, C.A.M. 1981, Persistent hypouraemic effect of long-term diflunisal administration. *Journal of Rheumatology*, **8:** 639–642.

Vane, J.R. 1971, Inhibition of prostaglandin synthesis as a mechanism of action for aspirin-like drugs. *Nature New Biology*, **231:** 232–235.

Vane, J.R. 1987, The evolution of non-steroidal anti-inflammatory drugs and their mechanism of action. *Drugs*, **33 (Suppl. 1):** 18–27.

Verbeeck, R.K., Boel, A., Buntinx, A. and De Schepper, P.J. 1980, Plasma protein binding and interaction studies with diflunisal, a new salicylate analgesic. *Biochemical Pharmacology*, **29:** 571–576.

Vestel, R.A. 1978, Drug use in the elderly: a review of problem and special considerations. *Drugs*, **16:** 358–382.

Vetto, A.A., Mannik, M., Zatarain-Rios, E. and Wener, M.H. 1990, Immune deposits in articular cartilage of patients with rheumatoid arthritis have a granular pattern not seen in osteoarthritis. *Rheumatology International*, **10:** 13–20.

Volans, G.N. 1974, Effects of food and exercise on the absorption of effervescent aspirin. *British Journal of Clinical Pharmacology*, **1:** 137–141.

Volpe, R. 1993, The management of subacute (DeQuervain's) thyroiditis. *Thyroid*, **3:** 253–255.

Von Rechenberg, H.K. 1961, *Butazolidin®. Phenylbutazon*. Stuttgart: Georg Thieme.

Voskuyl, A.E., Van De Laar, M.A., Moens, H.J. and Van der Korst, J.K. 1993, Extra-articular manifestations of rheumatoid arthritis: risk factors for serious gastrointestinal events. *Annals of the Rheumatic Diseases*, **52:** 771–775.

Walan, A., Bader, J.P., Classen, M., Lamers, C.B.H.W. and Piper, D.W. 1989, Effect of omeprazole and ranitidine on ulcer healing and relapse rates in patients with benign gastric ulcer. *New England Journal of Medicine*, **320:** 69–75.

Waldman, R.J., Hall, W.N., McGee, H. and Van Amburg, G. 1982, Aspirin as a risk factor in Reye's syndrome. *Journal of the American Medical Association*, **247:** 3089–3094.

Wallace, S.M. and Verbeeck, R.K. 1987, Plasma protein binding of drugs in the elderly. *Clinical Pharmacokinetics*, **12:** 41–72.

Wallace, S. and Whiting, B. 1976, Factors affecting drug binding in plasma of elderly patients. *British Journal of Clinical Pharmacology*, **3:** 327–330.

Walt, R., Katschinski, B., Logan, R., Ashley, J. and Langman, M. 1986, Rising frequency of ulcer perforation in elderly people in the United Kingdom. *Lancet*, **1:** 489–492.

Walter, E., Kaufmann, W. and Oster, P. 1981, Does chronic aspirin treatment increase blood pressure in man? *Klinische Wochenschrift*, **59:** 297–299.

Walters, J.S., Woodring, J.H., Stelling, C.B. and Rosenbaum, H.D. 1983, Salicylate-induced pulmonary edema. *Radiology*, **146:** 289–293.

Ward, J.R., Wilkens, R.F., Louie, J.S. and McAdam, L.P. 1978, Piroxicam and rheumatoid arthritis: a multicentre 14-week controlled double-blind study comparing piroxicam and aspirin. *Royal Society of Medicine International Congress Symposium Series I*, **1:** 31–39.

Warin, R.P. and Smith, R.J. 1982, Chronic urticaria: investigations with patch and challenge tests. *Contact Dermatitis*, **8:** 117–121.

Warrick, P., Rubin, A. and Gruber, C.M. 1974, Interactions in rats between the nonsteroidal anti-inflammatory drugs, aspirin and fenoprofen. *Proceedings of the Society for Experimental Biology and Medicine*, **147:** 599–607.

Watson, C.J., Deane, A.M., Doyle, P.T. and Bullock, K.N. 1990, Identifiable factors in postprostatectomy haemorrhage: the role of aspirin. *British Journal of Urology*, **66:** 85–87.

Webb, J., Downie, W.W., Dick, W.C. and Lee, P. 1973, Evaluation of digital joint circumference measurements in rheumatoid arthritis. *Scandinavian Journal of Rheumatology*, **2:** 127–131.

Webb, J., Whaley, K., MacSween, R.N.M., Nuki, G., Dick, W.C. and Buchanan, W.W. 1975, Liver disease in rheumatoid arthritis and Sjögren's syndrome: prospective study using biochemical and serological markers of hepatic dysfunction. *Annals of Rheumatic Diseases*, **34:** 70–80.

Weber, S., Rey, E., Pipeau, C., Lutfalla, G., Richard, M.-O., Daoud-El-Assaf, H., Olive, G. and Degeorges, M. 1983, Influence of aspirin on the haemodynamic effects of sublingual nitroglycerin. *Journal of Cardiovascular Pharmacology*, **5:** 874–877.

Weiler, J.M. 1992, Medical modifiers of sports injury: the use of non-steroidal anti-inflammatory drugs (NSAIDs) in sports soft-tissue injury. *Clinics in Sports Medicine*, **11:** 625–644.

Weissmann, G. 1991, Aspirin. *Scientific American*, **264:** 84–90.

Welch, R.W., Bentch, L.H. and Harris, S.C. 1978, Reduction of aspirin-induced gastrointestinal bleeding with cimetidine. *Gastroenterology*, **74:** 459–463.

Whaley, K., Goudie, R.B., Williamson, J., Nuki, G. Dick, W.C. and Buchanan, W.W. 1975a, Liver disease in Sjögren's syndrome and rheumatoid arthritis. *Lancet*, **1:** 861–863.

Whaley, K., Goudie, R.B., Williamson, J., Nuki, G., Dick, W.C. and Buchanan, W.W. 1975b, Liver disease in Sjögren's syndrome: prospective study using biochemical and serological markers of hepatic dysfunction. *Annals of the Rheumatic Diseases*, **34:** 70–80.

Whaley, K. and Webb, J. 1977, Liver and kidney disease in rheumatoid arthritis. *Clinical Rheumatic Diseases*, **3:** 527–547.

Whitehouse, M.W., Bluestone, R., Keppen, I. and Klinenberg, J.R. 1970, When is a drug inactive? Concerning the uricosuric activity of some anti-inflammatory drugs. *Journal of Pharmacy and Pharmacology*, **22:** 134–135.

Whitehouse, M.W., Kippen, I., Klinenberg, J.R., Schlosstein, L., Campion, D.S. and Bluestone, R. 1973, Increasing excretion of urate with displacing agents in man. *Annals of the New York Academy of Sciences*, **226:** 309–318.

Wilkens, R.F., Ward, J.R., Louie, J.S. and McAdam, L.P. 1982, Double-blind study comparing piroxicam and aspirin in the treatment of rheumatoid arthritis. *American Journal of Medicine*, **(Suppl.):** 23–26.

Wilkerson, R.D. and Glenn, J.M. 1977, Influence of non-steroidal anti-inflammatory drugs on cardiac toxicity. *American Heart Journal*, **94:** 454–455.

Williams, F.M., Ferner, R.E., Graham, M., Blain, P.G. and Alberti, K.G. 1990, The metabolic effects of aspirin in fasting and fed subjects: relevance to the aetiology of Reye's syndrome. *European Journal of Clinical Pharmacology*, **38:** 519–521.

Williams, H.J., Ward, J.R., Egger, M.J., Neuner, R., Brooks, R.H., Clegg, D.O., Field, E.H., Skosey, J.L, Alarcon, G.S. and Wilkens, R.F. 1993, Comparison of naproxen and acetaminophen in the two year study of treatment of osteoarthritis of the knee. *Arthritis and Rheumatism*, **36:** 1196–1206.

Williams, M.E., Weinblatt, M., Rosa, R.M., Griffin, V.L., Goldlust, M.B., Shang, S.F., Harrison, L.I. and Brown, R.S. 1986, Salsalate kinetics in patients with chronic renal failure undergoing hemodialysis. *Clinical Pharmacology and Therapeutics*, **39 (4):** 420–424.

Williams, R.C. Jr 1998, Acute rheumatic fever. *In:* J.H. Kleppel and P.A. Dieppe (eds), *Rheumatology*, 2nd edn, Vol. 2, pp. 1–10. London: Mosby.

Williams, R.L., Upton, R.A., Buskin, J.N. and Jones, R.M. 1981, Ketoprofen–aspirin interactions. *Clinical Pharmacology and Therapeutics*, **30:** 226–231.

Willis, J.V., Kendall, M.J. and Jack, D.B. 1980, A study of the effect of aspirin on the pharmacokinetics of oral and intravenous diclofenac sodium. *European Journal of Clinical Pharmacology*, **18:** 415–418.

Willis, T. 1992, *In:* T. Dring, C. Harper and J. Leigh (eds), *The London Practice of Physick*, 1684. Reprinted by the Classics of Medicine Library, New York.

Willoughby, J.M.T., Essignman, W.K., Weber, J.C.P. and Pinerva, R.F. 1986, Smoking and peptic ulcer in rheumatoid arthritis. *Clinical and Experimental Rheumatology*, **4:** 31–35.

Wilson, J.R. 1976, Aspirin hepatotoxicity in adults with rheumatoid arthritis. *Ohio State Medical Journal*, **72:** 577–583.

Windorfer, A., Kuenzer, W. and Urbanek, R. 1974, The influence of age on the activity of acetylsalicylic acid-esterase and protein-salicylate binding. *European Journal of Clinical Pharmacology*, **7 (3):** 227–231.

Wiseman, E.H., Chang, Y.H. and Hobbs, D.C. 1975, Interaction of sudoxicam and aspirin in animals and man. *Clinical Pharmacology and Therapeutics*, **18:** 441–448.

Woeber, K.A. and Ingbar, S.H. 1964, The effects of noncalorigenic congeners of salicylate on the peripheral metabolism of thyroxine. *Journal of Clinical Investigation*, **43:** 931–942.

Wohlgemuth, J. 1899, Über Aspirin (Acetylsalicylsäure). *Therapie Therapeutische Monatshäfte*, **3:** 276–278.

Wojtulewski, J.A., Gow, P.J., Walter, J., Grahame, R., Gibson, T. and Panayi, G.S. 1980, Clotrimazole in rheumatic arthritis. *Annals of the Rheumatic Diseases*, **39:** 469–472.

Wojtulewski, J.A., Walter, J. and Gray, J. 1978, Diflunisal compared with naproxen in the treatment of osteoarthritis of hip or knee – a double-blind trial. *Current Medical Research and Opinion*, **5:** 562–566.

Wolfe, M.M., Lichtenstein, D.R. and Singh, G. 1999, Gastrointestinal toxicity of nonsteroidal anti-inflammatory drugs. *New England Journal of Medicine*, **340:** 1888–1899.

Wolff, J. and Austen, F.K. 1958, Salicylates and thyroid function II. The effect on the thyroid–pituitary inter-relation. *Journal of Clinical Investigation*, **37:** 1144–1152.

Wollheim, F.A. 1996, Current pharmacological treatment of osteoarthritis. *Drugs*, **52 (Suppl. 3):** 27–38.

Wollheim, F.A. 2001, Approaches to rheumatoid arthritis in 2000. *Current Opinion in Rheumatology*, **13:** 193–201.

Wood, J.H. 1967, Effect of food on aspirin absorption. *Lancet*, **ii:** 212.

Woodford-Williams, E., Alvares, A.S., Webster, D., Landless, B. and Dixon, M.P. 1964, Serum protein patterns in normal and pathological ageing. *Gerontologia*, **10:** 86–99.

Wortmann, D.W., Kelsch, R.C., Kuhn, L., Sullivan, D.B. and Cassidy, J.T. 1980, Renal papillary necrosis in juvenile rheumatoid arthritis. *Journal of Pediatrics*, **40:** 97–37.

Yeomans N.D., Cook G.A. and Giraud, A.S. 1998, Selective COX-2 inhibitors: are they safe for the stomach? *Gastroenterology*, **115 (1):** 227–229.

Yeomans, N.D., Tulassay, Z. and Juhasz, L. 1998, A comparison of omeprazole with ranitidine for ulcers associated with nonsteroidal anti-inflammatory drugs. Acid Suppression Trial: Ranitidine versus Omeprazole for NSAID-associated Ulcer Treatment (ASTRONAUT) Study Group. *New England Journal of Medicine*, **338:** 719–726.

Yesair, D.W., Remington, L., Callahan, M. and Kensler, C.J. 1970, Comparative effects of salicylic acid, phenylbutazone, probenecid and other anions on the metabolism, distribution and excretion of indomethacin by rats. *Biochemical Pharmacology*, **19:** 1591–1600.

Yip, A.S., Chow, W.H., Tai, Y.T. and Cheung, K.L. 1990, Adverse effect of topical methylsalicylate ointment on warfarin anticoagulation: an unrecognized potential hazard. *Postgraduate Medical Journal*, **66:** 367–369.

Young, R.S., Torretti, D., Williams, R.H., Hendrikson, D. and Woods, M. 1984, Reye's syndrome associated with long-term aspirin therapy. *Journal of the American Medical Association*, **251:** 754–756.

Younossi, Z.M., Strum, W.B., Schatz, R.A., Teirstein, P.S., Cloutier, D.A. and Spinks, T.J. 1997, Effect of combined anticoagulation and low-dose aspirin treatment on upper gastrointestinal bleeding. *Digestive Diseases and Sciences*, **42:** 79–82.

Zambraski, E.J., Atkinson, D.C. and Dramond, J. Effects of salicylate as aspirin on renal prostaglandins and function in normal and sodium depleted dogs. *Journal of Pharmacology and Experimental Therapeutics*, **247:** 96–103.

Zeidler, H. 1992, Epidemiology and economics of NSAID-induced gastropathy. *Scandinavian Journal of Rheumatology*, **92 (Suppl.):** 3–8.

Zeitz, H.J. 1988, Bronchial asthma, nasal polyps, and aspirin sensitivity: Samter's syndrome. *Clinics of Chest Medicine*, **9:** 567–576.

Zeitz, J.H. and Jarmoszuk, I. 1985, Nasal polyps, bronchial asthma, and aspirin sensitivity: the Samter syndrome. *Comprehensive Therapeutics*, **11:** 21–26.

Zimmerman, H.J. 1974, Aspirin-induced hepatic injury. *Annals of Internal Medicine*, **80:** 103–105.

Zimmerman, H.J. 1981, Effects of aspirin and acetaminophen on the liver. *Archives of Internal Medicine*, **141:** 333–342.

Zipser, R.D. 1986, Role of renal prostaglandins and the effects of non-steroidal anti-inflammatory drugs in patients with liver disease. *American Journal of Medicine*, **81 (Suppl. 2B):** 95–103.

Zipser, R.D. and Henrich, W.L. 1986, Implications of nonsteroidal anti-inflammatory drug therapy. *American Journal of Medicine*, **80 (Suppl. 1A):** 78.

Zucker, P., Daum, F. and Cohen, M.I. 1975, Aspirin hepatitis. *American Journal of Diseases of Children*, **129:** 1433–1444.

Aspirin and NSAIDs in the Prevention of Cancer, Alzheimer's Disease and other Novel Therapeutic Actions

13

K.D. Rainsford

INTRODUCTION

Aspirin has been found to have potential therapeutic effects over the years, in a wide variety of conditions, that are not related to its conventional applications in the control of pain, inflammation, fever or thrombotic states. Many of these 'non-conventional' applications arose from either serendipitous observations (e.g. control of plasma glucose in diabetes mellitus, prevention of the decline in cognitive function in Alzheimer's disease) or the logical extension that inhibiting prostaglandin production that is otherwise upregulated in certain conditions (e.g. colorectal and other cancers) might lead to potentially beneficial effects of aspirin and related drugs. Here these potential therapeutic actions of aspirin and other related NSAIDs or analgesics are considered. Understanding the mechanisms of action has been a major factor in refining the development of therapeutic protocols with these drugs, as well as in giving biological credibility for their use in these non-conventional conditions. Furthermore, in some cases deeper understanding of the mechanisms of action of aspirin compared with other NSAIDs/analgesics has led to differentiation of prostaglandin-dependent from non-prostaglandin-dependent mechanisms. Recent advances in the understanding of the cellular and molecular biology of the control of cell growth and differentiation, apoptosis and angiogenesis have set the basis for further investigations on the mode of action of aspirin and related drugs in these processes, which has led to improved understanding of their potential as novel therapeutic (non-conventional) agents as well as giving insight into their actions in chronic inflammatory and neurodegenerative conditions.

PREVENTION OF COLORECTAL AND OTHER CANCERS

A substantial body of evidence from clinical, epidemiological and animal models and cellular systems is now available showing that aspirin and some other NSAIDs may prevent the development of colorectal cancer (Berkel *et al.*, 1996; Peleg *et al.*, 1996; Benamouzig *et al.*, 1997; Khan and Morrison, 1997; Ruffin *et al.*, 1997; Rosenberg *et al.*, 1998; Taketo, 1998; Bus *et al.*, 1999; Collet *et al.*, 1999; Muir and Logan, 1999; Shiff and Rigas, 1999a; 1999b; Vianio, 1999; Garcia-Rodriguez and Huerta-Arvarez, 2000; Langman *et al.*, 2000; see also Tables 13.1 and 13.2).

The data in Table 13.1, taken from a large number of trials with aspirin or other NSAIDs, show that there is reduced risk of developing colon cancer among users of these drugs. These studies are

TABLE 13.1

Epidemiological studies on the relation between intake of aspirin or other NSAIDs and occurrence of colorectal cancers.

Study design	P/R	Results	Reference: Authors
Population case–control study/colorectal cancer/715 cases and 727 controls	R	Aspirin use/colon and rectum/total: RR* = 0.57 (95% CI* 0.41–0.79)	Kune et al., 1988
Leisure World retired cohort (n = 13 987)	P	Aspirin use/RR = 1.5 (95% CI 1.1–2.2)	Paganini-Hill et al., 1989
Hospital case–control study/colorectal cancer/1326 cases and 4891 controls	R	NSAIDs/regular use not ended >1 year before: RR = 0.5 (95% CI 0.4–0.8)	Rosenberg et al., 1991
Nested case–control study within Cancer Prevention Study II/mortality/colon cancer/598 cases and 3058 controls	P	Aspirin use >16 times per month: RR = 0.6 (95% CI 0.4–0.89)	Thun et al., 1991
Case–control study with screening programme/polyps/147 cases and 329 controls	R	Aspirin: RR = 0.55 (95% CI 0.3–1.1) NSAIDs: RR = 0.56 (95% CI 0.3–1.2)	Logan et al., 1993
Intervention trial in polyp patients (n = 793)	P	Aspirin: RR = 0.52 (95% CI 0.31–0.89)	Greenberg et al., 1993
Case–control study of adenomatous polyps/157 cases and 480 controls	R	Daily NSAID use: RR = 0.36 (95% CI 0.20–0.63)	Martinez et al., 1995
Case–control study/colonic and rectal cancers/830 cases and 1662 controls/use of aspirin >2 times per day	R	Colon: RR = 0.34 (95% CI 0.14–0.82) Rectum: RR = 0.28 (95% CI 0.10–0.81)	Suh et al., 1993
Physicians' Health Study (cohort)/randomised trial/118 incident cases/males/colorectal cancer	P	Aspirin: RR = 1.15 (95% CI 0.8–1.85)	Gann et al., 1993
Cohort study/mortality/colon cancer (n = 950)/rectal cancer (n = 138)/aspirin use >16 times per month	P	Colon: RR = 0.63 (95% CI 0.44–0.89) Rectum: RR = 0.79 (95% CI 0.41–1.53)	Thun et al., 1993
Hospital case–control study/colorectal cancer/recent use (<4 years) 97 cases and 388 controls	R	Aspirin and NSAIDs, high dosage aspirin: RR = 0.25 (95% CI 0.09–0.73), NSAIDs: RR = 0.38 (95% CI 0.18–0.78)	Peleg et al., 1994
NHANES 1 cohort/12 years of follow-up/colorectal cancer/aspirin use in previous 30 days	P	RR = 0.85 (95% CI 0.63–1.15) Males aged <65 years: RR = 0.35 (95% CI 0.17–0.73)	Schreinemachers and Everson, 1994
Health Professionals' Follow-up Study (cohort)/251 incident cases of colorectal cancer	P	Regular aspirin use: RR = 0.68 (95% CI 0.52–0.92)	Giovannucci et al., 1994
Hospital case–control study/colorectal cancer/511 cases and 500 controls. Cohort study (Nurses Health Study)/331 cases, 551 651 person-years	R	NSAIDs/males: RR = 0.64 (95% CI 0.42–0.97), females: RR = 0.32 (95% CI 0.18–0.57), Aspirin: RR = 0.56 (95% CI 0.36–0.9) after 20 years	Muscat et al., 1994
Cohort study (Nurses' Health Study)/331 cases, 551 651 person-years	R	Aspirin: RR = 0.56 (95% CI 0.36–0.9) after 20 years	Giovannucci et al., 1995

Hospital-based colonoscopy case–control study (1992–1995) in which 210 patients with adenomas and 169 non-adenomas patients were compared	P	NSAID users OR (adjusted) = 0.56 (95% CI 0.34–0.92 cf. non-users). 71.8% had taken aspirin, 53.8% ibuprofen and remainder (<8.4%–1.9%) various NSAIDs. Regular users who stopped 1 yr previously had OR = 0.59 (CI 0.21–1.67)	Sandler et al., 1998
Non-concurrent cohort linkage study using Saskatchewan Prescription Drug Plan data from 1981–1995; 3844 colon cancer and 1971 rectal cancer cases	R	Rate ratios for both colon and rectal cancer decreased with dosage of NSAIDs in 6 months before diagnosis	Collett et al., 1999
Case–control study using UK General Practice Research Database; 1368 colon cancer and 593 rectal cancer cases (part of an overall study of GI cancers)	R	Reduction of Odds Ratios with increased number of NSAIDs prescriptions in colon cancer but only trend in rectal cancer in 13–24 month period before diagnosis	Langman et al., 2000

*RR, relative risk; OR, Odds Ratio; CI, confidence interval; NHANES 1, First National Health and Nutrition Examination Survey. P = Prospective, R = Retrospective studies. Modified from Berkel et al. (1996) and Taketo (1998b), with additional data from Collett et al. (1999) and Langman et al. (2000).

TABLE 13.2

Effects of aspirin and some other NSAIDs on the growth and proliferation of tumours induced in animal models.

Species/strain	Tumour-inducing agent	Study	Results/comments	Authors
Rats	Carcino-sarcoma cells	Osteolytic effects	Aspirin 500 mg/kg daily reduced osteolysis and hypercalcaemia	Powles et al., 1973
Mice	Transplantable fibrosarcoma; 3-methyl-cholanthrene (MC)	Induced fibrosarcomas	Aspirin and indomethacin suppressed growth	Plescia et al., 1975
Mice; CD$_{fl}$	Transplanted mast cell ascites (P815) or Lewis lung carcinomas		Aspirin 150 mg/kg twice daily inhibited tumour growth by 39–43% and amine concs of tumours. Similar effects with 3 mg/kg indomethacin, but 5 mg/kg reduced tumours by 80%	Hail, 1976
Mice; C3H	Transplantable fibrosarcoma; 3-methylcholanthrene	Induced fibrosarcomas	Aspirin 1 mg/ml in drinking water reduced tumour size by 41% with decrease in PGE$_2$ levels. Similar effect with 20 mg/ml indomethacin. Aspirin may have partly hydrolysed to salicylic acid, so effects could be partly due to latter	Lynch et al., 1978
Rats; F344	N-[4-(5-nitro-2-furyl)2-thiazolyl]-formanide (FANFF)	Bladder hyperplastic lesions	Aspirin 0.5% in diet inhibited lesions after 6 or 12 wks FANFT, reduced PGE$_2$ in bladder after 1–13 wks	Cohen et al., 1981
Rats; Sprague–Dawley	Dimethyl-hydrazine	^{5}H-thymidine, cell proliferation	Inhibition of PGE$_2$ production	Craven et al., 1982; 1983
Rats; male Fischer	FANFT	Bladder carcinoma and FANFT metabolism	Aspirin 0.5% in diet for 12 weeks followed by 56 wks control reduced numbers of rats with induced carcinomas to 37% from 87% in controls. FANFT and aspirin produced forestomach tumours, whereas FANFT alone did not	Murasaki et al., 1984

Species	Carcinogen	Tumour type	Findings	Reference
Rats; buffalo	N-methyl-N'-nitro-N-nitroso-guanidine (MNNG)	Gastric adeno-carcinomas Leiomyosarcomas	Treatment with aspirin 3.25 µg/ml in water and MNNG led to tumours, but treatments alone did not. Aspirin hydrolysis in water could be confounder	Cohen et al., 1984
Rats; male F344	FANFT alone and with sodium saccharin	Bladder carcinomas	Aspirin 0.5% in diet for 61 wks reduced incidence of bladder cancers to 20% and with saccharin 28% by aspirin. Part of effect could be due to aspirin inhibiting metabolism of FANFT	Sakata et al., 1986
Rats; male Wistar	N-Butyl-(4-hydroxybutyl-nitrosamine	Bladder carcinomas	Aspirin 0.1% in diet for 20 weeks reduced incidence of bladder cancer to 1/29 animals, cf. 8/29 in controls	Klan et al., 1993
Rats; male F344	Azoxymethane (AOM)	Colon carcinogenesis	Aspirin 200 and 400ppm for 52 wks in diet inhibited incidence to 53% and 47% from 78% and multiplicity (tumours/animal), 75% and 69% of controls respectively, coincident with reduction in tumour or mucosal PGE_2. Similar effects with piroxicam. Part of effect could be due to inhibition of metabolism of AOM as well as reduced PGE_2 production	Reddy et al., 1992; 1993
Rats; male Wistar	Dimethyl hydrazine (DH)	Colon tumours	Aspirin 30 and 60 but not 5mg/kg daily for 18 weeks reduced the number and incidence of tumours in a dose-related manner. Similar result with 5-amino salicylic acid 60mg/kg daily	Davis et al., 1992; Davies and Patterson, 1994
Mice; BDF (C57 bl/6xDBA/2F$_1$)	N-butyl-N-(4-hydroxybutyl) nitrosamine (BBN)	Bladder tumours	Aspirin 400 and 800mg/kg in diet failed to reduce induction of tumours, whereas sulindac 200 and 400mg/kg or ketoprofen 40 and 80mg/kg did	Rao et al., 1996

TABLE 13.2

continued

Species/strain	Tumour-inducing agent	Study	Results/comments	Authors
Rats; male Sprague–Dawley	DH	Measure of aberrant crypt foci (ACF)	Aspirin 50 mg/kg but not sodium salicylate 50 mg/kg or other NSAIDs daily reduced total ACFs and ACF formulation at 8 wks but not at 30 wks	Barnes *et al.*, 1997
Mice; female A/J	NNK	Lung tumours	Aspirin 294 mg/kg in diet and 5-lipoxyganase inhibitors A-79715 75 mg/kg and MK-886 25 mg/kg for 25 wks reduced multiplicity by 44, 75 and 52% respectively. Combination of A-79715 with aspirin reduced multiplicity by 87% (numbers of animals with tumours). Effects paralleled by reduced in plasma PGE_2 (aspirin and A-79715) and LTB_4 (A-79715, MK-886). Similar results in same model and system with aspirin 73–588 mg/kg in food reduced multiplicity, but no effect of paracetamol 1520 mg/kg	Rioux and Castonguay, 1998a; 1998b
Mice; female C57 BL/6J Min$^{+/+}$	Germline mutation induced adenomatous polyposis coli (APC)	Colon and other GI tumours, apoptosis and cytoskeletal β-catenin	Aspirin 200 mg/kg in diet for 10 wks reduced total tumours in small intestine by 44%. Reduction in majority of tumours in small intestine accounting for major part of effect. Aspirin increased enterocyte proliferation to normal, with increased apoptosis in small intestine and reduced β-catenin	Mahmoud *et al.*, 1998
Rats; male F344	Diethylnitrosamine + BBN + MMNG	General and liver tumours	Aspirin 0.3% and paracetamol 0.3% in diet for 66 wks reduced number of glutathione S-transferase placenta-	Uehara *et al.*, 1999

712 ■

| Mice; male 57BL/6J $min^{+/+}$ | Mutation of APC | Intestinal tumours | positive liver cells, while mixture of aspirin, paracetamol, dipyrone and ethenzyamide reduced liver tumours

Aspirin 400ppm in diet to 80 days of age had no effect although it reduced intestinal mucosal PGE_2 and 6-keto $PGF_{1\alpha}$ 6-keto $PGF_{1\alpha}$ production *ex vivo*. Indomethacin 9ppm reduced tumour load by 85% and inhibited PG production | Chiu *et al.*, 2000 |

from a variety of study designs and populations. An important aspect concerning dosage and timing has arisen from the studies of Collet *et al.* (1999) in their investigations of colonic and rectal cancers in patients from the Saskatchewan Prescription Drug Plan (1981–1995). These data showed that the risk of developing cancers in these sites was reduced progressively with increase in dosage and duration of use.

A large-scale pan-European prospective study supported by the European Union BIOMED1 programme is currently underway comparing 600 mg microencapsulated aspirin (Bayer) with identical placebo in subjects at risk with hereditary non-polyposis colon cancer (HNPCC) (Burn *et al.*, 1998). This study is of particular interest since this is one of the two commonest inherited colonic cancers (the other being familial adenomatous polyposis, FAP). The HNPCC phenotype, known as Lynch syndrome, is characterised by a high risk of developing colon cancer (80 per cent). A recent suggestion that mesalazine, the 5-aminosalicylate drug used in treatment of ulcerative colitis, could be useful in preventing colorectal cancer (Bus, *et al.*, 1999) is particularly interesting in view of the colonic disposition of this drug. More such prospective studies in which dosage and timing are controlled are now warranted.

A recent study from the Association pour la Prévention par l'aspirine du Cancer Colorectal was performed in 272 patients in 49 gastroenterology centres in France with a history of colorectal adenomas who received 160 or 300 mg/d lysine acetylsalicylate or placebo for 1 year and who were monitored by colonoscopy (Benamouzig *et al.*, 2003). The adenoma recurrence among 238 patients who completed the study was significantly reduced by the aspirin treatments in relation to adenoma size greater than 5 mm.

The potential for these drugs to prevent cancers in other sites is not yet so convincing, and there may be variability among different NSAIDs in their prophylactic activity (Berkel *et al.*, 1996; Castelao *et al.*, 2000; Coogan *et al.*, 2000; Langman *et al.*, 2000; Sharpe *et al.*, 2000; Tavani *et al.*, 2000). There have been some encouraging results from studies investigating the effects of aspirin in certain gastro-oesophageal cancers (Corley *et al.*, 2003). Thus, epidemiological data from the USA, National Health and Epidemiological Examination Survey (NHANES-I) and National Epidemiologic Follow-up Studies (NEFS) of 14 407 of 600 000 adult US residents showed that occasional users of aspirin had a 90 per cent decreased risk of developing oesophageal carcinoma (Funkouser and Sharp, 1995). A large multi-centre prospective study based on tumour registries in the State of Connecticut, a 15-county area of New Jersey, and a 3-county area of western Washington State of patients between the ages of 30 and 79 years who were diagnosed with one of four gastro-oesophageal cancers between 1993 and 1995 were compared with a randomly selected control group (Farrow *et al.*, 1998). After controlling for risk factors, current users of aspirin had decreased risks of developing oesophageal adenocarcinoma (odds ratio, OR = 0.40, confidence interval, CI = 0.24–0.58), oesophageal cell carcinoma (OR = 0.49, CI = 0.28–0.87) and non-cardia gastric adenocarcinoma (OR = 0.46, CI = 0.31–0.68), but not gastric cardia adenocarcinoma (OR = 0.80, CI = 0.54–1.19). The risk of developing one of these four carcinomas was also reduced in current users of non-aspirin NSAIDs, though the data for intake of these drugs and development of oesophageal squamous cell carcinomas (OR = 0.51, CI = 0.24–1.10) and oesophageal adenocarcinomas (OR = 0.81, CI = 0.51–1.30) were not so convincing in providing evidence of protection compared with users of aspirin or the non-aspirin NSAID users at risk of developing the other two cancers. Interestingly, the pattern of use in current users of aspirin was greater than that in non-aspirin NSAID use by age but not by gender in this study, and the use of aspirin was greater in males and of non-aspirin NSAIDs greater in females, so the age- and gender-based use of the drugs may affect the overall outcome if the tumour types have intrinsically different patterns depending on these factors. Still, these data are strongly suggestive that aspirin and other NSAIDs may have preventative action against upper gastrointestinal cancers.

Another epidemiological case–controlled study based on hospitals in Moscow, which included serological tests for *Helicobacter pylori* (a known potentially carcinogenic organism) among the variables assessed, showed that aspirin and other NSAIDs were associated with protection against stomach cancer (OR = 0.6, CI = 0.41–0.90) and that the predominant protective effect of aspirin was in non-cardia cancer, an effect also noted in the study of Farrow *et al.* (1998). In the Moscow study there were associations with age and level of education (OR = 0.49, CI = 0.31–0.77), while the risk was equally reduced in both men and women. The OR of aspirin-users without *H. pylori* was 0.81 (CI = 0.44–1.48), whereas it was reduced in individuals infected with *H. pylori*, the OR being 0.39 (CI = 0.19–0.77) in individuals adjusted for age and education, with similar results being related to cancer of the antrum. While these differences were claimed to be indicative of a trend for greater protection by aspirin in infected individuals, they were apparently not statistically significant. However, the trends were significant in individuals who had aspirin and non-aspirin NSAIDs (Zaridze *et al.*, 1999). Overall, these studies on gastro-oesophageal cancers are encouraging and are suggestive of a protective action of aspirin as well as other NSAIDs. Further studies are warranted in larger groups controlled for dose and duration of drug intake and *H. pylori* strain sub-types, as well as investigations on the mechanisms underlying this protection in patients at risk of different types of upper GI cancers.

In other epidemiological studies, NSAIDs of the acetic acid subtype were found to have greater protection than aspirin or salicylates or oxicams against bladder cancer (Castelao *et al.*, 2000). No benefits have been shown for aspirin intake and the development of ovarian cancer (Tavani *et al.*, 2000). Occurrence of breast cancer has been shown to be reduced among long-term NSAID users (Sharpe *et al.*, 2000).

It should be noted that NSAID (including aspirin) abusers, who consume analgesics in excess, have been shown to be at high risk of developing malignancies, especially in the GI and urinary tracts, breast and female genital tract but not the prostate (Bucher *et al.*, 1999). This underlies the importance of determining the effective dose ranges of NSAIDs in relation to the safety of long-term prophylaxis with these drugs.

Reduction in tumour growth and proliferation in laboratory animal models

An extensive number of studies has been reported investigating the effects of aspirin and other NSAIDs as an antiproliferative agent against tumours induced chemically, in tumour-bearing or gene-transfected animals (Kort *et al.*, 1986; Berket *et al.*, 1996; Chun-Hung *et al.*, 1997; Duperron *et al.*, 1997; Levy, 1997; Vainio, 1999; Table 13.2). Generally, the results have shown that dietary aspirin or i.p. aspirin reduces the numbers of tumours or their multiplicity (i.e. number of tumours per animal). The effects depend on the tumour type (including agent), the timing of aspirin or other NSAIDs in relation to induction if by a chemical carcinogen, and the sex and species of animal (Table 13.2). With chemical carcinogens there is the possibility that their cytochrome P_{450} or cyclo-oxygenase metabolism could be affected by co-administered NSAIDs (Levy, 1997). In those studies where the metabolism of the carcinogen was investigated it appeared that neither aspirin nor salicylic acid affected NNK metabolism (Duperron *et al.*, 1997), although aspirin may affect the metabolism of AOM (Reddy *et al.*, 1992) or FANFT (Sakata *et al.*, 1986) (Table 13.2). The contrasting effects of aspirin in the APC-mutant min[+/+] mouse (Table 13.2), where in one study it showed reduction in intestinal tumours in female mice (Mahmoud *et al.*, 1998) but not in male animals (Chiu *et al.*, 2000), are difficult to explain other than being due to sex, although in the latter study the dietary content of aspirin was relatively low.

The studies in tumours induced by carcinogens in the lungs, colon and bladder suggest that aspirin inhibits the proliferation of these cancers (Table 13.2), possibly coincident with effects on accelerating apoptosis.

In vitro studies

A substantial number of studies have shown that aspirin and other salicylates or NSAIDs directly inhibit the growth, differentiation and proliferation of tumour cells (Powles *et al.*, 1973; Arvind *et al.*, 1996; Elder and Paraskeva, 1996; Elder *et al.*, 1996; Ricchi *et al.*, 1997; Duffy *et al.*, 1998; Shinichi *et al.*, 2000; Smith *et al.*, 2000). There is evidence of differing effects of aspirin (100 µg/ml) or other NSAIDs on the growth and proliferation of cancer with different cell line types. In some studies relatively high concentrations of aspirin or salicylate have only been found to be effective in inhibiting growth, and thus human colorectal tumour cell lines were inhibited by salicylate with IC_{50}'s ranging from

1.65–7.38 mmol/l (Elder *et al.*, 1996). Salicylate caused all the cell lines to accumulate in the G_0–G_1 phase of the cell cycle as well as inducing apoptosis in carcinoma and *in vitro* transformed but not all adenoma cell lines. Duffy *et al.* (1998) found that aspirin and some other NSAIDs exhibited synergy with certain cancer chemotherapeutic agents, but the effect varied considerably with different cell lines and was independent of the cyclo-oxygenase inhibitory effects of the NSAIDs (as exemplified by the potent inhibitory effects of the PG-synthesis non-inhibitory drug, sulindac sulphone). Those NSAIDs that were more potent inhibitors were coincidentally inhibitors of the transport of the glutathione conjugates.

The mechanisms of action of NSAIDs have been considered to involve inhibition of prostaglandin production (especially that derived from the expression of the inflammogen-inducible COX-2), and prostaglandin-independent cell turnover – proliferation and apoptosis, DNA repair, signal transduction events (Ras, MAP kinase, Myc, NFκB/IκB), PPAR activation, angiogensis and induction of tumour immunity (Hanif *et al.*, 1996; Shiff and Rigas, 1999a; 1999b; Prescott, 2000; Prescott and Fitzpatrick, 2000; Fosslein, 2000). There is considerable evidence that some NSAIDs (e.g. the non-prostaglandin inhibitory forms of R(−)-flurbiprofen and sulindac sulphone) inhibit the development of chemically-induced or mutation/spontaneously induced tumours (Chiu *et al.*, 1997; Piazza *et al.*, 1997; Wechter *et al.*, 1997; 2000; Malkinson *et al.*, 1998; Skopińska-Rózewska *et al.*, 1998), suggesting that prevention or inhibition of tumour formation by some NSAIDs may be independent of prostaglandin inhibition. This has cast doubt on the role of inhibition of prostaglandin production in the anti-tumour effects of NSAIDs, even though increased production of prostaglandins of the E, F, I and D series has been shown in malignant tissues from patients with tumours (Stamford *et al.*, 1983; Smythies, 1988).

Moreover, it is well established that co-carcinogenic agents (e.g. phorbol esters) stimulate prostaglandins in target organs and that prostaglandins aid tumour development (Smythies, 1988). Recent extensive studies showing that selective COX-2 inhibitors, even highly selective drugs such as celecoxib, produced inhibitory effects in experimentally-induced or mutation-derived tumours in laboratory animal models (Oshima *et al.*, 1996; Yoshimi *et al.*, 1997; 1999; Matsunaga *et al.*, 1998; Taketo *et al.*, 1998a; 1998b; Fischer *et al.*, 1999; Nishimura *et al.*, 1999; Pentland *et al.*, 1999; Crew *et al.*, 2000., Fournier *et al.*, 2000; Harris *et al.*, 2000; Hussey and Tisdale, 2000; Masferrer *et al.*, 2000; Reddy *et al.*, 2000), combined with substantial evidence that there is a relationship between increased COX-2 expression and the development or progression of human carcinomas (Singer *et al.*, 1998; Tsujii *et al.*, 1998; Wolff *et al.*, 1998; Murata *et al.*, 1999; Sheehan *et al.*, 1999; Williams *et al.*, 1999; 2000) has argued strongly for the case for induced COX-2 derived prostaglandins being the main target for the action of both established and COX-2 selective NSAIDs in the prevention or control of cancer growth. The ever-increasing evidence for NSAIDs (both non-selective and selective COX-2 inhibitors) and non-PG synthesis inhibitory NSAIDs promoting or accelerating apoptosis and inhibiting angiogenesis (Hannif *et al.*, 1996; Tsujii *et al.*, 1998; Guerono and Diez, 1999; Elder *et al.*, 2000; Fosslien, 2000; Masferrer *et al.*, 2000; Shao *et al.*, 2000) suggests that these events may predominate and have a common basis in the action of aspirin and other NSAIDs. Differentiation between the PG-independent and dependent actions of aspirin and other NSAIDs on tumour growth and progression may not be clear cut, since at one level there are important cellular signals controlled by PGs, and at the other non-PG synthesis inhibitory NSAIDs may have more potent control of these cell signalling events that underlie tumourgenesis and metastasis. Recent studies showing that aspirin and salicylate inhibit tumour cell invasiveness induced in mice by Epstein–Barr virus by inhibiting the expression (induced by the latent membrane protein 1 derived from this virus) of a key matrix metalloproteinase (MMP-9) (Murono *et al.*, 2000; Jiang *et al.*, 2001) have also shown that aspirin 0.3–2.7 mmol/l inhibits the production of MMP-2, coincident with reduction in migration and invasion of SK-Hep-1 cells; production of MMP-2 was likewise decreased and the expression of E-cadherin increased in the colon tumour line, HT-29. Thus, MMP-2 and -9 and E-cadherin represent important sites for the actions of salicylates in controlling the metastatic process. As the inhibition of MMP-9 by salicylates occurs by inhibiting expression of the AP-1 and NFκB pathways, this suggests that one of the primary sites of the salicylates is in the control cancer proliferation is on these signal transduction pathways (Santini *et al.*, 1999; Murono *et al.*, 2000; Shao *et al.*, 2000).

ALZHEIMER'S DISEASE AND RELATED DEMENTIAS

There has been much interest in the potential use of NSAIDs as prophylactic or therapeutic treatments for Alzheimer's disease (AD) (Breitner *et al.*, 1994; 1995; Andersen *et al.*, 1995; Breitner, 1996; 1999; Hull *et al.*, 1999; McGeer and McGeer, 1998; 1999; Sugaya *et al.*, 2000). However, the debates about the utility of these drugs and their mechanisms of action rest on what is known about the clinical manifestations, diagnosis and molecular pathology of the disease (Bondareff, 1994; Lucca, 1999). As there are many facets and factors about these that are not yet clear, it is probably fair to say that we are only at an early stage in understanding how and when these drugs may be employed, what type of drug (e.g. COX-2 selective or non-selective) should be used, and how they work (Sugaya *et al.*, 2000).

Early epidemiological investigations suggested that patients with rheumatoid arthritis receiving NSAIDs had a lower incidence of AD than normal subjects (Breitner *et al.*, 1994; 1995; Breitner, 1996). This was followed by a trial in which indomethacin was claimed to reduce the deterioration in cognitive symptoms of AD (McGeer and McGeer, 1996). There followed a number of long-term longitudinal studies in human volunteers showing that use of NSAIDs, especially those available OTC (e.g. aspirin and ibuprofen), is associated with delay in the onset of symptoms of AD.

The case for using NSAIDs and other drugs (Evans, 2001) in preventing the onset and in therapy of AD is emphasised by recent data on the overall prevalence of this condition. Thus, Breteler and co-workers (1998a) showed that the prevalence of AD in a cohort study of the Rotterdam study population comprising 7983 subjects over the cycle of 55 years was 6.4 per cent and the incidence was 1 per 100 person-years. Both the prevalence and incidence increased markedly with age, with predominance in males.

Some insight into the clinical manifestation and pathogenesis of AD is instructive. AD is variously defined, but is probably simply stated to be a precipitous loss in cognitive function and memory. Definitions include evidence of an early age of onset and some relation to a reference of what can be considered to be normal memory and cognitive function for an individual (Hindmarch *et al.*, 1991; Cummings and Benson, 1992; Burns and Levy, 1994; Scadding and Gibbs, 1994). The symptoms of AD often develop slowly and insidiously, and the time of onset is difficult to establish. Part of the problem in diagnosing AD in an individual is that there are varying degrees of cognitive and memory loss in individuals, and these progress with age and at differing rates (Cummings and Benson, 1992). There is evidence of some loss of these functions, but these observations do not meet all the established criteria for AD. There may be difficulties in undertaking tasks such as calculations, or disorientation (especially loss of whereabouts in familiar locations) or common memory lapses. Mostly, potential for AD is only seen when these lapses or losses on memory or cognition achieve the stage of interfering with work or family life. The presence of the homozygous, $\epsilon4$ apolipoprotein E (ApoE) allele is claimed to be a predictor of development of AD in 95 per cent of patients with early dementia (de Knijf and van Duijn, 1998; Yamagata *et al.*, 2001). Varying combinations of apo-E4/E3, E3/E3, E3/E2, E4/E2 and E3/E2 alleles lead to a progressively lower risk of developing AD and this may be related to β-amyloid deposition (Corder *et al.*, 1995).

The pathogenesis of AD is complex, and some aspects are subject to debate. The original observation of Alois Alzheimer in 1907 of the presence of neuritic plaques and neurofibrillary tangles at autopsy (Alzheimer, 1907) has virtually gone unquestioned as one of the pathological hallmarks of the disease (Selko, 1994). Plaques stain for β-amyloid with intercalation dyes such as Congo Red, and give the characteristic bi-refringent appearance under polarised light. More than 30 proteins aggregate in neuritic plaques. Among the major ones is amyloid β4 protein (Aβ4), which is a pathognomic fragment of amyloid precursor protein (APP). Low-density lipoprotein (LDL)-related receptor is present with many of the proteins that it binds, including ApoE, which are internalised in neurones. There is debate about whether plaques are a consequence or a cause of the disease (Goedert *et al.*, 1994; Roses, 1994; Yankner, 1996). There are also views that place emphasis on the formation of neurofibrilliary tangles (NFTs) (Kosik, 1990; Goerdert *et al.*, 1992; Mandelkow *et al.*, 1992), which are hyperphosphorylated tau proteins that are components of microtubules in neurones. When tau proteins bind to themselves they form paired helical filaments that can no longer stabilise microtubules, and at this stage become hyperphosphorylated at various sites and form NFTs. Most now agree that Aβ4-containing plaques and NFTs are characteristic end-points or pathological consequences rather than primary causes of the disease.

Neuronal cell loss has also been identified in the cerebellum of patients with Alzheimer's disease (Sjöbeck and Englund, 2001). These authors found extensive neuronal loss, atrophy and gliosis in the vermis, and to a lesser extent neuronal loss in the inferior regions.

The reduction in the regional blood flow in the cerebral hemispheres has been observed in patients with AD (Nobili *et al.*, 2001). This may be related to well-recognised vascular factors that are considered to be risks in the development of AD and related dementias – for example, vascular dementia, which occurs in patients with a history of stroke (Breteler *et al.*, 1998b). The link to apo-E4 (which, as indicated above, is a high risk factor for the development of AD) and the HDL that are required for the removal of excess cholesterol from brain through the interaction with apo-E4 and heparin sulphate proteoglycans are pathological features that occur in subendothelial space of microvessels in the brain (Mulder and Terwel, 1998). Thus there is a strong link between abnormalities of cholesterol and lipid metabolism and the development of vascular changes that occur in AD. These vascular factors may also be important in patients with diabetes mellitus. The vascular components of AD, as well as abnormalities of platelet function that characterise cerebrovascular diseases and other conditions such as diabetes, may represent important sites for the actions of aspirin and related NSAIDs. This may also be a useful therapeutic strategy in vascular dementias (p. 720).

Inflammatory components of the pathogenesis of AD

The depositions of β-amyloid in those regions of the brain that are affected in AD are closely associated with a range of inflammatory reactions. These include activated complement proteins, cytokines, activated proteases (including metalloproteases), deposition of proteoglycans, and systemic increase in the acute phase proteins that represent the responses to activation by pro-inflammatory cytokines generated in response to amyloid activation (Lucca, 1999). Immunohistochemical staining for classical components of the complement pathway has shown that these are abundant in the brains of patients with AD in the regions where there are senile plaques. Evidence in these studies also shows that there are some late components of complement as well as the membrane attack complex (MAC). There is increased MAC content and degree of activation of microglia in the entorhinal cortex and superior frontal gyris of AD brains. There is evidence for increased consumption of complement proteins and C1q binding to β-amyloid.

A range of pro-inflammatory cytokines (IL-1, IL-6 and TNFα) has been shown to be increased in AD brain tissue. In particular, levels of IL-1 have been shown to be markedly increased in microglia and this has been correlated with increased numbers of neuritic plaques that have been observed in AD brains.

There is also evidence for increased oxyradicals and oxyradical damage in AD brains (McGeer and McGeer, 1998; Lucca, 1999). An elegant model has been proposed by Lucca (1999) to explain the complex interrelationship between apo-E, β-amyloid, and the induction of pro-inflammatory cytokines, oxyradicals and the activation of proteases that is thought to underlie the pathogenesis of AD.

The increase in pro-inflammatory cytokines may be related to levels of β-amyloid and the increased platelet activating factor (PAF), which would then be expected to cause increased prostaglandin production in local regions where there are accumulations of β-amyloid as well as increased production of cytokines. Increased production of IL1 and TNFα would be expected to lead to induction of COX-2, with consequent amplified production of prostaglandins. Increased production of COX-2 protein has been shown in a number of studies (Lukiw and Bizan, 1998; Lukiw *et al.*, 1998; Halliday, 2000a). Interestingly, increased expression of COX-1 characterised by increase in the production of messenger RNA and immunoreactive protein has been observed in AD brains. COX-1 immunoreactive microglia have been found to be associated with β-amyloid-bearing plaques and the fusiform cortex (Yermakova *et al.*, 1999). It would therefore appear that both COX-1 and COX-2 are important targets for actions of NSAIDs in AD. Additionally, intercellular signalling pathways, especially those involving the nuclear factor (NF) complex (Lukiw *et al.*, 1998) and cell cycle activation (Nagy *et al.*, 1998) may also represent important targets for the actions of these drugs.

Molecular and cellular aspects of the potential protective effects of aspirin and related drugs

Because there are a large number of inflammatory mediators that are produced in different areas of the brain associated with high level neurological function during Alzheimer's disease, the potential exists for aspirin and NSAIDs to act on the molecular and cellular processes underlying the development of these inflammatory reactions. In view of the current state of knowledge of aspirin and salicylate, it is possible to identify potential molecular targets as cyclo-oxygenase, oxyradical production, NFκB and other signal transduction pathways, leucocyte activation and release of certain enzymes, and certain other potential neuroprotective effects as yet not clearly defined but for which there have been indications from recent studies involving the activation of glutamate receptors. A strong case for the use of multiple antioxidants in the prevention and treatment of Alzheimer's disease has been recently highlighted by Prasad and co-workers (2000). Since salicylate is a potent antioxidant, it would be logical to include aspirin or salicylate in any multiple antioxidant therapy strategies. The potential for COX-2 modulation being a major part of the effects of NSAIDs in protection against the manifestations of Alzheimer's disease has been illustrated by studies using COX-2 inhibitors, e.g. nimesulide, which has been shown to have protective effects against excitotoxic insult in rats (Scali *et al.*, 2000). The protective effects of nimesulide might not only be due to its effects on COX-2 activity, but since this drug has well-known antioxidant and anticytokine properties the protective effects seen in this model system may be caused by other properties of the drug rather than COX-2 inhibitory activities. More specific antiprostaglandin effects can be seen in studies in the mouse model of Alzheimer's disease (involving transgenic transfer of a known human Alzheimer's disease gene), where it was found that chronically administered ibuprofen 375 ppm in feed from 10 months of age (when amyloid deposits first appear), and continuously fed for 6 months, resulted in reduction in interleukin-1β, and glial fibrillary acidic protein levels, and also marked reduction in the number of β-amyloid deposits (Lim *et al.*, 2000). Moreover, ibuprofen reduced the number of ubiquitin-labelled dystrophic neurites and the microglial levels of antiphosphotyrosine. Thus there would appear to be sound experimental evidence in support of ibuprofen having protective effects in Alzheimer's disease such as has been seen in recent clinical trials. Other NSAIDs may also suppress β-amyloid induced monocyte neurotoxicity (Dzenko *et al.*, 1997), but these types of studies have not been extended to the use of salicylates. As indicated in a subsequent section, there are a number of other activities of aspirin and NSAIDs that could be implicated in neuroprotection, and these may be relevant to the pathological manifestations of Alzheimer's disease.

The ultimate success with the use of NSAIDs in Alzheimer's disease will come from long-term studies in which the drug-induced changes in the pathological condition underlying the disease will be determined in relation to the improvement of cognitive and other psychological or psychiatric manifestations in prospective trials in which the drug intake has been quantified. One study in which the neuropathological findings and psychiatric assessments were undertaken retrospectively showed that NSAIDs did not result in any significant differences in the number of inflammatory glia, plaques or tangles, although there was an improvement in cognitive performance (Halliday *et al.*, 2000b). Unfortunately this *post-mortem* study was performed in relatively small numbers of subjects, and the intake of NSAIDs obviously has a variable quantitative basis by being a retrospective study. It may be that these studies indicate that longer-term usage of anti-inflammatory medications may be required in at-risk subjects in order to gain the full benefit for controlling the pathological as well as the cognitive changes underlying this condition (Austerman *et al.*, 1998). Moreover, the dose of aspirin and related drugs may need to be quantified for beneficial effects, as highlighted by the study of Stürmer *et al.* (1996), who showed that low-dose aspirin or short-term use of the drug only resulted in very minor improvement in cognitive function.

CHAPTER 13

Post-stroke dementia and ischaemic brain conditions

The potential for use of aspirin as an antiplatelet agent in controlling post-stroke dementia and other conditions where there may have been vascular injury has been indicated in a number of studies (van Kooten *et al.*, 1998; 1999). Van Kooten *et al.* (1999) showed that increased thromboxane A_2 biosynthesis (measured by the urinary excretion of 11-dehydro-TxB_2) was increased during the chronic phase of stroke and associated with a type of dementia. They found that patients receiving aspirin or an oral anticoagulant had clinical characteristics indicative of dementia that were comparable to control patients, thus indicating that there was some protection against the manifestations of dementia by aspirin treatment. In the review by van Kooten and co-workers (1998) of published studies in which the effects of aspirin on cognitive function were evaluated in subjects that had previously had stroke, four of six studies showed that there was some improvement with aspirin on cognitive function but in two other (largely observational) studies there was no benefit. The lack of appreciable effects of the drug in these population based studies may reflect their lack of sensitivity. In all these studies relatively few patients were studied, the maximum number being 3793 in one population based study. It may be necessary for studies to be performed in larger populations, where it is possible to control many of the variables.

Generalised neuroprotective effects

A number of animal model studies have shown that COX-2 production is increased in various areas of the brain following global ischaemia (Nogawa *et al.*, 1997; Nakayama *et al.*, 1998; Planas *et al.*, 1999). COX-2 production has been shown to be increased following treatment with N-methyl-D-aspartate (NMDA) primary cortical cells from post-natal and foetal mice (Hewett *et al.*, 2000). In this study the specific COX-2 inhibitor NS-398 but not the COX-1 inhibitor valeryl salicylate attenuated the NMDA-induced cell death. Aspirin 100 mM prior treatment did not prevent the NMDA mediated neuronal death but did in the presence of flurbiprofen. Other studies have indicated that aspirin does have neuroprotective effects against glutamate excitor toxicity (Riepe *et al.*, 1997; Gomes *et al.*, 1998) while salicylate, but not celecoxib, protects against dopaminergic neurotoxicity (Sairam *et al.*, 2003). Triflusal also has neuroprotective effects against excitatory injury (Acarin *et al.*, 2003). Aspirin also has protective effects against oxyradical-mediated against cerebrovascular ischaemia (Kuhn *et al.*, 1996). Recently a case report of a 64-year-old woman who presented with pain in the right orbit associated with the ophthalmic branch of the trigeminal nerve, which had produced pain that was unresponsive to non-aspirin NSAIDs, and who showed abnormal electroencephalogram and psychological symptoms was found to have markedly improved on low doses of aspirin for one month (Persegani *et al.*, 2001). There have been some studies investigating improvement in cognitive performance with aspirin (Austerman *et al.*, 1998; Peacock *et al.*, 1999) and while encouraging results were obtained this area requires more extensive investigation. Thus there may be considerable potential for aspirin focused on its antithrombotic effects in certain types of neurological conditions or neuronal injury resulting in dementia or neural cell loss.

COPPER SALICYLATES IN THE TREATMENT OF INFLAMMATORY DISEASES AND CANCER

Inflammatory diseases

The concept that regulation of copper (and also zinc) ion status during inflammation may confirm particular benefit therapeutically has been explored extensively by a considerable number of authors

(Sorenson, 1974; 1982a; 1982b; 1982c; Betts *et al.*, 1981; Milanino and Velo, 1981; Rainsford *et al.*, 1981; 1998; Crouch *et al.*, 1985; Milanino *et al.*, 1989; Berthon, 1992). In many inflammatory diseases, especially those that could be considered chronic, there is pronounced elevation of copper levels in the circulation. Because of this elevation it was suggested by Sorenson (1974) that anti-inflammatory compounds may act as chelators of copper ions, and this may coincidentally lead to their anti-inflammatory activity. He provided evidence that a number of copper chelates had anti-inflammatory activity and, since a number of anti-inflammatory compounds have potential chelating activity (Whitehouse, 1965), Sorenson's hypothesis would appear to be quite plausible. Over the years there have been a number of challenges to this general concept, but the original observations that copper chelates of anti-inflammatory drugs have greater potency as anti-inflammatory agents can still be shown with a considerable number of compounds (Sorenson, 1976; Williams *et al.*, 1976; Crouch *et al.*, 1985; Okuyama *et al.*, 1986; Beveridge, 1989; Korolkiewicz, 1989; Berthon, 1992; Sorenson, 1992a; 1992b; Zhiqiang *et al.*, 1998; Bouchaut *et al.*, 2000). In particular, in experimental animals in a variety of laboratory animal models, copper aspirinate has been shown to have marked anti-inflammatory activity in excess of that provided by an equimolar quantity of aspirin (Rainsford and Whitehouse, 1976a; Williams *et al.*, 1976; Korolkiewicz *et al.*, 1989; Sorenson, 1992b).

Interestingly, Hangarter and Lübke (1952) and Hangarter (1966) showed that a salicylate–copper complex (Permalon™) had pronounced anti-arthritic activity in patients with a variety of acute and chronic polyarthritic conditions. The copper–salicylate complex was given intravenously and contained 2 g of salicylic acid and 2.5 mg of copper (Hangarter and Lübke, 1951). In a summary of studies that had been performed with 160 patients with arthritic conditions who received 2–3 days intravenous infusion of 60–80 ml of Permalon™ mixture, which was shown to be very effective in controlling inflammation, Hangarter (1966) considered that the effect of Permalon™ was through actions on the reticuloendothelial system, especially on the increase in the production of antibodies and through antibacterial effects. A summary of the overall results that have been obtained by Hangarter and colleagues in their treatment of patients using copper salicylate (Permalon Therapy™) for the treatment of acute rheumatic fever, rheumatoid arthritis, cervical spinal shoulder and lumbar spinal syndromes and sciatica is shown in Table 13.3 (from Crouch *et al.*, 1985). More extensive studies have since shown that copper aspirinate, copper salicylate and copper di-isopropyl salicylate probably have anti-inflammatory actions due to a combination of effects that may involve these agents being superoxide dismutase

TABLE 13.3

Copper–salicylate (Permalon™) in the treatment of acute rheumatoid arthritis, cervical spine–shoulder, lumbar-spine syndromes and sciatica. (From Crouch *et al.*, 1985.)

Disease	Total no. of patients	Symptom-free	Clinical results		
			Improved	Slightly improved	Unchanged
Acute rheumatic fever	78	78 (100%)			
Cervical spine–shoulder and lumbar spine syndromes	160	95 (57%)	52 (32%)		13 (11%)
Sciatica without lumbar involvement	120	76 (63%)	38 (32%) .	6 (5%)	
Sciatica with lumbar involvement	160	95 (59%)	39 (24%)	10 (6%)	16 (11%)
Total	1140	744 (65%)	272 (24%)	16 (1%)	108 (10%)

mimetics, regulation of prostaglandin production and inhibition of nitric oxide synthetase (Sorenson, 1992a; Baquial and Sorenson, 1995).

The only other study in human subjects that appears to have been reported is that by Shackel *et al.* (1997), in which copper salicylate gel (containing 4.3 mg/g copper ion with 43.8 mg/g salicylate ion) was applied twice daily *topically* to the forearm of 93 patients with osteoarthritis of the hip or knee. The patients in this study appear to have been at various stages of disease. The authors observed that there were no differences in the pain rating of the copper salicylate treatment compared with controls that received placebo.

A possible interpretation of the differences in responses obtained in the uncontrolled trials by Hangarter and those in this trial by Shackel *et al.* (1997) may relate to the degree of inflammation in the patients with rheumatoid arthritis and related states studied by Hangarter, and the advanced state of osteoarthritis in the patients studied by Shackel and co-workers. Moreover, since the main effects of the copper-salicylate preparations have been in inflammatory states, it would appear that the major clinical actions of copper salicylate are in the control of chronic inflammation in RA rather than in pain responses in joint manifestations of OA.

An earlier claim by Sorenson (1974) that copper chelates of anti-inflammatory drugs such as aspirin may have anti-ulcer activity has also been the subject of much controversy. In a number of studies (Rainsford and Whitehouse, 1976b; Rainsford, 1982) these claims could not be entirely substantiated. It was suggested by Rainsford (1982) that copper complexes of aspirin and other salicylates could break down in the stomach and cause release of free copper, which would then become irritant to the mucosa. The apparent reduction in gastric ulcers seen in a number of studies (Williams *et al.*, 1976; Korolkiewicz *et al.*, 1986; Sorenson, 1992a) may have been due to the release of free copper ions causing mucosal irritation and subsequent production of prostaglandins, which would have anti-ulcer activity. Sorenson has challenged these concepts (Sorenson, 1984), and considered that there may have been technical issues that influenced the outcome of studies by Alich *et al.* (1983) and Rainsford and Whitehouse (1976b). However, the technical issues concerned the apparent inappropriate use of a suspending agent that might have complexing activity with copper, a claim that could not be substantiated experimentally. Thus the issue regarding whether copper aspirinate and other copper salicylates have anti-ulcer activity or lower ulcerogenic activity than non-steroidals still remains unresolved. It is at least possible to claim that when the anti-inflammatory activity of the copper complexes is related to that of the ulcerogenic activity of these compounds in comparison with that of the parent salicylate or other anti-inflammatory drugs, then this therapeutic ratio may be higher with the copper complexes.

Cancer and other conditions

A number of other pharmacological activities of copper complexes have also been investigated by Sorenson and other workers, among them potential anti-cancer activity (Leuthauser *et al.*, 1981; Willingham and Sorenson, 1986; Sorenson *et al.*, 1993), radiation protective or recovery activity (Sorenson, 1989; Sorenson *et al.*, 1992a; 1992b; Irving *et al.*, 1996), antidiabetic activity (Crouch *et al.*, 1985), anticonvulsant activity (Crouch *et al.*, 1985; Morgant *et al.*, 2000), antimutagenic activity (Crouch *et al.*, 1985), enhanced analgesic activity (Okuyama *et al.*, 1986) and inhibitory effects on development of ulcerative colitis (Keshavarzian *et al.*, 1992). Common pharmacological effects that may underlie all these diverse effects of copper salicylates and other copper complexes of anti-inflammatory drugs that reflect in part anti-inflammatory effects may centre on the superoxide dismutase-mimetic activity, and regulation of prostaglandin, nitric oxide and cytokine production as well as metabolic and direct cytotoxic effects of copper complexes (de Alvare *et al.*, 1976; Weser and Schubotz, 1981; Roch-Arvellier *et al.*, 1990; Berthon, 1992; Shuff *et al.*, 1992; Condorelli *et al.*, 1994; Booth *et al.*, 1997; Franco and Doria, 1997; Shen *et al.*, 1999).

Some aspects of the medicinal chemistry of copper salicylate complexes have already been discussed in Chapter 3. These studies highlight the potential for copper complexes to scavenge oxyradicals and control production of inflammatory medication that could be involved in tumour spreading and growth. Sorenson and co-workers have undertaken a considerable number of studies investigating the anti-tumour and radio-protective effects of copper salicylate complexes. Repeated intramuscular injection of 5 mg copper (II) (3,5-diisopropyl salicylate) or CuDIPS in 10 per cent Tween-80-saline vehicle caused a marked reduction in the size of Ehrlich carcinoma cells in mice, and also prolonged survival of these mice compared with those given the solvent (Tween-80-saline) i.m., CuSO$_4$ i.m., or DIPS i.m. alone (Leuthauser *et al.*, 1981). Thus this anti-tumour activity appears to be specific to the copper complex and not to the dissociated ligand or the copper ion (Leuthauser *et al.*, 1981). The authors of this study provided evidence that the superoxide dismutase-like activities were much lower in the tumour-bearing than in comparable normal muscle tissue, and attributed the beneficial effects of CuDIPS to its superoxide dismutase-like activity (Leuthauser *et al.*, 1981; Oberley *et al.*, 1982). They failed to consider the possibility of adrenocortical stimulation from local irritant actions of copper complex given by repeated intramuscular injection (Rainsford and Whitehouse, 1976a), which could itself be responsible for some effects contributing to anti-tumour activity. No signs of cellular infiltration of other inflammatory reactions were seen in the locus of the tumour in both control and CuDIPS-treated tumour-bearing animals (Oberley *et al.*, 1982), so it is presumed that there is little or no appreciable contribution of prostaglandin metabolism from these cells to the development of this cancerous condition. Also, it is assumed that no appreciable anti-inflammatory responses can be ascribed to CuDIPS in the development of its anti-tumour activity (Oberley *et al.*, 1982). The question arises about how CuDIPS can inhibit tumour growth by mimicking superoxide dismutase activity when there was no anti-tumour effect observed by superoxide dismutase itself, and when inhibition of superoxide dismutase activity to produce enhanced levels of cytotoxic superoxide anions (O$_2^{\bullet-}$) would seemingly have been a more useful therapeutic property. The lack of effects of superoxide dismutase might be due to its failure to penetrate lipid membranes, a feature that is possible with CuDIPS (Oberley *et al.*, 1982).

While there are obviously unresolved questions about the anti-tumour actions of CuDIPS, these interesting observations (Leuthauser *et al.*, 1981; Oberley *et al.*, 1982) should stimulate further research to determine the levels of superoxide dismutase-like activity and O$_2^{\bullet-}$ in tumour-bearing tissues of mice give CuDIPS, the influence of CuDIPS on the immunological defences, and the effects of this and related liposoluble copper–salicylate complexes on other presumed cytotoxic and 'differentiation'-controlling actions of both this and related metal complexes. At present, CuDIPS can only be regarded as an experimental drug. In the case of aspirin and other salicylates, these can of course be considered for therapeutic use in attempts to control tumours (as well as having benefits in relief of pain).

The relative toxicity of copper aspirinate was investigated by Sorenson *et al.* (1984). Although the growth rates of rats given 10 mg/kg of copper aspirinate were slightly lower than in control animals, there did not appear to be any significant changes either when viewed at necropsy or histopathologically. In this study the authors measured the copper and zinc concentrations in each of the major organs at periods of 1–5 months after treatment. In all except for increased levels of copper in the brain at 5 months, there did not appear to be any changes in the concentrations of both these metals. It would therefore appear that copper aspirinate has an acceptable toxicological profile and could be considered as a therapeutic agent.

THERAPEUTIC USES OF ASPIRIN IN PREGNANCY

There has been considerable interest in the potential for aspirin to be used for the following:

1. To prevent recurrent miscarriage, including that from pre-eclampsia
2. To prevent spontaneous abortion resulting from the anti-phospholipid antibodies that develop in systemic lupus erythematosus
3. To improve reproductive performance, especially in patients receiving *in vitro* fertilisation

4. To treat patent ductus arteriosus
5. To improve endometrial physiology.

Each of these claims must be balanced against the potential risk for teratogenic effects and the reduction in live birth weight that may occur, as is indicated by a large volume of experimental data from laboratory animal models (Beall and Klein, 1977; see also Chapter 8).

Safety compared with efficacy

Although careful evaluation of the potential risks of aspirin and other NSAIDs to cause abnormalities during pregnancy or abnormal functions has indicated that the relative risks are low (Østensen, 1994; CLASP Collaborative Group, 1995; Østensen and Østensen, 1996; Herrada et al., 2000; Nielsen et al., 2001; Østensen and Villiger, 2001), there is still much concern about the relative safety of aspirin in humans during pregnancy – including the potential to develop acute fatty liver in subjects in whom there is a mutation of the trifunctional protein of long-chain fatty acids in gestation (Baldwin, 2000). Optimisation of the NSAIDs and other antirheumatics that must necessarily be employed during pregnancy in patients with severe rheumatic disease, e.g. systemic lupus erythematosus, has been considered by Østensen (1994), and it has been considered that there is no evidence for any NSAID to have exhibited teratagenic effects in humans (Chapter 8). However, owing to the effects of these drugs on the inhibition of prostaglandin synthesis, which necessarily regulates the constriction of the ductus arteriosus *in utero*, and the potential for regulating hypertension in the neonate and labour, it is potentially possible for them to have both beneficial and untoward effects depending on how they are administered and, particularly, on the timing during pregnancy and the dosage of the drugs. The relative risks of an untoward event may also depend on the benefits that accrue in certain patients with rheumatic disease, e.g. systemic lupus erythematosus (Østensen, 1994; Østenson and Østensen, 1996).

Many of the assessments of safety and efficacy in different conditions have been made in relatively small groups, and only few prospective studies exist in relatively large populations. Among these are the CLASP Collaborative Group Study (1995), in which 4168 children were assessed at 12 months by questionnaires to general practitioners and 4365 were assessed at 18 months through direct questionnaires to parents. Among the outcome measures were: hospital visits during the first 18 months for congenital malformations, motor deficit, developmental delay, respiratory problems or bleeding problems; weight measurements; and the potential for delay in the development of certain established skills. The study was conducted in patient populations in the United Kingdom and Ottawa in Canada. The study was randomised involving 8914 mothers of children surviving to discharge following birth. There were no changes in (a) estimated gestational delivery (except in the patient group that survived at hospital discharge with this averaged about 38 weeks as expected), (b) the mean birth weight, and (c) any of the reasons accounting for a hospital visit at either 12 or 18 months. If anything there was a slight reduction in the number of congenital malformations when assessed at 12 months by the general practitioner or at 18 months by survey, but this did not achieve statistical significance. Thus this large randomised control trial in subjects who had received 60 mg aspirin compared with placebo would have included a significant number of patients who would likely have had pre-eclampsia, and there did not seem to be any measurable untoward effects.

Pre-eclampsia

Wallenburg (Wallenburg and Rotmans, 1988; 1997; Wallenburg, 2001) carefully evaluated the potential for aspirin given at low dosage to prevent pre-eclampsia and the complications that result therefrom during pregnancy in at-risk subjects. The initial studies were carried out in relatively small numbers, and over the years there have been variable interpretations as to the relative benefits of the aspirin therapy. There have even been occasional misleading statements on the Internet (www.aspirin-foundation.com/stories4.htm, accessed 18/9/01) giving the impression that aspirin may have a particularly high degree of benefit although the content of the story is usually more rational. An editorial by Barth (1998) in the *New England Journal of Medicine* highlighted the importance of timing of dosage and the other factors that should be considered in clinical trials of aspirin in pre-eclampsia.

A double-blind, randomised, placebo-controlled trial conducted in 240 pregnant women given aspirin 100 mg/day for various time periods showed that there was no effect of aspirin on blood pressure when taken on awakening, but this was evident at later times during the day, especially in individuals who had risks for hypertensive complications during pregnancy. The feature of this study is that the subjects received aspirin at 12 to 16 weeks of gestation, and it is probable that at this time those subjects at risk for developing hypertensive complications would show most marked responses to the drug. The authors emphasised in this study that doses of aspirin less than 80 mg/day did not affect placental thromboxane A_2 production, and the initiation of aspirin after 16 weeks' gestation could be a positive benefit, especially in subjects with pre-eclampsia (Hermida *et al.*, 1999).

The well-established precaution that aspirin (like other NSAIDs) should be only used with great caution and at the lowest effective dose during the first trimester of pregnancy (when) teratogenic effects might present a risk, as evidenced by studies in laboratory animal models – see Chapter 8) should perhaps be considered in the evaluation of safety versus efficacy in the use of aspirin for the prevention of pre-eclampsia and other conditions. An earlier study by Sibai *et al.* (1998) in patients with chronic hypertension who received low-dose aspirin indicated that women with elevated proteinuria at baseline were more likely to have premature delivery, and that this was dependent on the development of pre-eclampsia. The measurement of Doppler flow in the uterine arteries during the 17 to 23 week period in subjects who received 100 mg slow-release aspirin per day did not cause significant alteration in the incidence of pre-eclampsia or the delivery of a small-for-gestational-age baby. Although this negative result did highlight the relative safety of aspirin, it also had a benefit in that there was an reduction in the overall complications associated with utero-placental insufficiency. A more recent study by Goffinet *et al.* (2001) with an essentially similar study design also indicated that in subjects at risk for development of hypertension and pre-eclampsia there were no indicated benefits of aspirin taken at low dose.

Recent reviews and analysis of published data (Beaufils, 2000; Heyborne, 2000; Lee and Silver, 2000; Duley *et al.*, 2001) have highlighted the discrepancies between the various trials performed with aspirin, and overall there are considerations for having further clinical trials (Heyborne, 2000) on the use of early markers (Beaufils, 2000) and the potential for aspirin combinations, for example with heparin (Beaufils, 2000). In the detailed systematic review by Duley *et al.* (2001), a data analysis of the register of trials maintained by the Cochrane Pregnancy and Childbirth Group as well as in the Cochrane Control Trials Register and EMBSA suggest that the use of antiplatelet drugs was associated with a 15 per cent reduction in the risk of pre-eclampsia; this analysis covered 32 trials in 29 331 women. There was also an 8 per cent reduction in the risk of pre-term birth in 23 trials covering 28 268 women, and a 14 per cent reduction in foetal or neonatal death as evidenced in 13 selected trials covering 13 093 women. Small-for-gestational-age babies were reported in 25 trials covering 20 349 women. In spite of these small differences the relative risks were low, and in some cases did not achieve statistical significance.

Overall, therefore, it would appear that although there have been isolated reports highlighting the potential benefits of aspirin in pre-eclampsia (Chan *et al.*, 2001) much work needs to be undertaken in order fully to evaluate the potential risks and benefits, and especially the timing and dosage of administration of aspirin in relationship to other antihypertensive therapies and antiplatelet agents. Important considerations relating to concomitant disease as evidenced by the observations in patients with the antiphospholipid syndrome in systemic lupus erythematosus (Østenson, 1994; Østensen and Østensen, 1996; Østensen and Villiger, 2001) are important considerations for employing the antiplatelet effect of aspirin in these subjects, in whom inherent risks for development of lupus nephritis and other renal conditions attend the development of hypertensive responsiveness.

Recurrent miscarriage

Recent interest in the use of aspirin in prevention of recurrent miscarriage has been highlighted by the studies of Rai *et al.* (2000). This study evaluated the effects of low-dose aspirin (75 mg/day) on the live birth rate in women who had previously unexplained recurrent early miscarriage at less than 13 weeks' gestation, or in subjects with late pregnancy loss. While there was no apparent benefit to women with a history of recurrent early miscarriage, those who had suffered previous late miscarriage had an almost two-fold improvement in live birth rate. The reasons for these differential effects of aspirin are not apparent in this or previous studies.

Endometrial function and IVF treatment

Improvements in endometrial function in patients undergoing *in vitro* fertilisation have been studied for the responsiveness to aspirin administration (Harrington *et al.*, 2000). In the study by Rubinstein *et al.* (1999), treatment with a daily dose of 100 mg of aspirin compared with placebo approximately doubled the number of follicles, the number of oocytes retrieved, and the uterine and ovarian pulsality indices, and also resulted in the doubling of the pregnancy rate to 45 per cent and of the implantation rate. These studies have provided considerable encouragement for the use of aspirin, but the measurements for resistance in the ovarian vessels were subject to criticism (Sladkevicius (1999), with which the authors agreed in correspondence. In a later study, Hsieh *et al.* (2000) also showed significant improvements in endometrial function and a higher pregnancy rate, although this was less impressive than in the study by Rubinstein *et al.* (1999) (18.4 per cent compared to 9 per cent in controls).

Ductus arteriosus

A number of studies indicate that the prostaglandins, chiefly PGE_2, are primarily responsible for keeping the ductus arteriosus patent during growth of the foetus (Adeagbo *et al.*, 1982; Coceani and Olley, 1982). Use of aspirin and related NSAIDs in the final trimester, especially at 0.7 of gestation, may induce a risk of premature closure of the ductus and thus severely impair foetal circulation (Lines, 1977; Coceani and Olley, 1982). However, these drugs can be used therapeutically in the prematurely born infant to close the ductus (Coceani and Olley, 1982). A recent study by Grab *et al.* (2000) in which low-dose aspirin 100 mg/day was compared with placebo in a randomised double-blind trial indicated that when this drug was taken during the second and third trimester of pregnancy there were no alterations in utero-placental or foeto-placental haemodynamics, and there were no untoward outcomes in the development of moderate or severe constriction of the ductus arteriosus.

KAWASAKI DISEASE

Kawasaki disease, involving generalised vasculitis and various organo-pathies of unknown aetiology, was first described by Dr Tomasaku Kawasaki in 1967 in Japan as an acute febrile mucocutaneous lymph-node syndrome with accompanying fever, rash, conjunctivitis, inflammation of the oral mucosa, erythema, and swelling of the hands and feet with cervical adenitis (Onouchi and Kawasaki, 1999). Principal symptoms of this condition include fever that persists for 5 days or more, various peripheral inflammatory and oedematous conditions, and swelling of the lips, conjunctiva and tongue. There is also pronounced cervical lymphadenopathy. Following its initial description it was found to be a leading cause of acquired heart disease in children in both Japan and the USA, occurring with a higher frequency in boys than in girls and mostly in children under 5 years of age. While an aetiological agent has not been identified, much interest has focused on a microbial agent, perhaps centering on one of the enterotoxins produced by various strains of bacteria that lead to T-cell proliferation and cytokine release. While the condition is self-limiting, especially the febrile response, there is a high risk of mortality in individuals – presumably in those with MHC histocompatability associations.

The use of moderately low dosages of 30–50 mg/kg per day of aspirin in the acute phase of Kawasaki disease to control fever leads to shortening of the period of fever, but there are, as expected, risks of liver dysfunction and gastrointestinal haemorrhage associated with this drug. The use of intravenous gamma-globulin, although varying considerably with different formulations, is now a standard therapy, with reasonable outcomes in subjects with persistent coronary aneurisms. There are clear indications that 2 g/kg of intravenous gamma-globulin combined with at least 30–50 mg/day of aspirin provides maximum protection against the development of coronary abnormalities (Terai and Shulman, 1997). The recent application of the adenosine-5-diphosphate receptor platelet-activating inhibitor, ticlopidine, together with aspirin in successful application in a 7-month-old infant with Kawasaki disease complicated by thrombus of the giant coronary vessels indicated that it was not possible to use this combination successfully for thrombolytic therapy in this condition (O'Brien *et al.*, 2000). While there are few patients (relatively speaking) with Kawasaki disease worldwide, it is clear there is scope for improvement in the application of aspirin combined with other therapies in the treatment of this condition.

PERTURBED METABOLIC STATES

Salicylate, aspirin and some related drugs have long been known to have diverse (mostly inhibitory) effects on tissue and general body metabolism (Smith, 1966; Smith and Dawkins, 1971; Cohen, 1976; see Chapter 7). While interest has centered on the side or therapeutic effects of these drugs, it is possible that in some conditions the metabolic effects of the salicylates, including their capacity to inhibit prostaglandin production, could be exploited therapeutically. Here, interest has developed in the use of salicylates as part of the treatment of insulin diabetes, hyperthyroidism, Bartter's syndrome and various other conditions involving alteration to metabolism by the salicylates. Some speculative comments on possible uses in other diseases are also included in this section.

Diabetes mellitus

Salicylates have frequently been considered as possible antidiabetic agents, since Ebstein and Müller (1875) found that sodium salicylate 15 g daily reduced the urinary output of sugar in diabetic individuals. This observation was followed shortly afterwards by a series of papers from authors in Germany and other countries verifying that huge doses of salicylate (more than 5 g daily) were beneficial in diabetes (Ebstein, 1876; Ryba and Plumert, 1877; Foster, 1878; Williamson, 1901). No doubt the high doses of the drug also produced unpleasant side effects. This therapy never gained wide acceptance, even before the introduction of insulin (Smith and Dawkins, 1971). Although aspirin therapy for diabetes has been considered by several authors since the introduction of insulin, the interest has again waned (Smith and Dawkins, 1971). Part of the problem with using aspirin in insulin- and non-insulin-dependent diabetes is that the drug is regarded, along with many other drugs, as causing adverse effects on glucose homeostasis (Bressler and DeFronzo, 1997).

There have, however, been new insights into the mode of action of aspirin in stimulating insulin (and glucagon) secretion and enhancing glucose tolerance in diabetes (Giuliano *et al.*, 1978; 1981; Miccosi *et al.*, 1978; Tornvall and Allgen, 1980; McCarthy, 1981; Waitzman *et al.*, 1981; Ratzmann *et al.*, 1982a; 1982b; Seino *et al.*, 1982; Hundal *et al.*, 2002; Tran *et al.*, 2002). Moreover, a considerable number of studies (Gonzalez, 1980; Cotlier *et al.*, 1981; Davis *et al.*, 1997) have claimed that aspirin can

CHAPTER 13

also delay cataract formation that occurs in diabetic patients (Chew *et al.*, 1999) and, interestingly, in rheumatoid arthritics. Aspirin may also delay the progression of chronic renal failure in patients with diabetic nephropathy (Viberti *et al.*, 1994), and prevent cholesterol precipitation in subjects with gallstones (Adamek *et al.*, 1994; Bistran, 1994). Thus there has been considerable interest in the possibility of aspirin being used in the therapy of diabetes, and it is now regarded as part of the multifactorial interventions employed in the treatment of this disorder (Khan and Weir 1994; Lehman and Spinas, 2000).

The actions of aspirin in the control of glucose homeostasis in diabetic individuals are quite complex. Thus, Micossi *et al.* (1978) observed that treatment of insulin-dependent diabetics with 1 g/d aspirin for 3 days followed by 1 g on the fourth day caused a decrease in fasting serum glucose and the response to glucose (100 g), which was accompanied by an almost doubling of serum immunoreactive insulin. Reduction in serum glucose and increased insulin occurred in normal subjects and has been observed in Type 2 diabetics (Hundel *et al.*, 2002). Glucagon production had also increased markedly 30 minutes after the last dose of aspirin in diabetic compared with normal subjects, but it only caused a small increase that was evident over the entire time period. The serum free fatty acids and triglycerides were reduced in diabetic individuals, and only the former were slightly reduced in normal subjects. Similar effects of aspirin intake on glucose and insulin levels were observed by Giuliano *et al.* (1978) in normal subjects, and Chen and Robertson (1978) also observed that intravenous infusion of sodium salicylate to diabetic subjects caused glucose levels to return to normal and increase of insulin in the second phase of its secretion. Intravenous sodium salicylate was, however, shown to reduce plasma insulin in anaesthetised dogs (Vick-Mo *et al.*, 1978), so this contrasting action of salicylate may be related to the effects of anaesthetic.

Ratzman and co-workers (1982a; 1982b) showed that treatment with 3 g/d aspirin for 3 days followed by 1 g 1 hour prior to i.v. infusion of 8 or 16 mU/kg insulin did not alter the tissue sensitivity to this hormone in Type 1 diabetics although insulin resistance is overcome by the drug in Type 2 diabetics (Hundal *et al.*, 2002). Also, the mean C-peptide concentration was unaffected by the aspirin treatment. The metabolic clearance of insulin was slightly reduced by aspirin. In another study using the same treatment design, the levels of immunoreactive glucagon, growth hormone, non-esterified fatty acids and glycerol responses were unaffected by aspirin treatment. These studies confirmed the suggestion that the improvement of glucose tolerance by aspirin seems to be related to the biological effects of higher circulating insulin, but not to alterations in those endogenous hormones that negatively affect insulin responses.

Seino *et al.* (1982a; 1982b) also found that aspirin (1.5 g/d for 7 days with 1.5 g on the last day of dosage) enhanced insulin secretion in patients with mild insulin-dependent diabetes. This study also emphasised the requirement for maintaining aspirin dosage to achieve prolonged effects on insulin secretion and reduced glucose intolerance.

Enhancement of the effects of chlorpropamide (a drug used for treating diabetes insipidus) by salicylates has been shown to be due to the drug raising insulin levels and the lowering of cortisol (Richardson *et al.*, 1986).

Studies in normal rats have shown that aspirin 50–200 mg/kg reduces insulin levels (Arnold and Fernstrom, 1976; Kawashima *et al.*, 1980). Streptozotocin-diabetic rats given 25 mg/kg sodium salicylate in their drinking water for 1 day were found to have reduced glucagon and adrenaline responses to insulin hypoglycaemia (Patel and Skau, 1989). After 80–100 days continuous treatment with sodium salicylate the animals exhibited improvement in their diabetic state, as evidenced by body weight and plasma glucose. Part of the effects of salicylate may be due to reduced adrenaline-induced glycogenolysis (Miller *et al.*, 1985). Studies in the perfused rat pancreas preparation showed that 1–10 mmol/l sodium salicylate reduced insulin secretion, and the higher concentrations (5 and 10 mmol/l) reduced the basal but not glucose-stimulated glucagon secretion in the perfused rat pancreas (Garcia *et al.*, 1984). The authors of these studies suggested that the effects on insulin secretion were not related to cyclo-oxygenase inhibition (because salicylate is not a potent inhibitor of prostaglandin production), but other studies showed that salicylate inhibits COX activity (Evans *et al.*, 1983; Tran *et al.*, 2002). The results are, however, difficult to reconcile with the observations of increased insulin secretion in diabetic humans. The concentrations of salicylate employed in these studies were high (especially 5–10 mmol/l), and from these results may reflect toxic effects of the drugs.

However, it is known that salicylate or aspirin can induce hypoglycaemia in the alloxan-diabetic rats (and in diabetic individuals) and, conversely, also induce hyperglycaemia in normal rats and man (Smith and Dawkins, 1971). Interestingly, salicylamide and γ-resorcylic acid induce hyperglycaemia in the alloxan-diabetic rat, and benzoic, nicotinic and anthranilic acids induce hypoglycaemia (Smith and Dawkins, 1971). This suggests that the hypoglycaemic actions of salicylate reside primarily with the benzoic acid moiety.

Originally it was considered that the hypoglycaemic effects of salicylate were not due to effects of the drug on insulin release or actions (Smith and Dawkins, 1971), but more recent studies (Giuliano et al., 1978; 1981; Micossi et al., 1978; Tornvall and Allgen, 1980; Waitzman et al., 1980; Giuliano, 1981) indicate that this is most probable. It has been suggested (Guiliano et al., 1978; 1981) that aspirin affects insulin by inhibiting prostaglandin synthesis or affecting prostaglandin-regulated controls on glucose and arginine metabolism, the latter possibly being related to effects on NO metabolism (Nishio and Watanabe, 1998). The lack of effects of other NSAIDS (i.e. with their potent inhibitory actions on prostaglandin synthesis) on serum insulin levels in man (Micossi et al., 1978) would appear to mitigate against the major role of prostaglandins alone (Hirano et al., 1984) in the aspirin-induced release of insulin.

It is possible that salicylates reduce the production of glucose in diabetes by inhibiting intermediary metabolism and the uncoupling of oxidative phosphorylation in peripheral tissues (e.g. muscle fat deposits) that would normally act as sources for glucose (for review, see Smith and Dawkins, 1971). The suppression by the salicylates of the release of free fatty acids from adipose tissue is one important mechanism (Bizzi et al., 1965; Fang et al., 1968; Koyuncuoğlu et al., 1976; Vick-Mo et al., 1978; Balsare, 1981; Hundal et al., 2002). High toxic doses of salicylates may also induce metabolic ketosis (Mitchell et al., 1995), which is also evident in diabetics, so particular care is required concerning dosage of this drug. Salicylates may also exert actions on these control centres regulating sugar homeostasis in the brain and spinal cord. Gaitonde et al. (1967) found that direct infusion of sodium salicylate into the cerebral artery or ventricles induced hyperglycaemia in normal (i.e. non-diabetic) cats, but not when the drug was injected in other sites in the body. These authors found that sectioning of the mid-collicular region of the brainstem blocked the hyperglycaemic responses of salicylate, whereas severing of the lower spinal tracts was without effect (Gaitonde et al., 1967), and suggested that the site of action of salicylate might be in a region of the brain supplied by the vertebral artery. Such regional mechanisms could also be important in the hypoglycaemic responses to salicylate(s) in diabetes.

There have been attempts at modulating the activity of insulin by co-formulation with salicylates. The rationale behind this is that while there is no doubt that the recent biotechnological developments of human insulin will reduce the untoward immunological effects manifest with various preparations of insulin, it is possible that total dependence on the use of insulin by diabetics could be reduced by co-administration of aspirin.

A combination of sodium salicylate (50 mg) with an oral formulation of insulin (20 or 40 units) in capsules coated with a pH-dependent soluble polymer, Eudragit 5100, were found to have more potent and sustained hypoglycaemia than oral insulin in alloxan-diabetic rabbits (Hosny et al., 1998). A sodium salicylate thermo-reversible insulin liquid suppository has been found to produce lower cumulative ($AUC_{0\rightarrow 4h}$) plasma glucose levels in streptozotocin-diabetic rats than with insulin alone (Yun et al., 1999). The results suggest that oral or rectal diabetic therapy in insulin-dependent diabetes may benefit from the addition of salicylate, both from the direct actions of the drug on glucose metabolism and from the physicochemical consequences of the interaction of the drug with insulin. Likewise, the same argument could apply regarding the use of other hypoglycaemic agents (e.g. the sulphonylureas and biguanides).

Nowadays, the use of salicylates in diabetes is regarded more as a safety issue, i.e. in relation to glycaemic control (White et al., 1993), especially in non-insulin-dependent (Type 2) diabetes receiving sulphonyl-urea drugs (Krentz et al., 1994). However, the possibility that aspirin could delay the cataract formation so frequently encountered in diabetic patients is very important in the long-term therapy of this condition. Various authors (Cotlier and Sharma, 1981; Klein et al., 1987; Harding, 1992) have postulated that because salicylates compete with binding sites for tryptophan on albumin, aspirin will reduce (by the same mechanism) the binding of tryptophan (or its metabolite kynurenine) to lens

CHAPTER 13

proteins, so reducing the likelihood of tryptophan causing yellowing or browning of the lens, a visual manifestation of the cross-linking of lens proteins. Formation of advanced glycation end products leads to lens damage in streptozocin-diabetic rats (Heath *et al.*, 1996; Crabbe, 1998). In this model aspirin was found to have greater protection against formation of cataract than salicylate (Shastri *et al.*, 1998). This was related to the pronounced reduction in lens protein glycation by aspirin compared with salicylate. Aspirin also reduces the activity of aldose reductase, an enzyme implicated in the formation of advanced glycation end-products in diabetic cataracts (Gonzalez, 1980). Since the lens cannot oxidise excess glucose, it converts this sugar to sorbitol by the actions of aldose reductase. By inhibiting aldose reductase, aspirin prevents accumulation of toxic sorbitol (Gonzalez, 1980). The scavenging of oxyradicals by salicylates may also represent a mechanism for protection by aspirin (salicylate) against peroxidative damage of lipids occurring with cataract or other scar tissue damage (Beachy *et al.*, 1987; Riley and Harding, 1993). Protection against metal-ion catalysed oxidation of crystallin by aspirin has been observed in streptozotocin-diabetic rats (Jones and Hothersall, 1993). In well-controlled diabetic patients the plasma levels of 2,3-dihydroxybenzoic acid (a marker for oxidative stress) have been shown to be 23 per cent higher than in controls, thus supporting the concept that oxyradicals are involved in the pathogenses of chronic manifestations of diabetes (Ghiselli *et al.*, 1992). Salicylates may control the hydroxyl radicals that develop in the brains of hyperglycaemic rats (Li *et al.*, 1999). Aspirin may be superior to salicylate in therapeutic benefit in controlling these pathological manifestations (Pickup and Williams, 1997).

Aspirin may also prevent some of the vascular complications evident in diabetes, e.g. enhanced platelet aggregation from the supranormal thromboxane production (Jørgensen *et al.*, 1980; Halushka *et al.*, 1981; Krzywanek and Breddin, 1981) and erythrocyte adherence to endothelial cells (Wautier *et al.*, 1981), both of which could be important in the pathogenesis of retinopathy or other microangiographic changes in diabetes.

Thyroiditis and thyrotoxicosis

Salicylates mimic some effects of thyroxine, e.g. enhancement of body temperatures by uncoupling of oxidative phosphorylation (Austen *et al.*, 1958; Smith and Dawkins, 1971; Millhorn *et al.*, 1982). In contrast, they impair its action by: (a) displacing thyroxine (T_4) and the related thyrophenol, T_3 (3,5,3'-tri-iodothyronine) from their binding sites on plasma proteins, so reducing the levels of all active forms of this hormone in the blood and increasing their urinary excretion (Christensen, 1959; Good *et al.*, 1965; Larsen, 1972; Langer *et al.*, 1981; Barabetsky *et al.*, 1986; Kabadi and Danielson, 1987); (b) blocking the conversion of T_4, T_3 and reverse T_3 (3,5',3'-tri-iodothyronine (rT_3) to 3.3'di-iodothyronine by inhibiting 5'-iodothyronine deiodinase (Chopra and Solomon, 1980; Langer *et al.*, 1981); and (c) inhibiting the production of thyroid stimulating hormone (TSH)-regulated synthesis of thyroxine and related amino acids, and of thyroglobulin from the thyroid gland (Flock and Owen, 1965; Boeynaems *et al.*, 1975; Monarco *et al.*, 1977). By inhibiting the synthesis and enhancing elimination of thyroid hormones, the salicylates could be of value in the treatment of hyperthyroid states with the added benefit being that their minor thyroninomimetic action would enable some degree of tissue homeostasis to be retained. Thus Yamomoto and co-workers (1979) investigated the effects of aspirin 3–10 g daily compared with prednisolone 30–10 mg daily in the treatment of subacute thyroiditis with accompanying hyperthyroidism. With both treatments there was a marked improvement in clinical symptoms (fever, tenderness, struma, ESR) associated with reduced levels of TSH, T_4 and T_3 in the serum (Yamamoto *et al.*, 1979). While the effects of prednisolone were more immediate and pronounced than those of aspirin, the authors considered aspirin to be an effective treatment that exhibited none of the side effects evident with the steroid.

It is conceivable that higher doses of aspirin may be more effective in this condition and may also be useful adjuncts, with steroid and radioiodine, for treating the more advanced thyrotoxicosis state, thereby reducing the amounts and/or duration of treatments of these more toxic therapies.

Bartter's syndrome

In 1962, Bartter and co-workers described the eponymous syndrome in two patients who had pronounced retardation in growth with increasing circulating angiotensin-II, renin and aldosterone; and hypertrophy and hyperplasia of the renal juxtaglomerular apparatus (Bartter et al., 1962; 1977). Paradoxically, the blood pressure was normal but the pressor response to infusion of angiotensin-II was markedly reduced compared to that in normal individuals (Bartter et al., 1962). Since the original report, about 100 cases have been described in Negroes, Arabs, Caucasians and Orientals from a wide variety of geographical locations (Gill et al., 1976; Gill, 1980; Abdel-Ai et al., 1999), reflecting the apparent low incidence of this condition worldwide.

Since the initial description, the following additional abnormalities have been observed:

1. Impairment of pitressin-resistant urinary concentrating ability (Bartter et al., 1977; Gill, 1980)
2. Decreased pressor response to noradrenaline (Bartter, 1962; Gill, 1980)
3. Hypomagnesaemia (Bartter et al., 1962; Gill, 1980)
4. Hypouricaemia (Gill, 1980)
5. Increased output of prostaglandins E_2 and $F_{2\chi}$, 6-keto-prostaglandin $F_{1\alpha}$ (the hydration metabolite of PGI_2), but not of thromboxane B_2 (Gill et al., 1976; Bartter et al., 1977; Gill, 1980; Gullner et al., 1980)
6. A defect in both the first and second phases of platelet aggregation (Stoff et al., 1980).

While the aetiology of this syndrome has yet to be determined (Bartter et al., 1977; Baeler et al., 1980; Gill, 1980), there has been much interest in the possibility that the overproduction of prostaglandins by the kidney (Gill et al., 1976; Donker et al., 1977; Hsieb et al., 1978; Gill, 1980) and platelets (Stoff et al., 1980) could be a central feature of this condition. This could merely be a manifestation of increased turnover of membrane phospholipids in the hyperplastic state of the kidney evident in individuals with this syndrome. Other suggestions of a primary pathological cause include a defect reabsorption of sodium and chloride in the ascending tubules (Fichman et al., 1976; Gill and Bartter, 1978; Baeler et al., 1980; Carmine et al., 1982), or an imbalance of the three vasoactive hormonal systems, namely the kallikrein–kinin, renin–angiotensin and prostaglandin synthesis systems (McGiff, 1977).

While indomethacin has been the most extensively used drug in the treatment of Bartter's syndrome (Fichman et al., 1976; McGiff, 1977; Hsieb et al., 1978; Gullner et al., 1980), other inhibitors of prostaglandin synthesis (including aspirin) provide effective treatment (Norby et al., 1976; Littlewood et al., 1978; Ishikawa and Wada, 1979; Villa, 1980). In addition to inhibiting prostaglandin production, aspirin and indomethacin also reduce renin production and may have other actions contributing to their beneficial effects upon this condition (McGiff, 1977; Rado et al., 1978; Romero et al., 1978; Brooks et al., 1980).

Obesity

It has been suggested that salicylates could be used in the treatment of obesity as a consequence of the catabolic effects of these drugs. Obesity conditions are characterised by an abnormally high rate of metabolism of certain body constituents, and salicylates may attenuate this. This suggestion is entirely speculative, and is based on the known metabolic effects of salicylate itself.

It is well known that prolonged oral administration of high doses of aspirin or salicylate ($>200\,mg/kg$ per day) leads to significant weight loss in laboratory animals. Since salicylate itself exerts catabolic effects (Smith and Dawkins, 1971), the use of this drug in preference to its more ulcerogenic congener aspirin would seem obvious. Some of the catabolic effects of aspirin and salicylate arise from uncoupling of oxidative phosphorylation. Historically, the potent uncoupling agent 2,4-dinitrophenol was once considered for the treatment of obesity, but its toxicity and its loss of activity by extensive metabolism to amino derivatives have been major factors against the use of this compound. The mitochondrial uncoupling actions and lower toxicity of the salicylates would put them at a special advantage. While there are theoretical objections against the use of the salicylates because they inhibit fatty acid and triglyceride release from fat depots, this is overcome by the inhibition by these drugs of fatty acid synthesis. The genetically obese 'Zucher' strain of rats has hyperlipidaemia, hyperinsulinaemia and hyper-response platelets (Landgraf-Leurs *et al.*, 1981) and it may be possible to examine the effects of salicylates in this model.

Cystic fibrosis

Cystic fibrosis, or mucoviscidosis, is characterised by a hyperproduction of respiratory mucus (leading inevitably to marked respiratory distress) and a defect in sodium secretion (Harries, 1978; Davis and Di Sant'Agnese, 1980). The actions of salicylate or aspirin in inhibiting mucus synthesis (Rainsford, 1982) might also be beneficially exploited, perhaps in conjunction with the current therapy using a mucolytic agent such as *N*-acetylcysteine. The effect of aspirin on prostaglandin-mediated regulation of sodium secretion (Horton, 1973) as well as transcription of the cystic fibrosis transmembrane conductance protein (Tondelier *et al.*, 1999), might also be of some benefit.

IMMUNOLOGICAL EFFECTS

Prevention of allograft rejection

MacDonald and co-workers showed in 1970 that aspirin had beneficial effects against the hyperacute rejection of renal allografts in presensitised dogs (MacDonald *et al.*, 1970). They proposed that aspirin acts by its 'antiplatelet' activity (Thornell *et al.*, 1979), a concept that has since gained much support. It is also possible that the drug may exert beneficial effects inhibiting lymphocyte proliferation and other possible immunosuppressive effects (Gale, 1966; Opelz *et al.*, 1973; Smith *et al.*, 1975; Panush and Anthony, 1976; Gabourel *et al.*, 1977; Egorin *et al.*, 1978; Mobarok *et al.*, 1980). Aspirin may also protect ischaemically-injured kidneys from tissue damage (Matsuno *et al.*, 1997).

Multiple sclerosis

Lymphocyte adherence to measles-virus infected epithelial cells is substantially increased in patients with multiple sclerosis (MS) compared with normal individuals, and this is blocked both *in vitro* and *in vivo* by aspirin in a dose-dependent fashion (Dore-Duffy and Zurier, 1979a; 1979b; 1986; Offner *et al.*, 1982). The studies by Dore-Duffy and Zurier (1979a; 1979b) showed that lymphocytes from multiple sclerosis (MS) patients formed less avid E-rosettes with sheep red blood cells (SRBC), and they suggested that this and the adherence to measles-infected cells were prostaglandin-related mechanisms. Later studies by these authors implicated PGE_2-dependent cylic AMP production in lymphocytes as well as in monocytes from MS patients as a mechanism (Dore-Duffy and Zurier, 1986; 1991). The implication is that aspirin, by inhibiting PGE_2, would reverse the effects of PGE_2. although the drug can directly affect cyclic AMP production (Chapter 7).

Since there is a suggestion that disturbances of immunological functions underlie the development of MS (Berkow, 1977), the effects of aspirin in lymphocyte adherence (Dore-Duffy and Zurier, 1979), proliferation or other lymphocyte functions (Gale, 1966; Griswold and Ufeki, 1969; Opelz *et al.*, 1973; Smith *et al.*, 1975; Panush and Anthony, 1976; Gabourel *et al.*, 1977; Egorin, 1978; Cavallini *et al.*, 2001) may prove of some benefit in this condition. Likewise, aspirin or other salicylates may have some value in the treatment of autoimmune conditions by influencing lymphocyte functions (Gale, 1966; Griswold and Ufeki, 1969: Opelz *et al.*, 1973; Smith *et al.*, 1975; Panush and Anthony, 1976; Gabourel *et al.*, 1977; Egorin *et al.*, 1978;

Gerli *et al.*, 1998; Matasic *et al.*, 2000; Cavallini *et al.*, 2001). While the effect may not be expected to be as dramatic as that obtained with those immunoregulatory drugs currently used to treat these conditions, there may be specific benefits from incorporating salicylates into the therapeutic regimen used in such states. This is obviously one interesting area that requires more detailed investigation.

Antiviral and antiparasitic activities

Aspirin and salicylate will inhibit replication of several viruses (Kochamn *et al.*, 1965; Inglot, 1969; Dennis, 1979), but their efficacy compared to other antiviral agents is poor. The mechanism of action of these drugs may involve the inhibition of nucleic acid synthesis (as observed in mammalian tissues; Janakidevi and Smith, 1970), but has yet really to be determined. There may also be stimulatory effects of the drug on the immune system, including interferon-γ (IFN-γ) production, which may confer protection against cell invasion by viruses (Geber *et al.*, 1975). Combination of aspirin 10 mg/kg or ibuprofen 12 mg/kg with IFN-γ was found to reduce the numbers of viable *Listeria monocytogenes* in the spleen and liver of infected mice (Hockertz *et al.*, 1993; Hockertz and Heckenberger, 1996).

Growth of *Schistosoma mansoni* has been found to be affected by 4-iodoacetamidosalicylic acid, possibly as a consequence of it inhibiting lactate dehydrogenase and other metabolic enzymes (Tarrant and O'Hare, 1967). While sodium salicylate is ineffective against the Brazilian and Colombian strains of *Trypanosoma cruzi* (Duncanson *et al.*, 1978), it does appear to be effective against *T. rhodesiense* infections in mice (Vaughan, 1969).

OTHER POSSIBLE USES

Diarrhoea

Collier and coworkers, (Collier, 1974; Mennie *et al.*, 1975) suggested that aspirin could be used in the treatment of diarrhoea following radiation therapy or gastroenteric infections. The rationale for using aspirin is that it would inhibit the hyperproduction of prostaglandins in the gastrointestinal tract, which could be the cause of the diarrhoea (Collier, 1974). Several clinical reports attest to the efficacy of aspirin in treating diarrhoea from gastrointestinal infections (Editorial, 1980; Nalin, 1980; Stek, 1980). A bismuth sub-salicylate preparation (e.g. Pepto-bismol®, Norwich Eaton Pharmaceuticals) is currently used to treat diarrhoea, especially in children (Kruesi, 1980; Steinhoff *et al.*, 1980; Pickering *et al.*, 1981). There has been concern about the extensive bismuth and also possibly salicylate absorption, since originally it was believed that the drug was not absorbed from the gastrointestinal tract (Pickering *et al.*, 1981). Other NSAI drugs have also been used to treat diarrhoea (Buckhave and Rask Madsen, 1980).

Whilst there is no doubt about the efficacy of these drugs, there is some concern about the possibility of severe mucosal damage as a consequence of using these irritant drugs when the gastrointestinal tract is in a highly 'sensitised' state. One way around this might be to employ pro-drugs of salicylate acids that are non-irritant to the mucosa, but that generate an active prostaglandin synthesis inhibitor following absorption. Such drugs include meseclazone and various alkyl or glycoaryl esters of aspirin (see Chapter 3, and Rainsford and Willis, 1982).

Sickle cell anaemia

This inherited condition is frequently seen in Black African populations, where, as a consequence of the amino acid valine being substituted for glutamic acid on the sixth peptide of the β chain of haemoglobin (Berkow, 1977), abnormal sickle-cell haemoglobin (HbS) is produced.

On deoxygenation in peripheral tissues HbS becomes much less soluble, forming a semisolid gel that causes internal aggregation of red blood cells (RBCs), so causing the appearance of typical sickle-shaped cells (Berkow, 1977). The distorted RBCs are unable to pass through small capillaries, and plugs of these cells lead to thrombosis and infarction (Berkow, 1977). Individuals homozygous for this condition rarely survive beyond 40 years of age, and when younger frequently succumb to severe anaemia and an 'anplastic crisis' in which RBC production slows down during acute infarctions (Berkow, 1977). Interest in the possibility that aspirin could be used to treat this condition (see also Chapter 3) was aroused in 1973, when Klotz and Tam (1973) showed that aspirin acetylated HbS and coincidentally increased the oxygen affinity of HbS-bearing erythrocytes and even lysates thereof. The possibility was considered that there could be a shift in the oxygen dissociation curve for HbS resembling that observed following carbamylation of functional groups by cyanate, a compound that is used therapeutically in sickle cell anaemia (Klotz and Tam, 1973). Later studies (De Furia *et al.*, 1973) confirmed the acetylation of HbS by aspirin, but showed that the drug did not alter the oxygen dissociation curve or affinity of O_2 for HbS. It has been established that aspirin acetylates a variety of sites on the β chains of HbS, with most of the acetyl groups attached to the β-Lys-65, β-Lys-144 and β-Lys-90 residues. Rabinowitz *et al.* (1974) established from light-scattering studies that aspirin retarded sickling of HbS-bearing erythrocytes. Later in the search for more powerful sickling agents, Klotz's group (Walder *et al.*, 1977) found that acetyl-3,5-dibromosalicylic acid (i.e. 3.5 dibromo-aspirin) was an effective anti-sickling agent. This drug increased the minimum gelling concentration of HbS in parallel with a reduction in sickling (Walder *et al.*, 1977; 1979). The authors proposed that the two electron-withdrawing bromine atoms at the 3 and 5 positions of this molecule would give it greater acetylating potential than aspirin, and hence more potent anti-sickling activity (Walder *et al.*, 1977). Other halo-aspirins were also potent anti-sickling agents, but the acyl-(fluorophenyl)-salicylates were not (Vogt, 1980).

Another approach has been to employ reagents to cross-link peptide regions of HbS so as to prevent intermolecular interactions that occur during HbS gelation (Maugh, 1981). Various approaches have been adopted (Chaplain *et al.*, 1980). Some promising reagents include bis(3,4-dibromosalicyl) succinate and bis(3,5-dibromosalicyl) fumarate (Walder *et al.*, 1977). These salicylates, as well as aspirin, interfere with the binding of the Hb-oxygenation regulation, 2,3-diphosphoglycerate (Walder *et al.*, 1980; Wood *et al.*, 1981).

Earlier promising results were obtained in a clinical trial to show the effectiveness in controlling sickling with aspirin (Chaplain *et al.*, 1980; Osamo *et al.*, 1981). Since then some salicylates (i.e. of diaspirin or dihalo-aspirin) and related analogues have been studied as potential therapies for clinical use in the treatment of sickle-cell disease (Chapter 3). Overall however, clinical trials with diaspirin cross-lined haemoglobin preparations in blood transfusion and haemorrhagic shock have been disappointing (Przybelski *et al.*, 1999; Sloan *et al.*, 1999; Lamy *et al.*, 2000). The negative or poor outcomes in these studies may relate to the effects of haemoglobin in scavenging nitric oxide, participating in free-radical reactions, activation of the immune system and its neurotoxicity (Hess, 1996).

Thrombotic thrombocytopenic purpura

Thrombotic thrombocytopenic purpura (TTP) is a potentially fatal disease owing to abnormal aggregation of platelets and a protein, von Willebrand factor (vWF), in the small blood vessels of a number of organs, including spleen and kidneys (Fosang and Smith, 2001). Clots of proteins and platelets cause fragmentation of red cells, leading to serious neurological and renal abnormalities, anaemia, and a low platelet count (thrombocytopenia). It is accompanied by a reduced erythrocyte half-life, symptoms of lethargy, gastrointestinal distress, hepatomegaly, hypertension, proteinuria, haematuria and erythrocyte haemolysis (Berkow, 1977). The accumulation of vWF multimers in the blood in TTP originates from mutants in the zinc-proteinase, ADAMTS13, that cleaves these large aggregates, thus upsetting the balance of the interactions between platelets and between platelets and blood vessel walls (Fosang and Smith, 2001; Levy *et al.*, 2001).

The antiplatelet effects of aspirin have been employed for treatment of TTP (Birgens *et al.*, 1979; Peterson *et al.*, 1979). It is probably the acetylation of platelets that protects them as well as erythrocytes from destruction. It should be noted that administration of corticosteroids, plasmapheresis and splenectomy are also concomitant therapies employed in this condition (Berkow, 1977; Petersen *et al.*, 1979; Meyers, 1980).

Dermatological conditions

Salicylic acid has long been successfully used as a keratolytic agent to control callosities, hyperkeratosis and warts (Gross and Greenberg, 1948; Weirich, 1975). This drug has also been applied to control a whole variety of other conditions, including various forms of acne, athlete's foot (*Trichopyton* spp.) eczema, erysipelas, favus, and even dandruff (Gross and Greenberg, 1948; Elie and Durocher, 1983).

Elie and Durocher (1983) have shown that a 2 per cent salicylic acid lotion (in an alcohol base) was as effective as 0.05 per cent betamethasone-17,21 dipropionate (in the same base) in reducing the pruritus and scaling in erythematous squamous dermatosis of the scalp. The combination of the steroid and salicylic acid was noticeably superior to the individual drugs (Elie and Durocher, 1983), and also controlled the redness, which was not achieved with the separate drugs (Elie and Durocher, 1983). These results are of interest not only practically but also in illustrating the differing modes of action of these drugs. Other dermatological conditions (e.g. psoriasis) are also well controlled by such steroid–salicylic acid mixtures (Hillström *et al.*, 1982; Elie and Durocher, 1983), so there appear to be distinct advantages in this combination therapy. Salicylic acid appears to act on the stratum corneum by sloughing of corneocytes, perhaps reducing cell division (Roberts *et al.*, 1980). The drug itself does not influence cell division (Roberts *et al.*, 1980).

Salicylic acid and other salicylates exert powerful static effects against Gram-negative and Gram-positive bacteria, pathogenic yeasts, dermatophytes, moulds and other microbes (Gross and Greenberg, 1948; Weirich, 1975; Weirich *et al.*, 1975). This could represent an important part of the dermatological actions of this drug. The prevention of epithelial hyperplasia is also an important dermatological effect of salicylic acid (Weirich, 1975; Weirich *et al.*, 1975).

The potential for salicylate in modulating the skin carcinogenicity of UV light has been highlighted by studies in which the drug inhibited NFκB gene activation in cells lacking p53 (Paunesku *et al.*, 2000). This may be a useful property of salicylates in sunscreen preparations (Jiang *et al.*, 1997).

Iontophoretic application of salicylates to skin (Roberts *et al.*, 1998; Murakami *et al.*, 1999), the development of ion-pair systems (Megwa *et al.*, 2000) and ethanol/propylene glycol formulations (Levang *et al.*, 1999) have been explored as means of topical delivery of these drugs. Methyl salicylate has been employed as a stimulant to predict the absorption through the skin of sulphur mustard (Riviere *et al.*, 2001).

CHAPTER 13

MISCELLANEOUS

Salicylates have been suggested for the treatment of a variety of other conditions. Their mention here is not to say that they have proved efficacious or are to be considered the drug(s) of choice in a particular condition. However, this may give stimulus for further investigation.

Aspirin has been suggested as a treatment in endotoxic shock (Halushka *et al.*, 1981), and biliary dysfunction (Gross and Greenberg, 1948; McDonald *et al.*, 1970) – all conditions in which inhibition of the overproduction of prostaglandins appears to be of major significance. Stimulation of bile flow (choleresis) by salicylates may overcome biliary disorders, and this may be mediated by mechanisms other than those involving prostaglandins (e.g. metabolic formation, ATP productions) (Schmidt *et al.*, 1938; Bullock *et al.*, 1970; Pugh and Ruthishauser, 1978).

Salicylate has been shown to inhibit the attachment of the amoeba *Acanthamoeba castetellanii* to contact lenses (Tomlinson *et al.*, 2000), a procedure that may have practical application for reducing microbial infections in contact lens solutions.

Salicylates were reported in early literature (Gross and Greenberg, 1948) as being useful in venereal disease, schizophrenia, epilepsy and prostatitis. These drugs have also been reported to be useful in deodorising bowels, inducing hypnosis, and as a sedative (Gross and Greenberg, 1948). The actions of the salicylates in such states are probably weak, and derive essentially from the anti-inflammatory, antipyretic and/or analgesic actions, and not as a consequence of specific actions in these conditions.

The beneficial effects in some of these states (e.g. schizophrenia) is questionable. However, prostaglandin hyperproduction has been implicated in manifestations of schizophrenia (Horribin *et al.*, 1978; 1992; Langer *et al.*, 1981). This would appear to be an example of the possibility of predicting the use of aspirin and other salicylates, even though they have such diverse effects. The nature of their pathological consequences may be such that many common features are evident (e.g. inflammation, pain). Drugs used to control these manifestations are going to be effective therapies for control of symptoms, if not causes, regardless of the origin of the pathological state.

REFERENCES

Abdel-Al, Y.K., Badawi, M.H., Yaeesh, S.A., Habib, Y.Q., Al-Khuffash, F.A., Al-Ghanim, M.M. and Al-Najidi, A.K. 1999, Bartter's syndrome in Arabic children: review of 13 cases. *Pediatrics International*, **41:** 299–303.

Acarin, L., Gonzalez, B. and Castellano, B. 2002, Decrease of proinflammatory molecules correlates with neuroprotective effect of the fluorinated salicylate triflusal after postnatal excitotoxic damage. *Stroke* **33:** 2499–2505.

Adamek, H.E., Buttmann, A., Weber, J. and Riemann, J.F. 1994, Can aspirin prevent gallstone recurrence after successful extracorporeal shockwave lithotripsy? *Scandanavian Journal of Gastroenterology*, **29:** 355–359.

Adeagbo, A.S., Coceani, F. and Olley, P.M. 1982, The response of the lamb ductus venosus to prostaglandins and inhibitors of prostaglandin and thromboxane synthesis. *Circulation Research*, **51:** 580–586.

Alich, A.A., Welsh, V.J. and Wittmers, L.E. Jr 1983, Comparison of aspirin and copper aspirinate with respects to gastric mucosal damage in the rat. *Journal of Pharmaceutical Sciences*, **72:** 1457–1461.

Alzheimer, A. 1907, About a peculiar disease of the cerebral cortex. Allgemeine Zeitschrift für Psychiatrie und Psyschist. *Gerichtliche Medizin*, **64:** 146.

Andersen, K., Launer, L.J. Ott, A., Hoes, A.W., Breteler, M.M.B. and Hofman, A. 1995, Do nonsteroidal anti-inflammatory drugs decrease the risk for Alzheimer's disease? The Rotterdam Study. *Neurology*, **45:** 1441–1445.

Anthony, C.R. and Panush, R.S. 1978, Interaction of [14C]-acetylsalicylic acid with normal human peripheral blood lymphocytes. *Clinical and Experimental Immunology*, **31:** 482–489.

Arnold, M.A. and Fernstrom, J.D. 1976, Salicylate reduces serum insulin concentrations in the rat. *Life Sciences*, **19:** 813–818.

Arvind, P., Qiao, L., Papavassiliou, E., Goldin, E., Duceman, B. and Rigas, B. 1996, Aspirin and aspirin-like drugs induce HLA-DR expression in HT29 colon cancer cells. *International Journal of Oncology*, **8:** 1207–1211.

Austen, F.K., Rubini, M.E., Meroney, W.H. and Wolff, J. 1958, Salicylates and thyroid function. I. Depression of thyroid function. *Journal of Clinical Investigation*, **37:** 1131–1143.

Austermann, M., Grotemeyer, K.H., Evers, S., Rödding, D. and Husstedt, I.W. 1998, The influence of acetylsalicylic acid on cognitive processing: an event-related potentials study. *Psychopharmacology*, **138:** 369–374.

Baehler, R.W., Work, J., Kotchen, T.A., McMorrow, G. and Guthrie, G. 1980, Studies on the pathogenesis of Reye's syndrome. *American Journal of Medicine*, **69:** 933–983.

Baldwin, G.S. 2000, Do NSAIDs contribute to acute fatty liver of pregnancy? *Medical Hypothesis* **54:** 846–849.

Baquial, J.G. and Sorenson, J.R. 1995, Down-regulation of NADPH-diaphorase (nitric oxide synthase) may account for the pharmacological activities of Cu(II)2 (3,5-diisopropylsalicylate)4. *Journal of Inorganic Biochemistry*, **60:** 133–148.

Barabetsky, N.G., Chertow, B.S., Webb, M.D., Leonard, R.F. and Sivitz, W.I. 1986, Combined phenytoin and salicylate: effects on thyroid function tests. *Archives Internationales de Pharmacodynamie et de Therapie*, **284:** 166–176.

Barnes, C.J., Hardman, W.E., Cameron, I.L. and Lee, M. 1997, Aspirin, but not sodium salicylate, indomethacin, or nabumetone, reversibly represses 1,2-dimethylhydrazine–induces colonic aberrant crypt foci in rats. *Digestive Diseases and Sciences*, **42**: 920–926.

Barth, W. 1998, Low-dose aspirin for pre-eclampsia – the unresolved question. *New England Journal of Medicine* **338**: 756–757.

Bartter, F.C., Pronove, P., Gill, J.R. and MacCardle, R.C. 1962, Hyperplasia of the juxtaglomerular complex with hyperaldosteronism and hypokalemic alkalosis. *American Journal of Medicine*, **33**: 811–828.

Bartter, F.C., Gill, J.R. Jr and Frolich, J.C. 1977, Bartter's syndrome. *Advances in Nephrology from the Necker Hospital*, **7**: 191–198.

Beachy, N.A., Morris, S.M., Richards, R.D. and Varma, S.D. 1987, Photoperoxidation of lens lipids: inhibition by aspirin. *Photochemistry and Photobiology*, **45**: 677–678.

Beall, J.R. and Klein, M.F. 1977, Enhancement of aspirin-induced teratogenicity by food restriction in rats. *Toxicology and Applied Pharmacology* **39**: 489–495.

Beaufils, M. 2000, Aspirin and prevention of pre-eclampsia. *Revue de Medicine Interne*, **21** (**Suppl. 1**): 68S–74S.

Belsare, S.D. 1981, Studies on glucose tolerance during fat mobilization: influence of heparin and salicylate. *Indian Journal of Experimental Biology*, **19**: 88–89.

Benamouzig, R., Chaussade, S., Little, J., Munoz, N., Rautureau, J. and Couturier, D. 1997, Aspirin, NSAIDs and colorectal carcinogenesis. *Gastroenterologie Clinique et Biologique*, **21**: 188–196.

Benamouzig, R., Deyra, J., Martin, A., Girard, B., Jullian, E., Piednoir, B., Coururier, D., Coste, T., Little, J. and Chaussade, S., for the Association pour la Prévention par l'aspirine du Cancer Colorectal Study Group. 2003, Daily soluble aspirin and prevention of colorectal adenoma recurrence: One-year results of the APACC trial. *Gastroenterology*, **125**: 328–336.

Berkel, H.J., Holcombe, R.F., Middlebrooks, M., Kannan, K. 1996, Non-steroidal anti-inflammatory drugs and colorectal cancer. *Epidemiologic Reviews*, **18**: 205–214.

Berkow, R. (ed.) 1977, *The Merck Manual of Diagnosis and Therapy*. Rahway, NJ: Merck, Sharp & Dohme Research Laboratories Ltd.

Berthon, G. (ed.) 1992, *Handbook of Metal–Ligand Interactions in Biological Fluids*, Vols I and II. New York: Marcel Dekker.

Betts, H., Cleland, L. and Whitehouse, M.W. 1981, In vitro studies of anti-inflammatory copper complexes: some difficulties in their chemical interpretation. *In:* J.R.J. Sorenson (ed.), *Inflammatory Diseases and Copper*, pp. 553–562. New Jersey: Humana Press.

Beveridge, S.J. 1989, Copper therapy of inflammatory disorders: efficacy and biodistribution of topically applied copper complexes. *In:* R. Milanino, K.D. Rainsford and G.P. Velo (eds), *Copper and Zinc in Inflammation*, Vol. IV *Velo Inflammation and Drug Therapy Series*. Dordrecht: Kluwer Academic Publishers.

Bizzi, A., Garattini, S. and Veneroni, E. 1965, The action of salicylate in reducing plasma free fatty acids and its pharmacological consequences. *British Journal of Pharmacology and Chemotherapy*, **25**: 187–196.

Boeynaems, J.M., Van Sande, J. and Dumont, J.E. 1975, Blocking of dog thyroid secretion *in vitro* by inhibitors of prostaglandin synthesis. *Biochemical Pharmacology*, **24**: 1333–1337.

Bondareff, W. 1994, Subtypes of Alzheimer's disease. *In:* A. Burns and R. Levy (eds), *Dementia*, pp. 101–125. London: Chapman & Hall Medical.

Booth, B.L.K., Pitters, E., Mayer, B. and Sorenson, J.R.J. 1997, Down-regulation of porcine heart diaphorase by trimagnesium hexakis(3,5-di-isopropyl-salicylate, and down-regulation of nitric oxide synthase reactivity by $Mn_3(3,5\text{-DIPS})_6$ and $Cu(II)_2(3,5\text{-DIPS})_4$. *Metal-based Drugs*, **6**: 111–120.

Bouchaut, G.S., Brumas, V. and Berthon, G. 2000, Copper-ligand interactions and physiological free radical processes. Part 3. Influence of histidine, salicylic acid and anthranilic acid on copper-driven fenton chemistry in vitro. *Free Radical Research*, **32**: 451–461.

Breitner, J.C.S. 1996, The role of anti-inflammatory drugs in the prevention and treatment of Alzheimer's disease. *Annual Review of Medicine*, **47**: 401–411.

Breitner, J.C.S. 1999, Alzheimer's disease. *In: Abstracts of the International Conference on Inflammopharmacology and 6th Symposium on Side-Effects of Anti-Inflammatory Drugs. Chateau Elan, Braselton, Georgia, USA. 23–26 May.*

Breitner, J.C.S., Gau, B.A., Welsh, K.A., Plassman, B.L., McDonalds, W.M., Helms, M.J. and Anthony, J.C. 1994, Inverse association of anti-inflammatory treatments and Alzheimer's disease. *Neurology*, **44**: 227–232.

Breitner, J.C.S., Welsh, K.A., Helms, M.J., Gaskell, P.C., Gau, B.A., Roses, A.D., Pericakvance, M.A. and Saunders A.M. 1995, Delayed onset of Alzheimer's disease with nonsteroidal anti-inflammatory and histamine H_2 blocking drugs. *Neurobiology of Aging*, **16**: 523–530.

Bressler, P. and DeFronzo, R.A. 1997, Drug effects on glucose homeostasis. *In:* K.G.M.M. Alberti, P. Zimmert, R.A. DeFronzo and H. Keen (eds), *International Textbook of Diabetes Mellitus*, 2nd edn, pp. 230–250. Chichester: Wiley.

Breteler, M.M.B., Bots, M.L., Ott, A. and Hofman, A. 1998b, Risk factors for vascular disease and dementia. *Haemostasis*, **28**: 167–173.

Breteler, M.M.B., Ott, A. and Hofman, A. 1998a, The new epidemic: frequency of dementia in the Rotterdam study. *Haemostasis*, **28**: 117–123.

Brooks, P.M., Cossum, P.A. and Boyd, G.W. 1980, Rebound rise in renin concentrations after cessation of salicylates. *New England Journal of Medicine*, **303**: 562–564.

Bucher, C., Jordan, P., Nickeleit, V., Torhorst, J. and Mihatsch, M.J. 1999, Relative risk of malignant tumors in analgesic abusers. Effects of long-term intake of aspirin. *Clinical Nephrology*, **51**: 67–72.

Bukhave, K. and Rask-Madsen, J. 1980, An approach to evaluation of local intestinal PG production and clinical assessment of its inhibition by indomethacin in chronic diarrhea. *Advances in Prostaglandin Thromboxane Research*, **8**: 1627–1631.

Burke, V., Gracey, M. and Suharyono, S. 1980, Reduction by aspirin of intestinal fluid-loss in acute childhood gastroenteritis. *Lancet*, **1**: 1329–1330.

Burn, J., Alonso, A., Bertario, L. and Bishop, T. 1998, Concerted action polyp prevention, CAPP2: a trial of aspirin and/or resistant starch in people at risk of hereditary non-polyposis colon cancer (HNPCC). *In:* S.S. Baig (ed.), *Cancer Research Supported under BIOMED 1*, pp. 17–23. Brussels: IOS Press.

Burns, A. and Levy, R. (eds) 1994, *Dementia*. London: Chapman & Hall Medical.

Bus, P.J., Nagtegaal, I.D., Verspaget, H.W., Lamers, C.B.H.W., Geldof, H., Van Krieken J.H.J.M. and Griffioen, G. 1999, Mesalazine-induced apoptosis of colorectal cancer: on a verge of a new chemo-preventive era? *Alimentary Pharmacology and Therapeutics*, **13**: 1397–1402.

Carmine, Z., Ettore, B., Giuseppe, C. and Quirono, M. 1982, The renal tubular defect of Bartter's syndrome. *Nephron*, **32**: 140–148.

Castelao, J.E., Yuan, J.M, Gago-Dominguez, M., Yu, M.C. and Ross, R.K. 2000, Non-steroidal anti-inflammatory drugs and bladder cancer prevention. *British Journal of Cancer*, **82**: 1364–1369.

Cavallini, L., Francesconi, M.A., Zoccarato, F. and Alexandre, A. 2001, Involvement of nuclear factor-kappa B (NF-kappaB) activation in mitogen-induced salicylates acting as NF-kappaB inhibitors. *Biochemical Pharmacology*, **62**: 141–147.

Chan, W.S., Chunilal, S.D. and Ginsberg, J.S. 2001, Antithrombotic therapy during pregnancy. *Seminars in Perinatology*, **25**: 165–169.

Chaplin, H. Jr 1980, Hematuria in hemoglobin S disorders. *Archives of Internal Medicine*, **140 (12)**: 1573.

Chen, M. and Robertson, R.P. 1978, Restoration of the acute insulin response by sodium salicylate. *Diabetes*, **27**: 750–756.

Chew, E.Y., Benson, W.E., Remaley, N.A., Lindley, A.A., Burton, T.C., Csaky, K. and Williams, G.A. 1999, Results after lens extraction in patients with diabetic retinopathy: Early Treatment Diabetic Retinopathy Study Report Number 25. *Archives of Opthalmology*, **117**: 1600–1606.

Chiu C.-H., McEntee, M.F. and Whelan, J. 1997, Sulindac causes rapid regression of preexisting tumors in $-/+$ mice independent of prostaglandin biosynthesis. *Cancer Research*, **57**: 4267–4273.

Chiu, C.H., McEntee, M.F. and Whelan, J. 2000, Discordant effect of aspirin and indomethacin on intestinal tumor burden in Apc$^{-/+}$ mice. *Prostaglandins, Leukotrienes and Essential Fatty Acids*, **62**: 269–275.

Christensen, L.K. 1959, Thyroxine-releasing effect of salicylate and of 2,4-dinitrophenol. *Nature*, **183**: 1189–1190.

Chun-Hung, C., McEntee, M.F. and Whelan, J. 1997, Sulindac causes rapid regression of preexisting tumours in $-/+$ mice independent of prostaglandin biosynthesis. *Cancer Research*, **57**: 4267–4273.

CLASP Collaborative Group 1995, Low-dose aspirin in pregnancy and early childhood development: follow up of the collaborative low-dose aspirin study in pregnancy. *British Journal of Obstetrics and Gynaecology*, **102:** 861–868.

Coceani, F. and Olley, P.M. 1982, Prostaglandins in reproduction. *In:* H.J.M. Barnett, J. Hirsch and J.F. Mustard (eds), *Acetylsalicylic Acid: New Uses for an Old Drug*, pp. 109–122. New York: Raven Press.

Cohen, A., Geller, S.A., Horowitz, I., Toth, L.S. and Werther, J.L. 1984, Experimental models for gastric leiomyosarcoma. The effects of N-methyl-N'-nitro-N-nitrosoguanidine in combination with stress, aspirin, or sodium taurocholate. *Cancer*, **53:** 1088–1092.

Cohen, L.S. 1976, Clinical pharmacology of acetylsalicylic acid. *Seminars in Thrombosis and Hemostasis*, **2:** 146–175.

Cohen, S.A., Zenser, T.V., Murasaki, G., Fukushima, S., Mattammal, M.B., Rapp, N.S. and Davis, B.B. 1981, Aspirin inhibition of N-[4-(5-nitro-2-furyl)-2-thiazolyl]formamide-induced lesions of the urinary bladder correlated with inhibition of metabolism by bladder prostaglandin endoperoxide synthetase. *Cancer Research*, **41:** 3355–3359.

Collett, J.P., Sharpe, C., Belzile, E., Bivin, J.F., Hanley, J. and Abenhaim, L. 1999, Colorectal cancer prevention by non-steroidal anti-inflammatory drugs: effects of dosage and timing. *British Journal of Cancer*, **81:** 62–68.

Collier, H.O.J. 1974, Prostaglandins and gastrointestinal functions. *In:* H.J. Robinson and J.R. Vane (eds), *Prostaglandin Synthesis Inhibitors*, pp. 121–133. New York: Raven Press.

Condorelli, G., Costanzo, L.L., De Guidi, G., Giuffrida, S., Rizzarelli, E. and Vecchio, G. 1994, Inhibition of photohemolysis by copper(II) complexes with SOD-like activity. *Journal of Inorganic Biochemistry*, **54:** 257–265.

Coogan, P.F., Rosenberg, L., Palmer, J.R., Strom, B.L., Zauber, A.G., Stolley, P.D. and Shapiro, S. 2000, Nonsteroidal anti-inflammatory drugs and risk of digestive cancers at sites other than the large bowel. *Cancer Epidemiology, Biomarkers and Prevention*, **9:** 119–123.

Corder, E., Basun, H., Lannfelt, L., Viitanen, M. and Winblad, B. 1995, Apolipoprotein E-epsilon 4 gene dose. *Lancet*, **346:** 967–968.

Corley, D.A., Kerlikowske, K., Verma, R. and Buffler, P. 2003, Protective association of aspirin/NSAIDs and esophageal cancer: a systematic review and meta-analysis. *Gastroenterology*, **124:** 47–56

Cotlier, E. and Sharma, Y.R. 1981, Aspirin and senile cataracts in rheumatoid arthritis. *Lancet*, **1:** 338–339.

Crabbe, M.J. 1998, Cataract as a conformational disease – the Maillard reaction, alpha-crystallin and chemotherapy. *Cellular and Molecular Biology (Noisy-le-Grand)*, **44:** 1047–1050.

Craven, P.A. and DeRubertis, F.R., 1992, Effect of aspirin on 1,2-dimethylhydrazine-induced colonic carcinogenesis. *Carcinogenesis*, **13:** 541–546.

Craven, P.A., Saito, R. and DeRubertis, F.R. 1983, Role of prostaglandin synthesis in the modulation of proliferative activity of rat colonic epithelium. *Journal of Clinical Investigation*, **72:** 1365–1375.

Crew, T.E., Elder, D.J.E. and Paraskeva, C. 2000, A cyclooxygenase-2 (COX-2) selective non-steroidal anti-inflammatory drug enhances the growth inhibitory effect of butyrate in colorectal carcinoma cells expressing COX-2 protein: regulation of COX-2 by butyrate. *Carcinogenesis*, **21:** 66–77.

Crouch, R.K., Kensler, T.W., Oberley, L.W. and Sorenson, J.R.J. 1985, Possible medicinal uses of copper complexes. *In: Biological and Inorganic Copper Chemistry* (p. 139), K.D. Karlin and J. Zubieta (eds), New York: Adenine Press.

Cummings, J.L. and Benson, D.F. (eds) 1992, Cortical dementias: Alzheimer's disease and other cortical degenerations. *In: Dementia. A Clinical Approach*, 2nd edn, pp. 45–93. London: Butterworth-Heinemann.

Davies, A.E. and Patterson, F. 1994, Aspirin reduces the incidence of colonic carcinoma in the dimethylhydrazine rat animal model. *Australian and New Zealand Journal of Medicine*, **24:** 301–303.

Davis, A.W., Patterson, F. and Crouch, R. 1992, The effect of therapeutic drugs used in inflammatory bowel disease on the incidence and growth of colon cancer in the dimethyl hydrazine rat model. *British Journal of Cancer*, **66:** 777.

Davis, M.D., Kern, T.S. and Rand, L.I., 1997, Diabetic retinopathy. *In:* K.G.M.M. Alberti, P. Zimmert, R.A. DeFronzo and H. Keen (eds), *International Textbook of Diabetes Mellitus*, 2nd edn, pp. 1438–1442. Chichester: Wiley.

Davis, P.B. and Di Sant'Agnese, P.A. 1980, A review. Cystic fibrosis at forty – quo vadis? *Pediatric Research*, **14:** 83–87.

De Alvare, L.R., Goda, K. and Kimura, T. 1976, Mechanism of superoxide anion scavenging reaction by bis-(salicylato)copper(II) complex. *Biochemical and Biophysical Research Communications*, **69:** 687–694.

De Furia, F.G., Cerami, A., Bunn, H.F., Lu, Y.S. and Peterson, C.M. 1973, The effect of aspirin on sickling and oxygen affinity of erythrocytes. *Proceedings of the National Academy of Sciences USA*, **70 (12):** 3707–3710.

De Knijff, P. and Van Duijn, C.M. 1998, Role of APOE in dementia: a critical reappraisal. *Haemostasis*, **28:** 195–201.

Dennis, R.F. and Oh, J.O. 1979, Aspirin, cyclophosphamide, and dexamethasone effects on experimental secondary herpes simplex uveitis. *Archives of Opthalmology*, **97 (11):** 2170–2174.

Donker, A.J., De Jong, P.E., Van Eps, L.W., Brentjens, J.R., Bakker, K. and Doorenbos, H. 1977, Indomethacin in Bartter's syndrome: does the syndrome represent a state of hyperprostaglandinism? *Nephron*, **19:** 200–213.

Dore-Duffy, P. and Zurier, R.B. 1979a, E-rosette formation in normals and patients with multiple sclerosis: effect of prostaglandin and aspirin. *Clinical Immunological Immunopathology*, **13 (3):** 261–268.

Dore-Duffy, P. and Zurier, R.B. 1979b, Lymphocyte adherence in multiple sclerosis: effect of aspirin. *Journal of Clinical Investigation*, **63 (1):** 154–157.

Dore-Duffy, P. and Zurier, R.B. 1981, Lymphocyte adherence in multiple sclerosis. Role of monocytes and increased sensitivity of MS lymphocytes to prostaglandin E. *Clinical Immunological Immunopathology*, **19 (3):** 303–313.

Dore-Duffy, P. and Zurier, R.B. 1986, Lymphocyte adherence in multiple sclerosis: role of the cytoskeleton and prostaglandin E. *Prostaglandins and Leukotrienes in Medicine*, **23** (2–3): 277–287.

Duffy, C.P., Elliott, C.J., O'Connor, R.A., Heenan, M.M., Coyle, S., Cleary, I.M., Kavanagh, K., Verhaegen, S., O'Loughlin, C.M., NicAmhlaoibh, R. and Clynes, M. 1998, Enhancement of chemotherapeutic drug toxicity to human tumour cells in vitro by a subset of non-steroidal anti-inflammatory drugs (NSAIDs). *European Journal of Cancer*, **34 (8):** 1250–1259.

Duffy, C.P., Elliott, C.J., O'Connor, R.A., Heenan, M.M., Coyle, S., Cleary, I.M., Kavanagh, K., Verhaegen, S., O'Loughlin, C.M., Nicamhlaoibh, R. and Clynes, M. 1998, Enhancement of chemotherapeutic drug toxicity to human tumour cells in vitro by a subset of non-steroidal anti-inflammatory drugs (NSAIDs). *European Journal of Cancer*, **34:** 1250–1259.

Duley, L., Henderson-Smart, D., Knight, M. and King, J. 2001, Anti-platelet drugs for prevention of pre-eclampsia and its consequences: systematic review. *British Medical Journal*, **322:** 329–333.

Duncanson, F.P., Abelmann, W.H. and Pan, C. 1978, Growth of Trypanosoma cruizi *in vitro*, unaffected by sodium salicylate. *Annals of Tropical Medicine and Parasitology*, **72:** 577–578.

Duperron, C. and Castonguay, A. 1997, Chemopreventive efficacies of aspirin and sulindac against lung tumorigenesis in A/J mice. Carcinogenesis, **18:** 1001–1006.

Dzenko, K.A., Weltzien, R.B. and Pachter, J.S. 1997, Suppression of a β-induced monocyte neurotoxicity by anti-inflammatory compounds. *Journal of Neuroimmunology*, **80:** 6–12.

Ebstein, W. 1876, Zur Therapie des Diabetes mellitus, insbesondere über die Anwendung des salicylsauren Natron bei demselben. *Berliner Klinische Wochenschrift*, **13:** 337–340.

Ebstein, W. and Müller, J. 1875, Weitere Mitteilungen über die Behandlung des Diabetes Mellitus mit Carbolsäure nebst Bemerkungen über die Anwendung der Salicylsäure bei dieser Krankheit. *Berliner Klinische Wochenschrift*, **12:** 53–56.

Editorial 1980, Aspirin for childhood gastroenteritis. *Medical Journal of Australia*, **2:** 358.

Egorin, M.J., Felsted, R.L. and Bachur, N.R. 1978, Salicylate effects on the response of human lymphocytes to phytohemagglutinin isolectins. *Clinical Immunological Immunopathology*, **10:** 1–10.

Elder, D.J.E., Hague, A., Hicks, D.J. and Paraskeva, C. 1996, Differential growth inhibition by the aspirin metabolite salicylate in human colorectal tumor cell lines: enhanced apoptosis in carcinoma and *in vitro*-transformed adenoma relative to adenoma cell lines. *Cancer Research*, **56:** 2273–2276.

Elder, D.J.E., Halton, D.E., Crew, T.E. and Paraskeva, C. 2000, Apoptosis induction and cyclooxygenase-2 regulation in human colorectal, adenoma and carcinoma cell lines by the cyclooxygenase-2–selective non-steroidal anti-inflammatory drug NS-398. *International Journal of Cancer*, **86:** 553–560.

Elder, D.J.E. and Paraskeva, C. 1996, Are aspirin and other non-steroidal anti-inflammatory drugs effective in the prevention and treatment of colorectal cancer? *Lancet*, **348**: 485.

Elie, R., Durocher, L.P. and Kavalec, E.C. 1983, Effect of salicylic acid on the activity of betamethasone-17,21-dipropionate in the treatment of erythematous squamous dermatoses. *Journal of International Medical Research*, **11 (2)**: 108–112.

Evans, J. 2001, New paths to Alzheimer's drugs. *Chemistry in Britain*, **April**: 47–51.

Evans, M.H., Pace, C.S. and Clements, R.S. Jr. 1983, Endogenous prostaglandin synthesis and glucose-induced insulin secretion from the adult rat pancreatic islet. *Diabetes*, **32**: 509–515.

Fang, V., Foye, W.O., Robinson, S.M. and Jenkins, H.J. 1968, Hypoglycemic activity and chemical structure of the salicylates. *Journal of Pharmaceutical Sciences*, **57**: 2111–2116.

Farrow, D.C., Vaughan, T.L., Hansten, P.D., Stanford, J.L., Risch, H.A., Gammon, M.D., Chow, W.H., Dubrow, R., Ahsan, H., Mayne, S.T., Schoenberg, J.B., West, A.B., Rotterdam, H., Fraumeni, J.F. Jr and Blot, W.J. 1998, Use of aspirin and other nonsteroidal anti-inflammatory drugs and risk of esophageal and gastric cancer. *Cancer Epidemiology, Biomarkers and Prevention*, **7**: 97–102.

Fichman, M.P., Telfer, N., Zia, P., Speckart, P., Golub, M. and Rude, R. 1976, Role of prostaglandins in the pathogenesis of Bartter's syndrome. *American Journal of Medicine*, **60**: 785–797.

Fischer, S.M., Herng-Hsang, L.O., Gordon, G.B., Seibert, K., Kelloff, G., Lubet, R.A. and Conti, C.J. 1999, Chemopreventive activity of celecoxib, a specific cyclooxygenase-2 inhibitor, and indomethacin against ultraviolet light-induced skin carcinogenesis. *Molecular Carcinogenesis*, **25**: 231–240.

Flock, E.V. and Owen, C.A. 1965, Effect of salicylate on the metabolism of L-thyroxine. *Endocrinology*, **77**: 475–484.

Fosang, A.J. and Smith, P.J. 2001, To clot or not to clot. *Nature*, **413**: 475–478.

Fosslein, E. 2000, Molecular pathology of cyclooxygenase-2 in neoplasia. *Annals of Clinical and Laboratory Science*, **30**: 3–22.

Foster, B. 1878, Diabetic coma: acetonaemia. *British Medical Journal*, **1**: 79–81.

Fournier, D.B. and Gordon, G.B. 2000, COX-2 and colon cancer: potential targets for chemoprevention. *Journal of Cellular Biochemistry*, **34 (Suppl.)**: 97–102.

Franco, L. and Doria, D. 1997, Prostaglandins and nitric oxide in copper-complexes mediated protection against ethanol-induced gastric damage. *Pharmacological Research*, **36**: 395–399.

Funkhouser, E.M. and Sharp, G.B. 1995, Aspirin and reduced risk of esophageal carcinoma. *Cancer*, **76**: 1116–1119.

Gabourel, J.D., Davies, G.H. and Rittenberg, M.B. 1977, Effects of salicylate and phenobarbital on lymphocyte proliferation and function. *Clinical Immunological Immunopathology*, **7**: 53–61.

Gaitonde, B.B., Joglekar, S.N. and Shaligram, S.V. 1967, The hyperglycaemic action of sodium salicylate. *British Journal of Pharmacology*, **30 (3)**: 554–560.

Gale, G.R. 1966, Selective inhibition of deoxyribonucleic acid synthesis by salicylhydroxamic acid. *Proceedings of the Society for Experimental Biology and Medicine*, **122 (4)**: 1236–1240. .

Gann, P.H., Manson, J.E., Glynn, R.J., Buring, J.E. and Hennekins, C.H. 1993, Low-dose aspirin and incidence of colorectal tumours in a randomised trial. *Journal of the National Cancer Institute*, **85**: 1220–1224.

Garcia, C., Roncero, I., Tamarit-Rodriguez, J. and Tamarit, J. 1984, Effects of salicylate on insulin and glucagon secretion by the isolated and perfused rat pancreas. *Revue Espania Fisiologie* **40**: 477–482.

Garcia-Rodriguez, L.A. and Huerta-Alvarez, C. 2000, Reduced incidence of colorectal adenoma among long-term users non-steroidal anti-inflammatory drugs: a pooled analysis of published studies and a new population-based study. *Epidemiology*, **11**: 376–381.

Geber, W.F., Lefkowitz, S.S. and Hung, C.Y. 1975, Effect of ascorbic acid, sodium salicylate, and caffeine on the serum interferon level in response to viral infection. *Pharmacology*, **13**: 228–233.

Geling, N.G., Klimenko, T.V. and Milekhin, G.N, 1991, Indicators of prostaglandin synthase activity in thrombocytes of patients with schizophrenia with positive and negative disorders. *Zhournal Nevropatol Psikhiatr Im S S Korsokova*, **9**: 99–101.

Gerli, R., Paolucci, C., Gresele, P., Bistoni, O., Fiorucci, S., Muscat, C., Belia, S., Bertototto, A. and

Costantini, V. 1998, Salicylate inhibit adhesion and transmigration of T lymphocytes by preventing integrin activation induced by contact with endothelial cells. *Blood*, **92:** 2389–2398.

Ghiselli, A., Laurenti, O., De Mattia, G., Maiani, G. and Ferro-Luzzi, A. 1992, Salicylate hydroxylation as an early marker of *in vivo* oxidative stress in diabetic patients. *Free Radical Bioliogy and Medicine*, **13:** 621–626.

Gill, J.R. Jr 1980, Bartter's syndrome. *Annual Review of Medicine*, **31:** 405–419.

Gill, J.R. Jr and Bartter, F.C. 1978, Evidence for a prostaglandin-independent defect in chloride reabsorption in the loop of Henle as a proximal cause of Bartter's syndrome. *American Journal of Medicine*, **65:** 766–772.

Gill, J.R. Jr, Frolich, J.C., Bowden, R.E., Taylor, A.A., Keiser, H.R., Seyberth, H.W., Oates, J.A. and Bartter, F.C. 1976, Bartter's syndrome: a disorder characterised by high urinary prostaglandins and a dependence of hyperreninemia on prostaglandin synthesis. *American Journal of Medicine*, **61:** 43–51.

Giovannucci, E., Egan, K.M., Hunter, D.J., Stampfer, M.J., Colditz, G.A., Willett, W.C. and Speizer, F.E. 1995, Aspirin and the risk of colorectal cancer in women. *New England Journal of Medicine*, **333:** 609–614.

Giovannucci, E., Rimm, E.B., Stampfer, K.J., Colditz, G.A., Asherio, A. and Willett, W.C. 1994, Aspirin use and the risk for colorectal cancer and adenoma in male health professionals. *Annals of Internal Medicine*, **121:** 241–246.

Giuliano, D., Passariello, N., Torella, R.O., Cerciello, T., Varricchio, M. and Sgambato, S. 1981, Effects of acetylsalicylic acid on plasma glucose, free fatty acid, betahydroxybutyrate, glucagons and C-peptide responses to salbutamol in insulin-dependent diabetic subjects. *Acta Diabetologia Latino*, **18:** 27–36.

Giuliano, D., Torrella, R., Sinischalchie, N., Improto, L. and D'Onofrio, F. 1978, The effect of acetylsalicylic acid on insulin response to glucose and arginine in normal man. *Diabetologia*, **14:** 359–362.

Gliddon, A.E., Chung-A-On, K.O., Griffiths, K.D., Payne, J.W. and Davies, J.I. 1991, Observations on platelet prostaglandin metabolism in schizophrenia: the response to ADP. *Biochemical Society Transactions*, **19 (2):** 155S.

Goedert, M., Strittmatter, W.J. and Roses, A.D. 1994, Alzheimer's disease. Risky apolipoprotein in brain. *Nature*, **372:** 92–94.

Goffinet, F., Aboulker, D., Paris-Llado, J., Bucourt, M., Uzan, M., Papiernik, E. and Breart, G. 2001, Screening with a uterine Doppler in low-risk pregnant women followed by low-dose aspirin in women with abnormal results: a multicenter randomised controlled trial. *British Journal of Obstetrics and Gynaecology*, **108:** 510–518.

Gomes, I. 1998, Aspirin: a neuroprotective agent at high doses? *National Medical Journal of India*, **11 (1):** 14–17.

Gonzalez, E.R. 1980, Study to begin on whether aspirin can delay senile cataract formation. *Journal of American Medical Association*, **244 (23):** 2593–2594.

Good, B.P., Hetzel, B.S. and Hogg, B.M. 1965, Studies on the control of thyroid function in rats: effects of salicylate and related drugs. *Endocrinology*, **77:** 674–682.

Grab, D., Paulus, W.E., Erdmann, M., Terinde, R., Oberhoffer, R., Lang, D., Muche, R. and Kreienberg, R. 2000, Effects of low-dose aspirin on uterine and fetal blood flow during pregnancy; results of a randomised, placebo-controlled, double-blind trial. *Ultrasound in Obstetrics and Gynecology*, **15:** 19–27.

Greenberg, E.R., Baron, J.A., Freeman, D.H. Jr, Mandel, J.S. and Haile, R. 1993, Reduced risk of large bowel adenomas among aspirin users. The Polyp Prevention Study Group. *Journal of the National Cancer Institute*, **85:** 912–916.

Gross, M. and Greenberg, L.A. 1948, *The Salicylates. A Critical Bibliographic Review*. New Haven, CT: Hillhouse Press.

Guerono, M.T. and Diez, R.A. 1999, Induction of apoptosis by salicylates in B-cell chronic lymphocytic leukemia. *Blood*, **93:** 1123–1124.

Gullner, H.G., Smith, J.B. and Bartter, F.C. 1980, The principal metabolites of arachidonic acid are overproduced in Bartter's syndrome. *Advances in Prostaglandin Thromboxane Research*, **7:** 1185–1187.

Halliday, G., Robinson, S.R., Shepherd, C. and Kril, J. 2000a, Alzheimer's disease and inflammation: a review of cellular and therapeutic mechanisms. *Clinical and Experimental Pharmacology and Physiology*, **27:** 1–8.

Halliday, G.M., Shepherd, C.E., McCann, H., Reid, W.G.J., Grayson, D.A., Broe, A. and Kril, J.J.

2000, Effect of anti-inflammatory medications on neuropathological findings in Alzheimer disease. *Archives of Neurology*, **57**: 831–836.

Halushka, P.V., Rogers, R.C., Loadholt, C.B. and Colwell, J.A. 1981, Increased platelet thromboxane synthesis in diabetes mellitus. *Journal of Laboratory and Clinical Medicine*, **97**: 87–96.

Hangarter, W. 1966, Zur Behandlung des Ischiassyndroms mit Tropflinfusionen einer Kupfer-Natrium-Salizylat-Komplexverbindung (Permalon). *Zeitschrift für Rheumaforschung*, **25**: 289–296.

Hangarter, W. and Lübke, A. 1952, Über die Behandlung rheumatischer Erkrankungen mit einer Kupfer-Natrium-Salizylat-Komplex-verbindung (Permalon). *Deutsche Medizinsche Wochenschrift*, **77**: 870–872.

Hanif, R., Pittas, A., Feng, Y., Koutsos, M.I., Aiao, L., Stainano-Coico, L., Shiff, S.I. and Rigas, B. 1996, Effects of nonsteroidal anti-inflammatory drugs on proliferation and on induction of apoptosis in colon cancer cells by a prostaglandin-independent pathway. *Biochemical Pharmacology*, **52**: 237–245.

Harding, J.J. 1992, Pharmacological treatment strategies in age-related cataracts. *Drugs and Ageing*, **2**: 287–300.

Harries, J.T. 1978, Meconium in health and disease. *British Medical Bulletin*, **34 (1)**: 75–78.

Harrington, K., Kurdi, W., Aquilina, J., England, P. and Campbell, S. 2000, A prospective management study of slow-release aspirin in the palliation of uteroplacental insufficiency predicted by uterine Doppler at 20 weeks. *Ultrasound in Obstetrics and Gynecology* **15**: 13–18.

Harris, R.E., Alshafie, G.A., Abou-Issa, H. and Seibert, K. 2000, Chemoprevention of breast cancer in rats by celecoxib, a cyclooxygenase-2 inhibitor. *Cancer Research*, **60**: 2010–2013.

Heath, M.M., Rixon, K.C. and Harding, J.J. 1996, Glycation-induced inactivation of malate dehydrogenase protection by aspirin and a lens molecular chaperone, alpha-crystallin. *Biochimica et Biophysica Acta* **1315**: 176–184.

Hermida, R.C., Ayala, D.E., Fernandez, J.R., Mojon, A., Alonso, I., Silva, I., Ucieda, R., Codesido, J. and Iglesias, M. 1999, Administration time-dependent effects of aspirin in women at differing risk for pre-eclampsia. *Hypertension*, **34**: 1016–1023.

Herrada, J., Noble, L.S. and Kutteh, W.E. 2000, Safety of the treatment of antiphospholipid syndrome associated with recurrent pregnancy loss with enoxaparin and low dose aspirin. *Blood*, **96**: 4100 (Part 2).

Hess, J.R. 1996, Alternative oxygen carriers. *Current Opinion in Hematology*, **3**: 492–497.

Hewett, S.J., Uliasz, T.F., Vidwans, A.S. and Hewett, J.A. 2000, Cyclo-oxygenase-2 contributes to N-methyl-D-aspartate-mediated neuronal cell death in primary cortical cell culture. *Journal of Pharmacology and Experimental Therapeutics*, **293**: 417–425.

Heyborne, K.D. 2000, Pre-eclampsia prevention: lessons from the low-dose aspirin therapy trials. *American Journal of Obstetrics and Gynecology* **183**: 523–528.

Hial, V., Horakova, Z., Shaff, R.E. and Beaven, M.A. 1976, Alteration of tumor growth by aspirin and indomethacin: studies with two transplantable tumors in mice. *European Journal of Pharmacology*, **37**: 367–376.

Hindmarch, H. Hippius, G.K. and Wilcock, B. (eds) 1991, Scientific basis for therapeutic developments in Alzheimer's disease. *In: Dementia: Molecules, Methods and Measures*, pp. 61–66. Chichester: Wiley.

Hirano, T., Fukuyama, S., Nagano, S. and Takahashi, T. 1984, The effect of exogenous arachidonic acid on insulin secretion in isolated perifused hamster islets. *Endocrinologica Japonica*, **31**: 549–555.

Hockertz, S. and Heckenberger, R. 1996, Treatment of acute bacterial infection with a combination of acetylsalicylic acid/ibuprofen and interferon gamma. *Arzneimittel Forschung*, **46**: 1012–1015.

Hockertz, S., Paulini, I., Rogalla, K. and Schettler, T. 1993, Influence of acetylsalicylic acid on a Listeria monocytogenes infection. *Agents and Actions*, **40**: 119–123.

Horrobin, D.F. 1992, The relationship between schizophrenia and essential fatty acid and eicosanoid metabolism. *Prostaglandins, Leukotrienes and Essential Fatty Acids*, **46**: 71–77.

Horrobin, D.F., Ally, A.I., Karmali, R.A., Karmazyn, M., Manku, M.S. and Morgan, R.O. 1978, Prostaglandins and schizophrenia: further discussion of the evidence. *Psychological Medicine*, **8 (1)**: 43–48.

Horton, E.W. 1973, Prostaglandins at adrenergic nerve-endings. *British Medical Bulletin*, **29**: 148–151.

Hosny, E.A., Ghilzai, N.M., Al-Najar, T.A. and Elmazar, M.M. 1998, Hypoglycemic effect of oral insulin in diabetic rabbits using pH-dependent coated capsules containing sodium salicylate without and with sodium cholate. *Drug Development and Industrial Pharmacy*, **24**: 307–311.

Hsieh, B.S., Chen, W.Y.M., Yen, T.S. and Lai, M.S. 1978, Prostaglandin E excretion and inhibition of prostaglandin synthesis in Bartter's syndrome. *Taiwan Yi Xue Hui Zhi*, **77:** 831–838.

Hsieh, Y.Y., Tsai, H.D., Chang, C.C., Lo, H.Y. and Chen, C.L. 2000, Low-dose aspirin for infertile women with thin endometrium receiving intrauterine insemination: a prospective, randomised study. *Journal of Assisted Reproduction and Genetics*, **17:** 174–177.

Hull, M., Fiebich, B.L., Schumann, G., Lieb, K. and Bauer, J. 1999, Anti-inflammatory substances as new therapeutic option in Alzheimer's disease. *Drug Discovery Today*, **4:** 275–282.

Hundal, R.S., Petersen, K.F., Mayerson, A.B., Randhawa, P.S., Inzucchi, S., Shoelson, S.E. and Shulman, G.I. 2002, Mechanism by which high-dose aspirin improves glucose metabolism in type 2 diabetes. *J Clin Invest* **109:** 1321–1326.

Hussey, J.J.H. and Tisdale, M.J. 2000, Effect of the specific cyclo-oxygenase inhibitor, meloxicam, on tumour growth and cachexia in a murine model. *International Journal of Cancer*, **87:** 5–100.

Inglot, A.D. 1969, Comparison of the antiviral activity *in vitro* of some non-steroidal anti-inflammatory drugs. *Journal of General Virology*, **4:** 203–214.

Irving, K.J., Wear, M.A., Simmons, H., Tipton, L.G., Tipton, J.B., Maddox, K.M., Willingham, W.M. and Sorenson, J.R.J. 1996, An examination of the radioprotective and radiorecovery activities of Fe(III)(3,5-diisopropylsalicylate)$_3$ and Mn(III)$_2$(II)(μ_3-O)(3,5-diisopropylsalicylate)$_6$. *Inflammopharmacology*, **4:** 309–321.

Ishikawa, Y. and Wada, T. 1979, Escape from sodium and potassium retaining actions of aspirin-like drugs used in a patient with Bartter's syndrome. *Japanese Circulation Journal*, **43:** 757–767.

Jackson, G.E., Mkhonta-Gama, L., Voye, A. and Kelly, M. 2000, Design of copper-based anti-inflammatory drugs. *Journal of Inorganic Chemistry*, **79:** 147–152.

Janakidevi, K. and Smith, M.J.H. 1970, Effects of salicylate on the incorporation of orotic acid into nucleic acids of mouse tissues in vivo. *Journal of Pharmacy and Pharmacology*, **22 (1):** 51–55.

Jiang, M.-C., Liao, C.-F. and Lee, P.-H. 2001, Aspirin inhibits metalloproteinase-2 activity, increases E-cadeherin production, and inhibits *in vitro* invasion of tumor cells. *Biochemical and Biophysical Research Communications*, **282:** 671–677.

Jiang, R., Roberts, M.S., Frankerd, R.J. and Benson, H.A. 1997, Percutaneous absorption of sunscreen agents from liquid paraffin: self-association of octyl salicylate and effects on skin flux. *Journal of Pharmaceutical Sciences*, **86:** 791–796.

Jones, R.H. and Hothersall, J.S. 1993, Increased susceptibility to metal catalysed oxidation of diabetic lens beta L crystallin: possible protection by dietary supplementation with acetylsalicylic acid. *Experimental Eye Research*, **57:** 783–790.

Jorgensen, K.A., Mourits-Andersen, H.T., Ditzel, J. and Dyerberg, J. 1980, Acetylsalicylic acid and bleeding time in juvenile diabetes mellitus. *Thrombosis Research*, **20:** 611–615.

Kabadi, U.M. and Danielson, S. 1987, Misleading thyroid function tests and several homeostatic abnormalities induced by 'Disalcid' therapy. *Journal of the American Geratrics Society*, **35:** 255–257.

Kawashima, H., Monji, N. and Castro, A. 1980, Effects of calcium chloride on aspirin-induced hypoinsulinemia in rats. *Biochemical Pharmacology*, **29:** 1627–1630.

Keshavarzain, A., Haydek, J., Zabihi, R., Doria, M., D'Astice, M. and Sorenson, J.R.J. 1992, Agents capable of eliminating reactive oxygen species. Catalase, WR-2721, or Cu(II)$_2$(3,5-DIPS)$_4$ decrease experimental colitis. *Digestive Diseases and Sciences*, **37:** 1866–1873.

Khan, C.R. and Weir, G.C. (eds) 1994, *Joslin's Diabetes Mellitus*, 13th edn. Philadelphia: Lea & Febiger.

Khan, M.J. and Morrison, D.G. 1997, Chemoprevention for colorectal carcinoma. *Haematology–Oncology of North America*, **11:** 779.

Klan, R., Knispel, H.H. and Meier, T. 1993, Acetylsalicylic acid inhibition of n-butyl-(4-hydroxybutyl)nitrosamine-induced bladder carcinogenesis in rats. *Journal of Cancer Research in Clinical Oncology*, **119:** 482–485.

Klein, B.E., Klein, R. and Moss, S.E. 1987, Is aspirin use associated with lower rates of cataracts in diabetic individuals? *Diabetes Care*, **10:** 495–499.

Klotz, I.M. and Tam, J.W. 1973, Acetylation of sickle cell hemoglobin by aspirin. *Proceedings of the National Academy of Sciences USA*, **70 (5):** 1313–1315.

Kochman, M., Mastalerz, P. and Inglot, A.D. 1965, Inhibition of encephalomyocarditis virus replication by simple phosphonic and carboxylic acids. *Nature*, **207**: 888–890.

Korolkiewicz, Z., Hac, E., Gagalo, I., Gorczyca, P. and Lodzinska, A. 1989, The pharmacologic activity of complexes and mixtures with copper and salicylates or aminopyrine following oral dosing in rats. *Agents and Actions*, **26**: 355–359.

Kort, W.J., Hulsman, L.O.M., Van Schalwijk, W.P., Weijma, I.M., Zondervan, P.E. and Westbroek, D.L. 1986, Reductive effect of aspirin treatment on primary tumor growth and metastasis of implanted fibrosarcoma in rats. *Journal of the National Cancer Institute*, **76**: 711–720.

Kosik, K.S. 1990, Tau protein and Alzheimer's disease. *Current Opinion in Cell Biology*, **2**: 101–104.

Koyuncuoğlu, H., Öz, H., Geno, E., Sagduyu H., Aykoo, G., Sivas, A. and Uysal, M. 1976, The effects of sodium salicylate and flufenamic acid on the levels of some hormones and enzymes, and on lipid metabolism in rabbits. *Pharmacological Research Communications*, **8**: 267–278.

Krentz, A.J., Ferner, R.E. and Bailey, C.J. 1994, Comparative tolerability profiles of oral antidiabetic agents. *Drug Safety*, **11**: 223–241.

Kruesi, M. 1980, Bismuth preparations for diarrhoea. *Journal of the American Medical Association*, **244**: 1435.

Krzywanek, H.J. and Breddin, K. 1981, Platelet aggregation as a risk factor in diabetic subjects. *Hormone Metabolism Research*, **11 (Suppl.)**: 11–14.

Kuhn, W., Muller, T., Bttner, T. and Gerlach, M. 1996, Aspirin as a free radical scavenger: consequences for therapy of cerebrovascular ischemia. *Stroke*, **26**: 1959–1960.

Kune, G.A., Kune, S. and Watson, L.F. 1988, Colorectal cancer risk, chronic illnesses, operations, and medications: case–control results from the Melbourne Colorectal Cancer Study. *Cancer Research*, **48**: 4399–4404.

Lamy, M.L., Daily, E.K., Brichant, J.F., Larbuisson, R.P., Demeyere, R.H., Vandermeersch, E.A., Lehot, J.J., Parsloe, M.R., Berridge, J.C., Sinclair, C.J., Baron, J.F. and Przybelski, R.J. 2000, Randomised trial of diaspirin cross-linked haemoglobin solution as an alternative to blood transfusion after cardiac surgery: The DCLHb Cardiac Surgery Trial Collaborative Group. *Anesthesiology*, **92**: 646–656.

Landgraf-Leurs, M.M., Loy, A., Christea, C., Weber, P.C., Siess, W., Herberg, L.L. and Landgraf, R. 1981, Aggregation and thromboxane B2 formation in platelets and vascular prostacyclin production from genetically obese rats. *Prostaglandins*, **22**: 524–536.

Langman, M.J.S., Cheng, K.K., Gilman, E.A. and Lancashire, R.J. 2000, Effect of anti-inflammatory drugs on overall risk of common cancer: case–control study in General Practice Research Database. *British Medical Journal*, **320**: 1642–1646.

Larsen, P.R. 1972, Triiodothyronine: review of recent studies of its physiology and pathophysiology in man. *Metabolism*, **21 (11)**: 1073–1092.

Lee, R.M. and Silver, R.M. 2000, Recurrent pregnancy loss: summary and clinical recommendations. *Seminars in Reproductive Medicine*, **18**: 433–440.

Lehmann, R. and Spinas, G.A. 2000, Screening, diagnosis and management of diabetes mellitus and diabetic complications. *Therapeutische Umschrift*, **57**: 12–21.

Leuthauser, S.W.C., Oberley, L.W., Oberley, T.D., Sorenson, J.R.J. and Ramakrishna, K. 1981, Antitumor effect of a copper coordination compound with superoxide dismutase-like activity. *Journal of the National Cancer Institute*, **66**: 1077–1081.

Levang, A.K., Zhao, K. and Singh, J. 1999, Effect of ethanol/propylene glycol on the in vitro percutaneous absorption of aspirin, biophysical changes and macroscopic barrier properties of the skin. *International Journal of Pharmaceutics*, **181**: 255–263.

Levy, G.G., Nichols, W.C., Lian, E.C., Foroud, T. and McClintock, J.N., *et al.* 2001, Mutations in a member of the *ADAMTS* gene family cause thrombotic thrombocytopenic purpura. *Nature*, **413**: 488–494.

Levy, G.N. 1997, Prostaglandin H synthases, nonsteroidal anti-inflammatory drugs, and colon cancer. *FASEB Journal*, **11 (4)**: 234–247.

Li, P.A., Liu, G.-J., He, Q.P., Floyd, R.A. and Siesjö, B.K. 1999, Production of hydroxyl free radical by brain tissues in hyperglycemic rats subjected to transient forebrain ischemia. *Free Radical Biology and Medicine*, **27**: 1033–1040.

Lim, G.P., Yang, F., Chu, T., Chen, P., Beech, W., Teter, B., Tran, T., Ubeda, O., Ashe, K.H., Frautschy, S.A. and Cole, G.M. 2000, Ibuprofen suppresses plaque pathology and inflammation in a mouse model for Alzheimer's disease. *Journal of Neuroscience*, **20:** 5709–5714.

Littlewood, J.M., Lee, M.R. and Meadow, S.R. 1978, Treatment of Bartter's syndrome in early childhood with prostaglandin synthetase inhibitors. *Archives of Disease in Children*, **53:** 43–48.

Logan, R.F., Little, J., Hawtin, P.G. and Hardcastle, J.D. 1993, Effect of aspirin and non-steroidal anti-inflammatory drugs on colorectal adenomas: case–control study of subjects participating in the Nottingham Faecal Occult Blood Screening Programme. *British Medical Journal*, **307:** 285–289.

Luca, U. 1999, Nonsteroidal anti-inflammatory drugs and Alzheimer's Disease. *CNS Drugs*, **11:** 207–224.

Lukiw, W.J. and Bazan, N.G. 1998, Strong nuclear factor-κB-DNA binding parallels cyclooxygenase-2 gene transcription in aging and in sporadic Alzheimer's disease superior temporal lobe neocortex. *Journal of Neuroscience Research*, **53 (5): 583**–592.

Lynch, N.R., Castes, M., Astoin, M. and Salomon, J.S. 1978, Mechanisms of inhibition of tumour growth by aspirin and indomethacin. *British Journal of Cancer*, **38:** 503–512.

MacDonald, A., Busch, G.J., Alexander, J.L., Pheteplace, E.A., Menzoian, J. and Murray, J.E. 1970, Heparin and aspirin in the treatment of hyperacute rejection of renal allografts in presensitized dogs. *Transplantation*, **9:** 1–7.

Mahmoud, N.N., Dannenberg, A.J., Mestre, J., Bilinski, R.T., Churchill, M.R., Martucci, C., Newmark, H. and Bertagnolli, M.M. 1998, Aspirin prevents tumors in a murine model of familial adenomatous polyposis. *Surgery*, **124:** 225–231.

Malkinson, M.A., Koski, K.M., Dwyer-Nield, L.D., Rice, P.M., Rioux, N., Castonguay, A., Ahnen, D.J., Thompson, R.P. and Piazza, G.A. 1998, Inhibition of 4-(methylnitrosamino)-1-(3-pyridyl)-1-butanone-induced mouse lung tumour formation by FGN-1 (sulindac sulfone). *Carcinogenesis*, **19:** 1353–1356.

Mandelkow, E.M., Drewes, G., Biernat, J., Gustke, N., Van Lint, J., Vandenheede, J.R. and Mandelkow, E. 1992, Glycogen synthase kinase-3 and the Alzheimer-like state of microtubule-associated protein tau. *FEBS Letters*, **21:** 315–332.

Martinez, M.E., McPherson, R.S., Levin, B. and Annegers, J.F. 1995, Aspirin and other nonsteroidal anti-inflammatory drugs and risk of colorectal adenomatous polyps among endoscoped individuals. *Cancer Epidemiology Biomarkers and Prevalence*, **4:** 703–707.

Masferrer, J.L., Leahy, K.M., Koki, A.T., Zweiel, B.S., Settle, S, Woerner, B.M., Edwards, D.A, Flickinger, A.G., Moore, R.J. and Seibert, K. 2000, Antiangiogenic and antitumor activities of cyclooxygenase-2 inhibitors. *Cancer Research*, **60:** 1306–1311.

Matasic, R., Dietz, A.B. and Vuk-Pavlovic, S. 2000, Cyclooxygenase-independent inhibition of dendritic cell maturation by aspirin. *Immunology*, **101:** 53–60.

Matsunaga, K., Yoshimi, N., Yamada, Y., Shimizu, M., Kawabata, K., Ozawa, Y., Hara, A. and Mori, H. 1998, Inhibitory effects of nabumetone, a cyclooxygenase-2 inhibitor, and esculetin, a lipoxygenase inhibitor, on N-methyl-N-nitrosourea-induced mammary carcinogenesis in rats. *Japanese Journal of Cancer Research*, **89:** 496–501.

Matsuno, N., Muto, S., Uchiyama, M., Kozaki, K., Kageyama, T., Suzuki, T., Kozaki, M. and Nagao, T. 1997, Beneficial effect of aspirin and protease inhibitor as graft conditioning using machine perfusion in warm ischemically injured kidneys. *Transplantation Proceedings*, **29:** 3584–3585.

Maugh, T.H. II 1981, A new understanding of sickle cell emerges. *Science*, **211:** 265–267.

McCarthy, M.F. 1981, Maturity-onset diabetes mellitus – toward a physiological appropriate management. *Medical Hypotheses*, **7:** 1265–1285.

McGeer, E.G. and McGeer, P.L. 1998, The importance of inflammatory mechanisms in Alzheimer's disease. *Experimental Gerontology*, **33:** 371–378.

McGeer, P.L. and McGeer, E.G. 1996, Anti-inflammatory drugs in the fight against Alzheimer's disease. *Annals of the New York Academy of Sciences, USA*, **777:** 213–220.

McGeer, P.L. and McGeer, E.G. 1999, Inflammation of the brain in Alzheimer's disease: implications for therapy. *Journal of Leukocyte Biology*, **65:** 409–415.

McGiff, J.C. 1977, Bartter's syndrome results from an imbalance of vasoactive hormones. *Annals of Internal Medicine*, **87:** 369–372.

Megwa, S.A., Cross, S.E., Whitehouse, M.W., Benson, H.A. and Roberts, M.S. 2000, Effect of ion pairing with alkylamines on the in-vitro dermal penetration and local tissue disposition of salicylates. *Journal of Pharmacy and Pharmacology*, **52:** 929–940.

Mennie, A.T., Dalley, V.M., Dinneen, L.C. and Collier, H.O.J. 1975, Treatment of radiation-induced gastrointestinal distress with acetylsalicylate. *Lancet*, **15:** 942–943.

Micossi, P., Pontiroli, A.E., Baron, S.H., Tamayo, R.C., Lengel, F., Bevilacqua, M., Raggi, U., Norbiato, G. and Foa, P.P.1978, Aspirin stimulates insulin and glucagon secretion and increases glucose tolerance in normal and diabetic subjects. *Diabetes*, **27:** 1196–1204.

Milanino, R. and Velo, G.P. 1981, Multiple actions of copper in control of inflammation: studies in copper-deficient rats. *In:* K.D. Rainsford, K. Brune and M.W. Whitehouse, (eds), *Trace Elements in the Pathogenesis and Treatment of Inflammation*, pp. 209–230. Basel: Birkhäuser.

Milanino, R., Rainsford, K.D. and Velo, G.P. (eds) 1989, *Copper and Zinc in Inflammation*, Vol. IV. Dordrecht: Kluwer Academic Publishers.

Miller, J.D., Ganguli, S., Artal, R. and Sperling, M.A. 1985, Indomethacin and salicylate decrease epinephrine-induced glycogenolysis. *Metabolism*, **34:** 148–153.

Millhorn, D.E., Eldridge, F.L. and Waldrop, T.G. 1982, Effects of salicylate and 2,4-didinitrophenol on respiration and metabolism. *Journal of Applied Physiology*, **53:** 952.

Mitchell, G.A., Kassovska-Bratinova, S., Boukaftane, Y., Robert, M.F., Wang, S.P., Ashmarina, L., Lambert, M., Lapierre, P. and Potier, E. 1995, Medical aspects of ketone body metabolism. *Clinical and Investigative Medicine*, **18 (3):** 193–216.

Mobarok, A.T. and Morley, J. 1980, Actions of aspirin and dipyridamole on lymphocyte activation. *Agents and Actions*, **10:** 509–512.

Monaco, F., Davoli, C., Borghi, G. and Andreoli, M. 1977, Effects of 2-deoxy-glucose and acetylsalicylic acid on the thyroglobulin (TG) biosynthesis: inhibition of oligosaccharide chains formation. *Acta Endocrinologica Supplementum*, **212:** Abstract 291, 172.

Morgant, G., Dung, N.H., Daran, J.C., Viossat, B., Labouze, X., Roch-Arveiller, M., Greenaway, F.T., Cordes, W. and Sorenson, J.R. 2000, Low-temperature crystal structures of tetrakis-μ-3,5-diisopropylsalicylatobis-dimethylformamidodicopper(II) and tetrakis-μ-3,5-diisopropylsalicylatobis-diethyletheratodicopper(II) and their role in modulating polymorphonuclear leukocyte activity in overcoming seizures. *Journal of Inorganic Biochemistry*, **81:** 11–22.

Muir, K.R. and Logan, R.F.A. 1999, Aspirin, NSAIDs and colorectal cancer – what do the epidemiological studies show and what do they tell us about the modus operandi? *Apoptosis*, **4:** 389–396.

Mulder, M. and Terwel, D. 1998, Possible link between lipid metabolism and cerebral amyloid angiopathy in Alzheimer's disease: a role for high-density lipoproteins? *Haemostasis* **28:** 174–194.

Murakami, T., Ihara, C., Kiyonaka, G., Yumoto, R., Shigeki, S., Ikuta, Y. and Yata, N. 1999, Iontophoretic transdermal delivery of salicylic acid dissolved in ethanol-water mixture in rats. *Skin Pharmacology and Applied Skin Physiology*, **12:** 221–226.

Murasaki, G., Zenser, T.V., Davis, B.B. and Cohen, S.M. 1984, Inhibition by aspirin of N-[4-(5-nitro-2-furyl)-2-thiazolyl] formamide-induced bladder carcinogenesis and enhancement of forestomach carcinogenesis. *Carcinogenesis*, **5:** 53–55.

Murata, J., Sunao, Kawano, S., Tsuji, S., Tsuji, M., Sawaoka, H., Kimura, Y., Shiozaki, H. and Hori, M. 1999, Cyclooxygenase-2 overexpression enhances lymphatic invasion and metastasis in human gastric carcinoma. *American Journal of Gastroenterology*, **94:** 451–454.

Murono, S., Yoshizaki, T., Sato, H., Takeshita, H., Furukawa, M. and Pagano, J.S. 2000, Aspirin inhibits tumour cell invasiveness induced by Epstein–Barr virus latent membrane protein 1 through suppression of matrix metalloproteinase-9 expression. *Cancer Research*, **60:** 255–256.

Muscat, J.E., Stellman, S.D. and Wynder, E.L. 1994, Nonsteroidal anti-inflammatory drugs and colorectal cancer. *Cancer*, **74:** 1847–1854.

Myers, T.J., Wakem, C.J., Ball, E.D. and Tremont, S.J. 1980, Thrombotic thrombocytopenic purpura: combined treatment with plasmapheresis and antiplatelet agents. *Annals of Internal Medicine*, **92 (2 Pt 1):** 149–155.

Nagayama, M., Niwa, K., Nagayama, T., Ross, M.E. and Iadecola, C. 1999, The cyclooxygenase-2

CHAPTER 13

inhibitor NS-398 ameliorates ischemic brain injury in wild-type mice but not in mice with deletion of the inducible nitric oxide synthase gene. *Cerebral Blood Flow and Metabolism*, **19:** 1213–1219.

Nagy, Z.S., Esiri, M.M. and Smith, A.D. 1998, The cell division cycle and the pathophysiology of Alzheimer's disease. *Neuroscience*, **87:** 731–739.

Nakayama, M., Uchimura, K., Zhu, R.L., Nagayama, T., Rose, M.E., Stetler, R.A., Isakson, P.C., Chen, J. and Graham, S.H. 1998, Cyclooxygenase-2 inhibition prevents delayed death of CA1 hippocampal neurons following global ischemia. *Proceedings of the National Academy of Sciences, USA*, **95 (18):** 10954–10959.

Nalin, D.R. 1980, Aspirin and fluid losses in diarrhoea. *Lancet*, **2:** 793–794.

Nielsen, G.L., Sorenson, H.T., Larson, H. and Pedersen, L. 2001, Risk of adverse birth outcome and miscarriage in pregnant users of non-steroidal anti-inflammatory drugs: population based observational study and case–control study. *British Medical Journal*, **322:** 266–270.

Nishimura, G., Yanoma, S., Mizuno, H., Kawakami, K. and Tsukuda, M. 1999, A selective cyclooxygenase-2 inhibitor suppresses tumour growth in nude mouse xenografted with human head and neck squamous carcinoma cells. *Japanese Journal of Cancer Research*, **90:** 1152–1162.

Nishio, E. and Watanabe, Y. 1998, Aspirin and salicylate enhances the induction of inducible nitric oxide synthase in cultured rat smooth muscle cells. *Life Sciences*, **63 (6):** 429–439.

Nobili, F., Copello, Buffoni, F., Vitali, P., Girtler, N., Bordoni, C., Safaie-Semnani, E., Mariani, G. and Rodriguez, G. 2001, Regional cerebral blood flow and prognostic evaluation in Alzheimer's disease. *Dementia and Geriatric Cognitive Disorders*, **12:** 89–97.

Nogawa, S., Forster, C., Zhang, F., Nagayama, M., Ross, M.E., Iadecola, C. 1998, Interaction between inducible nitric oxide synthase and cyclooxygenase-2 after cerebral ischemia. *Proceedings of the National Academy of Sciences*, **95:** 10966–10971.

Nogawa, S., Zhang, F., Ross, M.E. and Iadecola, C. 1997, Cyclo-oxygenase-2 gene expression in neurons contributes to ischemic brain damage. *Journal of Neuroscience*, **17:** 2746–2755.

Norby, L., Flamenbaum, W., Lentz, R. and Ramwell, P. 1976, Prostaglandins and aspirin therapy in Bartter's syndrome. *Lancet*, **2:** 604–606.

O'Brien, M., Parness, I.A., Neufeld, E.J., Baker, A.L., Sundel, R.P. and Newburger, J.W. 2000, Ticlopidine plus aspirin for coronary thrombosis in Kawasaki disease. *Pediatrics* **105:** E641–E642.

Oberley, L.W., Leuthauser, S.W.G., Oberley, T.D., Sorenson, J.R.J. and Pasternak, R.F. 1982, Antitumor activities of compounds with superoxide dismutase activity. *In:* J.R.J. Sorenson, (ed.), *Inflammatory Diseases and Copper. The Metabolic and Therapeutic Roles of Copper and Other Essential Metalloelements in Humans*, pp. 423–430. New Jersey: Humana Press.

Offner, H., Danneskiold-Samsoe, B. and Dore-Duffy, P. 1982, Effect of prostaglandins and aspirin on active E-rosette formation in patients with multiple sclerosis. *Clinical Immunology and Immunepathology*, **22:** 159–167.

Okuyama, S., Hashimoto, S., Aihara, H., Willingham, W.M. and Sorenson, J.R. 1987, Copper complexes of non-steroidal anti-inflammatory agents: analgesic activity and possible opioid receptor activation. *Agents and Actions*, **21:** 130–144.

Onouchi, Z. and Kawasaki, T. 1999, Overview of pharmacological treatment of Kawasaki disease. *Drugs*, **58:** 813–822.

Opelz, G., Terasaki, P.I. and Hirata, A.A. 1973, Suppression of lymphocyte transformation by aspirin. *Lancet*, **2:** 478–480.

Osamo, N.O., Photiades, D.P. and Famodu, A.A. 1981, Therapeutic effect of aspirin in sickle cell anaemia. *Acta Haematologica*, **66:** 102–107.

Oshima, M., Dinchuk, J.E., Kargman, S.L., Oshima, H., Hancock, B., Kwong, E., Trzaskos, J.M., Evans, J.F. and Taketo, M.M. 1996, Suppression of intestinal polyposis in Apc$^{\Delta 716}$ knockout mice by inhibition of cyclooxygenase-2 (COX-2). *Cell*, **87:** 803–809.

Østensen, M. 1994, Optimization of antirheumatic drug treatment in pregnancy. *Clinical Pharmacokinetics*, **27:** 486–503.

Østensen, M. and Østensen, H. 1996, Safety of nonsteroidal anti-inflammatory drugs in pregnant patients with rheumatic disease. *Journal of Rheumatology*, **23:** 1045–1049.

Østensen, M. and Villiger, P.M. 2001, Nonsteroidal anti-inflammatory drugs in systemic lupus erythematosus. *Lupus*, **10:** 135–139.

Ota, S., Bamba, H. and Kato, A. 2000, Colorectal cancer and non-steroidal anti-inflammatory drugs. *Acta Pharmacologica Sinica*, **21:** 391–395.

Paganini-Hill, A., Chao, A., Ross, R.K. and Henderson, B.E. 1989, Aspirin use and chronic diseases: a cohort study of the elderly. *British Medical Journal*, **299:** 1247–1250.

Panush, R.S. and Anthony, C.R. 1976, Effects of acetylsalicylic acid on normal human peripheral blood lymphocytes. Inhibition of mitogen- and antigen-stimulated incorporation of tritiated thymidine. *Clinical and Experimental Immunology*, **23 (1):** 114–125.

Patel, D.G. and Skau, K.A. 1989, Effects of chronic sodium salicylate feeding on the impaired glucagon and epinephrine responses to insulin-induced hypoglycaemia in streptozotocin diabetic rats. *Diabetologia*, **32:** 61–66.

Paunesku, T., Chang-Liu, C.-M., Shearin-Jones, P., Watson, C., Milton, J., Oryhon, J., Salbego, D., Milosavijevic, A. and Woloschak., G.E. 2000, Identification of genes regulated by UV/salicylic acid. *International Journal of Radiation Biology*, **76:** 189–198.

Peacock, J.M., Rolsom, A.R., Knopman, D.S., Mosley, T.H., Goff, D.C. Jr and Szklo, M., for the Atherosclerosis Risk in Communities (ARIC) Study Investigators 1999, Association of nonsteroidal anti-inflammatory drugs and aspirin with cognitive performance in middle-aged adults. *Neuroepidemiology*, **18:** 134–143.

Peleg, I.I., Lubin, M.F., Cotsonis, G.A., Clark, W.S. and Wilcox, C.M. 1996, Long-term use of nonsteroidal anti-inflammatory drugs and other chemopreventors and risk of subsequent colorectal neoplasia. *Digestive Diseases and Sciences*, **41:** 1319–1326.

Peleg, I.I., Maibach, H.T., Brown, S.H. and Wilcox, C.M. 1994, Aspirin and nonsteroidal anti-inflammatory drug use and the risk of subsequent colorectal cancer. *Archives of Internal Medicine*, **154:** 394–399.

Pentland, A.P., Schoggins, J.W., Scott, G.A., Khan, K.N.M. and Han, R. 1999, Reduction of UV-induced skin tumours in hairless mice by selective COX-2 inhibition. *Carcinogenesis*, **20:** 1939–1944.

Persegani, C., Russo, P., Lugaresi, E., Nicolini, M. and Torlini, M. 2001, Neuroprotective effects of low-doses of aspirin. *Human Psychopharmacology. Clinical and Experimental*, **16:** 193–194.

Peterson, J., Amare, M., Henry, J.E. and Bone, R.C. 1979, Splenectomy and antiplatelet agents in thrombotic thrombocytopenic purpura. *American Journal of Medical Science*, **277 (1):** 75–89.

Piazza, G.A., Alberts, D.S., Hixson, L.J., Paranka, N.S., Li, H., Finn, T., Bogert, C., Guillen, J.S., Brendel, K., Gross, P.H., Sperl, G., Ritchie, J., Burt, R.W., Ellsworth, L., Ahnen, D.J. and Pamukcu, R. 1997, Sulindac sulfone inhibits azoxymethan-induced colon carcinogenesis in rats without reducing prostaglandin levels. *Cancer Research*, **57:** 2909–2915.

Pickering, L.K., Feldman, S., Ericsson, C.D. and Cleary, T.G. 1981, Absorption of salicylate and bismuth from a bismuth subsalicylate-containing compound (Pepto-Bismol). *Journal of Pediatrics*, **99 (4):** 654–656.

Pickup, J.C. and Williams, G. (eds)1997, *Textbook of Diabetes*, 2nd edn, Vol. 2. Oxford: Blackwell Science.

Planas, A.M., Soriano, M.A., Justicia, C. and Rodriguez-Farré, E. 1999, Induction of cyclooxygenase-2 in the rat brain after a mild episode of focal ischemia without tissue inflammation or neural cell damage. *Neuroscience Letters*, **275:** 141–144.

Plescia, O.J., Smith, A.H. and Grinwich, K. 1975, Subversion of immune system by tumor cells and role of prostaglandins. *Proceedings of the National Academy of Sciences USA*, **72:** 1848–1851.

Powles, T.J., Clark, S.A., Easty, G.C. and Neville, A.M. 1973, The inhibition by aspirin and indomethacin of osteolytic tumor deposits and hypercalcaemia in rats with Walker tumour, and its possible application to human breast cancer. *British Journal of Cancer*, **28:** 316–321.

Prasad, K.N., Hovland, A.R., Cole, W.C., Prasad, K.C., Bahreini, P., Edwards-Prasad, J. and Andreatta, C.P. 2000, Multiple antioxidants in the prevention and treatment of Alzheimer's disease: analysis of biologic rationale. *Clinical Neuropharmacology*, **23:** 2–13.

Prescott, S.M. 2000, Is cyclooxygenase-2 the alpha and the omega in cancer? *Journal of Clinical Investigation*, **105:** 1511–1513.

Prescott, S.M. and Fitzpatrick, F.A. 2000, Cyclooxygenase-2 and carcinogensis. *Biochimica et Biophysica Acta*, **1470:** 69–78.

Przybelski, R.J., Daily, E.K., Micheels, J., Sloan, E., Mols, P., Koenigsberg, M.D., Bickell, W.H., Thompson, D.R., Harviel, J.D. and Cohn, S.M. 1999, A safety assessment of diaspirin cross-linked hemoglobin (DCLHb) in the treatment of hemorrhagic, hypovolemic shock. *Prehospital and Disaster Medicine*, **14:** 251–264.

Pugh, P.M. and Rutishauser, S.C. 1978, Comparative effects of 2,4-dinitrophenol and sodium salicylate on bile secretion in the dog, cat, rabbit and guinea-pig. *General Pharmacology*, **9 (2):** 119–121.

Rabinowitz, I.N., Wolf, P.L. and Berman, S. 1974, Light scattering studies of retardation of sickling by aspirin-like drugs. *Research Communications in Chemical Pathology and Pharmacology*, **8 (2):** 417–420.

Rado, J.P., Simatupang, T., Boer, P. and Dorhout Mees, E.J. 1978, Pharmacologic studies in Bartter's syndrome: effect of DDAV and indomethacin on renal concentrating operation. Part II. *International Journal of Clinical Pharmacology and Biopharmacy*, **16:** 22–26.

Rai, R., Backos, M., Baxter, N., Chilcott, I. and Regan, L. 2000, Recurrent miscarriage – an aspirin a day? *Human Reproduction*, **15:** 2220–2223.

Rainsford, K.D. 1982, Development and therapeutic actions of oral copper complexes. *In:* J.R.J. Sorenson (ed.), *Inflammatory Diseases and Copper. The Metabolic and Therapeutic Roles of Copper and other Essential Metalloelements in Humans*, pp. 375–384. Clifton: Humana Press.

Rainsford, K.D. 1992, Gastric mucosal irritancy in rodents from copper complexes. *Experimental and Clinical Gastroenterology*, **2:** 65–77.

Rainsford, K.D., Brune, K. and Whitehouse, M.W. (eds) 1981, *Trace Elements in the Pathogenesis and Treatment of Inflammation*. Basel: Birkhäuser.

Rainsford, K.D., Milanino, R., Sorenson, J.R.J. and Velo, G.P. (eds) 1998, *Copper and Zinc in Inflammatory and Degenerative Diseases*. Dordrecht: Kluwer Academic Publishers.

Rainsford, K.D. and Whitehouse, M.W. 1976a, Concerning the merits of copper aspirin as a potential anti-inflammatory drug. *Journal of Pharmacy and Pharmacology*, **28:** 83–86.

Rainsford, K.D. and Whitehouse, M.W. 1976b, Gastric mucus effusion elicited by oral copper compounds: potential anti-ulcer activity. *Experientia*, **32:** 1172–1173.

Rainsford, K.D. and Willis, C. 1982, Relationship of gastric mucosal damage induced in pigs by anti-inflammatory drugs to their effects on prostaglandin production. *Digestive Diseases and Sciences*, **27:** 624–635.

Rao, K.V.N., Detrisac, C.J., Steele, V.E., Hawk, E.T., Kelloff, G.J. and McCormick, D.L. 1996, Differential activity of aspirin, ketoprofen and sulindac as cancer chemopreventive agents in the mouse urinary bladder. *Carcinogenesis*, **17:** 1435–1438.

Ratzmann, K.P., Besch, W., Witt, S. and Schulz, B. 1982b, Discrepant effect of the prostaglandin synthesis inhibitor acetylsalicylic acid on insulin and C-peptide. Response to glucose in man. *Hormones and Metabolism*, **14:** 508–512.

Ratzmann, K.P., Jahr, D., Besch, W., Heinke, P. and Schulz, B. 1982a, Effect of acetylsalicylic acid on insulin sensitivity in subjects with impaired glucose tolerance. *Acta Biologica Medica Germanica*, **41:** 929–933.

Reddy, B.S., Hirose, Y., Lubet, R., Steele, V., Kelloff, G., Paulson, S., Seibert, K. and Rao, C.V. 2000, Chemoprevention of colon cancer by specific cyclooxygenase-2 inhibitor, celecoxib, administered during different stages of carcinogenesis. Effect of the specific cyclooxygenase-2 inhibitor meloxicam on tumour growth and cachexia in a murine model. *Cancer Research*, **60:** 293–297.

Reddy, B.S., Rao, C.V., Rivenson, A. and Kelloff, G. 1993, Inhibitory effect of aspirin on azoxymethane-induced colon carcinogenesis in F344 rats. *Carcinogenesis*, **14:** 1493–1497.

Reddy, B.S., Tokumo, K., Kulkarni, N., Aliga, A. and Kelloff, G. 1992, Inhibition of colon carcinogenesis by prostaglandin synthesis inhibitors and related compounds. *Carcinogenesis*, **13:** 1019–1023.

Ricchi, P., Pignata, S., Di Popolo, A., Memoli, A., Apicella, A., Zarrilli, R. and Acquaviva, A.M. 1997, Effect of aspirin on cell proliferation and differentiation of colon adenocarcinoma Caco-2 cells. *International Journal of Cancer*, **73:** 880–884.

Richardson, T., Foster, J. and Mawer, G.E. 1986, Enhancement by sodium salicylate of the blood glucose lowering effect of chlorpropamide – drug interaction of summation of similar effects? *British Journal of Clinical Pharmacology*, **22:** 43–48.

Riepe, M.W., Kasischke, K. and Raupach, A. 1997, Acetylsalicylic acid increases tolerance against hypoxic and chemical hypoxia. *Stroke*, **28:** 2006–2011.

Riley, M.L. and Harding, J.J. 1993, The reaction of malondialdehyde with lens proteins and the protective effect of aspirin. *Biochemica et Biophysica Acta*, **1158:** 107–112.

Rioux, N. and Castonguay, A. 1998a, Inhibitors of lipoxygenase: a new class of cancer chemopreventive agents. *Carcinogenesis*, **19:** 1393–1400.

Rioux, N. and Castonguay, A. 1998b, Prevention of NNK-induced lung tumorigenesis in A/J mice by acetylsalicylic acid and NS-398. *Cancer Research*, **58:** 5354–5360.

Riviere, J.E., Smith, C.E., Budsabba, K., Brokks, J.D., Olajos, E.J., Salem, H. and Monteiro-Riviere, N.A. 2001, Use of methyl salicylate as a stimulant to predict the percutaneous absorption of sulfur mustard. *Journal of Applied Toxicology*, **21:** 91–99.

Roberts, D.L., Marshall, R. and Marks, R. 1980, Detection of the action of salicylic acid on the normal stratum corneum. *British Journal of Dermatology*, **103 (2):** 191–196.

Roberts, M.S., Lai, P.M. and Anissimov, Y.G. 1998, Epidermal iontophoresis: I. Development of the ionic mobility-pore model. *Pharmaceutical Research*, **15:** 1569–1578.

Roch-Arvellier, M., Revelant, V., Pham Huy, D., Maman, L., Fontagne, J., Sorenson, J.R.J. and Girooud, J.P. 1990,. Effects of some non-steroidal anti-inflammatory drug copper complexes on polymorphonuclear leukocyte oxidative metabolism. *Agents and Actions*, **31:** 65–71.

Romero, J.C., Dunlap, C.L. and Strong, C.G. 1976, The effect of indomethacin and other anti-inflammatory drugs on the renin-angiotensin system. *Journal of Clinical Investigation*, **58 (2):** 282–288.

Rosenberg, L., Louik, C. and Shapiro, S. 1998, Nonsteroidal anti-inflammatory drug use and reduced risk of large bowel carcinoma. *Cancer*, **82:** 2326–2333.

Rosenberg, L., Palmer, J.R., Zauber, A.G., Warshauer, M.E., Stolley, P.D. and Shapiro, S. 1991, A hypothesis: nonsteroidal anti-inflammatory drugs reduce the incidence of large bowel cancer. *Journal of the National Cancer Institute*, **83:** 355–358.

Roses, A.D. 1994, Apolipoprotein E affects the rate of Alzheimer disease expression: beta-amyloid burden is a secondary consequence dependent on APOE genotype and duration of disease. *Journal of Neuropathology and Experimental Neurology*, **53:** 427–428.

Rubinstein, M., Marazzi, A. and Polak de Fried, E. 1999, Low-dose aspirin treatment improves ovarian responsiveness, uterine and ovarian blood flow velocity, implantation, and pregnancy rates in patients undergoing in vitro fertilization: a prospective, randomized, double-blind placebo-controlled assay. *Fertility and Sterility*, **71:** 825–829.

Ruffin, M.T., Kirshnan, K., Rock, C.L., Normolle, D., Vaerten, M.A., Petersgolden, M., Crowell, J., Kellof, G., Boland, C.R. and Brenner, D.E. 1997, Suppression of human colorectal mucosal prostaglandins: determining the lowest effective aspirin dose. *Journal of the National Cancer Institute*, **89:** 1152–1160.

Ryba, J. and Plumert, A. 1877, Zur Behandlung des Diabetes Mellitus mit salicylsaurem Natron. *Prague Medizinische Wochenschrift*, **2:** 381–385, 401–406, 430–433.

Sairam, K., Saravanan, K.S., Banerjee, R. and Mohanakumar, K.P. 2003, Non-steroidal anti-inflammatory drug sodium salicylate, but not diclofenac or celecoxib, protects against 1-methyl-4-phenyl pyridinium-induced dopaminergic neurotoxicity in rats. *Brain Research*, **966:** 245–252.

Sakata, T., Hasegawa, R., Johansson, S.L., Zenser, T.V. and Cohen, S.M. 1986, Inhibition by aspirin of N-[4-(5-nitro-2-furyl)-2-thiazolyl]formamide initiation and sodium saccharin promotion of urinary bladder carcinogenesis in male F344 rats. *Cancer Research*, **46:** 3903–3906.

Sandler, R.S., Galanko, J.C., Murray, S.C., Helm, J.F. and Woosley, J.T. 1998, Aspirin and non-steroidal anti-inflammatory agents and risk for colorectal adenomas. *Gastroenterology*, **114:** 441–447.

Santini, G., Sciulli, M.G., Marinacci, R., Fusco, O., Spoletini, L., Pace, A., Ricciardulli, A., Natoli, C., Procopio, A., Maclouf, J. and Patrignani, P. 1999, Cycloxygenase-independent induction of $p21^{WAF-1/cip1}$, apoptosis and differentiation by L-745,337, a selective PGH synthase-2 inhibitor, and salicylate in HT-29 cells. *Apoptosis*, **4:** 151–162.

Sarraf, P., Mueller, E., Jones, D., King, F.J., DeAngelo, D.J., Partridge, J.B., Holden, S.A., Chen, L.B., Singer, S., Fletcher, C. and Spiegelman, B.M. 1998, Differentiation and reversal of malignant changes in colon cancer through PPAR. *Nature Medicine*, **4:** 1046–1052.

Scadding, J.W. and Gibbs, J. 1994, Neurobiological disease. *In:* R.L. Souhani and J. Roxham (eds), *Textbook of Medicine*, 2nd edn, pp. 858–982. Edinburgh: Churchill-Livingstone.

Scali, C., Prosperi, C., Vannaucchi, M.G., Pepeu, G. and Casamenti, F. 2000, Brain inflammatory reaction in an animal model of neuronal degeneration and its modulation by an anti-inflammatory drug: implication in Alzheimer's disease. *European Journal of Neuroscience*, **12:** 1900–1912.

Schmidt, C.R., Beazell, J.M., Atkinson, A.J. and Ivy, A.C. 1938, The effect of therapeutic agents on the volume and constituents of bile. *American Journal of Digestive Diseases*, **5:** 613–617.

Schreinemachers, D.M. and Everson, R.B. 1994, Aspirin use and lung, colon, and breast cancer incidence in a prospective study. *Epidemiology*, **5:** 138–146.

Seino, Y., Usami, M., Nakahara, H., Takemura, J., Nishi, S., Ishida, H., Ikeda, M. and Imura, H. 1982, Effect of acetylsalicylic acid on blood glucose and glucose regulatory hormones in mild diabetes. *Prostaglandins and Leukotrienes in Medicine*, **8:** 49–53.

Selko, D.J. 1994, Amyloid beta-protein precursor: new clues to the genesis of Alzheimer's disease. *Current Opinion in Neurobiology*, **4:** 708–716.

Shackel, N.A., Day, R.O., Kellett, B. and Brooks, P.M. 1997, Copper-salicylate gel for pain relief in osteoarthritis: a randomised controlled trial. *Medical Journal of Australia*, **167 (3):** 134–136.

Shao, J., Fujiwara, T., Kadowaki, Y., Fukazawa, T., Waku, T., Itoshima, T., Yamatsuji, T., Nishizaki, M., Roth, J.A. and Tanaka, N. 2000, Overexpression of the wild type p53 gene inhibits NF-κB activity and synergises with aspirin to induce apoptosis in human colon cancer cells. *Oncogene*, **19:** 726–736.

Sharpe, C.R., Collet, J.P., McNutt, M., Belzille, E., Boivin, J.F. and Hanley, J.A. 2000, Nested case–control study of the effects of non-steroidal anti-inflammatory drugs on breast cancer risk and stage. *British Journal of Cancer*, **83:** 112–120.

Shastri, G.V., Thomas, M., Victoria, A.J., Selvakumar, R., Kanagasabapathy, A.S., Thomas, K. and Lakshmi, K. 1998, Effect of aspirin and sodium salicylate on cataract development in diabetic rats. *Indian Journal of Experimental Biology*, **36:** 651–657.

Sheehan, K.M., Sheahan, K., O'Donoghue, D.P., MacSweeney, F., Conroy, R.M., Fitzgerald, D.J. and Murray, F.E. 1999, The relationship between cyclooxygenase-ω expression and colorectal cancer. *Journal of the American Medical Association*, **282:** 1254–1257.

Shen, Z.-Q., Ling, L., Wu, L.-O., Liu, W.-P. and Chen, Z.-H. 1999, Effects of copper-aspirin complex on plasma 6-keto-prostaglandin $F_{1\alpha}$ level and platelet cytosolic calcium in rabbits. *Platelets*, **10:** 345–348.

Shiff, S.J. and Rigas, B. 1999a, The role of cyclooxygenase inhibition in the antineoplastic effects of nonsteroidal anti-inflammatory drugs (NSAIDs). *Journal of Experimental Medicine*, **190:** 445–450.

Shiff, S.J. and Rigas, B. 1999b, Aspirin for cancer. *Nature Medicine*, **5:** 1348–1349.

Shuff, S.T., Chowdhary, P., Khan, M.F. and Sorenson, J.R.J. 1992, Stable superoxide dismutase (SOD)-mimetic ternary human serum albumin-Cu(II)(3,5-diisopropylsalicylate)2/Cu(II)2(3,5-diisopropylsalicylate)4 complexes in tissue distribution of the binary complex. *Biochemical Pharmacology*, **43 (7):** 1601–1612.

Sibai, B.M., Lindheimer, M., Hauth, J., Caritis, S., Vandorsten, P., Klebanoff, M., MacPherson, C., Landon, M., Miodovnik, M., Paul, R., Meis, P. and Dombrowski, M. 1998, Risk factors for preeclampsia, abruptio placentae, and adverse neonatal outcomes among women with chronic hypertension. *New England Journal of Medicine*, **339:** 667–671.

Singer, I.I., Kawka, D.W., Schloemann, S., Tessner, T., Riehl, T. and Stenson, W.F. 1998, Cyclooxygenase-2 induced in colonic epithelial cells in inflammatory bowel disease. *American Journal of Gastroenterology*, **115:** 297–306.

Sjöbeck, M. and Englund, E. 2001, Alzheimer's disease and the cerebellum: a morphologic study on neuronal and glial changes. *Dementia and Geriatric Cognitive Disorders*, **12:** 211–218.

Skopińska-Różewska, E., Piazza, G., Sommer, E., Pamukcu, R., Barcz, E., Filewska, M., Kupis, W., Caban, R., Rudziński, P., Bogdan, J., Mlekodaj, S. and Sikorska, E. 1998, Inhibition of angiogenesis by sulindac and its sulfone metabolite (FGN-1): A potential mechanism for their antineoplastic properties. *Tissue Reactions*, **10:** 85–89.

Sladkevicius, P. 1999, Is aspirin all it is cracked up to be – reproducibility of transvaginal color Doppler ultrasonography for ovarian and uterine vessels? *Fertility and Sterility*, **73:** 1069–1070.

Sloan, E.P., Koenigsberg, M., Gens, D., Cipolle, M., Runge, J., Mallory, M.N., Rodman, Jr, G., for the DCLHb Traumatic Hemorrhage Shock Study Group. 1999, Diaspirin cross-linked haemoglobin (DCLHb) in the treatment of severe traumatic hemorrhagic shock. *Journal of the American Medical Association*, **282**: 1857–1864.

Smith, M.J., Hoth, M. and Davis, K. 1975, Aspirin and lymphocyte transformation. *Annals of Internal Medicine*, **83**: 509–511.

Smith, M.J.H. 1966, Metabolic effects of salicylates. *In:* M.J.H. Smith and P.K. Smith (eds), *The Salicylates, A Critical Bibliographic Review*, pp. 49–105. New York: Wiley-Interscience.

Smith, M.J.H. and Dawkins, P.D. 1971, Salicylate and enzymes. *Journal of Pharmacy and Pharmacology*, **23**: 729–744.

Smith, M.L., Hawcroft, G. and Hull, M.A. 2000, The effect of non-steroidal anti-inflammatory drugs on human colorectal cancer cells: evidence of different mechanisms of action. *European Journal of Cancer*, **36**: 664–674.

Smythies, J.R. 1988, Cancer. *In:* P.B. Curtis Prior (ed.), *Prostaglandins. Biology and Chemistry of Prostaglandins and Related Eicosanoids*, pp. 631–636. Edinburgh: Churchill Livingstone.

Sorenson, J.R.J. 1974, The anti-inflammatory activity of some copper chelates. *In:* D.D. Hempell (ed.), *Trace Substances in Environmental Health*, Vol. VIII, pp. 305–311. Columbia, MI: University of Missouri.

Sorenson, J.R.J. 1976, Copper chelates as possible active forms of anti-arthritic agents. *Journal of Medicinal Chemistry*, **19**: 135–148.

Sorenson, J.R.J. 1982a, Copper complexes as the active metabolites of antiinflammatory agents. *In:* J.R.J. Sorenson (ed.), *Inflammatory Diseases and Copper. The Metabolic and Therapeutic Roles of Copper and Other Essential Metalloelements in Humans*, pp. 289–299. New Jersey: Humana Press.

Sorenson, J.R.J. 1982b, The anti-inflammatory activities of copper complexes. *In:* H. Sigel (ed.), *Metal Ions in Biological Systems*, Vol. 14, pp. 77–124. New York: Marcel Dekker.

Sorenson, J.R.J. (ed.) 1982c, *Inflammatory Diseases and Copper. The Metabolic and Therapeutic Roles of Copper and Other Essential Metalloelements in Humans*. New Jersey: Humana Press.

Sorensen, J.R.J. (ed.) 1982, The metabolic and therapeutic roles of copper and other metalloelements in humans. *In: Inflammatory Diseases and Copper*, pp. 289–299. Clifton, NJ: Humana Press.

Sorenson, J.R.J. 1984, The ulcerogenic potential of copper aspirinate seems to be more imaginary than real. *Journal of Pharmaceutical Sciences*, **73**: 1875–1877.

Sorenson, J.R.J. 1989, Copper complexes as 'radiation recovery' agents. *Chemistry in Britain*, pp. 169–172.

Sorenson, J.R.J. 1992a, Pharmacological activities of copper compounds. *In:* G. Berton (ed.), *Handbook on Metal-ligand Interactions in Biological Fluids*, Vol. II, Ch. 4.

Sorenson, J.R.J. 1992b, Copper-potentiation of non-steroidal anti-inflammatory drugs. *In:* G. Berton (ed.), *Handbook of Metal-ligand Interactions in Biological Fluids*, Vol. II, Ch. I.

Sorenson, J.R.J., Oberley, L.E., Crouch, R.K., Kensler, T.W., Kishore, V., Leuthauser, S.W.C., Oberley, T.D. and Pezeshk, A. 1983, Pharmacologic activities of copper compounds in chronic diseases. *Biological Trace Element Research*, **5**: 257–273.

Sorenson, J.R.J., Oberley, L.W., Oberley, T.D., Leuthauser, S.W.C., Ramakrishna, K., Vernino, L. and Kishore, V. 1982, Antineoplastic activities of some copper salicylates. *In:* D.D. Hemphill (ed.), *Trace Substances in Environmental Health*, Vol. XVI. Columbia, MI: University of Missouri.

Sorenson, J.R.J., Rolniak, T.M. and Chang, L.W. 1984, Preliminary chronic toxicity study of copper aspirinate. *Inorganica Chimica Acta*, **91**: L31–L34.

Sorenson, J.R.J., Sodergerg, L.S.F., Chang, L.W., Willingham, W.M., Baker, M.L., Barnett, J.B., Salari, H. and Bond, K. 1993, Copper-, iron-, manganese- and zinc-3,5-diisopropylsalicylate complexes increase survival of gamma-irradiated mice. *European Journal of Medicinal Chemistry*, **28**: 221–229.

Stamford, I.F., Carroll, M.A., Civier, A., Hensby, C.N. and Bennett, A. 1983, Identification of arachidonate metabolites in normal, benign and malignant human mammary tissues. *Journal of Pharmacy and Pharmacology*, **35**: 48–49.

Steinhoff, M.C., Douglas, R.G. Jr, Greenberg, H.B. and Callahan, D.R. 1980, Bismuth subsalicylate therapy of viral gastroenteritis. *Gastroenterology*, **78**: 1495–1499.

Stek, M. Jr 1980, Traveler's diarrhea in the Mediterranean basin. *Milanino Medicine*, **145**: 628–629.

CHAPTER 13

Stoff, J.S., Stemerman, M., Steer, M., Salzman, E. and Brown, R.S. 1980, A defect in platelet aggregation in Bartter's syndrome. *American Journal of Medicine*, **68:** 171–180.

Stürmer, T., Glynn, R.J., Field, T.S., Taylor, J.O. and Hennekens, C. 1996, Aspirin use and cognitive function in the elderly. *American Journal of Epidemiology*, **143:** 683–691.

Sugaya, K., Uz, T., Kumar, V. and Manev, H. 2000, New anti-inflammatory treatment strategy in Alzheimer's disease. *Japanese Journal of Pharmacology*, **82:** 85–94.

Suh, O., Mettlin, C. and Petrelli, N.J. 1993, Aspirin use, cancer, and polyps of the large bowel. *Cancer*, **72:** 1171–1177.

Taketo, M.M. 1998a, Cyclooxygenase-2 inhibitors in tumorigenesis (Part I). *Journal of the National Cancer Institute*, **90:** 1529–1534.

Taketo, M.M. 1998b, Cyclooxygenase-2 inhibitors in tumorigenesis (Part II). *Journal of the National Cancer Institute*, **90:** 1609–1620.

Tarrant, M.E. and O'Hare, J.P. 1967, Inactivation of *Schistosoma mansonii* lactate dehydrogenase by 4-iodoacetamide salicylic acid. *Biochemical Pharmacology*, **16:** 1421–1428.

Tavani, A., Gallus, S., Lavecchia, C., Conti, E., Montella, M. and Franceschi, S. 2000, Aspirin and ovarian cancer: an Italian case–control study. *Annals of Oncology*, **11:** 1171–1173.

Terai, M. and Shulman, S.T. 1997, Prevalence of coronary artery abnormalities in Kawasaki disease is highly dependent on gamma globulin dose but independent of salicylate dose. *Journal of Pediatrics*, **131:** 888–893.

Thornell, E., Jansson, R., Kral, J.G. and Svanik, J. 1979, Inhibition of prostaglandin synthesis as a treatment for biliary pain. *Lancet*, **1:** 584.

Thun, M.J., Namboordiri, M.M., Calle, E.E. *et al.*, 1993, Aspirin use and risk of fatal cancer. *Cancer Research*, **53:** 1322–1327.

Thun, M.J., Namboordiri, M.M. and Heath, C.W. Jr 1991, Aspirin use and reduced risk of fatal colon cancer, *New England Journal of Medicine*, **325:** 1593–1596.

Tomlinson, A., Simmons, P.A., Seal, D.V. and McFadyen, A.K. 2000, Salicylate inhibition of acanthamoeba attachment, to contact lenses – a model to reduce risk of infection. *Opthalmology*, **107:** 112–117.

Tondelier, D., Brouillard, F., Lipecka, J., Labarthe, R., Bali, M., Costa de Beauregard, M.-A., Torossi, T., Cougnon, M., Edelman, A. and Baudouin-Legros, M. 1999, Aspirin and some other nonsteroidal anti-inflammatory drugs inhibit cystic fibrosis transmembrane conductance regulator protein gene expression in T-84 cells. *Mediators of Inflammation*, **8:** 219–227.

Tornvall, G. and Allgen, L.G. 1980, Acute effects of acetylsalicylic acid on blood glucose and insulin in non-insulin dependent diabetes. *Acta Endocrinologica (Copenhagen)*, **239 (Suppl.):** 6–8.

Tran, P.O., Gleason, C.E. and Robertson, R.P. 2002, Inhibition of interleukin-1beta-induced COX-2 and EP3 gene expression by sodium salicylate enhances pancreatic islet beta-cell function. *Diabetes*, **51:** 1772–1778.

Tsujii, M., Kawano, S., Tsuji, S., Sawaoka, H., Hori, M. and Dubois, R.N. 1998, Cyclooxygenase regulates angiogenesis induced by colon cancer cells. *Cell*, **93:** 705–716.

Uehara, H., Otsuka, H. and Izumi, K. 1999, Modifying effects of a mixture of acetaminophen, aspirin, dipyrone and ethenzamide on a multiorgan initiation model and its carcinogenicity in male F344 rats. *Cancer Letters*, **135:** 83–90.

Vainio, H. 1999, Chemoprevention of cancer: a controversial and instructive story. *British Medical Bulletin*, **55:** 593–599.

Van Kooten, F., Ciabattoni, G., Koudstaal, P.J., Grobbee, D.E., Kluft, C. and Patrono, C. 1999, Increased thromboxane biosynthesis is associated with post-stroke dementia. *Stroke*, **30:** 1542–1547.

Van Kooten, F., Ciabattoni, G., Patrono, C. and Koudstaal, P.J. 1998, Role of platelet activation in dementia. *Haemostasis*, **28:** 202–208.

Vaughan, J.P. 1969, The effects of suramin and sodium salicylate on the morphological variation and virulence of two strains of Trypanosoma rhodesiense in mice. *Annals of Tropical Medicine and Parasitology*, **63:** 301–307.

Viberti, G., Wiseman, J.M., Pinto, J.R. and Messent, J. 1994, Diabetic Nephropathy. *In:* C.R. Khan and G.C. Weir (eds), *Joslin's Diabetes Mellitus*, 13th edn, pp. 363–371. Philadelphia: Lea & Febiger.

Vick-Mo, H., Hove, K. and Mjoes, O.D. 1978, Effects of sodium salicylate on plasma concentration and fatty acid turnover in dogs. *Acta Physiologica Scandinavica*, **103**: 113–119.

Villa, M.P., Zappulla, F., Salardi, S., Balsamo, A., Vecchi, F., Pirazzoli, P., Bernardi, F., Cicognani, A. and Cacciari, E. 1980, Aspirin and indomethacin therapy in 2 familial cases of Bartter's disease. Follow-up for growth and endocrine pattern. *Minerva Pediatrica*, **32**: 1259–1268.

Waitzman, M.B., Cornelius, L.M. and Evatt, B.L. 1980, Treatment of canine spontaneous diabetes mellitus with aspirin. *Metabolism Pediatrics and Ophthalmology*, **4**: 151–154.

Waitzman, M.B., Kaplan, H., Cornelius, L., Evatt, B. and Hunt, B. 1983, Aspirin and prostacyclin treatment of diabetic dogs. *Metabolism Pediatrics and Systematic Ophthalmology*, **7**: 153–158.

Walder, J.A., Walder, R.Y. and Arnone, A. 1980, Development of antisickling compounds that chemically modify haemoglobin S with 2,3-diphosphoglycerate binding site. *Journal of Molecular Biology*, **141**: 195–216.

Walder, J.A., Zaugg, R.H., Iwaoka, R.S., Watkin, W.G. and Klotz, I.M. 1977, Alternative aspirins as antisickling agents: acetyl-3,5-dibromosalicylic acid. *Proceedings of the National Academy of Sciences of the United States of America*, **74**: 5499–5503.

Walder, J.A., Zaugg, R.H., Walder, R.Y., Steele, J.M. and Klotz, I.M. 1979, Diaspirins that cross-link beta chains of hemoglobin: bis(3,5-dibromosalicyl) succinate and bis(3,5-dibromosalicyl) fumarate. *Biochemistry*, **18**: 4265–4270.

Wallenburg, H.C.S. 2001, Prevention of pre-eclampsia: status and perspectives 2000. *European Journal of Obstetrics and Gynecology and Reproductive Biology*, **94**: 13–22.

Wallenburg, H.C.S. and Rotmans, N. 1988, Prophylactic low-dose aspirin and dipyridamole in pregnancy. *Lancet*, **1**: 939.

Wautier, J.L., Paton, R.C., Wauter, M.P., Pintigny, D., Abadie, E., Passa, P. and Caen, J.P. 1981, Increased adhesion of erythrocytes to endothelial cells in diabetes mellitus and its relation to vascular complications. *New England Journal of Medicine*, **305**: 237–242.

Wechter, W.J., Kantoci, D., Murray, E.D. Jr, Quiggle, D.D., Leipold, D.D., Gibson, K.M. and McCracken, J.D. 1997, R-Flurbiprofen chemoprevention and treatment of intestinal adenomas in the APC$^{-/+}$ mouse model: implications for prophylaxis and treatment of colon cancer. *Cancer Research*, **57**: 4316–4324.

Wechter, W.J., Leipold, D.D., Murray, E.D. Jr, Quiggle, D, McCracken, J.D., Barrios, R.S. and Greenberg, N.M. 2000, E-7869 (R-Flurbiprofen) inhibits progression of prostate cancer in the TRAMP mouse. *Cancer Research*, **60**: 2203–2208.

Weirich, E.G. 1975, Dermatopharmacology of salicylic acid. I. Range of dermatotherapeutic effects of salicylic acid. *Dermatologica*, **151**: 268–273.

Weirich, E.G., Longauer, J.K. and Kirkwood, A.H. 1975, Dermatopharmacology of salicylic acid. II. Epidermal antihyperplastic effect of salicylic acid in animals. *Dermatologica*, **151**: 321–332.

Weser, U. and Schubotz, L.M. 1981, Catalytic reaction of copper complexes with superoxide. *In*: K.D. Rainsford, K. Brune and M.W. Whitehouse (eds), *Trace Elements in the Pathogenesis and Treatment of Inflammation*, pp. 103–120. Basel: Birkhäuser.

White, J.R. Jr, Hartman, J. and Campbell, R.K. 1993, Drug interactions in diabetic patients. The risk of losing glycemic control. *Postgraduate Medicine*, **93**: 131–132.

Whitehouse, M.W. 1965, Some biochemical and pharmacological properties of anti-inflammatory drugs. *Progress in Drug Research*, **8**: 321–429.

Williams, C.S., Mann, M. and Dubois, N.R. 1999, The role of cyclooxygenases in inflammation, cancer and development. *Oncogene*, **18**: 7908–7916.

Williams, C.S., Tsujii, M., Reese, J., Dey, S.D. and Du Bois, R.N. 2000, Host cyclooxygenase-2 modulates carcinoma growth. *Journal of Clinical Investigation*, **105**: 1589–1594.

Williams, D.A., Waltz, D.T. and Foye, W.O. 1976, Synthesis and biological evaluation of tetrakis-µ(acetylsalicylato)dicopper(II). *Journal of Pharmaceutical Sciences*, **65**: 126–128.

Williamson, R.T. 1901, On the treatment of glycosuria and diabetes mellitus with sodium salicylate. *British Medical Journal*, **1**: 760–762.

Willingham, W.M. and Sorenson, J.R.J. 1986, Physiologic role of copper complexes in antineoplasia. *Trace Elements in Medicine*, **3**: 119–152.

■ CHAPTER 13 ■

Wolff, H., Saukkonen, K., Anttila, S., Antti, K., Vainio, H. and Ristimäki, A. 1998, Expression of cyclooxygenase-2 in human lung carcinoma. *Cancer Research*, **58**: 4997–5001.

Wood, L.E., Haney, D.N., Patel, J.R., Clare, S.E., Shi, G.-Y., King, L.C. *et al.* 1981, Structural specificities in acylation of haemoglobin and sickle haemoglobin by diaspirins. *Journal of Biological Chemistry*, **256**: 7046–7052.

Yamagata, K., Urakami, K., Ikeda, J.I.Y., Adachi, H., Arai, H., Sasaki, K., Sato, K. and Nakashima, K. 2001, High expression of apolipoprotein E mRNA in the brains with sporadic Alzheimer's disease. *Dementia and Geriatric Cognitive Disorders*, **12**: 57–62.

Yankner, B.A. 1996, Mechanisms of neuronal degeneration in Alzheimer's disease. *Neuron*, **16**: 921–932.

Yermakova, A.V., Rollins, J., Callahan, L.M., Rogers, J. and O'Banion, M.K. 1999, Cyclooxygenase-1 in human Alzheimer and control brain: quantitative analysis of expression by microglia and CA3 hippocampus neurons. *Journal of Neuropathology and Experimental Neurology*, **58**: 1135–1146.

Yoshimi, J., Shimizu, M., Matsunaga, K., Yamada, Y., Fujii, K., Hara, A. and Mori, H. 1999, Chemopreventive effect of N-(2-cyclohexyloxy-4-nitrophenyl) methane sulfonamide (NS-398), a selective cycloxygenase-2 inhibitor, in rat colon carcinogensis induced by azoxymethane. *Japanese Journal of Cancer Research*, **90**: 406–412.

Yoshimi, N., Kawabata, K., Hara, A., Matsunaga, K., Yamada, Y. and Mori, H. 1997, Inhibitory effect of NS-398, a selective cyclooxygenase-2 inhibitor, on azoxymethane-induced aberrant crypt foci in colon carcinogenesis of F344 rats. *Japanese Journal of Cancer Research*, **88**: 1044–1051.

Yun, M., Choi, H., Jung, J. and Kim, C. 1999, Development of a thermo-reversible insulin liquid suppository with bioavailability of enhancement. *International Journal of Pharmaceutics*, **189**: 137–145.

Zaridze, D., Borisova, E., Maximovitch, D. and Chkhikvadze, V. 1999, Aspirin protects against gastric cancer: results of a case-control study from Moscow, Russia. *International Journal of Cancer*, **82**: 473–476.

Zhiqiang, S., Lei, W.Y., Li, L., Chen, Z.H. and Liu, W.P. 1998, Coordination of copper with aspirin improves its anti-inflammatory activity. *Inflammopharmacology*, **6**: 357–362.

Zurier, R.B. 1991, Role of prostaglandins E in inflammation and immune responses. *Advances in Prostaglandin Thromboxane Leukotriene Research*, **21B**: 947–953.

Index